Student Solutions Manual

Precalculus with Limits
A Graphing Approach

TEXAS EDITION

SIXTH EDITION

Ron Larson
The Pennsylvania State University,
The Behrend College

Australia • Brazil • Mexico • Singapore • United Kingdom • United States

ISBN-13: 978-1-285-86777-9
ISBN-10: 1-285-86777-7

Cengage Learning
200 First Stamford Place, 4th Floor
Stamford, CT 06902
USA

Cengage Learning is a leading provider of customized learning solutions with office locations around the globe, including Singapore, the United Kingdom, Australia, Mexico, Brazil, and Japan. Locate your local office at: **www.cengage.com/global**.

Cengage Learning products are represented in Canada by Nelson Education, Ltd.

To learn more about Cengage Learning Solutions, visit **www.cengage.com**.

Purchase any of our products at your local college store or at our preferred online store **www.cengagebrain.com**.

Printed in the United States of America
1 2 3 4 5 19 18 17 16 15

CONTENTS

CHAPTER 1

Section 1.1

1. (a) iii (b) i (c) v (d) ii (e) iv

3. parallel

5. Since $x = 3$ is a vertical line, all horizontal lines are perpendicular and have slope $m = 0$.

7. (a) $m = \frac{2}{3}$. Since the slope is positive, the line rises.

Matches L_2.

(b) m is undefined. The line is vertical. Matches L_3.
(c) $m = -2$. The line falls. Matches L_1.

9. Slope $= \frac{\text{rise}}{\text{run}} = \frac{3}{2}$

11.

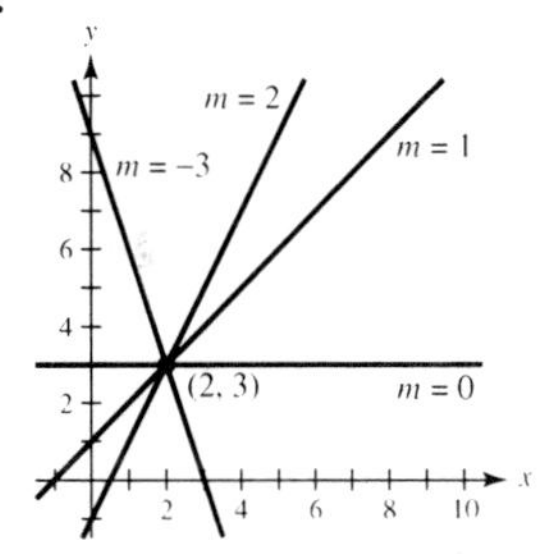

13. Slope $= \frac{0-(-10)}{-4-0} = \frac{10}{-4} = -\frac{5}{2}$

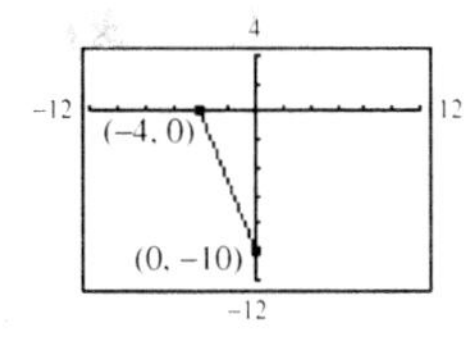

15. Slope is undefined.

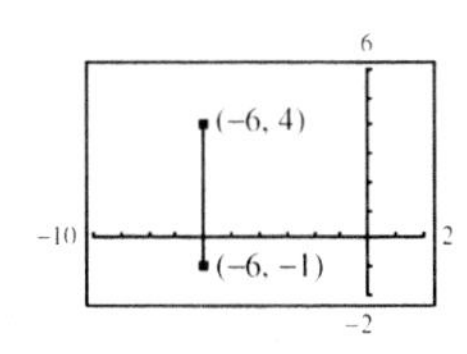

17. Since $m = 0$, y does not change. Three additional points are $(0, 1)$, $(3, 1)$, and $(-1, 1)$.

19. Since m is undefined, x does not change and the line is vertical. Three additional points are $(1, 1)$, $(1, 2)$, and $(1, 3)$.

21. Since $m = -2$, y decreases 2 for every unit increase in x. Three additional points are $(1, -11)$, $(2, -13)$, and $(3, -15)$.

23. Since $m = \frac{1}{2}$, y increase 1 for every increase of 2 in x.

Three additional points are $(9, -1)$, $(11, 0)$, and $(13, 1)$.

25. $m = 3$, $(0, -2)$

$$y + 2 = 3(x - 0)$$
$$y = 3x - 2 \Rightarrow 3x - y - 2 = 0$$

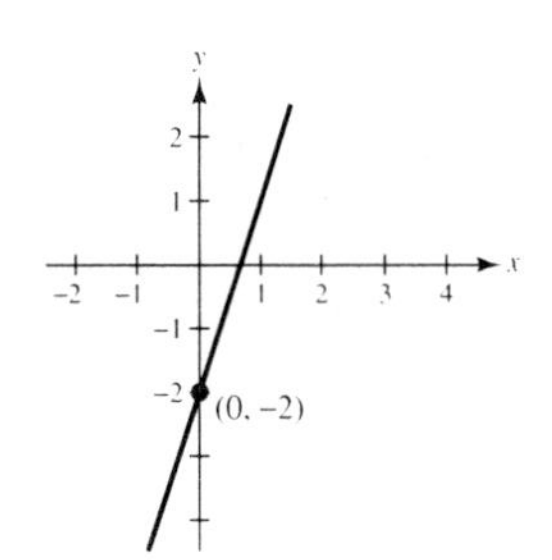

27. $m = -\frac{1}{2}$, $(2, -3)$

$$y - (-3) = -\frac{1}{2}(x - 2)$$
$$y + 3 = -\frac{1}{2}x + 1$$
$$2y + 4 = -x$$
$$x + 2y + 4 = 0$$

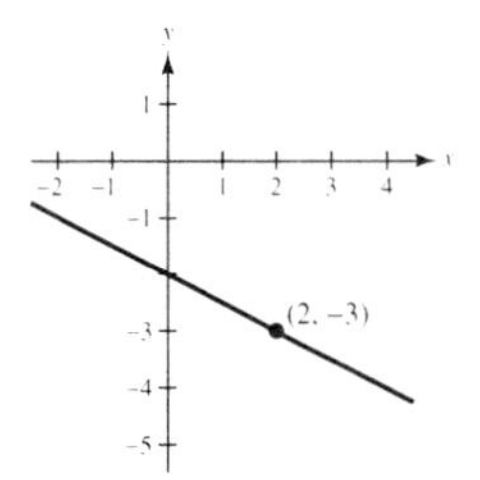

29. m is undefined, $(6, -1)$

$$x = 6$$
$x - 6 = 0$ vertical line

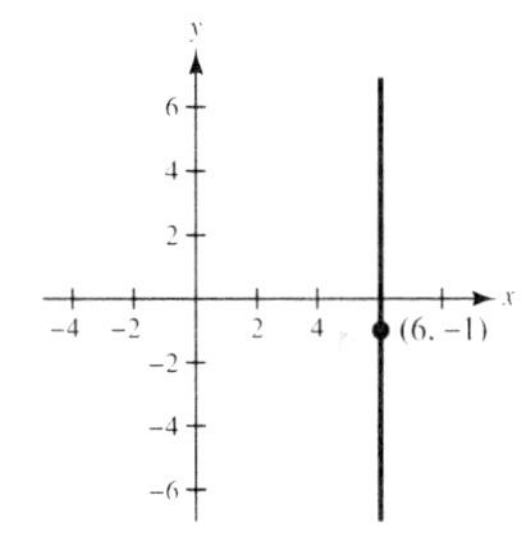

31. $m = 0, \left(-\frac{1}{2}, \frac{3}{2}\right)$

$$y - \frac{3}{2} = 0\left(x + \frac{1}{2}\right)$$

$$y - \frac{3}{2} = 0 \text{ horizontal line}$$

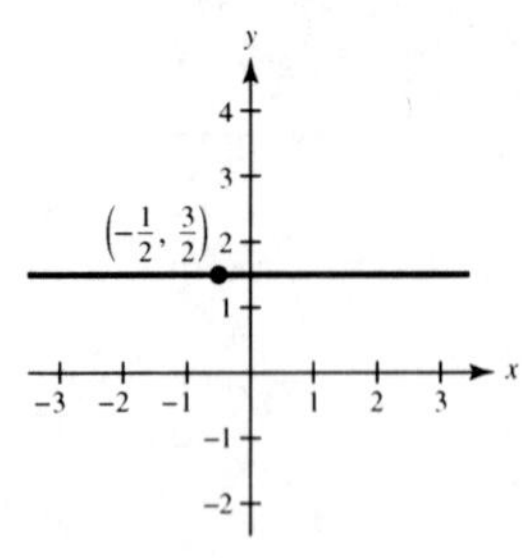

33. Begin by letting $x = 1$ correspond to 2001. Then using the points $(1, 1.6)$ and $(9, 5.2)$, you have

$$m = \frac{5.2 - 1.6}{9 - 1} = \frac{3.6}{8} = 0.45$$

$$y - 1.6 = 0.45(x - 1)$$
$$y = 0.45x + 1.15$$

When $x = 17$:
$$y = 0.45(17) + 1.15 = \$8.8 \text{ million}$$

35. $x - 2y = 4$

$$-2y = -x + 4$$
$$y = \frac{1}{2}x - 2$$

Slope: $\frac{1}{2}$

y-intercept: $(0, -2)$

The line passes through $(0, -2)$ and rises 1 unit for each horizontal increase of 2 units.

37. $2x - 5y + 10 = 0$

$$-5y = -2x - 10$$
$$y = \frac{2}{5}x + 2$$

Slope: $\frac{2}{5}$

y-intercept: $(0, 2)$

The line passes through $(0, 2)$ and rises 2 units for each horizontal increase of 5 units.

39. $x = -6$

Slope is undefined; no y-intercept.

The line is vertical and passes through $(-6, 0)$.

41. $3y + 2 = 0$

$$3y = -2$$
$$y = -\frac{2}{3}$$

Slope: 0

y-intercept: $\left(0, -\frac{2}{3}\right)$

The line is horizontal and passes through $\left(0, -\frac{2}{3}\right)$.

43. $5x - y + 3 = 0$

$$y = 5x + 3$$

(a) Slope: $m = 5$
y-intercept: $(0, 3)$

(b)

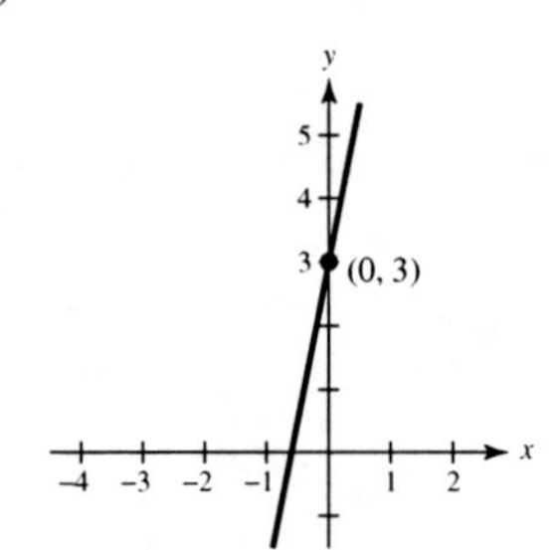

45. $5x - 2 = 0$

$$x = \frac{2}{5}$$

(a) Slope: undefined

No y-intercept

(b)

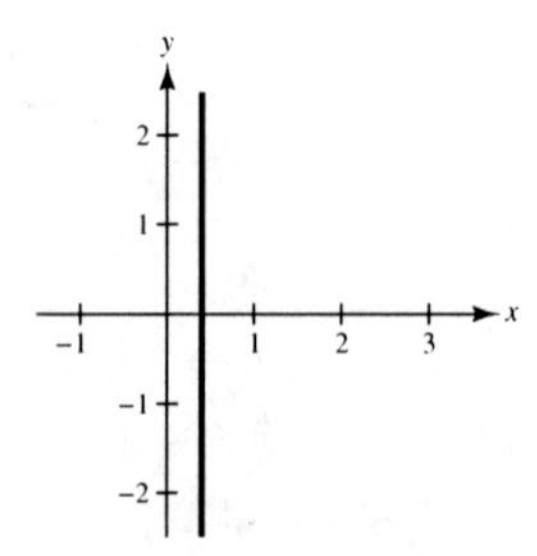

47. $3y + 5 = 0$

$$y = -\frac{5}{3}$$

(a) Slope: $m = 0$

y-intercept: $\left(0, -\frac{5}{3}\right)$

(b)

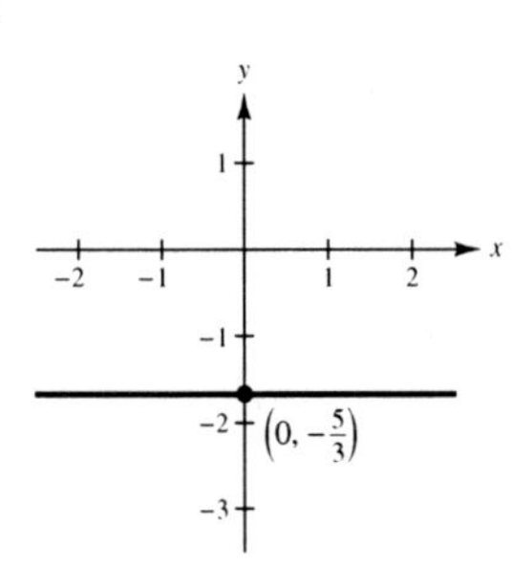

49. The slope is $\frac{-3-(-7)}{1-(-1)} = \frac{4}{2} = 2.$

$$y - (-3) = 2(x - 1)$$
$$y + 3 = 2x - 2$$
$$y = 2x - 5$$

51. $(5, -1), (-5, 5)$

$$y + 1 = \frac{5+1}{-5-5}(x - 5)$$
$$y = -\frac{3}{5}(x - 5) - 1$$
$$y = -\frac{3}{5}x + 2$$

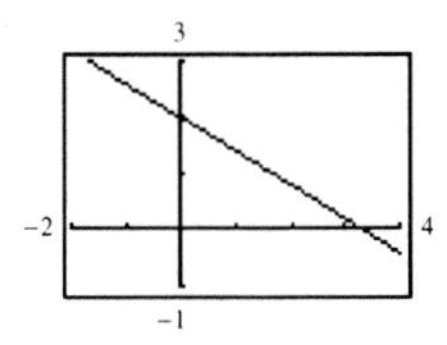

53. $(-8, 1), (-8, 7)$

Since both points have an x-coordinate of -8, the slope is undefined and the line is vertical.

$x + 8 = 0$

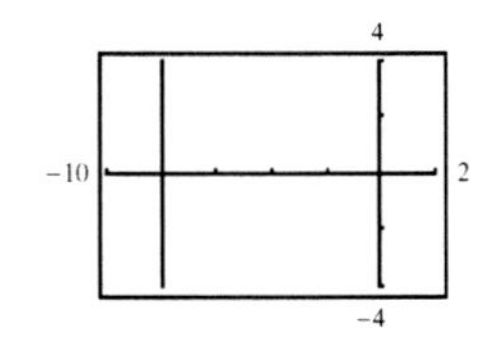

55. $\left(2, \frac{1}{2}\right), \left(\frac{1}{2}, \frac{5}{4}\right)$

$$y - \frac{1}{2} = \frac{\frac{5}{4} - \frac{1}{2}}{\frac{1}{2} - 2}(x - 2)$$
$$y = -\frac{1}{2}(x - 2) + \frac{1}{2}$$
$$y = -\frac{1}{2}x + \frac{3}{2}$$

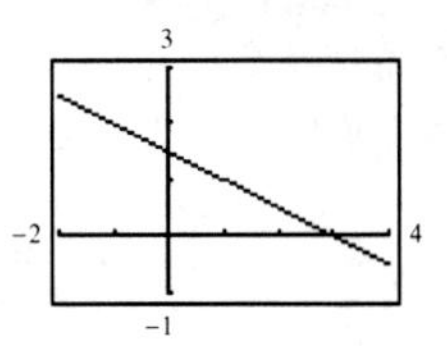

57. $\left(-\frac{1}{10}, -\frac{3}{5}\right), \left(\frac{9}{10}, -\frac{9}{5}\right)$

$$y + \frac{3}{5} = \frac{-\frac{9}{5} + \frac{3}{5}}{\frac{9}{10} + \frac{1}{10}}\left(x + \frac{1}{10}\right)$$
$$y + \frac{3}{5} = -\frac{6}{5}\left(x + \frac{1}{10}\right)$$
$$y = -\frac{6}{5}x - \frac{18}{25}$$

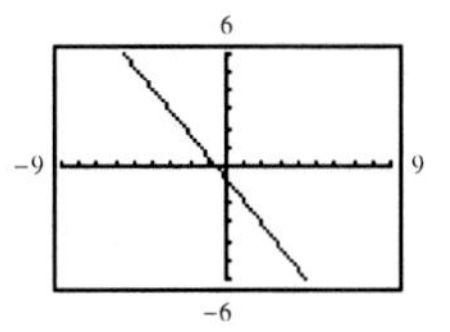

59. $(1, 0.6), (-2, -0.6)$

$$y - 0.6 = \frac{-0.6 - 0.6}{-2 - 1}(x - 1)$$
$$y = 0.4(x - 1) + 0.6$$
$$y = 0.4x + 0.2$$

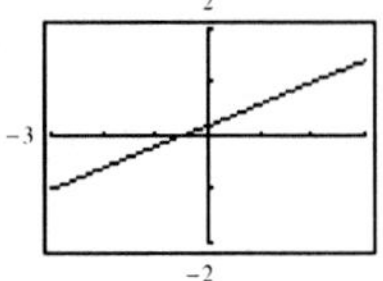

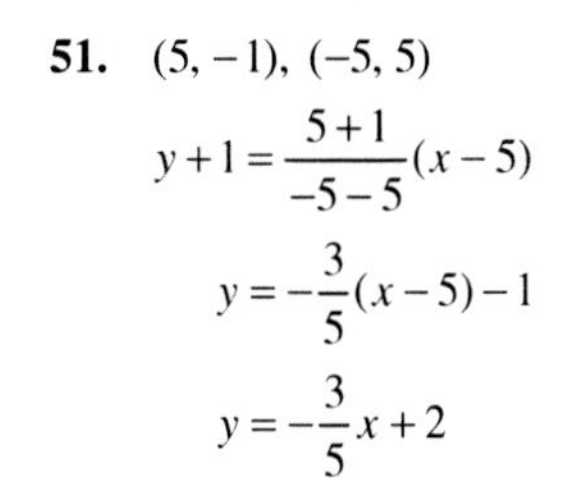

61.

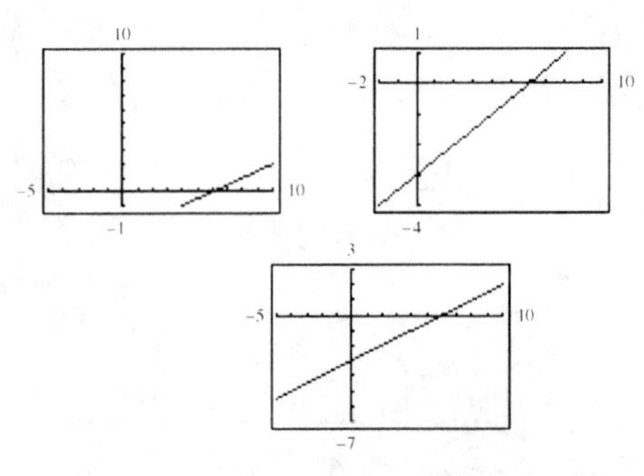

The first graph does not show both intercepts. The third graph is best because it shows both intercepts and gives the most accurate view of the slope by using a square setting.

63. L_1: $(0, -1), (5, 9)$

$$m_1 = \frac{9+1}{5-0} = 2$$

L_2: $(0, 3), (4, 1)$

$$m_2 = \frac{1-3}{4-0} = -\frac{1}{2} = -\frac{1}{m_1}$$

L_1 and L_2 are perpendicular.

65. L_1: $(3, 6), (-6, 0)$

$$m_1 = \frac{0-6}{-6-3} = \frac{2}{3}$$

L_2: $(0, -1), \left(5, \frac{7}{3}\right)$

$$m_2 = \frac{\frac{7}{3}+1}{5-0} = \frac{2}{3} = m_1$$

L_1 and L_2 are parallel.

67. $4x - 2y = 3$

$$y = 2x - \frac{3}{2}$$

Slope: $m = 2$

(a) Parallel slope: $m = 2$

$$y - 1 = 2(x - 2)$$
$$y = 2x - 3$$

(b) Perpendicular slope: $m = -\frac{1}{2}$

$$y - 1 = -\frac{1}{2}(x - 2)$$
$$y = -\frac{1}{2}x + 2$$

69. $3x + 4y = 7$

$$y = -\frac{3}{4}x + \frac{7}{4}$$

Slope: $m = -\frac{3}{4}$

(a) Parallel slope: $m = -\frac{3}{4}$

$$y - \frac{7}{8} = -\frac{3}{4}\left(x + \frac{2}{3}\right)$$
$$y = -\frac{3}{4}x + \frac{3}{8}$$

(b) Perpendicular slope: $m = \frac{4}{3}$

$$y - \frac{7}{8} = \frac{4}{3}\left(x + \frac{2}{3}\right)$$
$$y = \frac{4}{3}x + \frac{127}{72}$$

71. $6x + 2y = 9$

$$2y = -6x + 9$$
$$y = -3x + \frac{9}{2}$$

Slope: $m = -3$

(a) Parallel slope: $m = -3$

$$y + 1.4 = -3(x + 3.9)$$
$$y = -3x - 13.1$$

(b) Perpendicular slope: $m = \frac{1}{3}$

$$y + 1.4 = \frac{1}{3}(x + 3.9)$$
$$y = \frac{1}{3}x - \frac{1}{10}$$

73. $x - 4 = 0$ vertical line

Slope is undefined.

(a) $x - 3 = 0$ passes through $(3, -2)$ and is vertical.

(b) $y = -2$ passes through $(3, -2)$ and is horizontal.

75. $y + 2 = 0$

$y = -2$ horizontal line

Slope: $m = 0$

(a) $y = 1$ passes through $(-4, 1)$ and is horizontal.

(b) $x + 4 = 0$ passes through $(-4, 1)$ and is vertical.

77. The slope is 2 and $(-1, -1)$ line on the line. Hence,

$$y-(-1)=2(x-(-1))$$
$$y+1=2(x+1)$$
$$y=2x+1.$$

79. The slope of the given line is 2. Then y_2 has slope $-\frac{1}{2}$. Hence,

$$y-2=-\frac{1}{2}(x-(-2))$$
$$y-2=-\frac{1}{2}(x+2)$$
$$y=-\frac{1}{2}x+1.$$

81. The lines $y=\frac{1}{2}x$ and $y=-2x$ are perpendicular.

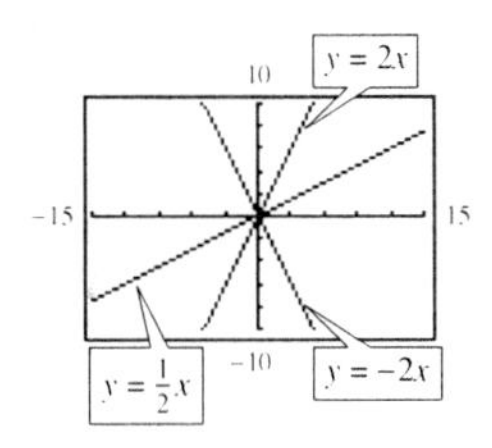

83. The lines $y=-\frac{1}{2}x$ and $y=-\frac{1}{2}x+3$ are parallel. Both are perpendicular to $y=2x-4$.

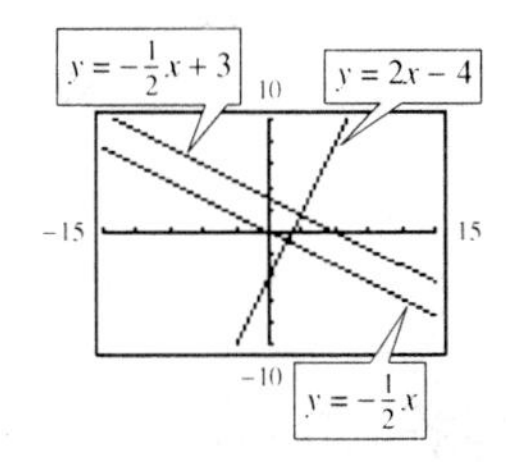

85. $\frac{\text{rise}}{\text{run}}=\frac{3}{4}=\frac{x}{\frac{1}{2}(32)}$

$$\frac{3}{4}=\frac{x}{16}$$
$$4x=48$$
$$x=12$$

The maximum height in the attic is 12 feet.

87.

(a)

Years	Slope
2000–2001	$1023-995=28$
2001–2002	$1247-1023=224$
2002–2003	$1211-1247=-36$
2003–2004	$1257-1211=46$
2004–2005	$1380-1257=123$
2005–2006	$1431-1380=51$
2006–2007	$1436-1431=5$
2007–2008	$1464-1436=28$

The greatest increase was \$224 million from 2001 to 2002.
The greatest decrease was \$36 million from 2002 to 2003.

(b) Using the points (0, 995) and (8, 1464), the slope is $m=\frac{1464-995}{8-0}=58.625$.

Then $y-995=58.625(x-0)$

$$y=58.625x+995$$

(c) There was an average increase in sales of about \$ 58.625 million per year from 2000 to 2008.

(d) When $x=10$: $y=58.625(10)+995$

$$y=\$1581.25 \text{ million}$$

Answers will vary.

89. $(9, 2540)$, $m=125$

$$V-2540=125(t-9)$$
$$V=125t+1415$$

91. $(9, 20{,}400)$, $m=-2000$

$$V-20{,}400=-2000(t-9)$$
$$V=-2000t+38{,}400$$

93. (a) $(0, 25{,}000)$, $(10, 2000)$

$$V-25{,}000=\frac{2000-25{,}000}{10-0}(t-0)$$
$$V-25{,}000=-2300t$$
$$V=-2300t+25{,}000$$

(b)

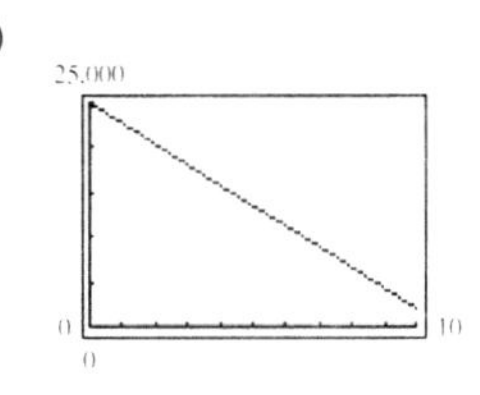

t	0	1	2	3	4	5
V	25,000	22,700	20,400	18,100	15,800	13,500

6	7	8	9	10
11,200	8900	6600	4300	2000

(c) $t = 0$: $V = -2300(0) + 25{,}000 = 25{,}000$
$t = 1$: $V = -2300(1) + 25{,}000 = 22{,}700$
etc.

95.

(a) $C = 36{,}500 + 9.25t + 18.50t$
$C = 36{,}500 + 27.75t$

(b) $R = tp$ (t hours at \$$p$ per hour)
$R = t(65)$
$R = 65t$

(c) $P = R - C$
$P = 65t - (36{,}500 + 27.75t)$
$P = 37.25t - 36{,}500$

(d) $P = 0$:
$37.25t - 36{,}500 = 0$
$37.25t = 36{,}500$
$t \approx 980$ hours

97.

(a) Using the points (1990, 75,365) and (2009, 87,163) the slope is

$$m = \frac{87{,}163 - 75{,}365}{2009 - 1990} = \frac{11{,}798}{19}.$$

The average annual increase in enrollment was about 621 students/year.

(b) 1995: $75{,}365 + 5(621) = 78{,}470$ students
2000: $75{,}365 + 10(621) = 81{,}575$ students
2005: $75{,}365 + 15(621) = 84{,}680$ students

(c) Using $m = 621$ and letting $x = 0$ correspond to 1990,

$y - 75{,}365 = 621(x - 0)$
$y = 621x + 75{,}365$

The slope is 621 and it determines the average increase in enrollment per year from 1990 to 2009.

99. False. The slopes are different:

$$\frac{4-2}{-1+8} = \frac{2}{7}$$
$$\frac{7+4}{-7-0} = -\frac{11}{7}$$

101. $\dfrac{x}{5} + \dfrac{y}{-3} = 1$

$-3x + 5y + 15 = 0$

a and b are the x- and y- intercepts.

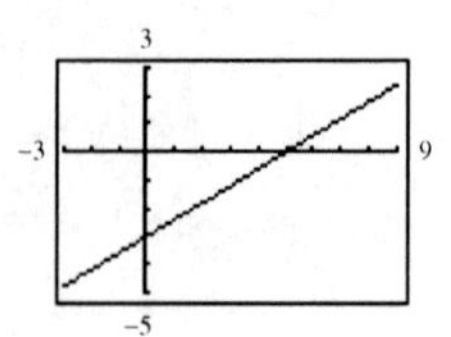

103. $\dfrac{x}{4} + \dfrac{y}{-\frac{2}{3}} = 1$

$-\dfrac{2}{3}x + 4y = \dfrac{-8}{3}$

$-2x + 12y = -8$

a and b are the x- and y-intercepts.

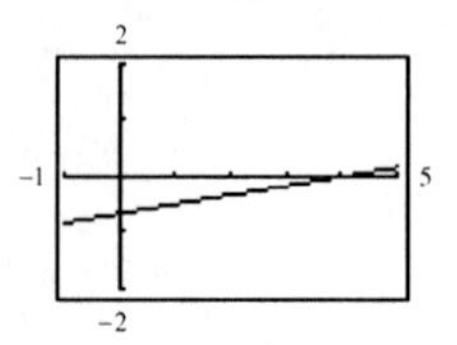

105. $\dfrac{x}{2} + \dfrac{y}{3} = 1$

$3x + 2y - 6 = 0$

107. $\dfrac{x}{-1/6} + \dfrac{y}{-2/3} = 1$

$-6x - \dfrac{3}{2}y = 1$

$12x + 3y + 2 = 0$

109. The slope is positive and the y-intercept is positive. Matches (a).

111. Both lines have positive slope, but their y-intercepts differ in sign. Matches (c).

113. No. The line $y = 2$ does not have and x-intercept.

115. Yes. Once a parallel line is established to the given line, there are an infinite number of distances away from that line, and thus an infinite number of parallel lines.

117. Yes. $x + 20$

119. No. The term $x^{-1} = \dfrac{1}{x}$ causes the expression to not be a polynomial.

121. No. This expression is not defined for $x = \pm 3$.

123. $x^2 - 6x - 27 = (x-9)(x+3)$

125. $2x^2 + 11x - 40 = (2x-5)(x+8)$

127. Answers will vary.

Section 1.2

1. domain, range, function

3. No. The input element $x = 3$ cannot be assigned to more than exactly one output element.

5. No. The domain of the function $f(x) = \sqrt{1+x}$ is $[-1, \infty)$ which does not include $x = -2$.

7. Yes, it does represent a function. Each domain value is matched with only one range value.

9. No, it does not represent a function. The domain values are each matched with three range values.

11. Yes, the relation represents y as a function of x. Each domain value is matched with only one range value.

13. (a) Each element of A is matched with exactly one element of B, so it does represent a function.
(b) The element 1 in A is matched with two elements, -2 and 1 of B, so it does not represent a function.
(c) Each element of A is matched with exactly one element of B, so it does represent a function.

15. Both are functions. For each year there is exactly one and only one average price of a name brand prescription and average price of a generic prescription.

17. $x^2 + y^2 = 4 \Rightarrow y = \pm\sqrt{4 - x^2}$

Thus, y *is not* a function of x. For instance, the values $y = 2$ and $y = -2$ both correspond to $x = 0$.

19. $y = \sqrt{x^2 - 1}$

This *is* a function of x.

21. $2x + 3y = 4 \Rightarrow y = \frac{1}{3}(4 - 2x)$

Thus, y *is* a function of x.

23. $y^2 = x^2 - 1 \Rightarrow y = \pm\sqrt{x^2 - 1}$

Thus, y *is not* a function of x. For instance, the values $y = \sqrt{3}$ and $y = -\sqrt{3}$ both correspond to $x = 2$.

25. $y = |4 - x|$

This *is* a functions of x.

27. $x = -7$ *does not* represent y as a function of x. All values of y correspond to $x = -7$.

29. $f(t) = 3t + 1$
(a) $f(2) = 3(2) + 1 = 7$
(b) $f(-4) = 3(-4) + 1 = -11$
(c) $f(t+2) = 3(t+2) + 1 = 3t + 7$

31. $h(t) = t^2 - 2t$
(a) $h(2) = 2^2 - 2(2) = 0$
(b) $h(1.5) = (1.5)^2 - 2(1.5) = -0.75$
(c) $h(x+2) = (x+2)^2 - 2(x+2) = x^2 + 2x$

33. $f(y) = 3 - \sqrt{y}$
(a) $f(4) = 3 - \sqrt{4} = 1$
(b) $f(0.25) = 3 - \sqrt{0.25} = 2.5$
(c) $f(4x^2) = 3 - \sqrt{4x^2} = 3 - 2|x|$

35. $q(x) = \dfrac{1}{x^2 - 9}$
(a) $q(-3) = \dfrac{1}{(-3)^2 - 9} = \dfrac{1}{9 - 9} = \dfrac{1}{0}$ undefined
(b) $q(2) = \dfrac{1}{(2)^2 - 9} = \dfrac{1}{4 - 9} = -\dfrac{1}{5}$
(c) $q(y+3) = \dfrac{1}{(y+3)^2 - 9} = \dfrac{1}{y^2 + 6y + 9 - 9} = \dfrac{1}{y^2 + 6y}$

37. $f(x) = \dfrac{|x|}{x}$
(a) $f(9) = \dfrac{|9|}{9} = 1$
(b) $f(-9) = \dfrac{|-9|}{-9} = -1$
(c) $f(t) = \dfrac{|t|}{t} = \begin{cases} 1, & t>0 \\ -1, & t<0 \end{cases}$

$f(0)$ is a undefined.

39. $f(x) = \begin{cases} 2x+1, & x < 0 \\ 2x+2, & x \ge 0 \end{cases}$
(a) $f(-1) = 2(-1) + 1 = -1$
(b) $f(0) = 2(0) + 2 = 2$
(c) $f(2) = 2(2) + 2 = 6$

41. $f(x) = \begin{cases} x^2+2, & x \le 1 \\ 2x^2+2, & x > 1 \end{cases}$
(a) $f(-2) = (-2)^2 + 2 = 6$
(b) $f(1) = (1)^2 + 2 = 3$
(c) $f(2) = 2(2)^2 + 2 = 10$

43. $f(x) = \begin{cases} x+2, & x < 0 \\ 4, & 0 \le x < 2 \\ x^2+1, & x \ge 2 \end{cases}$
(a) $f(-2) = (-2) + 2 = 0$
(b) $f(1) = 4$
(c) $f(4) = 4^2 + 1 = 17$

45. $f(x) = x^2$

$\{(-2, 4), (-1, 1), (0, 0), (1, 1), (2, 4)\}$

47. $f(x) = |x| + 2$

$\{(-2, 4), (-1, 3), (0, 2), (1, 3), (2, 4)\}$

49. $h(t) = \frac{1}{2}|t + 3|$

$h(-5) = \frac{1}{2}|-5 + 3| = \frac{1}{2}|-2| = \frac{1}{2}(2) = 1$

$h(-4) = \frac{1}{2}|-4 + 3| = \frac{1}{2}|-1| = \frac{1}{2}(1) = \frac{1}{2}$

$h(-3) = \frac{1}{2}|-3 + 3| = \frac{1}{2}|0| = 0$

$h(-2) = \frac{1}{2}|-2 + 3| = \frac{1}{2}|1| = \frac{1}{2}(1) = \frac{1}{2}$

$h(-1) = \frac{1}{2}|-1 + 3| = \frac{1}{2}|2| = \frac{1}{2}(2) = 1$

t	-5	-4	-3	-2	-1
$h(t)$	1	$\frac{1}{2}$	0	$\frac{1}{2}$	1

51. $f(x) = 15 - 3x = 0$

$3x = 15$

$x = 5$

53. $f(x) = \frac{3x - 4}{5} = 0$

$3x - 4 = 0$

$3x = 4$

$x = \frac{4}{3}$

55. $f(x) = 5x^2 + 2x - 1$

Since $f(x)$ is a polynomial, the domain is all real numbers x.

57. $h(t) = \frac{4}{t}$

Domain: All real numbers except $t = 0$

59. $f(x) = \sqrt[3]{x - 4}$

Domain: all real numbers x

61. $g(x) = \frac{1}{x} - \frac{3}{x + 2}$

Domain: All real numbers except $x = 0, x = -2$

63. $g(y) = \frac{y + 2}{\sqrt{y - 10}}$

$y - 10 > 0$

$y > 10$

Domain: all $y > 10$

65. $f(x) = \sqrt{4 - x^2}$

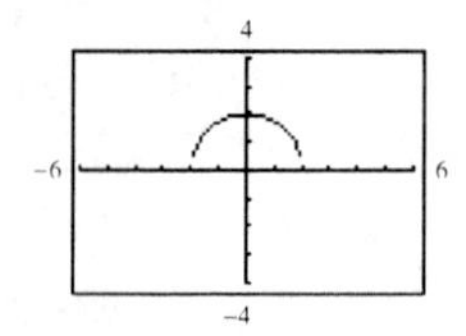

Domain: $[-2, 2]$

Range: $[0, 2]$

67. $g(x) = |2x + 3|$

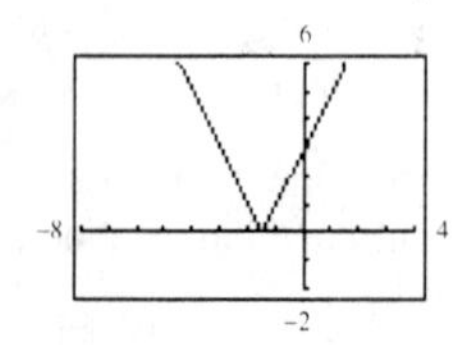

Domain: $(-\infty, \infty)$

Range: $[0, \infty)$

69. $A = \pi r^2, C = 2\pi r$

$r = \frac{C}{2\pi}$

$A = \pi\left(\frac{C}{2\pi}\right)^2 = \frac{C^2}{4\pi}$

71. (a) From the table, the maximum volume seems to be $1024\ \text{cm}^3$, corresponding to $x = 4$.

(b)

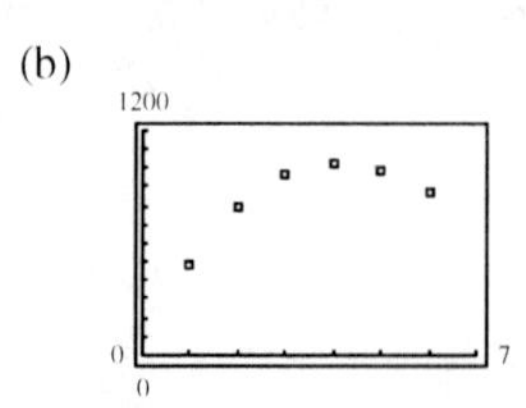

Yes, V is a function of x.

(c) $V = \text{length} \times \text{width} \times \text{height}$
$= (24 - 2x)(24 - 2x)x$
$= x(24 - 2x)^2 = 4x(12 - x)^2$

Domain: $0 < x < 12$

(d)

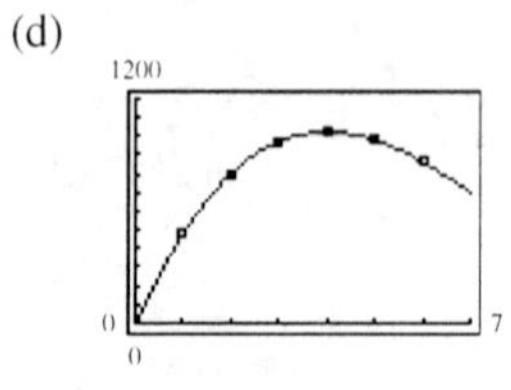

The function is a good fit. Answers will vary.

73. $A = l \cdot w = (2x)y = 2xy$

But $y = \sqrt{36 - x^2}$, so $A = 2x\sqrt{36 - x^2}$, $0 < x < 6$.

75. (a) Total Cost = Variable Costs + Fixed Costs

$C = 68.20x + 248{,}000$

(b) Revenue = Selling price × units sold

$R = 98.98x$

(c) Since $P = R - C$

$P = 98.98x - (68.20x + 248{,}000)$

$P = 30.78x - 248{,}000.$

77. (a) The independent variable is t and represents the year. The dependent variable is n and represents the numbers of miles traveled.

(b)

t	0	1	2	3	4	5	6
$n(t)$	581	645.26	699.04	742.34	775.16	797.5	809.36

t	7	8	9	10	11	12	13
$n(t)$	843.9	869.6	895.3	921	946.7	972.4	998.1

t	14	15	16	17
$n(t)$	1023.8	1049.5	1075.2	1100.9

(c) The model fits the data well.

79. No. If $x = 60$, $y = -0.004(60)^2 + 0.3(60) + 6$

$y = 9.6$ feet

Since the first baseman can only jump to a height of 8 feet, the throw will go over his head.

81. $f(x) = 2x$

$$\frac{f(x+c) - f(x)}{c} = \frac{2(x+c) - 2x}{c}$$

$$= \frac{2c}{c} = 2,\ c \neq 0$$

83. $f(x) = x^2 - x + 1,\ f(2) = 3$

$$\frac{f(2+h) - f(2)}{h} = \frac{(2+h)^2 - (2+h) + 1 - 3}{h}$$

$$= \frac{4 + 4h + h^2 - 2 - h + 1 - 3}{h}$$

$$= \frac{h^2 + 3h}{h} = h + 3,\ h \neq 0$$

85. $f(t) = \frac{1}{t},\ f(1) = 1$

$$\frac{f(t) - f(1)}{t - 1} = \frac{\frac{1}{t} - 1}{t - 1} = \frac{1 - t}{t(t-1)} = -\frac{1}{t},\ t \neq 1$$

87. False. The range of $f(x)$ is $(-1, \infty)$.

89. $f(x) = \sqrt{x} + 2$

Domain: $[0, \infty)$ or $x \geq 0$

Range: $[2, \infty)$ or $y \geq 2$

91. No. f is not the independent variable. Because the value of f depends on the value of x, x is the independent variable and f is the dependent variable.

93. $12 - \frac{4}{x+2} = \frac{12(x+2) - 4}{x+2} = \frac{12x + 20}{x+2}$

95. $$\frac{2x^3 + 11x^2 - 6x}{5x} \cdot \frac{x+10}{2x^2 + 5x - 3} = \frac{x(2x^2 + 11x - 6)(x+10)}{5x(2x-1)(x+3)}$$

$$= \frac{(2x-1)(x+6)(x+10)}{5(2x-1)(x+3)}$$

$$= \frac{(x+6)(x+10)}{5(x+3)},\ x \neq 0,\ \frac{1}{2}$$

Section 1.3

1. decreasing

3. Domain: $1 \leq x \leq 4$ or $[1, 4]$

5. If $f(2) \geq f(2)$ for all x in $(0, 3)$, then $(2, f(2))$ is a relative maximum of f.

7. Domain: all real numbers, $(-\infty, \infty)$

Range: $(-\infty, 1]$

$f(0) = 1$

9. Domain: $[-4, 4]$

Range: $[0, 4]$

$f(0) = 4$

11. 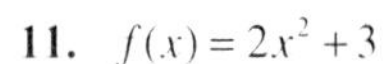

$f(x) = 2x^2 + 3$

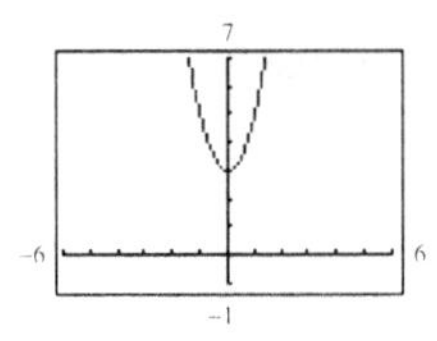

Domain: $(-\infty, \infty)$ Range: $[3, \infty)$

13. 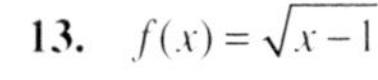

$f(x) = \sqrt{x - 1}$

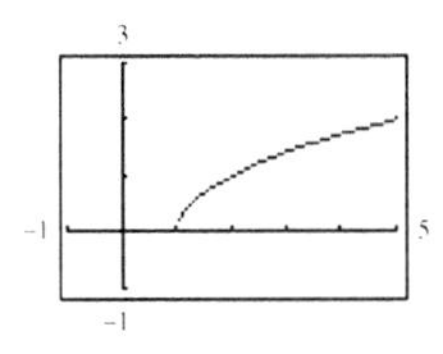

Domain: $x - 1 \geq 0 \Rightarrow x \geq 1$ or $[1, \infty)$

Range: $[0, \infty)$

15. $f(x) = |x + 3|$

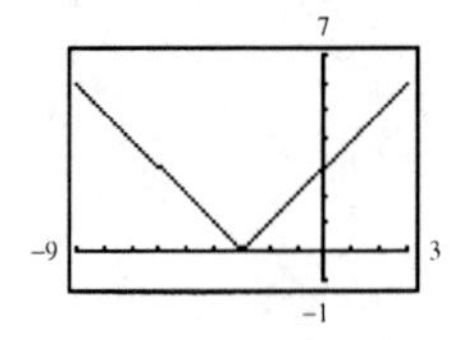

Domain: $(-\infty, \infty)$

Range: $[0, \infty)$

17. (a) Domain: $(-\infty, \infty)$

(b) Range: $[-2, \infty)$

(c) $f(x) = 0$ at $x = -1$ and $x = 3$.

(d) The values of $x = -1$ and $x = 3$ are the x-intercepts of the graph of f.

(e) $f(0) = -1$

(f) The value of $y = -1$ is the y-intercept of the graph of f.

(g) The value of f at $x = 1$ is $f(1) = -2$.

The coordinates of the point are $(1, -2)$.

(h) The value of f at $x = -1$ is $f(-1) = 0$.

The coordinates of the point are $(-1, 0)$.

(i) The coordinates of the point are $(-3, f(-3))$ or $(-3, 2)$.

19. $y = \frac{1}{2}x^2$

A vertical line intersects the graph just once, so y is a function of x. Graph $y_1 = \frac{1}{2}x^2$.

21. $x^2 + y^2 = 25$

A vertical line intersects the graph more than once, so y is not a function of x. Graph the circle as $y_1 = \sqrt{25 - x^2}$ and $y_2 = -\sqrt{25 - x^2}$.

23. $f(x) = \frac{3}{2}x$

f is increasing on $(-\infty, \infty)$.

25. $f(x) = x^3 - 3x^2 + 2$

f is increasing on $(-\infty, 0)$ and $(2, \infty)$.

f is decreasing on $(0, 2)$.

27. $f(x) = 3$

(a)

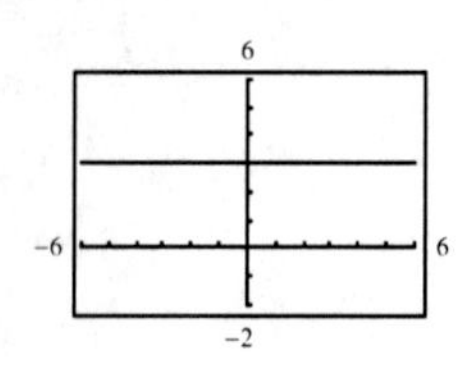

(b) f is constant on $(-\infty, \infty)$.

29. $f(x) = x^{2/3}$

(a)

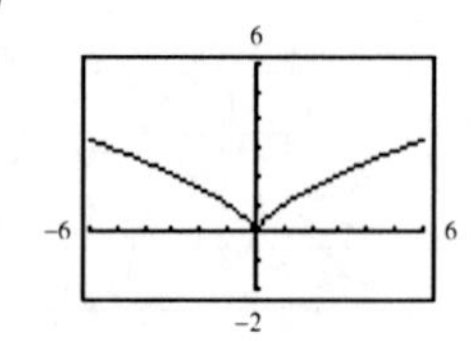

(b) Increasing on $(0, \infty)$

Decreasing on $(-\infty, 0)$

31. $f(x) = x\sqrt{x + 3}$

(a)

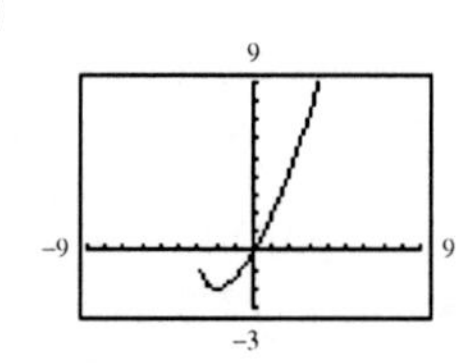

(b) Increasing on $(-2, \infty)$

Decreasing on $(-3, -2)$

33. $f(x) = |x + 1| + |x - 1|$

(a)

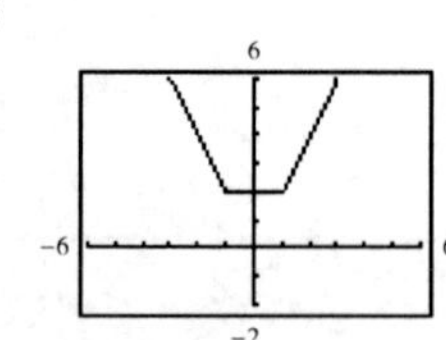

(b) Increasing on $(1, \infty)$, constant on $(-1, 1)$, decreasing on $(-\infty, -1)$

35. $f(x) = x^2 - 6x$

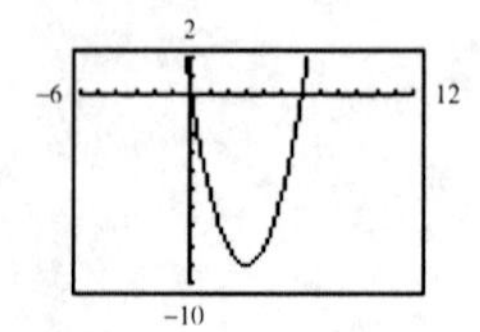

Relative minimum: $(3, -9)$

37. $y = 2x^3 + 3x^2 - 12x$

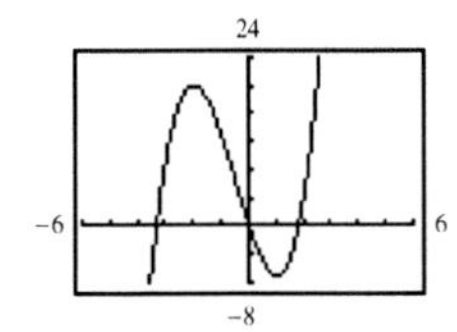

Relative minimum: $(1, -7)$

Relative maximum: $(-2, 20)$

39. $h(x) = (x-1)\sqrt{x}$

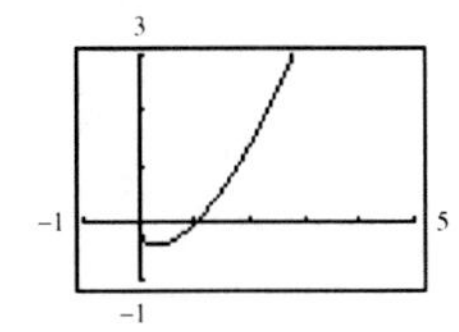

Relative minimum: $(0.33, -0.38)$

$(0, 0)$ is not a relative maximum because it occurs at the endpoint of the domain $[0, \infty)$.

41. $f(x) = x^2 - 4x - 5$

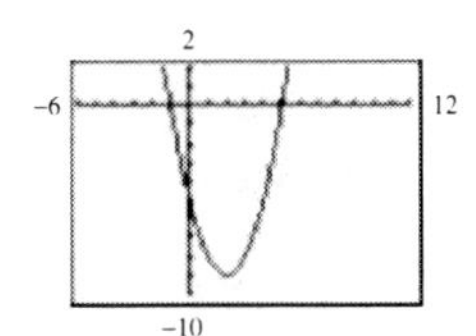

Relative minimum: $(2, -9)$

43. $f(x) = x^3 - 3x$

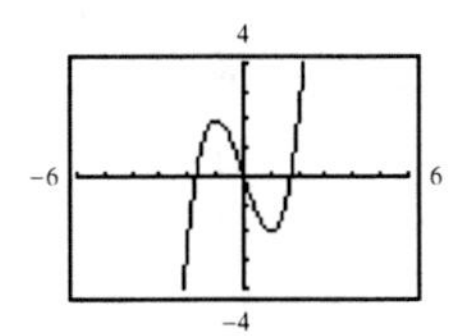

Relative minimum: $(1, -2)$

Relative maximum: $(-1, 2)$

45. $f(x) = 3x^2 - 6x + 1$

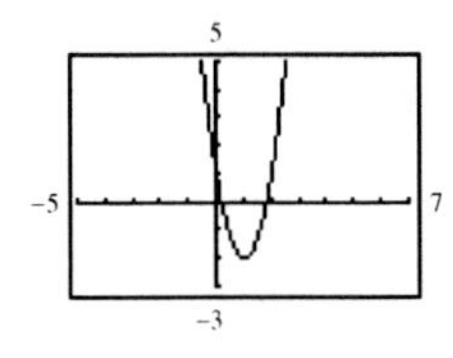

Relative minimum: $(1, -2)$

47. $f(x) = [\![x]\!] + 2$

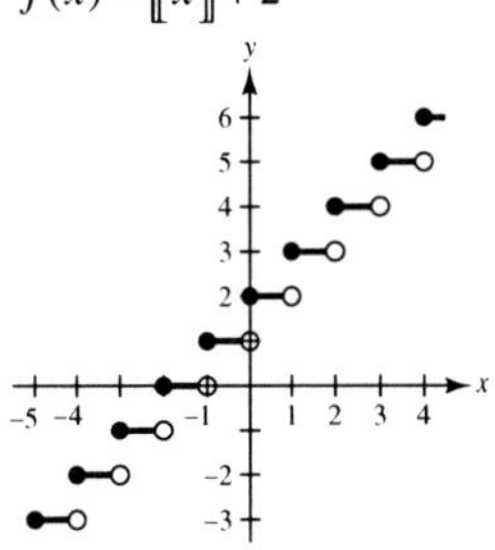

49. $f(x) = [\![x-1]\!] + 2$

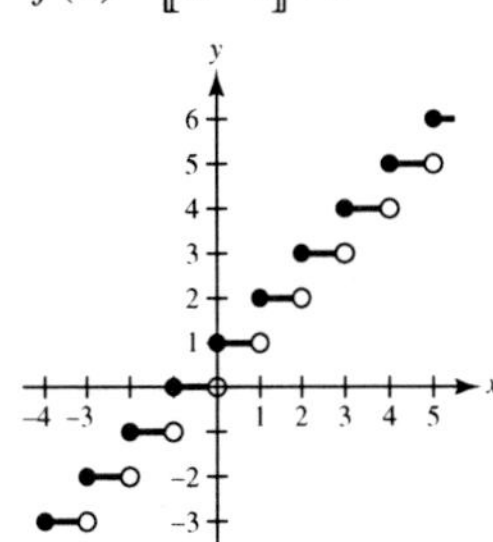

51. $f(x) = [\![2x]\!]$

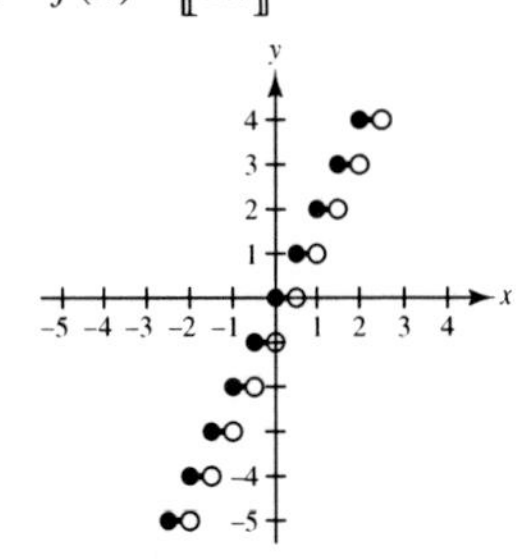

53. $s(x) = 2\left(\frac{1}{4}x - \left[\!\left[\frac{1}{4}x\right]\!\right]\right)$

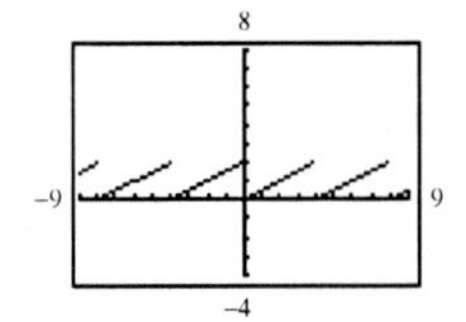

Domain: $(-\infty, \infty)$

Range: $[0, 2)$

Sawtooth pattern

55. $f(x) = \begin{cases} 2x+3, & x < 0 \\ 3-x, & x \geq 0 \end{cases}$

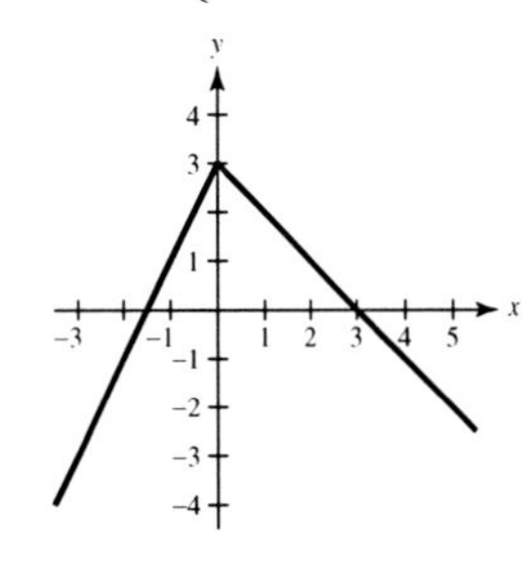

57. $f(x) = \begin{cases} \sqrt{x+4}, & x < 0 \\ \sqrt{4-x}, & x \geq 0 \end{cases}$

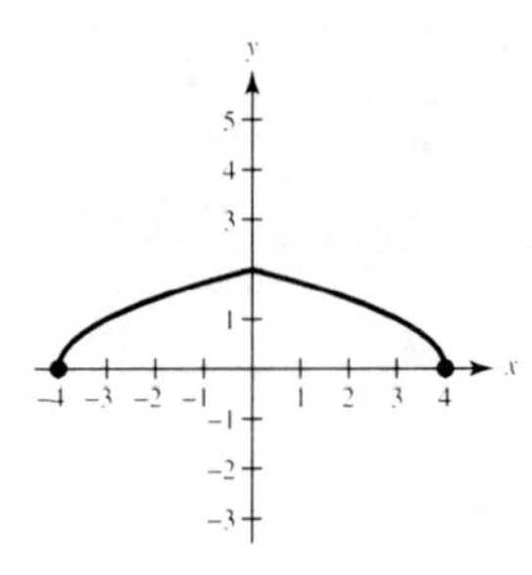

59. $f(x) = \begin{cases} x+3, & x \leq 0 \\ 3, & 0 < x \leq 2 \\ 2x-1, & x > 2 \end{cases}$

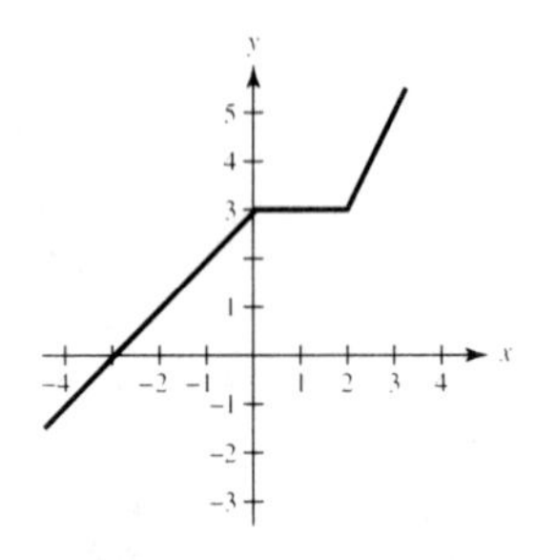

61. $f(x) = \begin{cases} 2x+1, & x \leq -1 \\ x^2-2, & x > -1 \end{cases}$

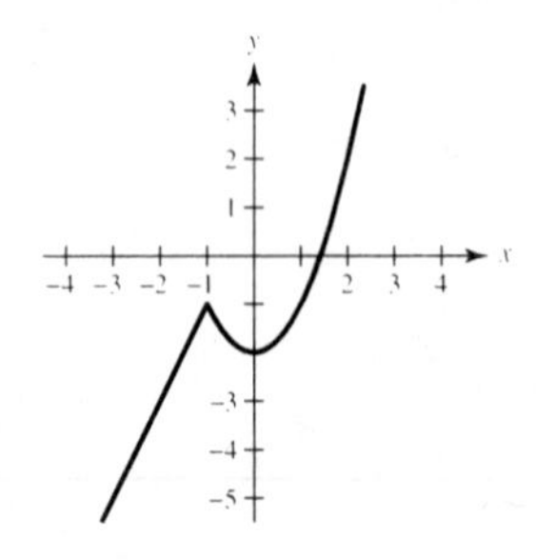

63. $f(x) = 5$ is even.

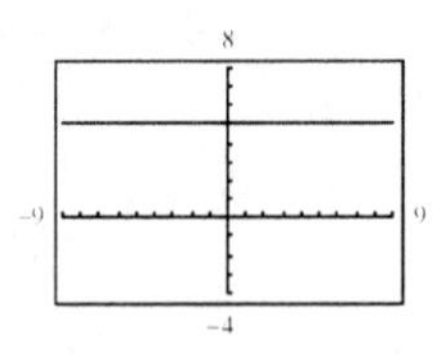

65. $f(x) = 3x - 2$ is neither even nor odd.

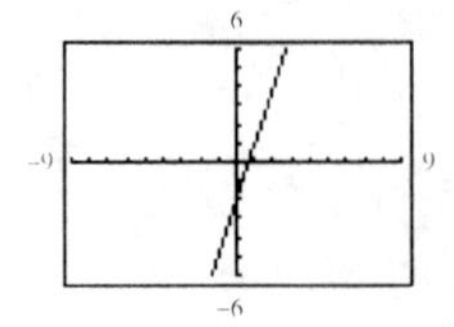

67. $h(x) = x^2 - 4$ is even.

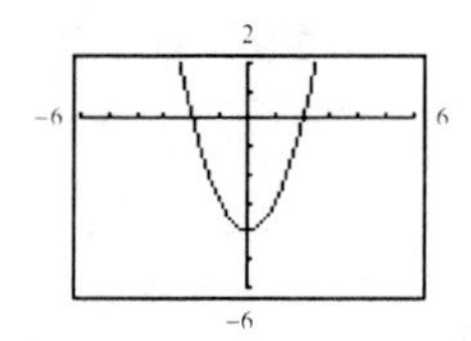

69. $f(x) = \sqrt{1-x}$ is neither even nor odd.

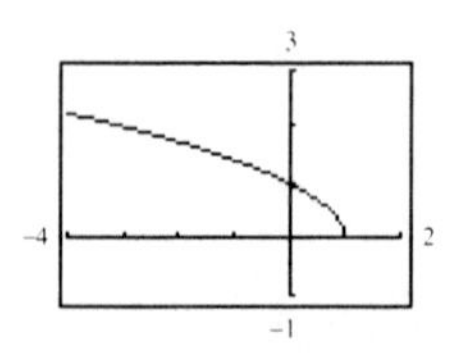

71. $f(x) = |x+2|$ is neither even nor odd.

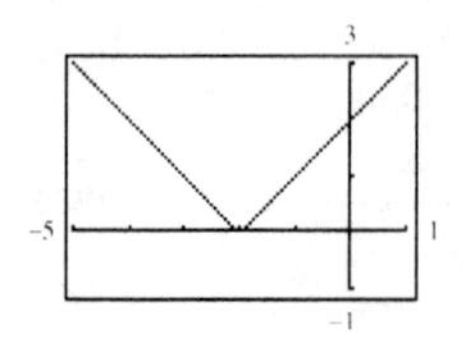

73. $\left(-\frac{3}{2}, 4\right)$

(a) If f is even, another point is $\left(\frac{3}{2}, 4\right)$.

(b) If f is odd, another point is $\left(\frac{3}{2}, -4\right)$.

75. $(4, 9)$

(a) If f is even, another point is $(-4, 9)$.

(b) If f is odd, another point is $(-4, -9)$.

77. $(x, -y)$

(a) If f is even, another point is $(-x, -y)$.

(b) If f is odd, another point is $(-x, y)$.

79. (a) $f(-t) = (-t)^2 + 2(-t) - 3$

$= t^2 - 2t - 3$

$\neq f(t) \neq -f(t)$

f is neither even nor odd.

(b)

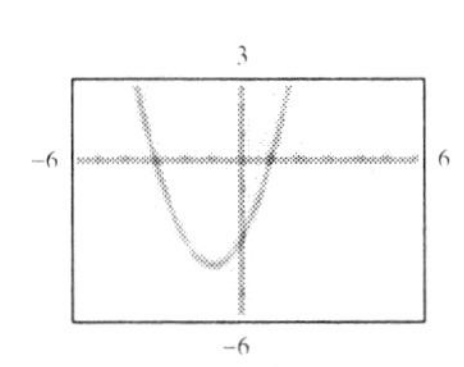

The graph is neither symmetric with respect to the origin nor with respect to the y-axis. So, f is neither even nor odd.

(c) Tables will vary. f is neither even nor odd.

81. (a) $g(-x) = (-x)^3 - 5(-x)$

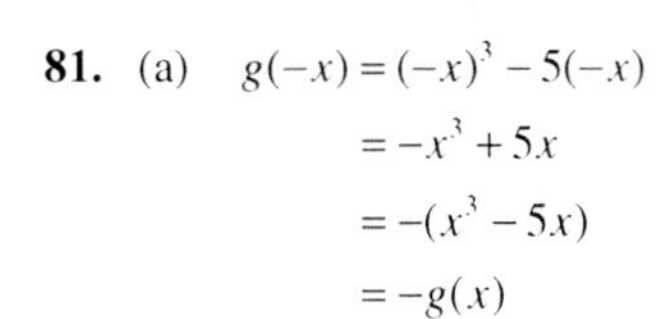

$= -x^3 + 5x$

$= -(x^3 - 5x)$

$= -g(x)$

g is odd.

(b)

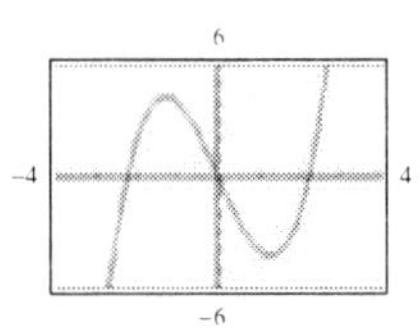

The graph is symmetric with respect to the origin. So, g is odd.

(c) Tables will vary. g is odd.

83. (a) $f(-x) = (-x)\sqrt{1-(-x)^2}$

$= -x\sqrt{1-x^2}$

$= -f(x)$

f is odd.

(b)

The graph is symmetric with respect to the orign. So, f is odd.

(c) Tables will vary. f is odd.

85. (a) $g(-s) = 4(-s)^{\frac{2}{3}}$

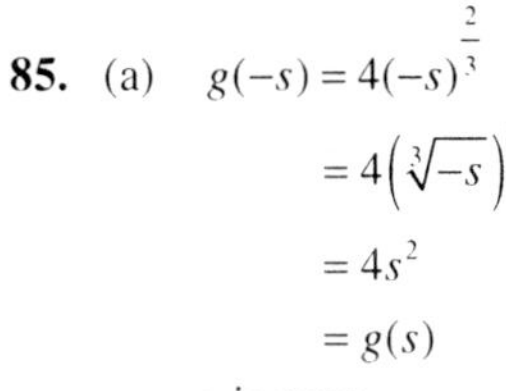

$= 4\left(\sqrt[3]{-s}\right)^2$

$= 4s^2$

$= g(s)$

g is even.

(b)

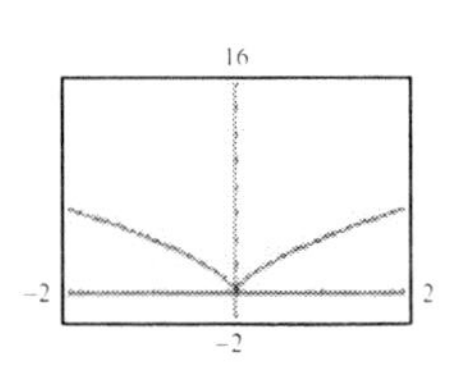

The graph is symmetric with respect to the y-axis. So, g is even.

(c) Tables will vary. g is even.

87. $f(x) = 4 - x \geq 0$

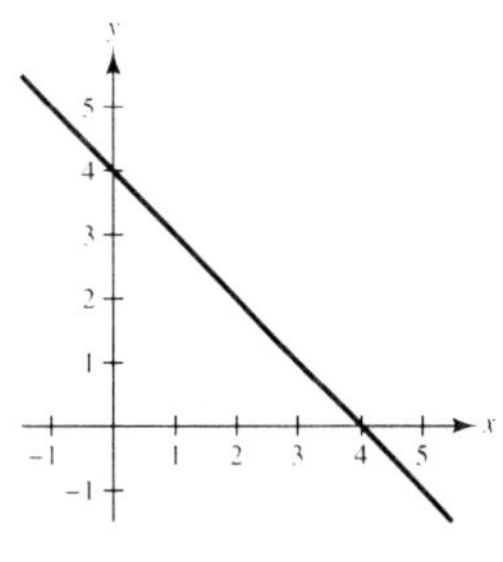

$f(x) \geq 0$

$4 - x \geq 0$

$4 \geq x$

$(-\infty, 4]$

89. $f(x) = x^2 - 9 \geq 0$

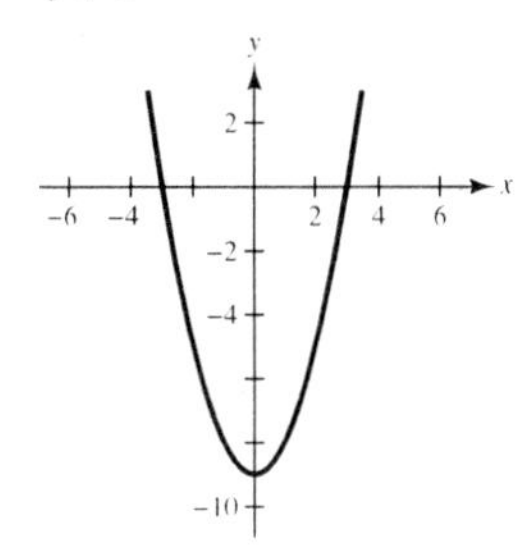

$f(x) \geq 0$

$x^2 - 9 \geq 0$

$x^2 \geq 9$

$x \geq 3$ or $x \leq -3$

$(-\infty, -3], [3, \infty)$

91. (a) C_2 is the appropriate model.

The cost of the first minute is \$1.05 and the cost increases \$0.08 when the next minute begins, and so on.

(b)

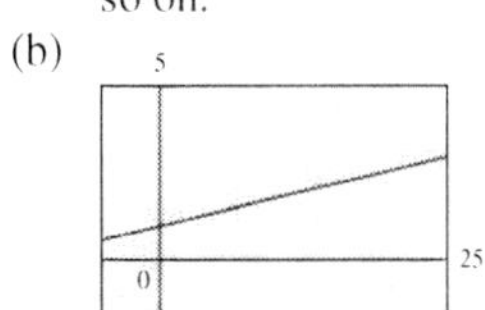

The cost of an 18-minute, 45-second call is

$C_2(18.75) = 1.05 - 0.08[\![-(18.75 - 1)]\!]$

$= \$2.49.$

93. $h = \text{top} - \text{bottom}$

$= (-x^2 + 4x - 1) - 2$

$= -x^2 + 4x - 3,\ 1 \leq x \leq 3$

95. (a)

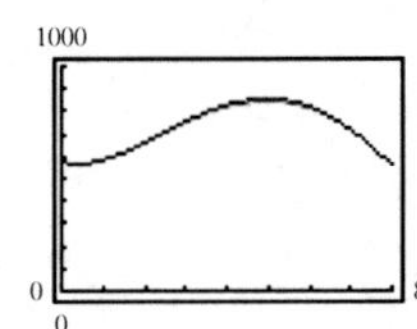

(b) The number of cooperative homes and condos was increasing from 2000 to 2005, and decreasing from 2005 to 2008.

(c) The maximum number of cooperative homes and condos was approximately 855 in 2005.

97. False. The domain of $f(x) = \sqrt{x^2}$ is the set of all real numbers.

99. c **101.** b **103.** a

105. Yes, $x = y^2 + 1$ defines x as a function of y. Any horizontal line can be drawn without intersecting the graph more than once.

107. Yes,
$$f(x) = \begin{cases} \vdots & \\ -1, & -1 \le x < 0 \\ 0, & 0 \le x < 1 \\ 1, & 1 \le x < 2 \\ \vdots & \end{cases}$$

109. f is an even function.

(a) $g(x) = -f(x)$ is even because $g(-x) = -f(-x) = -f(x) = g(x)$.

(b) $g(x) = f(-x)$ is even because $g(-x) = f(-(-x)) = f(x) = f(-x) = g(x)$.

(c) $g(x) = f(x) - 2$ is even because $g(-x) = f(-x) - 2 = f(x) - 2 = g(x)$.

(d) $g(x) = -f(x-2)$ is neither even nor odd because $g(-x) = -f(-x-2) = -f(x+2) \neq g(x)$ nor $-g(x)$.

111. $f(x) = a_{2n+1}x^{2n+1} + a_{2n-1}x^{2n-1} + \cdots + a_3x^3 + a_1x$

$$f(-x) = a_{2n+1}(-x)^{2n+1} + a_{2n-1}(-x)^{2n-1} + \cdots + a_3(-x)^3 + a_1(-x)$$
$$= -a_{2n+1}x^{2n+1} - a_{2n-1}x^{2n-1} - \cdots - a_3x^3 - a_1x = -f(x)$$

Therefore, $f(x)$ is odd.

113. $-2x^2 + 8x$

Terms: $-2x^2, 8x$

Coefficients: $-2, 8$

115. $\frac{x}{3} - 5x^2 + x^3$

Terms: $\frac{x}{3}, -5x^2, x^3$

Coefficients: $\frac{1}{3}, -5, 1$

117. $f(x) = -x^2 - x + 3$

(a) $f(4) = -(4)^2 - 4 + 3 = -17$

(b) $f(-2) = -(-2)^2 - (-2) + 3 = 1$

(c)
$$f(x-2) = -(x-2)^2 - (x-2) + 3$$
$$= -(x^2 - 4x + 4) - x + 2 + 3$$
$$= -x^2 + 3x + 1$$

119. $f(x) = x^2 - 2x + 9$

$$f(3+h) = (3+h)^2 - 2(3+h) + 9 = 9 + 6h + h^2 - 6 - 2h + 9$$
$$= h^2 + 4h + 12$$

$$f(3) = 3^2 - 2(3) + 9 = 12$$

$$\frac{f(3+h) - f(3)}{h} = \frac{(h^2 + 4h + 12) - 12}{h} = \frac{h(h+4)}{h} = h + 4, h \neq 0$$

Section 1.4

1. Horizontal shifts, vertical shifts, and reflections are rigid transformations.

3. $-f(x), f(-x)$

5.

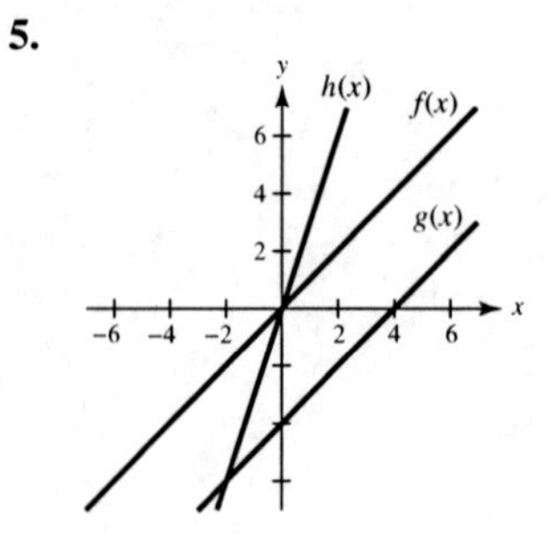

7.

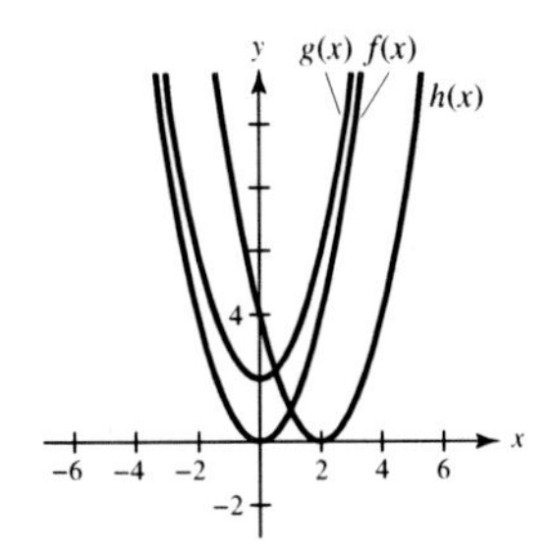

9.

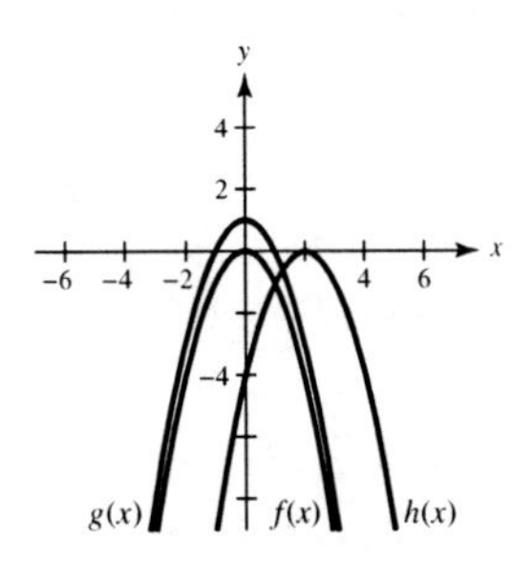

11.

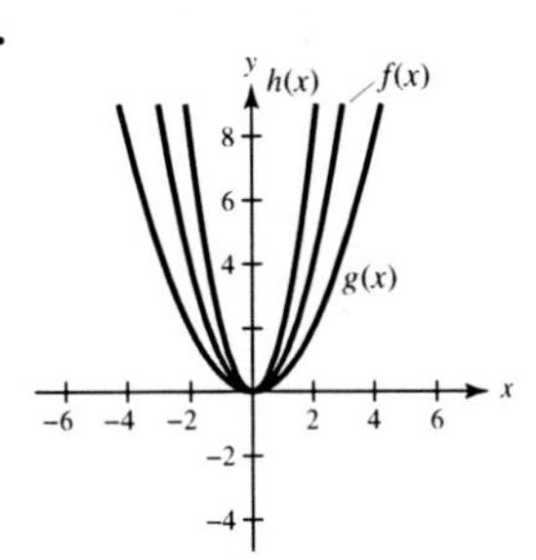

13.

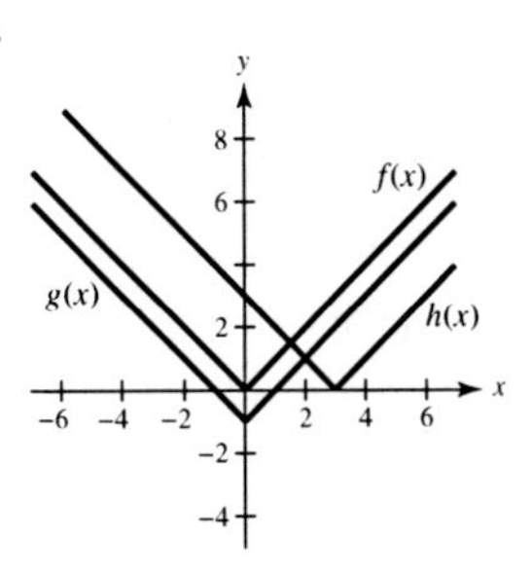

15.

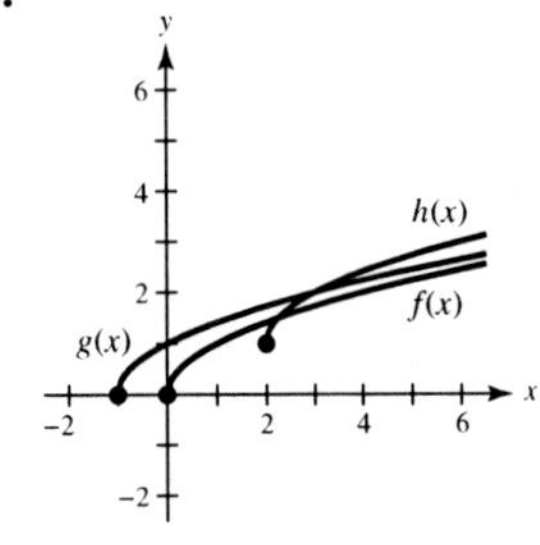

17.

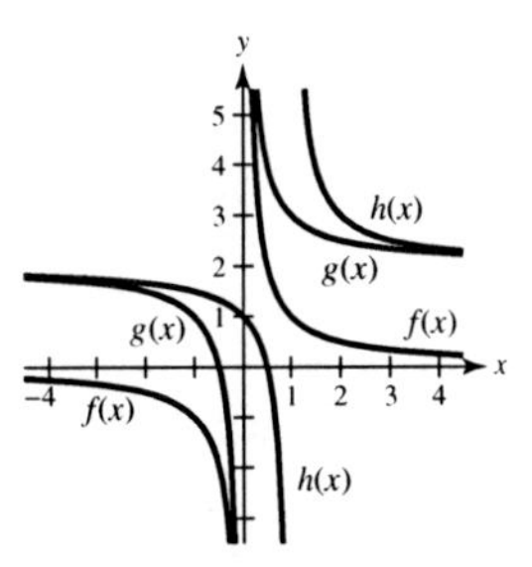

19. (a) $y = f(x) + 2$

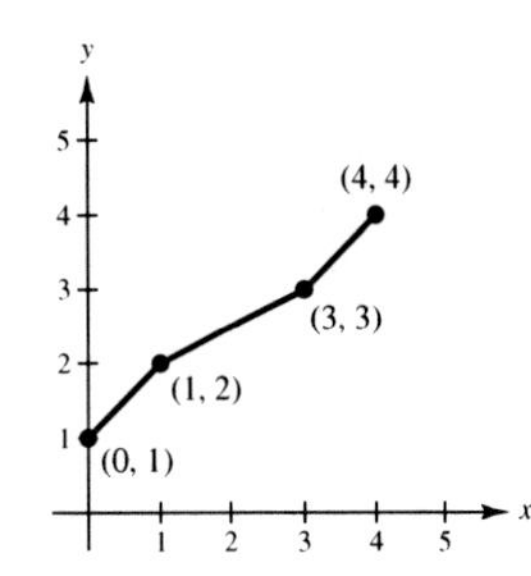

(b) $y = -f(x)$

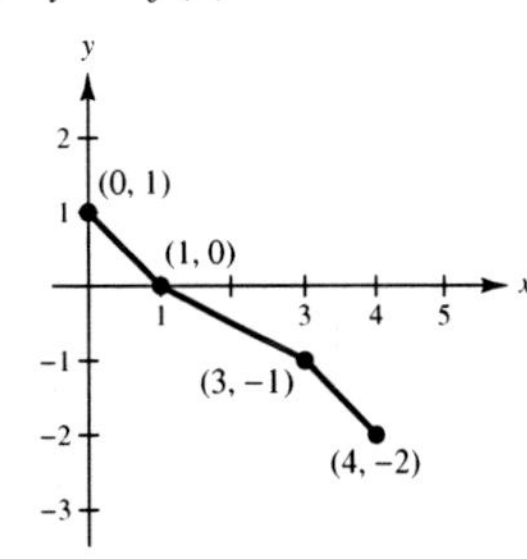

(c) $y = f(x - 2)$

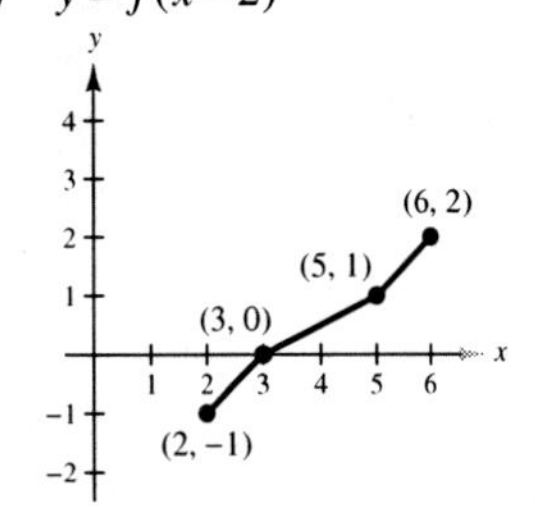

(d) $y = f(x + 3)$

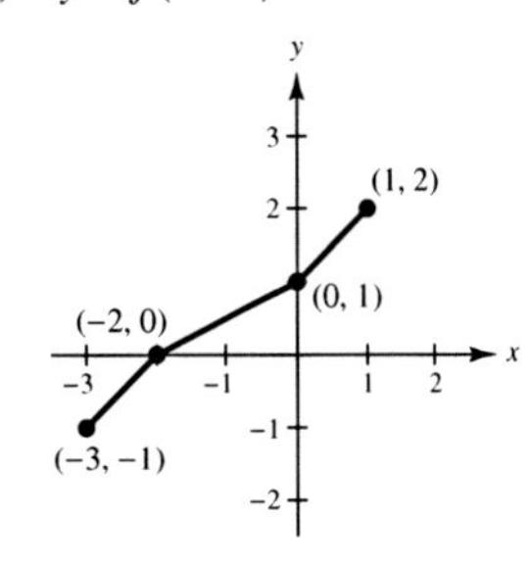

(e) $y = 2f(x)$

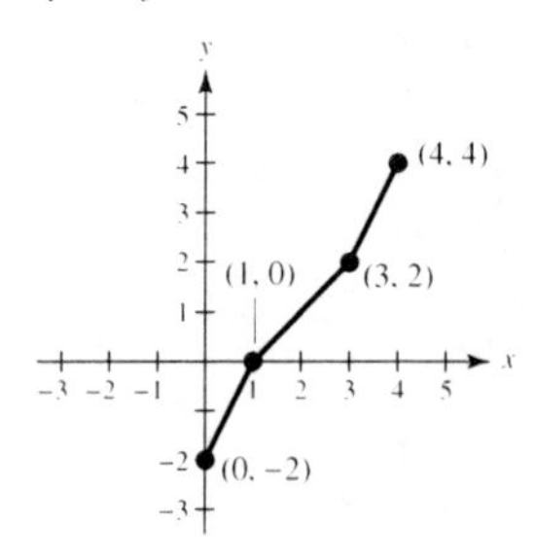

(f) $y = f(-x)$

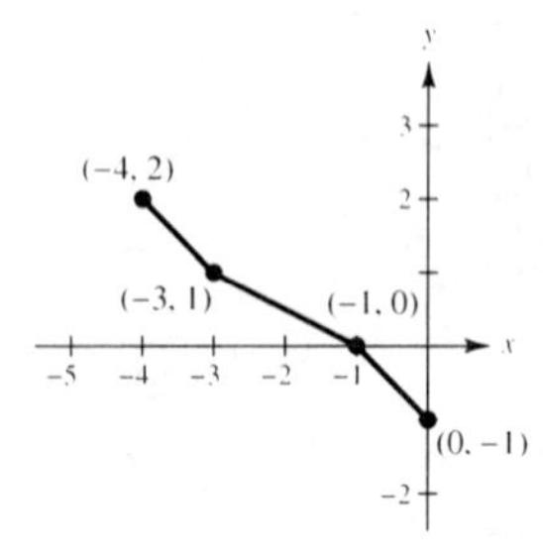

(g) Let $g(x) = f\left(\frac{1}{2}x\right)$. Then from the graph,

$$g(0) = f\left(\tfrac{1}{2}(0)\right) = f(0) = -1$$

$$g(2) = f\left(\tfrac{1}{2}(2)\right) = f(1) = 0$$

$$g(6) = f\left(\tfrac{1}{2}(6)\right) = f(3) = 1$$

$$g(8) = f\left(\tfrac{1}{2}(8)\right) = f(4) = 2$$

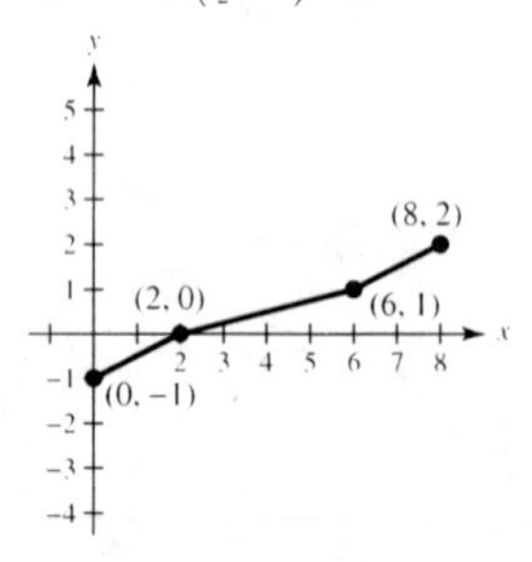

21. The graph of $f(x) = x^2$ should have been shifted one unit to the left instead of one unit to the right.

23. $y = \sqrt{x} + 2$ is $f(x) = \sqrt{x}$ shifted vertically upward two units.

25. $y = (x-4)^3$ is $f(x) = x^3$ shifted horizontally four units to the right.

27. $y = x^2 - 2$ is $f(x) = x^2$ shifted vertically two units downward.

29. Horizontal shift three units to left of $y = x$: $y = x + 3$ (or vertical shift three units upward)

31. Vertical shift one unit downward of $y = x^2$: $y = x^2 - 1$

33. Reflection in the x-axis and a vertical shift one unit upward of $y = \sqrt{x}$: $y = 1 - \sqrt{x}$

35. $y = -|x|$ is $f(x)$ reflected in the x-axis.

37. $y = (-x)^2$ is a reflection in the y-axis. In fact, $y = (-x)^2 = x^2$, therefore $y = (-x)^2$ is identical to $y = x^2$.

39. $y = \dfrac{1}{-x}$ is a reflection of $f(x) = \dfrac{1}{x}$ in the y-axis.

However, since $y = \dfrac{1}{-x} = -\dfrac{1}{x}$, either a reflection in the y-axis or a reflection in the x-axis produces the same graph.

41. $y = 4|x|$ is a vertical stretch of $f(x) = |x|$.

43. $g(x) = \dfrac{1}{4}x^3$ is a vertical shrink of $f(x) = x^3$.

45. $f(x) = \sqrt{4x}$ is a horizontal shrink of $f(x) = \sqrt{x}$. However, since $f(x) = \sqrt{4x} = 2\sqrt{x}$, it also can be described as a vertical stretch of $f(x) = \sqrt{x}$.

47. $f(x) = x^3 - 3x^2$

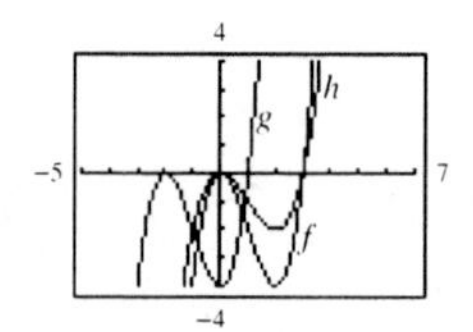

$g(x) = f(x+2) = (x+2)^3 - 3(x+2)^2$ is a horizontal shift two units to the left.

$h(x) = \dfrac{1}{2}f(x) = \dfrac{1}{2}(x^3 - 3x^2)$ is a vertical shrink.

49. $f(x) = x^3 - 3x^2$

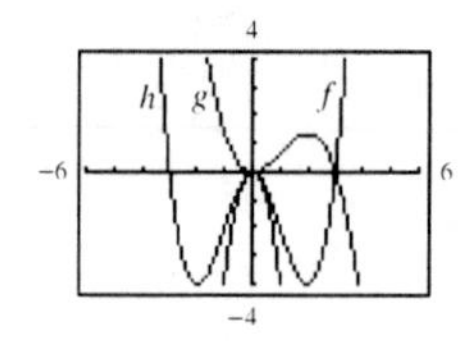

$g(x) = -\dfrac{1}{3}f(x) = -\dfrac{1}{3}(x^3 - 3x^2)$ is a reflection in the x-axis and a vertical shrink.

$h(x) = f(-x) = (-x)^3 - 3(-x)^2$ is a reflection in the y-axis.

51. (a) $f(x) = x^2$

(b) $g(x) = 2 - (x + 5)^2$ is obtained from f by a horizontal shift to the left five units, a reflection in the x-axis, and a vertical shift upward two units.

(c)

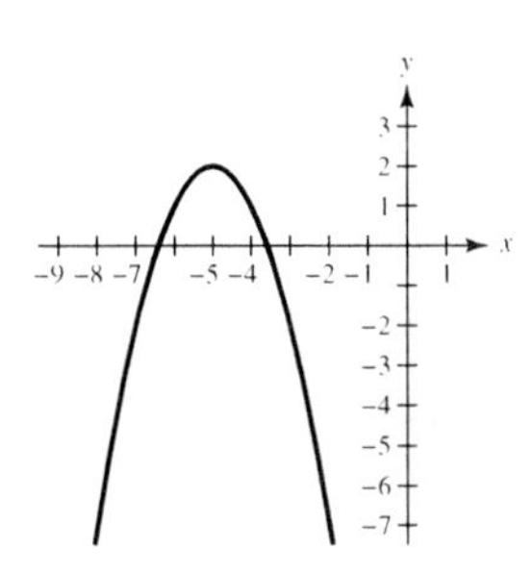

(d) $g(x) = 2 - f(x + 5)$

53. (a) $f(x) = x^2$

(b) $g(x) = 3 + 2(x - 4)^2$ is obtained from f by a horizontal shift four units to the right, a vertical stretch of 2, and a vertical shift upward three units.

(c)

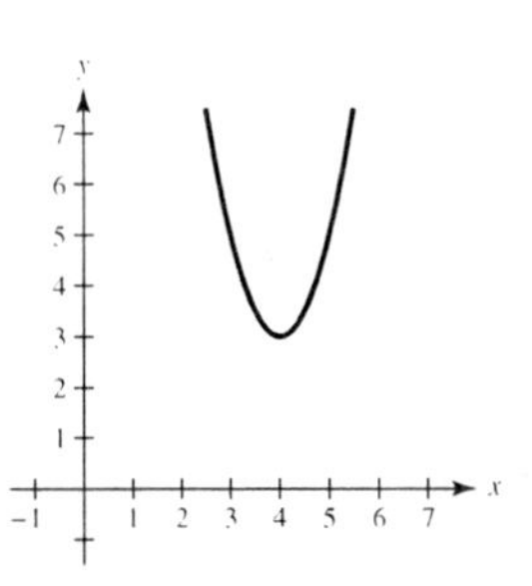

(d) $g(x) = 3 + 2f(x - 4)$

55. (a) $f(x) = x^3$

(b) $g(x) = 3(x - 2)^3$ is obtained from f by a horizontal shift two units to the right followed by a vertical stretch of 3.

(c)

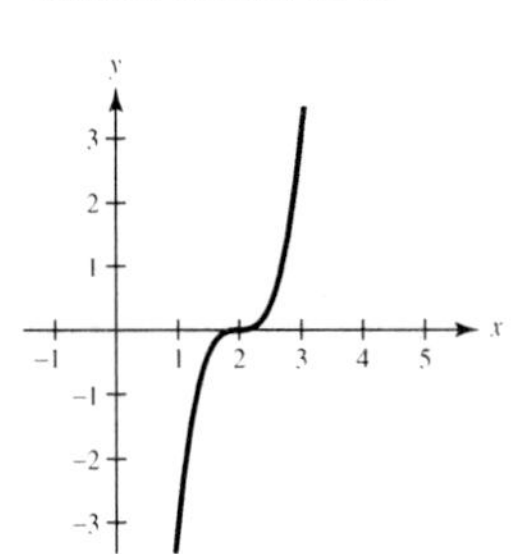

(d) $g(x) = 3f(x - 2)$

57. (a) $f(x) = x^3$

(b) $g(x) = (x - 1)^3 + 2$ is obtained from f by a horizontal shift one unit to the right and a vertical shift upward two units.

(c)

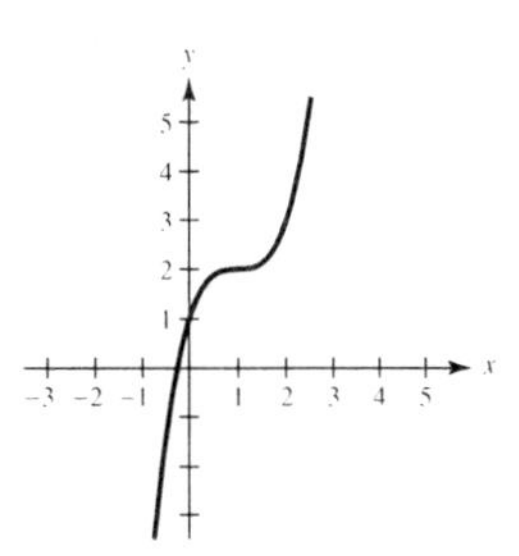

(d) $g(x) = f(x - 1) + 2$

59. (a) $f(x) = \dfrac{1}{x}$

(b) $g(x) = \dfrac{1}{x + 8} - 9$ is obtained from f by a horizontal shift eight units to the left and a vertical shift nine units downward.

(c)

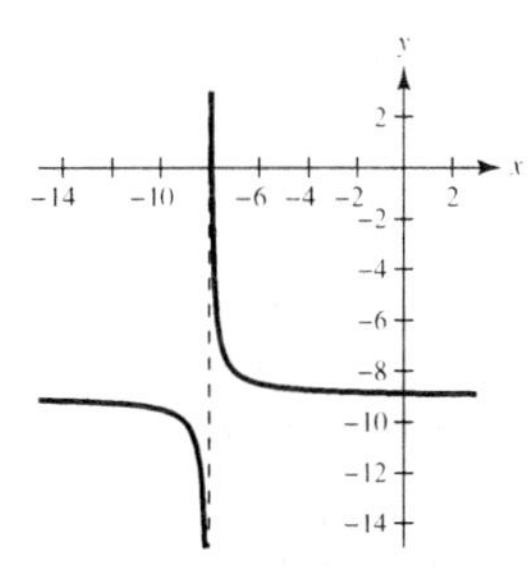

(d) $g(x) = f(x + 8) - 9$

61. (a) $f(x) = |x|$

(b) $g(x) = -2|x - 1| - 4$ is obtained from f by a horizontal shift one unit to the right, a vertical stretch of 2, a reflection in the x-axis, and a vertical shift downward four units.

(c)

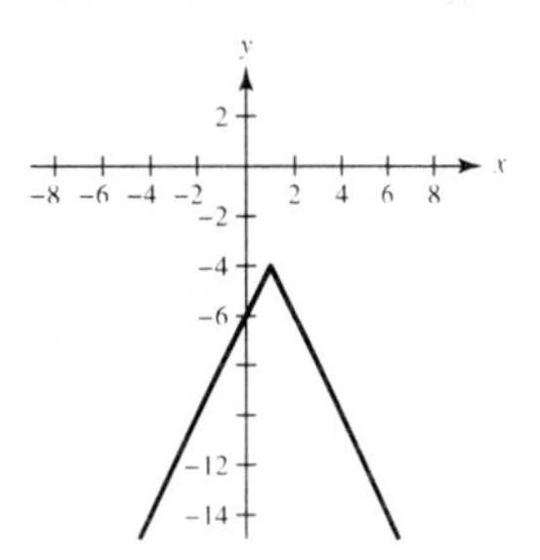

(d) $g(x) = -2f(x - 1) - 4$

63. (a) $f(x) = \sqrt{x}$

(b) $g(x) = -\frac{1}{2}\sqrt{x+3} - 1$ is obtained from f by a horizontal shift three units to the left, a vertical shrink, a reflection in the x-axis, and a vertical shift one unit downward.

(c)

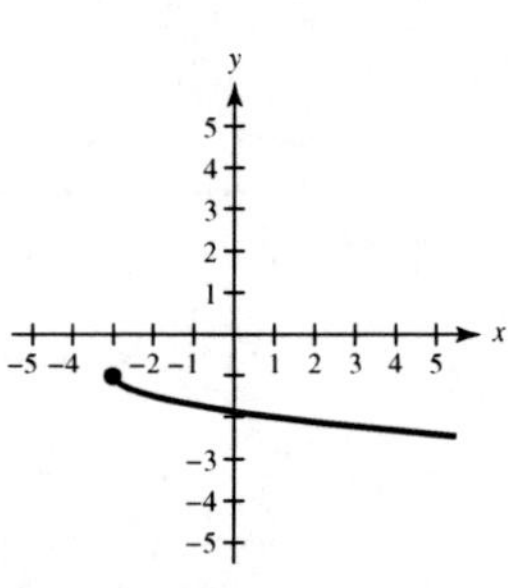

(d) $g(x) = -\frac{1}{2}f(x+3) - 1$

65. (a) $f(t)$ is a horizontal shift of 24.7 units to the right, a vertical shift of 183.4 units upward, a reflection in the t-axis (horizontal axis), and a vertical shrink.

(b)

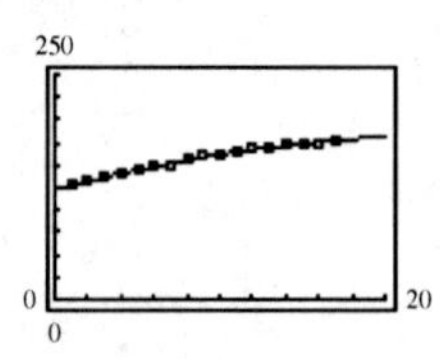

(c) $G(t) = F(t+10) = -0.099[(t+10) - 24.7]^2 + 183.4$

$= -0.099(t - 14.7)^2 + 183.4$

To make a horizontal shift 10 years backward (10 units left), add 10 to t.

67. False. $y = f(-x)$ is a reflection in the y-axis.

69. $y = f(-x)$ is a reflection in the y-axis, so the x-intercepts are $x = -2$ and $x = 3$.

71. $y = f(x) + 2$ is a vertical shift, so you cannot determine the x-intercepts.

73. The vertex is approximately at $(2, 1)$ and the graph opens upward. Matches (c).

75. The vertex is approximately $(2, -4)$ and the graph opens upward. Matches (c).

77. Answers will vary.

79. (a) Since $0 < a < 1$, $g(x) = ax^2$ will be a vertical shrink of $f(x) = x^2$.

(b) Since $a > 1$, $g(x) = ax^2$ will be a vertical stretch of $f(x) = x^2$.

81. Slope $L_1 : \frac{10+2}{2+2} = 3$

Slope $L_2 : \frac{9-3}{3+1} = \frac{3}{2}$

Neither parallel nor perpendicular

83. Domain: All $x \neq 9$

85. Domain:

$100 - x^2 \geq 0 \Rightarrow x^2 \leq 100 \Rightarrow -10 \leq x \leq 10$

Section 1.5

1. addition, subtraction, multiplication, division

3. $g(x)$

5. Since $(fg)(x) = 2x(x^2 + 1)$ and $f(x) = x^2 + 1$, $g(x) = 2x$, and $(fg)(x) = (gf)(x) = (2x)f(x)$.

7.

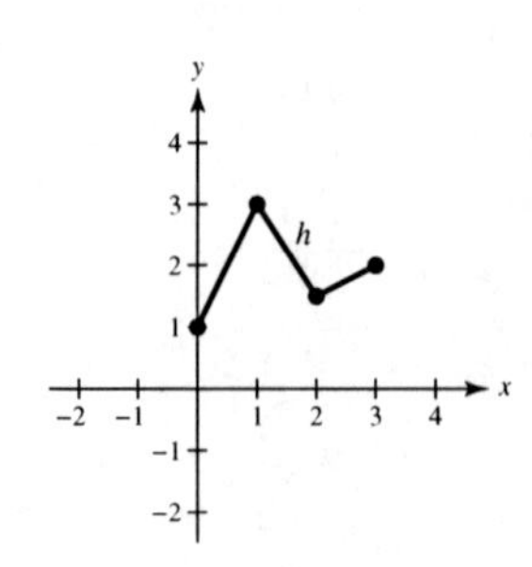

9.

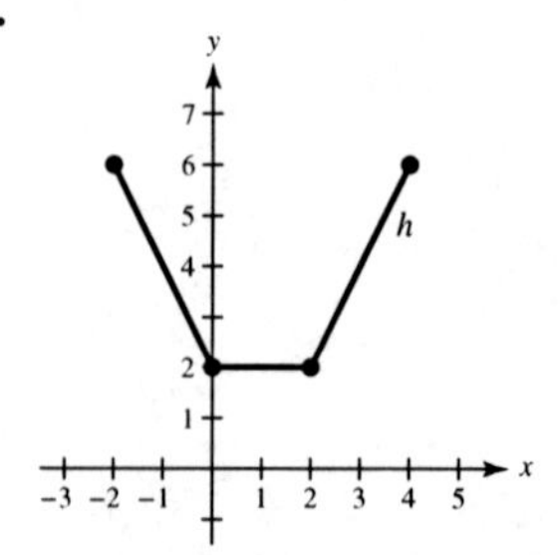

11. $f(x) = x + 3$, $g(x) = x - 3$

(a) $(f + g)(x) = f(x) + g(x) = (x+3) + (x-3) = 2x$

(b) $(f - g)(x) = f(x) - g(x) = (x+3) - (x-3) = 6$

(c) $(fg)(x) = f(x)g(x) = (x+3)(x-3) = x^2 - 9$

(d) $\left(\frac{f}{g}\right)(x) = \frac{f(x)}{g(x)} = \frac{x+3}{x-3}$

Domain: all $x \neq 3$

13. $f(x) = x^2,\ g(x) = 1 - x$

(a) $(f + g)(x) = f(x) + g(x) = x^2 + (1 - x) = x^2 - x + 1$

(b) $(f - g)(x) = f(x) - g(x) = x^2 - (1 - x) = x^2 + x - 1$

(c) $(fg)(x) = f(x) \cdot g(x) = x^2(1 - x) = x^2 - x^3$

(d) $\left(\dfrac{f}{g}\right)(x) = \dfrac{f(x)}{g(x)} = \dfrac{x^2}{1 - x}$

Domain: all $x \neq 1$

15. $f(x) = x^2 + 5,\ g(x) = \sqrt{1 - x}$

(a) $(f + g)(x) = x^2 + 5 + \sqrt{1 - x}$

(b) $(f - g)(x) = x^2 + 5 - \sqrt{1 - x}$

(c) $(fg)(x) = (x^2 + 5)\sqrt{1 - x}$

(d) $\left(\dfrac{f}{g}\right)(x) = \dfrac{x^2 + 5}{\sqrt{1 - x}}$

Domain: $x < 1$

17. $f(x) = \dfrac{1}{x},\ g(x) = \dfrac{1}{x^2}$

(a) $(f + g)(x) = \dfrac{1}{x} + \dfrac{1}{x^2} = \dfrac{x + 1}{x^2}$

(b) $(f - g)(x) = \dfrac{1}{x} - \dfrac{1}{x^2} = \dfrac{x - 1}{x^2}$

(c) $(fg)(x) = \dfrac{1}{x} \cdot \dfrac{1}{x^2} = \dfrac{1}{x^3}$

(d) $\left(\dfrac{f}{g}\right)(x) = \dfrac{\frac{1}{x}}{\frac{1}{x^2}} = x,\ x \neq 0$

Domain: $x \neq 0$

19. $(f + g)(3) = f(3) + g(3)$
$= (3^2 - 1) + (3 - 2)$
$= 8 + 1 = 9$

21. $(f - g)(0) = f(0) - g(0)$
$= (0 - 1) - (0 - 2)$
$= 1$

23. $(fg)(6) = f(6)g(6)$
$= (6^2 - 1)(6 - 2)$
$= (35)(4)$
$= 140$

25. $\left(\dfrac{f}{g}\right)(-5) = \dfrac{f(-5)}{g(-5)}$
$= \dfrac{(-5)^2 - 1}{-5 - 2}$
$= \dfrac{24}{-7}$
$= -\dfrac{24}{7}$

27. $(f - g)(2t) = f(2t) - g(2t)$
$= ((2t)^2 - 1) - (2t - 2)$
$= 4t^2 - 2t + 1$

29. $(fg)(-5t) = f(-5t)g(-5t)$
$= ((-5t)^2 - 1)(-5t - 2)$
$= (25t^2 - 1)(-5t - 2)$
$= -125t^3 - 50t^2 + 5t + 2$

31. $\left(\dfrac{f}{g}\right)(-t) = \dfrac{f(-t)}{g(-t)}$
$= \dfrac{(-t)^2 - 1}{-t - 2}$
$= \dfrac{t^2 - 1}{-t - 2} = \dfrac{1 - t^2}{t + 2},\ t \neq -2$

33.

35.

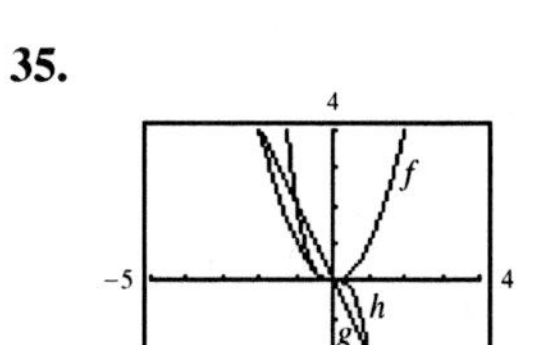

37. $f(x) = 3x,\ g(x) = -\dfrac{x^3}{10},\ (f + g)(x) = 3x - \dfrac{x^3}{10}$

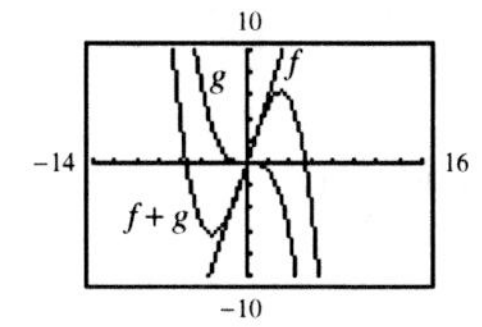

For $0 \le x \le 2$, $f(x)$ contributes more to the magnitude.

For $x > 6$, $g(x)$ contributes more to the magnitude.

39. $f(x) = 3x + 2,\ g(x) = -\sqrt{x+5},$
$(f+g)(x) = 3x + 2 - \sqrt{x+5}$

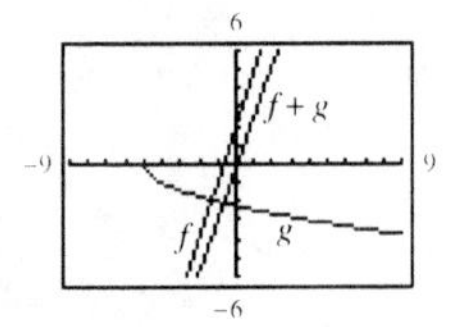

$f(x)$ contributes more to the magnitude in both intervals.

41. $f(x) = x^2,\ g(x) = x - 1$

(a) $(f \circ g)(x) = f(g(x)) = f(x-1) = (x-1)^2$

(b) $(g \circ f)(x) = g(f(x)) = g(x^2) = x^2 - 1$

(c) $(f \circ g)(0) = (0-1)^2 = 1$

43. $f(x) = 3x + 5,\ g(x) = 5 - x$

(a) $(f \circ g)(x) = f(g(x)) = f(5-x) = 3(5-x) + 5 = 20 - 3x$

(b) $(g \circ f)(x) = g(f(x)) = g(3x+5) = 5 - (3x+5) = -3x$

(c) $(f \circ g)(0) = 20$

45. (a) Domain of *f*: $x + 4 \geq 0$ or $x \geq -4$

(b) Domain of *g*: all real numbers

(c) $(f \circ g)(x) = f(g(x)) = f(x^2) = \sqrt{x^2+4}$

Domain: all real numbers

47. (a) Domain of *f*: all real numbers
(b) Domain of *g*: all $x \geq 0$
(c) $(f \circ g)(x) = f(g(x)) = f(\sqrt{x})$

$$= \left(\sqrt{x}\right)^2 + 1 = x + 1,\ x \geq 0$$

Domain: $x \geq 0$

49. (a) Domain of *f*: all $x \neq 0$
(b) Domain of *g*: all real numbers
(c) $(f \circ g)(x) = f(x+3) = \dfrac{1}{x+3}$

Domain: all $x \neq -3$

51. (a) Domain of *f*: all real numbers
(b) Domain of *g*: all real numbers
(c) $(f \circ g)(x) = f(g(x)) = f(3-x)$

$$= |(3-x) - 4| = |-x-1| = |x+1|$$

Domain: all real numbers

53. (a) Domain of *f*: all real numbers

(b) Domain of *g*: all $x \neq \pm 2$

(c) $(f \circ g)(x) = f(g(x)) = f\left(\dfrac{1}{x^2-4}\right) = \dfrac{1}{x^2-4} + 2$

Domain: $x \neq \pm 2$

55. (a) $(f \circ g)(x) = f(g(x)) = f(x^2) = \sqrt{x^2+4}$

Domain: all real numbers

$$(g \circ f)(x) = g(f(x)) = g\left(\sqrt{x+4}\right) = \left(\sqrt{x+4}\right)^2$$
$$= x + 4,\ x \geq -4$$

(b)

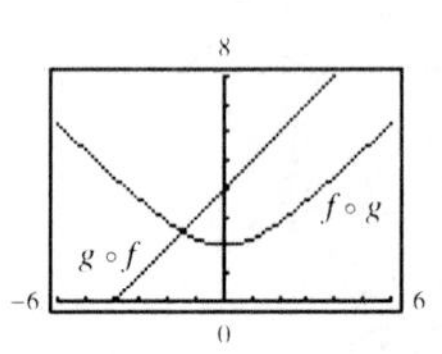

They are not equal.

57. (a) $(f \circ g)(x) = f(g(x)) = f(3x+9)$

$$= \frac{1}{3}(3x+9) - 3 = x$$

Domain: all real numbers

$$(g \circ f)(x) = g(f(x)) = g\left(\frac{1}{3}x - 3\right)$$
$$= 3\left(\frac{1}{3}x - 3\right) + 9 = x$$

(b)

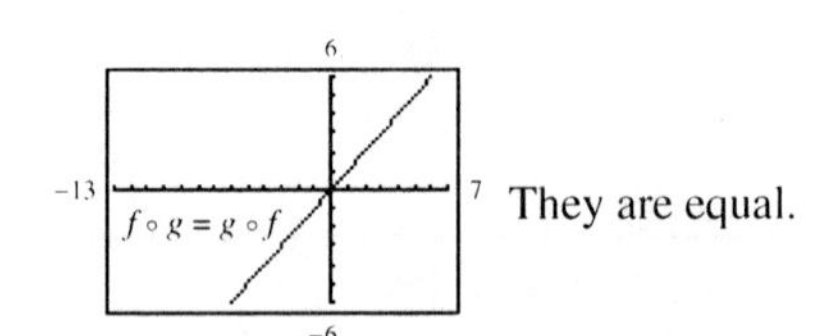

They are equal.

59. (a) $(f \circ g)(x) = f(g(x)) = f(x^6) = (x^6)^{2/3} = x^4$

Domain: all real numbers

$$(g \circ f)(x) = g(f(x)) = g(x^{2/3}) = (x^{2/3})^6 = x^4$$

(b)

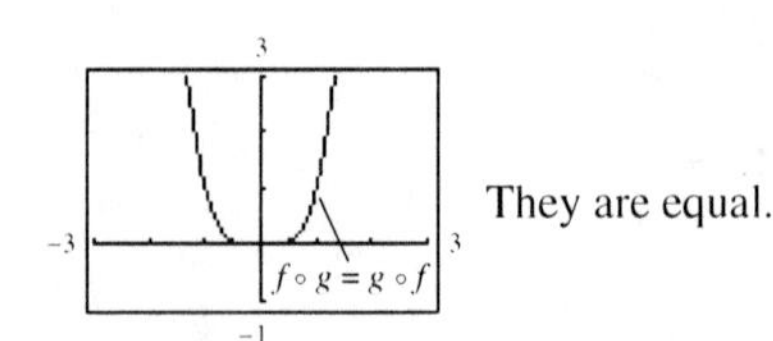

They are equal.

61. (a) $(f \circ g)(x) = f(g(x)) = f(4 - x) = 5(4 - x) + 4$
$= 24 - 5x$
$(g \circ f)(x) = g(f(x)) = g(5x + 4) = 4 - (5x + 4)$
$= -5x$

(b) They are not equal because $24 - 5x \neq -5x$.

(c)

x	$f(g(x))$	$g(f(x))$
0	24	0
1	19	–5
2	14	–10
3	9	–15

63. (a)
$(f \circ g)(x) = f(g(x)) = f(x^2 - 5) = \sqrt{(x^2 - 5) + 6} = \sqrt{x^2 + 1}$
$(g \circ f)(x) = g(f(x)) = g\left(\sqrt{x + 6}\right) = \left(\sqrt{x + 6}\right)^2 - 5$
$= (x + 6) - 5 = x + 1,\ x \geq -6$

(b) They are not equal because $\sqrt{x^2 + 1} \neq x + 1$.

(c)

x	$f(g(x))$	$g(f(x))$
0	1	1
–2	$\sqrt{5}$	–1
3	$\sqrt{10}$	4

65. (a) $(f \circ g)(x) = f(g(x)) = f(2x^3) = \left|2x^3\right|$
$(g \circ f)(x) = g(f(x)) = g(|x|) = 2|x|^3$

(b) They are equal because $\left|2x^3\right| = 2|x|^3$.

(c)

x	$f(g(x))$	$g(f(x))$
–1	2	2
0	0	0
1	2	2
2	16	16

67. (a) $(f + g)(3) = f(3) + g(3) = 2 + 1 = 3$

(b) $\left(\dfrac{f}{g}\right)(2) = \dfrac{f(2)}{g(2)} = \dfrac{0}{2} = 0$

69. (a) $(f \circ g)(2) = f(g(2)) = f(2) = 0$

(b) $(g \circ f)(2) = g(f(2)) = g(0) = 4$

71. Let $f(x) = x^2$ and $g(x) = 2x + 1$, then $(f \circ g)(x) = h(x)$. This is not a unique solution. Another possibility is $f(x) = (x + 1)^2$ and $g(x) = 2x$.

73. Let $f(x) = \sqrt[3]{x}$ and $g(x) = x^2 - 4$, then $(f \circ g)(x) = h(x)$. This answer is not unique. Other possibilities are

$f(x) = \sqrt[3]{x - 4}$ and $g(x) = x^2$ or
$f(x) = \sqrt[3]{-x}$ and $g(x) = 4 - x^2$ or
$f(x) = \sqrt[9]{x}$ and $g(x) = (x^2 - 4)^3$.

75. Let $f(x) = \dfrac{1}{x}$ and $g(x) = x + 2$, then $(f \circ g)(x) = h(x)$. This is not a unique solution. Other possibilities are

$f(x) = \dfrac{1}{x + 2}$ and $g(x) = x$ or $f(x) = \dfrac{1}{x + 1}$ and $g(x) = x + 1$.

77. Let $f(x) = x^2 + 2x$ and $g(x) = x + 4$, then $(f \circ g)(x) = h(x)$. This answer is not unique. Another possibility is $f(x) = x$ and $g(x) = (x + 4)^2 + 2(x + 4)$.

79. (a) $T(x) = R(x) + B(x) = \dfrac{3}{4}x + \dfrac{1}{15}x^2$

(b)

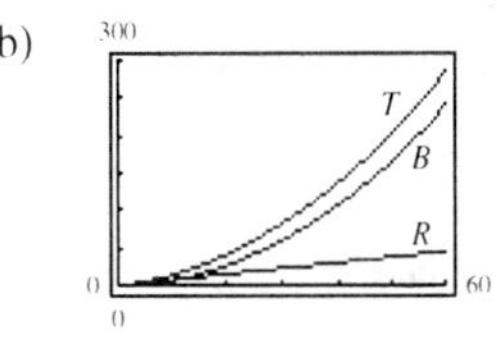

(c) $B(x)$ contributes more to $T(x)$ at higher speeds.

81. (a) $r(x) = \dfrac{x}{2}$

(b) $A(r) = \pi r^2$

(c) $(A \circ r)(x) = A(r(x))$
$= A\left(\dfrac{x}{2}\right) = \pi\left(\dfrac{x}{2}\right)^2 = \dfrac{1}{4}\pi x^2$

$A \circ r$ represents the area of the circular base of the tank with radius $\dfrac{x}{2}$.

83. (a) Since
$T = S_1 + S_2$, $T = (830 + 1.2t^2) + (390 + 75.4t)$
$T = 1.2t^2 + 75.4t + 1220$.

(b)

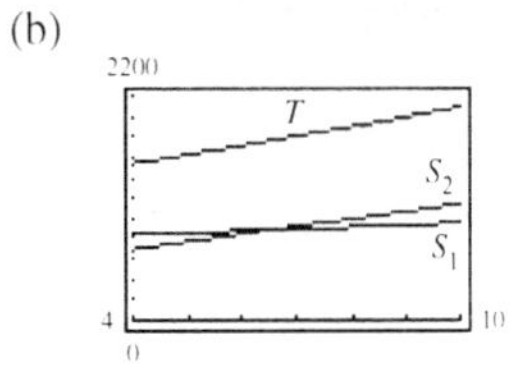

85. (a) $(N \circ T)(t) = N(T(t))$
$= N(2t + 1)$
$= 10(2t + 1)^2 - 20(2t + 1) + 600$
$= 40t^2 + 590$

$N \circ T$ represents the number of bacteria as a function of time.

(b) $(N \circ T)(6) = 10(13^2) - 20(13) + 600 = 2030$

At time $t = 6$, there are 2030 bacteria.

(c) $N = 800$ when $t \approx 2.3$ hours.

87. First, write the distance each plane is from point P. The plane that is 200 miles from point P is traveling at 450 miles per hour. Its distance is $200 - 450t$. Similarly, the other plane is $150 - 450t$ from point P.

So, the distance between the planes $s(t)$ can be found using the distance formula (or the Pythagorem Theorem): $s(t) = \sqrt{(200 - 450t)^2 + (150 - 450t)^2}$

$$s(t) = 50\sqrt{162t^2 - 126t + 25}$$

89. False. $g(x) = x - 3$

91. Let A, B, and C be the three siblings, in decreasing age. Then $A = 2B$ and $B = \frac{1}{2}C + 6$.

(a) $A = 2B = 2\left(\frac{1}{2}C + 6\right) = C + 12$

(b) If $A = 16$, then $B = 8$ and $C = 4$.

93. Let $f(x)$ and $g(x)$ be odd functions, and define $h(x) = f(x)g(x)$. Then

$$\begin{aligned} h(-x) &= f(-x)g(-x) \\ &= [-f(x)][-g(x)] \text{ since } f \text{ and } g \text{ are both odd} \\ &= f(x)g(x) = h(x). \end{aligned}$$

Thus, h is even.

Let $f(x)$ and $g(x)$ be even functions, and define $h(x) = f(x)g(x)$. Then

$$\begin{aligned} h(-x) &= f(-x)g(-x) \\ &= f(x)g(x) \text{ since } f \text{ and } g \text{ are both even} \\ &= h(x). \end{aligned}$$

Thus, h is even.

95. $g(-x) = \frac{1}{2}[f(-x) + f(-(-x))] = \frac{1}{2}[f(-x) + f(x)] = g(x)$, which shows that g is even.

$$\begin{aligned} h(-x) &= \frac{1}{2}[f(-x) - f(-(-x))] = \frac{1}{2}[f(-x) - f(x)] \\ &= -\frac{1}{2}[f(x) - f(-x)] = -h(x), \end{aligned}$$

which shows that h is odd.

97. (a) If $f(x) = x^2$ and $g(x) = \frac{1}{x-2}$, then

$$f(g(x)) = \left(\frac{1}{x-2}\right)^2 = \frac{1^2}{(x-2)^2} = \frac{1}{(x-2)^2} = h(x).$$

(b) If $f(x) = \frac{1}{x-2}$ and $g(x) = x^2$, then

$$f(g(x)) = \frac{1}{x^2 - 2} \neq h(x).$$

(c) If $f(x) = \frac{1}{x}$ and $g(x) = (x-2)^2$, then

$$f(g(x)) = \frac{1}{(x-2)^2} = h(x).$$

99. Three points on the graph of $y = -x^2 + x - 5$ are $(0, -5)$, $(1, -5)$, and $(2, -7)$.

101. Three points on the graph of $x^2 + y^2 = 24$ are $(\sqrt{24}, 0)$, $(-\sqrt{24}, 0)$, and $(0, \sqrt{24})$.

103. First $m = \frac{8-(-2)}{-3-(-4)} = \frac{10}{1} = 10$, and using the point $(-4, -2)$, $y - (-2) = 10(x - (-4))$

$$\begin{aligned} y + 2 &= 10x + 40 \\ y &= 10x + 38. \end{aligned}$$

105. First $m = \frac{4-(-1)}{\frac{1}{3} - \frac{3}{2}} = \frac{5}{-\frac{11}{6}} = -\frac{30}{11}$, and using the point

$\left(\frac{3}{2}, -1\right)$, $y - (-1) = -\frac{30}{11}\left(x - \frac{3}{2}\right)$

$$\begin{aligned} y + 1 &= -\frac{30}{11}x + \frac{45}{11} \\ y &= -\frac{30}{11}x + \frac{34}{11}. \end{aligned}$$

Section 1.6

1. inverse, f^{-1}

3. $y = x$

5. If a function is one-to-one, no two x-values in the domain can correspond to the same y-value in the range. Therefore, a horizontal line can intersect the graph at most once.

7. $f(x) = 6x$

$$f^{-1}(x) = \frac{1}{6}x$$

$$f\left(f^{-1}(x)\right) = f\left(\frac{1}{6}x\right) = 6\left(\frac{1}{6}x\right) = x$$

$$f^{-1}\left(f(x)\right) = f^{-1}(6x) = \frac{1}{6}(6x) = x$$

9. $f(x) = x + 7$

$f^{-1}(x) = x - 7$

$f(f^{-1}(x)) = f(x - 7) = (x - 7) + 7 = x$

$f^{-1}(f(x)) = f^{-1}(x + 7) = (x + 7) - 7 = x$

11. $f(x) = 2x + 1$

$f^{-1}(x) = \dfrac{x - 1}{2}$

$$f(f^{-1}(x)) = f\left(\frac{x-1}{2}\right) = 2\left(\frac{x-1}{2}\right) + 1$$

$$= (x - 1) + 1 = x$$

$$f^{-1}(f(x)) = f^{-1}(2x + 1) = \frac{(2x + 1) - 1}{2} = \frac{2x}{2} = x$$

13. $f(x) = \sqrt[3]{x}$

$f^{-1}(x) = x^3$

$f(f^{-1}(x)) = f(x^3) = \sqrt[3]{x^3} = x$

$f^{-1}(f(x)) = f^{-1}(\sqrt[3]{x}) = (\sqrt[3]{x})^3 = x$

15. The inverse is a line through $(-1, 0)$. Matches graph (c).

17. The inverse is half a parabola starting at $(1, 0)$. Matches graph (a).

19. $f(x) = x^3$, $g(x) = \sqrt[3]{x}$

$f(g(x)) = f(\sqrt[3]{x}) = (\sqrt[3]{x})^3 = x$

$g(f(x)) = g(x^3) = \sqrt[3]{x^3} = x$

Reflections in the line $y = x$

21. $f(x) = \sqrt{x - 4}$; $g(x) = x^2 + 4$, $x \ge 0$

$f(g(x)) = f(x^2 + 4)$, $x \ge 0$

$= \sqrt{(x^2 + 4) - 4} = x$

$g(f(x)) = g(\sqrt{x - 4})$

$= (\sqrt{x - 4})^2 + 4 = x$

Reflections in the line $y = x$

23. $f(x) = 1 - x^3$, $g(x) = \sqrt[3]{1 - x}$

$f(g(x)) = f(\sqrt[3]{1 - x}) = 1 - (\sqrt[3]{1 - x})^3 = 1 - (1 - x) = x$

$g(f(x)) = g(1 - x^3) = \sqrt[3]{1 - (1 - x^3)} = \sqrt[3]{x^3} = x$

Reflections in the line $y = x$

25. (a) $f(g(x)) = f\left(-\dfrac{2x + 6}{7}\right)$

$$= -\frac{7}{2}\left(-\frac{2x + 6}{7}\right) - 3 = x$$

$$g(f(x)) = g\left(-\frac{7}{2}x - 3\right)$$

$$= -\frac{2\left(-\frac{7}{2}x - 3\right) + 6}{7} = x$$

(b)

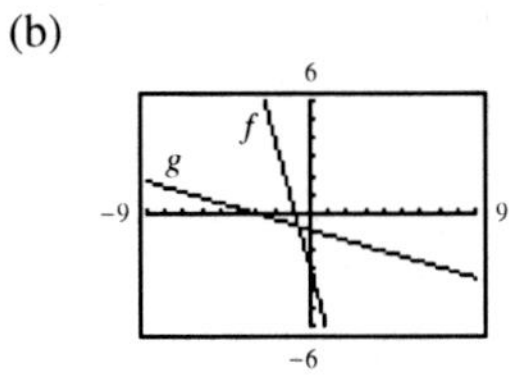

(c) $Y_1 = -\frac{7}{2}X - 3$

$Y_2 = -\frac{2X+6}{7}$

$Y_3 = Y_1(Y_2)$

$Y_4 = Y_2(Y_1)$

X	Y_3	Y_4
−4	−4	−4
−2	−2	−2
0	0	0
2	2	2
4	4	4

27. (a) $f(g(x)) = f\left(\sqrt[3]{x-5}\right) = \left(\sqrt[3]{x-5}\right)^3 + 5 = x$

$g(f(x)) = g(x^3+5) = \sqrt[3]{(x^3+5)-5} = x$

(b)

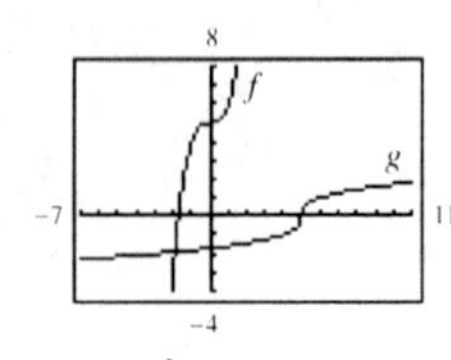

(c) $Y_1 = X^3 + 5$

$Y_2 = \sqrt[3]{X-5}$

$Y_3 = Y_1(Y_2)$

$Y_4 = Y_2(Y_1)$

X	Y_3	Y_4
−2	−2	−2
−1	−1	−1
0	0	0
1	1	1
4	4	4

29. (a) $f(g(x)) = f(8+x^2)$

$= -\sqrt{(8+x^2)-8}$

$= -\sqrt{x^2} = -(-x) = x,\ x \le 0$

$g(f(x)) = g(-\sqrt{x-8})$

$= 8 + (-\sqrt{x-8})^2$

$= 8 + (x-8) = x,\ x \ge 8$

(b)

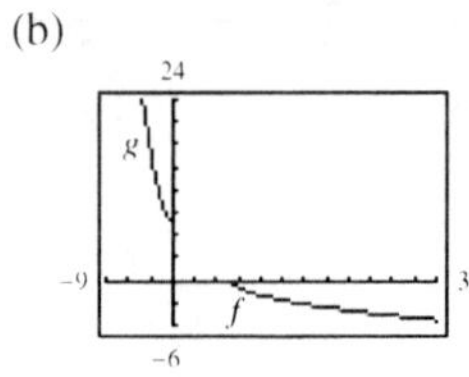

(c) $Y_1 = -\sqrt{x-8}$

$Y_2 = 8 + x^2,\ x = 0$

$Y_3 = Y_1(Y_2)$

$Y_4 = Y_2(Y_1)$

X	Y_3	Y_4
8	8	8
9	9	9
12	12	12
15	15	15
20	20	20

31. (a) $f(g(x)) = f\left(\frac{x}{2}\right)$

$= 2\left(\frac{x}{2}\right) = x$

$g(f(x)) = g(2x)$

$= \frac{2x}{2} = x$

(b)

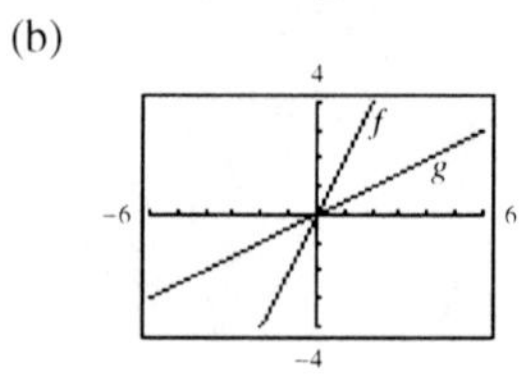

(c) $Y_1 = 2x$

$Y_2 = \frac{x}{2}$

$Y_3 = Y_1(Y_2)$

$Y_4 = Y_2(Y_1)$

X	Y_3	Y_4
−4	−4	−4
−2	−2	−2
0	0	0
2	2	2
4	4	4

33. (a) $f(g(x)) = f\left(-\frac{5x+1}{x-1}\right)$

$= \frac{\left(-\frac{5x+1}{x-1}\right) - 1}{\left(-\frac{5x+1}{x-1}\right) + 5} = \frac{\frac{6x}{x-1}}{\frac{6}{x-1}} = x,\ x \ne 1$

$g(f(x)) = g\left(\frac{x-1}{x+5}\right)$

$= -\frac{5\left(\frac{x-1}{x+5}\right) + 1}{\left(\frac{x-1}{x+5}\right) - 1} = \frac{\frac{6x}{x+5}}{\frac{6}{x+5}} = x,\ x \ne -5$

(b)

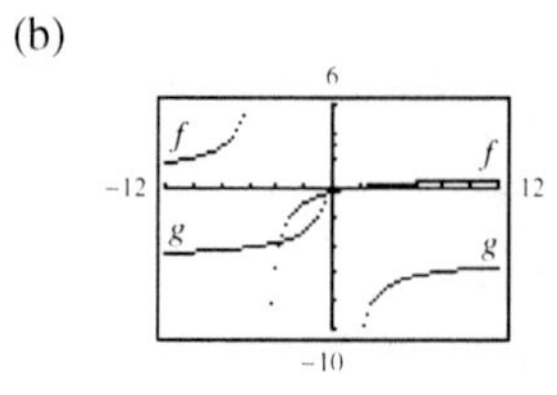

(c) $Y_1 = \frac{x-1}{x+5}$

$Y_2 = -\frac{5x+1}{x-1}$

$Y_3 = Y_1(Y_2)$

$Y_4 = Y_2(Y_1)$

X	Y_3	Y_4
−2	−2	−2
−1	−1	−1
0	0	0
1	1	1
2	2	2

35. Yes. No two elements, number of cans, in the domain correspond to the same element, price, in the range.

37. No. Both x-values, −3 and 0, in the domain correspond to the y-value 6 in the range.

39. Not a function.

41. It is the graph of a one-to-one function.

43. It is the graph of a one-to-one function.

45. $f(x) = 3 - \frac{1}{2}x$

f is one-to-one because a horizontal line will intersect the graph at most once.

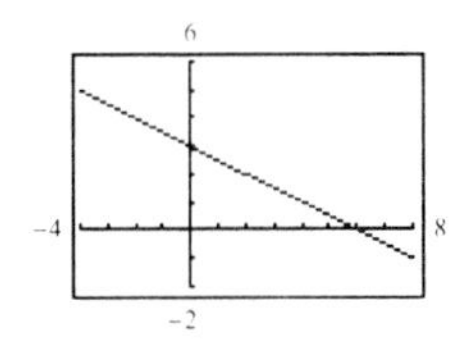

47. $h(x) = \frac{x^2}{x^2 + 1}$

h is not one-to-one because some horizontal lines intersect the graph twice.

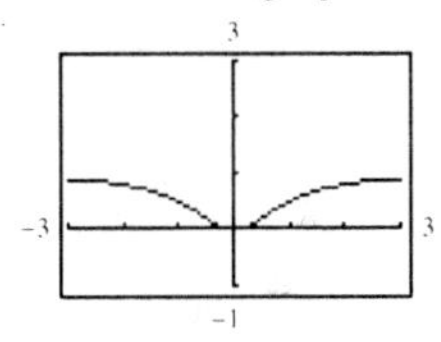

49. $h(x) = \sqrt{16 - x^2}$

h is not one-to-one because some horizontal lines intersect the graph twice.

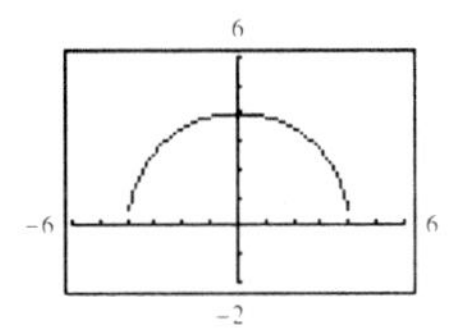

51. $f(x) = 10$

f is not one-to-one because the horizontal line $y = 10$ intersects the graph at every point on the graph.

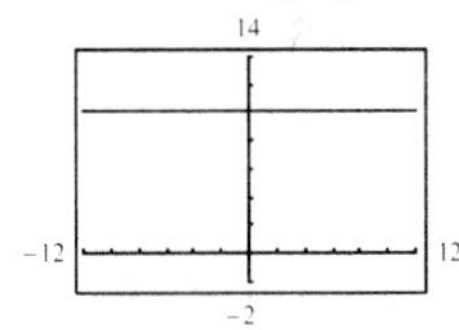

53. $g(x) = (x + 5)^3$

g is one-to-one because a horizontal line will intersect the graph at most once.

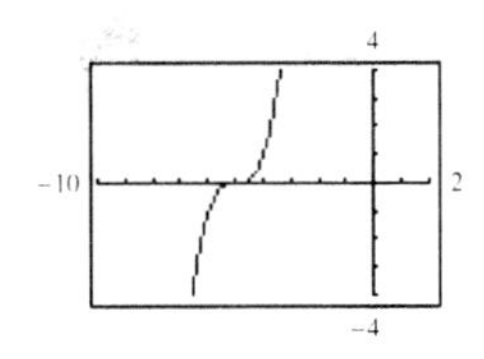

55. $h(x) = |x + 4| - |x - 4|$

h is not one-to-one because some horizontal lines intersect the graph more than once.

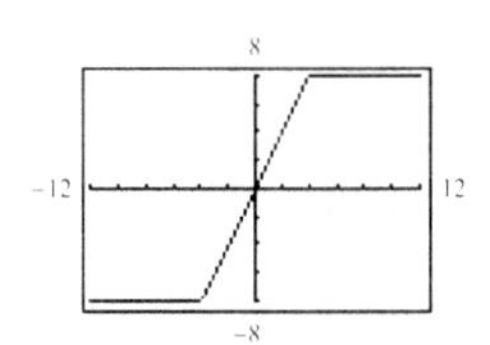

57.

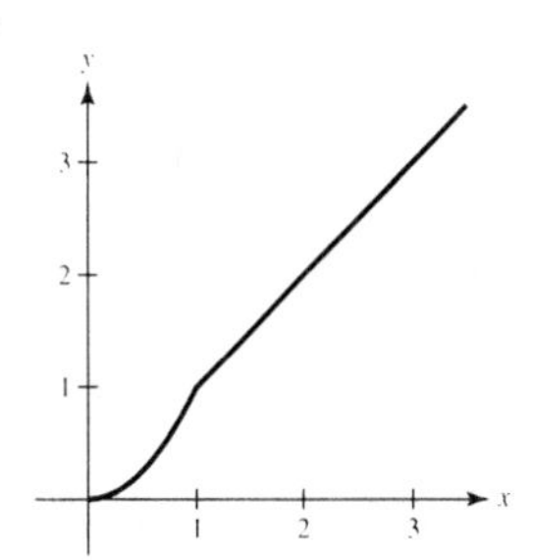

The graph of the function passes the Horizontal Line Test and does have an inverse function.

59. $f(x) = x^4$

f is not one-to-one.

For example, $f(2) = f(-2) = 16$.

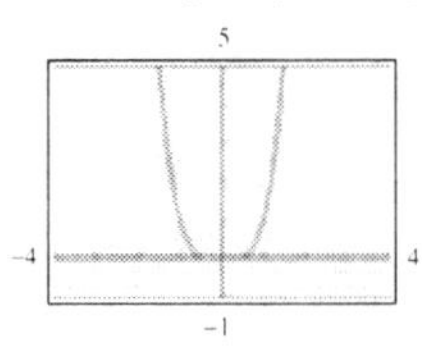

61.

$$f(x) = \frac{3x + 4}{5}$$

$$y = \frac{3x + 4}{5}$$

$$x = \frac{3y + 4}{5}$$

$$5x = 3y + 4$$

$$5x - 4 = 3y$$

$$\frac{5x - 4}{3} = y$$

$$f^{-1}(x) = \frac{5x - 4}{3}$$

f is one-to-one and has an inverse function.

63. $f(x) = \frac{1}{x^2}$ is not one-to-one.

For example, $f(1) = f(-1) = 1$.

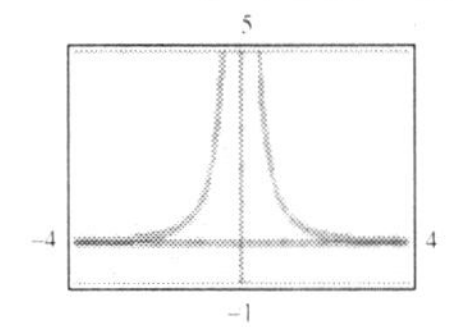

65. $f(x) = (x+3)^2,\ x \ge -3,\ y \ge 0$

$$y = (x+3)^2,\ x \ge -3,\ y \ge 0$$
$$x = (y+3)^2,\ y \ge -3,\ x \ge 0$$
$$\sqrt{x} = y+3,\ y \ge -3,\ x \ge 0$$
$$y = \sqrt{x} - 3,\ x \ge 0,\ y \ge -3$$

f is one-to-one and has an inverse function.

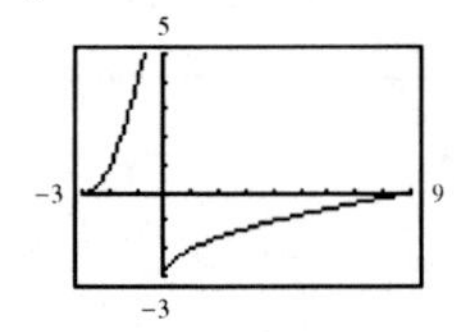

67. $f(x) = \sqrt{2x+3} \Rightarrow x \ge -\dfrac{3}{2},\ y \ge 0$

$$y = \sqrt{2x+3},\ x \ge -\frac{3}{2},\ y \ge 0$$
$$x = \sqrt{2y+3},\ y \ge -\frac{3}{2},\ x \ge 0$$
$$x^2 = 2y+3,\ x \ge 0,\ y \ge -\frac{3}{2}$$
$$y = \frac{x^2-3}{2},\ x \ge 0,\ y \ge -\frac{3}{2}$$

f is one-to-one and has an inverse function.

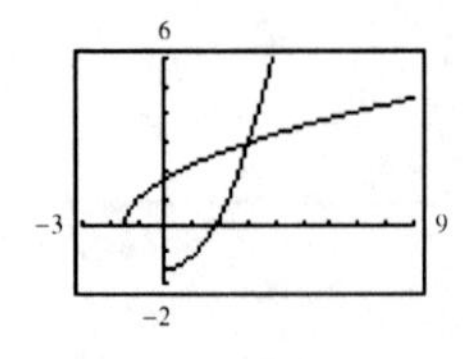

69. $f(x) = |x-2|,\ x \le 2,\ y \ge 0$

$$y = |x-2|$$
$$x = |y-2|,\ y \le 2,\ x \ge 0$$
$$x = -(y-2) \text{ since } y-2 \le 0.$$
$$x = -y+2$$
$$y = -x+2,\ x \ge 0,\ y \le 2$$

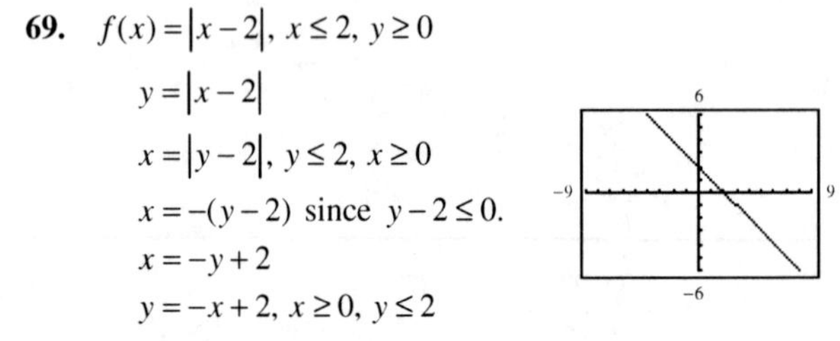

f is one-to-one and has an inverse function.

71. $f(x) = 2x-3$

$$y = 2x-3$$
$$x = 2y-3$$
$$y = \frac{x+3}{2}$$
$$f^{-1}(x) = \frac{x+3}{2}$$

Reflections in the line $y = x$

73. $f(x) = x^5$

$$y = x^5$$
$$x = y^5$$
$$y = \sqrt[5]{x}$$
$$f^{-1}(x) = \sqrt[5]{x}$$

Reflections in the line $y = x$

75. $f(x) = x^{3/5}$

$$y = x^{3/5}$$
$$x = y^{3/5}$$
$$y = x^{5/3}$$
$$f^{-1}(x) = x^{5/3}$$

Reflections in the line $y = x$

77. $f(x) = \sqrt{4-x^2},\ 0 \le x \le 2$

$$y = \sqrt{4-x^2}$$
$$x = \sqrt{4-y^2}$$
$$x^2 = 4-y^2$$
$$y^2 = 4-x^2$$
$$y = \sqrt{4-x^2}$$
$$f^{-1}(x) = \sqrt{4-x^2},\ 0 \le x \le 2$$

The graphs are the same.

79. $f(x) = \dfrac{4}{x}$

$$y = \frac{4}{x}$$
$$x = \frac{4}{y}$$
$$xy = 4$$
$$y = \frac{4}{x}$$
$$f^{-1}(x) = \frac{4}{x}$$

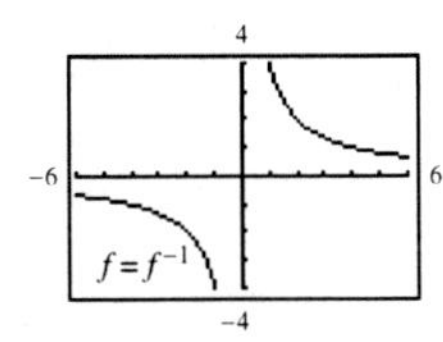

The graphs are the same.

81. If we let $f(x) = (x-2)^2$, $x \ge 2$, then *f* has an inverse function. [**Note:** We could also let $x \le 2$.]

$$y = (x-2)^2$$
$$x = (y-2)^2$$
$$\sqrt{x} = y - 2$$
$$\sqrt{x} + 2 = y$$
$$f^{-1}(x) = \sqrt{x} + 2$$

Domain of f: $x \ge 2$ Range of f: $y \ge 0$

Domain of f^{-1}: $x \ge 0$ Range of f^{-1}: $y \ge 2$

83. If we let $f(x) = |x+2|$, $x \ge -2$, then *f* has an inverse function. [**Note:** We could also let $x \le -2$.]

$$y = x + 2$$
$$x = y + 2$$
$$x - 2 = y$$
$$f^{-1}(x) = x - 2$$

Domain of *f*: $x \ge -2$

Domain of f^{-1}: $x \ge 0$

Range of *f*: $y \ge 0$

Range of f^{-1}: $y \ge -2$

85. If we let $f(x) = (x+3)^2$, $x \ge -3$ then *f* has an inverse function. [**Note:** We could also let $x \le -3$.]

$$y = (x+3)^2$$
$$x = (y+3)^2$$
$$\sqrt{x} = y + 3$$
$$y = \sqrt{x} - 3$$
$$f^{-1}(x) = \sqrt{x} - 3$$

Domain of f: $x \ge -3$ Range of f: $y \ge 0$

Domain of f^{-1}: $x \ge 0$ Range of f^{-1}: $y \ge -3$

87. If we let $f(x) = -2x^2 + 5$, $x \ge 0$, then *f* has an inverse function. [**Note:** We could also let $x \le -3$.]

$$y = -2x^2 + 5$$
$$x = -2y^2 + 5$$
$$x - 5 = -2y^2$$
$$y^2 = \frac{x-5}{-2} = \frac{5-x}{2}$$
$$y = \sqrt{\frac{(5-x)}{2}}$$
$$f^{-1}(x) = \sqrt{\frac{5-x}{2}}$$

Domain of f: $x \ge 0$ Range of f: $y \le 5$

Domain of f^{-1}: $x \le 5$ Range of f^{-1}: $y \ge 0$

89. If we let $f(x) = |x-4| + 1$, $x \ge 4$, then *f* has an inverse function. [**Note:** We could also let $x \le 4$.]

$$y = |x-4| + 1$$
$$y = x - 3 \text{ because } x \ge 4$$
$$x = y - 3$$
$$y = x + 3$$
$$f^{-1}(x) = x + 3$$

Domain of f: $x \ge 4$ Range of f: $y \ge 1$

Domain of f^{-1}: $x \ge 1$ Range of f^{-1}: $y \ge 4$

91.

x	$f(x)$
−2	−4
−1	−2
1	2
3	3

x	$f^{-1}(x)$
−4	−2
−2	−1
2	1
3	3

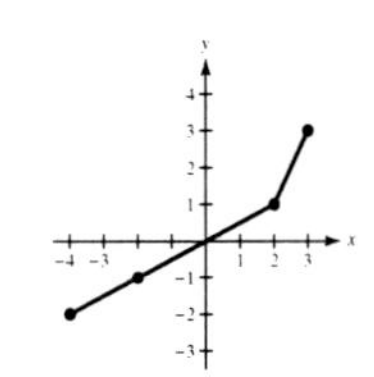

93. $f^{-1}(0) = \dfrac{1}{2}$ because $f\left(\dfrac{1}{2}\right) = 0$.

95. $(f \circ g)(2) = f(3) = -2$

97. $f^{-1}(g(0)) = f^{-1}(2) = 0$

99. $(g \circ f^{-1})(2) = g(0) = 2$

101. $f(x) = x^3 + x + 1$

(a) and (b)

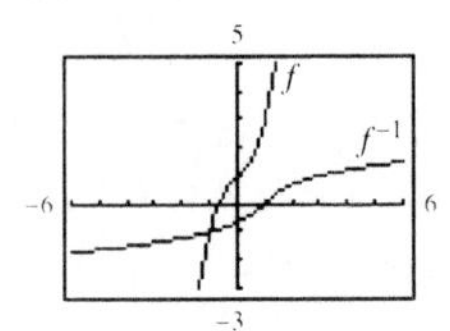

(c) The graph of the inverse relation is an inverse function since it satisfies the Vertical Line Test.

103. $g(x) = \dfrac{3x^2}{x^2 + 1}$

(a) and (b)

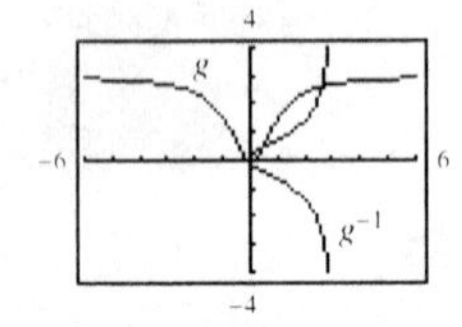

The graph of the inverse relation is not an inverse function since it does not satisfy the Vertical Line Test.

In Exercises 105 – 110, $f(x) = \frac{1}{8}x - 3,\ f^{-1}(x) = 8(x + 3),$ $g(x) = x^3,\ g^{-1}(x) = \sqrt[3]{x}.$

105. $(f^{-1} \circ g^{-1})(1) = f^{-1}(g^{-1}(1)) = f^{-1}\left(\sqrt[3]{1}\right)$

$$= 8\left(\sqrt[3]{1} + 3\right) = 8(1 + 3) = 32$$

107. $(f^{-1} \circ f^{-1})(6) = f^{-1}(f^{-1}(6)) = f^{-1}(8[6 + 3])$

$$= f^{-1}(72) = 8(72 + 3) = 600$$

109. $(fg)(x) = f(g(x)) = f\left(x^3\right) = \frac{1}{8}x^3 - 3$

Now find the inverse of $(f \circ g)(x) = \frac{1}{8}x^3 - 3$:

$$y = \frac{1}{8}x^3 - 3$$
$$x = \frac{1}{8}y^3 - 3$$
$$x + 3 = \frac{1}{8}y^3$$
$$8(x + 3) = y^3$$
$$\sqrt[3]{8(x + 3)} = y$$
$$(f \circ g)^{-1}(x) = 2\sqrt[3]{x + 3}$$

Note: $(f \circ g)^{-1} = g^{-1} \circ f^{-1}$

In Exercises 111 to 114,

$f(x) = x + 4,\ f^{-1}(x) = x - 4,\ g(x) = 2x - 5,\ g^{-1}(x) = \dfrac{x + 5}{2}.$

111. $(g^{-1} \circ f^{-1})(x) = g^{-1}(f^{-1}(x))$

$$= g^{-1}(x - 4)$$
$$= \frac{(x - 4) + 5}{2}$$
$$= \frac{x + 1}{2}$$

113. $(f \circ g)(x) = f(g(x)) = f(2x - 5) = (2x - 5) + 4 = 2x - 1.$

Now find the inverse function of $(f \circ g)(x) = 2x - 1.$

$$y = 2x - 1$$
$$x = 2y - 1$$
$$x + 1 = 2y$$
$$y = \frac{x + 1}{2}$$
$$(f \circ g)^{-1}(x) = \frac{x + 1}{2}$$

Note that $(f \circ g)^{-1}(x) = (g^{-1} \circ f^{-1})(x)$; see Exercise 111.

115. (a) Yes, *f* is one-to-one. For each European shoe size, there is exactly one U.S. shoe size.

(b) $f(11) = 45$

(c) $f^{-1}(43) = 10$ because $f(10) = 43.$

(d) $f(f^{-1}(41)) = f(8) = 41$

(e) $f^{-1}(f(13)) = f^{-1}(47) = 13$

117. (a) CALL ME LATER corresponds to numerical values: 3 1 12 12 0 13 5 0 12 1 20 5 18. Using *f* to encode,

$$f(3) = 19$$
$$f(1) = 9$$
$$f(12) = 64$$
$$f(12) = 64$$
$$f(0) = 4$$
$$f(13) = 69$$
$$f(5) = 29$$
$$f(0) = 4$$
$$f(12) = 64$$
$$f(1) = 9$$
$$f(20) = 104$$
$$f(5) = 29$$
$$f(18) = 94$$

(b) For $f(x) = 5x + 4$, $f^{-1}(x) = \dfrac{x - 4}{5}$.

Using f^{-1} to decode,
$$f^{-1}(119) = 23$$
$$f^{-1}(44) = 8$$
$$f^{-1}(9) = 1$$
$$f^{-1}(104) = 20$$
$$f^{-1}(4) = 0$$
$$f^{-1}(104) = 20$$
$$f^{-1}(49) = 9$$
$$f^{-1}(69) = 13$$
$$f^{-1}(29) = 5$$

Converting from numerical values to letters, the message is WHAT TIME.

119. False. $f(x) = x^2$ is even, but f^{-1} does not exist.

121. Yes. The inverse would give the time it took to complete n miles.

123. No. The function oscillates.

125. The graph of f^{-1} is a reflection of the graph of f in the line $y = x$.

127. (a) The function f will have an inverse function because no two temperatures in degrees Celsius will correspond to the same temperature in degrees Fahrenheit.

(b) $f^{-1}(50)$ would represent the temperature in degrees Celsius for a temperature of 50° Fahrenheit.

129. The constant function $f(x) = c$, whose graph is a horizontal line, would never have an inverse function.

131. We will show that $(f \circ g)^{-1}(x) = (g^{-1} \circ f^{-1})(x)$ for all x in their domains.

Let $y = (f \circ g)^{-1}(x) \Rightarrow (f \circ g)(y) = x$ then

$f(g(y)) = x \Rightarrow f^{-1}(x) = g(y)$.

Hence,

$(g^{-1} \circ f^{-1})(x) = g^{-1}(f^{-1}(x)) = g^{-1}(g(y)) = y = (f \circ g)^{-1}(x)$.

Thus, $g^{-1} \circ f^{-1} = (f \circ g)^{-1}$.

133. $\dfrac{27x^3}{3x^2} = 9x,\ x \neq 0$

135. $\dfrac{x^2 - 36}{6 - x} = \dfrac{(x - 6)(x + 6)}{-(x - 6)} = \dfrac{x + 6}{-1} = -x - 6,\ x \neq 6$

137. $x = 5$. No, it does not pass the Vertical Line Test.

139. $x^2 + y^2 = 9$

$y = \pm\sqrt{9 - x^2}$

No, y is not a function of x.

Section 1.7

1. positive

3. negative

5. (a)

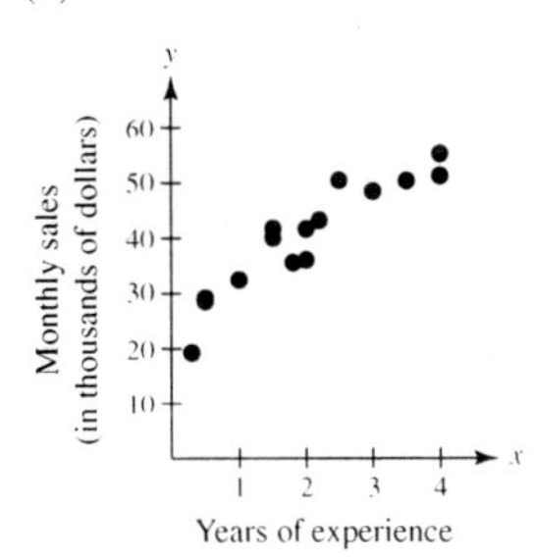

(b) Yes, the data appears somewhat linear. The more experience x corresponds to higher sales y.

7. Negative correlation

9. No correlation

11. (a) and (b)

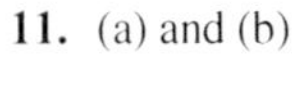

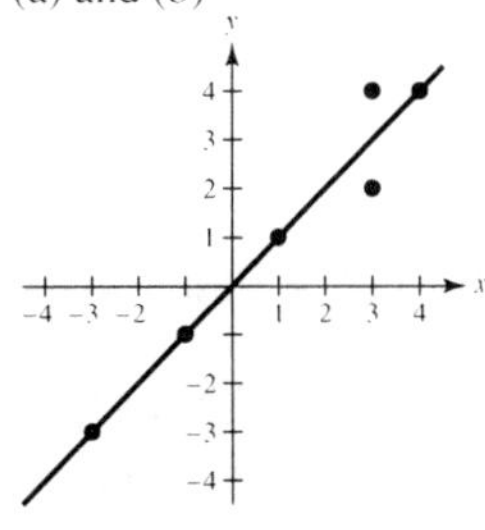

(c) $y = x$

13. (a) and (b)

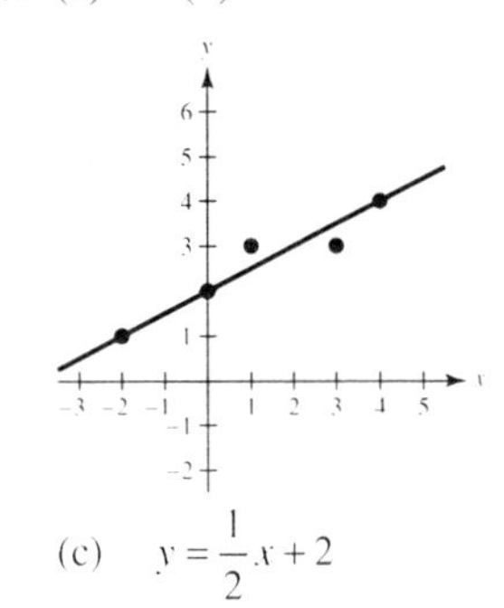

(c) $y = \dfrac{1}{2}x + 2$

15. $y = 0.46x + 1.6$

(a)

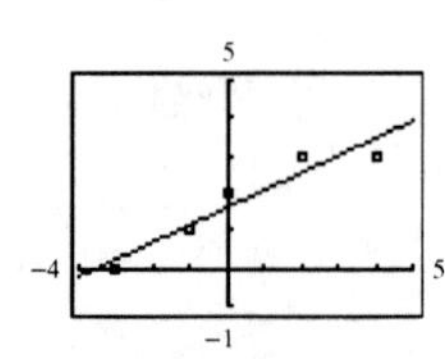

(b)

x	−3	−1	0	2	4
Linear equation	0.22	1.14	1.6	2.52	3.44
Given data	0	1	2	3	3

The model fits the data well.

17. (a)

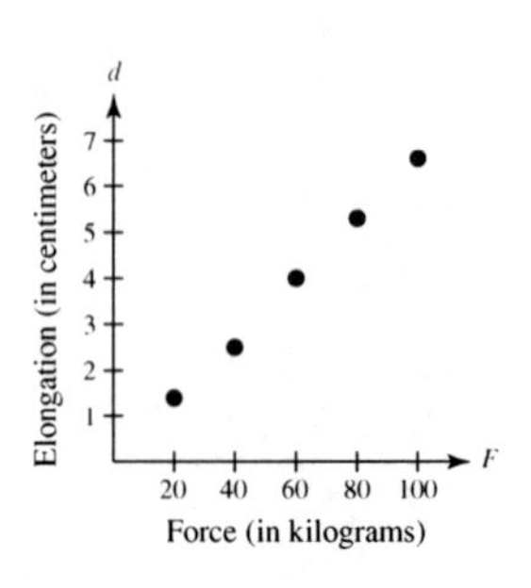

(b) $d = 0.07F - 0.3$

(c) $d = 0.066F$ or $F = 15.13d + 0.096$

(d) If $F = 55$, $d = 0.066(55) \approx 3.63\text{cm}$.

19. (a) and (c)

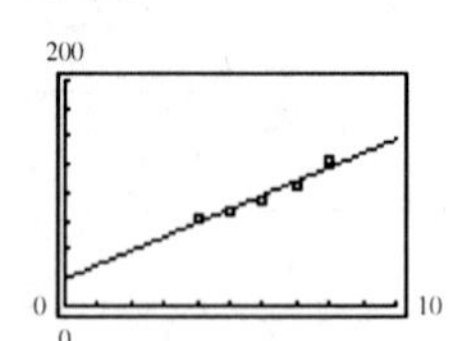

The model fits the data well.

(b) $T = 12.37t + 24.04$

(d) For 2010,
$t = 10$ and $T = 12.37(10) + 24.04 = \$147.74$ million.
For 2015,
$t = 15$ and $T = 12.37(15) + 24.04 = \$209.59$ million.

(e) 12.37; The slope represents the average annual increase in salaries (in millions of dollars).

21. (a) and (c)

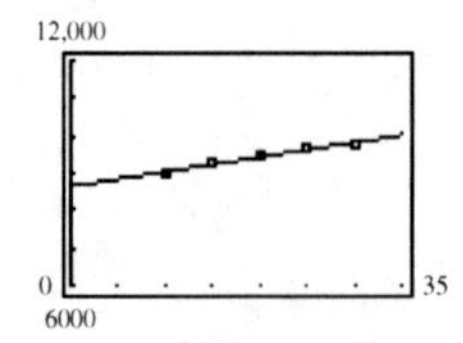

(b) $P = 38.98t + 8655.4$

(d)

Year	2010	2015	2020	2025	2030
Actual	9018	9256	9462	9637	9802
Model	9045.2	9240.1	9435	9629.9	9824.8

The model fits the data well.

(e) For 2050,
$t = 50$ and $P = 38.98(50) + 8655.4 = 10{,}604{,}400$ people.

Answers will vary.

23. (a) $y = 47.77x + 103.8$

Correlation coefficient: 0.81238

(b)

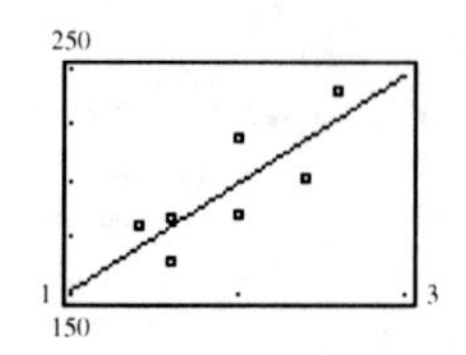

(c) The slope represents the increase in sales due to increased advertising.

(d) For \$1500, $x = 1.5$ and $y = 175.455$ or \$175,455.

25. (a) $T = -0.019t + 4.92$
$r \approx -0.886$

(b) The negative slope means that the winning times are generally decreasing over time.

(c)

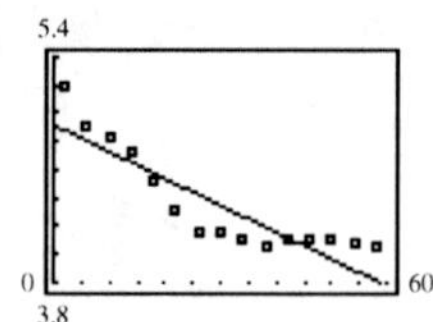

(d)

Year	1952	1956	1960	1964	1968
Actual	5.20	4.91	4.84	4.72	4.53
Model	4.88	4.81	4.73	4.65	4.58

Year	1972	1976	1980	1984	1988
Actual	4.32	4.16	4.15	4.12	4.06
Model	4.50	4.43	4.35	4.27	4.20

Year	1992	1996	2000	2004	2008
Actual	4.12	4.12	4.10	4.09	4.05
Model	4.12	4.05	3.97	3.89	3.82

The model does not fit the data well.

(e) The closer $|r|$ is to 1, the better the model fits the data.

(f) No. The winning times have leveled off in recent years, but the model values continue to decrease to unrealistic times.

27. True. To have positive correlation, the y-values tend to increase as x increases.

29. Answers will vary.

31. $f(x) = 2x^2 - 3x + 5$

(a) $f(-1) = 2 + 3 + 5 = 10$

(b) $f(w+2) = 2(w+2)^2 - 3(w+2) + 5$
$= 2w^2 + 5w + 7$

33. $6x + 1 = -9x - 8$
$15x = -9$
$x = -\frac{9}{15} = -\frac{3}{5}$

35. $8x^2 - 10x - 3 = 0$
$(4x+1)(2x-3) = 0$
$x = -\frac{1}{4}, \frac{3}{2}$

Chapter 1 Review

1.

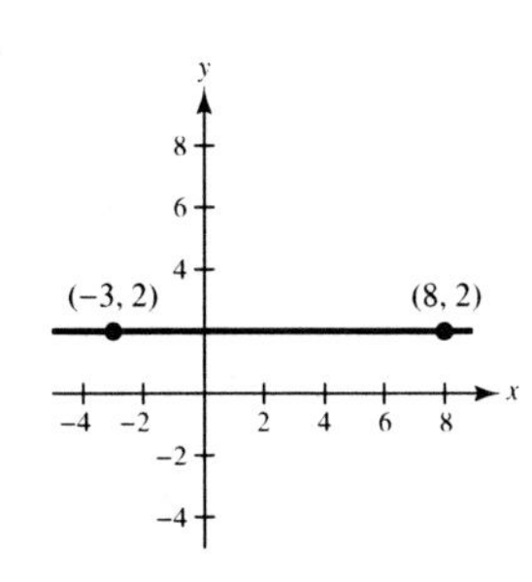

$m = \frac{2-2}{8-(-3)} = \frac{0}{11} = 0$

3.

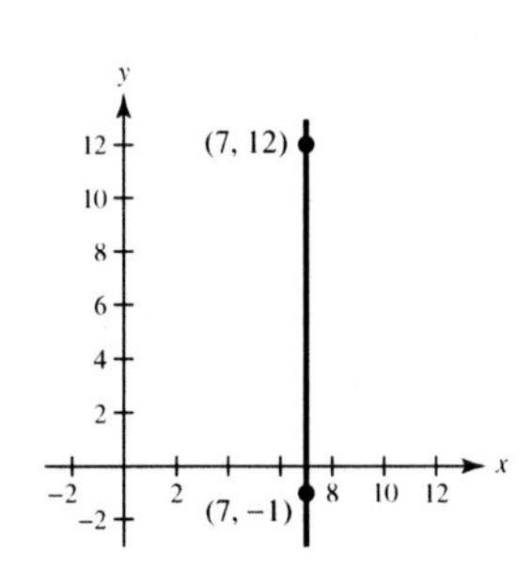

m is undefined.

5.

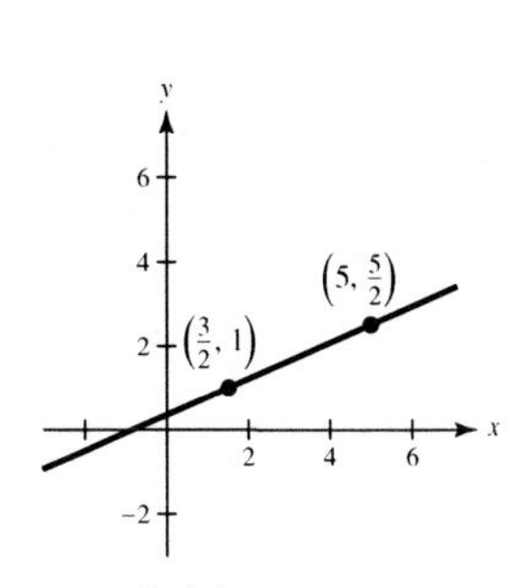

$m = \frac{(5/2) - 1}{5 - (3/2)} = \frac{3/2}{7/2} = \frac{3}{7}$

7.

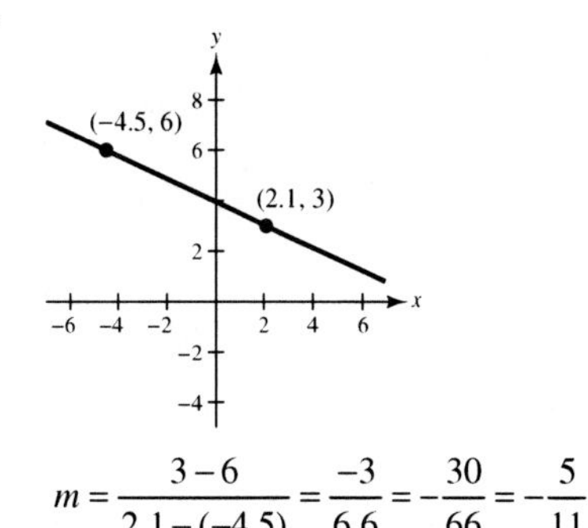

$m = \frac{3-6}{2.1-(-4.5)} = \frac{-3}{6.6} = -\frac{30}{66} = -\frac{5}{11}$

9. (a) $y + 1 = \frac{1}{4}(x-2)$
$4y + 4 = x - 2$
$-x + 4y + 6 = 0$

(b) Three additional points:
$(2+4, -1+1) = (6, 0)$
$(6+4, 0+1) = (10, 1)$
$(10+4, 1+1) = (14, 2)$
(other answers possible)

11. (a) $y + 5 = \frac{3}{2}(x-0)$
$2y + 10 = 3x$
$-3x + 2y + 10 = 0$

(b) Three additional points:
$(0+2, -5+3) = (2, -2)$
$(2+2, -2+3) = (4, 1)$
$(4+2, 1+3) = (6, 4)$
(other answers possible)

13. (a) $y - 6 = 0(x+2) = 0$
$y = 6$ (horizontal line)
$y - 6 = 0$

(b) Three additional points:
(0, 6), (1, 6), (−1, 6)
(other answers possible)

15. (a) m is undefined means that the line is vertical.
$x - 10 = 0$

(b) Three additional points: (10, 0), (10, 1), (10, 2)
(other answers possible)

17. $(2, -1), (4, -1)$

$m = \dfrac{-1-(-1)}{4-2} = \dfrac{0}{2} = 0$ (The line is horizontal.)

$y-(-1) = 0(x-2)$

$y = -1$

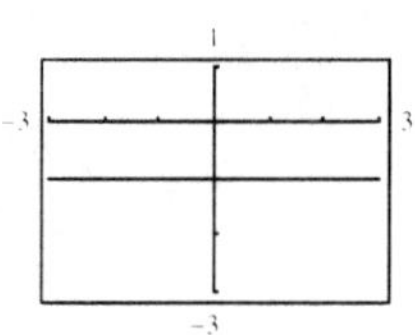

19. $\left(7, \dfrac{11}{3}\right), \left(9, \dfrac{11}{3}\right)$

$m = \dfrac{\frac{11}{3} - \frac{11}{3}}{9-7} = \dfrac{0}{2} = 0$ (The line is horizontal.)

$y - \dfrac{11}{3} = 0(x-7)$

$y = \dfrac{11}{3}$

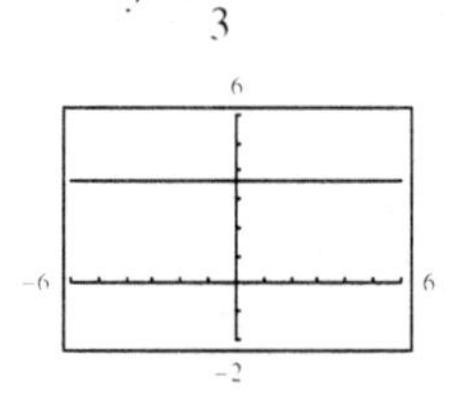

21. $(-1, 0), (6, 2)$

$m = \dfrac{2-0}{6-(-1)} = \dfrac{2}{7}$

$y - 0 = \dfrac{2}{7}(x+1)$

$y = \dfrac{2}{7}x + \dfrac{2}{7}$

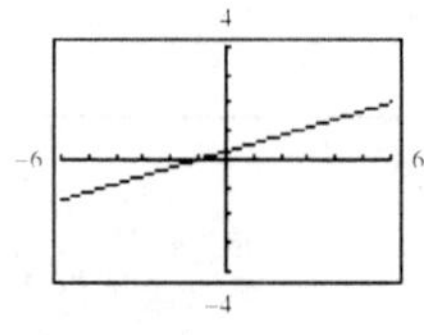

23. $(3, -1), (-3, 2)$

$m = \dfrac{2-(-1)}{-3-3} = \dfrac{3}{-6} = -\dfrac{1}{2}$

$y-(-1) = -\dfrac{1}{2}(x-3)$

$y = -\dfrac{1}{2}x + \dfrac{1}{2}$

For Exercise 25–28, $t = 0$ corresponds to 2010.

25. $(0, 12{,}500), m = 850$

$V - 12{,}500 = 850(t-0)$

$V = 850t + 12{,}500$

27. $(0, 625.50), m = 42.70$

$V - 625.50 = 42.70(t-0)$

$V = 42.70t + 625.50$

29. $(2, 160{,}000), (3, 185{,}000)$

$m = \dfrac{185{,}000 - 160{,}000}{3-2} = 25{,}000$

$S - 160{,}000 = 25{,}000(t-2)$

$S = 25{,}000t + 110{,}000$

For the fourth quarter let $t = 4$. Then we have $S = 25{,}000(4) + 110{,}000 = \$210{,}000$.

31. $5x - 4y = 8 \Rightarrow y = \dfrac{5}{4}x - 2$ and $m = \dfrac{5}{4}$

(a) Parallel slope: $m = \dfrac{5}{4}$

$y-(-2) = \dfrac{5}{4}(x-3)$

$4y + 8 = 5x - 15$

$0 = 5x - 4y - 23$

$y = \dfrac{5}{4}x - \dfrac{23}{4}$

(b) Perpendicular slope: $m = -\dfrac{4}{5}$

$y-(-2) = -\dfrac{4}{5}(x-3)$

$5y + 10 = -4x + 12$

$4x + 5y - 2 = 0$

$y = -\dfrac{4}{5}x + \dfrac{2}{5}$

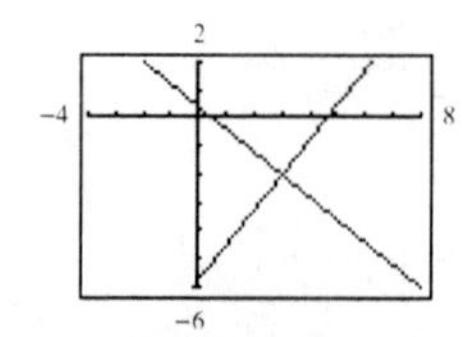

33. (a) Not a function. 20 is assigned two different values.
(b) Function

35. $16x^2 - y^2 = 0 \Rightarrow y = \pm 4x$
No, y is not a function of x. Some x-values correspond to two y-values. For example, $x = 1$ corresponds to $y = 4$ and $y = -4$.

37. $y = 2x - 3$
This is a function of x.

39. $y = \sqrt{1-x}$
This is a function of x.

41. $|y| = x + 2 \Rightarrow y = x + 2$ or $y = -(x + 2)$

Thus, y is not a function of x. Some x-values correspond to two y-values. For example, $x = 1$ corresponds to $y = 3$ and $y = -3$.

43. $f(x) = x^2 + 1$

(a) $f(1) = 1^2 + 1 = 2$

(b) $f(-3) = (-3)^2 + 1 = 10$

(c) $f(b^3) = (b^3)^2 + 1 = b^6 + 1$

(d) $f(x-1) = (x-1)^2 + 1 = x^2 - 2x + 2$

45. $h(x) = \begin{cases} 2x + 1, & x \le -1 \\ x^2 + 2, & x > -1 \end{cases}$

(a) $h(-2) = 2(-2) + 1 = -3$

(b) $h(-1) = 2(-1) + 1 = -1$

(c) $h(0) = 0^2 + 2 = 2$

(d) $h(2) = 2^2 + 2 = 6$

47. The domain of $f(x) = \dfrac{x-1}{x+2}$ is all real numbers $x \ne -2$.

49. $f(x) = \sqrt{25 - x^2}$

$$25 - x^2 \ge 0$$

$$(5 + x)(5 - x) \ge 0$$

The domain is $[-5, 5]$.

51. (a) $C(x) = 16{,}000 + 5.35x$

(b) $P(x) = R(x) - C(x)$

$= 8.20x - (16{,}000 + 5.35x)$

$= 2.85x - 16{,}000$

53. $f(x) = 2x^2 + 3x - 1$

$f(x + h) = 2(x + h)^2 + 3(x + h) - 1$

$= 2x^2 + 4xh + 2h^2 + 3x + 3h - 1$

$\dfrac{f(x+h) - f(x)}{h} =$

$\dfrac{(2x^2 + 4xh + 2h^2 + 3x + 3h - 1) - (2x^2 + 3x - 1)}{h}$

$= \dfrac{4xh + 2h^2 + 3h}{h}$

$= 4x + 2h + 3,\ h \ne 0$

55. Domain: all real numbers x

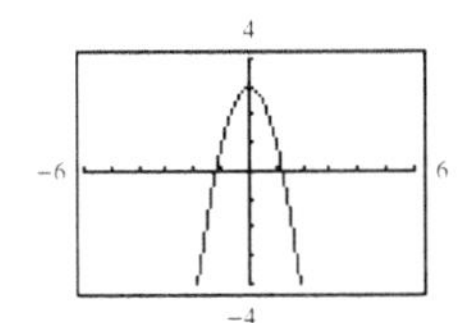

Range: $y \le 3$

57.

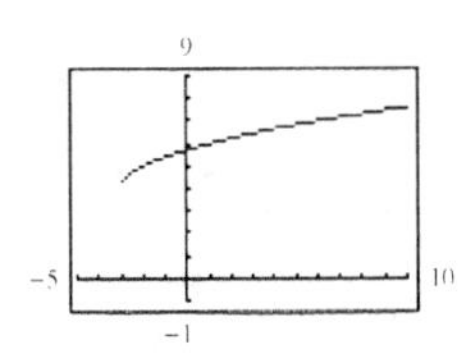

Domain: $[-3, \infty)$

Range: $[4, \infty)$

59.

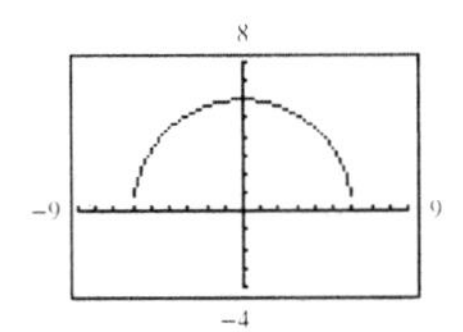

Domain: $36 - x^2 \ge 0 \Rightarrow x^2 \le 36 \Rightarrow -6 \le x \le 6$

Range: $0 \le y \le 6$

61.

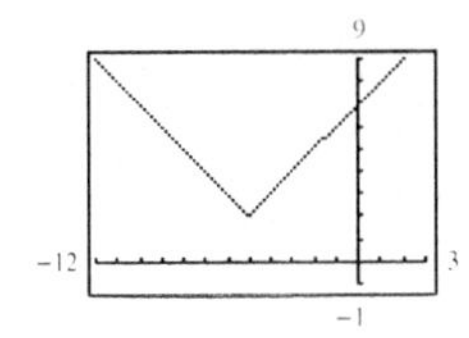

Domain: all real numbers x

Range: $[2, \infty)$

63. $y - 4x = x^2$

A vertical line intersects the graph just once, so y is a function of x. Solve for y and graph $y_1 = x^2 + 4x$.

65. $3x + y^2 - 2 = 0$

A vertical line intersects the graph more than once, so y is not a function of x. Solve for y and graph $y_1 = \sqrt{-3x + 2}$ and $y_2 = -\sqrt{-3x + 2}$.

67. $f(x) = x^3 - 3x$

(a)

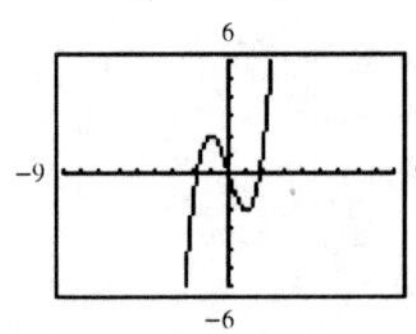

(b) Increasing on $(-\infty, -1)$ and $(1, \infty)$

Decreasing on $(-1, 1)$

69. $f(x) = x\sqrt{x-6}$

(a)

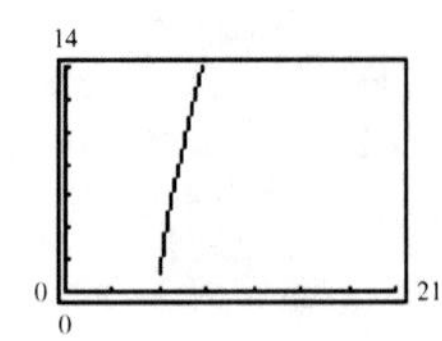

(b) Increasing on $(6, \infty)$

71. $f(x) = (x^2 - 4)^2$

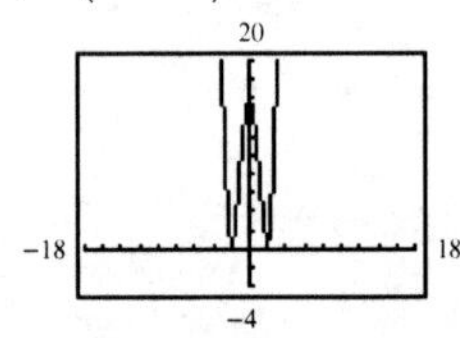

Relative minima: $(-2, 0)$ and $(2, 0)$

Relative maximum: $(0, 16)$

73. $h(x) = 4x^3 - x^4$

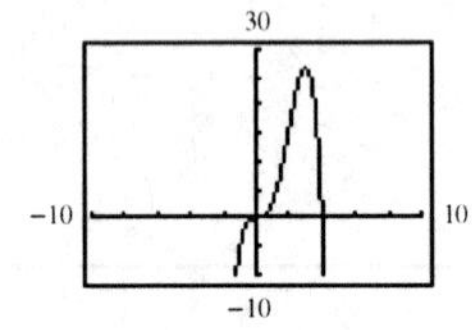

Relative maximum: $(3, 27)$

75. $f(x) = \begin{cases} 3x + 5, & x < 0 \\ x - 4, & x \geq 0 \end{cases}$

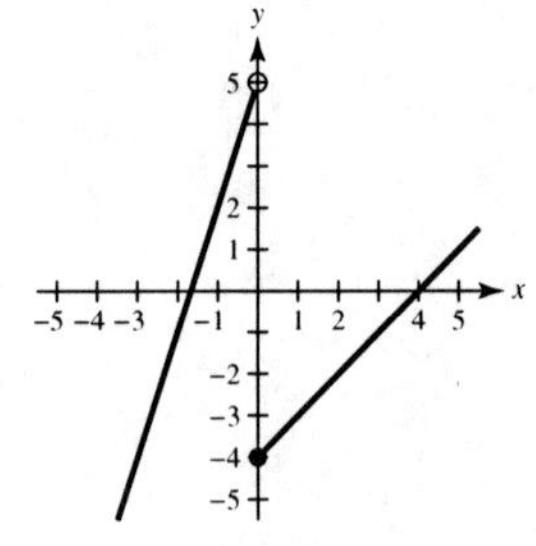

77. $f(x) = [\![x]\!] + 3$

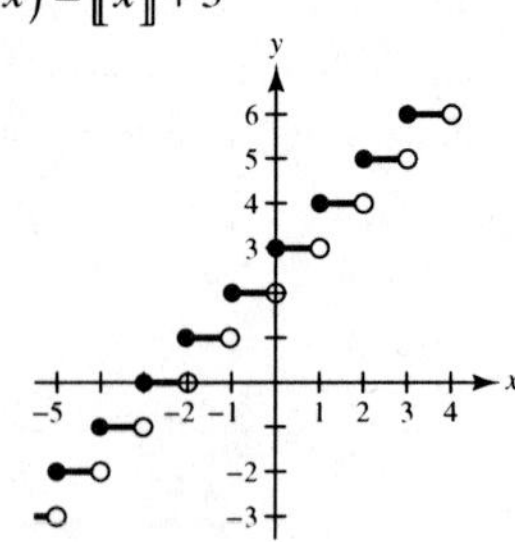

79. $f(-x) = (-x)^2 + 6$

$= x^2 + 6$

$= f(x)$

f is even.

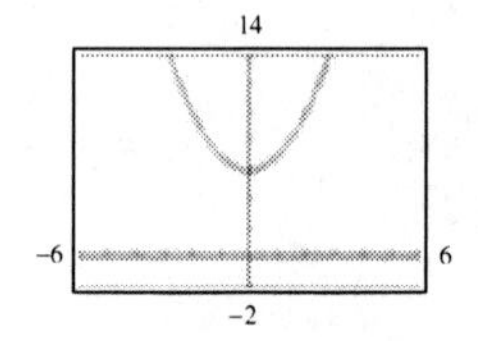

The graph is symmetric with respect to the y-axis. So, f is even.

81. $f(-x) = \left((-x)^2 - 8\right)^2$

$= (x^2 - 8)^2$

$= f(x)$

f is even.

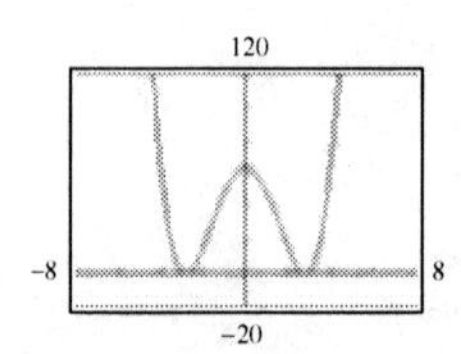

The graph is symmetric with respect to the y-axis. So, f is even.

83. $f(-x) = 3(-x)^{\frac{5}{2}} \neq f(x)$ and $f(-x) \neq -f(x)$

f is neither even nor odd.

(Note that the domain of f is $x \geq 0$.)

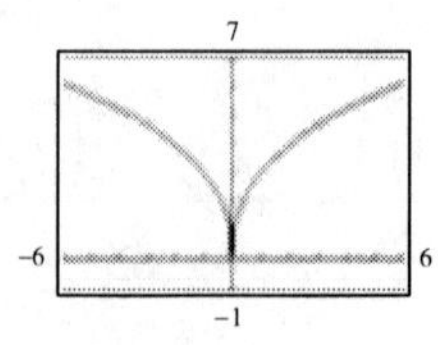

The graph is neither symmetric with respect to the origin nor with respect to the y-axis. So, f is neither even nor odd.

85. $f(-x) = \sqrt{4-(-x)^2}$
$= \sqrt{4-x^2}$
$= f(x)$

f is even.

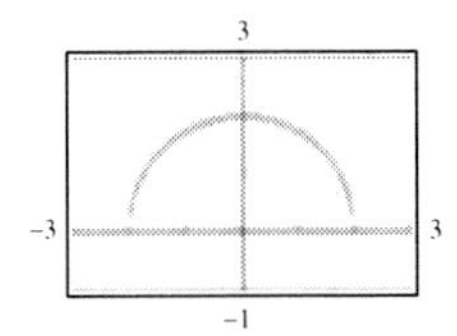

The graph is symmetric with respect to the y-axis. So, f is even.

87. Horizontal shift three units to the right of

$f(x) = \frac{1}{x}$: $y = \frac{1}{x-3}$

89. Horizontal shift two units to the right, followed by a vertical shift one unit upward of

$f(x) = x^2$: $y = (x-2)^2 + 1$

91. Vertical shift three units upward of

$f(x) = |x|$: $y = |x| + 3$

93.

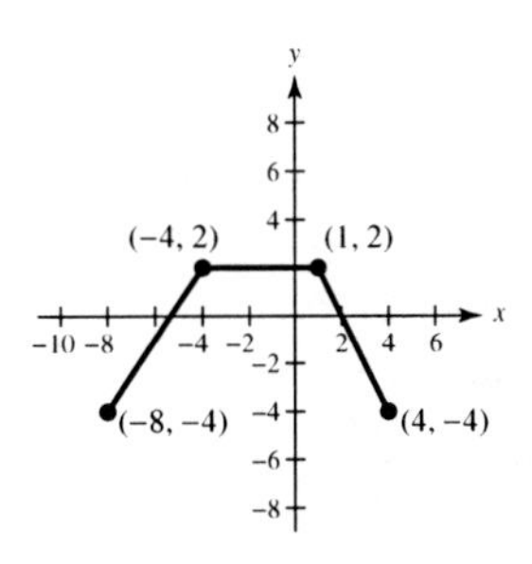

$y = f(-x)$ is a reflection in the y-axis.

95.

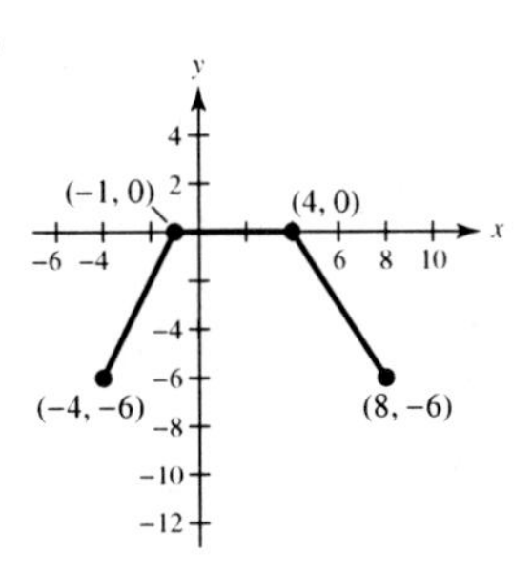

$y = f(x) - 2$ is a vertical shift two units downward.

97. $h(x) = \frac{1}{x} - 6$

(a) $f(x) = \frac{1}{x}$

(b) The graph of h is a vertical shift six units downward of f.

(c)

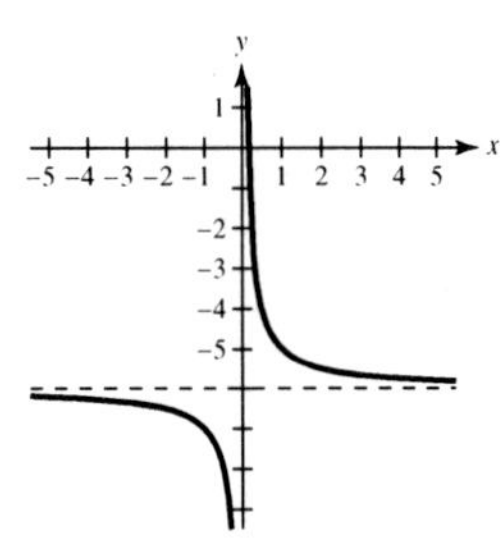

(d) $h(x) = f(x) - 6$

99. $h(x) = (x-2)^3 + 5$

(a) $f(x) = x^3$

(b) The graph of h is a horizontal shift of f two units to the right, followed by a vertical shift five units upward of f.

(c)

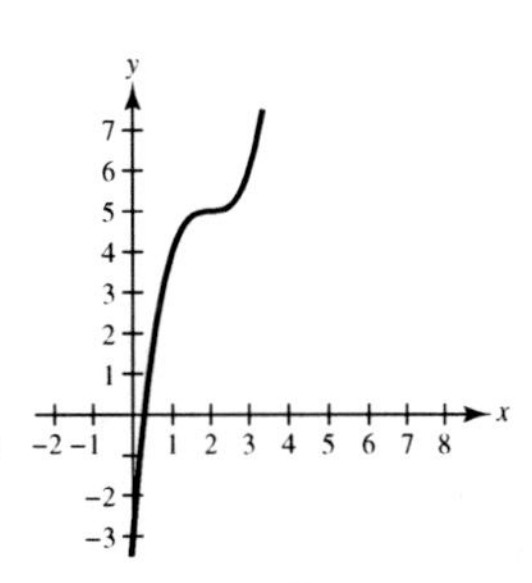

(d) $h(x) = (x-2)^3 + 5 = f(x-2) + 5$

101. $h(x) = -\sqrt{x} + 6$

(a) $f(x) = \sqrt{x}$

(b) The graph of h is a reflection in the x-axis and a vertical shift six units upward of f.

(c)

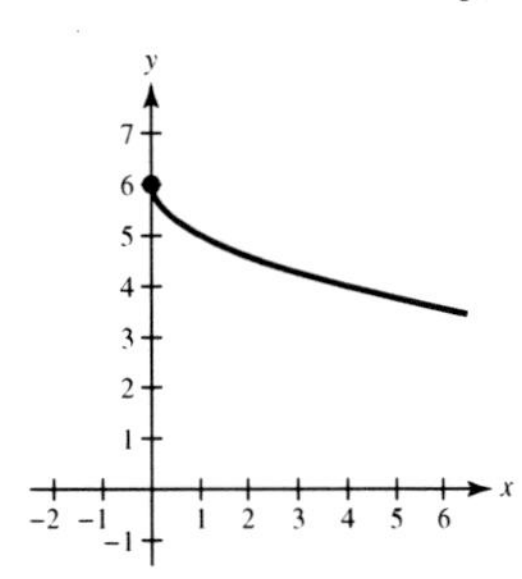

(d) $h(x) = -f(x) + 6$

103. $h(x) = |x| + 9$

(a) $f(x) = |x|$

(b) The graph of h is a vertical shift of f nine units upward.

(c)

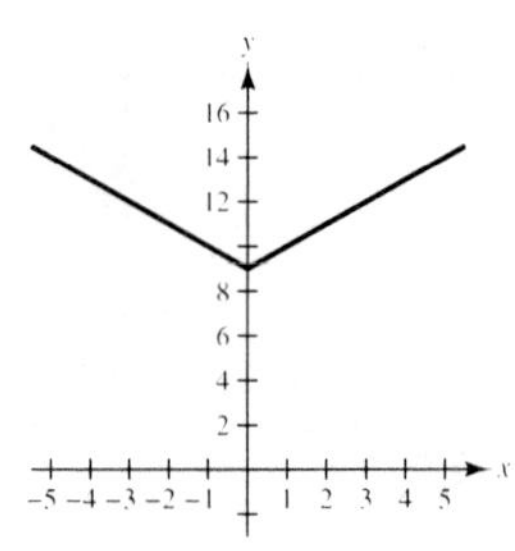

(d) $h(x) = |x| + 9$

$= f(x) + 9$

105. $h(x) = \dfrac{-2}{x+1} - 3$

(a) $f(x) = \dfrac{1}{x}$

(b) h is a horizontal shift one unit to the left, a reflection in the x-axis, a vertical stretch, followed by a vertical shift three units downward of f.

(c)

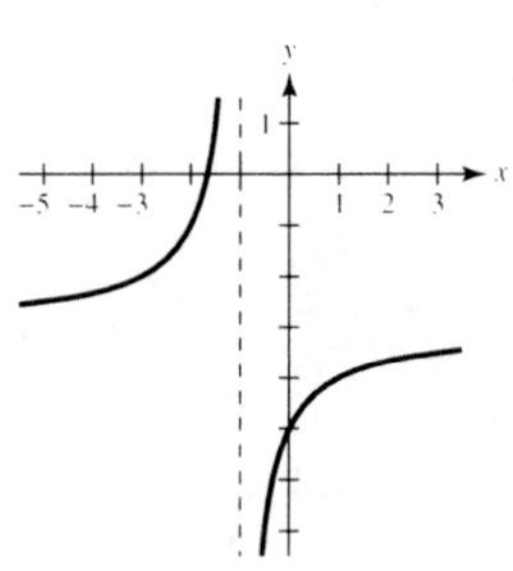

(d) $h(x) = -2f(x+1) - 3$

107. $(f - g)(4) = f(4) - g(4)$

$= [3 - 2(4)] - \sqrt{4}$

$= -5 - 2$

$= -7$

109. $(f + g)(25) = f(25) + g(25)$

$= -47 + 5$

$= -42$

111. $(fh)(1) = f(1)h(1) = (3 - 2(1))(3(1)^2 + 2)$

$= (1)(5) = 5$

113. $(h \circ g)(5) = h(g(5))$

$= h\left(\sqrt{5}\right)$

$= 3\left(\sqrt{5}\right)^2 + 2 = 17$

115. $(f \circ h)(-4) = f(h(-4))$

$= f(50)$

$= 3 - 2(50) = -97$

117. $f(x) = x^2, g(x) = x + 3$

$(f \circ g)(x) = f(x+3)$

$= (x+3)^2 = h(x)$

119. $f(x) = \sqrt{x}, g(x) = 4x + 2$

$(f \circ g)(x) = f(4x+2) = \sqrt{4x+2} = h(x)$

121. $f(x) = \dfrac{4}{x}, g(x) = x + 2$

$(f \circ g)(x) = f(x+2) = \dfrac{4}{x+2} = h(x)$

123.

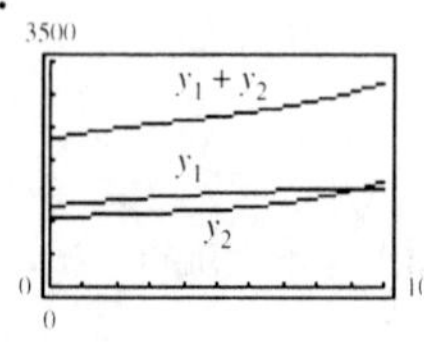

125. $f(x) = 6x$

$f^{-1}(x) = \dfrac{1}{6}x$

$f(f^{-1}(x)) = f\left(\dfrac{1}{6}x\right) = 6\left(\dfrac{1}{6}x\right) = x$

$f^{-1}(f(x)) = f^{-1}(6x) = \dfrac{1}{6}(6x) = x$

127. $f(x) = \dfrac{1}{2}x + 3$

$f^{-1}(x) = 2(x-3) = 2x - 6$

$f(f^{-1}(x)) = f(2(x-3))$

$= \dfrac{1}{2}(2(x-3)) + 3 = x - 3 + 3 = x$

$f^{-1}(f(x)) = f^{-1}\left(\dfrac{1}{2}x + 3\right)$

$= 2\left(\dfrac{1}{2}x + 3 - 3\right) = 2\left(\dfrac{1}{2}x\right) = x$

129. $f(x) = 3 - 4x, g(x) = \dfrac{3-x}{4}$

(a) $f(g(x)) = 3 - 4\left(\dfrac{3-x}{4}\right) = 3 - (3-x) = x$

$g(f(x)) = \dfrac{3 - (3-4x)}{4} = \dfrac{4x}{4} = x$

(b)

(c)

$Y_1 = 3 - 4X$

$Y_2 = \dfrac{3 - X}{4}$

$Y_3 = Y_1(Y_2)$

$Y_4 = Y_2(Y_1)$

X	Y_3	Y_4
−2	−2	−2
−1	−1	−1
0	0	0
1	1	1
2	2	2

131.

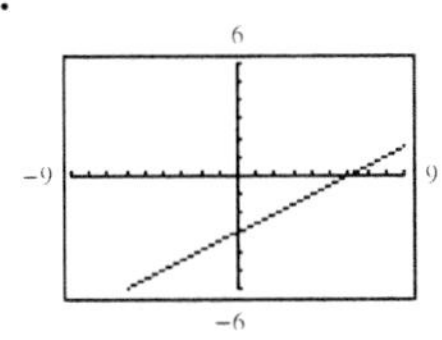

$f(x) = \dfrac{1}{2}x - 3$ passes the Horizontal Line Test, and hence is one-to-one and has an inverse function.

133.

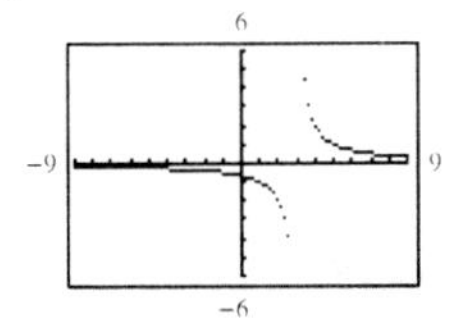

$h(t) = \dfrac{2}{t - 3}$ passes the Horizontal Line Test, and hence is one-to-one and has an inverse function.

135.
$$y = \frac{1}{2}x - 5$$
$$x = \frac{1}{2}y - 5$$
$$x + 5 = \frac{1}{2}y$$
$$y = 2(x + 5)$$
$$f^{-1}(x) = 2x + 10$$

137.
$$f(x) = 4x^3 - 3$$
$$y = 4x^3 - 3$$
$$x = 4y^3 - 3$$
$$x + 3 = 4y^3$$
$$\frac{x + 3}{4} = y^3$$
$$f^{-1}(x) = \sqrt[3]{\frac{x + 3}{4}}$$

139. $f(x) = \sqrt{x + 10}$
$$y = \sqrt{x + 10},\ x \ge -10,\ y \ge 0$$
$$x = \sqrt{y + 10},\ y \ge -10,\ x \ge 0$$
$$x^2 = y + 10$$
$$x^2 - 10 = y$$
$$f^{-1}(x) = x^2 - 10,\ x \ge 0$$

141. $f(x) = \dfrac{1}{4}x^2 + 1,\ x \ge 0$
$$y = \frac{1}{4}x^2 + 1$$
$$x = \frac{1}{4}y^2 + 1$$
$$x - 1 = \frac{1}{4}y^2$$
$$4(x - 1) = y^2$$
$$f^{-1}(x) = \sqrt{4(x - 1)} = 2\sqrt{x - 1}$$

The positive square root is chosen as f^{-1} since the domain of f is $[0, \infty)$.

143. Negative correlation

145. (a)

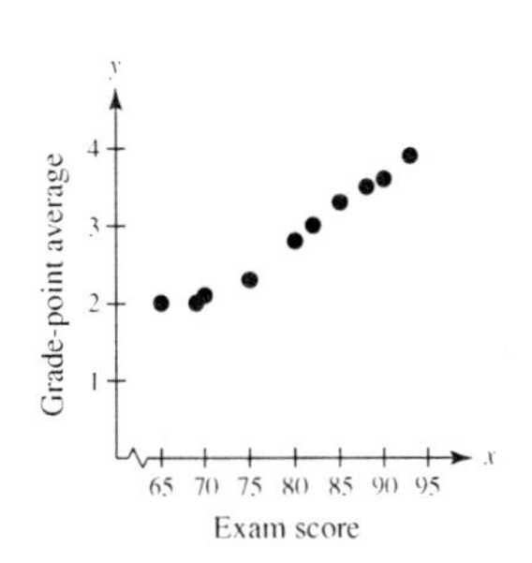

(b) Yes, the relationship is approximately linear. Higher entrance exam scores, x, are associated with higher grade-point averages, y.

147. (a)

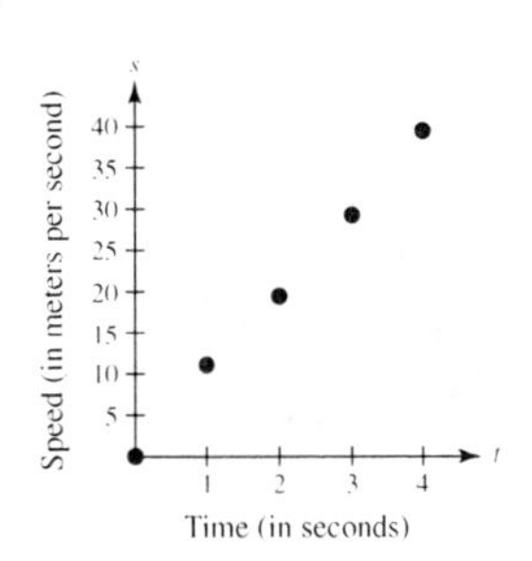

(b) $s \approx 10t$ (Approximations will vary.)

(c) $s = 9.7t + 0.4;\ r \approx 0.99933$

(d) For $t = 2.5$:
$$s = 9.7(2.5) + 0.4$$
$$= 24.25 + 0.4$$
$$= 24.65 \text{ m/sec}$$

149. False. $g(x) = -[(x - 6)^2 + 3] = -(x - 6)^2 - 3$ and $g(-1) = -52 \ne 28$

151. False. $f(x) = \dfrac{1}{x}$ or $f(x) = x$ satisfies $f = f^{-1}$.

Chapter 1 Test

1. $5x + 2y = 3$

$2y = -5x + 3$

$y = -\frac{5}{2}x + \frac{3}{2}$

$\text{Slope} = -\frac{5}{2}$

(a) Parallel line slope: $-\frac{5}{2}$

$y - 4 = -\frac{5}{2}(x - 0)$

$y = -\frac{5}{2}x + 4$

$5x + 2y - 8 = 0$

(b) Perpendicular line slope: $\frac{2}{5}$

$y - 4 = \frac{2}{5}(x - 0)$

$y = \frac{2}{5}x + 4$

$2x - 5y + 20 = 0$

2. $\text{Slope} = \frac{4-(-1)}{-3-2} = \frac{5}{-5} = -1$

$y + 1 = -1(x - 2)$

$y = -x + 1$

3. No. For some x there corresponds more than one value of y. For instance, if $x = 1$, $y = \pm\frac{1}{\sqrt{3}}$.

4. $f(x) = |x + 2| - 15$

(a) $f(-8) = |-8 + 2| - 15 = 6 - 15 = -9$

(b) $f(14) = |14 + 2| - 15 = 16 - 15 = 1$

(c) $f(t - 6) = |t - 6 + 2| - 15 = |t - 4| - 15$

5. $3 - x \geq 0 \Rightarrow$ domain is all $x \leq 3$.

6. Total Cost = Variable Costs + Fixed Costs

$C = 25.60x + 24{,}000$

Revenue = Price per unit × number of units

$R = 99.50x$

Profit = Revenue − Cost

$P = 99.50x - (25.60x + 24{,}000)$

$= 73.90x - 24{,}000$

7. $f(-x) = 2(-x)^3 - 3(-x)$

$= -2x^3 + 3x = -f(x)$

Odd

8. $f(-x) = 3(-x)^4 + 5(-x)^2$

$= 3x^4 + 5x^2 = f(x)$

Even

9. $h(x) = \frac{1}{4}x^4 - 2x^2 = \frac{1}{4}x^2(x^2 - 8)$

By graphing h, you see that the graph is increasing on $(-2, 0)$ and $(2, \infty)$ and decreasing on $(-\infty, -2)$ and $(0, 2)$.

10. $g(t) = |t + 2| - |t - 2|$

By graphing g, you see that the graph is increasing on $(-2, 2)$, and constant on $(-\infty, -2)$ and $(2, \infty)$.

11.

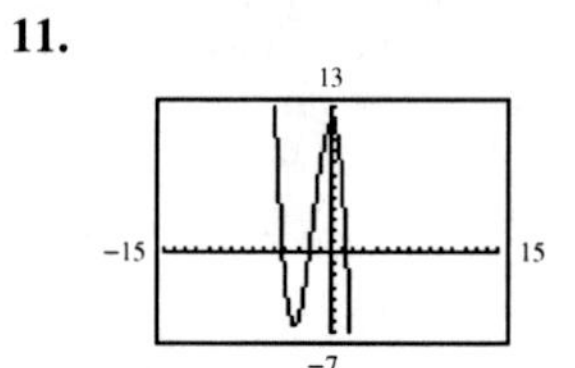

Relative minimum: $(-3.33, -6.52)$

Relative maximum: $(0, 12)$

12.

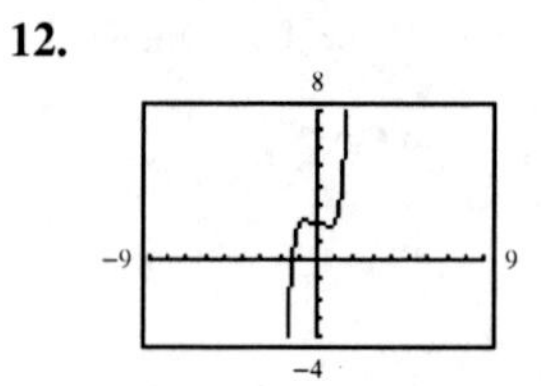

Relative minimum: $(0.77, 1.81)$

Relative maximum: $(-0.77, 2.19)$

13. (a) $f(x) = x^3$

(b) g is obtained from f by a horizontal shift five units to the right, a vertical stretch of 2, a reflection in the x-axis, and a vertical shift three units upward.

(c)

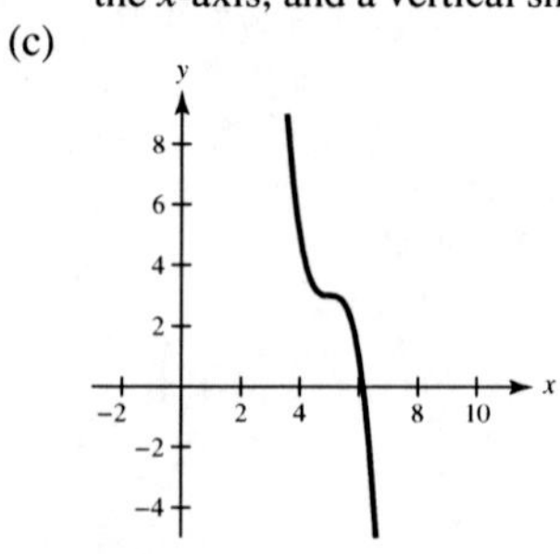

14. (a) $f(x) = \sqrt{x}$

(b) g is obtained from f by a reflection in the y-axis, and a horizontal shift seven units to the left.

(c)

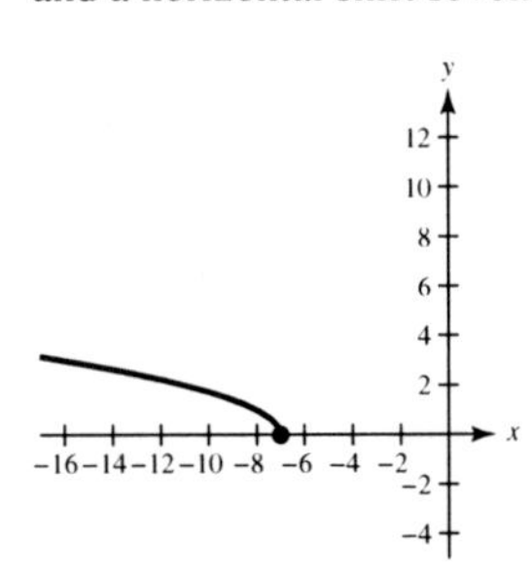

15. (a) $f(x) = |x|$

(b) g is obtained from f by a vertical stretch of 4 followed by a vertical shift seven units downward.

(c)

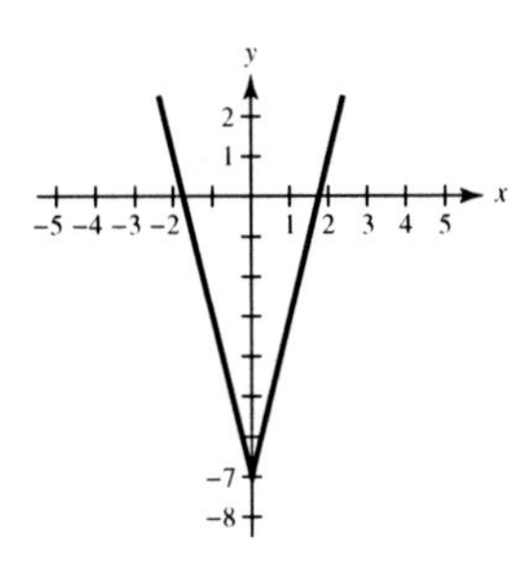

16. (a) $(f - g)(x) = x^2 - \sqrt{2 - x}$

Domain: $x \le 2$

(b) $\left(\dfrac{f}{g}\right)(x) = \dfrac{x^2}{\sqrt{2 - x}}$

Domain: $x < 2$

(c) $(f \circ g)(x) = f\left(\sqrt{2 - x}\right) = 2 - x$

Domain: $x \le 2$

(d) $(g \circ f)x = g\left(x^2\right) = \sqrt{2 - x^2}$

Domain: $-\sqrt{2} \le x \le \sqrt{2}$

17. $f(x) = x^3 + 8$

Yes, f is one-to-one and has an inverse function.

$$y = x^3 + 8$$
$$x = y^3 + 8$$
$$x - 8 = y^3$$
$$\sqrt[3]{x - 8} = y$$
$$f^{-1}(x) = \sqrt[3]{x - 8}$$

18. $f(x) = x^2 + 6$

No, f is not one-to-one, and does not have an inverse function.

19. $f(x) = \dfrac{3x\sqrt{x}}{8}$

Yes, f is one-to-one and has an inverse function.

$$y = \frac{3}{8}x^{3/2},\ x \ge 0,\ y \ge 0$$
$$x = \frac{3}{8}y^{3/2},\ y \ge 0,\ x \ge 0$$
$$\frac{8}{3}x = y^{3/2}$$
$$\left(\frac{8}{3}x\right)^{2/3} = y$$
$$f^{-1}(x) = \left(\frac{8}{3}x\right)^{2/3},\ x \ge 0$$

20. $C = 1.686t + 31.09$

Let $C = 50$ and solve for t.

$$50 = 1.686t + 31.09$$
$$18.91 = 1.686t$$
$$t \approx 11.21 \text{ or approximately 2012}$$

CHAPTER 2

Section 2.1

1. nonnegative integer, real

3. Yes. $f(x) = (x-2)^2 + 3$ is in the form $f(x) = a(x-h)^2 + k$. The vertex is $(2, 3)$.

5. $f(x) = (x-2)^2$ opens upward and has vertex $(2, 0)$. Matches graph (c).

7. $f(x) = x^2 + 3$ opens upward and has vertex $(0, 3)$. Matches graph (b).

9.

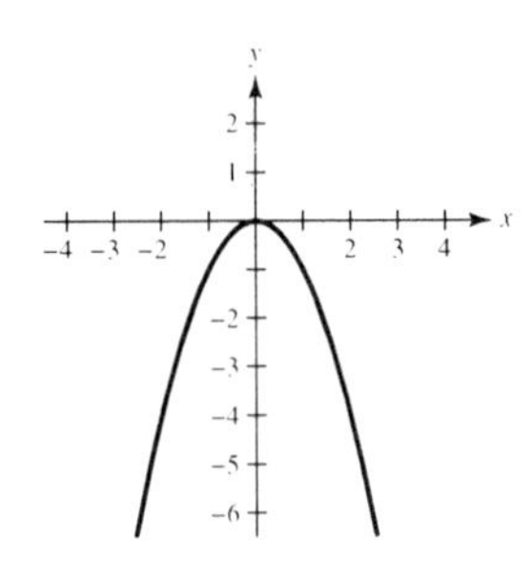

The graph of $y = -x^2$ is a reflection of $y = x^2$ in the x-axis.

11.

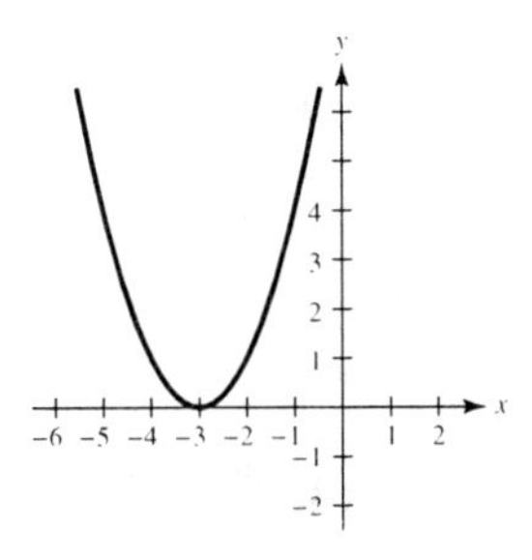

The graph of $y = (x+3)^2$ is a horizontal shift three units to the left of $y = x^2$.

13.

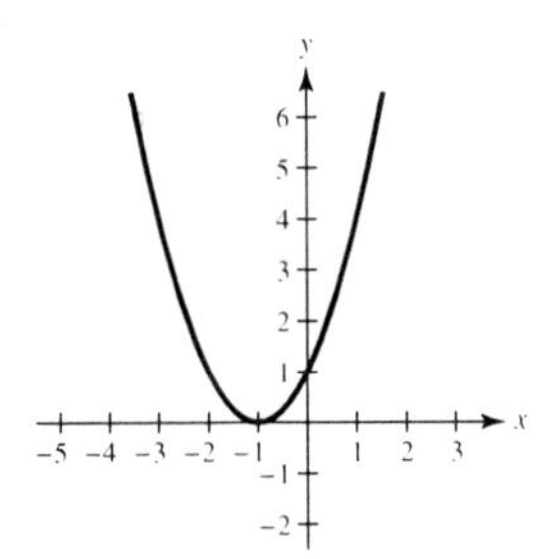

The graph of $y = (x+1)^2$ is a horizontal shift one unit to the left of $y = x^2$.

15.

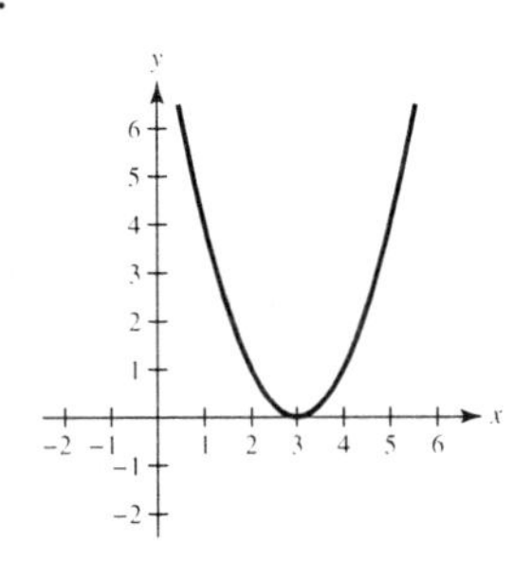

The graph of $y = (x-3)^2$ is a horizontal shift three units to the right of $y = x^2$.

17. $f(x) = 25 - x^2$
$= -x^2 + 25$
A parabola opening downward with vertex $(0, 25)$

19. $f(x) = \frac{1}{2}x^2 - 4$

A parabola opening upward with vertex $(0, -4)$

21. $f(x) = (x+4)^2 - 3$
A parabola opening upward with vertex $(-4, -3)$

23. $h(x) = x^2 - 8x + 16$
$= (x-4)^2$
A parabola opening upward with vertex $(4, 0)$

25. $f(x) = x^2 - x + \frac{5}{4}$

$$= (x^2 - x) + \frac{5}{4}$$

$$= \left(x^2 - x + \frac{1}{4}\right) + \frac{5}{4} - \frac{1}{4}$$

$$= \left(x - \frac{1}{2}\right)^2 + 1$$

A parabola opening upward with vertex $\left(\frac{1}{2}, 1\right)$

27. $f(x) = -x^2 + 2x + 5$
$= -(x^2 - 2x) + 5$
$= -(x^2 - 2x + 1) + 5 + 1$
$= -(x-1)^2 + 6$
A parabola opening downward with vertex $(1, 6)$

29. $h(x) = 4x^2 - 4x + 21$
$= 4(x^2 - x) + 21$
$= 4\left(x^2 - x + \frac{1}{4}\right) + 21 - 4\left(\frac{1}{4}\right)$
$= 4\left(x - \frac{1}{2}\right)^2 + 20$

A parabola opening upward with vertex $\left(\frac{1}{2}, 20\right)$

31. $f(x) = -(x^2 + 2x - 3)$
$= -(x^2 + 2x) + 3$
$= -(x^2 + 2x + 1) + 3 + 1$
$= -(x + 1)^2 + 4$
$-(x^2 + 2x - 3) = 0$
$-(x + 3)(x - 1) = 0$
$x + 3 = 0 \Rightarrow x = -3$
$x - 1 = 0 \Rightarrow x = 1$
A parabola opening downward with vertex $(-1, 4)$ and x-intercepts $(-3, 0)$ and $(1, 0)$.

33. $g(x) = x^2 + 8x + 11$
$= (x^2 + 8x) + 11$
$= (x^2 + 8x + 16) + 11 - 16$
$= (x + 4)^2 - 5$
$x^2 + 8x + 11 = 0$

$$x = \frac{-8 \pm \sqrt{8^2 - 4(1)(11)}}{2(1)} = \frac{-8 \pm \sqrt{64 - 44}}{2} = \frac{-8 \pm \sqrt{20}}{2} = \frac{-8 \pm 2\sqrt{5}}{2} = -4 \pm \sqrt{5}$$

A parabola opening upward with vertex $(-4, -5)$ and x-intercepts $(-4 \pm \sqrt{5}, 0)$.

35. $f(x) = -2x^2 + 16x - 31$
$= -2(x^2 - 8x) - 31$
$= -2(x^2 - 8x + 16) - 31 + 32$
$= -2(x - 4)^2 + 1$
$-2x^2 + 16x - 31 = 0$

$$x = \frac{-16 \pm \sqrt{16^2 - 4(-2)(-31)}}{2(-2)} = \frac{-16 \pm \sqrt{256 - 248}}{-4} = \frac{-16 \pm \sqrt{8}}{-4} = \frac{-16 \pm 2\sqrt{2}}{-4} = 4 \pm \frac{1}{2}\sqrt{2}$$

A parabola opening downward with vertex $(4, 1)$ and x-intercepts $\left(4 \pm \frac{1}{2}\sqrt{2}, 0\right)$

37. $(-1, 4)$ is the vertex.
$f(x) = a(x + 1)^2 + 4$

Since the graph passes through the point $(1, 0)$, we have:
$0 = a(1 + 1)^2 + 4$
$0 = 4a + 4$
$-1 = a$
Thus, $f(x) = -(x + 1)^2 + 4$. Note that $(-3, 0)$ is on the parabola.

39. $(-2,5)$ is the vertex.
$f(x) = a(x + 2)^2 + 5$
Since the graph passes through the point $(0, 9)$, we have:
$9 = a(0 + 2)^2 + 5$
$4 = 4a$
$1 = a$
Thus, $f(x) = (x + 2)^2 + 5$.

41. $(1, -2)$ is the vertex.
$f(x) = a(x - 1)^2 - 2$
Since the graph passes through the point $(-1, 14)$, we have:

$14 = a(-1 - 1)^2 - 2$
$14 = 4a - 2$
$16 = 4a$
$4 = a$
Thus, $f(x) = 4(x - 1)^2 - 2$.

43. $\left(\frac{1}{2}, 1\right)$ is the vertex.

$$f(x) = a\left(x - \frac{1}{2}\right)^2 + 1$$

Since the graph passes through the point $\left(-2, -\frac{21}{5}\right)$, we have:

$$-\frac{21}{5} = a\left(-2 - \frac{1}{2}\right)^2 + 1$$
$$-\frac{21}{5} = \frac{25}{4}a + 1$$
$$-\frac{26}{5} = \frac{25}{4}a$$
$$-\frac{104}{125} = a$$

Thus, $f(x) = -\frac{104}{125}\left(x - \frac{1}{2}\right)^2 + 1.$

45. $y = x^2 - 4x - 5$

x-intercepts: $(5, 0)$, $(-1, 0)$

$$0 = x^2 - 4x - 5$$
$$0 = (x - 5)(x + 1)$$
$$x = 5 \text{ or } x = -1$$

47. $y = x^2 + 8x + 16$

x-intercept: $(-4, 0)$

$$0 = x^2 + 8x + 16$$
$$0 = (x + 4)^2$$
$$x = -4$$

49. $y = x^2 - 4x$

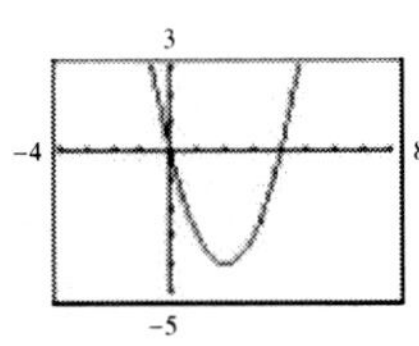

x-intercepts: $(0, 0)$, $(4, 0)$

$$0 = x^2 - 4x$$
$$0 = x(x - 4)$$
$$x = 0 \text{ or } x = 4$$

51. $y = 2x^2 - 7x - 30$

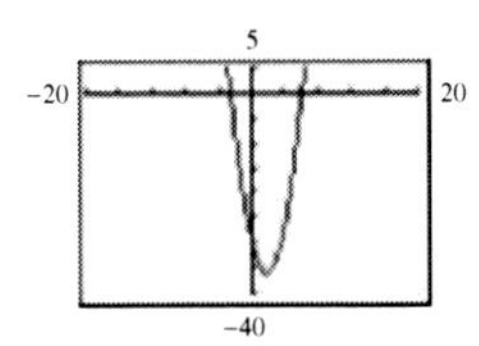

x-intercepts: $\left(-\frac{5}{2}, 0\right)$, $(6, 0)$

$$0 = 2x^2 - 7x - 30$$
$$0 = (2x + 5)(x - 6)$$
$$x = -\frac{5}{2} \text{ or } x = 6$$

53. $y = -\frac{1}{2}(x^2 - 6x - 7)$

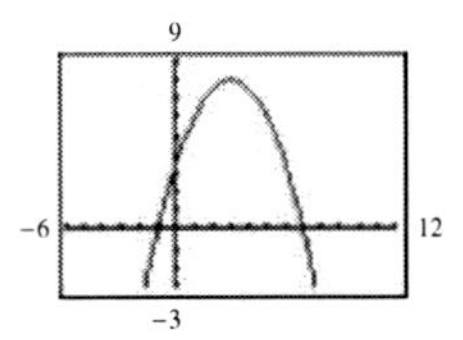

x-intercepts: $(-1, 0)$, $(7, 0)$

$$0 = -\frac{1}{2}(x^2 - 6x - 7)$$
$$0 = x^2 - 6x - 7$$
$$0 = (x + 1)(x - 7)$$
$$x = -1, 7$$

55. $f(x) = [x - (-1)](x - 3)$, opens upward

$$= (x + 1)(x - 3)$$
$$= x^2 - 2x - 3$$

$g(x) = -[x - (-1)](x - 3)$, opens downward

$$= -(x + 1)(x - 3)$$
$$= -(x^2 - 2x - 3)$$
$$= -x^2 + 2x + 3$$

Note: $f(x) = a(x + 1)(x - 3)$ has x-intercepts $(-1, 0)$ and $(3, 0)$ for all real numbers $a \neq 0$.

57. $f(x) = [x - (-3)]\left[x - \left(-\frac{1}{2}\right)\right](2)$, opens upward

$$= (x + 3)\left(x + \frac{1}{2}\right)(2)$$
$$= (x + 3)(2x + 1)$$
$$= 2x^2 + 7x + 3$$

$g(x) = -(2x^2 + 7x + 3)$, opens downward

$$= -2x^2 - 7x - 3$$

Note: $f(x) = a(x + 3)(2x + 1)$ has x-intercepts $(-3, 0)$ and $\left(-\frac{1}{2}, 0\right)$ for all real numbers $a \neq 0$.

59. Let $x =$ the first number and $y =$ the second number.

Then the sum is $x + y = 110 \Rightarrow y = 110 - x$.

The product is

$$P(x) = xy = x(110 - x) = 110x - x^2.$$
$$P(x) = -x^2 + 110x$$

$$= -(x^2 - 110x + 3025 - 3025)$$
$$= -[(x - 55)^2 - 3025]$$
$$= -(x - 55)^2 + 3025$$

The maximum value of the product occurs at the vertex of $P(x)$ and is 3025. This happens when $x = y = 55$.

61. Let x be the first number and y be the second number. Then $x + 2y = 24 \Rightarrow x = 24 - 2y$. The product is

$P = xy = (24 - 2y)y = 24y - 2y^2$.

Completing the square,

$$P = -2y^2 + 24y$$
$$= -2(y^2 - 12y + 36) + 72$$
$$= -2(y - 6)^2 + 72.$$

The maximum value of the product P occurs at the vertex of the parabola and equals 72. This happens when $y = 6$ and $x = 24 - 2(6) = 12$.

63. (a)

(b) Radius of semicircular ends of track: $r = \frac{1}{2}y$

Distance around two semicircular parts of track:

$$d = 2\pi r = 2\pi\left(\frac{1}{2}y\right) = \pi y$$

(c) Distance traveled around track in one lap:

$$d = \pi y + 2x = 200$$
$$\pi y = 200 - 2x$$
$$y = \frac{200 - 2x}{\pi}$$

(d) Area of rectangular region: $A = xy = x\left(\frac{200 - 2x}{\pi}\right)$

(e) The area is maximum when $x = 50$ and

$$y = \frac{200 - 2(50)}{\pi} = \frac{100}{\pi}.$$

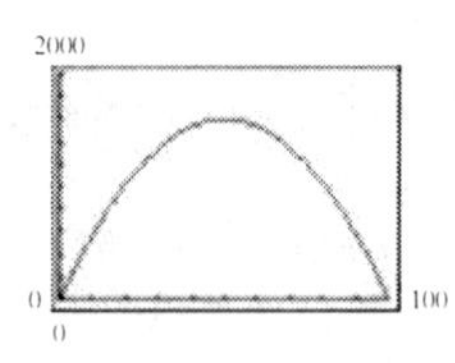

65. (a)

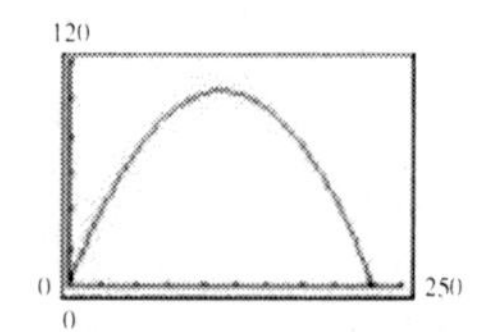

(b) When $x = 0$, $y = \frac{3}{2}$ feet.

(c) The vertex occurs at

$$x = \frac{-b}{2a} = \frac{-9/5}{2(-16/2025)} = \frac{3645}{32} \approx 113.9.$$

The maximum height is

$$y = \frac{-16}{2025}\left(\frac{3645}{32}\right)^2 + \frac{9}{5}\left(\frac{3645}{32}\right) + \frac{3}{2}$$
$$\approx 104.0 \text{ feet.}$$

(d) Using a graphing utility, the zero of y occurs at $x \approx 228.6$, or 228.6 feet from the punter.

67. (a)

$100 - 2x$

$x - 6$

$$A = lw$$
$$A = (100 - 2x)(x - 6)$$
$$A = -2x^2 + 112x - 600$$

(b) $Y_1 = -2x^2 + 112x - 600$

X	Y
25	950
26	960
27	966
28	968
29	966
30	960

The area is maximum when $x = 28$ inches.

69. $R(p) = -10p^2 + 1580p$

(a) When $p = \$50$, $R(50) = \$54{,}000$.
When $p = \$70$, $R(70) = \$61{,}600$.
When $p = \$90$, $R(90) = \$61{,}200$.

(b) The maximum R occurs at the vertex,

$$p = \frac{-b}{2a}$$
$$p = \frac{-1580}{2(-10)} = \$79$$

(c) When $p = \$79$, $R(79) = \$62{,}410$.

(d) Answers will vary.

71. (a)

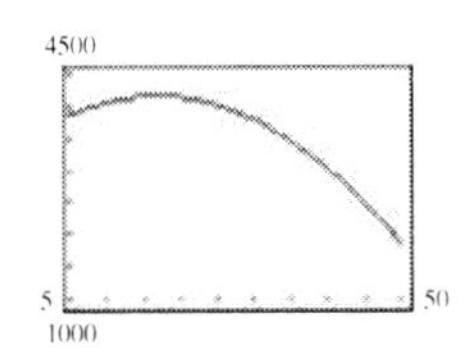

(b) Using the graph, during 1966 the maximum average annual consumption of cigarettes appears to have occurred and was 4155 cigarettes per person.

Yes, the warning had an effect because the maximum consumption occurred in 1966 and consumption decreased from then on.

(c) In 2000, $C(50) = 1852$ cigarettes per person.

$$\frac{1852}{365} \approx 5 \text{ cigarettes per day}$$

73. True.

$$-12x^2 - 1 = 0$$
$$12x^2 = -1, \text{ impossible}$$

75. The parabola opens downward and the vertex is $(-2, -4)$. Matches (c) and (d).

77. The graph of $f(x) = (x - z)^2$ would be a horizontal shift z units to the right of $g(x) = x^2$.

79. The graph of $f(x) = z(x - 3)^2$ would be a vertical stretch $(z > 1)$ and horizontal shift three units to the right of $g(x) = x^2$. The graph of $f(x) = z(x - 3)^2$ would be a vertical shrink $(0 < z < 1)$ and horizontal shift three units to the right of $g(x) = x^2$.

81. For $a < 0$, $f(x) = a\left(x + \frac{b}{2a}\right)^2 + \left(c - \frac{b^2}{4a}\right)$ is a maximum when $x = \frac{-b}{2a}$. In this case, the maximum value is $c - \frac{b^2}{4a}$. Hence,

$$25 = -75 - \frac{b^2}{4(-1)}$$
$$-100 = 300 - b^2$$
$$400 = b^2$$
$$b = \pm 20.$$

83. For $a > 0$, $f(x) = a\left(x + \frac{b}{2a}\right)^2 + \left(c - \frac{b^2}{4a}\right)$ is a minimum when $x = \frac{-b}{2a}$. In this case, the minimum value is $c - \frac{b^2}{4a}$. Hence,

$$10 = 26 - \frac{b^2}{4}$$
$$40 = 104 - b^2$$
$$b^2 = 64$$
$$b = \pm 8.$$

85. Let x = first number and y = second number.
Then $x + y = s$ or $y = s - x$.
The product is given by $P = xy$ or $P = x(s - x)$.

$$P = x(s - x)$$
$$P = sx - x^2$$

The maximum P occurs at the vertex when $x = \frac{-b}{2a}$.

$$x = \frac{-s}{2(-1)} = \frac{s}{2}$$

When $x = \frac{s}{2}$, $y = s - \frac{s}{2} = \frac{s}{2}$.

So, the numbers x and y are both $\frac{s}{2}$.

87. $y = ax^2 + bx - 4$

$(1, 0)$ on graph: $0 = a + b - 4$

$(4, 0)$ on graph: $0 = 16a + 4b - 4$

From the first equation, $b = 4 - a$.
Thus, $0 = 16a + 4(4 - a) - 4 = 12a + 12 \Rightarrow a = -1$ and hence $b = 5$, and $y = -x^2 + 5x - 4$.

89. $x + y = 8 \Rightarrow y = 8 - x$

$$-\frac{2}{3}x + 8 - x = 6$$
$$-\frac{5}{3}x + 8 = 6$$
$$-\frac{5}{3}x = -2$$
$$x = 1.2$$
$$y = 8 - 1.2 = 6.8$$

The point of intersection is $(1.2, 6.8)$.

91.
$$y = x + 3 = 9 - x^2$$
$$x^2 + x - 6 = 0$$
$$(x + 3)(x - 2) = 0$$
$$x = -3, \quad x = 2$$
$$y = -3 + 3 = 0$$
$$y = 2 + 3 = 5$$

Thus, $(-3, 0)$ and $(2, 5)$ are the points of intersection.

93. Answers will vary. (Make a Decision)

Section 2.2

1. continuous

3. (a) solution
 (b) $(x-a)$
 (c) $(a, 0)$

5. No. If f is an even-degree fourth-degree polynomial function, its left and right end behavior is either that it rises left and right or falls left and right.

7. Because f is a polynomial, it is a continuous on $[x_1, x_2]$ and $f(x_1) < 0$ and $f(x_2) > 0$. Then $f(x) = 0$ for some value of x in $[x_1, x_2]$.

9. $f(x) = -2x + 3$ is a line with y-intercept $(0, 3)$. Matches graph (f).

11. $f(x) = -2x^2 - 5x$ is a parabola with x-intercepts $(0, 0)$ and $\left(-\frac{5}{2}, 0\right)$ and opens downward. Matches graph (c).

13. $f(x) = -\frac{1}{4}x^4 + 3x^2$ has intercepts $(0, 0)$ and $(\pm 2\sqrt{3}, 0)$. Matches graph (e).

15. $f(x) = x^4 + 2x^3$ has intercepts $(0, 0)$ and $(-2, 0)$. Matches graph (g).

17. The graph of $f(x) = (x-2)^3$ is a horizontal shift two units to the right of $y = x^3$.

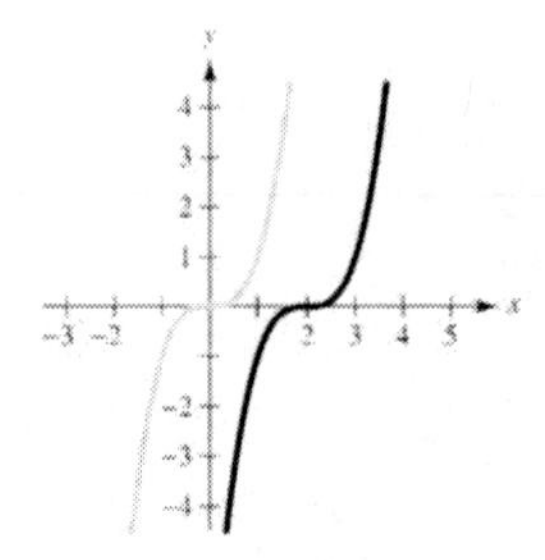

19. The graph of $f(x) = -x^3 + 1$ is a reflection in the x-axis and a vertical shift one unit upward of $y = x^3$.

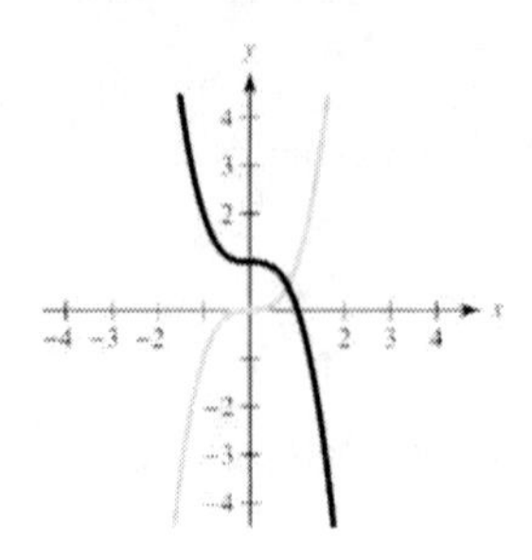

21. The graph of $f(x) = -(x-2)^3$ is a horizontal shift two units to the right and a reflection in the x-axis of $y = x^3$.

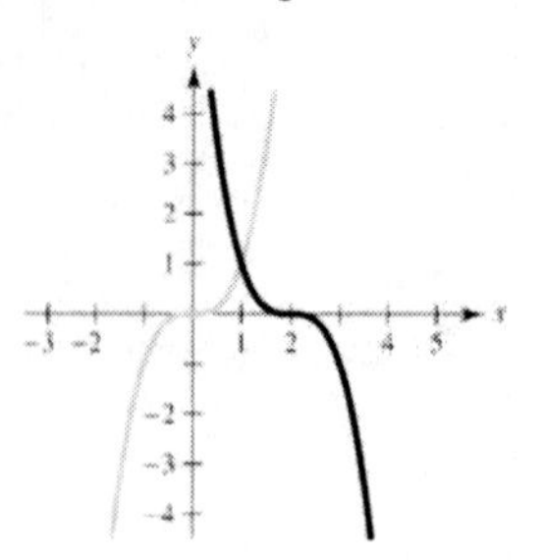

23.

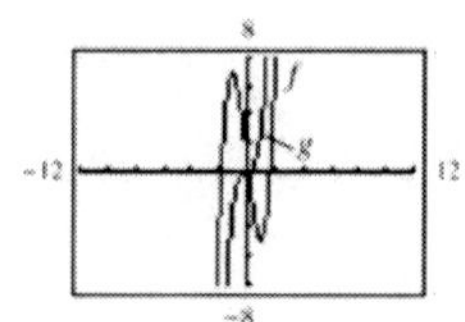

Yes, because both graphs have the same leading coefficient.

25.

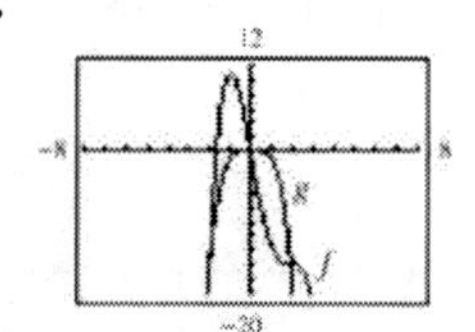

Yes, because both graphs have the same leading coefficient.

27.

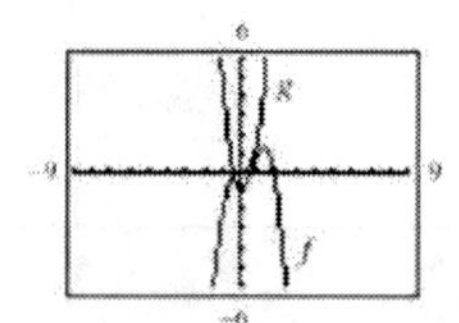

No, because the graphs have different leading coefficients.

29. $f(x) = 2x^4 - 3x + 1$

Degree: 4

Leading coefficient: 2

The degree is even and the leading coefficient is positive. The graph rises to the left and right.

31. $g(x) = 5 - \frac{7}{2}x - 3x^2$

Degree: 2

Leading coefficient: –3

The degree is even and the leading coefficient is negative. The graph falls to the left and right.

33. $f(x) = \frac{6x^5 - 2x^4 + 4x^2 - 5x}{3}$

Degree: 5

Leading coefficient: $\frac{6}{3} = 2$

The degree is odd and the leading coefficient is positive. The graph falls to the left and rises to the right.

35. $h(t) = -\frac{2}{3}\left(t^2 - 5t + 3\right)$

Degree: 2

Leading coefficient: $-\frac{2}{3}$

The degree is even and the leading coefficient is negative. The graph falls to the left and right.

37. (a) $f(x) = 3x^2 - 12x + 3$
$= 3\left(x^2 - 4x + 1\right) = 0$
$x = \frac{4 \pm \sqrt{16-4}}{2} = 2 \pm \sqrt{3}$

(b)

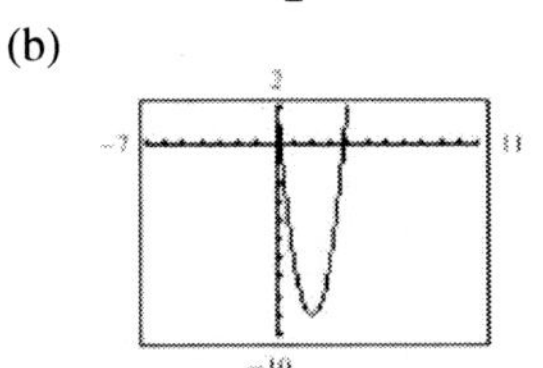

(c) $x \approx 3.732, 0.268$; the answers are approximately the same.

39. (a) $g(t) = \frac{1}{2}t^4 - \frac{1}{2}$
$= \frac{1}{2}(t+1)(t-1)(t^2+1) = 0$
$t = \pm 1$

(b)

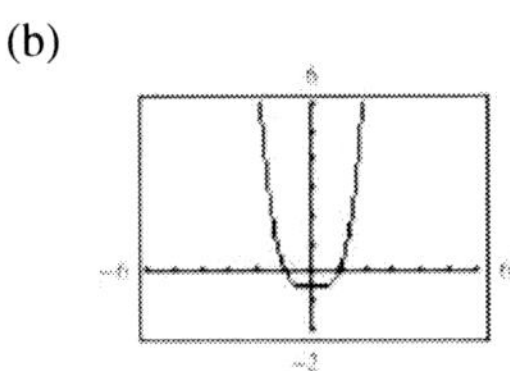

(c) $t = \pm 1$; the answers are the same.

41. (a) $f(x) = x^5 + x^3 - 6x$
$= x\left(x^4 + x^2 - 6\right)$
$= x\left(x^2 + 3\right)(x^2 - 2) = 0$
$x = 0, \pm\sqrt{2}$

(b)

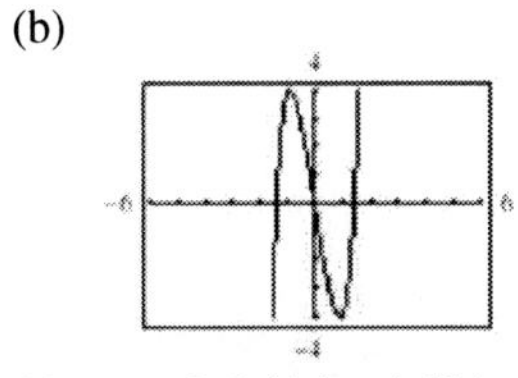

(c) $x = 0, 1.414, -1.414$; the answers are approximately the same.

43. (a) $f(x) = 2x^4 - 2x^2 - 40$
$= 2\left(x^4 - x^2 - 20\right)$
$= 2\left(x^2 + 4\right)\left(x + \sqrt{5}\right)\left(x - \sqrt{5}\right) = 0$
$x = \pm\sqrt{5}$

(b)

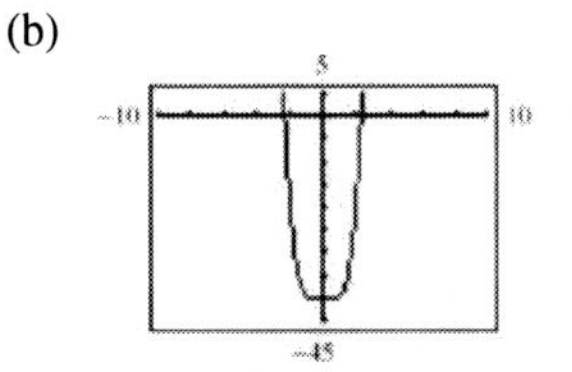

(c) $x = 2.236, -2.236$; the answers are approximately the same.

45. (a) $f(x) = x^3 - 4x^2 - 25x + 100$
$= x^2(x-4) - 25(x-4)$
$= (x^2 - 25)(x-4)$
$= (x-5)(x+5)(x-4) = 0$
$x = \pm 5, 4$

(b)

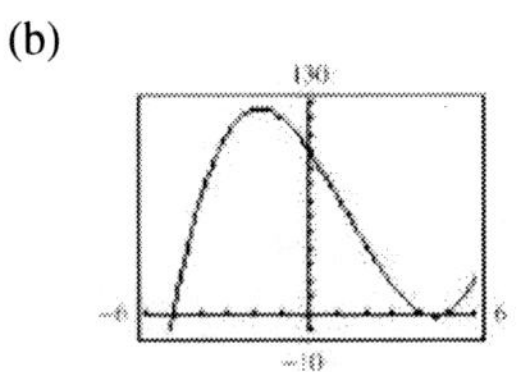

(c) $x = 4, 5, -5$; the answers are the same.

47. (a) $y = 4x^3 - 20x^2 + 25x$

$0 = 4x^3 - 20x^2 + 25x$

$0 = x(2x - 5)^2$

$x = 0, \frac{5}{2}$

(b)

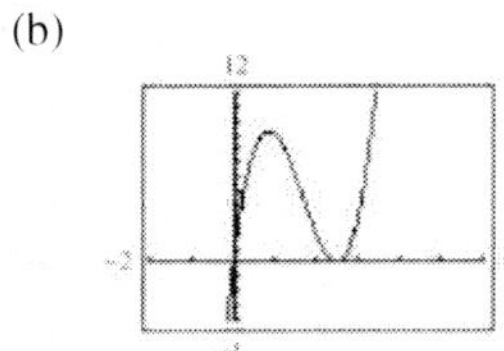

(c) $x = 0, \frac{5}{2}$; the answers are the same.

49. $f(x) = x^2 - 25$

$= (x + 5)(x - 5)$

$x = \pm 5$ (multiplicity 1)

51. $h(t) = t^2 - 6t + 9$

$= (t - 3)^2$

$t = 3$ (multiplicity 2)

53. $f(x) = x^2 + x - 2$

$= (x + 2)(x - 1)$

$x = -2, 1$ (multiplicity 1)

55. $f(t) = t^3 - 4t^2 + 4t$

$= t(t - 2)^2$

$t = 0$ (multiplicity 1), 2 (multiplicity 2)

57. $f(x) = \frac{1}{2}x^2 + \frac{5}{2}x - \frac{3}{2}$

$= \frac{1}{2}(x^2 + 5x - 3)$

$x = \frac{-5 \pm \sqrt{25 - 4(-3)}}{2} = -\frac{5}{2} \pm \frac{\sqrt{37}}{2}$

$\approx 0.5414, -5.5414$ (multiplicity 1)

59. $f(x) = 2x^4 - 6x^2 + 1$

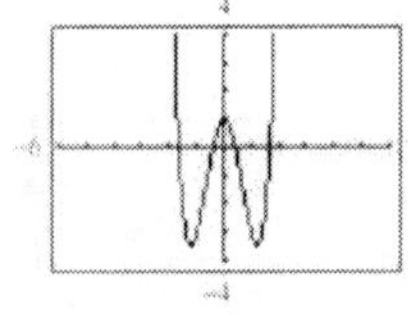

Zeros: $x \approx \pm 0.421, \pm 1.680$

Relative maximum: (0, 1)

Relative minima: (1.225, −3.5), (−1.225, −3.5)

61. $f(x) = x^5 + 3x^3 - x + 6$

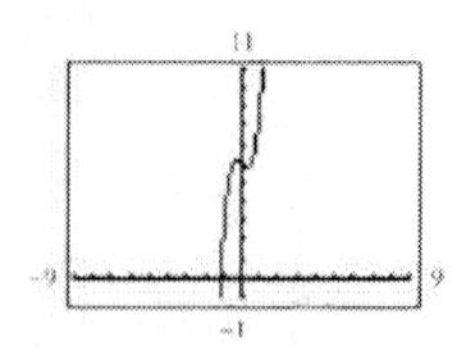

Zero: $x \approx -1.178$

Relative maximum: (−0.324, 6.218)

Relative minimum: (0.324, 5.782)

63. $f(x) = -2x^4 + 5x^2 - x - 1$

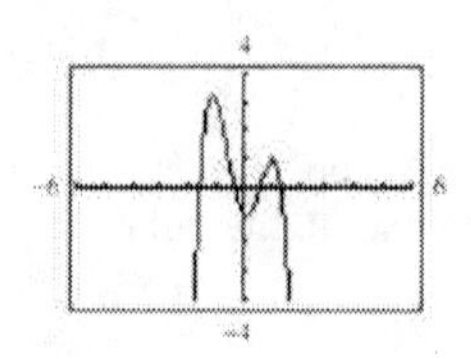

Zeros: −1.618, −0.366, 0.618, 1.366

Relative minimum: (0.101, −1.050)

Relative maxima: (−1.165, 3.267), (1.064, 1.033)

65. $f(x) = (x - 0)(x - 4) = x^2 - 4x$

Note: $f(x) = a(x - 0)(x - 4) = ax(x - 4)$ has zeros 0 and 4 for all nonzero real numbers a.

67. $f(x) = (x - 0)(x + 2)(x + 3) = x^3 + 5x^2 + 6x$

Note: $f(x) = ax(x + 2)(x + 3)$ has zeros 0, −2, and −3 for all nonzero real numbers a.

69. $f(x) = (x - 4)(x + 3)(x - 3)(x - 0)$

$= (x - 4)(x^2 - 9)x$

$= x^4 - 4x^3 - 9x^2 + 36x$

Note: $f(x) = a(x^4 - 4x^3 - 9x^2 + 36x)$ has zeros 4, −3, 3, and 0 for all nonzero real numbers a.

71. $f(x) = \left[x - \left(1 + \sqrt{3}\right)\right]\left[x - \left(1 - \sqrt{3}\right)\right]$

$= \left[(x - 1) - \sqrt{3}\right]\left[(x - 1) + \sqrt{3}\right]$

$= (x - 1)^2 - \left(\sqrt{3}\right)^2$

$= x^2 - 2x + 1 - 3$

$= x^2 - 2x - 2$

Note: $f(x) = a(x^2 - 2x - 2)$ has zeros $1 + \sqrt{3}$ and $1 - \sqrt{3}$ for all nonzero real numbers a.

73. $f(x) = (x-2)\left[x-\left(4+\sqrt{5}\right)\right]\left[x-\left(4-\sqrt{5}\right)\right]$

$= (x-2)\left[(x-4)-\sqrt{5}\right]\left[(x-4)+\sqrt{5}\right]$

$= (x-2)[(x-4)^2-5]$

$= x^3 - 10x^2 + 27x - 22$

Note: $f(x) = a(x-2)[(x-4)^2-5]$ has zeros $2,\ 4+\sqrt{5},$ and $4-\sqrt{5}$ for all nonzero real numbers a.

75. $f(x) = (x+2)^2(x+1) = x^3 + 5x^2 + 8x + 4$

Note: $f(x) = a(x+2)^2(x+1)$ has zeros $-2,\ -2,$ and -1 for all nonzero real numbers a.

77. $f(x) = (x+4)^2(x-3)^2$

$= x^4 + 2x^3 - 23x^2 - 24x + 144$

Note: $f(x) = a(x+4)^2(x-3)^2$ has zeros $-4,\ -4,\ 3,\ 3$ for all nonzero real numbers a.

79. $f(x) = -(x+1)^2(x+2)$

$= -x^3 - 4x^2 - 5x - 2$

Note: $f(x) = a(x+1)^2(x+2)^2,\ a < 0,$ has zeros $-1,\ -1,\ -2,$ rises to the left, and falls to the right.

81.

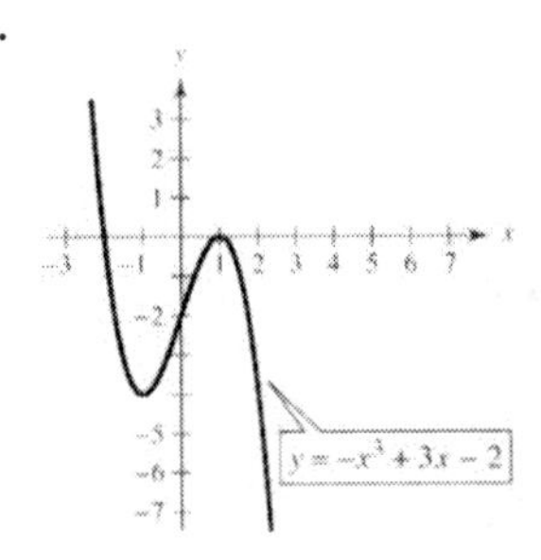

83.

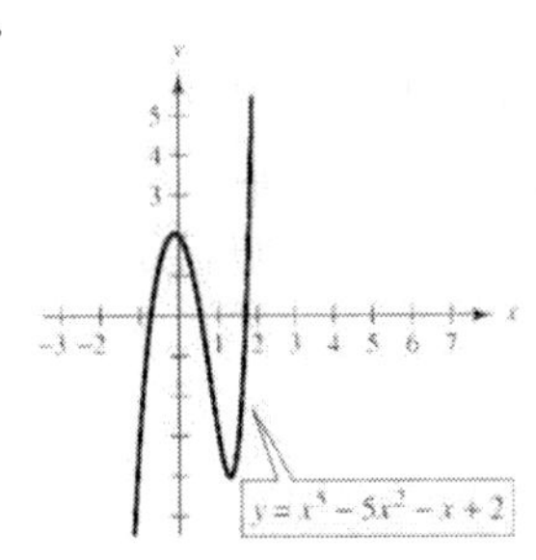

85. (a) The degree of f is odd and the leading coefficient is 1. The graph falls to the left and rises to the right.

(b) $f(x) = x^3 - 9x = x(x^2 - 9) = x(x-3)(x+3)$

Zeros: $0,\ 3,\ -3$

(c) and (d)

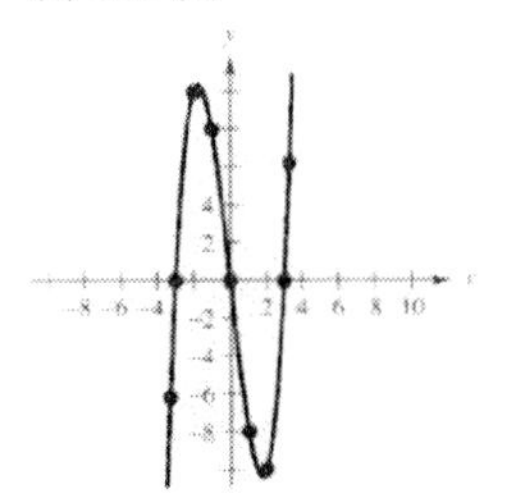

87. (a) The degree of f is odd and the leading coefficient is 1. The graph falls to the left and rises to the right.

(b) $f(x) = x^3 - 3x^2 = x^2(x-3)$

Zeros: $0,\ 3$

(c) and (d)

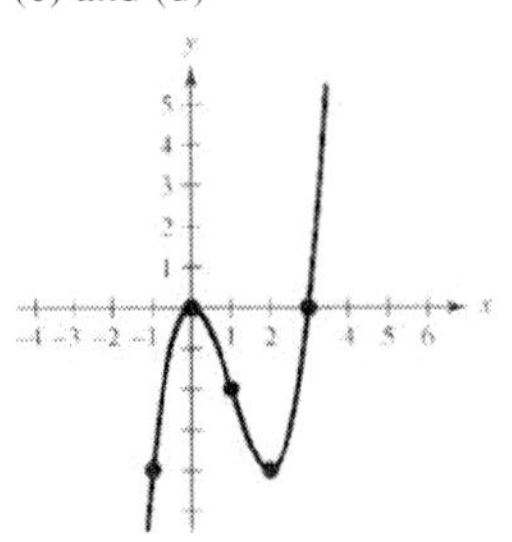

89. (a) The degree of f is even and the leading coefficient is -1. The graph falls to the left and falls to the right.

(b) $f(x) = -x^4 + 9x^2 - 20 = -(x^2-4)(x^2-5)$

Zeros: $\pm 2,\ \pm\sqrt{5}$

(c) and (d)

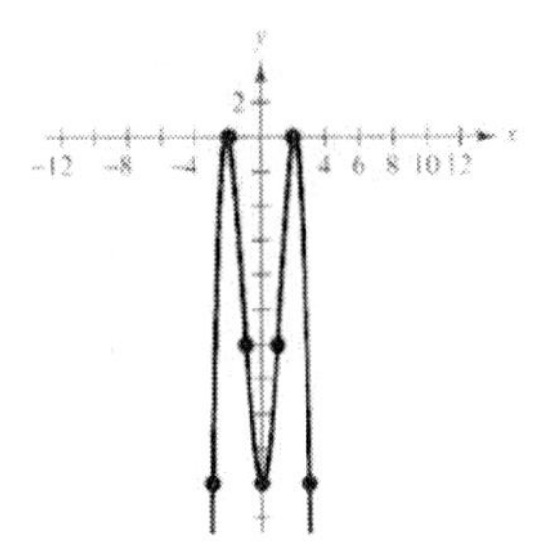

91. (a) The degree of f is odd and the leading coefficient is 1. The graph falls to the left and rises to the right.

(b) $f(x) = x^3 + 3x^2 - 9x - 27 = x^2(x+3) - 9(x+3)$
$= (x^2 - 9)(x+3)$
$= (x-3)(x+3)^2$

Zeros: 3, -3

(c) and (d)

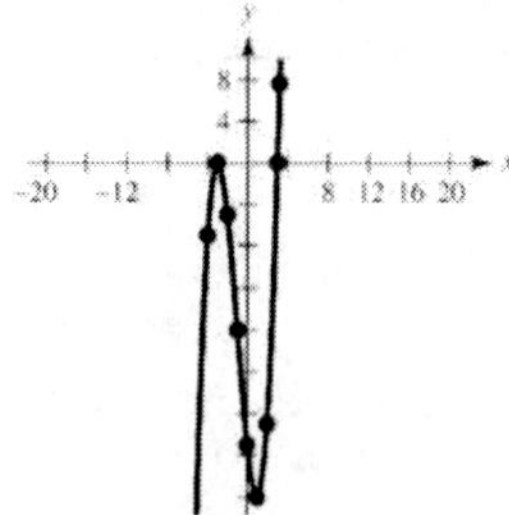

93. (a) The degree of g is even and the leading coefficient is $-\frac{1}{4}$. The graph falls to the left and falls to the right.

(b) $g(t) = -\frac{1}{4}(t^4 - 8t^2 + 16) = -\frac{1}{4}(t^2 - 4)^2$

Zeros: -2, -2, 2, 2

(c) and (d)

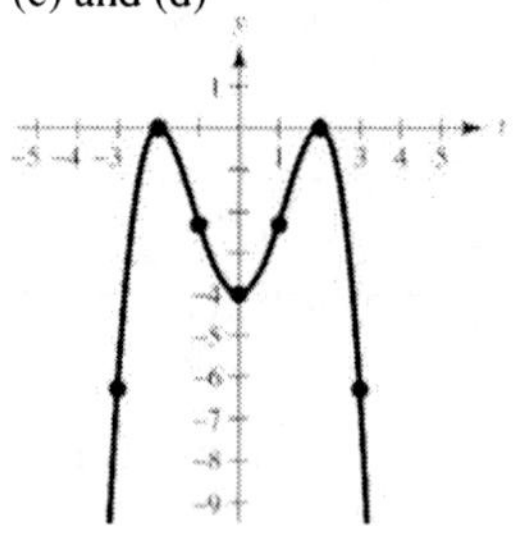

95. $f(x) = x^3 - 3x^2 + 3$

(a)

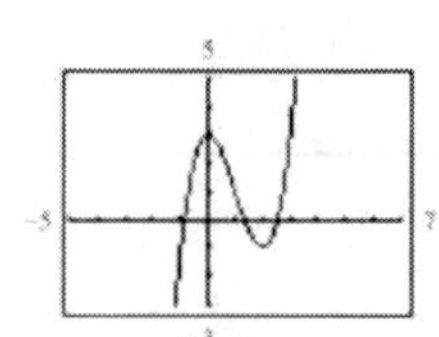

The function has three zeros. They are in the intervals $(-1, 0)$, $(1, 2)$ and $(2, 3)$.

(b) Zeros: -0.879, 1.347, 2.532

x	y
−0.9	−0.159
−0.89	−0.0813
−0.88	−0.0047
−0.87	0.0708
−0.86	0.14514
−0.85	0.21838
−0.84	0.2905

x	y
1.3	0.127
1.31	0.09979
1.32	0.07277
1.33	0.04594
1.34	0.0193
1.35	−0.0071
1.36	−0.0333

x	y
2.5	−0.125
2.51	−0.087
2.52	−0.0482
2.53	−0.0084
2.54	0.03226
2.55	0.07388
2.56	0.11642

97. $g(x) = 3x^4 + 4x^3 - 3$

(a)

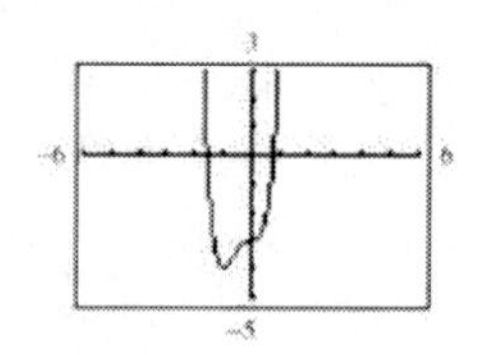

The function has two zeros. They are in the intervals $(-2, -1)$ and $(0, 1)$.

(b) Zeros: -1.585, 0.779

x	y_1
−1.6	0.2768
−1.59	0.09515
−1.58	−0.0812
−1.57	−0.2524
−1.56	−0.4184
−1.55	−0.5795
−1.54	−0.7356

x	y_1
0.75	−0.3633
0.76	−0.2432
0.77	−0.1193
0.78	0.00866
0.79	0.14066
0.80	0.2768
0.81	0.41717

99. $f(x) = x^4 - 3x^3 - 4x - 3$

(a)

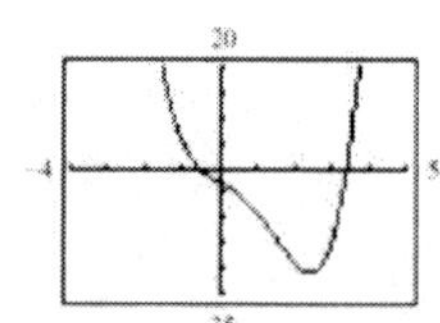

The function has two zeros. They are in the intervals $(-1, 0)$ and $(3, 4)$.

(b) Zeros: -0.578, 3.418

x	y_1
−0.61	0.2594
−0.60	0.1776
−0.59	0.09731
−0.58	0.0185
−0.57	−0.0589
−0.56	−0.1348
−0.55	−0.2094

x	y_1
3.39	−1.366
3.40	−0.8784
3.41	−0.3828
3.42	0.12071
3.43	0.63205
3.44	1.1513
3.45	1.6786

101. $f(x) = x^2(x+6)$

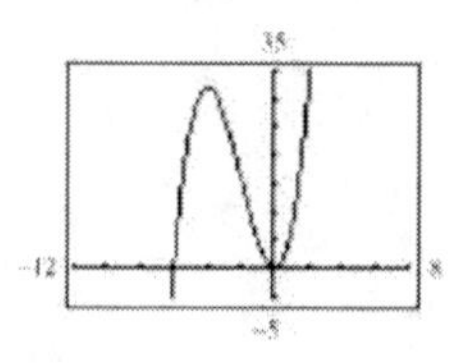

No symmetry
Two x-intercepts

103. $g(t) = -\frac{1}{2}(t-4)^2(t+4)^2$

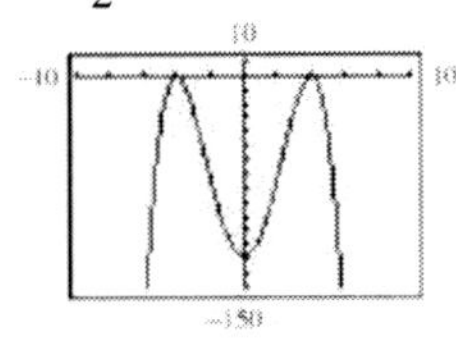

Symmetric with respect to the y-axis
Two x-intercepts

105. $f(x) = x^3 - 4x = x(x+2)(x-2)$

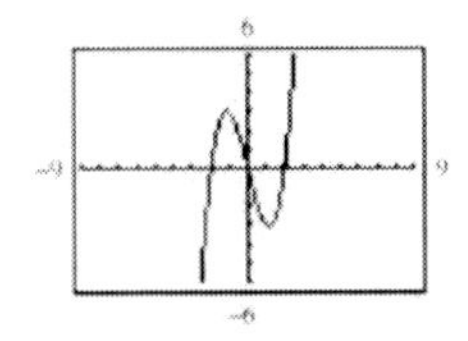

Symmetric with respect to the origin
Three x-intercepts

107. $g(x) = \frac{1}{5}(x+1)^2(x-3)(2x-9)$

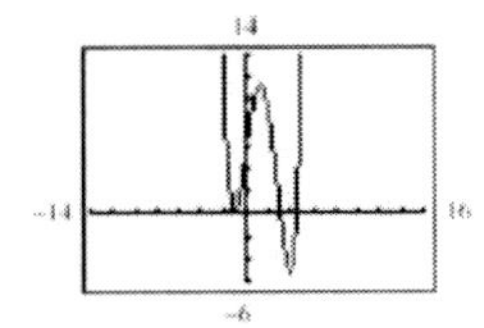

No symmetry
Three x-intercepts

109. (a) Volume = length × width × height

Because the box is made from a square, length = width.

Thus: Volume = $(\text{length})^2 \times$ height = $(36-2x)^2 x$

(b) Domain: $0 < 36 - 2x < 36$

$-36 < -2x < 0$

$18 > x > 0$

(c)

Height, x	Length and Width	Volume, V
1	$36-2(1)$	$1[36-2(1)]^2 = 1156$
2	$36-2(2)$	$2[36-2(2)]^2 = 2048$
3	$36-2(3)$	$3[36-2(3)]^2 = 2700$
4	$36-2(4)$	$4[36-2(4)]^2 = 3136$
5	$36-2(5)$	$5[36-2(5)]^2 = 3380$
6	$36-2(6)$	$6[36-2(6)]^2 = 3456$
7	$36-2(7)$	$7[36-2(7)]^2 = 3388$

Maximum volume is in the interval $5 < x < 7$.

(d)

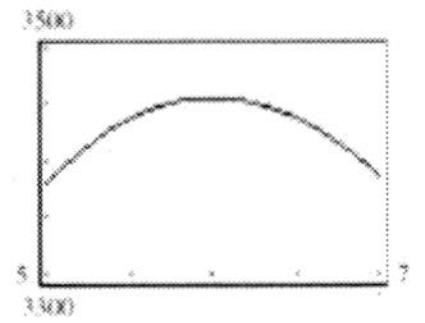

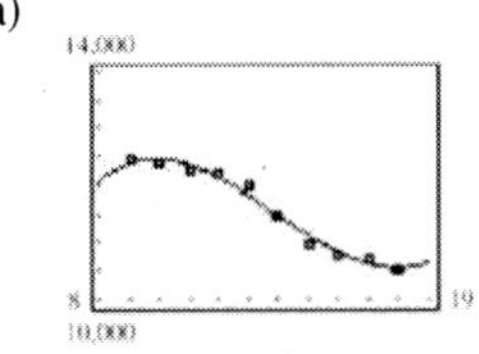

$x = 6$ when $V(x)$ is maximum.

111. The point of diminishing returns (where the graph changes from curving upward to curving downward) occurs when $x = 200$ The point is (200, 160) which corresponds to spending \$2,000,000 on advertising to obtain a revenue of \$160 million.

113. (a)

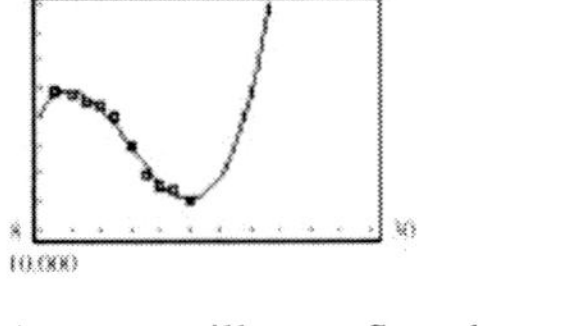

The model fits the data well.

(b)

Answers will vary. Sample answer: You could use the model to estimate production in 2010 because the result is somewhat reasonable, but you would not use the model to estimate the 2020 production because the result is unreasonably high.

(c)

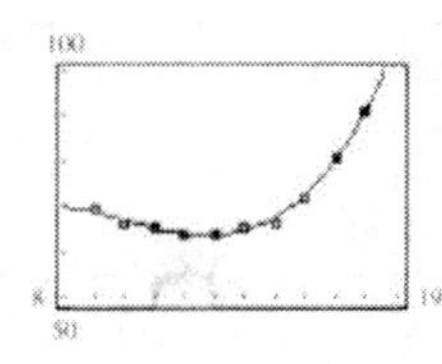

The model fits the data well.

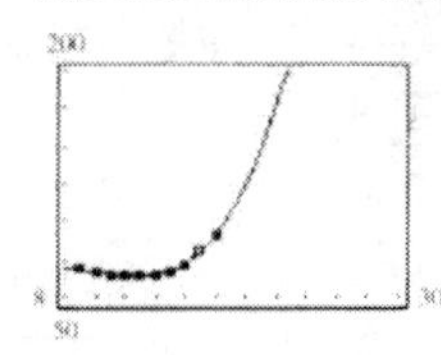

Answers will vary. Sample answer: You could use the model to estimate production in 2010 because the result is somewhat reasonable, but you would not use the model to estimate the 2020 production because the result is unreasonably high.

115. True. The degree is odd and the leading coefficient is -1.

117. False. The graph crosses the x-axis at $x = -3$ and $x = 0$.

119.

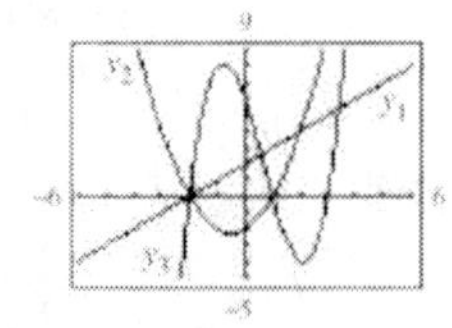

The graph of y_3 will fall to the left and rise to the right. It will have another x-intercept at $(3, 0)$ of odd multiplicity (crossing the x-axis).

121. $(f + g)(-4) = f(-4) + g(-4)$

$= -59 + 128 = 69$

123. $(f \circ g)\left(-\frac{4}{7}\right) = f\left(-\frac{4}{7}\right)g\left(-\frac{4}{7}\right) = (-11)\left(\frac{8 \cdot 16}{49}\right)$

$= -\frac{1408}{49} \approx -28.7347$

125. $(f \circ g)(-1) = f(g(-1)) = f(8) = 109$

127. $3(x - 5) < 4x - 7$

$3x - 15 < 4x - 7$

$-8 < x$

129. $\frac{5x - 2}{x - 7} \leq 4$

$$\frac{5x - 2}{x - 7} - 4 \leq 0$$

$$\frac{5x - 2 - 4(x - 7)}{x - 7} \leq 0$$

$$\frac{x + 26}{x - 7} \leq 0$$

$[x + 26 \geq 0 \text{ and } x - 7 < 0]$ or $[x + 26 \leq 0 \text{ and } x - 7 > 0]$

$[x \geq -26 \text{ and } x < 7]$ or $[x \leq -26 \text{ and } x > 7]$

$-26 \leq x < 7$ impossible

Section 2.3

1. $f(x)$ is the dividend, $d(x)$ is the divisor, $q(x)$ is the quotient, and $r(x)$ is the remainder.

3. constant term, leading coefficient

5. upper, lower

7. According to the Remainder Theorem, if you divide $f(x)$ by $x - 4$ and the remainder is 7, then $f(4) = 7$.

9. $$x + 3 \overline{)2x^2 + 10x + 12}$$ quotient $2x + 4$

$2x^2 + 6x$

$4x + 12$

$4x + 12$

0

$$\frac{2x^2 + 10x + 12}{x + 3} = 2x + 4, \; x \neq -3$$

11.

$$\begin{array}{r}
x^3+3x^2\qquad-1\\
x+2\overline{\big)x^4+5x^3+6x^2-x-2}\\
\underline{x^4+2x^3}\qquad\qquad\qquad\\
3x^3+6x^2\qquad\quad\\
\underline{3x^3+6x^2}\qquad\quad\\
-x-2\\
\underline{-x-2}\\
0
\end{array}$$

$$\frac{x^4+5x^3+6x^2-x-2}{x+2}=x^3+3x^2-1,\ x\neq-2$$

13.

$$\begin{array}{r}
x^2-\ 3x+1\\
4x+5\overline{\big)4x^3-7x^2-11x+5}\\
\underline{-4x^3+5x^2}\qquad\qquad\\
-12x^2-11x\quad\\
\underline{-12x^2-15x}\quad\\
4x+5\\
\underline{-\ 4x+5}\\
0
\end{array}$$

$$\frac{4x^3-7x^2-11x+5}{4x+5}=x^2-3x+1,\ x\neq-\frac{5}{4}$$

15.

$$\begin{array}{r}
7x^2\ \ -14x+28\\
x+2\overline{\big)7x^3+\ 0x^2\ +0x\ +\ 3}\\
\underline{7x^3+14x^2}\qquad\qquad\quad\\
-14x^2\qquad\qquad\quad\\
\underline{-14x^2-28x}\qquad\quad\\
28x+3\\
\underline{28x+56}\\
-53
\end{array}$$

$$\frac{7x^3+3}{x+2}=7x^2-14x+28-\frac{53}{x+2}$$

17.

$$\begin{array}{r}
3x+5\\
2x^2+0x+1\overline{\big)6x^3+10x^2+x+8}\\
\underline{6x^3+\ 0x^2\ +3x}\quad\\
10x^2-2x\ +8\\
\underline{10x^2+0x\ +5}\\
-2x+3
\end{array}$$

$$\frac{6x^3+10x^2+x+8}{2x^2+1}=3x+5-\frac{2x-3}{2x^2+1}$$

19.

$$\begin{array}{r}
x\\
x^2+1\overline{\big)x^3+0x^2+0x-9}\\
\underline{x^3\qquad\ +\ x}\quad\\
-x-9
\end{array}$$

$$\frac{x^3-9}{x^2+1}=x-\frac{x+9}{x^2+1}$$

21.

$$\begin{array}{r}
2x\\
x^2-2x+1\overline{\big)2x^3-4x^2-15x+5}\\
\underline{2x^3-4x^2+\ 2x}\quad\\
-17x+5
\end{array}$$

$$\frac{2x^3-4x^2-15x+5}{(x-1)^2}=2x-\frac{17x-5}{(x-1)^2}$$

23.

$$\begin{array}{r|rrrr}
5 & 3 & -17 & 15 & -25\\
 & & 15 & -10 & 25\\
\hline
 & 3 & -2 & 5 & 0
\end{array}$$

$$\frac{3x^3-17x^2+15x-25}{x-5}=3x^2-2x+5,\ x\neq5$$

25.

$$\begin{array}{r|rrrr}
3 & 6 & 7 & -1 & 26\\
 & & 18 & 75 & 222\\
\hline
 & 6 & 25 & 74 & 248
\end{array}$$

$$\frac{6x^3+7x^2-x+26}{x-3}=6x^2+25x+74+\frac{248}{x-3}$$

27.

$$\begin{array}{r|rrrr}
2 & 9 & -18 & -16 & 32\\
 & & 18 & 0 & -32\\
\hline
 & 9 & 0 & -16 & 0
\end{array}$$

$$\frac{9x^3-18x^2-16x+32}{x-2}=9x^2-16,\ x\neq2$$

29.

$$\begin{array}{r|rrrr}
-8 & 1 & 0 & 0 & 512\\
 & & -8 & 64 & -512\\
\hline
 & 1 & -8 & 64 & 0
\end{array}$$

$$\frac{x^3+512}{x+8}=x^2-8x+64,\ x\neq-8$$

31.

$$\begin{array}{r|rrrr}
-\frac{1}{2} & 4 & 16 & -23 & -15\\
 & & -2 & -7 & 15\\
\hline
 & 4 & 14 & -30 & 0
\end{array}$$

$$\frac{4x^3+16x^2-23x-15}{x+\frac{1}{2}}=4x^2+14x-30,\ x\neq-\frac{1}{2}$$

33. $y_2 = x - 2 + \dfrac{4}{x+2}$

$= \dfrac{(x-2)(x+2)+4}{x+2}$

$= \dfrac{x^2 - 4 + 4}{x+2}$

$= \dfrac{x^2}{x+2}$

$= y_1$

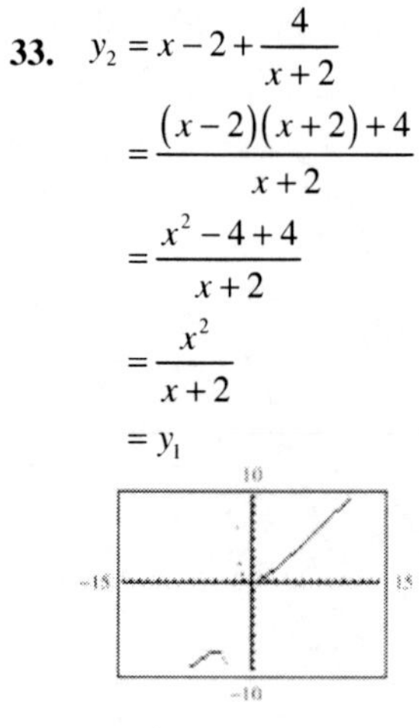

35. $y_2 = x^2 - 8 + \dfrac{39}{x^2+5}$

$= \dfrac{(x^2-8)(x^2+5)+39}{x^2+5}$

$= \dfrac{x^4 - 8x^2 + 5x^2 - 40 + 39}{x^2+5}$

$= \dfrac{x^4 - 3x^2 - 1}{x^2+5}$

$= y_1$

37. $f(x) = x^3 - x^2 - 14x + 11,\ k = 4$

$$\begin{array}{r|rrrr} 4 & 1 & -1 & -14 & 11 \\ & & 4 & 12 & -8 \\ \hline & 1 & 3 & -2 & 3 \end{array}$$

$f(x) = (x-4)(x^2 + 3x - 2) + 3$

$f(4) = (0)(26) + 3 = 3$

39.

$$\begin{array}{r|rrrr} \sqrt{2} & 1 & 3 & -2 & -14 \\ & & \sqrt{2} & 2+3\sqrt{2} & 6 \\ \hline & 1 & 3+\sqrt{2} & 3\sqrt{2} & -8 \end{array}$$

$f(x) = \left(x - \sqrt{2}\right)\left(x^2 + \left(3+\sqrt{2}\right)x + 3\sqrt{2}\right) - 8$

$f\left(\sqrt{2}\right) = 0\left(4 + 6\sqrt{2}\right) - 8 = -8$

41.

$$\begin{array}{r|rrrr} 1-\sqrt{3} & 4 & -6 & -12 & -4 \\ & & 4-4\sqrt{3} & 10-2\sqrt{3} & 4 \\ \hline & 4 & -2-4\sqrt{3} & -2-2\sqrt{3} & 0 \end{array}$$

$f(x) = (x - 1 + \sqrt{3})[4x^2 - (2 + 4\sqrt{3})x - (2 + 2\sqrt{3})]$

$f(1 - \sqrt{3}) = 0$

43. $f(x) = 2x^3 - 7x + 3$

(a)
$$\begin{array}{r|rrrr} 1 & 2 & 0 & -7 & 3 \\ & & 2 & 2 & -5 \\ \hline & 2 & 2 & -5 & -2 \end{array} \quad = f(1)$$

(b)
$$\begin{array}{r|rrrr} -2 & 2 & 0 & -7 & 3 \\ & & -4 & 8 & -2 \\ \hline & 2 & -4 & 1 & 1 \end{array} \quad = f(-2)$$

(c)
$$\begin{array}{r|rrrr} \frac{1}{2} & 2 & 0 & -7 & 3 \\ & & 1 & \frac{1}{2} & -\frac{13}{4} \\ \hline & 2 & 1 & -\frac{13}{2} & -\frac{1}{4} \end{array} \quad = f\left(\frac{1}{2}\right)$$

(d)
$$\begin{array}{r|rrrr} 2 & 2 & 0 & -7 & 3 \\ & & 4 & 8 & 2 \\ \hline & 2 & 4 & 1 & 5 \end{array} \quad = f(2)$$

45. $h(x) = x^3 - 5x^2 - 7x + 4$

(a)
$$\begin{array}{r|rrrr} 3 & 1 & -5 & -7 & 4 \\ & & 3 & -6 & -39 \\ \hline & 1 & -2 & -13 & -35 \end{array} \quad = h(3)$$

(b)
$$\begin{array}{r|rrrr} 2 & 1 & -5 & -7 & 4 \\ & & 2 & -6 & -26 \\ \hline & 1 & -3 & -13 & -22 \end{array} \quad = h(2)$$

(c)
$$\begin{array}{r|rrrr} -2 & 1 & -5 & -7 & 4 \\ & & -2 & 14 & -14 \\ \hline & 1 & -7 & 7 & -10 \end{array} \quad = h(-2)$$

(d)
$$\begin{array}{r|rrrr} -5 & 1 & -5 & -7 & 4 \\ & & -5 & 50 & -215 \\ \hline & 1 & -10 & 43 & -211 \end{array} \quad = h(-5)$$

47.

$$\begin{array}{r|rrrr} 2 & 1 & 0 & -7 & 6 \\ & & 2 & 4 & -6 \\ \hline & 1 & 2 & -3 & 0 \end{array}$$

$x^3 - 7x + 6 = (x-2)(x^2 + 2x - 3)$

$= (x-2)(x+3)(x-1)$

Zeros: 2, −3, 1

49.

$$\begin{array}{r|rrrr} \frac{1}{2} & 2 & -15 & 27 & -10 \\ & & 1 & -7 & 10 \\ \hline & 2 & -14 & 20 & 0 \end{array}$$

$2x^3 - 15x^2 + 27x - 10$

$= (x - \frac{1}{2})(2x^2 - 14x + 20)$

$= (2x-1)(x-2)(x-5)$

Zeros: $\frac{1}{2}$, 2, 5

51. (a)

$$\begin{array}{r|rrrr} -2 & 2 & 1 & -5 & 2 \\ & & -4 & 6 & -2 \\ \hline & 2 & -3 & 1 & 0 \end{array}$$

(b) $2x^2 - 3x + 1 = (2x - 1)(x - 1)$

Remaining factors: $(2x - 1)$, $(x - 1)$

(c) $f(x) = (x + 2)(2x - 1)(x - 1)$

(d) Real zeros: $-2, \frac{1}{2}, 1$

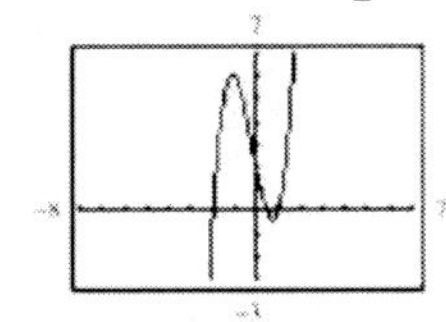

53. (a)

$$\begin{array}{r|rrrrr} 5 & 1 & -4 & -15 & 58 & -40 \\ & & 5 & 5 & -50 & 40 \\ \hline & 1 & 1 & -10 & 8 & 0 \end{array}$$

$$\begin{array}{r|rrrr} -4 & 1 & 1 & -10 & 8 \\ & & -4 & 12 & -8 \\ \hline & 1 & -3 & 2 & 0 \end{array}$$

(b) $x^2 - 3x + 2 = (x - 2)(x - 1)$

Remaining factors: $(x - 2)$, $(x - 1)$

(c) $f(x) = (x - 5)(x + 4)(x - 2)(x - 1)$

(d) Real zeros: $5, -4, 2, 1$

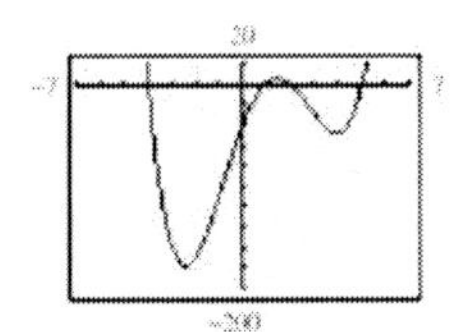

55. (a)

$$\begin{array}{r|rrrr} -\frac{1}{2} & 6 & 41 & -9 & -14 \\ & & -3 & -19 & 14 \\ \hline & 6 & 38 & -28 & 0 \end{array}$$

(b) $6x^2 + 38x - 28 = (3x - 2)(2x + 14)$

Remaining factors: $(3x - 2)$, $(x + 7)$

(c) $f(x) = (2x + 1)(3x - 2)(x + 7)$

(d) Real zeros: $-\frac{1}{2}, \frac{2}{3}, -7$

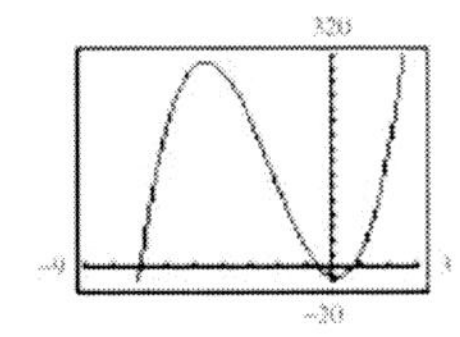

57. $f(x) = x^3 + 3x^2 - x - 3$

$p =$ factor of -3
$q =$ factor of 1

Possible rational zeros: $\pm 1, \pm 3$

$f(x) = x^2(x + 3) - (x + 3) = (x + 3)(x^2 - 1)$

Rational zeros: $\pm 1, -3$

59. $f(x) = 2x^4 - 17x^3 + 35x^2 + 9x - 45$

$p =$ factor of -45
$q =$ factor of 2

Possible rational zeros: $\pm 1, \pm 3, \pm 5, \pm 9, \pm 15, \pm 45, \pm\frac{1}{2}, \pm\frac{3}{2}, \pm\frac{5}{2}, \pm\frac{9}{2}, \pm\frac{15}{2}, \pm\frac{45}{2}$

Using synthetic division, -1, 3, and 5 are zeros.

$f(x) = (x + 1)(x - 3)(x - 5)(2x - 3)$

Rational zeros: $-1, 3, 5, \frac{3}{2}$

61. $f(x) = 2x^4 - x^3 + 6x^2 - x + 5$

4 variations in sign $\Rightarrow$ 4, 2, or 0 positive real zeros
$f(-x) = 2x^4 + x^3 + 6x^2 + x + 5$
0 variations in sign $\Rightarrow$ 0 negative real zeros

63. $g(x) = 4x^3 - 5x + 8$

2 variations in sign $\Rightarrow$ 2 or 0 positive real zeros
$g(-x) = -4x^3 + 5x + 8$
1 variation in sign $\Rightarrow$ 1 negative real zero

65. $f(x) = x^3 + x^2 - 4x - 4$

(a) $f(x)$ has 1 variation in sign $\Rightarrow$ 1 positive real zero.
$f(-x) = -x^3 + x^2 + 4x - 4$ has 2 variations in sign $\Rightarrow$ 2 or 0 negative real zeros.

(b) Possible rational zeros: $\pm 1, \pm 2, \pm 4$

(c)

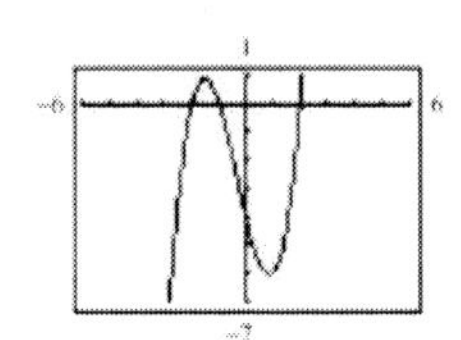

(d) Real zeros: $-2, -1, 2$

67. $f(x) = -2x^4 + 13x^3 - 21x^2 + 2x + 8$

(a) $f(x)$ has variations in sign $\Rightarrow$ 3 or 1 positive real zeros.

$f(-x) = -2x^4 - 13x^3 - 21x^2 - 2x + 8$ has 1 variation in sign $\Rightarrow$ 1 negative real zero.

(b) Possible rational zeros: $\pm\frac{1}{2}, \pm 1, \pm 2, \pm 4, \pm 8$

(c)

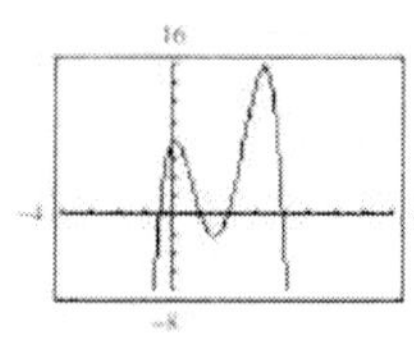

(d) Real zeros: $-\frac{1}{2}, 1, 2, 4$

69. $f(x) = 32x^3 - 52x^2 + 17x + 3$

(a) $f(x)$ has 2 variations in sign $\Rightarrow$ 2 or 0 positive real zeros.

$f(-x) = -32x^3 - 52x^2 - 17x + 3$ has 1 variation in sign $\Rightarrow$ 1 negative real zero.

(b) Possible rational zeros:

$$\pm\frac{1}{32}, \pm\frac{1}{16}, \pm\frac{1}{8}, \pm\frac{1}{4}, \pm\frac{1}{2}, \pm 1, \pm\frac{3}{32}, \pm\frac{3}{16}, \pm\frac{3}{8}, \pm\frac{3}{4}, \pm\frac{3}{2}, \pm 3$$

(c)

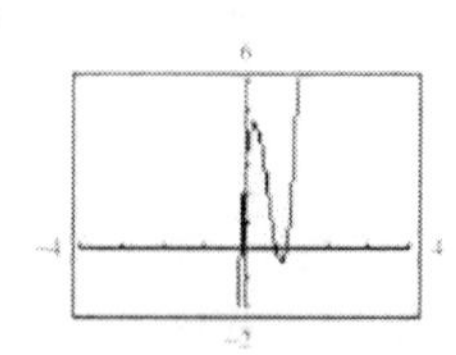

(d) Real zeros: $1, \frac{3}{4}, -\frac{1}{8}$

71. $f(x) = x^4 - 4x^3 + 15$

$$\begin{array}{r|rrrrr} 4 & 1 & -4 & 0 & 0 & 15 \\ & & 4 & 0 & 0 & 0 \\ \hline & 1 & 0 & 0 & 0 & 15 \end{array}$$

4 is an upper bound.

$$\begin{array}{r|rrrrr} -1 & 1 & -4 & 0 & 0 & 15 \\ & & -1 & 5 & -5 & 5 \\ \hline & 1 & -5 & 5 & -5 & 20 \end{array}$$

-1 is a lower bound.

Real zeros: 1.937, 3.705

73. $f(x) = x^4 - 4x^3 + 16x - 16$

$$\begin{array}{r|rrrrr} 5 & 1 & -4 & 0 & 16 & -16 \\ & & 25 & 105 & 525 & 2705 \\ \hline & 5 & 21 & 105 & 541 & 2689 \end{array}$$

5 is an upper bound.

$$\begin{array}{r|rrrrr} -3 & 1 & -4 & 0 & 16 & -16 \\ & & -3 & 21 & -63 & 141 \\ \hline & 1 & -7 & 21 & -47 & 125 \end{array}$$

-3 is a lower bound.

Real zeros: $-2, 2$

75. $P(x) = x^4 - \frac{25}{4}x^2 + 9$

$= \frac{1}{4}(4x^4 - 25x^2 + 36)$

$= \frac{1}{4}(4x^2 - 9)(x^2 - 4)$

$= \frac{1}{4}(2x + 3)(2x - 3)(x + 2)(x - 2)$

The rational zeros are $\pm\frac{3}{2}$ and ± 2.

77. $f(x) = x^3 - \frac{1}{4}x^2 - x + \frac{1}{4}$

$= \frac{1}{4}(4x^3 - x^2 - 4x + 1)$

$= \frac{1}{4}[x^2(4x - 1) - (4x - 1)]$

$= \frac{1}{4}(4x - 1)(x^2 - 1)$

$= \frac{1}{4}(4x - 1)(x + 1)(x - 1)$

The rational zeros are $\frac{1}{4}$ and ± 1.

79. $f(x) = x^3 - 1$

$= (x - 1)(x^2 + x + 1)$

Rational zeros: 1 $(x = 1)$

Irrational zeros: 0

Matches (d).

81. $f(x) = x^3 - x = x(x+1)(x-1)$

Rational zeros: $3\ (x = 0, \pm 1)$

Irrational zeros: 0

Matches (b).

83. $y = 2x^4 - 9x^3 + 5x^2 + 3x - 1$

Using the graph and synthetic division, $-\frac{1}{2}$ is a zero.

$-\frac{1}{2}$	2	−9	5	3	−1
		−1	5	−5	1
	2	−10	10	−2	0

$$y = \left(x + \frac{1}{2}\right)(2x^3 - 10x^2 + 10x - 2)$$

$x = 1$ is a zero of the cubic, so

$y = (2x+1)(x-1)(x^2 - 4x + 1).$

For the quadratic term, use the Quadratic Formula.

$$x = \frac{4 \pm \sqrt{16-4}}{2} = 2 \pm \sqrt{3}$$

The real zeros are $-\frac{1}{2}, 1, 2 \pm \sqrt{3}$.

85. $y = -2x^4 + 17x^3 - 3x^2 - 25x - 3$

Using the graph and synthetic division, -1 and $\frac{3}{2}$ are zeros.

$$y = -(x+1)(2x-3)(x^2 - 8x - 1)$$

For the quadratic term, use the Quadratic Formula.

$$x = \frac{8 \pm \sqrt{64+4}}{2} = 4 \pm \sqrt{17}$$

The real zeros are $-1, \frac{3}{2}, 4 \pm \sqrt{17}$.

87. $3x^4 - 14x^2 - 4x = 0$

$x(3x^3 - 14x - 4) = 0$

$x = 0$ is a real zero.

Possible ratoinal zeros: $\pm 1, \pm 2, \pm 4, \pm\frac{1}{3}, \pm\frac{2}{3}, \pm\frac{4}{3}$

−2	3	0	−14	−4
		−6	12	4
	3	−6	−2	0

$x = -2$ is a real zero.

Use the Quadratic Formula. $3x^2 - 6x - 2 = 0$

$$x = \frac{3 \pm \sqrt{15}}{3}$$

Real zeros: $x = 0, -2, \frac{3 \pm \sqrt{15}}{3}$

89. $z^4 - z^3 - 2z - 4 = 0$

Possible rational zeros: $\pm 1, \pm 2, \pm 4$

−1	1	−1	0	−2	−4
		−1	2	−2	4
	1	−2	2	−4	0

2	1	−2	2	−4
		2	0	4
	1	0	2	0

$z = -1$ and $z = 2$ are real zeros.

$$z^4 - z^3 - 2z - 4 = (z+1)(z-2)(z^2+2) = 0$$

The only real zeros are −1 and 2. You can verify this by graphing the function $f(z) = z^4 - z^3 - 2z - 4$.

91. $2y^4 + 7y^3 - 26y^2 + 23y - 6 = 0$

Possible rational zeros:

$\pm\frac{1}{2}, \pm 1, \pm\frac{3}{2}, \pm 2, \pm 3, \pm 6$

$$\begin{array}{r|rrrrr} \frac{1}{2} & 2 & 7 & -26 & 23 & -6 \\ & & 1 & 4 & -11 & 6 \\ \hline & 2 & 8 & -22 & 12 & 0 \end{array}$$

$$\begin{array}{r|rrrr} 1 & 2 & 8 & -22 & 12 \\ & & 2 & 10 & -12 \\ \hline & 2 & 10 & -12 & 0 \end{array}$$

$$\begin{array}{r|rrr} -6 & 2 & 10 & -12 \\ & & -12 & 12 \\ \hline & 2 & -2 & 0 \end{array}$$

$$\begin{array}{r|rr} 1 & 2 & -2 \\ & & 2 \\ \hline & 2 & 0 \end{array}$$

$x = \frac{1}{2}$, $x = 1$, $x = -6$, and $x = 1$ are real zeros.

$(y + 6)(y - 1)^2(2y - 1) = 0$

Real zeros: $x = -6, 1, \frac{1}{2}$

93. $4x^4 - 55x^2 - 45x + 36 = 0$

Possible rational zeros:

$\pm\frac{1}{2}, \pm\frac{3}{4}, \pm 1, \pm\frac{3}{2}, \pm 2, \pm\frac{9}{4}, \pm 3, \pm 4, \pm\frac{9}{2}, \pm 6, \pm 9, \pm 18$

$$\begin{array}{r|rrrrr} 4 & 4 & 0 & -55 & -45 & 36 \\ & & 16 & 64 & 36 & -36 \\ \hline & 4 & 16 & 9 & -9 & 0 \end{array}$$

$$\begin{array}{r|rrrr} -3 & 4 & 16 & 9 & -9 \\ & & -12 & -12 & 9 \\ \hline & 4 & 4 & -3 & 0 \end{array}$$

$$\begin{array}{r|rrr} \frac{1}{2} & 4 & 4 & -3 \\ & & 2 & 3 \\ \hline & 4 & 6 & 0 \end{array}$$

$$\begin{array}{r|rr} -\frac{3}{2} & 4 & 6 \\ & & -6 \\ \hline & 4 & 0 \end{array}$$

$x = 4$, $x = -3$, $x = \frac{1}{2}$, and $x = -\frac{3}{2}$ are real zeros.

$(x - 4)(x + 3)(2x - 1)(2x + 3) = 0$

Real zeros: $x = 4, -3, \frac{1}{2}, -\frac{3}{2}$

95. $8x^4 + 28x^3 + 9x^2 - 9x = 0$

$x(8x^3 + 28x^2 + 9x - 9) = 0$

$x = 0$ is a real zero.

Possible rational zeros:

$\pm 1, \pm 3, \pm 9, \pm\frac{1}{8}, \pm\frac{3}{8}, \pm\frac{9}{8}, \pm\frac{1}{2}, \pm\frac{3}{2}, \pm\frac{9}{2}, \pm\frac{1}{4}, \pm\frac{3}{4}, \pm\frac{9}{4}$

$$\begin{array}{r|rrrr} -3 & 8 & 28 & 9 & -9 \\ & & -24 & -12 & 9 \\ \hline & 8 & 4 & -3 & 0 \end{array}$$

$x = -3$ is a real zeros

Use the Quadratic Formula.

$8x^2 + 4x - 3 = 0 \qquad x = \frac{-\pm\sqrt{7}}{4}$

Real zeros: $x = 0, -3, \frac{-1 \pm \sqrt{7}}{4}$

97. $4x^5 + 12x^4 - 11x^3 - 42x^2 + 7x + 30 = 0$

Possible rational zeros:
$\pm 1, \pm 2, \pm 3, \pm 5, \pm 6, \pm 10, \pm 15, \pm 30,$
$\pm\frac{1}{2}, \pm\frac{1}{4}, \pm\frac{3}{2}, \pm\frac{3}{4}, \pm\frac{5}{2}, \pm\frac{5}{4}, \pm\frac{15}{2}, \pm\frac{15}{4}$

$$\begin{array}{r|rrrrrr} 1 & 4 & 12 & -11 & -42 & 7 & 30 \\ & & 4 & 16 & 5 & -37 & -30 \\ \hline & 4 & 16 & 5 & -37 & -30 & 0 \end{array}$$

$$\begin{array}{r|rrrrr} -1 & 4 & 16 & 5 & -37 & -30 \\ & & -4 & -12 & 7 & 30 \\ \hline & 4 & 12 & -7 & -30 & 0 \end{array}$$

$$\begin{array}{r|rrrr} -2 & 4 & 12 & -7 & -30 \\ & & -8 & -8 & 30 \\ \hline & 4 & 4 & -15 & 0 \end{array}$$

$$\begin{array}{r|rrr} \frac{3}{2} & 4 & 4 & -15 \\ & & 6 & 15 \\ \hline & 4 & 10 & 0 \end{array}$$

$$\begin{array}{r|rr} -\frac{5}{2} & 4 & 10 \\ & & -10 \\ \hline & 4 & 0 \end{array}$$

$x = 1$, $x = -1$, $x = -2$, $x = \frac{3}{2}$, and $x = -\frac{5}{2}$ are real zeros.

$(x - 1)(x + 1)(x + 2)(2x - 3)(2x + 5) = 0$

Real zeros: $x = 1, -1, -2, \frac{3}{2}, -\frac{5}{2}$.

99. $h(t) = t^3 - 2t^2 - 7t + 2$

(a) Zeros: –2, 3.732, 0.268

(b)
$$\begin{array}{r|rrrr} -2 & 1 & -2 & -7 & 2 \\ & & -2 & 8 & -2 \\ \hline & 1 & -4 & 1 & 0 \end{array} \quad t = -2 \text{ is a zero.}$$

(c) $h(t) = (t + 2)(t^2 - 4t + 1)$

$=(t + 2)\left[t - (\sqrt{3} + 2)\right]\left[t + (\sqrt{3} - 2)\right]$

101. $h(x) = x^5 - 7x^4 + 10x^3 + 14x^2 - 24x$

(a) $x = 0,\ 3,\ 4,\ \pm 1.414$

(b)
$$\begin{array}{r|rrrrr} 3 & 1 & -7 & 10 & 14 & -24 \\ & & 3 & -12 & -6 & 24 \\ \hline & 1 & -4 & -2 & 8 & 0 \end{array}$$

$x = 3$ is a zero.

$$\begin{array}{r|rrrr} 4 & 1 & -4 & -2 & 8 \\ & & 4 & 0 & -8 \\ \hline & 1 & 0 & -2 & 0 \end{array} \quad x = 4 \text{ is a zero.}$$

(c) $h(x) = x(x - 3)(x - 4)(x^2 - 2)$

$= x(x - 3)(x - 4)(x - \sqrt{2})(x + \sqrt{2})$

103. (a)

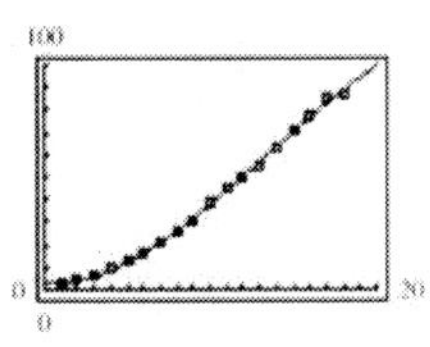

(b) The model fits the data well.

(c) $S = -0.0135t^3 + 0.545t^2 - 0.71t + 3.6$

$$\begin{array}{r|rrrr} 25 & -0.0135 & 0.545 & -0.71 & 3.6 \\ & & -0.3375 & 5.1875 & 111.9375 \\ \hline & -0.0135 & 0.2075 & 4.4775 & 115.5375 \end{array}$$

In 2015 (t = 25), the model predicts approximately 116 subscriptions per 100 people, obviously not a reasonable prediction because you cannot have more subscriptions than people.

105. (a) Combined length and width:

$4x + y = 120 \Rightarrow y = 120 - 4x$

$\text{Volume} = l \cdot w \cdot h = x^2 y$

$= x^2(120 - 4x)$

$= 4x^2(30 - x)$

(b)

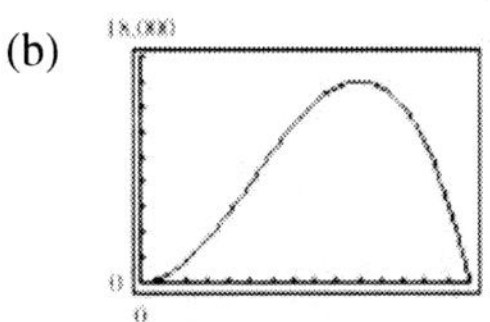

Dimension with maximum volume:
$20 \times 20 \times 40$

(c) $13{,}500 = 4x^2(30 - x)$

$4x^3 - 120x^2 + 13{,}500 = 0$

$x^3 - 30x^2 + 3375 = 0$

$$\begin{array}{r|rrrr} 15 & 1 & -30 & 0 & 3375 \\ & & 15 & -225 & -3375 \\ \hline & 1 & -15 & -225 & 0 \end{array}$$

$(x - 15)(x^2 - 15x - 225) = 0$

Using the Quadratic Formula, $x = 15$ or $\dfrac{15 \pm 15\sqrt{5}}{2}$.

The value of $\dfrac{15 - 15\sqrt{5}}{2}$ is not possible because it is negative.

107. False, $-\dfrac{4}{7}$ is a zero of *f*.

109. The zeros are 1, 1, and –2. The graph falls to the right.

$y = a(x - 1)^2(x + 2),\ a < 0$

Since $f(0) = -4,\ a = -2.$

$y = -2(x - 1)^2(x + 2)$

111. $f(x) = -(x + 1)(x - 1)(x + 2)(x - 2)$

113. (a) $\dfrac{x^2-1}{x-1}=x+1,\ x\neq 1$

(b) $\dfrac{x^3-1}{x-1}=x^2+x+1,\ x\neq 1$

(c) $\dfrac{x^4-1}{x-1}=x^3+x^2+x+1,\ x\neq 1$

In general,

$$\frac{x^n-1}{x-1}=x^{n-1}+x^{n-2}+\cdots+x+1,\ x\neq 1.$$

115. $9x^2-25=0$

$$(3x+5)(3x-5)=0$$

$$x=-\frac{5}{3},\ \frac{5}{3}$$

117. $2x^2+6x+3=0$

$$x=\frac{-6\pm\sqrt{6^2-4(2)(3)}}{2(2)}$$

$$=\frac{-6\pm\sqrt{12}}{4}$$

$$=\frac{-3\pm\sqrt{3}}{2}$$

$$x=-\frac{3}{2}+\frac{\sqrt{3}}{2},-\frac{3}{2}-\frac{\sqrt{3}}{2}$$

Section 2.4

1. (a) ii
(b) iii
(c) i

3. $(7+6i)+(8+5i)=(7+8)+(6+5)i$

$$=15+11i$$

The real part is 15 and the imaginary part is $11i$.

5. The additive inverse of $2-4i$ is $-2+4i$ so that $(2-4i)+(-2+4i)=0$.

7. $a+bi=-9+4i$

$a=-9$

$b=4$

9. $3a+(b+3)i=9+8i$

$3a=9 \qquad b+3=8$

$a=3 \qquad b=5$

11. $5+\sqrt{-16}=5+\sqrt{16(-1)}$

$$=5+4i$$

13. $-6=-6+0i$

15. $-5i+i^2=-5i-1=-1-5i$

17. $\left(\sqrt{-75}\right)^2=-75$

19. $\sqrt{-0.09}=\sqrt{0.09}i=0.3i$

21. $(4+i)-(7-2i)=(4-7)+(1+2)i$

$$=-3+3i$$

23. $(-1+8i)+(8-5i)=(-1+8)+(8-5)i$

$$=7+3i$$

25. $13i-(14-7i)=13i-14+7i=-14+20i$

27. $\left(\frac{3}{2}+\frac{5}{2}i\right)+\left(\frac{5}{3}+\frac{11}{3}i\right)=\left(\frac{3}{2}+\frac{5}{3}\right)+\left(\frac{5}{2}+\frac{11}{3}\right)i$

$$=\frac{9+10}{6}+\frac{15+22}{6}i$$

$$=\frac{19}{6}+\frac{37}{6}i$$

29. $(1.6+3.2i)+(-5.8+4.3i)=-4.2+7.5i$

31. $4(3+5i)=12+20i$

33. $(1+i)(3-2i)=3-2i+3i-2i^2$

$$=3+i+2$$

$$=5+i$$

35. $4i(8+5i)=32i+20i^2$

$$=32i+20(-1)$$

$$=-20+32i$$

37. $\left(\sqrt{14}+\sqrt{10}\,i\right)\left(\sqrt{14}-\sqrt{10}\,i\right)=14-10i^2=14+10=24$

39. $(6+7i)^2=36+42i+42i+49i^2$

$$=36+84i-49$$

$$=-13+84i$$

41. $(4+5i)^2-(4-5i)^2$

$$=\left[(4+5i)+(4-5i)\right]\left[(4+5i)-(4-5i)\right]=8(10i)=80i$$

43. $4-3i$ is the complex conjugate of $4+3i$.

$(4+3i)(4-3i)=16+9=25$

45. $-6+\sqrt{5}i$ is the complex conjugate of $-6-\sqrt{5}i$.

$\left(-6-\sqrt{5}i\right)\left(-6+\sqrt{5}i\right)=36+5=41$

47. $-\sqrt{20}i$ is the complex conjugate of $\sqrt{-20} = \sqrt{20}i$.

$$\left(\sqrt{20}i\right)\left(-\sqrt{20}i\right) = 20$$

49. $3 + \sqrt{2}i$ is the complex conjugate of

$$3 - \sqrt{-2} = 3 - \sqrt{2}\,i$$

$$(3 - \sqrt{2}i)(3 + \sqrt{2}i) = 9 + 2 = 11$$

51. $\dfrac{6}{i} = \dfrac{6}{i} \cdot \dfrac{-i}{-i} = \dfrac{-6i}{-i^2} = \dfrac{-6i}{1} = -6i$

53. $\dfrac{2}{4 - 5i} = \dfrac{2}{4 - 5i} \cdot \dfrac{4 + 5i}{4 + 5i} = \dfrac{8 + 10i}{16 + 25} = \dfrac{8}{41} + \dfrac{10}{41}i$

55.
$$\begin{aligned}\frac{2+i}{2-i} &= \frac{2+i}{2-i} \cdot \frac{2+i}{2+i}\\ &= \frac{4 + 4i + i^2}{4 + 1}\\ &= \frac{3 + 4i}{5} = \frac{3}{5} + \frac{4}{5}i\end{aligned}$$

57.
$$\begin{aligned}\frac{i}{(4 - 5i)^2} &= \frac{i}{16 - 25 - 40i}\\ &= \frac{i}{-9 - 40i} \cdot \frac{-9 + 40i}{-9 + 40i}\\ &= \frac{-40 - 9i}{81 + 40^2}\\ &= -\frac{40}{1681} - \frac{9}{1681}i\end{aligned}$$

59.
$$\begin{aligned}\frac{2}{1+i} - \frac{3}{1-i} &= \frac{2(1-i) - 3(1+i)}{(1+i)(1-i)}\\ &= \frac{2 - 2i - 3 - 3i}{1 + 1}\\ &= \frac{-1 - 5i}{2}\\ &= -\frac{1}{2} - \frac{5}{2}i\end{aligned}$$

61.
$$\begin{aligned}\frac{i}{3 - 2i} + \frac{2i}{3 + 8i} &= \frac{3i + 8i^2 + 6i - 4i^2}{(3 - 2i)(3 + 8i)}\\ &= \frac{-4 + 9i}{9 + 18i + 16}\\ &= \frac{-4 + 9i}{25 + 18i} \cdot \frac{25 - 18i}{25 - 18i}\\ &= \frac{-100 + 72i + 225i + 162}{25^2 + 18^2}\\ &= \frac{62 + 297i}{949}\\ &= \frac{62}{949} + \frac{297}{949}i\end{aligned}$$

63.
$$\begin{aligned}\sqrt{-18} - \sqrt{-54} &= 3\sqrt{2}i - 3\sqrt{6}i\\ &= 3\left(\sqrt{2} - \sqrt{6}\right)i\end{aligned}$$

65.
$$\begin{aligned}\left(-3 + \sqrt{-24}\right) + \left(7 - \sqrt{-44}\right) &= \left(-3 + 2\sqrt{6}i\right) + \left(7 - 2\sqrt{11}i\right)\\ &= 4 + \left(2\sqrt{6} - 2\sqrt{11}\right)i\\ &= 4 + 2\left(\sqrt{6} - \sqrt{11}\right)i\end{aligned}$$

67.
$$\begin{aligned}\sqrt{-6} \cdot \sqrt{-2} &= \left(\sqrt{6}i\right)\left(\sqrt{2}i\right)\\ &= \sqrt{12}\,i^2 = \left(2\sqrt{3}\right)(-1) = -2\sqrt{3}\end{aligned}$$

69. $\left(\sqrt{-10}\right)^2 = \left(\sqrt{10}i\right)^2 = 10i^2 = -10$

71.
$$\begin{aligned}\left(2 - \sqrt{-6}\right)^2 &= \left(2 - \sqrt{6}i\right)\left(2 - \sqrt{6}i\right)\\ &= 4 - 2\sqrt{6}i - 2\sqrt{6}i + 6i^2\\ &= 4 - 2\sqrt{6}i - 2\sqrt{6}i + 6(-1)\\ &= 4 - 6 - 4\sqrt{6}i\\ &= -2 - 4\sqrt{6}i\end{aligned}$$

73.
$$\begin{aligned}x^2 + 25 &= 0\\ x^2 &= -25\\ x &= \pm 5i\end{aligned}$$

75. $x^2 - 2x + 2 = 0;\ a = 1,\ b = -2,\ c = 2$

$$\begin{aligned}x &= \frac{-(-2) \pm \sqrt{(-2)^2 - 4(1)(2)}}{2(1)}\\ &= \frac{2 \pm \sqrt{-4}}{2}\\ &= \frac{2 \pm 2i}{2}\\ &= 1 \pm i\end{aligned}$$

77. $4x^2 + 16x + 17 = 0;\ a = 4,\ b = 16,\ c = 17$

$$\begin{aligned}x &= \frac{-16 \pm \sqrt{(16)^2 - 4(4)(17)}}{2(4)}\\ &= \frac{-16 \pm \sqrt{-16}}{8}\\ &= \frac{-16 \pm 4i}{8}\\ &= -2 \pm \frac{1}{2}i\end{aligned}$$

79. $16t^2 - 4t + 3 = 0;\ a = 16,\ b = -4,\ c = 3$

$$t = \frac{-(-4) \pm \sqrt{(-4)^2 - 4(16)(3)}}{2(16)} = \frac{4 \pm \sqrt{-176}}{32} = \frac{4 \pm 4\sqrt{11}i}{32} = \frac{1}{8} \pm \frac{\sqrt{11}}{8}i$$

81. $\frac{3}{2}x^2 - 6x + 9 = 0$ Multiply both sides by 2.

$3x^2 - 12x + 18 = 0;\ a = 3,\ b = -12,\ c = 18$

$$x = \frac{-(-12) \pm \sqrt{(-12)^2 - 4(3)(18)}}{2(3)} = \frac{12 \pm \sqrt{-72}}{6} = \frac{12 \pm 6\sqrt{2}i}{6} = 2 \pm \sqrt{2}i$$

83. $1.4x^2 - 2x - 10 = 0$ Multiply both sides by 5.

$7x^2 - 10x - 50 = 0;\ a = 7,\ b = -10,\ c = -50$

$$x = \frac{-(-10) \pm \sqrt{(-10)^2 - 4(7)(-50)}}{2(7)} = \frac{10 \pm \sqrt{1500}}{14} = \frac{10 \pm 10\sqrt{15}}{14} = \frac{5}{7} \pm \frac{5\sqrt{15}}{7}$$

85. $-6i^3 + i^2 = -6i^2 i + i^2 = -6(-1)i + (-1) = 6i - 1 = -1 + 6i$

87. $\left(\sqrt{-75}\right)^3 = \left(5\sqrt{3}i\right)^3 = 5^3\left(\sqrt{3}\right)^3 i^3 = 125\left(3\sqrt{3}\right)(-i)$

$= -375\sqrt{3}i$

89. $\frac{1}{i^3} = \frac{1}{i^3} \cdot \frac{i}{i} = \frac{i}{i^4} = \frac{i}{1} = i$

91. (a) $(2)^3 = 8$

(b) $\left(-1 + \sqrt{3}i\right)^3 = (-1)^3 + 3(-1)^2\left(\sqrt{3}i\right) + 3(-1)\left(\sqrt{3}i\right)^2 + \left(\sqrt{3}i\right)^3$

$= -1 + 3\sqrt{3}i - 9i^2 + 3\sqrt{3}i^3$

$= -1 + 3\sqrt{3}i + 9 - 3\sqrt{3}i$

$= 8$

(c) $\left(-1 - \sqrt{3}i\right)^3 = (-1)^3 + 3(-1)^2\left(-\sqrt{3}i\right) + 3(-1)\left(-\sqrt{3}i\right)^2 + \left(-\sqrt{3}i\right)^3$

$= -1 - 3\sqrt{3}i - 9i^2 - 3\sqrt{3}i^3$

$= -1 - 3\sqrt{3}i + 9 + 3\sqrt{3}i$

$= 8$

The three numbers are cube roots of 8.

93. (a) $i^{20} = (i^4)^5 = (1)^5 = 1$

(b) $i^{45} = (i^4)^{11} i = (1)^{11} i = i$

(c) $i^{67} = (i^4)^{16} i^3 = (1)^{16}(-i) = -i$

(d) $i^{114} = (i^4)^{28} i^2 = (1)^{28}(-1) = -1$

95. False. A real number $a + 0i = a$ is equal to its conjugate.

97. False. For example, $(1 + 2i) + (1 - 2i) = 2$, which is not an imaginary number.

99. True. Let $z_1 = a_1 + b_1 i$ and $z_2 = a_2 + b_2 i$. Then

$\overline{z_1 z_2} = \overline{(a_1 + b_1 i)(a_2 + b_2 i)}$

$= \overline{(a_1 a_2 - b_1 b_2) + (a_1 b_2 + b_1 a_2)i}$

$= (a_1 a_2 - b_1 b_2) - (a_1 b_2 + b_1 a_2)i$

$= (a_1 - b_1 i)\ (a_2 - b_2 i)$

$= \overline{a_1 + b_1 i}\ \overline{a_2 + b_2 i}$

$= \overline{z_1}\ \overline{z_2}.$

101. $\sqrt{-6}\sqrt{-6} = \sqrt{6}i\sqrt{6}i = 6i^2 = -6$

103. $(4x - 5)(4x + 5) = 16x^2 - 20x + 20x - 25 = 16x^2 - 25$

105. $\left(3x - \frac{1}{2}\right)(x + 4) = 3x^2 - \frac{1}{2}x + 12x - 2 = 3x^2 + \frac{23}{2}x - 2$

Section 2.5

1. Fundamental Theorem, Algebra

3. The Linear Factorization Theorem states that a polynomial function f of degree n, $n > 0$, has exactly n linear factors

$$f(x) = a(x - c_1)(x - c_2)\ldots(x - c_n).$$

5. $f(x) = x^3 + x$ has exactly 3 zeros. Matches (c).

7. $f(x) = x^5 + 9x^3$ has exactly 5 zeros. Matches (d).

9. $f(x) = x^2 + 25$

$x^2 + 25 = 0$

$x^2 = -25$

$x = \pm\sqrt{-25}$

$x = \pm 5i$

11. $f(x) = x^3 + 9x$

$$x^3 + 9x = 0$$
$$x(x^2 + 9) = 0$$
$$x = 0 \quad x^2 + 9 = 0$$
$$x^2 = -9$$
$$x = \pm\sqrt{-9}$$
$$x = \pm 3i$$

13. $f(x) = x^3 - 4x^2 + x - 4 = x^2(x-4) + 1(x-4) = (x-4)(x^2+1)$

Zeros: $4, \pm i$

The only real zero of $f(x)$ is $x = 4$. This corresponds to the x-intercept of $(4, 0)$ on the graph.

15. $f(x) = x^4 + 4x^2 + 4 = (x^2 + 2)^2$

Zeros: $\pm\sqrt{2}i, \ \pm\sqrt{2}i$

$f(x)$ has no real zeros and the graph of $f(x)$ has no x-intercepts.

17. $h(x) = x^2 - 4x + 1$

h has no rational zeros. By the Quadratic Formula, the zeros are

$$x = \frac{4 \pm \sqrt{16-4}}{2} = 2 \pm \sqrt{3}.$$

$$h(x) = \left[x - \left(2 + \sqrt{3}\right)\right]\left[x - \left(2 - \sqrt{3}\right)\right]$$
$$= (x - 2 - \sqrt{3})(x - 2 + \sqrt{3})$$

19. $f(x) = x^2 - 12x + 26$

f has no rational zeros. By the Quadratic Formula, the zeros are

$$x = \frac{12 \pm \sqrt{(-12)^2 - 4(26)}}{2} = 6 \pm \sqrt{10}.$$

$$f(x) = \left[x - (6 + \sqrt{10})\right]\left[x - (6 - \sqrt{10})\right]$$
$$= (x - 6 - \sqrt{10})(x - 6 + \sqrt{10})$$

21. $f(x) = x^2 + 25$

Zeros: $\pm 5i$

$f(x) = (x + 5i)(x - 5i)$

23. $f(x) = 16x^4 - 81$

$$= (4x^2 - 9)(4x^2 + 9)$$
$$= (2x - 3)(2x + 3)(2x + 3i)(2x - 3i)$$

Zeros: $\pm\frac{3}{2}, \ \pm\frac{3}{2}i$

25. $f(z) = z^2 - z + 56$

$$z = \frac{1 \pm \sqrt{1 - 4(56)}}{2}$$
$$= \frac{1 \pm \sqrt{-223}}{2}$$
$$= \frac{1}{2} \pm \frac{\sqrt{223}}{2}i$$

Zeros: $\frac{1}{2} \pm \frac{\sqrt{223}}{2}i$

$$f(z) = \left(z - \frac{1}{2} + \frac{\sqrt{223}i}{2}\right)\left(z - \frac{1}{2} - \frac{\sqrt{223}i}{2}\right)$$

27. $f(x) = x^4 + 10x^2 + 9$

$$= (x^2 + 1)(x^2 + 9)$$
$$= (x + i)(x - i)(x + 3i)(x - 3i)$$

The zeros of $f(x)$ are $x = \pm i$ and $x = \pm 3i$.

29. $f(x) = 3x^3 - 5x^2 + 48x - 80$

Using synthetic division, $\frac{5}{3}$ is a zero:

$\frac{5}{3}$	3	−5	48	−80
		5	0	80
	3	0	48	0

$$f(x) = \left(x - \frac{5}{3}\right)(3x^2 + 48)$$
$$= (3x - 5)(x^2 + 16)$$
$$= (3x - 5)(x + 4i)(x - 4i)$$

The zeros are $\frac{5}{3}, \ 4i, \ -4i$.

31. $f(t) = t^3 - 3t^2 - 15t + 125$

Possible rational zeros: $\pm 1, \ \pm 5, \ \pm 25, \ \pm 125$

−5	1	−3	−15	125
		−5	40	−125
	1	−8	25	0

By the Quadratic Formula, the zeros of $t^2 - 8t + 25$ are

$$t = \frac{8 \pm \sqrt{64 - 100}}{2} = 4 \pm 3i.$$

The zeros of $f(t)$ are $t = -5$ and $t = 4 \pm 3i$.

$$f(t) = \left[t - (-5)\right]\left[t - (4 + 3i)\right]\left[t - (4 - 3i)\right]$$
$$= (t + 5)(t - 4 - 3i)(t - 4 + 3i)$$

33. $f(x) = 5x^3 - 9x^2 + 28x + 6$

Possible rational zeros: $\pm 6, \pm \frac{6}{5}, \pm 3, \pm \frac{3}{5}, \pm 2, \pm \frac{2}{5}, \pm 1, \pm \frac{1}{5}$

$-\frac{1}{5}$	5	−9	28	6
		−1	2	−6
	5	−10	30	0

By the Quadratic Formula, the zeros of $5x^2 - 10x + 30$ are those of $x^2 - 2x + 6$:

$$x = \frac{2 \pm \sqrt{4 - 4(6)}}{2} = 1 \pm \sqrt{5}i$$

Zeros: $-\frac{1}{5}, 1 \pm \sqrt{5}i$

$$f(x) = 5\left(x + \frac{1}{5}\right)\left[x - \left(1 + \sqrt{5}i\right)\right]\left[x - \left(1 - \sqrt{5}i\right)\right]$$
$$= (5x + 1)\left(x - 1 - \sqrt{5}i\right)\left(x - 1 + \sqrt{5}i\right)$$

35. $g(x) = x^4 - 4x^3 + 8x^2 - 16x + 16$

Possible rational zeros: $\pm 1, \pm 2, \pm 4, \pm 8, \pm 16$

2	1	−4	8	−16	16
		2	−4	8	−16
2	1	−2	4	−8	0
		2	0	8	
	1	0	4	0	

$$g(x) = (x - 2)(x - 2)(x^2 + 4)$$
$$= (x - 2)^2(x + 2i)(x - 2i)$$

The zeros of g are $2, 2$, and $\pm 2i$.

37. (a) $f(x) = x^2 - 14x + 46$.

By the Quadric Formula,

$$x = \frac{14 \pm \sqrt{(-14)^2 - 4(46)}}{2} = 7 \pm \sqrt{3}.$$

The zeros are $7 + \sqrt{3}$ and $7 - \sqrt{3}$.

(b) $f(x) = \left[x - (7 + \sqrt{3})\right]\left[x - (7 - \sqrt{3})\right]$
$= (x - 7 - \sqrt{3})(x - 7 + \sqrt{3})$

(c) x-intercepts: $(7 + \sqrt{3}, 0)$ and $(7 - \sqrt{3}, 0)$

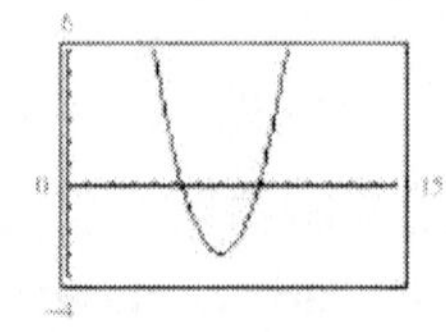

39. (a) $f(x) = 2x^3 - 3x^2 + 8x - 12$
$= (2x - 3)(x^2 + 4)$

The zeros are $\frac{3}{2}$ and $\pm 2i$.

(b) $f(x) = (2x - 3)(x + 2i)(x - 2i)$

(c) x-intercept: $\left(\frac{3}{2}, 0\right)$

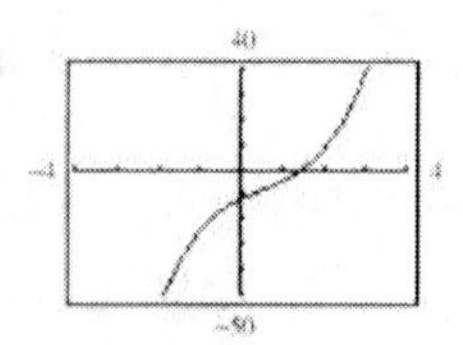

41. (a) $f(x) = x^3 - 11x + 150$
$= (x + 6)(x^2 - 6x + 25)$

Use the Quadratic Formula to find the zeros of $x^2 - 6x + 25$.

$$x = \frac{6 \pm \sqrt{(-6)^2 - 4(25)}}{2} = 3 \pm 4i.$$

The zeros are $-6, 3 + 4i$, and $3 - 4i$.

(b) $f(x) = (x + 6)(x - 3 + 4i)(x - 3 - 4i)$

(c) x-intercept: $(-6, 0)$

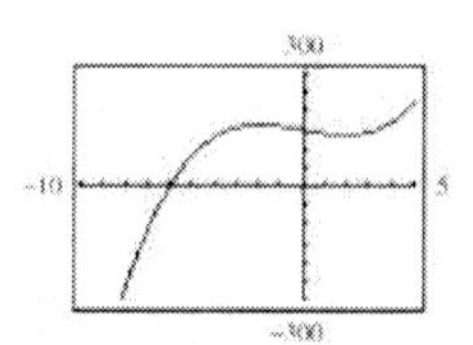

43. (a) $f(x) = x^4 + 25x^2 + 144$
$= (x^2 + 9)(x^2 + 16)$

The zeros are $\pm 3i, \pm 4i$.

(b) $f(x) = (x^2 + 9)(x^2 + 16)$
$= (x + 3i)(x - 3i)(x + 4i)(x - 4i)$

(c) No x-intercepts

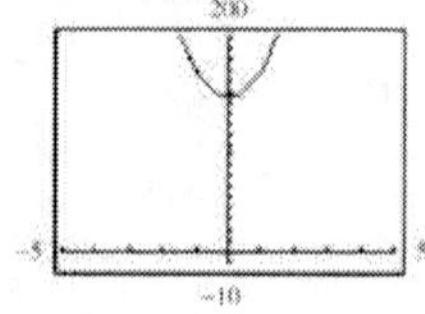

45. $f(x) = (x - 2)(x - i)(x + i)$
$= (x - 2)(x^2 + 1)$

Note that $f(x) = a(x^3 - 2x^2 + x - 2)$, where a is any nonzero real number, has zeros $2, \pm i$.

47. $f(x) = (x-2)^2(x-4-i)(x-4+i)$

$= (x-2)^2(x-8x+16+1)$

$= (x^2-4x+4)(x^2-8x+17)$

$= x^4-12x^3+53x^2-100x+68$

Note that $f(x) = a(x^4-12x^3+53x^2-100x+68)$, where a is any nonzero real number, has zeros $2,\ 2,\ 4 \pm i$.

49. Because $1+\sqrt{2}i$ is a zero, so is $1-\sqrt{2}i$.

$f(x) = (x-0)(x+5)(x-1-\sqrt{2}i)(x-1+\sqrt{2}i)$

$= (x^2+5x)(x^2-2x+1+2)$

$= (x^2+5x)(x^2-2x+3)$

$= x^4+3x^3-7x^2-15x$

Note that $f(x) = a(x^4+3x^3-7x^2+15x)$, where a is any nonzero real number, has zeros $0,\ -5,\ 1 \pm \sqrt{2}i$.

51. (a) $f(x) = a(x-1)(x+2)(x-2i)(x+2i)$

$= a(x-1)(x+2)(x^2+4)$

$f(1) = 10 = a(-2)(1)(5) \Rightarrow a = -1$

$f(x) = -(x-1)(x+2)(x-2i)(x+2i)$

(b) $f(x) = -(x-1)(x+2)(x^2+4)$

$= -(x^2+x-2)(x^2+4)$

$= -x^4-x^3-2x^2-4x+8$

53. (a) $f(x) = a(x+1)(x-2-\sqrt{5}i)(x-2+\sqrt{5}i)$

$= a(x+1)(x^2-4x+4+5)$

$= a(x+1)(x^2-4x+9)$

$f(-2) = 42 = a(-1)(4+8+9) \Rightarrow a = -2$

$f(x) = -2(x+1)(x-2-\sqrt{5}i)(x-2+\sqrt{5}i)$

(b) $f(x) = -2(x+1)(x^2-4x+9)$

$= -2x^3+6x^2-10x-18$

55. $f(x) = x^4-6x^2-7$

(a) $f(x) = (x^2-7)(x^2+1)$

(b) $f(x) = (x-\sqrt{7})(x+\sqrt{7})(x^2+1)$

(c) $f(x) = (x-\sqrt{7})(x+\sqrt{7})(x+i)(x-i)$

57. $f(x) = x^4-2x^3-3x^2+12x-18$

(a) $f(x) = (x^2-6)(x^2-2x+3)$

(b) $f(x) = (x+\sqrt{6})(x-\sqrt{6})(x^2-2x+3)$

(c) $f(x) = (x+\sqrt{6})(x-\sqrt{6})(x-1-\sqrt{2}i)(x-1+\sqrt{2}i)$

59. $f(x) = 2x^3+3x^2+50x+75$

Since is $5i$ a zero, so is $-5i$.

$$\begin{array}{r|rrrr} 5i & 2 & 3 & 50 & 75 \\ & & 10i & -50+15i & -75 \\ \hline & 2 & 3+10i & 15i & 0 \end{array}$$

$$\begin{array}{r|rrr} -5i & 2 & 3+10i & 15i \\ & & -10i & -15i \\ \hline & 2 & 3 & 0 \end{array}$$

The zero of $2x+3$ is $x = -\frac{3}{2}$. The zeros of f are $x = -\frac{3}{2}$ and $x = \pm 5i$.

Alternate solution

Since $x = \pm 5i$ are zeros of $f(x)$, $(x+5i)(x-5i) = x^2+25$ is a factor of $f(x)$. By long division we have:

$$\begin{array}{r|l} & \quad 2x+3 \\ x^2+0x+25 & 2x^3+3x^2+50x+75 \\ & \underline{2x^3+0x^2+50x} \\ & \qquad 3x^2+0x+75 \\ & \qquad \underline{3x^2+0x+75} \\ & \qquad\qquad 0 \end{array}$$

Thus, $f(x) = (x^2+25)(2x+3)$ and the zeros of f are $x = \pm 5i$ and $x = -\frac{3}{2}$.

61. $g(x) = x^3-7x^2-x+87$. Since $5+2i$ is a zero, so is $5-2i$.

$$\begin{array}{r|rrrr} 5+2i & 1 & -7 & -1 & 87 \\ & & 5+2i & -14+6i & -87 \\ \hline & 1 & -2+2i & -15+6i & 0 \end{array}$$

$$\begin{array}{r|rrr} 5-2i & 1 & -2+2i & -15+6i \\ & & 5-2i & 15-6i \\ \hline & 1 & 3 & 0 \end{array}$$

The zero of $x+3$ is $x = -3$.
The zeros of f are $-3,\ 5 \pm 2i$. .

63. $h(x)=3x^3-4x^2+8x+8$ Since $1-\sqrt{3}i$ is a zero, so is $1+\sqrt{3}i$.

$$\begin{array}{r|rrrr} 1-\sqrt{3}i & 3 & -4 & 8 & 8 \\ & & 3-3\sqrt{3}i & -10-2\sqrt{3}i & -8 \\ \hline & 3 & -1-3\sqrt{3}i & -2-2\sqrt{3}i & 0 \end{array}$$

$$\begin{array}{r|rrr} 1+\sqrt{3}i & 3 & -1-3\sqrt{3}i & -2-2\sqrt{3}i \\ & & 3+3\sqrt{3}i & 2+2\sqrt{3}i \\ \hline & 3 & 2 & 0 \end{array}$$

The zero of $3x+2$ is $x=-\frac{2}{3}$. The zeros of h are

$x=-\frac{2}{3}, 1\pm\sqrt{3}i.$

65. $h(x)=8x^3-14x^2+18x-9$. Since $\frac{1}{2}(1-\sqrt{5}i)$ is a zero, so is $\frac{1}{2}(1+\sqrt{5}i)$.

$$\begin{array}{r|rrrr} \frac{1}{2}\left(1-\sqrt{5}i\right) & 8 & -14 & 18 & -9 \\ & & 4-4\sqrt{5}i & -15+3\sqrt{5}i & 9 \\ \hline & 8 & -10-4\sqrt{5}i & 3+3\sqrt{5}i & 0 \end{array}$$

$$\begin{array}{r|rrr} \frac{1}{2}\left(1+\sqrt{5}i\right) & 8 & -10-4\sqrt{5}i & 3+3\sqrt{5}i \\ & & 4+4\sqrt{5}i & -3-3\sqrt{5}i \\ \hline & 8 & -6 & 0 \end{array}$$

The zero of $8x-6$ is $x=\frac{3}{4}$. The zeros of h are

$x=\frac{3}{4}, \frac{1}{2}(1\pm\sqrt{5}i).$

67. $f(x)=x^4+3x^3-5x^2-21x+22$

(a) The *root* feature yields the real roots 1 and 2, and the complex roots $-3\pm1.414i$.

(b) By synthetic division:

$$\begin{array}{r|rrrrr} 1 & 1 & 3 & -5 & -21 & 22 \\ & & 1 & 4 & -1 & -22 \\ \hline & 1 & 4 & -1 & -22 & 0 \end{array}$$

$$\begin{array}{r|rrrr} 2 & 1 & 4 & -1 & -22 \\ & & 2 & 12 & 22 \\ \hline & 1 & 6 & 11 & 0 \end{array}$$

The complex roots of $x^2+6x+11$ are

$$x=\frac{-6\pm\sqrt{6^2-4(11)}}{2}=-3\pm\sqrt{2}i.$$

69. $h(x)=8x^3-14x^2+18x-9$

(a) The *root* feature yields the real root 0.75, and the complex roots $0.5\pm1.118i$.

(b) By synthetic division:

$$\begin{array}{r|rrrr} \frac{3}{4} & 8 & -14 & 18 & -9 \\ & & 6 & -6 & 9 \\ \hline & 8 & -8 & 12 & 0 \end{array}$$

The complex roots of $8x^2-8x+12$ are

$$x=\frac{8\pm\sqrt{64-4(8)(12)}}{2(8)}=\frac{1}{2}\pm\frac{\sqrt{5}}{2}i.$$

71. To determine if the football reaches a height of 50 feet, set $h(t)=50$ and solve for t.

$$-16t^2+48t=50$$

$$-16t^2+48t-50=0$$

Using the Quadratic formula:

$$t=\frac{-(48)\pm\sqrt{48^2-4(-16)(-50)}}{2(-16)}$$

$$t=\frac{-(48)\pm\sqrt{-896}}{-32}$$

Because the discriminant is negative, the solutions are not real, therefore the football does not reach a height of 50 feet.

73. False, a third-degree polynomial must have at least one real zero.

75. Answers will vary.

77. $f(x)=x^2-7x-8$

$$=\left(x-\frac{7}{2}\right)^2-\frac{81}{4}$$

A parabola opening upward with vertex $\left(\frac{7}{2}, -\frac{81}{4}\right)$

79. $f(x)=6x^2+5x-6$

$$=6\left(x+\frac{5}{12}\right)^2-\frac{169}{24}$$

A parabola opening upward with vertex $\left(-\frac{5}{12}, -\frac{169}{24}\right)$

Section 2.6

1. rational functions

3. To determine the vertical asymptote(s) of the graph of $y = \frac{9}{x-3}$, find the real zeros of the denominator of the equation. (Assuming no common factors in the numerator and denominator)

5. $f(x) = \frac{1}{x-1}$

(a) Domain: all $x \neq 1$

(b)

x	$f(x)$
0.5	−2
0.9	−10
0.99	−100
0.999	−1000

x	$f(x)$
1.5	2
1.1	10
1.01	100
1.001	1000

x	$f(x)$
5	0.25
10	$0.\overline{1}$
100	$0.\overline{01}$
1000	$0.\overline{001}$

x	$f(x)$
−5	$-0.1\overline{6}$
−10	$-0.\overline{09}$
−100	$-0.\overline{0099}$
−1000	$-0.\overline{00099}$

(c) f approaches $-\infty$ from the left of 1 and ∞ from the right of 1.

7. $f(x) = \frac{3x}{|x-1|}$

(a) Domain: all $x \neq 1$

(b)

x	$f(x)$
0.5	3
0.9	27
0.99	297
0.999	2997

x	$f(x)$
1.5	9
1.1	33
1.01	303
1.001	3003

x	$f(x)$
5	3.75
10	$3.\overline{33}$
100	$3.\overline{03}$
1000	$3.\overline{003}$

x	$f(x)$
−5	−2.5
−10	−2.727
−100	−2.970
−1000	−2.997

(c) f approaches ∞ from both the left and the right of 1.

9. $f(x) = \frac{3x^2}{x^2 - 1}$

(a) Domain: all $x \neq 1$

(b)

x	$f(x)$
0.5	−1
0.9	−12.79
0.99	−147.8
0.999	−1498

x	$f(x)$
1.5	5.4
1.1	17.29
1.01	152.3
1.001	1502.3

x	$f(x)$
5	3.125
10	$3.\overline{03}$
100	$3.\overline{0003}$
1000	3

x	$f(x)$
−5	3.125
−10	$3.\overline{03}$
−100	$3.\overline{0003}$
−1000	3

(c) f approaches $-\infty$ from the left of 1, and ∞ from the right of 1. f approaches ∞ from the left of −1, and $-\infty$ from the right of −1.

11. $f(x) = \frac{2}{x+2}$

Vertical asymptote: $x = -2$
Horizontal asymptote: $y = 0$
Matches graph (a).

13. $f(x) = \frac{4x+1}{x}$

Vertical asymptote: $x = 0$
Horizontal asymptote: $y = 4$
Matches graph (c).

15. $f(x) = \frac{x-2}{x-4}$

Vertical asymptote: $x = 4$
Horizontal asymptote: $y = 1$
Matches graph (b).

17. $f(x) = \frac{1}{x^2}$

Vertical asymptote: $x = 0$
Horizontal asymptote: $y = 0$ or x-axis

19. $f(x) = \frac{2x^2}{x^2 + x - 6} = \frac{2x^2}{(x+3)(x-2)}$

Vertical asymptotes: $x = -3$, $x = 2$
Horizontal asymptote: $y = 2$

21. $f(x) = \frac{x(2+x)}{2x - x^2} = \frac{\cancel{x}(x+2)}{-\cancel{x}(x-2)} = -\frac{x+2}{x-2}, \ x \neq 0$

Vertical asymptote: $x = 2$
Horizontal asymptote: $y = -1$
Hole at $x = 0$

23. $f(x) = \dfrac{x^2 - 25}{x^2 + 5x} = \dfrac{\cancel{(x+5)}(x-5)}{x\cancel{(x+5)}} = \dfrac{x-5}{x},\ x \neq -5$

Vertical asymptote: $x = 0$

Horizontal asymptote: $y = 1$

Hole at $x = -5$

25. $f(x) = \dfrac{3x^2 + x - 5}{x^2 + 1}$

(a) Domain: all real numbers
(b) Continuous
(c) Vertical asymptote: none
Horizontal asymptote: $y = 3$

27. $f(x) = \dfrac{x^2 + 3x - 4}{-x^3 + 27} = \dfrac{(x+4)(x-1)}{-(x-3)(x^2+3x+9)}$

(a) Domain: all real numbers x except $x = 3$
(b) Not continuous at $x = 3$
(c) Vertical asymptote: $x = 3$
Horizontal asymptote: $y = 0$ or x-axis

29. $f(x) = \dfrac{x^2 - 16}{x - 4} = \dfrac{(x+4)\cancel{(x-4)}}{\cancel{x-4}} = x + 4,\ x \neq 4$

$g(x) = x + 4$

(a) Domain of f: all real x except $x = 4$
Domain of g: all real x
(b) Vertical asymptote:
f has none. g has none.
Hole: f has a hole at $x = 4$; g has none.
(c)

x	1	2	3	4	5	6	7
$f(x)$	5	6	7	Undef.	9	10	11
$g(x)$	5	6	7	8	9	10	11

(d)

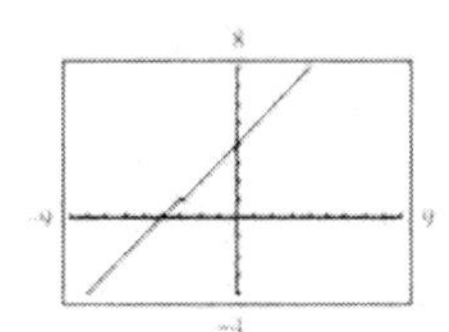

(e) Graphing utilities are limited in their resolution and therefore may not show a hole in a graph.

31. $f(x) = \dfrac{x^2 - 1}{x^2 - 2x - 3} = \dfrac{\cancel{(x+1)}(x-1)}{(x-3)\cancel{(x+1)}} = \dfrac{x-1}{x-3},\ x \neq -1$

$g(x) = \dfrac{x-1}{x-3}$

(a) Domain of f: all real x except $x = 3$ and $x = -1$
Domain of g: all real x except $x = 3$
(b) Vertical asymptote: f has a vertical asymptote at $x = 3$.
g has a vertical asymptote at $x = 3$.
Hole: f has a hole at $x = -1$; g has none.
(c)

x	-2	-1	0	1	2	3	4
$f(x)$	$\frac{3}{5}$	Undef.	$\frac{1}{3}$	0	-1	Undef.	3
$g(x)$	$\frac{3}{5}$	$\frac{1}{2}$	$\frac{1}{3}$	0	-1	Undef.	3

(d)

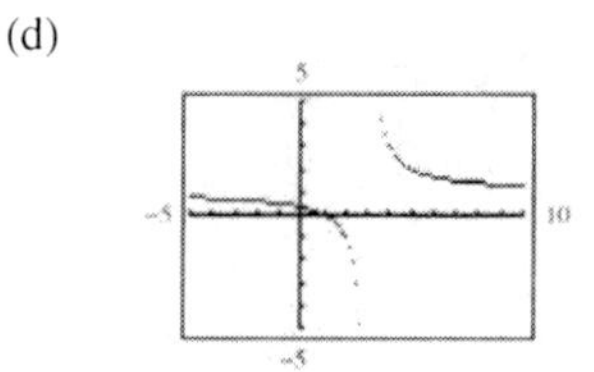

(e) Graphing utilities are limited in their resolution and therefore may not show a hole in a graph.

33. $f(x) = 4 - \dfrac{1}{x}$

As $x \to \pm\infty$, $f(x) \to 4$.

As $x \to \infty$, $f(x) \to 4$ but is less than 4.

As $x \to -\infty$, $f(x) \to 4$ but is greater than 4.

35. $f(x) = \dfrac{2x - 1}{x - 3}$

As $x \to \pm\infty$, $f(x) \to 2$.

As $x \to \infty$, $f(x) \to 2$ but is greater than 2.

As $x \to -\infty$, $f(x) \to 2$ but is less than 2.

37. $g(x) = \dfrac{x^2 - 4}{x + 3} = \dfrac{(x-2)(x+2)}{x+3}$

The zeros of g correspond to the zeros of the numerator and are $x = \pm 2$.

39. $f(x) = 1 - \dfrac{2}{x - 5} = \dfrac{x - 7}{x - 5}$

The zero of f corresponds to the zero of the numerator and is $x = 7$.

41. $g(x) = \dfrac{x^2 - 2x - 3}{x^2 + 1} = \dfrac{(x-3)(x+1)}{x^2+1} = 0$

Zeros: $x = -1, 3$

43. $f(x) = \dfrac{2x^2 - 5x + 2}{2x^2 - 7x + 3} = \dfrac{(2x-1)(x-2)}{(2x-1)(x-3)} = \dfrac{x-2}{x-3},\ x \neq \dfrac{1}{2}$

Zero: $x = 2$ ($x = \dfrac{1}{2}$ is not in the domain.)

45. $C = \dfrac{255p}{100 - p},\ 0 \le p < 100$

(a) find $C(10) = \dfrac{255(10)}{100 - 10} = \$28.3$ million

$C(40) = \dfrac{255(40)}{100 - 40} = \$170$ million

$C(75) = \dfrac{255(75)}{100 - 75} = \$765$ million

(b)

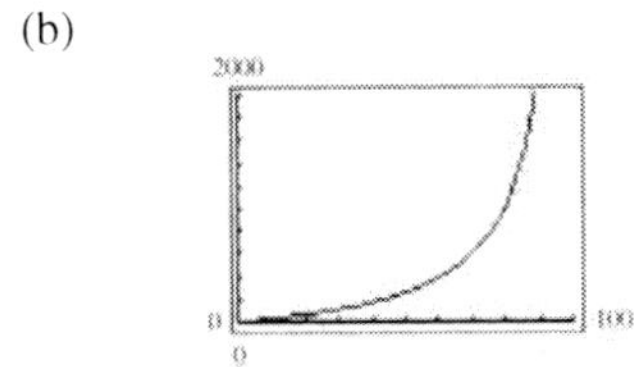

(c) No. The function is undefined at $p = 100\,\%$.

47.

(a)

M	200	400	600	800	1000	1200	1400	1600	1800	2000
t	0.472	0.596	0.710	0.817	0.916	1.009	1.096	1.178	1.255	1.328

The greater the mass, the more time required per oscillation. The model is a good fit to the actual data.

(b) You can find M corresponding to $t = 1.056$ by finding the point of intersection of

$$t = \frac{38M + 16{,}965}{10(M + 500)} \text{ and } t = 1.056.$$

If you do this, you obtain $M \approx 1306$ grams.

49. (a) The model fits the data well.

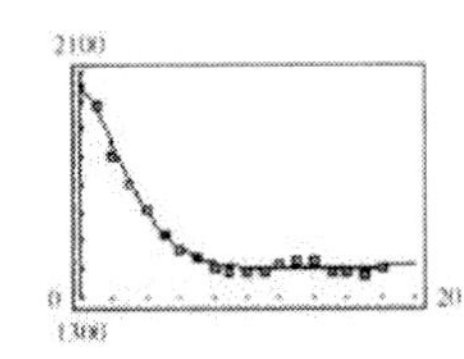

(a) $N = \dfrac{77.095t^2 - 216.04t + 2050}{0.052t^2 - 0.8t + 1}$

$2009: N(19) \approx 1{,}412{,}000$

$2010: N(20) \approx 1{,}414{,}000$

$2011: N(11) \approx 1{,}416{,}000$

Answers will vary.

(b) Horizontal asymptote: $N = 1482.6$ (approximately) Answers will vary.

51. False. For example, $f(x) = \dfrac{1}{x^2 + 1}$ has no vertical asymptote.

53. No. If $x = c$ is also a zero of the denominator of f, then f is undefined at $x = c$, and the graph of f may have a hole of vertical asymptote at $x = c$.

55.

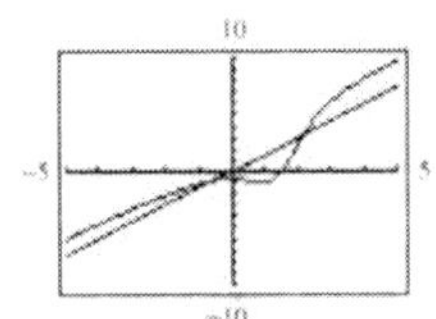

The graphs of $y_1 = \dfrac{3x^3 - 5x^2 + 4x - 5}{2x^2 - 6x + 7}$ and $y_2 = \dfrac{3x^3}{2x^2}$ are approximately the same graph as $x \to \infty$ and $x \to -\infty$.

Therefore as $x \to \pm\infty$, the graph of a rational function $y = \dfrac{a_n x^n + \ldots a_1 x + a_0}{b_m x^m + \ldots b_1 x + b_0}$ appears to be very close to the graph of $y = \dfrac{a_n x^n}{b_m x^m}$.

57. $y - 2 = \dfrac{-1 - 2}{0 - 3}(x - 3) = 1(x - 3)$

$y = x - 1$

$y - x + 1 = 0$

59. $y - 7 = \dfrac{10 - 7}{3 - 2}(x - 2) = 3(x - 2)$

$y = 3x + 1$

$3x - y + 1 = 0$

61.

$$\begin{array}{r} x + 9 \\ x - 4 \overline{)\, x^2 + 5x + 6} \\ \underline{x^2 - 4x} \\ 9x + 6 \\ \underline{9x - 36} \\ 42 \end{array}$$

$$\frac{x^2 + 5x + 6}{x - 4} = x + 9 + \frac{42}{x - 4}$$

63.

$$\begin{array}{r} 2x^2 - 9 \\ x^2 + 5 \overline{)\, 2x^4 + 0x^3 + x^2 + 0x - 11} \\ \underline{2x^4 + \quad 10x^2} \\ -9x^2 - \quad 11 \\ \underline{-9x^2 - \quad 45} \\ 34 \end{array}$$

$$\frac{2x^4 + x^2 - 11}{x^2 + 5} = 2x^2 - 9 + \frac{34}{x^2 + 5}$$

Section 2.7

1. slant, asymptote

3. Yes. Because the numerator's degree is exactly 1 greater than that of the denominator, the graph of f has a slant asymptote.

5.

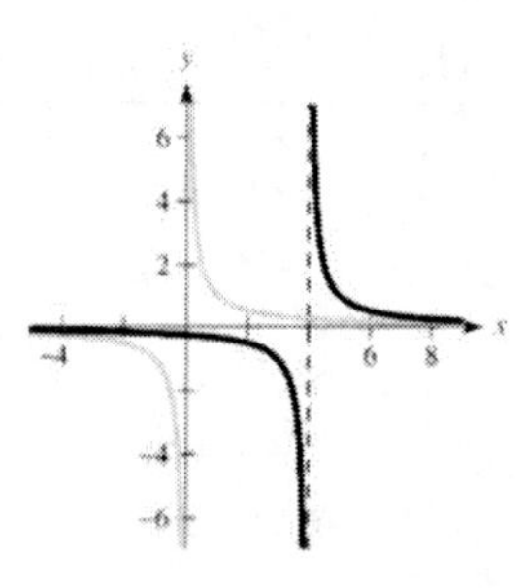

The graph of $g = \frac{1}{x-4}$ is a horizontal shift four units to the right of the graph of $f(x) = \frac{1}{x}$.

7.

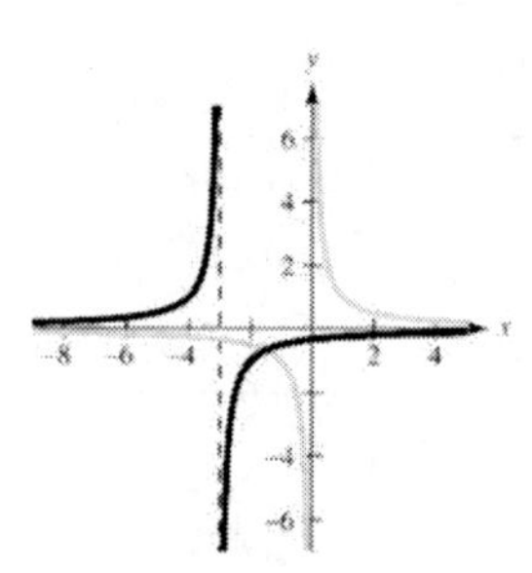

The graph of $g(x) = \frac{-1}{x+3}$ is a reflection in the x-axis and a horizontal shift three units to the left of the graph of $f(x) = \frac{1}{x}$.

9. $g(x) = \frac{2}{x} + 1$

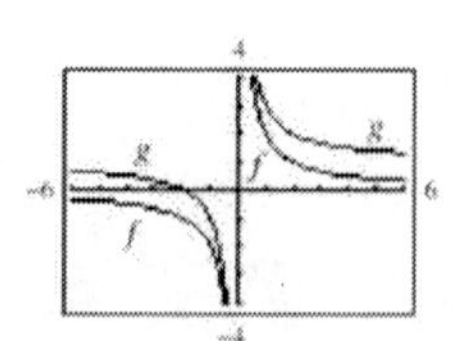

Vertical shift one unit upward

11. $g(x) = -\frac{2}{x}$

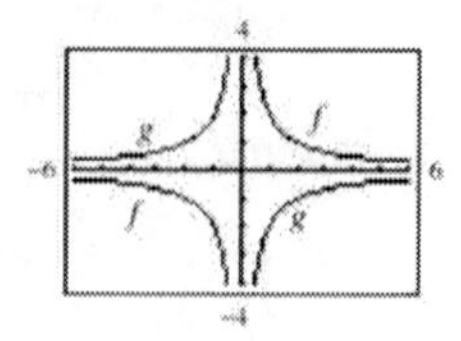

Reflection in the x-axis

13. $g(x) = \frac{2}{x^2} - 2$

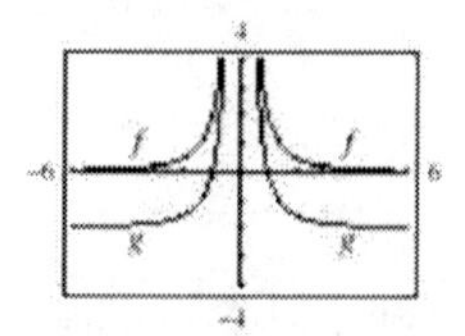

Vertical shift two units downward

15. $g(x) = \frac{2}{(x-2)^2}$

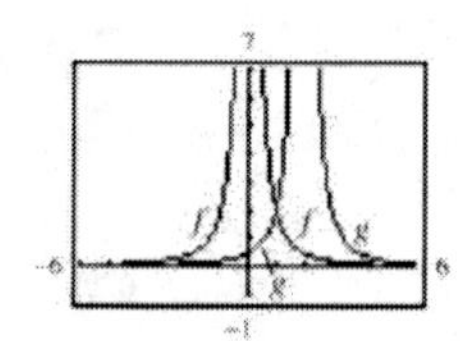

Horizontal shift two units to the right

17. $f(x) = \frac{1}{x+2}$

y-intercept: $\left(0, \frac{1}{2}\right)$

Vertical asymptote: $x = -2$
Horizontal asymptote: $y = 0$

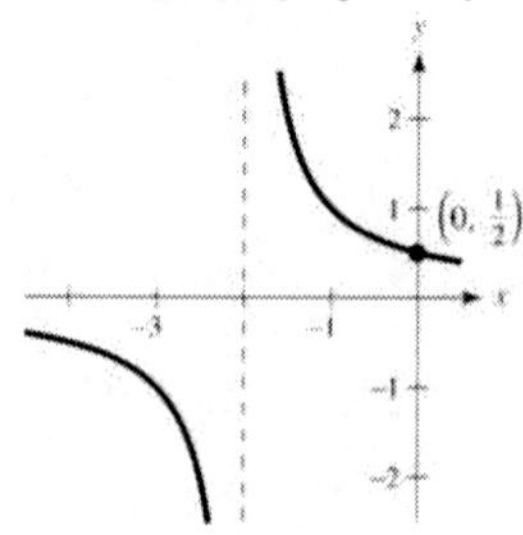

x	-4	-3	-1	0	1
y	$-\frac{1}{2}$	-1	1	$\frac{1}{2}$	$\frac{1}{3}$

19. $C(x) = \dfrac{5 + 2x}{1 + x} = \dfrac{2x + 5}{x + 1}$

x-intercept: $\left(-\dfrac{5}{2},\ 0\right)$

y-intercept: $(0,\ 5)$

Vertical asymptote: $x = -1$

Horizontal asymptote: $y = 2$

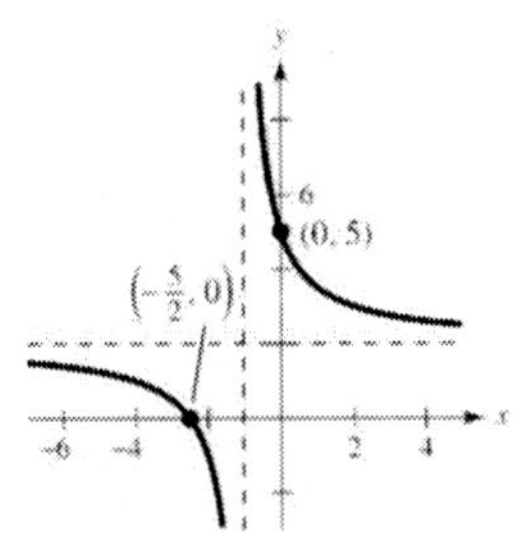

x	-4	-3	-2	0	1	2
$C(x)$	1	$\dfrac{1}{2}$	-1	5	$\dfrac{7}{2}$	3

21. $f(t) = \dfrac{1 - 2t}{t} = -\dfrac{2t - 1}{t}$

t-intercept: $\left(\dfrac{1}{2},\ 0\right)$

Vertical asymptote: $t = 0$

Horizontal asymptote: $y = -2$

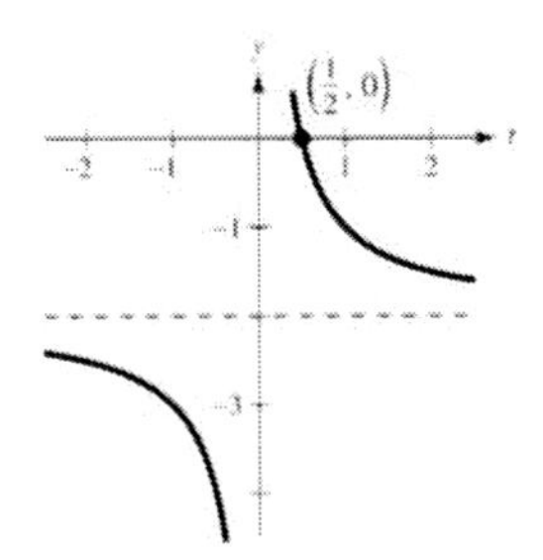

x	-2	-1	$\dfrac{1}{2}$	1	2
y	$-\dfrac{5}{2}$	-3	0	-1	$-\dfrac{3}{2}$

23. $f(x) = \dfrac{x^2}{x^2 - 4}$

Intercept: $(0,\ 0)$

Vertical asymptotes: $x = 2,\ x = -2$

Horizontal asymptote: $y = 1$

y-axis symmetry

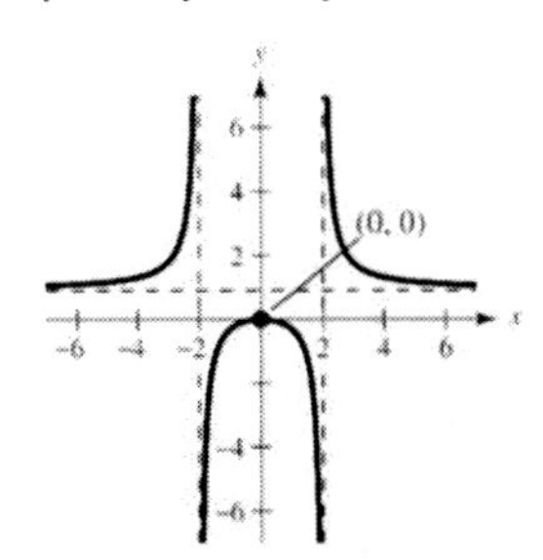

x	-4	-1	0	-1	4
y	$\dfrac{4}{3}$	$-\dfrac{1}{3}$	0	$-\dfrac{1}{3}$	$\dfrac{4}{3}$

25. $g(x) = \dfrac{4(x + 1)}{x(x - 4)}$

Intercept: $(-1,\ 0)$

Vertical asymptotes: $x = 0$ and $x = 4$

Horizontal asymptote: $y = 0$

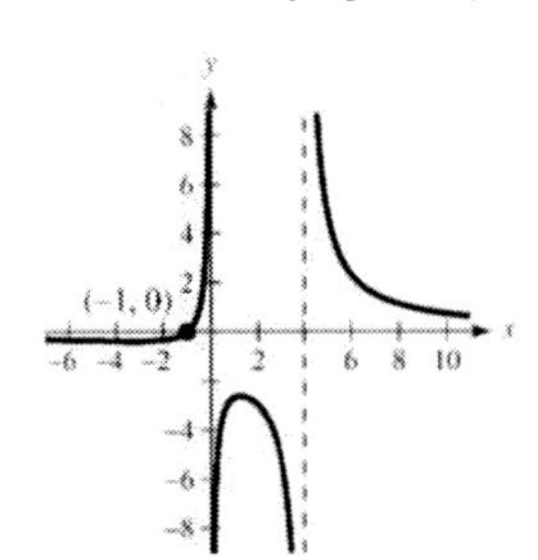

x	-2	-1	1	2	3	5	6
y	$-\dfrac{1}{3}$	0	$-\dfrac{8}{3}$	-3	$-\dfrac{16}{3}$	$\dfrac{24}{5}$	$\dfrac{7}{3}$

27. $f(x) = \dfrac{3x}{x^2 - x - 2} = \dfrac{3x}{(x+1)(x-2)}$

Intercept: $(0, 0)$

Vertical asymptotes: $x = -1, 2$

Horizontal asymptote: $y = 0$

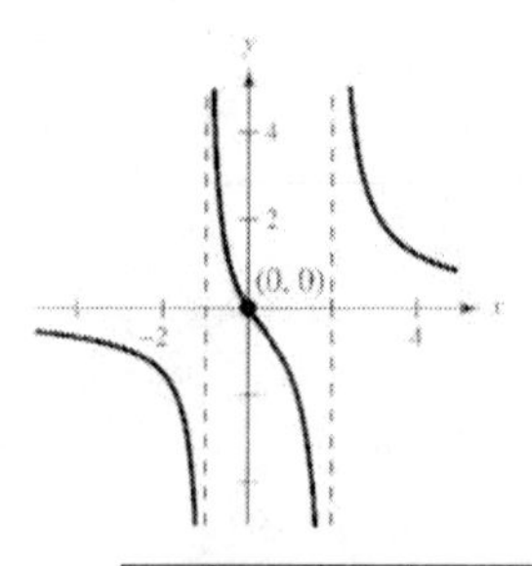

x	-3	0	1	3	4
y	$-\dfrac{9}{10}$	0	$-\dfrac{3}{2}$	$\dfrac{9}{4}$	$\dfrac{6}{5}$

29. $f(x) = \dfrac{x^2 + 3x}{x^2 + x - 6} = \dfrac{x(x+3)}{(x-2)(x+3)} = \dfrac{x}{x-2},$

$x \neq -3$

Intercept: $(0, 0)$

Vertical asymptote: $x = 2$

(There is a hole at $x = -3$.)

Horizontal asymptote: $y = 1$

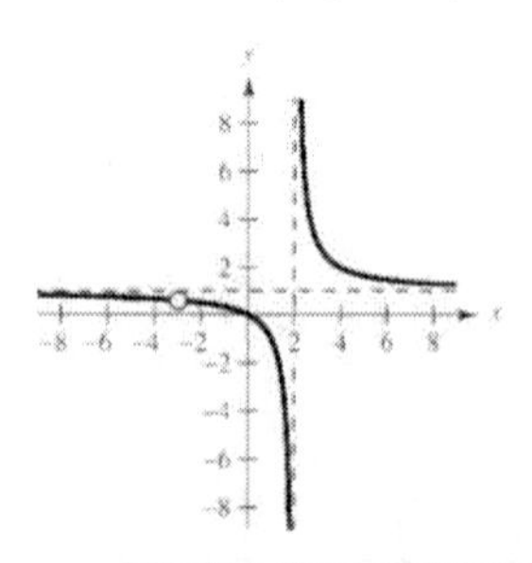

x	-2	-1	0	1	2	3
y	$\dfrac{1}{2}$	$\dfrac{1}{3}$	0	-1	Undef.	3

31. $f(x) = \dfrac{x^2 - 1}{x + 1} = \dfrac{(x+1)(x-1)}{x+1} = x - 1,$

$x \neq -1$

The graph is a line, with a hole at $x = -1$.

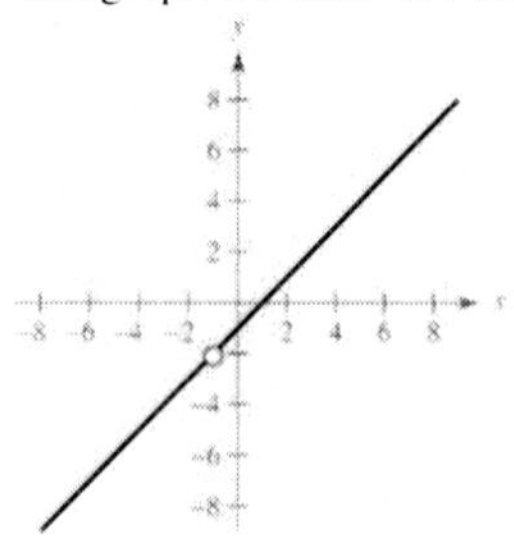

33. $f(x) = \dfrac{2 + x}{1 - x} = -\dfrac{x + 2}{x - 1}$

Vertical asymptote: $x = 1$

Horizontal asymptote: $y = -1$

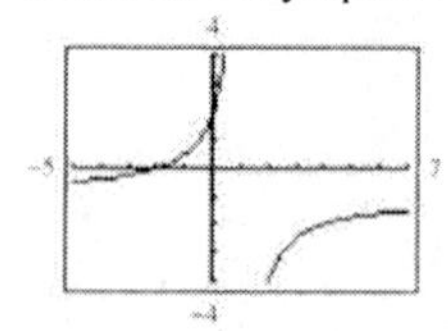

Domain: $x \neq 1$ or $(-\infty, 1) \cup (1, \infty)$

35. $f(t) = \dfrac{3t + 1}{t}$

Vertical asymptote: $t = 0$

Horizontal asymptote: $y = 3$

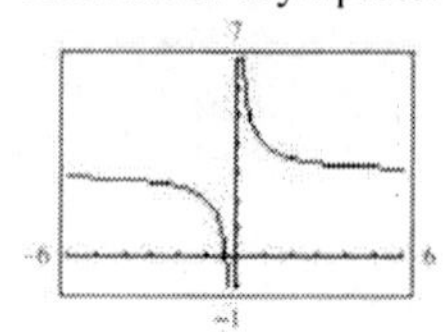

Domain: $t \neq 0$ or $(-\infty, 0) \cup (0, \infty)$

37. $h(t) = \dfrac{4}{t^2 + 1}$

Domain: all real numbers or $(-\infty, \infty)$

Horizontal asymptote: $y = 0$

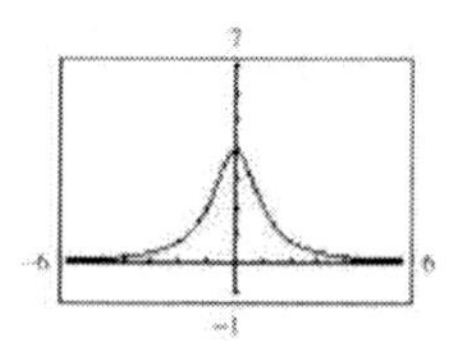

39. $f(x) = \dfrac{x + 1}{x^2 - x - 6} = \dfrac{x + 1}{(x-3)(x+2)}$

Domain: all real numbers except $x = 3, -2$

Vertical asymptotes: $x = 3, x = -2$

Horizontal asymptote: $y = 0$

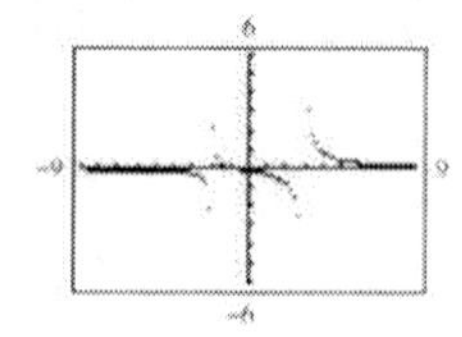

41. $f(x) = \dfrac{20x}{x^2+1} - \dfrac{1}{x} = \dfrac{19x^2-1}{x(x^2+1)}$

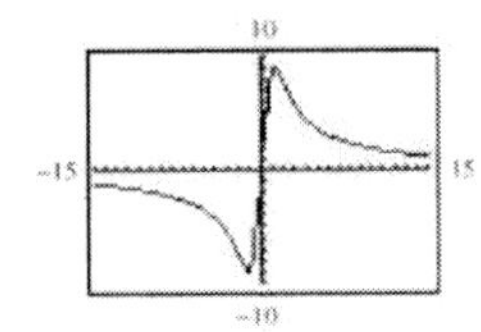

Domain: all real numbers except 0, or $(-\infty, 0) \cup (0, \infty)$
Vertical asymptote: $x = 0$
Horizontal asymptote: $y = 0$

43. $h(x) = \dfrac{6x}{\sqrt{x^2+1}}$

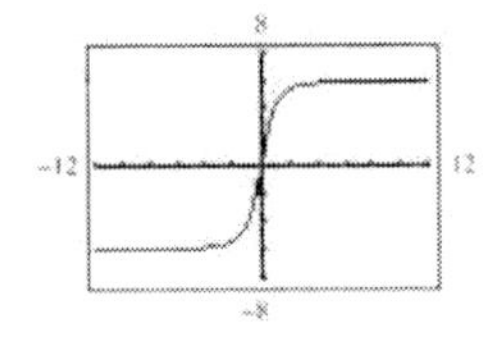

There are two horizontal asymptotes, $y = \pm 6$.

45. $g(x) = \dfrac{4|x-2|}{x+1}$

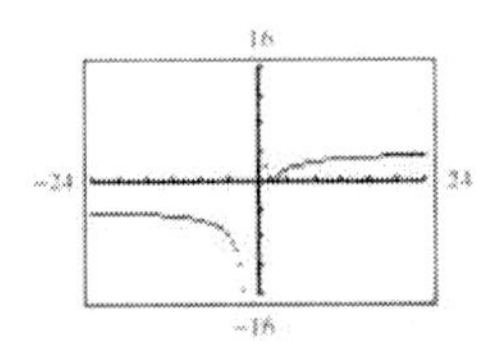

There are two horizontal asymptotes, $y = \pm 4$ and one vertical asymptote, $x = -1$.

47. $f(x) = \dfrac{4(x-1)^2}{x^2-4x+5}$

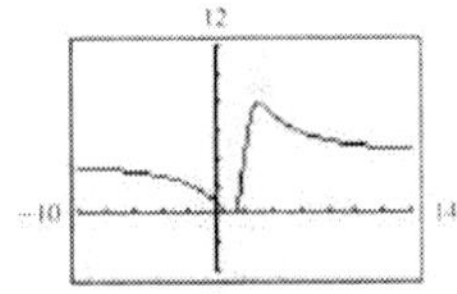

The graph crosses its horizontal asymptote, $y = 4$.

49. $f(x) = \dfrac{2x^2+1}{x} = 2x + \dfrac{1}{x}$

Vertical asymptote: $x = 0$
Slant asymptote: $y = 2x$
Origin symmetry

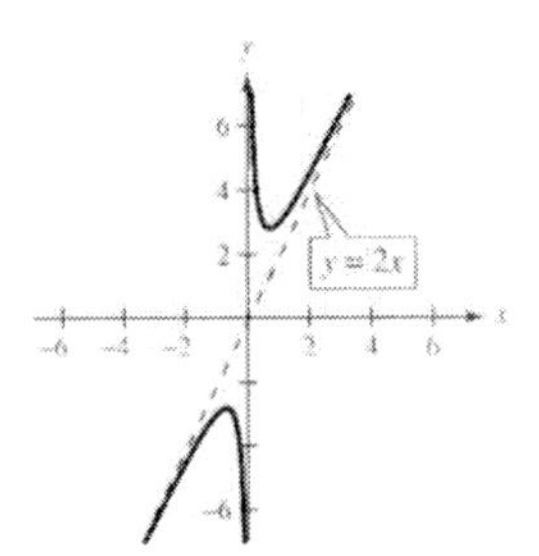

51. $h(x) = \dfrac{x^2}{x-1} = x + 1 + \dfrac{1}{x-1}$

Intercept: (0, 0)
Vertical asymptote: $x = 1$
Slant asymptote: $y = x + 1$

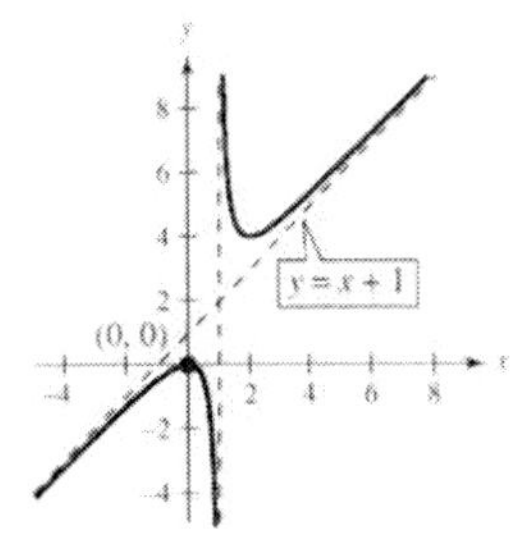

53. $g(x) = \dfrac{x^3}{2x^2-8} = \dfrac{1}{2}x + \dfrac{4x}{2x^2-8}$

Intercept: (0, 0)
Vertical asymptotes: $x = \pm 2$
Slant asymptote: $y = \dfrac{1}{2}x$
Origin symmetry

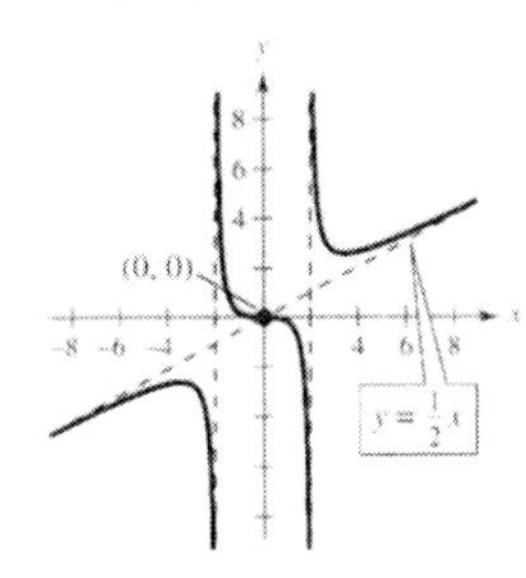

55. $f(x) = \dfrac{x^3 + 2x^2 + 4}{2x^2 + 1} = \dfrac{x}{2} + 1 + \dfrac{3 - \frac{x}{2}}{2x^2 + 1}$

Intercepts: $(-2.594, 0)$, $(0, 4)$

Slant asymptote: $y = \dfrac{x}{2} + 1$

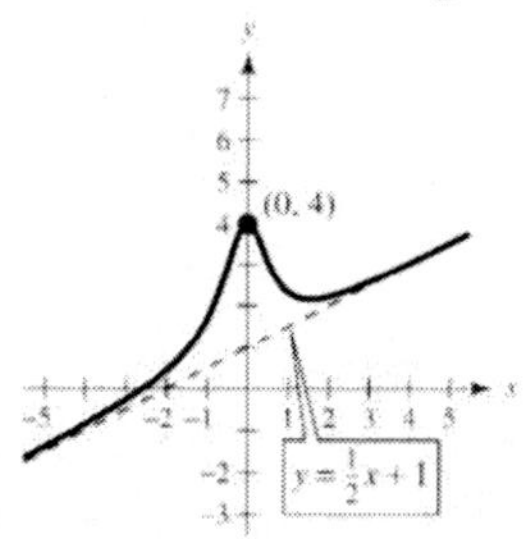

57. $y = \dfrac{x+1}{x-3}$

x-intercept: $(-1, 0)$

$0 = \dfrac{x+1}{x-3}$

$0 = x + 1$

$-1 = x$

59. $y = \dfrac{1}{x} - x$

x-intercepts: $(\pm 1, 0)$

$0 = \dfrac{1}{x} - x$

$x = \dfrac{1}{x}$

$x^2 = 1$

$x = \pm 1$

61. $y = \dfrac{2x^2 + x}{x+1} = 2x - 1 + \dfrac{1}{x+1}$

Domain: all real numbers except $x = -1$
Vertical asymptote: $x = -1$
Slant asymptote: $y = 2x - 1$

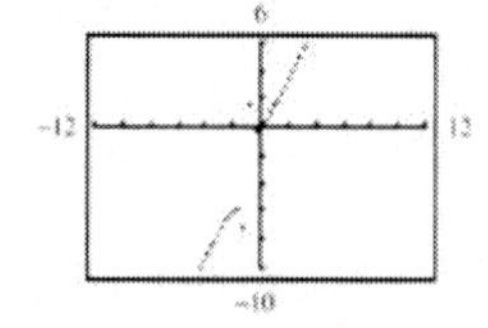

63. $y = \dfrac{1 + 3x^2 - x^3}{x^2} = \dfrac{1}{x^2} + 3 - x = -x + 3 + \dfrac{1}{x^2}$

Domain: all real numbers except 0
or $(-\infty, 0) \cup (0, \infty)$
Vertical asymptote: $x = 0$
Slant asymptote: $y = -x + 3$

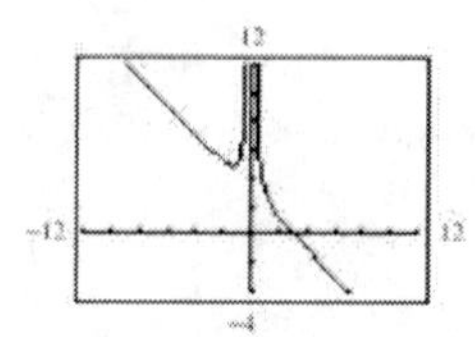

65. $f(x) = \dfrac{x^2 - 5x + 4}{x^2 - 4} = \dfrac{(x-4)(x-1)}{(x-2)(x+2)}$

Vertical asymptotes: $x = 2$, $x = -2$
Horizontal asymptote: $y = 1$
No slant asymptotes, no holes

67. $f(x) = \dfrac{2x^2 - 5x + 2}{2x^2 - x - 6} = \dfrac{(2x-1)(x-2)}{(2x+3)(x-2)} = \dfrac{2x-1}{2x+3}$,

$x \neq 2$

Vertical asymptote: $x = -\dfrac{3}{2}$
Horizontal asymptote: $y = 1$
No slant asymptotes
Hole at $x = 2$, $\left(2, \dfrac{3}{7}\right)$

69. $f(x) = \dfrac{2x^3 - x^2 - 2x + 1}{x^2 + 3x + 2}$

$= \dfrac{(x-1)(x+1)(2x-1)}{(x+1)(x+2)}$

$= \dfrac{(x-1)(2x-1)}{x+2}$, $x \neq -1$

Long division gives

$f(x) = \dfrac{2x^2 - 3x + 1}{x+2} = 2x - 7 + \dfrac{15}{x+2}$.

Vertical asymptote: $x = -2$
No horizontal asymptote
Slant asymptote: $y = 2x - 7$
Hole at $x = -1$, $(-1, 6)$

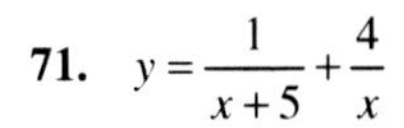

71. $y = \dfrac{1}{x+5} + \dfrac{4}{x}$

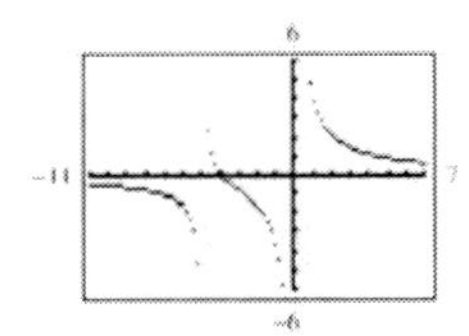

x-intercept: (–4, 0)

$$0 = \frac{1}{x+5} + \frac{4}{x}$$
$$-\frac{4}{x} = \frac{1}{x+5}$$
$$-4(x+5) = x$$
$$-4x - 20 = x$$
$$-5x = 20$$
$$x = -4$$

73. $y = \dfrac{1}{x+2} + \dfrac{2}{x+4}$

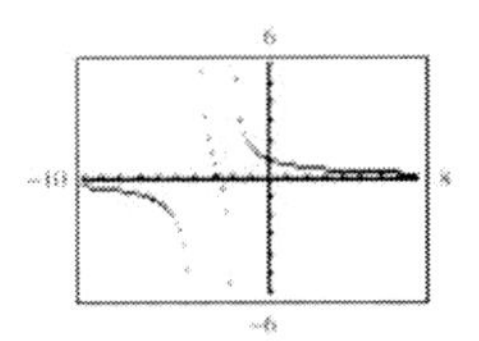

x-intercept: $\left(-\dfrac{8}{3}, 0\right)$

$$\frac{1}{x+2} + \frac{2}{x+4} = 0$$
$$\frac{1}{x+2} = \frac{-2}{x+4}$$
$$x + 4 = -2x - 4$$
$$3x = -8$$
$$x = -\frac{8}{3}$$

75. $y = x - \dfrac{6}{x-1}$

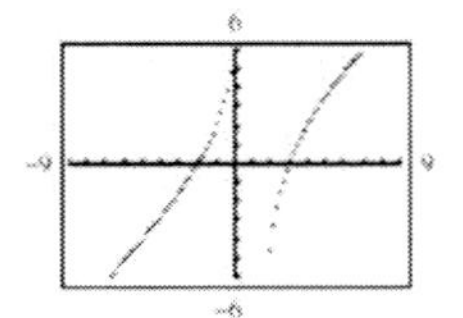

x-intercepts: $(-2, 0)$, $(3, 0)$

$$0 = x - \frac{6}{x-1}$$
$$\frac{6}{x-1} = x$$
$$6 = x(x-1)$$
$$0 = x^2 - x - 6$$
$$0 = (x+2)(x-3)$$
$$x = -2,\ x = 3$$

77. $y = x + 2 - \dfrac{1}{x+1}$

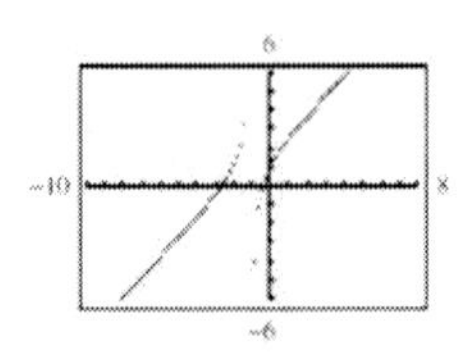

x-intercepts: $(-2.618, 0)$, $(-0.382, 0)$

$$x + 2 = \frac{1}{x+2}$$
$$x^2 + 3x + 2 = 1$$
$$x^2 + 3x + 1 = 0$$
$$x = \frac{-3 \pm \sqrt{9-4}}{2}$$
$$= -\frac{3}{2} \pm \frac{\sqrt{5}}{2}$$
$$\approx -2.618, -0.382$$

79. $y = x + 1 + \dfrac{2}{x-1}$

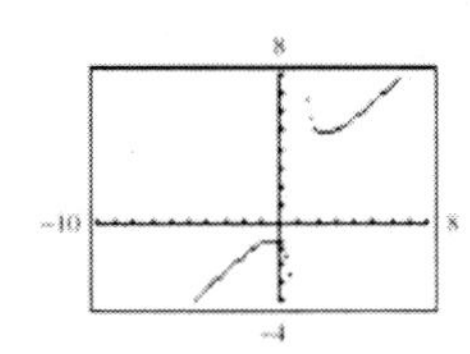

No x-intercepts

$$x + 1 + \frac{2}{x-1} = 0$$
$$\frac{2}{x-1} = -x - 1$$
$$2 = -x^2 + 1$$
$$x^2 + 1 = 0$$

No real zeros

81. $y = x + 3 - \dfrac{2}{2x - 1}$

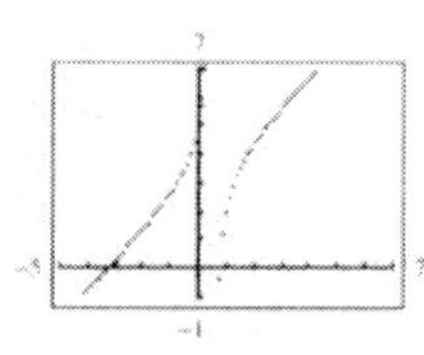

x-intercepts: $(0.766, 0)$, $(-3.266, 0)$

$$x + 3 - \frac{2}{2x - 1} = 0$$

$$x + 3 = \frac{2}{2x - 1}$$

$$2x^2 + 5x - 3 = 2$$

$$2x^2 + 5x - 5 = 0$$

$$x = \frac{-5 \pm \sqrt{25 - 4(2)(-5)}}{4}$$

$$= \frac{-5 \pm \sqrt{65}}{4}$$

$$\approx 0.766, -3.266$$

83.

(a) $0.25(50) + 0.75(x) = C(50 + x)$

$$\frac{12.5 + 0.75x}{50 + x} = C$$

$$\frac{50 + 3x}{200 + 4x} = C$$

$$C = \frac{3x + 50}{4(x + 50)}$$

(b) Domain: $x \geq 0$ and $x \leq 1000 - 50 = 950$

Thus, $0 \leq x \leq 950$.

(c)

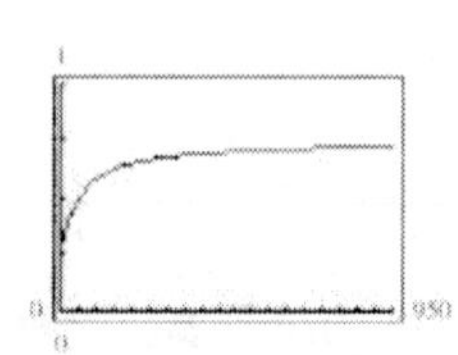

As the tank fills, the rate that the concentration is increasing slows down. It approaches the horizontal asymptote $C = \dfrac{3}{4} = 0.75$. When the tank is full $(x = 950)$, the concentration is $C = 0.725$.

85.

(a) $A = xy$ and

$$(x - 2)(y - 4) = 30$$

$$y - 4 = \frac{30}{x - 2}$$

$$y = 4 + \frac{30}{x - 2} = \frac{4x + 22}{x - 2}$$

Thus, $A = xy = x\left(\dfrac{4x + 22}{x - 2}\right) = \dfrac{2x(2x + 11)}{x - 2}$.

(b) Domain: Since the margins on the left and right are each 1 inch, $x > 2$, or $(2, \infty)$.

(c)

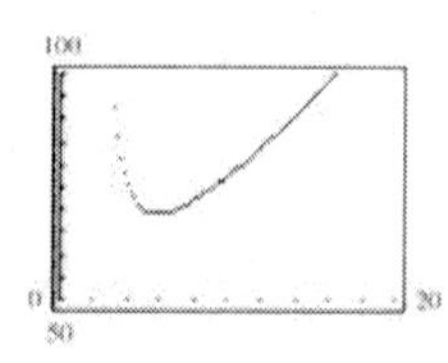

The area is minimum when $x \approx 5.87$ in. and $y \approx 11.75$ in.

87. $C = 100\left(\dfrac{200}{x^2} + \dfrac{x}{x + 30}\right)$, $1 \leq x$

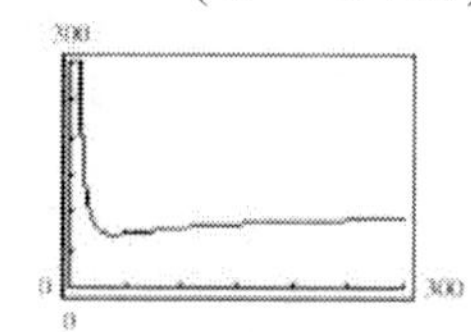

The minimum occurs when $x = 40.4 \approx 40$.

89.

$$C = \frac{3t^2 + t}{t^3 + 50}, \quad 0 \leq t$$

(a) The horizontal asymptote is the t-axis, or $C = 0$. This indicates that the chemical eventually dissipates.

(b)

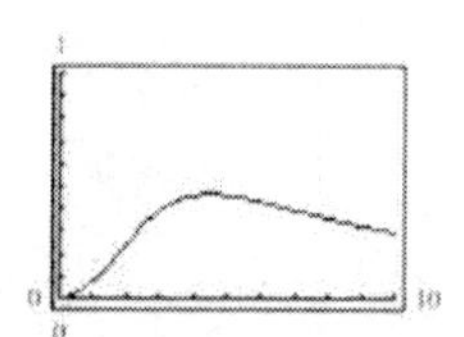

The maximum occurs when $t \approx 4.5$.

(c) Graph C together with $y = 0.345$. The graphs intersect at $t \approx 2.65$ and $t \approx 8.32$. $C < 0.345$ when $0 \leq t \leq 2.65$ hours and when $t > 8.32$ hours.

91.

(a) $A = -0.2182t + 5.665$

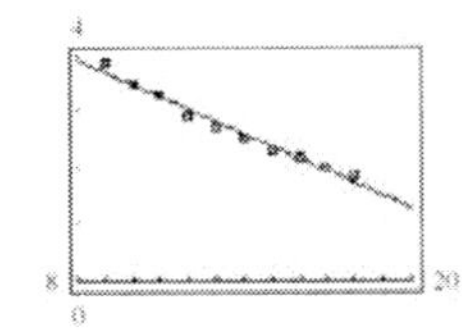

(b) $A = \dfrac{1}{0.0302t - 0.020}$

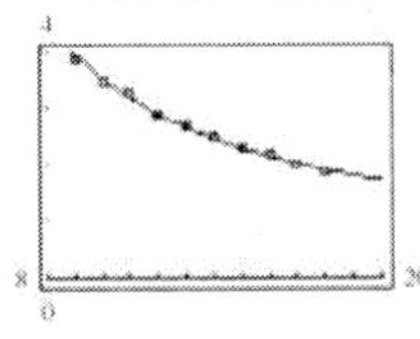

(c)

Year	1999	2000	2001	2002
Original data, A	3.9	3.5	3.3	2.9
Model from (a), A	3.7	3.5	3.3	3.0
Model from (b), A	4.0	3.5	3.2	2.9

Year	2003	2004	2005	2006
Original data, A	2.7	2.5	2.3	2.2
Model from (a), A	2.8	2.6	2.4	2.2
Model from (b), A	2.7	2.5	2.3	2.2

Year	2007	2008
Original data, A	2.0	1.9
Model from (a), A	2.0	1.7
Model from (b), A	2.0	1.9

Answers will vary.

93. False. The graph of a rational function is continuous when the polynomial in the denominator has no real zeros.

95. $h(x) = \dfrac{6 - 2x}{3 - x} = \dfrac{2(3 - x)}{3 - x} = 2,\ x \neq 3$

Since $h(x)$ is not reduced and $(3 - x)$ is a factor of both the numerator and the denominator, $x = 3$ is not a horizontal asymptote.

There is a hole in the graph at $x = 3$.

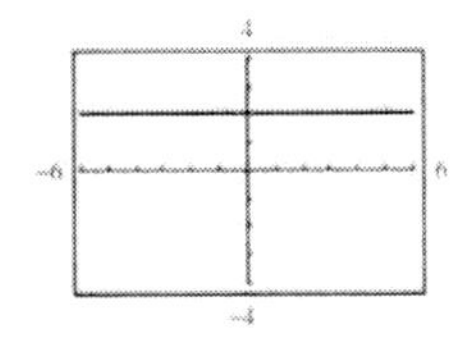

97. *Horizontal asymptotes:*

If the degree of the numerator is greater than the degree of the denominator, then there is no horizontal asymptote. If the degree of the numerator is less than the degree of the denominator, then there is a horizontal asymptote at $y = 0$.

If the degree of the numerator is equal to the degree of the denominator, then there is a horizontal asymptote at the line given by the ratio of the leading coefficients.

Vertical asymptotes:

Set the denominator equal to zero and solve.

Slant asymptotes:

If there is no horizontal asymptote and the degree of the numerator is exactly one greater than the degree of the denominator, then divide the numerator by the denominator. The slant asymptote is the result, not including the remainder.

99. $\left(\dfrac{x}{8}\right)^{-3} = \left(\dfrac{8}{x}\right)^{3} = \dfrac{512}{x^3}$

101. $\dfrac{3^{7/6}}{3^{1/6}} = 3^{6/6} = 3$

103.

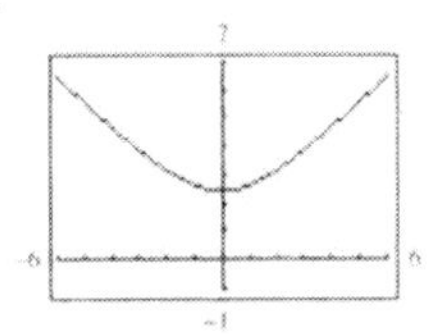

Domain: all x

Range: $y \geq \sqrt{6}$

105.

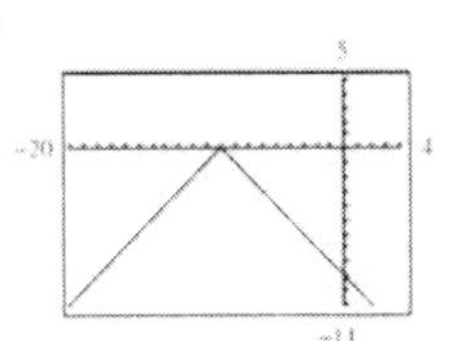

Domain: all x

Range: $y \leq 0$

107. Answers will vary.

Section 2.8

1. quadratic

3. Quadratic

5. Linear

7. Neither

9. (a)

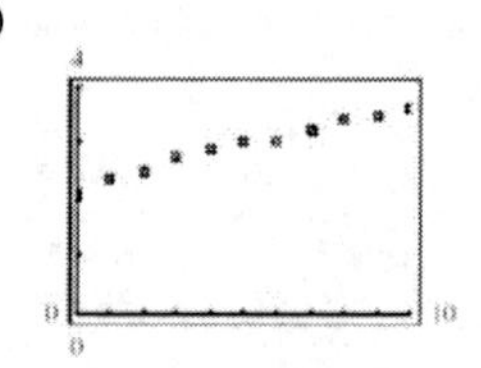

(b) Linear model is better.

(c) $y = 0.14x + 2.2$, linear

(d)

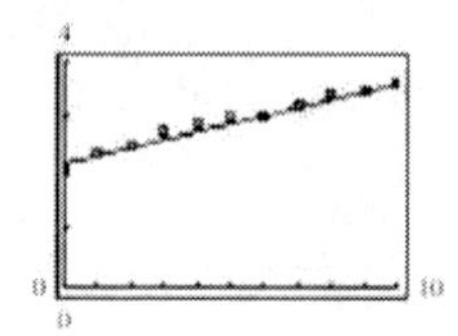

(e)

x	0	1	2	3	4
y	2.1	2.4	2.5	2.8	2.9
Model	2.2	2.4	2.5	2.6	2.8

x	5	6	7	8	9	10
y	3.0	3.0	3.2	3.4	3.5	3.6
Model	2.9	3.0	3.2	3.4	3.5	3.6

11. (a)

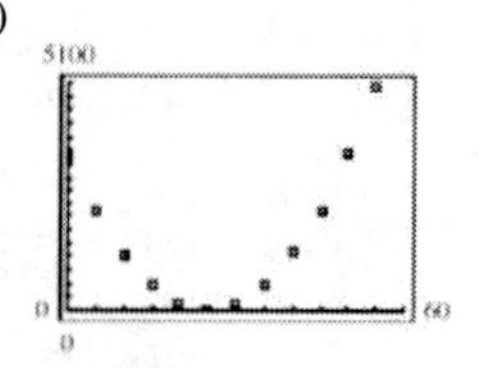

(b) Quadratic model is better.

(c) $y = 5.55x^2 - 277.5x + 3478$

(d)

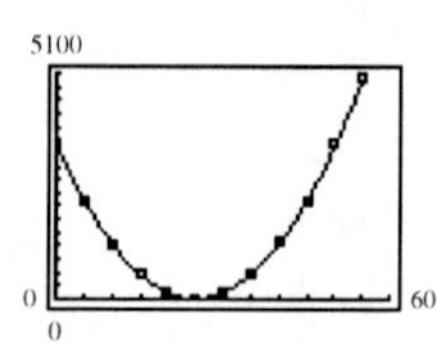

(e)

x	0	5	10	15	20	25
y	3480	2235	1250	565	150	12
Model	3478	2229	1258	564	148	9

x	30	35	40	45	50	55
y	145	575	1275	2225	3500	5010
Model	148	564	1258	2229	3478	5004

13. (a)

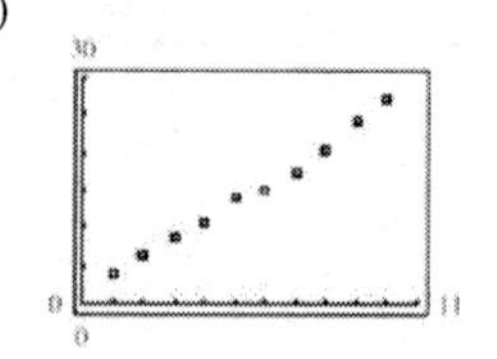

(b) Linear model is better.

(c) $y = 2.48x + 1.1$

(d)

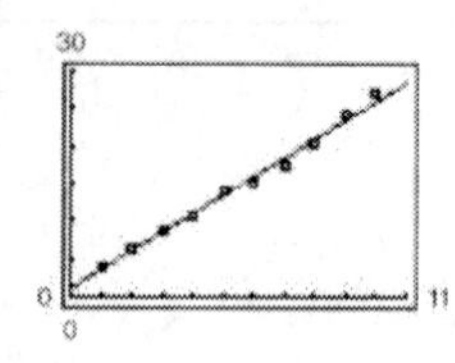

(e)

x	1	2	3	4	5
Actual, y	4.0	6.5	8.8	10.6	13.9
Model, y	3.6	6.1	8.5	11.0	13.5

x	6	7	8	9	10
Actual, y	15.0	17.5	20.1	24.0	27.1
Model, y	16.0	18.5	20.9	23.4	25.9

15. (a)

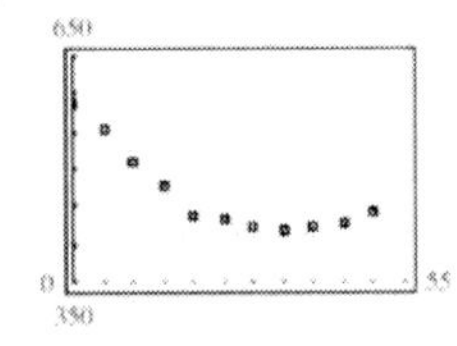

(b) Quadratic is better.

(c) $y = 0.14x^2 - 9.9x + 591$

(d)

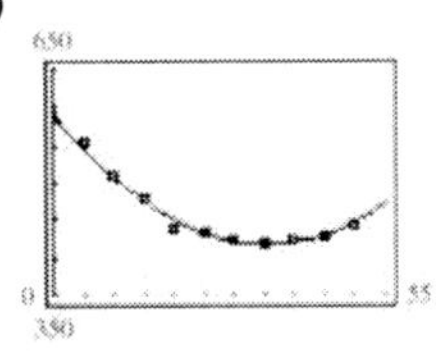

(e)

x	0	5	10	15	20	25
Actual, y	587	551	512	478	436	430
Model, y	591	545	506	474	449	431

x	30	35	40	45	50
Actual, y	424	420	423	429	444
Model, y	420	416	419	429	446

17. (a)

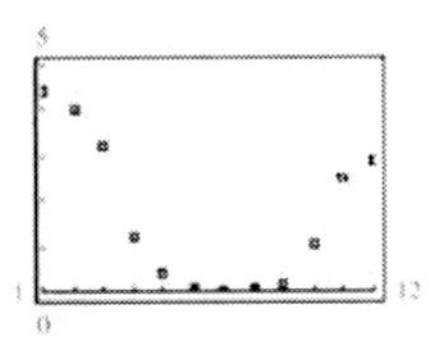

(b) $P = 0.1323t^2 - 1.893t + 6.85$

(c)

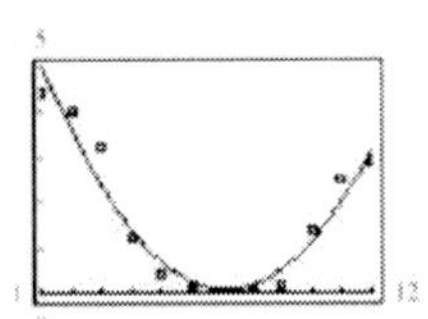

(d) The model's minimum is $H \approx 0.1$ at $t = 7.4$. This corresponds to July.

19. (a)

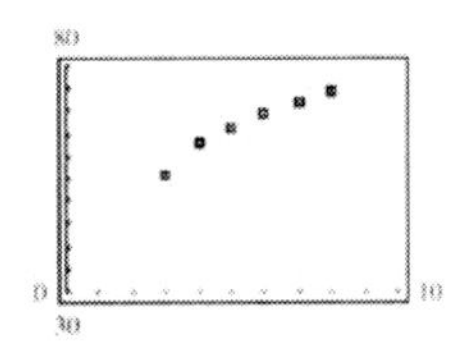

(b) $P = -0.5638t^2 + 9.690t + 32.17$

(c)

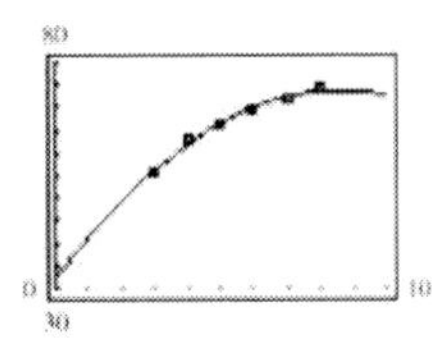

(d) To determine when the percent P of the U.S. population who used the Internet falls below 60%, set P = 60. Using the Quadratic Formula, solve for t.

$$-0.5638t^2 + 9.690t + 32.17 = 60$$

$$-0.5638t^2 + 9.690t - 27.83 = 0$$

$$t = \frac{-(9.690) \pm \sqrt{(9.690)^2 - 4(-0.5638)(-27.83)}}{2(-0.5638)}$$

$$t = \frac{-9.690 \pm \sqrt{156.658316}}{-1.1276}$$

$t \approx 13.54$ or 2014

Therefore after 2014, the percent of the U.S. population who use the Internet will fall below 60%.

This is not a good model for predicting future years. In fact, by 2021, the model gives negative values for the percentage.

21. (a)

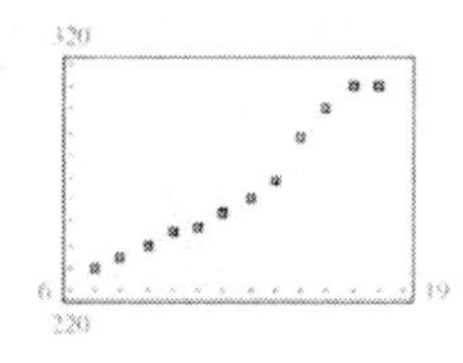

(b) $T = 7.97t + 166.1$

$r^2 \approx 0.9469$

(c)

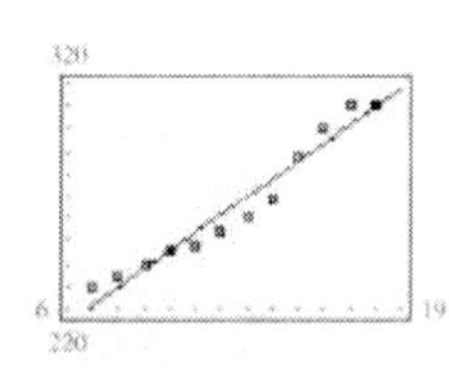

(d) $T = 0.459t^2 - 3.51t + 232.4$

$r^2 \approx 0.9763$

(e)

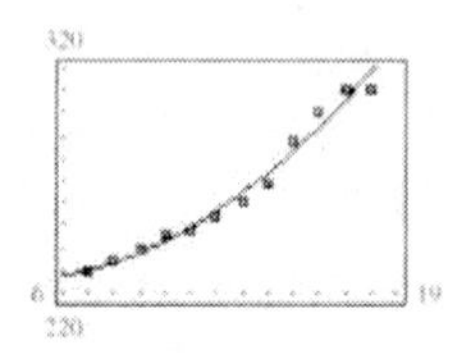

(f) The quadratic model is a better fit. Answers will vary.

(g) To determine when the number of televisions, T, in homes reaches 350 million, set T for the model equal to 350 and solve for t.

Linear model:

$$7.97t + 166.1 = 350$$
$$7.97t = 183.9$$
$$t \approx 23.1 \text{ or } 2014$$

Quadratic model:

$$0.459t^2 - 3.51t + 232.4 = 350$$
$$0.459t^2 - 3.51t - 117.6 = 0$$

Using the Quadratic Formula,

$$t = \frac{-(-3.51) \pm \sqrt{(-3.51)^2 - 4(0.459)(-117.6)}}{2(0.459)}$$

$$t = \frac{3.51 \pm \sqrt{228.2337}}{0.918}$$

$$t \approx 20.3 \text{ or } 2011$$

23. True. A quadratic model with a positive leading coefficient opens upward. So, its vertex is at its lowest point.

25. The model is above all data points.

27. (a) $(f \circ g)(x) = f(x^2 + 3) = 2(x^2 + 3) - 1 = 2x^2 + 5$

(b) $(g \circ f)(x) = g(2x - 1) = (2x - 1)^2 + 3 = 4x^2 - 4x + 4$

29. (a) $(f \circ g)(x) = f(\sqrt[3]{x + 1}) = x + 1 - 1 = x$

(b) $(g \circ f)(x) = g(x^3 - 1) = \sqrt[3]{x^3 - 1 + 1} = x$

31. f is one-to-one.

$$y = 2x + 5$$
$$x = 2y + 5$$
$$2y = x - 5$$
$$y = \frac{(x - 5)}{2} \Rightarrow f^{-1}(x) = \frac{x - 5}{2}$$

33. f is one-to-one on $[0, \infty)$.

$$y = x^2 + 5, x \ge 0$$
$$x = y^2 + 5, y \ge 0$$
$$y^2 = x - 5$$
$$y = \sqrt{x - 5} \Rightarrow f^{-1}(x) = \sqrt{x - 5}, x \ge 5$$

35. For $1 - 3i$, the complex conjugate is $1 + 3i$.

$$(1 - 3i)(1 + 3i) = 1 - 9i^2$$
$$= 1 + 9$$
$$= 10$$

37. For $-5i$, the complex conjugate is $5i$.

$$(-5i)(5i) = -25i^2$$
$$= 25$$

Chapter 2 Review Exercises

1.

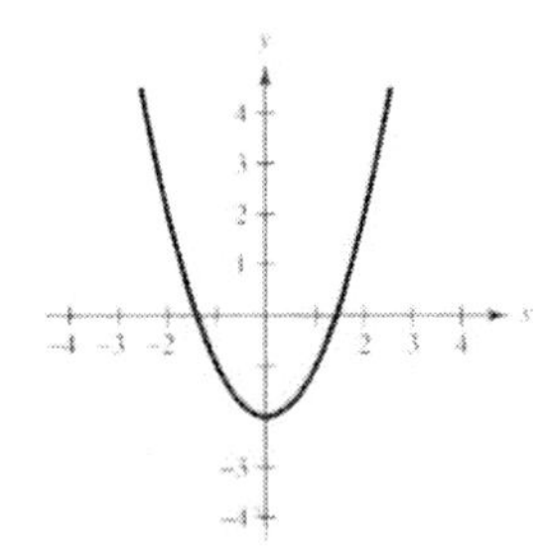

The graph of $y = x^2 - 2$ is a vertical shift two units downward of $y = x^2$.

3.

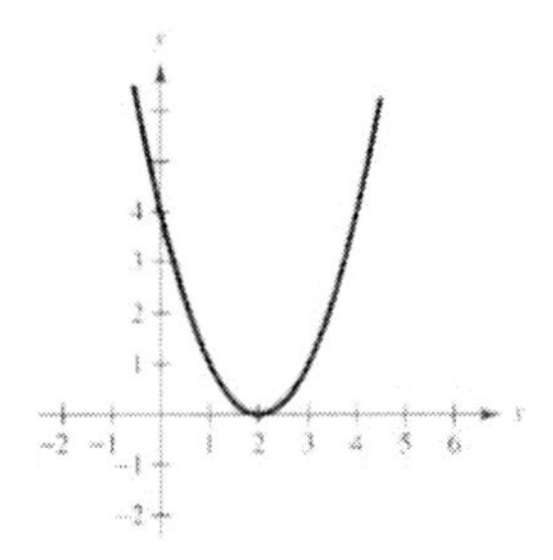

The graph of $y = (x - 2)^2$ is a horizontal shift two units to the right of $y = x^2$.

5.

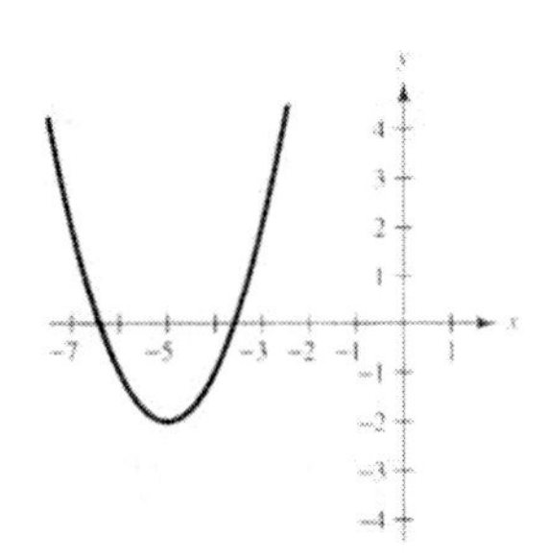

The graph of $y = (x + 5)^2 - 2$ is a horizontal shift five units to the left and a vertical shift two units downward of $y = x^2$.

7. The graph of $f(x) = (x + 3/2)^2 + 1$ is a parabola opening upward with vertex $\left(-\frac{3}{2}, 1\right)$, and no x-intercepts.

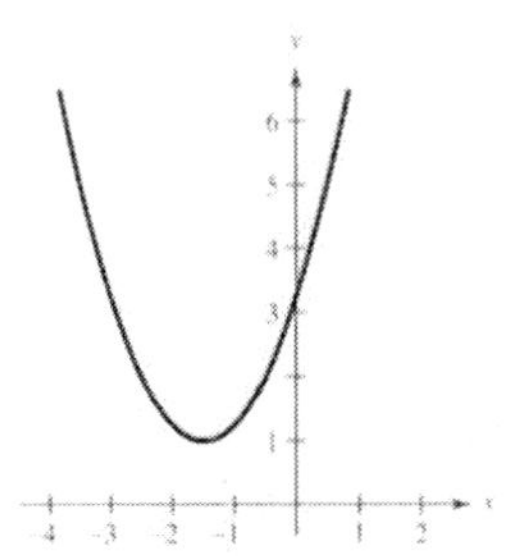

9.

$$f(x) = \frac{1}{3}(x^2 + 5x - 4)$$

$$= \frac{1}{3}\left(x^2 + 5x + \frac{25}{4} - \frac{25}{4}\right) - \frac{4}{3}$$

$$= \frac{1}{3}\left(x + \frac{5}{2}\right)^2 - \frac{41}{12}$$

The graph of f is a parabola opening upward with vertex $\left(-\frac{5}{2}, -\frac{41}{12}\right)$.

x-intercepts: $\frac{1}{3}\left(x + \frac{5}{2}\right)^2 - \frac{41}{12} = 0.$

or

$$\frac{1}{3}\left(x^2 + 5x - 4\right) = 0$$

$$x^2 + 5x - 4 = 0$$

Use Quadratic formula.

$$x = \frac{-5 \pm \sqrt{41}}{2}$$

$$\left(\frac{-5 + \sqrt{41}}{2}, 0\right), \left(\frac{-5 - \sqrt{41}}{2}, 0\right)$$

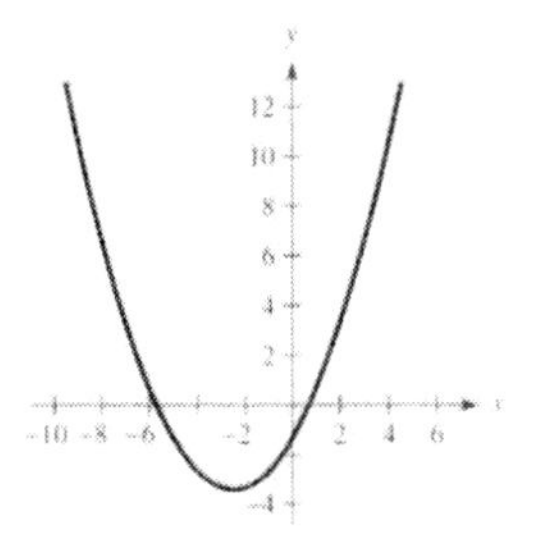

11. Vertex: $(1, -4) \Rightarrow f(x) = a(x-1)^2 - 4$

Point: $(2, -3) \Rightarrow -3 = a(2-1)^2 - 4$

$1 = a$

Thus, $f(x) = (x-1)^2 - 4$.

13.

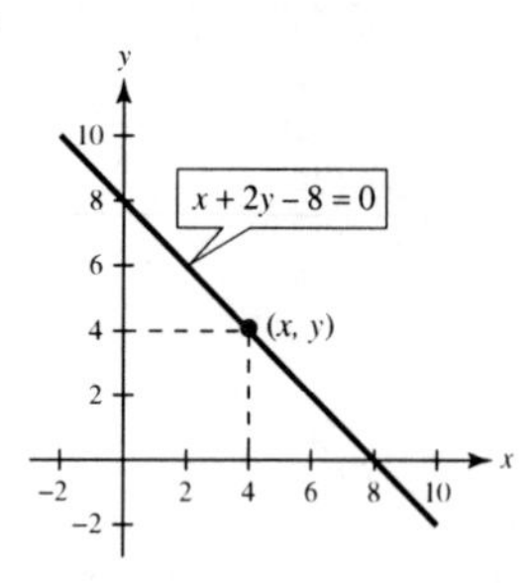

(a) $A = xy$

If $x + 2y - 8 = 0$, then $y = \dfrac{8-x}{2}$.

$A = x\left(\dfrac{8-x}{2}\right)$

$A = 4x - \dfrac{1}{2}x^2, \ 0 < x < 8$

(b)

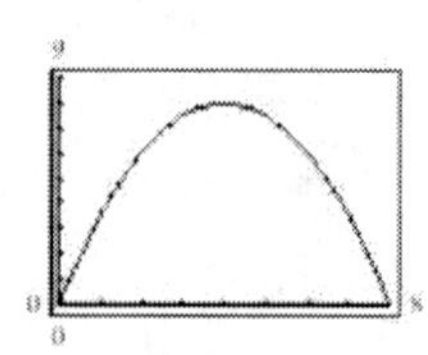

Using the graph, when $x = 4$, the area is a maximum.

When $x = 4$, $y = \dfrac{8-4}{2} = 2$.

(c)
$$
\begin{aligned}
A &= -\frac{1}{2}x^2 + 4x \\
&= -\frac{1}{2}(x^2 - 8x) \\
&= -\frac{1}{2}(x^2 - 8x + 16 - 16) \\
&= -\frac{1}{2}(x-4)^2 + 8
\end{aligned}
$$

The graph of A is a parabola opening downward, with vertex $(4, 8)$. Therefore, when $x = 4$, the maximum area is 8 square units.

Yes, graphically and algebraically the same dimensions result.

15.

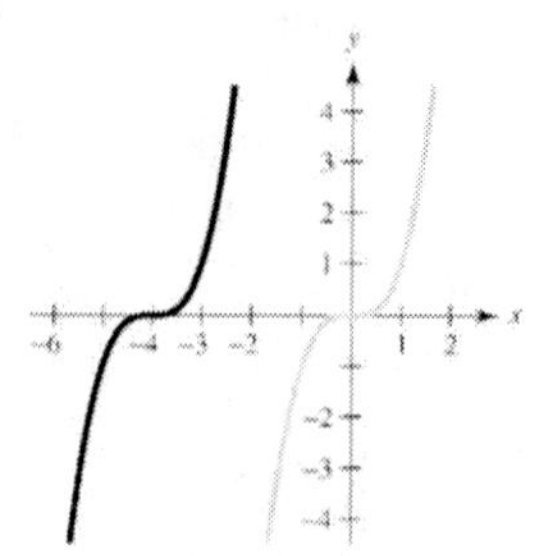

The graph of $f(x) = (x+4)^3$ is a horizontal shift four units to the left of $y = x^3$.

17.

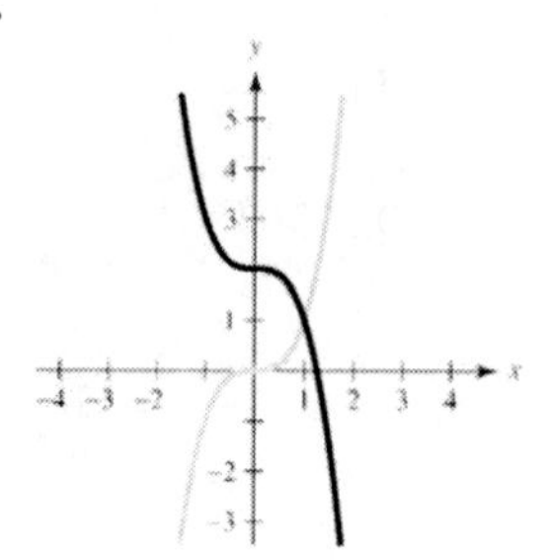

The graph if $f(x) = -x^3 + 2$ is a reflection in the x-axis and a vertical shift two units upward of $y = x^3$.

19.

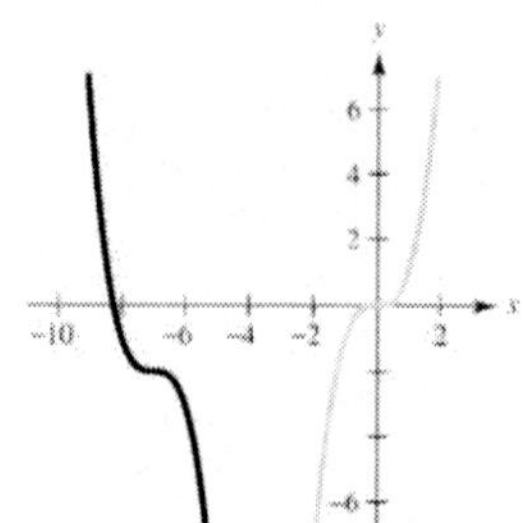

The graph of $f(x) = -(x+7)^3 - 2$ is a horizontal shift seven units to the left, a reflection in the x-axis, and a vertical shift two units downward of $y = x^3$.

21. $f(x) = \dfrac{1}{2}x^3 - 2x + 1; \ g(x) = \dfrac{1}{2}x^3$

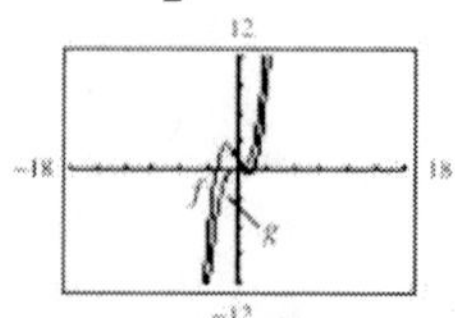

The graphs have the same end behavior. Both functions are of the same degree and have positive leading coefficients.

23. $f(x) = -x^2 + 6x + 9$

The degree is even and the leading coefficient is negative. The graph falls to the left and right.

25. $f(x) = \frac{3}{4}\left(x^4 + 3x^2 + 2\right)$

The degree is even and the leading coefficient is positive. The graph rises to the left and right.

27. (a) $x^4 - x^3 - 2x^2 = x^2\left(x^2 - x - 2\right)$

$= x^2(x-2)(x+1) = 0$

Zeros: $x = -1, 0, 0, 2$

(b)

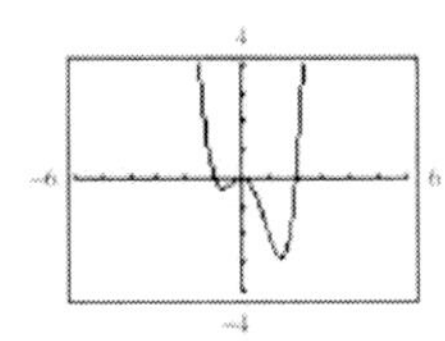

(c) Zeros: $x = -1, 0, 0, 2$; the same

29. (a) $t^3 - 3t = t\left(t^2 - 3\right) = t\left(t + \sqrt{3}\right)\left(t - \sqrt{3}\right) = 0$

Zeros: $t = 0, \pm\sqrt{3}$

(b)

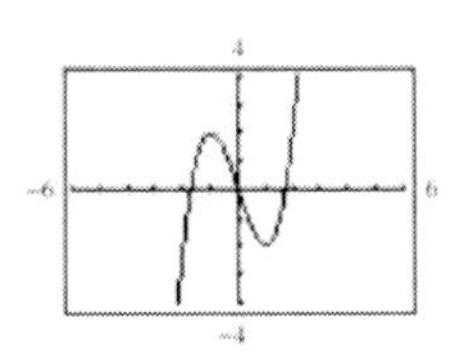

(c) Zeros: $t = 0, \pm 1.732$; the same

31. (a) $x(x+3)^2 = 0$

Zeros: $x = 0, -3, -3$

(b)

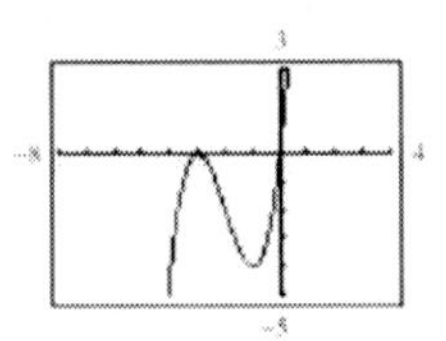

(c) Zeros: $x = -3, -3, 0$; the same

33. $f(x) = (x+2)(x-1)^2(x-5)$

$= x^4 - 5x^3 - 3x^2 + 17x - 10$

35. $f(x) = (x-3)\left(x - 2 + \sqrt{3}\right)\left(x - 2 - \sqrt{3}\right)$

$= x^3 - 7x^2 + 13x - 3$

37. (a) The degree of f is even and the leading coefficient is 1. The graph rises to the left and rises to the right.

(b) $f(x) = x^4 - 2x^3 - 12x^2 + 18x + 27$

$= (x^4 - 12x^2 + 27) - (2x^3 - 18x)$

$= (x^2 - 9)(x^2 - 3) - 2x(x^2 - 9)$

$= (x^2 - 9)(x^2 - 3 - 2x)$

$= (x+3)(x-3)(x^2 - 2x - 3)$

$= (x+3)(x-3)(x-3)(x+1)$

Zeros: $-3, 3, 3, -1$

(c) and (d)

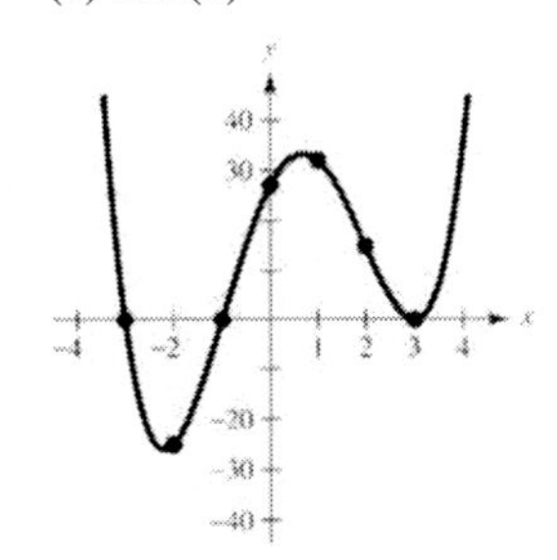

39. $f(x) = x^3 + 2x^2 - x - 1$

(a) $f(-3) < 0, f(-2) > 0 \Rightarrow$ zero in $(-3, -2)$

$f(-1) > 0, f(0) < 0 \Rightarrow$ zero in $(-1, 0)$

$f(0) < 0, f(1) > 0 \Rightarrow$ zero in $(0, 1)$

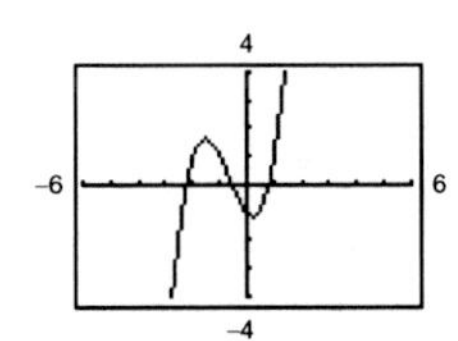

(b) Zeros: $-2.247, -0.555, 0.802$

41. $f(x) = x^4 - 6x^2 - 4$

(a) $f(-3) > 0, f(-2) < 0 \Rightarrow$ zero in $(-3, -2)$

$f(2) < 0, f(3) > 0 \Rightarrow$ zero in $(2, 3)$

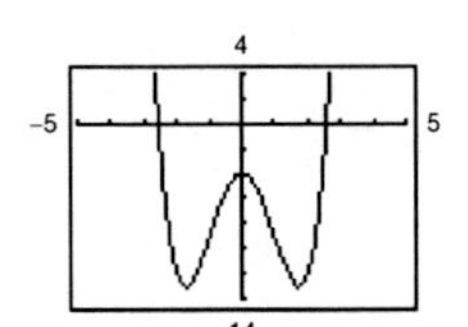

(b) Zeros: ± 2.570

43.

$$\begin{array}{r} 8x + 5 \\ 3x - 2 \overline{\big)\,24x^2 - x - 8} \\ \underline{24x^2 - 16x} \\ 15x - 8 \\ \underline{15x - 10} \\ 2 \end{array}$$

Thus, $\frac{24x^2 - x - 8}{3x - 2} = 8x + 5 + \frac{2}{3x - 2}$.

45.

$$\begin{array}{r} x^2-2 \\ x^2-1\overline{)x^4-3x^2+2} \\ \underline{x^4-x^2} \\ -2x^2+2 \\ \underline{-2x^2+2} \\ 0 \end{array}$$

Thus, $\dfrac{x^4-3x^2+2}{x^2-1}=x^2-2,\ (x\pm 1).$

47.

$$\begin{array}{r} 5x+2 \\ x^2-3x+1\overline{)5x^3-13x^2-\ x+2} \\ \underline{5x^3-15x^2+5x} \\ 2x^2-6x+2 \\ \underline{2x^2-6x+2} \\ 0 \end{array}$$

Thus, $\dfrac{5x^3-13x^2-x+2}{x^2-3x+1}=5x+2,$

$\left(x\neq\dfrac{1}{2}(3\pm\sqrt{5})\right).$

49.

$$\begin{array}{r} 3x^2+5x+8 \\ 2x^2+0x-1\overline{)6x^4+10x^3+13x^2-5x+2} \\ \underline{6x^4+\ 0x^3\ -3x^2} \\ 10x^3+16x^2-5x \\ \underline{10x^3+\ 0x^2-5x} \\ 16x^2-0+2 \\ \underline{16x^2+0-8} \\ 10 \end{array}$$

Thus,

$\dfrac{6x^4+10x^3+13x^2-5x+2}{2x^2-1}=3x^2+5x+8+\dfrac{10}{2x^2-1}.$

51.

$$\begin{array}{r|rrrrr} -2 & 0.25 & -4 & 0 & 0 & 0 \\ & & -\frac{1}{2} & 9 & -18 & 36 \\ \hline & \frac{1}{4} & -\frac{9}{2} & 9 & -18 & 36 \end{array}$$

Thus, $\dfrac{0.25x^4-4x^3}{x+2}=\dfrac{1}{4}x^3-\dfrac{9}{2}x^2+9x-18+\dfrac{36}{x+2}.$

53.

$$\begin{array}{r|rrrrr} \frac{2}{3} & 6 & -4 & -27 & 18 & 0 \\ & & 4 & 0 & -18 & 0 \\ \hline & 6 & 0 & -27 & 0 & 0 \end{array}$$

Thus, $\dfrac{6x^4-4x^3-27x^2+18x}{x-(2/3)}=6x^3-27x,\ x\neq\dfrac{2}{3}.$

55.

$$\begin{array}{r|rrrr} 4 & 3 & -10 & 12 & -22 \\ & & 12 & 8 & 80 \\ \hline & 3 & 2 & 20 & 58 \end{array}$$

Thus, $\dfrac{3x^3-10x^2+12x-22}{x-4}=3x^2+2x+20+\dfrac{58}{x-4}.$

57. (a)

$$\begin{array}{r|rrrrr} -3 & 1 & 10 & -24 & 20 & 44 \\ & & -3 & -21 & 135 & -465 \\ \hline & 1 & 7 & -45 & 155 & -421=f(-3) \end{array}$$

(b)

$$\begin{array}{r|rrrrr} -2 & 1 & 10 & -24 & 20 & 44 \\ & & -2 & -16 & 80 & -200 \\ \hline & 1 & 8 & -40 & 100 & -156=f(-2) \end{array}$$

59. $f(x)=x^3+4x^2-25x-28$

(a)

$$\begin{array}{r|rrrr} 4 & 1 & 4 & -25 & -28 \\ & & 4 & 32 & 28 \\ \hline & 1 & 8 & 7 & 0 \end{array}$$

$(x-4)$ is a factor.

(b) $x^2+8x+7=(x+1)(x+7)$

Remaining factors: $(x+1)$, $(x+7)$

(c) $f(x)=(x-4)(x+1)(x+7)$

(d) Zeros: 4, -1, -7

61. $f(x)=x^4-4x^3-7x^2+22x+24$

(a)

$$\begin{array}{r|rrrrr} -2 & 1 & -4 & -7 & 22 & 24 \\ & & -2 & 12 & -10 & -24 \\ \hline & 1 & -6 & 5 & 12 & 0 \end{array}$$

$(x+2)$ is a factor.

$$\begin{array}{r|rrrr} 3 & 1 & -6 & 5 & 12 \\ & & 3 & -9 & -12 \\ \hline & 1 & -3 & -4 & 0 \end{array}$$

$(x-3)$ is a factor.

(b) $x^2-3x-4=(x-4)(x+1)$

Remaining factors: $(x-4)$, $(x+1)$

(c) $f(x)=(x+2)(x-3)(x-4)(x+1)$

(d) Zeros: -2, 3, 4, -1

63. $f(x)=4x^3-11x^2+10x-3$

Possible rational zeros: $\pm 1,\ \pm 3,\ \pm\frac{3}{2},\ \pm\frac{3}{4},\ \pm\frac{1}{2},\ \pm\frac{1}{4}$

Zeros: 1, 1, $\dfrac{3}{4}$

65. $g(x)=5x^3-6x+9$ has two variation in sign $\Rightarrow 0$ or 2 positive real zeros.

$g(-x)=-5x^3+6x+9$ has one variation in sign $\Rightarrow 1$ negative real zero.

67.
$$\begin{array}{r|rrrr} 1 & 4 & -3 & 4 & -3 \\ & & 4 & 1 & 5 \\ \hline & 4 & 1 & 5 & 2 \end{array}$$

All entries positive; $x = 1$ is upper bound.

$$\begin{array}{r|rrrr} -\frac{1}{4} & 4 & -3 & 4 & -3 \\ & & -1 & 1 & -\frac{5}{4} \\ \hline & 4 & -4 & 5 & -\frac{17}{4} \end{array}$$

Alternating signs; $x = -\frac{1}{4}$ is lower bound.

Real zero: $x = \dfrac{3}{4}$

69. $f(x) = 6x^3 + 31x^2 - 18x - 10$

Possible rational zeros:

$\pm 1, \pm 2, \pm 5, \pm 10, \pm\frac{1}{2}, \pm\frac{5}{2}, \pm\frac{1}{3}, \pm\frac{2}{3}, \pm\frac{5}{3}, \pm\frac{10}{3}, \pm\frac{1}{6}, \pm\frac{5}{6}$

Using synthetic division and a graph check $x = \frac{5}{6}$.

$$\begin{array}{r|rrrr} \frac{5}{6} & 6 & 31 & -18 & -10 \\ & & 5 & 30 & 10 \\ \hline & 6 & 36 & 12 & 0 \end{array}$$

$x = \frac{5}{6}$ is a real zero.

Rewrite in polynomial form and use the Quadratic Formula:

$$6x^2 + 36x + 12 = 0$$
$$6(x^2 + 6x + 2) = 0$$
$$x^2 + 6x + 2 = 0$$
$$x = \frac{-6 \pm \sqrt{(6)^2 - 4(1)(2)}}{2(1)}$$
$$x = \frac{-6 \pm \sqrt{28}}{2} = \frac{-6 \pm 2\sqrt{7}}{2} = -3 \pm \sqrt{7}$$

Real zeros: $x = \frac{5}{6},\ -3 \pm \sqrt{7}$

71. $f(x) = 6x^4 - 25x^3 + 14x^2 + 27x - 18$

Possible rational zeros:

$\pm 1, \pm 2, \pm 3, \pm 6, \pm 9, \pm 18, \pm\frac{1}{2}, \pm\frac{3}{2}, \pm\frac{9}{2}, \pm\frac{1}{3}, \pm\frac{2}{3}, \pm\frac{1}{6}$

Use a graphing utility to see that $x = -1$ and $x = 3$ are probably zeros.

$$\begin{array}{r|rrrrr} -1 & 6 & -25 & 14 & 27 & -18 \\ & & -6 & 31 & -45 & 18 \\ \hline & 6 & -31 & 45 & -18 & 0 \end{array}$$

$$\begin{array}{r|rrrr} 3 & 6 & -31 & 45 & -18 \\ & & 18 & -39 & 18 \\ \hline & 6 & -13 & 6 & 0 \end{array}$$

$$6x^4 - 25x^3 + 14x^2 + 27x - 18 = (x + 1)(x - 3)(6x^2 - 13x + 6)$$
$$= (x + 1)(x - 3)(3x - 2)(2x - 3)$$

Zeros: $x = -1, 3, \dfrac{2}{3}, \dfrac{3}{2}$

73. $6 + \sqrt{-25} = 6 + 5i$

75. $-2i^2 + 7i = 2 + 7i$

77. $(7 + 5i) + (-4 + 2i) = (7 - 4) + (5i + 2i)$
$= 3 + 7i$

79. $5i(13 - 8i) = 65i - 40i^2 = 40 + 65i$

81. $(10 - 8i)(2 - 3i) = 20 - 30i - 16i + 24i^2$
$= -4 - 46i$

83. $(3 + 7i)^2 + (3 - 7i)^2 = (9 + 42i - 49) + (9 - 42i - 49)$
$= -80$

85. $\left(\sqrt{-16} + 3\right)\left(\sqrt{-25} - 2\right) = (4i + 3)(5i - 2)$
$= -20 - 8i + 15i - 6$
$= -26 + 7i$

87. $\sqrt{-9} + 3 + \sqrt{-36} = 3i + 3 + 6i$
$= 3 + 9i$

89. $\dfrac{6 + i}{i} = \dfrac{6 + i}{i} \cdot \dfrac{-i}{-i} = \dfrac{-6i - i^2}{-i^2}$
$= \dfrac{-6i + 1}{1} = 1 - 6i$

91. $\dfrac{3 + 2i}{5 + i} \cdot \dfrac{5 - i}{5 - i} = \dfrac{15 + 10i - 3i + 2}{25 + 1}$
$= \dfrac{17}{26} + \dfrac{7}{26}i$

93.
$$x^2 + 16 = 0$$
$$x^2 = -16$$
$$x = \pm\sqrt{-16}$$
$$x = \pm 4i$$

95.
$$x^2 + 3x + 6 = 0$$
$$x = \frac{-3 \pm \sqrt{3^2 - 4(1)(6)}}{2(1)}$$
$$x = \frac{-3 \pm \sqrt{-15}}{2}$$
$$x = -\frac{3}{2} \pm \frac{\sqrt{15}}{2}i$$

97.
$$3x^2 - 5x + 6 = 0$$
$$x = \frac{-(-5) \pm \sqrt{(-5)^2 - 4(3)(6)}}{2(3)}$$
$$x = \frac{5 \pm \sqrt{-47}}{6}$$
$$x = \frac{5}{6} \pm \frac{\sqrt{47}}{6}i$$

99. $x^2 + 6x + 9 = 0$

$$(x+3)^2 = 0$$
$$x + 3 = 0$$
$$x = -3$$

101. $x^3 + 16x = 0$

$$x(x^2 + 16) = 0$$
$$x = 0 \quad x^2 + 16 = 0$$
$$x^2 = -16$$
$$x = \pm 4i$$

103. $f(x) = 3x(x-2)^2$

Zeros: 0, 2, 2

105. $h(x) = x^3 - 7x^2 + 18x - 24$

$$\begin{array}{r|rrrr} 4 & 1 & -7 & 18 & -24 \\ & & 4 & -12 & 24 \\ \hline & 1 & -3 & 6 & 0 \end{array}$$

$x = 4$ is a zero. Applying the Quadratic Formula on $x^2 - 3x + 6$,

$$x = \frac{3 \pm \sqrt{9 - 4(6)}}{2} = \frac{3}{2} \pm \frac{\sqrt{15}}{2}i.$$

Zeros: $4, \frac{3}{2} + \frac{\sqrt{15}}{2}i, \frac{3}{2} - \frac{\sqrt{15}}{2}i$

$$h(x) = (x-4)\left(x - \frac{3 + \sqrt{15}i}{2}\right)\left(x - \frac{3 - \sqrt{15}i}{2}\right)$$

107. $f(x) = 2x^4 - 5x^3 + 10x - 12$

$$\begin{array}{r|rrrrr} 2 & 2 & -5 & 0 & 10 & -12 \\ & & 4 & -2 & -4 & 12 \\ \hline & 2 & -1 & -2 & 6 & 0 \end{array}$$

$x = 2$ is a zero.

$$\begin{array}{r|rrrr} -\frac{3}{2} & 2 & -1 & -2 & 6 \\ & & -3 & 6 & -6 \\ \hline & 2 & -4 & 4 & 0 \end{array}$$

$x = -\frac{3}{2}$ is a zero.

$$f(x) = (x-2)\left(x + \frac{3}{2}\right)(2x^2 - 4x + 4)$$
$$= (x-2)(2x+3)(x^2 - 2x + 2)$$

By the Quadratic Formula, applied to $x^2 - 2x + 2$,

$$x = \frac{2 \pm \sqrt{4 - 4(2)}}{2} = 1 \pm i.$$

Zeros: $2, -\frac{3}{2}, 1 \pm i$

$$f(x) = (x-2)(2x+3)(x-1+i)(x-1-i)$$

109. $f(x) = x^5 + x^4 + 5x^3 + 5x^2$

$$= x^2(x^3 + x^2 + 5x + 5)$$
$$= x^2\left[x^2(x+1) + 5(x+1)\right]$$
$$= x^2(x+1)(x^2+5)$$
$$= x^2(x+1)(x+\sqrt{5}i)(x-\sqrt{5}i)$$

Zeros: $0, 0, -1, \pm\sqrt{5}i$

111. $f(x) = x^3 - 4x^2 + 6x - 4$

(a) $x^3 - 4x^2 + 6x - 4 = (x-2)(x^2 - 2x + 2)$

By the Quadratic Formula for $x^2 - 2x + 2$,

$$x = \frac{2 \pm \sqrt{(-2)^2 - 4(2)}}{2} = 1 \pm i$$

Zeros: $2, 1+i, 1-i$

(b) $f(x) = (x-2)(x-1-i)(x-1+i)$

(c) x-intercept: $(2, 0)$

113. (a) $f(x) = -3x^3 - 19x^2 - 4x + 12$

$$= -(x+1)(3x^2 + 16x - 12)$$

$$\begin{array}{r|rrrr} -1 & -3 & -19 & -4 & 12 \\ & & 3 & 16 & -12 \\ \hline & -3 & -16 & 12 & 0 \end{array}$$

$$3x^2 + 16x - 12 = 0$$
$$(3x-2)(x+6) = 0$$
$$3x - 2 = 0 \Rightarrow x = \frac{2}{3}$$
$$x + 6 = 0 \Rightarrow x = -6$$

Zeros: $-1, \frac{2}{3}, -6$

(b) $f(x) = -(x+1)(3x-2)(x+6)$

(c) x-intercepts: $(-1, 0), (-6, 0), \left(\frac{2}{3}, 0\right)$

115. $f(x) = x^4 + 34x^2 + 225$

(a) $x^4 + 34x^2 + 225 = (x^2 + 9)(x^2 + 25)$

Zeros: $\pm 3i, \pm 5i$

(b) $(x+3i)(x-3i)(x+5i)(x-5i)$

(c) No x-intercepts

117. Since $5i$ is a zero, so is $-5i$.

$$f(x) = (x-4)(x+2)(x-5i)(x+5i)$$
$$= (x^2 - 2x - 8)(x^2 + 25)$$
$$= x^4 - 2x^3 + 17x^2 - 50x - 200$$

119. Since $-3+5i$ is a zero, so is $-3-5i$.

$$\begin{aligned} f(x) &= (x-1)(x+4)(x+3-5i)(x+3+5i) \\ &= (x^2+3x-4)((x+3)^2+25) \\ &= (x^2+3x-4)(x^2+6x+34) \\ &= x^4+9x^3+48x^2+78x-136 \end{aligned}$$

121. $f(x)=x^4-2x^3+8x^2-18x-9$

(a) $f(x)=(x^2+9)(x^2-2x-1)$

(b) For the quadratic

$x^2-2x-1, x=\dfrac{2\pm\sqrt{(-2)^2-4(-1)}}{2}=1\pm\sqrt{2}.$

$f(x)=\left(x^2+9\right)\left(x-1+\sqrt{2}\right)\left(x-1-\sqrt{2}\right)$

(c) $f(x)=\left(x+3i\right)\left(x-3i\right)\left(x-1+\sqrt{2}\right)\left(x-1-\sqrt{2}\right)$

123. Zeros: $-2i$, 2i

$(x+2i)(x-2i)=x^2+4$ is a factor.

$f(x)=(x^2+4)(x+3)$

Zeros: $\pm 2i, -3$

125. $f(x)=\dfrac{2-x}{x+3}$

(a) Domain: all $x\neq -3$

(b) Not continuous

(c) Horizontal asymptote: $y=-1$

Vertical asymptote: $x=-3$

127. $f(x)=\dfrac{2}{x^2-3x-18}=\dfrac{2}{(x-6)(x+3)}$

(a) Domain: all $x\neq 6, -3$

(b) Not continuous

(c) Horizontal asymptote: $y=0$

Vertical asymptotes: $x=6, x=-3$

129. $f(x)=\dfrac{7+x}{7-x}$

(a) Domain: all $x\neq 7$

(b) Not continuous

(c) Horizontal asymptote: $y=-1$

Vertical asymptote: $x=7$

131. $f(x)=\dfrac{4x^2}{2x^2-3}$

(a) Domain: all $x\neq \pm\sqrt{\dfrac{3}{2}}=\pm\dfrac{\sqrt{6}}{2}$

(b) Not continuous

(c) Horizontal asymptote: $y=2$

Vertical asymptote: $x=\pm\sqrt{\dfrac{3}{2}}=\pm\dfrac{\sqrt{6}}{2}$

133. $f(x)=\dfrac{2x-10}{x^2-2x-15}=\dfrac{2(x-5)}{(x-5)(x+3)}=\dfrac{2}{x+3}, x\neq 5$

(a) Domain: all $x\neq 5, -3$

(b) Not continuous

(c) Vertical asymptote: $x=-3$

(There is a hole at $x=5$.)

Horizontal asymptote: $y=0$

135. $f(x)=\dfrac{x-2}{|x|+2}$

(a) Domain: all real numbers

(b) Continuous

(c) No vertical asymptotes

Horizontal asymptotes: $y=1, y=-1$

137. $C=\dfrac{528p}{100-p}, 0\leq p<100$

(a) When $p=25$, $C=\dfrac{528(25)}{100-25}=\$176$ million.

When $p=50$, C$=\dfrac{528(50)}{100-50}=\$528$ million.

When $p=75$, C$=\dfrac{528(75)}{100-75}=\$1584$ million.

(b) 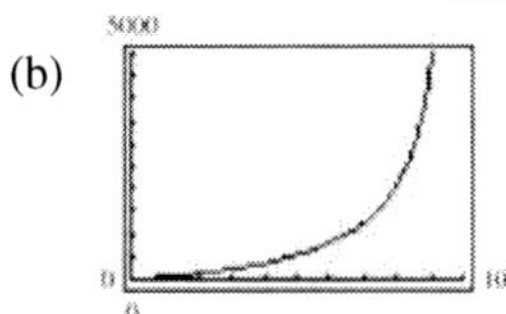

Answers will vary.

(c) No. As $p\to 100$, C approaches infinity.

139. $f(x)=\dfrac{x^2-5x+4}{x^2-1}$

$=\dfrac{(x-4)(x-1)}{(x-1)(x+1)}$

$=\dfrac{x-4}{x+1}, x\neq 1$

Vertical asymptote: $x=-1$

Horizontal asymptote: $y=1$

No slant asymptotes

Hole at $x=1$

141. $f(x) = \dfrac{3x^2 + 5x - 2}{x + 1}$

$= \dfrac{(3x - 1)(x + 2)}{x + 1}$

Vertical asymptote: $x = -1$

Horizontal asymptote: none

Long division gives:

$$\begin{array}{r} 3x + 2 \\ x + 1 \overline{)\, 3x^2 + 5x - 2} \\ \underline{3x^2 + 3x} \\ 2x - 2 \\ \underline{2x + 2} \\ -4 \end{array}$$

Slant asymptote: $y = 3x + 2$

143. $f(x) = \dfrac{2x - 1}{x - 5}$

Intercepts: $\left(0, \dfrac{1}{5}\right), \left(\dfrac{1}{2}, 0\right)$

Vertical asymptote: $x = 5$

Horizontal asymptote: $y = 2$

x	-4	-1	0	$\frac{1}{2}$	1	6	8
y	1	$\frac{1}{2}$	$\frac{1}{5}$	0	$-\frac{1}{4}$	11	5

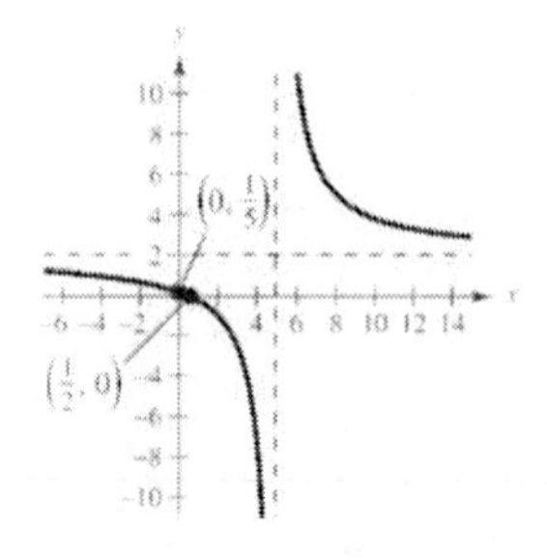

145. $f(x) = \dfrac{2x^2}{x^2 - 4}$

Intercept: $(0, 0)$

y-axis symmetry

Vertical asymptotes: $x = \pm 2$

Horizontal asymptote: $y = 2$

x	-6	-4	-1	0	1	4	6
y	$\frac{9}{4}$	$\frac{8}{3}$	$-\frac{2}{3}$	0	$-\frac{2}{3}$	$\frac{8}{3}$	$\frac{9}{4}$

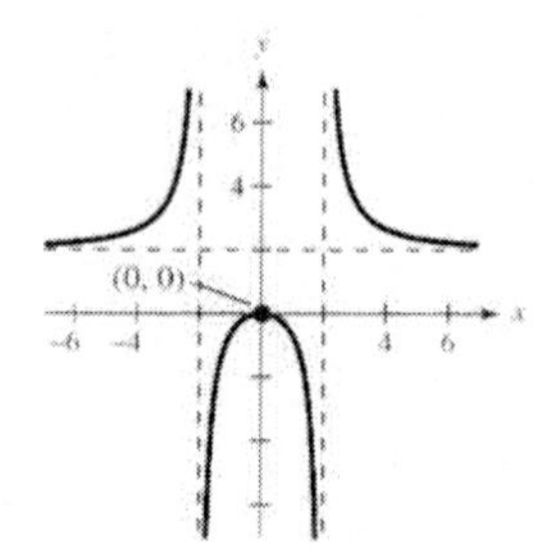

147. $f(x) = \dfrac{2}{(x + 1)^2}$

Intercept: $(0, 2)$

Horizontal asymptote: $y = 0$

Vertical asymptote: $x = -1$

x	-4	-3	-2	0	1	2
y	$\frac{2}{9}$	$\frac{1}{2}$	2	2	$\frac{1}{2}$	$\frac{2}{9}$

149. $f(x) = \dfrac{2x^3}{x^2+1} = 2x - \dfrac{2x}{x^2+1}$

Intercept: $(0, 0)$
Origin symmetry
Slant asymptote: $y = 2x$

x	-2	-1	0	1	2
y	$-\frac{16}{5}$	-1	0	1	$\frac{16}{5}$

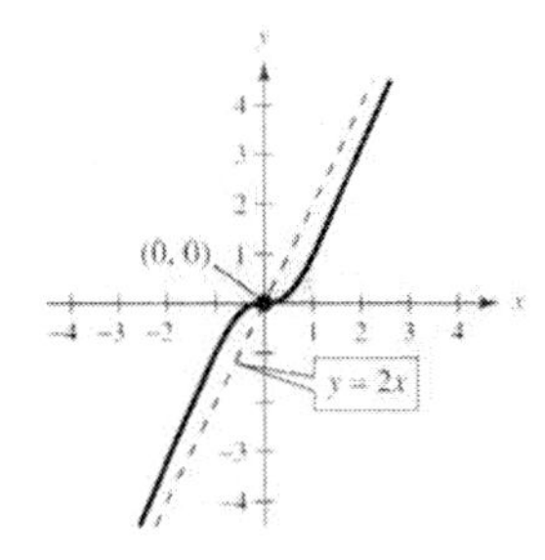

151. $f(x) = \dfrac{x^2 - x + 1}{x - 3} = x + 2 + \dfrac{7}{x-3}$

Intercept: $\left(0, -\frac{1}{3}\right)$

Vertical asymptote: $x = 3$
Slant asymptote: $y = x + 2$

x	-4	0	2	4	5
y	-3	$-\frac{1}{3}$	-3	13	10.5

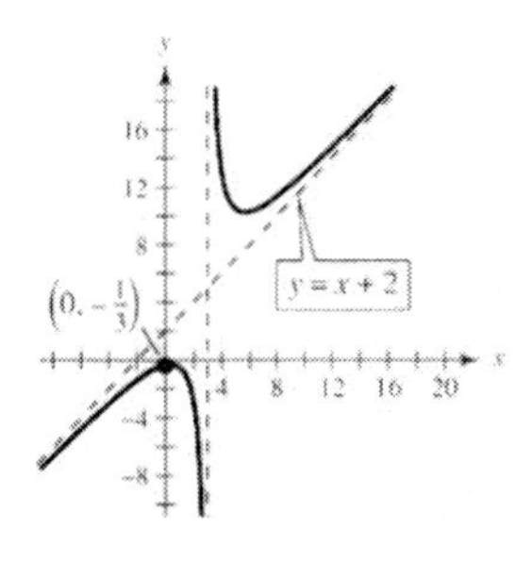

153. (a)

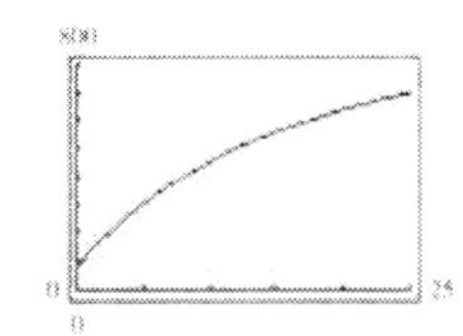

5 years: $N(5) = \dfrac{20(4+3(5))}{1+0.05(5)} = 304$ thousand fish

10 years: $N(10) = \dfrac{20(4+3(10))}{1+0.05(10)} = 453.\overline{3}$ thousand fish

25 years: $N(25) = \dfrac{20(4+3(25))}{1+0.05(25)} = 702.\overline{2}$ thousand fish

(b) The maximum number of fish is $N = 1{,}200{,}000$. The graph of N has a horizontal asymptote at $N = 1200$ or 1,200,000 fish.

155. Quadratic model

157. Linear model

159. (a)

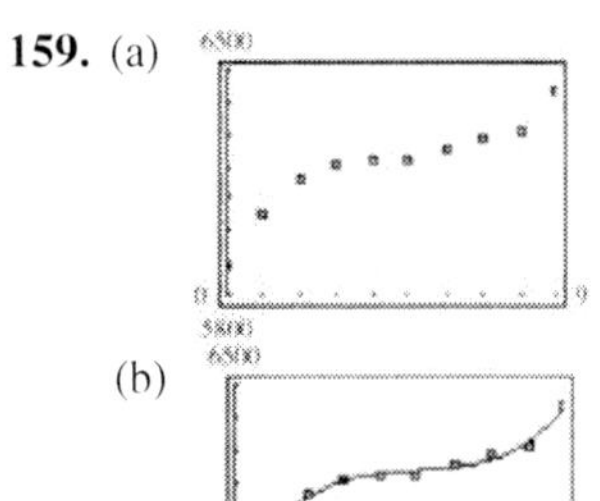

(b)

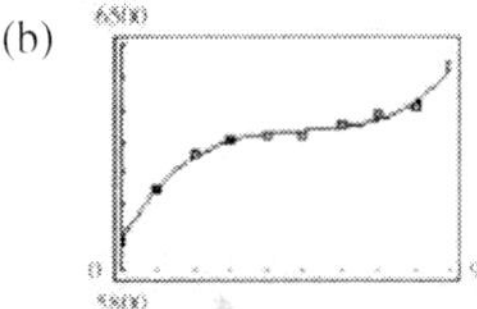

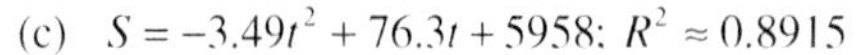

(c) $S = -3.49t^2 + 76.3t + 5958$; $R^2 \approx 0.8915$

(d)

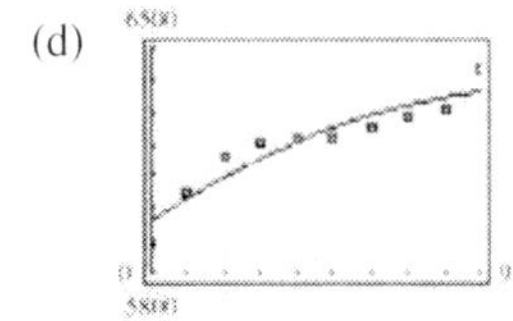

(e) The cubic model is a better fit because it more closely follows the pattern of the data.

(f) For 2012, let $t = 12$.

$$S(12) = 2.520(12)^3 - 37.51(12)^2 + 192.4(12) + 5895 = 7157 \text{ stations}$$

161. False. The degree of the numerator is two more than the degree of the denominator.

163. False. $(1+i)+(1-i) = 2$, a real number

165. Not every rational function has a vertical asymptote. For example,

$$y = \frac{x}{x^2+1}.$$

167. The error is $\sqrt{-4} \neq 4i$. In fact,

$$-i\left(\sqrt{-4} - 1\right) = -i(2i - 1) = 2 + i.$$

Chapter 2 Test

1. $y = x^2 + 4x + 3 = x^2 + 4x + 4 - 1 = (x+2)^2 - 1$

Vertex: $(-2, -1)$

$x = 0 \Rightarrow y = 3$

$y = 0 \Rightarrow x^2 + 4x + 3 = 0 \Rightarrow (x+3)(x+1) = 0 \Rightarrow x = -1, -3$

Intercepts: $(0, 3)$, $(-1, 0)$, $(-3, 0)$

2. Let $y = a(x-h)^2 + k$. The vertex $(3, -6)$ implies that $y = a(x-3)^2 - 6$. For $(0, 3)$ you obtain $3 = a(0-3)^2 - 6 = 9a - 6 \Rightarrow a = 1$.

Thus, $y = (x-3)^2 - 6 = x^2 - 6x + 3$.

3. $f(x) = 4x^3 + 4x^2 + x = x(4x^2 + 4x + 1) = x(2x+1)^2$

Zeros: 0 (multiplicity 1)

$-\frac{1}{2}$ (multiplicity 2)

4. $f(x) = -x^3 + 7x + 6$

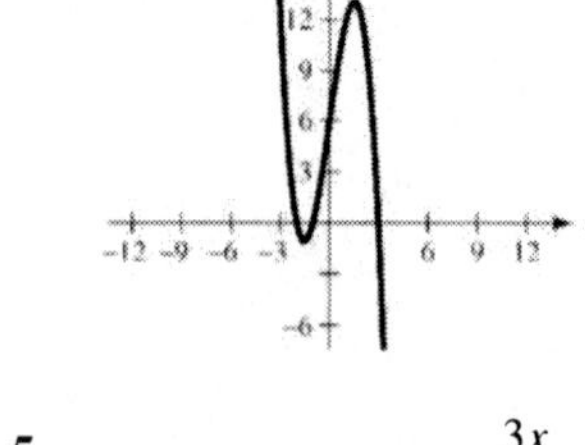

5.

$$\begin{array}{r} 3x \\ x^2+1 \overline{\big) 3x^3 + 0x^2 + 4x - 1} \\ \underline{3x^3 \quad\;\; + 3x} \\ x - 1 \end{array}$$

$3x + \dfrac{x-1}{x^2+1}$

6.

2	2	0	−5	0	−3
		4	8	6	12
	2	4	3	6	9

$2x^3 + 4x^2 + 3x + 6 + \dfrac{9}{x-2}$

7.

−2	3	0	−6	5	−1
		−6	12	−12	14
	3	−6	6	−7	13

$f(-2) = 13$

8. Possible rational zeros:

$\pm 24, \pm 12, \pm 8, \pm 6, \pm 4, \pm 3, \pm 2, \pm 1, \pm\frac{3}{2}, \pm\frac{1}{2}$

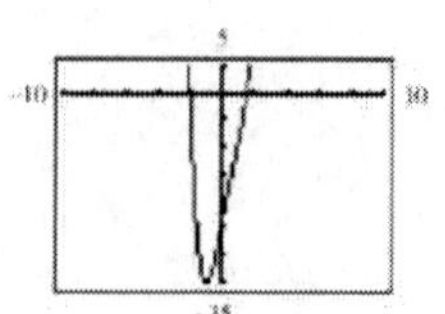

Rational zeros: $-2, \frac{3}{2}$

9. Possible rational zeros: $\pm 2, \pm 1, \pm\frac{2}{3}, \pm\frac{1}{3}$

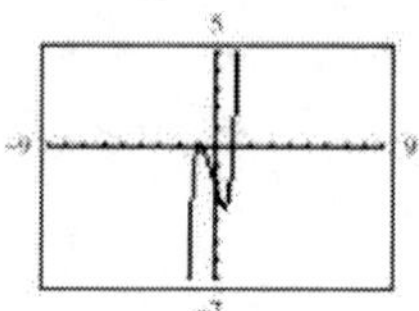

Rational zeros: $\pm 1, -\frac{2}{3}$

10. $f(x) = x^3 - 7x^2 + 11x + 19$

$= (x+1)(x^2 - 8x + 19)$

For the quadratic,

$x = \dfrac{8 \pm \sqrt{64 - 4(19)}}{2} = 4 \pm \sqrt{3}i.$

Zeros: $-1, 4 \pm \sqrt{3}\,i$

$f(x) = (x+1)(x - 4 + \sqrt{3}i)(x - 4 - \sqrt{3}i)$

11. $(-8 - 3i) + (-1 - 15i) = -9 - 18i$

12. $\left(10 + \sqrt{-20}\right) - \left(4 - \sqrt{-14}\right) = 6 + 2\sqrt{5}i + \sqrt{14}i = 6 + \left(2\sqrt{5} + \sqrt{14}\right)i$

13. $(2+i)(6-i) = 12 + 6i - 2i + 1 = 13 + 4i$

14. $(4+3i)^2 - (5+i)^2 = (16 + 24i - 9) - (25 + 10i - 1) = -17 + 14i$

15. $\dfrac{8+5i}{6-i} \cdot \dfrac{6+i}{6+i} = \dfrac{48 + 30i + 8i - 5}{36+1} = \dfrac{43}{37} + \dfrac{38}{37}i$

16. $\dfrac{5i}{2+i} \cdot \dfrac{2-i}{2-i} = \dfrac{10i + 5}{4+1} = 1 + 2i$

17. $\dfrac{(2i-1)}{(3i+2)} \cdot \dfrac{2-3i}{2-3i} = \dfrac{6 - 2 + 4i + 3i}{4+9}$

$= \dfrac{4}{13} + \dfrac{7}{13}i$

18. $x^2 + 75 = 0$

$$x^2 = -75$$
$$x = \pm\sqrt{-75}$$
$$= \pm 5\sqrt{3}i$$

19. $x^2 - 2x + 8 = 0$

$$x = \frac{-(-2) \pm \sqrt{(-2)^2 - 4(1)(8)}}{2(1)}$$
$$x = \frac{2 \pm \sqrt{-28}}{2}$$
$$x = \frac{2 \pm 2\sqrt{7}i}{2}$$
$$x = 1 \pm \sqrt{7}i$$

20.

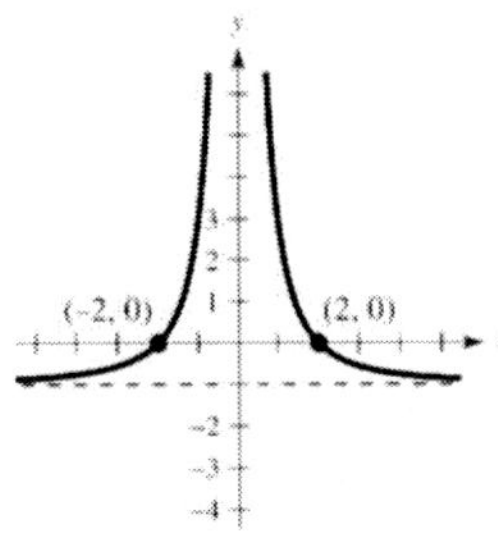

Vertical asymptote: $x = 0$
Intercepts: $(2, 0)$, $(-2, 0)$
Symmetry: y-axis
Horizontal asymptote: $y = -1$

21. $g(x) = \dfrac{x^2 + 2}{x - 1} = x + 1 + \dfrac{3}{x - 1}$

Vertical asymptote: $x = 1$
Intercept: $(0, -2)$
Slant asymptote: $y = x + 1$

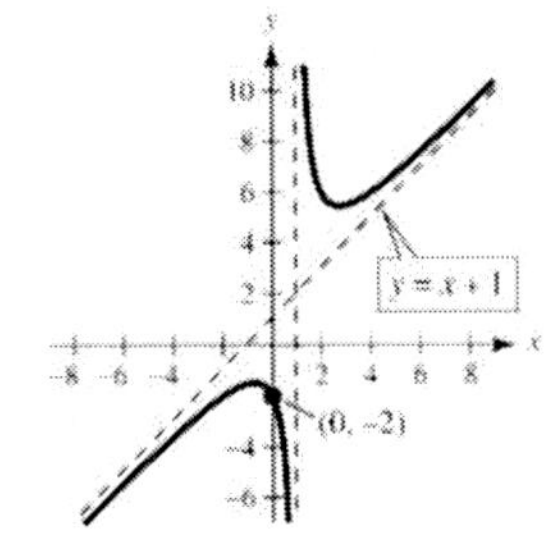

22. $f(x) = \dfrac{2x^2 + 9}{5x^2 + 2}$

Horizontal asymptote: $y = \dfrac{2}{5}$

y-axis symmetry

Intercept: $\left(0, \dfrac{9}{2}\right)$

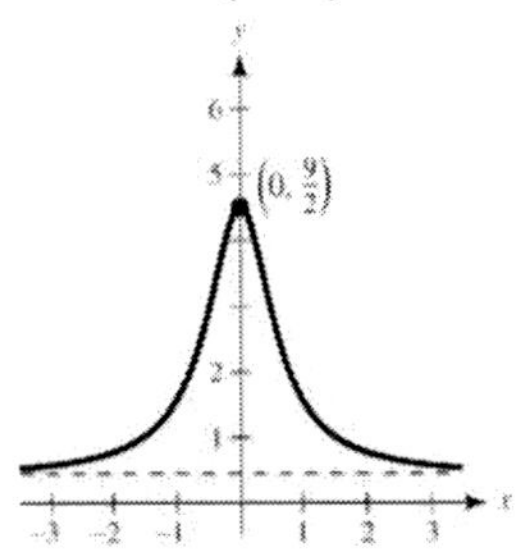

23. (a)

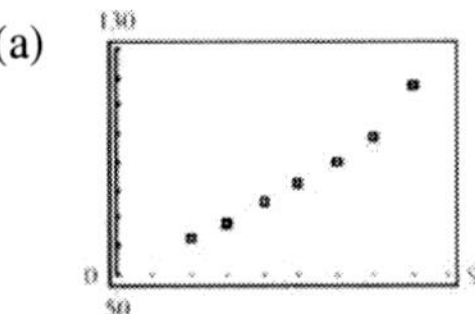

(b) $A = 0.861t^2 + 0.03t + 60.0$

(c)

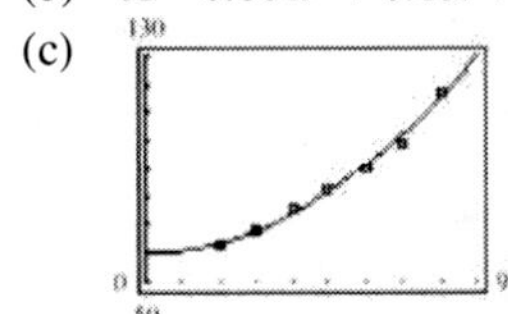

The model fits the data well.

(d) For 2010, let $t = 10$.

$$A(10) = 0.861(10)^2 + 0.03(10) + 60.0$$
$$= \$146.4 \text{ million}$$

For 2012, let $t = 12$.

$$A(12) = 0.861(12)^2 + 0.03(12) + 60.0$$
$$\approx \$184.3 \text{ billion}$$

(e) Answers will vary.

CHAPTER 3

Section 3.1

1. transcendental

3. The graph of $f(x+1)$ is a horizontal shift one unit to the left of $f(x) = 5^x$.

5. $(3.4)^{6.8} \approx 4112.033$

7. $5^{-\pi} \approx 0.006$

9. $g(x) = 5^x$

x	-2	-1	0	1	2
y	$\frac{1}{25}$	$\frac{1}{5}$	1	5	25

Asymptote: $y = 0$

Intercept: $(0, 1)$

Increasing

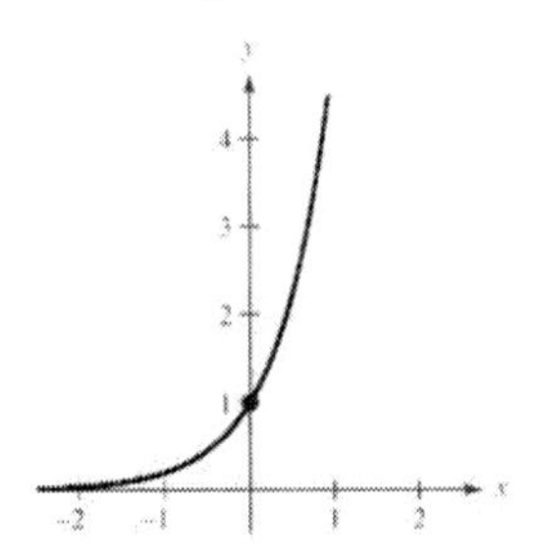

11. $f(x) = \left(\frac{1}{5}\right)^x = 5^{-x}$

x	-2	-1	0	1	2
y	25	5	1	$\frac{1}{5}$	$\frac{1}{25}$

Asymptote: $y = 0$

Intercept: $(0, 1)$

Decreasing

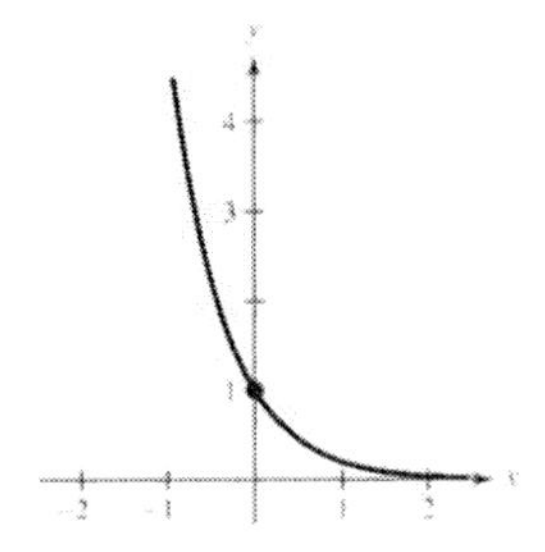

13. $h(x) = 3^x$

x	-2	-1	0	1	2
y	$\frac{1}{9}$	$\frac{1}{3}$	1	3	9

Asymptote: $y = 0$
Intercept: $(0, 1)$
Increasing

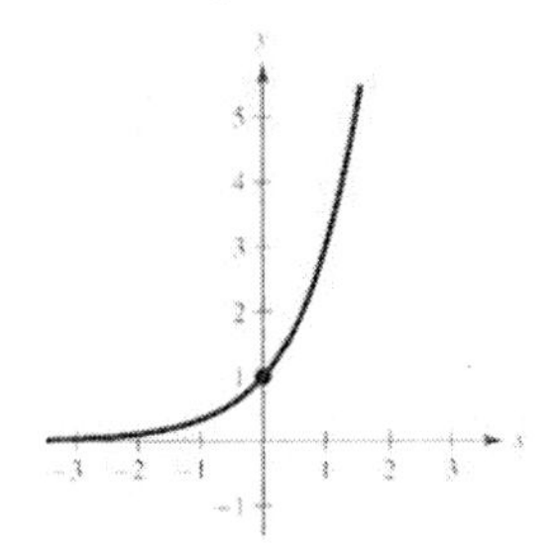

15. $g(x) = 3^{-x}$

x	-2	-1	0	1	2
y	9	3	1	$\frac{1}{3}$	$\frac{1}{9}$

Asymptote: $y = 0$
Intercept: $(0, 1)$
Decreasing

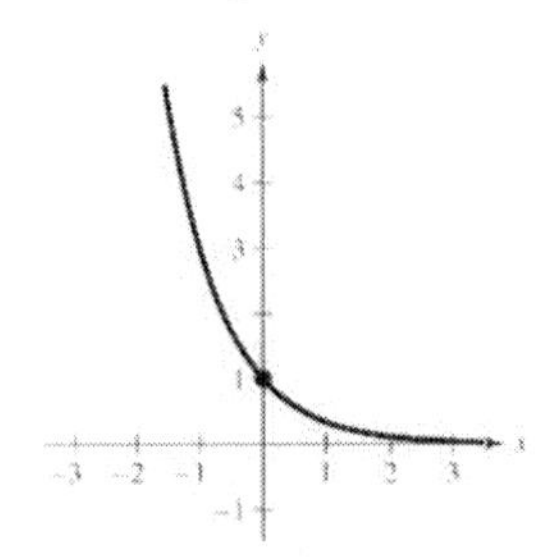

17. $f(x) = 2^{x-2}$ rises to the right.
Asymptote: $y = 0$
Intercept: $\left(0, \frac{1}{4}\right)$
Matches graph (d).

19. $f(x) = 2^x - 4$ rises to the right.
Asymptote: $y = -4$
Intercept: $(0, -3)$
Matches graph (c).

21. The graph of $g(x) = 3^{x-5}$ is a horizontal shift five units to the right of $f(x) = 3^x$.

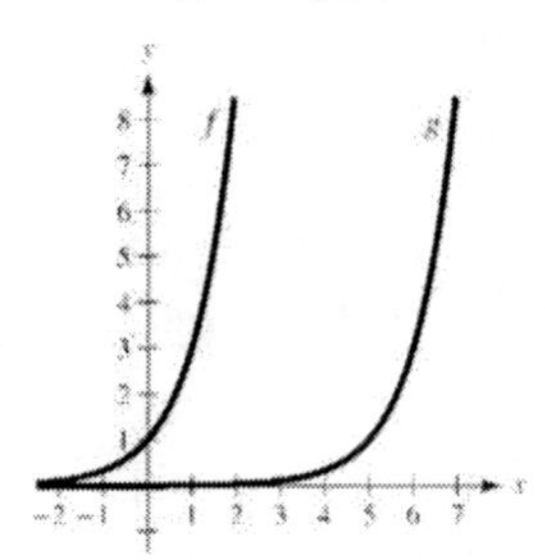

23. The graph of $g(x) = -\left(\frac{3}{5}\right)^{x+4}$ is a reflection in the x-axis and a horizontal shift four units to the left of $f(x) = \left(\frac{3}{5}\right)^x$.

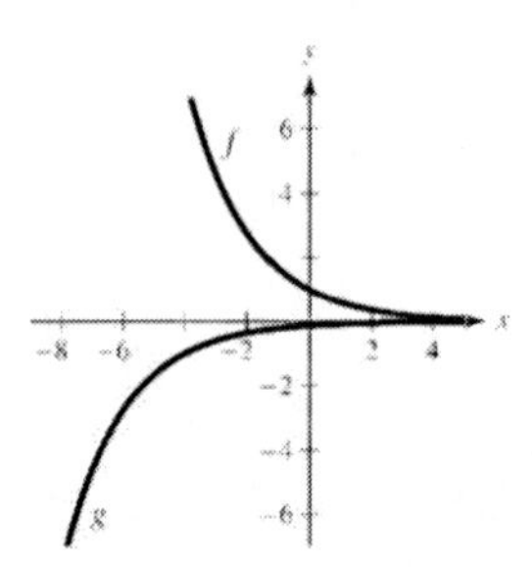

25. The graph of $g(x) = 4^{x-2} - 3$ is a horizontal shift two units right and a vertical shift three units downward of $f(x) = 4x$.

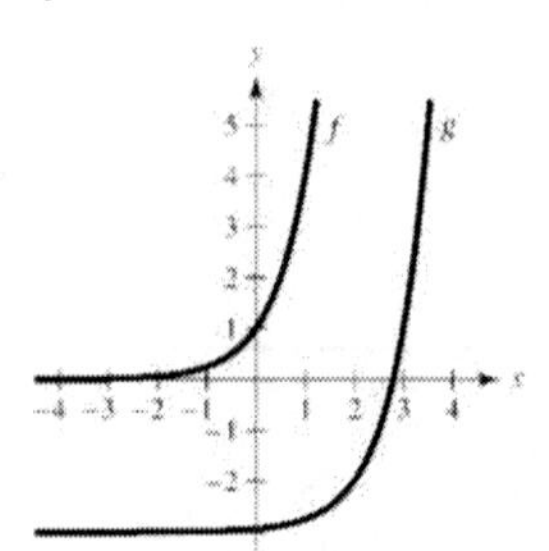

27. (a)

(b) $e^2 \approx 7.3891$

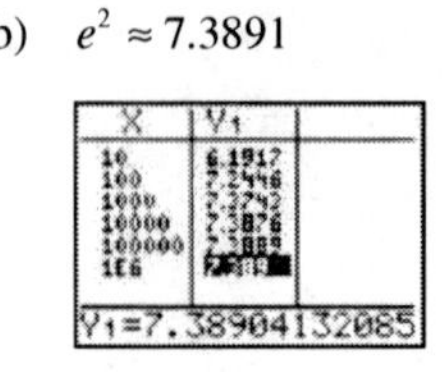

29. $e^{9.2} \approx 9897.129$

31. $50e^{4(0.02)} \approx 54.164$

33. $f(x) = \left(\frac{5}{2}\right)^x$

x	-2	-1	0	1	2
$f(x)$	0.16	0.4	1	2.5	6.25

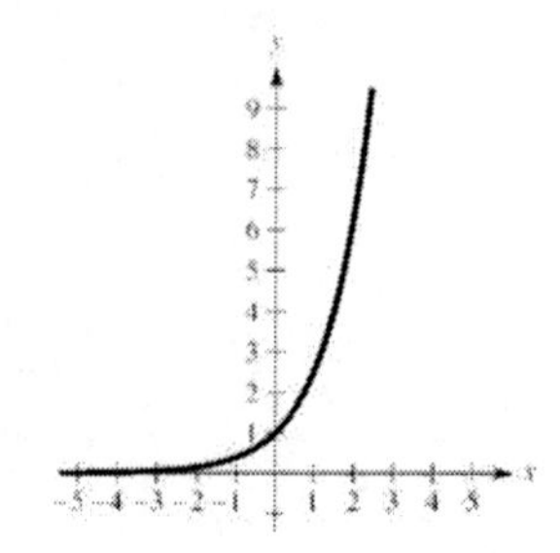

Asymptote: $y = 0$

35. $f(x) = 6^x$

x	-2	-1	0	1	2
$f(x)$	0.03	0.17	1	6	36

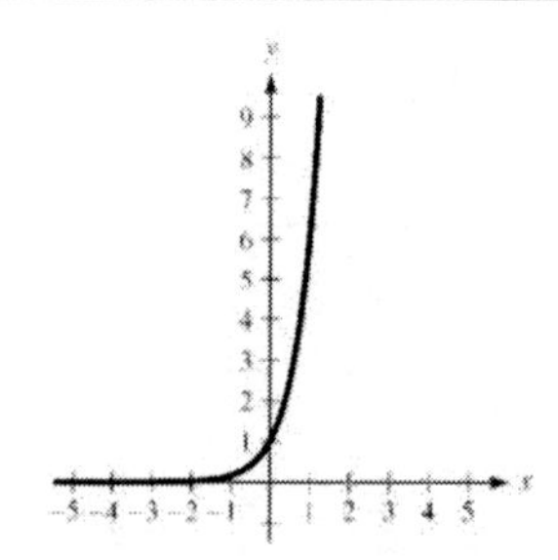

Asymptote: $y = 0$

37. $f(x) = 3^{x+2}$

x	-3	-2	-1	0	1
$f(x)$	0.33	1	3	9	27

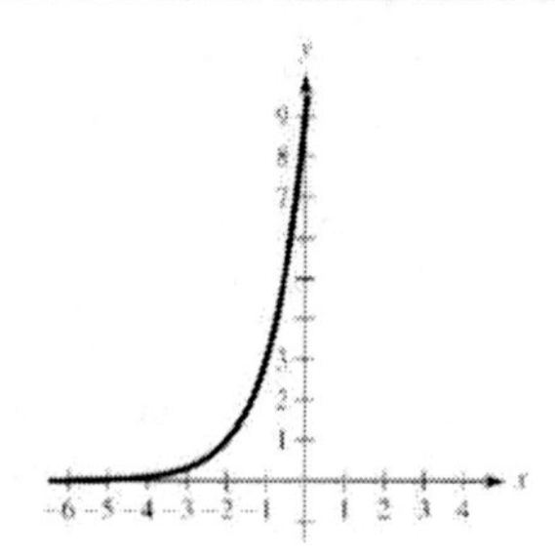

Asymptote: $y = 0$

39. $y = 3^{x-2} + 1$

x	-1	0	1	2	3	4
y	1.04	1.11	1.33	2	4	10

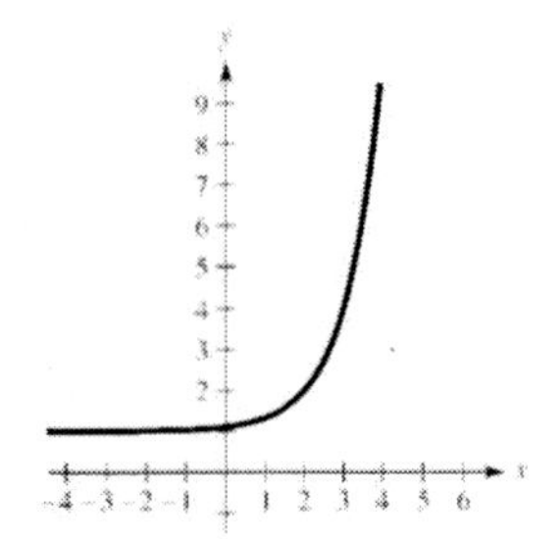

Asymptote: $y = 1$

41. $f(x) = e^{-x}$

x	-2	-1	0	1	2
$f(x)$	7.39	2.72	1	0.37	0.14

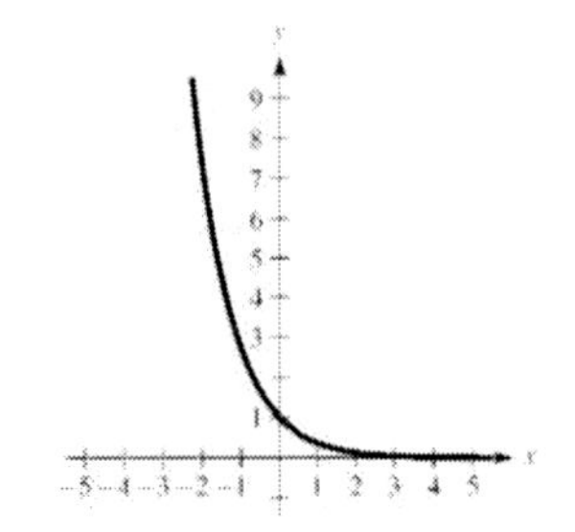

Asymptote: $y = 0$

43. $f(x) = 3e^{x+4}$

x	-6	-5	-4	-3	-2
$f(x)$	0.41	1.10	3	8.15	22.17

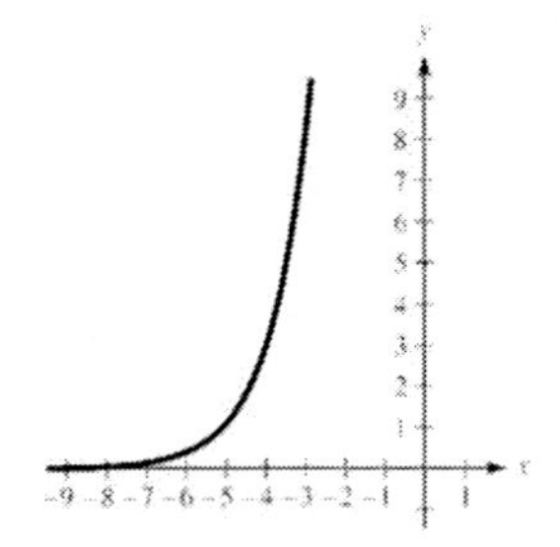

Asymptote: $y = 0$

45. $f(x) = 2 + e^{x-5}$

x	3	4	5	6	7
$f(x)$	2.14	2.37	3	4.72	9.39

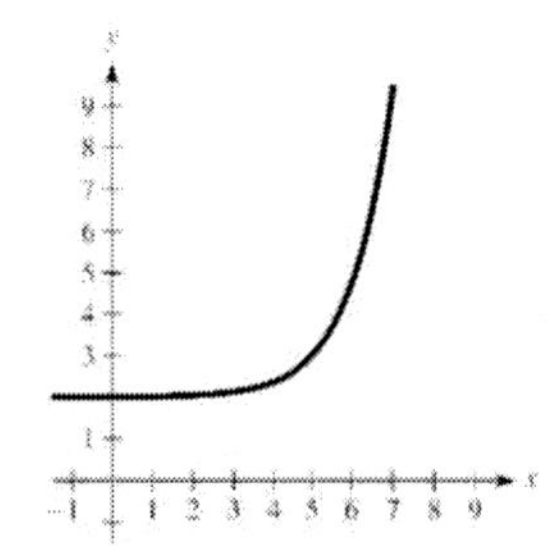

Asymptote: $y = 2$

47. $s(t) = 2e^{0.12t}$

t	-2	-1	0	1	2
$s(t)$	1.57	1.77	2	2.26	2.54

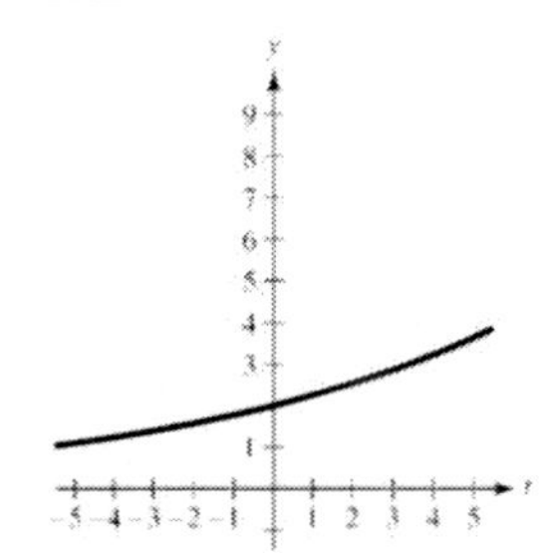

Asymptote: $y = 0$

49. $f(x) = \dfrac{8}{1 + e^{-0.5x}}$

(a)

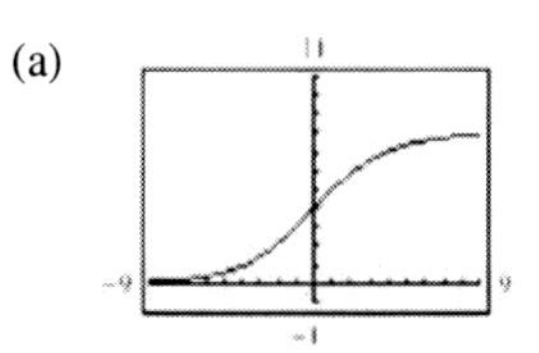

(b)

x	-30	-20	-10	0	10	20	30
$f(x)$	≈ 0	≈ 0	0.05	4	7.95	≈ 8	≈ 8

Horizontal asymptotes: $y = 0$, $y = 8$

51. $f(x) = \dfrac{-6}{2 - e^{0.2x}}$

(a)

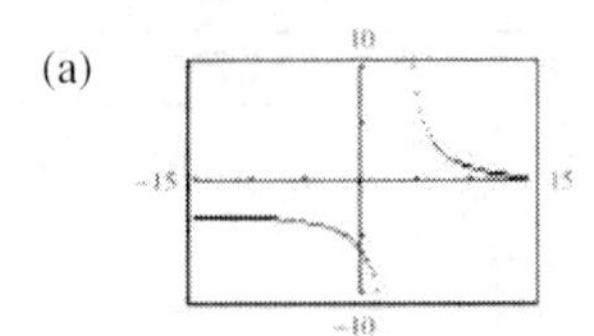

(b)

x	-20	-10	0	3	3.4	3.46
$f(x)$	-3.03	-3.22	-6	-34	-230	-2617

x	3.47	4	5	10	20
$f(x)$	3516	26.6	8.4	1.11	0.11

Horizontal asymptotes: $y = -3$, $y = 0$

Vertical asymptote: $x \approx 3.47$

53.

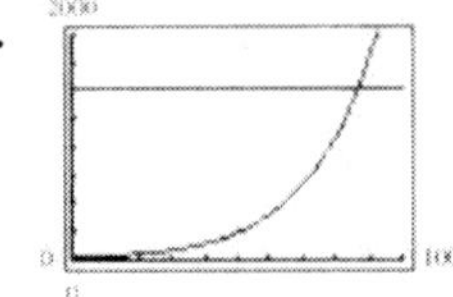

Intersection: $(86.350, 1500)$

55. $f(x) = x^2 e^{-x}$

(a)

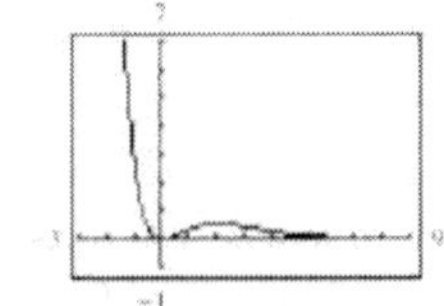

(b) Decreasing on $(-\infty, 0)$, $(2, \infty)$
Increasing on $(0, 2)$

(c) Relative maximum: $(2, 4e^{-2}) \approx (2, 0.541)$
Relative minimum: $(0, 0)$

57. $P = 2500$, $r = 2\% = 0.02$, $t = 10$

Compounded n times per year:

$$A = P\left(1 + \frac{r}{n}\right)^{nt} = 2500\left(1 + \frac{0.02}{n}\right)^{10n}$$

Compounded continuously:
$A = Pe^{rt} = 2500e^{(0.02)(10)}$

n	1	2	4	12
A	\$3047.49	\$3050.48	\$3051.99	\$3053.00

n	365	Continuous
A	\$3053.49	\$3053.51

59. $P = 2500$, $r = 4\% = 0.04$, $t = 20$

Compounded n times per year:

$$A = P\left(1 + \frac{r}{n}\right)^{nt} = 2500\left(1 + \frac{0.04}{n}\right)^{20n}$$

Compounded continuously:
$A = Pe^{rt} = 2500e^{(0.04)(20)}$

n	1	2	4	12
A	\$5477.81	\$5520.10	\$5541.79	\$5556.46

n	365	Continuous
A	\$5563.61	\$5563.85

61. $P = 12{,}000$, $r = 4\% = 0.04$, $A = pe^{rt} = 12{,}000e^{0.04t}$

t	1	10	20	30	40	50
A	\$12,489.73	\$17,901.90	\$26,706.49	\$39,841.40	\$59,436.39	\$88,668.67

t	1	10	20	30	40	50
A	\$12,742.04	\$21,865.43	\$39,841.40	\$75,595.77	\$132,278.12	\$241,026.44

63. $P = 12{,}000$, $r = 3.5\% = 0.035$, $A = pe^{rt} = 12{,}000e^{0.035t}$

t	1	10	20	30	40	50
A	\$12,427.44	\$17,028.81	\$24,165.03	\$34,291.81	\$48,662.40	\$69,055.23

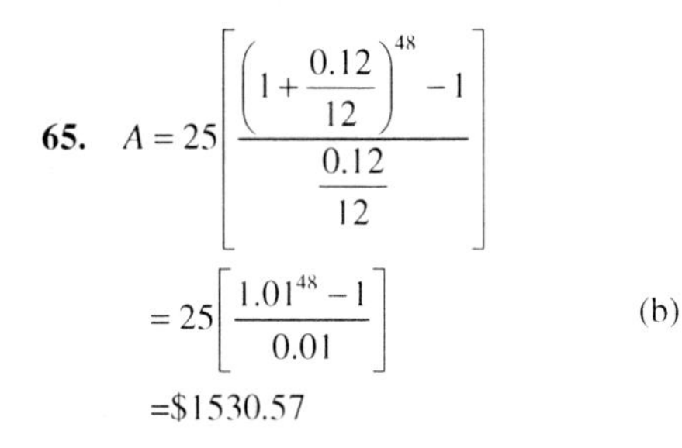

65. $A = 25\left[\dfrac{\left(1+\dfrac{0.12}{12}\right)^{48}-1}{\dfrac{0.12}{12}}\right]$

$= 25\left[\dfrac{1.01^{48}-1}{0.01}\right]$

$= \$1530.57$

67. $A = 200\left[\dfrac{\left(1+\dfrac{0.06}{12}\right)^{72}-1}{\dfrac{0.06}{12}}\right]$

$= \$17,281.77$

69. (a) $y_1 = 500(1+0.07)^x$

$y_2 = 500\left(1+\dfrac{0.07}{4}\right)^{4x}$

$y_3 = 500e^{0.07x}$

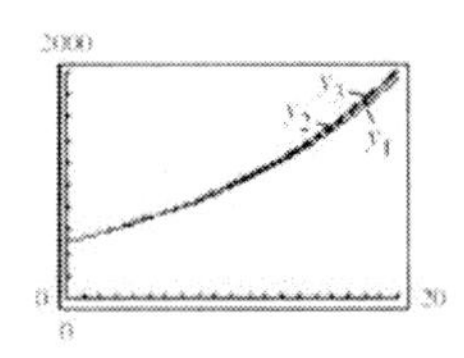

(b) y_3 has the highest return.

After 20 years:

$y_2 - y_1 = 2003.20 - 1934.84 = \$68.36$

$y_3 - y_2 = 2027.60 - 2003.20 = \$24.40$

$y_3 - y_1 = 2027.60 - 1934.84 = \$92.76$

71. $Q = 10\left(\dfrac{1}{2}\right)^{\frac{t}{5715}}$

(a) When $t = 0$, $Q = 10\left(\dfrac{1}{2}\right)^{\frac{0}{5715}} = 10(1) = 10$ grams.

(b) When $t = 2000$,

$Q = 10\left(\frac{1}{2}\right)^{2000/5715} \approx 7.85$ grams.

(c)

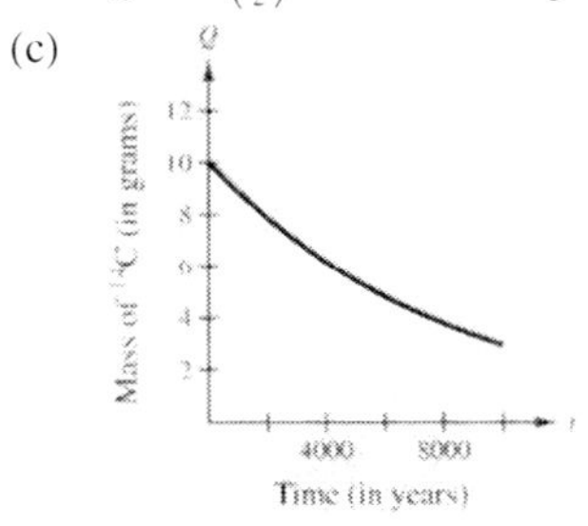

73. (a)

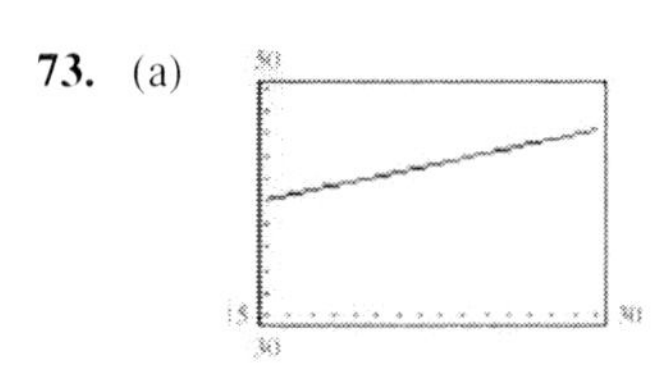

(b)

Year	2023	2024	2025	2026	2027	2028	2029	2030
P	43.4	43.8	44.2	44.7	45.1	45.5	46.0	46.4

Year	2015	2016	2017	2018	2019	2020	2021	2022
P	40.1	40.5	40.9	41.3	41.7	42.1	42.5	43.0

(c) $P = 34.706e^{0.0097t} = 50$

$e^{0.0097t} = 1.441$

$0.0097t = \ln(1.441)$

$t \approx 37.7$ or 2037

(Answers will vary.)

75. True. $f(x) = 1^x$ is not an exponential function.

77. The graph decreases for all x and has a positive y-intercept. Matches (d).

79.

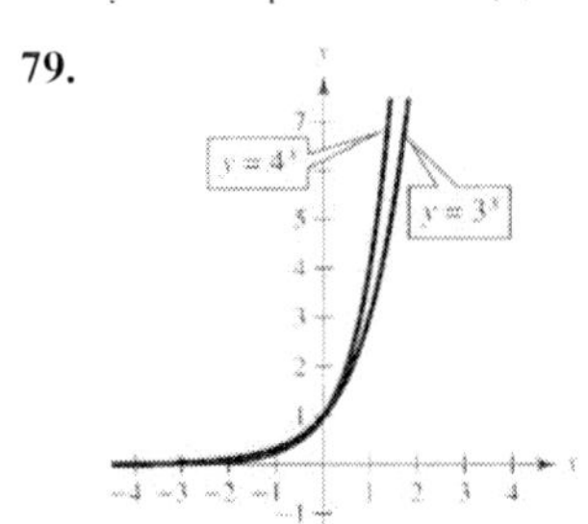

The solution set to $3^x < 4^x$ is $x > 0$.

81. $e^{\pi} \approx 23.14$, $\pi^{e} \approx 22.46$

$e^{\pi} > \pi^{e}$

83. $5^{-3} = 0.008$, $3^{-5} \approx 0.0041$

$5^{-3} > 3^{-5}$

85. f has an inverse because f is one-to-one.

$y = 5x - 7$

$x = 5y - 7$

$x + 7 = 5y$

$f^{-1}(x) = \dfrac{1}{5}(x+7)$

87. f has an inverse because f is one-to-one.

$y = \sqrt[3]{x+8}$

$x = \sqrt[3]{y+8}$

$x^3 = y + 8$

$x^3 - 8 = y$

$f^{-1}(x) = x^3 - 8$

89. Answers will vary.

Section 3.2

1. logarithmic function

3. $a^{\log_a x} = x$

5. If $\log_a b = c$, then $a^c = b$.

7. $\log_4 64 = 3 \Rightarrow 4^3 = 64$

9. $\log_7 \frac{1}{49} = -2 \Rightarrow 7^{-2} = \frac{1}{49}$

11. $\log_{32} 4 = \frac{2}{5} \Rightarrow 32^{2/5} = 4$

13. $\log_2 \sqrt{2} = \frac{1}{2} \Rightarrow 2^{1/2} = \sqrt{2}$

15. $5^3 = 125 \Rightarrow \log_5 125 = 3$

17. $81^{1/4} = 3 \Rightarrow \log_{81} 3 = \dfrac{1}{4}$

19. $6^{-2} = \frac{1}{36} \Rightarrow \log_6 \frac{1}{36} = -2$

21. $g^a = 4 \Rightarrow \log_g 4 = a$

23. $\log_2 16 = \log_2 2^4 = 4$

25. $$\begin{aligned} g\left(\frac{1}{1000}\right) &= \log_{10}\left(\frac{1}{1000}\right) \\ &= \log_{10}(10^{-3}) \\ &= -3 \end{aligned}$$

27. $\log_{10} 345 \approx 2.538$

29. $6\log_{10} 14.8 \approx 7.022$

31. $$\begin{aligned} \log_7 x &= \log_7 9 \\ x &= 9 \end{aligned}$$

33. $$\begin{aligned} \log_4 4^2 &= x \\ 2 &= x \end{aligned}$$

35. $$\begin{aligned} \log_8 x &= \log_8 10^{-1} \\ x &= 10^{-1} = \frac{1}{10} \end{aligned}$$

37. $\log_4 4^{3x} = (3x)\log_4 4 = 3x$

39. $$\begin{aligned} 3\log_2\left(\tfrac{1}{2}\right) &= 3\log_2(2^{-1}) \\ &= 3(-1) = -3 \end{aligned}$$

41. $f(x) = 3^x$ and $g(x) = \log_3 x$ are inverse functions of each other.

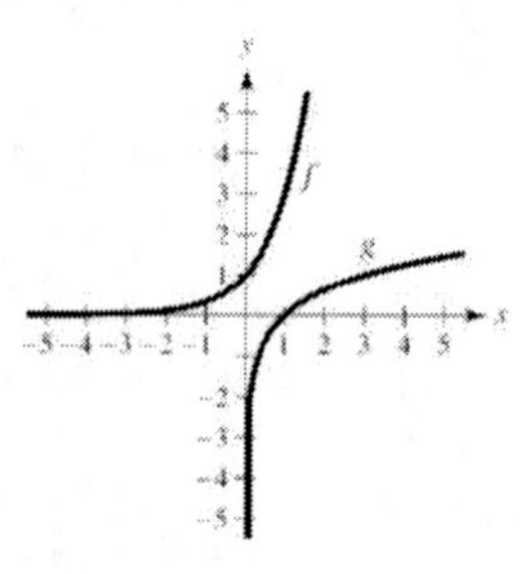

43. $f(x) = 15^x$ and $g(x) = \log_{15} x$ are inverse functions of each other.

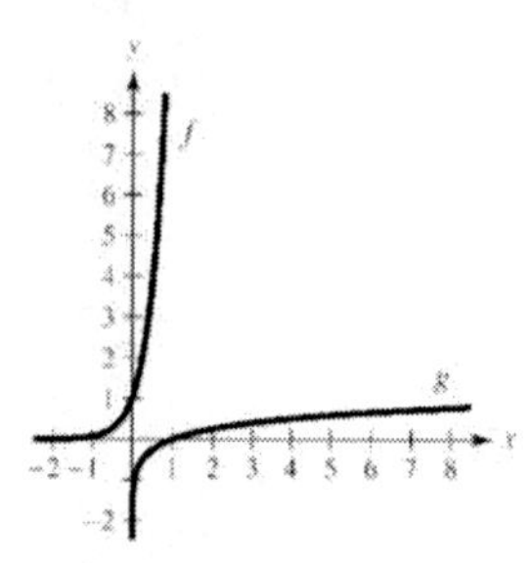

45. $y = \log_2(x+2)$

Domain: $x + 2 > 0 \Rightarrow x > -2$
Vertical asymptote: $x = -2$

$$\begin{aligned} \log_2(x+2) &= 0 \\ x + 2 &= 1 \\ x &= -1 \end{aligned}$$

x-intercept: $(-1, 0)$

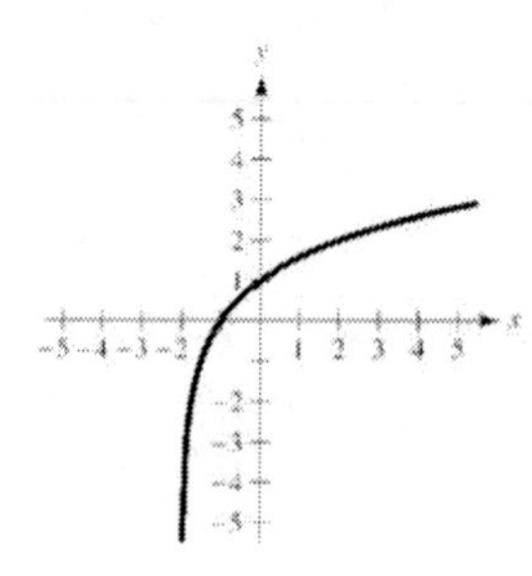

47. $y = 1 + \log_2 x$

Domain: $x > 0$
Vertical asymptote: $x = 0$

$$1 + \log_2 x = 0$$
$$\log_2 x = -1$$
$$x = 2^{-1} = \frac{1}{2}$$

x-intercept: $\left(\frac{1}{2}, 0\right)$

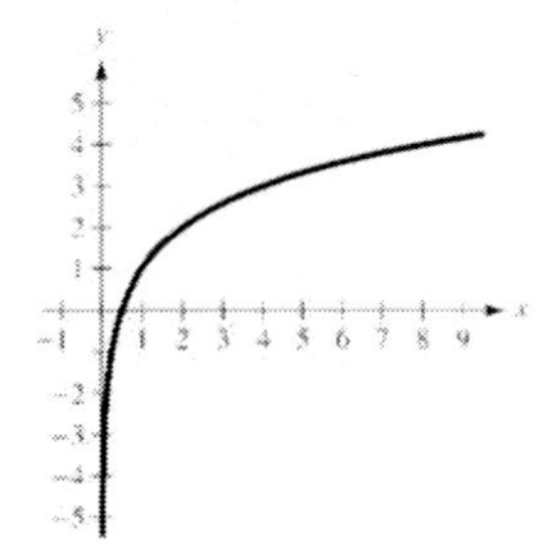

49. $y = 1 + \log_2(x - 2)$

Domain: $x - 2 > 0 \Rightarrow x > 2$
Vertical asymptote: $x = 2$

$$1 + \log_2(x - 2) = 0$$
$$\log_2(x - 2) = -1$$
$$x - 2 = 2^{-1} = \tfrac{1}{2}$$
$$x = \tfrac{5}{2}$$

x-intercept: $\left(\tfrac{5}{2}, 0\right)$

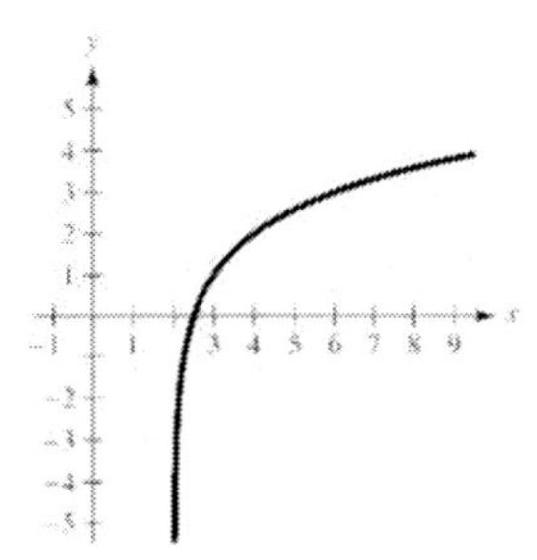

51. $f(x) = \log_3 x + 2$

Asymptote: $x = 0$
Point on graph: $(1, 2)$
Matches graph (b).

53. $f(x) = -\log_3(x + 2)$

Asymptote: $x = -2$
Point on graph: $(-1, 0)$
Matches graph (d).

55. The graph of $g(x) = -\log_{10} x$ is a reflection in the x-axis of the graph of $f(x) = \log_{10} x$.

57. The graph of $g(x) = 4 - \log_2 x$ is a reflection in the x-axis followed by a vertical shift four units upward of the graph of $f(x) = \log_2 x$.

59. The graph of $g(x) = -2 + \log_8(x + 3)$ is a horizontal shift three units to the left and a vertical shift two units downward of the graph of $f(x) = \log_8 x$.

61. $\ln 1 = 0 \Rightarrow e^0 = 1$

63. $\ln e = 1 \Rightarrow e^1 = e$

65. $\ln \sqrt{e} = \frac{1}{2} \Rightarrow e^{1/2} = \sqrt{e}$

67. $\ln 9 = 2.1972\ldots \Rightarrow e^{2.1972\ldots} = 9$

69. $e^3 = 20.0855\ldots \Rightarrow \ln 20.0855\ldots = 3$

71. $e^{1.3} = 3.6692\ldots \Rightarrow \ln 3.6692\ldots = 1.3$

73. $\sqrt[3]{e} = 1.3956\ldots \Rightarrow \ln 1.3956\ldots = \frac{1}{3}$

75. $\sqrt{e^3} = 4.4816\ldots \Rightarrow e^{3/2} = 4.4816\ldots \Rightarrow \ln 4.4816\ldots = \frac{3}{2}$

77. $\ln \sqrt{42} \approx 1.869$

79. $-\ln\left(\frac{1}{2}\right) \approx 0.693$

81. $\ln e^2 = 2$
(Inverse Property)

83. $e^{\ln 1.8} = 1.8$
(Inverse Property)

85. $e \ln 1 = e(0) = 0$

87. $\ln e^{\ln e} = 1 \; (\ln e = 1)$

89. $f(x) = \ln(x - 1)$

Domain: $x > 1$
The domain is $(1, \infty)$.
Vertical asymptote: $x - 1 = 0 \Rightarrow x = 1$

x-intercept:

$$\ln(x - 1) = 0$$
$$x - 1 = e^0$$
$$x - 1 = 1$$
$$x = 2$$

The x-intercept is $(2, 0)$.

$y = \ln(x-1) \Rightarrow e^y + 1 = x$

x	1.02	1.14	2	8.39
y	–4	–2	0	2

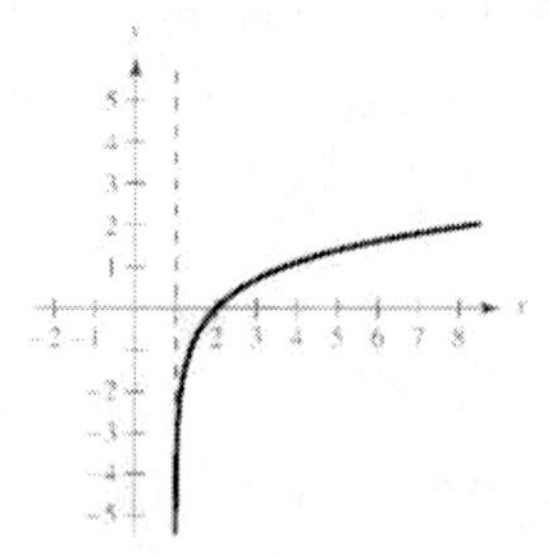

91. $g(x) = \ln(-x)$

Domain: $-x > 0 \Rightarrow x < 0$
The domain is $(-\infty, 0)$.

Vertical asymptote: $-x = 0 \Rightarrow x = 0$
x-intercept:

$$0 = \ln(-x)$$
$$e^0 = -x$$
$$-1 = x$$

The x-intercept is $(-1, 0)$.

$y = \ln(-x) \Rightarrow e^y = -x$

x	–0.14	–0.37	–1	–2.72	–7.39
y	–2	–1	0	1	2

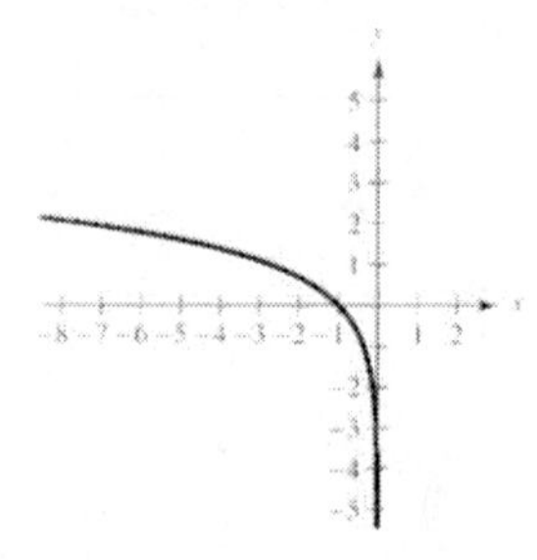

93. $g(x) = \ln(x+3)$ is a horizontal shift three units to the left.

95. $g(x) = \ln x - 5$ is a vertical shift five units downward.

97. $g(x) = \ln(x-1) + 2$ is a horizontal shift one unit to the right and a vertical shift two units upward.

99. $f(x) = \frac{x}{2} - \ln\frac{x}{4}$

(a)

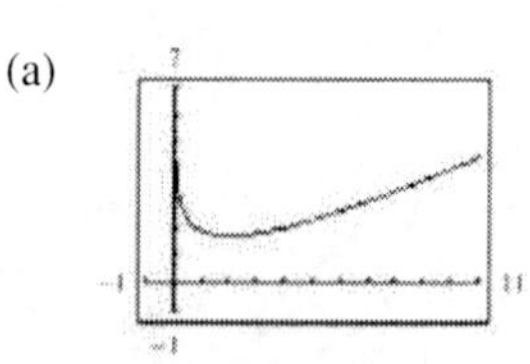

(b) Domain: $(0, \infty)$

(c) Increasing on $(2, \infty)$

Decreasing on $(0, 2)$

(d) Relative minimum: $(2, 1.693)$

101. $h(x) = 4x \ln x$

(a)

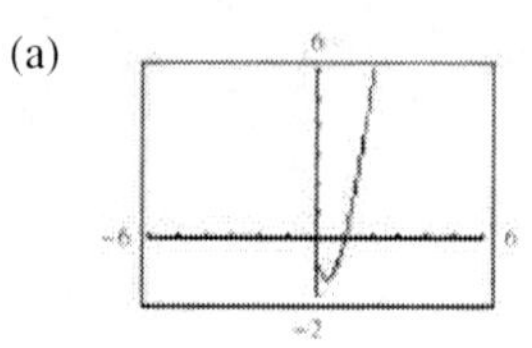

(b) Domain: $(0, \infty)$

(c) Increasing on $(0.368, \infty)$

Decreasing on $(0, 0.368)$

(d) Relative minimum: $(0.368, -1.472)$

103. $f(x) = \ln\left(\frac{x+2}{x-1}\right)$

(a)

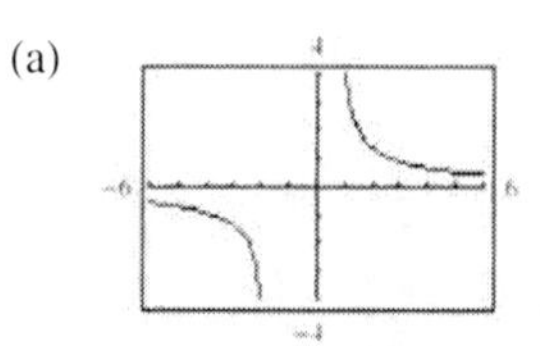

(b) $\frac{x+2}{x-1} > 0$; Critical numbers: $1, -2$

Test intervals: $(-\infty, -2), (-2, 1), (1, \infty)$

Testing these three intervals, we see that the domain is $(-\infty, -2) \cup (1, \infty)$.

(c) The graph is decreasing on $(-\infty, -2)$ and $(1, \infty)$.

(d) There are no relative maximum or minimum values.

105. $f(x) = \ln\left(\frac{x^2}{10}\right)$

(a)

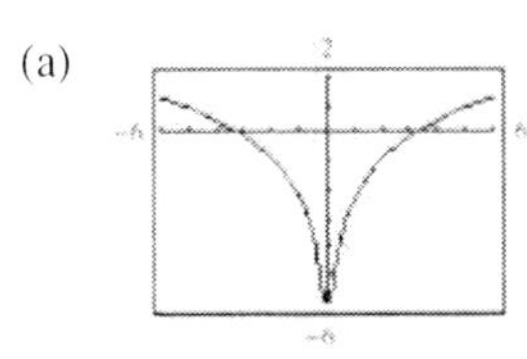

(b) $\frac{x^2}{10} > 0 \Rightarrow x \neq 0$; Domain: all $x \neq 0$

(c) The graph is increasing on $(0, \infty)$ and decreasing on $(-\infty, 0)$.

(d) There are no relative maximum or minimum values.

107. $f(x) = \sqrt{\ln x}$

(a)

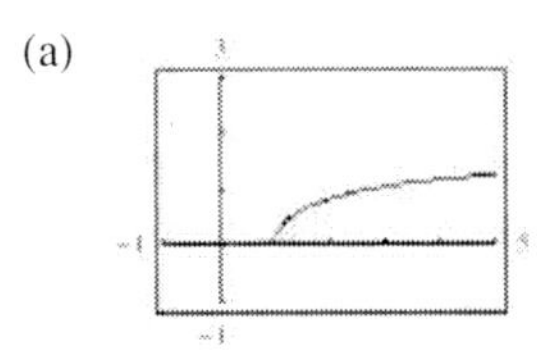

(b) $\ln x \geq 0 \Rightarrow x \geq 1$; Domain: $x \geq 1$

(c) The graph is increasing on $(1, \infty)$.

(d) Relative minimum: $(1, 0)$

109. $f(t) = 80 - 17\log_{10}(t+1),\ 0 \leq t \leq 12$

(a) $f(0) = 80 - 17\log_{10}(0+1) = 80$

(b) $f(4) = 80 - 17\log_{10}(4+1) \approx 68.1$

(c) $f(10) = 80 - 17\log_{10}(10+1) \approx 62.3$

(d)

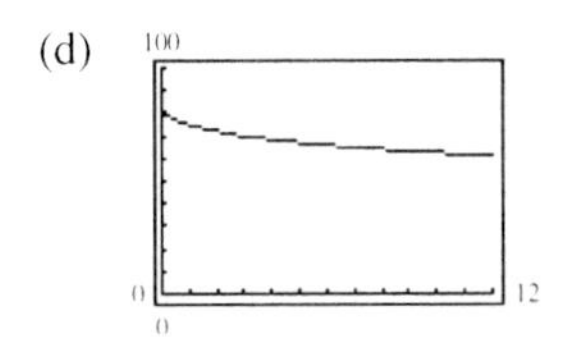

111. $t = \frac{\ln K}{0.055}$

(a)

K	1	2	4	6	8	10	12
t	0	12.6	25.2	32.6	37.8	41.9	45.2

As the amount increases, the time increases, but at a lesser rate.

(b)

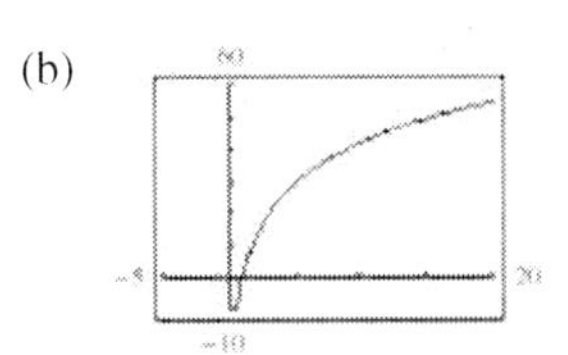

113. $t = 16.625\ln\left(\frac{x}{x - 750}\right),\ x > 750$

(a) $16.625\ln\left(\frac{897.72}{897.72 - 750}\right) \approx 30$ years

$16.625\ln\left(\frac{1659.24}{1659.24 - 750}\right) \approx 10$ years

(b) $(897.72)(30)(12) = 323{,}179.20$

$(1659.24)(10)(12) = 199{,}108.80$

Interest for 30-year loan is

$323{,}179.20 - 150{,}000 = \$173{,}179.20$.

Interest for 10-year loan is

$199{,}108.80 - 150{,}000 = \$49{,}108.80$.

115. False. You would reflect $y = 6^x$ in the line $y = x$.

117. $5 = \log_b 32$

$b^5 = 32 = 2^5$

$b = 2$

119. $2 = \log_b\left(\frac{1}{16}\right)$

$b^2 = \frac{1}{16} = \left(\frac{1}{4}\right)^2$

$b = \frac{1}{4}$

121. The vertical asymptote is to the right of the y-axis, and the graph increases. Matches (b).

123. $f(x) = \log_a x$ is the inverse function of $g(x) = a^x$, where $a > 0,\ a \neq 1$.

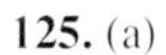

125. (a)

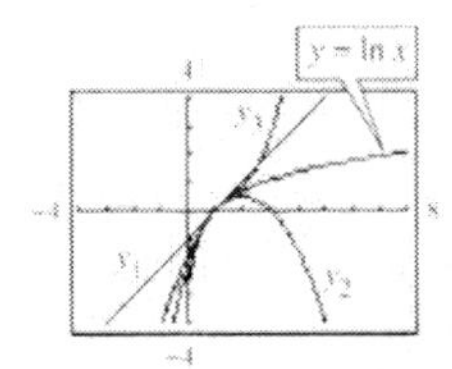

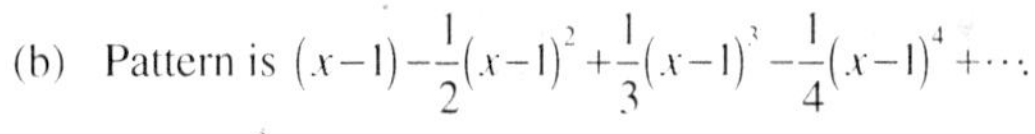

(b) Pattern is $(x-1) - \frac{1}{2}(x-1)^2 + \frac{1}{3}(x-1)^3 - \frac{1}{4}(x-1)^4 + \cdots$.

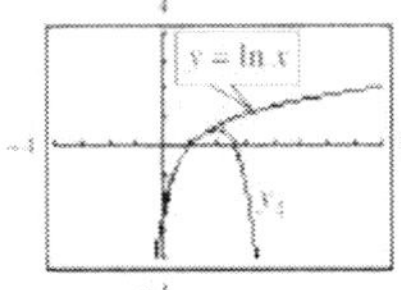

As you use more terms, the graph better approximates the graph of $\ln x$ on the interval $(0, 2)$.

127. $f(x) = \frac{\ln x}{x}$

(a)

x	1	5	10	10^2	10^4	10^6
$f(x)$	0	0.322	0.230	0.046	0.00092	0.0000138

(b) As x increases without bound, $f(x)$ approaches 0.

129. $x^2 + 2x - 3 = (x+3)(x-1)$

131. $12x^2 + 5x - 3 = (4x+3)(3x-1)$

133. $16x^2 - 25 = (4x+5)(4x-5)$

135. $2x^3 + x^2 - 45x = x(2x^2 + x - 45)$

$= x(2x-9)(x+5)$

137. $(f+g)(2) = f(2) + g(2) = [3(2)+2] + [2^3 - 1] = 8 + 7 = 15$

139. The graphs of $y_1 = 5x - 7$ and $y_2 = x + 4$ intersect when $x = 2.75$ or $\frac{11}{4}$.

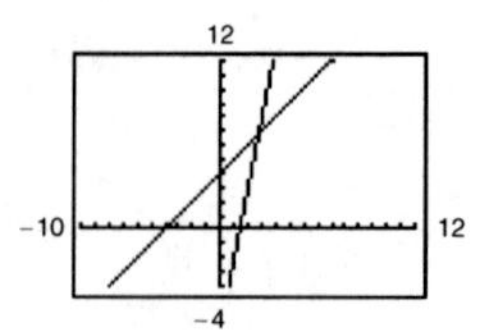

141. The graphs of $y_1 = \sqrt{3x-2}$ and $y_2 = 9$ intersect when $x \approx 27.667$ or $\frac{83}{3}$.

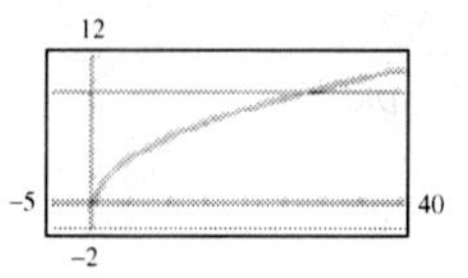

Section 3.3

1. change-of-base

3. Using the change of base formula:
$\log_3 24 = \frac{\ln 24}{\ln 3}$

5. (a) $\log_5 x = \frac{\log_{10} x}{\log_{10} 5}$

(b) $\log_5 x = \frac{\ln x}{\ln 5}$

7. (a) $\log_{1/5} x = \frac{\log_{10} x}{\log_{10} 1/5} = \frac{\log_{10} x}{-\log_{10} 5}$

(b) $\log_{1/5} x = \frac{\ln x}{\ln 1/5} = \frac{\ln x}{-\ln 5}$

9. (a) $\log_a\left(\frac{3}{10}\right) = \frac{\log_{10}\left(\frac{3}{10}\right)}{\log_{10} a}$

(b) $\log_a\left(\frac{3}{10}\right) = \frac{\ln\left(\frac{3}{10}\right)}{\ln a}$

11. (a) $\log_{2.6} x = \frac{\log_{10} x}{\log_{10} 2.6}$

(b) $\log_{2.6} x = \frac{\ln x}{\ln 2.6}$

13. $\log_3 7 = \frac{\ln 7}{\ln 3} \approx 1.771$

15. $\log_{1/2} 4 = \frac{\ln 4}{\ln(1/2)} = -2$

17. $\log_6 0.9 = \frac{\ln 0.9}{\ln 6} \approx -0.059$

19. $\log_{15} 1460 = \frac{\ln 1460}{\ln 15} \approx 2.691$

21. $\ln 20 = \ln(4 \cdot 5)$

$= \ln 4 + \ln 5$

23. $\ln\frac{25}{4} = \ln 25 - \ln 4$

$= \ln 5^2 - \ln 4$

$= 2\ln 5 - \ln 4$

25. $\log_b 25 = \log_b 5^2$

$= 2\log_b 5$

$\approx 2(0.8271) \approx 1.6542$

27. $\log_b \sqrt{3} = \frac{1}{2}\log_b 3$

$\approx \frac{1}{2}(0.5646)$

≈ 0.2823

29. $f(x) = \log_3(x+2) = \dfrac{\ln(x+2)}{\ln 3}$

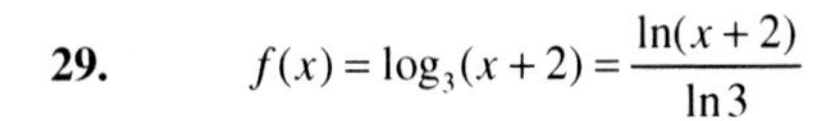

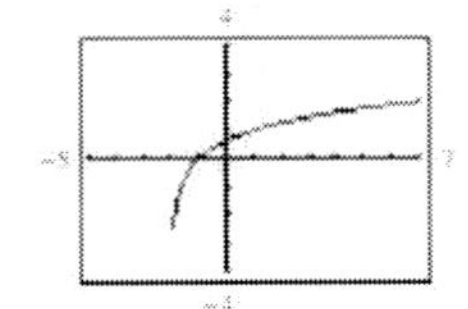

31. $f(x) = \log_{1/2}(x-2) = \dfrac{\ln(x-2)}{\ln(1/2)} = \dfrac{\ln(x-2)}{-\ln 2}$

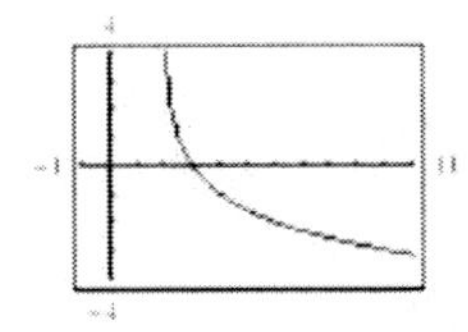

33. $f(x) = \log_{1/4}(x^2) = \dfrac{\ln x^2}{\ln(1/4)} = \dfrac{\ln x^2}{-\ln 4}$

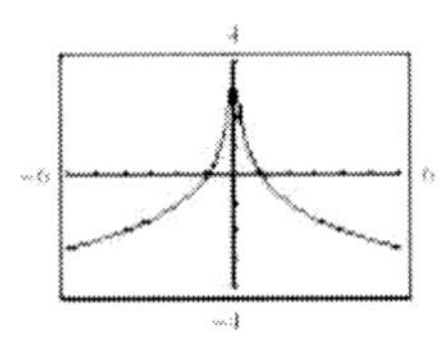

35. $f(x) = \log_{1/2}\left(\dfrac{x}{2}\right) = \dfrac{\ln(x/2)}{\ln(1/2)} = \dfrac{\ln(x/2)}{-\ln 2}$

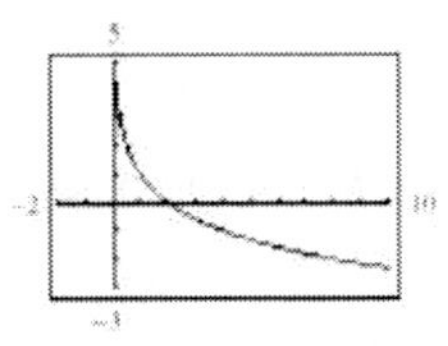

37. $\log_4 8 = \log_4 2^3 = 3\log_4 2$
$= 3\log_4 4^{1/2} = 3\left(\frac{1}{2}\right)\log_4 4$
$= \frac{3}{2}$

39. $\log_2(4^2 \cdot 3^4) = \log_2 4^2 + \log_2 3^4$
$= 2\log_2 4 + 4\log_2 3$
$= 2\log_2 2^2 + 4\log_2 3$
$= 4\log_2 2 + 4\log_2 3$
$= 4 + 4\log_2 3$

41. $\ln(5e^6) = \ln 5 + \ln e^6 = \ln 5 + 6 = 6 + \ln 5$

43. $\ln\dfrac{6}{e^2} = \ln 6 - \ln e^2$
$= \ln 6 - 2\ln e = \ln 6 - 2$

45. $\log_5 \frac{1}{250} = \log_5 1 - \log_5 250 = 0 - \log_5(125 \cdot 2)$
$= -\log_5(5^3 \cdot 2) = -[\log_5 5^3 + \log_5 2]$
$= -[3\log_5 5 + \log_5 2] = -3 - \log_5 2$

47. $\log_{10} 5x = \log_{10} 5 + \log_{10} x$

49. $\log_{10}\left(\dfrac{t}{8}\right) = \log_{10} t - \log_{10} 8$

51. $\log_8 x^4 = 4\log_8 x$

53. $\ln\sqrt{z} = \ln z^{1/2} = \dfrac{1}{2}\ln z$

55. $\ln xyz = \ln x + \ln y + \ln z$

57. $\log_6 ab^3c^2 = \log_6 a + \log_6 b^3 + \log_6 c^2$
$= \log_6 a + 3\log_6 b + 2\log_6 c$

59. $\ln\sqrt[3]{\dfrac{x}{y}} = \dfrac{1}{3}\ln\dfrac{x}{y}$
$= \dfrac{1}{3}\left[\ln x - \ln y\right]$
$= \dfrac{1}{3}\ln x - \dfrac{1}{3}\ln y$

61. $\ln\left(\dfrac{x^2-1}{x^3}\right) = \ln\left(\dfrac{(x+1)(x-1)}{x^3}\right)$
$= \ln(x+1) + \ln(x-1) - \ln x^3$
$= \ln(x+1) + \ln(x-1) - 3\ln x,\ x > 1$

63. $\ln\left(\dfrac{x^4\sqrt{y}}{z^5}\right) = \ln x^4\sqrt{y} - \ln z^5$
$= \ln x^4 + \ln\sqrt{y} - \ln z^5$
$= 4\ln x + \dfrac{1}{2}\ln y - 5\ln z$

65. $y_1 = \ln\left[x^3(x+4)\right]$

$y_2 = 3\ln x + \ln(x+4)$

(a)

(b)

x	0.5	1	1.5	2	3	10
y_1	−0.5754	1.6094	2.9211	3.8712	5.2417	9.5468
y_2	−0.5754	1.6094	2.9211	3.8712	5.2417	9.5468

(c) The graphs and table suggest that

$y_1 = y_2$ for $x > 0$. In fact,

$$y_1 = \ln\left[x^3(x+4)\right] = \ln x^3 + \ln(x+4)$$
$$= 3\ln x + \ln(x+4) = y_2.$$

67. $y_1 = \left(\dfrac{x^4}{x-2}\right)$, $y_2 = 4\ln x - \ln(x-2)$

(a)

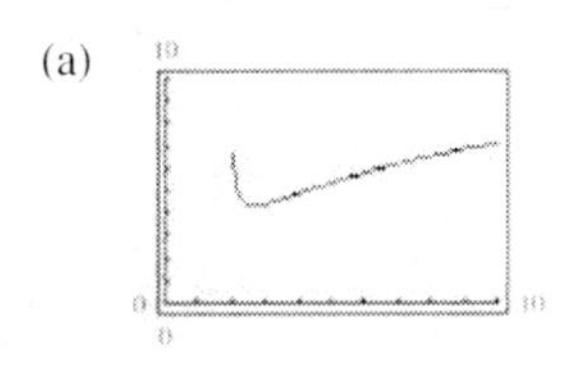

(b)

x	0	3	4	5
y_1	Error	4.39	4.85	5.34
y_2	Error	4.39	4.85	5.34

x	6	7	8
y_1	5.78	6.17	6.53
y_2	5.78	6.17	6.53

(c) The graphs and table suggest that $y_1 = y_2$.

In fact,

$$y_1 = \ln\left(\frac{x^4}{x-2}\right)$$
$$= \ln x^4 - \ln(x-2)$$
$$= 4\ln x - \ln(x-2)$$
$$= y_2.$$

69. $\ln x + \ln 4 = \ln 4x$

71. $\log_4 z - \log_4 y = \log_4 \dfrac{z}{y}$

73. $2\log_2(x+3) = \log_2(x+3)^2$

75. $\frac{1}{2}\ln(x^2+4) = \ln(x^2+4)^{1/2}$

$$= \ln\sqrt{x^2+4}$$

77. $\ln x - 3\ln(x+1) = \ln x - \ln(x+1)^3$

$$= \ln\frac{x}{(x+1)^3}$$

79. $\ln(x-2) - (x+2) = \ln\left(\dfrac{x-2}{x+2}\right)$

81. $\ln x - 2[\ln(x+2) + \ln(x-2)] = \ln x - 2\ln[(x+2)(x-2)]$

$$= \ln x - 2\ln(x^2-4)$$
$$= \ln x - \ln(x^2-4)^2$$
$$= \ln\frac{x}{(x^2-4)^2}$$

83. $\frac{1}{3}\left[2\ln(x+3) + \ln x(x^2-1)\right] = \frac{1}{3}\left[\ln(x+3)^2 + \ln x - \ln(x^2-1)\right]$

$$= \frac{1}{3}\left[\ln\left[x(x+3)^2\right] - \ln(x^2-1)\right]$$
$$= \frac{1}{3}\ln\frac{x(x+3)^2}{x^2-1}$$
$$= \ln\sqrt[3]{\frac{x(x+3)^2}{x^2-1}}$$

85. $y_1 = 2\left[\ln 8 - \ln(x^2+1)\right]$

$$y_2 = \ln\left[\frac{64}{(x^2+1)^2}\right]$$

(a)

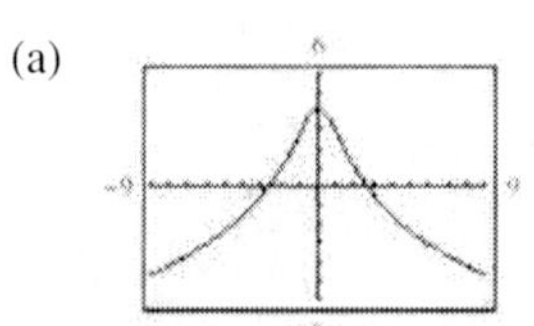

(b)

x	−8	−4	−2	0
y_1	−4.1899	−1.5075	0.9400	4.1589
y_2	−4.1899	−1.5075	0.9400	4.1589

x	2	4	8
y_1	0.9400	−1.5075	−4.1899
y_2	0.9400	−1.5075	−4.1899

(c) The graphs and table suggest that $y_1 = y_2$. In fact,

$$y_1 = 2\left[\ln 8 - \ln(x^2+1)\right]$$
$$= 2\ln\frac{8}{x^2+1} = \ln\frac{64}{(x^2+1)^2} = y_2.$$

87. $y_1 = \ln x + \frac{1}{2}\ln(x+1)$, $y_2 = \ln\left(x\sqrt{x+1}\right)$, $x > 0$

(a)

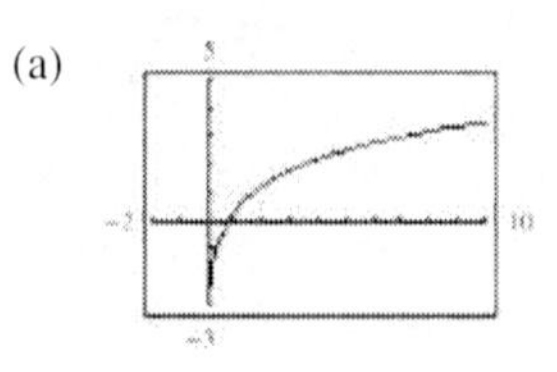

(b)

x	0	1	2	5	10
y_1	ERROR	0.34657	1.2425	2.5053	3.5015
y_2	ERROR	0.34657	1.2425	2.5053	3.5015

(c) The graphs and table suggest that $y_1 = y_2$. In fact,

$$y_1 = \ln x + \tfrac{1}{2}\ln(x+1) = \ln x + \ln(x+1)^{1/2}$$
$$= \ln\left[x\sqrt{x+1}\right] = y_2.$$

89. $y_1 = \ln x^2$

$y_2 = 2\ln x$

(a)

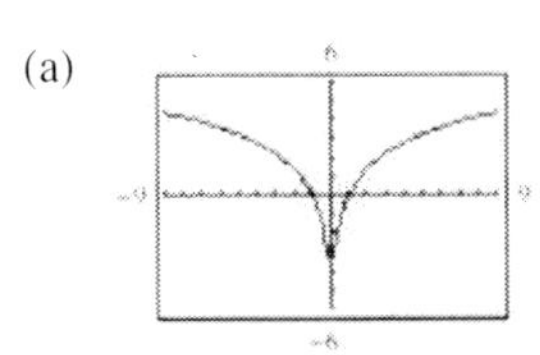

(The domain of y_2 is $x > 0$.)

(b)

x	−8	−4	1	2	4
y_1	4.1589	2.7726	0	1.3863	2.7726
y_2	Error	Error	0	1.3863	2.7726

(c) The graphs and table suggest that $y_1 = y_2$ for $x > 0$. The functions are not equivalent because the domains are different.

91. $y_1 = \ln(x-2) + \ln(x+2)$, $y_2 = \ln(x^2-4)$

(a)

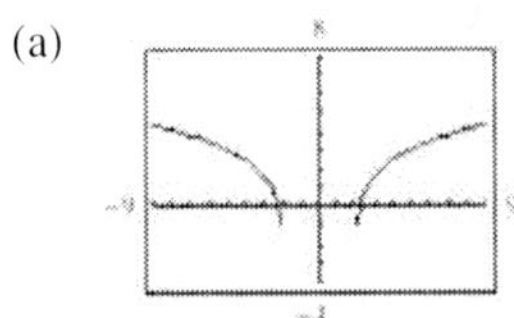

(b)

x	−4	−3	0	3
y_1	Error	Error	Error	1.61
y_2	2.48	1.61	Error	1.61

x	4	5	6
y_1	2.48	3.04	3.47
y_2	2.48	3.04	3.47

(c) The graphs and table suggest that $y_1 = y_2$ for $x > -2$.

Using the properties of logarithms,

$$y_1 = \ln(x-2) + \ln(x-2)$$
$$= \ln[(x-2)(x+2)]$$
$$= \ln(x^2-4)$$
$$= y_2 \text{ For } x > -2.$$

The domain of y_1 is all real x such that $x > -2$, however the domain of y_2 is all real x such that $x < -2$ or $x > 2$.

93. $\log_3 9 = 2\log_3 3 = 2$

95. $\log_4 16^{3.4} = 3.4\log_4(4^2) = 6.8\log_4 4 = 6.8$

97. $\log_2(-4)$ is undefined. -4 is not in the domain of $\log_2 x$.

99. $\log_5 75 - \log_5 3 = \log_5 \frac{75}{3} = \log_5 25 = \log_5 5^2 = 2$

101. $\ln e^3 - \ln e^7 = 3 - 7 = -4$

103. $2\ln e^4 = 2(4)\ln e = 8$

105. $\ln\left(\dfrac{1}{\sqrt{e}}\right) = \ln(1) - \ln e^{1/2} = 0 - \dfrac{1}{2}\ln e = -\dfrac{1}{2}$

107. $\beta = 10\log_{10}\left(\dfrac{I}{10^{-12}}\right)$

(a)
$$\beta = 10\left[\log_{10} I - \log_{10} 10^{-12}\right]$$
$$= 10\left[\log_{10} I - (-12)\right]$$
$$= 10\left[\log_{10} I + 12\right]$$
$$= 120 + 10\log_{10} I$$

(b)

I	10^{-4}	10^{-6}	10^{-8}	10^{-10}	10^{-12}	10^{-14}
β	80	60	40	20	0	−20

109. (a)

(b) $T - 21 = 54.4(0.964)^t$

$T = 21 + 54.4(0.964)^t$

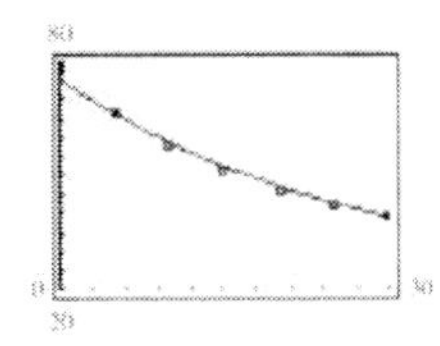

The data $(t, T-21)$ fits the model

$T - 21 = 54.4(0.964)^t$.

The model

$T = 21 + 54.4(0.964)^t$

fits the original data.

(c) $\ln(T - 21) = -0.0372t + 3.9971$, linear model

$$T - 21 = e^{-0.0372t+3.9971}$$
$$T = 21 + 54.4e^{-0.0372t}$$
$$= 21 + 54.4(0.964)^t$$

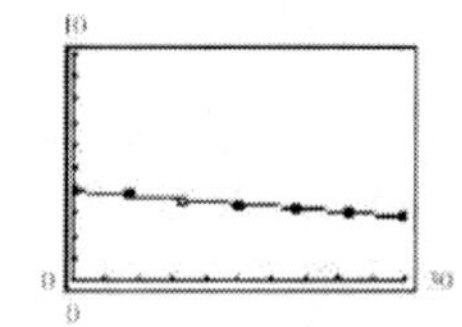

(d)

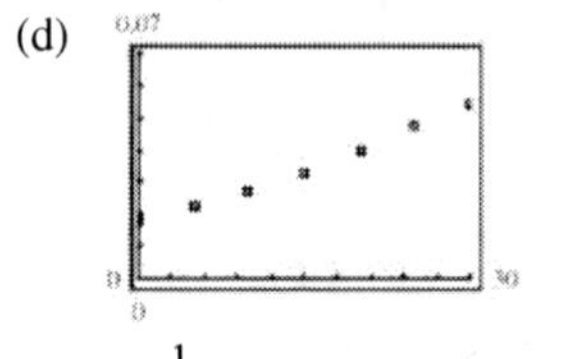

$$\frac{1}{T-21} = 0.00121t + 0.01615, \text{ linear model}$$

$$T - 21 = \frac{1}{0.00121t + 0.01615}$$
$$T = 21 + \frac{1}{0.00121t + 0.01615}$$

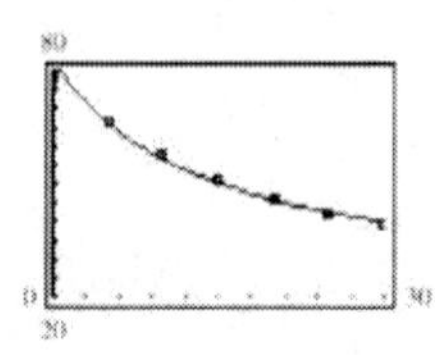

111. True. By the Product Property: $\ln(uv) = \ln u + \ln v$

113. False. $\sqrt{\ln x} \neq \frac{1}{2}\ln x$

In fact, $\ln x^{1/2} = \frac{1}{2}\ln x$.

115. True. In fact, if $\ln x < 0$, then $0 < x < 1$.

117. The error is an improper use of the Quotient Property of logarithms.

$$\ln\frac{x^2}{\sqrt{x^2+4}} = \ln x^2 - \ln\sqrt{x^2+4}$$
$$= 2\ln x - \ln(x^2+4)^{1/2}$$
$$= 2\ln x - \tfrac{1}{2}\ln(x^2+4)$$

119. The natural logarithms of 14 team integers between 1 and 20 can be approximated using $\ln 2$, $\ln 3$, and $\ln 5$.

$\ln 1 = 0$, $\ln 2 \approx 0.6931$, $\ln 3 \approx 1.0986$, $\ln 5 \approx 1.6094$

$\ln 2 \approx 0.6931$

$\ln 3 \approx 1.0986$

$\ln 4 = \ln 2 + \ln 2 \approx 0.6931 + 0.6931 = 1.3862$

$\ln 5 \approx 1.6094$

$\ln 6 = \ln 2 + \ln 3 \approx 0.6931 + 1.0986 = 1.7917$

$\ln 8 = \ln 2^3 = 3\ln 2 \approx 3(0.6931) = 2.0793$

$\ln 9 = \ln 3^2 = 2\ln 3 \approx 2(1.0986) = 2.1972$

$\ln 10 = \ln 5 + \ln 2 \approx 1.6094 + 0.6931 = 2.3025$

$\ln 12 = \ln 2^2 + \ln 3 = 2\ln 2 + \ln 3 \approx 2(0.6931) + 1.0986 = 2.4848$

$\ln 15 = \ln 5 + \ln 3 \approx 1.6094 + 1.0986 = 2.7080$

$\ln 16 = \ln 2^4 = 4\ln 2 \approx 4(0.6931) = 2.7724$

$\ln 18 = \ln 3^2 + \ln 2 = 2\ln 3 + \ln 2 \approx 2(1.0986) + 0.6931 = 2.8903$

$\ln 20 = \ln 5 + \ln 2^2 = \ln 5 + 2\ln 2 \approx 1.6094 + 2(0.6931) = 2.9956$

121. No. The domains are not the same.

The domain of $y_1 = \ln[x(x-2)]$ is $(-\infty, 0)$, $(2, \infty)$. This can be found by solving the quadratic inequality, $x(x-2) > 0$.

The domain of $y_2 = \ln x + \ln(x-2)$ is $(2, \infty)$. This can be found by the intersection of the intervals $(0, \infty)$ and $(2, \infty)$, the domains of each term respectively.

123. $\dfrac{24xy^{-2}}{16x^{-3}y} = \dfrac{24xx^3}{16yy^2} = \dfrac{3x^4}{2y^3}$

125.
$$(18x^3y^4)^{-3}(18x^3y^4)^4 = \frac{(18x^3y^4)^4}{(18x^3y^4)^3}$$
$$= 18x^3y^4$$

127.
$$x^2 - 6x + 2 = 0$$
$$x = \frac{6 \pm \sqrt{36-4(2)}}{2} = 3 \pm \sqrt{7}$$

129.
$$x^4 - 19x^2 + 48 = 0$$
$$(x^2-16)(x^2-3) = 0$$
$$(x-4)(x+4)\left(x-\sqrt{3}\right)\left(x+\sqrt{3}\right) = 0$$
$$x = \pm 4, \ \pm\sqrt{3}$$

Section 3.4

1. (a) $x = y$
(b) $x = y$
(c) x
(d) x

3. $\ln e^7 = 7$ because of the Inverse Property.

5. To solve $3 + \ln x = 10$, isolate the logarithmic term by subtracting 3 from each side of the equation.

7. $4^{2x-7} = 64$

(a) $x = 5$
$4^{2(5)-7} = 4^3 = 64$
Yes, $x = 5$ is a solution.

(b) $x = 2$
$4^{2(2)-7} = 4^{-3} = \frac{1}{64} \neq 64$
No, $x = 2$ is not a solution.

9. $3e^{x+2} = 75$

(a) $x = -2 + e^{25}$
$3e^{(-2+e^{25})+2} = 3e^{e^{25}} \neq 75$
No, $x = -2 + e^{25}$ is not a solution.

(b) $x = -2 + \ln 25$
$3e^{(-2+\ln 25)+2} = 3e^{\ln 25} = 3(25) = 75$
Yes, $x = -2 + \ln 25$ is a solution.

(c) $x \approx 1.2189$
$3e^{1.2189+2} = 3e^{3.2189} \approx 75$
Yes, $x \approx 1.2189$ is an approximate solution.

11. $\log_4(3x) = 3$

$$4^3 = 3x$$
$$x = \frac{64}{3} \approx 21.333$$

(a) Yes, $x \approx 21.3560$ is an approximate solution.
(b) No, $x = -4$ is not a solution.
(c) Yes, $x = \frac{64}{3}$ is a solution.

13. $\ln(x - 1) = 3.8$

(a) $x = 1 + e^{3.8}$
$\ln(1 + e^{3.8} - 1) = \ln e^{3.8} = 3.8$
Yes, $x = 1 + e^{3.8}$ is a solution.

(b) $x \approx 45.7012$
$\ln(45.7012 - 1) = \ln(44.7012) \approx 3.8$
Yes, $x \approx 45.7012$ is an approximate solution.

(c) $x = 1 + \ln 3.8$
$\ln(1 + \ln 3.8 - 1) = \ln(\ln 3.8) \approx 0.289$
No, $x = 1 + \ln 3.8$ is not a solution.

15. $f(x) = 2^x$, $g(x) = 8$

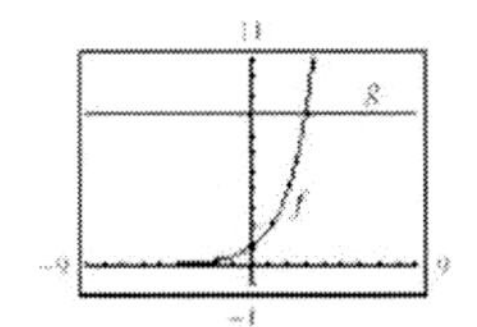

Point of intersection: $(3, 8)$

Algebraically:
$$2^x = 8$$
$$2^x = 2^3$$
$$x = 3$$

17. $f(x) = 5^{x-2} - 15$

$g(x) = 10$

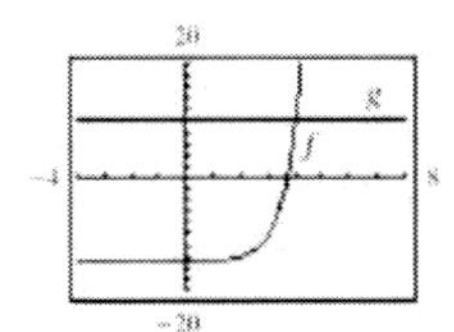

Point of intersection: $(4, 10)$

Algebraically:
$$5^{x-2} - 15 = 10$$
$$5^{x-2} = 25 = 5^2$$
$$x - 2 = 2$$
$$x = 4$$

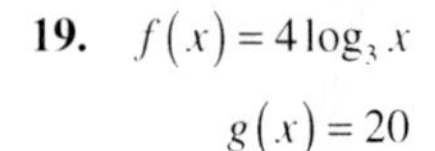

19. $f(x) = 4\log_3 x$

$g(x) = 20$

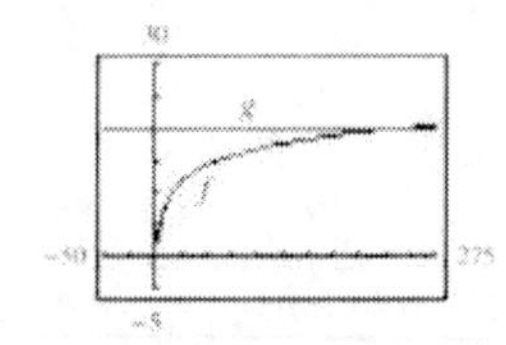

Point of intersection: $(243, 20)$

Algebraically: $4\log_3 x = 20$

$\log_3 x = 5$

$x = 3^5 = 243$

21. $f(x) = \ln e^{x+1}$

$g(x) = 2x + 5$

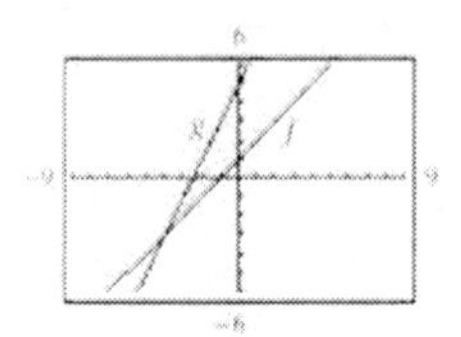

Point of intersection: $(-4, -3)$

Algebraically: $\ln e^{x+1} = 2x + 5$

$x + 1 = 2x + 5$

$-4 = x$

23. $4^x = 16$

$4^x = 4^2$

$x = 2$

25. $5^x = \dfrac{1}{625}$

$5^x = \dfrac{1}{5^4} = 5^{-4}$

$x = -4$

27. $\left(\dfrac{1}{8}\right)^x = 64$

$8^{-x} = 8^2$

$-x = 2$

$x = -2$

29. $\left(\dfrac{2}{3}\right)^x = \dfrac{81}{16}$

$\left(\dfrac{3}{2}\right)^{-x} = \left(\dfrac{3}{2}\right)^4$

$-x = 4$

$x = -4$

31. $e^x = 14$

$\ln e^x = \ln 14$

$x = \ln 14$

33. $6(10^x) = 216$

$10^x = 36$

$\log_{10} 10^x = \log_{10} 36$

$x = \log_{10} 36 \approx 1.5563$

35. $2^{x+3} = 256$

$2^x \cdot 2^3 = 256$

$2^x = 32$

$x = 5$

Alternate solution:

$2^{x+3} = 2^8$

$x + 3 = 8$

$x = 5$

37. $\ln x - \ln 5 = 0$

$\ln x = \ln 5$

$x = 5$

39. $\ln x = -9$

$e^{\ln x} = e^{-9}$

$x = e^{-9} = e^{1\backslash 9}$

41. $\log_x 625 = 4$

$x^4 = 625$

$x^4 = 5^4$

$x = 5$

43. $\log_{10} x = -1$

$x = 10^{-1}$

$x = \dfrac{1}{10}$

45. $\ln(2x - 1) = 5$

$2x - 1 = e^5$

$x = \dfrac{1 + e^5}{2} \approx 74.707$

47. $\ln e^{x^2} = x^2 \ln e = x^2$

49. $e^{\ln x^2} = x^2$

51. $-1 + \ln e^{2x} = -1 + 2x = 2x - 1$

53. $5 + e^{\ln(x^2+1)} = 5 + x^2 + 1$

$= x^2 + 6$

55.
$$8^{3x} = 360$$
$$\ln 8^{3x} = \ln 360$$
$$3x \ln 8 = \ln 360$$
$$3x = \frac{\ln 360}{\ln 8}$$
$$x = \frac{1}{3}\frac{\ln 360}{\ln 8}$$
$$x \approx 0.944$$

57.
$$5^{-t/2} = 0.20 = \frac{1}{5}$$
$$-\frac{t}{2}\ln 5 = \ln\left(\frac{1}{5}\right)$$
$$-\frac{t}{2}\ln 5 = -\ln 5$$
$$\frac{t}{2} = 1$$
$$t = 2$$

59.
$$250e^{0.02x} = 10{,}000$$
$$e^{0.02x} = 40$$
$$0.02x = \ln 40$$
$$x = \frac{\ln 40}{0.02}$$
$$x \approx 184.444$$

61.
$$500e^{-x} = 300$$
$$e^{-x} = \frac{3}{5}$$
$$-x = \ln\frac{3}{5}$$
$$x = -\ln\frac{3}{5} = \ln\frac{5}{3} \approx 0.511$$

63.
$$7 - 2e^x = 5$$
$$-2e^x = -2$$
$$e^x = 1$$
$$x = \ln 1 = 0$$

65.
$$5(2^{3-x}) - 13 = 100$$
$$5(2^{3-x}) = 113$$
$$2^{3-x} = \frac{113}{5}$$
$$\ln 2^{3-x} = \ln\left(\frac{113}{5}\right)$$
$$3 - x = \frac{\ln(113/5)}{\ln 2}$$
$$x = 3 - \frac{\ln(113/5)}{\ln 2}$$
$$x \approx -1.498$$

67.
$$\left(1 + \frac{0.10}{12}\right)^{12t} = 2$$
$$\left(\frac{12.1}{12}\right)^{12t} = 2$$
$$(12t)\ln\left(\frac{12.1}{12}\right) = \ln 2$$
$$t = \frac{1}{12}\frac{\ln 2}{\ln(12.1/12)}$$
$$t \approx 6.960$$

69.
$$5000\left[\frac{(1 + 0.005)^x}{0.005}\right] = 250{,}000$$
$$5000(1.005)^x = 1250$$
$$1.005^x = 0.25$$
$$x\ln(1.005) = \ln 0.25$$
$$x = \frac{\ln 0.25}{\ln(1.005)}$$
$$x \approx -277.951$$

71.
$$e^{2x} - 4e^x - 5 = 0$$
$$(e^x - 5)(e^x + 1) = 0$$
$$e^x = 5 \text{ or } e^x = -1$$
$$x = \ln 5 \approx 1.609$$
$(e^x = -1$ is impossible.)

73.
$$e^x = e^{x^2 - 2}$$
$$x = x^2 - 2$$
$$x^2 - x - 2 = 0$$
$$(x - 2)(x + 1) = 0$$
$$x = 2,\ -1$$

75.
$$e^{x^2 - 3x} = e^{x - 2}$$
$$x^2 - 3x = x - 2$$
$$x^2 - 4x + 2 = 0$$
$$x = \frac{4 \pm \sqrt{16 - 8}}{2}$$
$$x = 2 \pm \sqrt{2}$$
$$x \approx 3.414,\ 0.586$$

77.
$$\frac{400}{1 + e^{-x}} = 350$$
$$1 + e^{-x} = \frac{400}{350} = \frac{8}{7}$$
$$e^{-x} = \frac{1}{7}$$
$$-x = \ln\left(\frac{1}{7}\right) = -\ln 7$$
$$x = \ln 7 \approx 1.946$$

79. $\dfrac{40}{1-5e^{-0.01x}} = 200$

$$1-5e^{-0.01x} = \frac{40}{200} = \frac{1}{5}$$

$$5e^{-0.01x} = \frac{4}{5}$$

$$e^{-0.01x} = \frac{4}{25}$$

$$-0.01x = \ln\left(\frac{4}{25}\right)$$

$$x = \frac{\ln(4/25)}{-0.01}$$

$$x \approx 183.258$$

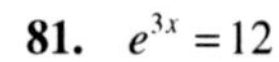

81. $e^{3x} = 12$

(a)

x	0.6	0.7	0.8	0.9	1.0
e^{3x}	6.05	8.17	11.02	14.88	20.09

Using the table, a solution to the equation must lie in (0.8, 0.9).

(b)

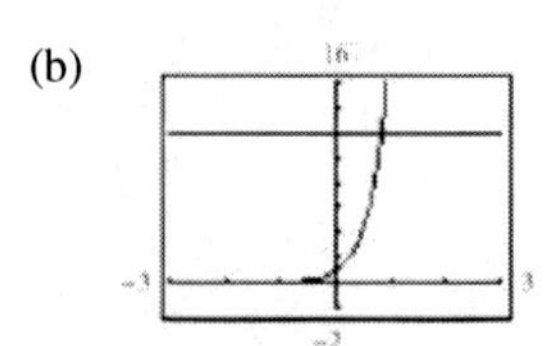

(c)

$$e^{3x} = 12$$

$$\ln e^{3x} = \ln 12$$

$$3x = \ln 12$$

$$x = \tfrac{1}{3}\ln 12 \approx 0.828$$

83. $\left(1+\dfrac{0.065}{365}\right)^{365t} = 4$

$$1.000178^{365t} = 4$$

The zero of $y = 1.000178^{365t} - 4$ is $t \approx 21.330$.

85. $\dfrac{3000}{2+e^{2x}} = 2$

The zero of $y = \dfrac{3000}{2+e^{2x}} - 2$ is $x \approx 3.656$.

87. $g(x) = 6e^{1-x} - 25$

Zero at $x \approx -0.427$

89. $g(t) = e^{0.09t} - 3$

Zero at $t \approx 12.207$

91. $\ln x = -3$

$$x = e^{-3} \approx 0.050$$

93. $\ln 4x = 2.1$

$$4x = e^{2.1}$$

$$x = \frac{1}{4}e^{2.1} \approx 2.042$$

95. $\log_5(3x+2) = \log_5(6-x)$

$$3x+2 = 6-x$$

$$4x = 4$$

$$x = 1$$

97. $-2+2\ln 3x = 17$

$$2\ln 3x = 19$$

$$\ln 3x = \frac{19}{2}$$

$$3x = e^{19/2}$$

$$x = \frac{1}{3}e^{19/2}$$

$$x \approx 4453.242$$

99. $7\log_4(0.6x) = 12$

$$\log_4(0.6x) = \frac{12}{7}$$

$$4^{12/7} = 0.6x = \frac{3}{5}x$$

$$x = \frac{5}{3}4^{12/7}$$

$$\approx 17.945$$

101. $\log_{10}(z-3) = 2$

$$10^{\log_{10}(z-3)} = 10^2$$

$$z-3 = 100$$

$$z = 103$$

103. $\ln\sqrt{x+2} = 1$

$$\sqrt{x+2} = e^1$$

$$x+2 = e^2$$

$$x = e^2 - 2 \approx 5.389$$

105. $\ln(x+1)^2 = 2$

$$e^{\ln(x+1)^2} = e^2$$
$$(x+1)^2 = e^2$$

$x+1 = e$ or $x+1 = -e$

$x = e-1 \approx 1.718$

or

$x = -e-1 \approx -3.718$

107. $\log_4 x - \log_4(x-1) = \dfrac{1}{2}$

$$\log_4\left(\frac{x}{x-1}\right) = \frac{1}{2}$$
$$4^{\log_4(x/x-1)} = 4^{1/2}$$
$$\frac{x}{x-1} = 2$$
$$x = 2(x-1)$$
$$x = 2x-2$$
$$2 = x$$

109. $\ln(x+5) = \ln(x-1) - \ln(x+1)$

$$\ln(x+5) = \ln\left(\frac{x-1}{x+1}\right)$$
$$x+5 = \frac{x-1}{x+1}$$
$$(x+5)(x+1) = x-1$$
$$x^2+6x+5 = x-1$$
$$x^2+5x+6 = 0$$
$$(x+2)(x+3) = 0$$

$x = -2$ or $x = -3$

Both of these solutions are extraneous, so the equation has no solution.

111. $\log_{10} 8x - \log_{10}\left(1+\sqrt{x}\right) = 2$

$$\log_{10}\frac{8x}{1+\sqrt{x}} = 2$$
$$\frac{8x}{1+\sqrt{x}} = 10^2$$
$$8x = 100 + 100\sqrt{x}$$
$$8x - 100\sqrt{x} - 100 = 0$$
$$2x - 25\sqrt{x} - 25 = 0$$
$$\sqrt{x} = \frac{25 \pm \sqrt{25^2 - 4(2)(-25)}}{4}$$
$$= \frac{25 \pm 5\sqrt{33}}{4}$$

Choosing the positive value, we have $\sqrt{x} \approx 13.431$ and $x \approx 180.384$.

113. $y = \log_a x,\ a^y = x$

$$\log_e a^y = \log_e x$$
$$y\log_e a = \log_e x$$
$$y = \frac{\log_e x}{\log_e a}$$
$$\log_a x = \frac{\log_e x}{\log_e a} = \frac{\ln x}{\ln a}$$

115. $\ln 2x = 2.4$

(a)

x	2	3	4	5	6
$\ln 2x$	1.39	1.79	2.08	2.30	2.48

Using the table, a solution must fall within the interval (5, 6).

(b)

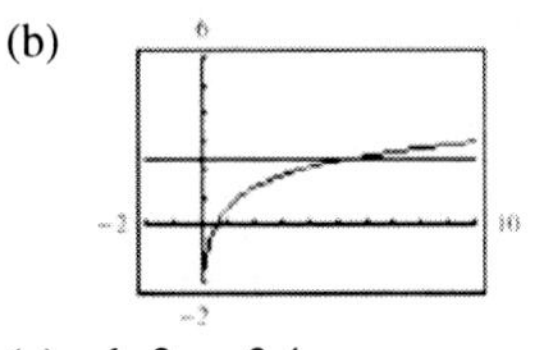

(c) $\ln 2x = 2.4$

$$e^{\ln 2x} = e^{2.4}$$
$$2x = e^{2.4}$$
$$x = \frac{1}{2}e^{2.4} \approx 5.512$$

117. $6\log_3(0.5x) = 11$

(a)

x	12	13	14	15	16
$6\log_3(0.5x)$	9.79	10.22	10.63	11.00	11.36

Using the table, a solution must fall within the interval $(14, 15)$.

(b)

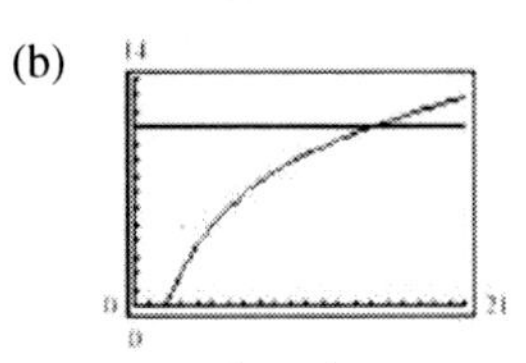

(c) $6\log_3(0.5x) = 11$

$$\log_3(0.5x) = \frac{11}{6}$$
$$3^{\log_3(0.5x)} = 3^{11/6}$$
$$0.5x = 3^{11/6}$$
$$x = 2(3^{11/6}) \approx 14.988$$

119. $\log_{10} x = x^3 - 3$

Graphing $y = \log_{10} x - x^3 + 3$, you obtain two zeros, $x \approx 1.469$ and $x \approx 0.001$.

121. $\ln x + \ln(x - 2) = 1$

Graphing $y = \ln x + \ln(x - 2) - 1$, you obtain one zero, $x \approx 2.928$.

123. $\ln(x - 3) + \ln(x + 3) = 1$

Graphing $y = \ln(x - 3) + \ln(x + 3) - 1$, you obtain one zero, $x \approx 3.423$.

125. $y_1 = 7$

$y_2 = 2^{x-1} - 5$

Intersection: $(4.585, 7)$

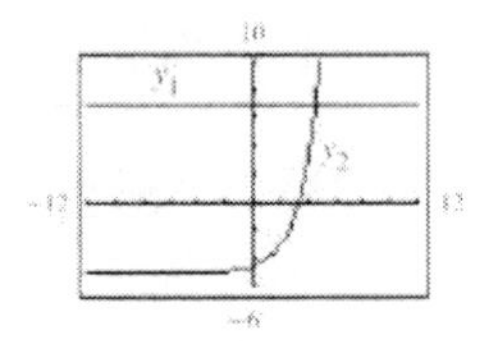

127. $y_1 = 80$

$y_2 = 4e^{-0.2x}$

Intersection: $(-14.979, 80)$

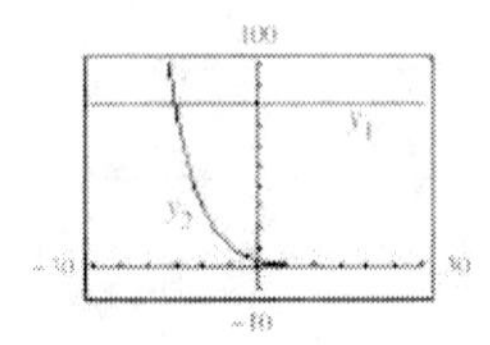

129. $y_1 = 3.25$

$y_2 = \frac{1}{2}\ln(x + 2)$

Intersection: $(663.142, 3.25)$

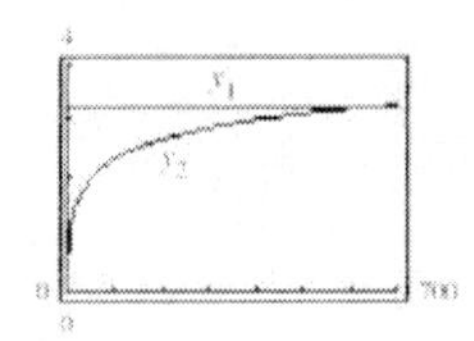

131. $2x^2e^{2x} + 2xe^{2x} = 0$

$$(2x^2 + 2x)e^{2x} = 0$$

$$2x^2 + 2x = 0 \quad (\text{since } e^{2x} \neq 0)$$

$$2x(x + 1) = 0$$

$$x = 0, -1$$

133. $-xe^{-x} + e^{-x} = 0$

$$(-x + 1)e^{-x} = 0$$

$$-x + 1 = 0 \quad (\text{since } e^{-x} \neq 0)$$

$$x = 1$$

135. $2x\ln x + x = 0$

$$x(2\ln x + 1) = 0$$

$$2\ln x + 1 = 0 \quad (\text{since } x > 0)$$

$$\ln x = -\tfrac{1}{2}$$

$$x = e^{-1/2} \approx 0.607$$

137. $\dfrac{1 + \ln x}{2} = 0$

$$1 + \ln x = 0$$

$$\ln x = -1$$

$$x = e^{-1} = \frac{1}{e} \approx 0.368$$

139. Exponential growth

$$y = ae^{bx}$$

$$\frac{y}{a} = e^{bx}$$

$$\ln\left(\frac{y}{a}\right) = \ln e^{bx}$$

$$\ln\left(\frac{y}{a}\right) = bx$$

$$\frac{1}{b}\ln\left(\frac{y}{a}\right) = x$$

$$\text{or } x = \frac{1}{b}(\ln y - \ln a)$$

141. Gaussian model

$$y = ae^{-(x-b)^2/c}$$

$$\frac{y}{a} = e^{-(x-b)^2/c}$$

$$\ln\left(\frac{y}{a}\right) = \ln e^{-(x-b)^2/c}$$

$$\ln\left(\frac{y}{a}\right) = -\frac{(x-b)^2}{c}$$

$$c\ln\left(\frac{y}{a}\right) = -(x-b)^2$$

$$c(\ln y - \ln a) = -(x-b)^2$$

$$-c(\ln y - \ln a) = (x-b)^2$$

$$c(\ln a - \ln y) = (x-b)^2$$

$$\pm\sqrt{c(\ln a - \ln y)} = x - b$$

$$b \pm \sqrt{c(\ln a - \ln y)} = x$$

143. (a)

$$\begin{aligned}2000 &= 1000e^{0.075t}\\ 2 &= e^{0.075t}\\ \ln 2 &= 0.075t\\ t &= \frac{\ln 2}{0.075} \approx 9.24 \text{ years}\end{aligned}$$

(b)

$$\begin{aligned}3000 &= 1000e^{0.075t}\\ 3 &= e^{0.075t}\\ \ln 3 &= 0.075t\\ t &= \frac{\ln 3}{0.075} \approx 14.65 \text{ years}\end{aligned}$$

145. (a)

$$\begin{aligned}2000 &= 1000e^{0.025t}\\ 2 &= e^{0.025t}\\ \ln 2 &= 0.025t\\ t &= \frac{\ln 2}{0.025} \approx 27.73 \text{ years}\end{aligned}$$

(b)

$$\begin{aligned}3000 &= 1000e^{0.025t}\\ 3 &= e^{0.025t}\\ \ln 3 &= 0.025t\\ t &= \frac{\ln 3}{0.025} \approx 43.94 \text{ years}\end{aligned}$$

147. $P = 5000\left(1 - \frac{4}{4 + e^{-0.002x}}\right)$

(a) When $P = \$300$:

$$\begin{aligned}300 &= 5000\left(1 - \frac{4}{4 + e^{-0.002x}}\right)\\ \frac{3}{50} &= 1 - \frac{4}{4 + e^{-0.002x}}\\ -\frac{47}{50} &= -\frac{4}{4 + e^{-0.002x}}\\ 47(4 + e^{-0.002x}) &= 4(50)\\ 188 + 47e^{-0.002x} &= 200\\ 47e^{-0.002x} &= 12\\ e^{-0.002x} &= \frac{12}{47}\\ \ln e^{-0.002x} &= \ln\left(\frac{12}{47}\right)\\ -0.002x &= \ln\left(\frac{12}{47}\right)\\ x &= -\frac{1}{0.002}\ln\left(\frac{12}{47}\right) \approx 6.82 \text{ units}\end{aligned}$$

(b) When $P = \$250$:

$$\begin{aligned}250 &= 5000\left(1 - \frac{4}{4 + e^{-0.002x}}\right)\\ \frac{1}{20} &= 1 - \frac{4}{4 + e^{-0.002x}}\\ -\frac{19}{20} &= -\frac{4}{4 + e^{-0.002x}}\\ 19\left(4 + e^{-0.002x}\right) &= 4(20)\\ 76 + 19e^{-0.002x} &= 80\\ 19e^{-0.002x} &= 4\\ e^{-0.002x} &= \tfrac{4}{19}\\ \ln e^{-0.002x} &= \ln\left(\tfrac{4}{19}\right)\\ -0.002x &= \ln\left(\tfrac{4}{19}\right)\\ x &= -\frac{1}{0.002x}\ln\left(\tfrac{4}{19}\right) \approx 779 \text{ units}\end{aligned}$$

149. $y = 13{,}107 - 2077.6\ln t$

To determine when there were about 7100 commercial banks, let $y = 7100$ and solve for t.

$$\begin{aligned}7100 &= 13{,}107 - 2077.6\ln t\\ -6007 &= -2077.6\ln t\\ \frac{6007}{2077.6} &= \ln t\\ e^{\frac{6007}{2077.6}} &= e^{\ln t}\\ e^{\frac{6007}{2077.6}} &= t\\ 18.0 &\approx t\end{aligned}$$

or 2008

151. $T = 20\left[1 + 7\left(2^{-h}\right)\right]$

(a)

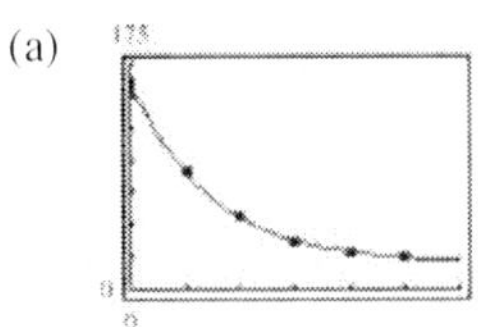

(b) We see a horizontal asymptote at $y = 20$. This represents the room temperature.

(c) $100 = 20[1 + 7(2^{-h})]$

$$5 = 1 + 7(2^{-h})$$
$$4 = 7(2^{-h})$$
$$\frac{4}{7} = 2^{-h}$$
$$\ln\left(\frac{4}{7}\right) = \ln 2^{-h}$$
$$\ln\left(\frac{4}{7}\right) = -h \ln 2$$
$$\frac{\ln(4/7)}{-\ln 2} = h$$
$$h \approx 0.81 \text{ hour}$$

153. False. The equation $e^x = 0$ has no solutions.

155. The error is that both sides of the equation should be divided by 2, before taking the natural log.

$$2e^x = 10$$
$$e^x = 5$$
$$\ln e^x = \ln 5$$
$$x = \ln 5$$

157. To solve $5^x = 34$, the Inverse Property should be used. First, take the natural log of both sides, then solve for x.

$$5^x = 34$$
$$\ln 5^x = \ln 34$$
$$x \ln 5 = \ln 34$$
$$x = \frac{\ln 34}{\ln 5}$$

159. Yes. The doubling time is given by

$$2P = Pe^{rt}$$
$$2 = e^{rt}$$
$$\ln 2 = rt$$
$$t = \frac{\ln 2}{r}.$$

The time to quadruple is given by

$$4P = Pe^{rt}$$
$$4 = e^{rt}$$
$$\ln 4 = rt$$
$$t = \frac{\ln 4}{r} = \frac{\ln 2^2}{r} = \frac{2 \ln 2}{r} = 2\left[\frac{\ln 2}{r}\right]$$

which is twice as long.

161. $f(x) = 3x^3 - 4$

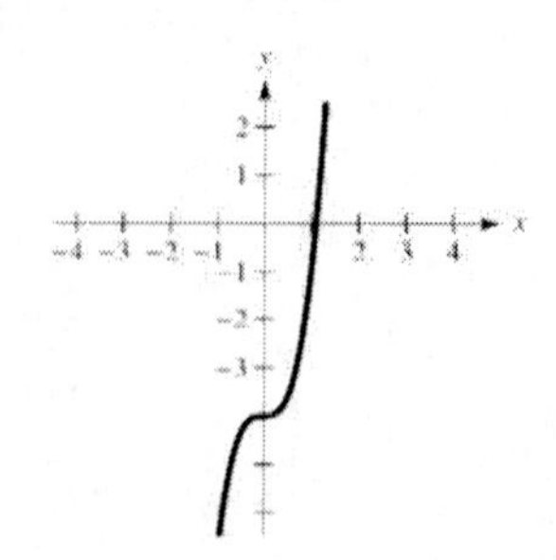

163. $f(x) = |x| + 9$

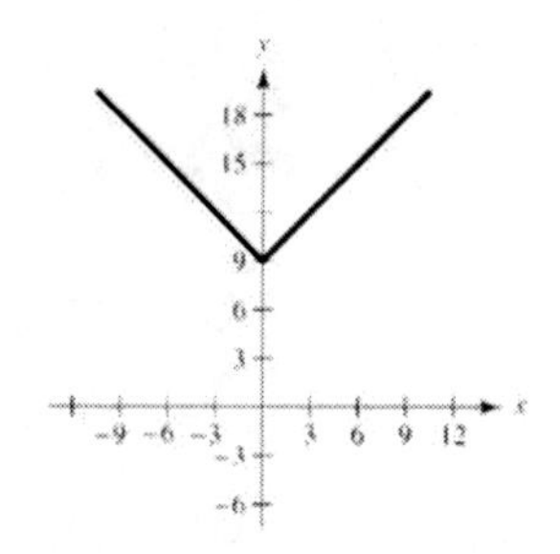

165. $f(x) = \begin{cases} 2x, & x < 0 \\ -x^2 + 4, & x \ge 0 \end{cases}$

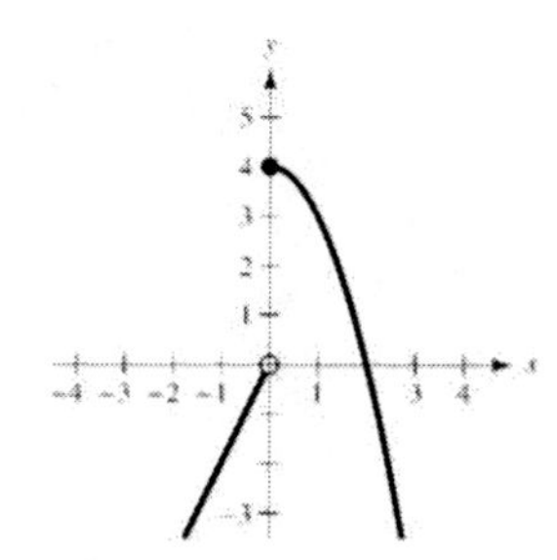

Section 3.5

1. (a) Exponential growth model: $y = ae^{bx}$, $b > 0$

(b) Exponential decay model: $y = ae^{-bx}$, $b > 0$

(c) Logistic growth model: $y = \dfrac{a}{1 + be^{-rx}}$

(d) Gaussian model: $y = ae^{-(x-b)^2/c}$

(e) Natural logarithmic model: $y = a + b\ln x$

(f) Common logarithmic modal: $y = a + b\log_{10} x$

3. sigmoidal

5. The model $y = 120e^{-0.25x}$ would represent an exponential decay model because the exponent's coefficient is negative.

7. $y = 2e^{x/4}$

This is an exponential growth model.

Matches graph (c).

9. $y = 6 + \log_{10}(x + 2)$

This is a logarithmic model, and contains $(-1, 6)$.

Matches graph (b).

11. $y = \ln(x + 1)$

This is a logarithmic model.

Matches graph (d).

13. Since $A = 10{,}000e^{0.035t}$, the time to double is given by

$$20{,}000 = 10{,}000e^{0.035t}$$
$$2 = e^{0.035t}$$
$$\ln 2 = 0.035t$$
$$t = \frac{\ln 2}{0.035} \approx 19.8 \text{ years.}$$

Amount after 10 years:
$A = 10{,}000e^{0.035(10)} \approx \$14{,}190.68$

15. Since $A = 7500e^{rt}$ and $A = 15{,}000$ when $t = 21$, we have the following.

$$15{,}000 = 7500e^{21r}$$
$$2 = e^{21r}$$
$$\ln 2 = 21r$$
$$r = \frac{\ln 2}{21} \approx 0.033 = 3.3\%$$

Amount after 10 years:
$A = 7500e^{0.033(10)} \approx \$10{,}432.26$

17. Since $A = 5000e^{rt}$ and $A = 5665.74$ when $t = 10$, we have the following.

$$5665.74 = 5000e^{10r}$$
$$\frac{5665.74}{5000} = e^{10r}$$
$$\ln\left(\frac{5665.74}{5000}\right) = 10r$$
$$r = \frac{1}{10}\ln\left(\frac{5665.74}{5000}\right)$$
$$\approx 0.0125 = 1.25\%$$

The time to double is given by
$$10{,}000 = 5000e^{0.0125t}$$
$$2 = e^{0.0125t}$$
$$\ln 2 = 0.0125t$$
$$t = \frac{\ln 2}{0.0125} \approx 55.5 \text{ years.}$$

19. Since $A = Pe^{0.045t}$ and $A = 100{,}000$ when $t = 10$, we have the following.

$$100{,}000 = Pe^{0.045(10)}$$
$$\frac{100{,}000}{e^{0.45}} = P \approx \$63{,}762.82$$

The time to double is given by

$$127{,}525.64 = 63{,}762.82e^{0.045t}$$
$$2 = e^{0.045t}$$
$$\ln 2 = 0.045t$$
$$t = \frac{\ln 2}{0.045} \approx 15.4 \text{ years.}$$

21. $3P = Pe^{rt}$

$$3 = e^{rt}$$
$$\ln 3 = rt$$
$$\frac{\ln 3}{r} = t$$

r	2%	4%	6%	8%	10%	12%
$t = \dfrac{\ln 3}{r}$	54.93	27.47	18.31	13.73	10.99	9.16

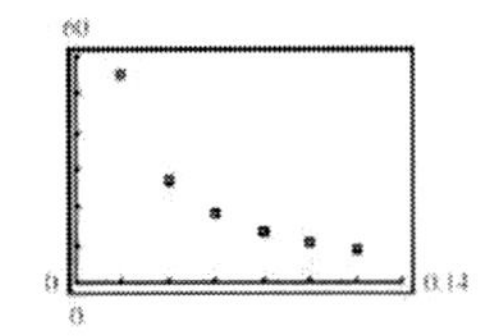

23.

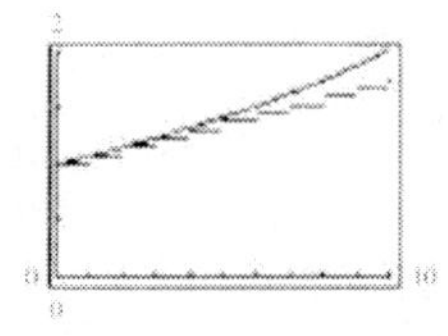

$A = 1 + 0.075[\![t]\!]$

$A = e^{0.07t}$

Continuous compounding results in faster growth.

25. $\frac{1}{2}C = Ce^{k(1599)}$

$$\frac{1}{2} = e^{1599k}$$
$$k = \frac{\ln(1/2)}{1599}$$
$$y = Ce^{kt}$$
$$= 10e^{[\ln(1/2)/1599](1000)}$$
$$\approx 6.48g$$

27. $\frac{1}{2}C = Ce^{k(5700)}$

$$\frac{1}{2} = e^{5700k}$$
$$k = \frac{\ln\left(\frac{1}{2}\right)}{5700}$$
$$y = Ce^{kt}$$
$$y = 3e^{\left[\ln(1/2)/5700\right]1000}$$
$$y \approx 2.66$$

29. $y = ae^{bx}$

$$1 = ae^{b(0)} \Rightarrow 1 = a$$
$$10 = e^{b(3)}$$
$$\ln 10 = 3b$$
$$\frac{\ln 10}{3} = b \Rightarrow b \approx 0.7675$$

Thus, $y = e^{0.7675x}$.

31. $(0, 4) \Rightarrow a = 4$

$$(5, 1) \Rightarrow 1 = 4e^{b(5)} \Rightarrow b = \frac{1}{5}\ln\left(\frac{1}{4}\right)$$
$$= -\frac{1}{5}\ln 4 \approx -0.2773$$

Thus, $y = 4e^{-0.2773x}$.

33. (a) According to the model, $P = 333.68e^{-0.0099t}$, the population was decreasing because the coefficient of the exponent is negative.

(b) For 2000, let $t = 0$.

$P = 333.68e^{-0.0099(0)} \approx 333{,}680$ people

For 2005, let $t = 5$.

$P = 333.68e^{-0.0099(5)} \approx 317{,}565$ people

For 2008, let $t = 8$.

$P = 333.68e^{-0.0099(8)} \approx 308{,}272$ people

(c) To find when the population is expected to be 290,000, let $P = 290$, and solve for t.

$$290 = 333.68e^{-0.0099t}$$
$$\frac{290}{333.68} = e^{-0.0099t}$$
$$\ln\left(\frac{290}{333.68}\right) = -0.0099t$$
$$\frac{\ln\left(\frac{290}{333.68}\right)}{-0.0099} = t$$
$$14.17 \approx t$$

or 2014

35. $P = 1155.4e^{kt}$

(a) In 2002 ($t = 2$), the population was 1,200,000 ($P = 1200$).

$$1200 = 1155.4e^{k(2)}$$
$$\frac{1200}{1155.4} = e^{2k}$$
$$\ln\left(\frac{1200}{1155.4}\right) = 2k$$
$$\frac{1}{2}\ln\left(\frac{1200}{1155.4}\right) = k$$
$$0.0189 \approx k$$
$$P = 1155.4e^{0.0189t}$$

(b) For 2015, let $t = 15$ and find P.

$P = 1155.4e^{0.0189(15)}$

$P \approx 1534.104$ or 1,534,104 people

37. $y = Ce^{kt}$

$$\frac{1}{2}C = Ce^{5700k}$$
$$\frac{1}{2} = e^{5700k}$$
$$\ln\left(\tfrac{1}{2}\right) = 5700k$$
$$\frac{\ln\left(\frac{1}{2}\right)}{5700} = k$$

The ancient charcoal has 15% as much radioactive carbon as it did originally.

$$0.15C = Ce^{(\ln(1/2)/5700)t}$$
$$0.15 = e^{(\ln(1/2)/5700)t}$$
$$\ln(0.15) = (\ln(1/2)/5700)t$$
$$\frac{\ln(0.15)}{(\ln(1/2)/5700)} = t$$
$$15{,}600.7 \approx t$$

or about 15,601 years ago

39. (a) Use the points (0, 49,200) and (2, 32,590), where t is the time since purchased and V is the value.

$$m = \frac{49{,}200 - 32{,}590}{0 - 2} = -8305$$

The V-intercept is 49,200, so the linear model is $V = -8305t + 49{,}200$.

(b) Using the points (0, 49,200) and (2, 32,590), first find a.

$$V = ae^{kt}$$
$$49{,}200 = ae^{k(0)}$$
$$49{,}200 = a$$

So, $V = 49{,}200e^{kt}$

Next, find k, using the second point.

$$32{,}590 = 49{,}200e^{k(2)}$$
$$\frac{32{,}590}{49{,}200} = e^{2k}$$
$$\ln\left(\frac{32{,}590}{49{,}200}\right) = 2k$$
$$\frac{1}{2}\ln\left(\frac{32{,}590}{49{,}200}\right) = k$$
$$-0.2059 \approx k$$
$$V = 49{,}200e^{-0.2059t}$$

(c)

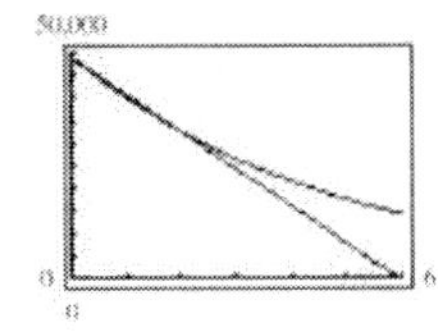

(d) The exponential model has a greater depreciation rate the first year.

(e) Using the graph, the value of the sedan with the linear model is greater for $0 < t < 2$, and the value of the sedan with the exponential model is greater for $t \geq 2$.

41. $y = 0.0266e^{-(x-100)^2/450}$, $70 \leq x \leq 115$

(a)

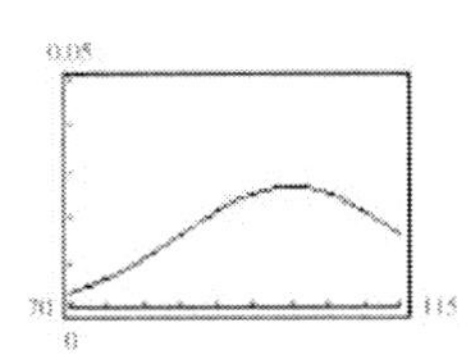

(b) Maximum point is $x = 100$, the average IQ score.

43. $p(t) = \dfrac{1000}{1 + 9e^{-0.1656t}}$

(a) $p(5) = \dfrac{1000}{1 + 9e^{-0.1656(5)}} \approx 203$ animals

(b) $500 = \dfrac{1000}{1 + 9e^{-0.1656t}}$

$$1 + 9e^{-0.1656}t = 2$$
$$9e^{-0.1656t} = 1$$
$$e^{-0.1656t} = \frac{1}{9}$$
$$t = \frac{-\ln(1/9)}{0.1656} \approx 13 \text{ months}$$

(c)

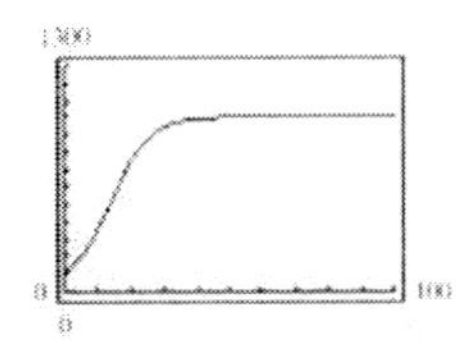

The horizontal asymptotes are $p = 0$ and $p = 1000$. The population will approach 1000 as time increases.

45. $R = \log_{10}\left(\dfrac{I}{I_0}\right) = \log_{10} I \Rightarrow I = 10^R$

(a) $I = 10^7 = 10{,}000{,}000$

(b) $I = 10^{8.1} \approx 125{,}892{,}541$

(c) $I = 10^{6.1} \approx 1{,}258{,}925$

47. $\beta(I) = 10\log_{10}(I/I_0)$, where $I_0 = 10^{-12}$ watt per square meter.

(a) $\beta(10^{-10}) = 10 \cdot \log_{10}\left(\frac{10^{-10}}{10^{-12}}\right) = 10\log_{10}10^2 = 20$ decibels

(b) $\beta(10^{-5}) = 10 \cdot \log_{10}\left(\frac{10^{-5}}{10^{-12}}\right) = 10\log_{10}10^7 = 70$ decibels

(c) $\beta(10^{0}) = 10 \cdot \log_{10}\left(\frac{10^{0}}{10^{-12}}\right) = 10\log_{10}10^{12} = 120$ decibels

49. $$\beta = 10\log_{10}\left(\frac{I}{I_0}\right)$$

$$10^{\beta/10} = \frac{I}{I_0}$$

$$I = I_0 10^{\beta/10}$$

$$\%\text{ decrease} = \frac{I_0 10^{8.8} - I_0 10^{7.2}}{I_0 10^{8.8}} \times 100 = 97.5\%$$

51. $\text{pH} = -\log_{10}[\text{H}^+] = -\log_{10}[2.3 \times 10^{-5}] \approx 4.64$

53. $$\text{pH} = -\log_{10}[\text{H}^+]$$

$$-\text{pH} = \log_{10}[\text{H}^+]$$

$$10^{-\text{pH}} = [\text{H}^+]$$

$$\frac{\text{Hydrogen ion concentration of grape}}{\text{Hydrogen ion concentration of backing soda}} = \frac{10^{-3.5}}{10^{-8.0}} = 10^{4.5} \approx 31{,}623 \text{ times}$$

55. $$u = 120{,}000\left[\frac{0.075t}{1 - \left(\frac{1}{1 + 0.075/12}\right)^{12t}} - 1\right]$$

(a)

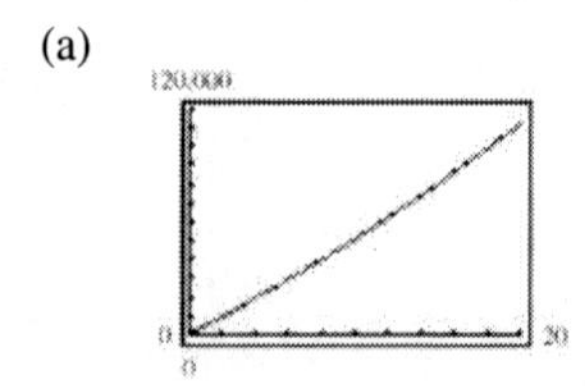

(b) From the graph, when $u = 120{,}000$, $t \approx 21.2$ years. Yes, a mortgage of approximately 37.6 years will result in about \$240,000 of interest.

57. $t = -10\ln\left(\frac{T - 70}{98.6 - 70}\right)$

At 9:00 A.M. we have

$t = -10\ln[(85.7 - 70)/(98.6 - 70)] \approx 6$ hours.

Thus, we can conclude that the person died 6 hours before 9 A.M., or 3:00 A.M.

59. False. The domain could be all real numbers.

61. The graph of a Gaussian model will not have an x-intercept because no x-value will yield a negative y-value.

63. A Gaussian model will have the maximum value occur at the average value of the independent variable.

65. An exponential growth model will have a graph which shows a steadily increasing rate of growth.

67. $4x - 3y - 9 = 0 \Rightarrow y = \frac{1}{3}(4x - 9)$

Slope: $\frac{4}{3}$

Matches (a).

Intercepts: $(0, -3)$, $\left(\frac{9}{4}, 0\right)$

69. $y = 25 - 2.25x$

Slope: -2.25

Matches (d).

Intercepts: $(0, 25)$, $\left(\frac{100}{9}, 0\right)$

71. $f(x) = 2x^3 - 3x^2 + x - 1$

The graph falls to the left and rises to the right.

73. $g(x) = -1.6x^5 + 4x^2 - 2$

The graph rises to the left and falls to the right.

75.
$$\begin{array}{r|rrrr} 4 & 2 & -8 & 3 & -9 \\ & & 8 & 0 & 12 \\ \hline & 2 & 0 & 3 & 3 \end{array}$$

$$\frac{2x^3 - 8x^2 + 3x - 9}{x - 4} = 2x^2 + 3 + \frac{3}{x - 4}$$

77. Answers will vary.

Section 3.6

1. $y = ax^b$

3. A scatter plot can be created to provide an idea of what type of model will best fit a set of data.

5. Logarithmic model

7. Quadratic model

9. Exponential model

11. Quadratic model

13. 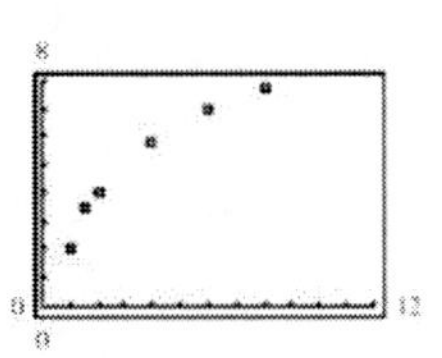

Logarithmic model

15. 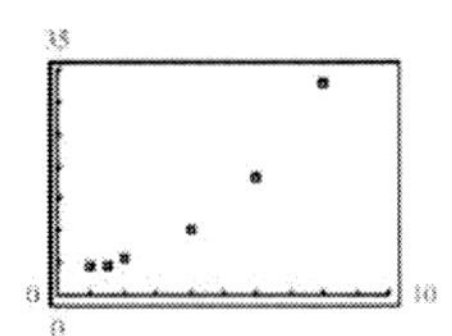

Exponential model

17. 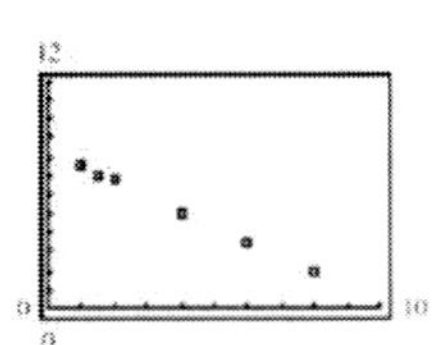

Linear model

19. $y = 4.752(1.2607)^x$

Coefficient of determination: 0.96773

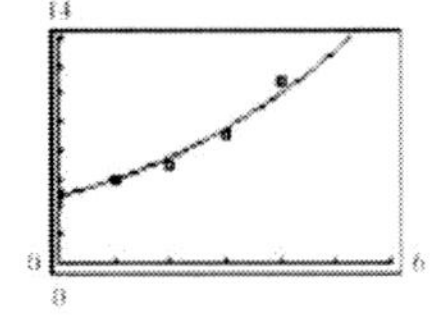

21. $y = 8.463(0.7775)^x$

Coefficient of determination: 0.86639

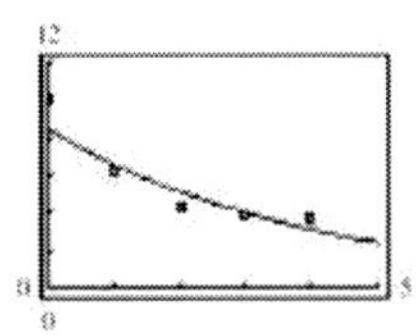

23. $y = 2.083 + 1.257 \ln x$

Coefficient of determination: 0.98672

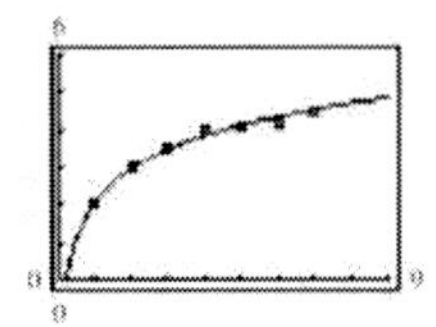

25. $y = 9.826 - 4.097 \ln x$

Coefficient of determination: 0.93704

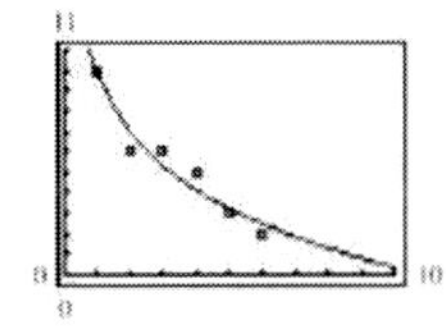

27. $y = 1.985x^{0.760}$

Coefficient of determination: 0.99686

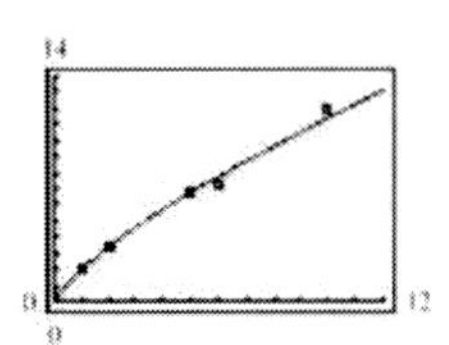

29. $y = 16.103x^{-3.174}$

Coefficient of determination: 0.88161

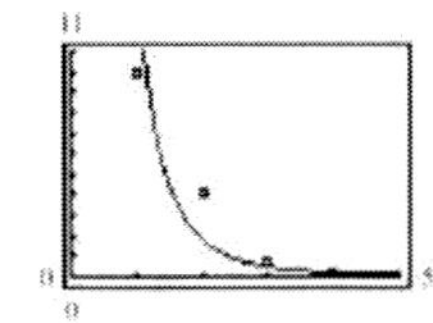

31. (a) Exponential model:

$S = 1876.645(1.1980)^t$

$r^2 = 0.9996$

Power model: $S = 1905.844t^{0.6018}$

$r^2 = 0.9559$

(b) Exponential model:

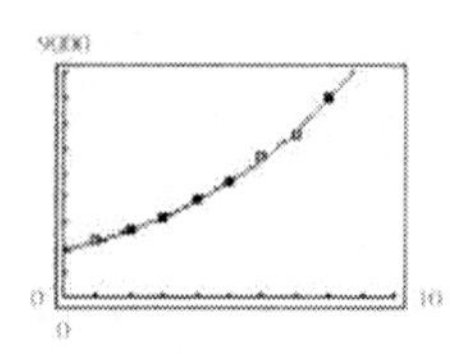

Power model:

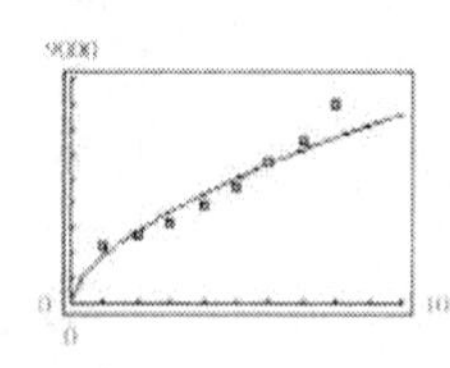

(c) The exponential model with $r^2 \approx 0.9996$ is a better fit, as compared to the power model with $r^2 \approx 0.9559$.

33. (a) Linear model:

$p = 2.89t + 252.9,\ r^2 \approx 0.9987$

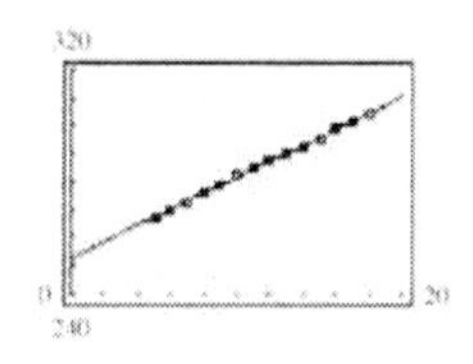

Power model:

(b) $p = 222.94t^{0.1048},\ r^2 \approx 0.9850$

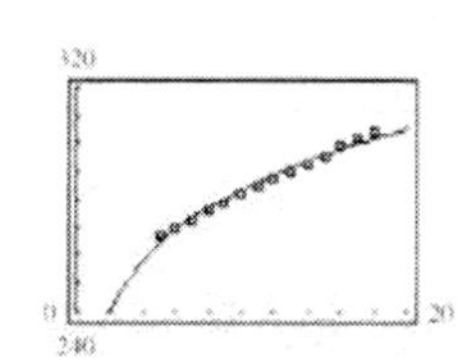

Exponential model:

(c) $p = 254.445(1.0102)^t,\ r^2 \approx 0.9972$

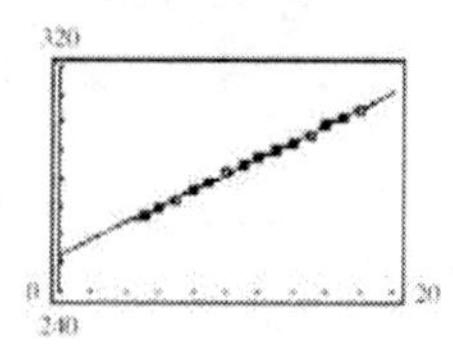

Logarithmic model:

(d) $p = 29.813\ln t + 215.36,\ r^2 \approx 0.9803$

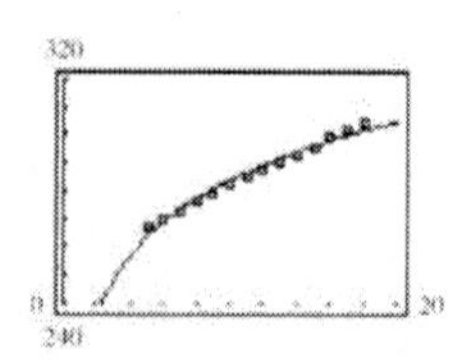

(e) The linear model from part (a) is the best fit because its coefficient of determination, $r^2 \approx 0.9987$, is closest to 1.

(f) Linear model:

Year	2009	2010	2011	2012	2013	2014
Population (in millions)	307.8	310.7	313.6	316.5	319.4	322.3

Power model:

Year	2009	2010	2011	2012	2013	2014
Population (in millions)	303.5	305.2	306.7	308.2	309.7	311.1

Exponential model:

Year	2009	2010	2011	2012	2013	2014
Population (in millions)	308.6	311.7	314.9	318.1	321.3	324.6

Logarithmic model:

Year	2009	2010	2011	2012	2013	2014
Population (in millions)	303.1	304.7	306.1	307.5	308.8	310.1

(g) and (h) Answers will vary.

35. (a) $N = 1315.584(1.0644)^t$

(b) $N = 1315.584\, e^{(\ln 1.0644)t}$

$= 1315.584e^{0.0624t}$

(c) For 2009, let $t = 9$ and find N.

$N = 1315.584e^{(0.0624)(9)} \approx 2307$ stores

Answers will vary.

37. (a) $P = \dfrac{91.3686}{1 + 765.5440e^{-0.2547x}}$

(b)

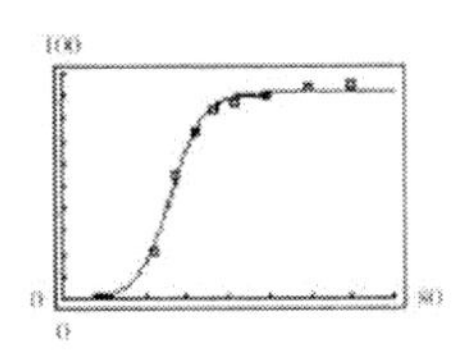

The model fits the data well.

39. True.

41. Answers will vary.

43. $2x + 5y = 10$

$5y = -2x + 10$

$y = -\frac{2}{5}x + 2$

Slope: $-\frac{2}{5}$

y-intercept: $(0, 2)$

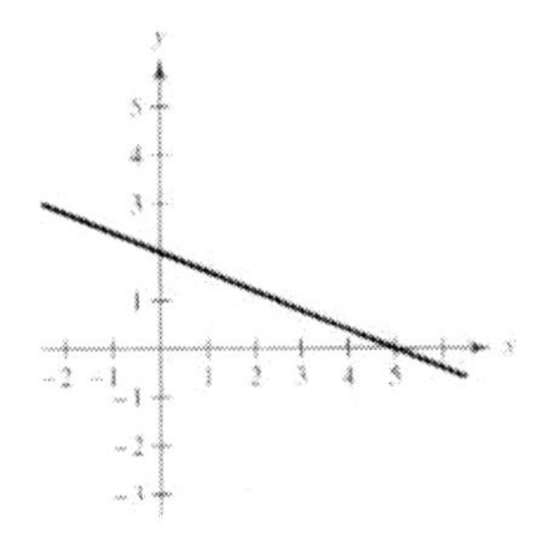

45. $1.2x + 3.5y = 10.5$

$35y = -12x + 105$

$y = -\frac{12}{35}x + \frac{105}{35}$

$= -\frac{12}{35}x + 3$

Slope: $-\frac{12}{35}$

y-intercept: $(0, 3)$

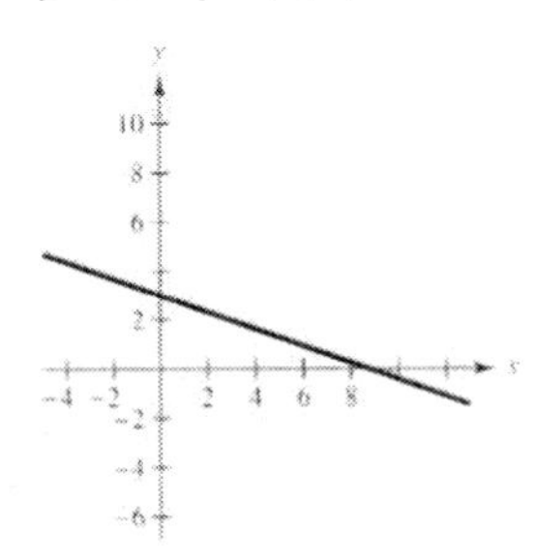

47. $f(x) = a(x - h)^2 + k$

$f(x) = a(x + 1)^2 + 2$ Vertex: $(-1, 2)$

$1 = a(0 + 1)^2 + 2$ Point: $(0, 1)$

$1 = a + 2$

$-1 = a$

$f(x) = -(x + 1)^2 + 2$

49. $f(x) = a(x - h)^2 + k$

$f(x) = a(x - 3)^2 + 2$ Vertex: $(3, 2)$

$0 = a(4 - 3)^2 + 2$ Point: $(4, 0)$

$0 = a + 2$

$-2 = a$

$f(x) = -2(x - 3)^2 + 2$

Chapter 3 Review Exercises

1. $(1.45)^{2\pi} \approx 10.3254$

3. $60^{2(-1.1)} = 60^{-2.2}$
≈ 0.0001

5. $f(x) = 4^x$

Intercept: $(0, 1)$

Horizontal asymptote: x-axis

Increasing on: $(-\infty, \infty)$

Matches graph (c).

7. $f(x) = -4^x$

Intercept: $(0, -1)$

Horizontal asymptote: x-axis

Decreasing on: $(-\infty, \infty)$

Matches graph (b).

9. $f(x) = 6^x$

Intercept: $(0, 1)$

Horizontal asymptote: x-axis or $y = 0$

Increasing on: $(-\infty, \infty)$

x	-2	-1	0	1	2
y	$\frac{1}{36}$	$\frac{1}{6}$	1	6	36

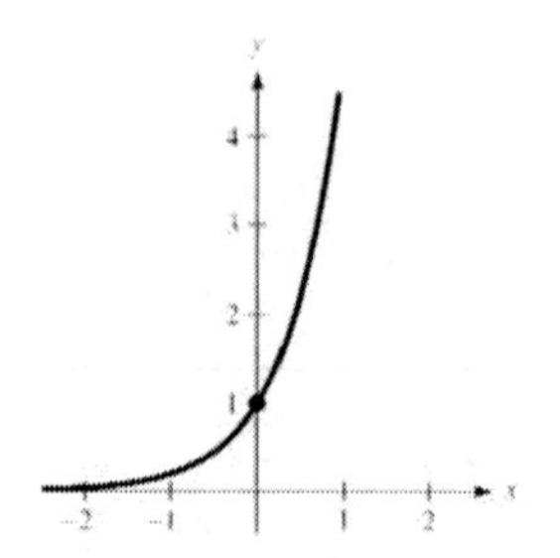

11. $g(x) = 6^{-x}$

Intercept: $(0, 1)$

Horizontal asymptote: x-axis or $y = 0$

Decreasing on: $(-\infty, \infty)$

x	-2	-1	0	1	2
y	36	6	1	$\frac{1}{6}$	$\frac{1}{36}$

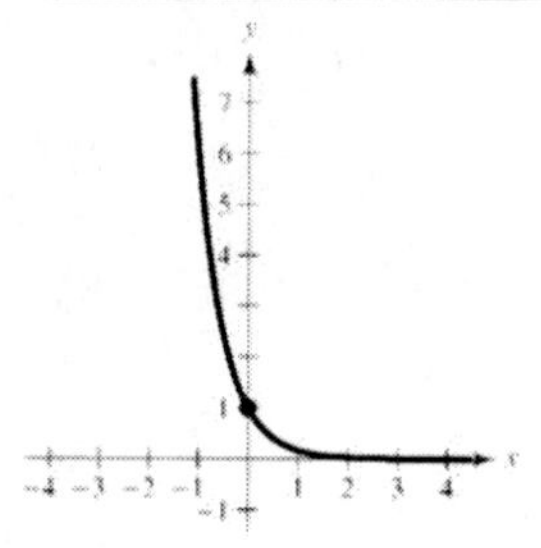

13. $h(x) = e^{x-1}$

x	0	1	2	3	4
$h(x)$	0.37	1	2.72	7.39	20.09

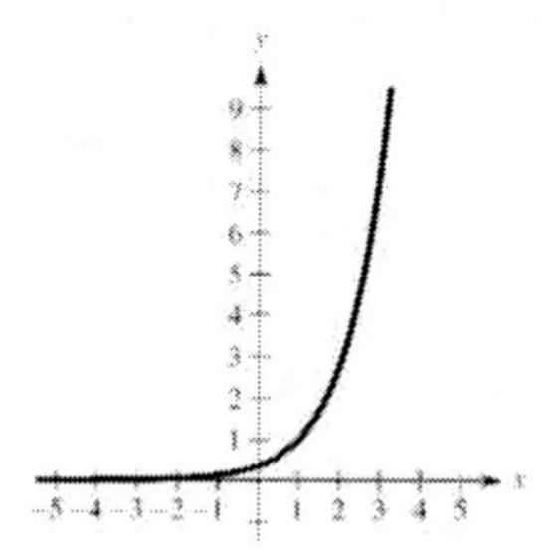

Horizontal asymptote: $y = 0$

15. $h(x) = -e^x$

x	-2	-1	0	1	2
$h(x)$	-0.14	-0.37	-1	-2.72	-7.39

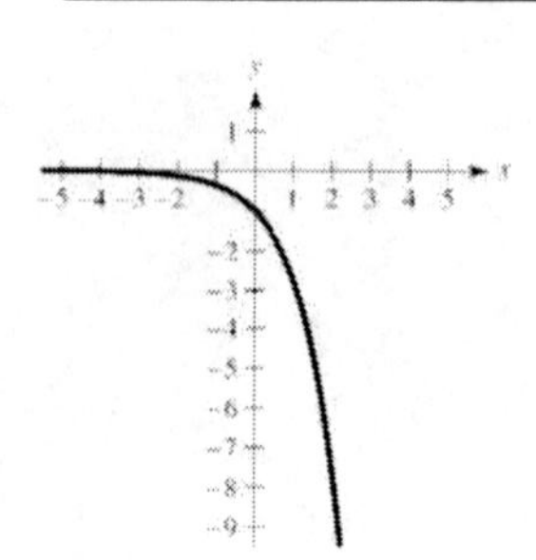

Horizontal asymptote: $y = 0$

17. $f(x) = 4e^{-0.5x}$

x	-1	0	1	2	3	4
$f(x)$	6.59	4	2.43	1.47	0.89	0.54

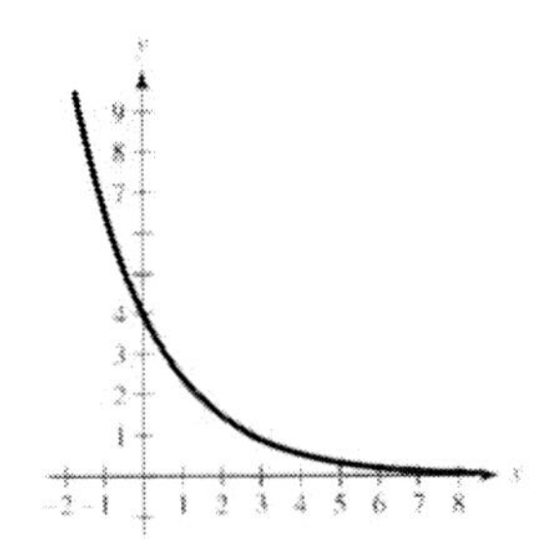

Horizontal asymptote: $y = 0$

19. $A = Pe^{rt} = 10{,}000e^{0.08t}$

t	1	10	20	30	40	50
A	\$10,832.87	\$22,255.41	\$49,530.32	\$110,231.76	\$245,325.30	\$545,981.50

21. 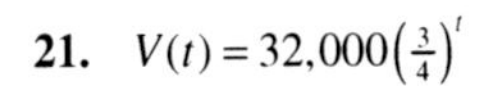
$V(t) = 32{,}000\left(\frac{3}{4}\right)^t$

(a)

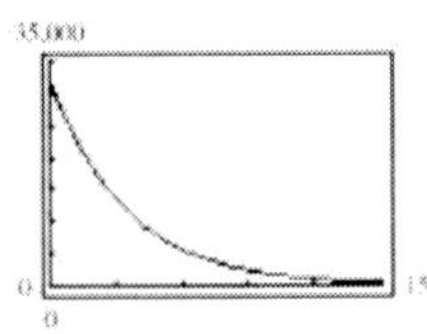

(b) After 2 years, $V(2) = 32{,}000\left(\frac{3}{4}\right)^2 = \$18{,}000.$

(c) The SUV depreciates most rapidly when it is first sold. Answers will vary.

23. $\log_5 125 = 3$

$5^3 = 125$

25. $\log_{64} 2 = \dfrac{1}{6}$

$64^{1/6} = 2$

27. $4^3 = 64$

$\log_4 64 = 3$

29. $125^{\frac{2}{3}} = 25$

$\log_{125} 25 = \dfrac{2}{3}$

31. $\left(\dfrac{1}{2}\right)^{-3} = 8$

$\log_{1/2} 8 = -3$

33. $\log_6 216 = \log_6 6^3$

$= 3\log_6 6$

$= 3$

35. $\log_4\left(\dfrac{1}{4}\right) = \log_4\left(4^{-1}\right)$

$= -\log_4 4$

$= -1$

37. $g(x) = -\log_2 x + 5 = 5 - \dfrac{\ln x}{\ln 2}$

Domain: $x > 0$

Vertical asymptote: $x = 0$

x-intercept: $(32, 0)$

x	1	2	3	4	5
y	5	4	3.42	3	2.68

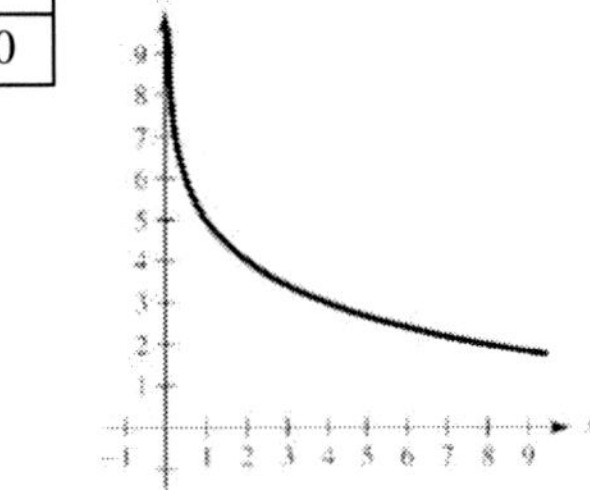

39. $f(x) = \log_2(x-1) + 6 = 6 + \dfrac{\ln(x-1)}{\ln(2)}$

Domain: $x > 1$

Vertical asymptote: $x = 1$

x-intercept: $(1.016, 0)$

x	2	3	4	5	6
y	6	7	7.59	8	8.32

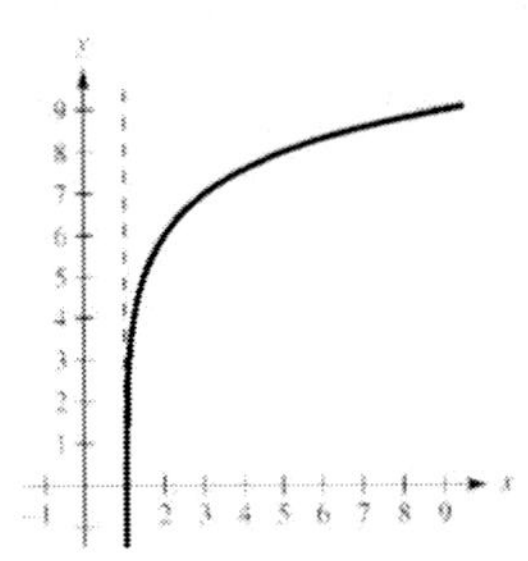

41. $\ln(21.5) \approx 3.068$

43. $\ln\sqrt{6} \approx 0.896$

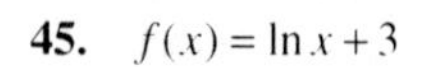

45. $f(x) = \ln x + 3$

Domain: $(0, \infty)$

Vertical asymptote: $x = 0$

x-intercept: $(0.05, 0)$

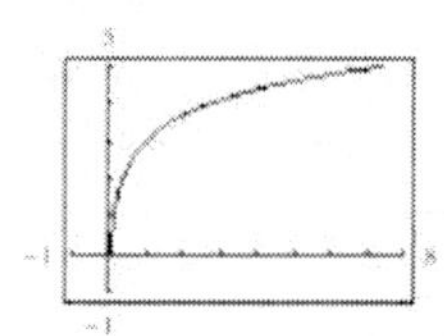

47. $h(x) = \frac{1}{2}\ln x$

Domain: $x > 0$

Vertical asymptote: $x = 0$

x-intercept: $(1, 0)$

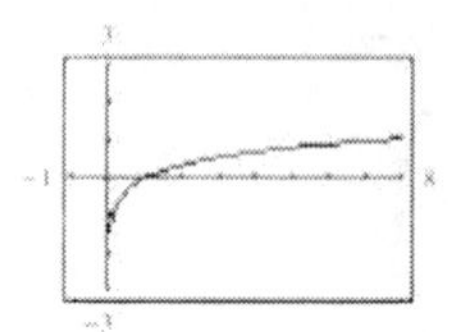

49. $t = 50\log_{10}\dfrac{18{,}000}{18{,}000 - h}$

(a) $0 \le h < 18{,}000$

(b)

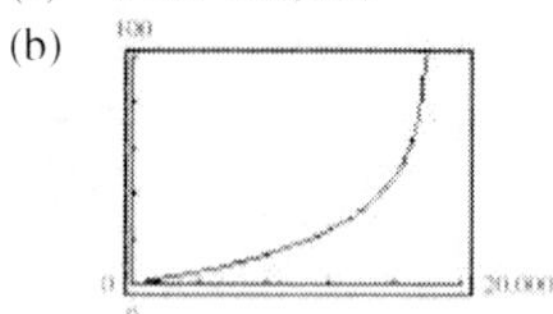

Vertical asymptote: $h = 18{,}000$

(c) The plane climbs at a faster rate as it approaches its absolute ceiling.

(d) If $h = 4000$, $t = 50\log_{10}\dfrac{18{,}000}{18{,}000 - 4000} \approx 5.46$ minutes.

51. $\log_4 9 = \dfrac{\log_{10} 9}{\log_{10} 4} \approx 1.585$

$\log_4 9 = \dfrac{\ln 9}{\ln 4} \approx 1.585$

53. $\log_{14} 364 = \dfrac{\log_{10} 364}{\log_{10} 14} \approx 2.235$

$\log_{14} 364 = \dfrac{\ln 364}{\ln 14} \approx 2.235$

55. $f(x) = \log_2(x-1) = \dfrac{\ln(x-1)}{\ln 2}$

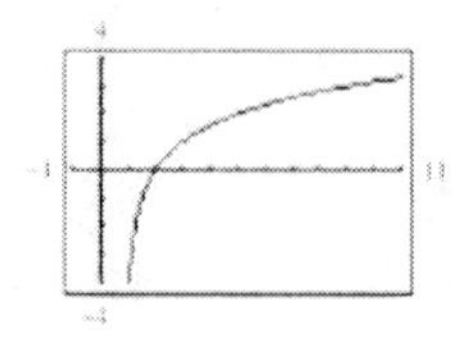

57. $f(x) = -\log_{1/2}(x+2) = -\dfrac{\ln(x+2)}{\ln(1/2)} = \dfrac{\ln(x+2)}{\ln 2}$

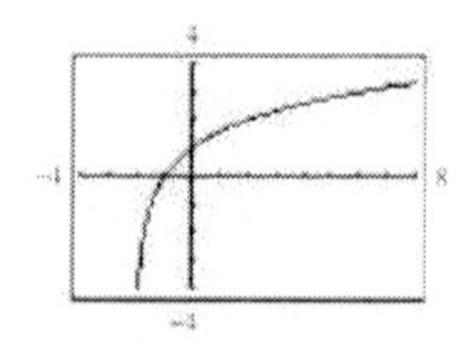

59. $\log_b 9 = \log_b 3^2$

$= 2\log_b 3$

$= 2(0.5646)$

$= 1.1292$

61. $\log_b \sqrt{5} = \log_b 5^{1/2}$

$= \frac{1}{2}\log_b 5$

$= \frac{1}{2}(0.8271)$

$= 0.41355$

63. $\ln(5e^{-2}) = \ln 5 + \ln e^{-2}$

$= \ln 5 - 2\ln e$

$= \ln 5 - 2$

65. $\log_{10} 200 = \log_{10}(2 \cdot 100)$

$= \log_{10} 2 + \log_{10} 10^2$

$= \log_{10} 2 + 2$

67. $\log_5 5x^2 = \log_5 5 + \log_5 x^2 = 1 + 2\log_5 x$

69. $\log_{10}\dfrac{5\sqrt{y}}{x^2} = \log_{10} 5\sqrt{y} - \log_{10} x^2$

$= \log_{10} 5 + \log_{10}\sqrt{y} - \log_{10} x^2$

$= \log_{10} 5 + \frac{1}{2}\log_{10} y - 2\log_{10} x$

71. $\ln\left(\dfrac{x+3}{xy}\right) = \ln(x+3) - \ln(xy)$

$= \ln(x+3) - \ln x - \ln y$

73. $\log_2 9 + \log_2 x = \log_2(9x)$

75. $\frac{1}{2}\ln(2x-1) - 2\ln(x+1) = \ln\sqrt{2x-1} - \ln(x+1)^2$

$= \ln\dfrac{\sqrt{2x-1}}{(x+1)^2}$

77. $\ln 3 + \frac{1}{3}\ln(4-x^2) - \ln x = \ln\left[\dfrac{3(4-x^2)^{1/3}}{x}\right] = \ln\dfrac{3\sqrt[3]{4-x^2}}{x}$

79. $s = 25 - \dfrac{13\ln(h/12)}{\ln 3}$

(a)

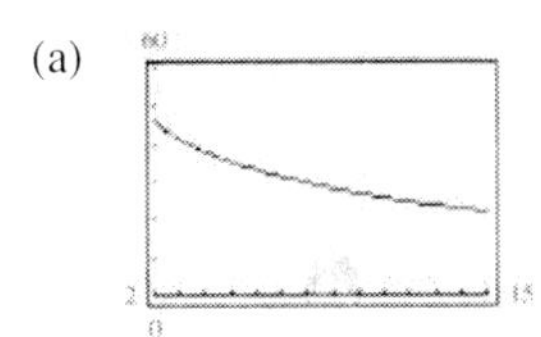

(b)

h	4	6	8	10	12	14
s	38	33.2	29.8	27.2	25	23.2

(c) As the depth increases, the number of miles of roads cleared decreases.

81. $10^x = 10{,}000$

$10^x = 10^4$

$x = 4$

83. $6^x = \dfrac{1}{216} = \dfrac{1}{6^3} = 6^{-3} \Rightarrow x = -3$

85. $2^{x+1} = \dfrac{1}{16}$

$2^{x+1} = 2^{-4}$

$x + 1 = -4$

$x = -5$

87. $\log_8 x = 4$

$8^4 = x$

$4096 = x$

89. $\log_2(x-1) = 3$

$2^3 = x - 1$

$x = 9$

91. $\ln x = 4$

$x = e^4 \approx 54.598$

93. $\ln(x-1) = 2$

$e^2 = x - 1$

$x = 1 + e^2$

95. $3e^{-5x} = 132$

$e^{-5x} = 44$

$-5x = \ln 44$

$x = -\dfrac{\ln 44}{5} \approx -0.757$

97. $2^x + 13 = 35$

$2^x = 22$

$x\ln 2 = \ln 22$

$x = \dfrac{\ln 22}{\ln 2} \approx 4.459$

99. $-4(5^x) = -68$

$5^x = 17$

$x\ln 5 = \ln 17$

$x = \dfrac{\ln 17}{\ln 5} \approx 1.760$

101. $2e^{x-3} - 1 = 4$

$2e^{x-3} = 5$

$e^{x-3} = \dfrac{5}{2}$

$x - 3 = \ln\left(\dfrac{5}{2}\right)$

$x = 3 + \ln\left(\dfrac{5}{2}\right) \approx 3.916$

103. $e^{2x} - 7e^x + 10 = 0$

$(e^x - 5)(e^x - 2) = 0$

$e^x = 5 \Rightarrow x = \ln 5 \approx 1.609$

$e^x = 2 \Rightarrow x = \ln 2 \approx 0.693$

105. $\ln 3x = 6.4$

$e^{6.4} = 3x$

$\dfrac{1}{3}e^{6.4} = x$

$200.615 \approx x$

107. $\ln x - \ln 5 = 2$

$\ln\left(\dfrac{x}{5}\right) = 2$

$e^2 = \dfrac{x}{5}$

$5e^2 = x$

$36.945 \approx x$

109. $\ln\sqrt{x+1} = 2$

$\dfrac{1}{2}\ln(x+1) = 2$

$\ln(x+1) = 4$

$x + 1 = e^4$

$x = e^4 - 1$

≈ 53.598

111. $\log_4(x-1) = \log_4(x-2) - \log_4(x+2)$

$$\log_4(x-1) = \log_4\left(\frac{x-2}{x+2}\right)$$
$$x-1 = \frac{x-2}{x+2}$$
$$(x-1)(x+2) = x-2$$
$$x^2 + x - 2 = x - 2$$
$$x^2 = 0$$
$$x = 0 \quad \text{(extraneous)}$$

No solution

113. $\log_{10}(1-x) = -1$

$$10^{-1} = 1 - x$$
$$x = 1 - 10^{-1} = 0.9$$

115. $xe^x + e^x = 0$

$$(x+1)e^x = 0$$
$$x + 1 = 0 \quad (\text{since } e^x \neq 0)$$
$$x = -1$$

117. $x\ln x + x = 0$

$$x(\ln x + 1) = 0$$
$$\ln x + 1 = 0 \quad (\text{since } x > 0)$$
$$\ln x = -1$$
$$x = e^{-1} = \frac{1}{e} \approx 0.368$$

119. To find the time for a \$7550 deposit to double earning 6.9% interest compounded continuously, set $A = 15{,}100$ and solve for t using the formula $A = pe^{rt}$.

$$15{,}100 = 7550\, e^{0.069t}$$
$$2 = e^{0.069t}$$
$$\ln 2 = 0.069t$$
$$\frac{\ln 2}{0.069} = t$$
$$10.05 \approx t$$

About 10.05 years

121. $y = 3e^{-2x/3}$

Decreasing exponential
Matches graph (e).

123. $y = \ln(x+3)$

Logarithmic function shifted to left
Matches graph (f).

125. $y = 2e^{-(x+4)^2/3}$

Gaussian model
Matches graph (a).

127. $P = 6707.7e^{kt}$

Use the point (t, P), or $(18, 9222)$, to find k,

$$9222 = 6707.7e^{k(18)}$$
$$\frac{9222}{6707.7} = e^{1.8k}$$
$$\ln\left(\frac{9222}{6707.7}\right) = 18k$$
$$\frac{1}{18}\ln\left(\frac{9222}{6707.7}\right) = k$$
$$0.01769 \approx k$$
$$P = 6707.7e^{0.01769t}$$

For 2020, let $t = 30$, and find P.

$$P = 6707.7e^{0.01769(30)}$$
$$\approx 11{,}403.908 \text{ or } 11{,}403{,}908 \text{ people}$$

129. $N = \dfrac{62}{1 + 5.4e^{-0.24t}}$

(a) To find the number of weeks, t, to read 40 words per minute, set $N = 40$ and solve for t.

$$40 = \frac{62}{1 + 5.4e^{-0.24t}}$$
$$1 + 5.4e^{-0.24t} = 1.55$$
$$5.4e^{-0.24t} = 0.55$$
$$e^{-0.24t} = 0.10185$$
$$-0.24t = \ln(0.10185)$$
$$t = \frac{\ln(0.10185)}{-0.24}$$
$$t \approx 9.52 \text{ weeks}$$

(b) Set $N = 60$ and find t.

$$60 = \frac{62}{1 + 5.4e^{-0.24t}}$$
$$1 + 5.4e^{-0.24t} = 1.0\overline{3}$$
$$5.4e^{-0.24t} = 0.0\overline{3}$$
$$e^{-0.24t} = \frac{0.0\overline{3}}{5.4}$$
$$-0.24t = \ln\left(\frac{0.0\overline{3}}{5.4}\right)$$
$$t = \frac{\ln\left(\frac{0.0\overline{3}}{5.4}\right)}{-0.24}$$
$$t \approx 21.20 \text{ weeks}$$

131. Logistic model

133. (a) Linear model: $N = 41.5t + 2722$

$r^2 \approx 0.9785$

Exponential model: $N = 2728(1.0142)^t$

$r^2 \approx 0.9818$

Power model: $N = 2727.6t^{0.0497}$

$r^2 \approx 0.8398$

(b) Linear model:

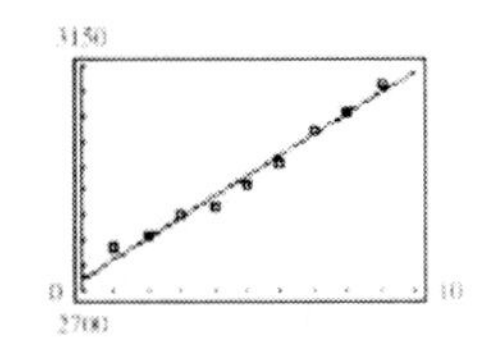

Exponential model:

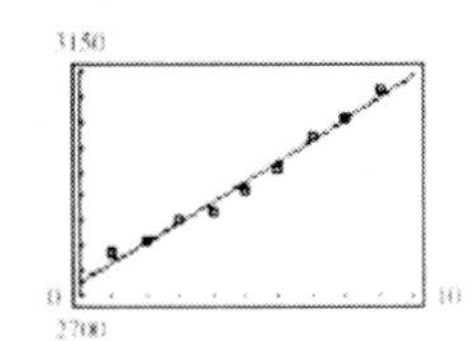

Power model:

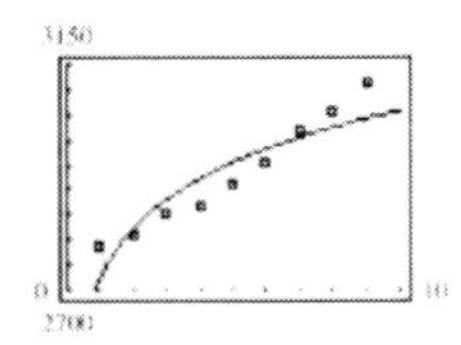

(c) The exponential model with $r^2 \approx 0.9818$ is the best fit because the coefficient of determination is closest to 1.

(d) Find N when $t = 10$, using the exponential model.

$N = 2728(1.0142)^{10}$

≈ 3141.09 or 3,141,090 female athletes

(e) Find t when $N = 3720$ using the exponential model.

$$3720 = 2728(1.0142)^t$$
$$1.\overline{36} = 1.0142^t$$
$$\ln 1.\overline{36} = \ln 1.0142^t$$
$$\ln 1.\overline{36} = t \ln 1.0142$$
$$\frac{\ln 1.\overline{36}}{\ln 1.0142} = t$$
$$22 \approx t$$

The school year 2021 – 2022

135. $e^{x-1} = e^x \cdot e^{-1} = \dfrac{e^x}{e}$

True (by properties of exponents).

137. False. The domain of $f(x) = \ln(x)$ is $x > 0$.

139. Since $1 < \sqrt{2} < 2$, then $2^1 < 2^{\sqrt{2}} < 2^2 \Rightarrow 2 < 2^{\sqrt{2}} < 4$.

Chapter 3 Test

1. $f(x) = 10^{-x}$

x	−2	−1	0	1	2
$f(x)$	100	10	1	0.1	0.01

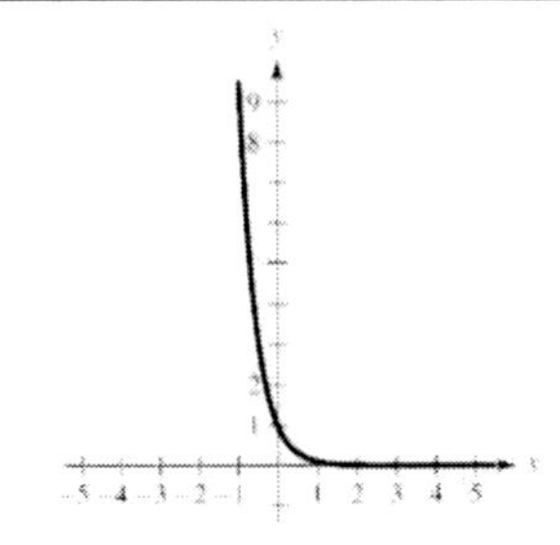

Horizontal asymptote: $y = 0$

Intercept: $(0, 1)$

2. $f(x) = -6^{x-2}$

x	0	2	3	4
$f(x)$	−0.03	−1	−6	−36

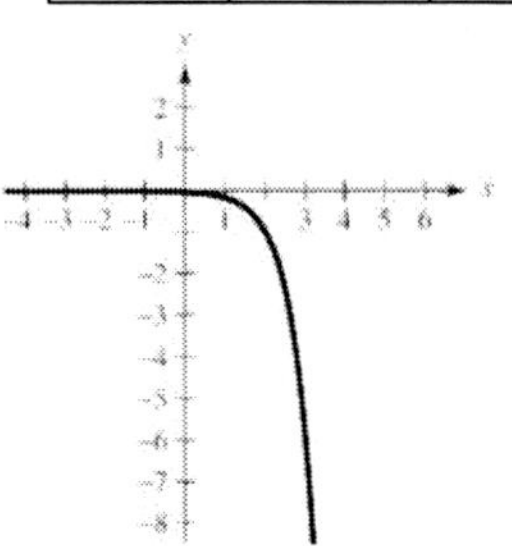

Horizontal asymptote: $y = 0$

Intercept: $\left(0, -\dfrac{1}{36}\right)$

3. $f(x) = 1 - e^{2x}$

x	-2	-1	0	1	2
$f(x)$	0.9817	0.8647	0	-6.3891	-53.5982

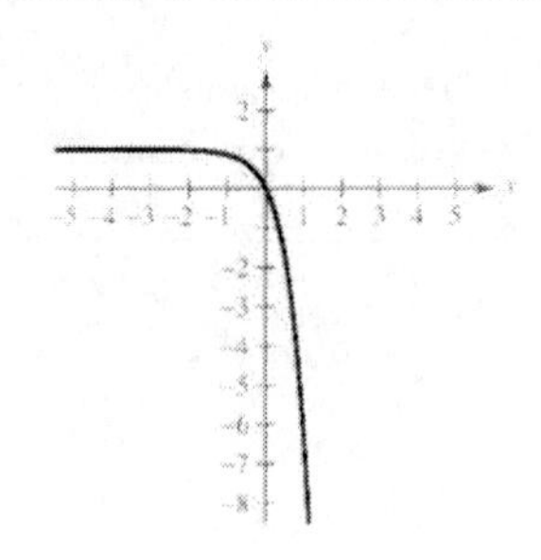

Horizontal asymptote: $y = 1$

Intercept: $(0, 0)$

4. $\log_7 7^{-0.89} = -0.89 \log_7 7$
$= -0.89$

5. $4.6 \ln e^2 = 4.6(2) \ln e = 9.2$

6. $5 - \log_{10} 1000 = 5 - \log_{10} 10^3$
$= 5 - 3 = 2$

7. $f(x) = -\log_{10} x - 6$

Domain: $x > 0$
Vertical asymptote: $x = 0$
x-intercept: $\left(10^{-6}, 0\right) \approx (0, 0)$

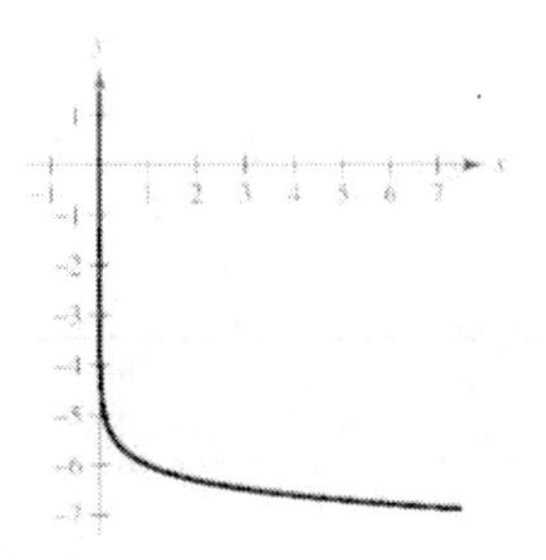

8. $f(x) = \ln(x - 4)$

Domain: $x > 4$
Vertical asymptote: $x = 4$
x-intercept: $(5, 0)$

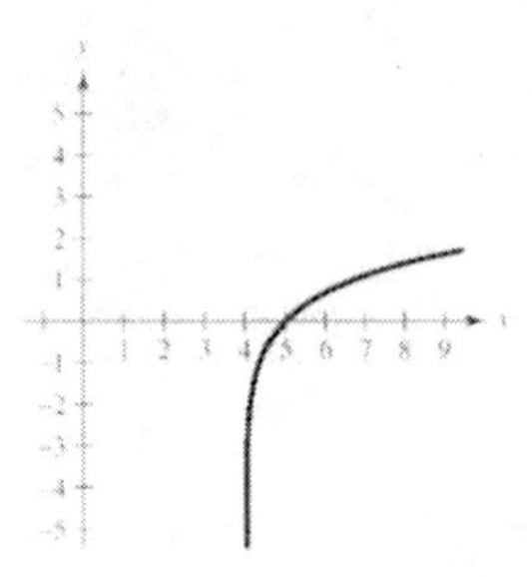

9. $f(x) = 1 + \ln(x + 6)$

Domain: $x > -6$
Vertical asymptote: $x = -6$
x-intercept: $(-5.632, 0)$

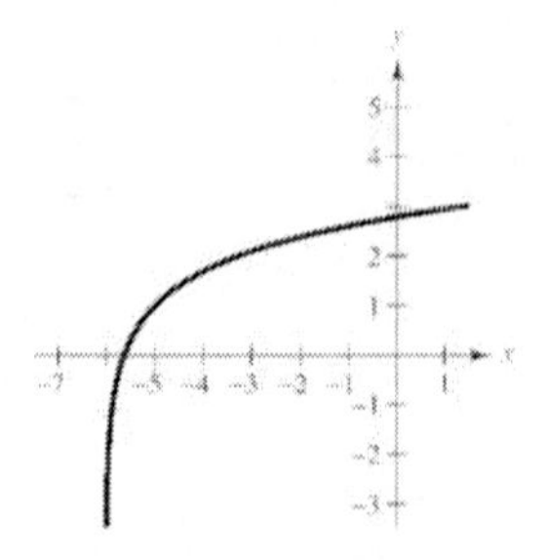

10. $\log_7 44 = \dfrac{\ln 44}{\ln 7} \approx 1.945$

11. $\log_{2/5}(0.9) = \dfrac{\ln(0.9)}{\ln(2/5)} \approx 0.115$

12. $\log_{12} 64 = \dfrac{\ln 64}{\ln 12} \approx 1.674$

13. $\log_2 3a^4 = \log_2 3 + \log_2 a^4 = \log_2 3 + 4\log_2 a$

14. $\ln \dfrac{5\sqrt{x}}{6} = \ln 5 + \ln\sqrt{x} - \ln 6$
$= \ln 5 + \dfrac{1}{2}\ln x - \ln 6$

15. $\ln \dfrac{x\sqrt{x+1}}{2e^4} = \ln x + \ln\sqrt{x+1} - \ln(2) - \ln e^4$
$= \ln x + \dfrac{1}{2}\ln(x+1) - \ln 2 - 4$

16. $\log_3 13 + \log_3 y = \log_3(13y)$

17. $4 \ln x - 4 \ln y = \ln x^4 - \ln y^4 = \ln\left(\dfrac{x^4}{y^4}\right) = \ln\left(\dfrac{x}{y}\right)^4$

18. $\ln x - \ln(x+2) + \ln(2x-3) = \ln\left[\dfrac{x(2x-3)}{x+2}\right]$

19. $3^x = 81 = 3^4$
$x = 4$

20. $5^{2x} = 2500$

$2x \ln 5 = \ln 2500$

$x = \dfrac{1}{2} \dfrac{\ln 2500}{\ln 5} \approx 2.431$

21. $\log_7 x = 3$

$$7^3 = x$$
$$x = 343$$

22. $\log_{10}(x-4) = 5$

$$10^5 = x - 4$$
$$x = 10^5 + 4$$
$$= 100{,}004$$

23. $\dfrac{1025}{8+e^{4x}} = 5$

$$1025 = 40 + 5e^{4x}$$
$$985 = 5e^{4x}$$
$$e^{4x} = 197$$
$$4x = \ln(197)$$
$$x = \frac{1}{4}\ln(197) \approx 1.321$$

24. $-xe^{-x} + e^{-x} = 0$

$$e^{-x}(1-x) = 0$$
$$x = 1$$

25. $\log_{10} x - \log_{10}(8-5x) = 2$

$$\log_{10}\left(\frac{x}{8-5x}\right) = 2$$
$$10^2 = \frac{x}{8-5x}$$
$$800 - 500x = x$$
$$800 = 501x$$
$$x = \frac{800}{501} \approx 1.597$$

26. $2x\ln x - x = 0$

$$2\ln x = 1,\ (x \neq 0)$$
$$\ln x = \frac{1}{2}$$
$$x = e^{1/2} \approx 1.649$$

27. $\frac{1}{2} = 1e^{k(22)}$ (half-life is 22 years)

$$\ln\tfrac{1}{2} = 22k$$
$$k = \tfrac{1}{22}\ln\tfrac{1}{2} = -\tfrac{1}{22}\ln 2 \approx -0.03151$$
$$A = e^{-0.03151(19)} \approx 0.54953 \text{ or } 55\% \text{ remains}$$

28. (a) Logarithmic model: $R = 200.7\ln t + 57.835$

Exponential model: $R = 115.47(1.227)^t$

Power model: $R = 119.22t^{0.6703}$

(b) Logarithmic model:

Exponential model:

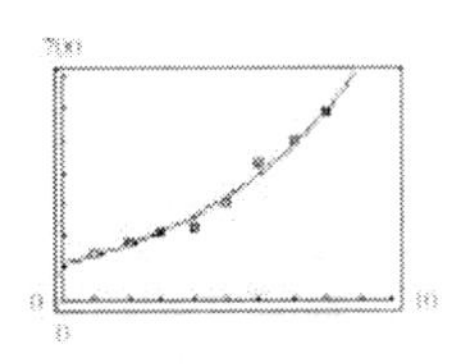

Power model:

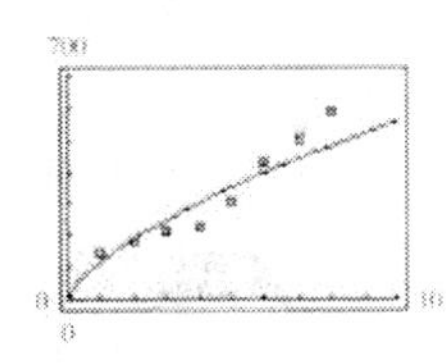

(c) The exponential model is the best fit. Using the exponential model, let $t = 15$ and find R.

$$R = 115.47(1.227)^{15}$$
$$R \approx 2483.92 \text{ or } \$2{,}483{,}920{,}000$$

Cumulative Test for Chapters 1–3

1. (a) Slope $= \dfrac{8-4}{-5-(-1)} = \dfrac{4}{-4} = -1$

$$y - 8 = -1(x - (-5)) = -x - 5$$
$$y = -x + 3$$

(b) Three additional points: (0, 3), (1, 2), (2, 1)

2. (a) $y - 1 = -2\left(x + \dfrac{1}{2}\right)$

$$y - 1 = -2x - 1$$
$$y = -2x$$

(b) Three additional points: (0, 0), (1, −2), (2, −4)

3. (a) Vertical line: $x = -\frac{3}{7}$ or $x + \frac{3}{7} = 0$

(b) Three additional points: $\left(-\frac{3}{7}, 0\right), \left(-\frac{3}{7}, 1\right), \left(-\frac{3}{7}, 2\right)$

4. $f(x) = \dfrac{x}{x-2}$

(a) $f(5) = \dfrac{5}{5-2} = \dfrac{5}{3}$

(b) $f(2)$ is undefined (division by 0).

(c) $f(5+4s) = \dfrac{5+4s}{(5+4s)-2} = \dfrac{5+4s}{3+4s}$

5. $f(x) = \begin{cases} 3x-8, & x < 0 \\ x^2+4, & x \geq 0 \end{cases}$

(a) $f(-8) = 3(-8) - 8 = -32$

(b) $f(0) = 0^2 + 4 = 4$

(c) $f(4) = 4^2 + 4 = 20$

6. No, for some x values there corresponds two values of y.

7.

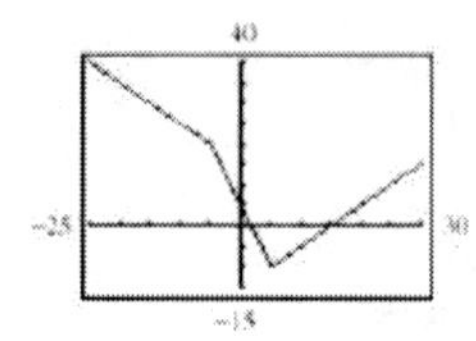

Decreasing on $(-\infty, 5)$
Increasing on $(5, \infty)$

8. $f(x) = \sqrt{x}$

(a) The graph of $r(x) = \frac{1}{4}\sqrt{x}$ is a vertical shrink of the graph of $f(x) = \sqrt{x}$.

(b) The graph of $h(x) = \sqrt{x} - 3$ is a vertical shift three units downward of the graph of $f(x) = \sqrt{x}$.

(c) The graph of $g(x) = -\sqrt{x+3}$ is a horizontal shift three units to the left and a reflection in the x-axis of the graph of $f(x) = \sqrt{x}$.

9. $(f+g)(-4) = f(-4) + g(-4)$

$= \left[-(-4)^2 + 3(-4) - 10\right] + \left[4(-4) + 1\right]$

$= -38 - 15 = -53$

10. $(g-f)\left(\frac{3}{4}\right) = \left[4\left(\frac{3}{4}\right) + 1\right] - \left[-\left(\frac{3}{4}\right)^2 + 3\left(\frac{3}{4}\right) - 10\right]$

$= 4 - (-8.3125) = 12.3125 = \dfrac{197}{16}$

11. $(g \circ f)(-2) = g(f(-2)) = g(-20) = 4(-20) + 1 = -79$

12. $(fg)(-1) = f(-1)g(-1) = (-14)(-3) = 42$

13. Yes, $h(x) = 5x - 2$ has an inverse function.

$y = 5x - 2$

$x = 5y - 2$

$x + 2 = 5y$

$\dfrac{x+2}{5} = y$

$h^{-1}(x) = \dfrac{x+2}{5}$

14. $f(x) = -(x-2)^2 + 5$

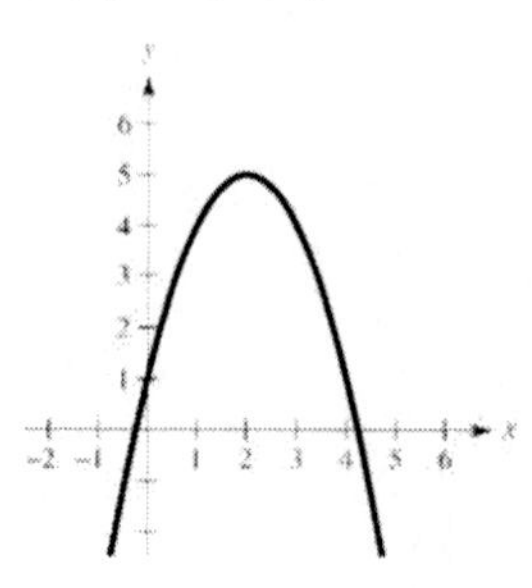

15. $f(x) = x^2 - 6x + 5$

$f(x) = (x-3)^2 - 4$

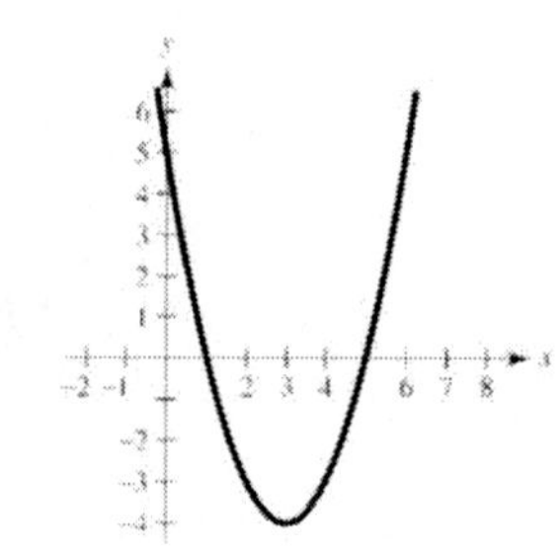

16. $f(x) = x^3 + 2x^2 - 9x - 18$

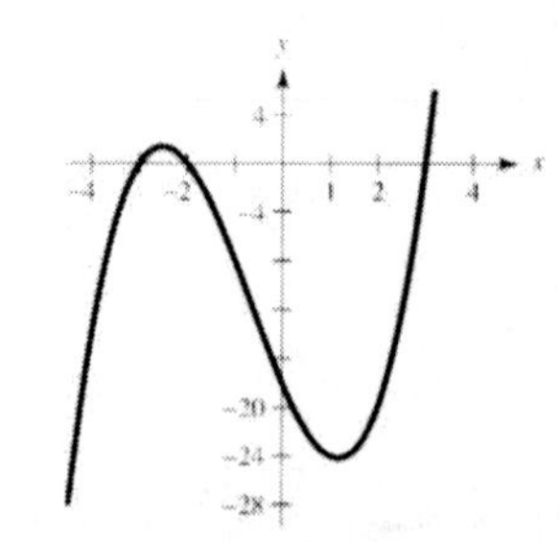

17. $x^3 + 2x^2 + 4x + 8 = (x+2)(x^2+4)$

$x + 2 = 0 \Rightarrow x = -2$

$x^2 + 4 = 0$

$x^2 = -4$

$x = \pm\sqrt{-4} \Rightarrow \pm 2i$

Zeros: $-2, \ \pm 2i$

18. Using a graphing utility, $x \approx 1.424$.

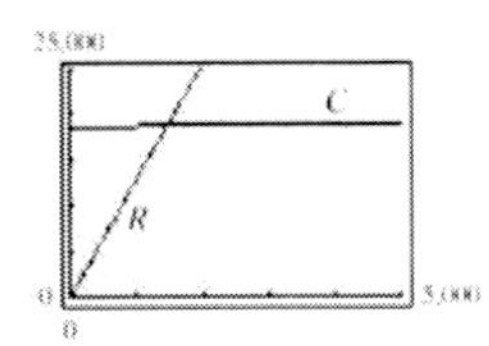

19.
$$\begin{array}{r} 4x+2 \\ x+3\overline{)4x^2+14x-9} \\ \underline{4x^2+12x} \\ 2x-9 \\ \underline{2x+6} \\ -15 \end{array}$$

$$\frac{4x^2+14x-9}{x+3} = 4x+2-\frac{15}{x+3}$$

20.
$$\begin{array}{c|cccc} 6 & 2 & -5 & 6 & -20 \\ & & 12 & 42 & 288 \\ \hline & 2 & 7 & 48 & 268 \end{array}$$

$$\frac{2x^3-5x^2+6x-20}{x-6} = 2x^2+7x+48+\frac{268}{x-6}$$

21. $(-5+4i)(-5-4i) = 25-16i^2$
$= 25-16(-1)$
$= 25+16 = 41$

22. $f(x) = (x-0)(x+3)\left[x-\left(1+\sqrt{5}i\right)\right]\left[x-\left(1-\sqrt{5}i\right)\right]$
$= x(x+3)[(x-1)^2+5]$
$= (x^2+3x)(x^2-2x+6)$
$= x^4+x^3+18x$

23. $f(x) = \dfrac{2x}{x-3}$

Vertical asymptote: $x = 3$
Horizontal asymptote: $y = 2$

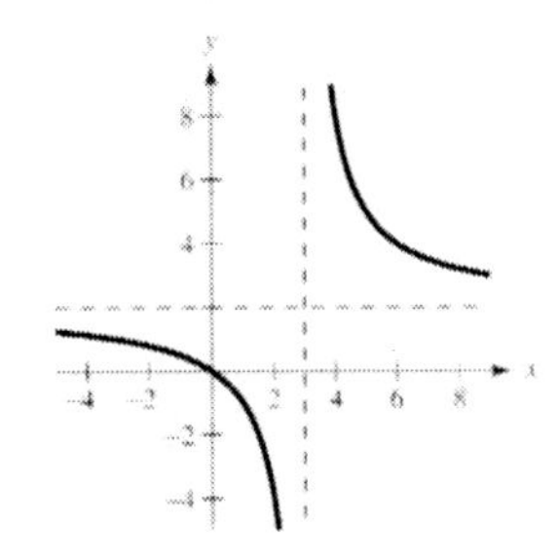

24. $f(x) = \dfrac{5x}{x^2+x-6} = \dfrac{5x}{(x+3)(x-2)}$

Vertical asymptote: $x = -3, 2$
Horizontal asymptote: $y = 0$

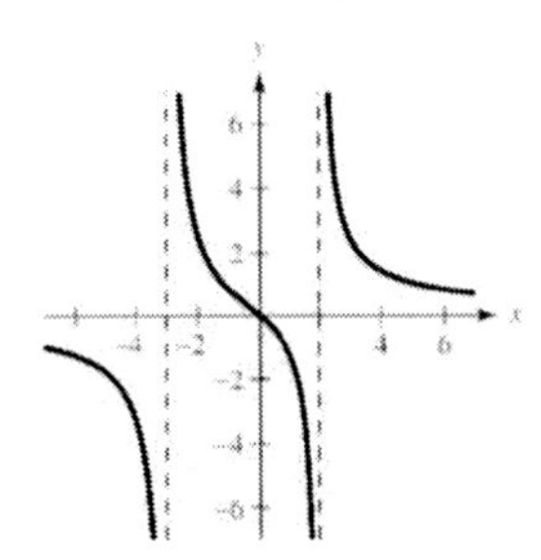

25. $f(x) = \dfrac{x^2-3x+8}{x-2} = x-1+\dfrac{6}{x-2}$

Vertical asymptote: $x = 2$
Slant asymptote: $y = x-1$

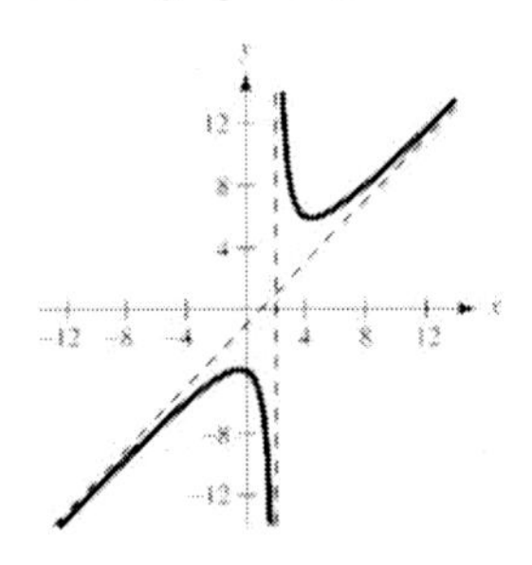

26. $(1.85)^{3.1} \approx 6.733$

27. $58^{\sqrt{5}} \approx 8772.934$

28. $e^{-\frac{8}{5}} \approx 0.202$

29. $4e^{2.56} \approx 51.743$

30. $f(x) = -3^{x+4}-5$

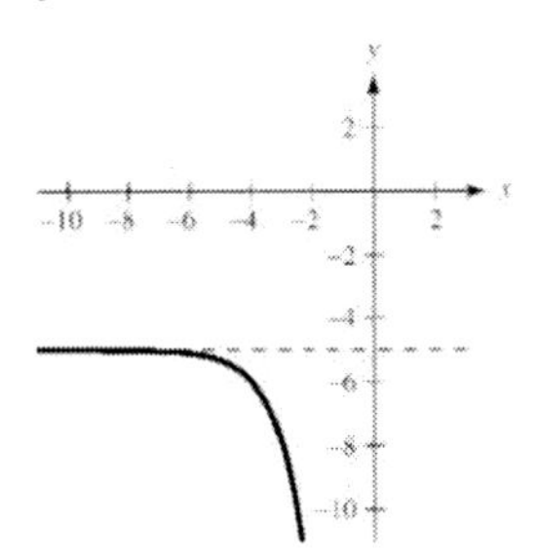

31. $f(x) = -\left(\frac{1}{2}\right)^{-x}-3$

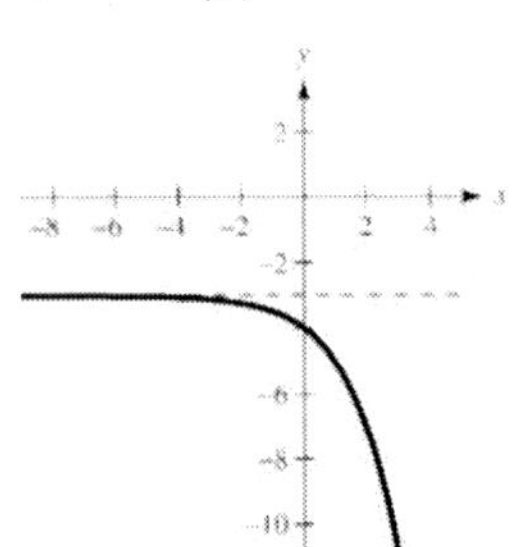

32. 
$f(x) = 4 + \log_{10}(x - 3)$

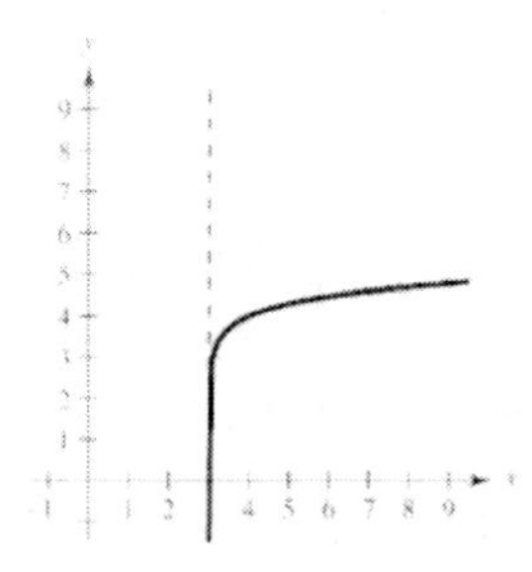

33. $f(x) = \ln(4 - x)$

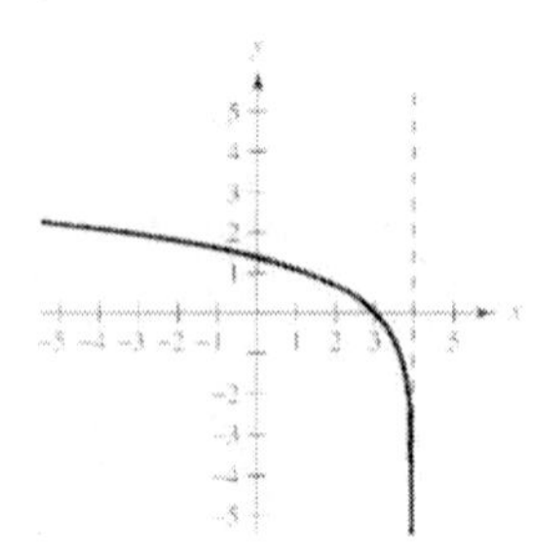

34. $\log_5 16 = \dfrac{\ln 16}{\ln 5} \approx 1.723$

35. $\log_9 6.8 = \dfrac{\ln 6.8}{\ln 9} \approx 0.872$

36. $\log_2\left(\dfrac{3}{2}\right) = \dfrac{\ln\left(\frac{3}{2}\right)}{\ln 2} \approx 0.585$

37. $\ln\left(\dfrac{x^2 - 4}{x^2 + 1}\right) = \ln[(x - 2)(x + 2)] - \ln(x^2 + 1)$

$= \ln(x - 2) + \ln(x + 2) - \ln(x^2 + 1)$

38. $2 \ln x - \ln(x - 1) + \ln(x + 1) = \ln x^2\left(\dfrac{x + 1}{x - 1}\right)$

39. $6e^{2x} = 72$

$e^{2x} = 12$

$2x = \ln 12$

$x = \dfrac{1}{2}\ln 12 \approx 1.242$

40. $4^{x-5} + 21 = 30$

$4^{x-5} = 9$

$(x - 5)\ln 4 = \ln 9$

$x - 5 = \dfrac{\ln 9}{\ln 4}$

$x = 5 + \dfrac{\ln 9}{\ln 4} \approx 6.585$

41. $\log_2 x + \log_2 5 = 6$

$\log_2 5x = 6$

$5x = 2^6 = 64$

$x = \dfrac{64}{5} = 12.8$

42. $250e^{0.05x} = 500{,}000$

$e^{0.05x} = 2000$

$0.05x = \ln 2000$

$x = 20 \ln 2000$

≈ 152.018

43. $2x^2e^{2x} - 2xe^{2x} = 0$

$(2x^2 - 2x)e^{2x} = 0$

$2x^2 - 2x = 0$

$2x(x - 1) = 0$

$x = 0, 1$

44. $\ln(2x - 5) - \ln x = 1$

$\ln\dfrac{2x - 5}{x} = 1$

$e = \dfrac{2x - 5}{x}$

$ex = 2x - 5$

$x(e - 2) = -5$

$x = \dfrac{-5}{e - 2} < 0$

No solution because $\ln\left(\dfrac{-5}{e - 2}\right)$ does not exist.

45. (a) Let x and y be the lengths of the sides.
$2x + 2y = 546 \Rightarrow y = 273 - x$
$A = xy = x(273 - x)$

(b)

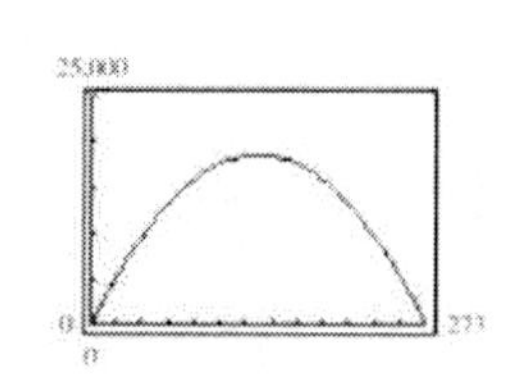

Domain: $0 < x < 273$

(c) If $A = 15{,}000$, then $x = 76.23$ or 196.77.
Dimensions in feet:
76.23×196.77 or 196.77×76.23

46. (a) Quadratic model: $y = 0.0178t^2 - 0.130t + 1.26$

$r^2 \approx 0.9778$

Exponential model: $y = 1.002(1.025)^t$

$r^2 \approx 0.4009$

Power model: $y = 1.041t^{0.0564}$

$r^2 \approx 0.1686$

(b) Quadratic model:

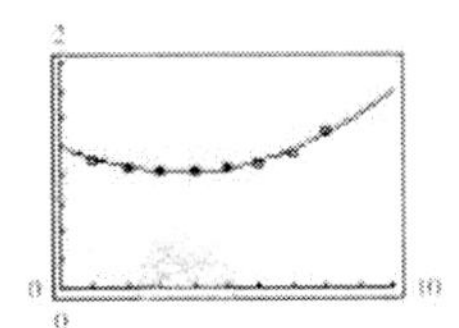

Exponential model:

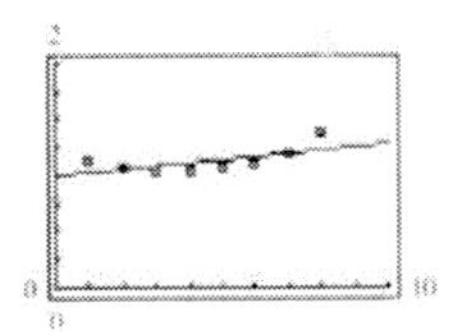

Power model:

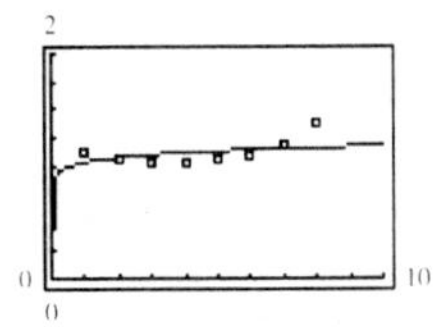

(c) The quadratic model is the best fit because its coefficient of determination, $r^2 \approx 0.9778$, is closest to 1.

(d) Using the quadratic model, let $t = 10$ and find y.

$y = 0.0178(10)^2 - 0.130(10) + 1.26$

$y \approx \$1.74$

Answers will vary.

CHAPTER 4

Section 4.1

1. Trigonometry

3. standard position

5. radian

7. One-half revolution of a circle is equal to $180°$ or π radians.

9. The angles $315°$ and $-225°$ are not coterminal. See figure.

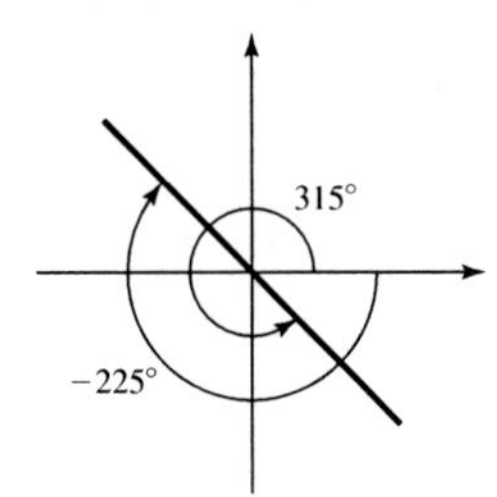

11.

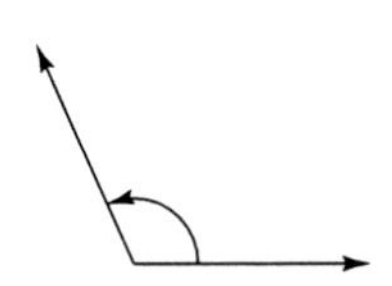

The angle shown is approximately 2 radians.

13. (a) Since $0 < \dfrac{\pi}{6} < \dfrac{\pi}{2}$, $\dfrac{\pi}{6}$ lies in Quadrant I.

(b) Since $\pi < \dfrac{5\pi}{4} < \dfrac{3\pi}{2}$, $\dfrac{5\pi}{4}$ lies in Quadrant III.

15. (a) Since $\dfrac{3\pi}{2} < \dfrac{7\pi}{4} < 2\pi$, $\dfrac{7\pi}{4}$ lies in Quadrant IV.

(b) Since $\dfrac{5\pi}{2} < \dfrac{11\pi}{4} < 3\pi$, $\dfrac{11\pi}{4}$ lies in Quadrant II.

17. (a) Since $-\dfrac{\pi}{2} < -1 < 0$, -1 lies in Quadrant IV.

(b) Since $-\pi < -2 < -\dfrac{\pi}{2}$, -2 lies in Quadrant III.

19. (a) $\dfrac{3\pi}{2}$

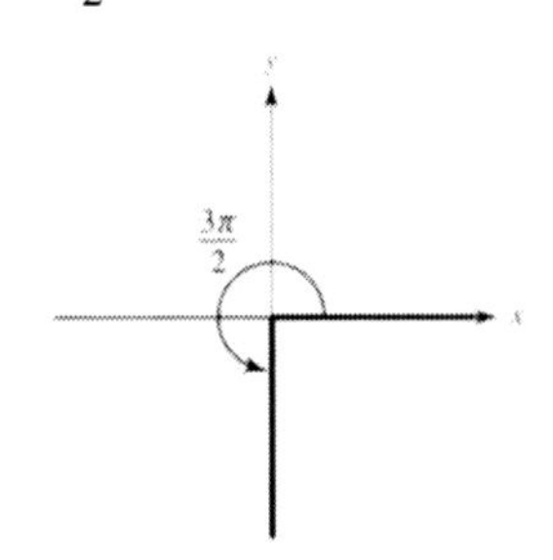

(b) $-\dfrac{7\pi}{2}$

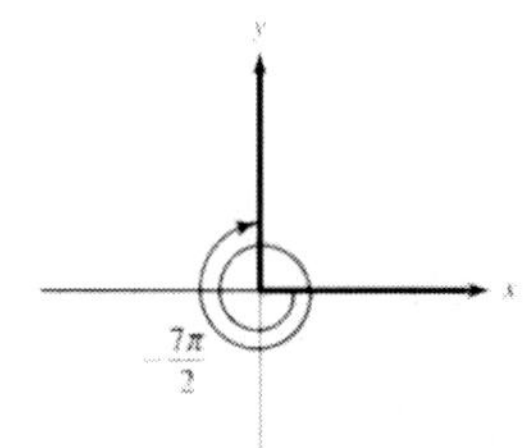

21. (a) $-\dfrac{7\pi}{4}$

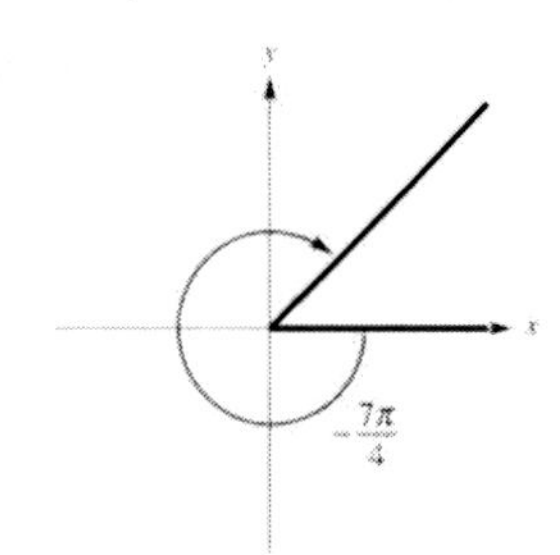

(b) $-\dfrac{5\pi}{2}$

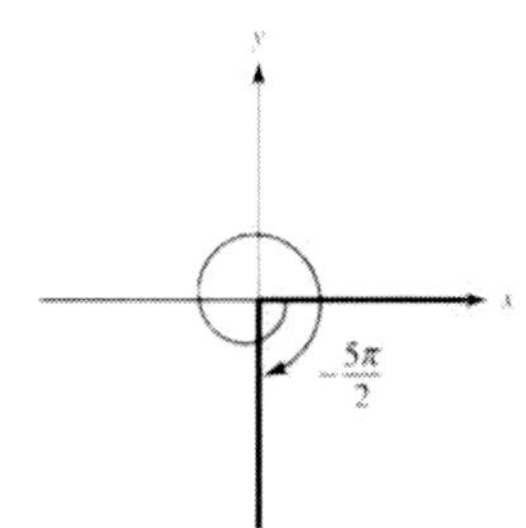

23. (a) 4

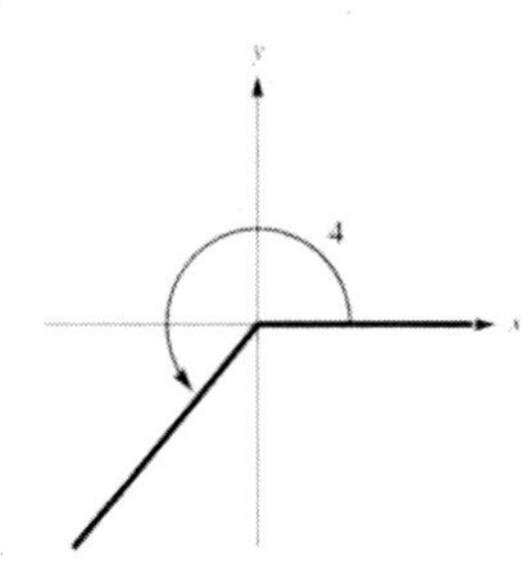

(b) -3

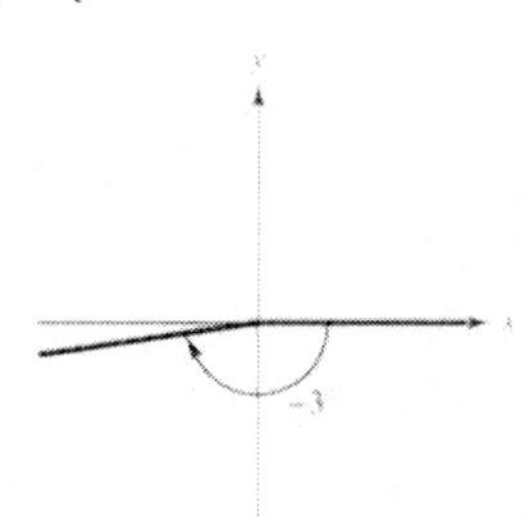

25. (a) Coterminal angles for $\frac{\pi}{6}$:

$$\frac{\pi}{6} + 2\pi = \frac{13\pi}{6}$$

$$\frac{\pi}{6} - 2\pi = -\frac{11\pi}{6}$$

(b) Coterminal angles for $\frac{2\pi}{3}$:

$$\frac{2\pi}{3} + 2\pi = \frac{8\pi}{3}$$

$$\frac{2\pi}{3} - 2\pi = -\frac{4\pi}{3}$$

27. (a) Coterminal angles for $-\frac{9\pi}{4}$:

$$-\frac{9\pi}{4} + 2\pi = -\frac{\pi}{4}$$

$$-\frac{9\pi}{4} + 4\pi = \frac{7\pi}{4}$$

(b) Coterminal angles for $-\frac{2\pi}{15}$:

$$-\frac{2\pi}{15} + 2\pi = \frac{28\pi}{15}$$

$$-\frac{2\pi}{15} - 2\pi = -\frac{32\pi}{15}$$

29.

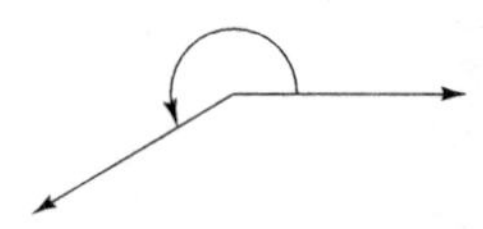

The angle shown is approximately $210°$.

31. (a) Since $0° < 55° < 90°$, $55°$ lies in Quadrant I.
(b) Since $180° < 215° < 270°$, $215°$ lies in Quadrant III.

33. (a) Since $90° < 150° < 180°$, $150°$ lies in Quadrant II.
(b) Since $270° < 282° < 360°$, $282°$ lies in Quadrant IV.

35. (a) Since $-180° < -132° 50' < -90°$, $-132° 50'$ lies in Quadrant III.
(b) Since $-360° < -336° 30' < -270°$, $-336° 30'$ lies in Quadrant I.

37. (a) $45°$

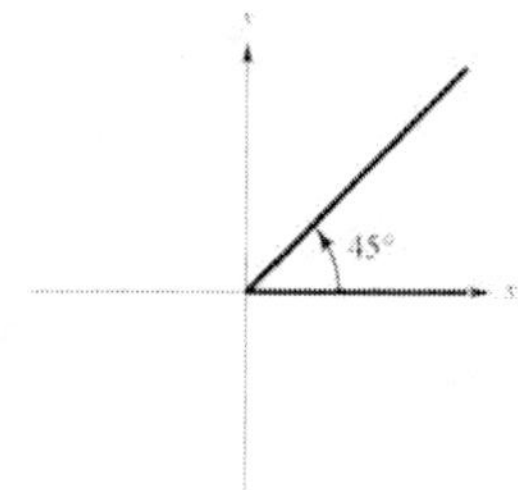

(b) $90°$

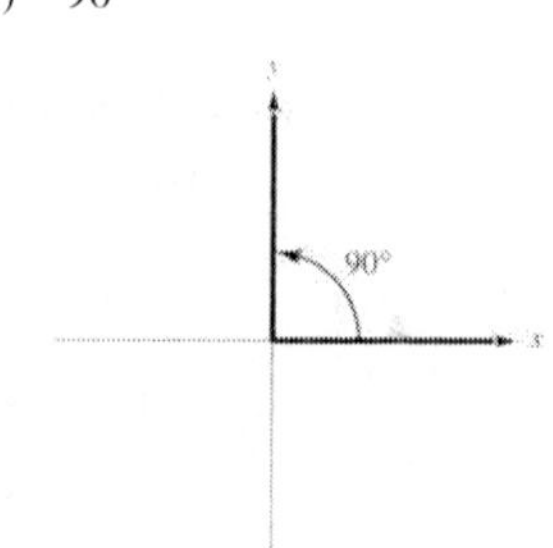

39. (a) $30°$

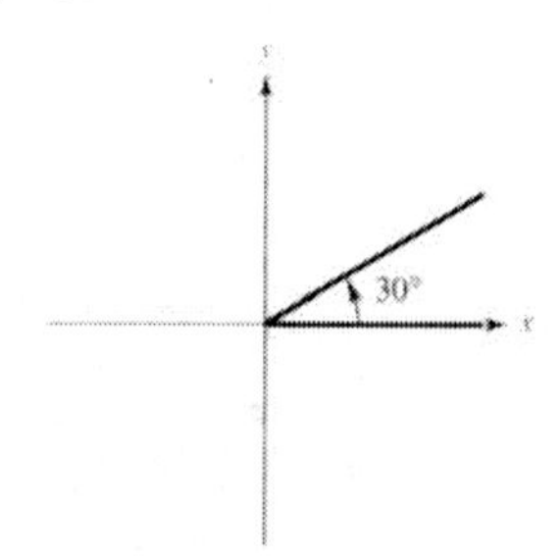

(b) $150°$

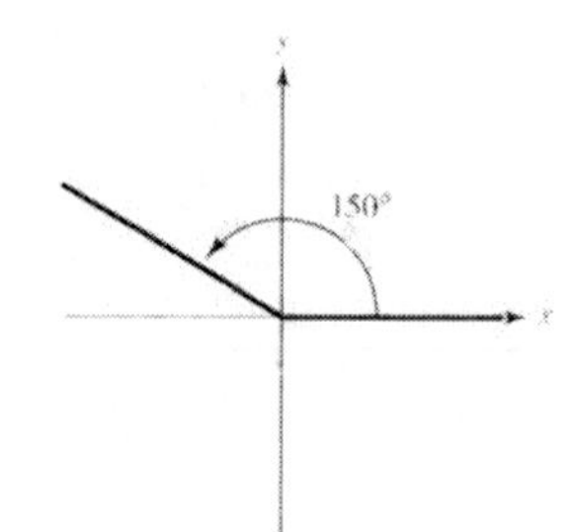

41. (a) $405°$

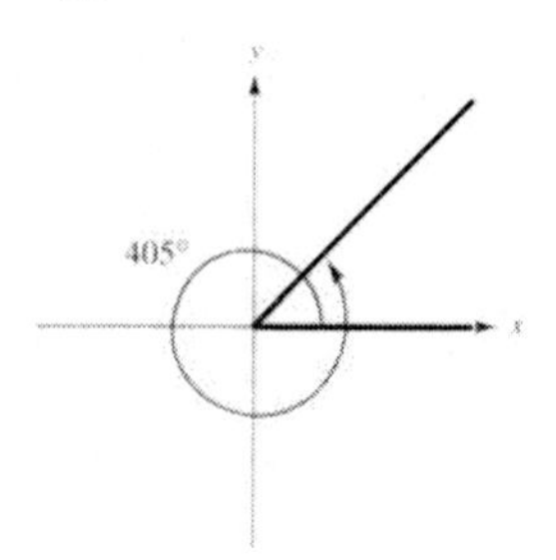

(b) 780°

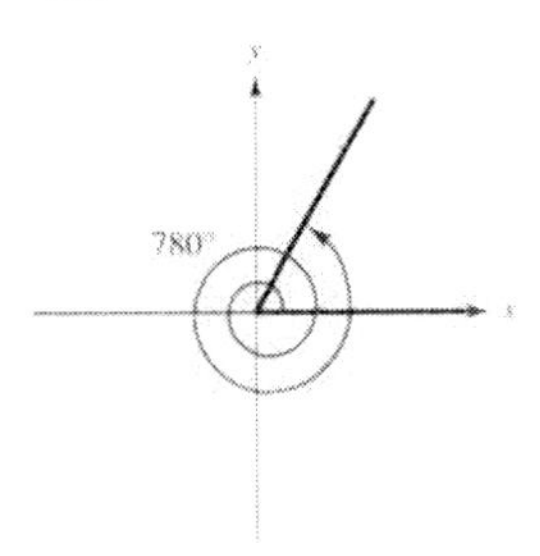

43. (a) Coterminal angles for $52°$:
$52° + 360° = 412°$
$52° - 360° = -308°$

(b) Coterminal angles for $-36°$:
$-36° + 360° = 324°$
$-36° - 360° = -396°$

45. (a) Coterminal angles for $300°$:
$300° + 360° = 660°$
$300° - 360° = -60°$

(b) Coterminal angles for $230°$:
$230° + 360° = 590°$
$230° - 360° = -130°$

47. (a) $30° = 30°\left(\frac{\pi}{180°}\right) = \frac{\pi}{6}$

(b) $150° = 150°\left(\frac{\pi}{180°}\right) = \frac{5\pi}{6}$

49. (a) $-20° = -20°\left(\frac{\pi}{180°}\right) = -\frac{\pi}{9}$

(b) $-240° = -240°\left(\frac{\pi}{180°}\right) = -\frac{4\pi}{3}$

51. (a) $\frac{3\pi}{2} = \frac{3\pi}{2}\left(\frac{180°}{\pi}\right) = 270°$

(b) $-\frac{7\pi}{6} = -\frac{7\pi}{6}\left(\frac{180°}{\pi}\right) = -210°$

53. (a) $\frac{7\pi}{3} = \frac{7\pi}{3}\left(\frac{180°}{\pi}\right) = 420°$

(b) $-\frac{13\pi}{60} = -\frac{13\pi}{60}\left(\frac{180°}{\pi}\right) = -39°$

55. $115° = 115°\left(\frac{\pi}{180°}\right) \approx 2.007$ radians

57. $-216.35° = -216.35°\left(\frac{\pi}{180°}\right) \approx -3.776$ radians

59. $-0.78° = -0.78°\left(\frac{\pi}{180°}\right) \approx -0.014$ radian

61. $\frac{\pi}{7} = \frac{\pi}{7}\left(\frac{180°}{\pi}\right) \approx 25.714°$

63. $6.5\pi = 6.5\pi\left(\frac{180°}{\pi}\right) = 1170°$

65. $-2 = -2\left(\frac{180°}{\pi}\right) \approx -114.592°$

67. $64°\ 45' = 64° + \left(\frac{45}{60}\right)^{\circ} = 64.75°$

69. $85°\ 18'\ 30'' = 85° + \left(\frac{18}{60}\right)^{\circ} + \left(\frac{30}{3600}\right)^{\circ} \approx 85.308°$

71. $-125°\ 36'' = -125° - \left(\frac{36}{3600}\right)^{\circ} = -125.01°$

73. $280.6° = 280° + 0.6(60)' = 280°\ 36'$

75. $-345.12° = -345°\ 7'\ 12''$

77. $-0.355 = -0.355\left(\frac{180°}{\pi}\right)$
$\approx -20.34° = -20°\ 20'\ 24''$

79. Complement: $90° - 24° = 66°$
Supplement: $180° - 24° = 156°$

81. Complement: $90° - 87° = 3°$
Supplement: $180° - 87° = 93°$

83. Complement: $\frac{\pi}{2} - \frac{\pi}{3} = \frac{\pi}{6}$

Supplement: $\pi - \frac{\pi}{3} = \frac{2\pi}{3}$

85. Complement: $\frac{\pi}{2} - \frac{\pi}{6} = \frac{\pi}{3}$

Supplement: $\pi - \frac{\pi}{6} = \frac{5\pi}{6}$

87. $s = r\theta$
$6 = 5\theta$
$\theta = \frac{6}{5}$ radians

89. $s = r\theta$
$8 = 15\theta$
$\theta = \frac{8}{15}$ radian

91. $s = r\theta$
$35 = 14.5\theta$
$\theta = \frac{70}{29} \approx 2.414$ radians

93. $s = r\theta$, θ in radians

$$s = 14(180)\left(\frac{\pi}{180}\right) = 14\pi \approx 43.982 \text{ inches}$$

95. $s = r\theta$, θ in radians

$$s = 27\left(\frac{2\pi}{3}\right) = 18\pi \text{ meters} \approx 56.55 \text{ meters}$$

97. $r = \dfrac{s}{\theta} = \dfrac{36}{\pi/2} = \dfrac{72}{\pi}$ feet ≈ 22.92 feet

99. $r = \dfrac{s}{\theta} = \dfrac{82}{135°(\pi/180°)} = \dfrac{328}{3\pi}$ miles ≈ 34.80 miles

101. The angle between Omaha and Dallas:

$$\theta = 41°\ 15'\ 50'' - 32°\ 47'\ 39''$$
$$= 8°\ 28'\ 11'' \approx 0.1478 \text{ radian}$$
$$s = \theta r = (0.1478)(4000) \approx 591.2 \text{ miles}$$

103. $\theta = \dfrac{s}{r} = \dfrac{450}{6378} \approx 0.07056$ radian $\approx 4.04°$

$\approx 4°\ 2'\ 33.02''$

105. $\theta = \dfrac{s}{r} = \dfrac{24}{5} = 4.8$ rad $\approx 275.02°$

107. Linear speed $= \dfrac{s}{t} = \dfrac{r\theta}{t} = \dfrac{(6400 + 1250)2\pi}{110} \approx 436.967$ km/min

109. (a) $\dfrac{\text{Revolutions}}{\text{Second}} = \dfrac{2400}{60} = 40$ rev/sec

Angular speed $= (2\pi)(40) = 80\pi$ rad/sec

(b) Radius of saw blade $= \dfrac{7.5}{2} = 3.75$ in.

Radius in feet $= \dfrac{3.75}{12} = 0.3125$ ft

Speed $= \dfrac{s}{t} = \dfrac{r\theta}{t} = r\dfrac{\theta}{t} = r(\text{angular speed})$

$= 0.3125(80\pi) = 78.54$ ft/sec

111. (a) $200 \le \dfrac{\text{revolutions}}{\text{minute}} \le 500$

$2\pi(200) \le$ angular speed $\le 2\pi(500)$

400π rad/min $\le$ angular speed $\le 1000\pi$ rad/min

(b) Speed $= \dfrac{s}{t} = \dfrac{r\theta}{t} = r\dfrac{\theta}{t} = r(\text{angular speed}) = 6(\text{angular speed})$

$6(400\pi) \le$ linear speed $\le 6(1000\pi)$

For the outermost track, 6000π cm/min

113. False, 1 radian $= \left(\dfrac{180}{\pi}\right)^{\circ} \approx 57.3°$, so one radian is much larger than one degree.

115. True: $\dfrac{2\pi}{3} + \dfrac{\pi}{4} + \dfrac{\pi}{12} = \dfrac{8\pi + 3\pi + \pi}{12} = \pi = 180°$

117. $A = \dfrac{1}{2}r^2\theta = \dfrac{1}{2}(10)^2 \cdot \dfrac{\pi}{3} = \dfrac{50}{3}\pi$ square meters

119. $A = \dfrac{1}{2}r^2\theta$, $s = r\theta$

(a) $\theta = 0.8 \Rightarrow A = \dfrac{1}{2}r^2(0.8) = 0.4r^2$ Domain: $r > 0$

$s = r\theta = r(0.8)$ Domain: $r > 0$

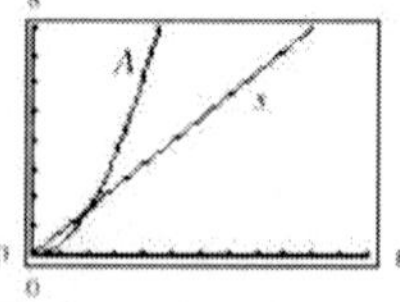

The area function changes more rapidly for $r > 1$ because it is quadratic and the arc length function is linear.

(b) $r = 10 \Rightarrow A = \dfrac{1}{2}(10^2)\theta = 50\theta$ Domain: $0 < \theta < 2\pi$

$s = r\theta = 10\theta$ Domain: $0 < \theta < 2\pi$

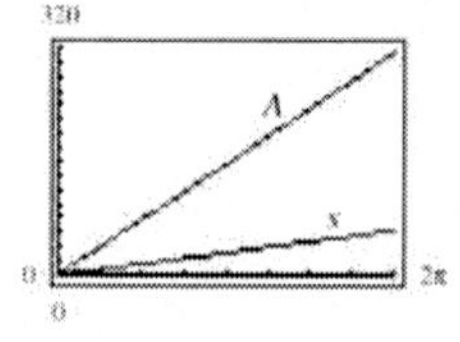

121. Answers will vary.

123. The graph of g is a horizontal shift one unit to the right of $f(x) = x^3$.

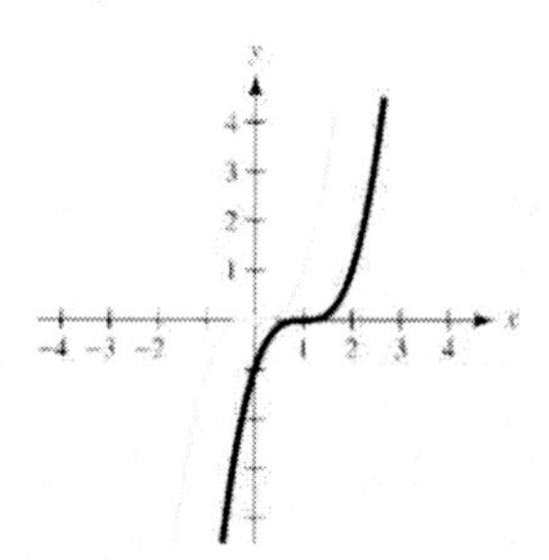

125. The graph of g is a reflection in the x-axis and a vertical shift two units upward of $f(x) = x^3$.

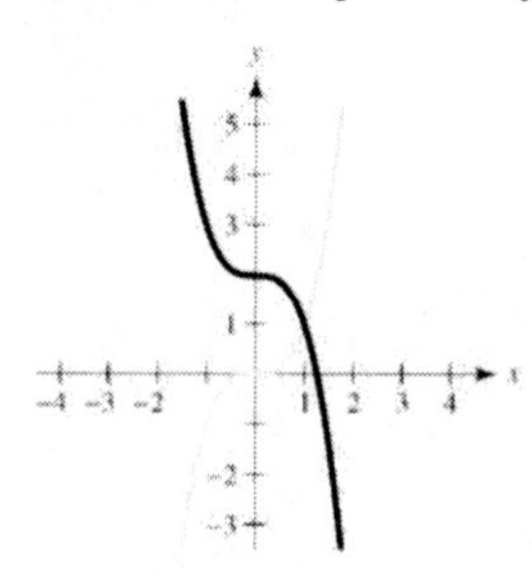

127. The graph of g is a horizontal shift one unit to the left and a vertical shift three units downward of $f(x) = x^3$.

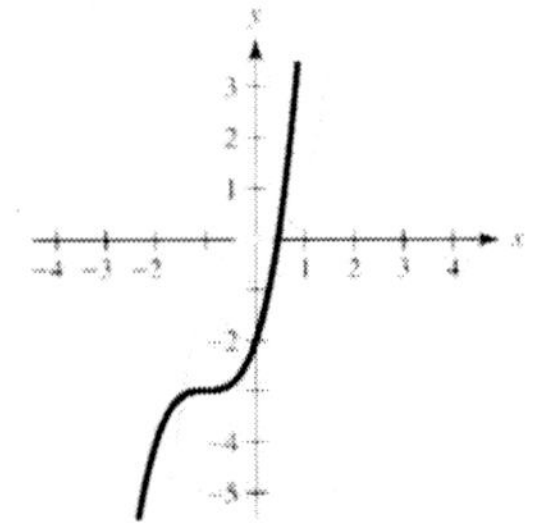

Section 4.2

1. unit circle

3. odd, even

5. The cosine and secant functions are even; the sine, tangent, cosecant, and cotangent functions are odd.

7. The unit circle can be divided into eight equal arcs using t-values that are multiples of $\frac{\pi}{4}$.

9. $\sin\theta = y = \frac{15}{17}$

$\cos\theta = x = -\frac{8}{17}$

$\tan\theta = \frac{y}{x} = -\frac{15}{8}$

$\cot\theta = \frac{x}{y} = -\frac{8}{15}$

$\sec\theta = \frac{1}{x} = -\frac{17}{8}$

$\csc\theta = \frac{1}{y} = \frac{17}{15}$

11. $\sin\theta = y = -\frac{5}{13}$

$\cos\theta = x = \frac{12}{13}$

$\tan\theta = \frac{y}{x} = -\frac{5}{12}$

$\cot\theta = \frac{x}{y} = -\frac{12}{5}$

$\sec\theta = \frac{1}{x} = \frac{13}{12}$

$\csc\theta = \frac{1}{y} = -\frac{13}{5}$

13. $t = \frac{\pi}{4}$ corresponds to $\left(\frac{\sqrt{2}}{2}, \frac{\sqrt{2}}{2}\right)$.

15. $t = \frac{7\pi}{6}$ corresponds to $\left(-\frac{\sqrt{3}}{2}, -\frac{1}{2}\right)$.

17. $t = \frac{2\pi}{3}$ corresponds to $\left(-\frac{1}{2}, \frac{\sqrt{3}}{2}\right)$.

19. $t = \frac{3\pi}{2}$ corresponds to $(0, -1)$.

21. $t = -\frac{7\pi}{4}$ corresponds to $\left(\frac{\sqrt{2}}{2}, \frac{\sqrt{2}}{2}\right)$.

23. $t = \frac{\pi}{4}$ corresponds to $\left(\frac{\sqrt{2}}{2}, \frac{\sqrt{2}}{2}\right)$.

$\sin t = y = \frac{\sqrt{2}}{2}$

$\cos t = x = \frac{\sqrt{2}}{2}$

$\tan t = \frac{y}{x} = 1$

25. $t = -\frac{7\pi}{4}$ corresponds to $\left(\frac{\sqrt{2}}{2}, \frac{\sqrt{2}}{2}\right)$.

$\sin t = y = \frac{\sqrt{2}}{2}$

$\cos t = x = \frac{\sqrt{2}}{2}$

$\tan t = \frac{y}{x} = 1$

27. $t = \frac{2\pi}{3}$ corresponds to $\left(-\frac{1}{2}, \frac{\sqrt{3}}{2}\right)$.

$\sin t = y = \frac{\sqrt{3}}{2}$

$\cos t = x = -\frac{1}{2}$

$\tan t = \frac{y}{x} = -\sqrt{3}$

29. $t = -\frac{5\pi}{3}$ corresponds to $\left(\frac{1}{2}, \frac{\sqrt{3}}{2}\right)$.

$\sin t = y = \frac{\sqrt{3}}{2}$

$\cos t = x = \frac{1}{2}$

$\tan t = \frac{y}{x} = \sqrt{3}$

31. $t = -\frac{\pi}{6}$ corresponds to $\left(\frac{\sqrt{3}}{2}, -\frac{1}{2}\right)$.

$\sin t = y = -\frac{1}{2}$

$\cos t = x = \frac{\sqrt{3}}{2}$

$\tan t = \frac{y}{x} = -\frac{\sqrt{3}}{3}$

33. $t = \frac{3\pi}{4}$ corresponds to $\left(-\frac{\sqrt{2}}{2}, \frac{\sqrt{2}}{2}\right)$.

$\sin t = y = \frac{\sqrt{2}}{2}$

$\cos t = x = -\frac{\sqrt{2}}{2}$

$\tan t = \frac{y}{x} = -1$

$\csc t = \frac{1}{y} = \sqrt{2}$

$\sec t = \frac{1}{x} = -\sqrt{2}$

$\cot t = \frac{x}{y} = -1$

35. $t = \frac{\pi}{2}$ corresponds to $(0, 1)$.

$\sin t = y = 1$

$\cos t = x = 0$

$\tan t = \frac{y}{x}$ is undefined.

$\csc t = \frac{1}{y} = 1$

$\sec t = \frac{1}{x}$ is undefined.

$\cot t = \frac{x}{y} = 0$

37. $t = -\frac{4\pi}{3}$ corresponds to $\left(-\frac{1}{2}, \frac{\sqrt{3}}{2}\right)$.

$\sin t = y = \frac{\sqrt{3}}{2}$

$\cos t = x = -\frac{1}{2}$

$\tan t = \frac{y}{x} = -\sqrt{3}$

$\cot t = \frac{x}{y} = -\frac{\sqrt{3}}{3}$

$\sec t = \frac{1}{x} = -2$

$\csc t = \frac{1}{y} = \frac{2\sqrt{3}}{3}$

39. $\sin 5\pi = \sin \pi = 0$

41. Because $\frac{8\pi}{3} = 2\pi + \frac{2\pi}{3}$:

$$\cos\frac{8\pi}{3} = \cos\left(2\pi + \frac{2\pi}{3}\right) = \cos\frac{2\pi}{3} = -\frac{1}{2}$$

43. Because $-\frac{13\pi}{6} = -2\pi - \frac{\pi}{6}$:

$$\cos\left(-\frac{13\pi}{6}\right) = \cos\left(-\frac{\pi}{6}\right) = \cos\left(\frac{11\pi}{6}\right) = \frac{\sqrt{3}}{2}$$

45. Because $-\frac{9\pi}{4} = -2\pi - \frac{\pi}{4}$

$$\sin\left(-\frac{9\pi}{4}\right) = \sin\left(-\frac{\pi}{4}\right) = -\frac{\sqrt{2}}{2}$$

47. $\sin t = \frac{1}{3}$

(a) $\sin(-t) = -\sin t = -\frac{1}{3}$

(b) $\csc(-t) = -\csc t = -3$

49. $\cos(-t) = -\frac{1}{5}$

(a) $\cos t = \cos(-t) = -\frac{1}{5}$

(b) $\sec(-t) = \frac{1}{\cos(-t)} = -5$

51. $\sin t = \frac{4}{5}$

(a) $\sin(\pi - t) = \sin t = \frac{4}{5}$

(b) $\sin(t + \pi) = -\sin t = -\frac{4}{5}$

53. $\sin\frac{\pi}{7} \approx 0.4339$

55. $\cos\frac{11\pi}{5} \approx 0.8090$

57. $\csc 1.3 \approx 1.0378$

59. $\cos(-1.7) \approx -0.1288$

61. $\csc 0.8 = \frac{1}{\sin 0.8} \approx 1.3940$

63. $\sec 22.8 = \frac{1}{\cos 22.8} \approx -1.4486$

65. $\cot 2.5 = \frac{1}{\tan 2.5} \approx -1.3386$

67. $\csc(-1.5) = \frac{1}{\sin(-1.5)} \approx -1.0025$

69. $\sec(-4.3) = \frac{1}{\cos(-4.3)} \approx -2.4950$

71. (a) $\sin 5 \approx -1$

(b) $\cos 2 \approx -0.4$

73. (a) $\sin t = 0.25$

$t \approx 0.25$ or 2.89

(b) $\cos t = -0.25$

$t \approx 1.82$ or 4.46

75. $I = 5e^{-2t}\sin t$

$I(0.7) = 5e^{-1.4}\sin 0.7$

≈ 0.79 ampere

77. $y(t) = \frac{1}{4}\cos 6t$

(a) $y(0) = \frac{1}{4}\cos 0 = 0.2500$ ft

(b) $y\left(\frac{1}{4}\right) = \frac{1}{4}\cos\frac{3}{2} \approx 0.0177$ ft

(c) $y\left(\frac{1}{2}\right) = \frac{1}{4}\cos 3 \approx -0.2475$ ft

79. False. $\sin\left(\frac{-4\pi}{3}\right) = \frac{\sqrt{3}}{2} > 0$

81. True, because a is coterminal with $a - 6\pi$, $\tan a = \tan(a - 6\pi)$.

83. True. The values are the same.

85. (a) The points (x_1, y_1) and (x_2, y_2) are symmetric about the origin.

(b) Because of the symmetry of the points, you can make the conjecture that $\sin(t_1 + \pi) = -\sin t_1$.

(c) Because of the symmetry of the points, you can make the conjecture that $\cos(t_1 + \pi) = -\cos t_1$.

87. $\cos\theta = x = \cos(-\theta)$

$\sec\theta = \frac{1}{\cos\theta} = \frac{1}{\cos(-\theta)} = \sec(-\theta)$

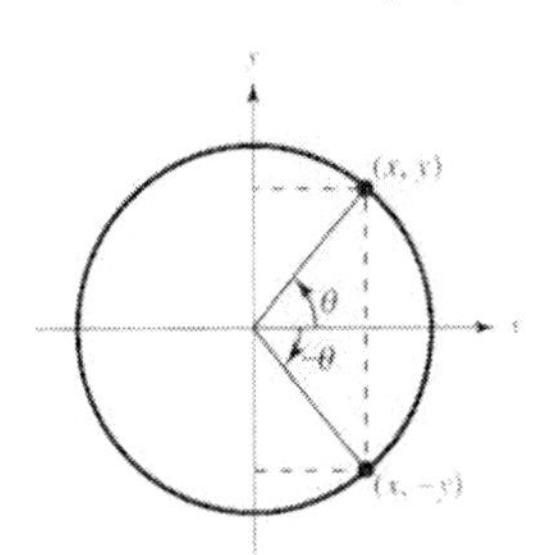

$\sin\theta = y = -\sin(-\theta)$

$\csc\theta = \frac{1}{y} = -\csc(-\theta)$

$\tan\theta = \frac{y}{x} = -\tan(-\theta)$

$\cot\theta = \frac{x}{y} = -\cot(-\theta)$

89. $f(t) = \sin t$ and $g(t) = \tan t$

Both f and g are odd functions.

$h(t) = f(t)g(t) = \sin t \tan t$

$h(-t) = \sin(-t)\tan(-t)$

$= (-\sin t)(-\tan t)$

$= \sin t \tan t = h(t)$

The function $h(t) = f(t)g(t)$ is even.

91. $f(x) = \dfrac{2x}{x-3}$

Asymptotes: $x = 3,\ y = 2$

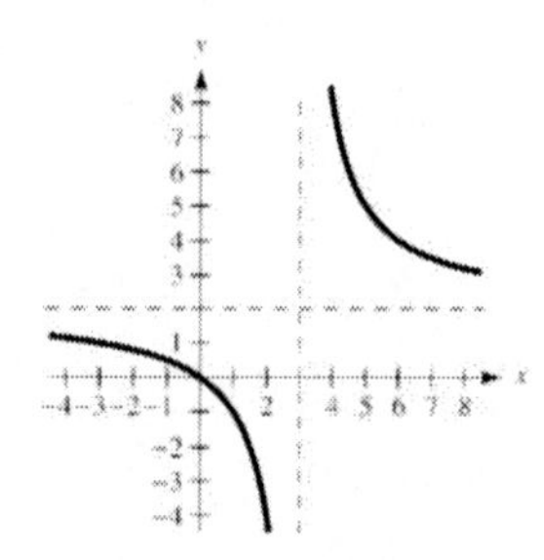

93. $f(x) = \dfrac{x^2+3x-10}{2x^2-8} = \dfrac{(x-2)(x+5)}{2(x-2)(x+2)}$

$= \dfrac{x+5}{2(x+2)},\quad x \neq 2$

Asymptotes: $x = -2,\ y = \dfrac{1}{2}$

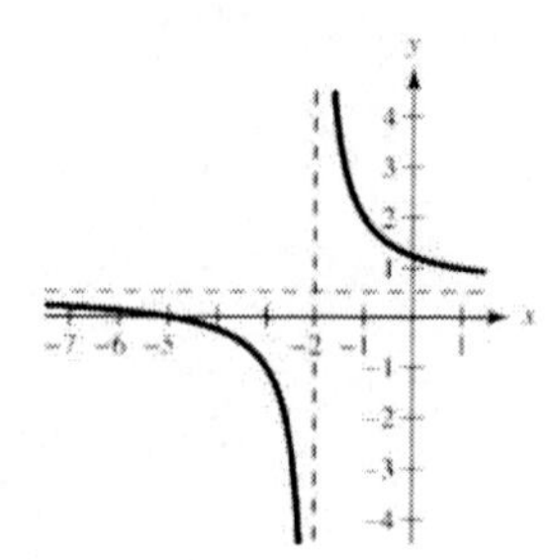

Section 4.3

1. (a) iii
(b) vi
(c) ii
(d) v
(e) i
(f) iv

3. elevation, depression

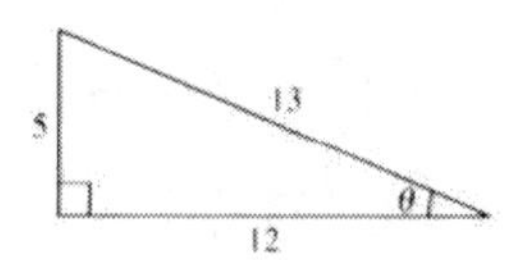

Figure for Exercise 4–6

5. The side adjacent to θ has length 12.

7.

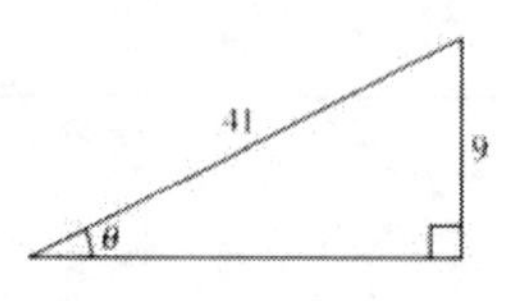

$\text{adj} = \sqrt{41^2 - 9^2} = \sqrt{1600} = 40$

$\sin\theta = \dfrac{\text{opp}}{\text{hyp}} = \dfrac{9}{41}$

$\cos\theta = \dfrac{\text{adj}}{\text{hyp}} = \dfrac{40}{41}$

$\tan\theta = \dfrac{\text{opp}}{\text{adj}} = \dfrac{9}{40}$

$\cot\theta = \dfrac{\text{adj}}{\text{opp}} = \dfrac{40}{9}$

$\sec\theta = \dfrac{\text{hyp}}{\text{adj}} = \dfrac{41}{40}$

$\csc\theta = \dfrac{\text{hyp}}{\text{opp}} = \dfrac{41}{9}$

9.

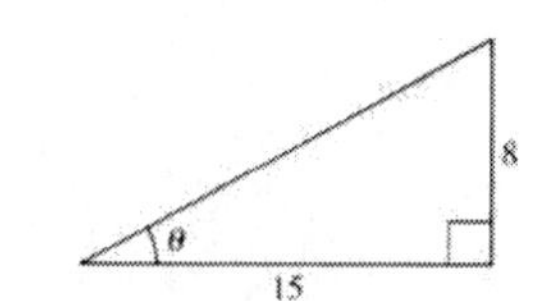

$\text{hyp} = \sqrt{8^2 + 15^2} = 17$

$\sin\theta = \dfrac{\text{opp}}{\text{hyp}} = \dfrac{8}{17}$

$\cos\theta = \dfrac{\text{adj}}{\text{hyp}} = \dfrac{15}{17}$

$\tan\theta = \dfrac{\text{opp}}{\text{adj}} = \dfrac{8}{15}$

$\csc\theta = \dfrac{\text{hyp}}{\text{opp}} = \dfrac{17}{8}$

$\sec\theta = \dfrac{\text{hyp}}{\text{adj}} = \dfrac{17}{15}$

$\cot\theta = \dfrac{\text{adj}}{\text{opp}} = \dfrac{15}{8}$

11.

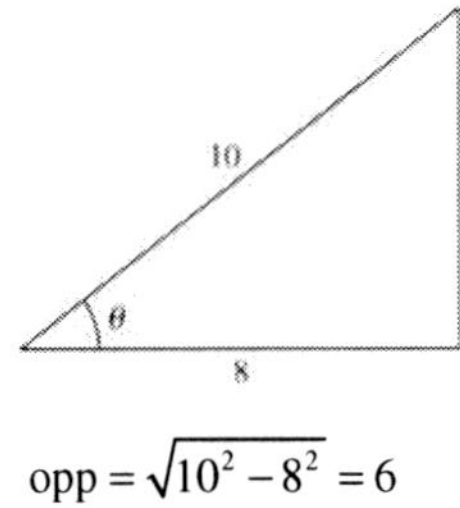

$\text{opp} = \sqrt{10^2 - 8^2} = 6$

$\sin\theta = \frac{\text{opp}}{\text{hyp}} = \frac{6}{10} = \frac{3}{5}$

$\cos\theta = \frac{\text{adj}}{\text{hyp}} = \frac{8}{10} = \frac{4}{5}$

$\tan\theta = \frac{\text{opp}}{\text{adj}} = \frac{6}{8} = \frac{3}{4}$

$\csc\theta = \frac{\text{hyp}}{\text{opp}} = \frac{10}{6} = \frac{5}{3}$

$\sec\theta = \frac{\text{hyp}}{\text{adj}} = \frac{10}{8} = \frac{5}{4}$

$\cot\theta = \frac{\text{adj}}{\text{opp}} = \frac{8}{6} = \frac{4}{3}$

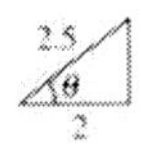

$\text{opp} = \sqrt{2.5^2 - 2^2} = 1.5$

$\sin\theta = \frac{\text{opp}}{\text{hyp}} = \frac{1.5}{2.5} = \frac{3}{5}$

$\cos\theta = \frac{\text{adj}}{\text{hyp}} = \frac{2}{2.5} = \frac{4}{5}$

$\tan\theta = \frac{\text{opp}}{\text{adj}} = \frac{1.5}{2} = \frac{3}{4}$

$\csc\theta = \frac{\text{hyp}}{\text{opp}} = \frac{2.5}{1.5} = \frac{5}{3}$

$\sec\theta = \frac{\text{hyp}}{\text{adj}} = \frac{2.5}{2} = \frac{5}{4}$

$\cot\theta = \frac{\text{adj}}{\text{opp}} = \frac{2}{1.5} = \frac{4}{3}$

The function values are the same since the triangles are similar and the corresponding sides are proportional.

13.

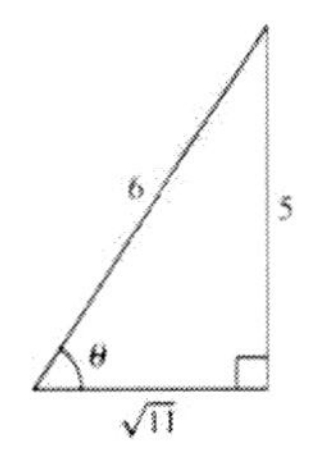

Given: $\sin\theta = \frac{5}{6} = \frac{\text{opp}}{\text{hyp}}$

$5^2 + (\text{adj})^2 = 6^2$

$\text{adj} = \sqrt{11}$

$\cos\theta = \frac{\text{adj}}{\text{hyp}} = \frac{\sqrt{11}}{6}$

$\tan\theta = \frac{\text{opp}}{\text{adj}} = \frac{5}{\sqrt{11}} = \frac{5\sqrt{11}}{11}$

$\cot\theta = \frac{\text{adj}}{\text{opp}} = \frac{\sqrt{11}}{5}$

$\sec\theta = \frac{\text{hyp}}{\text{adj}} = \frac{6}{\sqrt{11}} = \frac{6\sqrt{11}}{11}$

$\csc\theta = \frac{\text{hyp}}{\text{opp}} = \frac{6}{5}$

15.

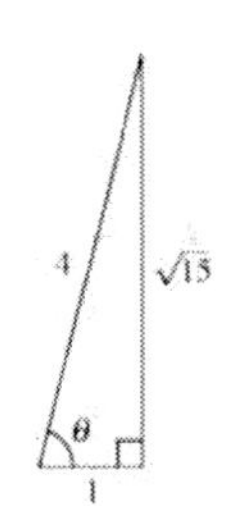

Given: $\sec\theta = 4 = \frac{4}{1} = \frac{\text{hyp}}{\text{adj}}$

$(\text{opp})^2 + 1^2 = 4^2$

$\text{opp} = \sqrt{15}$

$\sin\theta = \frac{\text{opp}}{\text{hyp}} = \frac{\sqrt{15}}{4}$

$\cos\theta = \frac{\text{adj}}{\text{hyp}} = \frac{1}{4}$

$\tan\theta = \frac{\text{opp}}{\text{adj}} = \sqrt{15}$

$\cot\theta = \frac{\text{adj}}{\text{opp}} = \frac{1}{\sqrt{15}} = \frac{\sqrt{15}}{15}$

$\csc\theta = \frac{\text{hyp}}{\text{opp}} = \frac{4}{\sqrt{15}} = \frac{4\sqrt{15}}{15}$

17.

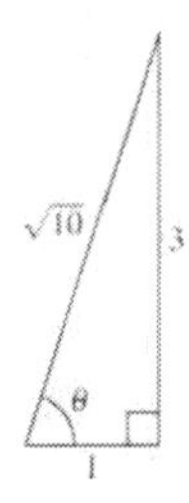

Given: $\tan\theta = 3 = \frac{3}{1} = \frac{\text{opp}}{\text{adj}}$

$$3^2 + 1^2 = (\text{hyp})^2$$
$$\text{hyp} = \sqrt{10}$$
$$\sin\theta = \frac{\text{opp}}{\text{hyp}} = \frac{3\sqrt{10}}{10}$$
$$\cos\theta = \frac{\text{adj}}{\text{hyp}} = \frac{\sqrt{10}}{10}$$
$$\sec\theta = \frac{\text{hyp}}{\text{adj}} = \sqrt{10}$$
$$\cot\theta = \frac{\text{adj}}{\text{opp}} = \frac{1}{3}$$
$$\csc\theta = \frac{\text{hyp}}{\text{opp}} = \frac{\sqrt{10}}{3}$$

19.

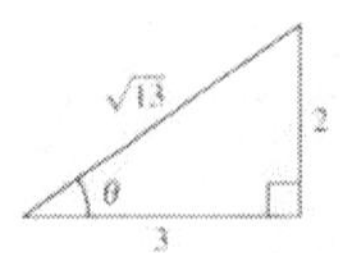

Given: $\cot\theta = \frac{3}{2} = \frac{\text{adj}}{\text{opp}}$

$$3^2 + 2^2 = \text{hyp}^2$$
$$13 = \text{hyp}^2$$
$$\sqrt{13} = \text{hyp}$$
$$\sin\theta = \frac{\text{opp}}{\text{hyp}} = \frac{2}{\sqrt{13}} = \frac{2\sqrt{13}}{13}$$
$$\cos\theta = \frac{\text{adj}}{\text{hyp}} = \frac{3}{\sqrt{13}} = \frac{3\sqrt{13}}{13}$$
$$\tan\theta = \frac{\text{opp}}{\text{adj}} = \frac{2}{3}$$
$$\sec\theta = \frac{\text{hyp}}{\text{adj}} = \frac{\sqrt{13}}{3}$$
$$\csc\theta = \frac{\text{hyp}}{\text{opp}} = \frac{\sqrt{13}}{2}$$

	Function	θ*(deg)*	θ*(rad)*	*Function Value*
21.	sin	30°	$\frac{\pi}{6}$	$\frac{1}{2}$
23.	tan	60°	$\frac{\pi}{3}$	$\sqrt{3}$
25.	cot	60°	$\frac{\pi}{3}$	$\frac{\sqrt{3}}{3}$
27.	cos	30°	$\frac{\pi}{6}$	$\frac{\sqrt{3}}{2}$
29.	cot	45°	$\frac{\pi}{4}$	1

31. (a) $\sin 10° \approx 0.1736$

(b) $\cos 80° \approx 0.1736$

33. (a) $\sec 42°\,12' = \sec 42.2° = \dfrac{1}{\cos 42.2°} \approx 1.3499$

(b) $\csc 48°\,7' = \dfrac{1}{\sin\left(48 + \frac{7}{60}\right)°} \approx 1.3432$

35. Make sure that your calculator is in radian mode.

(a) $\cot\frac{\pi}{16} = \dfrac{1}{\tan(\pi/16)} \approx 5.0273$

(b) $\tan\frac{\pi}{8} \approx 0.4142$

37. $\sin\theta = \dfrac{1}{\csc\theta}$

39. $\tan\theta = \dfrac{1}{\cot\theta}$

41. $\sec\theta = \dfrac{1}{\cos\theta}$

43. $\tan\theta = \dfrac{\sin\theta}{\cos\theta}$

45. $\sin^2\theta + \cos^2\theta = 1$

47. $\sin(90° - \theta) = \cos\theta$

49. $\tan(90° - \theta) = \cot\theta$

51. $\sec(90° - \theta) = \csc\theta$

53. $\sin 60° = \dfrac{\sqrt{3}}{2}, \ \cos 60° = \dfrac{1}{2}$

(a) $\tan 60° = \dfrac{\sin 60°}{\cos 60°} = \sqrt{3}$

(b) $\sin 30° = \cos 60° = \dfrac{1}{2}$

(c) $\cos 30° = \sin 60° = \dfrac{\sqrt{3}}{2}$

(d) $\cot 60° = \dfrac{\cos 60°}{\sin 60°} = \dfrac{1}{\sqrt{3}} = \dfrac{\sqrt{3}}{3}$

55. $\csc\theta = 3,\ \sec\theta = \dfrac{3\sqrt{2}}{4}$

(a) $\sin\theta = \dfrac{1}{\csc\theta} = \dfrac{1}{3}$

(b) $\cos\theta = \dfrac{1}{\sec\theta} = \dfrac{2\sqrt{2}}{3}$

(c) $\tan\theta = \dfrac{\sin\theta}{\cos\theta} = \dfrac{1/3}{\left(2\sqrt{2}\right)/3} = \dfrac{\sqrt{2}}{4}$

(d) $\sec(90° - \theta) = \csc\theta = 3$

57. $\cos\alpha = \dfrac{1}{3}$

(a) $\sec\alpha = \dfrac{1}{\cos\alpha} = 3$

(b) $\sin^2\alpha + \cos^2\alpha = 1$

$$\sin^2\alpha + \left(\tfrac{1}{3}\right)^2 = 1$$

$$\sin^2\alpha = \frac{8}{9}$$

$$\sin\alpha = \frac{2\sqrt{2}}{3}$$

(c) $\cot\alpha = \dfrac{\cos\alpha}{\sin\alpha}$

$$= \frac{\frac{1}{3}}{\frac{2\sqrt{2}}{3}}$$

$$= \frac{1}{2\sqrt{2}}$$

$$= \frac{\sqrt{2}}{4}$$

(d) $\sin(90° - \alpha) = \cos\alpha = \dfrac{1}{3}$

59. $\tan\theta\cot\theta = \tan\theta\left(\dfrac{1}{\tan\theta}\right) = 1$

61. $\tan\theta\cos\theta = \left(\dfrac{\sin\theta}{\cos\theta}\right)\cos\theta = \sin\theta$

63. $(1 + \cos\theta)(1 - \cos\theta) = 1 - \cos^2\theta$

$$= \left(\sin^2\theta + \cos^2\theta\right) - \cos^2\theta$$

$$= \sin^2\theta$$

65. $\dfrac{\sin\theta}{\cos\theta} + \dfrac{\cos\theta}{\sin\theta} = \dfrac{\sin^2\theta + \cos^2\theta}{\sin\theta\cos\theta}$

$$= \frac{1}{\sin\theta\cos\theta}$$

$$= \frac{1}{\sin\theta}\cdot\frac{1}{\cos\theta} = \csc\theta\sec\theta$$

67. (a) $\sin\theta = \dfrac{1}{2} \Rightarrow \theta = 30° = \dfrac{\pi}{6}$

(b) $\csc\theta = 2 \Rightarrow \theta = 30° = \dfrac{\pi}{6}$

69. (a) $\sec\theta = 2 \Rightarrow \theta = 60° = \dfrac{\pi}{3}$

(b) $\cot\theta = 1 \Rightarrow \theta = 45° = \dfrac{\pi}{4}$

71. (a) $\csc\theta = \dfrac{2\sqrt{3}}{3} \Rightarrow \theta = 60° = \dfrac{\pi}{3}$

(b) $\sin\theta = \dfrac{\sqrt{2}}{2} \Rightarrow \theta = 45° = \dfrac{\pi}{4}$

73. $\tan 30° = \dfrac{y}{105} \Rightarrow y = 105\tan 30° = 105\cdot\dfrac{\sqrt{3}}{3} = 35\sqrt{3}$

$\cos 30° = \dfrac{105}{r} \Rightarrow r = \dfrac{105}{\cos 30°} = \dfrac{105}{\sqrt{3}/2} = \dfrac{210}{\sqrt{3}} = 70\sqrt{3}$

75. $\cos 60° = \dfrac{x}{16} \Rightarrow x = 16\cos 60° = 16\left(\dfrac{1}{2}\right) = 8$

$\sin 60° = \dfrac{y}{16} \Rightarrow y = 16\sin 60° = 16\left(\dfrac{\sqrt{3}}{2}\right) = 8\sqrt{3}$

77. (a)

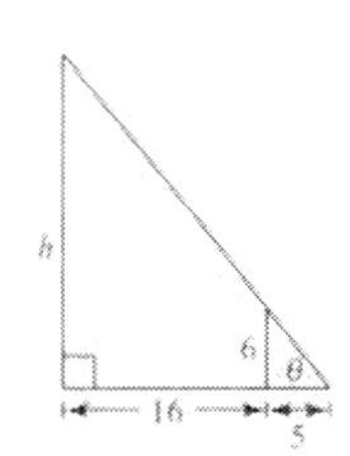

(b) $\tan\theta = \dfrac{6}{5}$ and $\tan\theta = \dfrac{h}{21}$

Thus, $\dfrac{6}{5} = \dfrac{h}{21}$.

(c) $h = \dfrac{6(21)}{5} = 25.2$ feet

79. (a) $\tan\theta = \dfrac{50}{50} = 1 \Rightarrow \theta = 45°$

(b) $L^2 = 50^2 + 50^2 = 2\cdot 50^2 \Rightarrow L = 50\sqrt{2}$ feet

(c) $\dfrac{50\sqrt{2}}{6} = \dfrac{25\sqrt{2}}{3}$ ft/sec rate down the zip line

$\dfrac{50}{6} = \dfrac{25}{3}$ ft/sec vertical rate

81. $\tan\theta = \dfrac{\text{opp}}{\text{adj}}$

$\tan 58° = \dfrac{w}{100}$

$w = 100\tan 58° \approx 160$ feet

83. (a)

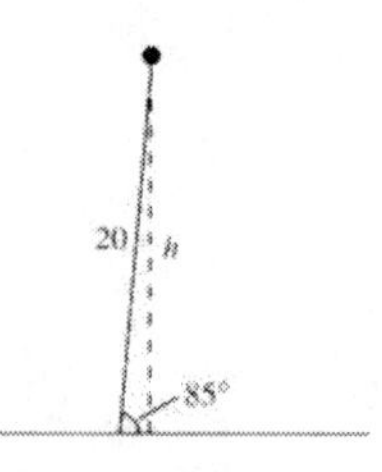

(b) $\sin 85° = \dfrac{h}{20}$

(c) $h = 20\sin 85°$

≈ 19.9 meters

(d) As the breeze becomes stronger and the angle the balloon makes with the ground decreases, the side of the triangle labeled h will decrease in height.

(e)

Angle, θ	80°	70°	60°	50°
Height	19.7	18.8	17.3	15.3

Angle, θ	40°	30°	20°	10°
Height	12.9	10.0	6.8	3.5

(f) As the angle the balloon makes with the ground approaches 0°, the height h of the balloon approaches 0 meters.

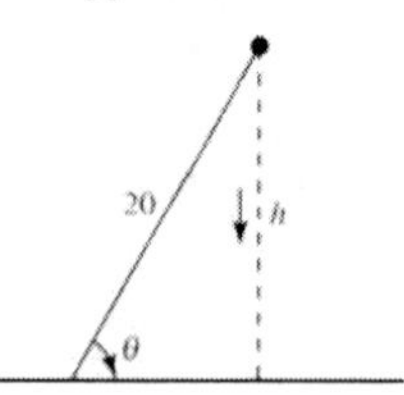

85. False

$\sin 45° + \cos 45° = \dfrac{\sqrt{2}}{2} + \dfrac{\sqrt{2}}{2} = \sqrt{2} \neq 1$

87. Yes, with the Pythagorean Theorem.

89. (a)

θ	0°	20°	40°	60°	80°
$\sin\theta$	0	0.3420	0.6428	0.8660	0.9848
$\cos\theta$	1	0.9397	0.7660	0.5000	0.1736
$\tan\theta$	0	0.3640	0.8391	1.7321	5.6713

(b) Sine and tangent are increasing; cosine is decreasing.

(c) In each case, $\tan\theta = \dfrac{\sin\theta}{\cos\theta}$.

91. $f(x) = e^{3x}$

x	−1	0	1	2
$f(x)$	0.05	1	20.09	403.43

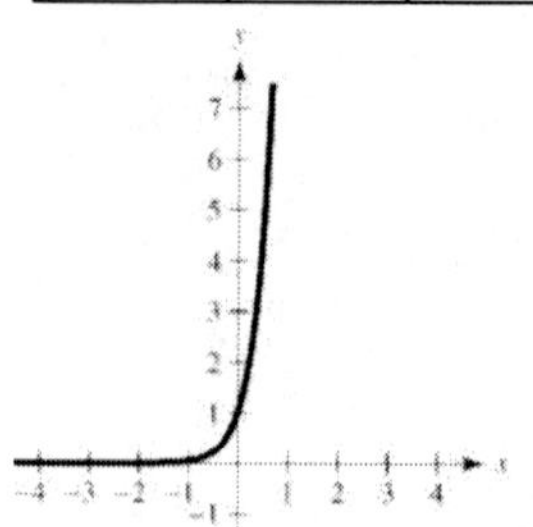

Horizontal asymptote: $y = 0$

93. $f(x) = 2 + e^x$

x	−1	0	1	2
$f(x)$	2.05	3	22.09	405.43

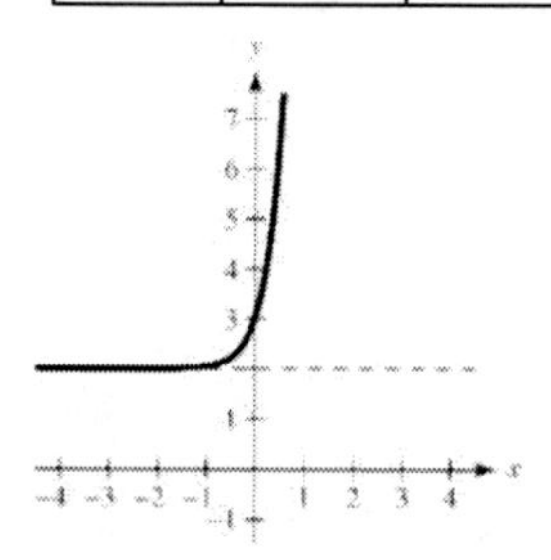

Horizontal asymptote: $y = 2$

95. $f(x) = \log_3 x = \dfrac{\ln x}{\ln 3}$

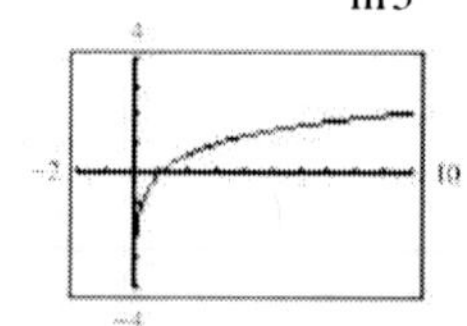

Domain: $(0, \infty)$

Vertical asymptote: $x = 0$

x-intercept: $(1, 0)$

97. $f(x) = -\log_3 x = -\dfrac{\ln x}{\ln 3}$

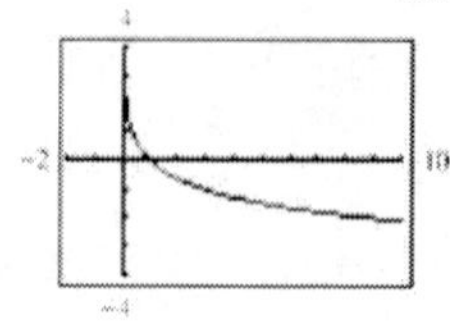

Domain: $(0, \infty)$

Vertical asymptote: $x = 0$

x-intercept: $(1, 0)$

Section 4.4

1. $\frac{y}{r}$

3. $\frac{y}{x}$

5. $\cos\theta$

7. reference

9. $\sin\theta = 0$ at the quadrant angles $\theta = 0$ and π.

11. (a) $(x, y) = (4, 3)$

$r = \sqrt{16+9} = 5$

$\sin\theta = \frac{y}{r} = \frac{3}{5}$

$\cos\theta = \frac{x}{r} = \frac{4}{5}$

$\tan\theta = \frac{y}{x} = \frac{3}{4}$

$\csc\theta = \frac{r}{y} = \frac{5}{3}$

$\sec\theta = \frac{r}{x} = \frac{5}{4}$

$\cot\theta = \frac{x}{y} = \frac{4}{3}$

(b) $(x, y) = (-8, -15)$

$r = \sqrt{64+225} = 17$

$\sin\theta = \frac{y}{r} = -\frac{15}{17}$

$\cos\theta = \frac{x}{r} = -\frac{8}{17}$

$\tan\theta = \frac{y}{x} = \frac{15}{8}$

$\csc\theta = \frac{r}{y} = -\frac{17}{15}$

$\sec\theta = \frac{r}{x} = -\frac{17}{8}$

$\cot\theta = \frac{x}{y} = \frac{8}{15}$

13. (a) $(x, y) = (-\sqrt{3}, -1)$

$r = \sqrt{3+1} = 2$

$\sin\theta = \frac{y}{r} = -\frac{1}{2}$

$\cos\theta = \frac{x}{r} = \frac{-\sqrt{3}}{2}$

$\tan\theta = \frac{y}{x} = \frac{\sqrt{3}}{3}$

$\csc\theta = \frac{r}{y} = -2$

$\sec\theta = \frac{r}{x} = \frac{-2\sqrt{3}}{3}$

$\cot\theta = \frac{x}{y} = \sqrt{3}$

(b) $(x, y) = (-2, 2)$

$r = \sqrt{4+4} = 2\sqrt{2}$

$\sin\theta = \frac{y}{r} = \frac{\sqrt{2}}{2}$

$\cos\theta = \frac{x}{r} = -\frac{\sqrt{2}}{2}$

$\tan\theta = \frac{y}{x} = -1$

$\csc\theta = \frac{r}{y} = \sqrt{2}$

$\sec\theta = \frac{r}{x} = -\sqrt{2}$

$\cot\theta = \frac{x}{y} = -1$

15. $(x, y) = (7, 24)$

$r = \sqrt{49+576} = 25$

$\sin\theta = \frac{y}{r} = \frac{24}{25}$

$\cos\theta = \frac{x}{r} = \frac{7}{25}$

$\tan\theta = \frac{y}{x} = \frac{24}{7}$

$\csc\theta = \frac{r}{y} = \frac{25}{24}$

$\sec\theta = \frac{r}{x} = \frac{25}{7}$

$\cot\theta = \frac{x}{y} = \frac{7}{24}$

17. $(x, y) = (5, -12)$

$r = \sqrt{5^2 + (-12)^2} = \sqrt{25 + 144} = \sqrt{169} = 13$

$\sin\theta = \frac{y}{r} = -\frac{12}{13}$

$\cos\theta = \frac{x}{r} = \frac{5}{13}$

$\tan\theta = \frac{y}{x} = -\frac{12}{5}$

$\csc\theta = \frac{r}{y} = -\frac{13}{12}$

$\sec\theta = \frac{r}{x} = \frac{13}{5}$

$\cot\theta = \frac{x}{y} = -\frac{5}{12}$

19. $(x, y) = (-4, 10)$

$r = \sqrt{16 + 100} = 2\sqrt{29}$

$\sin\theta = \frac{y}{r} = \frac{5\sqrt{29}}{29}$

$\cos\theta = \frac{x}{r} = -\frac{2\sqrt{29}}{29}$

$\tan\theta = \frac{y}{x} = -\frac{5}{2}$

$\csc\theta = \frac{r}{y} = \frac{\sqrt{29}}{5}$

$\sec\theta = \frac{r}{x} = -\frac{\sqrt{29}}{2}$

$\cot\theta = \frac{x}{y} = -\frac{2}{5}$

21. $(x, y) = (-10, 8)$

$r = \sqrt{(-10)^2 + 8^2} = \sqrt{164} = 2\sqrt{41}$

$\sin\theta = \frac{y}{r} = \frac{8}{2\sqrt{41}} = \frac{4\sqrt{41}}{41}$

$\cos\theta = \frac{x}{r} = \frac{-10}{2\sqrt{41}} = -\frac{5\sqrt{41}}{41}$

$\tan\theta = \frac{y}{x} = \frac{8}{-10} = -\frac{4}{5}$

$\csc\theta = \frac{r}{y} = \frac{\sqrt{41}}{4}$

$\sec\theta = \frac{r}{x} = -\frac{\sqrt{41}}{5}$

$\cot\theta = \frac{x}{y} = -\frac{5}{4}$

23. $\sin\theta < 0 \Rightarrow \theta$ lies in Quadrant III or Quadrant IV.
$\cos\theta < 0 \Rightarrow \theta$ lies in Quadrant II or Quadrant III.
$\sin\theta < 0$ and $\cos\theta < 0 \Rightarrow \theta$ lies in Quadrant III.

25. $\cot\theta > 0 \Rightarrow \theta$ lies in Quadrant I or Quadrant III.
$\cos\theta > 0 \Rightarrow \theta$ lies in Quadrant I or Quadrant IV.
$\cot\theta > 0$ and $\cos\theta > 0 \Rightarrow \theta$ lies in Quadrant I.

27. $\sin\theta = \frac{y}{r} = \frac{3}{5} \Rightarrow x^2 = 25 - 9 = 16$

θ in Quadrant II $\Rightarrow x = -4$

$\sin\theta = \frac{y}{r} = \frac{3}{5}$

$\cos\theta = \frac{x}{r} = -\frac{4}{5}$

$\tan\theta = \frac{y}{x} = -\frac{3}{4}$

$\csc\theta = \frac{r}{y} = \frac{5}{3}$

$\sec\theta = \frac{r}{x} = -\frac{5}{4}$

$\cot\theta = \frac{x}{y} = -\frac{4}{3}$

29. $\sin\theta < 0 \Rightarrow y < 0$

$\tan\theta = \frac{y}{x} = \frac{-15}{8} \Rightarrow r = 17$

$\sin\theta = \frac{y}{r} = -\frac{15}{17}$

$\cos\theta = \frac{x}{r} = \frac{8}{17}$

$\cot\theta = \frac{x}{y} = -\frac{8}{15}$

$\csc\theta = \frac{r}{y} = -\frac{17}{15}$

$\sec\theta = \frac{r}{x} = \frac{17}{8}$

31. $\sec\theta = \frac{r}{x} = \frac{2}{-1} \Rightarrow y^2 = 4 - 1 = 3$

$\sin\theta \geq 0 \Rightarrow y = \sqrt{3}$

$\sin\theta = \frac{y}{r} = \frac{\sqrt{3}}{2}$

$\cos\theta = \frac{x}{r} = -\frac{1}{2}$

$\tan\theta = \frac{y}{x} = -\sqrt{3}$

$\csc\theta = \frac{r}{y} = \frac{2\sqrt{3}}{3}$

$\sec\theta = \frac{r}{x} = -2$

$\cot\theta = \frac{x}{y} = -\frac{\sqrt{3}}{3}$

33. $\cot\theta$ is undefined $\Rightarrow \theta = n\pi$.

$\frac{\pi}{2} \le \theta \le \frac{3\pi}{2} \Rightarrow \theta = \pi,\ y = 0,\ x = -r$

$\sin\theta = \frac{y}{r} = 0$

$\cos\theta = \frac{x}{r} = \frac{-r}{r} = -1$

$\tan\theta = \frac{y}{x} = \frac{0}{x} = 0$

$\csc\theta = \frac{r}{y}$ is undefined.

$\sec\theta = \frac{r}{x} = -1$

$\cot\theta$ is undefined.

35. To find a point on the terminal side of θ, use any point on the line $y = -x$ that lies in Quadrant II. $(-1,\ 1)$ is one such point.

$x = -1,\ y = 1,\ r = \sqrt{2}$

$\sin\theta = \frac{y}{r} = \frac{1}{\sqrt{2}} = \frac{\sqrt{2}}{2}$

$\cos\theta = \frac{x}{r} = -\frac{1}{\sqrt{2}} = -\frac{\sqrt{2}}{2}$

$\tan\theta = \frac{y}{x} = -1$

$\csc\theta = \frac{r}{y} = \sqrt{2}$

$\sec\theta = \frac{r}{x} = -\sqrt{2}$

$\cot\theta = \frac{x}{y} = -1$

37. To find a point on the terminal side of θ, use any point on the line $y = 2x$ that lies in Quadrant III. $(-1,\ -2)$ is one such point.

$x = -1,\ y = -2,\ r = \sqrt{5}$

$\sin\theta = \frac{y}{r} = -\frac{2}{\sqrt{5}} = -\frac{2\sqrt{5}}{5}$

$\cos\theta = \frac{x}{r} = -\frac{1}{\sqrt{5}} = -\frac{\sqrt{5}}{5}$

$\tan\theta = \frac{y}{x} = \frac{-2}{-1} = 2$

$\csc\theta = \frac{r}{y} = \frac{\sqrt{5}}{-2} = -\frac{\sqrt{5}}{2}$

$\sec\theta = \frac{r}{x} = \frac{\sqrt{5}}{-1} = -\sqrt{5}$

$\cot\theta = \frac{x}{y} = \frac{-1}{-2} = \frac{1}{2}$

39. $(x,\ y) = (-1,\ 0)$

$\sec\pi = \frac{r}{x} = \frac{1}{-1} = -1$

41. $(x,\ y) = (0,\ -1)$

$\cot\left(\frac{3\pi}{2}\right) = \frac{x}{y} = \frac{0}{-1} = 0$

43. $(x,\ y) = (1,\ 0)$

$\sec 0 = \frac{r}{x} = \frac{1}{1} = 1$

45. $(x,\ y) = (-1,\ 0)$

$\cot\pi = \frac{x}{y} = -\frac{1}{0} \Rightarrow$ undefined

47. $\theta = 120°$

$\theta' = 180° - 120° = 60°$

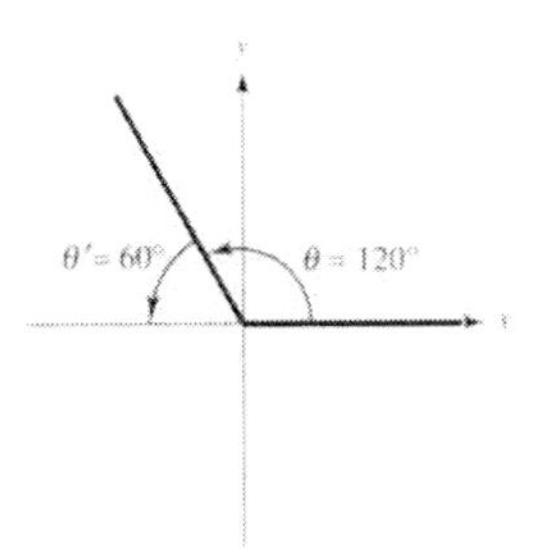

49. $\theta = 150°$

$\theta' = 180° - 150° = 30°$

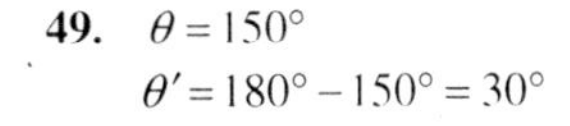

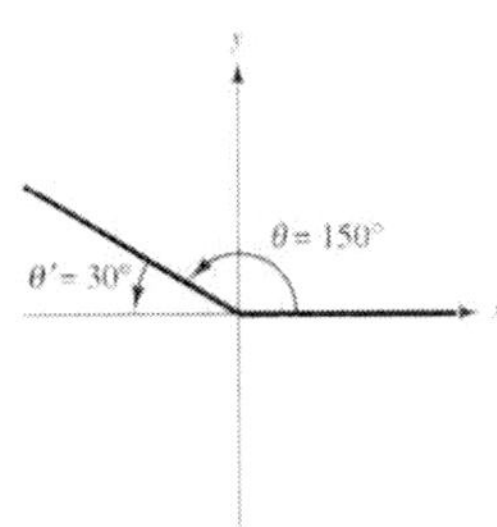

51. $\theta = -45°$ is coterminal with $315°$.

$\theta' = 360° - 315° = 45°$

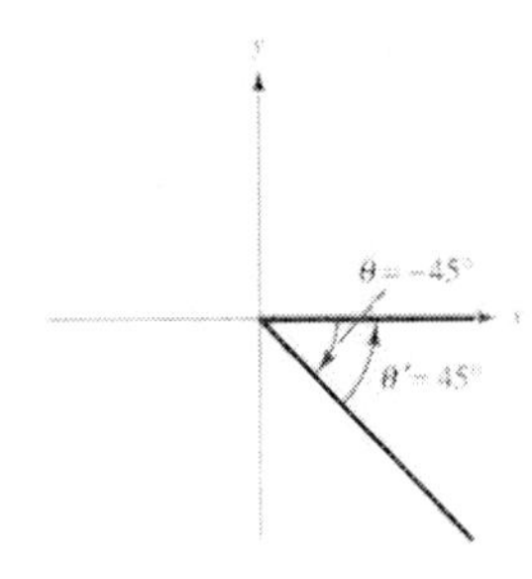

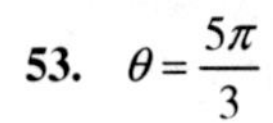

53. $\theta = \dfrac{5\pi}{3}$

$\theta' = 2\pi - \dfrac{5\pi}{3} = \dfrac{\pi}{3}$

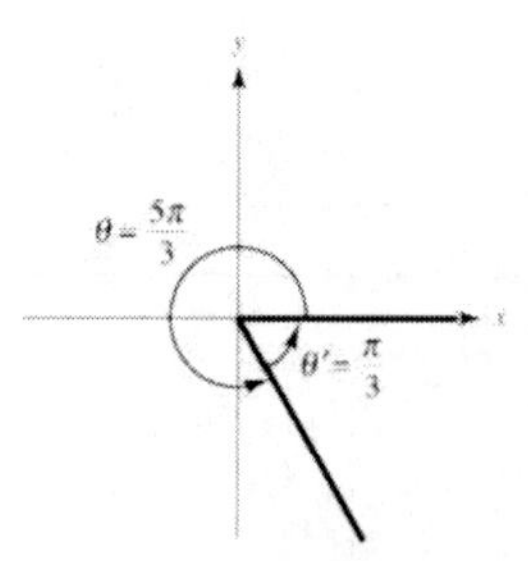

55. $\theta = -\dfrac{5\pi}{6}$ is coterminal with $\dfrac{7\pi}{6}$.

$\theta' = \dfrac{7\pi}{6} - \pi = \dfrac{\pi}{6}$

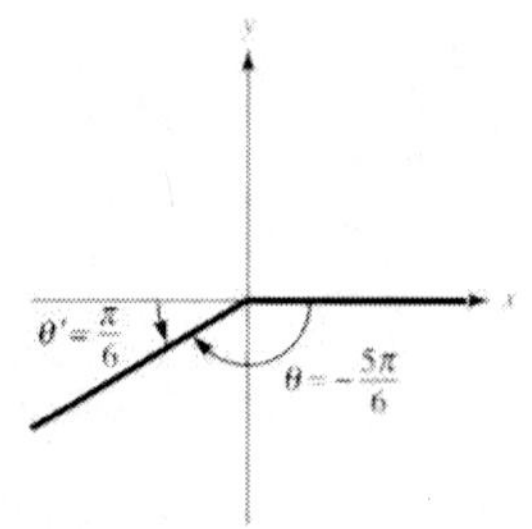

57. $\theta = \dfrac{11\pi}{6}$

$\theta' = 2\pi - \dfrac{11\pi}{6} = \dfrac{\pi}{6}$

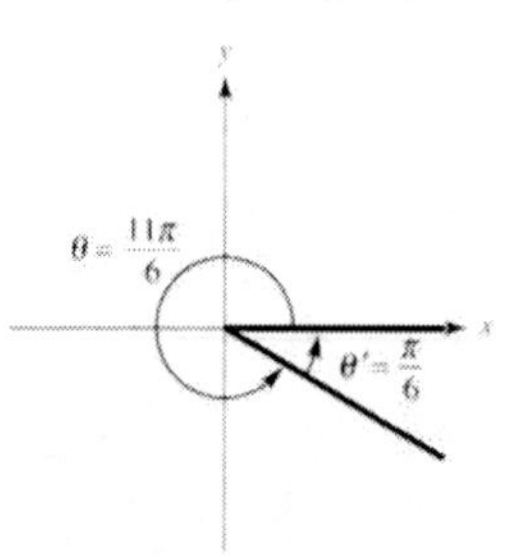

59. $\theta = 208°$

$\theta' = 208° - 180° = 28°$

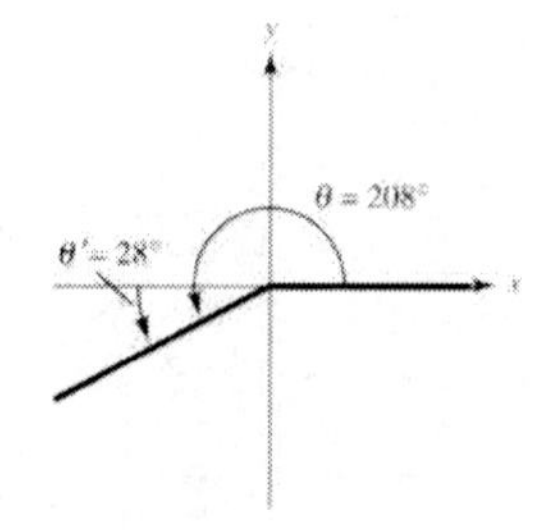

61. $\theta = -292°$

$\theta' = 360° - 292° = 68°$

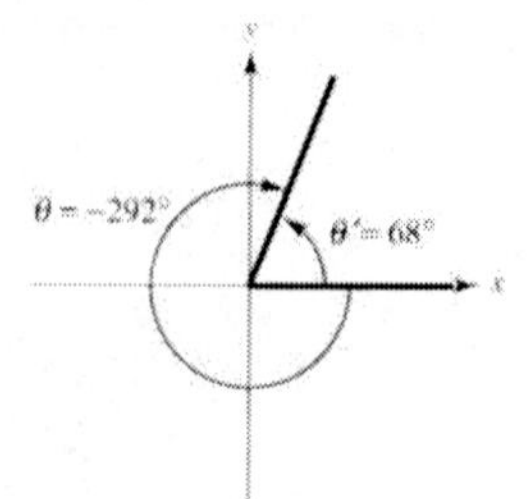

63. $\theta = \dfrac{11\pi}{5}$ is coterminal with $\dfrac{\pi}{5}$.

$\theta' = \dfrac{\pi}{5}$

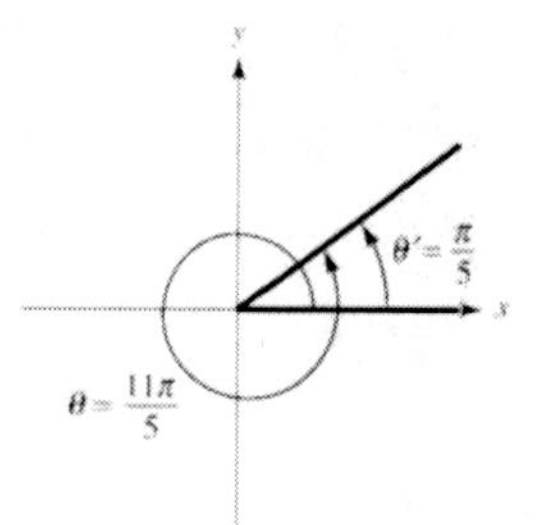

65. $\theta = -1.8$ lies in Quadrant III.

$\theta' = \pi - 1.8 \approx 1.342$

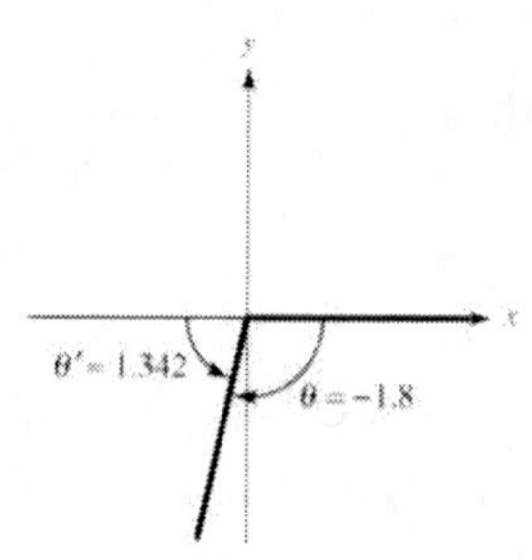

67. $\theta = 225°, \theta' = 360° - 225° = 45°$, Quadrant III

$\sin 225° = -\sin 45° = -\dfrac{\sqrt{2}}{2}$

$\cos 225° = -\cos 45° = -\dfrac{\sqrt{2}}{2}$

$\tan 225° = \tan 45° = 1$

69. $\theta = -750°$ is coterminal with $330°$, Quadrant IV.

$\theta' = 360° - 330° = 30°$

$\sin(-750°) = -\sin 30° = -\dfrac{1}{2}$

$\cos(-750°) = \cos 30° = \dfrac{\sqrt{3}}{2}$

$\tan(-750°) = -\tan 30° = -\dfrac{\sqrt{3}}{3}$

71. $\theta = \frac{5\pi}{3}$, Quadrant IV

$$\theta' = 2\pi - \frac{5\pi}{3} = \frac{\pi}{3}$$

$$\sin\frac{5\pi}{3} = -\sin\frac{\pi}{3} = -\frac{\sqrt{3}}{2}$$

$$\cos\frac{5\pi}{3} = \cos\frac{\pi}{3} = \frac{1}{2}$$

$$\tan\frac{5\pi}{3} = -\tan\frac{\pi}{3} = -\sqrt{3}$$

73. $\theta = -\frac{\pi}{6}$ is coterminal with $\frac{5\pi}{6}$, Quadrant IV.

$$\theta' = 2\pi - \frac{5\pi}{6} = \frac{7\pi}{6}$$

$$\sin\left(-\frac{\pi}{6}\right) = \sin\frac{7\pi}{6} = -\frac{1}{2}$$

$$\cos\left(-\frac{\pi}{6}\right) = -\cos\frac{7\pi}{6} = \frac{\sqrt{3}}{2}$$

$$\tan\left(-\frac{\pi}{6}\right) = -\tan\frac{7\pi}{6} = -\frac{\sqrt{3}}{3}$$

75. $\theta = \frac{11\pi}{4}$ is coterminal with $\frac{3\pi}{4}$, Quadrant II.

$$\theta' = \pi - \frac{3\pi}{4} = \frac{\pi}{4}$$

$$\sin\frac{11\pi}{4} = \sin\frac{\pi}{4} = \frac{\sqrt{2}}{2}$$

$$\cos\frac{11\pi}{4} = -\cos\frac{\pi}{4} = -\frac{\sqrt{2}}{2}$$

$$\tan\frac{11\pi}{4} = -\tan\frac{\pi}{4} = -1$$

77. $\theta = -\frac{17\pi}{6}$ is coterminal with $\frac{7\pi}{6}$, Quadrant III.

$$\theta' = \frac{7\pi}{6} - \pi = \frac{\pi}{6}$$

$$\sin\left(-\frac{17\pi}{6}\right) = -\sin\left(\frac{\pi}{6}\right) = -\frac{1}{2}$$

$$\cos\left(-\frac{17\pi}{6}\right) = -\cos\left(\frac{\pi}{6}\right) = -\frac{\sqrt{3}}{2}$$

$$\tan\left(-\frac{17\pi}{6}\right) = \tan\left(\frac{\pi}{6}\right) = \frac{\sqrt{3}}{3}$$

79. $\sin\theta = -\frac{3}{5}$

$$\sin^2\theta + \cos^2\theta = 1$$

$$\cos^2\theta = 1 - \sin^2\theta$$

$$\cos^2\theta = 1 - \left(-\frac{3}{5}\right)^2$$

$$\cos^2\theta = 1 - \frac{9}{25}$$

$$\cos^2\theta = \frac{16}{25}$$

$\cos\theta > 0$ in Quadrant IV.

$$\cos\theta = \frac{4}{5}$$

81. $\csc\theta = -2$

$$1 + \cot^2\theta = \csc^2\theta$$

$$\cot^2\theta = \csc^2\theta - 1$$

$$\cot^2\theta = (-2)^2 - 1$$

$$\cot^2\theta = 3$$

$\cot\theta < 0$ in Quadrant IV.

$$\cot\theta = -\sqrt{3}$$

83. $\sec\theta = -\frac{9}{4}$

$$1 + \tan^2\theta = \sec^2\theta$$

$$\tan^2\theta = \sec^2\theta - 1$$

$$\tan^2\theta = \left(-\frac{9}{4}\right)^2 - 1$$

$$\tan^2\theta = \frac{65}{16}$$

$\tan\theta > 0$ in Quadrant III.

$$\tan\theta = \frac{\sqrt{65}}{4}$$

85. $\sin\theta = \frac{2}{5}$ and $\cos\theta < 0 \Rightarrow \theta$ is in Quadrant II.

$$\cos\theta = -\sqrt{1 - \sin^2\theta} = -\sqrt{1 - \frac{4}{25}} = -\frac{\sqrt{21}}{5}$$

$$\tan\theta = \frac{\sin\theta}{\cos\theta} = \frac{2/5}{-\sqrt{21}/5} = \frac{-2}{\sqrt{21}} = -\frac{2\sqrt{21}}{21}$$

$$\csc\theta = \frac{1}{\sin\theta} = \frac{5}{2}$$

$$\sec\theta = \frac{1}{\cos\theta} = \frac{-5}{\sqrt{21}} = -\frac{5\sqrt{21}}{21}$$

$$\cot\theta = \frac{1}{\tan\theta} = -\frac{\sqrt{21}}{2}$$

87. $\tan\theta = -4$ and $\cos\theta < 0 \Rightarrow \theta$ is in Quadrant II.

$$\sec\theta = -\sqrt{1+\tan^2\theta} = -\sqrt{1+16} = -\sqrt{17}$$

$$\cos\theta = \frac{1}{\sec\theta} = \frac{-1}{\sqrt{17}} = -\frac{\sqrt{17}}{17}$$

$$\sin\theta = \tan\theta\cos\theta = (-4)\left(-\frac{\sqrt{17}}{17}\right) = \frac{4\sqrt{17}}{17}$$

$$\csc\theta = \frac{1}{\sin\theta} = \frac{17}{4\sqrt{17}} = \frac{\sqrt{17}}{4}$$

$$\cot\theta = \frac{1}{\tan\theta} = -\frac{1}{4}$$

89. $\csc\theta = -\frac{3}{2}$ and $\tan\theta < 0 \Rightarrow \theta$ is in Quadrant IV.

$$\sin\theta = \frac{1}{\csc\theta} = -\frac{2}{3}$$

$$\cos\theta = \sqrt{1-\sin^2\theta} = \sqrt{1-\frac{4}{9}} = \frac{\sqrt{5}}{3}$$

$$\sec\theta = \frac{1}{\cos\theta} = \frac{3}{\sqrt{5}} = \frac{3\sqrt{5}}{5}$$

$$\tan\theta = \frac{\sin\theta}{\cos\theta} = \frac{-2/3}{\sqrt{5}/3} = \frac{-2}{\sqrt{5}} = -\frac{2\sqrt{5}}{5}$$

$$\cot\theta = \frac{1}{\tan\theta} = -\frac{\sqrt{5}}{2}$$

91. $\sin 10° \approx 0.1736$

93. $\tan 245° \approx 2.1445$

95. $\cos(-110°) \approx -0.3420$

97. $\sec(-280°) = \frac{1}{\cos(-280°)}$

≈ 5.7588

99. $\tan\left(\frac{2\pi}{9}\right) \approx 0.8391$

101. $\csc\left(-\frac{8\pi}{9}\right) = \frac{1}{\sin\left(-\frac{8\pi}{9}\right)}$

≈ -2.9238

103. (a) $\sin\theta = \frac{1}{2} \Rightarrow$ reference angle is $30°$ or $\frac{\pi}{6}$ and θ is in Quadrant I or Quadrant II.

Values in degrees: $30°, 150°$

Values in radians: $\frac{\pi}{6}, \frac{5\pi}{6}$

(b) $\sin\theta = -\frac{1}{2} \Rightarrow$ reference angle is $30°$ or $\frac{\pi}{6}$ and θ is in Quadrant III or Quadrant IV.

Values in degrees: $210°, 330°$

Values in radians: $\frac{7\pi}{6}, \frac{11\pi}{6}$

105. (a) $\csc\theta = \frac{2\sqrt{3}}{3} \Rightarrow$ reference angle is $60°$ or $\frac{\pi}{3}$ and θ is in Quadrant I or Quadrant II.

Values in degrees: $60°, 120°$

Values in radians: $\frac{\pi}{3}, \frac{2\pi}{3}$

(b) $\cot\theta = -1 \Rightarrow$ reference angle is $45°$ or $\frac{\pi}{4}$ and θ is in Quadrant II or Quadrant IV.

Values in degrees: $135°, 315°$

Values in radians: $\frac{3\pi}{4}, \frac{7\pi}{4}$

107. (a) $\sec\theta = -\frac{2\sqrt{3}}{3} \Rightarrow$ reference angle is $\frac{\pi}{6}$ or $30°$, and θ is in Quadrant II or Quadrant III.

Values in degrees: $150°, 210°$

Values in radians: $\frac{5\pi}{6}, \frac{7\pi}{6}$

(b) $\cos\theta = -\frac{1}{2} \Rightarrow$ reference angle is $\frac{\pi}{3}$ or $60°$, and θ is in Quadrant II or Quadrant III.

Values in degrees: $120°, 240°$

Values in radians: $\frac{2\pi}{3}, \frac{4\pi}{3}$

109. (a) $f(\theta)+g(\theta)=\sin 30°+\cos 30°=\frac{1}{2}+\frac{\sqrt{3}}{2}=\frac{1+\sqrt{3}}{2}$

(b) $\cos 30°-\sin 30°=\frac{\sqrt{3}-1}{2}$

(c) $[\cos 30°]^2=\left(\frac{\sqrt{3}}{2}\right)^2=\frac{3}{4}$

(d) $\sin 30°\cos 30°=\left(\frac{1}{2}\right)\left(\frac{\sqrt{3}}{2}\right)=\frac{\sqrt{3}}{4}$

(e) $\sin(2\cdot 30°)=\sin 60°=\frac{\sqrt{3}}{2}$

(f) $\cos(-30°)=\cos 30°=\frac{\sqrt{3}}{2}$

111. (a) $f(\theta)+g(\theta)=\sin 315°+\cos 315°=-\frac{\sqrt{2}}{2}+\frac{\sqrt{2}}{2}=0$

(b) $\cos 315°-\sin 315°=\frac{\sqrt{2}}{2}-\left(\frac{-\sqrt{2}}{2}\right)=\sqrt{2}$

(c) $[\cos 315°]^2=\left(\frac{\sqrt{2}}{2}\right)^2=\frac{1}{2}$

(d) $\sin 315°\cos 315°=\left(\frac{-\sqrt{2}}{2}\right)\left(\frac{\sqrt{2}}{2}\right)=-\frac{1}{2}$

(e) $\sin(2\cdot 315°)=\sin\ 630°=-1$

(f) $\cos(-315°)=\cos(315°)=\frac{\sqrt{2}}{2}$

113. (a) $f(\theta)+g(\theta)=\sin 150°+\cos 150°=\frac{1}{2}+\frac{-\sqrt{3}}{2}=\frac{1-\sqrt{3}}{2}$

(b) $\cos 150°-\sin 150°=\frac{-\sqrt{3}}{2}-\frac{1}{2}=\frac{-1-\sqrt{3}}{2}$

(c) $[\cos 150°]^2=\left(\frac{-\sqrt{3}}{2}\right)^2=\frac{3}{4}$

(d) $\sin 150°\cos 150°=\frac{1}{2}\cdot\frac{-\sqrt{3}}{2}=-\frac{\sqrt{3}}{4}$

(e) $\sin(2\cdot 150°)=\sin 300°=-\frac{\sqrt{3}}{2}$

(f) $\cos(-150°)=\cos(150°)=-\frac{\sqrt{3}}{2}$

115. (a) $f(\theta)+g(\theta)=\sin\frac{7\pi}{6}+\cos\frac{7\pi}{6}=-\frac{1}{2}-\frac{\sqrt{3}}{2}=\frac{-1-\sqrt{3}}{2}$

(b) $\cos\frac{7\pi}{6}-\sin\frac{7\pi}{6}=\frac{-\sqrt{3}}{2}-\left(-\frac{1}{2}\right)=\frac{1-\sqrt{3}}{2}$

(c) $\left[\cos\frac{7\pi}{6}\right]^2=\left(\frac{-\sqrt{3}}{2}\right)^2=\frac{3}{4}$

(d) $\sin\frac{7\pi}{6}\cos\frac{7\pi}{6}=\left(-\frac{1}{2}\right)\left(-\frac{\sqrt{3}}{2}\right)=\frac{\sqrt{3}}{4}$

(e) $\sin\left(2\cdot\frac{7\pi}{6}\right)=\sin\frac{7\pi}{3}=\frac{\sqrt{3}}{2}$

(f) $\cos\left(\frac{-7\pi}{6}\right)=\cos\left(\frac{7\pi}{6}\right)=-\frac{\sqrt{3}}{2}$

117. (a) $f(\theta)+g(\theta)=\sin\frac{4\pi}{3}+\cos\frac{4\pi}{3}=\frac{-\sqrt{3}}{2}-\frac{1}{2}=\frac{-1-\sqrt{3}}{2}$

(b) $\cos\frac{4\pi}{3}-\sin\frac{4\pi}{3}=-\frac{1}{2}-\left(\frac{-\sqrt{3}}{2}\right)=\frac{\sqrt{3}-1}{2}$

(c) $\left[\cos\frac{4\pi}{3}\right]^2=\left(-\frac{1}{2}\right)^2=\frac{1}{4}$

(d) $\sin\frac{4\pi}{3}\cos\frac{4\pi}{3}=\left(\frac{-\sqrt{3}}{2}\right)\left(-\frac{1}{2}\right)=\frac{\sqrt{3}}{4}$

(e) $\sin\left(2\cdot\frac{4\pi}{3}\right)=\sin\frac{8\pi}{3}=\frac{\sqrt{3}}{2}$

(f) $\cos\left(\frac{-4\pi}{3}\right)=\cos\left(\frac{4\pi}{3}\right)=-\frac{1}{2}$

119. (a) $f(\theta)+g(\theta)=\sin 270°+\cos 270°=-1+0=-1$

(b) $\cos 270°-\sin 270°=0-(-1)=1$

(c) $[\cos 270°]^2=0^2=0$

(d) $\sin 270°\cos 270°=(-1)(0)=0$

(e) $\sin(2\cdot 270°)=\sin 540°=0$

(f) $\cos(-270°)=\cos(270°)=0$

121. (a) $f(\theta)+g(\theta)=\sin\frac{7\pi}{2}+\cos\frac{7\pi}{2}=-1+0=-1$

(b) $\cos\frac{7\pi}{2}-\sin\frac{7\pi}{2}=0-(-1)=1$

(c) $\left[\cos\frac{7\pi}{2}\right]^2=0^2=0$

(d) $\sin\frac{7\pi}{2}\cos\frac{7\pi}{2}=(-1)(0)=0$

(e) $\sin\left(2\cdot\frac{7\pi}{2}\right)=\sin 7\pi=0$

(f) $\cos\left(\frac{-7\pi}{2}\right)=\cos\left(\frac{7\pi}{2}\right)=0$

123. $T = 76.35 + 15.95\cos\left(\frac{\pi t}{6} - \frac{7\pi}{6}\right)$

(a) January $(t = 1)$: $T(1) = 76.35 + 15.95\cos(-\pi)$
$= 60.4° \text{ F}$

(b) July $(t = 7)$: $T(7) = 76.35 + 15.95\cos(0)$
$= 92.3° \text{ F}$

(c) October $(t = 10)$: $T(10) = 76.35 + 15.95\cos\left(\frac{\pi}{2}\right)$
$= 76.35° \text{ F}$

125. $\sin\theta = \frac{6}{d} \Rightarrow d = \frac{6}{\sin\theta}$

(a) $\theta = 30°$

$d = \frac{6}{\sin 30°} = \frac{6}{(1/2)} = 12 \text{ miles}$

(b) $\theta = 90°$

$d = \frac{6}{\sin 90°} = \frac{6}{1} = 6 \text{ miles}$

(c) $\theta = 120°$

$d = \frac{6}{\sin 120°} \approx 6.9 \text{ miles}$

127. True. Since $0 < \cos\theta < 1$ for $0 < \theta < \pi/2$, or Quadrant I, then $\sin\theta < \frac{\sin\theta}{\cos\theta} = \tan\theta$.

129. False. If $90° < \theta < 180°$, $\sin\theta > 0$.

131. (a)

θ	$0°$	$20°$	$40°$	$60°$	$80°$
$\sin\theta$	0	0.3420	0.6428	0.8660	0.9848
$\sin(180° - \theta)$	0	0.3420	0.6428	0.8660	0.9848

(b) It appears that $\sin\theta = \sin(180° - \theta)$.

133. If your classmate used a calculator and found $\tan\left(\frac{\pi}{2}\right) = 0.0274224385$, then the calculator was in the wrong mode, degrees instead of radians.

135.
$$3x - 7 = 14$$
$$3x = 21$$
$$x = 7$$

137. $x^2 - 2x - 5 = 0$

$$x = \frac{2 \pm \sqrt{4 + 20}}{2} = 1 \pm \sqrt{6}$$
$$x \approx 3.449, -1.449$$

139.
$$\frac{3}{x-1} = \frac{x+2}{9}$$
$$27 = (x-1)(x+2)$$
$$x^2 + x - 29 = 0$$
$$x = \frac{-1 \pm \sqrt{1 + 4(29)}}{2} = \frac{-1 \pm \sqrt{117}}{2}$$
$$x \approx -5.908, 4.908$$

Section 4.5

1. amplitude

3. $\frac{2\pi}{b}$

5. The period of $y = \sin x$ is 2π.

7. The constant d of $y = \sin x + d$ is a vertical shift of d units.

9. $f(x) = \sin x$

(a) x-intercepts:
$(-2\pi, 0), (-\pi, 0), (0, 0), (\pi, 0), (2\pi, 0)$

(b) y-intercept: $(0, 0)$

(c) Increasing on: $\left(-2\pi, -\frac{3\pi}{2}\right), \left(-\frac{\pi}{2}, \frac{\pi}{2}\right), \left(\frac{3\pi}{2}, 2\pi\right)$

Decreasing on: $\left(-\frac{3\pi}{2}, -\frac{\pi}{2}\right), \left(\frac{\pi}{2}, \frac{3\pi}{2}\right)$

(d) Relative maxima: $\left(-\frac{3\pi}{2}, 1\right), \left(\frac{\pi}{2}, 1\right)$

Relative minima: $\left(-\frac{\pi}{2}, -1\right), \left(\frac{3\pi}{2}, -1\right)$

11. $y = 3\sin 2x$

Period: $\frac{2\pi}{2} = \pi$

Amplitude: $|3| = 3$

13. $y = \frac{5}{2}\cos\frac{x}{2}$

Period: $\frac{2\pi}{1/2} = 4\pi$

Amplitude: $\left|\frac{5}{2}\right| = \frac{5}{2}$

15. $y = \frac{2}{3}\sin \pi x$

Period: $\frac{2\pi}{\pi} = 2$

Amplitude: $\left|\frac{2}{3}\right| = \frac{2}{3}$

17. $y = -2\sin x$

Period: $\frac{2\pi}{1} = 2\pi$

Amplitude: $|-2| = 2$

19. $y = \frac{1}{4}\cos\frac{2x}{3}$

Period: $\frac{2\pi}{2/3} = 3\pi$

Amplitude: $\left|\frac{1}{4}\right| = \frac{1}{4}$

21. $f(x) = \sin x$

$g(x) = \sin(x - \pi)$

The graph of g is a horizontal shift to the right π units of the graph of f (a phase shift).

23. $f(x) = \cos 2x$

$g(x) = -\cos 2x$

The graph of g is a reflection in the x-axis of the graph of f.

25. $f(x) = \cos x$

$g(x) = -5\cos x$

The graph of g has five times the amplitude of f, and is reflected in the x-axis.

27. $f(x) = \sin 2x$

$g(x) = 3 + \sin 2x$

The graph of g is a vertical shift three units upward of the graph of f.

29. The graph of g has twice the amplitude as the graph of f.

31. The graph of g is a horizontal shift π units to the right of the graph of f.

33. $f(x) = \sin x$

Period: 2π

Amplitude: 1

$g(x) = -4\sin x$

Period: 2π

Amplitude: $|-4| = 4$

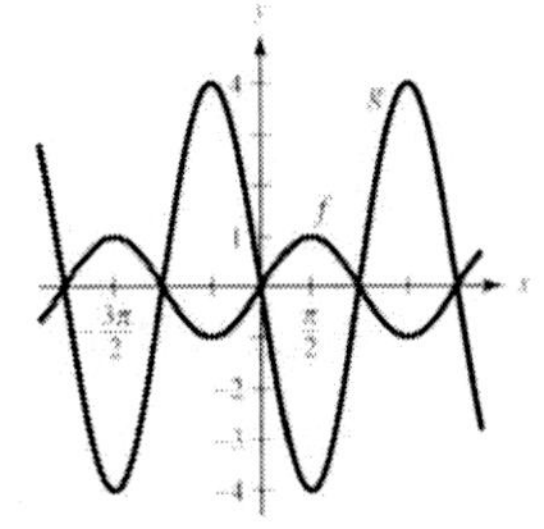

35. $f(x) = \cos x$

Period: 2π

Amplitude: 1

$g(x) = 1 + \cos x$ is the graph of f shifted vertically one unit upward.

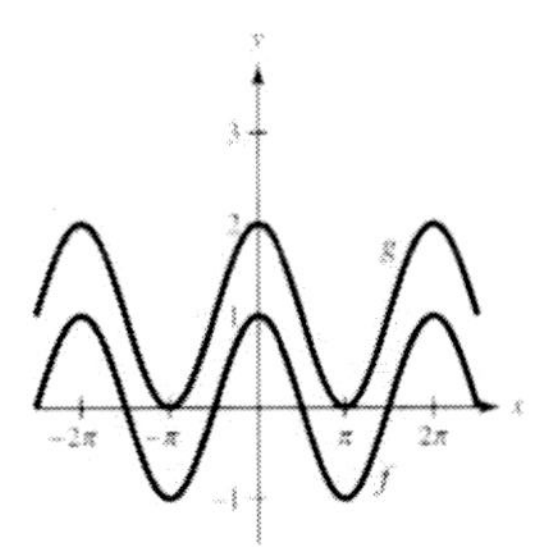

37. $f(x) = -\frac{1}{2}\sin\frac{x}{2}$

Period: 4π

Amplitude: $\frac{1}{2}$

$g(x) = 3 - \frac{1}{2}\sin\frac{x}{2}$ is the graph of f shifted vertically three units upward.

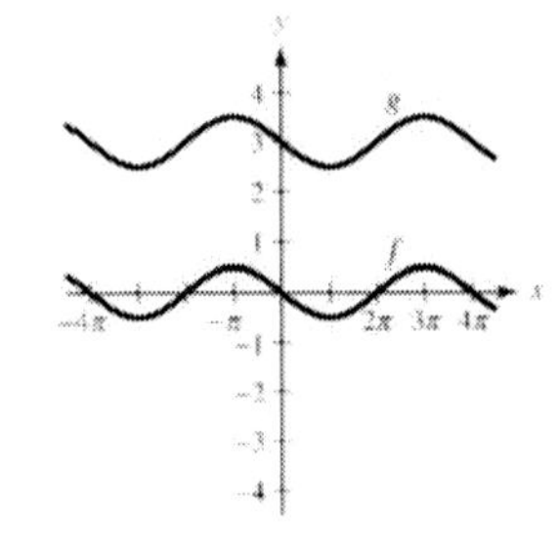

39. $f(x) = \sin x,\ g(x) = \cos\left(x - \dfrac{\pi}{2}\right)$

Period: 2π
Amplitude: 1

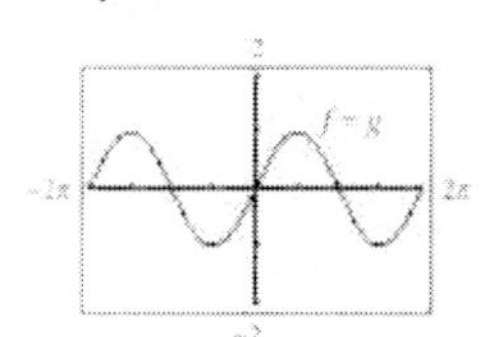

Conjecture: $\sin x = \cos\left(x - \dfrac{\pi}{2}\right)$

41. $f(x) = \cos x$

$g(x) = -\sin\left(x - \dfrac{\pi}{2}\right) = \sin\left(\dfrac{\pi}{2} - x\right) = \cos x$

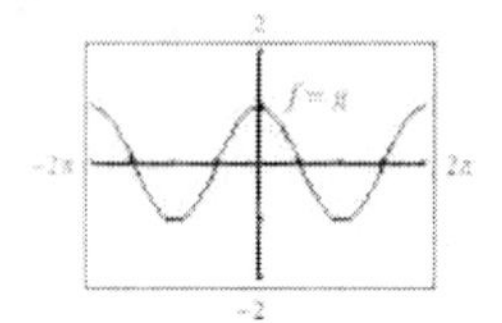

Conjecture: $\cos x = -\sin\left(x - \dfrac{\pi}{2}\right)$

43. $y = 3\sin x$

Period: 2π

Amplitude: 3

Key points:

$(0, 0),\ \left(\dfrac{\pi}{2}, 3\right),\ (\pi, 0),\ \left(\dfrac{3\pi}{2}, -3\right),\ (2\pi, 0)$

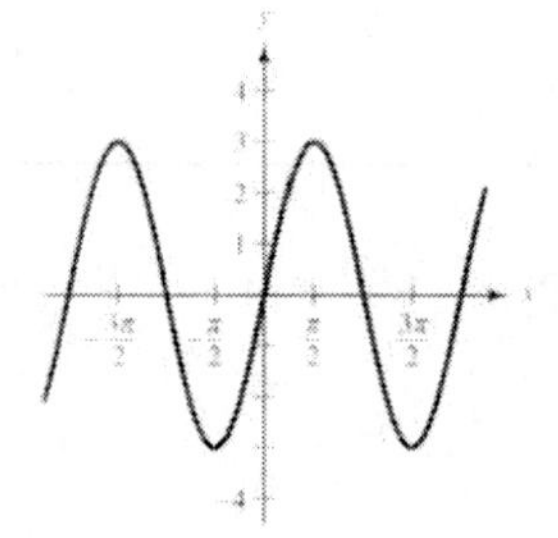

45. $y = \dfrac{1}{4}\cos x$

Period: 2π

Amplitude: $\dfrac{1}{4}$

Key points:

$\left(0, \dfrac{1}{4}\right),\ \left(\dfrac{\pi}{2}, 0\right),\ \left(\pi, -\dfrac{1}{4}\right),\ \left(\dfrac{3\pi}{2}, 0\right),\ \left(2\pi, \dfrac{1}{4}\right)$

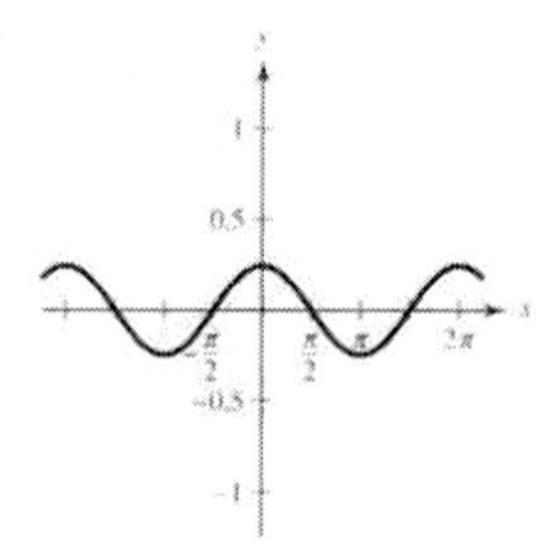

47. $y = \cos\dfrac{x}{2}$

Period: 4π

Amplitude: 1

Key points: $(0, 1), (\pi, 0), (2\pi, -1), (3\pi, 0), (4\pi, 1)$

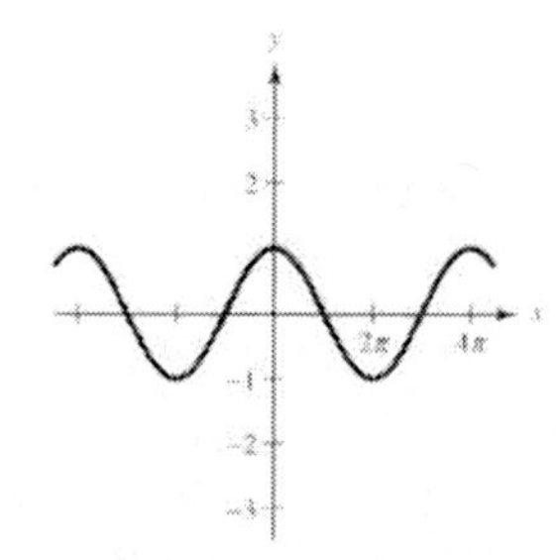

49. $y = \sin\left(x - \dfrac{\pi}{4}\right);\ a = 1,\ b = 1,\ c = \dfrac{\pi}{4}$

Period: 2π
Amplitude: 1

Shift: Set $x - \dfrac{\pi}{4} = 0$ and $x - \dfrac{\pi}{4} = 2\pi$

$x = \dfrac{\pi}{4} \qquad x = \dfrac{9\pi}{4}$

Key points:

$\left(\frac{\pi}{4}, 0\right), \left(\frac{3\pi}{4}, 1\right), \left(\frac{5\pi}{4}, 0\right), \left(\frac{7\pi}{4}, -1\right), \left(\frac{9\pi}{4}, 0\right)$

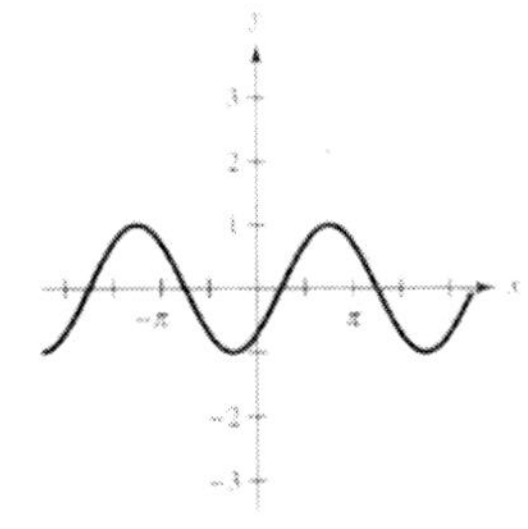

51. $y = -8\cos(x + \pi)$

Period: 2π

Amplitude: 8

Key points:

$(-\pi, -8), \left(-\frac{\pi}{2}, 0\right), (0, 8), \left(\frac{\pi}{2}, 0\right), (\pi, -8)$

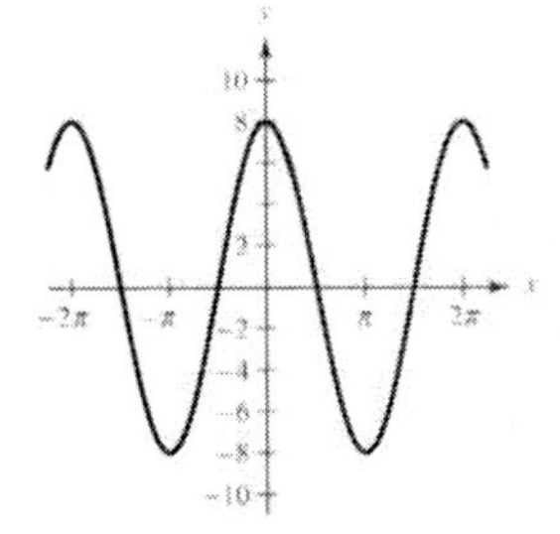

53. $y = 1 - \sin\frac{2\pi x}{3}$

Period: 3

Amplitude: 1

Vertical shift one unit upward and a reflection in the x-axis of $f(x) = \sin\frac{2\pi x}{3}$.

Key points: $(0, 1), \left(\frac{3}{4}, 0\right), \left(\frac{3}{2}, 1\right), \left(\frac{9}{4}, 2\right), (3, 1)$

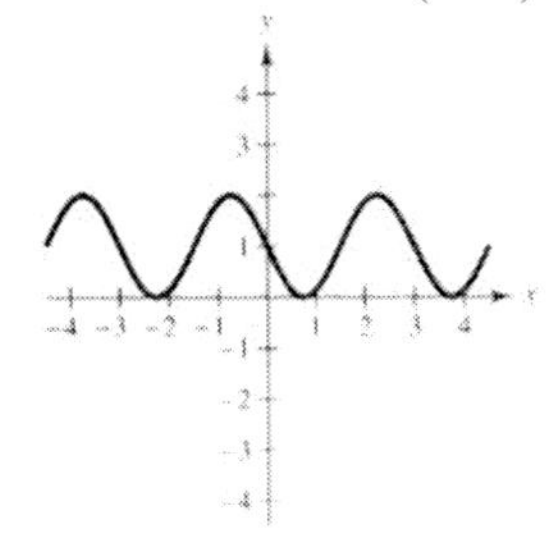

55. $y = \cos\left(\frac{x}{2} - \frac{\pi}{4}\right)$

Period: 4π

Amplitude: $\frac{2}{3}$

Shift: $\frac{x}{2} - \frac{\pi}{4} = 0$ and $\frac{x}{2} - \frac{\pi}{4} = 2\pi$

$x = \frac{\pi}{2}$ $\quad x = \frac{9\pi}{2}$

Key points:

$\left(\frac{\pi}{2}, \frac{2}{3}\right), \left(\frac{3\pi}{2}, 0\right), \left(\frac{5\pi}{2}, -\frac{2}{3}\right), \left(\frac{7\pi}{2}, 0\right), \left(\frac{9\pi}{2}, \frac{2}{3}\right)$

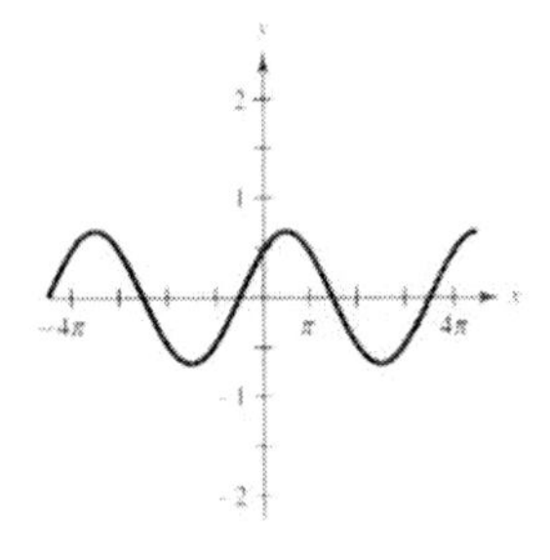

57. $y = -2\sin\frac{2\pi x}{3}$

Amplitude: 2

Period: $\frac{2\pi}{2\pi/3} = 3$

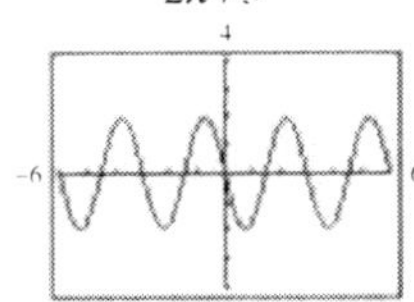

59. $y = -4 + 5\cos\frac{\pi t}{12}$

Amplitude: 5

Period: $\frac{2\pi}{\pi/12} = 24$

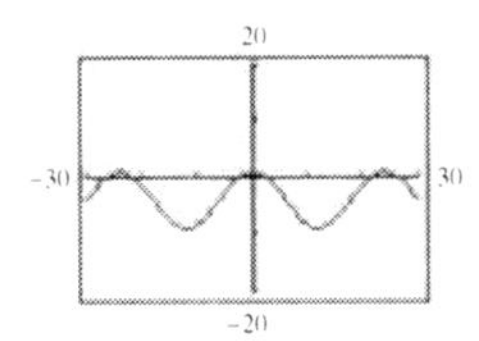

61. 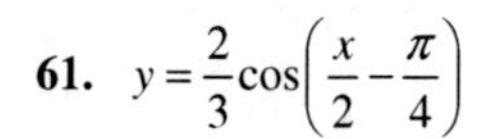
$y = \frac{2}{3}\cos\left(\frac{x}{2} - \frac{\pi}{4}\right)$

Amplitude: $\frac{2}{3}$

Period: $\frac{2\pi}{1/2} = 4\pi$

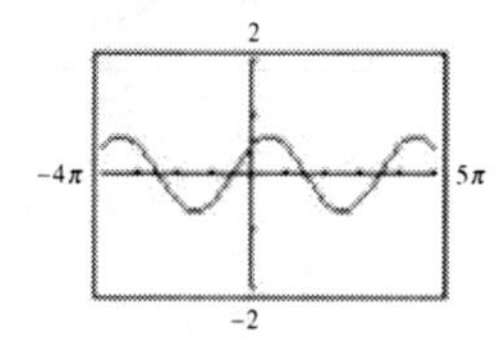

63. $y = -2\sin(4x + \pi)$

Amplitude: 2

Period: $\frac{\pi}{2}$

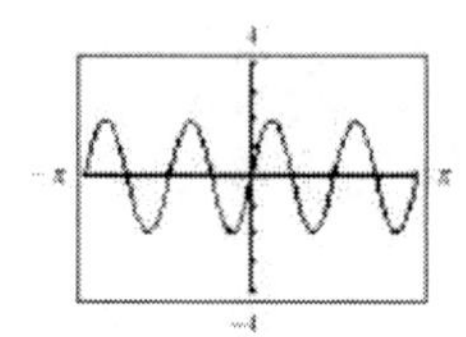

65. $y = \cos\left(2\pi x - \frac{\pi}{2}\right) + 1$

Amplitude: 1
Period: 1

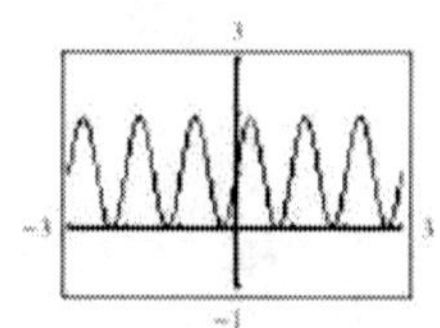

67. $y = 5\sin(\pi - 2x) + 10$

Amplitude: 5
Period: π

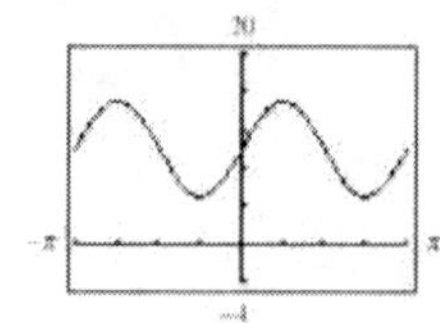

69. $y = \frac{1}{100}\sin 120\pi t$

Amplitude: $\frac{1}{100}$

Period: $\frac{1}{60}$

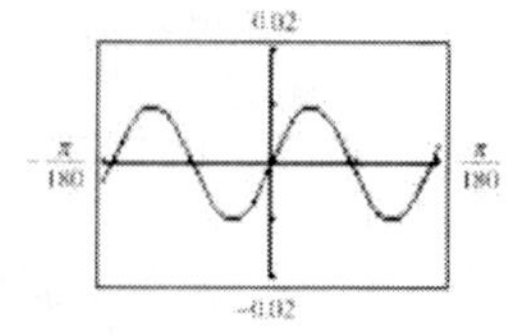

71. $f(x) = a\cos x + d$

Amplitude: $\frac{1}{2}[8 - 0] = 4$

Since $f(x)$ is the graph of $g(x) = 4\cos x$ reflected about the x-axis and shifted vertically four units upward, we have $a = -4$ and $d = 4$. Thus, $f(x) = -4\cos x + 4$
$= 4 - 4\cos x.$

73. $f(x) = a\cos x + d$

Amplitude: $\frac{1}{2}[7 - (-5)] = 6$

Since $f(x)$ is the graph of $g(x) = 6\cos x$ reflected about the x-axis and shifted vertically one unit upward, we have $a = -6$ and $d = 1$. Thus,

$f(x) = -6\cos x + 1.$

75. $f(x) = a\sin(bx - c)$

Amplitude: $|a| = 3$

Since the graph is reflected about the x-axis, we have $a = -3$.

Period: $\frac{2\pi}{b} = \pi \Rightarrow b = 2$

Phase shift: $c = 0$

Thus, $f(x) = -3\sin 2x.$

77. $f(x) = a\sin(bx - c)$

Amplitude: $a = 1$

Period: $2\pi \Rightarrow b = 1$

Phase shift: $bx - c = 0$ when $x = \frac{\pi}{4}.$

$$(1)\left(\frac{\pi}{4}\right) - c = 0 \Rightarrow c = \frac{\pi}{4}$$

Thus, $f(x) = \sin\left(x - \frac{\pi}{4}\right).$

79. $y_1 = \sin x$

$y_2 = -\frac{1}{2}$

In the interval $[-2\pi, 2\pi]$, $\sin x = -\frac{1}{2}$ when

$x = -\frac{5\pi}{6}, -\frac{\pi}{6}, \frac{7\pi}{6}, \frac{11\pi}{6}$.

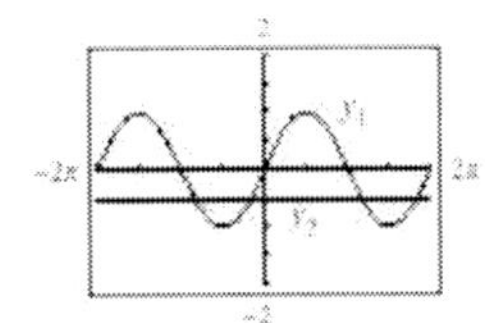

81. $v = 0.85\sin\frac{\pi t}{3}$

(a) 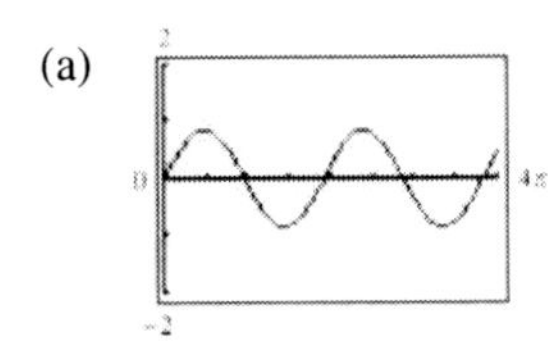

(b) Time for one cycle = one period $= \frac{2\pi}{\pi/3} = 6\text{ sec}$

(c) Cycles per min $= \frac{60}{6} = 10$ cycles per min

(d) The period would decrease.

83. $h = 25\sin\frac{\pi}{15}(t - 75) + 30$

(a) 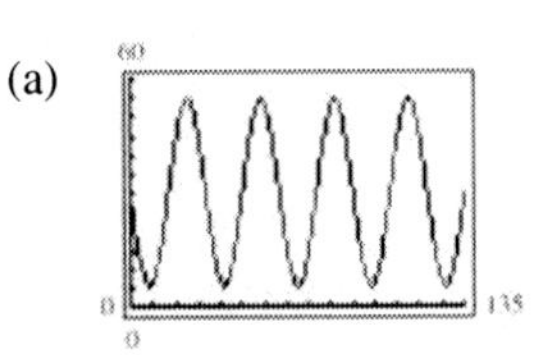

(b) Minimum: $30 - 25 = 5$ feet
Maximum: $30 + 25 = 55$ feet

85. $C = 30.3 + 21.6\sin\left(\frac{2\pi t}{365} + 10.9\right)$

(a) Period: $\frac{2\pi}{b} = \frac{2\pi}{(2\pi/365)} = 365$ days

This is to be expected: 365 days = 1 year.

(b) The constant 30.3 gallons is the average daily fuel consumption.

(c) 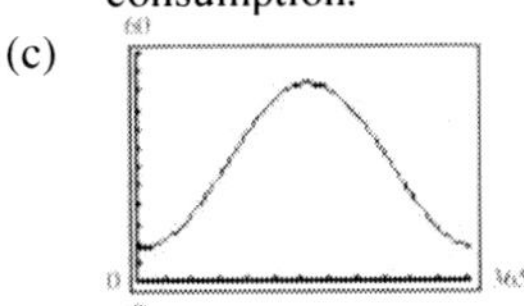

Consumption exceeds 40 gallons/day when $124 \le x \le 252$. (Graph C together with $y = 40$.) (Beginning of May through part of September)

87. (a) and (c)

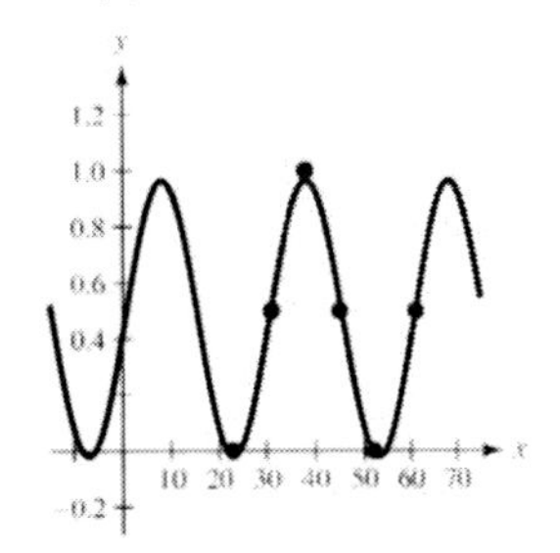

The model fits the data well.

(b) $y = 0.493\sin(0.209x - 0.114) + 0.472$

(d) Period $= \frac{2\pi}{0.209} \approx 30$ days

(e) June 12, 2013 ($t = 538$): $y \approx 0.129$ or 12.9%

89. True. The period of $f(x) = \sin x$ is 2π, the period of $g(x) = \sin(x + 2\pi)$ is also 2π, and g is a horizontal shift of 2π to the left which is equivalent to a horizontal shift of 2π to the right.

91. False. The amplitude is $\frac{1}{2}$ that of $y = \cos x$.

93.

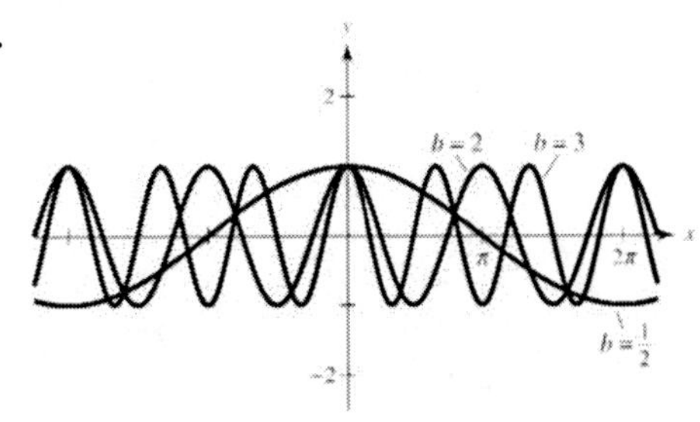

The value of b affects the period of the graph.

$b = \frac{1}{2} \rightarrow \frac{1}{2}$ cycle

$b = 2 \rightarrow 2$ cycles

$b = 3 \rightarrow 3$ cycles

95. The graph passes through (0, 0) and has period π. Matches (e).

97. The period is 4π and the amplitude is 1. Since (0, 1) and (π, 0) are on the graph, matches (c).

99. (a) $h(x) = \cos^2 x$ is even.

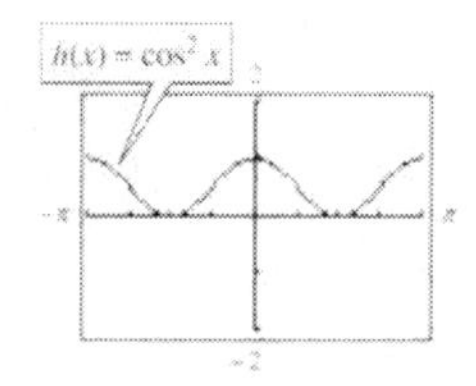

(b) $h(x) = \sin^2 x$ is even.

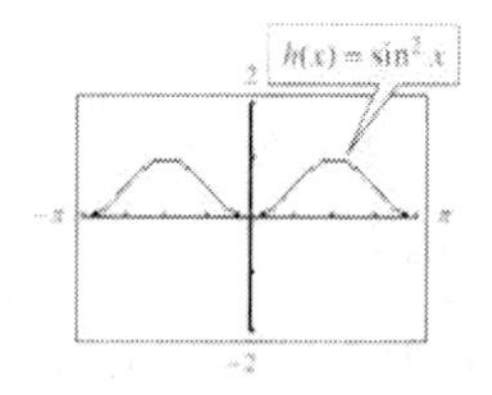

(c) $h(x) = \sin x \cos x$ is odd.

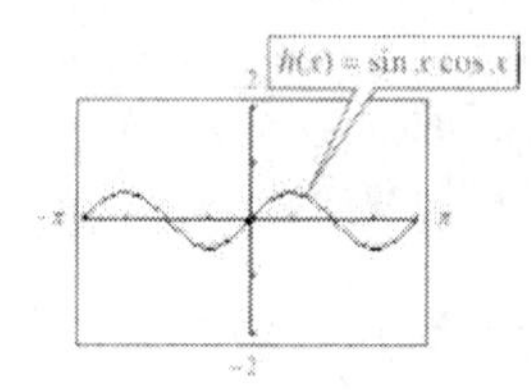

101. (a)

x	-1	-0.1	-0.01	-0.001
$\frac{\sin x}{x}$	0.8415	0.9983	1.0	1.0

x	0	0.001	0.01	0.1	1
$\frac{\sin x}{x}$	Undef.	1.0	1.0	0.9983	0.8415

(b)

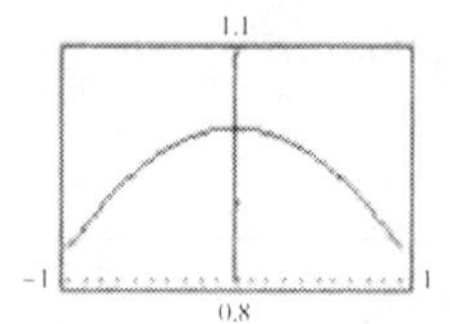

As $x \to 0$, $f(x) = \dfrac{\sin x}{x}$ approaches 1.

(c) As x approaches 0, $\dfrac{\sin x}{x}$ approaches 1.

103. (a)

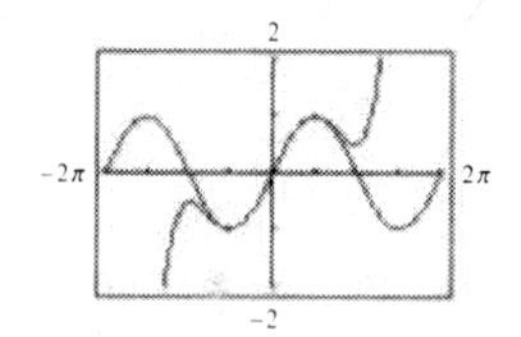

The polynomial function is a good approximation of the sine function when x is close to 0.

(b)

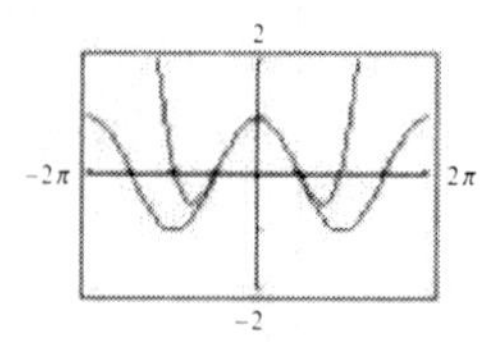

The polynomial function is a good approximation of the cosine function when x is close to 0.

(c) Next term for sine approximation: $-\dfrac{x^7}{7!}$

Next term for cosine approximation: $-\dfrac{x^6}{6!}$

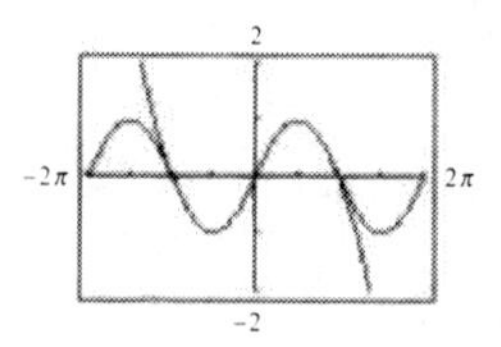

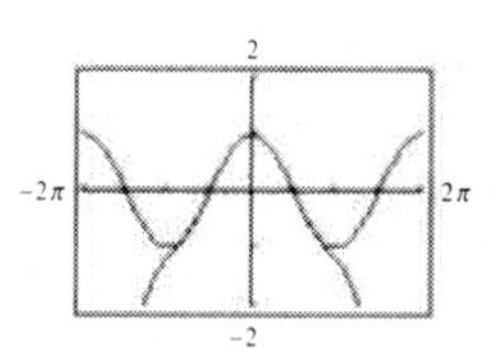

The accuracy increased.

105.

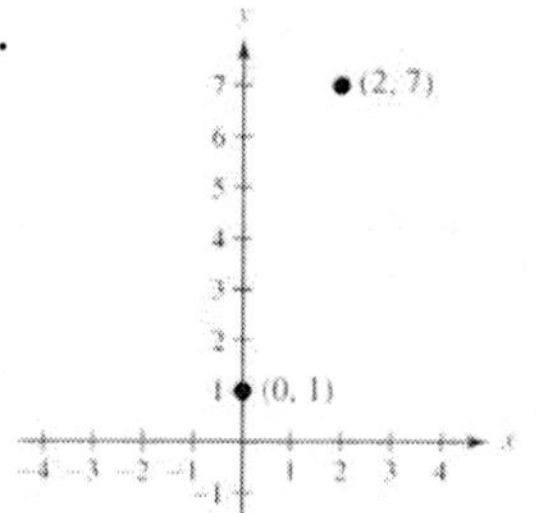

Slope $= \dfrac{7-1}{2-0} = 3$

107. $8.5 = 8.5\left(\dfrac{180°}{\pi}\right) \approx 487.014°$

109. Answers will vary. (Make a Decision)

Section 4.6

1. vertical

3. The tangent and cotangent functions have a period of π and a range of all real numbers.

5. $f(x) = \tan x$

(a) x-intercepts: $(-2\pi, 0), (-\pi, 0), (0, 0), (\pi, 0), (2\pi, 0)$

(b) y-intercept: $(0, 0)$

(c) Increasing on:

$$\left(-2\pi, -\frac{3\pi}{2}\right), \left(-\frac{3\pi}{2}, -\frac{\pi}{2}\right), \left(-\frac{\pi}{2}, \frac{\pi}{2}\right), \left(\frac{\pi}{2}, \frac{3\pi}{2}\right), \left(\frac{3\pi}{2}, 2\pi\right)$$

Never decreasing

(d) No relative extrema

(e) Vertical asymptotes: $x = -\frac{3\pi}{2}, -\frac{\pi}{2}, \frac{\pi}{2}, \frac{3\pi}{2}$

7. $f(x) = \sec x$

(a) No x-intercepts

(b) y-intercept: $(0, 1)$

(c) Increasing on:

$$\left(-2\pi, -\frac{3\pi}{2}\right), \left(-\frac{3\pi}{2}, -\pi\right), \left(0, \frac{\pi}{2}\right), \left(\frac{\pi}{2}, \pi\right)$$

Decreasing on

$$\left(-\pi, -\frac{\pi}{2}\right), \left(-\frac{\pi}{2}, 0\right), \left(\pi, \frac{3\pi}{2}\right), \left(\frac{3\pi}{2}, 2\pi\right)$$

(d) Relative minima: $(-2\pi, 1), (0, 1), (\pi, 1)$

Relative minima: $(-\pi, -1), (\pi, -1)$

(e) Vertical asymptotes: $x = -\frac{3\pi}{2}, -\frac{\pi}{2}, \frac{\pi}{2}, \frac{3\pi}{2}$

9. $y = \frac{1}{2}\tan x$

Period: π

Two consecutive asymptotes: $x = -\frac{\pi}{2}$ and $x = \frac{\pi}{2}$

x	$-\frac{\pi}{4}$	0	$\frac{\pi}{4}$
y	$-\frac{1}{2}$	0	$\frac{1}{2}$

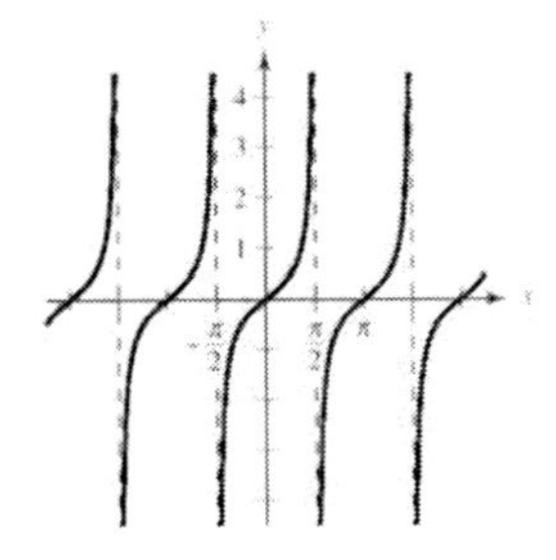

11. $y = -2\tan 2x$

Period: $\frac{\pi}{2}$

Two consecutive asymptotes: $2x = -\frac{\pi}{2} \Rightarrow x = -\frac{\pi}{4}$

$2x = \frac{\pi}{2} \Rightarrow x = \frac{\pi}{4}$

x	$-\frac{\pi}{8}$	0	$\frac{\pi}{8}$
y	2	0	-2

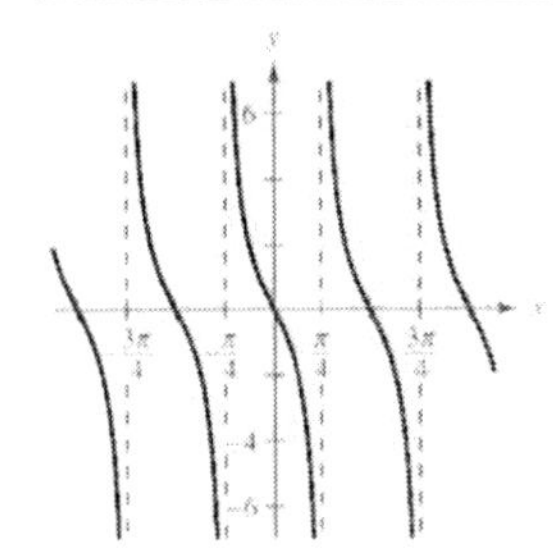

13. $y = \frac{1}{2}\cot\frac{x}{2}$

Period: $\frac{\pi}{1/2} = 2\pi$

Two consecutive asymptotes:

$\frac{x}{2} = 0 \Rightarrow x = 0$

$\frac{x}{2} = \pi \Rightarrow x = 2\pi$

x	$\frac{\pi}{2}$	π	$\frac{3\pi}{2}$
y	$\frac{1}{2}$	0	$-\frac{1}{2}$

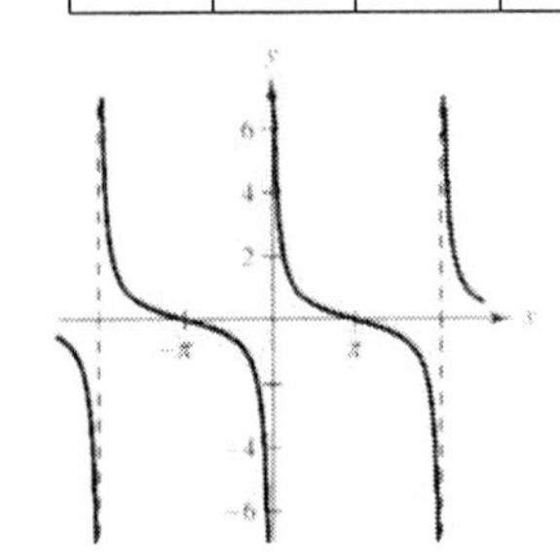

15. $y = 3\csc\frac{x}{2}$

Period: $\frac{2\pi}{1/2} = 4\pi$

Two consecutive asymptotes:

$\frac{x}{2} = 0 \Rightarrow x = 0$

$\frac{x}{2} = \pi \Rightarrow x = 2\pi$

x	$\frac{\pi}{2}$	π	$\frac{3\pi}{2}$
y	4.243	3	4.243

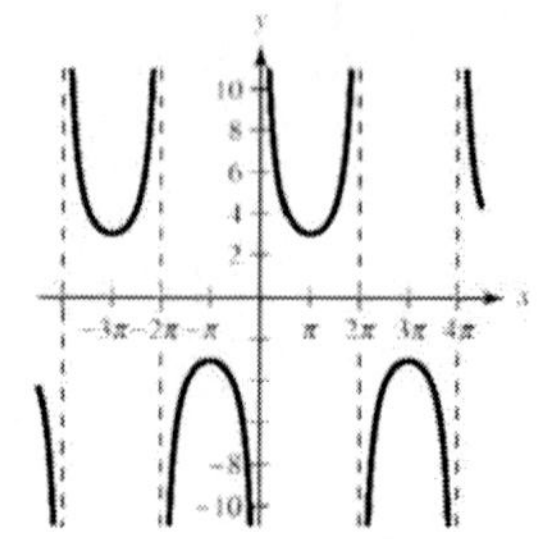

17. $y = -\frac{1}{2}\sec x$

Period: 2π

Two consecutive asymptotes:

$x = -\frac{\pi}{2},\ x = \frac{\pi}{2}$

x	$-\frac{\pi}{4}$	0	$\frac{\pi}{4}$
y	–0.707	–0.5	–0.707

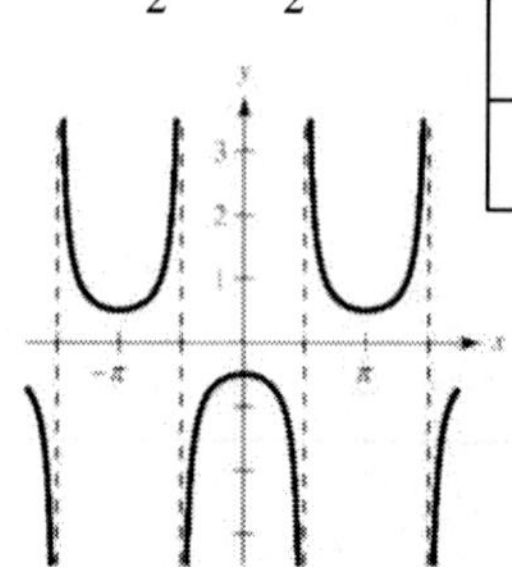

19. $y = \sec \pi x - 3$

Period: 2

Shift graph of sec πx down three units.

Two consecutive asymptotes:

$x = -\frac{1}{2},\ x = \frac{1}{2}$

x	–0.25	0	0.25
y	–1.586	–2	–1.586

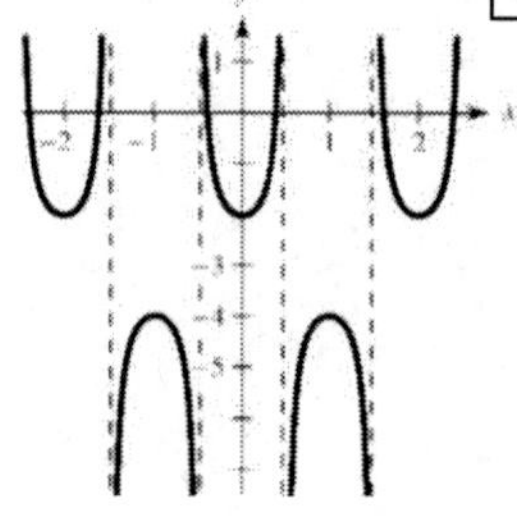

21. $y = 2\tan\frac{\pi x}{4}$

Period: $\frac{\pi}{\pi/4} = 4$

Two consecutive asymptotes: $\frac{\pi x}{4} = -\frac{\pi}{2},\ x = -2$

$\frac{\pi x}{4} = \frac{\pi}{2} \Rightarrow x = 2$

x	–1	0	1
y	–2	0	2

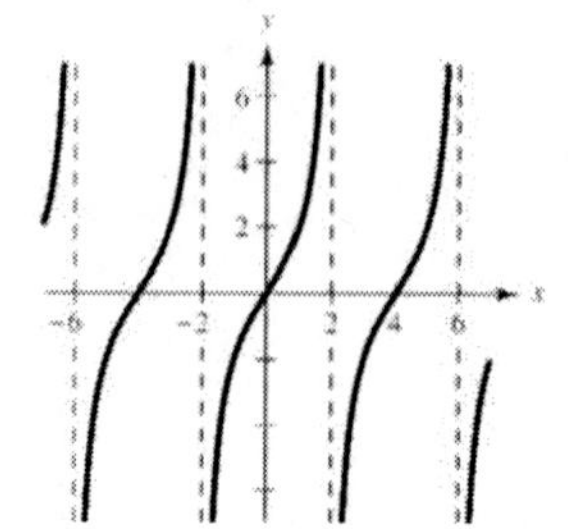

23. $y = \frac{1}{2}\sec(2x - \pi)$

Period: $\frac{2\pi}{2} = \pi$

Two consecutive asymptotes: $x = \pm\frac{\pi}{4}$

x	$-\frac{\pi}{8}$	0	$\frac{\pi}{8}$
y	–0.707	–0.5	–0.707

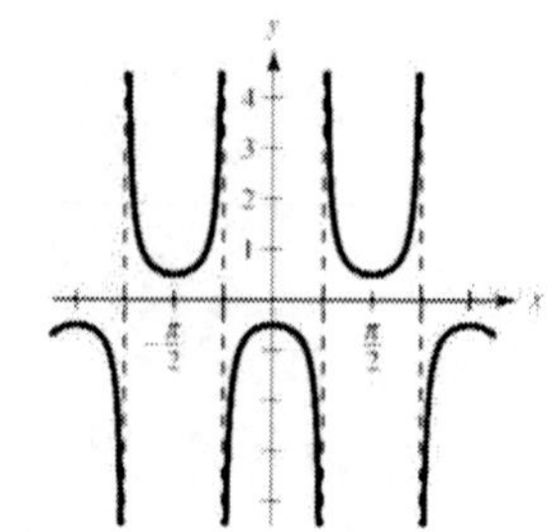

25. $y = \csc(\pi - x)$

Period: 2π
Two consecutive asymptotes:

$\pi - x = 0$ and $\pi - x = 2\pi$
$x = \pi$ $\qquad x = -\pi$

x	$\frac{\pi}{4}$	$\frac{\pi}{2}$	$\frac{3\pi}{4}$
y	1.414	1	1.414

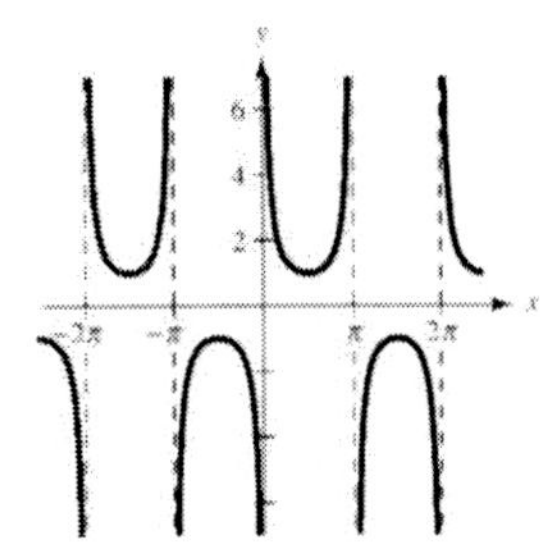

27. $y = 2\cot\left(x - \frac{\pi}{2}\right)$

Period: π
Two consecutive asymptotes:

$x - \frac{\pi}{2} = 0 \Rightarrow x = \frac{\pi}{2}$

$x - \frac{\pi}{2} = \pi \Rightarrow x = \frac{3\pi}{2}$

x	$\frac{3\pi}{4}$	π	$\frac{5\pi}{4}$
y	2	0	-2

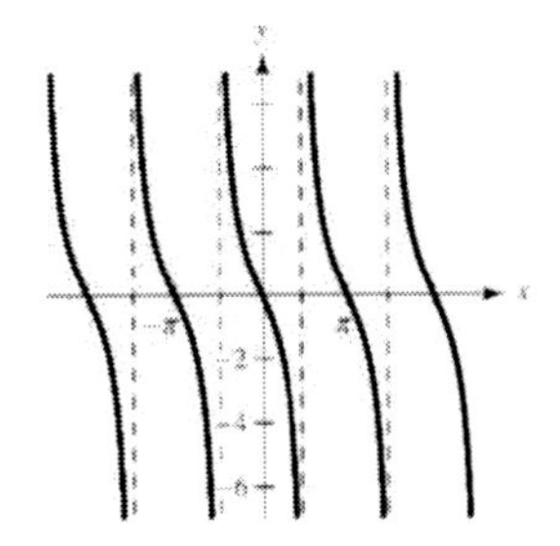

29. $y = 2\csc 3x = \frac{2}{\sin(3x)}$

Period: $\frac{2\pi}{3}$

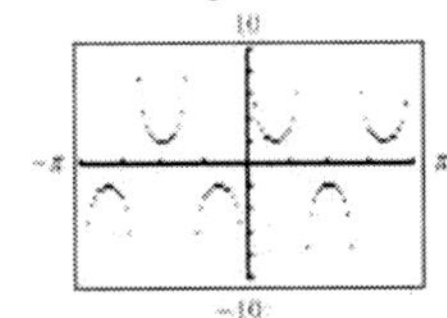

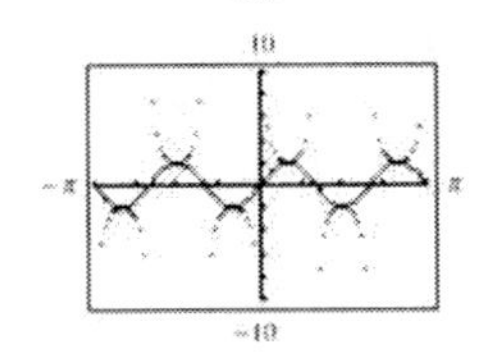

31. $y = -2\sec 4x$
$= \frac{-2}{\cos 4x}$

Period: $\frac{\pi}{2}$

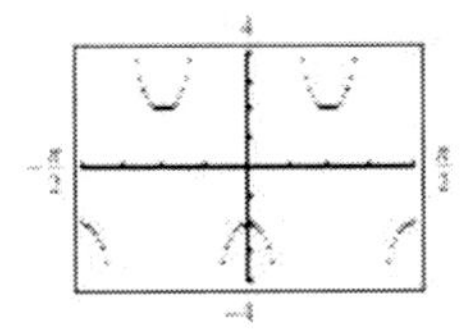

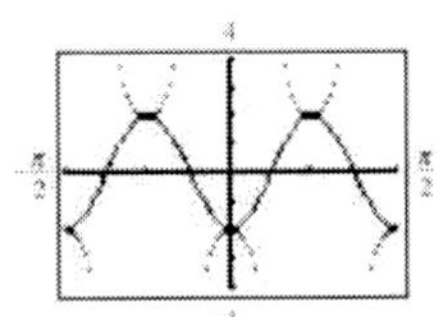

33. $y = \frac{1}{3}\sec\left(\frac{\pi x}{2} + \frac{\pi}{2}\right)$
$= \frac{1}{3\cos\left(\frac{\pi x}{2} + \frac{\pi}{2}\right)}$

Period: 4

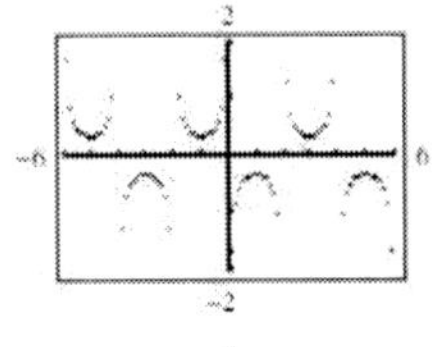

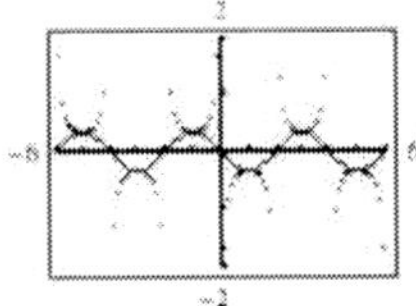

35. $\tan x = 1$

The solutions appear to be:

$$x = -\frac{7\pi}{4},\ -\frac{3\pi}{4},\ \frac{\pi}{4},\ \frac{5\pi}{4}$$

(or in decimal form: −5.498, −2.356, 0.785, 3.927)

37. $\sec x = -2$

The solutions appear to be:

$$x = \pm\frac{2\pi}{3},\ \pm\frac{4\pi}{3}$$

(or in decimal form: −4.819, −2.094, 2.094, 4.189)

39. $\tan x = \sqrt{3}$

The solutions appear to be:

$$x = -\frac{5\pi}{3},\ -\frac{2\pi}{3},\ \frac{\pi}{3},\ \frac{4\pi}{3}$$

(or in decimal form: −5.236, −2.094, 1.047, 4.189)

41. The graph of $f(x) = \sec x$ has y-axis symmetry. Thus, the function is even.

43. The function

$$f(x) = \csc 2x = \frac{1}{\sin 2x}$$

has origin symmetry. Thus, the function is odd.

45. $f(x) = \tan\left(x - \frac{\pi}{2}\right)$

$$\begin{aligned} f(-x) &= \tan\left(-x - \frac{\pi}{2}\right) \\ &= \tan\left(-\left(x + \frac{\pi}{2}\right)\right) \\ &= \frac{\sin\left(-\left(x + \frac{\pi}{2}\right)\right)}{\cos\left(-\left(x + \frac{\pi}{2}\right)\right)} \\ &= -\frac{\sin\left(x + \frac{\pi}{2}\right)}{\cos\left(x + \frac{\pi}{2}\right)} \\ &= -\tan\left(x + \frac{\pi}{2}\right) \\ &= -\tan\left(x - \frac{\pi}{2}\right) \end{aligned}$$

Thus, the function is odd.

47. $y_1 = \sin x \csc x$ and $y_2 = 1$

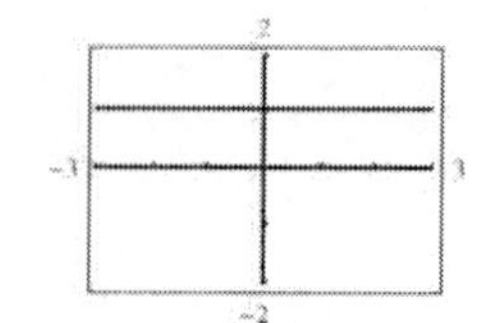

Not equivalent because y_1 is not defined at 0.

$$\sin x \csc x = \sin x\left(\frac{1}{\sin x}\right) = 1,\ \sin x \neq 0$$

49. $y_1 = \dfrac{\cos x}{\sin x}$ and $y_2 = \cot x = \dfrac{1}{\tan x}$

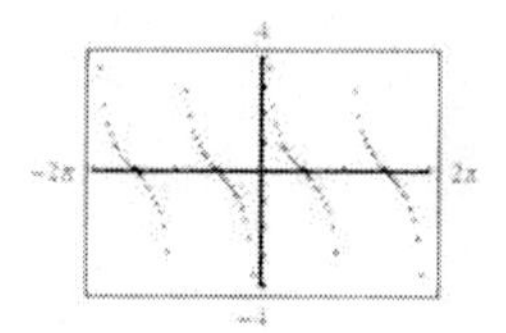

Equivalent

$$\cot x = \frac{\cos x}{\sin x}$$

51. $f(x) = x\cos x$

As $x \to 0, f(x) \to 0.$

Odd function

$$f\left(\frac{3\pi}{2}\right) = 0$$

Matches graph (d).

53. $g(x) = |x|\sin x$

As $x \to 0, g(x) \to 0.$

Odd function

$$g(2\pi) = 0$$

Matches graph (b).

55. $f(x) = e^{-x}\cos x$

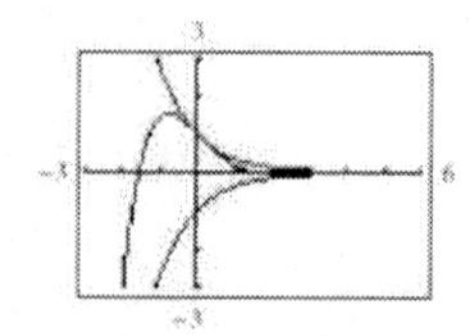

Since the damping factor is e^{-x}

$$-e^{-x} \le e^{-x}\cos x \le e^{-x}.$$

$y = \pm e^{-x}$ touches $y = e^{-x}\cos x$ at $x = n\pi$.

$y = e^{-x}\cos x$ has x-intercepts at $\cos x = 0$ or $x = \dfrac{\pi}{2} + n\pi$.

57. $y = e^{-x^2/4}\cos x$

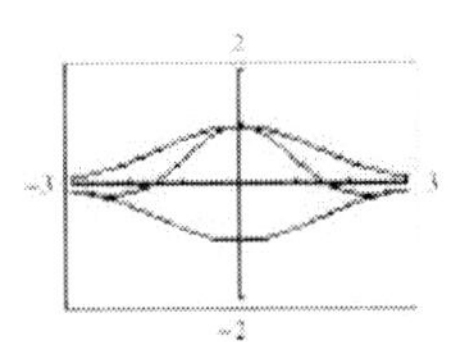

Since $e^{-x^2/4}$ is the damping factor,
$-e^{-x^2/4} \le e^{-x^2/4}\cos x \le e^{-x^2/4}$.

$y = \pm e^{-x^2/4}$ touches $y = e^{-x^2/4}\cos x$ at $x = n\pi$.

$y = e^{-x^2/4}\cos x$ has x-intercepts at $x = \dfrac{\pi}{2} + n\pi$.

59. $f(x) = \tan x$

(a) As $x \to \dfrac{\pi^+}{2}$, $f(x) \to -\infty$.

(b) As $x \to \dfrac{\pi^-}{2}$, $f(x) \to \infty$.

(c) As $x \to -\dfrac{\pi^+}{2}$, $f(x) \to -\infty$.

(d) As $x \to -\dfrac{\pi^-}{2}$, $f(x) \to \infty$.

61. $f(x) = \cot x$

(a) As $x \to 0^+$, $f(x) \to \infty$.

(b) As $x \to 0^-$, $f(x) \to -\infty$.

(c) As $x \to \pi^+$, $f(x) \to \infty$.

(d) As $x \to \pi^-$, $f(x) \to -\infty$.

63. $\tan x = \dfrac{5}{d}$

$$d = \frac{5}{\tan x} = 5\cot x$$

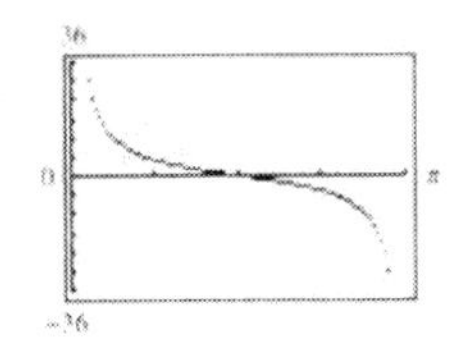

65. (a)

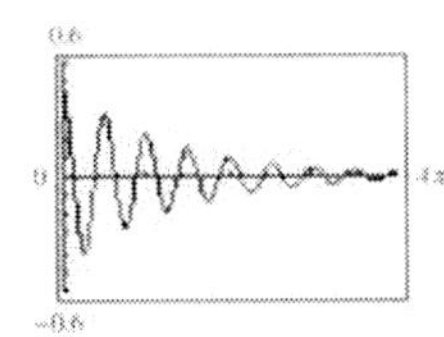

(b) The displacement function is not periodic, but damped. It approaches 0 as t increases.

67. (a) Yes. For each t there corresponds one and only one value of y.

(b) One way to determine the frequency is to note that the time between the first and second maximum points is $t = 0.7622 - 0 = 0.7622$. Thus, the frequency is approximately $(0.7622)^{-1} = 1.3$ oscillations per second.

(c) $y = 12(0.221)^t\cos(8.2t)$

To obtain this model, first fit an exponential model $y = ab^t$ to the data points $(0, 12)$, $(0.7622, 3.76)$, and $(1.5476, 1.16)$. This yields $y = 12(0.2210)^t$. Using $\dfrac{2\pi}{0.7622} \approx 8.2$ for the cosine term, you obtain the model above.

(d) $\ln 0.221 \approx -1.51 \Rightarrow y = 12e^{-1.51t}\cos(8.2t)$

(e)

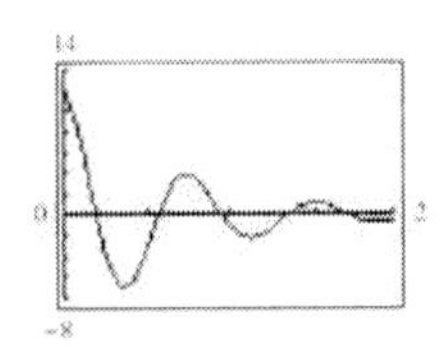

69. True. For $x \to -\infty$, $2^x \to 0$.

71. True.

$$y = \sec x = \frac{1}{\cos x}$$

If the reciprocal of $y = \sin x$ is translated $\pi/2$ units to the left, then

$$y = \frac{1}{\sin\left(x + \frac{\pi}{2}\right)} = \frac{1}{\cos x} = \sec x.$$

73. $f(x) = \tan\dfrac{\pi x}{2}$, $g(x) = \dfrac{1}{2}\sec\dfrac{\pi x}{2}$

(a)

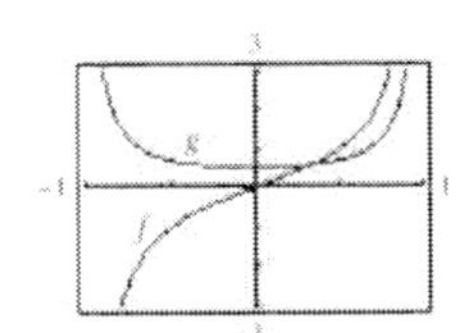

(b) The interval in which $f < g$ is $\left(-1, \dfrac{1}{3}\right)$.

(c) The interval in which $2f < 2g$ is $\left(-1, \dfrac{1}{3}\right)$, which is the same interval as part (b).

75. (a)

x	-1	-0.1	-0.01	-0.001
$\dfrac{\tan x}{x}$	1.5574	1.0033	1.0	1.0

x	0	0.001	0.01	0.1	1
$\dfrac{\tan x}{x}$	Undef.	1.0	1.0	1.0033	1.5574

(b)

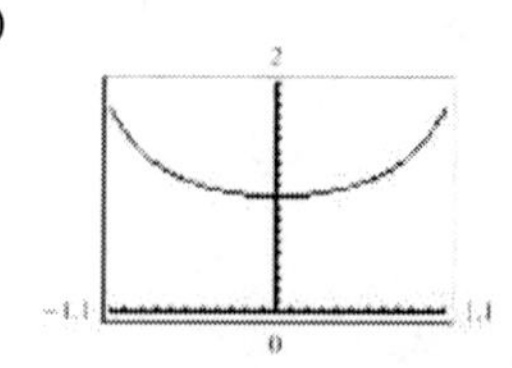

As $x \to 0, f(x) = \dfrac{\tan x}{x} \to 1.$

(c) The ratio approaches 1 as x approaches 0.

77.

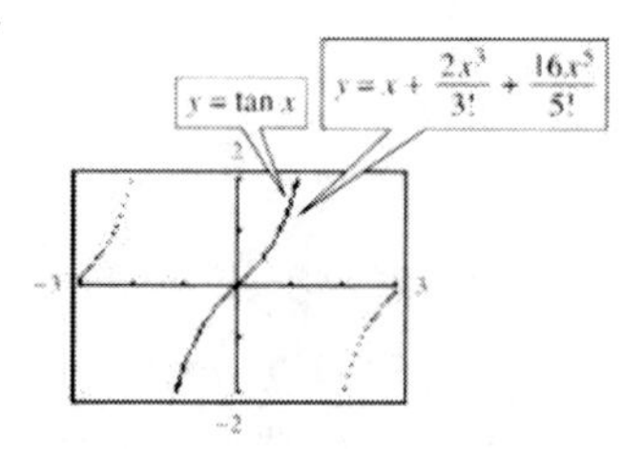

The graphs of $y_1 = \tan x$ and $y_2 = x + \dfrac{2x^3}{3!} + \dfrac{16x^5}{5!}$ are similar on the interval $\left(-\dfrac{\pi}{2}, \dfrac{\pi}{2}\right)$.

79. Distributive Property

81. Additive Identity Property

83. Not one-to-one

85. $y = \sqrt{3x-14},\ x \ge \dfrac{14}{3},\ y \ge 0$

$x = \sqrt{3y-14},\ y \ge \dfrac{14}{3},\ x \ge 0$

$x^2 = 3y - 14$

$y = \dfrac{1}{3}\left(x^2 + 14\right)$

$f^{-1}(x) = \dfrac{1}{3}(x^2 + 14),\ x \ge 0$

87. $y = x^2 + 3x - 4$

Domain: $(-\infty, \infty)$

x-intercepts:

$x^2 + 3x - 4 = 0$

$(x+4)(x-1) = 0$

$x + 4 = 0 \Rightarrow x = -4$

$x - 1 = 0 \Rightarrow x = 1$

$x = -4 \qquad x = 1$

$(-4, 0), (1, 0)$

y- intercept: $y = 0^2 + 3(0) - 4 = -4$

$(0, -4)$

No asymptotes

89. $f|x| = 3^{x+1} + 2$

Domain: $(-\infty, \infty)$

x-intercept: none

y-intercept: $y = 3^{0+1} + 2 = 3 + 2 = 5$

$(0, 5)$

Horizontal asymptote: $y = 2$

Section 4.7

1. $y = \sin^{-1} x,\ -1 \le x \le 1$

3. The inverse sine function can be denoted as $\sin^{-1} x$ or $\arcsin x$.

5. (a) $y = \arcsin\dfrac{1}{2} \Rightarrow \sin y = \dfrac{1}{2}$ for

$-\dfrac{\pi}{2} \le y \le \dfrac{\pi}{2} \Rightarrow y = \dfrac{\pi}{6}$

(b) $y = \arcsin 0 \Rightarrow \sin y = 0$ for $-\dfrac{\pi}{2} \le y \le \dfrac{\pi}{2} \Rightarrow y = 0$

7. (a) $\arcsin 1 = \dfrac{\pi}{2}$ because $\sin\dfrac{\pi}{2} = 1$ and $-\dfrac{\pi}{2} \le \dfrac{\pi}{2} \le \dfrac{\pi}{2}$.

(b) $\arccos 1 = 0$ because $\cos 0 = 1$ and $0 \le 1 \le \pi$.

9. (a) $\arctan\dfrac{\sqrt{3}}{3} = \dfrac{\pi}{6}$ because $\tan\dfrac{\pi}{6} = \dfrac{\sqrt{3}}{3}$ and $-\dfrac{\pi}{2} < \dfrac{\pi}{6} < \dfrac{\pi}{2}$.

(b) $\arctan(-1) = -\dfrac{\pi}{4}$ because $\tan\left(-\dfrac{\pi}{4}\right) = -1$ and $-\dfrac{\pi}{2} < -\dfrac{\pi}{4} < \dfrac{\pi}{2}$.

11. (a) $y = \arctan\left(-\sqrt{3}\right) \Rightarrow \tan y = -\sqrt{3}$ for

$-\dfrac{\pi}{2} < y < \dfrac{\pi}{2} \Rightarrow y = -\dfrac{\pi}{3}$

(b) $y = \arctan\sqrt{3} \Rightarrow \tan y = \sqrt{3} \Rightarrow y = \dfrac{\pi}{3}$

13. (a) $y = \sin^{-1}\frac{\sqrt{3}}{2} \Rightarrow \sin y = \frac{\sqrt{3}}{2}$ for

$-\frac{\pi}{2} \le y \le \frac{\pi}{2} \Rightarrow y = \frac{\pi}{3}$

(b) $y = \tan^{-1}\left(\frac{-\sqrt{3}}{3}\right) \Rightarrow \tan y = \frac{-\sqrt{3}}{3} \Rightarrow y = -\frac{\pi}{6}$

15. (a)

x	−1.0	−0.8	−0.6	−0.4	−0.2
y	−1.5708	−0.9273	−0.6435	−0.4115	−0.2014

x	0	0.2	0.4	0.6	0.8	1
y	0	0.2014	0.4115	0.6435	0.9273	1.5708

(b)

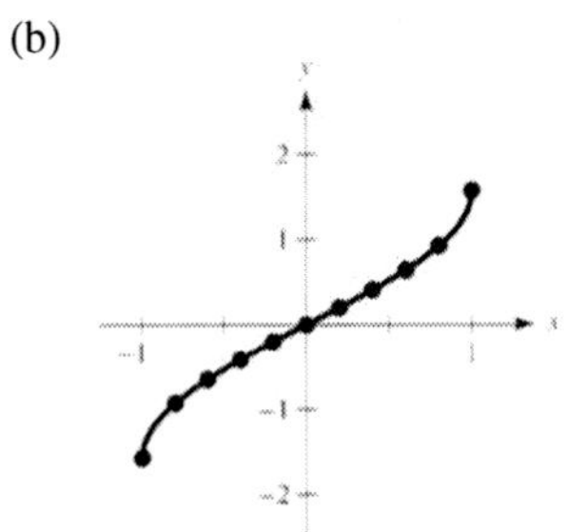

(c)

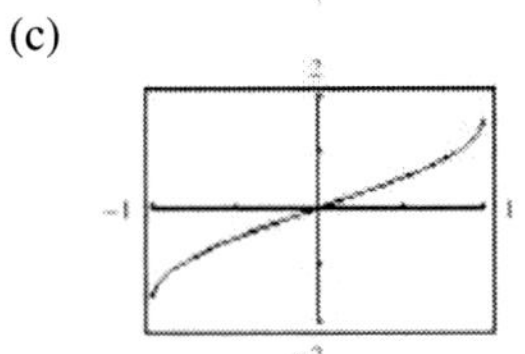

The graphs are the same.

(d) Intercept: $(0, 0)$; symmetric to the origin

17. $y = \arctan x$

$x = -\sqrt{3} \Rightarrow y = -\frac{\pi}{3}\left(\tan\left(-\frac{\pi}{3}\right) = -\sqrt{3}\right)$

$y = -\frac{\pi}{6} \Rightarrow x = -\frac{\sqrt{3}}{3}\left(\tan\left(-\frac{\pi}{6}\right) = -\frac{\sqrt{3}}{3}\right)$

$y = \frac{\pi}{4} \Rightarrow x = 1\left(\tan\frac{\pi}{4} = 1\right)$

19. $\arcsin 0.45 \approx 0.47$

21. $\tan^{-1} 15 \approx 1.50$

23. $\cos^{-1}(0.75) \approx 0.72$

25. $\arcsin(-0.75) \approx -0.85$

27. $\arctan(-6) \approx -1.41$

29. $\sin^{-1}\left(\frac{3}{4}\right) \approx 0.85$

31. $\arctan\left(\frac{7}{2}\right) \approx 1.29$

33. $g(x) = \arcsin(x + 3)$ is a horizontal shift three units to the left of $f(x) = \arcsin x$.

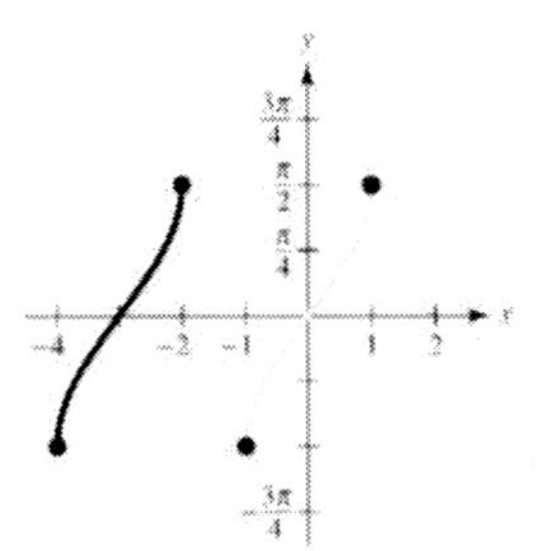

35. $g(x) = \arcsin(-x)$ is a reflection about the y-axis of $f(x) = \arcsin x$.

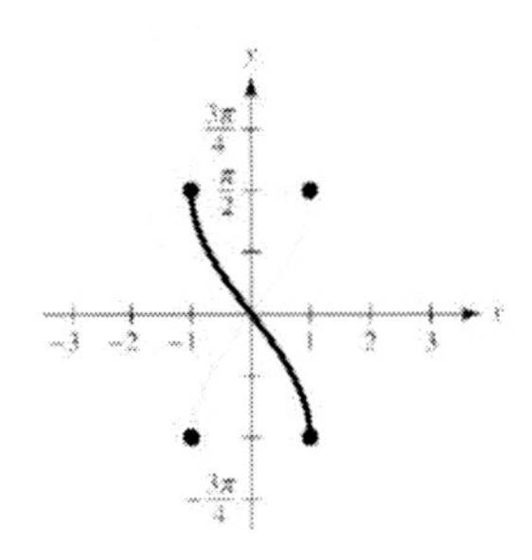

37. $g(x) = \arccos(x + \pi)$ is a horizontal shift π units to the left of $f(x) = \arccos x$.

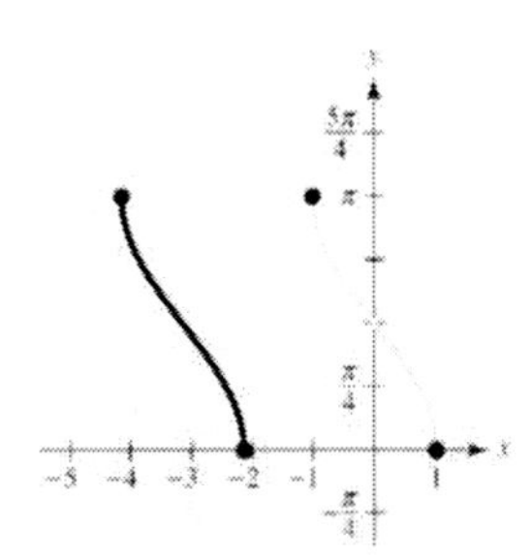

39. $g(x) = \arccos(-x - 2) = \arccos(-(x + 2))$ is a reflection about the y-axis and a horizontal shift two units to the left of $f(x) = \arccos x$.

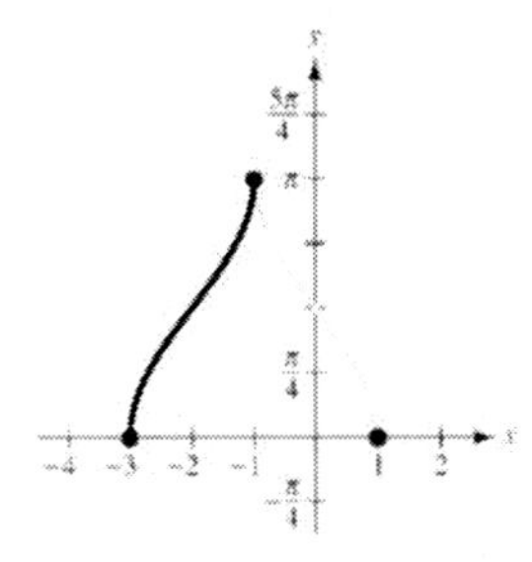

41. $g(x) = \arctan(x+1)$ is a horizontal shift one unit to the left of $f(x) = \arctan x$.

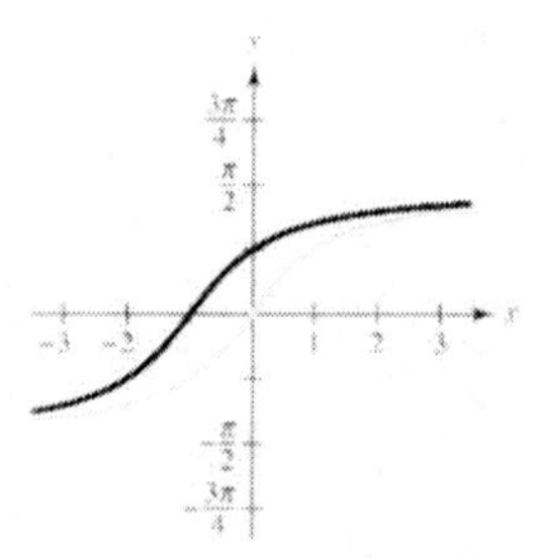

43. $g(x) = \arctan(-x)$ is a reflection about the y-axis of $f(x) = \arctan x$.

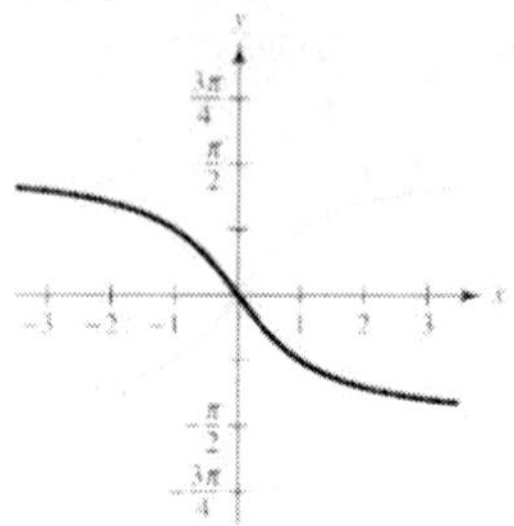

45. $\tan\theta = \dfrac{x}{8}$

$\theta = \arctan\dfrac{x}{8}$

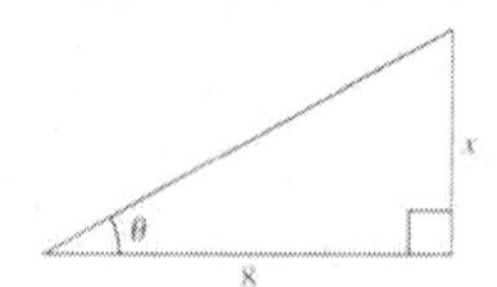

47. $\sin\theta = \dfrac{x+2}{5}$

$\theta = \arcsin\left(\dfrac{x+2}{5}\right)$

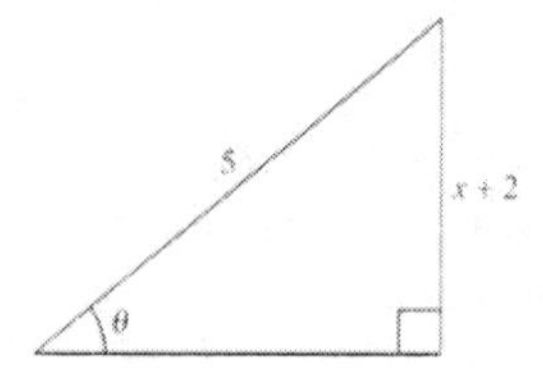

49. Let y be the third side. Then

$y^2 = 2^2 - x^2 = 4 - x^2 \Rightarrow y = \sqrt{4-x^2}$

$\sin\theta = \dfrac{x}{2} \Rightarrow \theta = \arcsin\dfrac{x}{2}$

$\cos\theta = \dfrac{\sqrt{4-x^2}}{2} \Rightarrow \theta = \arccos\dfrac{\sqrt{4-x^2}}{2}$

$\tan\theta = \dfrac{x}{\sqrt{4-x^2}} \Rightarrow \theta = \arctan\dfrac{x}{\sqrt{4-x^2}}$

51. Let y be the hypotenuse. Then

$y^2 = (x+1)^2 + 2^2 = x^2 + 2x + 5 \Rightarrow y = \sqrt{x^2 + 2x + 5}.$

$\sin\theta = \dfrac{x+1}{\sqrt{x^2+2x+5}} \Rightarrow \theta = \arcsin\dfrac{x+1}{\sqrt{x^2+2x+5}}$

$\cos\theta = \dfrac{2}{\sqrt{x^2+2x+5}} \Rightarrow \theta = \arccos\dfrac{2}{\sqrt{x^2+2x+5}}$

$\tan\theta = \dfrac{x+1}{2} \Rightarrow \theta = \arctan\dfrac{x+1}{2}$

53. $\sin(\arcsin 0.7) = 0.7$

55. $\cos[\arccos(-0.3)] = -0.3$

57. $\arcsin(\sin 3\pi) = \arcsin(0) = 0$

Note: 3π is not in the range of the arcsine function.

59. $\arctan\left(\tan\dfrac{11\pi}{6}\right) = \arctan\left(-\dfrac{\sqrt{3}}{3}\right) = -\dfrac{\pi}{6}$

61. $\sin^{-1}\left(\sin\dfrac{5\pi}{2}\right) = \sin^{-1} 1 = \dfrac{\pi}{2}$

63. $\sin^{-1}\left(\tan\dfrac{5\pi}{4}\right) = \sin^{-1} 1 = \dfrac{\pi}{2}$

65. Let $y = \arctan\dfrac{4}{3}$. Then

$\tan y = \dfrac{4}{3},\ 0 < y < \dfrac{\pi}{2},$ and

$\sin y = \dfrac{4}{5}.$

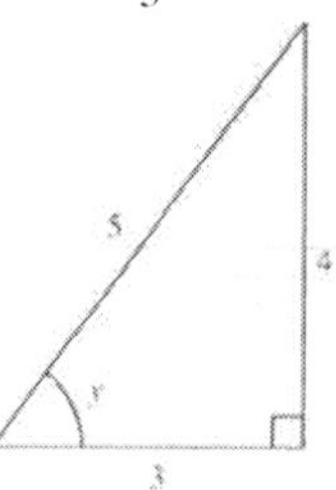

67. Let $y = \arcsin\dfrac{24}{25}$. Then

$\sin y = \dfrac{24}{25},$ and $\cos y = \dfrac{7}{25}.$

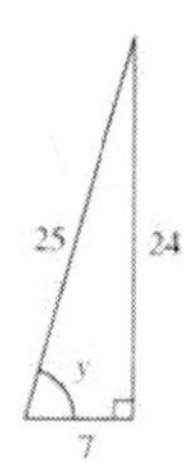

69. Let $y = \arctan\left(-\frac{3}{5}\right)$. Then,

$\tan y = -\frac{3}{5}, \ -\frac{\pi}{2} < y < 0$ and $\sec y = \frac{\sqrt{34}}{5}$.

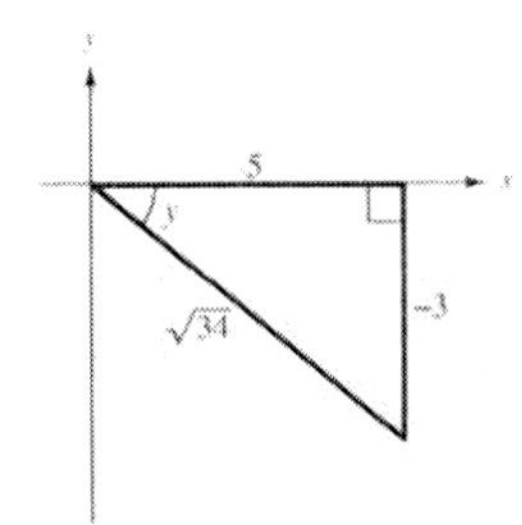

71. Let $y = \arccos\left(-\frac{2}{3}\right)$. Then, $\cos y = -\frac{2}{3}, \ \frac{\pi}{2} < y < \pi$ and $\sin y = \frac{\sqrt{5}}{3}$.

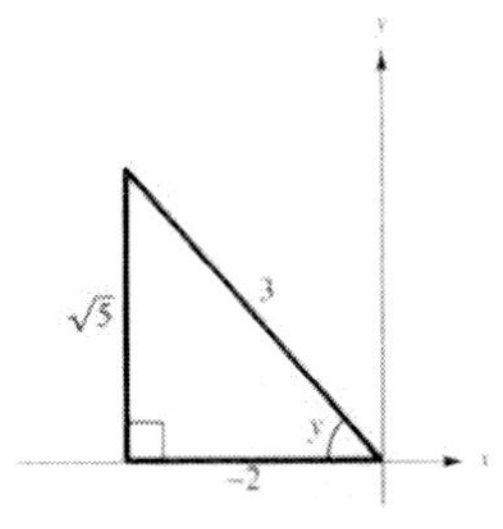

73. Let $y = \arctan x$. Then,

$\tan y = x$ and $\cot y = \frac{1}{x}$.

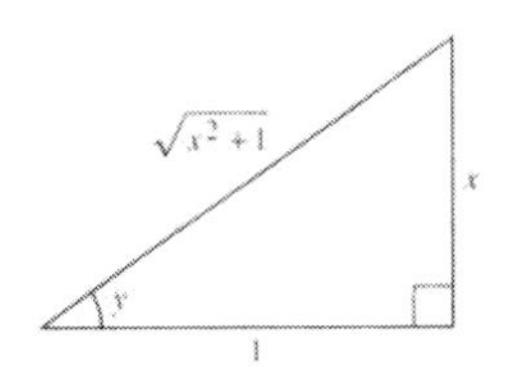

75. Let $y = \arccos(x + 2)$, $\cos y = x + 2$.

Opposite side: $\sqrt{1 - (x + 2)^2}$

$\sin y = \frac{\sqrt{1 - (x + 2)^2}}{1} = \sqrt{-x^2 - 4x - 3}$

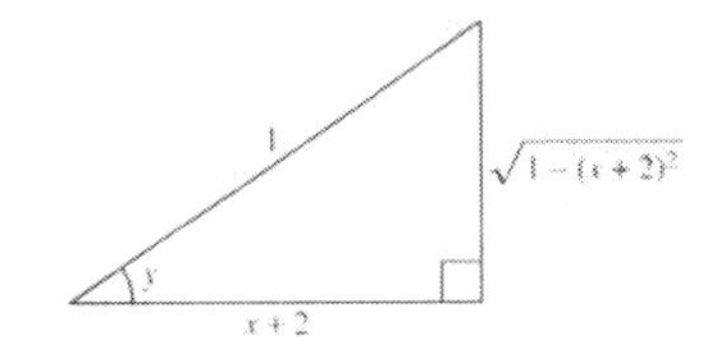

77. Let $y = \arccos \frac{x}{5}$. Then $\cos y = \frac{x}{5}$, and $\tan y = \frac{\sqrt{25 - x^2}}{x}$.

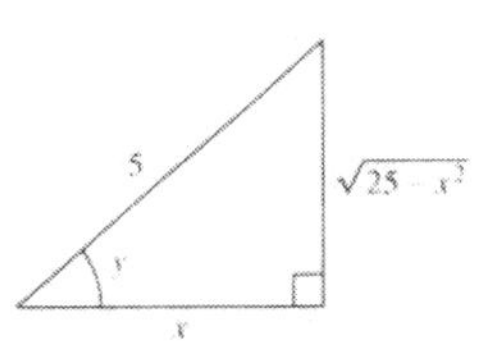

79. Let $y = \arctan \frac{x}{\sqrt{7}}$. Then $\tan y = \frac{x}{\sqrt{7}}$ and $\csc y = \frac{\sqrt{7 + x^2}}{x}$.

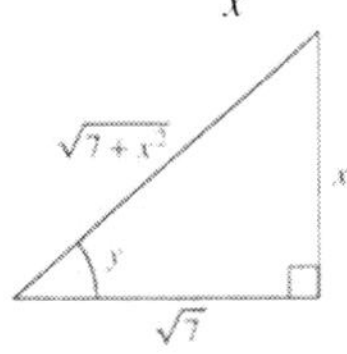

81. Let $y = \arctan \frac{14}{x}$. Then $\tan y = \frac{14}{x}$ and

$\sin y = \frac{14}{\sqrt{196 + x^2}}$. Thus, $y = \arcsin\left(\frac{14}{\sqrt{196 + x^2}}\right)$.

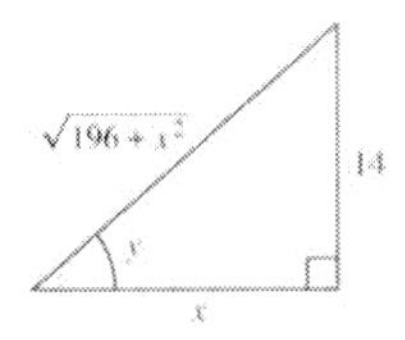

83. Let $y = \arccos \frac{3}{\sqrt{x^2 - 2x + 10}}$. Then,

$\cos y = \frac{3}{\sqrt{x^2 - 2x + 10}} = \frac{3}{\sqrt{(x - 1)^2 + 9}}$

and $\sin y = \frac{|x - 1|}{\sqrt{(x - 1)^2 + 9}}$. Thus,

$y = \arcsin \frac{|x - 1|}{\sqrt{(x - 1)^2 + 9}} = \arcsin \frac{|x - 1|}{\sqrt{x^2 - 2x + 10}}$

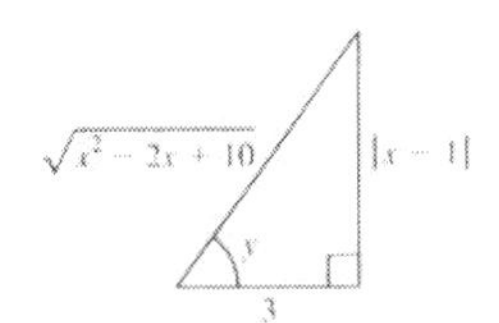

85. $y = 2\arccos x$

Domain: $-1 \le x \le 1$

Range: $0 \le y \le 2\pi$

Vertical stretch of $f(x) = \arccos x$

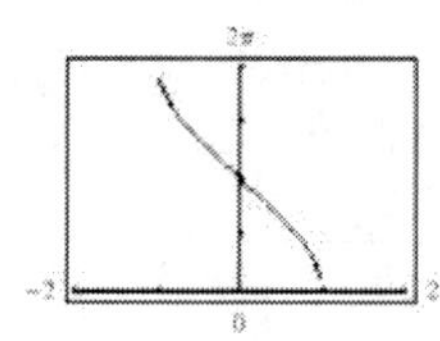

87. $f(x) = \arcsin(x - 2)$

Domain: $1 \le x \le 3$

Range. $-\frac{\pi}{2} \le y \le \frac{\pi}{2}$

Horizontal shift two units to the right of $f(x) = \arcsin x$

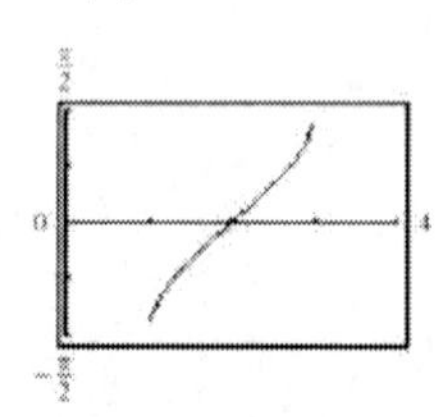

89. $f(x) = \arctan 2x$

Domain: all real numbers

Range: $-\frac{\pi}{2} < y < \frac{\pi}{2}$

Horizontal shrink of $y = \arctan x$

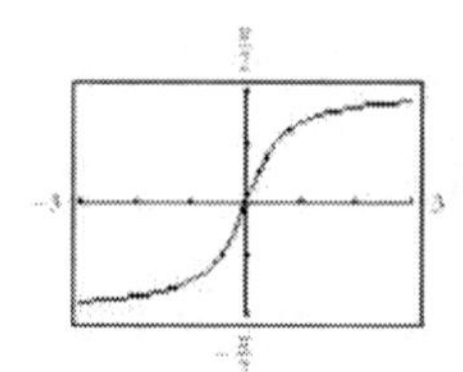

91. $f(t) = 3\cos 2t + 3\sin 2t$

$$= \sqrt{3^2 + 3^2}\sin\left(2t + \arctan\frac{3}{3}\right)$$

$$= 3\sqrt{2}\sin(2t + \arctan 1)$$

$$= 3\sqrt{2}\sin\left(2t + \frac{\pi}{4}\right)$$

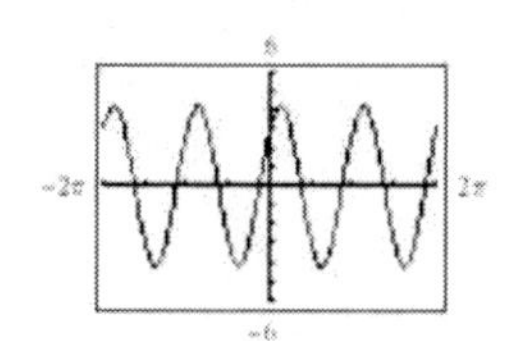

The two forms are equivalent.

93. As $x \to 1^-$, $\arcsin x \to \frac{\pi}{2}$.

95. As $x \to \infty$, $\arctan x \to \frac{\pi}{2}$.

97. As $x \to -1^+$, $\arccos x \to \pi$.

99. (a)

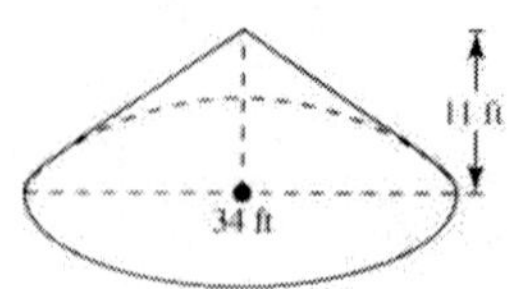

(b) $\tan\theta = \frac{11}{17} \Rightarrow \theta \approx 0.5743$ radian or $32.9°$

(c) $\tan(0.5743) = \frac{h}{20} \Rightarrow h = 20\tan(0.5743)$

≈ 12.94 feet

101. $\beta = \arctan\frac{3x}{x^2 + 4}$, $x > 0$

(a)

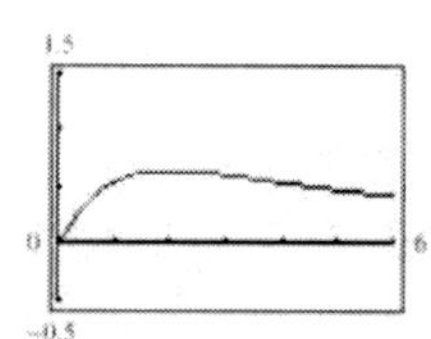

(b) β is maximum when $x = 2$ feet.

(c) The graph has a horizontal asymptote at $\beta = 0$. As the camera moves further from the picture, the angle subtended by the camera approaches 0.

103. (a) $\tan\theta = \frac{x}{20}$

$$\theta = \arctan\frac{x}{20}$$

(b) When $x = 5$,

$$\theta = \arctan\frac{5}{20} \approx 14.0°, \ (0.24 \text{ rad}).$$

When $x = 12$,

$$\theta = \arctan\frac{12}{20} \approx 31.0°, \ (0.54 \text{ rad}).$$

105. False. $\tan\frac{5\pi}{4} = 1 \Rightarrow \arctan 1 = \frac{\pi}{4}$ because the range of $\arctan x$ is $\left(-\frac{\pi}{2}, \frac{\pi}{2}\right)$

107. $y = \text{arccot}\, x$ if and only if $\cot y = x$, $-\infty < x < \infty$ and $0 < y < \pi$.

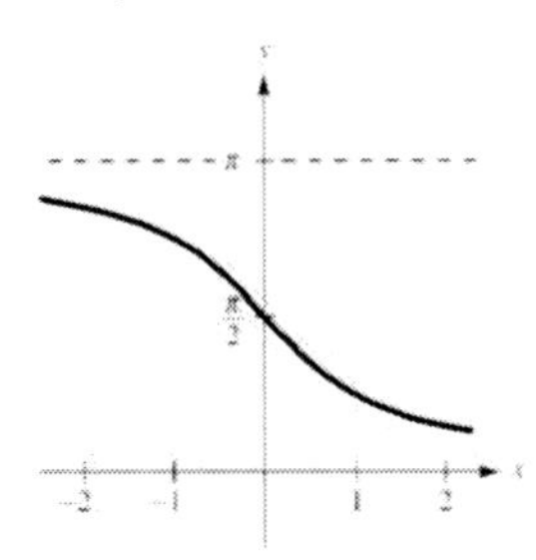

109. $y = \text{arccsc}\, x$ if and only if $\csc y = x$.

Domain: $(-\infty, -1] \cup [1, \infty)$

Range: $\left[-\frac{\pi}{2}, 0\right) \cup \left(0, \frac{\pi}{2}\right]$

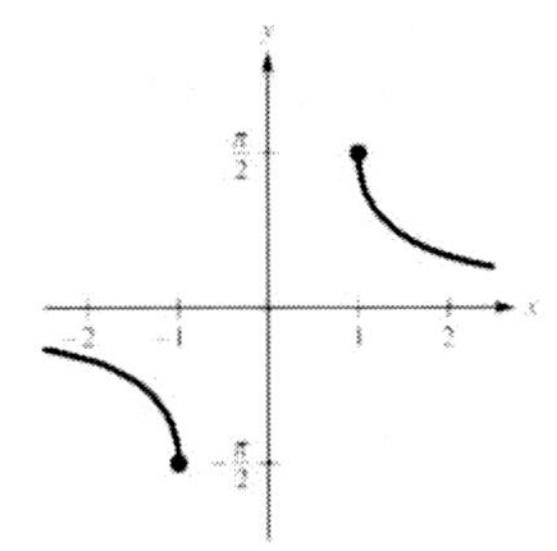

111. $y = \text{arcsec}\sqrt{2} \Rightarrow \sec y = \sqrt{2}$ and

$$0 \le y < \frac{\pi}{2} \cup \frac{\pi}{2} < y \le \pi \Rightarrow y = \frac{\pi}{4}$$

113. $y = \text{arccot}\left(-\sqrt{3}\right) \Rightarrow \cot y = -\sqrt{3}$ and

$$0 < y < \pi \Rightarrow y = \frac{5\pi}{6}$$

115. Let $y = \arcsin(-x)$. Then,

$$\sin y = -x$$
$$-\sin y = x$$
$$\sin(-y) = x$$
$$-y = \arcsin x$$
$$y = -\arcsin x.$$

Therefore, $\arcsin(-x) = -\arcsin x$.

117. $\arcsin x + \arccos x = \frac{\pi}{2}$

Let $\alpha = \arcsin x \Rightarrow \sin \alpha = x$.
Let $\beta = \arccos x \Rightarrow \cos \beta = x$.
Hence, $\sin a = \cos \beta \Rightarrow a$ and β are complementary angles $\Rightarrow a + \beta = \frac{\pi}{2} \Rightarrow \arcsin x + \arccos x = \frac{\pi}{2}$.

119. $\frac{4}{4\sqrt{2}} = \frac{1}{\sqrt{2}} = \frac{\sqrt{2}}{2}$

121. $\frac{2\sqrt{3}}{6} = \frac{\sqrt{3}}{3}$

123. $\sin \theta = \frac{5}{6}$

Adjacent side: $\sqrt{6^2 - 5^2} = \sqrt{11}$

$$\cos \theta = \frac{\sqrt{11}}{6}$$

$$\tan \theta = \frac{5}{\sqrt{11}} = \frac{5\sqrt{11}}{11}$$

$$\csc \theta = \frac{6}{5}$$

$$\sec \theta = \frac{6}{\sqrt{11}} = \frac{6\sqrt{11}}{11}$$

$$\cot \theta \frac{\sqrt{11}}{5}$$

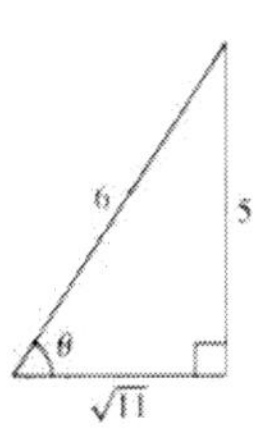

125. $\sin \theta = \frac{3}{4}$

Adjacent side: $\sqrt{16 - 9} = \sqrt{7}$

$$\cos \theta = \frac{\sqrt{7}}{4}$$

$$\tan \theta = \frac{3}{\sqrt{7}} = \frac{3\sqrt{7}}{7}$$

$$\csc \theta = \frac{4}{3}$$

$$\sec \theta = \frac{4}{\sqrt{7}} = \frac{4\sqrt{7}}{7}$$

$$\cot \theta = \frac{\sqrt{7}}{3}$$

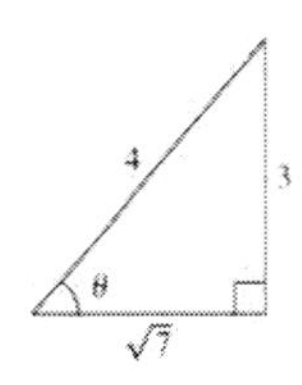

Section 4.8

1. harmonic motion

3. No. N 20° E means a direction first of due north, then 20° east of north.

5. Given: $A = 30°, b = 10$

$B = 90° - 30° = 60°$

$\tan A = \frac{a}{b} \Rightarrow a = b\tan A = 10\tan 30° \approx 5.77$

$\cos A = \frac{b}{c} \Rightarrow c = \frac{b}{\cos A} = \frac{10}{\cos 30°} \approx 11.55$

7. Given: $B = 71°, b = 14$

$\tan B = \frac{b}{a} \Rightarrow a = \frac{b}{\tan B} = \frac{14}{\tan 71°} \approx 4.82$

$\sin B = \frac{b}{c} \Rightarrow c = \frac{b}{\sin B} = \frac{14}{\sin 71°} \approx 14.81$

$A = 90° - 71° = 19°$

9. Given: $a = 6, b = 12$

$c^2 = a^2 + b^2 \Rightarrow c = \sqrt{36 + 144} \approx 13.42$

$\tan A = \frac{a}{b} = \frac{6}{12} = \frac{1}{2} \Rightarrow A = \arctan\frac{1}{2} \approx 26.57°$

$B = 90° - 26.57° = 63.43°$

11. Given: $b = 16, c = 54$

$a = \sqrt{c^2 - b^2} = \sqrt{2660} \approx 51.58$

$\cos A = \frac{b}{c} = \frac{16}{54} \Rightarrow A = \arccos\left(\frac{16}{54}\right) \approx 72.76°$

$B = 90° - 72.76° = 17.24°$

13. $A = 12°15', c = 430.5$

$B = 90° - 12°15' = 77°45'$

$\sin 12°15' = \frac{a}{430.5}$

$a = 430.5\sin 12°15' \approx 91.34$

$\cos 12°15' = \frac{b}{430.5}$

$b = 430.5\cos 12°15' \approx 420.70$

15. $\tan\theta = \frac{h}{b/2}$

$h = \frac{1}{2}b\tan\theta$

$h = \frac{1}{2}(8)\tan 52° \approx 5.12$ in.

17. $\tan\theta = \frac{h}{b/2}$

$h = \frac{1}{2}b\tan\theta$

$h = \frac{1}{2}(18.5)\tan 41.6° \approx 8.21$ ft

19. $\sin 80° = \frac{h}{20}$

$h = 20\sin 80°$

≈ 19.70 feet

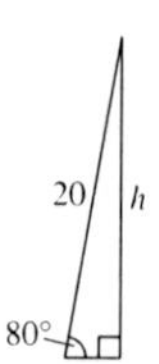

21. $\sin 66° = \frac{h}{120}$

$120\sin 66° = h$

$109.63 \text{ feet} \approx h$

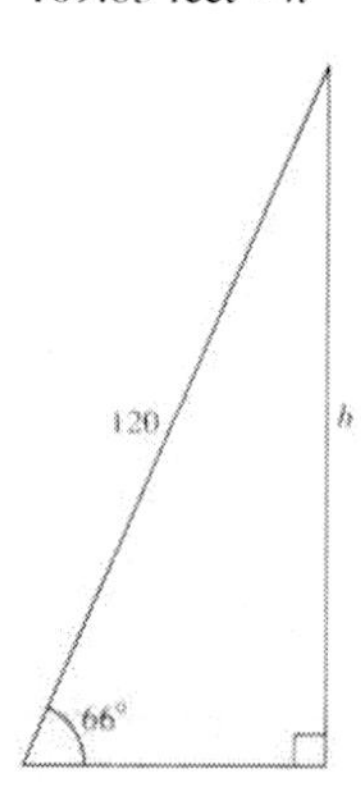

23. $\sin 31.5° = \frac{x}{4000}$

$x = 4000\sin 31.5°$

≈ 2089.99 feet

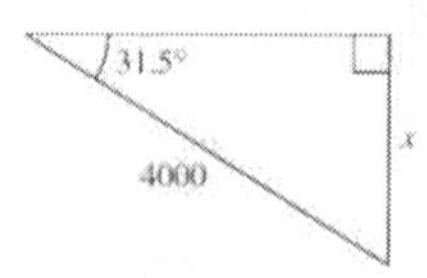

25. (a)

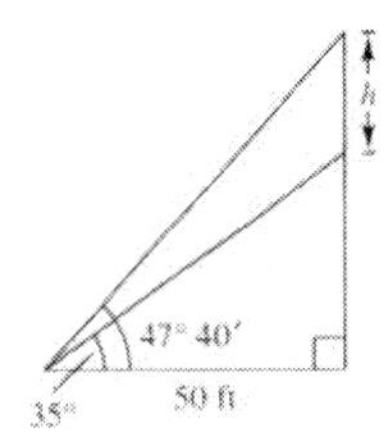

(b) Let the height of the church $= x$ and the height of the church and steeple $= y$. Then:

$$\tan 35° = \frac{x}{50} \text{ and } \tan 47°40' = \frac{y}{50}$$
$$x = 50\tan 35° \text{ and } y = 50\tan 47°40'$$
$$h = y - x = 50(\tan 47°40' - \tan 35°)$$

(c) $h \approx 19.9$ feet

27. (a)

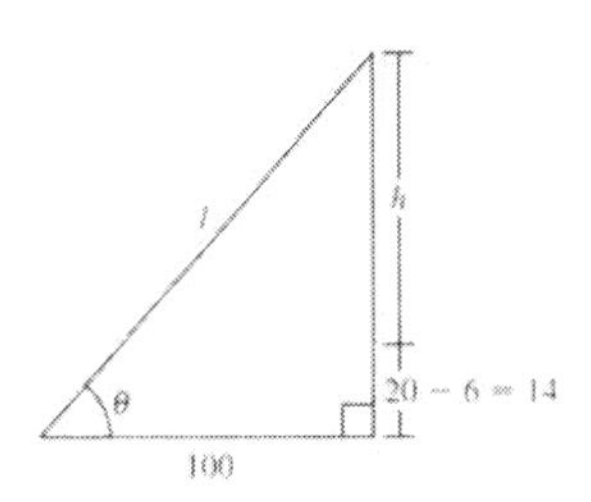

$$l^2 = (h+14)^2 + 100^2$$
$$l = \sqrt{h^2 + 28h + 10{,}196}$$

(b) $\cos\theta = \dfrac{100}{h}$

$$\theta = \arccos\left(\frac{100}{l}\right)$$

(c) If $\theta = 35°$, then $\cos 35° = \dfrac{100}{l}$.

$$l = \frac{100}{\cos 35°}$$
$$l \approx 122.08 \text{ feet}$$

When $l = 122.08$ feet, using part (a):

$$122.08^2 = (h+14)^2 + 100^2$$
$$4903.5264 = (h+14)^2$$
$$70.025 = h + 14$$
$$56.03 \text{ feet} \approx h$$

29.

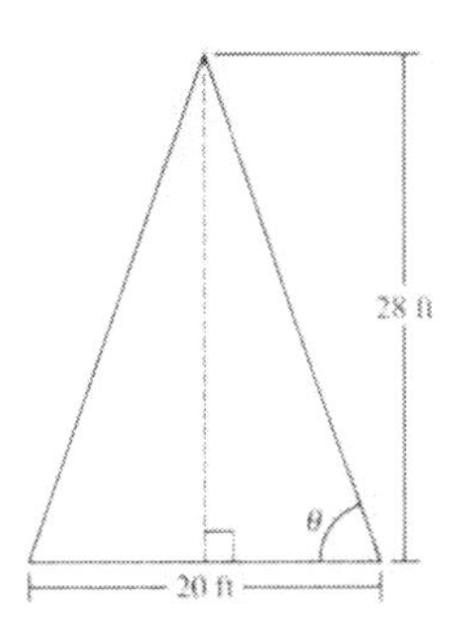

$$\tan\theta = \frac{28}{10}$$
$$\theta = \tan^{-1}\left(\frac{28}{10}\right)$$
$$\theta \approx 70.35°$$

31. $\sin\alpha = \dfrac{4000}{16{,}500} \Rightarrow \alpha \approx 14.03°$

$\theta \approx 90° - \alpha \approx 75.97°$

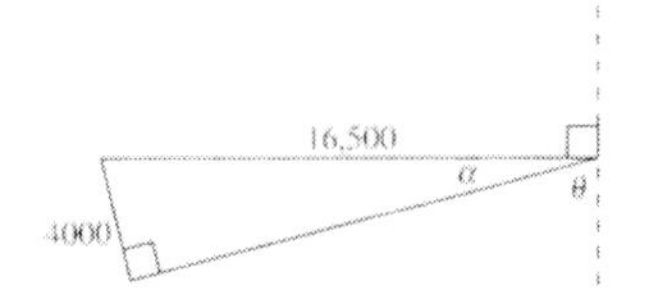

33. Since the airplane speed is $\left(275\dfrac{\text{ft}}{\text{sec}}\right)\left(60\dfrac{\text{sec}}{\text{min}}\right) = 16{,}500\dfrac{\text{ft}}{\text{min}}$, after one minute its distance traveled is 16,500 feet.

$$\sin 18° = \frac{a}{16{,}500}$$
$$a = 16{,}500\sin 18°$$
$$\approx 5099 \text{ ft}$$

35. $\sin 9.5° = \dfrac{x}{4}$

$$x = 4\sin 9.5°$$
$$\approx 0.66 \text{ mile}$$

37. $90° - 29° = 61°$

$(20)(6) = 120$ nautical miles

$\sin 61° = \frac{a}{120} \Rightarrow a = 120\sin 61° \approx 104.95$ nautical miles south

$\cos 61° = \frac{b}{120} \Rightarrow b = 120\cos 61° \approx 58.18$ nautical miles west

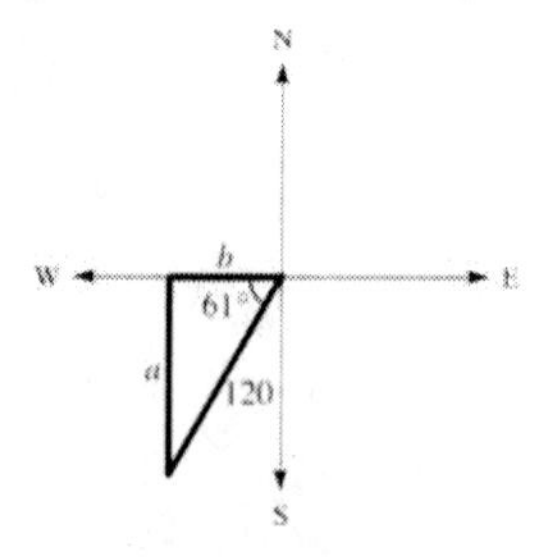

39. $\theta = 32°,\ \phi = 68°$ *Note* : *ABC* forms a right triangle.

(a) $\alpha = 90° - 32° = 58°$

Bearing from *A* to *C*: N 58° E

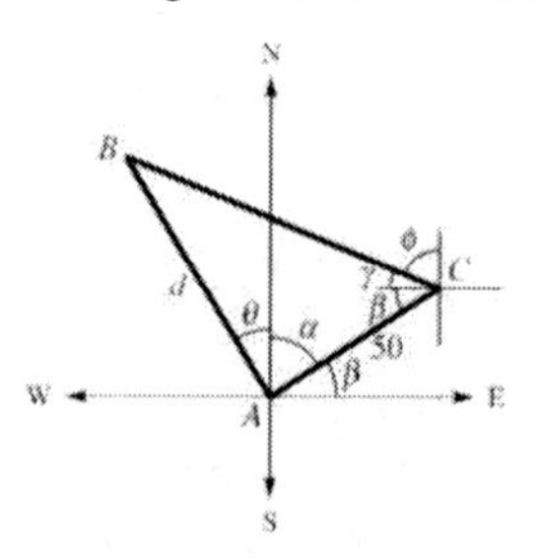

(b) $\beta = \theta = 32°$
$\gamma = 90° - \phi = 22°$
$C = \beta + \gamma = 54°$

$\tan C = \frac{d}{50} \Rightarrow \tan 54° = \frac{d}{50} \Rightarrow d \approx 68.82$ m

41. $\tan\theta = \frac{45}{30} = \frac{3}{2} \Rightarrow \theta \approx 56.31°$

Bearing: N 56.31°W

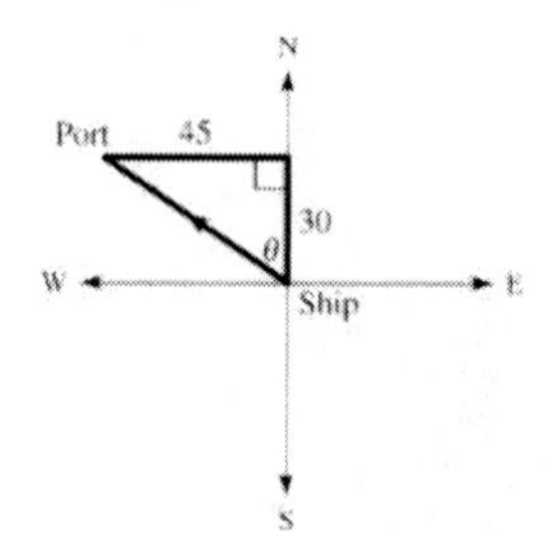

43. $\tan 6.5° = \frac{350}{d} \Rightarrow d \approx 3071.91$ ft

$\tan 4° = \frac{350}{D} \Rightarrow D \approx 5005.23$ ft

Distance between ships: $D - d \approx 1933.3$ ft

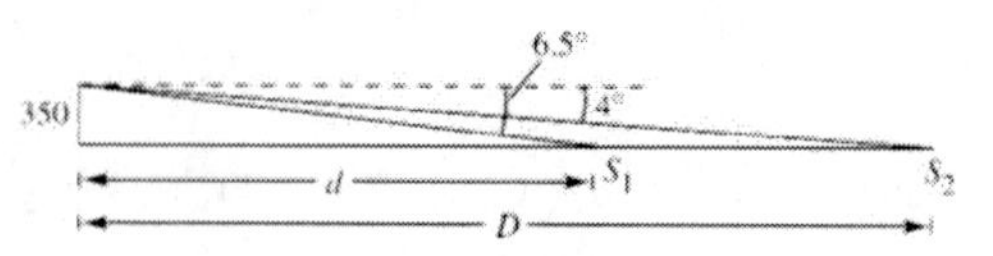

45. $\tan 57° = \frac{a}{x} \Rightarrow x = a\cot 57°$

$\tan 16° = \frac{a}{x + (55/6)}$

$\tan 16° = \frac{a}{a\cot 57° + (55/6)}$

$\cot 16° = \frac{a\cot 57° + (55/6)}{a}$

$a\cot 16° - a\cot 57° = \frac{55}{6}$

$\Rightarrow a \approx 3.23$ miles $\approx 17{,}054$ feet

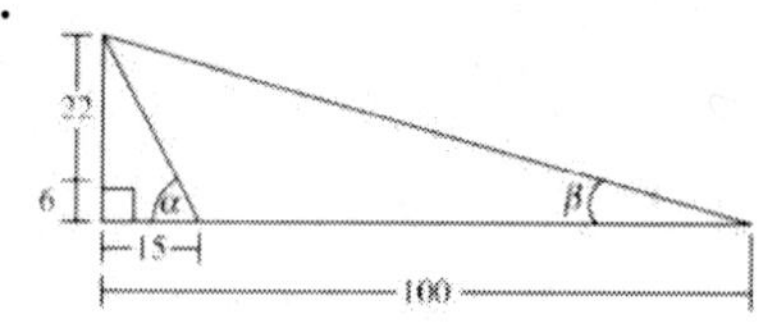

47.

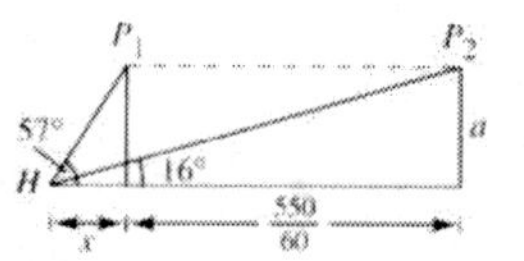

(a) $\tan\alpha = \frac{28}{15}$

$\alpha = \tan^{-1}\left(\frac{28}{15}\right)$

$\alpha \approx 61.82°$

$\tan\beta = \frac{28}{100}$

$\beta = \tan^{-1}\left(\frac{28}{100}\right)$

$\beta \approx 15.64°$

(b)

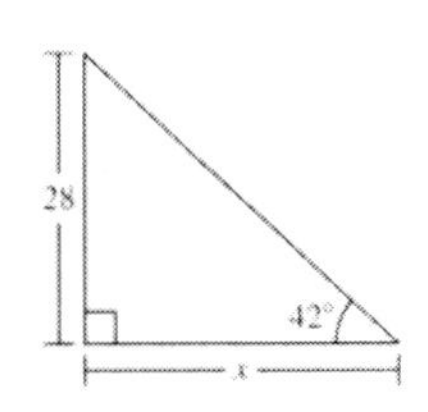

$$\tan 42° = \frac{28}{x}$$

$$x = \frac{28}{\tan 42°}$$

$$x \approx 31.10 \text{ feet}$$

49. $L_1 : 3x - 2y = 5 \Rightarrow y = \frac{3}{2}x - \frac{5}{2} \Rightarrow m_1 = \frac{3}{2}$

$L_2 : x + y = 1 \Rightarrow y = -x + 1 \Rightarrow m_2 = -1$

$$\tan\alpha = \left|\frac{-1-(3/2)}{1+(-1)(3/2)}\right| = \left|\frac{-5/2}{-1/2}\right| = 5$$

$$\alpha = \arctan 5 \approx 78.7°$$

51. The diagonal of the base has a length of

$\sqrt{a^2 + a^2} = \sqrt{2}a$. Now, we have:

$$\tan\theta = \frac{a}{\sqrt{2}a} = \frac{1}{\sqrt{2}}$$

$$\theta = \arctan\frac{1}{\sqrt{2}} \approx 35.3°$$

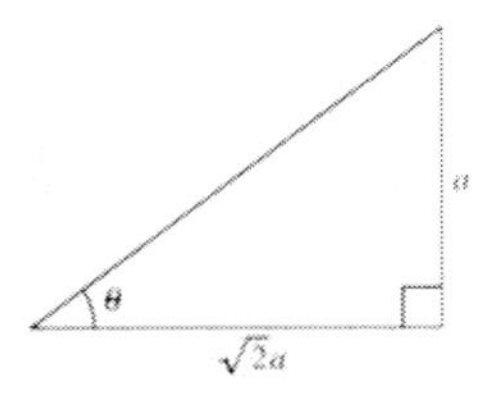

53. $\cos 30° = \frac{b}{r}$

$$b = \cos 30° r$$

$$b = \frac{\sqrt{3}r}{2}$$

$$y = 2b$$

$$= 2\left(\frac{\sqrt{3}r}{2}\right) = \sqrt{3}r$$

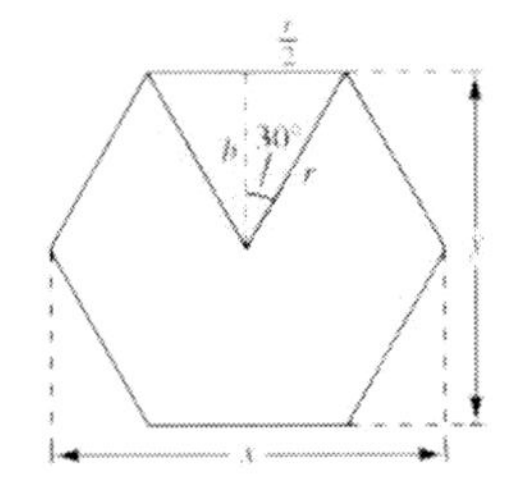

55. $d = 0$ when $t = 0$, $a = 8$, period $= 2$

Use $d = a\sin\omega t$ since $d = 0$ when $t = 0$.

Period: $\frac{2\pi}{\omega} = 2 \Rightarrow \omega = \pi$

Thus, $d = 8\sin\pi t$.

57. $d = 3$ when $t = 0$, $a = 3$, period $= 1.5$

Use $d = a\cos\omega t$ since $d = 3$ when $t = 0$.

Period: $\frac{2\pi}{\omega} = 1.5 \Rightarrow \omega = \frac{4}{3}\pi$

Thus, $d = 3\cos\frac{4}{3}\pi t$.

59. $d = 4\cos 8\pi t$

(a) Maximum displacement = amplitude = 4

(b) Frequency: $\frac{\omega}{2\pi} = \frac{8\pi}{2\pi}$

$= 4$ cycles per unit of time

(c) When $t = 5$, $d = 4\cos(8\pi(5)) = 4$.

(d) Last positive value of t for which $d = 0$:

$$4\cos 8\pi t = 0$$

$$\cos 8\pi t = 0$$

$$8\pi t = \frac{\pi}{2}$$

$$t = \frac{\pi}{2} \cdot \frac{1}{8\pi} = \frac{1}{16}$$

61. $d = \frac{1}{16}\sin 140\pi t$

(a) Maximum displacement: amplitude $= \frac{1}{16}$

(b) Frequency: $\frac{\omega}{2\pi} = \frac{140\pi}{2\pi}$

$= 70$ cycles per unit of time

(c) When $t = 5$, $d = \frac{1}{16}(\sin 140\pi(0)) = 0$.

(d) Least positive value of t for which $d = 0$:

$$\frac{1}{16}\sin 140\pi t = 0$$

$$\sin 140\pi t = 0$$

$$140\pi t = \pi$$

$$t = \frac{\pi}{140\pi} = \frac{1}{140}$$

63. $d = a\sin\omega t$

$$\text{Period} = \frac{2\pi}{\omega} = \frac{1}{\text{frequency}}$$

$$\frac{2\pi}{\omega} = \frac{1}{264}$$

$$\omega = 2\pi(264) = 528\pi$$

65. $y = \frac{1}{4}\cos 16t,\ t > 0$

(a)

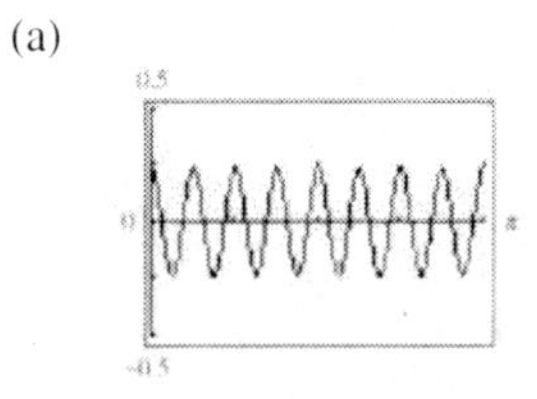

(b) Period: $\frac{2\pi}{16} = \frac{\pi}{8}$ seconds

(c) $\frac{1}{4}\cos 16t = 0$ when

$$16t = \frac{\pi}{2} \Rightarrow t = \frac{\pi}{32} \text{ seconds.}$$

67. (a) and (b)

Base 1	Base 2	Altitude	Area
8	$8+16\cos 10°$	8 sin 10°	22.1
8	$8+16\cos 20°$	8 sin 20°	42.5
8	$8+16\cos 30°$	8 sin 30°	59.7
8	$8+16\cos 40°$	8 sin 40°	72.7
8	$8+16\cos 50°$	8 sin 50°	80.5
8	$8+16\cos 60°$	8 sin 60°	83.1
8	$8+16\cos 70°$	8 sin 70°	80.7

Maximum ≈ 83.1 square feet

(c) $A = \frac{1}{2}(b_1 + b_2)h$

$$= \tfrac{1}{2}[8 + 8 + 16\cos\theta]8\sin\theta$$

$$= 64(1 + \cos\theta)\sin\theta$$

(d) Maximum area is approximately 83.1 square feet for $\theta = 60°$.

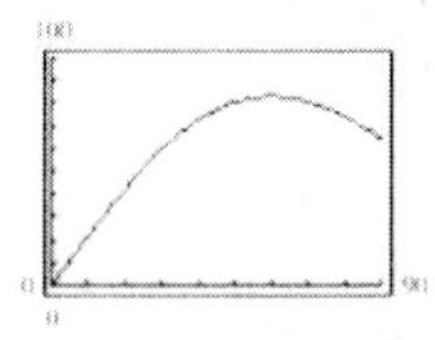

69. (a)

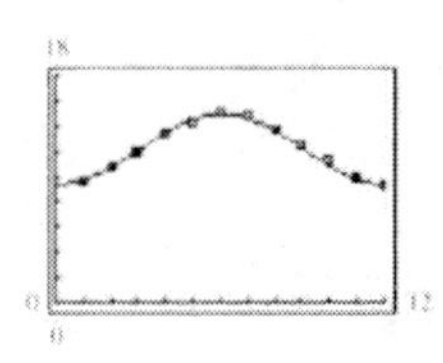

(b) $\text{period} = \frac{2\pi}{n} = \frac{2\pi}{\pi/6} = 12$

The period is what you expect as the model examines the number of hours of daylight over one year (12 months).

(c) $\text{Amplitude} = |2.77| = 2.77$

The amplitude represents the maximum displacement from the average number of hours of daylight.

71. True Simple harmonic motion is given by $a\sin\omega t$ or $a\cos\omega t$, when a is a real number not another function of t.

73. False The amplitude of simple harmonic motion $a\sin\omega t$ or $a\cos\omega t$ is given by a.

75.

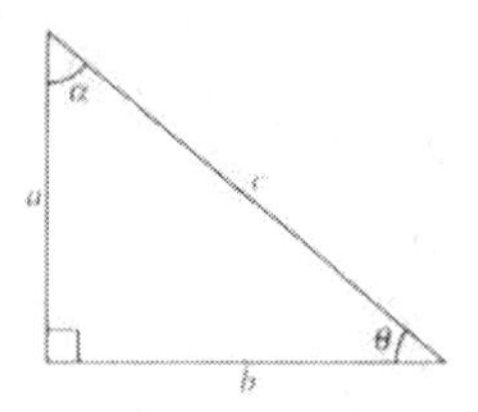

Answers will vary but a possible combination follows.

Given θ and a:

$$\alpha = 90° - \theta$$

$$\tan\theta = \frac{a}{b} \Rightarrow b = \frac{a}{\tan\theta}$$

$$\sin\theta = \frac{a}{c} \Rightarrow c = \frac{a}{\sin\theta}$$

Given θ and b:

$$\alpha = 90° - \theta$$

$$\tan\theta = \frac{a}{b} \Rightarrow a = b\tan\theta$$

$$\cos\theta = \frac{b}{c} \Rightarrow c = \frac{b}{\cos\theta}$$

Given a and b:

$$\tan\theta = \frac{a}{b} \Rightarrow \theta = \tan^{-1}\left(\frac{a}{b}\right)$$

$$\alpha = 90° - \theta$$

$$a^2 + b^2 = c^2 \Rightarrow c = \sqrt{a^2 + b^2}$$

Given a and c:

$$\sin\theta = \frac{a}{c} \Rightarrow \theta = \sin^{-1}\left(\frac{a}{c}\right)$$

$$\alpha = 90° - \theta$$

$$a^2 + b^2 = c^2 \Rightarrow b = \sqrt{c^2 - a^2}$$

Given b and c:

$$\cos\theta = \frac{b}{c} \Rightarrow \theta = \cos^{-1}\left(\frac{b}{c}\right)$$

$\alpha = 90° - \theta$

$a^2 + b^2 = c^2 \Rightarrow a = \sqrt{c^2 - b^2}$

Given θ and c:

$\alpha = 90° - \theta$

$\sin\theta = \frac{a}{c} \Rightarrow a = c\sin\theta$

$\cos\theta = \frac{b}{c} \Rightarrow b = c\cos\theta$

77. $y - 2 = 4(x + 1)$

$4x - y + 6 = 0$

79. Slope $= \frac{6-2}{-2-3} = -\frac{4}{5}$

$y - 2 = -\frac{4}{5}(x - 3)$

$5y - 10 = -4x + 12$

$4x + 5y - 12 = 0$

81. Domain: $(-\infty, \infty)$

83. Domain: $(-\infty, \infty)$

Chapter 4 Review Exercises

1. 40° or 0.7 radian

3. (a)

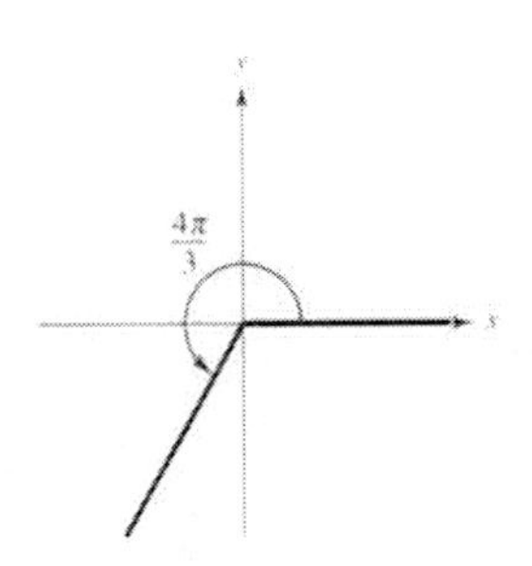

(b) Quadrant III

(c) $\frac{4\pi}{3} + 2\pi = \frac{10\pi}{3}$

$\frac{4\pi}{3} - 2\pi = -\frac{2\pi}{3}$

5. (a)

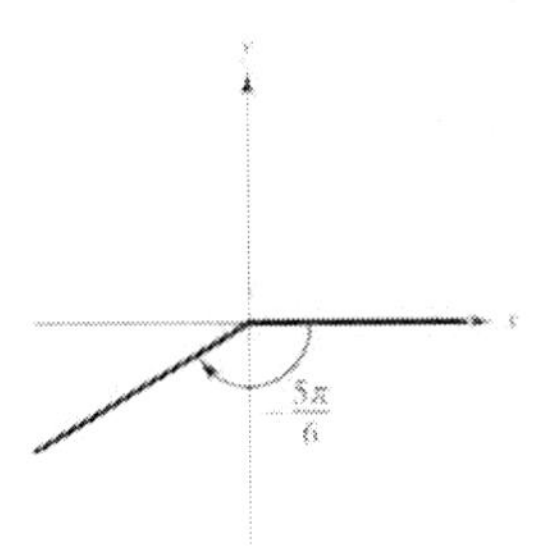

(b) Quadrant III

(c) $-\frac{5\pi}{6} + 2\pi = \frac{7\pi}{6}$

$-\frac{5\pi}{6} - 2\pi = -\frac{17\pi}{6}$

7. (a)

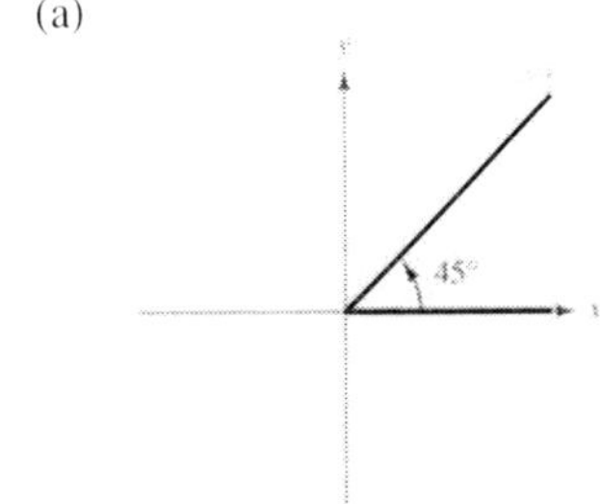

(b) Quadrant I

(c) $45° + 360° = 405°$

$45° - 360° = -315°$

9. (a)

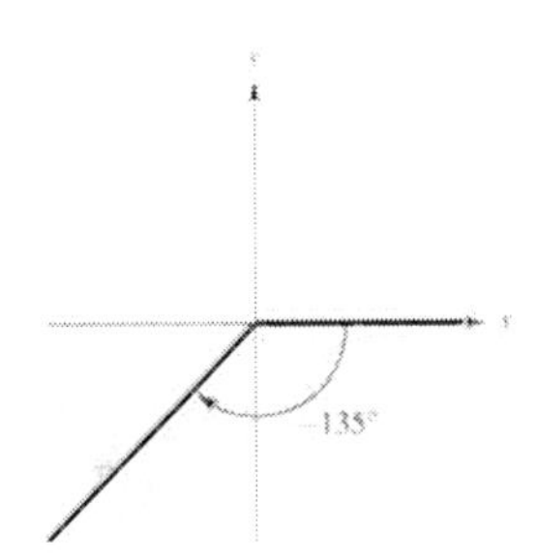

(b) Quadrant III

(c) $-135° + 360° = 225°$

$-135° - 360° = -495°$

11. $135°16'45'' = \left(135 + \frac{16}{60} + \frac{45}{3600}\right)^\circ \approx 135.279°$

13. $6°34'19'' = \left(6 + \frac{34}{60} + \frac{19}{3600}\right)^\circ \approx 6.572°$

15. $135.29° = 135° + (0.29)(60)' = 135°17'24''$

17. $-85.36° = -[85 + 0.36(60')] = -85°21'36''$

19. $94° = 94° \cdot \frac{\pi \text{ rad}}{180°} = \frac{47\pi}{90} \text{rad} \approx 1.641 \text{ rad}$

21. $415° = 415° \cdot \frac{\pi \text{ rad}}{180°} = \frac{83\pi}{36} \text{rad} \approx 7.243 \text{ rad}$

23. $\frac{5\pi}{7} = \frac{5\pi}{7}\left(\frac{180°}{\pi}\right) \approx 128.571°$

25. $-3.5 = -3.5\left(\frac{180°}{\pi}\right) \approx -200.535°$

27. Complement: $\frac{\pi}{2} - \frac{\pi}{8} = \frac{3\pi}{8}$

Supplement: $\pi - \frac{\pi}{8} = \frac{7\pi}{8}$

29. Complement: $\frac{\pi}{2} - \frac{3\pi}{10} = \frac{\pi}{5}$

Supplement: $\pi - \frac{3\pi}{10} = \frac{7\pi}{10}$

31. Complement of $5°$: $90° - 5° = 85°$
Supplement of $5°$: $180° - 5° = 175°$

33. Complement of $157°$: Not possible; $157°$ is greater than $90°$.
Supplenent of $157°: 180° - 157° = 23°$

35.
$$s = r\theta$$
$$25 = 12\theta$$
$$\theta = \frac{25}{12} \approx 2.083 \text{ radians}$$

37.
$$s = r\theta$$
$$s = 20(138°)\frac{\pi}{180°}$$
$$s = \frac{46\pi}{3} \approx 48.171 \text{ m}$$

39. In one revolution, the arc length traveled is $s = 2\pi r = 2\pi(6) = 12\pi \text{cm}$. The time required for one revolution is $t = \frac{1}{500}$ minutes.

Linear speed $= \frac{s}{t} = \frac{12\pi}{1/500} = 6000$ cm/min

41. $t = \frac{7\pi}{4}$ corresponds to $\left(\frac{\sqrt{2}}{2}, -\frac{\sqrt{2}}{2}\right)$.

43. $t = \frac{5\pi}{6}$ corresponds to $\left(-\frac{\sqrt{3}}{2}, \frac{1}{2}\right)$.

45. $t = -\frac{2\pi}{3}$ corresponds to $\left(-\frac{1}{2}, -\frac{\sqrt{3}}{2}\right)$.

47. $t = -\frac{5\pi}{4}$ corresponds to $\left(-\frac{\sqrt{2}}{2}, \frac{\sqrt{2}}{2}\right)$.

49. $t = \frac{\pi}{4}$ corresponds to $\left(\frac{\sqrt{2}}{2}, \frac{\sqrt{2}}{2}\right)$.

$$\sin\frac{\pi}{4} = y = \frac{\sqrt{2}}{2}$$
$$\cos\frac{\pi}{4} = x = \frac{\sqrt{2}}{2}$$
$$\tan\frac{\pi}{4} = \frac{y}{x} = 1$$
$$\cot\frac{\pi}{4} = \frac{x}{y} = 1$$
$$\sec\frac{\pi}{4} = \frac{1}{x} = \sqrt{2}$$
$$\csc\frac{\pi}{4} = \frac{1}{y} = \sqrt{2}$$

51. $t = 2\pi$ corresponds to $(1, 0)$.

$$\sin 2\pi = y = 0$$
$$\cos 2\pi = x = 1$$
$$\tan 2\pi = \frac{y}{x} = 0$$
$$\cot 2\pi = \frac{x}{y}, \text{ undefined}$$
$$\sec 2\pi = \frac{1}{x} = 1$$
$$\csc 2\pi = \frac{1}{y}, \text{ undefined}$$

53. $t = -\frac{11\pi}{6}$ corresponds to $\left(\frac{\sqrt{3}}{2}, \frac{1}{2}\right)$.

$$\sin\left(-\frac{11\pi}{6}\right) = y = \frac{1}{2}$$
$$\cos\left(-\frac{11\pi}{6}\right) = x = \frac{\sqrt{3}}{2}$$
$$\tan\left(-\frac{11\pi}{6}\right) = \frac{y}{x} = \frac{\sqrt{3}}{3}$$
$$\cot\left(-\frac{11\pi}{6}\right) = \frac{x}{y} = \sqrt{3}$$
$$\sec\left(-\frac{11\pi}{6}\right) = \frac{1}{x} = \frac{2\sqrt{3}}{3}$$
$$\csc\left(-\frac{11\pi}{6}\right) = \frac{1}{y} = 2$$

55. $t = -\frac{\pi}{2}$ corresponds to $(0, -1)$.

$\sin\left(-\frac{\pi}{2}\right) = y = -1$

$\cos\left(-\frac{\pi}{2}\right) = x = 0$

$\tan\left(-\frac{\pi}{2}\right) = \frac{y}{x}$, undefined

$\cot\left(-\frac{\pi}{2}\right) = \frac{x}{y} = 0$

$\sec\left(-\frac{\pi}{2}\right) = \frac{1}{x}$, undefined

$\csc\left(-\frac{\pi}{2}\right) = \frac{1}{y} = -1$

57. $\cos 4\pi = \cos 0 = 1$

59. $\sin\left(-\frac{17\pi}{6}\right) = \sin\left(\frac{7\pi}{6}\right) = -\frac{1}{2}$

61. $\sin t = \frac{3}{5}$

(a) $\sin(-t) = -\sin t = -\frac{3}{5}$

(b) $\csc(-t) = \frac{1}{\sin(-t)} = -\frac{5}{3}$

63. $\sin(-t) = -\frac{2}{3}$

(a) $\sin t = -\sin(-t) = -\left(-\frac{2}{3}\right) = \frac{2}{3}$

(b) $\csc t = \frac{1}{\sin t} = \frac{3}{2}$

65. $\cot 2.3 = \frac{1}{\tan 2.3} \approx -0.8935$

67. $\cos\frac{5\pi}{3} = \frac{1}{2}$

69. The opposite side is $\sqrt{9^2 - 4^2} = \sqrt{81 - 16} = \sqrt{65}$.

$\sin\theta = \frac{\text{opp}}{\text{hyp}} = \frac{\sqrt{65}}{9}$

$\cos\theta = \frac{\text{adj}}{\text{hyp}} = \frac{4}{9}$

$\tan\theta = \frac{\text{opp}}{\text{adj}} = \frac{\sqrt{65}}{4}$

$\csc\theta = \frac{1}{\sin\theta} = \frac{9}{\sqrt{65}} = \frac{9\sqrt{65}}{65}$

$\sec\theta = \frac{1}{\cos\theta} = \frac{9}{4}$

$\cot\theta = \frac{1}{\tan\theta} = \frac{4}{\sqrt{65}} = \frac{4\sqrt{65}}{65}$

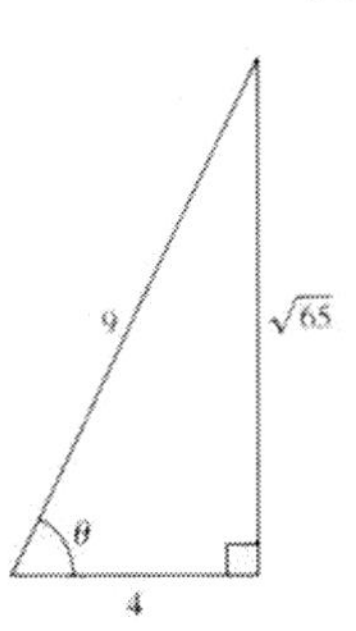

71. The hypotenuse is $\sqrt{12^2 + 10^2} = \sqrt{244} = 2\sqrt{61}$.

$\sin\theta = \frac{\text{opp}}{\text{hyp}} = \frac{10}{2\sqrt{61}} = \frac{5}{\sqrt{61}} = \frac{5\sqrt{61}}{61}$

$\cos\theta = \frac{\text{adj}}{\text{hyp}} = \frac{12}{2\sqrt{61}} = \frac{6}{\sqrt{61}} = \frac{6\sqrt{61}}{61}$

$\tan\theta = \frac{\text{opp}}{\text{adj}} = \frac{10}{12} = \frac{5}{6}$

$\csc\theta = \frac{1}{\sin\theta} = \frac{\sqrt{61}}{5}$

$\sec\theta = \frac{1}{\cos\theta} = \frac{\sqrt{61}}{6}$

$\cot\theta = \frac{1}{\tan\theta} = \frac{6}{5}$

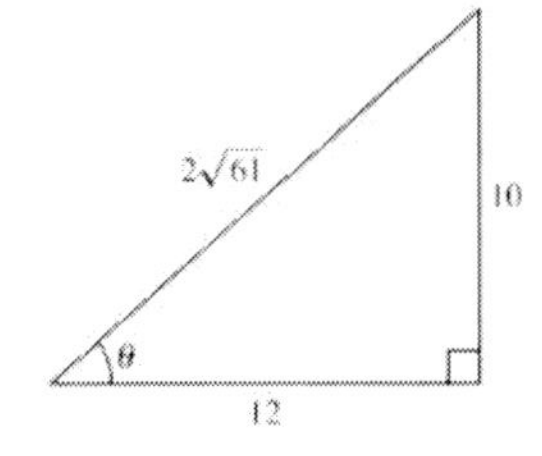

73. Given: $\sin\theta = \frac{7}{24}$

$\cos\theta = \frac{\sqrt{527}}{24}$

$\tan\theta = \frac{7}{\sqrt{527}} = \frac{7\sqrt{527}}{527}$

$\cot\theta = \frac{\sqrt{527}}{7}$

$\sec\theta = \frac{24}{\sqrt{527}} = \frac{24\sqrt{527}}{527}$

$\csc\theta = \frac{24}{7}$

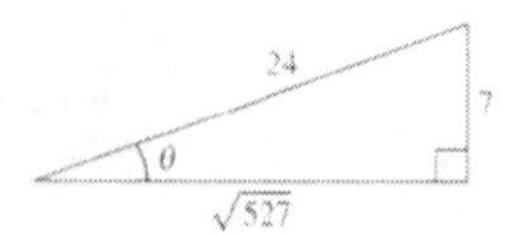

75. Given: $\tan\theta = \frac{1}{4}$

$\sin\theta = \frac{1}{\sqrt{17}} = \frac{\sqrt{17}}{17}$

$\cos\theta = \frac{4}{\sqrt{17}} = \frac{4\sqrt{17}}{17}$

$\cot\theta = 4$

$\sec\theta = \frac{\sqrt{17}}{4}$

$\csc\theta = \sqrt{17}$

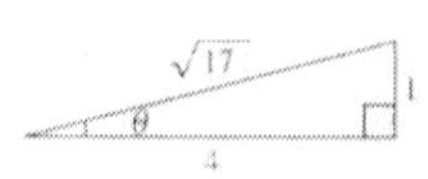

77. (a) $\cos 84° \approx 0.1045$
(b) $\sin 6° \approx 0.1045$

79. (a) $\cos\frac{\pi}{4} \approx 0.7071$
(b) $\sec\frac{\pi}{4} \approx 1.4142$

81. $\csc\theta\tan\theta = \frac{1}{\sin\theta}\cdot\frac{\sin\theta}{\cos\theta} = \frac{1}{\cos\theta} = \sec\theta$

83. $\tan 62° = \frac{w}{125}$

$x = 125\tan 62°$
≈ 235 feet

85. $x = 12,\ y = 16,\ r = \sqrt{144+256} = \sqrt{400} = 20$

$\sin\theta = \frac{y}{r} = \frac{4}{5}$

$\cos\theta = \frac{x}{r} = \frac{3}{5}$

$\tan\theta = \frac{y}{x} = \frac{4}{3}$

$\csc\theta = \frac{r}{y} = \frac{5}{4}$

$\sec\theta = \frac{r}{x} = \frac{5}{3}$

$\cot\theta = \frac{x}{y} = \frac{3}{4}$

87. $x = -7,\ y = 2,\ r = \sqrt{49+4} = \sqrt{53}$

$\sin\theta = \frac{y}{r} = \frac{2}{\sqrt{53}} = \frac{2\sqrt{53}}{53}$

$\cos\theta = \frac{x}{r} = -\frac{7}{\sqrt{53}} = -\frac{7\sqrt{53}}{53}$

$\tan\theta = \frac{y}{x} = -\frac{2}{7}$

$\csc\theta = \frac{\sqrt{53}}{2}$

$\sec\theta = -\frac{\sqrt{53}}{7}$

$\cot\theta = -\frac{7}{2}$

89. $x = \frac{2}{3}$ and $y = \frac{5}{8},\ r = \sqrt{\left(\frac{2}{3}\right)^2 + \left(\frac{5}{8}\right)^2} = \frac{\sqrt{481}}{24}$

$\sin\theta = \frac{y}{r} = \frac{5/8}{\sqrt{481}/24} = \frac{15\sqrt{481}}{481}$

$\cos\theta = \frac{x}{r} = \frac{2/3}{\sqrt{481}/24} = \frac{16\sqrt{481}}{481}$

$\tan\theta = \frac{y}{x} = \frac{5/8}{2/3} = \frac{15}{16}$

$\csc\theta = \frac{r}{y} = \frac{\sqrt{481}}{15}$

$\sec\theta = \frac{r}{x} = \frac{\sqrt{481}}{16}$

$\cot\theta = \frac{x}{y} = \frac{16}{15}$

91. $\sec\theta = \frac{6}{5}$, $\tan\theta < 0 \Rightarrow \theta$ is in Quadrand IV.

$r = 6, x = 5, y = -\sqrt{36-25} = -\sqrt{11}$

$\sin\theta = \frac{y}{r} = -\frac{\sqrt{11}}{6}$

$\cos\theta = \frac{x}{r} = \frac{5}{6}$

$\tan\theta = \frac{y}{x} = -\frac{\sqrt{11}}{5}$

$\csc\theta = -\frac{6\sqrt{11}}{11}$

$\sec\theta = \frac{6}{5}$

$\cot\theta = -\frac{5\sqrt{11}}{11}$

93. $\sin\theta = \frac{3}{8}$, $\cos\theta < 0 \Rightarrow \theta$ is in Quadrant II.

$y = 3, r = 8, x = -\sqrt{55}$

$\sin\theta = \frac{y}{r} = \frac{3}{8}$

$\cos\theta = \frac{x}{r} = -\frac{\sqrt{55}}{8}$

$\tan\theta = \frac{y}{x} = -\frac{3}{\sqrt{55}} = -\frac{3\sqrt{55}}{55}$

$\csc\theta = \frac{8}{3}$

$\sec\theta = -\frac{8}{\sqrt{55}} = -\frac{8\sqrt{55}}{55}$

$\cot\theta = -\frac{\sqrt{55}}{3}$

95. $\theta = 330°$

Reference angle:
$\theta' = 360° - 330° = 30°$

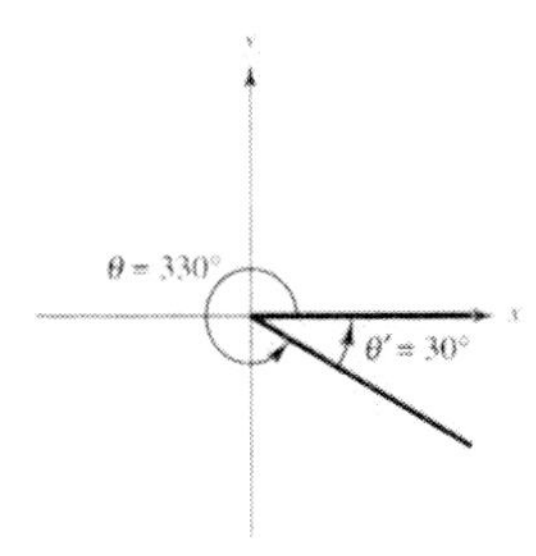

97. $\theta = \frac{5\pi}{4}$

Reference angle: $\theta' = \frac{5\pi}{4} - \pi = \frac{\pi}{4}$

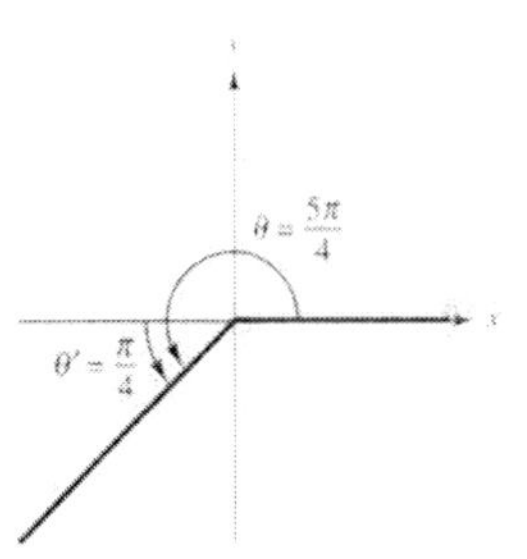

99. $\theta = 264°$

Reference angle:

$\theta' = 264° - 180° = 84°$

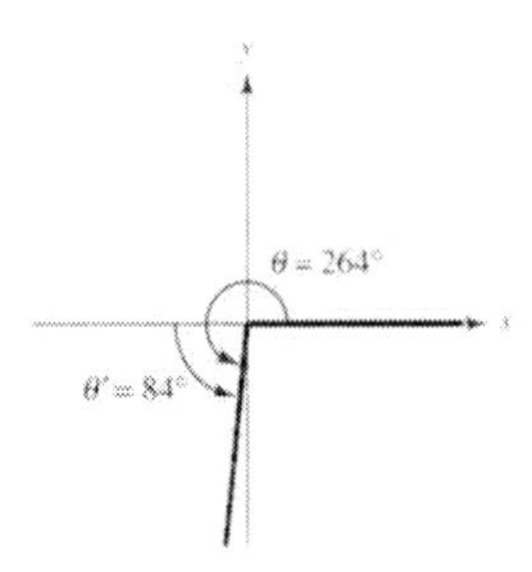

101. $\theta = -\frac{6\pi}{5}$ is coterminal with $\frac{4\pi}{5}$.

Reference angle:

$\theta' = \pi - \frac{4\pi}{5} = \frac{\pi}{5}$

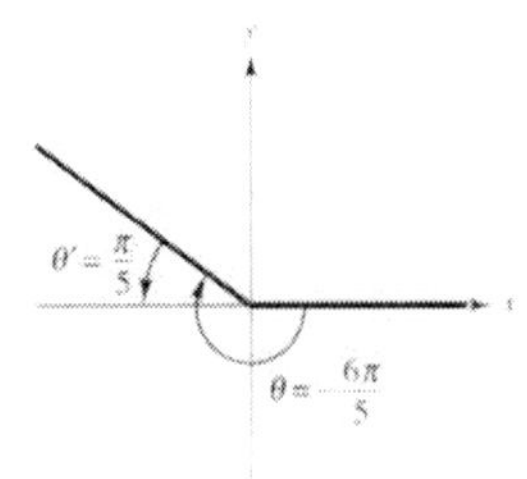

103. $240°$ is Quadrant III with reference angle $60°$.

$\sin 240° = -\sin 60° = -\frac{\sqrt{3}}{2}$

$\cos 240° = -\cos 60° = -\frac{1}{2}$

$\tan 240° = \frac{-\sqrt{3}/2}{-1/2} = \sqrt{3}$

105. $-210°$ is coterminal with $150°$ in Quadrant II with reference angle $30°$.

$$\sin(-210°) = \sin(30°) = \frac{1}{2}$$

$$\cos(-210°) = -\cos(30°) = -\frac{\sqrt{3}}{2}$$

$$\tan(-210°) = \frac{1/2}{-\sqrt{3}/2} = -\frac{1}{\sqrt{3}} = -\frac{\sqrt{3}}{3}$$

107. $-9\pi/4$ is coterminal with $7\pi/4$ in Quadrant IV with reference angle $\pi/4$.

$$\sin\left(-\frac{9\pi}{4}\right) = -\sin\left(\frac{\pi}{4}\right) = -\frac{\sqrt{2}}{2}$$

$$\cos\left(-\frac{9\pi}{4}\right) = \cos\left(\frac{\pi}{4}\right) = \frac{\sqrt{2}}{2}$$

$$\tan\left(-\frac{9\pi}{4}\right) = \frac{-\sqrt{2}/2}{\sqrt{2}/2} = -1$$

109. $\sin(4\pi) = \sin(0) = 0$

$\cos(4\pi) = 1$

$\tan(4\pi) = 0$

111. $\tan 33° \approx 0.6494$

113. $\sec\frac{12\pi}{5} = \frac{1}{\cos\left(\frac{12\pi}{5}\right)} \approx 3.2361$

115. $f(x) = 3\sin x$

Amplitude: 3

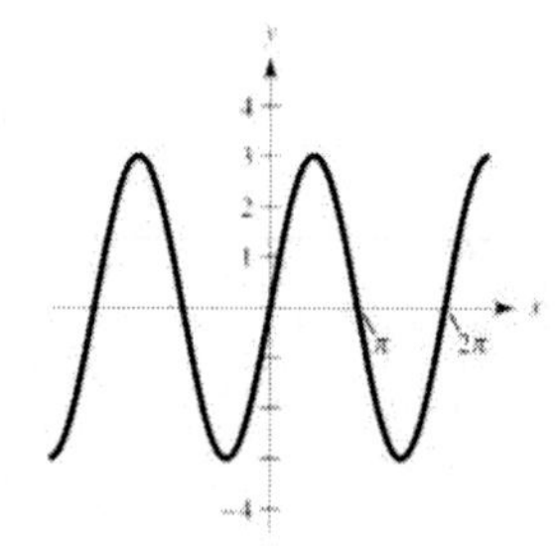

117. $f(x) = \frac{1}{4}\cos x$

Amplitude: $\frac{1}{4}$

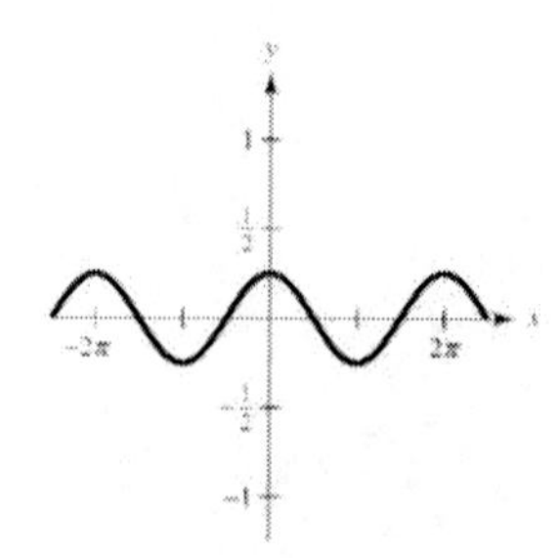

119. Period: $\frac{2\pi}{\pi} = 2$

Amplitude: 5

121. Period: $\frac{2\pi}{2} = \pi$

Amplitude: 3.4

123. $y = 3\cos 2\pi x$

Amplitude: 3

Period: $\frac{2\pi}{2\pi} = 1$

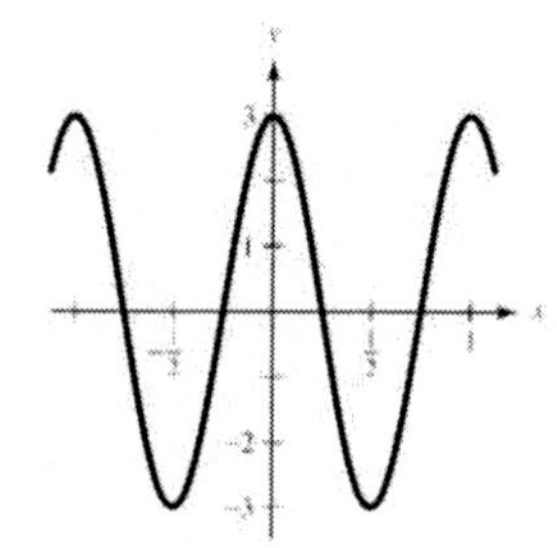

125. $f(x) = 5\sin\frac{2x}{5}$

Amplitude: 5

Period: $\frac{2\pi}{2/5} = 5\pi$

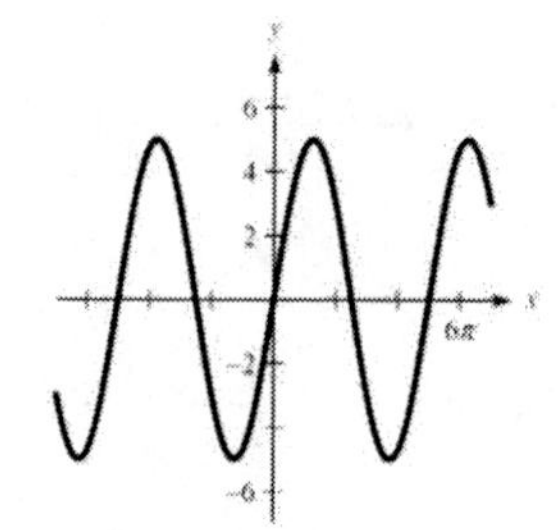

127. $f(x) = -\frac{5}{2}\cos\left(\frac{x}{4}\right)$

Amplitude: $\frac{5}{2}$

Period: $\frac{2\pi}{1/4} = 8\pi$

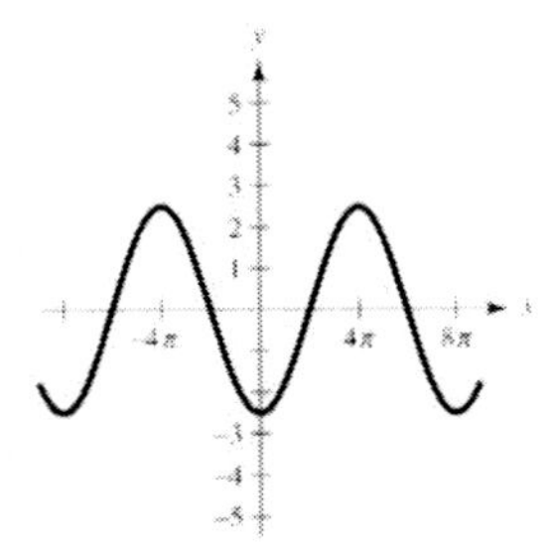

129. $f(x) = \frac{5}{2}\sin(x - \pi)$

Amplitude: $\frac{5}{2}$

Period: 2π

Shift:

$x - \pi = 0$ and $x - \pi = 2\pi$

$x = \pi$ $\quad$ $x = 3\pi$

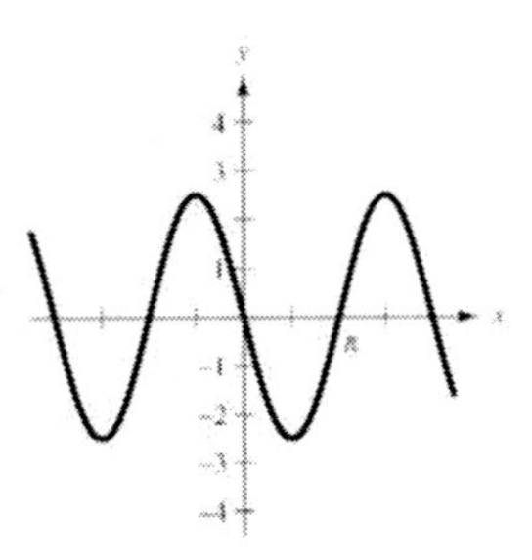

131. $f(x) = 2 - \cos\frac{\pi x}{2}$

Amplitude: 1

Period: $\frac{2\pi}{(\pi/2)} = 4$

Reflection in x-axis

Vertical shift upward two units

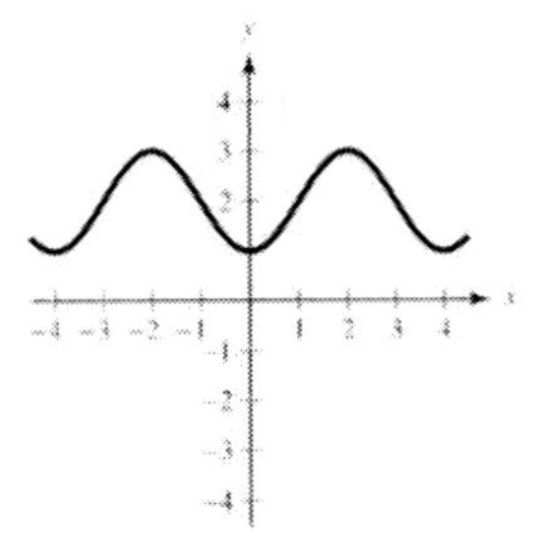

133. $f(x) = -3\cos\left(\frac{x}{2} - \frac{\pi}{4}\right)$

Amplitude: 3

Period: $\frac{2\pi}{1/2} = 4\pi$

Reflection in x-axis

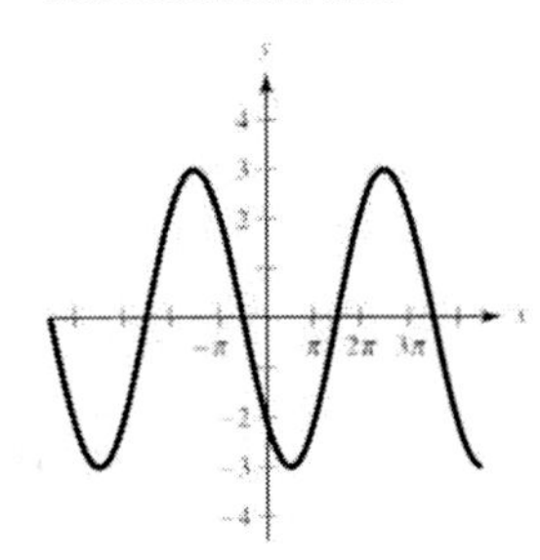

135. $f(x) = a\cos(bx - c)$

Amplitude: 2

Period: 2π

Shift: $\frac{\pi}{4}$

$f(x) = -2\cos\left(x - \frac{\pi}{4}\right)$

137. $f(x) = a\cos(bx - c)$

Amplitude: 4

Period: π

Shift: $\frac{\pi}{2}$

$f(x) = -4\cos\left(2x - \frac{\pi}{2}\right)$

139. $S = 48.4 - 6.1\cos\frac{\pi t}{6}$

Maximum sales: $t = 6$

(June)

Minimum sales: $t = 12$

(December)

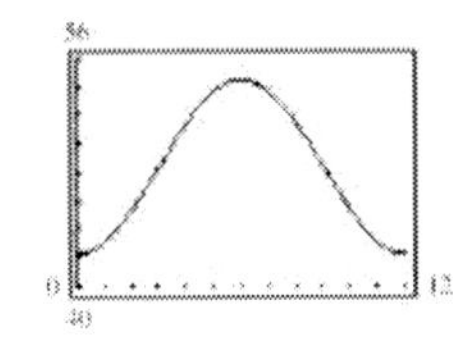

141. $f(x) = -\tan\dfrac{\pi x}{4}$

Period: $\dfrac{\pi}{(\pi/4)} = 4$

Two consecutive asymptotes:

$x = -2,\ x = 2$

Reflected in x-axis

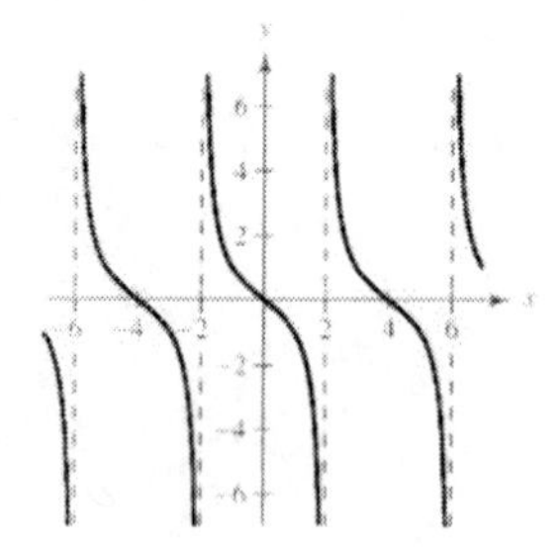

143. $f(x) = \dfrac{1}{4}\tan\left(x - \dfrac{\pi}{2}\right)$

Period: $\dfrac{\pi}{1} = \pi$

Two consecutive asymptotes:

$x = 0,\ x = \pi$

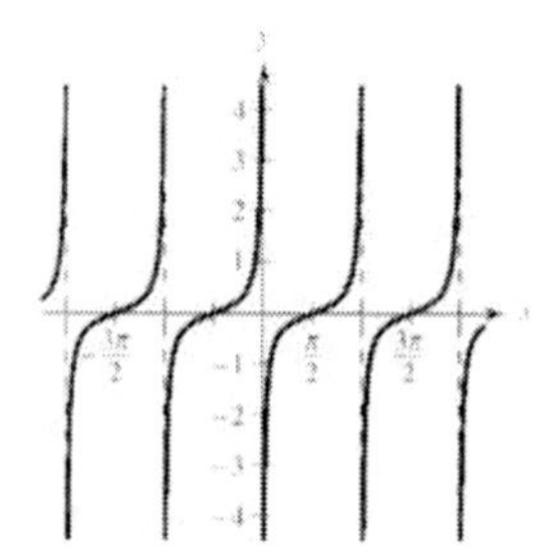

145. $f(x) = 3\cot\dfrac{x}{2}$

Period: $\dfrac{\pi}{1/2} = 2\pi$

Two consecutive asymptotes:

$\dfrac{x}{2} = 0 \Rightarrow x = 0$

$\dfrac{x}{2} = \pi \Rightarrow x = 2\pi$

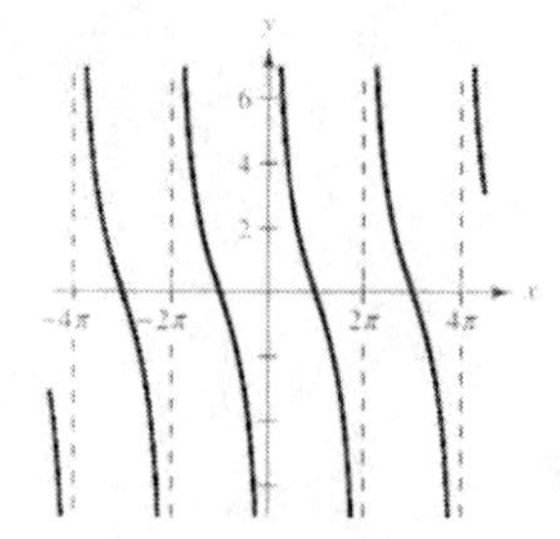

147. $f(x) = \dfrac{1}{2}\cot\left(x - \dfrac{\pi}{2}\right)$

Period: π

Two consecutive asymptotes:

$x - \dfrac{\pi}{2} = 0 \Rightarrow x = \dfrac{\pi}{2}$

$x - \dfrac{\pi}{2} = \pi \Rightarrow x = \dfrac{3\pi}{2}$

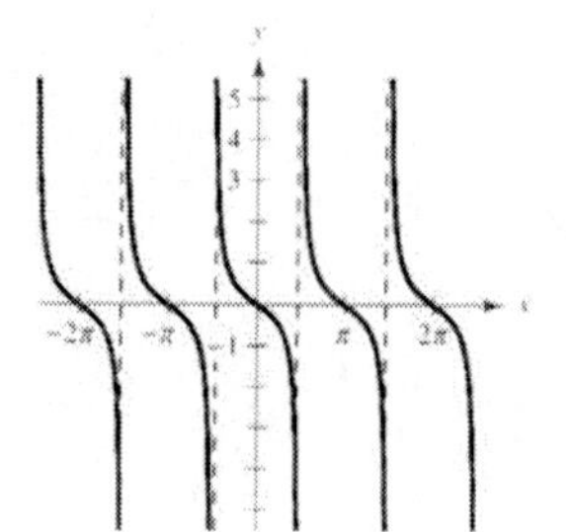

149. $f(x) = \dfrac{1}{4}\sec x$

Period: 2π

Two consecutive asymptotes:

$x = \dfrac{\pi}{2},\ x = \dfrac{3\pi}{2}$

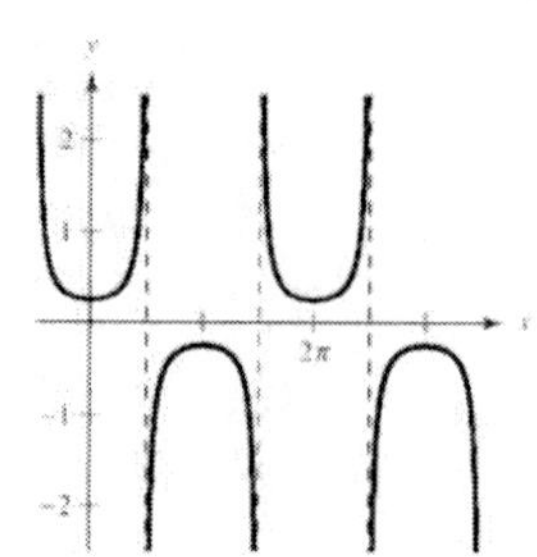

151. $f(x) = \dfrac{1}{4}\csc 2x$

Period: $\dfrac{2\pi}{2} = \pi$

Two consecutive asymptotes:

$2x = 0 \Rightarrow x = 0$

$2x = \pi \Rightarrow x = \dfrac{\pi}{2}$

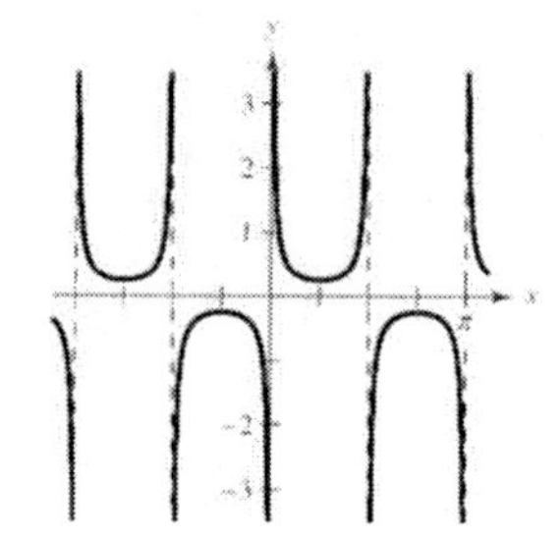

153. $f(x) = \sec\left(x - \frac{\pi}{4}\right)$

Secant function shifted $\frac{\pi}{4}$ to right

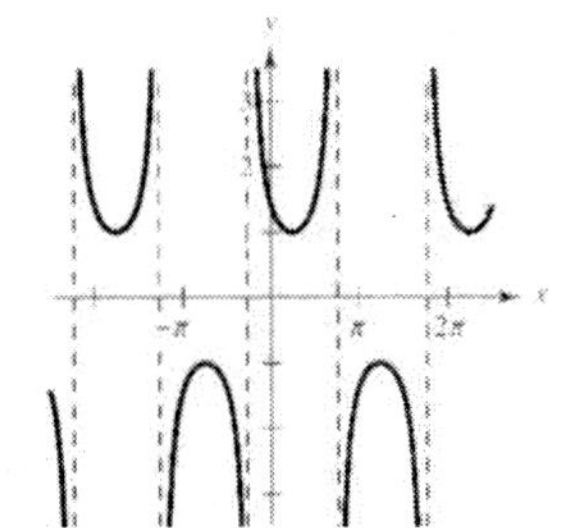

155. $f(x) = \frac{1}{4}\tan\frac{\pi x}{2},\ g(x) = \frac{1}{4}\cot\frac{\pi x}{2}$

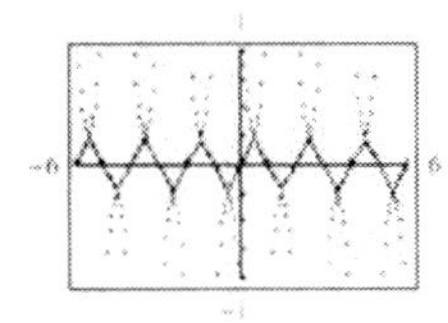

Answers will vary.

157. $f(x) = 4\cot(2x - \pi),\ g(x) = 4\tan(2x - \pi)$

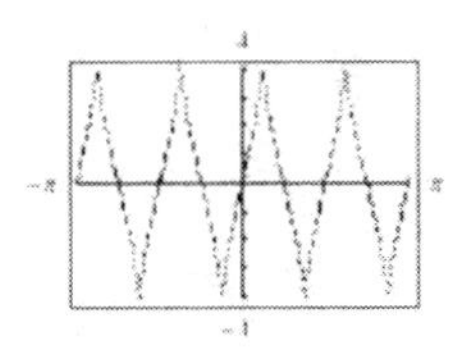

Answers will vary.

159. $f(x) = 2\sec(x - \pi),\ g(x) = 2\cos(x - \pi)$

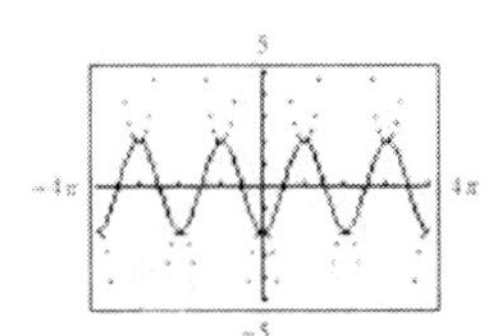

Answers will vary.

161. $f(x) = \csc\left(3x - \frac{\pi}{2}\right),\ g(x) = \sin\left(3x - \frac{\pi}{2}\right)$

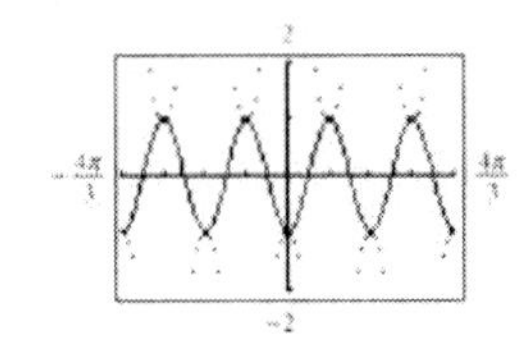

Answers will vary.

163. $f(x) = e^{x}\sin 2x$

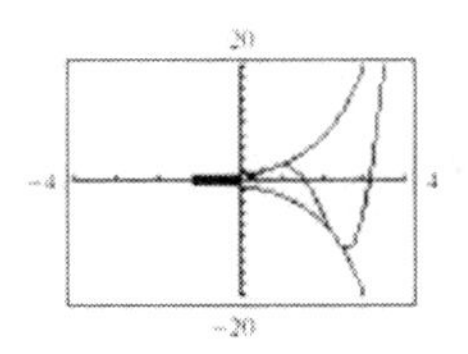

Since the damping factor is e^{-x}, $-e^{-x} \le e^{-x}\sin 2x \le e^{-x}$.

$y = \pm e^{-x}$ touches $y = e^{-x}\sin 2x$ at $x = \frac{\pi}{4} + n\frac{\pi}{2}$.

$f(x) = e^{-x}\sin 2x$ has x-intercepts at $\sin 2x = 0$ or $x = \frac{n\pi}{2}$.

165. $f(x) = 2x\cos x$

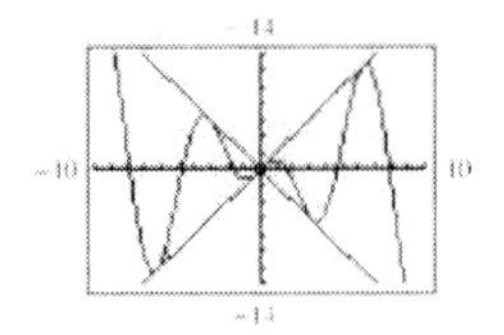

Since the damping factor is $2x$,

$-2x \le 2x\cos x \le 2x$.

$y = \pm 2x$ touches $y = 2x\cos x$ at $x = n\pi$.

$y = 2x\cos x$ has x- intercepts at $\cos x = 0$

or $x = \frac{\pi}{2} + n\pi$.

167. (a) $\arcsin(-1) = -\frac{\pi}{2}$ because $\sin\left(-\frac{\pi}{2}\right) = -1$.

(b) $\arcsin(0) = 0$ because $\sin 0 = 0$.

169. (a) $\arccos\left(\frac{\sqrt{2}}{2}\right) = \frac{\pi}{4}$ because $\cos\frac{\pi}{4} = \frac{\sqrt{2}}{2}$.

(b) $\arccos\left(-\frac{\sqrt{3}}{2}\right) = \frac{5\pi}{6}$ because $\cos\frac{5\pi}{6} = -\frac{\sqrt{3}}{2}$.

171. $\arccos(0.42) \approx 1.14$

173. $\sin^{-1}(-0.94) \approx -1.22$

175. $\arctan(-12) \approx -1.49$

177. $\tan^{-1}(0.81) \approx 0.68$

179. $\sin\theta = \frac{x}{16} \Rightarrow \theta = \arcsin\left(\frac{x}{16}\right)$

181. Let $y = \arcsin(x-1)$. Then

$$\sin y = (x-1) = \frac{x-1}{1} \text{ and}$$

$$\sec y = \frac{1}{\sqrt{-x^2+2x}} = \frac{\sqrt{-x^2+2x}}{-x^2+2x}.$$

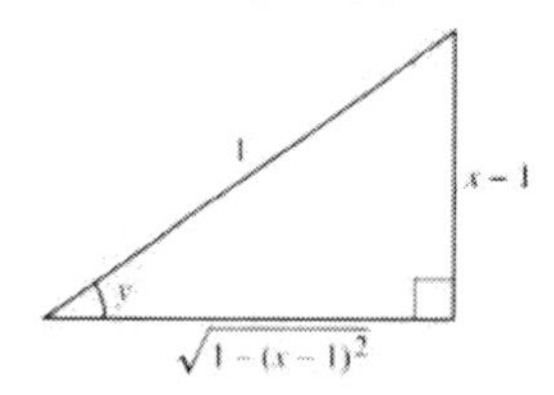

183. Let $y = \arccos\frac{x^2}{4-x^2}$. Then $\cos y = \frac{x^2}{4-x^2}$ and

$$\sin y = \frac{\sqrt{(4-x^2)^2-(x^2)^2}}{4-x^2} = \frac{\sqrt{16-2x^2}}{4-x} = \frac{2\sqrt{4-2x^2}}{4-x^2}.$$

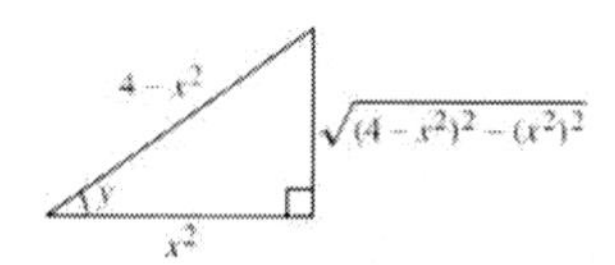

185. $\sin(1°10') = \frac{a}{3.5}$

$$a = 3.5\sin(1°10') = 3.5\sin\left(\frac{7°}{6}\right) \approx 0.0713 \text{ or 7 meters}$$

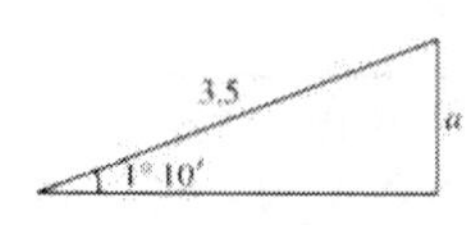

187. $\tan 14° = \frac{y}{37{,}000} \Rightarrow y = 37{,}000\tan 14° \approx 9225.1$ feet

$\tan 58° = \frac{x+y}{37{,}000} \Rightarrow x+y = 37{,}000\tan 58° \approx 59{,}212$ feet

$x = 59{,}212.4 - 9225.1 \approx 49{,}987.2$ feet

The towns are approximately 50,000 feet apart or 9.47 miles.

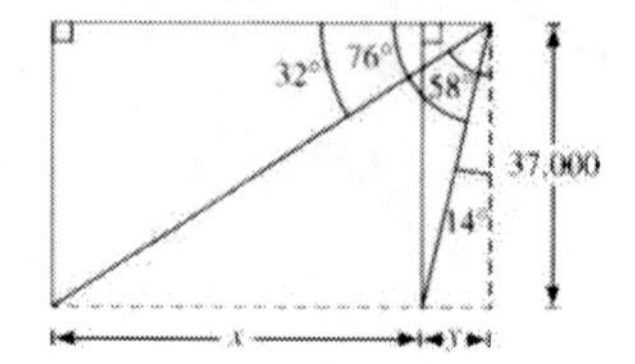

189. Use cosine model with amplitude 3 feet.

Period: 15 seconds

$$d = 3\cos\left(\frac{2\pi}{15}t\right)$$

191. False. $y = \sin\theta$ is a function, but it is not one-to-one.

193. False. The sine or cosine functions are used to model simple harmonic motion.

195. $\tan\theta = \frac{0.672s^2}{3000}$

(a)

s	10	20	30	40	50	60
θ	1.28°	5.12°	11.40°	19.72°	29.25°	38.88°

(b) θ increases at an increasing rate. The function is not linear.

197. (a)

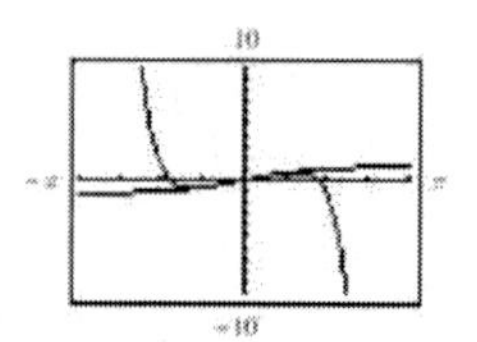

The polynomial function is a good approximation for the arctangent function when x is close to 0.

(b) Next term: $\frac{x^9}{9}$

$$\arctan x \approx x - \frac{x^3}{3} + \frac{x^5}{5} - \frac{x^7}{7} + \frac{x^9}{9}$$

The accuracy of the approximation increases as more terms are added.

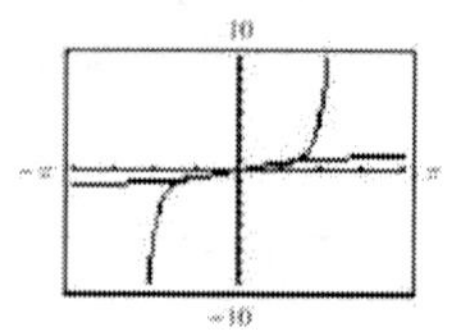

Chapter 4 Test

1. (a)

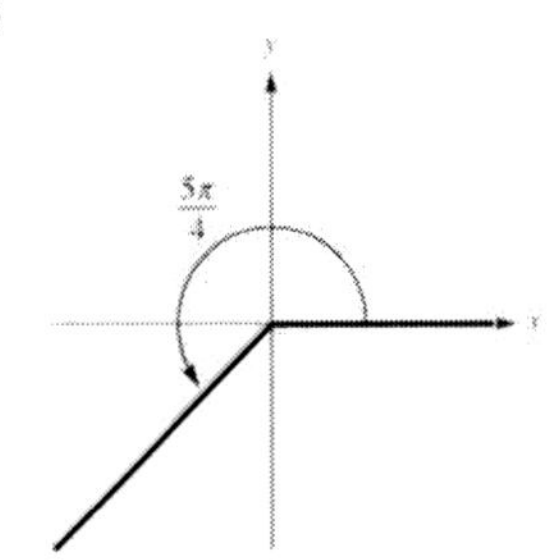

(b) $\frac{5\pi}{4} + 2\pi = \frac{13\pi}{4}$; $\frac{5\pi}{4} - 2\pi = -\frac{3\pi}{4}$

(c) $\frac{5\pi}{4} \cdot \frac{180}{\pi} = 225°$

2. angular speed $= \frac{\theta}{\text{time}}$

$$\frac{100 \text{ kilometers}}{\text{hour}} \Rightarrow \frac{100{,}000 \text{ meters}}{3600 \text{ seconds}} \Rightarrow \frac{250}{9} \text{ m/sec}$$

Since $s = \theta r$, $\theta = \frac{s}{r}$.

$$\theta = \frac{\frac{250}{9}\frac{\text{m}}{\text{sec}}}{0.625 \text{ m}} = 44\frac{4}{9} \text{ rad/sec}.$$

3. $x = -1$, $y = 4$, $r = \sqrt{1+16} = \sqrt{17}$

$\sin\theta = \frac{4}{\sqrt{17}} = \frac{4\sqrt{17}}{17}$

$\cos\theta = -\frac{1}{\sqrt{17}} = -\frac{\sqrt{17}}{17}$

$\tan\theta = -4$

$\csc\theta = \frac{\sqrt{17}}{4}$

$\sec\theta = -\sqrt{17}$

$\cot\theta = -\frac{1}{4}$

4. $\tan\theta = \frac{7}{2}$

$\tan^2\theta + 1 = \sec^2\theta \Rightarrow \sec\theta = \sqrt{\frac{49}{4} + 1} = \frac{\sqrt{53}}{2}$

$\cos\theta = \frac{2}{\sqrt{53}} = \frac{2\sqrt{53}}{53}$

$\sin\theta = \tan\theta\cos\theta = \frac{7\sqrt{53}}{53}$

$\csc\theta = \frac{53}{7\sqrt{53}} = \frac{\sqrt{53}}{7}$

$\cot\theta = \frac{2}{7}$

$\sec\theta = \frac{\sqrt{53}}{2}$

5. $\theta = 255°$

$\theta' = 255° - 180° = 75°$

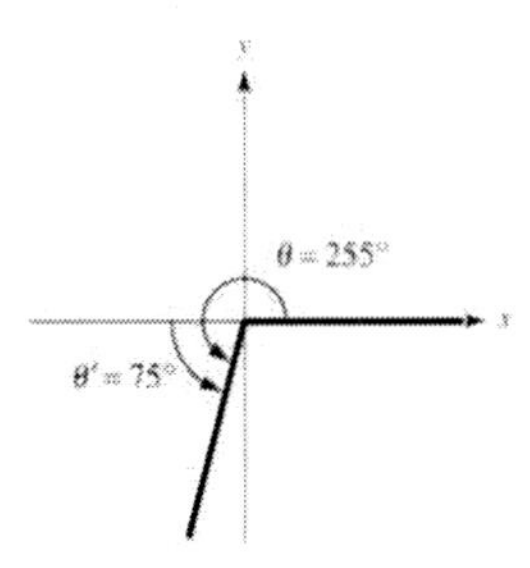

6. $\sec\theta = \frac{1}{\cos\theta} < 0 \Rightarrow$ Quadrants II or III

$\tan\theta > 0 \Rightarrow$ Quadrants I or III

Hence, Quadrant III.

7. $\cos\theta = -\frac{\sqrt{2}}{2}$

$\theta = 135°, 225°$

8. $\csc\theta = \frac{1}{\sin\theta} = 1.030 \Rightarrow \sin\theta = \frac{1}{1.030}$ and θ in Quadrant I or II.

Using a calculator, $\theta = 1.33, 1.81$ radians.

9. $\cos\theta = -\frac{3}{5}$, $\sin\theta > 0$, Quadrant II

$\sin\theta = \frac{4}{5}$

$\sec = -\frac{5}{3}$

$\cot\theta = -\frac{3}{4}$

$\tan\theta = -\frac{4}{3}$

$\csc\theta = \frac{5}{4}$

10. $g(x) = -2\sin\left(x - \frac{\pi}{4}\right)$

Period: $\frac{2\pi}{1} = 2\pi$

Amplitude: 2, shifted $\pi/4$ to the right

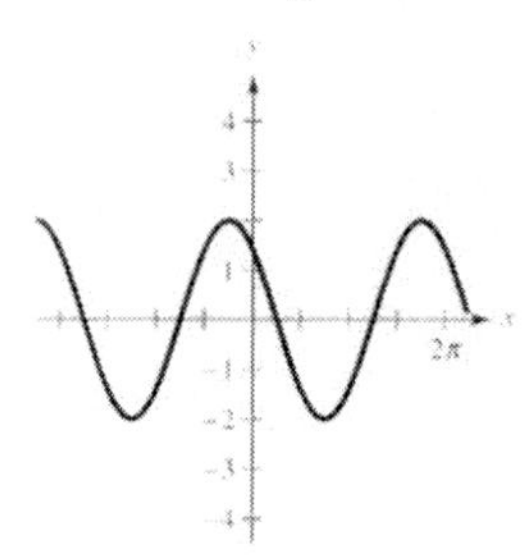

11. $f(x) = \frac{1}{2}\tan 4x$

Period: $\frac{\pi}{4}$

Two consecutive asymptotes:

$x = -\frac{\pi}{8}$, $x = \frac{\pi}{8}$

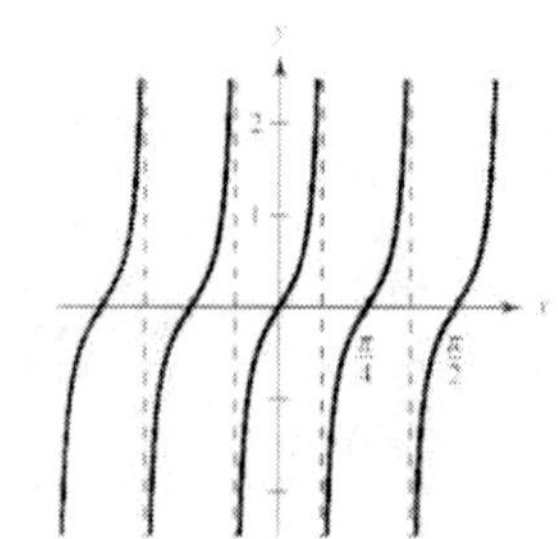

12. $f(x) = \frac{1}{2}\sec(x - \pi)$ is $y = \frac{1}{2}\sec x$ shifted π to the right.

Period: 2π

Two consecutive asymptotes:

$$x - \pi = -\frac{\pi}{2} \Rightarrow x = \frac{\pi}{2}$$
$$x - \pi = \frac{\pi}{2} \Rightarrow x = \frac{3\pi}{2}$$

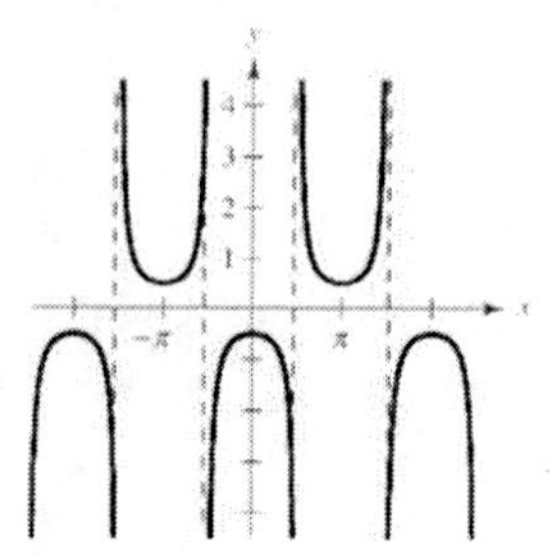

13. $f(x) = 2\cos(\pi - 2x) + 3 = 2\cos(2x - \pi) + 3$

Amplitude: 2

Shifted $\frac{\pi}{2}$ to the right, period: π

Shifted vertically upward 3

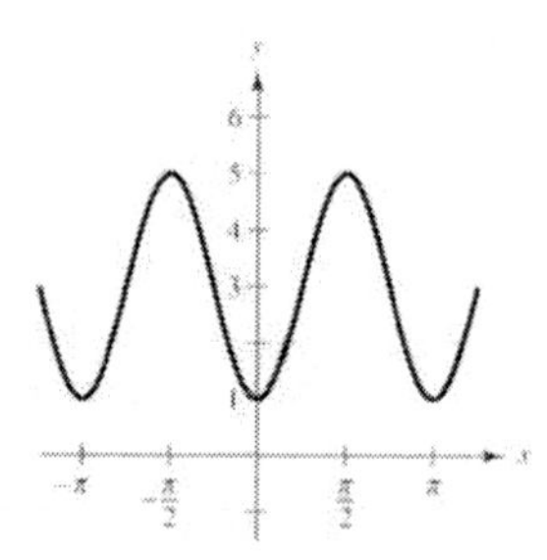

14. $f(x) = 2\csc\left(x + \frac{\pi}{2}\right)$

Period: π

Shifted $\frac{\pi}{2}$ to the left

Two consecutive asymptotes:

$$x + \frac{\pi}{2} = 0 \Rightarrow x = -\frac{\pi}{2}$$
$$x + \frac{\pi}{2} = \pi \Rightarrow x = \frac{\pi}{2}$$

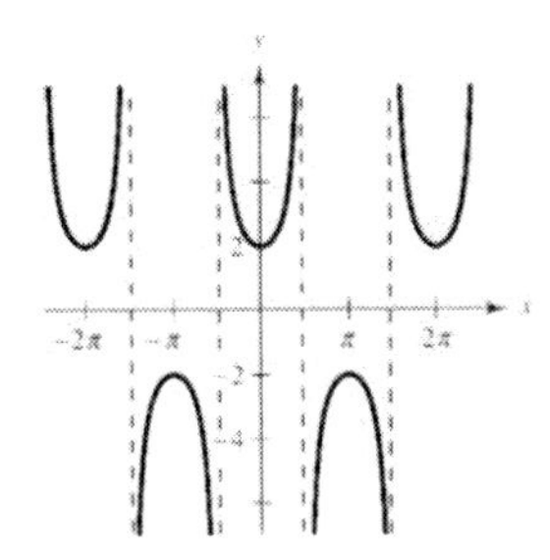

15. $f(x) = 2\cot\left(x - \dfrac{\pi}{2}\right)$

Period: π

Two consecutive asymptotes:

$x - \dfrac{\pi}{2} = 0 \Rightarrow x = \dfrac{\pi}{2};$

$x - \dfrac{\pi}{2} = \pi \Rightarrow x = \dfrac{3\pi}{2}$

16.

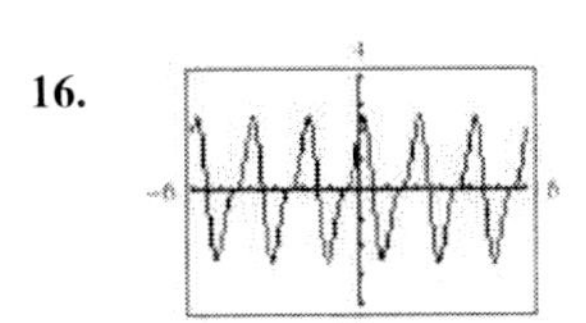

Period is 2.

17.

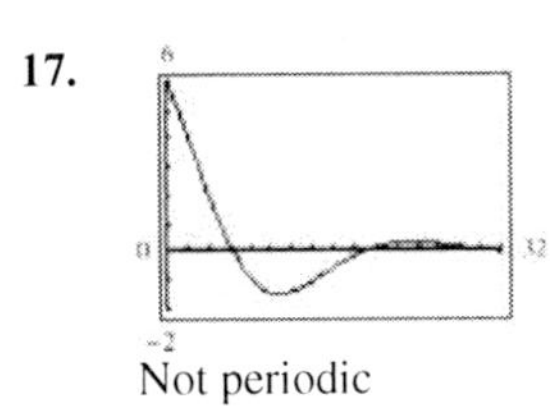

Not periodic

18. Amplitude: 2

Reflected in x-axis $\Rightarrow a = -2$

Period $= 4\pi \Rightarrow 4\pi = \dfrac{2\pi}{b} \Rightarrow b = \dfrac{1}{2}$

Horizontal shift $\dfrac{\pi}{4}$ units to the right

$y = -2\sin\left(\dfrac{x}{2} - \dfrac{\pi}{4}\right)$

19. Let $u = \arccos\dfrac{2}{3} \Rightarrow \cos u = \dfrac{2}{3}$.

Then $\tan\left(\arccos\dfrac{2}{3}\right) = \tan u = \dfrac{\sqrt{5}}{2}$.

20. $f(x) = 2\arcsin\left(\dfrac{1}{2}x\right)$

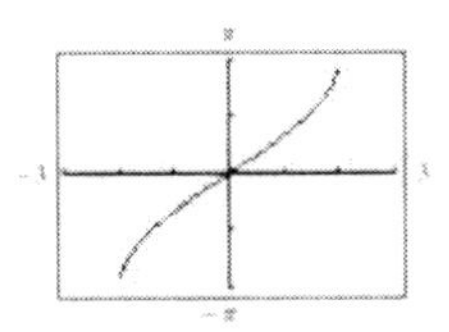

21. $f(x) = 2\arccos x$

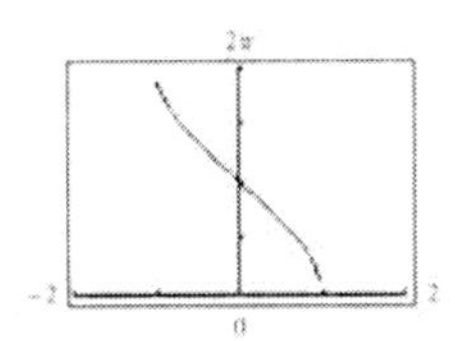

22. $f(x) = \arctan\left(\dfrac{x}{2}\right)$

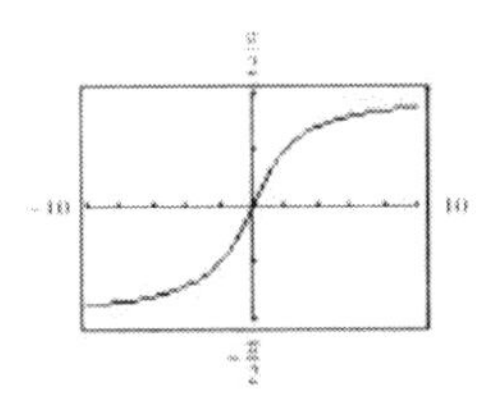

23.

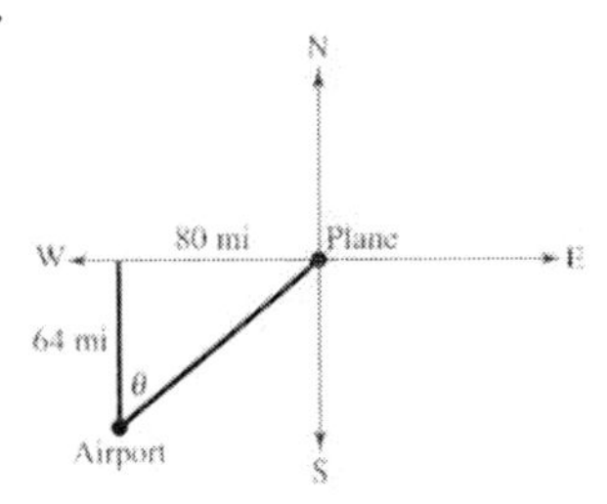

$\tan\theta = \dfrac{80}{96}$

$\theta = \arctan\dfrac{80}{64} \approx 51.34°$

Bearing: S 51.34° W

CHAPTER 5

Section 5.1

1. (a) $\sin u = \dfrac{1}{\csc u}$, matches (iii).

(b) $\cos u = \dfrac{1}{\sec u}$, matches (i).

(c) $\tan u = \dfrac{1}{\cot u}$, matches (ii).

3. $\sin u$

5. $\cos u$

7. $\sin x = \dfrac{1}{2}$, $\cos x = \dfrac{\sqrt{3}}{2}$, x is in Quadrant I.

$$\tan x = \frac{1/2}{\sqrt{3}/2} = \frac{1}{\sqrt{3}} = \frac{\sqrt{3}}{3}$$

$$\cot x = \sqrt{3}$$

$$\csc x = 2$$

$$\sec x = \frac{2}{\sqrt{3}} = \frac{2\sqrt{3}}{3}$$

9. $\sec\theta = \sqrt{2}$, $\sin\theta = -\dfrac{\sqrt{2}}{2} \Rightarrow \theta$ is in Quadrant IV.

$$\cos\theta = \frac{1}{\sec\theta} = \frac{1}{\sqrt{2}} = \frac{\sqrt{2}}{2}$$

$$\tan\theta = \frac{\sin\theta}{\cos\theta} = \frac{-\sqrt{2}/2}{\sqrt{2}/2} = -1$$

$$\cot\theta = \frac{1}{\tan\theta} = -1$$

$$\csc\theta = -\sqrt{2}$$

11. $\tan x = \dfrac{7}{24}$, $\sec x = \dfrac{-25}{24} \Rightarrow x$ is in Quadrant III.

$$\cot x = \frac{24}{7}$$

$$\cos x = -\frac{24}{25}$$

$$\sin x = -\sqrt{1-\cos^2 x} = -\frac{7}{25}$$

$$\csc x = \frac{1}{\sin x} = -\frac{25}{7}$$

13. $\sec\phi = -\dfrac{17}{15}$, $\sin\phi = \dfrac{8}{17}$, ϕ is in Quadrant II.

$$\cos\phi = -\frac{15}{17}$$

$$\csc\phi = \frac{17}{8}$$

$$\tan\phi = \frac{8/17}{-15/17} = -\frac{8}{15}$$

$$\cot\phi = -\frac{15}{8}$$

15. $\sin(-x) = -\sin x = -\dfrac{2}{3} \Rightarrow \sin x = \dfrac{2}{3}$

$\sin x = \dfrac{2}{3}$, $\tan x = -\dfrac{2\sqrt{5}}{5} \Rightarrow x$ is in Quadrant II.

$$\cos x = -\sqrt{1-\sin^2 x} = -\sqrt{1-\frac{4}{9}} = -\frac{\sqrt{5}}{3}$$

$$\cot x = \frac{1}{\tan x} = -\frac{\sqrt{5}}{2}$$

$$\sec x = \frac{1}{\cos x} = -\frac{3\sqrt{5}}{5}$$

$$\csc x = \frac{1}{\sin x} = \frac{3}{2}$$

17. $\tan\theta = 2$, $\sin\theta < 0 \Rightarrow \theta$ is in Quadrant III.

$$\sec\theta = -\sqrt{\tan^2\theta + 1} = -\sqrt{5}$$

$$\cos\theta = \frac{1}{\sec\theta} = \frac{1}{-\sqrt{5}} = -\frac{\sqrt{5}}{5}$$

$$\cot\theta = \frac{1}{\tan\theta} = \frac{1}{2}$$

$$\sin\theta = -\sqrt{1-\cos^2\theta}$$

$$= -\sqrt{1-\frac{1}{5}} = -\sqrt{\frac{4}{5}} = \frac{-2}{\sqrt{5}} = -\frac{2\sqrt{5}}{5}$$

$$\csc\theta = \frac{1}{\sin\theta} = -\frac{\sqrt{5}}{2}$$

19. $\csc\theta$ is undefined and $\cos\theta < 0 \Rightarrow \theta = \pi$.

$\sin\theta = 0$

$\cos\theta = -1$

$\tan\theta = 0$

$\cot\theta$ is undefined.

$\sec\theta = -1$

21. $\sec x \cos x = \dfrac{1}{\cos x} \cdot \cos x = 1$

Matches (d).

23. $\cot^2 x - \csc^2 x = \cot^2 x - (1 + \cot^2 x) = -1$

Matches (b).

25. $\dfrac{\sin(-x)}{\cos(-x)} = \dfrac{-\sin x}{\cos x} = -\tan x$

Matches (e).

27. $\sin x \sec x = \sin x\left(\dfrac{1}{\cos x}\right) = \tan x$

Matches (b).

29. $\sec^4 x - \tan^4 x = (\sec^2 x + \tan^2 x)(\sec^2 x - \tan^2 x)$

$= (\sec^2 x + \tan^2 x)(1)$

$= \sec^2 x + \tan^2 x$

Matches (f).

31. $\dfrac{\sec^2 x - 1}{\sin^2 x} = \dfrac{\tan^2 x}{\sin^2 x} = \dfrac{\sin^2 x}{\cos^2 x} \cdot \dfrac{1}{\sin^2 x} = \sec^2 x$

Matches (e).

33. $\cot x \sin x = \dfrac{\cos x}{\sin x} \sin x = \cos x$

35. $\sin\phi(\csc\phi - \sin\phi) = \sin\phi\csc\phi - \sin^2\phi$

$= \sin\phi \cdot \dfrac{1}{\sin\phi} - \sin^2\phi$

$= 1 - \sin^2\phi$

$= \cos^2\phi$

37. $\dfrac{\csc x}{\cot x} = \dfrac{1}{\sin x} \cdot \dfrac{\sin x}{\cos x} = \dfrac{1}{\cos x} = \sec x$

39. $\sec\alpha \dfrac{\sin\alpha}{\tan\alpha} = \dfrac{1}{\cos\alpha}(\sin\alpha)\cot\alpha$

$= \dfrac{1}{\cos\alpha}(\sin\alpha)\left(\dfrac{\cos\alpha}{\sin\alpha}\right) = 1$

41. $\sin\left(\dfrac{\pi}{2} - x\right)\csc x = \cos x \cdot \dfrac{1}{\sin x} = \cot x$

43. $\dfrac{\cos^2 y}{1 - \sin y} = \dfrac{1 - \sin^2 y}{1 - \sin y}$

$= \dfrac{(1 + \sin y)(1 - \sin y)}{1 - \sin y}$

$= 1 + \sin y$

45. $\cot^2 x - \cot^2 x \cos^2 x = \cot^2 x(1 - \cos^2 x)$

$= \dfrac{\cos^2 x}{\sin^2 x}\sin^2 x = \cos^2 x$

47. $\dfrac{\cos^2 x - 4}{\cos x - 2} = \dfrac{(\cos x + 2)(\cos x - 2)}{\cos x - 2} = \cos x + 2$

49. $\tan^4 x + 2\tan^2 x + 1 = (\tan^2 x + 1)^2$

$= (\sec^2 x)^2 = \sec^4 x$

51. $\sin^4 x - \cos^4 x = (\sin^2 x + \cos^2 x)(\sin^2 x - \cos^2 x)$

$= (1)(\sin^2 x - \cos^2 x) = \sin^2 x - \cos^2 x$

53. $\csc^3 x - \csc^2 x - \csc x + 1 = \csc^2 x(\csc x - 1) - (\csc x - 1)$

$= (\csc^2 x - 1)(\csc x - 1)$

$= \cot^2 x(\csc x - 1)$

55. $(\sin x + \cos x)^2 = \sin^2 x + 2\sin x \cos x + \cos^2 x$

$= (\sin^2 x + \cos^2 x) + 2\sin x \cos x$

$= 1 + 2\sin x \cos x$

57. $(\csc x + 1)(\csc x - 1) = \csc^2 x - 1 = \cot^2 x$

59. $\dfrac{1}{1 + \cos x} + \dfrac{1}{1 - \cos x} = \dfrac{1 - \cos x + 1 + \cos x}{(1 + \cos x)(1 - \cos x)}$

$= \dfrac{2}{1 - \cos^2 x}$

$= \dfrac{2}{\sin^2 x}$

$= 2\csc^2 x$

61. $\tan x - \dfrac{\sec^2 x}{\tan x} = \dfrac{\tan^2 x - \sec^2 x}{\tan x} = \dfrac{-1}{\tan x} = -\cot x$

63. $\tan x + \dfrac{\cos x}{1 + \sin x} = \dfrac{\sin x}{\cos x} + \dfrac{\cos x}{1 + \sin x}$

$= \dfrac{\sin x + \sin^2 x + \cos^2 x}{\cos x(1 + \sin x)}$

$= \dfrac{\sin x + 1}{\cos x(1 + \sin x)}$

$= \dfrac{1}{\cos x}$

$= \sec x$

65. $\dfrac{\sin^2 y}{1 - \cos y} = \dfrac{1 - \cos^2 y}{1 - \cos y}$

$= \dfrac{(1 + \cos y)(1 - \cos y)}{1 - \cos y}$

$= 1 + \cos y$

67. $\dfrac{\sin x}{\tan x} = \sin x\left(\dfrac{\cos x}{\sin x}\right)$

$= \cos x$

69. $\dfrac{3}{\sec x - \tan x} \cdot \dfrac{\sec x + \tan x}{\sec x + \tan x} = \dfrac{3(\sec x + \tan x)}{\sec^2 x - \tan^2 x}$

$= \dfrac{3(\sec x + \tan x)}{1}$

$= 3(\sec x + \tan x)$

71. $y_1 = \cos\left(\dfrac{\pi}{2} - x\right),\ y_2 = \sin x$

x	0.2	0.4	0.6	0.8	1.0	1.2	1.4
y_1	0.1987	0.3894	0.5646	0.7174	0.8415	0.9320	0.9854
y_2	0.1987	0.3894	0.5646	0.7174	0.8415	0.9320	0.9854

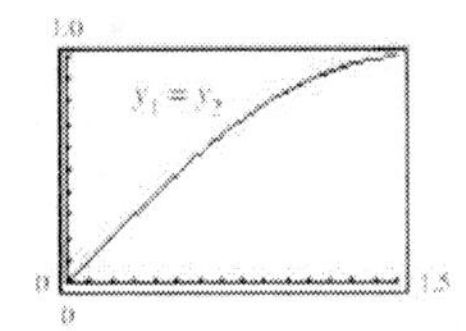

Conjecture: $y_1 = y_2$

73. $y_1 = \dfrac{\cos x}{1 - \sin x},\ y_2 = \dfrac{1 + \sin x}{\cos x}$

x	0.2	0.4	0.6	0.8	1.0	1.2	1.4
y_1	1.2230	1.5085	1.8958	2.4650	3.4082	5.3319	11.6814
y_2	1.2230	1.5085	1.8958	2.4650	3.4082	5.3319	11.6814

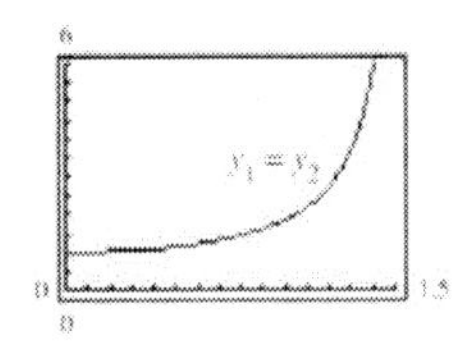

Conjecture: $y_1 = y_2$

75. $y_1 = \cos x \cot x + \sin x = \csc x$

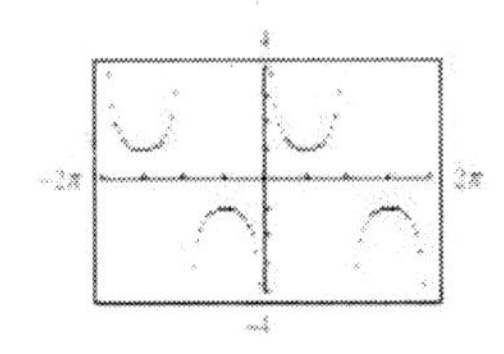

77. $y_1 = \sec x - \dfrac{\cos x}{1 + \sin x} = \tan x$

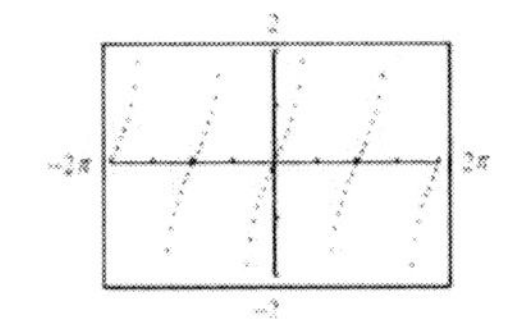

79. $x = 5\sin\theta,\ 0 < \theta < \dfrac{\pi}{2}$

$\sqrt{25 - x^2} = \sqrt{25 - (5\sin\theta)^2}$

$= \sqrt{25 - 25\sin^2\theta}$

$= \sqrt{25(1 - \sin^2\theta)}$

$= \sqrt{25\cos^2\theta}$

$= 5\cos\theta$

81. $x = 3\sec\theta,\ 0 < \theta < \dfrac{\pi}{2}$

$\sqrt{x^2 - 9} = \sqrt{(3\sec\theta)^2 - 9}$

$= \sqrt{9\sec^2\theta - 9}$

$= \sqrt{9(\sec^2\theta - 1)}$

$= \sqrt{9\tan^2\theta}$

$= 3\tan\theta$

83. $x = 3\sin\theta,\ 0 < \theta < \dfrac{\pi}{2}$

$\sqrt{9 - x^2} = \sqrt{9 - 9\sin^2\theta}$

$= \sqrt{9\cos^2\theta} = 3\cos\theta$

85. $2x = 3\tan\theta,\ 0 < \theta < \dfrac{\pi}{2}$

$\sqrt{4x^2 + 9} = \sqrt{9\tan^2\theta + 9}$

$= \sqrt{9\sec^2\theta} = 3\sec\theta$

87. $4x = 3\sec\theta,\ 0 < \theta < \dfrac{\pi}{2}$

$\sqrt{16x^2 - 9} = \sqrt{9\sec^2\theta - 9}$

$= \sqrt{9\tan^2\theta} = 3\tan\theta$

89. $x = \sqrt{2}\sin\theta,\ 0 < \theta < \dfrac{\pi}{2}$

$\sqrt{2 - x^2} = \sqrt{2 - 2\sin^2\theta}$

$= \sqrt{2\cos^2\theta} = \sqrt{2}\cos\theta$

91. $\sin\theta = \sqrt{1 - \cos^2\theta}$

Let $y_1 = \sin x$ and $y_2 = \sqrt{1 - \cos^2 x},\ 0 \le x < 2\pi.$

$y_1 = y_2$ for $0 \le x \le \pi$, so we have

$\sin\theta = \sqrt{1 - \cos^2\theta}$ for $0 \le \theta \le \pi$.

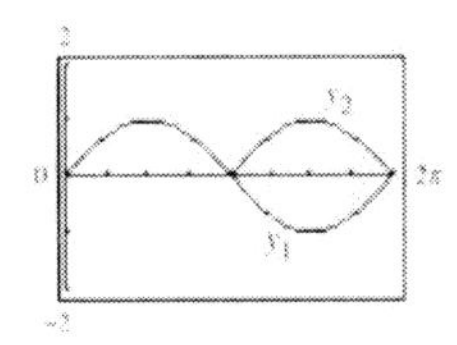

93. $\sec\theta = \sqrt{1+\tan^2\theta}$

Let $y_1 = \dfrac{1}{\cos x}$ and $y_2 = \sqrt{1+\tan^2 x}$, $0 \le x < 2\pi$.

$y_1 = y_2$ for $0 \le x < \dfrac{\pi}{2}$ and $\dfrac{3\pi}{2} < x < 2\pi$, so we have $\sec\theta = \sqrt{1+\tan^2\theta}$ for $0 \le \theta < \dfrac{\pi}{2}$ and $\dfrac{3\pi}{2} < \theta < 2\pi$.

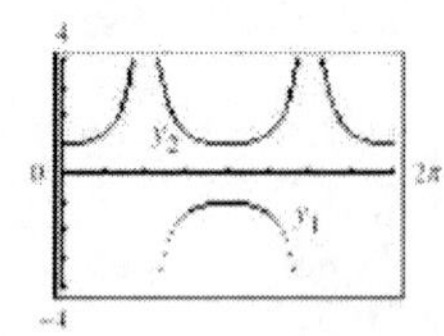

95. $\ln|\cos\theta| - \ln|\sin\theta| = \ln\left|\dfrac{\cos\theta}{\sin\theta}\right| = \ln|\cot\theta|$

97. $\ln(1+\sin x) - \ln|\sec x| = \ln\left|\dfrac{1+\sin x}{\sec x}\right|$

$= \ln|\cos x(1+\sin x)|$

99. $\ln|\sec x| + \ln|\sin x| = \ln|\sec x \cdot \sin x|$

$= \ln\left|\dfrac{1}{\cos x}\cdot \sin x\right|$

$= \ln|\tan x|$

101. Let $\theta = \dfrac{7\pi}{6}$. Then

$\cos\theta = \cos\dfrac{7\pi}{6} = -\dfrac{\sqrt{3}}{2} \ne \sqrt{1-\sin^2\theta} = \dfrac{\sqrt{3}}{2}$.

103. Let $\theta = \dfrac{5\pi}{3}$. Then

$\sin\theta = \sin\dfrac{5\pi}{3} = -\dfrac{\sqrt{3}}{2} \ne \sqrt{1-\cos^2\theta} = \dfrac{\sqrt{3}}{2}$.

105. Let $\theta = \dfrac{7\pi}{4}$. Then

$\csc\theta = \csc\dfrac{7\pi}{4} = -\sqrt{2} \ne \sqrt{1+\cos^2\theta} = \sqrt{2}$.

107. (a) $\csc^2 132^\circ - \cot^2 132^\circ \approx 1.8107 - 0.8107 = 1$

(b) $\csc^2\dfrac{2\pi}{7} - \cot^2\dfrac{2\pi}{7} \approx 1.6360 - 0.6360 = 1$

109. $\cos\left(\dfrac{\pi}{2}-\theta\right) = \sin\theta$

(a) $\theta = 80^\circ$

$\cos(90^\circ - 80^\circ) = \sin 80^\circ$

$0.9848 = 0.9848$

(b) $\theta = 0.8$

$\cos\left(\dfrac{\pi}{2} - 0.8\right) = \sin 0.8$

$0.7174 = 0.7174$

111. $\mu W\cos\theta = W\sin\theta$

$\mu = \dfrac{W\sin\theta}{W\cos\theta} = \tan\theta$

113. $\sec x\tan x - \sin x = \dfrac{1}{\cos x}\dfrac{\sin x}{\cos x} - \sin x$

$= \sin x(\sec^2 x - 1)$

$= \sin x \cdot \tan^2 x$

115. False.

$\cos 0 \sec\dfrac{\pi}{4} \ne 1$

117. As $x \to \dfrac{\pi^-}{2}$, $\sin x \to 1$ and $\csc x \to 1$.

119. As $x \to \dfrac{\pi^-}{2}$, $\tan x \to \infty$ and $\cot x \to 0$.

121. $\sin\theta$

$\cos\theta = \pm\sqrt{1-\sin^2\theta}$

$\tan\theta = \dfrac{\sin\theta}{\cos\theta} = \pm\dfrac{\sin\theta}{\sqrt{1-\sin^2\theta}}$

$\csc\theta = \dfrac{1}{\sin\theta}$

$\sec\theta = \pm\dfrac{1}{\sqrt{1-\sin^2\theta}}$

$\cot\theta = \pm\dfrac{\sqrt{1-\sin^2\theta}}{\sin\theta}$

The sign + or − depends on the choice of θ.

123. $\dfrac{1}{x+5} + \dfrac{x}{x-8} = \dfrac{x-8+x(x+5)}{(x+5)(x-8)}$

$= \dfrac{x^2+6x-8}{(x+5)(x-8)}$

125. $\dfrac{2x}{x^2-4}-\dfrac{7}{x+4}=\dfrac{2x(x+4)-7(x+2)(x-2)}{(x+2)(x-2)(x+4)}$

$=\dfrac{2x^2+8x-7x^2+28}{(x+2)(x-2)(x+4)}$

$=\dfrac{-5x^2+8x+28}{(x+2)(x-2)(x+4)}$

127. $f(x)=\dfrac{1}{2}\sin \pi x$

Period: $\dfrac{2\pi}{\pi}=2$

Amplitude: $\dfrac{1}{2}$

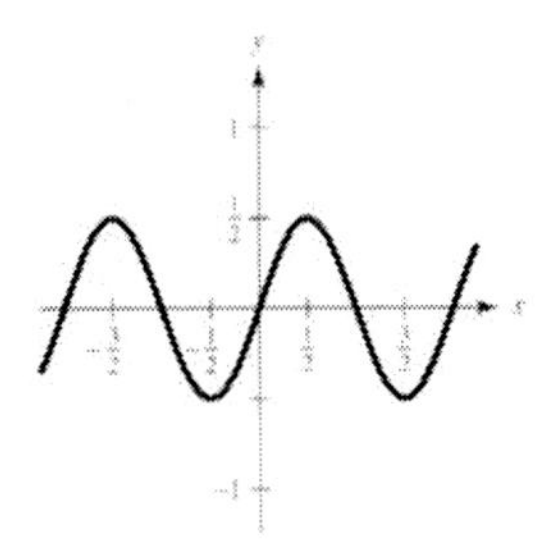

129. $f(x)=\dfrac{1}{2}\cot\left(x+\dfrac{\pi}{4}\right)$

Period: π

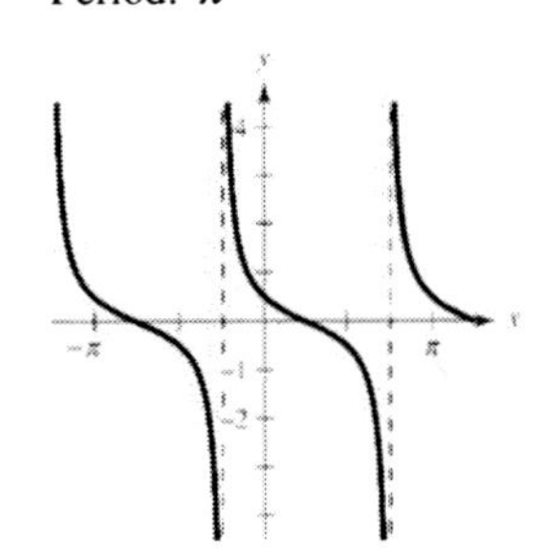

Section 5.2

1. $\cot u$

3. $\tan u$

5. $\cos^2 u$

7. $-\sin u$

9. No. Algebraic techniques must be used to produce a valid proof to verify a trigonometric identity.

11. $\sin t \csc t=\sin t\left(\dfrac{1}{\sin t}\right)=1$

13. $\dfrac{\csc^2 x}{\cot x}=\dfrac{1}{\sin^2 x}\cdot\dfrac{\sin x}{\cos x}=\dfrac{1}{\sin x\cdot\cos x}$

$=\csc x\cdot\sec x$

15. $\cos^2\beta-\sin^2\beta=(1-\sin^2\beta)-\sin^2\beta$

$=1-2\sin^2\beta$

17. $\tan^2\theta+6=(\tan^2\theta+1)+5$

$=\sec^2\theta+5$

19. $(1+\sin x)(1-\sin x)=1-\sin^2 x=\cos^2 x$

21. $\dfrac{1}{\sec x\tan x}=\cos x\cdot\dfrac{\cos x}{\sin x}$

$=\dfrac{\cos^2 x}{\sin x}$

$=\dfrac{1-\sin^2 x}{\sin x}$

$=\dfrac{1}{\sin x}-\sin x$

$=\csc x-\sin x$

x	0.2	0.4	0.6	0.8	1.0	1.2	1.4
y_1	4.8348	2.1785	1.2064	0.6767	0.3469	0.1409	0.0293
y_2	4.8348	2.1785	1.2064	0.6767	0.3469	0.1409	0.0293

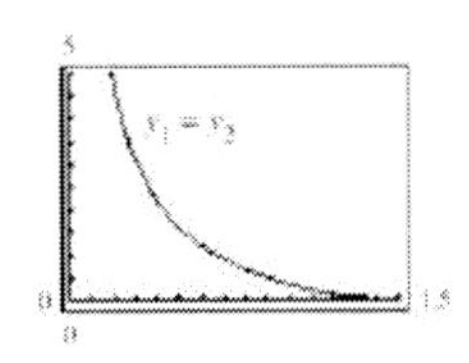

23. $\csc x - \sin x = \dfrac{1}{\sin x} - \sin x$

$= \dfrac{1 - \sin^2 x}{\sin x}$

$= \dfrac{\cos^2 x}{\sin x}$

$= \cos x \cdot \dfrac{\cos x}{\sin x}$

$= \cos x \cdot \cot x$

x	0.2	0.4	0.6	0.8	1.0	1.2	1.4
y_1	4.8348	2.1785	1.2064	0.6767	0.3469	0.1409	0.0293
y_2	4.8348	2.1785	1.2064	0.6767	0.3469	0.1409	0.0293

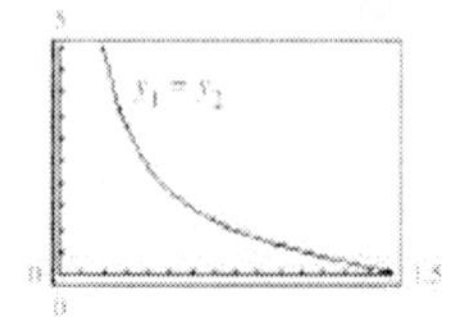

25. $\sin x + \cos x \cot x = \sin x + \cos x \dfrac{\cos x}{\sin x}$

$= \dfrac{\sin^2 x + \cos^2 x}{\sin x}$

$= \dfrac{1}{\sin x}$

$= \csc x$

x	0.2	0.4	0.6	0.8	1.0	1.2	1.4
y_1	5.0335	2.5679	1.7710	1.3940	1.1884	1.0729	1.0148
y_2	5.0335	2.5679	1.7710	1.3940	1.1884	1.0729	1.0148

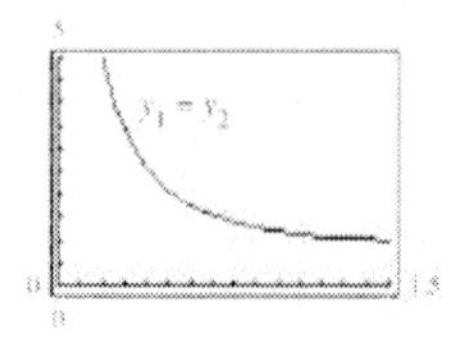

27. $\dfrac{1}{\tan x} + \dfrac{1}{\cot x} = \dfrac{\cot x + \tan x}{\tan x \cdot \cot x}$

$= \cot x + \tan x$

x	0.2	0.4	0.6	0.8	1.0	1.2	1.4
y_1	5.1359	2.7880	2.1458	2.0009	2.1995	2.9609	5.9704
y_2	5.1359	2.7880	2.1458	2.0009	2.1995	2.9609	5.9704

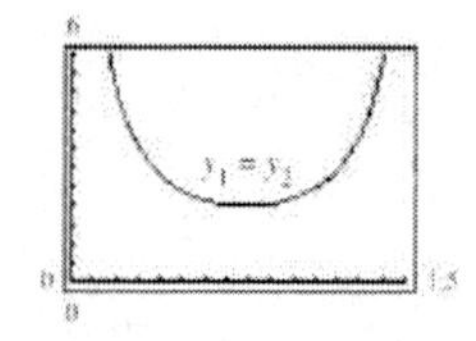

29. The error is in line 1: $\cot(-x) = -\cot x$.

31. $\sec^4 x - 2\sec^2 x + 1 = \left(\sec^2 x - 1\right)^2$

$= \left(\tan^2 x\right)^2$

$= \tan^4 x$

33. $\sin^{1/2} x \cos x - \sin^{5/2} x \cos x = \sin^{1/2} x \cos x\left(1 - \sin^2 x\right)$

$= \sin^{1/2} x \cos x \cdot \cos^2 x$

$= \cos^3 x\sqrt{\sin x}$

35. $\cot\left(\dfrac{\pi}{2} - x\right)\csc x = \tan x \csc x$

$= \dfrac{\sin x}{\cos x} \cdot \dfrac{1}{\sin x}$

$= \dfrac{1}{\cos x} = \sec x$

37. $\dfrac{\csc(-x)}{\sec(-x)} = \dfrac{1/\sin(-x)}{1/\cos(-x)}$

$= \dfrac{\cos(-x)}{\sin(-x)}$

$= \dfrac{\cos x}{-\sin x}$

$= -\cot x$

39. $\dfrac{\cos x - \cos y}{\sin x + \sin y} + \dfrac{\sin x - \sin y}{\cos x + \cos y}$

$= \dfrac{(\cos x + \cos y)(\cos x - \cos y) + (\sin x + \sin y)(\sin x - \sin y)}{(\sin x + \sin y)(\cos x + \cos y)}$

$= \dfrac{\cos^2 x - \cos^2 y + \sin^2 x - \sin^2 y}{(\sin x + \sin y)(\cos x + \cos y)}$

$= \dfrac{1 - 1}{(\sin x + \sin y)(\cos x + \cos y)}$

$= 0$

41. $\dfrac{\cos\theta}{1 - \sin\theta} = \dfrac{\cos\theta}{1 - \sin\theta} \cdot \dfrac{1 + \sin\theta}{1 + \sin\theta}$

$$= \frac{\cos\theta(1+\sin\theta)}{1-\sin^2\theta}$$
$$= \frac{\cos\theta(1+\sin\theta)}{\cos^2\theta}$$
$$= \frac{1+\sin\theta}{\cos\theta}$$
$$= \frac{1}{\cos\theta} + \frac{\sin\theta}{\cos\theta}$$
$$= \sec\theta + \tan\theta$$

43. $\sin^2\left(\frac{\pi}{2}-x\right)+\sin^2 x = \cos^2 x + \sin^2 x = 1$

45. $\sin x \csc\left(\frac{\pi}{2}-x\right) = \sin x \sec x$
$$= \sin x\left(\frac{1}{\cos x}\right)$$
$$= \tan x$$

47. $2\sec^2 x - 2\sec^2 x\sin^2 x - \sin^2 x - \cos^2 x$
$$= 2\sec^2 x(1-\sin^2 x) - (\sin^2 x + \cos^2 x)$$
$$= 2\sec^2 x\left(\cos^2 x\right) - 1$$
$$= 2\cdot\frac{1}{\cos^2 x}\cdot\cos^2 x - 1$$
$$= 2 - 1 = 1$$

49. $\frac{\cot x\tan x}{\sin x} = \frac{1}{\sin x} = \csc x$

51. $\csc\theta\tan\theta = \frac{1}{\sin\theta}\cdot\frac{\sin\theta}{\cos\theta}$
$$= \frac{1}{\cos\theta}$$
$$= \sec\theta$$

53. $1 - \frac{\sin^2\theta}{1-\cos\theta} = \frac{1-\cos\theta-\sin^2\theta}{1-\cos\theta}$

$$\sin 2u = 2\sin u\cos u = 2\left(-\frac{1}{\sqrt{37}}\right)\left(\frac{6}{\sqrt{37}}\right) = -\frac{12}{37}$$
$$\cos 2u = \cos^2 u - \sin^2 u = \frac{36}{37} - \frac{1}{37} = \frac{35}{37}$$
$$\tan 2u = \frac{2\tan u}{1-\tan^2 u} = \frac{2(-1/6)}{1-(-1/6)^2} = \frac{-2/6}{35/36} = -\frac{12}{35}$$

55. $\frac{\sin\beta}{1-\cos\beta}\cdot\frac{1+\cos\beta}{1+\cos\beta} = \frac{\sin\beta(1+\cos\beta)}{1-\cos^2\beta}$
$$= \frac{\sin\beta(1+\cos\beta)}{\sin^2\beta}$$
$$= \frac{1+\cos\beta}{\sin\beta}$$

57. $\frac{\tan^3\alpha - 1}{\tan\alpha - 1} = \frac{(\tan\alpha-1)(\tan^2\alpha+\tan\alpha+1)}{\tan\alpha-1} = \tan^2\alpha+\tan\alpha+1$

59. It appears that $y_1 = 1$. Analytically,
$$\frac{1}{\cot x+1}+\frac{1}{\tan x+1} = \frac{\tan x+1+\cot x+1}{(\cot x+1)(\tan x+1)}$$
$$= \frac{\tan x+\cot x+2}{\cot x\tan x+\cot x+\tan x+1}$$
$$= \frac{\tan x+\cot x+2}{\tan x+\cot x+2}$$
$$= 1.$$

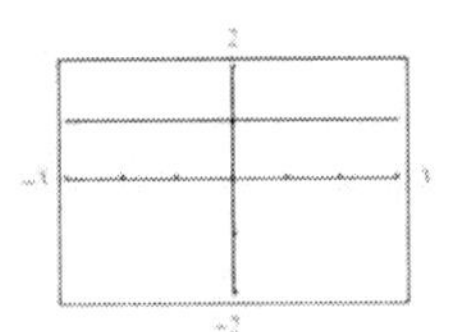

61. It appears that $y_1 = \sin x$. Analytically,
$$\frac{1}{\sin x} - \frac{\cos^2 x}{\sin x} = \frac{1-\cos^2 x}{\sin x} = \frac{\sin^2 x}{\sin x} = \sin x.$$

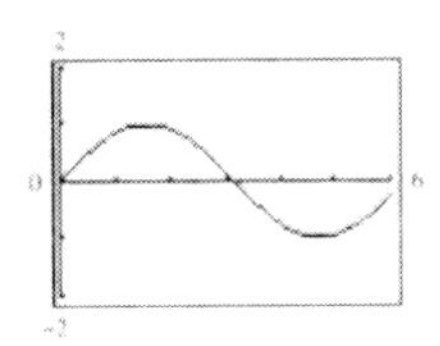

63. $\ln|\cot\theta| = \ln\left|\frac{\cos\theta}{\sin\theta}\right|$
$$= \ln\frac{|\cos\theta|}{|\sin\theta|}$$
$$= \ln|\cos\theta| - \ln|\sin\theta|$$

65. $\sin^2 35° + \sin^2 55° = \cos^2(90° - 35°) + \sin^2 55°$
$$= \cos^2 55° + \sin^2 55° = 1$$

67. $\cos^2 20° + \cos^2 52° + \cos^2 38° + \cos^2 70°$
$$= \cos^2 20° + \cos^2 52° + \sin^2(90° - 38°) + \sin^2(90° - 70°)$$
$$= \cos^2 20° + \cos^2 52° + \sin^2 52° + \sin^2 20°$$
$$= (\cos^2 20° + \sin^2 20°) + (\cos^2 52° + \sin^2 52°)$$
$$= 1 + 1 = 2$$

69. $\tan^5 x = \tan^3 x\cdot\tan^2 x$
$$= \tan^3 x(\sec^2 x - 1)$$
$$= \tan^3 x\sec^2 x - \tan^3 x$$

71. $(\sin^2 x - \sin^4 x)\cos x = \sin^2 x(1-\sin^2 x)\cos x$
$$= \sin^2 x\cdot\cos^2 x\cdot\cos x$$
$$= \cos^3 x\sin^2 x$$

73. Let $\theta = \sin^{-1} x \Rightarrow \sin\theta = x = \frac{x}{1}$.

From the diagram,

$\tan(\sin^{-1} x) = \tan\theta = \frac{x}{\sqrt{1-x^2}}$.

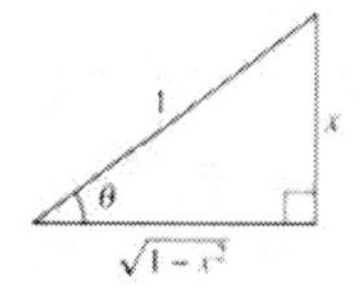

75. Let $\theta = \sin^{-1}\frac{x-1}{4} \Rightarrow \sin\theta = \frac{x-1}{4}$.

From the diagram,

$\tan\left(\sin^{-1}\frac{x-1}{4}\right) = \tan\theta = \frac{x-1}{\sqrt{16-(x-1)^2}}$.

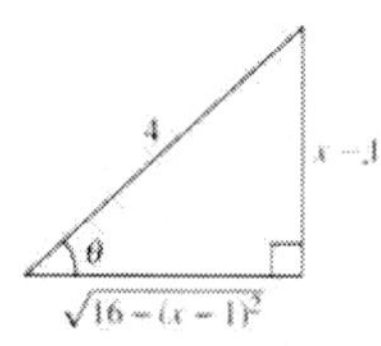

77. (a) $\frac{h\sin(90° - \theta)}{\sin\theta} = \frac{h\cos\theta}{\sin\theta} = h\cot\theta$

(b)

θ	15°	30°	45°	60°	75°	90°
s	18.66	8.66	5	2.89	1.34	0

(c) Maximum: 15°
Minimum: 90°

(d) Noon

79. True. For instance, $(\sec^2\theta - 1)/\sec^2\theta = \sin^2\theta$ was verified two different ways on page 358 of the text.

81. False. Just because the equation is true for one value of θ, you cannot conclude that the equation is an identity. For example,

$\sin^2\frac{\pi}{4} + \cos^2\frac{\pi}{4} = 1 \neq 1 + \tan^2\frac{\pi}{4}$.

83. (a)
$$\frac{\sin x}{1+\cos x} = \frac{\sin x}{1+\cos x}\cdot\frac{1-\cos x}{1-\cos x} = \frac{\sin x(1-\cos x)}{1-\cos^2 x} = \frac{\sin x(1-\cos x)}{\sin^2 x} = \frac{1-\cos x}{\sin x}$$

(b) Not true for $x = 0$ because $(1-\cos x)/\sin x$ is not defined for $x = 0$.

85.
$$\sqrt{a^2-u^2} = \sqrt{a^2 - a^2\sin^2\theta} = \sqrt{a^2(1-\sin^2\theta)} = \sqrt{a^2\cos^2\theta} = a\cos\theta$$

87.
$$\sqrt{a^2+u^2} = \sqrt{a^2 + a^2\tan^2\theta} = \sqrt{a^2(1+\tan^2\theta)} = \sqrt{a^2\sec^2\theta} = a\sec\theta$$

89. $\sqrt{\tan^2 x} = |\tan x|$

Let $x = \frac{3\pi}{4}$. Then, $\sqrt{\tan^2 x} = \sqrt{(-1)^2} = 1 \neq \tan\left(\frac{3\pi}{4}\right) = -1$.

91. $|\tan\theta| = \sqrt{\sec^2\theta - 1}$

Let $\theta = \frac{3\pi}{4}$. Then,

$\tan\frac{3\pi}{4} = -1 \neq \sqrt{\left[\sec(3\pi/4)\right]^2 - 1} = \sqrt{2-1} = \sqrt{1} = 1$.

93. When n is even, $\cos\left[\frac{(2n+1)\pi}{2}\right] = \cos\frac{\pi}{2} = 0$.

When n is odd, $\cos\left[\frac{(2n+1)\pi}{2}\right] = \cos\frac{3\pi}{2} = 0$.

Thus, $\cos\left[\frac{(2n+1)\pi}{2}\right] = 0$ for all n.

95. $f(x) = 2^x + 3$

x	−4	−2	0	2	3
y	3.0625	3.25	4	7	11

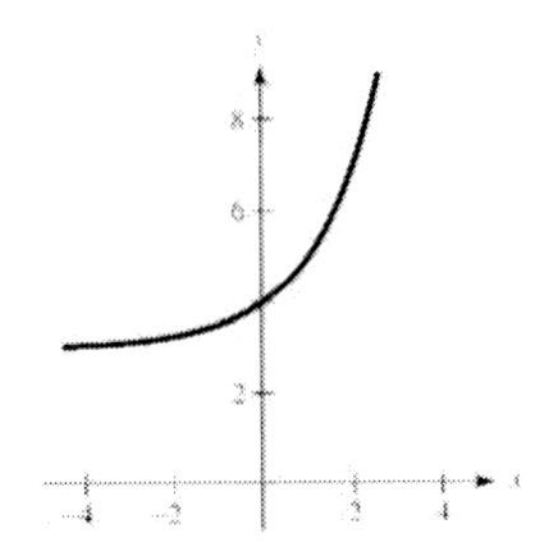

Horizontal asymptote: $y = 3$ as $x \to -\infty$

97. $f(x) = 2^{-x} + 1$

x	−3	−2	0	2	4
y	9	5	2	1.25	1.0625

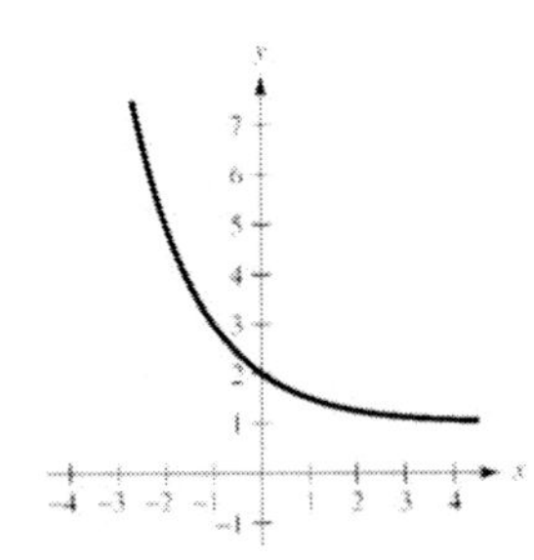

Horizontal asymptote: $y = 1$ as $x \to -\infty$

Section 5.3

1. general

3. No. The solutions to $\cos x = 0$ are $x = \frac{\pi}{2} + n\pi$.

5. $2\cos x - 1 = 0$

(a) $x = \frac{\pi}{3}$: $2\cos\frac{\pi}{3} - 1 = 2\left(\frac{1}{2}\right) - 1 = 0$

(b) $x = \frac{5\pi}{3}$: $2\cos\frac{5\pi}{3} - 1 = 2\left(\frac{1}{2}\right) - 1 = 0$

7. $3\tan^2 2x - 1 = 0$

(a) $x = \frac{\pi}{12}$: $3\left[\tan\left(\frac{2\pi}{12}\right)\right]^2 - 1 = 3\tan^2\frac{\pi}{6} - 1 = 3\left(\frac{1}{\sqrt{3}}\right)^2 - 1 = 0$

(b) $x = \frac{5\pi}{12}$: $3\left[\tan\left(\frac{10\pi}{12}\right)\right]^2 - 1 = 3\tan^2\frac{5\pi}{6} - 1$

$= 3\left(-\frac{1}{\sqrt{3}}\right)^2 - 1 = 0$

9. $2\sin^2 x - \sin x - 1 = 0$

(a) $x = \frac{\pi}{2}$: $2\sin^2\left(\frac{\pi}{2}\right) - \sin\left(\frac{\pi}{2}\right) - 1 = 2 - 1 - 1 = 0$

(b) $x = \frac{7\pi}{6}$: $2\sin^2\left(\frac{7\pi}{6}\right) - \sin\left(\frac{7\pi}{6}\right) - 1 = 2\left(\frac{1}{4}\right) - \left(-\frac{1}{2}\right) - 1 = 0$

11. $\sin x = \frac{1}{2}$

$x = 30°, 150°$

13. $\cos x = -\frac{1}{2}$

$x = 120°, 240°$

15. $\tan x = 1$

$x = 45°, 225°$

17. $\cos x = -\frac{\sqrt{3}}{2}$

$x = \frac{5\pi}{6}, \frac{7\pi}{6}$

19. $\cot x = -1$

$x = \frac{3\pi}{4}, \frac{7\pi}{4}$

21. $\tan x = -\frac{\sqrt{3}}{3}$

$x = \frac{5\pi}{6}, \frac{11\pi}{6}$

23. $\csc x = -2 \Rightarrow \sin x = -\frac{1}{2}$

$x = \frac{7\pi}{6}, \frac{11\pi}{6}$

25. $\cot x = \sqrt{3} \Rightarrow \tan x = \frac{\sqrt{3}}{3}$

$x = \frac{\pi}{6}, \frac{7\pi}{6}$

27. $\tan x = -1$

$x = \frac{3\pi}{4}, \frac{7\pi}{4}$

29. $2\sin x + 1 = 0$

$$2\sin x = -1$$
$$\sin x = -\frac{1}{2}$$
$$x = \frac{7\pi}{6} + 2n\pi$$
$$x = \frac{11\pi}{6} + 2n\pi$$

31. $\sqrt{3}\sec x - 2 = 0$

$$\sqrt{3}\sec x = 2$$
$$\sec x = \frac{2}{\sqrt{3}}$$
$$\cos x = \frac{\sqrt{3}}{2}$$
$$x = \frac{\pi}{6} + 2n\pi$$
$$\text{or } x = \frac{11\pi}{6} + 2n\pi$$

33. $\sec^2 x - 4 = 0$

$$\sec^2 x = \frac{4}{3}$$
$$\sec x = \pm\frac{2}{\sqrt{3}}$$
$$\cos x = \pm\frac{\sqrt{3}}{2}$$
$$x = \frac{\pi}{6} + n\pi \text{ or } x = \frac{5\pi}{6} + n\pi$$

35. $4\cos^2 x - 1 = 0$

$$\cos^2 x = \frac{1}{4}$$
$$\cos x = \pm\frac{1}{2}$$
$$x = \frac{\pi}{3} + n\pi \text{ or } x = \frac{2\pi}{3} + n\pi$$

37. $\tan x + \sqrt{3} = 0$

$$\tan x = -\sqrt{3}$$
$$x = \frac{2\pi}{3}, \frac{5\pi}{3}$$

39. $\csc^2 x - 2 = 0$

$$\csc^2 x = 2$$
$$\csc x = \pm\sqrt{2}$$
$$\sin x = \pm\frac{1}{\sqrt{2}}$$
$$x = \frac{\pi}{4}, \frac{3\pi}{4}, \frac{5\pi}{4}, \frac{7\pi}{4}$$

41. $\cos^3 x = \cos x$

$$\cos^3 x - \cos x = 0$$
$$\cos x\,(\cos^2 x - 1) = 0$$

$\cos x = 0$ or $\cos^2 x - 1 = 0$

$x = \frac{\pi}{2}, \frac{3\pi}{2}$ $\quad$ $\cos x = \pm 1$

$x = 0, \pi$

43. $\sec^2 x - \sec x - 2 = 0$

$(\sec x - 2)(\sec x + 1) = 0$

$\sec x - 2 = 0$ or $\sec x + 1 = 0$

$\sec x = 2$ $\quad$ $\sec x = -1$

$x = \frac{\pi}{3}, \frac{5\pi}{3}$ $\quad$ $x = \pi$

45. $2\sin x + \csc x = 0$

$$2\sin x + \frac{1}{\sin x} = 0$$
$$2\sin^2 x + 1 = 0$$

Since $2\sin^2 x + 1 > 0$, there are no solutions.

47. $2\sec^2 x + \tan^2 x - 3 = 0$

$$2\sec^2 x + \sec^2 x - 1 - 3 = 0$$
$$3\sec^2 x - 4 = 0$$
$$\sec^2 x = \frac{4}{3}$$
$$\sec x = \pm\frac{2}{\sqrt{3}} = \pm\frac{2\sqrt{3}}{3}$$
$$x = \frac{\pi}{6}, \frac{5\pi}{6}, \frac{7\pi}{6}, \frac{11\pi}{6}$$

49. $2\sin^2 x + 3\sin x + 1 = 0$

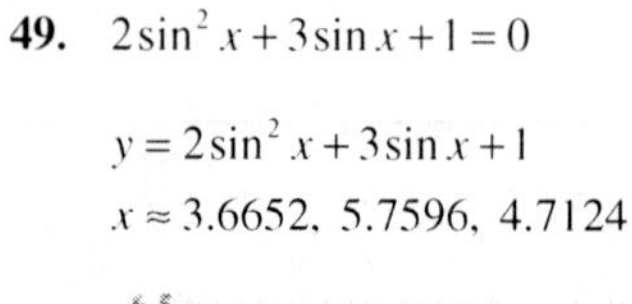

$y = 2\sin^2 x + 3\sin x + 1$

$x \approx 3.6652, 5.7596, 4.7124$

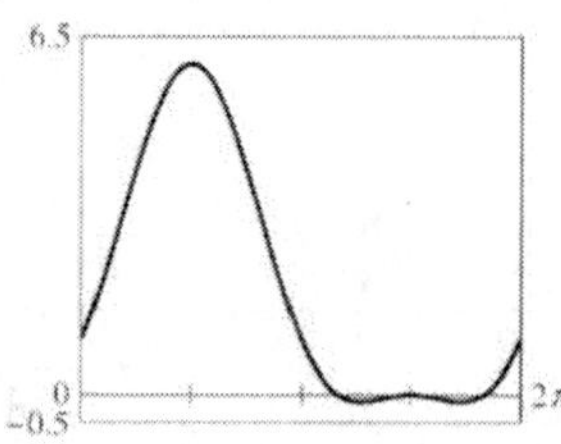

51. $4\sin^2 x = 2\cos x + 1$

$y = 4\sin^2 x - 2\cos x - 1$

$x \approx 0.8614,\ 5.4218$

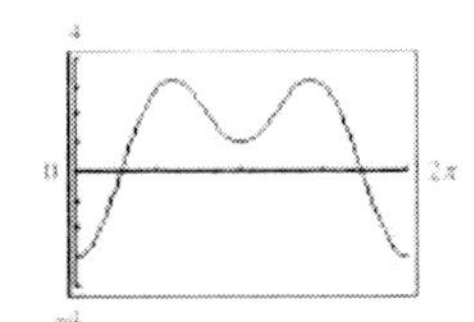

53. $\csc x + \cot x = 1$

$y = \dfrac{1}{\sin x} + \dfrac{\cos x}{\sin x} - 1$

$x \approx 1.5708\left(\dfrac{\pi}{2}\right)$

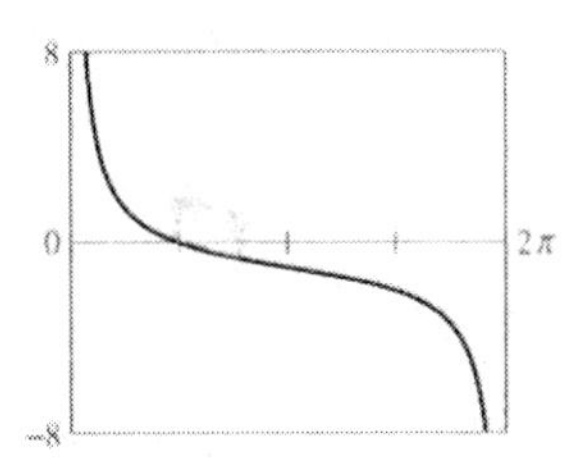

55. $\dfrac{\cos x \cot x}{1 - \sin x} = 3$

$y = \dfrac{\cos x}{(1 - \sin x)\tan x} - 3$

$x \approx 0.5236,\ 2.6180$

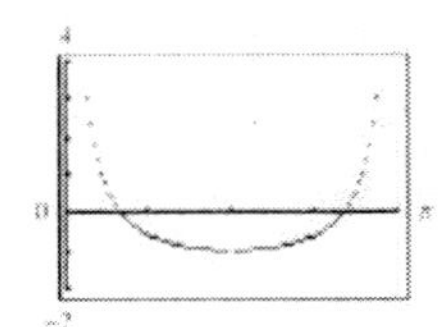

57. (a)

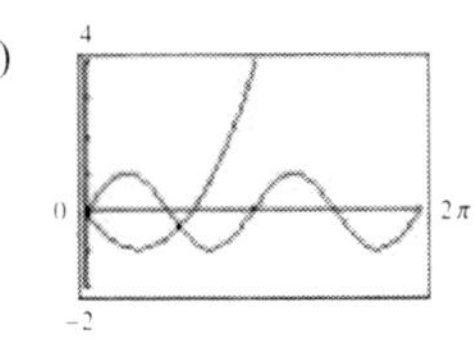

(b) $\sin 2x = x^2 - 2x$

(c) Points of intersection: $(0, 0)$, $(1.7757, -0.3984)$

59. (a)

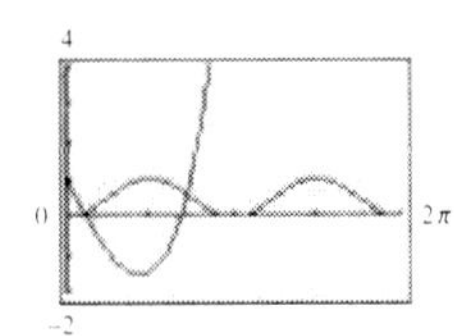

(b) $\sin^2 x = e^x - 4x$

(c) Points of intersection:
$(0.3194, 0.0986)$, $(2.2680, 0.5878)$

61. $\cos\dfrac{x}{4} = 0$

$\dfrac{x}{4} = \dfrac{\pi}{2} + 2n\pi$ or $\dfrac{x}{4} = \dfrac{3\pi}{2} + 2n\pi$

$x = 2\pi + 8n\pi$ or $x = 6\pi + 8n\pi$

Combining, $x = 2\pi + 4n\pi$.

63. $\sin 4x = 1$

$4x = \dfrac{\pi}{2} + 2n\pi$

$x = \dfrac{\pi}{8} + \dfrac{n\pi}{2}$

65. $\sin 2x = -\dfrac{\sqrt{3}}{2}$

$2x = \dfrac{4\pi}{3} + 2n\pi$ or $2x = \dfrac{5\pi}{3} + 2n\pi$

$x = \dfrac{2\pi}{3} + n\pi$ $\qquad x = \dfrac{5\pi}{6} + n\pi$

67. $\cos\dfrac{x}{2} = \dfrac{\sqrt{2}}{2}$

$\dfrac{x}{2} = \dfrac{\pi}{4} + 2n\pi$ or $\dfrac{x}{2} = \dfrac{7\pi}{4} + 2n\pi$

$x = \dfrac{\pi}{2} + 4n\pi$ $\qquad x = \dfrac{7\pi}{2} + 4n\pi$

69. $y = \sin\dfrac{\pi x}{2} + 1$

From the graph in the textbook we see that the curve has x-intercepts at $x = -1$ and at $x = 3$.

71. $y = \tan^2\left(\dfrac{\pi x}{6}\right) - 3$

From the graph in the textbook, we see that the curve has x-intercepts at $x = \pm 2$.

73. $2\cos x - \sin x = 0$

Graph $y_1 = 2\cos x - \sin x$ and estimate the zeros.

$x \approx 1.1071,\ 4.2487$

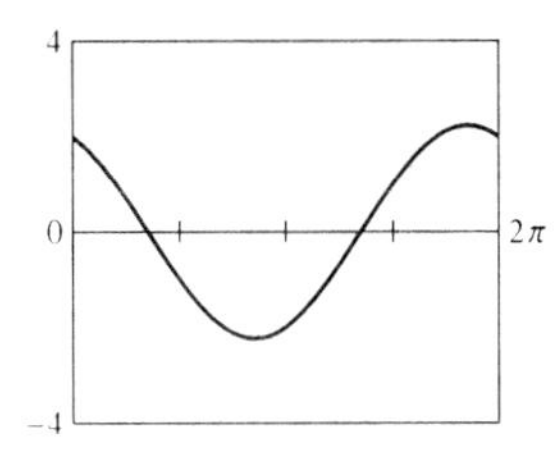

75. $x\tan x - 1 = 0$

Graph $y_1 = x\tan x - 1$ and estimate the zeros.

$x \approx 0.8603, 3.4256$

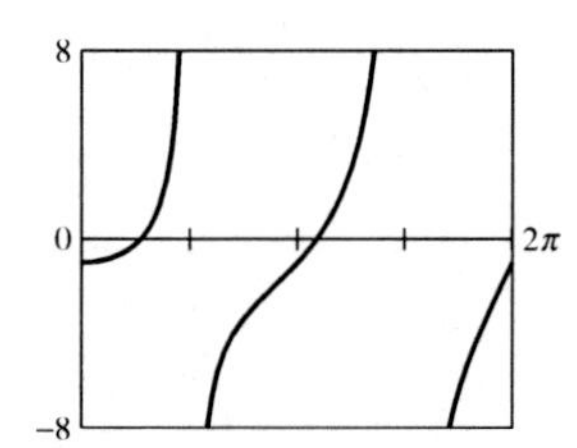

77. $\sec^2 x + 0.5\tan x - 1 = 0$

Graph $y_1 = \dfrac{1}{(\cos x)^2} + 0.5\tan x - 1$ and estimate the zeros.

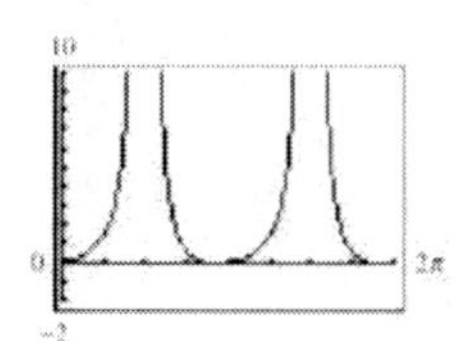

$x = 0, x \approx 2.6779, 3.1416, 5.8195$

79. $12\sin^2 x - 13\sin x + 3 = 0$

Graph $y_1 = 12\sin^2 x - 13\sin x + 3$ and estimate the zeros.

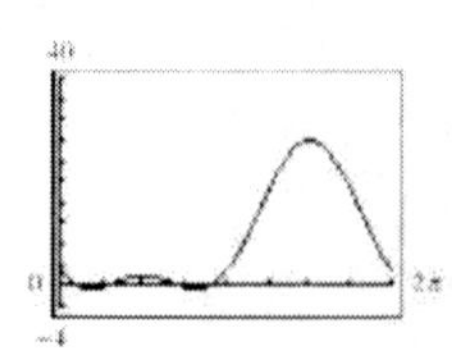

$x \approx 0.3398, 0.8481, 2.2935, 2.8018$

81. $\tan^2 x - 6\tan x + 5 = 0$

$(\tan x - 1)(\tan x - 5) = 0$

$\tan x - 1 = 0$ $\qquad$ $\tan x - 5 = 0$

$\tan x = 1$ $\qquad$ $\tan x = 5$

$x = \pi/4,\ 5\pi/4$ $\qquad$ $x = \arctan 5 \approx 1.37$

$x = \pi + \arctan 5 \approx 4.51$

83. $3\tan^2 x + 5\tan x - 4 = 0,\ \left[-\dfrac{\pi}{2}, \dfrac{\pi}{2}\right]$

$x \approx -1.154, 0.534$

85. $4\cos^2 x - 2\sin x + 1 = 0,\ \left[-\dfrac{\pi}{2}, \dfrac{\pi}{2}\right]$

$x \approx 1.110$

87. $f(x) = \sin 2x$

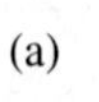

(a)

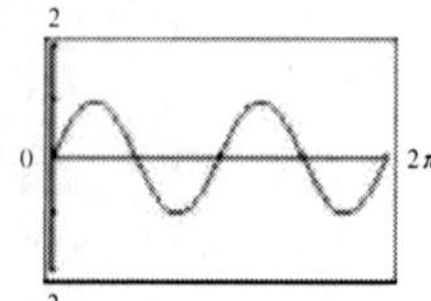

Maxima: $(0.7854, 1), (3.9270, 1)$

Minima: $(2.3562, -1), (5.4978, -1)$

(b) $2\cos 2x = 0$

$$\cos 2x = 0$$

$$2x = \frac{\pi}{2} + n\pi$$

$$x = \frac{\pi}{4} + \frac{n\pi}{2}$$

The zeros are 0.7854, 2.3562, 3.9270, and 5.4978.

89. $f(x) = \sin^2 x + \cos x$

(a)

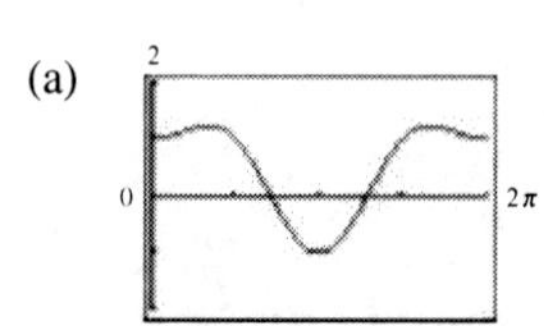

Maxima: $(1.0472, 1.25), (5.2360, 1.25)$

Minima: $(0, 1), (3.1416, -1), (6.2832, 1)$

(b) $2\sin x\cos x - \sin x = 0$

$$\sin x(2\cos x - 1) = 0$$

$$\sin x = 0 \Rightarrow x = n\pi$$

$$\cos x = \frac{1}{2} \Rightarrow x = \frac{\pi}{3} + 2n\pi,\ \frac{5\pi}{3} + 2n\pi$$

The zeros are 0, 1.0472, 3.1416, 5.2360, and 6.2832.

91. $f(x) = \sin x + \cos x$

(a)

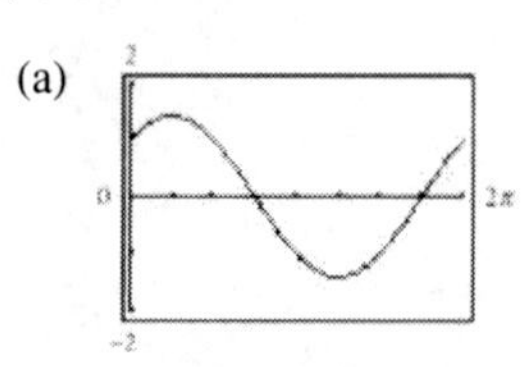

Maximum: $(0.7854, 1.4142)$

Minimum: $(3.9270, -1.4142)$

(b) $\cos x - \sin x = 0$

$$\cos x = \sin x$$

$$1 = \frac{\sin x}{\cos x}$$

$$\tan x = 1$$

$$x = \frac{\pi}{4} + 2n\pi,\ \frac{5\pi}{4} + 2n\pi$$

The zeros are 0.7854 and 3.9270.

93. $f(x) = \tan\frac{\pi x}{4}$

$\tan 0 = 0,$ but 0 is not positive. By graphing

$y = \tan\frac{\pi x}{4} - x,$

you see that the smallest positive fixed point is $x = 1$.

95. $S = 74.50 - 43.75\cos\frac{\pi t}{6}$

t	1	2	3	4	5	6	7	8	9	10	11	12
S	36.6	52.6	74.5	96.4	112.4	118.3	112.4	96.4	74.5	52.6	36.6	30.8

$S > 100$ for $t = 5, 6, 7$ (May, June, July)

97. $f(x) = \cos\frac{1}{x}$

(a) The domain of $f(x)$ is all real numbers except 0.
(b) The graph has y-axis symmetry and a horizontal asymptote at $y = 1$.
(c) As $x \to 0$, $f(x)$ oscillates between -1 and 1.
(d) There are an infinite number of solution in the interval $[-1, 1]$.

$$\frac{1}{x} = \frac{\pi}{2} + n\pi = \frac{\pi + 2n\pi}{2} \Rightarrow x = \frac{2}{\pi(2n+1)}$$

(e) The greatest solution appears to occur at $x \approx 0.6366$.

99.

$$y = \frac{1}{12}(\cos 8t - 3\sin 8t)$$

$$\frac{1}{12}(\cos 8t - 3\sin 8t) = 0$$

$$\cos 8t = 3\sin 8t$$

$$\frac{1}{3} = \tan 8t$$

$$8t = 0.32175 + n\pi$$

$$t = 0.04 + \frac{n\pi}{8}$$

In the interval $0 \le t \le 1$, $t \approx 0.04$, 0.43, and 0.83 second.

101.

$$r = \frac{1}{32}v_0^2 \sin 2\theta$$

$$300 = \frac{1}{32}(100)^2 \sin 2\theta$$

$$\sin 2\theta = 0.96$$

$2\theta \approx 1.287$ or $2\theta \approx \pi - 1.287 \approx 1.855$

$\theta \approx 0.6435 \approx 37°$ or $\theta \approx 0.9275 \approx 53°$

103. $A = 2x\cos x,\ 0 \le x \le \frac{\pi}{2}$

(a)

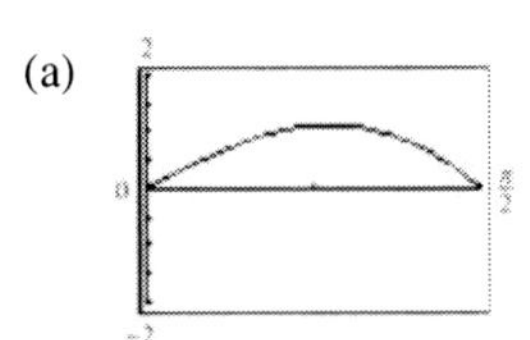

The maximum area of $A \approx 1.12$ occurs when $x \approx 0.86$.

(b) $A \ge 1$ for $0.6 < x < 1.1$

105. $y_1 = 2\sin x$

$y_2 = 3x + 1$

From the graph we see that there is only one point of intersection.

107. False. $\sin x - x = 0$ has one solution, $x = 0$.

109. False. The equation has no solution because $-1 \le \sin x \le 1$.

111. Answers will vary.

113. $124° = 124°\left(\frac{\pi}{180°}\right) \approx 2.164$ radians

115. $-0.41° = -0.41°\left(\frac{\pi}{180°}\right) \approx -0.007$ radian

117. Answers will vary. (Make a Decision).

Section 5.4

1. $\sin u \cos v - \cos u \sin v$

3. $\frac{\tan u + \tan v}{1 - \tan u \tan v}$

5. $\cos u \cos v + \sin u \sin v$

7. Sample answer:
$\sin 105° = \sin(45° + 60°) = \sin 45° \cos 60° + \cos 45° \sin 60°$

9. (a) $\cos(240° - 0°) = \cos(240°) = -\frac{1}{2}$

(b) $\cos(240°) - \cos 0° = -\frac{1}{2} - 1 = -\frac{3}{2}$

11. (a) $\cos\left(\frac{\pi}{4}+\frac{\pi}{3}\right) = \cos\frac{\pi}{4}\cos\frac{\pi}{3} - \sin\frac{\pi}{4}\sin\frac{\pi}{3}$

$= \frac{\sqrt{2}}{2}\left(\frac{1}{2}\right) - \frac{\sqrt{2}}{2}\left(\frac{\sqrt{3}}{2}\right)$

$= \frac{\sqrt{2}-\sqrt{6}}{4}$

(b) $\cos\frac{\pi}{4} + \cos\frac{\pi}{3} = \frac{\sqrt{2}}{2} + \frac{1}{2} = \frac{\sqrt{2}+1}{2}$

13. (a) $\sin 105° = \sin(135° - 30°)$

$= \sin 135° \cos 30° - \cos 135° - \sin 30°$

$= \left(\frac{\sqrt{2}}{2}\right)\left(\frac{\sqrt{3}}{2}\right) - \left(\frac{\sqrt{3}}{2}\right)\left(\frac{1}{2}\right)$

$= \frac{\sqrt{6}}{4} + \frac{\sqrt{2}}{4} = \frac{\sqrt{6}+\sqrt{2}}{4}$

(b) $\sin 135° - \sin 30° = \frac{\sqrt{2}}{2} - \frac{1}{2} = \frac{\sqrt{2}-1}{2}$

15. $\sin 105° = \sin(60° + 45°)$

$= \sin 60° \cos 45° + \sin 45° \cos 60°$

$= \frac{\sqrt{3}}{2}\cdot\frac{\sqrt{2}}{2} + \frac{\sqrt{2}}{2}\cdot\frac{1}{2}$

$= \frac{\sqrt{2}}{4}(\sqrt{3}+1)$

$\cos 105° = \cos(60° + 45°)$

$= \cos 60° \cos 45° - \sin 60° \sin 45°$

$= \frac{1}{2}\cdot\frac{\sqrt{2}}{2} - \frac{\sqrt{3}}{2}\cdot\frac{1}{2}$

$= \frac{\sqrt{2}}{4}(1-\sqrt{3})$

$\tan 105° = \tan(60° + 45°)$

$= \frac{\tan 60° + \tan 45°}{1 - \tan 60° \tan 45°}$

$= \frac{\sqrt{3}+1}{1-\sqrt{3}} = \frac{\sqrt{3}+1}{1-\sqrt{3}}\cdot\frac{1+\sqrt{3}}{1+\sqrt{3}}$

$= \frac{4+2\sqrt{3}}{-2} = -2-\sqrt{3}$

17. $\sin 195° = \sin(225° - 30°)$

$= \sin 225° \cos 30° - \sin 30° \cos 225°$

$= -\sin 45° \cos 30° + \sin 30° \cos 45°$

$= -\frac{\sqrt{2}}{2}\cdot\frac{\sqrt{3}}{2} + \frac{1}{2}\cdot\frac{\sqrt{2}}{2} = \frac{\sqrt{2}}{4}(1-\sqrt{3})$

$\cos 195° = \cos(225° - 30°)$

$= \cos 225° \cos 30° + \sin 225° \sin 30°$

$= -\cos 45° \cos 30° - \sin 45° \cos 30°$

$= -\frac{\sqrt{2}}{2}\cdot\frac{\sqrt{3}}{2} - \frac{\sqrt{2}}{2}\cdot\frac{1}{2} = -\frac{\sqrt{2}}{4}\left(\sqrt{3}+1\right)$

$\tan 195° = \tan(225° - 30°)$

$= \frac{\tan 225° - \tan 30°}{1 + \tan 225° \tan 30°}$

$= \frac{\tan 45° - \tan 30°}{1 + \tan 45° \tan 30°}$

$= \frac{1-(\sqrt{3}/3)}{1+(\sqrt{3}/3)} = \frac{3-\sqrt{3}}{3+\sqrt{3}}\cdot\frac{3-\sqrt{3}}{3-\sqrt{3}}$

$= \frac{12-6\sqrt{3}}{6} = 2-\sqrt{3}$

19. $\sin\frac{11\pi}{12} = \sin\left(\frac{3\pi}{4}+\frac{\pi}{6}\right)$

$= \sin\frac{3\pi}{4}\cos\frac{\pi}{6} + \sin\frac{\pi}{6}\cos\frac{3\pi}{4}$

$= \frac{\sqrt{2}}{2}\cdot\frac{\sqrt{3}}{2} + \frac{1}{2}\left(-\frac{\sqrt{2}}{2}\right) = \frac{\sqrt{2}}{4}(\sqrt{3}-1)$

$\cos\frac{11\pi}{12} = \cos\left(\frac{3\pi}{4}+\frac{\pi}{6}\right)$

$= \cos\frac{3\pi}{4}\cos\frac{\pi}{6} - \sin\frac{3\pi}{4}\sin\frac{\pi}{6}$

$= -\frac{\sqrt{2}}{2}\cdot\frac{\sqrt{3}}{2} - \frac{\sqrt{2}}{2}\cdot\frac{1}{2}$

$= -\frac{\sqrt{2}}{4}(\sqrt{3}+1)$

$\tan\frac{11\pi}{12} = \tan\left(\frac{3\pi}{4}+\frac{\pi}{6}\right)$

$= \frac{\tan(3\pi/4) + \tan(\pi/6)}{1 - \tan(3\pi/4)\tan(\pi/6)}$

$= \frac{-1+(\sqrt{3}/3)}{1-(-1)(\sqrt{3}/3)}$

$= \frac{-3+\sqrt{3}}{3+\sqrt{3}}\cdot\frac{3-\sqrt{3}}{3-\sqrt{3}}$

$= \frac{-12+6\sqrt{3}}{6} = -2+\sqrt{3}$

21. $-\dfrac{\pi}{12} = \dfrac{\pi}{6} - \dfrac{\pi}{4}$

$$\sin\left(-\frac{\pi}{12}\right) = \sin\left(\frac{\pi}{6} - \frac{\pi}{4}\right)$$
$$= \sin\frac{\pi}{6}\cos\frac{\pi}{4} - \sin\frac{\pi}{4}\cos\frac{\pi}{6}$$
$$= \frac{1}{2}\cdot\frac{\sqrt{2}}{2} - \frac{\sqrt{2}}{2}\cdot\frac{\sqrt{3}}{2} = \frac{\sqrt{2}}{4}(1-\sqrt{3})$$

$$\cos\left(-\frac{\pi}{12}\right) = \cos\left(\frac{\pi}{6} - \frac{\pi}{4}\right)$$
$$= \cos\frac{\pi}{6}\cos\frac{\pi}{4} + \sin\frac{\pi}{6}\sin\frac{\pi}{4}$$
$$= \frac{\sqrt{3}}{2}\cdot\frac{\sqrt{2}}{2} + \frac{1}{2}\cdot\frac{\sqrt{2}}{2} = \frac{\sqrt{2}}{4}(\sqrt{3}+1)$$

$$\tan\left(-\frac{\pi}{12}\right) = \tan\left(\frac{\pi}{6} - \frac{\pi}{4}\right)$$
$$= \frac{\tan(\pi/6) - \tan(\pi/4)}{1 + \tan(\pi/6)\tan(\pi/4)}$$
$$= \frac{\left(\sqrt{3}/3\right) - 1}{1 + \left(\sqrt{3}/3\right)} = \frac{\sqrt{3}-3}{\sqrt{3}+3}\cdot\frac{\sqrt{3}-3}{\sqrt{3}-3}$$
$$= \frac{12 - 6\sqrt{3}}{-6} = -2 + \sqrt{3}$$

23. $\sin 75° = \sin(30° + 45°)$

$$= \sin 30°\cos 45° + \sin 45°\cos 30°$$
$$= \frac{1}{2}\cdot\frac{\sqrt{2}}{2} + \frac{\sqrt{2}}{2}\cdot\frac{\sqrt{3}}{2} = \frac{\sqrt{2}}{4}\left(1+\sqrt{3}\right)$$

$$\cos 75° = \cos(30° + 45°)$$
$$= \cos 30°\cos 45° - \sin 30°\sin 45°$$
$$= \frac{\sqrt{3}}{2}\cdot\frac{\sqrt{2}}{2} - \frac{1}{2}\cdot\frac{\sqrt{2}}{2} = \frac{\sqrt{2}}{4}\left(\sqrt{3}-1\right)$$

$$\tan 75° = \tan(30° + 45°)$$
$$= \frac{\tan 30° + \tan 45°}{1 - \tan 30°\tan 45°}$$
$$= \frac{\left(\sqrt{3}/3\right) + 1}{1 - \left(\sqrt{3}/3\right)} = \frac{\sqrt{3}+3}{3-\sqrt{3}}\cdot\frac{3+\sqrt{3}}{3+\sqrt{3}}$$
$$= \frac{6\sqrt{3}+12}{6} = \sqrt{3} + 2$$

25. $\sin(-285°) = \sin(45° - 330°)$

$$= \sin 45°\cos 330° - \cos 45°\sin 330°$$
$$= \left(\frac{\sqrt{2}}{2}\right)\left(\frac{\sqrt{3}}{2}\right) - \left(\frac{\sqrt{2}}{2}\right)\left(-\frac{1}{2}\right)$$
$$= \frac{\sqrt{6}}{4} + \frac{\sqrt{2}}{4} = \frac{\sqrt{6}+\sqrt{2}}{4}$$

$$\cos(-285°) = \cos(45° - 330°)$$
$$= \cos 45°\cos 330° + \sin 45°\sin 330°$$
$$= \left(\frac{\sqrt{2}}{2}\right)\left(\frac{\sqrt{3}}{2}\right) + \left(\frac{\sqrt{2}}{2}\right)\left(-\frac{1}{2}\right)$$
$$= \frac{\sqrt{6}}{4} - \frac{\sqrt{2}}{4} = \frac{\sqrt{6}-\sqrt{2}}{4}$$

$$\tan(-285°) = \tan(45° - 330°)$$
$$= \frac{\tan 45° - \tan 330°}{1 + \tan 45°\tan 330°}$$
$$= \frac{1 - \left(\frac{-\sqrt{3}}{3}\right)}{1 + (1)\left(\frac{-\sqrt{3}}{3}\right)}$$
$$= \frac{\frac{3+\sqrt{3}}{3}}{\frac{3-\sqrt{3}}{3}}$$
$$= \frac{3+\sqrt{3}}{3-\sqrt{3}}\cdot\frac{3+\sqrt{3}}{3+\sqrt{3}}$$
$$= \frac{12+6\sqrt{3}}{6} = 2 + \sqrt{3}$$

27. $\dfrac{13\pi}{12} = \dfrac{3\pi}{4} + \dfrac{\pi}{3}$

$$\sin\frac{13\pi}{12} = \sin\left(\frac{3\pi}{4} + \frac{\pi}{3}\right) = \sin\frac{3\pi}{4}\cos\frac{\pi}{3} + \sin\frac{\pi}{3}\cos\frac{3\pi}{4}$$
$$= \frac{\sqrt{2}}{2}\cdot\frac{1}{2} + \frac{\sqrt{3}}{2}\left(-\frac{\sqrt{2}}{2}\right) = \frac{\sqrt{2}-\sqrt{6}}{4}$$

$$\cos\frac{13\pi}{12} = \cos\left(\frac{3\pi}{4} + \frac{\pi}{3}\right) = \cos\frac{3\pi}{4}\cos\frac{\pi}{3} - \sin\frac{3\pi}{4}\sin\frac{\pi}{3}$$
$$= \left(\frac{\sqrt{2}}{-2}\right)\left(\frac{1}{2}\right) - \left(\frac{\sqrt{2}}{2}\right)\left(\frac{\sqrt{3}}{2}\right) = -\frac{\sqrt{6}+\sqrt{2}}{4}$$

$$\tan\frac{13\pi}{12} = \tan\left(\frac{3\pi}{4} + \frac{\pi}{3}\right) = \frac{\tan(3\pi/4) + \tan(\pi/3)}{1 - \tan(3\pi/4)\tan(\pi/3)}$$
$$= \frac{(-1)+\sqrt{3}}{1-(-1)\sqrt{3}} = \frac{\sqrt{3}-1}{\sqrt{3}+1} = 2 - \sqrt{3}$$

29. $-\dfrac{7\pi}{12} = \dfrac{\pi}{6} - \dfrac{3\pi}{4}$

$$\sin\left(-\frac{7\pi}{12}\right) = \sin\left(\frac{\pi}{6} - \frac{3\pi}{4}\right) = \sin\frac{\pi}{6}\cos\frac{3\pi}{4} - \sin\frac{3\pi}{4}\cos\frac{\pi}{6}$$
$$= \frac{1}{2}\left(-\frac{\sqrt{2}}{2}\right) - \frac{\sqrt{2}}{2}\left(\frac{\sqrt{3}}{2}\right) = -\frac{\sqrt{2}+\sqrt{6}}{4}$$

$$\cos\left(-\frac{7\pi}{12}\right) = \cos\left(\frac{\pi}{6} - \frac{3\pi}{4}\right) = \cos\frac{\pi}{6}\cos\frac{3\pi}{4} + \sin\frac{\pi}{6}\sin\frac{3\pi}{4}$$
$$= \frac{\sqrt{3}}{2}\left(-\frac{\sqrt{2}}{2}\right) + \frac{1}{2}\cdot\frac{\sqrt{2}}{2} = \frac{\sqrt{2}-\sqrt{6}}{4}$$

$$\tan\left(-\frac{7\pi}{12}\right) = \tan\left(\frac{\pi}{6} - \frac{3\pi}{4}\right) = \frac{\tan(\pi/6) - \tan(3\pi/4)}{1 + \tan(\pi/6)\tan(3\pi/4)}$$
$$= \frac{(\sqrt{3}/3) - (-1)}{1 + (\sqrt{3}/3)(-1)} = \frac{3+\sqrt{3}}{3-\sqrt{3}} = 2 + \sqrt{3}$$

31. $\cos 60°\cos 10° - \sin 60°\sin 10° = \cos(60° + 10°)$

$$= \cos 70°$$

33. $\dfrac{\tan 325° - \tan 116°}{1 + \tan 325°\tan 116°} = \tan(325° - 116°)$

$$= \tan 209°$$

35. $\sin 3.5\cos 1.2 - \cos 3.5\sin 1.2 = \sin(3.5 - 1.2) = \sin 2.3$

37. $\cos\dfrac{\pi}{9}\cos\dfrac{\pi}{7} - \sin\dfrac{\pi}{9}\sin\dfrac{\pi}{7} = \cos\left(\dfrac{\pi}{9} + \dfrac{\pi}{7}\right) = \cos\left(\dfrac{16\pi}{63}\right)$

39. $\sin\dfrac{\pi}{12}\cos\dfrac{\pi}{4} + \cos\dfrac{\pi}{12}\sin\dfrac{\pi}{4} = \sin\left(\dfrac{\pi}{12} + \dfrac{\pi}{4}\right)$

$$= \sin\left(\frac{\pi}{3}\right)$$
$$= \frac{\sqrt{3}}{2}$$

41. $\sin 120°\cos 60° - \cos 120°\sin 60° = \sin(120° - 60°)$

$$= \sin(60°)$$
$$= \frac{\sqrt{3}}{2}$$

43. $\dfrac{\tan\dfrac{5\pi}{6} - \tan\dfrac{\pi}{6}}{1 + \tan\dfrac{5\pi}{6}\tan\dfrac{\pi}{6}} = \tan\left(\dfrac{5\pi}{6} - \dfrac{\pi}{6}\right)$

$$= \tan\left(\frac{2\pi}{3}\right)$$
$$= -\sqrt{3}$$

45. $y_1 = \sin\left(\dfrac{\pi}{6} + x\right)$

$$= \sin\frac{\pi}{6}\cos x + \sin x \cdot \cos\frac{\pi}{6}$$
$$= \frac{1}{2}\cos x + \frac{\sqrt{3}}{2}\sin x$$
$$= \frac{1}{2}\left(\cos x + \sqrt{3}\sin x\right) = y_2$$

x	0.2	0.4	0.6	0.8	1.0	1.2	1.4
y_1	0.6621	0.7978	0.9017	0.9696	0.9989	0.9883	0.9364
y_2	0.6621	0.7978	0.9017	0.9696	0.9989	0.9883	0.9384

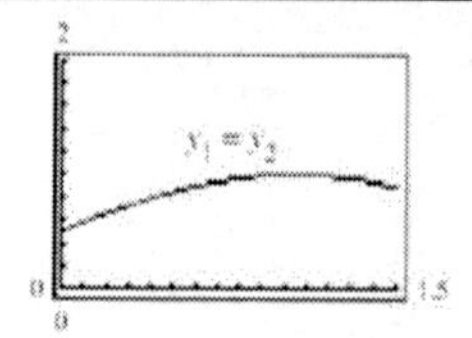

47. $y_1 = \cos(x + \pi)\cos(x - \pi)$

$$= (\cos x \cdot \cos\pi - \sin x \cdot \sin\pi)[\cos x\cos\pi + \sin x\sin\pi]$$
$$= [-\cos x][-\cos x] = \cos^2 x = y_2$$

x	0.2	0.4	0.6	0.8	1.0	1.2	1.4
y_1	0.9605	0.8484	0.6812	0.4854	0.2919	0.1313	0.0289
y_2	0.9605	0.8484	0.6812	0.4854	0.2919	0.1313	0.0289

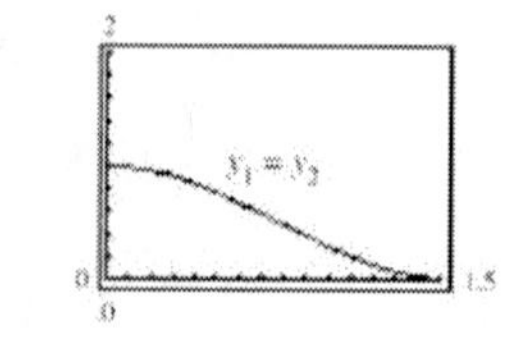

49. $\sin(u + v) = \sin u\cos v + \cos u\sin v$

$$= \left(\frac{5}{13}\right)\left(\frac{-4}{5}\right) + \left(\frac{-12}{13}\right)\left(\frac{3}{5}\right)$$
$$= \frac{-20}{65} - \frac{36}{65} = -\frac{56}{65}$$

51. $\tan(u + v) = \dfrac{\tan u + \tan v}{1 - \tan u\tan v}$

$$= \frac{\left(\frac{-5}{12}\right) + \left(-\frac{3}{4}\right)}{1 - \left(-\frac{5}{12}\right)\left(-\frac{3}{4}\right)}$$
$$= \frac{-\frac{7}{6}}{1 - \frac{15}{48}} = \frac{-\frac{7}{6}}{\frac{33}{48}}$$
$$= -\frac{7}{6}\cdot\frac{48}{33} = -\frac{56}{33}$$

53. $\cos(u+v) = \cos u \cos v - \sin u \sin v$

$$= \left(\frac{-15}{17}\right)\left(\frac{-3}{5}\right) - \left(\frac{-8}{17}\right)\left(\frac{-4}{5}\right)$$

$$= \frac{45}{85} - \frac{32}{85} = \frac{13}{85}$$

55. $\sin(v-u) = \sin v \cos u - \cos v \sin u$

$$= \left(\frac{-4}{5}\right)\left(\frac{-15}{17}\right) - \left(\frac{-3}{5}\right)\left(\frac{-8}{17}\right)$$

$$= \frac{60}{85} - \frac{24}{85} = \frac{36}{85}$$

57. $\sin(\arcsin x + \arccos x) = \sin(\arcsin x)\cos(\arccos x) + \sin(\arccos x)\cos(\arcsin x)$

$$= x \cdot x + \sqrt{1-x^2} \cdot \sqrt{1-x^2}$$

$$= x^2 + 1 - x^2$$

$$= 1$$

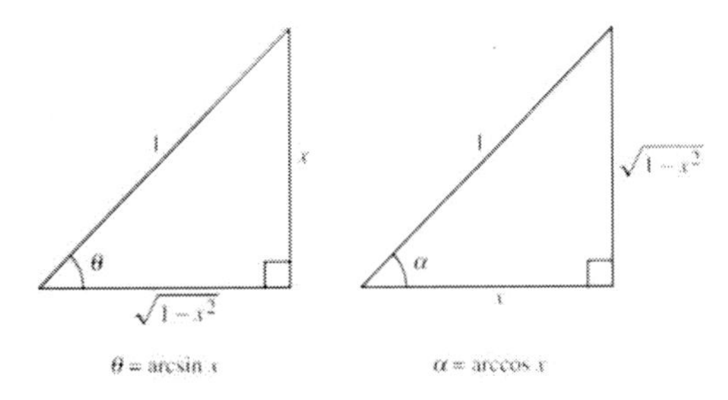

$\theta = \arcsin x$ $\alpha = \arccos x$

59. Let: $u = \arctan 2x$ and $v = \arccos x$

$\tan u = 2x$ $\cos v = x$

$$\sin(\arctan 2x - \arccos x) = \sin(u - v)$$

$$= \sin u \cos v - \cos u \sin v$$

$$= \frac{2x}{\sqrt{4x^2+1}}(x) - \frac{1}{\sqrt{4x^2+1}}\left(\sqrt{1-x^2}\right)$$

$$= \frac{2x^2 - \sqrt{1-x^2}}{\sqrt{4x^2+1}}$$

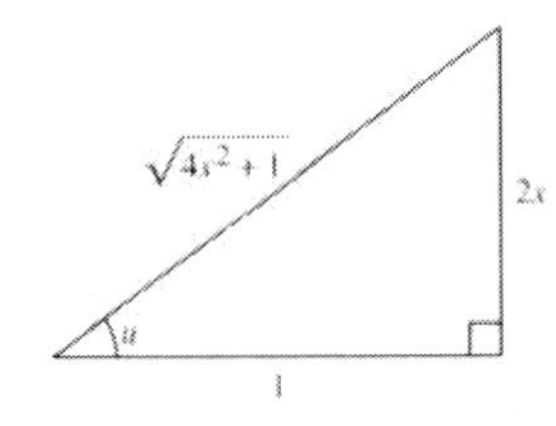

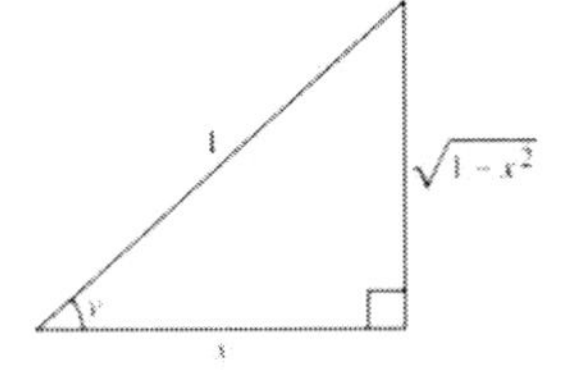

61. $\sin^{-1}(-1) = -\dfrac{\pi}{2}$

$$\sin\left(\frac{\pi}{2} + \sin^{-1}(-1)\right) = \sin\left(\frac{\pi}{2} - \frac{\pi}{2}\right) = \sin 0 = 0$$

63. $\sin^{-1} 1 = \dfrac{\pi}{2}$

$$\cos(\pi + \sin^{-1} 1) = \cos\left(\pi + \frac{\pi}{2}\right) = \cos\frac{3\pi}{2} = 0$$

65. $\sin^{-1} 1 = \dfrac{\pi}{2}$ because $\sin\dfrac{\pi}{2} = 1$.

$\cos^{-1} 1 = 0$ because $\cos 0 = 1$.

$$\sin(\sin^{-1} 1 + \cos^{-1} 1) = \sin\left(\frac{\pi}{2} + 0\right) = 1$$

67. $\sin(\sin^{-1} 0 - \cos^{-1} 0)$

$\sin^{-1} 0 = 0$ and $\cos^{-1} 0 = \dfrac{\pi}{2}$

$$\sin\left(0 - \frac{\pi}{2}\right) = \sin\left(-\frac{\pi}{2}\right) = -1$$

69. Let $\theta = \cos^{-1}\frac{3}{5} \Rightarrow \cos\theta = \frac{3}{5}$.

Let $\phi = \sin^{-1}\frac{5}{13} \Rightarrow \sin\phi = \frac{5}{13}$.

$$\sin\left(\cos^{-1}\tfrac{3}{5} - \sin^{-1}\tfrac{5}{13}\right) = \sin(\theta - \phi)$$

$$= \sin\theta\cos\phi - \cos\theta\sin\phi$$

$$= \left(\tfrac{4}{5}\right)\left(\tfrac{12}{13}\right) - \left(\tfrac{3}{5}\right)\left(\tfrac{5}{13}\right) = \tfrac{33}{65}$$

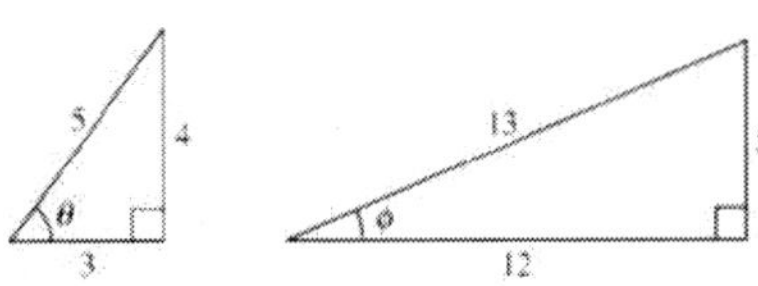

71. Let $\theta = \tan^{-1}\frac{3}{4} \Rightarrow \tan\theta = \frac{3}{4}$.

Let $\phi = \sin^{-1}\frac{3}{5} \Rightarrow \sin\phi = \frac{3}{5}$.

$$\sin\left(\tan^{-1}\tfrac{3}{4} + \sin^{-1}\tfrac{3}{5}\right) = \sin(\theta + \phi)$$

$$= \sin\theta\cos\phi + \sin\phi\cos\theta$$

$$= \tfrac{3}{5}\left(\tfrac{4}{5}\right) + \tfrac{3}{5}\left(\tfrac{4}{5}\right) = \tfrac{24}{25}$$

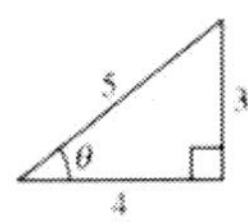

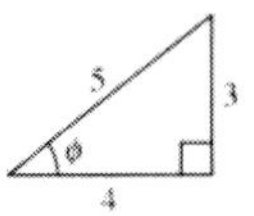

Note: $\theta = \phi$

73. $\sin\left(\dfrac{\pi}{2} + x\right) = \sin\dfrac{\pi}{2}\cos x + \sin x \cos\dfrac{\pi}{2}$

$$= (1)\cos x + 0 = \cos x$$

75. $\tan(x+\pi)-\tan(\pi-x)=\dfrac{\tan x+\tan\pi}{1-\tan x\cdot\tan\pi}-\dfrac{\tan\pi-\tan x}{1+\tan\pi\tan x}$

$=\dfrac{\tan x}{1}-\left(-\dfrac{\tan x}{1}\right)=2\tan x$

77. $\sin(x+y)+\sin(x-y)=\sin x\cos y+\sin y\cos x$

$+\sin x\cos y-\sin y\cos x=2\sin x\cos y$

79. $\cos(x+y)\cos(x-y)=[\cos x\cos y-\sin x\sin y]$

$[\cos x\cos y+\sin x\sin y]$

$=\cos^2 x\cos^2 y-\sin^2 x\sin^2 y=\cos^2 x(1-\sin^2 y)-\sin^2 x\sin^2 y$

$=\cos^2 x-\sin^2 y(\cos^2 x+\sin^2 x)=\cos^2 x-\sin^2 y$

81.
$$\sin\left(x+\frac{\pi}{3}\right)+\sin\left(x-\frac{\pi}{3}\right)=1$$
$$\sin x\cos\frac{\pi}{3}+\cos x\sin\frac{\pi}{3}+\sin x\cos\frac{\pi}{3}-\cos x\sin\frac{\pi}{3}=1$$
$$2\sin x(0.5)=1$$
$$\sin x=1$$
$$x=\frac{\pi}{2}$$

83.
$$\tan(x+\pi)+2\sin(x+\pi)=0$$
$$\frac{\tan x+\tan\pi}{1-\tan x\tan\pi}+2(\sin x\cos\pi+\cos x\sin\pi)=0$$
$$\frac{\tan x+0}{1-\tan x(0)}+2[\sin x(-1)+\cos x(0)]=0$$
$$\frac{\tan x}{1}-2\sin x=0$$
$$\frac{\sin x}{\cos x}=2\sin x$$
$$\sin x=2\sin x\cos x$$
$$\sin x(1-2\cos x)=0$$
$$\sin x=0 \quad\text{or}\quad \cos x=\frac{1}{2}$$
$$x=0,\ \pi \qquad x=\frac{\pi}{3},\ \frac{5\pi}{3}$$

85. Graph $y_1=\cos\left(x+\dfrac{\pi}{4}\right)+\cos\left(x-\dfrac{\pi}{4}\right)$ and $y_2=1$, and find the point(s) of intersection.

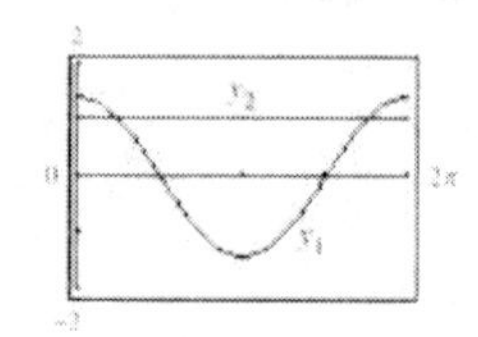

$x\approx 0.7854,\ 5.4978$

87. Graph $y_1=\sin\left(x+\dfrac{\pi}{2}\right)+\cos^2 x$ and approximate the zeros.

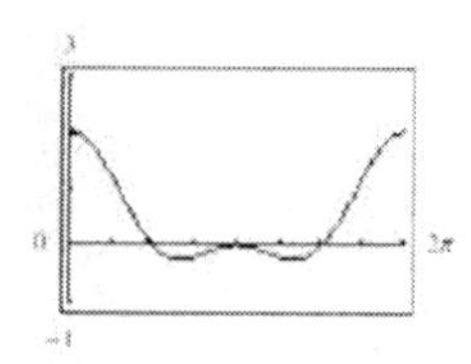

$x=\dfrac{\pi}{2},\ \pi,\ \dfrac{3\pi}{2}$

89. $y_1+y_2=A\cos 2\pi\left(\dfrac{t}{T}-\dfrac{x}{\lambda}\right)+A\cos 2\pi\left(\dfrac{t}{T}+\dfrac{x}{\lambda}\right)$

$=A\left[\cos\left(\dfrac{2\pi t}{T}\right)\cos\left(\dfrac{2\pi x}{\lambda}\right)+\sin\left(\dfrac{2\pi t}{T}\right)\sin\left(\dfrac{2\pi x}{\lambda}\right)\right]$

$+A\left[\cos\left(\dfrac{2\pi t}{T}\right)\cos\left(\dfrac{2\pi x}{\lambda}\right)-\sin\left(\dfrac{2\pi t}{T}\right)\sin\left(\dfrac{2\pi x}{\lambda}\right)\right]$

$=2A\cos\left(\dfrac{2\pi t}{T}\right)\cos\left(\dfrac{2\pi x}{\lambda}\right)$

91. False. $\cos(u\pm v)=\cos u\cos v\mp\sin u\sin v$

93. $\cos(n\pi+\theta)=\cos n\pi\cos\theta-\sin n\pi\sin\theta$

$=(-1)^n(\cos\theta)-(0)(\sin\theta)$

$=(-1)^n(\cos\theta)$, where n is an integer.

95. $C=\arctan\dfrac{b}{a}\Rightarrow\tan C=\dfrac{b}{a}\Rightarrow\ \sin C=\dfrac{b}{\sqrt{a^2+b^2}},$

$\cos C=\dfrac{a}{\sqrt{a^2+b^2}}$

$\sqrt{a^2+b^2}\sin(B\theta+C)$

$=\sqrt{a^2+b^2}\left(\sin B\theta\cdot\dfrac{a}{\sqrt{a^2+b^2}}+\dfrac{b}{\sqrt{a^2+b^2}}\cdot\cos B\theta\right)$

$=a\sin B\theta+b\cos B\theta$

97. $\sin\theta+\cos\theta$: $a=1,\ b=1,\ B=1$

(a) $C=\arctan\dfrac{b}{a}=\arctan 1=\dfrac{\pi}{4}$

$\sin\theta+\cos\theta=\sqrt{a^2+b^2}\sin(B\theta+C)$

$=\sqrt{2}\sin\left(\theta+\dfrac{\pi}{4}\right)$

(b) $C=\arctan\dfrac{a}{b}=\arctan 1=\dfrac{\pi}{4}$

$\sin\theta+\cos\theta=\sqrt{a^2+b^2}\cos(B\theta-C)$

$=\sqrt{2}\cos\left(\theta-\dfrac{\pi}{4}\right)$

99. $12\sin 3\theta + 5\cos 3\theta;\ a = 12,\ b = 5,\ B = 3$

(a) $C = \arctan\frac{b}{a} = \arctan\frac{5}{12} \approx 0.3948$

$$12\sin 3\theta + 5\cos 3\theta = \sqrt{a^2+b^2}\,\sin(B\theta + C)$$
$$\approx 13\sin(3\theta + 0.3948)$$

(b) $C = \arctan\frac{a}{b} = \arctan\frac{12}{5} \approx 1.1760$

$$12\sin 3\theta + 5\cos 3\theta = \sqrt{a^2+b^2}\,\cos(B\theta - C)$$
$$\approx 13\cos(3\theta - 1.1760)$$

101. $C = \arctan\frac{b}{a} = \frac{\pi}{2} \Rightarrow a = 0$

$\sqrt{a^2+b^2} = 2 \Rightarrow b = 2$

$B = 1$

$$2\sin\left(\theta + \frac{\pi}{2}\right) = (0)(\sin\theta) + (2)(\cos\theta) = 2\cos\theta$$

103.
$$\frac{\sin(x+h)-\sin x}{h} = \frac{\sin x\cos h + \cos x\sin h - \sin x}{h}$$
$$= \frac{\sin x(\cos h - 1) + \cos x\sin h}{h}$$
$$= \frac{\cos x\sin h}{h} - \frac{\sin x(1-\cos h)}{h}$$

105. From the figure, it appears that $u + v = w$. Assume that u, v, and w are all in Quadrant I. From the figure:

$$\tan u = \frac{s}{3s} = \frac{1}{3}$$
$$\tan v = \frac{s}{2s} = \frac{1}{2}$$
$$\tan w = \frac{s}{s} = 1$$

$$\tan(u+v) = \frac{\tan u + \tan v}{1 - \tan u\tan v} = \frac{(1/3)+(1/2)}{1-(1/3)(1/2)} = \frac{5/6}{1-(1/6)} = 1 = \tan w$$

Thus, $\tan(u+v) = \tan w$. Because u, v, and w are all in Quadrant I, we have $\arctan[\tan(u+v)] = \arctan[\tan w]$

$u + v = w$.

107. $\cos(u+v+w)$

$$= \cos[(u+v)+w]$$
$$= \cos(u+v)\cos w - \sin(u+v)\sin w$$
$$= (\cos u\cos v - \sin u\sin v)\cos w - (\sin u\cos v + \cos u\sin v)\sin w$$
$$= \cos u\cos v\cos w - \sin u\sin v\cos w - \sin u\cos v\sin w - \cos u\sin v\sin w$$

109. $x = 0:\ y = -\frac{1}{2}(0-10) + 14 = 5 + 14 = 19$

y-intercept: $(0, 19)$

$y = 0$: $0 = -\frac{1}{2}(x-10) + 14$

$= -\frac{1}{2}x + 19 \Rightarrow x = 38$

x-intercept: $(38, 0)$

111. $x = 0:\ |2(0) - 9| - 5 = 9 - 5 = 4$

y-intercept: $(0, 4)$

$y = 0$: $|2x - 9| = 5 \Rightarrow x = 7, 2$

x-intercepts: $(2, 0)$, $(7, 0)$

Section 5.5

1. $\dfrac{1+\cos 2u}{2}$

3. $-2\sin\left(\dfrac{u+v}{2}\right)\sin\left(\dfrac{u-v}{2}\right)$

5. $\tan\dfrac{u}{2}$

7. (a) $\sin 2u = 2\sin u\cos u$; matches (ii).

(b) $\cos 2u = 1-2\sin^2 u$; matches (i).

(c) $\tan 2u = \dfrac{2\tan u}{1-\tan^2 u}$; matches (iii).

9.

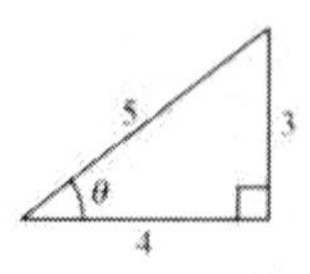

(a) $\sin\theta = \dfrac{3}{5}$

(b) $\cos\theta = \dfrac{4}{5}$

(c) $\tan\theta = \dfrac{3}{4}$

(d) $\sin 2\theta = 2\sin\theta\cos\theta$

$$= 2\left(\frac{3}{5}\right)\left(\frac{4}{5}\right)$$

$$= \frac{24}{25}$$

(e) $\cos 2\theta = \cos^2\theta - \sin^2\theta$

$$= \left(\frac{4}{5}\right)^2 - \left(\frac{3}{5}\right)^2$$

$$= \frac{7}{25}$$

(f) $\sec 2\theta = \dfrac{1}{\cos 2\theta}$

$$= \frac{25}{7}$$

(g) $\csc 2\theta = \dfrac{1}{\sin 2\theta}$

$$= \frac{25}{24}$$

(h) $\cot 2\theta = \dfrac{1}{\tan 2\theta}$

$$= \frac{1-\tan^2\theta}{2\tan\theta}$$

$$= \frac{1-\left(\frac{3}{4}\right)^2}{2\left(\frac{3}{4}\right)} = \frac{\frac{7}{16}}{\frac{3}{2}}$$

$$= \frac{7}{16}\cdot\frac{2}{3} = \frac{7}{24}$$

11. $\sin 2x - \sin x = 0$

Solutions: 0, 1.047, 3.142, 5.236

Analytically:

$$\sin 2x - \sin x = 0$$

$$2\sin x\cos x - \sin x = 0$$

$$\sin x(2\cos x - 1) = 0$$

$\sin x = 0$ or $2\cos x - 1 = 0$

$x = 0,\ \pi$ $\qquad \cos x = \dfrac{1}{2}$

$x = \dfrac{\pi}{3},\ \dfrac{5\pi}{3}$

$x = 0,\ \dfrac{\pi}{3}, \pi,\ \dfrac{5\pi}{3}$

13. $4\sin x\cos x = 1$

$x \approx 0.2618,\ 1.3090,\ 3.4034,\ 4.4506$

Analytically:

$4\sin x\cos x = 1$

$$2\sin(2x) = 1$$

$$\sin(2x) = \frac{1}{2}$$

$$2x = \frac{\pi}{6},\ \frac{5\pi}{6},\ \frac{13\pi}{6},\ \frac{17\pi}{6}$$

$$x = \frac{\pi}{12},\ \frac{5\pi}{12},\ \frac{13\pi}{12},\ \frac{17\pi}{12}$$

15. $\cos 2x - \cos x = 0$

$x \approx 0,\ 2.094,\ 4.189,$ (6.283 not in interval)

Analytically:

$$\cos 2x - \cos x = 0$$

$$2\cos^2 x - 1 - \cos x = 0$$

$$(2\cos x + 1)(\cos x - 1) = 0$$

$\cos x = -\dfrac{1}{2},\qquad \cos x = 1$

$x = \dfrac{2\pi}{3},\ \dfrac{4\pi}{3},\ 0,$ (2π not in interval)

17. $\sin 4x = -2\sin 2x$

$x \approx 0, 1.571, 3.142, 4.712$

$$\sin 4x = -2\sin 2x$$
$$\sin 4x + 2\sin 2x = 0$$
$$2\sin 2x\cos 2x + 2\sin 2x = 0$$
$$2\sin 2x(\cos 2x + 1) = 0$$

$2\sin 2x = 0$ or $\cos 2x + 1 = 0$

$\sin 2x = 0$ $\qquad \cos 2x = -1$

$2x = n\pi$ $\qquad 2x = \pi + 2n\pi$

$x = \dfrac{n}{2}\pi$ $\qquad x = \dfrac{\pi}{2} + n\pi$

$x = 0, \dfrac{\pi}{2}, \pi, \dfrac{3\pi}{2}$ $\qquad x = \dfrac{\pi}{2}, \dfrac{3\pi}{2}$

19. $\cos 2x + \sin x = 0$

$x \approx 1.5708, 3.6652, 5.7596$

Algebraically:

$$\cos 2x + \sin x = 0$$
$$1 - 2\sin^2 x + \sin x = 0$$
$$2\sin^2 x - \sin x - 1 = 0$$

$(2\sin x + 1)(\sin x - 1) = 0$

$\sin x = -\dfrac{1}{2} \Rightarrow x = \dfrac{7\pi}{6}, \dfrac{11\pi}{6}$

$\sin x = 1 \quad \Rightarrow x = \dfrac{\pi}{2}$

21. $\sin u = \dfrac{3}{5},\ 0 < u < \dfrac{\pi}{2} \Rightarrow \cos u = \dfrac{4}{5}$

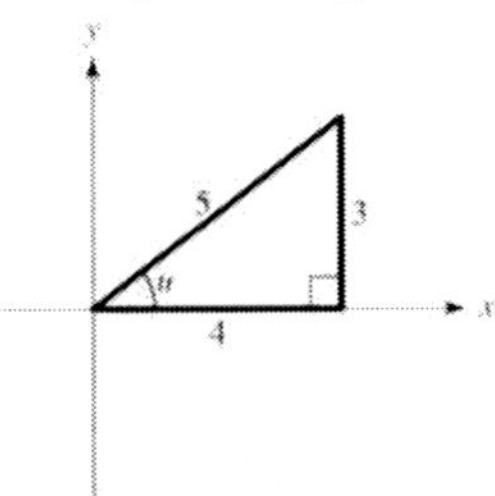

$\sin 2u = 2\sin u\cos u = 2\cdot\dfrac{3}{5}\cdot\dfrac{4}{5} = \dfrac{24}{25}$

$\cos 2u = \cos^2 u - \sin^2 u = \dfrac{16}{25} - \dfrac{9}{25} = \dfrac{7}{25}$

$\tan 2u = \dfrac{2\tan u}{1 - \tan^2 u} = \dfrac{2(3/4)}{1-(9/16)} = \dfrac{24}{7}$

23. $\tan u = \dfrac{1}{2},\ \pi < u < \dfrac{3\pi}{2} \Rightarrow \sin u = -\dfrac{1}{\sqrt{5}}$ and

$\cos u = -\dfrac{2}{\sqrt{5}}$

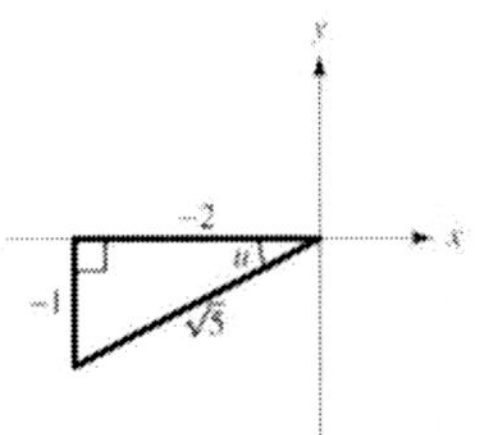

$\sin 2u = 2\sin u\cos u = 2\left(-\dfrac{1}{\sqrt{5}}\right)\left(-\dfrac{2}{\sqrt{5}}\right) = \dfrac{4}{5}$

$\cos 2u = \cos^2 u - \sin^2 u = \left(-\dfrac{2}{\sqrt{5}}\right)^2 - \left(-\dfrac{1}{\sqrt{5}}\right)^2 = \dfrac{3}{5}$

$\tan 2u = \dfrac{2\tan u}{1-\tan^2 u} = \dfrac{2(1/2)}{1-(1/4)} = \dfrac{4}{3}$

25. $\sec u = -2,\ \dfrac{\pi}{2} < u < \pi$

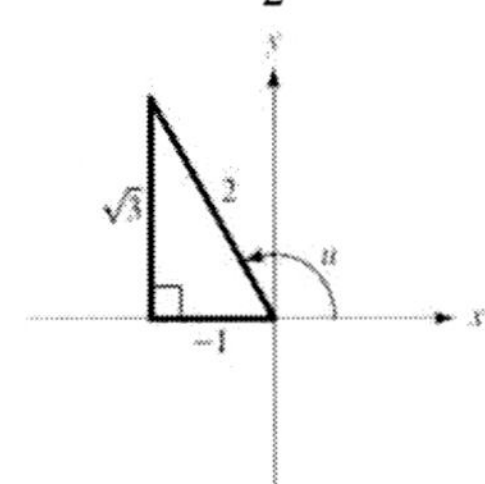

$$\sin 2u = 2\sin u\cos u$$
$$= 2\left(\frac{\sqrt{3}}{2}\right)\left(-\frac{1}{2}\right)$$
$$= -\frac{\sqrt{3}}{2}$$

$$\cos 2u = \cos^2 u - \sin^2 u$$
$$= \left(-\frac{1}{2}\right)^2 - \left(\frac{\sqrt{3}}{2}\right)^2$$
$$= \frac{1}{4} - \frac{3}{4} = -\frac{1}{2}$$

$$\tan 2u = \frac{2\tan u}{1-\tan^2 u}$$
$$= \frac{2\left(-\frac{\sqrt{3}}{1}\right)}{1-\left(-\frac{\sqrt{3}}{1}\right)^2}$$
$$= \frac{-2\sqrt{3}}{1-3}$$
$$= \sqrt{3}$$

27. $8\sin x\cos x = 4(2\sin x\cos x) = 4\sin 2x$

29. $$4 - 8\sin^2 x = 4\left(1 - 2\sin^2 x\right) = 4\cos 2x$$

31. $$\begin{aligned}\cos^4 x &= \left(\cos^2 x\right)\left(\cos^2 x\right)\\ &= \left(\frac{1+\cos 2x}{2}\right)\left(\frac{1+\cos 2x}{2}\right)\\ &= \frac{1+2\cos 2x+\cos^2 2x}{4}\\ &= \frac{1+2\cos 2x+\left(1+\cos 4x\right)/2}{4}\\ &= \frac{2+4\cos 2x+1+\cos 4x}{8}\\ &= \frac{3+4\cos 2x+\cos 4x}{8}\\ &= \frac{1}{8}(3+4\cos 2x+\cos 4x)\end{aligned}$$

33. $$\begin{aligned}\left(\sin^2 x\right)\left(\cos^2 x\right) &= \left(\frac{1-\cos 2x}{2}\right)\left(\frac{1+\cos 2x}{2}\right)\\ &= \frac{1-\cos^2 2x}{4}\\ &= \frac{1}{4}\left(1-\frac{1+\cos 4x}{2}\right)\\ &= \frac{1}{8}(2-1-\cos 4x)\\ &= \frac{1}{8}(1-\cos 4x)\end{aligned}$$

35. $$\begin{aligned}\sin^2 x\cos^4 x &= \sin^2 x\cos^2 x\cos^2 x\\ &= \left(\frac{1-\cos 2x}{2}\right)\left(\frac{1+\cos 2x}{2}\right)\left(\frac{1+\cos 2x}{2}\right)\\ &= \frac{1}{8}(1-\cos 2x)(1+\cos 2x)(1+\cos 2x)\\ &= \frac{1}{8}\left(1-\cos^2 2x\right)(1+\cos 2x)\\ &= \frac{1}{8}\left(1+\cos 2x-\cos^2 2x-\cos^3 2x\right)\\ &= \frac{1}{8}\left[1+\cos 2x-\left(\frac{1+\cos 4x}{2}\right)-\cos 2x\left(\frac{1+\cos 4x}{2}\right)\right]\\ &= \frac{1}{16}\left[2+2\cos 2x-1-\cos 4x-\cos 2x-\cos 2x\cos 4x\right]\\ &= \frac{1}{16}\left[1+\cos 2x-\cos 4x-\left(\frac{1}{2}\cos 2x+\frac{1}{2}\cos 6x\right)\right]\\ &= \frac{1}{32}(2+2\cos 2x-2\cos 4x-\cos 2x-\cos 6x)\\ &= \frac{1}{32}(2+\cos 2x-2\cos 4x-\cos 6x)\end{aligned}$$

37. $$\sin^2 2x = \frac{1-\cos 4x}{2} = \frac{1}{2}-\frac{1}{2}\cos 4x = \frac{1}{2}(1-\cos 4x)$$

39. $$\cos^2\frac{x}{2} = \frac{1+\cos x}{2} = \frac{1}{2}+\frac{1}{2}\cos x = \frac{1}{2}(1+\cos x)$$

41. $$\tan^2 2x = \frac{1-\cos 4x}{1+\cos 4x}$$

43. $$\begin{aligned}\sin^4\frac{x}{2} &= \left(\sin^2\frac{x}{2}\right)\left(\sin^2\frac{x}{2}\right)\\ &= \left(\frac{1-\cos x}{2}\right)\left(\frac{1-\cos x}{2}\right)\\ &= \frac{1}{4}\left(1-2\cos x+\cos^2 x\right)\\ &= \frac{1}{4}\left(1-2\cos x+\frac{1+\cos 2x}{2}\right)\\ &= \frac{1}{8}(2-4\cos x+1+\cos 2x)\\ &= \frac{1}{8}(3-4\cos x+\cos 2x)\end{aligned}$$

45. (a) $\cos\frac{\theta}{2} = \sqrt{\frac{1+\cos\theta}{2}} = \sqrt{\frac{1+(15/17)}{2}} = \sqrt{\frac{16}{17}} = \frac{4}{\sqrt{17}} = \frac{4\sqrt{17}}{17}$

(b) $\sin\frac{\theta}{2} = \sqrt{\frac{1-\cos\theta}{2}} = \sqrt{\frac{1-(15/17)}{2}} = \frac{\sqrt{17}}{17}$

(c) $\tan\frac{\theta}{2} = \frac{\sin\theta}{1+\cos\theta} = \frac{8/17}{1+(15/17)} = \frac{8}{12} = \frac{1}{4}$

(d) $\sec\frac{\theta}{2} = \frac{1}{\cos(\theta/2)} = \frac{1}{\sqrt{(1+\cos\theta)/2}} = \frac{\sqrt{2}}{\sqrt{1+(15/17)}} = \frac{\sqrt{17}}{4}$

(e) $\csc\frac{\theta}{2} = \frac{1}{\sin(\theta/2)} = \frac{1}{\sqrt{(1-\cos\theta)/2}} = \frac{1}{\sqrt{1-(15/17)/2}} = \frac{1}{1/\sqrt{17}} = \sqrt{17}$

(f) $\cot\frac{\theta}{2} = \frac{1+\cos\theta}{\sin\theta} = \frac{1+(15/17)}{8/17} = 4$

(g) $2\sin\frac{\theta}{2}\cos\frac{\theta}{2} = 2\left(\frac{1}{\sqrt{17}}\right)\left(\frac{4\sqrt{17}}{17}\right) = \frac{8}{17}, \ (=\sin\theta)$

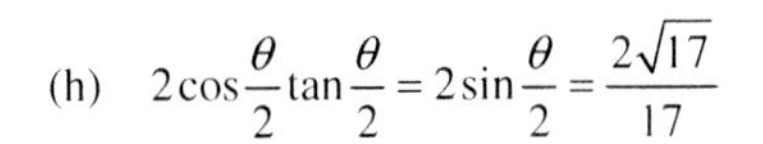

(h) $2\cos\frac{\theta}{2}\tan\frac{\theta}{2} = 2\sin\frac{\theta}{2} = \frac{2\sqrt{17}}{17}$

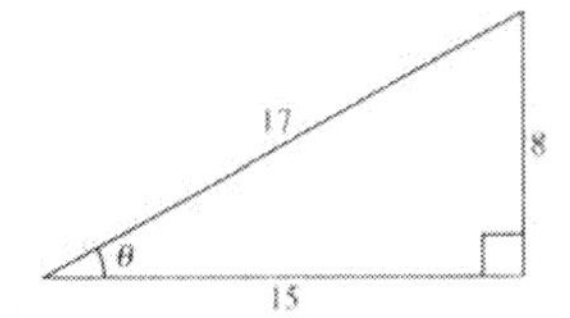

$\sin\theta = \frac{8}{17}$

$\cos\theta = \frac{15}{17}$

47. $\sin 75° = \sin\left(\frac{150}{2}\right)° = \sqrt{\frac{1-\cos 150°}{2}}$

$= \sqrt{\frac{1+\frac{\sqrt{3}}{2}}{2}}$

$= \sqrt{\frac{\frac{2+\sqrt{3}}{2}}{2}}$

$= \frac{\sqrt{2+\sqrt{3}}}{2}$

$\cos 75° = \cos\left(\frac{150}{2}\right)° = \sqrt{\frac{1+\cos 150°}{2}}$

$= \sqrt{\frac{1-\frac{\sqrt{3}}{2}}{2}}$

$= \sqrt{\frac{\frac{2-\sqrt{3}}{2}}{2}} = \frac{\sqrt{2-\sqrt{3}}}{2}$

$\tan 75° = \tan\left(\frac{150}{2}\right)° = \frac{1-\cos 150°}{\sin 150°}$

$= \frac{1+\frac{\sqrt{3}}{2}}{\frac{1}{2}}$

$= \frac{\frac{2+\sqrt{3}}{2}}{\frac{1}{2}} = 2+\sqrt{3}$

49. $\sin 67°30' = \sin\left(\frac{135}{2}\right)° = \sqrt{\frac{1-\cos 135°}{2}}$

$= \sqrt{\frac{1+\frac{\sqrt{2}}{2}}{2}}$

$= \sqrt{\frac{\frac{2+\sqrt{2}}{2}}{2}} = \frac{\sqrt{2+\sqrt{2}}}{2}$

$\cos 67°30' = \cos\left(\frac{135}{2}\right)° = \sqrt{\frac{1+\cos 135°}{2}}$

$= \sqrt{\frac{1-\frac{\sqrt{2}}{2}}{2}}$

$= \sqrt{\frac{\frac{2-\sqrt{2}}{2}}{2}}$

$= \frac{\sqrt{2-\sqrt{2}}}{2}$

$\tan 67°30' = \tan\left(\frac{135}{2}\right)°$

$= \frac{1-\cos 135°}{\sin 135°}$

$= \frac{1+\frac{\sqrt{2}}{2}}{\frac{\sqrt{2}}{2}}$

$= \frac{\frac{2+\sqrt{2}}{2}}{\frac{\sqrt{2}}{2}}$

$= \frac{2+\sqrt{2}}{\sqrt{2}}$

$= \frac{2\sqrt{2}+2}{2} = \sqrt{2}+1$

51. $\sin\frac{\pi}{8} = \sin\left[\frac{1}{2}\left(\frac{\pi}{4}\right)\right] = \sqrt{\frac{1-\cos \pi/4}{2}} = \frac{1}{2}\sqrt{2-\sqrt{2}}$

$\cos\frac{\pi}{8} = \cos\left[\frac{1}{2}\left(\frac{\pi}{4}\right)\right] = \sqrt{\frac{1+\cos(\pi/4)}{2}} = \frac{1}{2}\sqrt{2+\sqrt{2}}$

$\tan\frac{\pi}{8} = \tan\left[\frac{1}{2}\left(\frac{\pi}{4}\right)\right] = \frac{\sin(\pi/4)}{1+\cos(\pi/4)} = \frac{\sqrt{2}/2}{1+\left(\sqrt{2}/2\right)} = \sqrt{2}-1$

53. $\sin\frac{3\pi}{8} = \sin\left(\frac{1}{2}\cdot\frac{3\pi}{4}\right) = \sqrt{\frac{1-\cos(3\pi/4)}{2}} = \sqrt{\frac{1+\left(\sqrt{2}/2\right)}{2}}$

$= \frac{1}{2}\sqrt{2+\sqrt{2}}$

$\cos\frac{3\pi}{8} = \cos\left(\frac{1}{2}\cdot\frac{3\pi}{4}\right) = \sqrt{\frac{1+\cos(3\pi/4)}{2}} = \sqrt{\frac{1-\left(\sqrt{2}/2\right)}{2}}$

$= \frac{1}{2}\sqrt{2-\sqrt{2}}$

$\tan\frac{3\pi}{8} = \tan\left(\frac{1}{2}\cdot\frac{3\pi}{4}\right) = \frac{\sin(3\pi/4)}{1+\cos(3\pi/4)} = \frac{\sqrt{2}/2}{1-\left(\sqrt{2}/2\right)} = \frac{\sqrt{2}}{2-\sqrt{2}}$

$= \sqrt{2}+1$

55. $\sin u = \frac{5}{13}, \frac{\pi}{2} < u < \pi \Rightarrow \cos u = -\frac{12}{13}$

$$\sin\left(\frac{u}{2}\right) = \sqrt{\frac{1-\cos u}{2}} = \sqrt{\frac{1+(12/13)}{2}} = \frac{5\sqrt{26}}{26}$$

$$\cos\left(\frac{u}{2}\right) = \sqrt{\frac{1+\cos u}{2}} = \sqrt{\frac{1-(12/13)}{2}} = \frac{\sqrt{26}}{26}$$

$$\tan\left(\frac{u}{2}\right) = \frac{\sin u}{1+\cos u} = \frac{5/13}{1-(12/13)} = \frac{5}{1} = 5$$

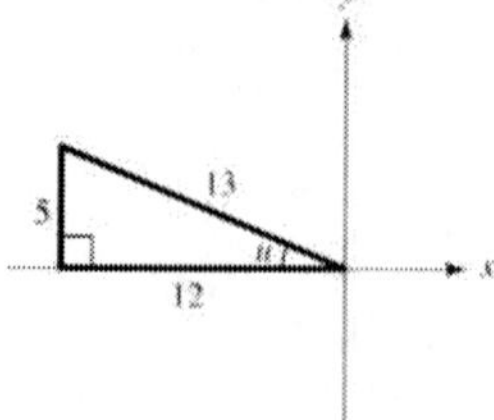

57. $\tan u = -\frac{8}{5}, \frac{3\pi}{2} < u < 2\pi$

$$\sin u = -\frac{8}{\sqrt{89}}, \cos u = \frac{5}{\sqrt{89}}$$

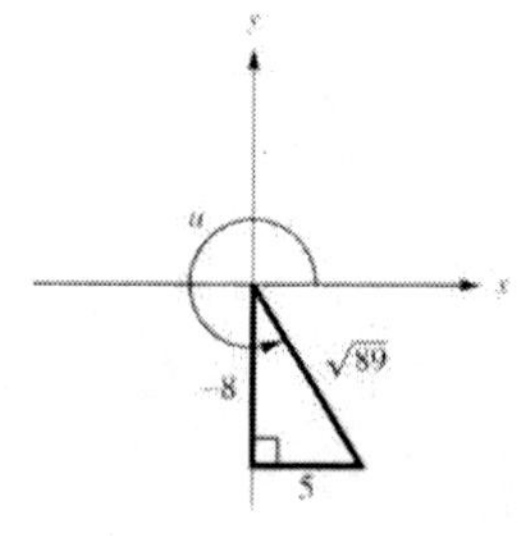

$$\sin\left(\frac{u}{2}\right) = \sqrt{\frac{1-\cos u}{2}} = \sqrt{\frac{1-\left(5/\sqrt{89}\right)}{2}} = \sqrt{\frac{\sqrt{89}-5}{2\sqrt{89}}}$$

$$= \sqrt{\frac{89-5\sqrt{89}}{178}}$$

$$\cos\left(\frac{u}{2}\right) = -\sqrt{\frac{1+\cos u}{2}} = -\sqrt{\frac{1+\left(5/\sqrt{89}\right)}{2}} = -\sqrt{\frac{\sqrt{89}+5}{2\sqrt{89}}}$$

$$= -\sqrt{\frac{89+5\sqrt{89}}{178}}$$

$$\tan\left(\frac{u}{2}\right) = \frac{1-\cos u}{\sin u} = \frac{1-\left(5/\sqrt{89}\right)}{-8/\sqrt{89}} = \frac{5-\sqrt{89}}{8}$$

59. $\csc u = -\frac{5}{3}, \pi < u < \frac{3\pi}{2}$

$$\sin u = -\frac{3}{5}, \cos u = -\frac{4}{5}$$

$$\sin\left(\frac{u}{2}\right) = \sqrt{\frac{1-\cos u}{2}} = \sqrt{\frac{1+(4/5)}{2}} = \frac{3}{\sqrt{10}} = \frac{3\sqrt{10}}{10}$$

$$\cos\left(\frac{u}{2}\right) = -\sqrt{\frac{1+\cos u}{2}} = -\sqrt{\frac{1-(4/5)}{2}} = \frac{-1}{\sqrt{10}} = -\frac{\sqrt{10}}{10}$$

$$\tan\left(\frac{u}{2}\right) = \frac{1-\cos u}{\sin u} = \frac{1+(4/5)}{-3/5} = -3$$

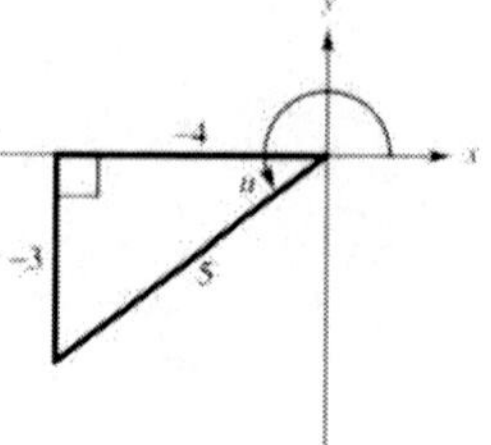

61. $\sqrt{\frac{1-\cos 6x}{2}} = |\sin 3x|$

63. $-\sqrt{\frac{1-\cos 8x}{1+\cos 8x}} = -\frac{\sqrt{(1-\cos 8x)/2}}{\sqrt{(1+\cos 8x)/2}}$

$$= -\left|\frac{\sin 4x}{\cos 4x}\right| = -|\tan 4x|$$

65.

$$\sin\frac{x}{2} + \cos x = 0$$

$$\sin\frac{x}{2} + \cos 2\left(\frac{x}{2}\right) = 0$$

$$\sin\frac{x}{2} + 1 - 2\sin^2\frac{x}{2} = 0$$

$$2\sin^2\frac{x}{2} - \sin\frac{x}{2} - 1 = 0$$

$$\left(2\sin\frac{x}{2} + 1\right)\left(\sin\frac{x}{2} - 1\right) = 0$$

$2\sin\frac{x}{2} + 1 = 0$ $\sin\frac{x}{2} - 1 = 0$

$\sin\frac{x}{2} = \frac{-1}{2}$ $\sin\frac{x}{2} = 1$

$\frac{x}{2} = \frac{7\pi}{6}, \frac{11\pi}{6}$ $\frac{x}{2} = \frac{\pi}{2}$

$x = \frac{7\pi}{3}, \frac{11\pi}{6}$ $x = \pi$

Only $x = \pi$ is in the interval $[0, 2\pi)$.

67.
$$\cos\frac{x}{2} - \sin x = 0$$
$$\pm\sqrt{\frac{1+\cos x}{2}} = \sin x$$
$$\frac{1+\cos x}{2} = \sin^2 x$$
$$1+\cos x = 2\sin^2 x$$
$$1+\cos x = 2 - 2\cos^2 x$$
$$2\cos^2 x + \cos x - 1 = 0$$
$$(2\cos x - 1)(\cos x + 1) = 0$$
$2\cos x - 1 = 0$ or $\cos x + 1 = 0$

$\cos x = \frac{1}{2}$, $x = \frac{\pi}{3}, \frac{5\pi}{3}$

$\cos x = -1$, $x = \pi$

69. $6\sin\frac{\pi}{3}\cos\frac{\pi}{3} = 6\cdot\frac{1}{2}\left[\sin\left(\frac{\pi}{3}+\frac{\pi}{3}\right) + \sin\left(\frac{\pi}{3}-\frac{\pi}{3}\right)\right]$
$$= 3\left(\sin\frac{2\pi}{3} + \sin 0\right) = 3\sin\frac{2\pi}{3}$$

71. $\sin 5\theta\cos 3\theta = \frac{1}{2}\left[\sin(5\theta+3\theta) + \sin(5\theta-3\theta)\right]$
$$= \frac{1}{2}(\sin 8\theta + \sin 2\theta)$$

73. $10\cos 75^\circ\cos 15^\circ = 5\left[\cos(75^\circ - 15^\circ) + \cos(75^\circ + 15^\circ)\right]$
$$= 5(\cos 60^\circ + \cos 90^\circ)$$

75. $\sin(x+y)\sin(x-y)$
$$= \frac{1}{2}\left[\cos((x+y)-(x-y)) - \cos((x+y)+(x-y))\right]$$
$$= \frac{1}{2}(\cos 2y - \cos 2x)$$

77. $\sin 5\theta - \sin\theta = 2\cos\left(\frac{5\theta+\theta}{2}\right)\sin\left(\frac{5\theta-\theta}{2}\right)$
$$= 2\cos 3\theta\sin 2\theta$$

79. $\cos 6x + \cos 2x = 2\cos\left(\frac{6x+2x}{2}\right)\cos\left(\frac{6x-2x}{2}\right)$
$$= 2\cos 4x\cos 2x$$

81. $\sin(\alpha+\beta) - \sin(\alpha-\beta)$
$$= 2\cos\left(\frac{\alpha+\beta+\alpha-\beta}{2}\right)\sin\left(\frac{\alpha+\beta-\alpha+\beta}{2}\right)$$
$$= 2\cos\alpha\sin\beta$$

83. $\cos\left(\theta+\frac{\pi}{2}\right) - \cos\left(\theta-\frac{\pi}{2}\right)$
$$= -2\sin\left(\frac{\theta+(\pi/2)+\theta-(\pi/2)}{2}\right)\sin\left(\frac{\theta+(\pi/2)-\theta+(\pi/2)}{2}\right)$$
$$= -2\sin\theta\sin\frac{\pi}{2} = -2\sin\theta$$

85. $\sin 195^\circ + \sin 105^\circ = 2\sin\left(\frac{195^\circ+105^\circ}{2}\right)\cos\left(\frac{195^\circ-105^\circ}{2}\right)$
$$= 2\sin(150^\circ)\cos(45^\circ)$$
$$= 2\left(\frac{1}{2}\right)\left(\frac{\sqrt{2}}{2}\right) = \frac{\sqrt{2}}{2}$$

87. $\cos\frac{5\pi}{12} + \cos\frac{\pi}{12}$
$$= 2\cos\left(\frac{(5\pi/12)+(\pi/12)}{2}\right)\cos\left(\frac{(5\pi/12)-(\pi/12)}{2}\right)$$
$$= 2\cos\left(\frac{\pi}{4}\right)\cos\left(\frac{\pi}{6}\right) = 2\left(\frac{\sqrt{2}}{2}\right)\left(\frac{\sqrt{3}}{2}\right) = \frac{2\sqrt{6}}{4} = \frac{\sqrt{6}}{2}$$

89. $\sin 6x + \sin 2x = 0$
$$2\sin\left(\frac{6x+2x}{2}\right)\cos\left(\frac{6x-2x}{2}\right) = 0$$
$$\sin 4x\cos 2x = 0$$
$\sin 4x = 0$ or $\cos 2x = 0$

$4x = n\pi$, $x = \frac{n\pi}{4}$

$2x = \frac{\pi}{2} + n\pi$, $x = \frac{\pi}{4} + \frac{n\pi}{2}$

$$x = 0, \frac{\pi}{4}, \frac{\pi}{2}, \frac{3\pi}{4}, \pi, \frac{5\pi}{4}, \frac{3\pi}{2}, \frac{7\pi}{4}$$

91.
$$\frac{\cos 2x}{\sin 3x - \sin x} - 1 = 0$$
$$\frac{\cos 2x}{\sin 3x - \sin x} = 1$$
$$\frac{\cos 2x}{2\cos 2x\sin x} = 1$$
$$2\sin x = 1$$
$$\sin x = \frac{1}{2}$$
$$x = \frac{\pi}{6}, \frac{5\pi}{6}$$

Figure for Exercises 93–96.

93. $\sin^2 \alpha = \left(\frac{5}{13}\right)^2 = \frac{25}{169}$

$\sin^2 \alpha = 1 - \cos^2 \alpha = 1 - \left(\frac{12}{13}\right)^2 = 1 - \frac{144}{169} = \frac{25}{169}$

95. $\sin \alpha \cos \beta = \left(\frac{5}{13}\right)\left(\frac{4}{5}\right) = \frac{4}{13}$

$$\sin \alpha \cos \beta = \cos\left(\frac{\pi}{2} - \alpha\right)\sin\left(\frac{\pi}{2} - \beta\right) = \left(\frac{5}{13}\right)\left(\frac{4}{5}\right) = \frac{4}{13}$$

97.
$$\csc 2\theta = \frac{1}{\sin 2\theta} = \frac{1}{2\sin\theta\cos\theta} = \frac{1}{\sin\theta}\cdot\frac{1}{2\cos\theta} = \frac{\csc\theta}{2\cos\theta}$$

99.
$$\cos^2 2\alpha - \sin^2 2\alpha = \cos\left[2(2\alpha)\right] = \cos 4\alpha$$

101.
$$(\sin x + \cos x)^2 = \sin^2 x + 2\sin x\cos x + \cos^2 x = (\sin^2 x + \cos^2 x) + 2\sin x\cos x = 1 + \sin 2x$$

103.
$$1 + \cos 10y = 1 + \cos^2 5y - \sin^2 5y = 1 + \cos^2 5y - (1 - \cos^2 5y) = 2\cos^2 5y$$

105.
$$\sec\frac{u}{2} = \frac{1}{\cos(u/2)} = \pm\sqrt{\frac{2}{1+\cos u}} = \pm\sqrt{\frac{2\sin u}{\sin u(1+\cos u)}} = \pm\sqrt{\frac{2\sin u}{\sin u + \sin u\cos u}} = \pm\sqrt{\frac{(2\sin u)/(\cos u)}{(\sin u)/(\cos u) + (\sin u\cos u)/(\cos u)}} = \pm\sqrt{\frac{2\tan u}{\tan u + \sin u}}$$

107.
$$\begin{aligned}\cos 3\beta &= \cos(2\beta + \beta)\\ &= \cos 2\beta\cos\beta - \sin 2\beta\sin\beta\\ &= (\cos^2\beta - \sin^2\beta)\cos\beta - 2\sin\beta\cos\beta\sin\beta\\ &= \cos^3\beta - \sin^2\beta\cos\beta - 2\sin^2\beta\cos\beta\\ &= \cos^3\beta - 3\sin^2\beta\cos\beta\end{aligned}$$

109. Case 1:

$$\frac{\sin x + \sin y}{\cos x + \cos y} = \frac{2\sin\left(\frac{x+y}{2}\right)\cos\left(\frac{x-y}{2}\right)}{2\cos\left(\frac{x+y}{2}\right)\cos\left(\frac{x-y}{2}\right)} = \frac{\sin\left(\frac{x+y}{2}\right)}{\cos\left(\frac{x+y}{2}\right)} = \tan\left(\frac{x+y}{2}\right)$$

Case 2:

$$\frac{\sin x - \sin y}{\cos x + \cos y} = \frac{2\cos\left(\frac{x+y}{2}\right)\sin\left(\frac{x-y}{2}\right)}{2\cos\left(\frac{x+y}{2}\right)\cos\left(\frac{x-y}{2}\right)} = \frac{\sin\left(\frac{x-y}{2}\right)}{\cos\left(\frac{x-y}{2}\right)} = \tan\left(\frac{x-y}{2}\right)$$

111. $\sin^2 x = \frac{1 - \cos 2x}{2} = \frac{1}{2} - \frac{\cos 2x}{2}$

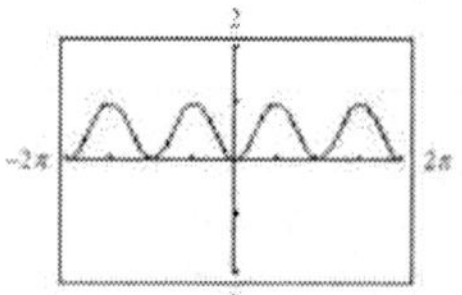

113. $f(x) = \cos^4 x = \frac{1}{8}(3 + 4\cos 2x + \cos 4x)$

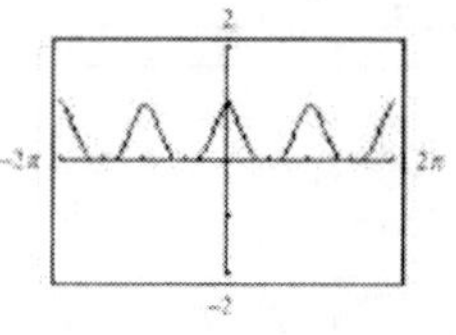

115. Let $\theta = \arcsin x$.

$$\sin(2\arcsin x) = 2\sin(\arcsin x)\cos(\arcsin x) = 2x\sqrt{1 - x^2}$$

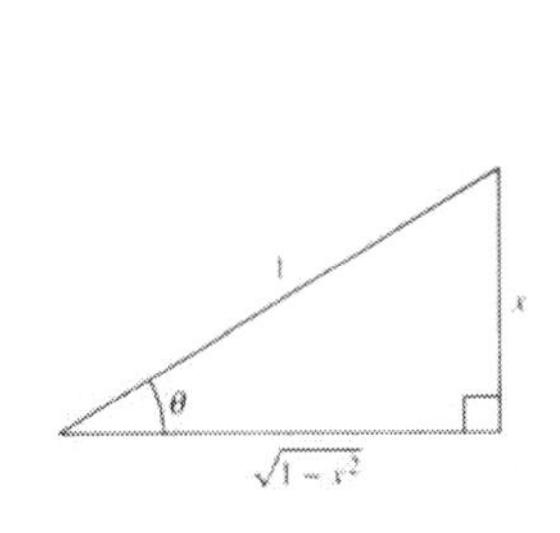

Figure for 115 and 117.

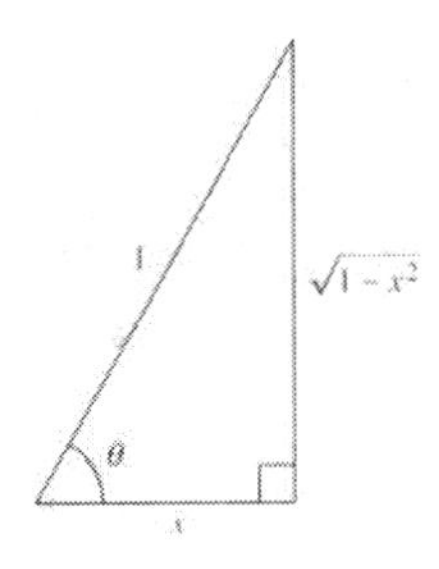

Figure for 116 and 118.

117. Let $\theta = \arcsin x$.

$$\cos(2\arcsin x) = 1 - 2\sin^2(\arcsin x)$$
$$= 1 - 2x^2$$

119. Let $\theta = \arctan x$.

$$\cos(2\arctan x) = 1 - 2\sin^2(\arctan x)$$
$$= 1 - 2\left(\frac{x}{\sqrt{1+x^2}}\right)^2$$
$$= 1 - \frac{2x^2}{1+x^2}$$
$$= \frac{1-x^2}{1+x^2}$$

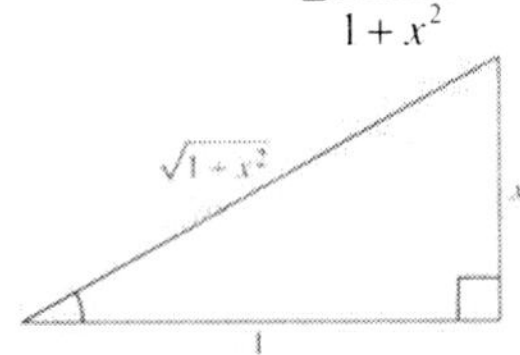

Figure for 119 and 120.

121. (a) $f(x) = \sin 2x - \sin x = 0$

$$2\sin x\cos x - \sin x = 0$$
$$\sin x(2\cos x - 1) = 0$$
$$\sin x = 0 \Rightarrow x = 0,\ \pi,\ 2\pi$$
$$\cos x = \frac{1}{2} \Rightarrow x = \frac{\pi}{3},\ \frac{5\pi}{3}$$

(b)
$$2\cos 2x - \cos x = 0$$
$$2(2\cos^2 x - 1) - \cos x = 0$$
$$4\cos^2 x - \cos x - 2 = 0$$
$$\cos x = \frac{1 \pm \sqrt{1+32}}{8} = \frac{1 \pm \sqrt{33}}{8}$$
$$x = \arccos\left(\frac{1 \pm \sqrt{33}}{8}\right)$$
$$x = 2\pi - \arccos\left(\frac{1 \pm \sqrt{33}}{8}\right)$$
$$x \approx 0.5678,\ 2.2057,\ 4.0775,\ 5.7154$$

123. $r = \frac{1}{32}v_0^{\,2}\sin 2\theta$

Let $r = 130$ feet and $v_0 = 75$ feet per second, and find θ.

$$130 = \frac{1}{32}(75)^2\sin 2\theta$$
$$4160 = 5625\sin 2\theta$$
$$\frac{4160}{5625} = \sin 2\theta$$
$$\sin^{-1}\left(\frac{4160}{5625}\right) = 2\theta$$
$$47.69° \approx 2\theta \qquad (0° < \theta < 90°)$$
$$23.85° = \theta$$

125. $\frac{x}{2} = 2r\sin^2\frac{\theta}{2},\ x = 4r\left[\sin\frac{\theta}{2}\right]^2 = 4r\frac{1-\cos\theta}{2}$

$$= 2r(1 - \cos\theta)$$

127. False. If $x = \pi$, $\sin\frac{x}{2} = \sin\frac{\pi}{2} = 1$, whereas

$$-\sqrt{\frac{1-\cos\pi}{2}} = -1.$$

129. $f(x) = 2\sin x\left[2\cos^2\left(\frac{x}{2}\right) - 1\right]$

(a)

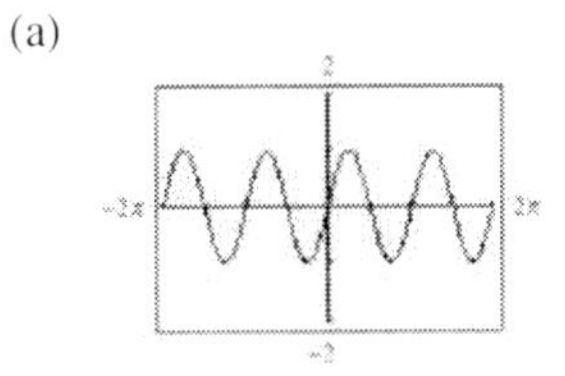

(b) The graph appears to be that of $y = \sin 2x$.

(c) $2\sin x\left[2\cos^2\left(\frac{x}{2}\right) - 1\right] = 2\sin x\left[2\frac{1+\cos x}{2} - 1\right]$

$$= 2\sin x\left[\cos x\right] = \sin 2x$$

131. (a)

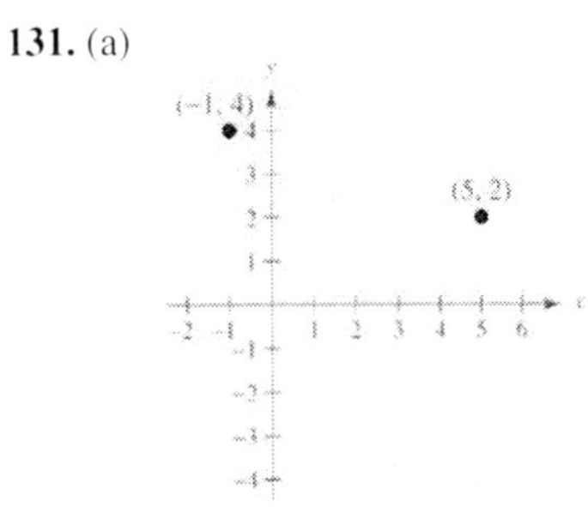

(b) Distance: $\sqrt{(5+1)^2 + (2-4)^2} = \sqrt{40} = 2\sqrt{10}$

(c) Midpoint: $\left(\frac{-1+5}{2},\ \frac{4+2}{2}\right) = (2,\ 3)$

133. (a)

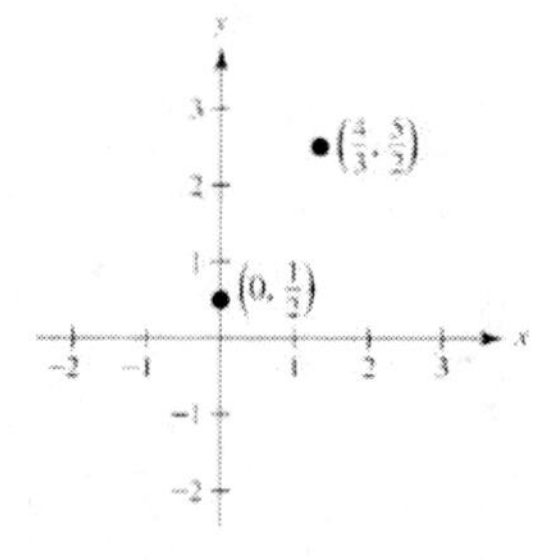

(b) Distance:

$$\sqrt{\left(\frac{4}{3}\right)^2 + \left(\frac{5}{2} - \frac{1}{2}\right)^2} = \sqrt{\frac{16}{9} + 4}$$

$$= \frac{\sqrt{52}}{3} = \frac{2\sqrt{13}}{3}$$

(c) Midpoint: $\left(\frac{0 + (4/3)}{2}, \frac{(1/2) + (5/2)}{2}\right) = \left(\frac{2}{3}, \frac{3}{2}\right)$

135. (a) Complement: $90° - 55° = 35°$

Supplement: $180° - 55° = 125°$

(b) Complement: None

Supplement: $180° - 162° = 18°$

137. (a) Complement: $\frac{\pi}{2} - \frac{\pi}{18} = \frac{8\pi}{18} = \frac{4\pi}{9}$

Supplement: $\pi - \frac{\pi}{18} = \frac{17\pi}{18}$

(b) Complement: $\frac{\pi}{2} - \frac{9\pi}{20} = \frac{\pi}{20}$

Supplement: $\pi - \frac{9\pi}{20} = \frac{11\pi}{20}$

139. $s = r\theta \Rightarrow \theta = \frac{s}{r} = \frac{7}{15} \approx 0.467$ rad

Chapter 5 Review Exercises

1. $\frac{1}{\cos x} = \sec x$

3. $\frac{1}{\sec x} = \cos x$

5. $\sqrt{1 - \cos^2 x} = |\sin x|$

7. $\csc\left(\frac{\pi}{2} - x\right) = \sec x$

9. $\sec(-x) = \sec x$

11. $\sin x = \frac{4}{5}$, $\cos x = \frac{3}{5}$, Quadrant I

$\tan x = \frac{\sin x}{\cos x} = \frac{4}{3}$

$\cot x = \frac{3}{4}$

$\sec x = \frac{5}{3}$

$\csc x = \frac{5}{4}$

13. $\sin\left(\frac{\pi}{2} - x\right) = \cos x = \frac{1}{\sqrt{2}} = \frac{\sqrt{2}}{2}$,

$\sin x = -\frac{1}{\sqrt{2}} = -\frac{\sqrt{2}}{2}$, Quadrant IV

$\tan x = -1$

$\cot x = -1$

$\sec x = \sqrt{2}$

$\csc x = -\sqrt{2}$

15. $\frac{1}{\tan^2 x + 1} = \frac{1}{\sec^2 x} = \cos^2 x$

17. $\frac{\sin^2\theta + \cos^2\theta}{\sin\theta} = \frac{1}{\sin\theta}$

$= \csc\theta$

19. $\tan^2\theta(\csc^2\theta - 1) = \tan^2\theta(\cot^2\theta)$

$= \tan^2\theta\left(\frac{1}{\tan^2\theta}\right) = 1$

21. $\tan\left(\frac{\pi}{2} - x\right)\sec x = \cot x \sec x$

$= \frac{\cos x}{\sin x} \cdot \frac{1}{\cos x} = \frac{1}{\sin x} = \csc x$

23. $\frac{\sin^2\alpha - \cos^2\alpha}{\sin^2\alpha - \sin\alpha\cos\alpha} = \frac{(\sin\alpha + \cos\alpha)(\sin\alpha - \cos\alpha)}{\sin\alpha(\sin\alpha - \cos\alpha)}$

$= \frac{\sin\alpha + \cos\alpha}{\sin\alpha} = 1 + \cot\alpha$

25. $\sin^{-1/2} x \cos x = \frac{\cos x}{\sin^{1/2} x}$

$= \frac{\cos x}{\sqrt{\sin x}} \cdot \frac{\sqrt{\sin x}}{\sqrt{\sin x}}$

$= \frac{\cos x}{\sin x}\sqrt{\sin x} = \cot x\sqrt{\sin x}$

27. $\cos x(\tan^2 x + 1) = \cos x \sec^2 x$

$= \frac{1}{\sec x}\sec^2 x = \sec x$

29. $\sin^3\theta + \sin\theta\cos^2\theta = \sin\theta(\sin^2\theta + \cos^2\theta)$
$= \sin\theta$

31. $\sin^5 x\cos^2 x = \sin^4 x\cos^2 x\sin x$
$= (1-\cos^2 x)^2\cos^2 x\sin x$
$= (1-2\cos^2 x + \cos^4 x)\cos^2 x\sin x$
$= (\cos^2 x - 2\cos^4 x + \cos^6 x)\sin x$

33. $\sqrt{\dfrac{1-\sin\theta}{1+\sin\theta}} = \sqrt{\dfrac{1-\sin\theta}{1+\sin\theta}\cdot\dfrac{1-\sin\theta}{1-\sin\theta}} = \sqrt{\dfrac{(1-\sin\theta)^2}{1-\sin^2\theta}}$
$= \sqrt{\dfrac{(1-\sin\theta)^2}{\cos^2\theta}} = \dfrac{|(1-\sin\theta)|}{|\cos\theta|} = \dfrac{1-\sin\theta}{|\cos\theta|}$

Note: We can drop the absolute value on $1-\sin\theta$ since it is always nonnegative.

35. $\dfrac{\csc(-x)}{\sec(-x)} = -\dfrac{\csc x}{\sec x} = -\dfrac{\cos x}{\sin x} = -\cot x$

37. $\csc^2\left(\dfrac{\pi}{2} - x\right) - 1 = \sec^2 x - 1 = \tan^2 x$

39. $2\sin x - 1 = 0$
$\sin x = \dfrac{1}{2}$
$x = \dfrac{\pi}{6} + 2n\pi$
$x = \dfrac{5\pi}{6} + 2n\pi$

41. $\sin x = \sqrt{3} - \sin x$
$2\sin x = \sqrt{3}$
$\sin x = \dfrac{\sqrt{3}}{2}$
$x = \dfrac{\pi}{3} + 2n\pi$
$x = \dfrac{2\pi}{3} + 2n\pi$

43. $3\sqrt{3}\tan x = 3$
$\tan x = \dfrac{1}{\sqrt{3}}$
$x = \dfrac{\pi}{6} + n\pi$

45. $3\csc^2 x = 4$
$\csc^2 x = \dfrac{4}{3}$
$\sin^2 x = \dfrac{3}{4}$
$\sin x = \pm\dfrac{\sqrt{3}}{2}$
$x = \dfrac{\pi}{3} + n\pi$
$x = \dfrac{2\pi}{3} + n\pi$

47. $4\cos^2 x - 3 = 0$
$\cos^2 x = \dfrac{3}{4}$
$\cos x = \pm\dfrac{\sqrt{3}}{2}$
$x = \dfrac{\pi}{6} + n\pi$
$x = \dfrac{5\pi}{6} + n\pi$

49. $\sin x - \tan x = 0$
$\sin x - \dfrac{\sin x}{\cos x} = 0$
$\sin x\cos x - \sin x = 0$
$\sin x(\cos x - 1) = 0$
$\sin x = 0$ or $\cos x - 1 = 0$
$x = n\pi$ $\quad\cos x = 1$
$x = 2n\pi$

51. $2\cos^2 x - \cos x - 1 = 0$
$(2\cos x + 1)(\cos x - 1) = 0$
$2\cos x + 1 = 0$ or $\cos x - 1 = 0$
$\cos x = -\dfrac{1}{2}$ $\quad\cos x = 1$
$x = \dfrac{2\pi}{3}, \dfrac{4\pi}{3}$ $\quad x = 0$

53. $\cos^2 x + \sin x = 1$
$1 - \sin^2 + \sin x = 1$
$\sin x(\sin x - 1) = 0$
$\sin x = 0$ or $\sin x = 1$
$x = 0, \pi$ $\quad x = \dfrac{\pi}{2}$

55. $2\sin 2x = \sqrt{2}$

$$\sin 2x = \frac{\sqrt{2}}{2}$$

$$2x = \frac{\pi}{4}, \frac{3\pi}{4}, \frac{9\pi}{4}, \frac{11\pi}{4}$$

$$x = \frac{\pi}{8}, \frac{3\pi}{8}, \frac{9\pi}{8}, \frac{11\pi}{8}$$

57. $\cos 4x(\cos x - 1) = 0$

$\cos 4x = 0$ or $\cos x - 1 = 0$

$$4x = \frac{\pi}{2}, \frac{3\pi}{2}, \frac{5\pi}{2}, \frac{7\pi}{2}, \frac{9\pi}{2}, \frac{11\pi}{2}, \frac{13\pi}{2}, \frac{15\pi}{2}$$

or $\cos x = 1$

$$x = \frac{\pi}{8}, \frac{3\pi}{8}, \frac{5\pi}{8}, \frac{7\pi}{8}, \frac{9\pi}{8}, \frac{11\pi}{8}, \frac{13\pi}{8}, \frac{15\pi}{8}, 0$$

59. $2\sin 2x - 1 = 0$

$$\sin 2x = \frac{1}{2}$$

$$2x = \frac{\pi}{6} + 2n\pi \quad \text{or} \quad 2x = \frac{5\pi}{6} + 2n\pi$$

$$x = \frac{\pi}{12} + n\pi \quad \text{or} \quad x = \frac{5\pi}{12} + n\pi$$

61. $2\sin^2 3x - 1 = 0$

$$\sin^2 3x = \frac{1}{2}$$

$$\sin 3x = \pm\frac{\sqrt{2}}{2}$$

$$3x = \frac{\pi}{4} + \frac{n\pi}{2}$$

$$x = \frac{\pi}{12} + \frac{n\pi}{6}$$

63. $\sin^2 x - 2\sin x = 0$

$\sin x(\sin x - 2) = 0$

$\sin x = 0$ or $\sin x = 2$ (impossible)

$x = 0, \pi$

65. $\tan^2 x + \tan x - 12 = 0$

$(\tan x + 4)(\tan x - 3) = 0$

$\tan x + 4 = 0$ or $\tan x - 3 = 0$

$\tan x = -4$ $\tan x = 3$

$x = \pi + \arctan(-4) \approx 1.82$ $x = \arctan 3 \approx 1.25$

$x = 2\pi + \arctan(-4) \approx 4.96$ $x = \pi + \arctan 3 \approx 4.39$

67. $\sin 285° = \sin(315° - 30°)$

$$= \sin 315° \cos 30° - \cos 315° \sin 30°$$

$$= \left(-\frac{\sqrt{2}}{2}\right)\left(\frac{\sqrt{3}}{2}\right) - \left(\frac{\sqrt{2}}{2}\right)\left(\frac{1}{2}\right) = -\frac{\sqrt{6}+\sqrt{2}}{4}$$

$\cos 285° = \cos(315° - 30°) = \cos 315° \cos 30° + \sin 315° \sin 30°$

$$= \left(\frac{\sqrt{2}}{2}\right)\left(\frac{\sqrt{3}}{2}\right) + \left(-\frac{\sqrt{2}}{2}\right)\left(\frac{1}{2}\right) = \frac{\sqrt{6}-\sqrt{2}}{4}$$

$$\tan 285° = -\frac{\sqrt{6}+\sqrt{2}}{\sqrt{6}-\sqrt{2}} = -2 - \sqrt{3}$$

69. $\sin\frac{31\pi}{12} = \sin\left(\frac{11\pi}{6} + \frac{3\pi}{4}\right) = \sin\frac{11\pi}{6}\cos\frac{3\pi}{4} + \sin\frac{3\pi}{4}\cos\frac{11\pi}{6}$

$$= \left(-\frac{1}{2}\right)\left(-\frac{\sqrt{2}}{2}\right) + \left(\frac{\sqrt{2}}{2}\right)\left(\frac{\sqrt{3}}{2}\right) = \frac{\sqrt{2}+\sqrt{6}}{4}$$

$\cos\frac{31\pi}{12} = \cos\left(\frac{11\pi}{6} + \frac{3\pi}{4}\right) = \cos\frac{11\pi}{6}\cos\frac{3\pi}{4} - \sin\frac{11\pi}{6}\sin\frac{3\pi}{4}$

$$= \left(\frac{\sqrt{3}}{2}\right)\left(-\frac{\sqrt{2}}{2}\right) - \left(-\frac{1}{2}\right)\left(\frac{\sqrt{2}}{2}\right) = \frac{\sqrt{2}-\sqrt{6}}{4}$$

$$\tan\frac{31\pi}{12} = \frac{\sin(31\pi/12)}{\cos(31\pi/12)} = \frac{\sqrt{2}+\sqrt{6}}{\sqrt{2}-\sqrt{2}} = -2 - \sqrt{3}$$

71. $\sin 130° \cos 50° + \cos 130° \sin 50° = \sin(130° + 50°)$

$$= \sin 180° = 0$$

73. $\dfrac{\tan 25° + \tan 50°}{1 - \tan 25° \tan 50°} = \tan(25° + 50°) = \tan 75°$

75. $\sin(u + v) = \sin u \cos v + \cos u \sin v$

$$= \left(\frac{4}{5}\right)\left(-\frac{7}{25}\right) + \left(\frac{-3}{5}\right)\left(\frac{24}{25}\right)$$

$$= -\frac{28}{125} - \frac{72}{125} = -\frac{100}{125} = -\frac{4}{5}$$

77. $\tan(u - v) = \dfrac{\tan u - \tan v}{1 + \tan u + \tan v}$

$$= \frac{\left(-\frac{4}{3}\right) - \left(-\frac{24}{7}\right)}{1 + \left(-\frac{4}{3}\right)\left(-\frac{24}{7}\right)}$$

$$= \frac{\frac{44}{21}}{\frac{39}{7}} = \frac{44}{117}$$

79. $\cos(u+v) = \cos u + \cos v - \sin u \sin v$

$$= \left(-\frac{3}{5}\right)\left(-\frac{7}{25}\right) - \left(\frac{4}{5}\right)\left(\frac{24}{25}\right)$$

$$= \frac{21}{125} - \frac{96}{125}$$

$$= -\frac{75}{125} = -\frac{3}{5}$$

81. $\cos\left(x + \frac{\pi}{2}\right) = \cos x \cos\frac{\pi}{2} - \sin x \sin\frac{\pi}{2}$

$$= (\cos x)(0) - (\sin x)(1) = -\sin x$$

83. $\cot\left(\frac{\pi}{2} - x\right) = \frac{\cos[(\pi/2) - x]}{\sin[(\pi/2) - x]}$

$$= \frac{\cos(\pi/2)\cos x + \sin(\pi/2)\sin x}{\sin(\pi/2)\cos x - \sin x \cos(\pi/2)}$$

$$= \frac{\sin x}{\cos x} = \tan x$$

85. $\cos 3x = \cos(2x + x)$

$$= \cos 2x \cos x - \sin 2x \sin x$$

$$= (\cos^2 x - \sin^2 x)\cos x - 2\sin x \cos x \sin x$$

$$= \cos^3 x - 3\sin^2 x \cos x$$

$$= \cos^3 x - 3\cos x(1 - \cos^2 x)$$

$$= \cos^3 x - 3\cos x + 3\cos^3 x$$

$$= 4\cos^3 x - 3\cos x$$

87. $\sin\left(x + \frac{\pi}{2}\right) - \sin\left(x - \frac{\pi}{2}\right) = \sqrt{2}$

$$2\cos x \sin\frac{\pi}{2} = \sqrt{2} \quad \text{(Sum-to-Product)}$$

$$\cos x = \frac{\sqrt{2}}{2}$$

$$x = \frac{\pi}{4}, \frac{7\pi}{4}$$

89. $\sin u = \frac{5}{7}, \quad 0 < u < \frac{\pi}{2}$

$$\sin 2u = 2\sin u \cos u$$

$$= 2\left(\frac{5}{7}\right)\left(\frac{2\sqrt{6}}{7}\right)$$

$$= \frac{20\sqrt{6}}{49}$$

$$\cos 2u = \cos^2 u - \sin^2 u$$

$$= \left(\frac{2\sqrt{6}}{7}\right)^2 - \left(\frac{5}{7}\right)^2$$

$$= \frac{24}{49} - \frac{25}{49} = -\frac{1}{49}$$

$$\tan 2u = \frac{2\tan u}{1 - \tan^2 u}$$

$$= \frac{2\left(\frac{5}{2\sqrt{6}}\right)}{1 - \left(\frac{5}{2\sqrt{6}}\right)^2}$$

$$= \frac{\frac{5}{\sqrt{6}}}{1 - \frac{25}{24}}$$

$$= \frac{\frac{5}{\sqrt{6}}}{-\frac{1}{24}}$$

$$= \frac{5}{\sqrt{6}}\left(-\frac{24}{1}\right)$$

$$= -\frac{120}{\sqrt{6}}$$

$$= \frac{-120\sqrt{6}}{6}$$

$$= -20\sqrt{6}$$

91. $\tan u = -\frac{2}{9}, \frac{\pi}{2} < u < \pi$

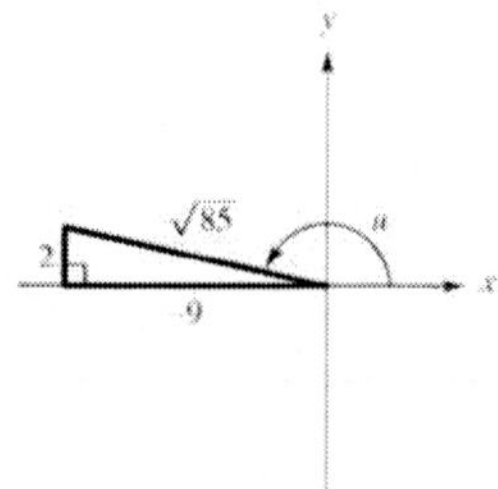

$\sec^2 u = \tan^2 u + 1 = \frac{4}{81} + 1 = \frac{85}{81} \Rightarrow \sec u = -\frac{\sqrt{85}}{9}$

$\cos u = \frac{-9\sqrt{85}}{85}, \quad \sin u = (\tan u)(\cos u) = \frac{2\sqrt{85}}{85}$

$\sin 2u = 2\sin u \cos u$

$= 2\left(\frac{2\sqrt{85}}{85}\right)\left(\frac{-9\sqrt{85}}{85}\right) = -\frac{36}{85}$

$\cos 2u = 1 - 2\sin^2 u = 1 - 2\left(\frac{4}{85}\right) = \frac{77}{85}$

$\tan 2u = \frac{\sin 2u}{\cos 2u} = -\frac{36}{77}$

93. $6\sin x\cos x = 3[2\sin x\cos x] = 3\sin x2x$

95. $1 - 4\sin^2 x\cos^2 x = 1 - (2\sin x\cos x)^2$

$= 1 - \sin^2 2x = \cos^2 2x$

97. $r = \frac{1}{32}v_0^2 \sin 2\theta$

$100 = \frac{1}{32}(80)^2 \sin 2\theta$

$\sin 2\theta = 0.5$

$2\theta = 30^\circ$ or $2\theta = 180^\circ - 30^\circ = 150^\circ$

$\theta = 15^\circ \qquad \theta = 75^\circ$

99. $\sin^6 x = \left(\frac{1-\cos 2x}{2}\right)^3 = \frac{1}{8}(1 - 3\cos 2x + 3\cos^2 2x - \cos^3 2x)$

$= \frac{1}{8}\left[1 - 3\cos 2x + 3\left(\frac{1+\cos 4x}{2}\right) - \cos 2x\left(\frac{1+\cos 4x}{2}\right)\right]$

$= \frac{1}{8}\left(1 - 3\cos 2x + \frac{3}{2} + \frac{3}{2}\cos 4x - \frac{1}{2}\cos 2x - \frac{1}{2}\cos 2x\cos 4x\right)$

$= \frac{1}{16}\left(5 - 7\cos 2x + 3\cos 4x - \frac{1}{2}[\cos 2x + \cos 6x]\right)$

$= \frac{1}{32}(10 - 15\cos 2x + 6\cos 4x - \cos 6x)$

101. $\cos^4 2x = \left(\frac{1+\cos 4x}{2}\right)^2$

$= \frac{1}{4}(1 + 2\cos 4x + \cos^2 4x)$

$= \frac{1}{4}\left(1 + 2\cos 4x + \frac{1+\cos 8x}{2}\right)$

$= \frac{1}{8}(2 + 4\cos 4x + 1 + \cos 8x)$

$= \frac{1}{8}(3 + 4\cos 4x + \cos 8x)$

103. $\tan^2 4x = \frac{1-\cos 8x}{1+\cos 8x}$

105. $\sin 15^\circ = \sin\left(\frac{30}{2}\right)^\circ = \sqrt{\frac{1-\cos 30^\circ}{2}}$

$= \sqrt{\frac{1-\frac{\sqrt{3}}{2}}{2}}$

$= \sqrt{\frac{2-\sqrt{3}}{2}}$

$\cos 15^\circ = \cos\left(\frac{30}{2}\right)^\circ = \sqrt{\frac{1+\cos 30^\circ}{2}}$

$= \sqrt{\frac{1+\frac{\sqrt{3}}{2}}{2}}$

$= \sqrt{\frac{2+\sqrt{3}}{2}}$

$\tan 15^\circ = \tan\left(\frac{30}{2}\right)^\circ = \frac{\sin 30^\circ}{1+\cos 30^\circ}$

$= \frac{\frac{1}{2}}{1+\frac{\sqrt{3}}{2}}$

$= \frac{\frac{1}{2}}{\frac{2+\sqrt{3}}{2}}$

$= \frac{1}{2+\sqrt{3}}$

$= 2 - \sqrt{3}$

107. $\sin\left(\frac{7\pi}{8}\right) = \sin\left(\frac{1}{2}\cdot\frac{7\pi}{4}\right) = \sqrt{\frac{1-\cos(7\pi/4)}{2}}$

$= \sqrt{\frac{1-(\sqrt{2}/2)}{2}} = \sqrt{\frac{2-\sqrt{2}}{2}}$

$\cos\left(\frac{7\pi}{8}\right) = \cos\left(\frac{1}{2}\cdot\frac{7\pi}{4}\right) = -\sqrt{\frac{1+\cos(7\pi/4)}{2}}$

$= -\sqrt{\frac{1+(\sqrt{2}/2)}{2}} = -\frac{\sqrt{2+\sqrt{2}}}{2}$

$\tan\left(\frac{7\pi}{8}\right) = \tan\left(\frac{1}{2}\cdot\frac{7\pi}{4}\right) = \frac{\sin(7\pi/4)}{1+\cos(7\pi/4)}$

$= \frac{-\sqrt{2}/2}{1+(\sqrt{2}/2)} = \frac{-\sqrt{2}}{2+\sqrt{2}} = 1-\sqrt{2}$

109. $\sin u = \frac{3}{5},\ 0 < u < \frac{\pi}{2} \Rightarrow \cos u = \frac{4}{5}$

$\sin\left(\frac{u}{2}\right) = \sqrt{\frac{1-\cos u}{2}}$

$= \sqrt{\frac{1-(4/5)}{2}} = \frac{1}{\sqrt{10}} = \frac{\sqrt{10}}{10}$

$\cos\left(\frac{u}{2}\right) = \sqrt{\frac{1+\cos u}{2}}$

$= \sqrt{\frac{1+(4/5)}{2}} = \frac{3}{\sqrt{10}} = \frac{3\sqrt{10}}{10}$

$\tan\left(\frac{u}{2}\right) = \frac{1-\cos u}{\sin u} = \frac{1-(4/5)}{3/5} = \frac{1}{3}$

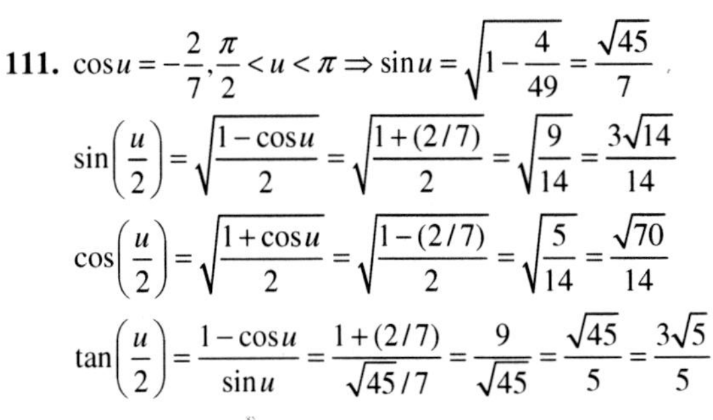

111. $\cos u = -\frac{2}{7},\ \frac{\pi}{2} < u < \pi \Rightarrow \sin u = \sqrt{1-\frac{4}{49}} = \frac{\sqrt{45}}{7}$

$\sin\left(\frac{u}{2}\right) = \sqrt{\frac{1-\cos u}{2}} = \sqrt{\frac{1+(2/7)}{2}} = \sqrt{\frac{9}{14}} = \frac{3\sqrt{14}}{14}$

$\cos\left(\frac{u}{2}\right) = \sqrt{\frac{1+\cos u}{2}} = \sqrt{\frac{1-(2/7)}{2}} = \sqrt{\frac{5}{14}} = \frac{\sqrt{70}}{14}$

$\tan\left(\frac{u}{2}\right) = \frac{1-\cos u}{\sin u} = \frac{1+(2/7)}{\sqrt{45}/7} = \frac{9}{\sqrt{45}} = \frac{\sqrt{45}}{5} = \frac{3\sqrt{5}}{5}$

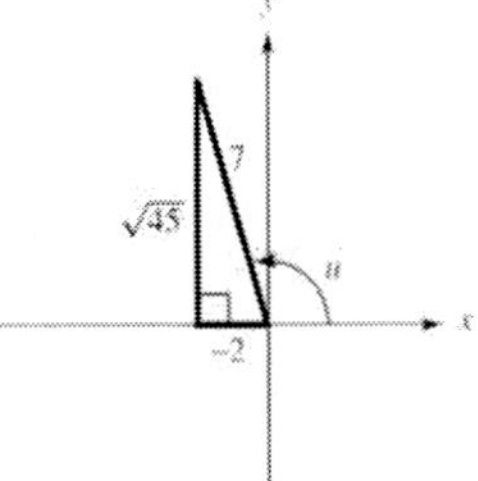

113. $-\sqrt{\frac{1+\cos 8x}{2}} = -|\cos 4x|$

115. $\frac{\sin 10x}{1+\cos 10x} = \tan 5x$

117. Volume V of the trough will be area A of the isosceles triangle times the length l of the trough.

$V = A\cdot l,\ A = \frac{1}{2}bh$

$\cos\frac{\theta}{2} = \frac{h}{0.5} \Rightarrow h = 0.5\cos\frac{\theta}{2}$

$\sin\frac{\theta}{2} = \frac{b/2}{0.5} \Rightarrow \frac{b}{2} = 0.5\sin\frac{\theta}{2}$

$A = 0.5\sin\frac{\theta}{2}0.5\cos\frac{\theta}{2} = (0.5)^2\sin\frac{\theta}{2}\cos\frac{\theta}{2} = 0.25\sin\frac{\theta}{2}\cos\frac{\theta}{2}\ \text{m}^2$

$V = (0.25)(4)\sin\frac{\theta}{2}\cos\frac{\theta}{2} = \sin\frac{\theta}{2}\cos\frac{\theta}{2}\ \text{m}^3$

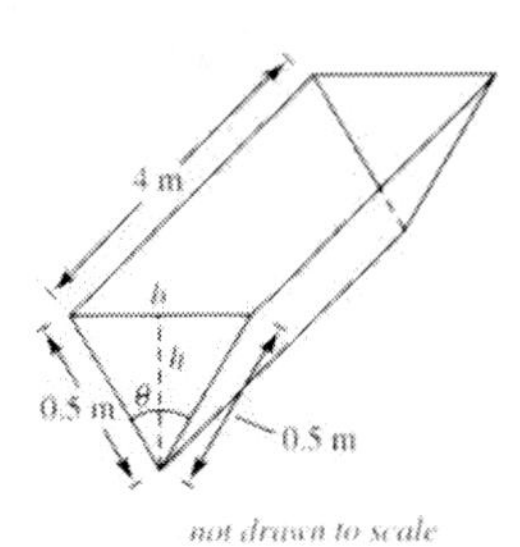

119. $6\sin\frac{\pi}{4}\cos\frac{\pi}{4} = 6\left[\frac{1}{2}\sin\left(\frac{\pi}{4}+\frac{\pi}{4}\right)+\sin\left(\frac{\pi}{4}+\frac{\pi}{4}\right)\right]$

$= 3\left(\sin\frac{\pi}{2}+\sin 0\right) = 3$

121. $\sin 5\alpha \sin 4\alpha = \frac{1}{2}\left[\cos(5\alpha-4\alpha)-\cos(5\alpha+4\alpha)\right]$

$= \frac{1}{2}(\cos\alpha - \cos 9\alpha)$

123. $\cos 5\theta + \cos 4\theta = 2\cos\left(\frac{9\theta}{2}\right)\cos\left(\frac{\theta}{2}\right)$

125. $\sin\left(x+\frac{\pi}{4}\right)-\sin\left(x-\frac{\pi}{4}\right) = 2\cos x \sin x\frac{\pi}{4} = \sqrt{2}\cos x$

127. $y = 1.5\sin 8t - 0.5\cos 8t$

$a = \frac{3}{2},\ b = -\frac{1}{2},\ B = 8,\ C = \arctan\left(-\frac{1/2}{3/2}\right)$

$y = \sqrt{\left(\frac{3}{2}\right)^2+\left(\frac{1}{2}\right)^2}\sin\left(8t+\arctan\left(-\frac{1}{3}\right)\right)$

$y = \frac{1}{2}\sqrt{10}\sin\left(8t+\arctan\left(-\frac{1}{3}\right)\right)$

129. The amplitude is $\frac{\sqrt{10}}{2}$.

131. False, if $\frac{\pi}{2} < \theta < \pi \Rightarrow \frac{\pi}{4} < \frac{\theta}{2} < \frac{\pi}{2}$,

which is in Quadrant I $\Rightarrow \cos\left(\frac{\theta}{2}\right) > 0$.

133. True.

$4\sin(-x)\cos(-x) = 4(-\sin x)(\cos x)$

$= -4\sin x\cos x$

$= -2(2\sin x\cos x) = -2\sin 2x$

135. Answers will vary. See page 350.

137. No. A trigonometric equation with an infinite number of solutions does not have to be an identity. For example, the general solution of $\sin x = \frac{1}{2}$ is $x = \frac{\pi}{6} + 2n\pi$ and $x = \frac{5\pi}{6} + 2n\pi$, which produces an infinite number of solutions.

139. $y_1 = \sec^2\left(\frac{\pi}{2} - x\right) = \csc^2 x$

$y_2 = \cot^2 t$

$\csc^2 x = \cot^2 x + 1$

Let $y_3 = y_2 + 1 = \cot^2 x + 1 = y_1$.

Chapter 5 Test

1. $\tan\theta = \frac{3}{2},\ \cos\ \theta < 0 \Rightarrow \theta$ in Quadrant III

$\sec^2\theta = \tan^2\theta + 1 = \frac{9}{4} + 1 = \frac{13}{4} \Rightarrow \sec\theta = -\frac{\sqrt{13}}{2}$

$\cos\theta = -\frac{2}{\sqrt{13}} = -\frac{2\sqrt{13}}{13}$

$\sin\theta = \tan\theta\cos\theta = \frac{3}{2}\left(-\frac{2\sqrt{13}}{13}\right) = -\frac{3\sqrt{13}}{13}$

$\csc\theta = -\frac{13}{3\sqrt{13}} = -\frac{\sqrt{13}}{3}$

$\cot\theta = \frac{2}{3}$

2. $\csc^2\beta(1-\cos^2\beta) = \frac{1}{\sin^2\beta}\cdot\sin^2\beta = 1$

3. $\frac{\sec^4 x - \tan^4 x}{\sec^2 x + \tan^2 x} = \frac{[(\sec^2 x) + (\tan^2 x)][\sec^2 x - \tan^2 x]}{\sec^2 x + \tan^2 x}$

$= \sec^2 x - \tan^2 x = 1$

4. $\frac{\cos\theta}{\sin\theta} + \frac{\sin\theta}{\cos\theta} = \frac{\cos^2\theta + \sin^2\theta}{\sin\theta\cos\theta} = \frac{1}{\sin\theta\cos\theta} = \csc\theta\sec\theta$

5. Since $\tan^2\theta = \sec^2\theta - 1$ for all θ, then

$\tan\ \theta = -\sqrt{\sec^2\theta - 1}$ in Quadrants II and IV.

Thus, $\frac{\pi}{2} < \theta \le \pi$ and $\frac{3\pi}{2} < \theta < 2\pi$.

6. Conjecture: $y_1 = y_2$

Algebraically,

$$y_1 = \sin x + \cos x \cot x = \sin x + \frac{\cos^2 x}{\sin x}$$
$$= \frac{\sin^2 x + \cos^2 x}{\sin x}$$
$$= \frac{1}{\sin x} = \csc x.$$

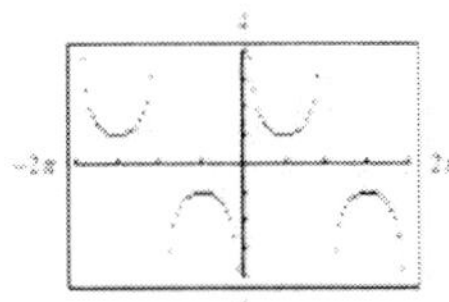

7. $\sin\theta \cdot \sec\theta = \sin\theta = \sin\theta \frac{1}{\cos\theta} = \tan\theta$

8. $\sec^2 x \tan^2 x + \sec^2 x = \sec^2 x(\tan^2 x + 1) = \sec^4 x$

9. $$\frac{\csc\alpha + \sec\alpha}{\sin\alpha + \cos\alpha} = \frac{\frac{1}{\sin\alpha} + \frac{1}{\cos\alpha}}{\sin\alpha + \cos\alpha} = \frac{\frac{\cos\alpha + \sin\alpha}{\sin\alpha \cdot \cos\alpha}}{(\sin\alpha + \cos\alpha)}$$
$$= \frac{1}{\sin\alpha\cos\alpha} = \frac{\cos^2\alpha + \sin^2\alpha}{\sin\alpha\cos\alpha}$$
$$= \frac{\cos\alpha}{\sin\alpha} + \frac{\sin\alpha}{\cos\alpha} = \cot\alpha + \tan\alpha$$

10. $$\cos\left(x + \frac{\pi}{2}\right) = \cos x \cos\frac{\pi}{2} - \sin x \sin x\frac{\pi}{2}$$
$$= 0 - \sin x = -\sin x$$

11. $$\sin(n\pi + \theta) = \sin n\pi \cos\theta + \cos n\pi \sin\theta$$
$$= 0 + (-1)^n \sin\theta$$
$$= (-1)^n \sin\theta$$

12. $$(\sin x + \cos x)^2 = \sin^2 x + \cos^2 x + 2\sin x\cos x$$
$$= 1 + \sin 2x$$

13. $$\tan 255^\circ = \tan(225^\circ + 30^\circ)$$
$$= \frac{\tan 225^\circ + \tan 30^\circ}{1 - \tan 225^\circ \tan 30^\circ}$$
$$= \frac{1 + \frac{\sqrt{3}}{3}}{1 - (1)\left(\frac{\sqrt{3}}{3}\right)}$$
$$= \frac{\frac{3+\sqrt{3}}{3}}{\frac{3-\sqrt{3}}{3}}$$
$$= \frac{3+\sqrt{3}}{3} \cdot \frac{3}{3-\sqrt{3}}$$
$$= \frac{3+\sqrt{3}}{3-\sqrt{3}}$$
$$= \frac{3+\sqrt{3}}{3-\sqrt{3}} \cdot \frac{3+\sqrt{3}}{3+\sqrt{3}}$$
$$= \frac{9 + 6\sqrt{3} + 3}{9 - 3}$$
$$= \frac{12 + 6\sqrt{3}}{6} = 2 + \sqrt{3}$$

14. $$\sin^4 x \tan^2 x = \frac{\sin^6 x}{\cos^2 x}$$
$$= \frac{1}{32} \cdot \frac{(10 - 15\cos 2x + 6\cos 4x - \cos 6x)}{(1 + \cos 2x)/2}$$
$$= \frac{1}{16}\left(\frac{10 - 15\cos 2x + 6\cos 4x - \cos 6x}{1 + \cos 2x}\right)$$

15. $$\frac{\sin 4\theta}{1 + \cos 4\theta} = \tan\frac{4\theta}{2} = \tan 2\theta$$

16. $$4\cos 2\theta \sin 4\theta = 4\left(\frac{1}{4}\right)[\sin(2\theta + 4\theta) - \sin(2\theta - 4\theta)]$$
$$= 2[\sin 6\theta - \sin(-2\theta)]$$
$$= 2(\sin 6\theta + \sin 2\theta)$$

17. $$\sin 3\theta - \sin 4\theta = 2\cos\left(\frac{3\theta + 4\theta}{2}\right)\sin\left(\frac{3\theta - 4\theta}{2}\right)$$
$$= 2\cos\left(\frac{7\theta}{2}\right)\sin\left(\frac{-\theta}{2}\right)$$
$$= -2\cos\frac{7\theta}{2}\sin\frac{\theta}{2}$$

18.
$$\tan^2 x + \tan x = 0$$
$$\tan x(\tan x + 1) = 0$$
$$\tan x = 0 \Rightarrow x = 0,\ \pi$$
$$\tan x + 1 = 0 \Rightarrow \tan x = -1 \Rightarrow x = \frac{3\pi}{4}, \frac{7\pi}{4}$$

19.
$$\sin 2\alpha - \cos\alpha = 0$$
$$2\sin\alpha\cos\alpha - \cos\alpha = 0$$
$$\cos\alpha(2\sin\alpha - 1) = 0$$
$$\cos\alpha = 0 \Rightarrow \alpha = \frac{\pi}{2},\ \frac{3\pi}{2}$$
$$2\sin\alpha - 1 = 0 \Rightarrow \sin\alpha = \frac{1}{2} \Rightarrow \alpha = \frac{\pi}{6},\ \frac{5\pi}{6}$$

20.
$$4\cos^2 x - 3 = 0$$
$$\cos^2 x = \frac{3}{4}$$
$$\cos x = \pm\frac{\sqrt{3}}{2}$$
$$x = \frac{\pi}{6},\ \frac{5\pi}{6},\ \frac{7\pi}{6},\ \frac{11\pi}{6}$$

21.
$$\csc^2 x - \csc x - 2 = 0$$
$$(\csc x - 2)(\csc x + 1) = 0$$
$$\csc x - 2 = 0 \Rightarrow \csc x = 2 \Rightarrow \sin x = \frac{1}{2} \Rightarrow x = \frac{\pi}{6},\ \frac{5\pi}{6}$$
$$\csc x + 1 = 0 \Rightarrow \csc x = -1 \Rightarrow \sin x = -1 \Rightarrow x = \frac{3\pi}{2}$$

22.
$$3\cos x - x = 0$$
$$x \approx 1.170,\ -2.663,\ -2.938$$

23.
$$\sin 2u = 2\sin u\cos u = 2\frac{2}{\sqrt{5}}\cdot\frac{1}{\sqrt{5}} = \frac{4}{5}$$
$$\tan 2u = \frac{2\tan u}{1 - \tan^2 u} = \frac{2(2)}{1 - 2^2} = -\frac{4}{3}$$
$$\cos 2u = 1 - 2\sin^2 u = 1 - 2\left(\frac{2}{\sqrt{5}}\right)^2 = -\frac{3}{5}$$

24.
$$n = \frac{\sin[(\theta/2) + (\alpha/2)]}{\sin(\theta/2)}$$
$$\frac{3}{2} = \frac{\sin\left[(\theta/2) + 30^\circ\right]}{\sin(\theta/2)}$$
$$3\sin\frac{\theta}{2} = 2\left[\sin\frac{\theta}{2}\cos 30^\circ + \cos\frac{\theta}{2}\sin 30^\circ\right]$$
$$3\sin\frac{\theta}{2} = \sqrt{3}\sin\frac{\theta}{2} + \cos\frac{\theta}{2}$$
$$\left(3 - \sqrt{3}\right)\sin\frac{\theta}{2} = \cos\frac{\theta}{2}$$
$$\tan\frac{\theta}{2} = \frac{1}{3 - \sqrt{3}}$$
$$\frac{\theta}{2} = \arctan\left(\frac{1}{3 - \sqrt{3}}\right) \approx 38.26^\circ$$
$$\theta \approx 76.52^\circ$$

CHAPTER 6

Section 6.1

1. oblique

3. $\frac{1}{2}bc\ \sin A;\ \frac{1}{2}ab\ \sin C;\ \frac{1}{2}ac\ \sin B$

5. The two cases AAS (two angles and one side) or ASA (angle side angle) and SSA (two sides and an angle opposite) can be solved using the Law of Sines.

7. Given: $A = 25°,\ B = 60°,\ a = 12$

$C = 180° - 25° - 60° = 95°$

$$b = \frac{a}{\sin A}(\sin B) = \frac{12}{\sin 25°}(\sin 60°) \approx 24.59 \text{ in.}$$

$$c = \frac{a}{\sin A}(\sin C) = \frac{12}{\sin 25°}(\sin 95°) \approx 28.29 \text{ in.}$$

9. Given: $B = 15°,\ C = 125°,\ c = 20$

$A = 180° - 15° - 125° = 40°$

$$a = \frac{c}{\sin C}(\sin A) = \frac{20}{\sin 125°}(\sin 40°) \approx 15.69 \text{ cm}$$

$$b = \frac{c}{\sin C}(\sin B) = \frac{20}{\sin 125°}(\sin 15°) \approx 6.32 \text{ cm}$$

11. Given: $A = 80°\ 15',\ B = 25°\ 30',\ b = 2.8$

$C = 180° - 80°\ 15' - 25°\ 30' = 74°\ 15'$

$$a = \frac{b}{\sin B}(\sin A) = \frac{2.8}{\sin 25°\ 30'}(\sin 80°\ 15') \approx 6.41 \text{ km}$$

$$c = \frac{b}{\sin B}(\sin C) = \frac{2.8}{\sin 25°\ 30'}(\sin 74°\ 15') \approx 6.26 \text{ km}$$

13. Given: $A = 36°,\ a = 8,\ b = 5$

$$\sin B = \frac{b\ \sin A}{a} = \frac{5\ \sin(36°)}{8} \approx 0.3674 \Rightarrow B \approx 21.6°$$

$C = 180° - A - B \approx 180° - 36° - 21.6° = 122.4°$

$$c = \frac{a}{\sin A}(\sin C) = \frac{8}{\sin(36°)}\sin(122.4°) \approx 11.49$$

15. Given: $A = 102.4°,\ C = 16.7°,\ a = 21.6$

$B = 180° - A - C = 180° - 102.4° - 16.7° = 60.9°$

$$b = \frac{a}{\sin A}(\sin B) = \frac{21.6}{\sin 102.4°}(\sin 60.9°) \approx 19.32$$

$$c = \frac{a}{\sin A}(\sin C) = \frac{21.6}{\sin 102.4°}(\sin 16.7°) \approx 6.36$$

17. Given: $A = 110°15',\ a = 48,\ b = 16$

$$\sin B = \frac{b\ \sin A}{a} = \frac{16\ \sin 110°\ 15'}{48} \approx 0.31273 \Rightarrow B \approx 18°\ 13'$$

$C = 180° - A - B \approx 180° - 110°\ 15' - 18°\ 13' = 51°\ 32'$

$$c = \frac{a}{\sin A}(\sin C) = \frac{48}{\sin 110°\ 15'}(\sin 51°\ 32') \approx 40.06$$

19. Given: $A = 110°,\ a = 125,\ b = 100$

$$\sin B = \frac{b\ \sin A}{a} = \frac{100\ \sin 110°}{125} \approx 0.75175 \Rightarrow B \approx 48.74°$$

$C = 180° - A - B \approx 21.26°$

$$c = \frac{a}{\sin A}(\sin C) = \frac{125\ \sin 21.26°}{\sin 110°} \approx 48.23$$

21. Given: $B = 28°,\ C = 104°,\ a = 3\frac{5}{8}$

$A = 180° - B - C = 180° - 28° - 104° = 48°$

$$b = \frac{a}{\sin A}(\sin B) = \frac{\frac{29}{8}}{\sin 48°}\sin 28° \approx 2.29$$

$$c = \frac{a}{\sin A}(\sin C) = \frac{\frac{29}{8}}{\sin 48°}\sin 104° \approx 4.73$$

23. Given: $B = 10°,\ C = 135°,\ c = 45$

$A = 180° - B - C = 180° - 10° - 135° = 35°$

$$a = \frac{c}{\sin C}(\sin A) = \frac{45}{\sin 135°}\sin 35° \approx 36.50$$

$$b = \frac{c}{\sin C}(\sin B) = \frac{45}{\sin 135°}\sin 10° \approx 11.05$$

25. Given: $C = 85°\ 20',\ a = 35,\ c = 50$

$$\sin A = a\frac{\sin C}{c} = 35\frac{\sin 85°\ 20'}{50} \approx 0.6978$$

$A \approx 44°\ 14'$

$B = 180° - C - A = 180° - 85°\ 20' - 44°\ 14' = 50°\ 26'$

$$b = \frac{c}{\sin C}(\sin B) = \frac{50}{\sin 85°\ 20'}\sin 50°\ 26' \approx 38.67$$

27. Given: $A = 76°,\ a = 18,\ b = 20$

$$\sin B = \frac{b\ \sin A}{a} = \frac{20\ \sin 76°}{18} \approx 1.078$$

No solution

29. Given: $A = 58°$, $a = 11.4$, $b = 12.8$

$$\sin B = \frac{b \sin A}{a} = \frac{12.8 \sin 58°}{11.4} \approx 0.9522 \Rightarrow B \approx 72.21° \text{ or } 107.79°$$

Case 1

$B \approx 72.21°$

$C = 180° - 58° - 72.21° = 49.79°$

$$c = \frac{a}{\sin A}(\sin C) = \frac{11.4}{\sin 58°}(\sin 49.79°) \approx 10.27$$

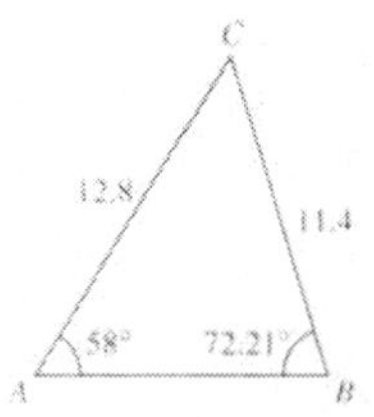

Case 2

$B \approx 107.79°$

$C \approx 180° - 58° - 107.79° = 14.21°$

$$c = \frac{a}{\sin A}(\sin C) = \frac{11.4}{\sin 58°}(\sin 14.21°) \approx 3.30$$

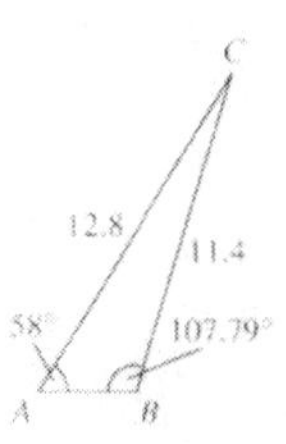

31. Given: $A = 36°$ and $a = 5$

(a) One solution: If $b \le a = 5$ or if $b = \dfrac{5}{\sin 36°}$ (right triangle)

(b) Two solutions: If $a = 5 < b < \dfrac{5}{\sin 36°}$

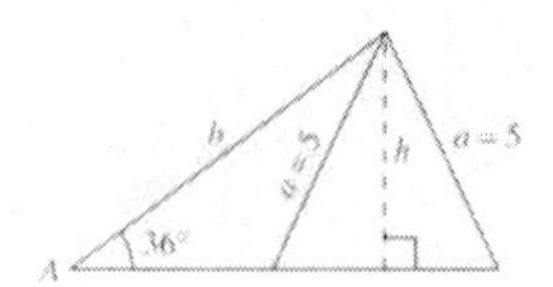

(c) No solution: If $b > \dfrac{5}{\sin 36°}$

33. Given: $A = 10°$ and $a = 10.8$

(a) One solution: If $b \le a = 10.8$ or $b = \dfrac{10.8}{\sin 10°}$ (right triangle)

(b) Two solutions: If $a = 10.8 < b < \dfrac{10.8}{\sin 10°}$

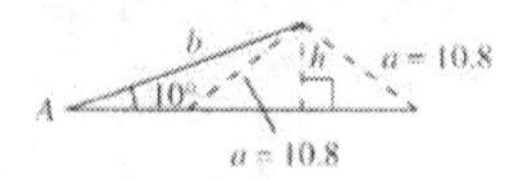

(c) No solution: If $b > \dfrac{10.8}{\sin 10°}$

35. $\text{Area} = \frac{1}{2}ab \sin C$

$= \frac{1}{2}(6)(10) \sin(110°)$

≈ 28.2 square units

37. $\text{Area} = \frac{1}{2}ac \sin A$

$= \frac{1}{2}(67)(85) \sin(38° \ 45')$

≈ 1782.3 square units

39. $\text{Area} = \frac{1}{2}ac \sin B$

$= \frac{1}{2}(103)(58) \sin 75° \ 15'$

≈ 2888.6 square units

41. Angle $CAB = 70°$

Angle $B = 20° + 14° = 34°$

(a)

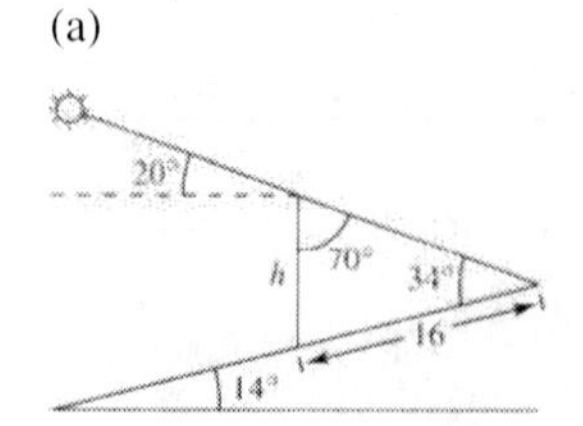

(b) $\dfrac{16}{\sin 70°} = \dfrac{h}{\sin 34°}$

(c) $h = \dfrac{16 \sin 34°}{\sin 70°} \approx 9.52$ meters

43. $\sin A = \dfrac{a \sin B}{b} = \dfrac{500 \sin(46°)}{720} \approx 0.4995$

$A \approx 29.97°$

$\angle ACD = 90° - 29.97° \approx 60°$

Bearing: S 60° W or (240° in plane navigation)

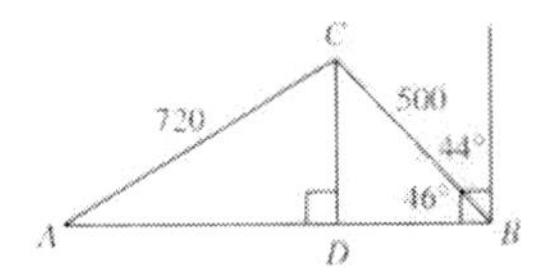

45. $\angle ACD = 65°$

$\angle ADC = 180° - 65° - 15° = 100°$

$\angle CDB = 180° - 100° = 80°$

$\angle B = 180° - 80° - 70° = 30°$

$a = \dfrac{b}{\sin B}(\sin A) = \dfrac{30}{\sin 30°}(\sin 15°) \approx 15.53 \text{ km}$

$c = \dfrac{b}{\sin B}(\sin C) = \dfrac{30}{\sin 30°}(\sin 135°) \approx 42.43 \text{ km}$

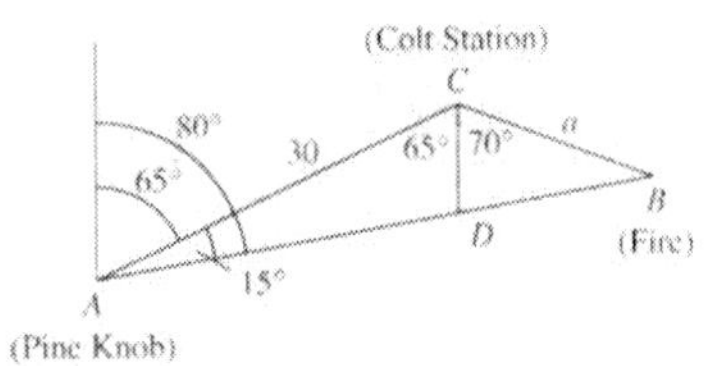

47. $\dfrac{\sin(42° - \theta)}{10} = \dfrac{\sin 48°}{17}$

$\sin(42° - \theta) = \dfrac{10}{17}\sin 48° \approx 0.4371$

$42° - \theta \approx 25.919$

$\theta \approx 16.1°$

49. (a) $\sin\alpha = \dfrac{5.45}{58.36} \approx 0.0934$

$\alpha \approx 5.36°$

(b) $\dfrac{d}{\sin\beta} = \dfrac{58.36}{\sin\theta} \Rightarrow \sin\beta = \dfrac{d \sin\theta}{58.36}$

$\beta = \sin^{-1}\left[\dfrac{d \sin\theta}{58.36}\right]$

(c) $\theta + \beta + 90° + 5.36° = 180° \Rightarrow \beta = 84.64° - \theta$

$d = \sin\beta\left(\dfrac{58.36}{\sin\theta}\right) = \sin(84.64° - \theta)\dfrac{58.36}{\sin\theta}$

(d)

θ	10°	20°	30°	40°	50°	60°
d	324.1	154.2	95.2	63.8	43.3	28.1

51. False. If just the three angles are known, the triangle cannot be solved.

53. False. The cases that give two angles and a side do have unique solutions, they are AAS and ASA.

55. Answers will vary. $A = 36°$, $a = 5$

(a) $b = 4$ one solution

(b) $b = 7$ two solutions $[h = b \sin A < a < b]$

(c) $b = 10$ no solution $[a < h = b \sin A]$

57. $\tan\theta = \dfrac{\sin\theta}{\cos\theta} = -\dfrac{12}{5}$

$\sec\theta = \dfrac{13}{5}$

$\cot\theta = -\dfrac{5}{12}$

$\csc\theta = -\dfrac{13}{12}$

59. $6 \sin 8\theta \cos 3\theta = 6\left(\dfrac{1}{2}\right)[\sin(8\theta + 3\theta) + \sin(8\theta - 3\theta)]$

$= 3(\sin 11\theta + \sin 5\theta)$

61. $3 \cos\dfrac{\pi}{6} \sin\dfrac{5\pi}{3} = 3\left(\dfrac{1}{2}\right)\left[\sin\left(\dfrac{\pi}{6} + \dfrac{5\pi}{3}\right) - \sin\left(\dfrac{\pi}{6} - \dfrac{5\pi}{3}\right)\right]$

$= \dfrac{3}{2}\left[\sin\left(\dfrac{11\pi}{6}\right) - \sin\left(-\dfrac{3\pi}{2}\right)\right]$

$= \dfrac{3}{2}\left[\sin\left(\dfrac{11\pi}{6}\right) + \sin\left(\dfrac{3\pi}{2}\right)\right]$

Section 6.2

1. $c^2 = a^2 + b^2 - 2ab\cos C$

3. No. ASA, two angles and the included side, would use the Law of Sines.

5. Yes. SSS, three sides, would use the Law of Cosines.

7. Given: $a = 12,\ b = 16,\ c = 18$

$$\cos A = \frac{b^2 + c^2 - a^2}{2bc} = \frac{16^2 + 18^2 - 12^2}{2(16)(18)}$$
$$\approx 0.75694 \Rightarrow A \approx 40.80°$$
$$\sin B = \frac{b\sin A}{a} \approx 0.8712 \Rightarrow B \approx 60.61°$$
$$C \approx 180° - 60.61° - 40.80° = 78.59°$$

9. Given: $a = 8.5,\ b = 9.2,\ c = 10.8$

$$\cos A = \frac{b^2 + c^2 - a^2}{2bc} = \frac{9.2^2 + 10.8^2 - 8.5^2}{2(9.2)(10.8)} \approx 0.6493 \Rightarrow A \approx 49.51°$$
$$\sin B = \frac{b\ \sin A}{a} \approx \frac{9.2\ \sin 49.51°}{8.5} \approx 0.82315 \Rightarrow B \approx 55.40°$$
$$C \approx 180° - 55.40° - 49.51° = 75.09°$$

11. Given: $a = 10,\ c = 15,\ B = 20°$

$$b^2 = a^2 + c^2 - 2ac\cos B = 100 + 225 - 2(10)(15)\cos 20°$$
$$\approx 43.0922 \Rightarrow b \approx 6.56 \text{ mm}$$
$$\cos A = \frac{b^2 + c^2 - a^2}{2bc} \approx \frac{43.0922 + 225 - 100}{2(6.56)(15)}$$
$$\approx 0.8541 \Rightarrow A \approx 31.40°$$
$$C \approx 180° - 20° - 31.40° = 128.60°$$

13. Given: $a = 6,\ b = 8,\ c = 12$

$$\cos A = \frac{b^2 + c^2 - a^2}{2bc} = \frac{64 + 144 - 36}{2(8)(12)} \approx 0.8958 \Rightarrow A \approx 26.4°$$
$$\sin B = \frac{b\sin A}{a} \approx \frac{8\sin 26.4°}{6} \approx 0.5928 \Rightarrow B \approx 36.3°$$
$$C \approx 180° - 26.4° - 36.3° = 117.3°$$

15. Given: $A = 50°,\ b = 15,\ c = 30$

$$a^2 = b^2 + c^2 - 2bc\cos A = 225 + 900 - 2(15)(30)\cos 50°$$
$$\approx 546.49 \Rightarrow a \approx 23.38$$
$$\cos B = \frac{a^2 + c^2 - b^2}{2ac} \approx \frac{546.49 + 900 - 225}{2(23.4)(30)}$$
$$\approx 0.8708 \Rightarrow B \approx 29.4°$$
$$C = 180° - A - B \approx 180° - 50° - 29.5° = 100.6°$$

17. Given: $a = 9,\ b = 12,\ c = 15$

$$\cos C = \frac{a^2 + b^2 - c^2}{2ab} = \frac{81 + 144 - 225}{2(9)(12)} = 0 \Rightarrow C = 90°$$
$$\sin A = \frac{9}{15} = \frac{3}{5} \Rightarrow A \approx 36.9°$$
$$B \approx 180° - 90° - 36.9° = 53.1°$$

19. Given: $a = 75.4,\ b = 48,\ c = 48$

$$\cos A = \frac{b^2 + c^2 - a^2}{2bc} = \frac{48^2 + 48^2 - 75.4^2}{2(48)(48)} \approx -0.2338 \Rightarrow A \approx 103.5°$$
$$\sin B = \frac{b\ \sin A}{a} \approx \frac{48\ \sin(103.5°)}{75.4} \approx 0.6190 \Rightarrow B \approx 38.2°$$

$C = B \approx 38.2°$ (Because of roundoff error, $A + B + C \neq 180°$.)

21. Given: $B = 8°15' = 8.25°,\ a = 26,\ c = 18$

$$b^2 = a^2 + c^2 - 2ac\ \cos B = 26^2 + 18^2 - 2(26)(18)\ \cos(8.25°)$$
$$\approx 73.6863 \Rightarrow b \approx 8.58$$
$$\sin C = \frac{c\ \sin B}{b} \approx \frac{18\ \sin(8.25°)}{8.58} \approx 0.3 \Rightarrow C \approx 17.51° \approx 17°\ 31'$$
$$A = 180° - B - C \approx 180° - 8.25° - 17.51° = 154.24° \approx 154°\ 14'$$

23. Given: $B = 75°20' \approx 75.33°,\ a = 6.2,\ c = 9.5$

$$b^2 = a^2 + c^2 - 2ac\ \cos B$$
$$= 6.2^2 + 9.5^2 - 2(6.2)(9.5)\cos(75.33°)$$
$$= 98.86$$
$$b \approx 9.94$$
$$\sin C = \frac{c\sin B}{b} = \frac{9.5\sin(75.33°)}{9.94}$$
$$\approx 0.9246 \Rightarrow C \approx 67°33'53''$$
$$A = 180° - B - C \approx 37°6'7''$$

25. $d^2 = 4^2 + 8^2 - 2(4)(8)\cos 30°$

$$\approx 24.57 \Rightarrow d \approx 4.96$$
$$2\phi = 360° - 2\theta \Rightarrow \phi = 150°$$
$$c^2 = 4^2 + 8^2 - 2(4)(8)\cos 150° \approx 135.43$$
$$c \approx 11.64$$

27. $\cos\phi = \dfrac{10^2 + 14^2 - 20^2}{2(10)(14)}$

$\phi \approx 111.8°$

$2\theta = 360° - 2(111.80°)$

$\theta = 68.2°$

$d^2 = 10^2 + 14^2 - 2(10)(14)\ \cos 68.2°$

$d \approx 13.86$

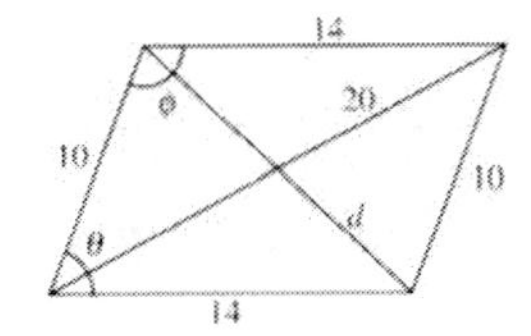

29. $\cos\alpha = \dfrac{15^2 + 12.5^2 - 10^2}{2(15)(12.5)} = 0.75 \Rightarrow \alpha \approx 41.41°$

$\cos\beta = \dfrac{15^2 + 10^2 - 12.5^2}{2(15)(10)} = 0.5625 \Rightarrow \beta \approx 55.77°$

$\delta = 180° - 41.41° - 55.77° \approx 82.82°$

$\mu = 180° - \delta \approx 97.18°$

$b^2 = 12.5^2 + 10^2 - 2(12.5)(10)\ \cos(97.18°) \approx 287.50$

$b \approx 16.96$

$\sin\omega = \dfrac{10}{16.96}\sin\mu \approx 0.585 \Rightarrow \omega \approx 35.8°$

$\sin\epsilon = \dfrac{12.5}{16.99}\sin\mu \approx 0.731 \Rightarrow \epsilon \approx 47°$

$\theta = \alpha + \omega \approx 77.2°$

$\phi = \beta + \epsilon \approx 102.8°$

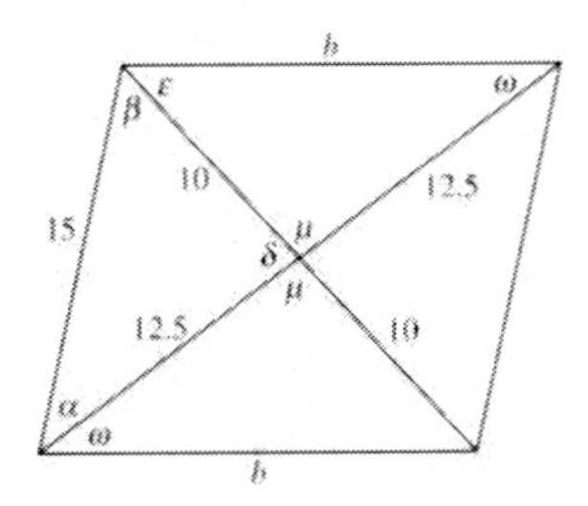

31. Given: $a = 8,\ c = 5,\ B = 40°$

Given two sides and included angle, use the Law of Cosines.

$b^2 = a^2 + c^2 - 2ac\ \cos B = 64 + 25 - 2(8)(5)\ \cos 40°$

$\approx 27.7164 \Rightarrow b \approx 5.26$

$\cos A = \dfrac{b^2 + c^2 - a^2}{2bc} \approx \dfrac{(5.26)^2 + 25 - 64}{2(5.26)(5)}$

$\approx -0.2154 \Rightarrow A \approx 102.44°$

$C \approx 180° - 102.44° - 40° \approx 37.56°$

33. Given: $A = 24°,\ a = 4,\ b = 18$

Given two sides and an angle opposite one of them, use the Law of Sines.

$h = b\sin A = 18\sin 24° \approx 7.32$

Because $a < h$, no triangle is formed.

35. Given: $A = 42°,\ B = 35°,\ c = 1.2$

Given two angles and a side, use the Law of Sines.

$C = 180° - 42° - 35° = 103°$

$a = \dfrac{c\sin A}{\sin C} = \dfrac{1.2\sin 42°}{\sin 103°} \approx 0.82$

$b = \dfrac{c\sin B}{\sin C} = \dfrac{1.2\sin 35°}{\sin 103°} \approx 0.71$

37. Given: $a = 12,\ b = 24,\ c = 18$

$s = \dfrac{a+b+c}{2} = 27$

$\text{Area} = \sqrt{s(s-a)(s-b)(s-c)}$

$= \sqrt{27(15)(3)(9)}$

≈ 104.57 square inches

39. Given: $a = 5,\ b = 8,\ c = 10$

$s = \dfrac{a+b+c}{2} = \dfrac{23}{2} = 11.5$

$\text{Area} = \sqrt{s(s-a)(s-b)(s-c)}$

$= \sqrt{11.5(6.5)(3.5)(1.5)}$

≈ 19.81 square units

41. Given: $a = 1.24,\ b = 2.45,\ c = 1.25$

$s = \dfrac{a+b+c}{2} = 2.47$

$\text{Area} = \sqrt{s(s-a)(s-b)(s-c)}$

$= \sqrt{2.47(1.23)(0.02)(1.22)}$

≈ 0.27 square feet

43. Given: $a = 3.5,\ b = 10.2,\ c = 9$

$s = \dfrac{a+b+c}{2} = 11.35$

$\text{Area} = \sqrt{s(s-a)(s-b)(s-c)}$

$= \sqrt{11.35(7.85)(1.15)(2.35)}$

≈ 15.52 square units

45. Given: $a = 10.59,\ b = 6.65,\ c = 12.31$

$s = \dfrac{a+b+c}{2} = 14.775$

$\text{Area} = \sqrt{s(s-a)(s-b)(s-c}$

$= \sqrt{14.775(4.185)(8.125)(2.465)}$

≈ 35.19 square units

47. Angle at $B = 180° - 80° = 100°$

$b^2 = 240^2 + 380^2 - 2(240)(380)\cos 100°$

$\approx 233{,}673.4 \Rightarrow b \approx 483.4$ meters

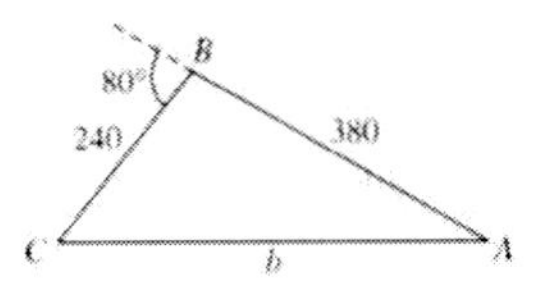

49.

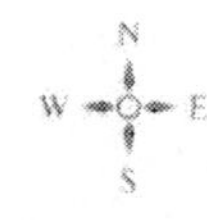

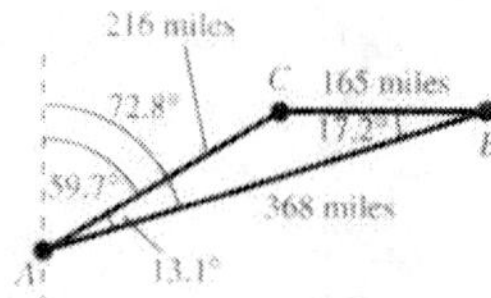

$a = 165,\ b = 216,\ c = 368$

$$\cos B = \frac{165^2 + 368^2 - 216^2}{2(165)(368)} \approx 0.9551$$

$B \approx 17.2°$

$$\cos A = \frac{216^2 + 368^2 - 165^2}{2(216)(368)} \approx 0.9741$$

$A \approx 13.1°$

(a) Bearing of Minneapolis (C) from Phoenix (A)
N (90° − 17.2° − 13.1°) E
N 59.7° E

(b) Bearing of Albany (B) from Phoenix (A)
N (90° − 17.2°) E
N 72.8° E

51. The largest angle is across from the largest side.

$$\cos C = \frac{650^2 + 575^2 - 725^2}{2(650)(575)}$$

$c \approx 72.3°$

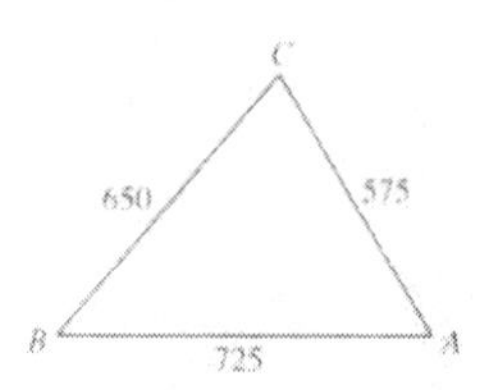

53. $\overline{RS} = \sqrt{8^2 + 10^2} = \sqrt{164} = 2\sqrt{41} \approx 12.8$ feet

$\overline{PQ} = \frac{1}{2}\sqrt{16^2 + 10^2} = \frac{1}{2}\sqrt{356} = \sqrt{89} \approx 9.4$ feet

$\tan P = \frac{10}{16}$

$P = \arctan\frac{5}{8} \approx 32.0°$

$\overline{QS} = \sqrt{8^2 + 9.4^2 - 2(8)(9.4)\cos 32°}$

$\approx \sqrt{24.81} \approx 5.0$ feet

55. $s = \frac{a+b+c}{2} = \frac{145 + 257 + 290}{2} = 346$

Area $= \sqrt{s(s-a)(s-b)(s-c)}$

$= \sqrt{346(201)(89)(56)} \approx 18{,}617.7$ square feet

57. (a) $7^2 = 1.5^2 + x^2 - 2(1.5)(x)\cos\theta$

$49 = 2.25 + x^2 - 3x\cos\theta$

(b) $x^2 - 3x\cos\theta = 46.75$

$$x^2 - 3x\cos\theta + \left(\frac{3\cos\theta}{2}\right)^2 = 46.75 + \left(\frac{3\cos\theta}{2}\right)^2$$

$$\left[x - \frac{3\cos\theta}{2}\right]^2 = \frac{187}{4} + \frac{9\cos^2\theta}{4}$$

$$x - \frac{3\cos\theta}{2} = \pm\sqrt{\frac{187 + 9\cos^2\theta}{4}}$$

Choosing the positive values of x, we have

$$x = \frac{1}{2}\left(3\cos\theta + \sqrt{9\cos^2\theta + 187}\right).$$

(c)

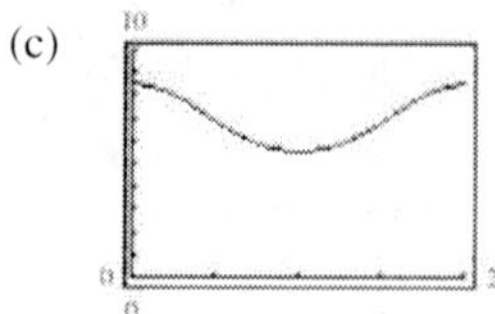

(d) Note that $x = 8.5$ when $\theta = 0$ and $\theta = 2\pi$, and $x = 5.5$ when $\theta = \pi$. Thus, the distance is $2(8.5 - 5.5) = 2(3) = 6$ inches.

59. True. The third side is found by the Law of Cosines. The other angles are determined by the Law of Sines.

61. $$\frac{1}{2}bc(1 + \cos A) = \frac{1}{2}bc\left[1 + \frac{b^2 + c^2 - a^2}{2bc}\right]$$

$$= \frac{1}{2}bc\left[\frac{2bc + b^2 + c^2 - a^2}{2bc}\right]$$

$$= \frac{1}{4}\left[(b+c)^2 - a^2\right]$$

$$= \frac{1}{4}\left[(b+c) + a\right]\left[(b+c) - a\right]$$

$$= \frac{b+c+a}{2} \cdot \frac{b+c-a}{2}$$

$$= \frac{a+b+c}{2} \cdot \frac{-a+b+c}{2}$$

63. Given: $a = 12,\ b = 30,\ A = 20°$

$a^2 = b^2 + c^2 - 2bc\ \cos A$

$12^2 = 30^2 + c^2 - 2(30)(c)\ \cos 20°$

$c^2 - (60\ \cos 20°)c + 756 = 0$

Solving this quadratic equation, $c \approx 21.97,\ 34.41$.

For $c = 21.97$,

$$\cos B = \frac{a^2 + c^2 - b^2}{2ac} \approx \frac{12^2 + 21.97^2 - 30^2}{2(12)(21.97)} \approx -0.5184 \Rightarrow B \approx 121.2°$$

$C \approx 180° - 121.2° - 20° = 38.8°$.

For $c = 34.41$,

$$\cos B = \frac{a^2 + c^2 - b^2}{2ac} \approx \frac{12^2 + 34.41^2 - 30^2}{2(12)(34.41)} \approx 0.5183 \Rightarrow B \approx 58.8°$$

$C \approx 180° - 58.8° - 20° = 101.2°$.

Using the Law of Sines,

$$\sin B = \frac{b\ \sin A}{a} = \frac{30\ \sin 20°}{12}$$

$\approx 0.8551 \Rightarrow B \approx 58.8°$ or $121.2°$.

For $B = 58.8°$, $C = 180° - 58.8° - 20° = 101.2°$ and

$$c = \frac{a\ \sin C}{\sin A} \approx 34.42.$$

For $B = 121.2°$, $C = 180° - 121.2° - 20° = 38.8°$ and

$$c = \frac{a\ \sin C}{\sin A} \approx 21.98.$$

This gives the same result as using the Law of Cosines. An advantage of using the Law of Cosines is that it is easier to choose the correct value to avoid the ambiguous case. Its disadvantage is that there are more computations. The opposite is true for the Law of Sines.

65. Since $0 < C < 180°$, $\cos\left(\frac{C}{2}\right) = \sqrt{\frac{1 + \cos C}{2}}$.

Hence,

$$\cos\left(\frac{C}{2}\right) = \sqrt{\frac{1 + \left(a^2 + b^2 - c^2\right)/2ab}{2}} = \sqrt{\frac{2ab + a^2 + b^2 - c^2}{4ab}}.$$

On the other hand,

$$s(s - c) = \frac{1}{2}(a + b + c)\left(\frac{1}{2}(a + b + c) - c\right)$$
$$= \frac{1}{2}(a + b + c)\frac{1}{2}(a + b - c)$$
$$= \frac{1}{4}((a + b)^2 - c^2)$$
$$= \frac{1}{4}(a^2 + b^2 + 2ab - c^2).$$

Thus, $\sqrt{\frac{s(s - c)}{ab}} = \sqrt{\frac{a^2 + b^2 + 2ab - c^2}{4ab}}$ and we have verified that $\cos\left(\frac{C}{2}\right) = \sqrt{\frac{s(s - c)}{ab}}$.

67. $\arcsin(-1) = -\frac{\pi}{2}$ because $\sin\left(-\frac{\pi}{2}\right) = -1$.

69. $\tan^{-1}\left(\sqrt{3}\right) = \frac{\pi}{3}$ because $\tan\left(\frac{\pi}{3}\right) = \sqrt{3}$.

Section 6.3

1. directed line segment

3. magnitude

5. standard position

7. resultant

9. Two directed line segments that have the same magnitude and direction are equivalent.

11. $\mathbf{u} = \langle 6 - 2,\ 5 - 4\rangle = \langle 4,\ 1\rangle = \mathbf{v}$

13. Initial point: $(0,\ 0)$

Terminal point: $(1,\ 3)$

$\mathbf{v} = \langle 1 - 0,\ 3 - 0\rangle = \langle 1,\ 3\rangle$

$\|\mathbf{v}\| = \sqrt{(1)^2 + (3)^2} = \sqrt{10} \approx 3.16$

15. Initial point: $(2,\ 2)$

Terminal point: $(-1,\ 4)$

$\mathbf{v} = \langle -1 - 2,\ 4 - 2\rangle = \langle -3,\ 2\rangle$

$\|\mathbf{v}\| = \sqrt{(-3)^2 + 2^2} = \sqrt{13} \approx 3.61$

17. Initial point: $(3,\ -2)$

Terminal point: $(3,\ 3)$

$\mathbf{v} = \langle 3 - 3,\ 3 - (-2)\rangle = \langle 0,\ 5\rangle$

$\|\mathbf{v}\| = 5$

19. Initial point: $(-3,\ -5)$

Terminal point: $(5,\ 1)$

$\mathbf{v} = \langle 5 - (-3),\ 1 - (-5)\rangle = \langle 8,\ 6\rangle$

$\|\mathbf{v}\| = \sqrt{(8)^2 + (6)^2} = \sqrt{64 + 36} = \sqrt{100} = 10$

21. Initial point: $\left(\frac{2}{5}, 1\right)$

Terminal point: $\left(1, \frac{2}{5}\right)$

$$\mathbf{v} = \left\langle 1 - \tfrac{2}{5}, \tfrac{2}{5} - 1 \right\rangle = \left\langle \tfrac{3}{5}, -\tfrac{3}{5} \right\rangle$$

$$\|\mathbf{v}\| = \sqrt{\left(\tfrac{3}{5}\right)^2 + \left(-\tfrac{3}{5}\right)^2} = \sqrt{\tfrac{18}{25}} = \tfrac{3}{5}\sqrt{2} \approx 0.85$$

23. Initial point: $\left(-\frac{2}{3}, -1\right)$

Terminal point: $\left(\frac{1}{2}, \frac{4}{5}\right)$

$$\mathbf{v} = \left\langle \tfrac{1}{2} - \left(-\tfrac{2}{3}\right), \tfrac{4}{5} - (-1) \right\rangle = \left\langle \tfrac{7}{6}, \tfrac{9}{5} \right\rangle$$

$$\|\mathbf{v}\| = \sqrt{\left(\frac{7}{6}\right)^2 + \left(\frac{9}{5}\right)^2} = \frac{\sqrt{4141}}{30} \approx 2.1450$$

25. $-\mathbf{v}$

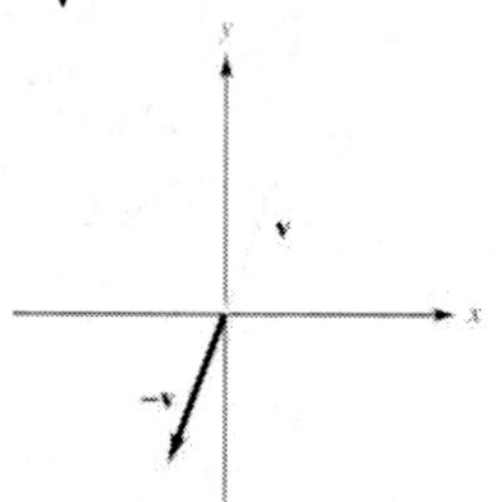

27. $\mathbf{u} + \mathbf{v}$

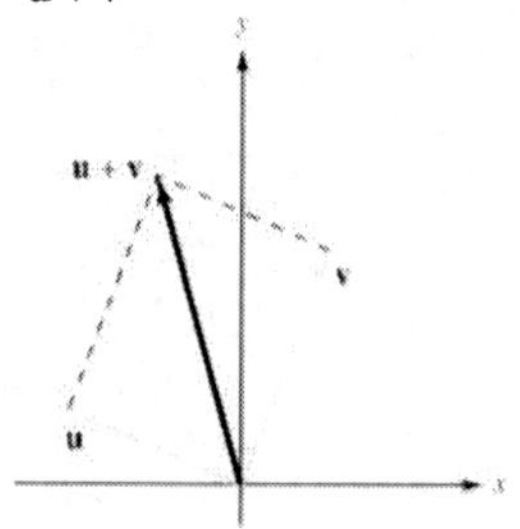

29. $\mathbf{u} + 2\mathbf{v}$

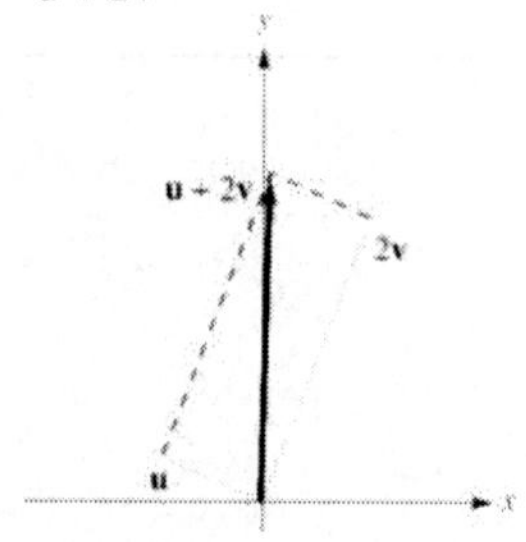

31.

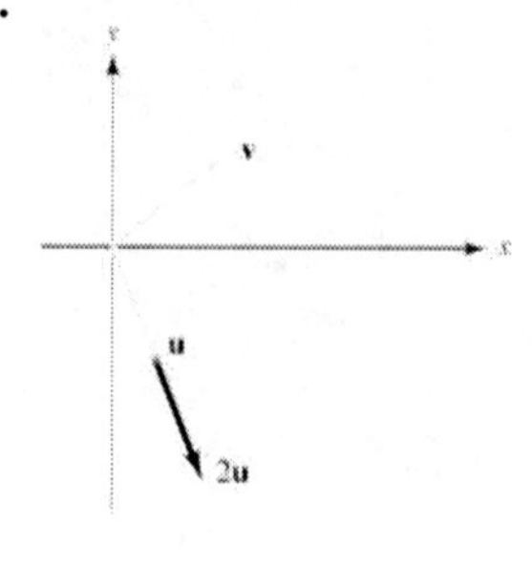

33.

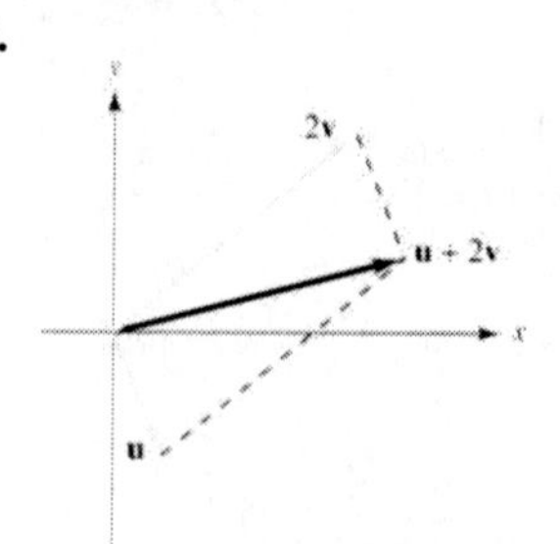

35.

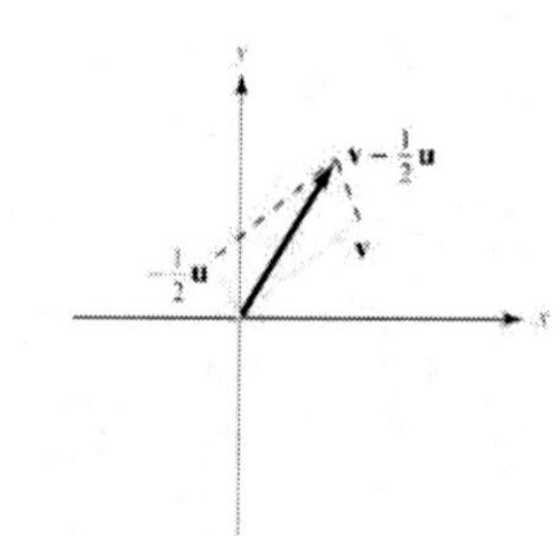

37. $\mathbf{u} = \langle 4, 2 \rangle$, $\mathbf{v} = \langle 7, 1 \rangle$

(a) $\mathbf{u} + \mathbf{v} = \langle 11, 3 \rangle$

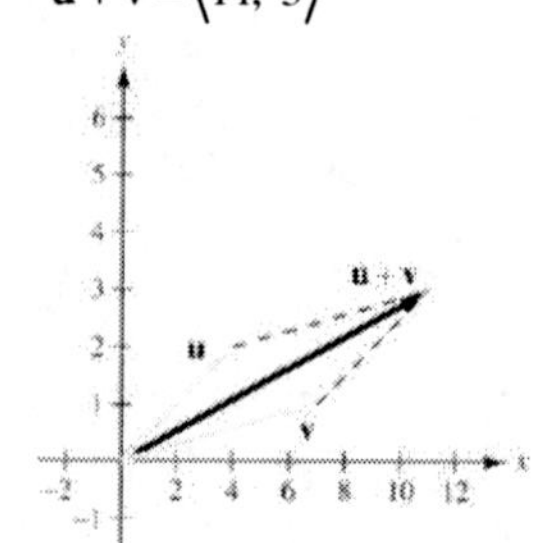

(b) $\mathbf{u} - \mathbf{v} = \langle -3, 1 \rangle$

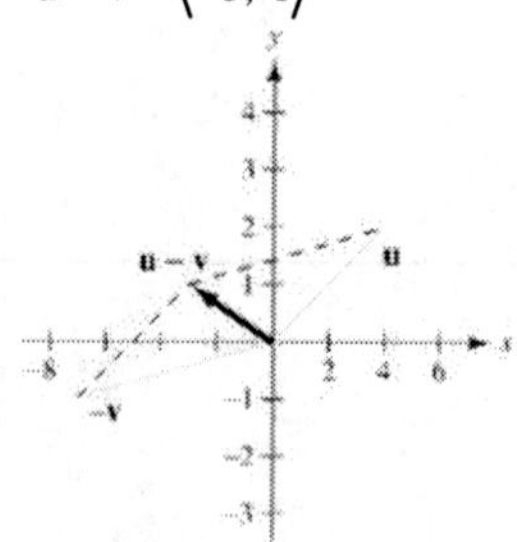

(c) $2\mathbf{u} - 3\mathbf{v} = \langle 8, 4 \rangle - \langle 21, 3 \rangle = \langle -13, 1 \rangle$

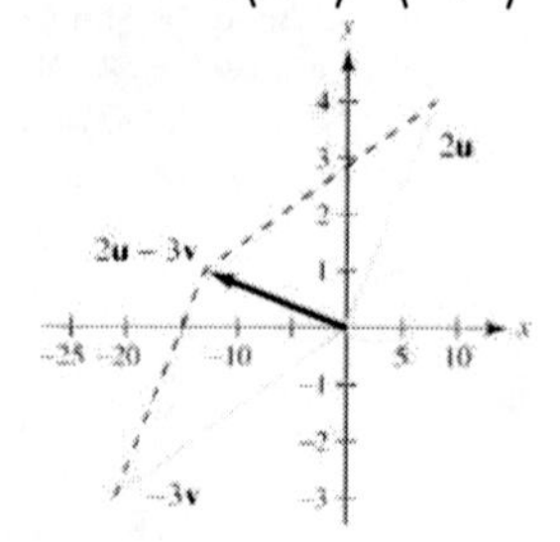

(d) $\mathbf{v} + 4\mathbf{u} = \langle 7, 1\rangle + \langle 16, 8\rangle = \langle 23, 9\rangle$

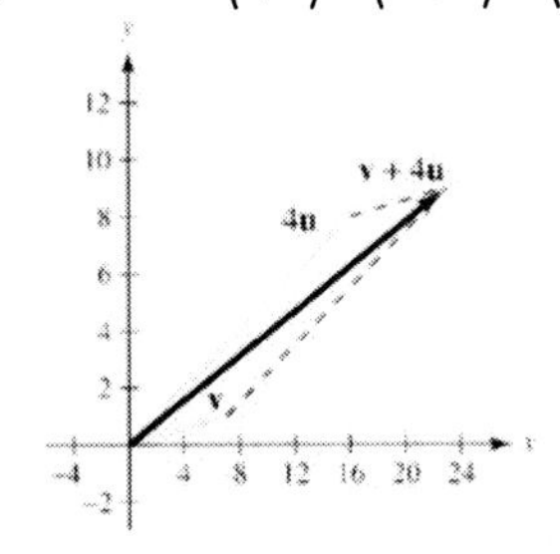

39. $\mathbf{u} = \langle -6, -8\rangle, \mathbf{v} = \langle 2, 4\rangle$

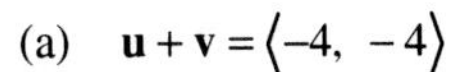
(a) $\mathbf{u} + \mathbf{v} = \langle -4, -4\rangle$

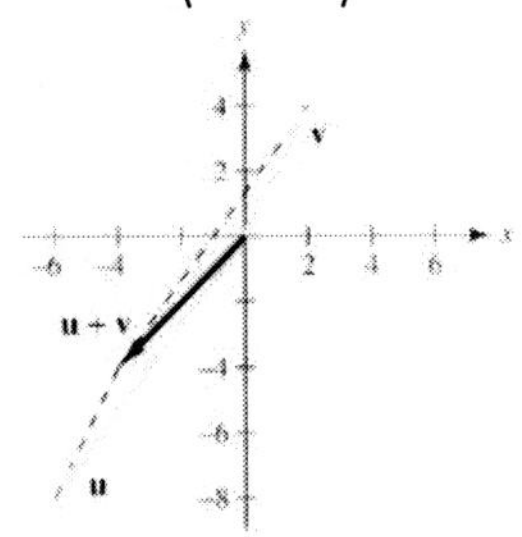

(b) $\mathbf{u} - \mathbf{v} = \langle -8, -12\rangle$

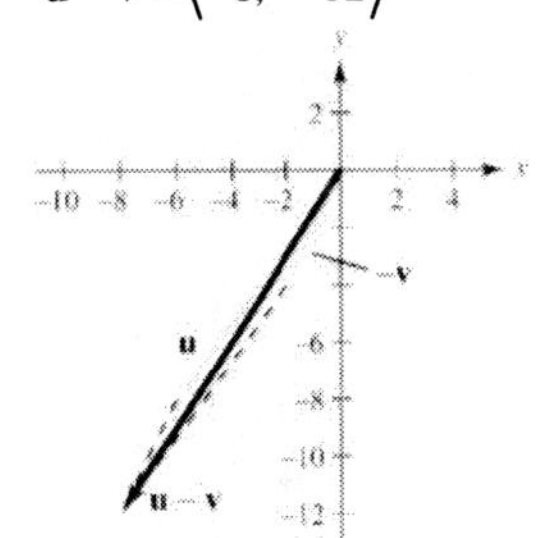

(c) $2\mathbf{u} - 3\mathbf{v} = \langle -12, -16\rangle - \langle 6, 12\rangle = \langle -18, -28\rangle$

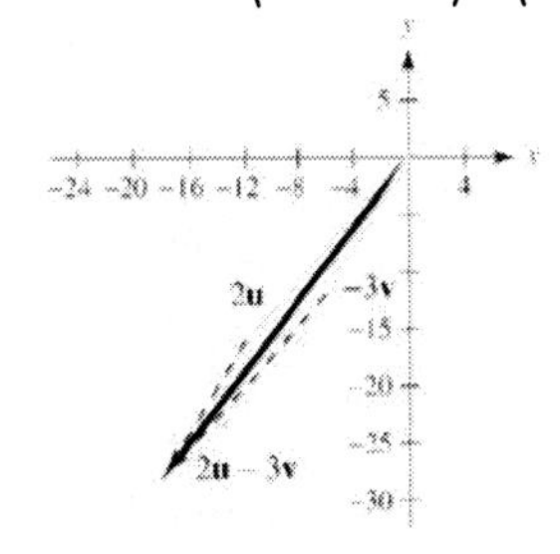

(d) $\mathbf{v} + 4\mathbf{u} = \langle 2, 4\rangle + 4\langle -6, -8\rangle = \langle -22, -28\rangle$

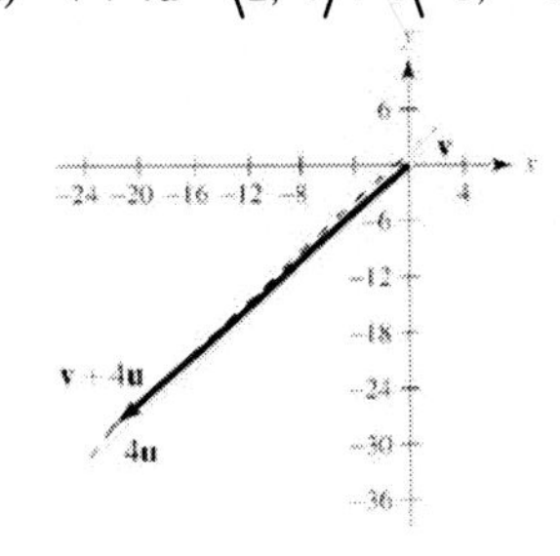

41. $\mathbf{u} = \mathbf{i} + \mathbf{j}, \mathbf{v} = 2\mathbf{i} - 3\mathbf{j}$

(a) $\mathbf{u} + \mathbf{v} = 3\mathbf{i} - 2\mathbf{j}$

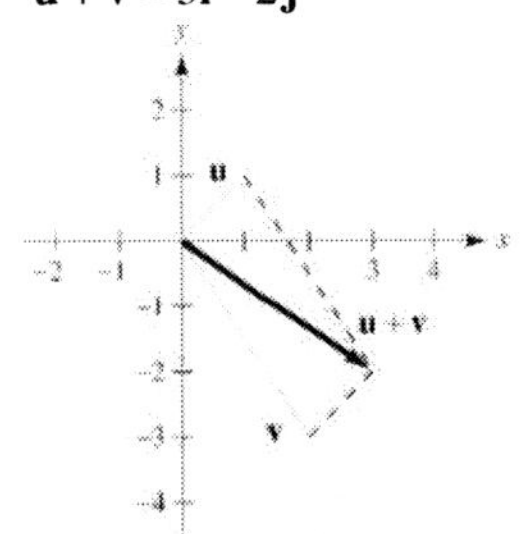

(b) $\mathbf{u} - \mathbf{v} = -\mathbf{i} + 4\mathbf{j}$

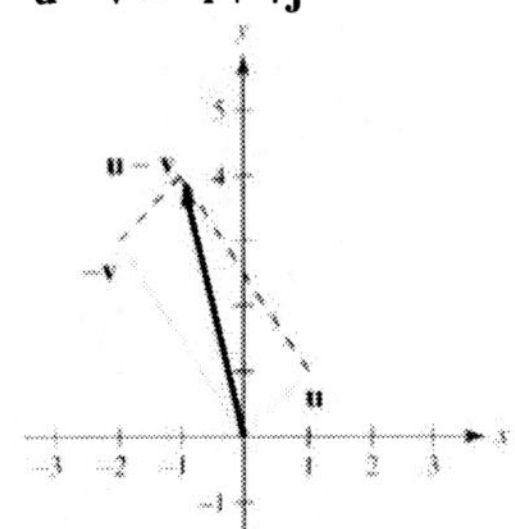

(c) $2\mathbf{u} - 3\mathbf{v} = (2\mathbf{i} + 2\mathbf{j}) - (6\mathbf{i} - 9\mathbf{j}) = -4\mathbf{i} + 11\mathbf{j}$

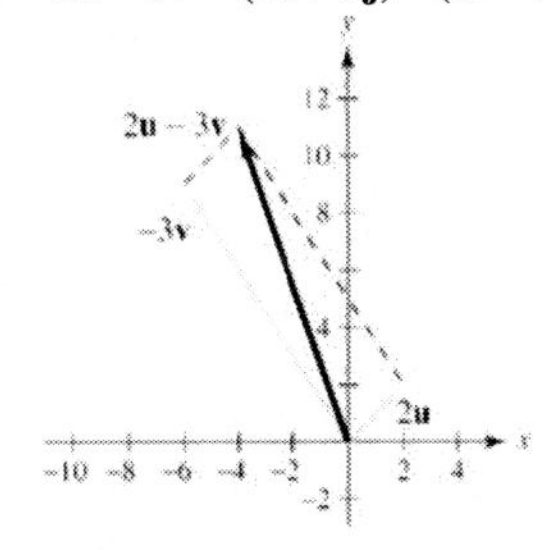

(d) $\mathbf{v} + 4\mathbf{u} = (2\mathbf{i} - 3\mathbf{j}) + (4\mathbf{i} + 4\mathbf{j}) = 6\mathbf{i} + \mathbf{j}$

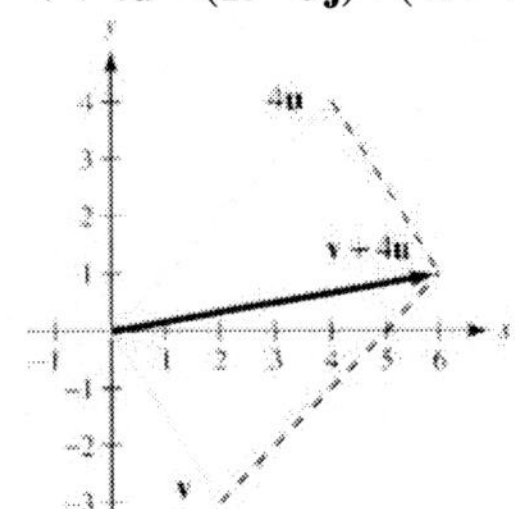

43. $\mathbf{w} = \mathbf{u} + \mathbf{v}$

45. $\mathbf{u} = \mathbf{w} - \mathbf{v}$

47. $\mathbf{u} = \langle 6, 0\rangle$

$\|\mathbf{u}\| = 6$

Unit vector: $\frac{1}{6}\langle 6, 0\rangle = \langle 1, 0\rangle$

49. $\mathbf{v} = \langle -1, 1 \rangle$

$\|\mathbf{v}\| = \sqrt{2}$

$$\text{Unit vector} = \frac{1}{\|\mathbf{v}\|}\mathbf{v} = \left\langle -\frac{1}{\sqrt{2}}, \frac{1}{\sqrt{2}} \right\rangle = \left\langle -\frac{\sqrt{2}}{2}, \frac{\sqrt{2}}{2} \right\rangle$$

51. $\|\mathbf{v}\| = \|\langle -24, -7 \rangle\| = \sqrt{(-24)^2 + (-7)^2} = 25$

Unit vector: $\frac{1}{25}\langle -24, -7 \rangle = \langle -\frac{24}{25}, -\frac{7}{25} \rangle$

53. $\mathbf{u} = \frac{1}{\|\mathbf{v}\|}\mathbf{v} = \frac{1}{\sqrt{16+9}}(4\mathbf{i} - 3\mathbf{j}) = \frac{1}{5}(4\mathbf{i} - 3\mathbf{j}) = \frac{4}{5}\mathbf{i} - \frac{3}{5}\mathbf{j}$

55. $\mathbf{w} = 2\mathbf{j}$

$\mathbf{u} = \frac{1}{2}(2\mathbf{j}) = \mathbf{j}$

57.
$$8\left(\frac{1}{\|\mathbf{u}\|}\mathbf{u}\right) = 8\left(\frac{1}{\sqrt{5^2 + 6^2}}\langle 5, 6 \rangle\right) = \frac{8}{\sqrt{61}}\langle 5, 6 \rangle = \left\langle \frac{40\sqrt{61}}{61}, \frac{48\sqrt{61}}{61} \right\rangle = \frac{40\sqrt{61}}{61}\mathbf{i} + \frac{48\sqrt{61}}{61}\mathbf{j}$$

59.
$$7\left(\frac{1}{\|\mathbf{u}\|}\mathbf{u}\right) = 7\left(\frac{1}{\sqrt{3^2 + 4^2}}\langle 3, 4 \rangle\right) = \frac{7}{5}\langle 3, 4 \rangle = \left\langle \frac{21}{5}, \frac{28}{5} \right\rangle = \frac{21}{5}\mathbf{i} + \frac{28}{5}\mathbf{j}$$

61.
$$8\left(\frac{1}{\|\mathbf{u}\|}\mathbf{u}\right) = 8\left(\frac{1}{2}\langle -2, 0 \rangle\right) = 4\langle -2, 0 \rangle = \langle -8, 0 \rangle = -8\mathbf{i}$$

63. $\mathbf{v} = \langle 4 - (-3), 5 - 1 \rangle = \langle 7, 4 \rangle = 7\mathbf{i} + 4\mathbf{j}$

65. $\mathbf{v} = \langle 2 - (-1), 3 - (-5) \rangle = \langle 3, 8 \rangle = 3\mathbf{i} + 8\mathbf{j}$

67.
$$\mathbf{v} = \frac{3}{2}\mathbf{u} = \frac{3}{2}(2\mathbf{i} - \mathbf{j}) = 3\mathbf{i} - \frac{3}{2}\mathbf{j} = \langle 3, -\frac{3}{2} \rangle$$

69.
$$\mathbf{v} = \mathbf{u} + 2\mathbf{w} = (2\mathbf{i} - \mathbf{j}) + 2(\mathbf{i} + 2\mathbf{j}) = 4\mathbf{i} + 3\mathbf{j} = \langle 4, 3 \rangle$$

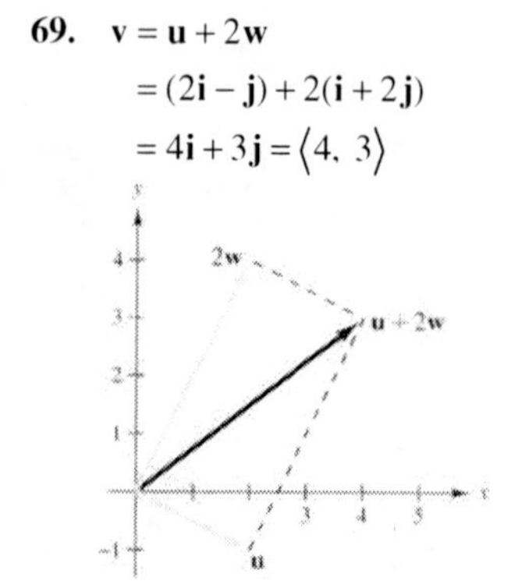

71.
$$\mathbf{v} = \frac{1}{2}(3\mathbf{u} + \mathbf{w}) = \frac{1}{2}(3\langle 2, -1 \rangle + \langle 1, 2 \rangle) = \langle \frac{7}{2}, -\frac{1}{2} \rangle$$

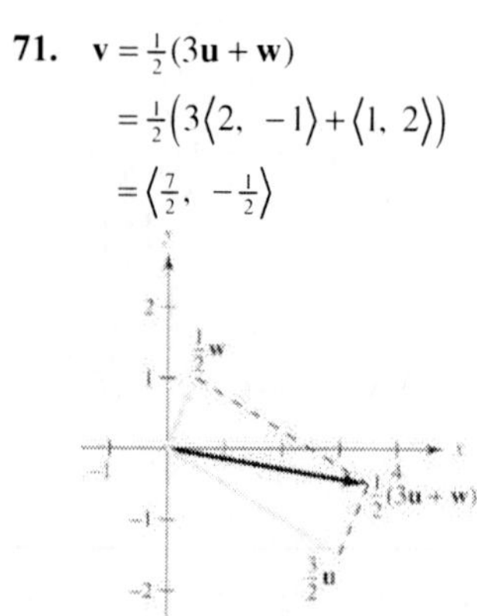

73. $\mathbf{v} = 5(\cos 30°\mathbf{i} + \sin 30°\mathbf{j})$

$\|\mathbf{v}\| = 5, \ \theta = 30°$

75. $\mathbf{v} = 6\mathbf{i} - 6\mathbf{j}$

$\|\mathbf{v}\| = \sqrt{6^2 + (-6)^2} = \sqrt{72} = 6\sqrt{2}$

$\tan\theta = -\frac{6}{6} = -1$

Since $\mathbf{v}$ lies in Quadrant IV, $\theta = 315°$.

77. $\mathbf{v} = -2\mathbf{i} + 5\mathbf{j}$

$\|\mathbf{v}\| = \sqrt{(-2)^2 + 5^2} = \sqrt{29}$

$\tan\theta = -\frac{5}{2}$

Since $\mathbf{v}$ lies in Quadrant II, $\theta \approx 111.8°$.

79. $\mathbf{v} = \langle 3\cos 0°, 3\sin 0° \rangle$

$= \langle 3, 0 \rangle$

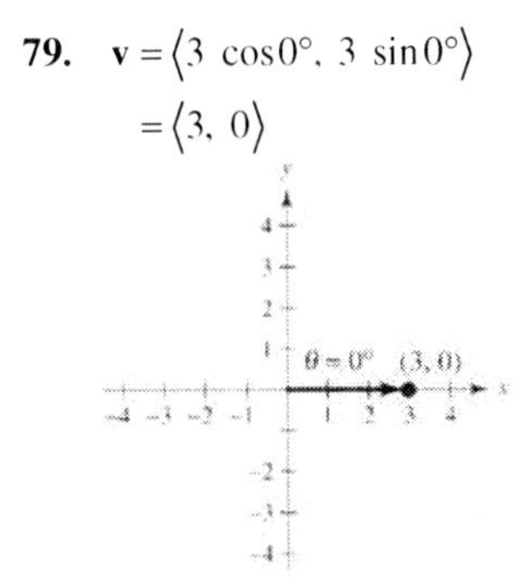

81. $\mathbf{v} = \left\langle \frac{7}{2}\cos 150°, \frac{7}{2}\sin 150° \right\rangle$

$= \left\langle \frac{7}{2}\cdot\left(\frac{-\sqrt{3}}{2}\right), \frac{7}{2}\left(\frac{1}{2}\right) \right\rangle$

$= \left\langle \frac{-7\sqrt{3}}{4}, \frac{7}{4} \right\rangle$

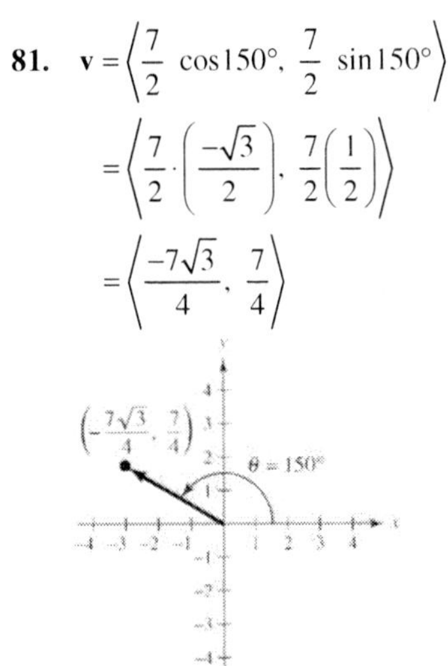

83. $\mathbf{v} = \langle 3\sqrt{2}\cos 150°, 3\sqrt{2}\sin 150° \rangle$

$= \left\langle -\frac{3\sqrt{6}}{2}, \frac{3\sqrt{2}}{2} \right\rangle$

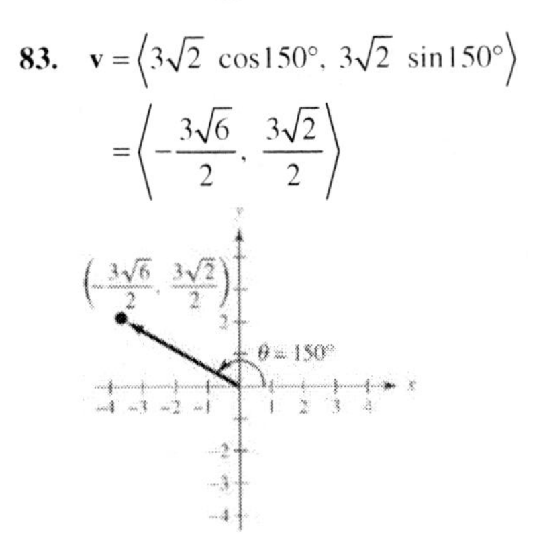

85. $\mathbf{v} = 2\left(\frac{1}{\sqrt{3^2 + 1^2}}\right)(\mathbf{i} + 3\mathbf{j})$

$= \frac{2}{\sqrt{10}}(\mathbf{i} + 3\mathbf{j})$

$= \frac{\sqrt{10}}{5}\mathbf{i} + \frac{3\sqrt{10}}{5}\mathbf{j}$

$= \left\langle \frac{\sqrt{10}}{5}, \frac{3\sqrt{10}}{5} \right\rangle$

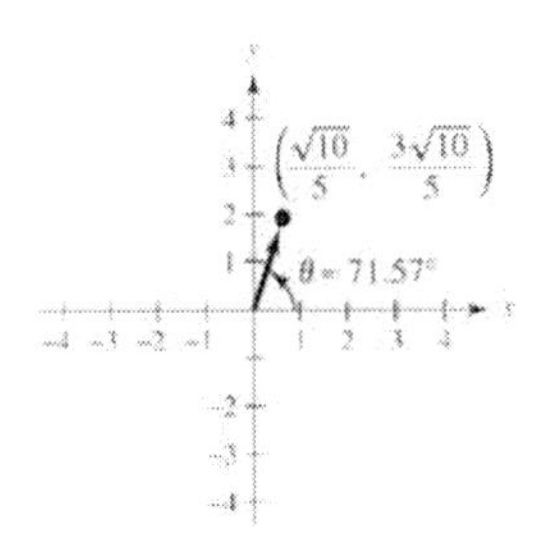

87. $\mathbf{u} = \langle 5\cos 60°, 5\sin 60° \rangle = \left\langle \frac{5}{2}, \frac{5\sqrt{3}}{2} \right\rangle$

$\mathbf{v} = \langle 5\cos 90°, 5\sin 90° \rangle = \langle 0, 5 \rangle$

$\mathbf{u} + \mathbf{v} = \left\langle \frac{5}{2}, \frac{5\sqrt{3}}{2} \right\rangle + \langle 0, 5 \rangle = \left\langle \frac{5}{2}, 5 + \frac{5}{2}\sqrt{3} \right\rangle$

89. $\mathbf{u} = \langle 20\cos 45°, 20\sin 45° \rangle = \langle 10\sqrt{2}, 10\sqrt{2} \rangle$

$\mathbf{v} = \langle 50\cos 150°, 50\sin 150° \rangle = \langle -25\sqrt{3}, 25 \rangle$

$\mathbf{u} + \mathbf{v} = \langle 10\sqrt{2} - 25\sqrt{3}, 10\sqrt{2} + 25 \rangle$

91. $\mathbf{v} = \mathbf{i} + \mathbf{j}$

$\mathbf{w} = 2(\mathbf{i} - \mathbf{j})$

$\mathbf{u} = \mathbf{v} - \mathbf{w} = -\mathbf{i} + 3\mathbf{j}$

$\|\mathbf{v}\| = \sqrt{2}$

$\|\mathbf{w}\| = 2\sqrt{2}$

$\|\mathbf{v} - \mathbf{w}\| = \sqrt{10}$

$\cos\alpha = \frac{\|\mathbf{v}\|^2 + \|\mathbf{w}\|^2 - \|\mathbf{v} - \mathbf{w}\|^2}{2\|\mathbf{v}\|\|\mathbf{w}\|} = \frac{2 + 8 - 10}{2\sqrt{2}\cdot 2\sqrt{2}} = 0$

$\alpha = 90°$

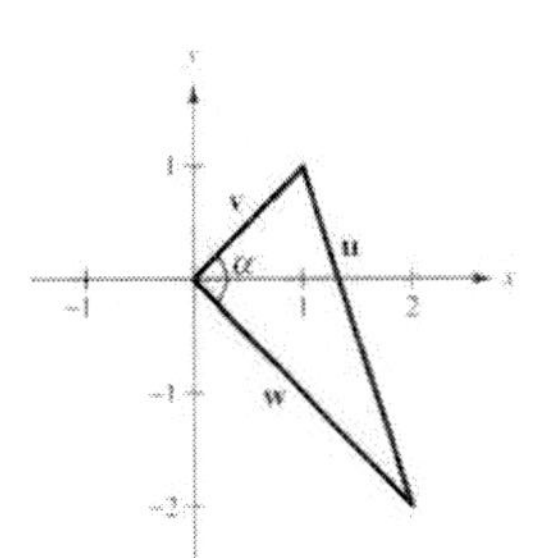

93. $\mathbf{u} = 400\cos 25°\mathbf{i} + 400\sin 25°\mathbf{j}$

$\mathbf{v} = 300\cos 70°\mathbf{i} + 300\sin 70°\mathbf{j}$

$\mathbf{u} + \mathbf{v} \approx 465.13\mathbf{i} + 450.96\mathbf{j}$

$\|\mathbf{u} + \mathbf{v}\| \approx \sqrt{(465.13)^2 + (450.96)^2} \approx 647.85$

$\alpha = \arctan\left(\frac{450.96}{465.13}\right) \approx 44.11°$

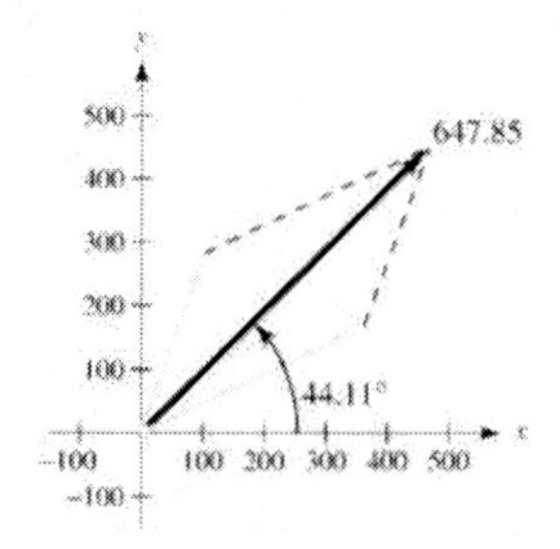

95. Force One: $\mathbf{u} = 45\mathbf{i}$

Force Two: $\mathbf{v} = 60\cos\theta\mathbf{i} + 60\sin\theta\mathbf{j}$

Resultant Force: $\mathbf{u} + \mathbf{v} = (45 + 60\cos\theta)\mathbf{i} + 60\sin\theta\mathbf{j}$

$$\|\mathbf{u} + \mathbf{v}\| = \sqrt{(45 + 60\cos\theta)^2 + (60\sin\theta)^2} = 90$$

$$2025 + 5400\cos\theta + 3600 = 8100$$

$$5400\cos\theta = 2475$$

$$\cos\theta = \frac{2475}{5400} \approx 0.4583$$

$$\theta \approx 62.7°$$

97. Horizontal component of velocity:
$70\cos 40° \approx 53.62$ ft/sec

Vertical component of velocity:
$70\sin 40° \approx 45.0$ ft/sec

99. Rope $\overline{AC}$: $\mathbf{u} = 10\mathbf{i} - 24\mathbf{j}$

The vector lies in Quadrant IV and its reference angle is $\arctan\left(\frac{12}{5}\right)$.

$$\mathbf{u} = \|\mathbf{u}\|\left[\cos\left(\arctan\frac{12}{5}\right)\mathbf{i} - \sin\left(\arctan\frac{12}{5}\right)\mathbf{j}\right]$$

Rope $\overline{BC}$: $\mathbf{v} = -20\mathbf{i} - 24\mathbf{j}$

The vector lies in Quadrant III and its reference angle is $\arctan\left(\frac{6}{5}\right)$.

$$\mathbf{v} = \|\mathbf{v}\|\left[-\cos\left(\arctan\frac{6}{5}\right)\mathbf{i} - \sin\left(\arctan\frac{6}{5}\right)\mathbf{j}\right]$$

Resultant: $\mathbf{u} + \mathbf{v} = -5000\mathbf{j}$

$$\|\mathbf{u}\|\cos\left(\arctan\frac{12}{5}\right) - \|\mathbf{v}\|\cos\left(\arctan\frac{6}{5}\right) = 0$$

$$-\|\mathbf{u}\|\sin\left(\arctan\frac{12}{5}\right) - \|\mathbf{v}\|\sin\left(\arctan\frac{6}{5}\right) = -5000$$

Solving this system of equations yields:

$T_{AC} = \|\mathbf{u}\| \approx 3611.1$ pounds

$T_{BC} = \|\mathbf{v}\| \approx 2169.5$ pounds

101. (a) Tow line 1: $\mathbf{u} = \|\mathbf{u}\|(\cos\theta\mathbf{i} + \sin\theta\mathbf{j})$

Tow line 2: $\mathbf{v} = \|\mathbf{u}\|(\cos(-\theta)\mathbf{i} + \sin(-\theta)\mathbf{j})$

Resultant:

$$\mathbf{u} + \mathbf{v} = 6000\mathbf{i} = \left[\|\mathbf{u}\|\cos\theta + \|\mathbf{u}\|\cos(-\theta)\right]\mathbf{i} \Rightarrow$$

$$6000 = 2\|\mathbf{u}\|\cos\theta \Rightarrow \|\mathbf{u}\| \approx 3000\sec\theta$$

$T = \|\mathbf{u}\| = 3000\sec\theta$

Domain: $0° \le \theta < 90°$

(b)

θ	10°	20°	30°	40°	50°	60°
T	3046.3	3192.5	3464.1	3916.2	4667.2	6000.0

(c)

(d) The tension increases because the component in the direction of the motion of the barge decreases.

103. (a) $\mathbf{u} = 220\mathbf{i}$, $\mathbf{v} = 150\cos 30°\mathbf{i} + 150\sin 30°\mathbf{j}$

$$\mathbf{u} + \mathbf{v} = \left(220 + 75\sqrt{3}\right)\mathbf{i} + 75\mathbf{j}$$

$$\|\mathbf{u} + \mathbf{v}\| = \sqrt{\left(220 + 75\sqrt{3}\right)^2 + 75^2} \approx 357.85 \text{ newtons}$$

$$\tan\theta = \frac{75}{220 + 75\sqrt{3}} \Rightarrow \theta \approx 12.1°$$

(b) $\mathbf{u} + \mathbf{v} = 220\mathbf{i} + (150\cos\theta\mathbf{i} + 150\sin\theta\mathbf{j})$

$$M = \|\mathbf{u} + \mathbf{v}\| = \sqrt{220^2 + 150^2(\cos^2\theta + \sin^2\theta) + 2(220)(150)\cos\theta}$$

$$= \sqrt{70{,}900 + 66{,}000\cos\theta} = 10\sqrt{709 + 660\cos\theta}$$

$$\alpha = \arctan\left(\frac{15\sin\theta}{22 + 15\cos\theta}\right)$$

(c)

θ	0°	30°	60°	90°	120°	150°	180°
M	370.0	357.9	322.3	266.3	194.7	117.2	70.0
α	0°	12.1°	23.8°	34.3°	41.9°	39.8°	0°

(d)

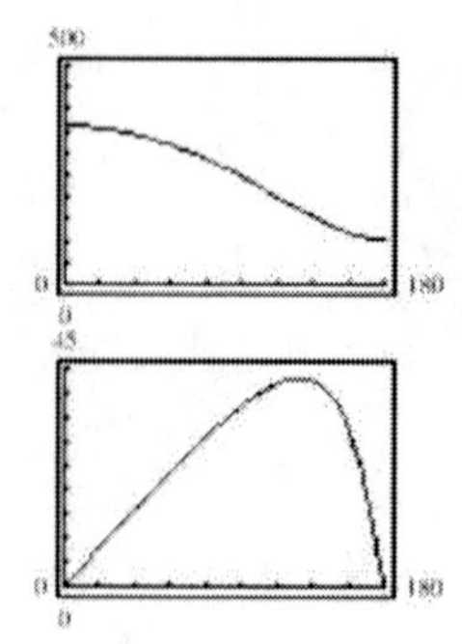

(e) For increasing θ, the two vectors tend to work against each other, resulting in a decrease in the magnitude of the resultant.

105. Airspeed: $\mathbf{v} = 860(\cos 302°\mathbf{i} + \sin 302°\mathbf{j})$

Groundspeed: $\mathbf{u} = 800(\cos 310°\mathbf{i} + \sin 310°\mathbf{j})$

$\mathbf{w} + \mathbf{v} = \mathbf{u}$

$$\mathbf{w} = \mathbf{u} - \mathbf{v} = 800(\cos 310°\mathbf{i} + \sin 310°\mathbf{j}) - 860(\cos 302°\mathbf{i} + \sin 302°\mathbf{j})$$
$$\approx 58.50\mathbf{i} + 116.49\mathbf{j}$$

$$\|\mathbf{w}\| = \sqrt{58.50^2 + 116.49^2} \approx 130.35 \text{ km/hr}$$

$$\theta = \arctan\left(\frac{116.49}{58.50}\right) \approx 63.3°$$

Direction: N 26.7° E

107. True by definition.

109. True. In fact, $a = b = 0$.

111. True. **a** and **d** are parallel, and pointing in opposite directions.

113. True. $\mathbf{a} - \mathbf{b} = \mathbf{c}$ and $\mathbf{u} = -\mathbf{b}$

115. True. $\mathbf{a} + \mathbf{w} = 2\mathbf{a} = -2\mathbf{d}$

117. False. $\mathbf{u} - \mathbf{v} = 2\mathbf{u}$ and $-2(\mathbf{b} + \mathbf{t}) = -2(-2\mathbf{u}) = 4\mathbf{u}$

119. (a) The angle between them is 0°.
(b) The angle between them is 180°.
(c) No. At most it can be equal to the sum when the angle between them is 0°.

121. Let $\mathbf{v} = (\cos\theta)\mathbf{i} + (\sin\theta)\mathbf{j}$.

$$\|\mathbf{v}\| = \sqrt{\cos^2\theta + \sin^2\theta} = \sqrt{1} = 1$$

Therefore, **v** is a unit vector for any value of θ.

123. Answers will vary. Possible answers:
(a) To add two vectors geometrically, first position them (without changing their lengths or directions) so that the initial point of the second vector **v** coincides with the terminal point of the first vector **u**. Then, the sum $\mathbf{u} + \mathbf{v}$ is the vector formed by joining the initial point of the first vector **u** with the terminal point of the second vector **v**.
(b) The product of a vector **v** and a scalar k is the vector that is $|k|$ times as long as **v**. If k is positive, then kv has the same direction as **v**, and if k is negative, then $k\mathbf{v}$ has the opposite direction as **v**.

125. $\mathbf{u} = \langle 5-1,\ 2-6\rangle = \langle 4,\ -4\rangle$

$\mathbf{v} = \langle 9-4,\ 4-5\rangle = \langle 5,\ -1\rangle$

$\mathbf{u} - \mathbf{v} = \langle -1,\ -3\rangle$

$\mathbf{v} - \mathbf{u} = \langle 1,\ 3\rangle$

127. $$\left(\frac{6x^4}{7y^{-2}}\right)\left(14x^{-1}y^5\right) = \frac{12x^4y^5y^2}{x}$$
$$= 12x^3y^7,\ x \neq 0,\ y \neq 0$$

129. $$(18x)^0(4xy)^2(3x^{-1}) = \frac{16x^2y^2(3)}{x}$$
$$= 48xy^2,\ x \neq 0$$

131. $(2.1\times10^9)(3.4\times10^{-4}) = 7.14\times10^5$

133. $\cos x(\cos x + 1) = 0$

$$\cos x = 0 \Rightarrow x = \frac{\pi}{2} + n\pi$$
$$\cos x = -1 \Rightarrow x = \pi + 2n\pi$$

135. $3\sec x + 4 = 10$

$$\sec x = 2$$
$$\cos x = \frac{1}{2}$$
$$x = \frac{\pi}{3} + 2n\pi,\ \frac{5\pi}{3} + 2n\pi$$

Section 6.4

1. Yes. Let $\mathbf{u} = \langle u_1,\ u_2\rangle$ and $\mathbf{v} = \langle v_1,\ v_2\rangle$.

$$\mathbf{u}\cdot\mathbf{v} = u_1v_1 + u_2v_2 = v_1u_1 + v_2u_2 = \mathbf{v}\cdot\mathbf{u}$$

3. The dot product of two vectors is a scalar.

5. $$\left(\frac{\mathbf{u}\cdot\mathbf{v}}{\|\mathbf{v}\|^2}\right)\mathbf{v}$$

7. $\mathbf{u}\cdot\mathbf{v} = \langle 6,\ 3\rangle\cdot\langle 2,\ -4\rangle = 6(2) + 3(-4) = 0$

9. $\mathbf{u}\cdot\mathbf{v} = \langle 5,\ 1\rangle\cdot\langle 3,\ -1\rangle = 5(3) + 1(-1) = 14$

11. $\mathbf{u} = \langle 2,\ 2\rangle$

$\mathbf{u}\cdot\mathbf{u} = 2(2) + 2(2) = 8$, scalar

13. $\mathbf{u} = \langle 2,\ 2\rangle,\ \mathbf{v} = \langle -3,\ 4\rangle$

$\mathbf{u}\cdot 2\mathbf{v} = 2\mathbf{u}\cdot\mathbf{v} = 4(-3) + 4(4) = 4$, scalar

15. $(3\mathbf{w}\cdot\mathbf{v})\mathbf{u} = \left(3\langle 1, -4\rangle\cdot\langle -3, 4\rangle\right)\langle 2, 2\rangle$
$= \left(3(-3)+(-12)(4)\right)\langle 2, 2\rangle$
$= -57\langle 2, 2\rangle$
$= \langle -114, -114\rangle$, vector

17. $\mathbf{u} = \langle -5, 12\rangle$
$\|\mathbf{u}\| = \sqrt{\mathbf{u}\cdot\mathbf{u}} = \sqrt{(-5)^2+12^2} = 13$

19. $\mathbf{u} = 20\mathbf{i}+25\mathbf{j}$
$\|\mathbf{u}\| = \sqrt{\mathbf{u}\cdot\mathbf{u}} = \sqrt{(20)^2+(25)^2} = \sqrt{1025} = 5\sqrt{41}$

21. $\mathbf{u} = -4\mathbf{j}$
$\|\mathbf{u}\| = \sqrt{\mathbf{u}\cdot\mathbf{u}} = \sqrt{(-4)(-4)} = 4$

23. $\mathbf{u} = \langle -1, 0\rangle,\ \mathbf{v} = \langle 0, 2\rangle$
$$\cos\theta = \frac{\mathbf{u}\cdot\mathbf{v}}{\|\mathbf{u}\|\ \|\mathbf{v}\|} = \frac{0}{(1)(2)} = 0 \Rightarrow \theta = 90°$$

25. $\mathbf{u} = 3\mathbf{i}+4\mathbf{j},\ \mathbf{v} = -2\mathbf{i}+3\mathbf{j}$
$$\cos\theta = \frac{\mathbf{u}\cdot\mathbf{v}}{\|\mathbf{u}\|\ \|\mathbf{v}\|} = \frac{-6+12}{(5)\left(\sqrt{3}\right)} = \frac{6}{5\sqrt{13}}$$
$$\theta = \arccos\left(\frac{6}{5\sqrt{13}}\right) \approx 70.56°$$

27. $\mathbf{u} = 2\mathbf{i},\ \mathbf{v} = -3\mathbf{j}$
$$\cos\theta = \frac{\mathbf{u}\cdot\mathbf{v}}{\|\mathbf{u}\|\ \|\mathbf{v}\|} = \frac{0}{(2)(3)} = 0 \Rightarrow \theta = 90°$$

29. $$\mathbf{u} = \left(\cos\frac{\pi}{3}\right)\mathbf{i}+\left(\sin\frac{\pi}{3}\right)\mathbf{j} = \frac{1}{2}\mathbf{i}+\frac{\sqrt{3}}{2}\mathbf{j}$$
$$\mathbf{v} = \left(\cos\frac{3\pi}{4}\right)\mathbf{i}+\left(\sin\frac{3\pi}{4}\right)\mathbf{j} = \frac{\sqrt{2}}{2}\mathbf{i}+\frac{\sqrt{2}}{2}\mathbf{j}$$
$\|\mathbf{u}\| = \|\mathbf{v}\| = 1$
$$\cos\theta = \frac{\mathbf{u}\cdot\mathbf{v}}{\|\mathbf{u}\|\ \|\mathbf{v}\|} = \mathbf{u}\cdot\mathbf{v} = \left(\frac{1}{2}\right)\left(-\frac{\sqrt{2}}{2}\right)+\left(\frac{\sqrt{3}}{2}\right)\left(\frac{\sqrt{2}}{2}\right) = \frac{-\sqrt{2}+\sqrt{6}}{4}$$
$$\theta = \arccos\left(\frac{-\sqrt{2}+\sqrt{6}}{4}\right) = 75° = \frac{5\pi}{12}$$

31. $\mathbf{u} = 2\mathbf{i}-4\mathbf{j},\ \mathbf{v} = 3\mathbf{i}-5\mathbf{j}$
$$\cos\theta = \frac{\mathbf{u}\cdot\mathbf{v}}{\|\mathbf{u}\|\ \|\mathbf{v}\|} = \frac{6+20}{\sqrt{20}\sqrt{34}}$$
$$= \frac{13\sqrt{170}}{170} \Rightarrow \theta \approx 4.40°$$

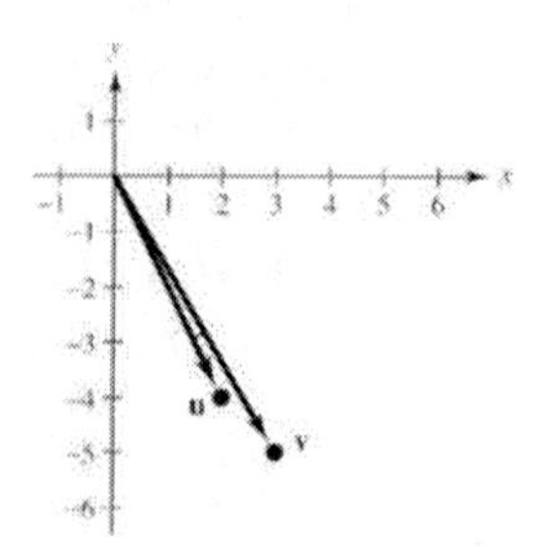

33. $\mathbf{u} = 6\mathbf{i}-2\mathbf{j},\ \mathbf{v} = 8\mathbf{i}-5\mathbf{j}$
$$\cos\theta = \frac{\mathbf{u}\cdot\mathbf{v}}{\|\mathbf{u}\|\ \|\mathbf{v}\|} = \frac{48+10}{\sqrt{40}\sqrt{89}}$$
$$= \frac{29\sqrt{890}}{890} \Rightarrow \theta \approx 13.57°$$

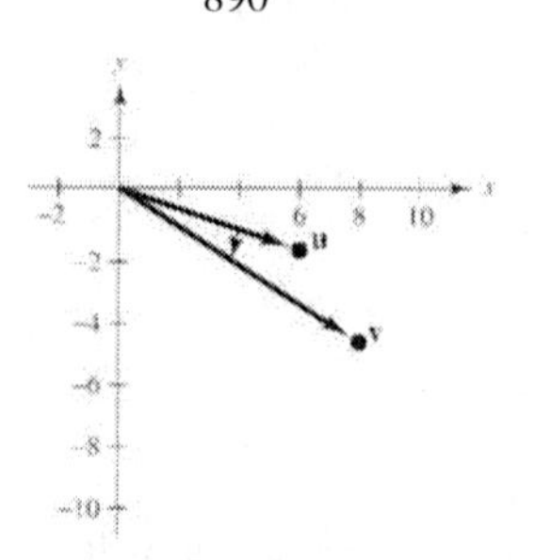

35. $P = (1, 2),\ Q = (3, 4),\ R = (2, 5)$
$\overrightarrow{PQ} = \langle 2, 2\rangle,\ \overrightarrow{PR} = \langle 1, 3\rangle,\ \overrightarrow{QR} = \langle -1, 1\rangle$
$$\cos\alpha = \frac{\overrightarrow{PQ}\cdot\overrightarrow{PR}}{\left\|\overrightarrow{PQ}\right\|\ \left\|\overrightarrow{PR}\right\|} = \frac{8}{\left(2\sqrt{2}\right)\left(\sqrt{10}\right)} \Rightarrow \alpha = \arccos\frac{2}{\sqrt{5}} \approx 26.57°$$
$$\cos\beta = \frac{\overrightarrow{PQ}\cdot\overrightarrow{QR}}{\left\|\overrightarrow{PQ}\right\|\ \left\|\overrightarrow{QR}\right\|} = 0 \Rightarrow \beta = 90°$$
$\gamma \approx 180° - 26.6° - 90° = 63.43°$

37. $P = (-3, 0),\ Q = (2, 2),\ R = (0, 6)$
$\overrightarrow{PQ} = \langle 5, 2\rangle,\ \overrightarrow{QR} = \langle -2, 4\rangle,\ \overrightarrow{PR} = \langle 3, 6\rangle,$
$\overrightarrow{QP} = \langle -5, -2\rangle$
$$\cos\alpha = \frac{\overrightarrow{PQ}\cdot\overrightarrow{PR}}{\left\|\overrightarrow{PQ}\right\|\ \left\|\overrightarrow{PR}\right\|} = \frac{27}{\left(\sqrt{29}\right)\left(\sqrt{45}\right)} \Rightarrow \alpha \approx 41.63°$$
$$\cos\beta = \frac{\overrightarrow{QR}\cdot\overrightarrow{QP}}{\left\|\overrightarrow{QR}\right\|\ \left\|\overrightarrow{QP}\right\|} = \frac{2}{\left(\sqrt{20}\right)\left(\sqrt{29}\right)} \Rightarrow \beta \approx 85.24°$$
$\phi = 180° - 41.6° - 85.2° \approx 53.13°$

39. $\mathbf{u}\cdot\mathbf{v} = \|\mathbf{u}\|\ \|\mathbf{v}\|\ \cos\theta$
$$= (9)(36)\ \cos\frac{3\pi}{4}$$
$$= 324\left(-\frac{\sqrt{2}}{2}\right)$$
$$= -162\sqrt{2}$$

41. $\|\mathbf{u}\| = 4,\ \|\mathbf{v}\| = 10,\ \theta = \dfrac{2\pi}{3}$

$$\begin{aligned}\mathbf{u}\cdot\mathbf{v} &= \|\mathbf{u}\|\ \|\mathbf{v}\|\ \cos\theta\\ &= (4)(10)\cos\frac{2\pi}{3}\\ &= 40\left(-\frac{1}{2}\right)\\ &= -20\end{aligned}$$

43. $$\begin{aligned}\mathbf{u}\cdot\mathbf{v} &= \langle 10,\ -6\rangle\cdot\langle 9,\ 15\rangle\\ &= (10)(9) + (-6)(15)\\ &= 90 - 90\\ &= 0\end{aligned}$$

u and **v** are orthogonal.

45. $$\begin{aligned}\mathbf{u}\cdot\mathbf{v} &= \langle 0,\ 1\rangle\cdot\langle 1,\ -1\rangle\\ &= (0)(1) + (1)(-1)\\ &= 0 - 1 = -1 \neq 0\end{aligned}$$

u and **v** are not orthogonal.

47. $\mathbf{u} = \langle 10,\ 20\rangle$ and $\mathbf{v} = \langle -5,\ 10\rangle$

$\mathbf{u} \neq k\mathbf{v} \Rightarrow$ Not parallel

$$\begin{aligned}\mathbf{u}\cdot\mathbf{v} &= \langle 10,\ 20\rangle\cdot\langle -5,\ 10\rangle\\ &= (10)(-5) + (20)(10)\\ &= -50 + 200 = 150 \neq 0 \Rightarrow \text{Not orthogonal}\end{aligned}$$

49. $\mathbf{u} = -\dfrac{3}{5}\mathbf{i} + \dfrac{7}{10}\mathbf{i}$ and $\mathbf{v} = 12\mathbf{i} - 14\mathbf{j}$

$$\mathbf{u} = k\mathbf{v} \Rightarrow -\frac{3}{5}\mathbf{i} + \frac{7}{10}\mathbf{j} = \left(-\frac{1}{20}\right)(12\mathbf{i} - 14\mathbf{j}) \Rightarrow k = -\frac{1}{20}$$

u and **v** are parallel.

51. $\mathbf{u}\cdot\mathbf{v} = \langle 2,\ -k\rangle\cdot\langle 3,\ 2\rangle = 6 - 2k = 0 \Rightarrow k = 3$

53. $\mathbf{u}\cdot\mathbf{v} = \langle 1,\ 4\rangle\cdot\langle 2k,\ -5\rangle = 2k - 20 = 0 \Rightarrow k = 10$

55. $\mathbf{u}\cdot\mathbf{v} = \langle -3k,\ 2\rangle\cdot\langle -6,\ 0\rangle = 18k = 0 \Rightarrow k = 0$

57. $\mathbf{u} = \langle 3,\ 4\rangle,\ \mathbf{v} = \langle 8,\ 2\rangle$

$$\begin{aligned}\mathbf{w}_1 = \text{proj}_{\mathbf{v}}\mathbf{u} &= \left(\frac{\mathbf{u}\cdot\mathbf{v}}{\|\mathbf{v}\|^2}\right)\mathbf{v}\\ &= \left(\frac{32}{68}\right)\mathbf{v} = \frac{8}{17}\langle 8,\ 2\rangle = \frac{16}{17}\langle 4,\ 1\rangle\end{aligned}$$

$$\mathbf{w}_2 = \mathbf{u} - \mathbf{w}_1 = \langle 3,\ 4\rangle - \frac{16}{17}\langle 4,\ 1\rangle = \frac{13}{17}\langle -1,\ 4\rangle$$

$$\mathbf{u} = \mathbf{w}_1 + \mathbf{w}_2 = \frac{16}{17}\langle 4,\ 1\rangle + \frac{13}{17}\langle -1,\ 4\rangle$$

59. $\mathbf{u} = \langle 0,\ 3\rangle,\ \mathbf{v} = \langle 2,\ 15\rangle$

$$\mathbf{w}_1 = \text{proj}_{\mathbf{v}}\mathbf{u} = \left(\frac{\mathbf{u}\cdot\mathbf{v}}{\|\mathbf{v}\|^2}\right)\mathbf{v} = \frac{45}{229}\langle 2,\ 15\rangle$$

$$\begin{aligned}\mathbf{w}_2 = \mathbf{u} - \mathbf{w}_1 &= \langle 0,\ 3\rangle - \frac{45}{229}\langle 2,\ 15\rangle\\ &= \left\langle -\frac{90}{229},\ \frac{12}{229}\right\rangle = \frac{6}{229}\langle -15,\ 2\rangle\end{aligned}$$

$$\mathbf{u} = \mathbf{w}_1 + \mathbf{w}_2 = \frac{45}{229}\langle 2,\ 15\rangle + \frac{6}{229}\langle -15,\ 2\rangle$$

61. $\text{proj}_{\mathbf{v}}\mathbf{u} = \mathbf{u}$ since they are parallel.

$$\begin{aligned}\text{proj}_{\mathbf{v}}\mathbf{u} &= \frac{\mathbf{u}\cdot\mathbf{v}}{\|\mathbf{v}\|^2}\mathbf{v}\\ &= \frac{18+8}{36+16}\mathbf{v} = \frac{26}{52}\langle 6,\ 4\rangle = \langle 3,\ 2\rangle = \mathbf{u}\end{aligned}$$

63. $\text{proj}_{\mathbf{v}}\mathbf{u} = 0$ since they are perpendicular.

$$\text{proj}_{\mathbf{v}}\mathbf{u} = \frac{\mathbf{u}\cdot\mathbf{v}}{\|\mathbf{v}\|^2}\mathbf{v} = 0,\ \text{since } \mathbf{u}\cdot\mathbf{v} = 0.$$

65. $\mathbf{u} = \langle 2,\ 6\rangle$

For **v** to be orthogonal to **u**, $\mathbf{u}\cdot\mathbf{v}$ must equal 0.

Two possibilities: $\langle 6,\ -2\rangle$ and $\langle -6,\ 2\rangle$

67. $\mathbf{u} = \dfrac{1}{2}\mathbf{i} - \dfrac{3}{4}\mathbf{j}$

For **v** to be orthogonal to **u**, $\mathbf{u}\cdot\mathbf{v}$ must equal 0.

Two possibilities: $\dfrac{3}{4}\mathbf{i} + \dfrac{1}{2}\mathbf{j}$ and $-\dfrac{3}{4}\mathbf{i} - \dfrac{1}{2}\mathbf{j}$

69. $W = \|\text{proj}_{\overrightarrow{PQ}}\mathbf{v}\|\ \|\overrightarrow{PQ}\|$, where $\overrightarrow{PQ} = \langle 4,\ 7\rangle$ and $\mathbf{v} = \langle 1,\ 4\rangle$

$$\text{proj}_{\overrightarrow{PQ}}\mathbf{v} = \left(\frac{\mathbf{v}\cdot\overrightarrow{PQ}}{\|\overrightarrow{PQ}\|^2}\right)\overrightarrow{PQ} = \left(\frac{32}{65}\right)\langle 4,\ 7\rangle$$

$$W = \|\text{proj}_{\overrightarrow{PQ}}\mathbf{v}\|\ \|\overrightarrow{PQ}\| = \left(\frac{32\sqrt{65}}{65}\right)\left(\sqrt{65}\right) = 32$$

71. $\mathbf{u} = \langle 1225,\ 2445\rangle$ and $\mathbf{v} = \langle 12.20,\ 8.50\rangle$

(a) $$\begin{aligned}\mathbf{u}\cdot\mathbf{v} &= (1225)(12.20) + (2445)(8.50)\\ &= \$35{,}727.50\end{aligned}$$

$\mathbf{u}\cdot\mathbf{v}$ is the total amount paid to both levels of employees at the temp agency.

(b) To increase wages by 2%, multiply **v** by 1.02 or $1.02\mathbf{v} = \langle 12.444,\ 8.67\rangle$.

73. (a) $\mathbf{F} = -30{,}000\mathbf{j}$, Gravitational force

$\mathbf{v} = \langle \cos(d^\circ), \sin(d^\circ) \rangle$

$$\mathbf{w}_1 = \text{proj}_{\mathbf{v}}\mathbf{F} = \left(\frac{\mathbf{F}\cdot\mathbf{v}}{\|\mathbf{v}\|^2}\right)\mathbf{v} = (\mathbf{F}\cdot\mathbf{v})\mathbf{v} = -30{,}000\ \sin(d^\circ)\langle \cos d^\circ, \sin d^\circ \rangle$$

$$= \langle -30{,}000\ \sin d^\circ\ \cos d^\circ, -30{,}000\ \sin^2 d^\circ \rangle$$

Force needed: $30{,}000\ \sin(d^\circ)$

(b)

d	0°	1°	2°	3°	4°	5°
Force	0	523.6	047.0	570.1	2092.7	2614.7

d	6°	7°	8°	9°	10°
Force	3135.9	3656.1	4157.2	4693.0	5209.4

(c) $\mathbf{w}_2 = \mathbf{F} - \mathbf{w}_1 = -30{,}000\mathbf{j} + 2614.7(\cos(5^\circ)\mathbf{i} + \sin(5^\circ)\mathbf{j})$

$\approx 2604.75\mathbf{i} - 29{,}772.11\mathbf{j}$

$\|\mathbf{w}_2\| \approx 29{,}885.8$ pounds

75. (a) $\mathbf{F} = 250\langle \cos 30^\circ, \sin 30^\circ \rangle$

$\overrightarrow{PQ} = d\langle 1, 0 \rangle$

$$\mathbf{w} = \mathbf{F}\cdot\overrightarrow{PQ} = 250\frac{\sqrt{3}}{2}d$$

$$= 125\sqrt{3}d$$

(b)

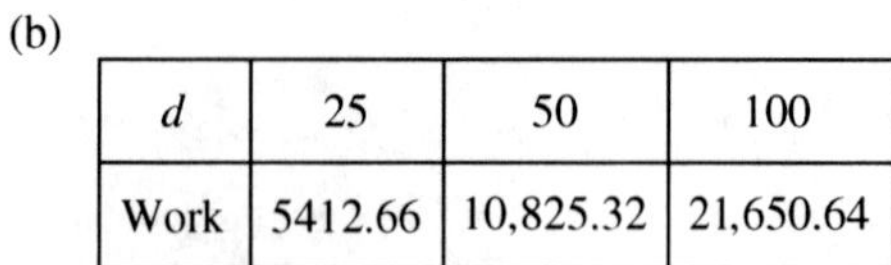

d	25	50	100
Work	5412.66	10,825.32	21,650.64

77. Work $= (\cos 35^\circ)(15{,}691)(800)$

$= 10{,}282{,}651.78$ newton-meters

79. True. $\mathbf{u}\cdot\mathbf{v} = 0$

81. $\mathbf{u}\cdot\mathbf{v} = 0 \Rightarrow$ they are orthogonal (unit vectors).

83. If $\mathbf{u}$ is a unit vector, then $\mathbf{u}\cdot\mathbf{u} = 1$. The angle between $\mathbf{u}$ and itself is 0. So, since $\cos\theta = \dfrac{\mathbf{u}\cdot\mathbf{u}}{\|\mathbf{u}\|\ \|\mathbf{u}\|}$ and if

$\theta = 0$, $\cos = 1 \Rightarrow \dfrac{\mathbf{u}\cdot\mathbf{u}}{\|\mathbf{u}\|\ \|\mathbf{u}\|} = 1$.

85. (a) $\text{proj}_{\mathbf{v}}\mathbf{u} = \mathbf{u} \Rightarrow \mathbf{u}$ and $\mathbf{v}$ are parallel.

(b) $\text{proj}_{\mathbf{v}}\mathbf{u} = \mathbf{0} \Rightarrow \mathbf{u}$ and $\mathbf{v}$ are orthogonal.

87. Use the Law of Cosines on the triangle:

$$\|\mathbf{u}-\mathbf{v}\|^2 = \|\mathbf{u}\|^2 + \|\mathbf{v}\|^2 - 2\|\mathbf{u}\|\ \|\mathbf{v}\|\ \cos\theta$$

$$= \|\mathbf{u}\|^2 + \|\mathbf{v}\|^2 - 2\mathbf{u}\cdot\mathbf{v}$$

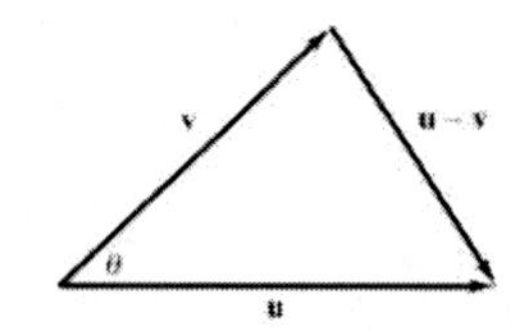

89. From trigonometry, $x = \cos\theta$ and $y = \sin\theta$.

Thus, $\mathbf{u} = \langle x, y \rangle = \cos\theta\mathbf{i} + \sin\theta\mathbf{j}$.

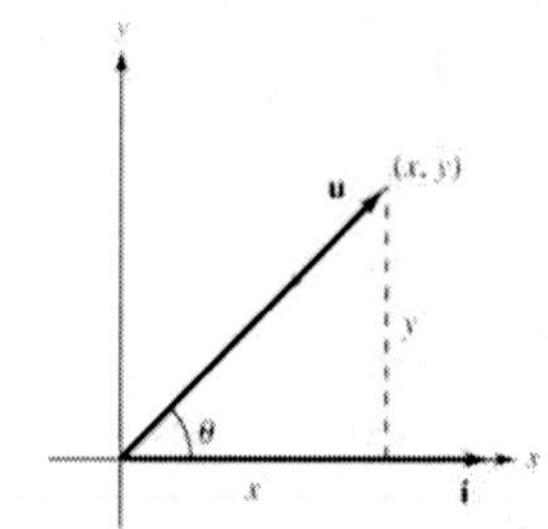

91. $g(x) = f(x-4)$ is a horizontal shift of f four units to the right.

93. $g(x) = f(x) + 6$ is a vertical shift of f six units upward.

95. $3i(4-5i) = 12i + 15 = 15 + 12i$

97. $(1+3i)(1-3i) = 1 - (3i)^2 = 1 + 9 = 10$

99. $$\frac{3}{1+i} + \frac{2}{2-3i} = \frac{3}{1+i}\cdot\frac{1-i}{1-i} + \frac{2}{2-3i}\cdot\frac{2+3i}{2+3i}$$

$$= \frac{3-3i}{2} + \frac{4+6i}{13}$$

$$= \frac{39-39i+8+12i}{26}$$

$$= \frac{47}{26} - \frac{27}{26}i$$

Section 6.5

1. absolute value

3. nth root

5. In the trigonometric form $z = r\ (\cos\theta + i\ \sin\theta)$, r is the distance from the origin to the point (a, b).

7. $|4i| = 4$

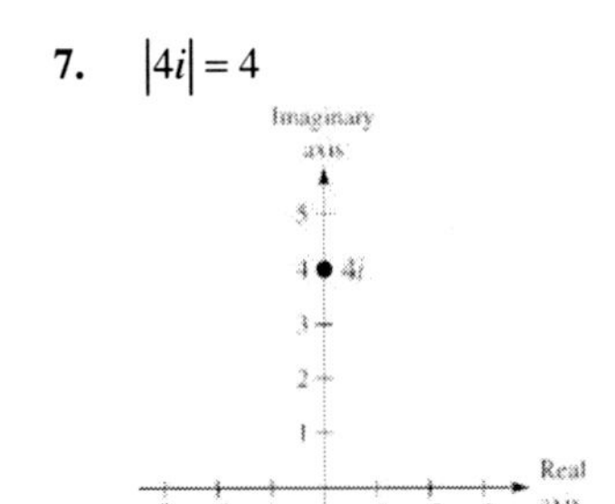

9. $|-5| = 5$

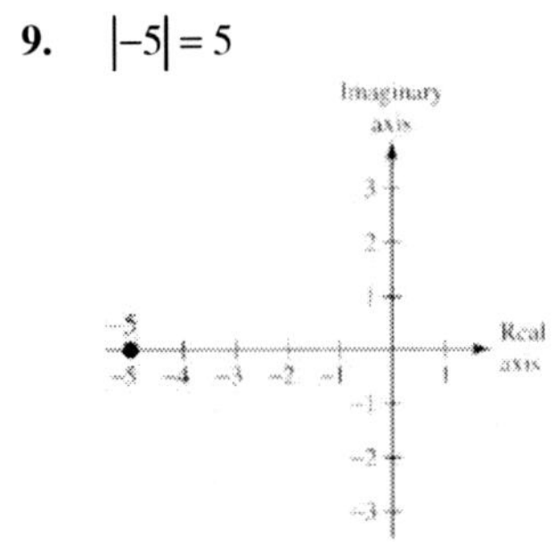

11. $|-4+4i| = \sqrt{(-4)^2 + (4)^2}$
$= \sqrt{32} = 4\sqrt{2}$

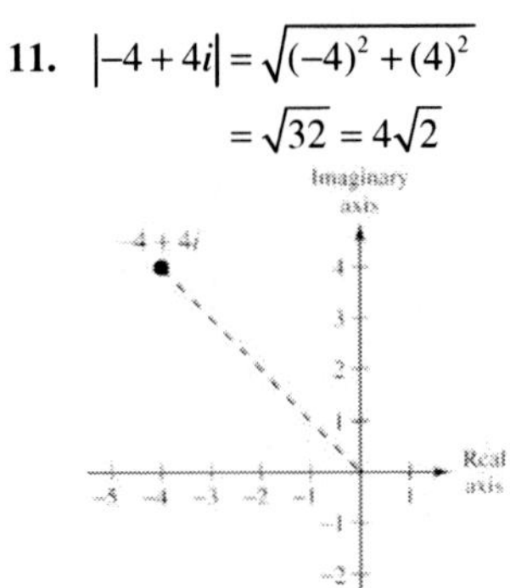

13. $|9+7i| = \sqrt{(9)^2 + (7)^2}$
$= \sqrt{81+49} = \sqrt{130}$

15. $z = 2i$
$r = \sqrt{(0)^2 + (2)^2} = \sqrt{4} = 2$
$\tan\theta = \dfrac{2}{0}$, undefined $\Rightarrow \theta = \pi/2$
$z = 2(\cos\pi/2 + i\sin\pi/2)$

17. $z = -4$
$r = \sqrt{(-4)^2 + (0)^2} = \sqrt{16} = 4$
$\tan\theta = \dfrac{0}{-4} = 0 \Rightarrow \theta = \pi$
$z = 4(\cos\pi + i\sin\pi)$

19. $z = -3 - 3i$
$r = \sqrt{(-3)^2 + (-3)^2} = \sqrt{9+9} = \sqrt{18} = 3\sqrt{2}$
$\tan\theta = \dfrac{-3}{-3} = 1 \Rightarrow \theta = \dfrac{5\pi}{4}$
$z = 3\sqrt{2}\left(\cos\dfrac{5\pi}{4} + i\sin\dfrac{5\pi}{4}\right)$

21. $z = \sqrt{3} - i$
$r = \sqrt{\left(\sqrt{3}\right)^2 + (-1)^2} = 2$
$\tan\theta = \dfrac{-1}{\sqrt{3}} \Rightarrow \theta = \dfrac{11\pi}{6}$
$z = 2\left(\cos\dfrac{11\pi}{6} + i\sin\dfrac{11\pi}{6}\right)$

23. $z = -8i$

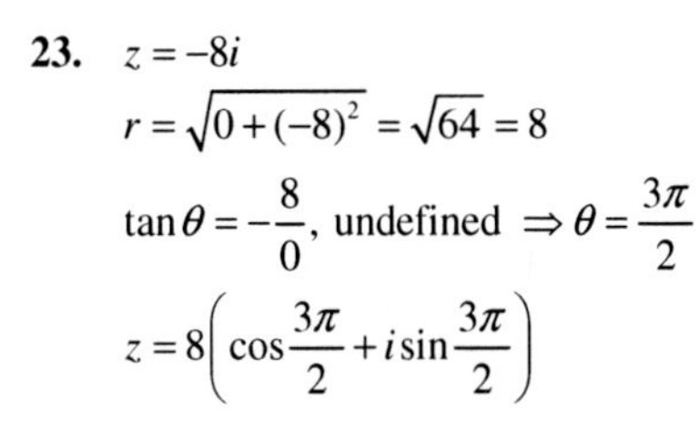

$r = \sqrt{0 + (-8)^2} = \sqrt{64} = 8$
$\tan\theta = -\dfrac{8}{0}$, undefined $\Rightarrow \theta = \dfrac{3\pi}{2}$
$z = 8\left(\cos\dfrac{3\pi}{2} + i\sin\dfrac{3\pi}{2}\right)$

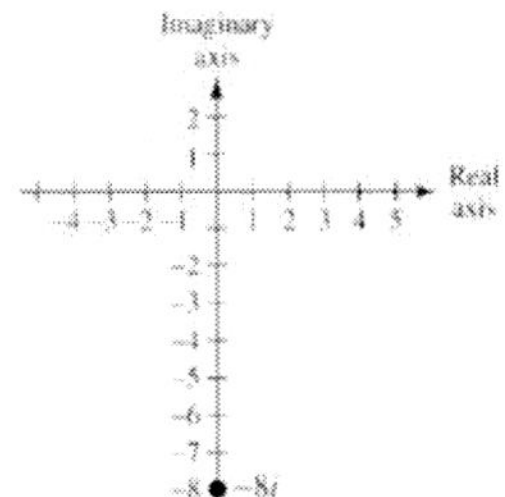

25. $z = -5i$

$r = \sqrt{(0)^2 + (-5)^2} = \sqrt{25} = 5$

$\tan\theta = -\frac{5}{0}$, undefined $\Rightarrow \theta = \frac{3\pi}{2}$

$z = 5\left(\cos\frac{3\pi}{2} + i\sin\frac{3\pi}{2}\right)$

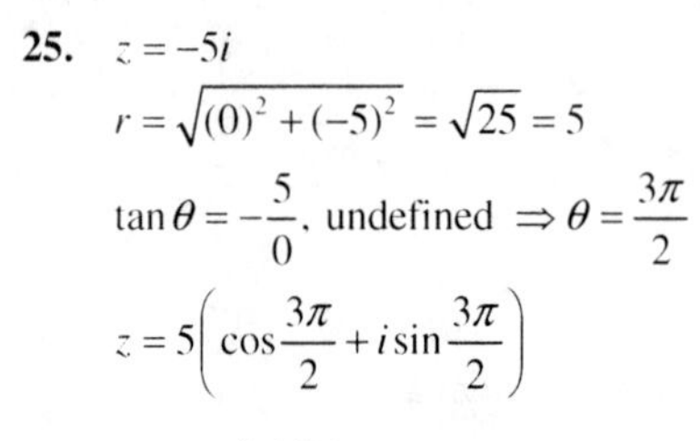

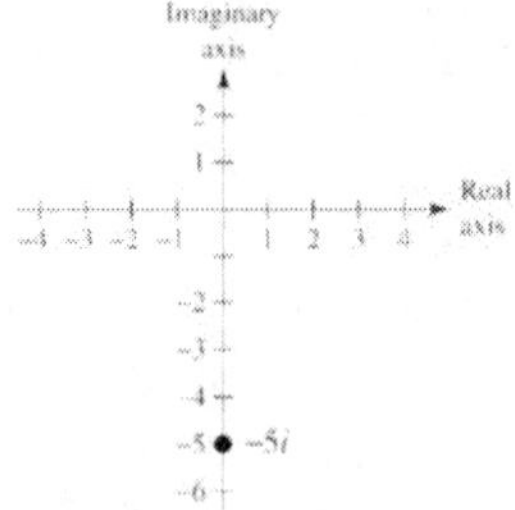

27. $z = 5 - 5i$

$r = \sqrt{5^2 + (-5)^2} = \sqrt{50} = 5\sqrt{2}$

$\tan\theta = -\frac{5}{5} = -1 \Rightarrow \theta = \frac{7\pi}{4}$

$z = 5\sqrt{2}\left(\cos\frac{7\pi}{4} + i\sin\frac{7\pi}{4}\right)$

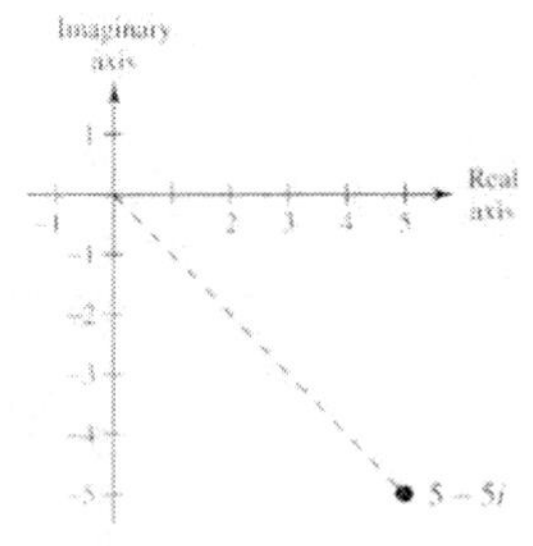

29. $z = \sqrt{3} + i$

$r = \sqrt{\left(\sqrt{3}\right)^2 + 1^2} = \sqrt{4} = 2$

$\tan\theta = \frac{1}{\sqrt{3}} = \frac{\sqrt{3}}{3} \Rightarrow \theta = \frac{\pi}{6}$

$z = 2\left(\cos\frac{\pi}{6} + i\sin\frac{\pi}{6}\right)$

31. $z = 1 + i$

$r = \sqrt{(1)^2 + (1)^2} = \sqrt{2}$

$\tan\theta = \frac{1}{1} = 1 \Rightarrow \theta = \pi/4$

$z = \sqrt{2}\left(\cos\frac{\pi}{4} + i\sin\frac{\pi}{4}\right)$

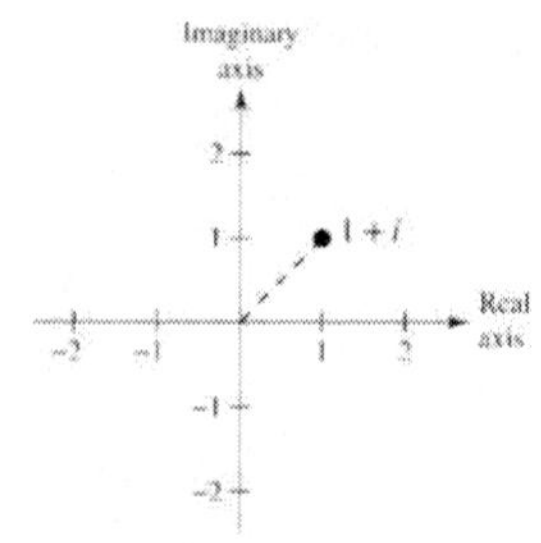

33. $z = -2\left(1 + \sqrt{3}i\right)$

$r = \sqrt{(-2)^2 + \left(-2\sqrt{3}\right)^2} = \sqrt{16} = 4$

$\tan\theta = \frac{\sqrt{3}}{1} = \sqrt{3} \Rightarrow \theta = \frac{4\pi}{3}$

$z = 4\left(\cos\frac{4\pi}{3} + i\sin\frac{4\pi}{3}\right)$

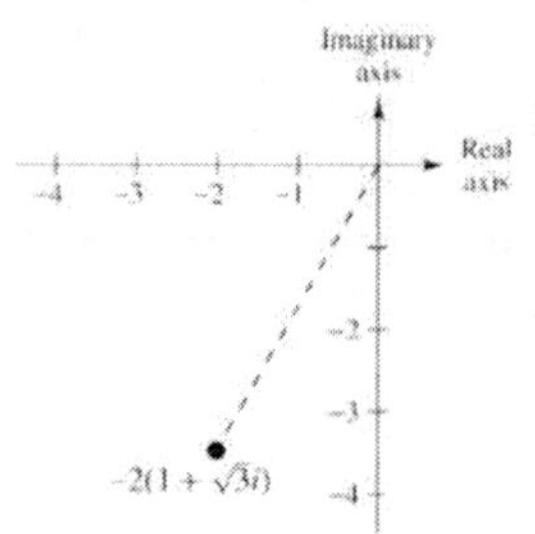

35. $z = -7 + 4i$

$r = \sqrt{49 + 16} = \sqrt{65}$

$\tan\theta = \frac{4}{-7} \Rightarrow \theta \approx 2.62$ radians or $150.26°$

$z = \sqrt{65}\left(\cos 150.26° + i\sin 150.26°\right)$

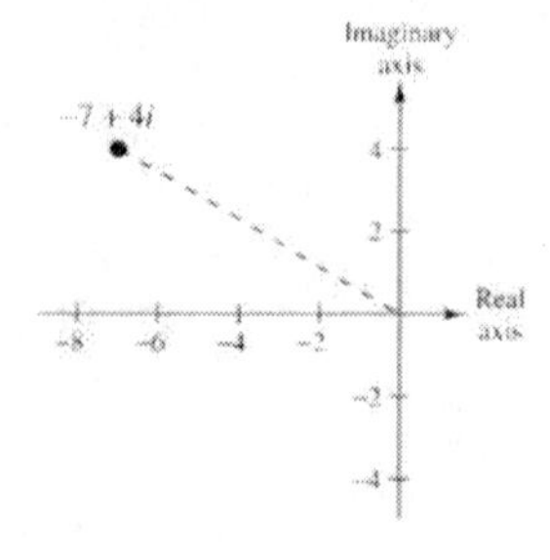

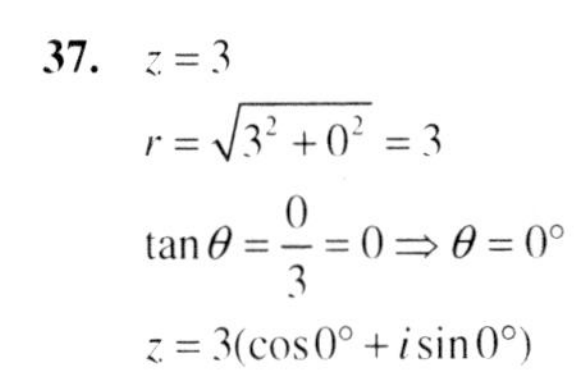

37. $z = 3$

$r = \sqrt{3^2 + 0^2} = 3$

$\tan\theta = \dfrac{0}{3} = 0 \Rightarrow \theta = 0°$

$z = 3(\cos 0° + i\sin 0°)$

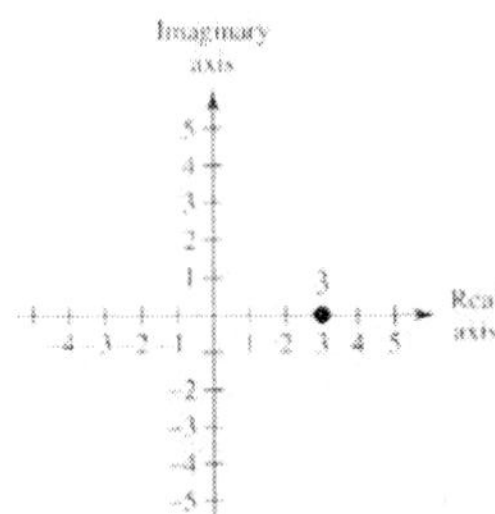

39. $z = 3 + \sqrt{3}i$

$r = \sqrt{9+3} = \sqrt{12} = 2\sqrt{3}$

$\tan\theta = \dfrac{\sqrt{3}}{3} \Rightarrow \theta = \dfrac{\pi}{6}$ or $30°$

$z = 2\sqrt{3}\left(\cos\dfrac{\pi}{6} + i\sin\dfrac{\pi}{6}\right)$

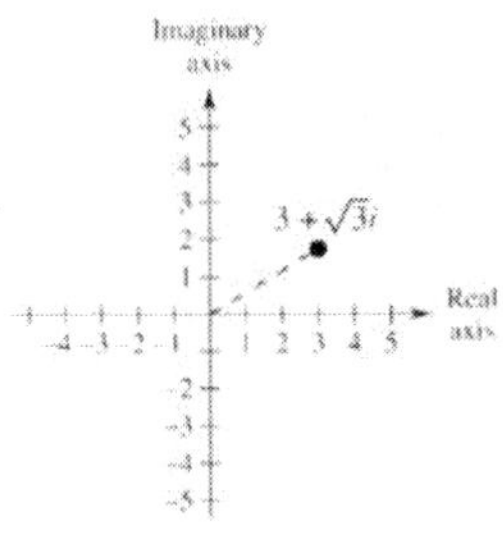

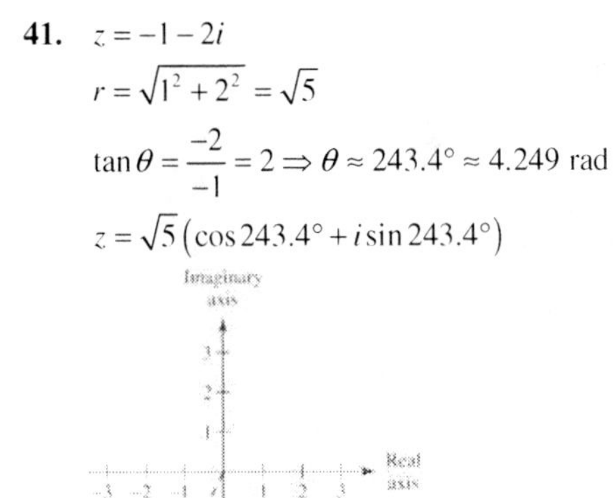

41. $z = -1 - 2i$

$r = \sqrt{1^2 + 2^2} = \sqrt{5}$

$\tan\theta = \dfrac{-2}{-1} = 2 \Rightarrow \theta \approx 243.4° \approx 4.249$ rad

$z = \sqrt{5}\left(\cos 243.4° + i\sin 243.4°\right)$

43. $z = 5 + 2i$

$r = \sqrt{25+4} = \sqrt{29} \approx 5.385$

$\tan\theta = \frac{2}{5} \Rightarrow \theta \approx 21.80° \approx 0.381$ rad

$z = \sqrt{29}\left(\cos 21.80° + i\sin 21.80°\right)$

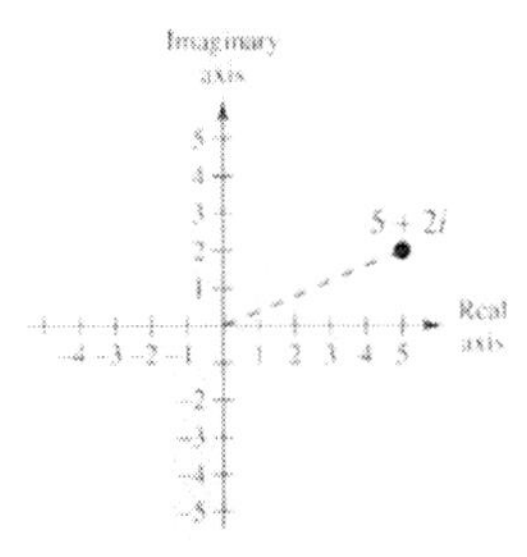

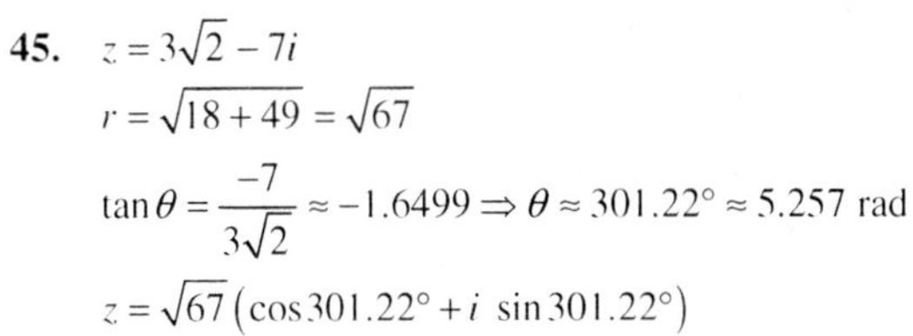

45. $z = 3\sqrt{2} - 7i$

$r = \sqrt{18+49} = \sqrt{67}$

$\tan\theta = \dfrac{-7}{3\sqrt{2}} \approx -1.6499 \Rightarrow \theta \approx 301.22° \approx 5.257$ rad

$z = \sqrt{67}\left(\cos 301.22° + i\ \sin 301.22°\right)$

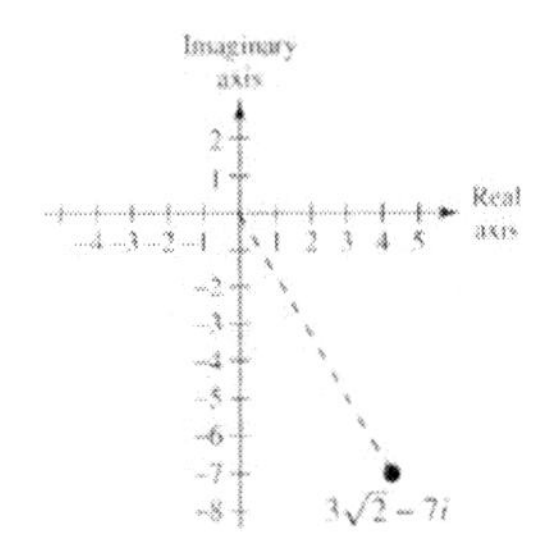

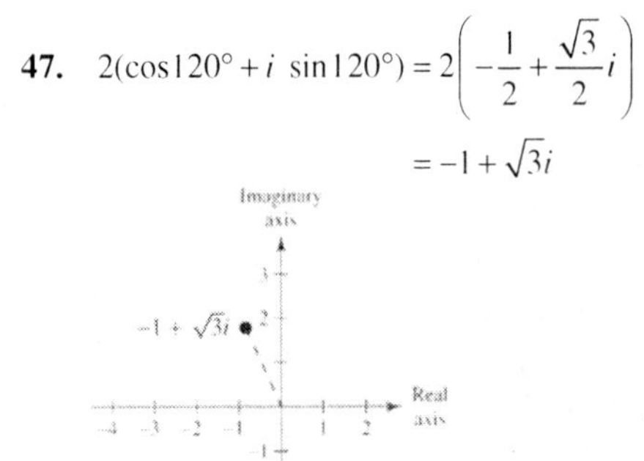

47. $2(\cos 120° + i\ \sin 120°) = 2\left(-\dfrac{1}{2} + \dfrac{\sqrt{3}}{2}i\right)$

$= -1 + \sqrt{3}i$

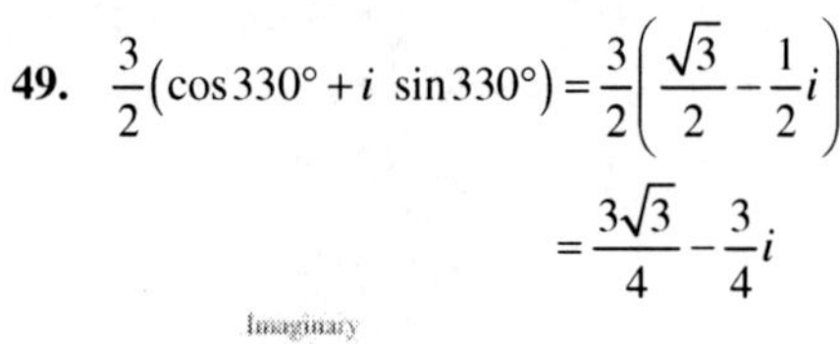

49. $\dfrac{3}{2}(\cos 330° + i\ \sin 330°) = \dfrac{3}{2}\left(\dfrac{\sqrt{3}}{2} - \dfrac{1}{2}i\right)$

$$= \frac{3\sqrt{3}}{4} - \frac{3}{4}i$$

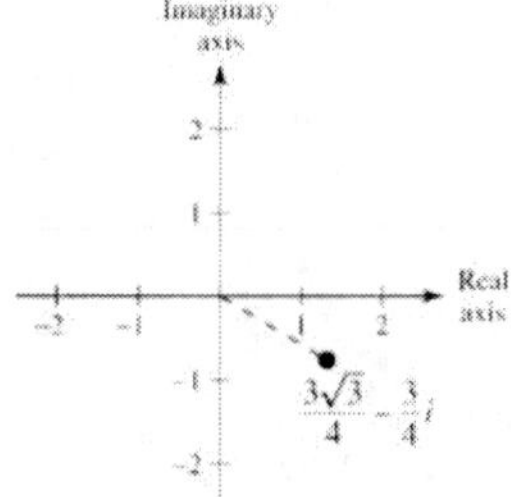

51. $3.75\left(\cos\dfrac{3\pi}{4} + i\ \sin\dfrac{3\pi}{4}\right) = -\dfrac{15\sqrt{2}}{8} + \dfrac{15\sqrt{2}}{8}i$

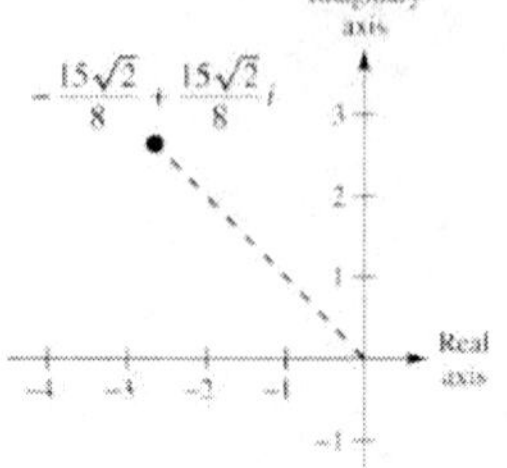

53. $6\left(\cos\dfrac{\pi}{3} + i\ \sin\dfrac{\pi}{3}\right) = 6\left(\dfrac{1}{2} + \dfrac{\sqrt{3}}{2}i\right) = 3 + 3\sqrt{3}i$

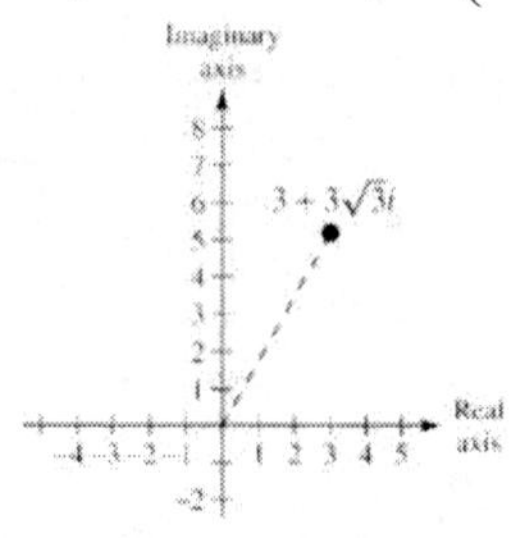

55. $4\left(\cos\dfrac{3\pi}{2} + i\ \sin\dfrac{3\pi}{2}\right) = 4(0 - i) = -4i$

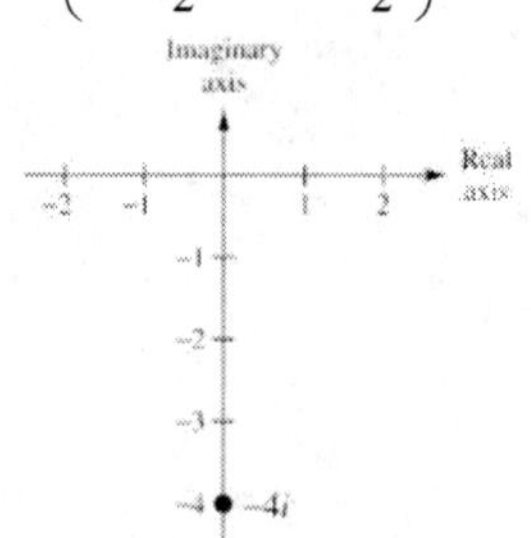

57. $3\left[\cos(18°\ 45') + i\ \sin(18°\ 45')\right] \approx 2.8408 + 0.9643i$

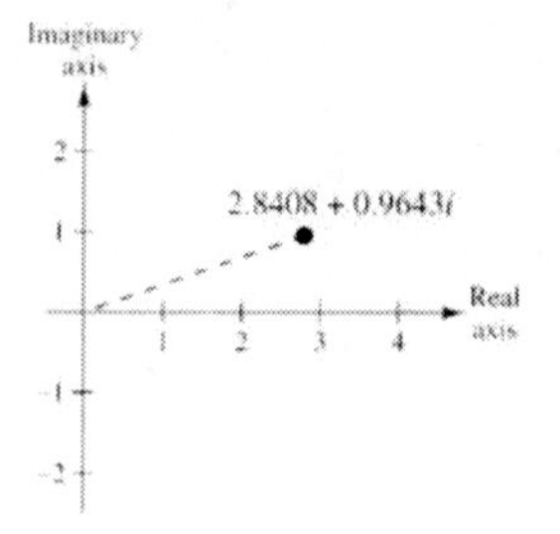

59. $5\left(\cos\dfrac{\pi}{9} + i\ \sin\dfrac{\pi}{9}\right) \approx 4.6985 + 1.7101i$

61. $9(\cos 58° + i\ \sin 58°) \approx 4.7693 + 7.6324i$

63. $z = \dfrac{\sqrt{2}}{2}(1 + i)$

$$z^2 = \frac{\sqrt{2}}{2}(1+i)\frac{\sqrt{2}}{2}(1+i) = \frac{1}{2}(2i) = i$$

$$z^3 = z^2 z = i\left[\frac{\sqrt{2}}{2}(1+i)\right] = \frac{\sqrt{2}}{2}(i-1)$$

$$z^4 = z^3 z = \frac{\sqrt{2}}{2}(i-1)\frac{\sqrt{2}}{2}(1+i) = \frac{1}{2}(-2) = -1$$

The absolute value of each power is 1.

65. $\left[2\left(\cos\dfrac{\pi}{2} + i\ \sin\dfrac{\pi}{2}\right)\right]\left[5\left(\cos\dfrac{3\pi}{2} + i\ \sin\dfrac{3\pi}{2}\right)\right]$

$$= (2)(5)\left[\cos\left(\frac{\pi}{2} + \frac{3\pi}{2}\right) + i\ \sin\left(\frac{\pi}{2} + \frac{3\pi}{2}\right)\right]$$

$$= 10(\cos 2\pi + i\ \sin 2\pi)$$

$$= 10(\cos 0 + i\ \sin 0)$$

67. $\left[\dfrac{2}{3}\left(\cos\dfrac{4\pi}{3} + i\ \sin\dfrac{4\pi}{3}\right)\right]\left[9\left(\cos\dfrac{5\pi}{3} + i\ \sin\dfrac{5\pi}{3}\right)\right]$

$$= \left(\frac{2}{3}\right)(9)\left[\cos\left(\frac{4\pi}{3} + \frac{5\pi}{3}\right) + i\ \sin\left(\frac{4\pi}{3} + \frac{5\pi}{3}\right)\right]$$

$$= 6(\cos 3\pi + i\ \sin 3\pi)$$

$$= 6(\cos \pi + i\ \sin \pi)$$

69. $\left[\frac{5}{3}(\cos 140° + i\ \sin 140°)\right]\left[\frac{2}{3}(\cos 60° + i\ \sin 60°)\right]$
$= \left(\frac{5}{3}\right)\left(\frac{2}{3}\right)\left[\cos(140° + 60°) + i\ \sin(140° + 60°)\right]$
$= \frac{10}{9}(\cos 200° + i\ \sin 200°)$

71. $\left[\frac{11}{20}(\cos 290° + i\ \sin 290°)\right]\left[\frac{2}{5}(\cos 200° + i\ \sin 200°)\right]$
$= \left(\frac{11}{20}\right)\left(\frac{2}{5}\right)\left[\cos(290° + 200°) + i\ \sin(290° + 200°)\right]$
$= \frac{11}{50}(\cos 490° + i\ \sin 490°)$
$= \frac{11}{50}(\cos 130° + i\ \sin 130°)$

73. $\frac{\cos 50° + i\ \sin 50°}{\cos 20° + i\ \sin 20°} = \cos(50° - 20°) + i\ \sin(50° - 20°)$
$= \cos 30° + i\ \sin 30°$

75. $\frac{2(\cos 120° + i\ \sin 120°)}{4(\cos 40° + i\ \sin 40°)}$
$= \frac{1}{2}\left[\cos(120° - 40°) + i\ \sin(120° - 40°)\right]$
$= \frac{1}{2}(\cos 80° + i\ \sin 80°)$

77. $\frac{18(\cos 54° + i\ \sin 54°)}{3(\cos 102° + i\ \sin 102°)}$
$= 6(\cos(54° - 102°) + i\ \sin(54° - 102°))$
$= 6(\cos(-48°) + i\ \sin(-48°))$
$= 6(\cos 312° + i\ \sin 312°)$

79. (a) $2 - 2i = 2\sqrt{2}\left(\cos\frac{7\pi}{4} + i\ \sin\frac{7\pi}{4}\right)$
$1 + i = \sqrt{2}\left(\cos\frac{\pi}{4} + i\ \sin\frac{\pi}{4}\right)$

(b) $(2 - 2i)(1 + i)$
$= 2\sqrt{2}\left(\cos\frac{7\pi}{4} + i\ \sin\frac{7\pi}{4}\right)\sqrt{2}\left(\cos\frac{\pi}{4} + i\ \sin\frac{\pi}{4}\right)$
$= 4(\cos 2\pi + i\ \sin 2\pi) = 4$

(c) $(2 - 2i)(1 + i) = 2 + 2 = 4$

81. (a) $2 + 2i = 2\sqrt{2}(\cos 45° + i\ \sin 45°)$
$1 - i = \sqrt{2}\left[\cos(-45°) + i\ \sin(-45°)\right]$

(b)

$(2 + 2i)(1 - i)$
$= \left[2\sqrt{2}(\cos 45° + i\ \sin 45°)\right]\left[\sqrt{2}(\cos(-45°) + i\ \sin(-45°))\right]$
$= 4(\cos 0° + i\ \sin 0°) = 4$

(c) $(2 + 2i)(1 - i) = 2 - 2i + 2i - 2i^2 = 2 + 2 = 4$

83. (a) $-2i = 2\left[\cos(-90°) + i\ \sin(-90°)\right]$
$1 + i = \sqrt{2}(\cos 45° + i\ \sin 45°)$

(b) $-2i(1 + i)$
$= 2[\cos(-90°) + i\ \sin(-90°)]\left[\sqrt{2}(\cos 45° + i\ \sin 45°)\right]$
$= 2\sqrt{2}\left[\cos(-45°) + i\ \sin(-45°)\right]$
$= 2\sqrt{2}\left[\frac{1}{\sqrt{2}} - \frac{1}{\sqrt{2}}i\right] = 2 - 2i$

(c) $-2i(1 + i) = -2i - 2i^2 = -2i + 2 = 2 - 2i$

85. (a) $-2i = 2\left(\cos\frac{3\pi}{2} + i\ \sin\frac{3\pi}{2}\right)$
$\sqrt{3} - i = 2\left(\cos\frac{11\pi}{6} + i\ \sin\frac{11\pi}{6}\right)$

(b) $-2i\left(\sqrt{3} - i\right)$
$= 2\left(\cos\frac{3\pi}{2} + i\ \sin\frac{3\pi}{2}\right)2\left(\cos\frac{11\pi}{6} + i\ \sin\frac{11\pi}{6}\right)$
$= 4\left(\cos\frac{20\pi}{6} + i\ \sin\frac{20\pi}{6}\right)$
$= 4\left(-\frac{1}{2} - \frac{\sqrt{3}}{2}i\right)$
$= -2 - 2\sqrt{3}i$

(c) $-2i\left(\sqrt{3} - i\right) = -2\sqrt{3}i - 2$

87. (a) $2 = 2(\cos 0 + i\ \sin 0)$
$1 - i = \sqrt{2}\left(\cos\frac{7\pi}{4} + i\ \sin\frac{7\pi}{4}\right)$

(b) $2(1 - i) = 2\sqrt{2}\left(\cos\frac{7\pi}{4} + i\ \sin\frac{7\pi}{4}\right)$
$= 2\sqrt{2}\left(\frac{\sqrt{2}}{2} - \frac{\sqrt{2}}{2}i\right)$
$= 2(1 - i) = 2 - 2i$

(c) $2(1 - i) = 2 - 2i$

89. (a) $3+4i = 5(\cos 0.927 + i\ \sin 0.927)$

$1-\sqrt{3}i = 2\left(\cos\frac{5\pi}{3} + i\sin\frac{5\pi}{3}\right)$

(b) $\frac{5(\cos 0.927 + i\ \sin 0.927)}{2(\cos 5\pi/3 + i\ \sin 5\pi/3)}$

$= \frac{5}{2}\left[\cos\left(0.927 - \frac{5\pi}{3}\right) + i\ \sin\left(0.927 - \frac{5\pi}{3}\right)\right]$

$= \frac{5}{2}\left[\cos 1.974 + i\ \sin 1.974\right]$

$\approx \left(\frac{3}{4} - \sqrt{3}\right) + \left(\frac{3\sqrt{3}}{4} + 1\right)i$

(c) $\frac{3+4i}{1-\sqrt{3}i} \cdot \frac{1+\sqrt{3}i}{1+\sqrt{3}i} = \frac{3+3\sqrt{3}i+4i+4\sqrt{3}i}{1-3i^2}$

$= \frac{\left(3-4\sqrt{3}\right)+\left(4+3\sqrt{3}\right)i}{4}$

$= \left(\frac{3}{4} - \sqrt{3}\right) + \left(1 + \frac{3\sqrt{3}}{4}\right)i$

91. (a) $5 = 5(\cos 0 + i\ \sin 0)$

$2+2i = 2\sqrt{2}\left(\cos\frac{\pi}{4} + i\ \sin\frac{\pi}{4}\right)$

(b) $\frac{5}{2+2i} = \frac{5(\cos 0 + i\ \sin 0)}{2\sqrt{2}\left(\cos\frac{\pi}{4} + i\ \sin\frac{\pi}{4}\right)}$

$= \frac{5}{2\sqrt{2}}\left(\cos\left(-\frac{\pi}{4}\right) + i\ \sin\left(-\frac{\pi}{4}\right)\right)$

$= \frac{5}{2\sqrt{2}}\left(\frac{\sqrt{2}}{2} - \frac{\sqrt{2}}{2}i\right) = \frac{5}{4} - \frac{5}{4}i$

(c) $\frac{5}{2+2i} \cdot \frac{2-2i}{2-2i} = \frac{5(2-2i)}{4+4} = \frac{5}{4} - \frac{5}{4}i$

93. (a) $4i = 4\left(\cos\frac{\pi}{2} + i\ \sin\frac{\pi}{2}\right)$

$-1+i = \sqrt{2}\left(\cos\frac{3\pi}{4} + i\ \sin\frac{3\pi}{4}\right)$

(b) $\frac{4i}{-1+i} = \frac{4}{\sqrt{2}}\left(\cos\left(\frac{\pi}{2} - \frac{3\pi}{4}\right) + i\ \sin\left(\frac{\pi}{2} - \frac{3\pi}{4}\right)\right)$

$= \frac{4}{\sqrt{2}}\left(\cos\left(-\frac{\pi}{4}\right) + i\ \sin\left(-\frac{\pi}{4}\right)\right)$

$= \frac{4}{\sqrt{2}}\left(\frac{\sqrt{2}}{2} - \frac{\sqrt{2}}{2}i\right) = 2-2i$

(c) $\frac{4i}{-1+i} \cdot \frac{-1-i}{-1-i} = \frac{-4i+4}{2} = 2-2i$

95. Let $z = x+iy$.

$|z| = 2 \Rightarrow 2 = \sqrt{x^2+y^2} \Rightarrow 4 = x^2+y^2$

Circle of radius 2

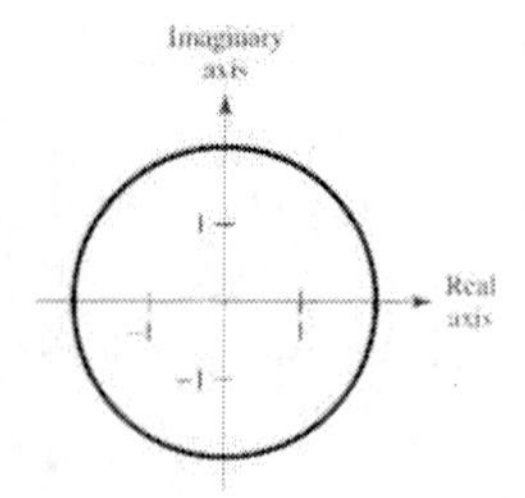

97. Let $z = x+iy$.

$|z| = 4 \Rightarrow 4 = \sqrt{x^2+y^2} \Rightarrow x^2+y^2 = 16$

Circle of radius 4

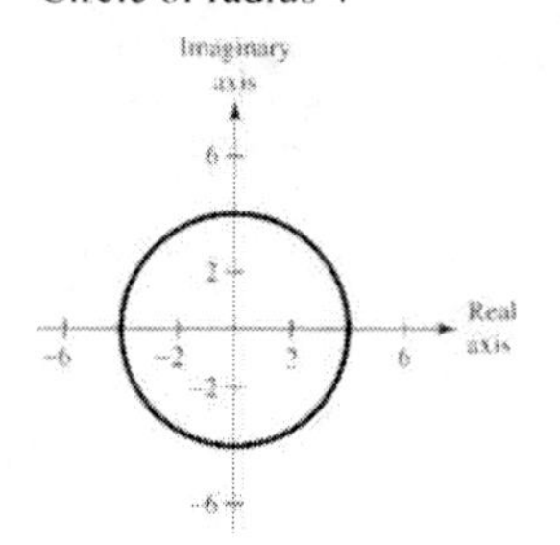

99. Let $z = x+iy$.

$|z| = 7 \Rightarrow 7 = \sqrt{x^2+y^2} \Rightarrow x^2+y^2 = 49$

Circle of radius 7

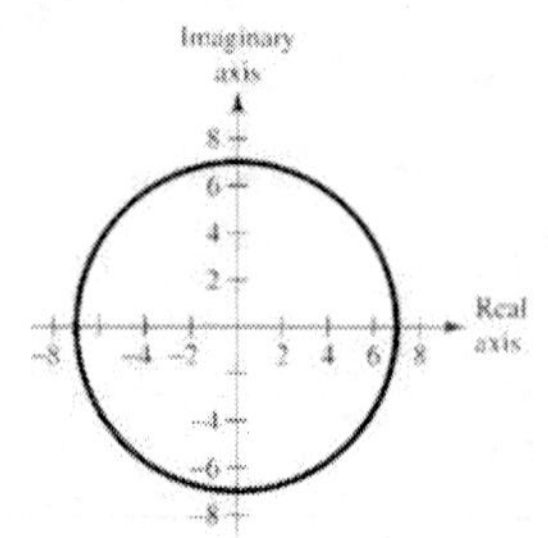

101. $\theta = \frac{\pi}{6}$

Let $z = x+iy$ such that:

$\tan\frac{\pi}{6} = \frac{y}{x} \Rightarrow \frac{y}{x} = \frac{1}{\sqrt{3}} \Rightarrow y = \frac{1}{\sqrt{3}}x$

Line

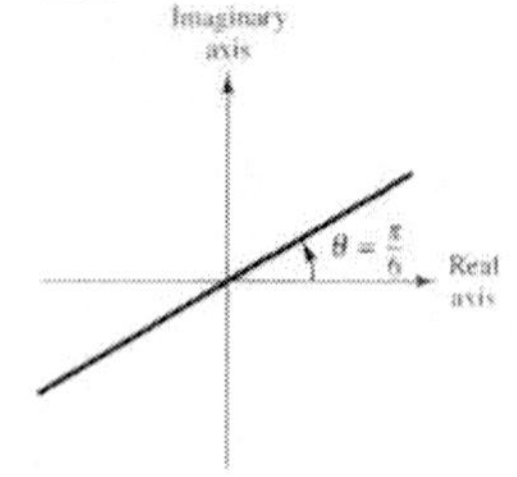

103. $\theta = \frac{\pi}{3}$

Let $z = x + iy$ such that:

$\tan\frac{\pi}{3} = \frac{y}{x}$

$\frac{y}{x} = \sqrt{3}$

$y = \sqrt{3}x$

Line

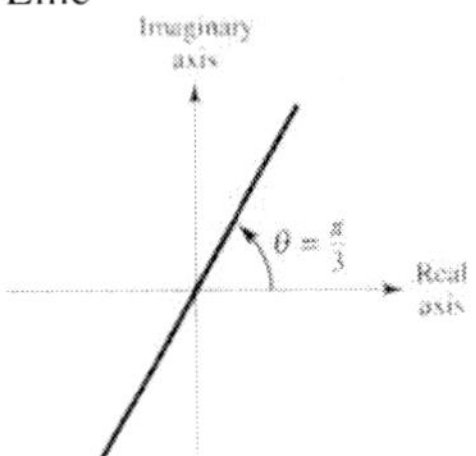

105. $\theta = \frac{2\pi}{3}$

Let $z = x + iy$ such that:

$\tan\frac{2\pi}{3} = \frac{y}{x} \Rightarrow -\sqrt{3} = \frac{y}{x} \Rightarrow y = -\sqrt{3}x$

Line

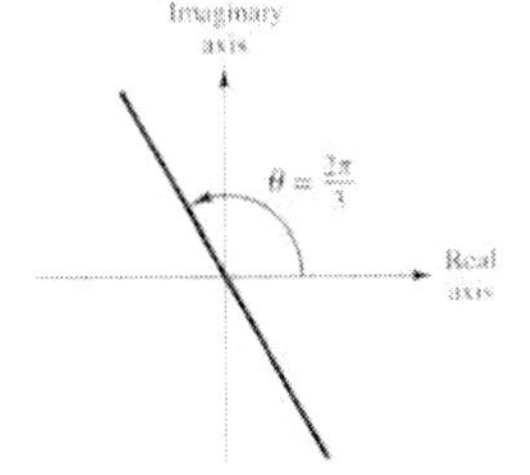

107. $(1+i)^3 = \left[\sqrt{2}\left(\cos\frac{\pi}{4} + i\sin\frac{\pi}{4}\right)\right]^3$

$= \left(\sqrt{2}\right)^3\left(\cos\frac{3\pi}{4} + i\sin\frac{3\pi}{4}\right)$

$= 2\sqrt{2}\left(-\frac{\sqrt{2}}{2} + \frac{\sqrt{2}}{2}i\right)$

$= -2 + 2i$

109. $(-1+i)^6 = \left[\sqrt{2}\left(\cos\frac{3\pi}{4} + i\sin\frac{3\pi}{4}\right)\right]^6$

$= \left(\sqrt{2}\right)^6\left(\cos\frac{18\pi}{4} + i\sin\frac{18\pi}{4}\right)$

$= 8\left(\cos\frac{9\pi}{2} + i\sin\frac{9\pi}{2}\right)$

$= 8\left[\cos\left(4\pi + \frac{\pi}{2}\right) + i\sin\left(4\pi + \frac{\pi}{2}\right)\right]$

$= 8\left(\cos\frac{\pi}{2} + i\sin\frac{\pi}{2}\right)$

$= 8(0 + i) = 8i$

111. $2\left(\sqrt{3} + i\right)^5 = 2\left[2\left(\cos\frac{\pi}{6} + i\sin\frac{\pi}{6}\right)\right]^5$

$= 2\left[2^5\left(\cos\frac{5\pi}{6} + i\sin\frac{5\pi}{6}\right)\right]$

$= 64\left(-\frac{\sqrt{3}}{2} + \frac{1}{2}i\right)$

$= -32\sqrt{3} + 32i$

113. $\left[5\left(\cos 20^\circ + i\sin 20^\circ\right)\right]^3 = 5^3\left(\cos 60^\circ + i\sin 60^\circ\right)$

$= \frac{125}{2} + \frac{125\sqrt{3}}{2}i$

115. $\left(\cos\frac{5\pi}{4} + i\sin\frac{5\pi}{4}\right)^{10}$

$= \cos\frac{25\pi}{2} + i\sin\frac{25\pi}{2}$

$= \cos\left(12\pi + \frac{\pi}{2}\right) + i\sin\left(12\pi + \frac{\pi}{2}\right)$

$= \cos\frac{\pi}{2} + i\sin\frac{\pi}{2} = i$

117. $\left[2(\cos 1.25 + i\sin 1.25)\right]^4 = 2^4(\cos 5 + i\sin 5)$

$\approx 4.5386 - 15.3428i$

119. $\left[2(\cos\pi + i\sin\pi)\right]^8 = 2^8(\cos 8\pi + i\sin 8\pi)$

$= 256(1) = 256$

121. $(3-2i)^5 = \left[\sqrt{13}\left(\cos(5.6952) + i\sin(5.6952)\right)\right]^5$

$= 169\sqrt{13}\left[\cos(28.476) + i\sin(28.476)\right]$

$\approx -597 - 122i$

123. $\left[4(\cos 10^\circ + i\sin 10^\circ)\right]^6 = 4^6(\cos 60^\circ + i\sin 60^\circ)$

$= 4096\left(\frac{1}{2} + \frac{\sqrt{3}}{2}i\right)$

$= 2048 + 2048\sqrt{3}i$

125. $\left[3\left(\cos\frac{\pi}{8} + i\sin\frac{\pi}{8}\right)\right]^2 = 3^2\left(\cos\frac{\pi}{4} + i\sin\frac{\pi}{4}\right)$

$= 9\left(\frac{\sqrt{2}}{2} + i\frac{\sqrt{2}}{2}\right)$

$= \frac{9}{2}\sqrt{2} + \frac{9}{2}\sqrt{2}i$

127. $\left[-\frac{1}{2}\left(1 + \sqrt{3}i\right)\right]^6 = \left[\cos\frac{4\pi}{3} + i\sin\frac{4\pi}{3}\right]^6$

$= \cos 8\pi + i\sin 8\pi$

$= 1$

129. $2i = 2\left(\cos\frac{\pi}{2} + i\ \sin\frac{\pi}{2}\right)$

Square roots:

$\sqrt{2}\left(\cos\frac{\pi}{4} + i\ \sin\frac{\pi}{4}\right) = 1 + i$

$\sqrt{2}\left(\cos\frac{5\pi}{4} + i\ \sin\frac{5\pi}{4}\right) = -1 - i$

131. $-3i = 3\left(\cos\frac{3\pi}{2} + i\ \sin\frac{3\pi}{2}\right)$

Square roots:

$\sqrt{3}\left(\cos\frac{3\pi}{4} + i\ \sin\frac{3\pi}{4}\right) = -\frac{\sqrt{6}}{2} + \frac{\sqrt{6}}{2}i$

$\sqrt{3}\left(\cos\frac{7\pi}{4} + i\ \sin\frac{7\pi}{4}\right) = \frac{\sqrt{6}}{2} - \frac{\sqrt{6}}{2}i$

133. $2 - 2i = 2\sqrt{2}\left(\cos\frac{7\pi}{4} + i\ \sin\frac{7\pi}{4}\right)$

Square roots:

$8^{1/4}\left(\cos\frac{7\pi}{8} + i\ \sin\frac{7\pi}{8}\right) \approx -1.554 + 0.644i$

$8^{1/4}\left(\cos\frac{15\pi}{8} + i\ \sin\frac{15\pi}{8}\right) \approx 1.554 - 0.644i$

135. $1 + \sqrt{3}i = 2\left(\cos\frac{\pi}{3} + i\ \sin\frac{\pi}{3}\right)$

Square roots:

$\sqrt{2}\left(\cos\frac{\pi}{6} + i\ \sin\frac{\pi}{6}\right) = \frac{\sqrt{6}}{2} + \frac{\sqrt{2}}{2}i$

$\sqrt{2}\left(\cos\frac{7\pi}{6} + i\ \sin\frac{7\pi}{6}\right) = -\frac{\sqrt{6}}{2} - \frac{\sqrt{2}}{2}i$

137. (a) Square roots of $5(\cos 120° + i\ \sin 120°)$:

$\sqrt{5}\left[\cos\left(\frac{120° + 360°k}{2}\right) + i\ \sin\left(\frac{120° + 360°k}{2}\right)\right],\ k = 0, 1$

$k = 0:\ \sqrt{5}(\cos 60° + i\ \sin 60°)$

$k = 1:\ \sqrt{5}(\cos 240° + i\ \sin 240°)$

(b)

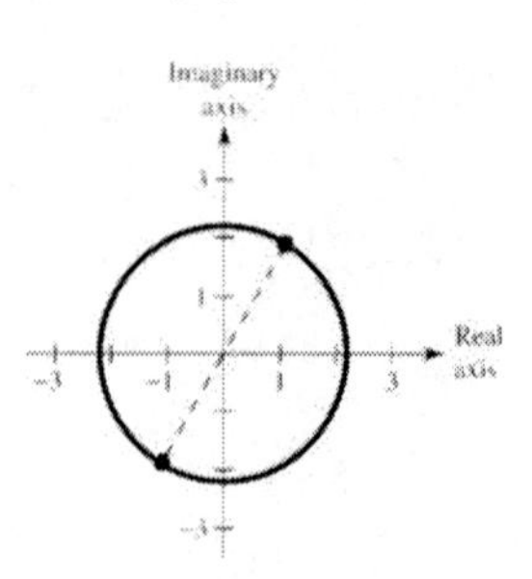

(c) $\frac{\sqrt{5}}{2} + \frac{\sqrt{15}}{2}i,\ -\frac{\sqrt{5}}{2} - \frac{\sqrt{15}}{2}i$

139. (a) Fourth roots of $8\left(\cos\frac{2\pi}{3} + i\ \sin\frac{2\pi}{3}\right)$:

$\sqrt[4]{8}\left[\cos\left(\frac{(2\pi/3) + 2k\pi}{4}\right) + i\sin\left(\frac{(2\pi/3) + 2k\pi}{4}\right)\right]$

$k = 0, 1, 2, 3$

$k = 0:\ \sqrt[4]{8}(\cos\pi/6 + i\ \sin\pi/6)$

$k = 1:\ \sqrt[4]{8}(\cos 2\pi/3 + i\ \sin 2\pi/3)$

$k = 2:\ \sqrt[4]{8}(\cos 7\pi/6 + i\ \sin 7\pi/6)$

$k = 3:\ \sqrt[4]{8}(\cos 5\pi/3 + i\ \sin 5\pi/3)$

(b)

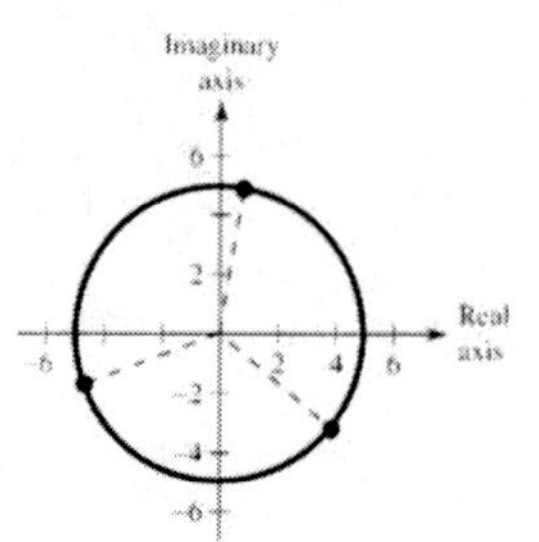

(c) $1.4565 + 0.8409i,\ -0.8409 + 1.4565i,$
$-1.4565 - 0.8409i,\ 0.8409 - 1.4565i$

141. (a) Cube roots of $-25i = 25\left(\cos\frac{3\pi}{2} + i\ \sin\frac{3\pi}{2}\right)$:

$\sqrt[3]{25}\left[\cos\left(\frac{\frac{3\pi}{2} + 2k\pi}{3}\right) + i\ \sin\left(\frac{\frac{3\pi}{2} + 2k\pi}{3}\right)\right]$

$k = 0, 1, 2$

$k = 0:\ \sqrt[3]{25}[\cos\pi/2 + i\ \sin\pi/2]$

$k = 1:\ \sqrt[3]{25}[\cos 7\pi/6 + i\ \sin 7\pi/6]$

$k = 2:\ \sqrt[3]{25}[\cos 11\pi/6 + i\ \sin 11\pi/6]$

(b)

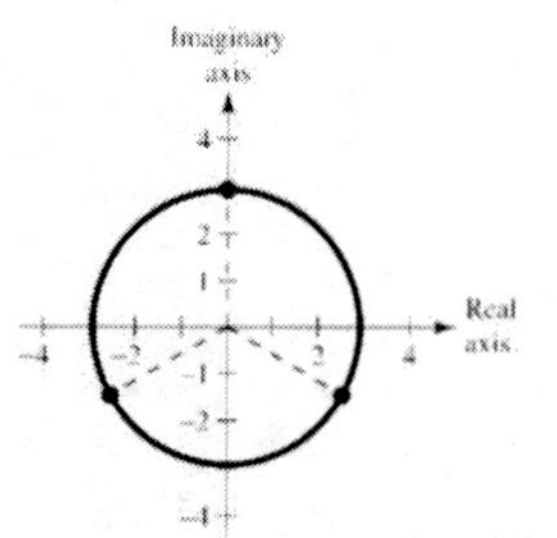

(c) $2.9240i,\ -2.5323 - 1.4620i,\ 2.5323 - 1.4620i$

143. (a) Cube roots of

$$-\frac{125}{2}\left(1+\sqrt{3}i\right)=125\left(\cos\frac{4\pi}{3}+i\sin\frac{4\pi}{3}\right):$$

$$\sqrt[3]{125}\left[\cos\left(\frac{(4\pi/3)+2k\pi}{3}\right)+i\sin\left(\frac{(4\pi/3)+2k\pi}{3}\right)\right]$$

$k=0,\ 1,\ 2$

$k=0:\ 5\left(\cos\frac{4\pi}{9}+i\sin\frac{4\pi}{9}\right)$

$k=1:\ 5\left(\cos\frac{10\pi}{9}+i\sin\frac{10\pi}{9}\right)$

$k=2:\ 5\left(\cos\frac{16\pi}{9}+i\sin\frac{16\pi}{9}\right)$

(b)

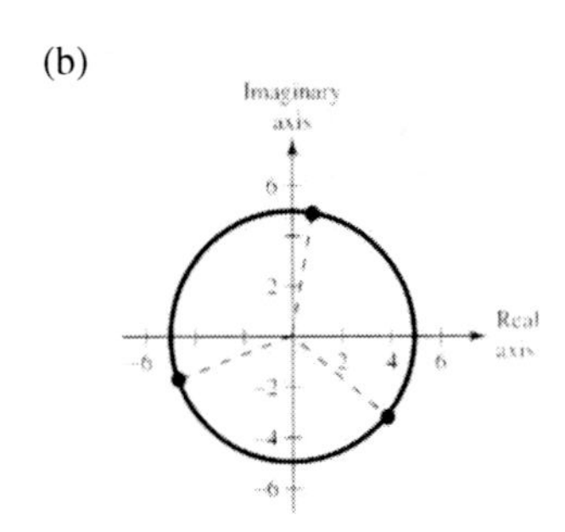

(c) $0.8682+4.9240i,\ -4.6985-1.7101i,$
$3.8302-3.2139i$

145. (a) Cube roots of $64i=64\left(\cos\frac{\pi}{2}+i\sin\frac{\pi}{2}\right)$:

$$(64)^{1/3}\left[\cos\left(\frac{(\pi/2)+2k\pi}{3}\right)+i\sin\left(\frac{(\pi/2)+2k\pi}{3}\right)\right],\ k=0,\ 1,\ 2$$

$k=0:\ 4\left(\cos\frac{\pi}{6}+i\sin\frac{\pi}{6}\right)$

$k=1:\ 4\left(\cos\frac{5\pi}{6}+i\sin\frac{5\pi}{6}\right)$

$k=2:\ 4\left(\cos\frac{9\pi}{6}+i\sin\frac{9\pi}{6}\right)=4\left(\cos\frac{3\pi}{2}+i\sin\frac{3\pi}{2}\right)$

(b)

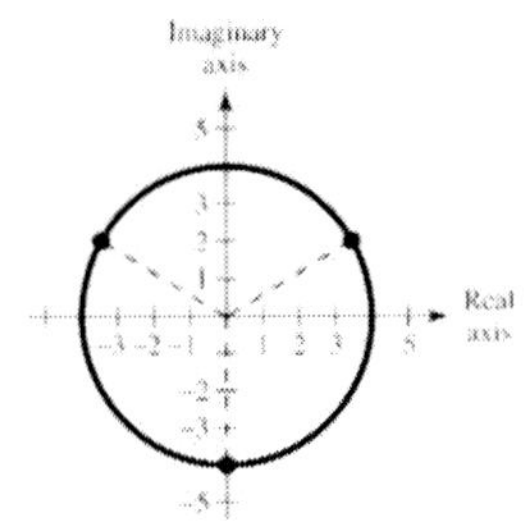

(c) $2\sqrt{3}+2i,\ -2\sqrt{3}+2i,\ -4i$

147. (a) Fifth roots of $1=\cos 0+i\sin 0$:

$\cos\frac{2k\pi}{5}+i\sin\frac{2k\pi}{5},\ k=0,\ 1,\ 2,\ 3,\ 4$

$k=0:\ \cos 0+i\sin 0$

$k=1:\ \cos\frac{2\pi}{5}+i\sin\frac{2\pi}{5}$

$k=2:\ \cos\frac{4\pi}{5}+i\sin\frac{4\pi}{5}$

$k=3:\ \cos\frac{6\pi}{5}+i\sin\frac{6\pi}{5}$

$k=4:\ \cos\frac{8\pi}{5}+i\sin\frac{8\pi}{5}$

(b)

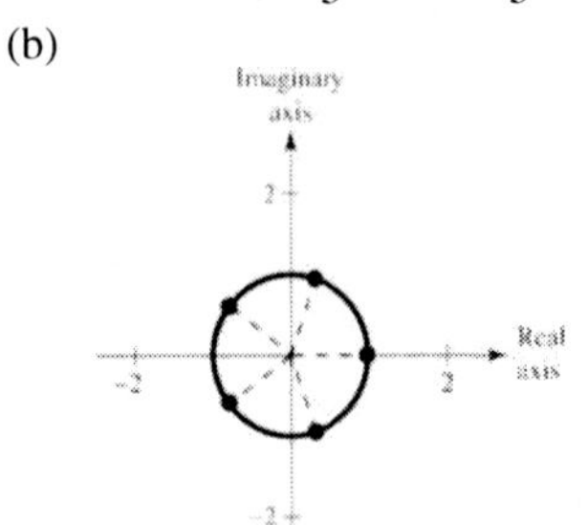

(c) $1,\ 0.3090+0.9511i,\ -0.8090+0.5878i,$
$-0.8090-0.5878i,\ 0.3090-0.9511i$

149. (a) Cube root of $-125=125(\cos 180°+i\sin 180°)$

$$\sqrt[3]{125}\left[\cos\left(\frac{180°+360k}{3}\right)+i\sin\left(\frac{180°+360k}{3}\right)\right],$$

$k=0,\ 1,\ 2$

$k=0:\ 5(\cos 60°+i\sin 60°)$

$k=1:\ 5(\cos 180°+i\sin 180°)$

$k=2:\ 5(\cos 300°+i\sin 300°)$

(b)

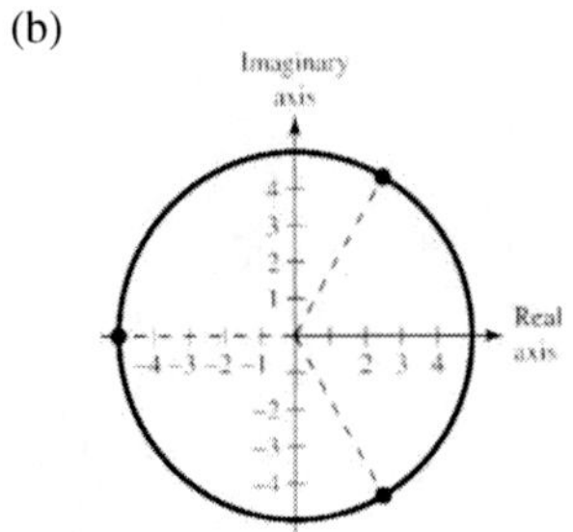

(c) $\frac{5}{2}+\frac{5\sqrt{3}}{2}i,\ -5,\ \frac{5}{2}-\frac{5\sqrt{3}}{2}i$

151. (a) Fifth roots of

$$128(-1+i) = 128\sqrt{2}\left(\cos\frac{3\pi}{4} + i\,\sin\frac{3\pi}{4}\right):$$

$$\left(128\sqrt{2}\right)^{1/5}\left[\cos\left(\frac{3\pi/4 + 2k\pi}{5}\right) + i\,\sin\left(\frac{3\pi/4 + 2k\pi}{5}\right)\right]$$

$k = 0, 1, 2, 3, 4$

$k = 0: 2\sqrt{2}\left(\cos\frac{3\pi}{20} + i\,\sin\frac{3\pi}{20}\right)$

$k = 1: 2\sqrt{2}\left(\cos\frac{11\pi}{20} + i\,\sin\frac{11\pi}{20}\right)$

$k = 2: 2\sqrt{2}\left(\cos\frac{19\pi}{20} + i\,\sin\frac{19\pi}{20}\right)$

$k = 3: 2\sqrt{2}\left(\cos\frac{27\pi}{20} + i\,\sin\frac{27\pi}{20}\right)$

$k = 4: 2\sqrt{2}\left(\cos\frac{7\pi}{4} + i\,\sin\frac{7\pi}{4}\right)$

(b)

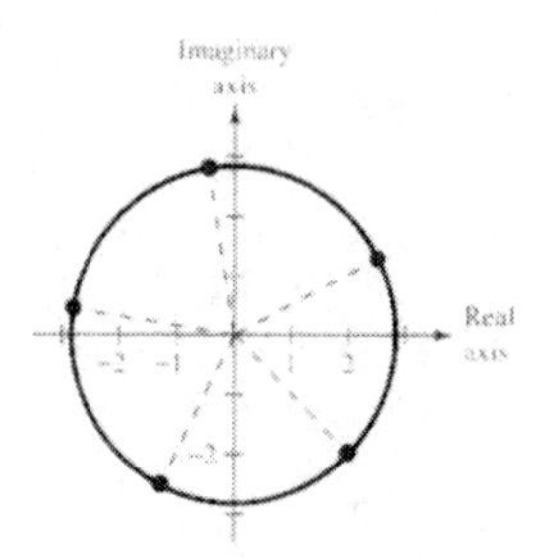

(c) $2.52 + 1.28i, -0.44 + 2.79i, -2.79 + 0.44i, -1.28 - 2.52i, 2 - 2i$

153. $x^4 - i = 0$

$x^4 = i$

The solutions are the fourth roots of

$i = \cos\frac{\pi}{2} + i\,\sin\frac{\pi}{2}.$

$$\sqrt[4]{1}\left[\cos\left(\frac{(\pi/2) + 2k\pi}{4}\right) + i\,\sin\left(\frac{(\pi/2) + 2k\pi}{4}\right)\right]$$

$k = 0, 1, 2, 3$

$k = 0: \cos\frac{\pi}{8} + i\,\sin\frac{\pi}{8}$

$k = 1: \cos\frac{5\pi}{8} + i\,\sin\frac{5\pi}{8}$

$k = 2: \cos\frac{9\pi}{8} + i\,\sin\frac{9\pi}{8}$

$k = 3: \cos\frac{13\pi}{8} + i\,\sin\frac{13\pi}{8}$

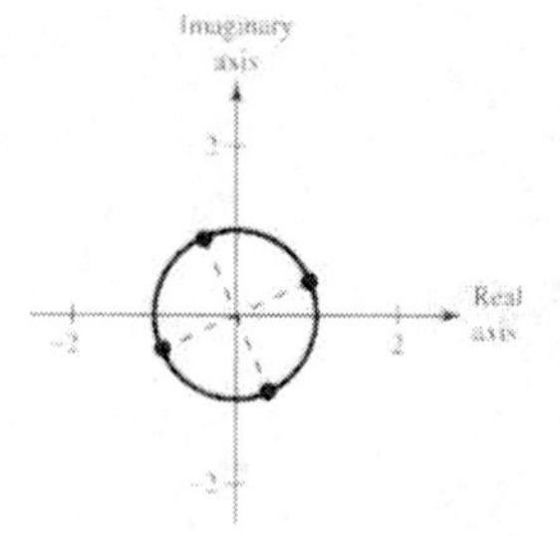

155. $x^5 + 243 = 0$

$x^5 = -243$

The solutions are the fifth roots of

$-243 = 243[\cos\pi + i\sin\pi]:$

$$\sqrt[5]{243}\left[\cos\left(\frac{\pi + 2k\pi}{5}\right) + i\sin\left(\frac{\pi + 2k\pi}{5}\right)\right],\ k = 0, 1, 2, 3, 4$$

$k = 0: 3\left(\cos\frac{\pi}{5} + i\sin\frac{\pi}{5}\right)$

$k = 1: 3\left(\cos\frac{3\pi}{5} + i\sin\frac{3\pi}{5}\right)$

$k = 2: 3\left(\cos\frac{5\pi}{5} + i\sin\frac{5\pi}{5}\right) = 3(\cos\pi + i\sin\pi)$

$k = 3: 3\left(\cos\frac{7\pi}{5} + i\sin\frac{7\pi}{5}\right)$

$k = 4: 3\left(\cos\frac{9\pi}{5} + i\sin\frac{9\pi}{5}\right)$

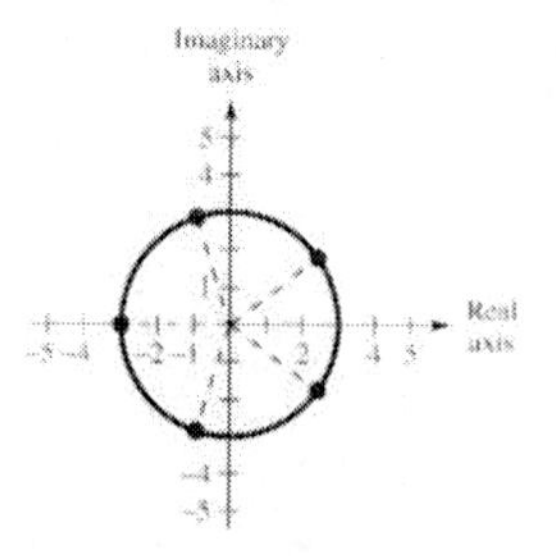

157. $x^4 + 16i = 0$

$x^4 = -16i$

The solutions are the fourth roots of

$$-16i = 16\left[\cos\frac{3\pi}{2} + i\sin\frac{3\pi}{2}\right]:$$

$$\sqrt[4]{16}\left[\cos\left(\frac{(3\pi/2) + 2k\pi}{4}\right) + i\sin\left(\frac{(3\pi/2) + 2k\pi}{4}\right)\right],$$

$k = 0, 1, 2, 3$

$k = 0: 2\left[\cos\left(\frac{3\pi}{8}\right) + i\sin\left(\frac{3\pi}{8}\right)\right]$

$k = 1: 2\left[\cos\left(\frac{7\pi}{8}\right) + i\sin\left(\frac{7\pi}{8}\right)\right]$

$k = 2: 2\left[\cos\left(\frac{11\pi}{8}\right) + i\sin\left(\frac{11\pi}{8}\right)\right]$

$k = 3: 2\left[\cos\left(\frac{15\pi}{8}\right) + i\sin\left(\frac{15\pi}{8}\right)\right]$

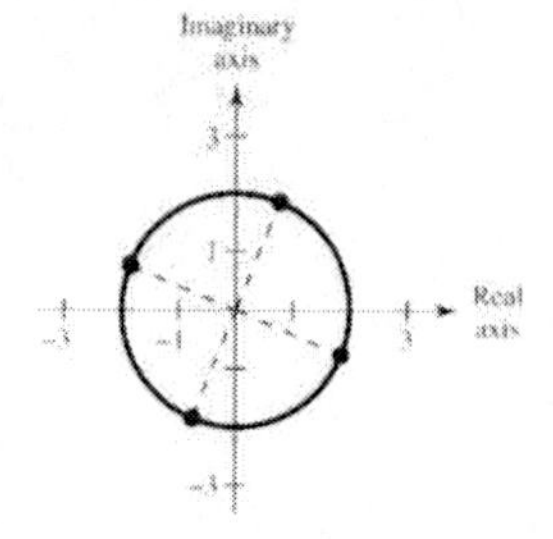

159. $x^3 - (1 - i) = 0$

$x^3 = 1 - i$

The solutions are the cube roots of $1 - i = \sqrt{2}(\cos 315° + i\sin 315°)$:

$$\sqrt[3]{\sqrt{2}}\left[\cos\left(\frac{315° + 360°k}{3}\right) + i\sin\left(\frac{315° + 360°}{3}\right)\right],\ k = 0, 1, 2$$

$k = 0: \sqrt[6]{2}(\cos 105° + i\sin 105°)$

$k = 1: \sqrt[6]{2}(\cos 225° + i\sin 225°)$

$k = 2: \sqrt[6]{2}(\cos 345° + i\sin 345°)$

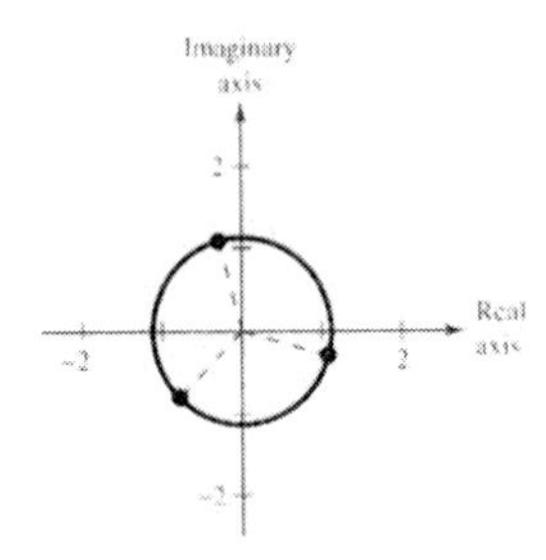

161. $E = I \cdot Z$

$= (10 + 2i)(4 + 3i)$

$= (40 + 6) + (30 + 8)i$

$= 34 + 38i$

163. $Z = \dfrac{E}{I}$

$= \dfrac{5 + 5i}{2 + 4i} \cdot \dfrac{2 - 4i}{2 - 4i}$

$= \dfrac{(10 + 20) + (10 - 20)i}{4 + 16}$

$= \dfrac{30 - 10i}{20}$

$= \dfrac{3}{2} - \dfrac{1}{2}i$

165. $I = \dfrac{E}{Z}$

$= \dfrac{12 + 24i}{12 + 20i} \cdot \dfrac{12 - 20i}{12 - 20i}$

$= \dfrac{(144 + 480) + (288 - 240)i}{144 + 400}$

$= \dfrac{624 + 48i}{544}$

$= \dfrac{39}{34} + \dfrac{3}{34}i$

167. True. $\left[\dfrac{1}{2}(1 - \sqrt{3}i)\right]^9 = \left[\dfrac{1}{2} - \dfrac{\sqrt{3}}{2}i\right]^9 = -1$

169. True. $z_1 z_2 = r_1 r_2\left[\cos(\theta_1 + \theta_2) + i\sin(\theta_1 + \theta_2)\right] = 0$ if and only if $r_1 = 0$ and or $r_2 = 0$.

171. $\dfrac{z_1}{z_2} = \dfrac{r_1(\cos\theta_1 + i\sin\theta_1)}{r_2(\cos\theta_2 + i\sin\theta_2)} \cdot \dfrac{\cos\theta_2 - i\sin\theta_2}{\cos\theta_2 - i\sin\theta_2}$

$= \dfrac{r_1}{r_2(\cos^2\theta_2 + \sin^2\theta_2)}[\cos\theta_1\cos\theta_2 + \sin\theta_1\sin\theta_2$

$+ i(\sin\theta_1\cos\theta_2 - \sin\theta_2\cos\theta_1)]$

$= \dfrac{r_1}{r_2}[\cos(\theta_1 - \theta_2) + i\sin(\theta_1 - \theta_2)]$

173. (a) $z\bar{z} = [r(\cos\theta + i\sin\theta)][r(\cos(-\theta) + i\sin(-\theta))]$

$= r^2[\cos(\theta - \theta) + i\sin(\theta - \theta)]$

$= r^2[\cos 0 + i\sin 0]$

$= r^2$

(b) $\dfrac{z}{\bar{z}} = \dfrac{r(\cos\theta + i\sin\theta)}{r[\cos(-\theta) + i\sin(-\theta)]}$

$= \dfrac{r}{r}[\cos(\theta - (-\theta)) + i\sin(\theta - (-\theta))]$

$= \cos 2\theta + i\sin 2\theta$

175. Let $a = 0$ and $b = \pi$ in Euler's Formula:

$e^{a+bi} = e^a(\cos b + i\sin b)$

$e^{0+\pi i} = e^0(\cos\pi + i\sin\pi)$

$e^{\pi i} = -1$

$e^{\pi i} + 1 = 0$

177. $d = 16\cos\left(\dfrac{\pi}{4}t\right)$

Maximum displacement: 16

Lowest possible t-value: $\dfrac{\pi}{4}t = \dfrac{\pi}{2} \Rightarrow t = 2$

179. $d = \dfrac{1}{8}\cos(12\pi t)$

Maximum displacement: $\dfrac{1}{8}$

Lowest possible t-value: $12\pi t = \dfrac{\pi}{2} \Rightarrow t = \dfrac{1}{24}$

Chapter 6 Review Exercises

1. Given: $A = 32°,\ B = 50°,\ a = 16$

$$C = 180° - 32° - 50° = 98°$$

$$b = \frac{a \sin B}{\sin A} = \frac{16 \sin 50°}{\sin 32°} \approx 23.13$$

$$c = \frac{a \sin C}{\sin A} = \frac{16 \sin 98°}{\sin 32°} \approx 29.90$$

3. Given: $B = 25°,\ C = 105°,\ c = 25$

$$A = 180° - 25° - 105° = 50°$$

$$b = \frac{c \sin B}{\sin C} = \frac{25 \sin 25°}{\sin 105°} \approx 10.94$$

$$a = \frac{c \sin A}{\sin C} = \frac{25 \sin 50°}{\sin 105°} \approx 19.83$$

5. Given:

$A = 60°\ 15' = 60.25°,\ B = 45°\ 30' = 45.5°,\ b = 4.8$

$$C = 180° - 60.25° - 45.5° = 74.25° = 74°\ 15'$$

$$a = \frac{b \sin A}{\sin B} = \frac{4.8 \sin 60.25°}{\sin 45.5°} \approx 5.84$$

$$c = \frac{b \sin C}{\sin B} = \frac{4.8 \sin 74.25°}{\sin 45.5°} \approx 6.48$$

7. Given: $A = 75°,\ a = 2.5,\ b = 16.5$

$$\sin B = \frac{b \sin A}{a} = \frac{16.5 \sin 75°}{2.5} \approx 6.375 \Rightarrow \text{no triangle formed}$$

No solution

9. Given: $B = 115°,\ a = 9,\ b = 14.5$

$$\sin A = \frac{a\ \sin B}{b} = \frac{9\ \sin 115°}{14.5} \approx 0.5625 \Rightarrow A \approx 34.2°$$

$$C \approx 180° - 115° - 34.2° = 30.8°$$

$$c = \frac{b}{\sin B}(\sin C) \approx \frac{14.5}{\sin 115°}(\sin 30.8°) \approx 8.18$$

11. Given: $A = 33°,\ b = 7,\ c = 10$

$$\text{Area} = \frac{1}{2} bc\ \sin A$$

$$= \frac{1}{2}(7)(10) \sin 33°$$

$$\approx 19.06 \text{ square units}$$

13. Given: $C = 122°,\ b = 18,\ a = 29$

$$\text{Area} = \frac{1}{2} ab\ \sin C$$

$$= \frac{1}{2}(29)(18) \sin 122°$$

$$\approx 221.34 \text{ square units}$$

15. $\sin 28° = \dfrac{h}{75}$

$$h = 75 \sin 28° \approx 35.21 \text{ feet}$$

$$\cos 28° = \frac{x}{75}$$

$$x = 75 \cos 28° \approx 66.22 \text{ feet}$$

$$\tan 45° = \frac{H}{x}$$

$$H = x \tan 45° \approx 66.22 \text{ feet}$$

Height of tree: $H - h \approx 31$ feet

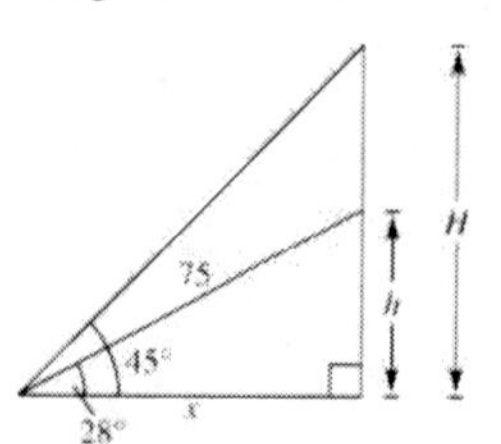

17. Given: $a = 8,\ b = 14,\ c = 17$

$$\cos C = \frac{a^2 + b^2 - c^2}{2ab}$$

$$= \frac{64 + 196 - 289}{2(8)(14)} \approx -0.1295 \Rightarrow C \approx 97.44°$$

$$\cos B = \frac{a^2 + c^2 - b^2}{2ac}$$

$$= \frac{64 + 289 - 196}{2(8)(17)} \approx 0.5772 \Rightarrow B \approx 54.75°$$

$$A = 180° - B - C = 180° - 54.75° - 97.44° = 27.81°$$

19. Given: $a = 9,\ b = 12,\ c = 20$

$$\cos\ C = \frac{a^2 + b^2 - c^2}{2ab}$$

$$= \frac{81 + 144 - 400}{2(9)(12)}$$

$$\approx -0.8102 \Rightarrow C \approx 144.1°$$

$$\sin A = \frac{a \sin C}{c}$$

$$= \frac{9 \sin(144.1°)}{20}$$

$$\approx 0.264 \Rightarrow A \approx 15.3$$

$$B = 180° - 144.1° - 15.3° = 20.6°$$

21. Given: $a = 6.5$, $b = 10.2$, $c = 16$

$$\cos A = \frac{b^2 + c^2 - a^2}{2bc} = \frac{10.2^2 + 16^2 - 6.5^2}{2(10.2)(16)} \approx 0.97 \Rightarrow A \approx 13.19°$$

$$\cos B = \frac{a^2 + c^2 - b^2}{2ac} = \frac{6.5^2 + 16^2 - 10.2^2}{2(6.5)(16)} \approx 0.93 \Rightarrow B \approx 20.98°$$

$$C = 180° - A - B \approx 145.83°$$

23. Given: $C = 65°$, $a = 25$, $b = 12$

$$c^2 = a^2 + b^2 - 2ab\cos C = 25^2 + 12^2 - 2(25)(12)\cos 65°$$

$$\approx 515.4290 \Rightarrow c \approx 22.70$$

$$\sin A = \frac{a\sin C}{c} = \frac{25\sin 65°}{22.70} \approx 0.998 \Rightarrow A \approx 86.38°$$

$$B = 180° - A - C \approx 28.62°$$

25. $a^2 = 5^2 + 8^2 - 2(5)(8)\cos 152° \approx 159.6 \Rightarrow a \approx 12.63$ ft

$b^2 = 5^2 + 8^2 - 2(5)(8)\cos 28° \approx 18.36 \Rightarrow b \approx 4.285$ ft

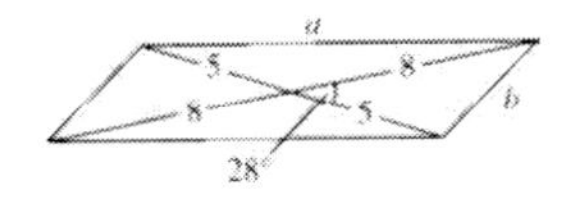

27. Given: $a = 3$, $b = 6$, $c = 8$

$$s = \frac{a+b+c}{2} = \frac{3+6+8}{2} = 8.5$$

$$\text{Area} = \sqrt{s(s-a)(s-b)(s-c)}$$

$$= \sqrt{8.5(8.5-3)(8.5-6)(8.5-8)}$$

$$= \sqrt{58.4375} \approx 7.64 \text{ square units}$$

29. Given: $a = 64.8$, $b = 49.2$, $c = 24.1$

$$s = \frac{a+b+c}{2} = \frac{64.8 + 49.2 + 24.1}{2} = 69.05$$

$$\text{Area} = \sqrt{s(s-a)(s-b)(s-c)}$$

$$= \sqrt{69.05(4.25)(19.85)(44.95)}$$

$$\approx 511.7 \text{ square units}$$

31. Initial point: $(-5, 4)$

Terminal point: $(2, -1)$

$$\mathbf{v} = \langle 2 - (-5), -1 - 4 \rangle = \langle 7, -5 \rangle$$

$$\|\mathbf{v}\| = \sqrt{7^2 + (-5)^2} = \sqrt{74}$$

33. Initial point: $(0, 10)$

Terminal point: $(7, 3)$

$$\mathbf{v} = \langle 7 - 0, 3 - 10 \rangle = \langle 7, -7 \rangle$$

$$\|\mathbf{v}\| = \sqrt{7^2 + (-7)^2} = \sqrt{98} = 7\sqrt{2}$$

35. (a) $\mathbf{u} + \mathbf{v} = \langle -1, -3 \rangle + \langle -3, 6 \rangle = \langle -4, 3 \rangle$

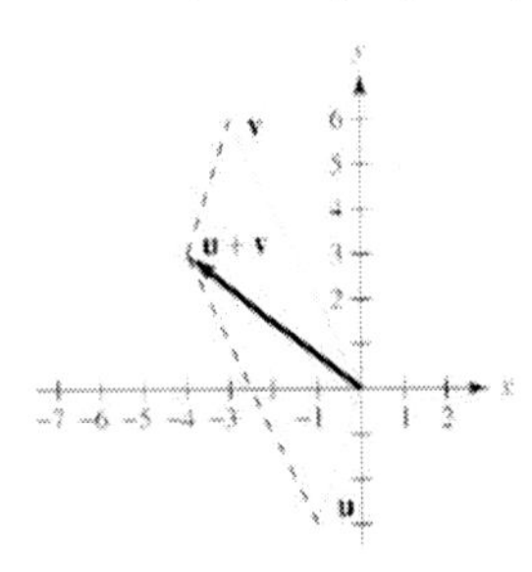

(b) $\mathbf{u} - \mathbf{v} = \langle 2, -9 \rangle$

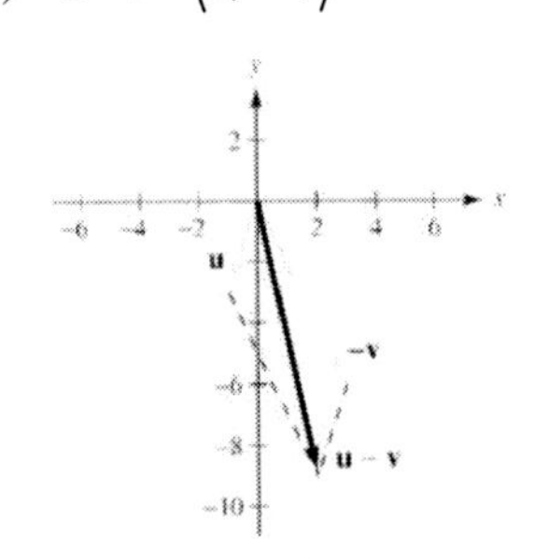

(c) $3\mathbf{u} = \langle -3, -9 \rangle$

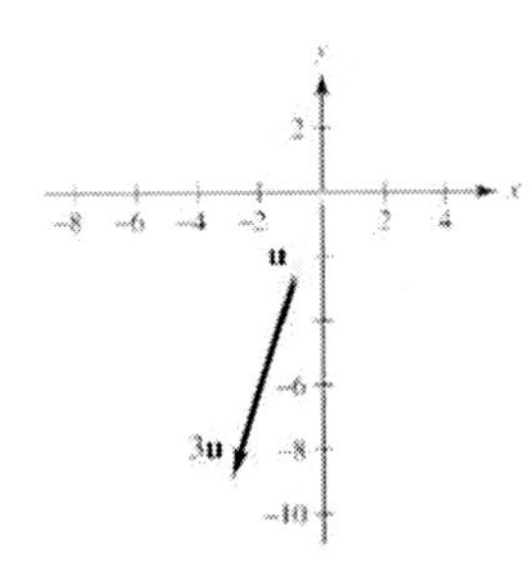

(d) $2\mathbf{v} + 5\mathbf{u} = \langle -6, 12 \rangle + \langle -5, -15 \rangle = \langle -11, -3 \rangle$

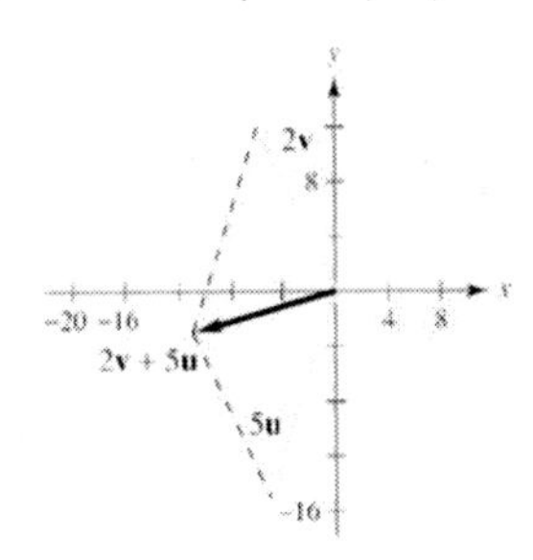

37. (a) $\mathbf{u} + \mathbf{v} = \langle -5, 2 \rangle + \langle 4, 4 \rangle = \langle -1, 6 \rangle$

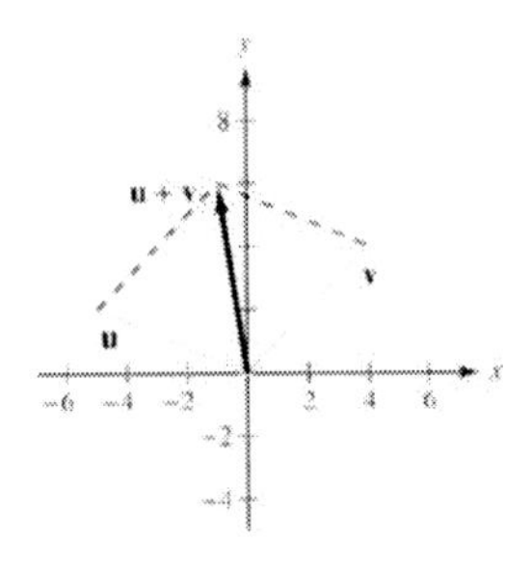

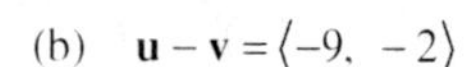
(b) $\mathbf{u} - \mathbf{v} = \langle -9, -2 \rangle$

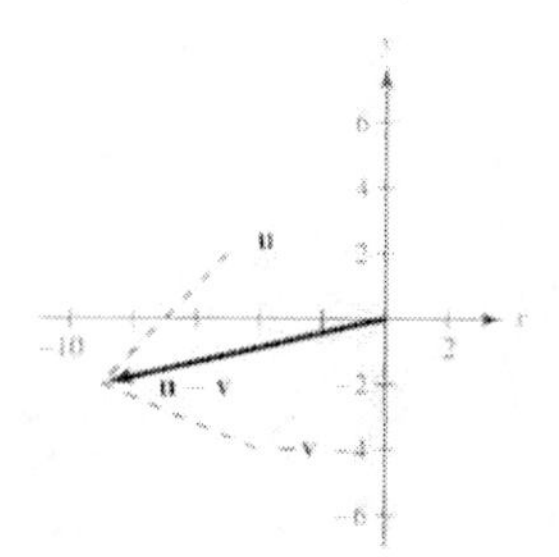

(c) $3\mathbf{u} = \langle -15, 6 \rangle$

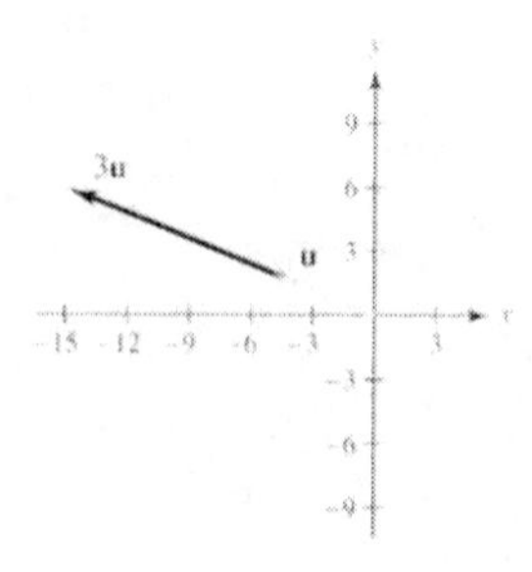

(d) $2\mathbf{v} + 5\mathbf{u} = \langle 8, 8 \rangle + \langle -25, 10 \rangle = \langle -17, 18 \rangle$

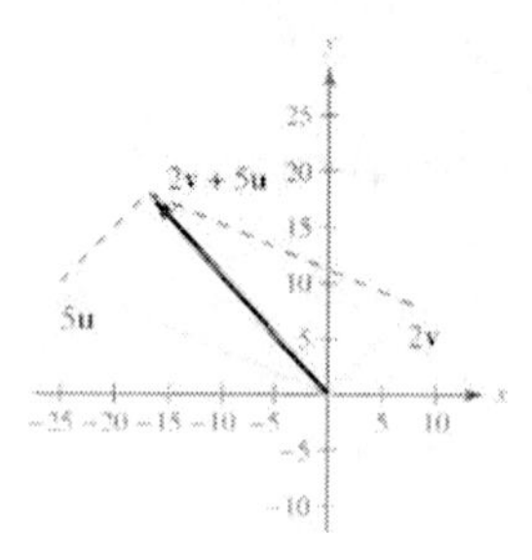

39. (a) $\mathbf{u} + \mathbf{v} = (2\mathbf{i} - \mathbf{j}) + (5\mathbf{i} + 3\mathbf{j}) = 7\mathbf{i} + 2\mathbf{j}$

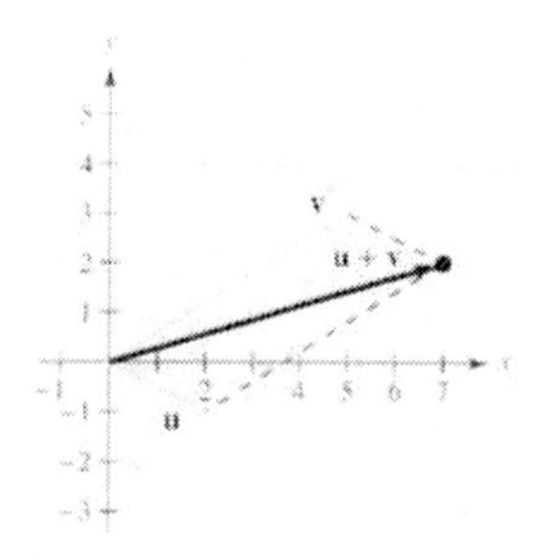

(b) $\mathbf{u} - \mathbf{v} = -3\mathbf{i} - 4\mathbf{j}$

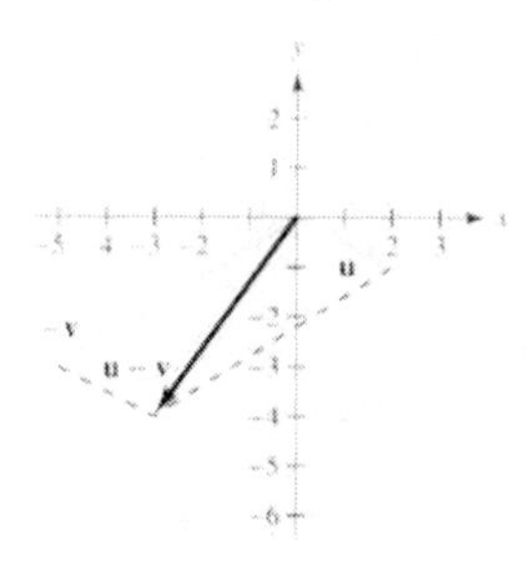

(c) $3\mathbf{u} = 6\mathbf{i} - 3\mathbf{j}$

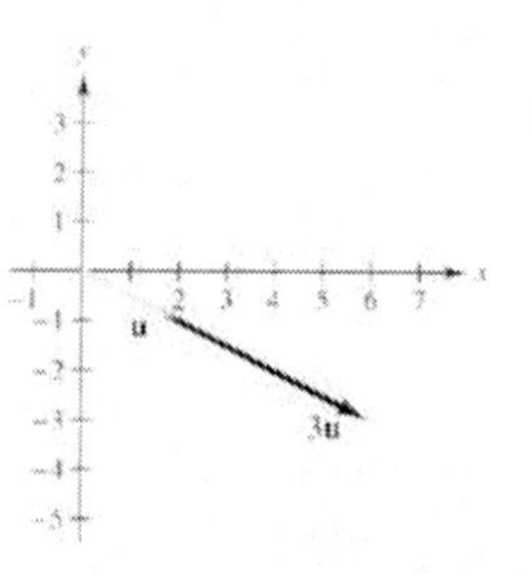

(d) $2\mathbf{v} + 5\mathbf{u} = (10\mathbf{i} + 6\mathbf{j}) + (10\mathbf{i} - 5\mathbf{j}) = 20\mathbf{i} + \mathbf{j}$

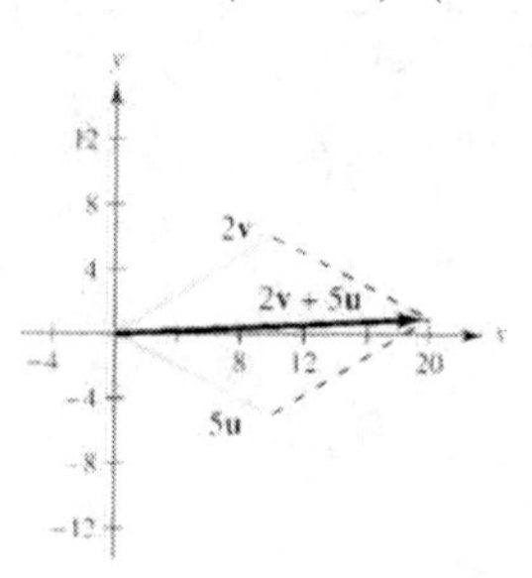

41. $3\mathbf{v} = 3(10\mathbf{i} + 3\mathbf{j}) = 30\mathbf{i} + 9\mathbf{j} = \langle 30, 9 \rangle$

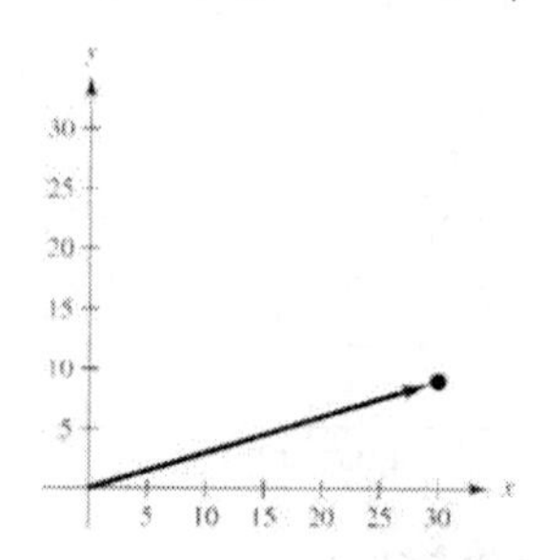

43. $\mathbf{w} = 4\mathbf{u} + 5\mathbf{v} = 4(6\mathbf{i} - 5\mathbf{j}) + 5(10\mathbf{i} - 3\mathbf{j})$
$= 74\mathbf{i} - 5\mathbf{j}$

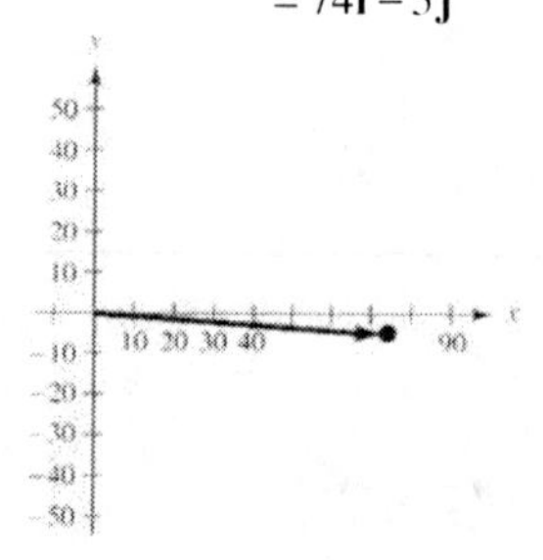

45. $\|\mathbf{u}\| = 6$

Unit vector: $\dfrac{1}{6}\langle 0, -6 \rangle = \langle 0, -1 \rangle$

47. $\|\mathbf{v}\| = \sqrt{5^2 + (-2)^2} = \sqrt{29}$

Unit vector: $\dfrac{1}{\sqrt{29}}\langle 5, -2 \rangle = \left\langle \dfrac{5}{\sqrt{29}}, \dfrac{-2}{\sqrt{29}} \right\rangle$

49. $\mathbf{u} = \langle 1 - (-8), -5 - 3 \rangle = \langle 9, -8 \rangle = 9\mathbf{i} - 8\mathbf{j}$

51. $\mathbf{v} = 7(\cos 60°\mathbf{i} + \sin 60°\mathbf{j})$

$\|\mathbf{v}\| = 7$

$\theta = 60°$

53. $\mathbf{v} = 5\mathbf{i} + 4\mathbf{j}$

$\|\mathbf{v}\| = \sqrt{5^2 + 4^2} = \sqrt{41}$

$\tan\theta = \dfrac{4}{5} \Rightarrow \theta \approx 38.7°$

55. $\mathbf{v} = -3\mathbf{i} - 3\mathbf{j}$

$\|\mathbf{v}\| = \sqrt{(-3)^2 + (-3)^2} = 3\sqrt{2}$

$\tan\theta = \dfrac{-3}{-3} = 1 \Rightarrow \theta \approx 225°$

57. Force One: $\mathbf{u} = 85\mathbf{i}$

Force Two: $\mathbf{v} = 50\ \cos 15°\mathbf{i} + 50\ \sin 15°\mathbf{j}$

Resultant Force:

$\mathbf{u} + \mathbf{v} = (85 + \cos 15°)\mathbf{i} + (50\sin 15°)\mathbf{j}$

$\|\mathbf{u} + \mathbf{v}\| = \sqrt{(85 + 50\cos 15°)^2 + (50\sin 15°)^2}$

$= \sqrt{85^2 + 8500\ \cos 15° + 50^2}$

$= 133.92$ lb

$\tan\theta = \dfrac{50\sin 15°}{85 + 50\cos 15°} \Rightarrow \theta \approx 5.5°$ from the 85-pound force.

59. By symmetry, the magnitudes of the tensions are equal.

$\mathbf{T} = \|\mathbf{T}\|(\cos 120°\mathbf{i} + \sin 120°\mathbf{j})$

$\|\mathbf{T}\|\sin 120° = \dfrac{1}{2}(200) \Rightarrow \|\mathbf{T}\| = \dfrac{100}{\sqrt{3}/2} = \dfrac{200}{\sqrt{3}} \approx 115.5$ lb

61. $\mathbf{u} \cdot \mathbf{v} = \langle 0, -2\rangle \cdot \langle 1, 10\rangle = 0 - 20 = -20$

63. $\mathbf{u} \cdot \mathbf{v} = \langle 6, -1\rangle \cdot \langle 2, 5\rangle = 6(2) + (-1)(5) = 7$

65. $\mathbf{u} \cdot \mathbf{u} = \langle -3, -4\rangle \cdot \langle -3, -4\rangle$

$= 9 + 16 = 25 = \|\mathbf{u}\|^2$

67. $4\mathbf{u} \cdot \mathbf{v} = 4\langle -3, -4\rangle \cdot \langle 2, 1\rangle = 4(-6, -4) = -40$

69. $\mathbf{u} = \langle 2\sqrt{2}, -4\rangle,\ \mathbf{v} = \langle -\sqrt{2}, 1\rangle$

$\cos\theta = \dfrac{\mathbf{u} \cdot \mathbf{v}}{\|\mathbf{u}\|\|\mathbf{v}\|} = \dfrac{-8}{(\sqrt{24})(\sqrt{3})} \Rightarrow \theta \approx 160.5°$

71. $\mathbf{u} = \cos\dfrac{7\pi}{4}\mathbf{i} + \sin\dfrac{7\pi}{4}\mathbf{j} = \left\langle \dfrac{1}{\sqrt{2}}, -\dfrac{1}{\sqrt{2}} \right\rangle$

$\mathbf{v} = \cos\dfrac{5\pi}{6}\mathbf{i} + \sin\dfrac{5\pi}{6}\mathbf{j} = \left\langle -\dfrac{\sqrt{3}}{2}, \dfrac{1}{2} \right\rangle$

$\cos\theta = \dfrac{\mathbf{u} \cdot \mathbf{v}}{\|\mathbf{u}\|\|\mathbf{v}\|} = \dfrac{\left(-\sqrt{3}/(2\sqrt{2})\right) - \left(1/(2\sqrt{2})\right)}{(1)(1)}$

$\theta \approx 165°$ or $\dfrac{11\pi}{12}$

73. $\cos\theta = \dfrac{\mathbf{u} \cdot \mathbf{v}}{\|\mathbf{u}\|\|\mathbf{v}\|} = 0 \Rightarrow \theta = 90°$

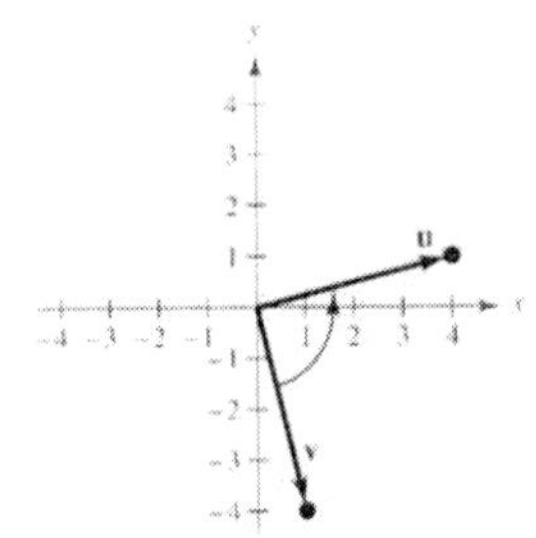

75. $\cos\theta = \dfrac{\mathbf{u} \cdot \mathbf{v}}{\|\mathbf{u}\|\|\mathbf{v}\|} = \dfrac{70 - 15}{\sqrt{74}\sqrt{109}}$

$\approx 0.612 \Rightarrow \theta \approx 52.2°$

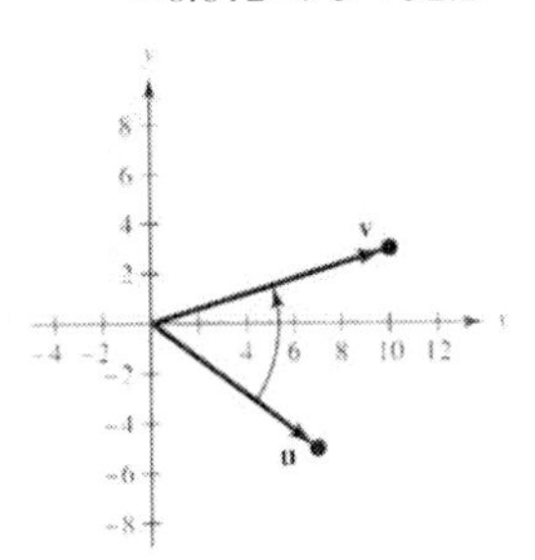

77. $\mathbf{u} = \langle 39, -12\rangle,\ \mathbf{v} = \langle -26, 8\rangle$

$\mathbf{u} \cdot \mathbf{v} = 39(-26) + (-12)(8)$

$= -1110 \neq 0 \Rightarrow \mathbf{u}$ and $\mathbf{v}$ are not orthogonal.

$\mathbf{v} = -\frac{2}{3}\mathbf{u} \Rightarrow \mathbf{u}$ and $\mathbf{v}$ parallel.

79. $\mathbf{u} = \langle 8, 5\rangle,\ \mathbf{v} = \langle -2, 4\rangle$

$\mathbf{u} \cdot \mathbf{v} = 8(-2) + (5)(4)$

$= 4 \neq 0 \Rightarrow \mathbf{u}$ and $\mathbf{v}$ are not orthogonal.

$\mathbf{u} \neq k\mathbf{v} \Rightarrow \mathbf{u}$ and $\mathbf{v}$ are not parallel.

Neither

81. $\langle 1, -k\rangle \cdot \langle 1, 2\rangle = 1 - 2k = 0 \Rightarrow k = \frac{1}{2}$

83. $\langle k, -1\rangle \cdot \langle 2, -2\rangle = 2k + 2 = 0 \Rightarrow k = -1$

85. $\mathbf{u} = \langle -4, 3\rangle, \mathbf{v} = \langle -8, -2\rangle$

$$\text{proj}_{\mathbf{v}}\mathbf{u} = \left(\frac{\mathbf{u}\cdot\mathbf{v}}{\|\mathbf{v}\|^2}\right)\mathbf{v} = \left(\frac{26}{68}\right)\langle -8, -2\rangle = -\frac{13}{17}\langle 4, 1\rangle$$

$$\mathbf{u} - \text{proj}_{\mathbf{v}}\mathbf{u} = \langle -4, 3\rangle - \left\langle \frac{-52}{17}, \frac{-13}{17}\right\rangle = \left\langle -\frac{16}{17}, \frac{64}{17}\right\rangle$$

$$\mathbf{u} = \left\langle -\frac{52}{17}, -\frac{13}{17}\right\rangle + \left\langle -\frac{16}{17}, -\frac{64}{17}\right\rangle$$

87. $\mathbf{u} = \langle 2, 7\rangle, \mathbf{v} = \langle 1, -1\rangle$

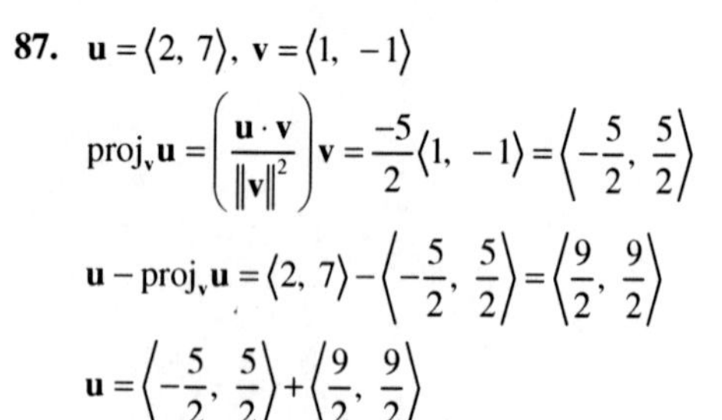

$$\text{proj}_{\mathbf{v}}\mathbf{u} = \left(\frac{\mathbf{u}\cdot\mathbf{v}}{\|\mathbf{v}\|^2}\right)\mathbf{v} = \frac{-5}{2}\langle 1, -1\rangle = \left\langle -\frac{5}{2}, \frac{5}{2}\right\rangle$$

$$\mathbf{u} - \text{proj}_{\mathbf{v}}\mathbf{u} = \langle 2, 7\rangle - \left\langle -\frac{5}{2}, \frac{5}{2}\right\rangle = \left\langle \frac{9}{2}, \frac{9}{2}\right\rangle$$

$$\mathbf{u} = \left\langle -\frac{5}{2}, \frac{5}{2}\right\rangle + \left\langle \frac{9}{2}, \frac{9}{2}\right\rangle$$

89. 48 inches = 4 feet

Work $= 18{,}000(4) = 72{,}000$ ft · lb

91. $z = 7i$

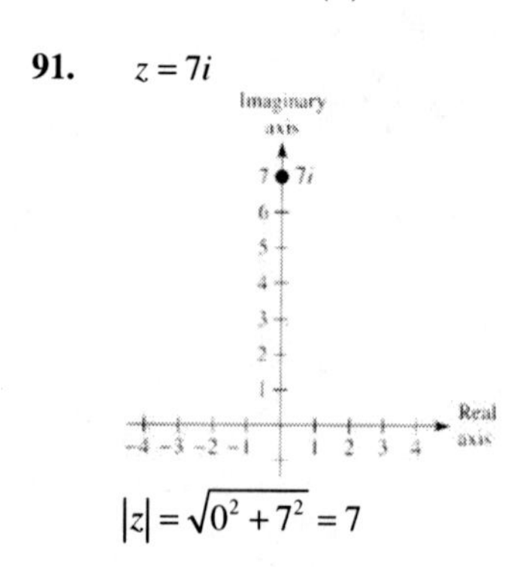

$$|z| = \sqrt{0^2 + 7^2} = 7$$

93. $z = 5 + 3i$

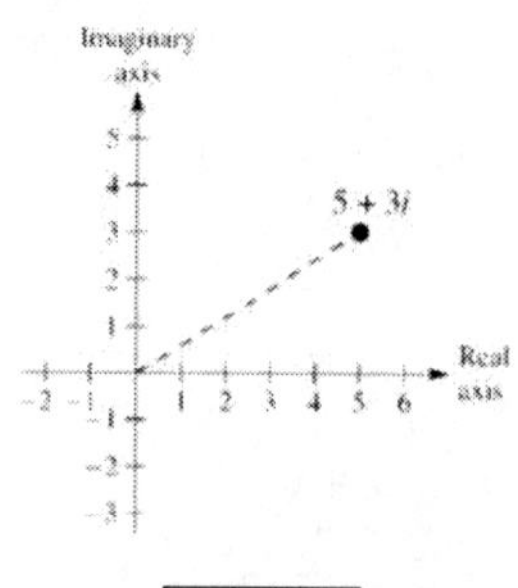

$$|z| = \sqrt{(5)^2 + (3)^2} = \sqrt{34}$$

95. $z = 2 - 2i$

$$|z| = \sqrt{4+4} = 2\sqrt{2}$$

$$\theta = \frac{7\pi}{4}$$

$$z = 2\sqrt{2}\left(\cos\frac{7\pi}{4} + i\sin\frac{7\pi}{4}\right)$$

97. $z = -\sqrt{3} - i$

$$|z| = \sqrt{3+1} = 2$$

$$\theta = \frac{7\pi}{6}$$

$$z = 2\left(\cos\frac{7\pi}{6} + i\sin\frac{7\pi}{6}\right)$$

99. $$\left[\frac{5}{2}\left(\cos\frac{\pi}{2} + i\sin\frac{\pi}{2}\right)\right]\left[4\left(\cos\frac{\pi}{4} + i\sin\frac{\pi}{4}\right)\right] = 10\left[\cos\frac{3\pi}{4} + i\sin\frac{3\pi}{4}\right]$$

101. $$\frac{20(\cos 320° + i\sin 320°)}{5(\cos 80° + i\sin 80°)} = 4[\cos 240° + i\sin 240°]$$

103. (a) $$2 - 2i = 2\sqrt{2}\left(\cos\frac{7\pi}{4} + i\sin\frac{7\pi}{4}\right)$$

$$3 + 3i = 3\sqrt{2}\left(\cos\frac{\pi}{4} + i\sin\frac{\pi}{4}\right)$$

(b) $$2\sqrt{2}\left(\cos\frac{7\pi}{4} + i\sin\frac{7\pi}{4}\right)3\sqrt{2}\left(\cos\frac{\pi}{4} + i\sin\frac{\pi}{4}\right) = 12(\cos 2\pi + i\sin 2\pi) = 12$$

(c) $(2 - 2i)(3 + 3i) = 6 + 6 = 12$

105. (a) $$3 - 3i = 3\sqrt{2}\left(\cos\frac{7\pi}{4} + i\sin\frac{7\pi}{4}\right)$$

$$2 + 2i = 2\sqrt{2}\left(\cos\frac{\pi}{4} + i\sin\frac{\pi}{4}\right)$$

(b) $$\frac{3\sqrt{2}\left(\cos\frac{7\pi}{4} + i\sin\frac{7\pi}{4}\right)}{2\sqrt{2}\left(\cos\frac{\pi}{4} + i\sin\frac{\pi}{4}\right)} = \frac{3}{2}\left(\cos\frac{3\pi}{2} + i\sin\frac{3\pi}{2}\right) = \frac{3}{2}(-i) = -\frac{3}{2}i$$

(c) $$\frac{3 - 3i}{2 + 2i}\cdot\frac{(2 - 2i)}{(2 - 2i)} = \frac{6 - 12i - 6}{8} = -\frac{12i}{8} = -\frac{3}{2}i$$

107. $$\left[5\left(\cos\frac{\pi}{12} + i\sin\frac{\pi}{12}\right)\right]^4 = 5^4\left(\frac{4\pi}{12} + i\sin\frac{4\pi}{12}\right)$$

$$= 625\left(\cos\frac{\pi}{3} + i\sin\frac{\pi}{3}\right)$$

$$= 625\left(\frac{1}{2} + \frac{\sqrt{3}}{2}i\right)$$

$$= \frac{625}{2} + \frac{625\sqrt{3}}{2}i$$

109. $(2+3i)^6 \approx \left[\sqrt{13}\left(\cos 56.3^\circ + i\ \sin 56.3^\circ\right)\right]^6$

$= 13^3\left(\cos 337.9^\circ + i\ \sin 337.9^\circ\right)$

$\approx 13^3\left(0.9263 - 0.3769i\right)$

$\approx 2035 - 828i$

111. $-\sqrt{3} + i = 2\left(\cos\frac{5\pi}{6} + i\ \sin\frac{5\pi}{6}\right)$

Square roots:

$\sqrt{2}\left(\cos\frac{5\pi}{12} + i\ \sin\frac{5\pi}{12}\right) \approx 0.3660 + 1.3660i$

$\sqrt{2}\left(\cos\frac{17\pi}{12} + i\ \sin\frac{17\pi}{12}\right) \approx -0.3660 - 1.3660i$

113. $-2i = 2\left(\cos\frac{3\pi}{2} + i\ \sin\frac{3\pi}{2}\right)$

Square roots:

$\sqrt{2}\left(\cos\frac{3\pi}{4} + i\ \sin\frac{3\pi}{4}\right) = \sqrt{2}\left(-\frac{\sqrt{2}}{2} + i\frac{\sqrt{2}}{2}\right) = -1 + i$

$\sqrt{2}\left(\cos\frac{7\pi}{4} + i\ \sin\frac{7\pi}{4}\right) = 1 - i$

115. (a) Sixth roots of $-729i = 729\left(\cos\frac{3\pi}{2} + i\sin\frac{3\pi}{2}\right)$:

$\sqrt[6]{729}\left(\cos\frac{(3\pi/2) + 2k\pi}{6} + i\ \sin\frac{(3\pi/2) + 2k\pi}{6}\right),$

$k = 0, 1, 2, 3, 4, 5$

$k = 0:\ 3\left(\cos\frac{\pi}{4} + i\sin\frac{\pi}{4}\right)$

$k = 1:\ 3\left(\cos\frac{7\pi}{12} + i\sin\frac{7\pi}{12}\right)$

$k = 2:\ 3\left(\cos\frac{11\pi}{12} + i\sin\frac{11\pi}{12}\right)$

$k = 3:\ 3\left(\cos\frac{5\pi}{4} + i\sin\frac{5\pi}{4}\right)$

$k = 4:\ 3\left(\cos\frac{19\pi}{12} + i\sin\frac{19\pi}{12}\right)$

$k = 5:\ 3\left(\cos\frac{23\pi}{12} + i\sin\frac{23\pi}{12}\right)$

(b)

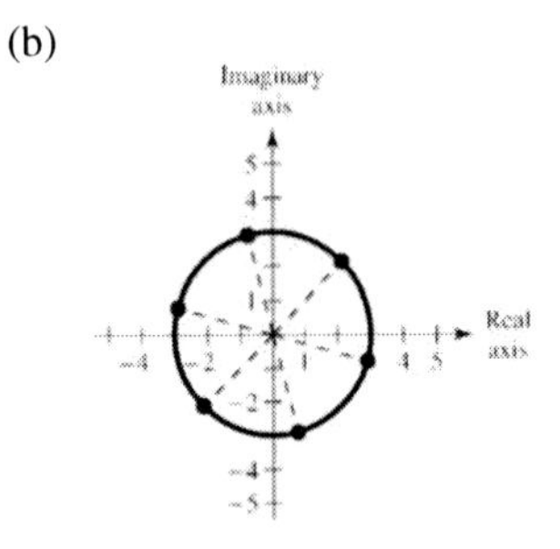

(c) $\frac{3\sqrt{2}}{2} + \frac{3\sqrt{2}}{2}i,\ -0.7765 + 2.898i,\ -2.898 + 0.7765i,$

$-\frac{3\sqrt{2}}{2} - \frac{3\sqrt{2}}{2}i,\ 0.7765 - 2.898i,\ 2.898 - 0.7765i$

$i,\ 0.7765 - 2.898i,\ 2.898 - 0.07765i$

117. (a) Cube roots of $8 = 8(\cos 0 + i\sin 0)$:

$\sqrt[3]{8}\left(\cos\left(\frac{2\pi k}{3}\right) + i\ \sin\frac{2\pi k}{3}\right),\ k = 0, 1, 2$

$k = 0:\ 2\left(\cos 0 + i\ \sin 0\right)$

$k = 1:\ 2\left(\cos\frac{2\pi}{3} + i\sin\frac{2\pi}{3}\right)$

$k = 2:\ 2\left(\cos\frac{4\pi}{3} + i\sin\frac{4\pi}{3}\right)$

(b)

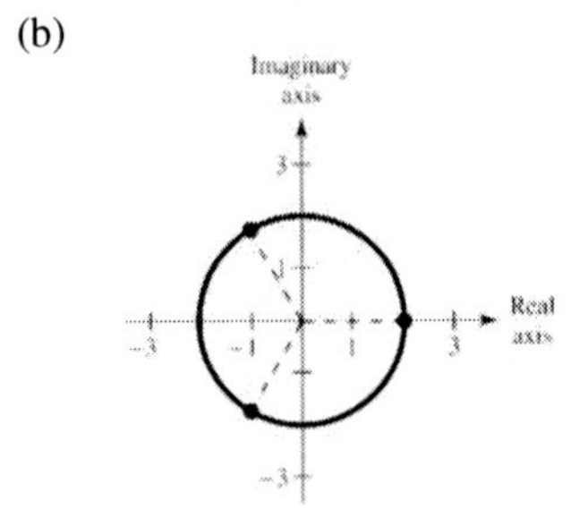

(c) $2,\ -1 + \sqrt{3}i,\ -1 - \sqrt{3}i$

119. $x^4 + 256 = 0$

$x^4 = -256 = 256(\cos\pi + i\sin\pi)$

$\sqrt[4]{-256} = 4\left[\cos\left(\frac{\pi + 2\pi k}{4}\right) + i\ \sin\left(\frac{\pi + 2\pi k}{4}\right)\right]$

$k = 0, 1, 2, 3$

$k = 0:\ 4\left(\cos\frac{\pi}{4} + i\sin\frac{\pi}{4}\right) = \frac{4\sqrt{2}}{2} + \frac{4\sqrt{2}}{2}i = 2\sqrt{2} + 2\sqrt{2}i$

$k = 1:\ 4\left(\cos\frac{3\pi}{4} + i\sin\frac{3\pi}{4}\right) = -\frac{4\sqrt{2}}{2} + \frac{4\sqrt{2}}{2}i = -2\sqrt{2} + 2\sqrt{2}i$

$k = 2:\ 4\left(\cos\frac{5\pi}{4} + i\sin\frac{5\pi}{4}\right) = -\frac{4\sqrt{2}}{2} - \frac{4\sqrt{2}}{2}i = -2\sqrt{2} - 2\sqrt{2}i$

$k = 3:\ 4\left(\cos\frac{7\pi}{4} + i\sin\frac{7\pi}{4}\right) = \frac{4\sqrt{2}}{2} - \frac{4\sqrt{2}}{2}i = 2\sqrt{2} - 2\sqrt{2}i$

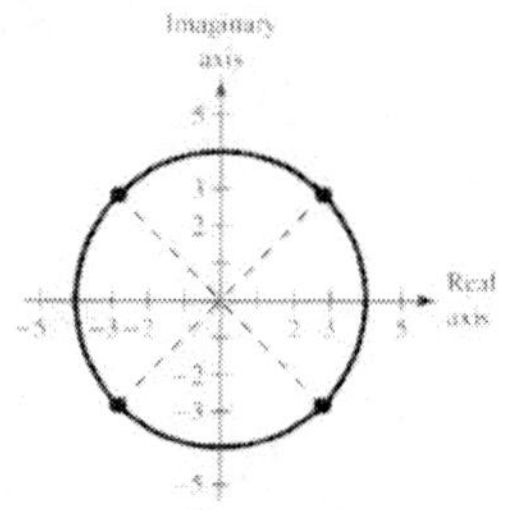

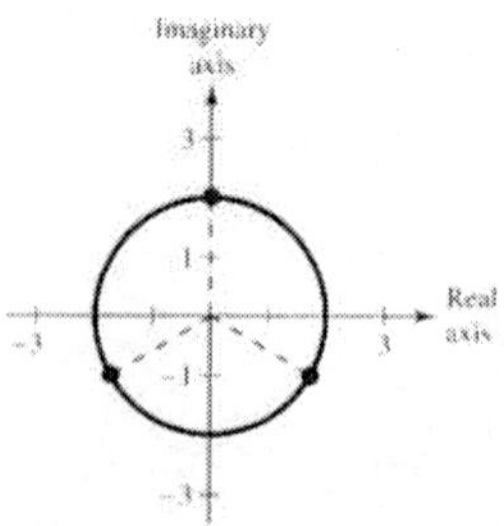

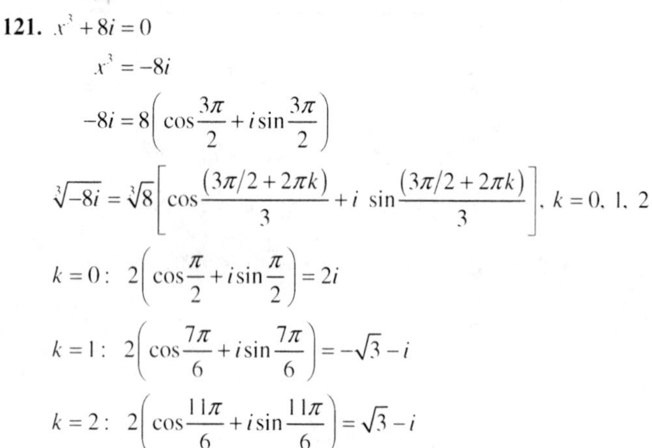

121. $x^3 + 8i = 0$

$$x^3 = -8i$$

$$-8i = 8\left(\cos\frac{3\pi}{2} + i\sin\frac{3\pi}{2}\right)$$

$$\sqrt[3]{-8i} = \sqrt[3]{8}\left[\cos\frac{(3\pi/2 + 2\pi k)}{3} + i\,\sin\frac{(3\pi/2 + 2\pi k)}{3}\right],\ k = 0, 1, 2$$

$$k = 0:\ 2\left(\cos\frac{\pi}{2} + i\sin\frac{\pi}{2}\right) = 2i$$

$$k = 1:\ 2\left(\cos\frac{7\pi}{6} + i\sin\frac{7\pi}{6}\right) = -\sqrt{3} - i$$

$$k = 2:\ 2\left(\cos\frac{11\pi}{6} + i\sin\frac{11\pi}{6}\right) = \sqrt{3} - i$$

123. True. $\sin 90°$ is defined in the Law of Sines.

Chapter 6 Test

1. $A = 36°,\ B = 98°,\ c = 16$

$$C = 180° - 36° - 98° = 46°$$

$$a = \frac{c}{\sin C}\sin A \approx 13.07$$

$$b = \frac{a}{\sin A}\sin B \approx 22.03$$

2. $a = 2,\ b = 4,\ c = 5$

$$\cos C = \frac{a^2 + b^2 - c^2}{2ab} = \frac{4 + 16 - 25}{2(2)(4)} = -0.3125 \Rightarrow C \approx 108.21°$$

$$\cos B = \frac{a^2 + c^2 - b^2}{2ac} = \frac{4 + 25 - 16}{2(2)(5)} = 0.65 \Rightarrow B \approx 49.46°$$

$$\cos A = \frac{b^2 + c^2 - a^2}{2bc} = \frac{16 + 25 - 4}{2(4)(5)} = 0.925 \Rightarrow A \approx 22.33°$$

3. $A = 35°,\ b = 8,\ c = 12$

$$a^2 = b^2 + c^2 - 2bc \cdot \cos A$$

$$= 64 + 144 - 2(8)(12)\cos 35° \approx 50.7228$$

$$a \approx 7.12$$

$$\sin B = \frac{\sin A}{a}b \approx 0.6443 \Rightarrow B \approx 40.11°$$

$$C = 180° - A - B \approx 104.89°$$

4. $A = 25°,\ b = 28,\ a = 18$

$$\sin B = \frac{\sin A}{a}b = \frac{(\sin 25°)28}{18} \approx 0.6574$$

$B_1 \approx 41.10°$ or $B_2 \approx 180° - 41.1° \approx 138.90°$

Case 1: $B_1 = 41.10°$

$$C = 180° - A - B_1 \approx 113.90°$$

$$c = \frac{a}{\sin A}\sin C \approx 38.94$$

Case 2: $B_2 = 138.90°$

$$C = 180° - A - B_2 \approx 16.10°$$

$$c = \frac{a}{\sin A}\sin C \approx 11.81$$

5. No triangle possible $(5.2 \le 10.1)$

6. $\sin B = \dfrac{\sin A}{a} b = \dfrac{\sin 150°}{9.4} 4.8$

$\approx 0.2553 \Rightarrow B \approx 14.8°$

$C = 180° - A - B = 15.2°$

$c = \dfrac{a}{\sin A} \sin C \approx 4.9$

7. Law of Cosines:

$a^2 = b^2 + c^2 - 2bc\cos\theta$

$= 480^2 + 565^2 - 2(480)(565)\cos 80°$

$= 455{,}438.2 \Rightarrow a \approx 674.9$ ft

8. $s = \dfrac{a+b+c}{2} = \dfrac{55+85+100}{2} = 120$

$A = \sqrt{s(s-a)(s-b)(s-c)}$

$= \sqrt{120(65)(35)(20)}$

≈ 2336.7 square meters

9. $\mathbf{w} = \langle 4 - (-8), 1 - (-12)\rangle = \langle 12, 13\rangle$

$\|\mathbf{w}\| = \sqrt{12^2 + 13^2} = \sqrt{313} \approx 17.69$

10. $\mathbf{u} = \langle 0, -4\rangle, \mathbf{v} = \langle 4, 6\rangle$

(a) $2\mathbf{v} + \mathbf{u} = 2\langle 4, 6\rangle + \langle 0, -4\rangle = \langle 8, 12\rangle + \langle 0, -4\rangle = \langle 8, 8\rangle$

(b) $\mathbf{u} - 3\mathbf{v} = \langle 0, -4\rangle - 3\langle 4, 6\rangle$

$= \langle 0, -4\rangle - \langle 12, 18\rangle$

$= \langle -12, -22\rangle$

(c) $5\mathbf{u} - \mathbf{v} = 5\langle 0, -4\rangle - \langle 4, 6\rangle$

$= \langle 0, -20\rangle - \langle 4, 6\rangle$

$= \langle -4, -26\rangle$

11. $\mathbf{u} = \langle -5, 2\rangle, \mathbf{v} = \langle -1, -10\rangle$

(a) $2\mathbf{v} + \mathbf{u} = 2\langle -1, -10\rangle + \langle -5, 2\rangle$

$= \langle -2, -20\rangle + \langle -5, -2\rangle$

$= \langle -7, -18\rangle$

(b) $\mathbf{u} - 3\mathbf{v} = \langle -5, 2\rangle - 3\langle -1, -10\rangle$

$= \langle -5, 2\rangle - \langle -3, -30\rangle$

$= \langle -2, 32\rangle$

(c) $5\mathbf{u} - \mathbf{v} = 5\langle -5, 2\rangle - \langle -1, -10\rangle$

$= \langle -25, 10\rangle - \langle -1, -10\rangle$

$= \langle -24, 20\rangle$

12. (a) $2\mathbf{v} + \mathbf{u} = 2(6\mathbf{i} + 9\mathbf{j}) + (\mathbf{i} - \mathbf{j}) = 13\mathbf{i} + 17\mathbf{j}$

(b) $\mathbf{u} - 3\mathbf{v} = (\mathbf{i} - \mathbf{j})) - 3(6\mathbf{i} + 9\mathbf{j}) = -17\mathbf{i} - 28\mathbf{j}$

(c) $5\mathbf{u} - \mathbf{v} = 5(\mathbf{i} - \mathbf{j}) - (6\mathbf{i} + 9\mathbf{j}) = -\mathbf{i} - 14\mathbf{j}$

13. (a) $2\mathbf{v} + \mathbf{u} = 2(-\mathbf{i} - 2\mathbf{j}) + (2\mathbf{i} + 3\mathbf{j}) = -\mathbf{j}$

(b) $\mathbf{u} - 3\mathbf{v} = (2\mathbf{i} + 3\mathbf{j}) - 3(-\mathbf{i} - 2\mathbf{j}) = 5\mathbf{i} + 9\mathbf{j}$

(c) $5\mathbf{u} - \mathbf{v} = 5(2\mathbf{i} + 3\mathbf{j}) - (-\mathbf{i} - 2\mathbf{j}) = 11\mathbf{i} + 17\mathbf{j}$

14. $\mathbf{v} = 6\mathbf{i} - 4\mathbf{j}$

$\mathbf{u} = \dfrac{\mathbf{v}}{\|\mathbf{v}\|}$

$\|\mathbf{v}\| = \sqrt{(6)^2 + (4)^2} = \sqrt{52} = 2\sqrt{13}$

$\mathbf{u} = \dfrac{6}{2\sqrt{13}}\mathbf{i} - \dfrac{4}{2\sqrt{13}}\mathbf{j}$

$\mathbf{u} = \dfrac{3\sqrt{13}}{13}\mathbf{i} - \dfrac{2\sqrt{13}}{13}\mathbf{j}$

15. $12\dfrac{\langle 3, -5\rangle}{\|\langle 3, -5\rangle\|} = \dfrac{12}{\sqrt{34}}\langle 3, -5\rangle = \left\langle \dfrac{36}{\sqrt{34}}, \dfrac{-60}{\sqrt{34}}\right\rangle$

$= \left\langle \dfrac{18\sqrt{34}}{17}, \dfrac{-30\sqrt{34}}{17}\right\rangle$

16. $250(\cos 45°\mathbf{i} + \sin 45°\mathbf{j})$, first force

$130(\cos(-60°)\mathbf{i} + \sin(-60°)\mathbf{j})$, second force

Resultant: $\left[250\left(\dfrac{\sqrt{2}}{2}\right) + 130\left(\dfrac{1}{2}\right)\right]\mathbf{i} + \left[250\left(\dfrac{\sqrt{2}}{2}\right) + 130\left(-\dfrac{\sqrt{3}}{2}\right)\right]\mathbf{j}$

$= (125\sqrt{2} + 65)\mathbf{i} + (125\sqrt{2} - 65\sqrt{3})\mathbf{j}$

Magnitude: $\sqrt{(125\sqrt{2} + 65)^2 + (125\sqrt{2} - 65\sqrt{3})^2} \approx 250.15$

Direction: $\theta = \arctan\left(\dfrac{125\sqrt{2} - 65\sqrt{3}}{125\sqrt{2} + 65}\right) \Rightarrow \theta \approx 14.9°$

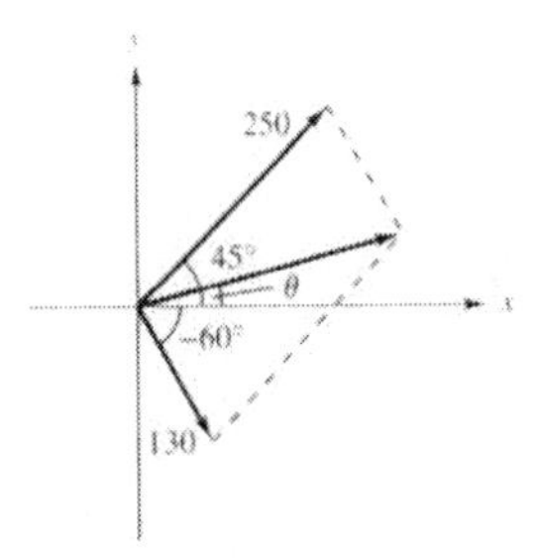

17. $\mathbf{u} = \langle -9, 4\rangle, \mathbf{v} = \langle 1, 2\rangle$

$\mathbf{u} \cdot \mathbf{v} = \langle -9, 4\rangle \cdot \langle 1, 2\rangle$

$= (-9)(1) + (4)(2)$

$= -9 + 8 = -1$

18. $\cos\theta = \dfrac{\mathbf{u} \cdot \mathbf{v}}{\|\mathbf{u}\|\|\mathbf{v}\|} = \dfrac{-8}{\sqrt{53(4)}} \Rightarrow \theta \approx 105.9°$

19. $\mathbf{u} = \langle 9, -12 \rangle,\ \mathbf{v} = \langle -4, -3 \rangle$

$$\mathbf{u} \cdot \mathbf{v} = \langle 9,\ -12 \rangle \cdot \langle -4,\ -3 \rangle$$
$$= (9)(-4) + (-12)(-3)$$
$$= -36 + 36 = 0$$

Yes, **u** and **v** are orthogonal since $\mathbf{u} \cdot \mathbf{v} = 0$.

20. $\text{proj}_{\mathbf{v}}\mathbf{u} = \frac{-37}{26}\langle -5,\ -1 \rangle = \left\langle \frac{185}{26},\ \frac{37}{26} \right\rangle = \mathbf{w}_1$

$$\mathbf{w}_2 = \mathbf{u} - \mathbf{w}_1 = \langle 6,\ 7 \rangle - \left\langle \frac{185}{26},\ \frac{37}{26} \right\rangle = \left\langle \frac{-29}{26},\ \frac{145}{26} \right\rangle$$

$\mathbf{u} = \mathbf{w}_1 + \mathbf{w}_2$

21. $|z| = 2\sqrt{2}$

$$z = 2\sqrt{2}\left(\cos\frac{3\pi}{4} + i\sin\frac{3\pi}{4} \right)$$

22. $100(\cos 240° + i\sin 240°) = -50 - 50\sqrt{3i}$

23. $\left[3\left(\cos\frac{5\pi}{6} + i\sin\frac{5\pi}{6} \right) \right]^8 = 3^8\left(\cos\frac{40\pi}{6} + i\sin\frac{40\pi}{6} \right)$

$$= 3^8\left(-\frac{1}{2} + \frac{\sqrt{3}}{2}i \right)$$
$$= -3280.5 + 3280.5\sqrt{3i}$$
$$= -\frac{6561}{2} + \frac{6561}{2}\sqrt{3i}$$

24. $(3 - 3i)^6 = \left[3\sqrt{2}\left(\cos\frac{7\pi}{4} + i\sin\frac{7\pi}{4} \right) \right]^6$

$$= 5832\left(\cos\frac{42\pi}{4} + i\sin\frac{42\pi}{4} \right) = 5832i$$

25. $128\left(1 + \sqrt{3i}\right) = 256\left(\frac{1}{2} + \frac{\sqrt{3}}{2}i \right) = 256\left(\cos\frac{\pi}{3} + i\sin\frac{\pi}{3} \right)$

Fourth roots: $\sqrt[4]{256}\left(\cos\frac{\pi/3 + 2\pi k}{4} + i\sin\frac{(\pi/3) + 2\pi k}{4} \right)$

$k = 0,\ 1,\ 2,\ 3$

Four roots are: $4\left(\cos\frac{\pi}{12} + i\sin\frac{\pi}{12} \right) \approx 3.8637 + 1.0353i$

$$4\left(\cos\frac{7\pi}{12} + i\sin\frac{7\pi}{12} \right) \approx -1.0353 + 3.8637i$$
$$4\left(\cos\frac{13\pi}{12} + i\sin\frac{13\pi}{12} \right) \approx -3.8637 - 1.0353i$$
$$4\left(\cos\frac{19\pi}{12} + i\sin\frac{19\pi}{12} \right) \approx 1.0353 - 3.8637i$$

26. $x^4 = 625i$

Fourth roots of $625i = 625\left(\cos\frac{\pi}{2} + i\sin\frac{\pi}{2} \right)$

$$\sqrt[4]{625}\left(\cos\left(\frac{(\pi/2) + 2\pi k}{4} \right) + i\sin\left(\frac{(\pi/2) + 2\pi k}{4} \right) \right)$$

$k = 0,\ 1,\ 2,\ 3$

Four roots are: $5\left(\cos\frac{\pi}{8} + i\sin\frac{\pi}{8} \right)$

$$5\left(\cos\frac{5\pi}{8} + i\sin\frac{5\pi}{8} \right)$$
$$5\left(\cos\frac{9\pi}{8} + i\sin\frac{9\pi}{8} \right)$$
$$5\left(\cos\frac{13\pi}{8} + i\sin\frac{13\pi}{8} \right)$$

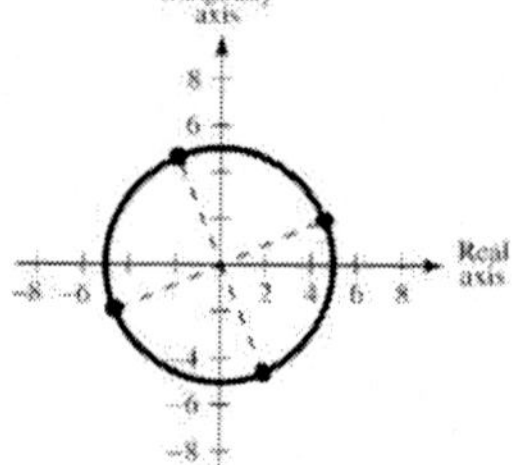

Chapters 4–6 Cumulative Test

1. $\theta = -150°$

(a)

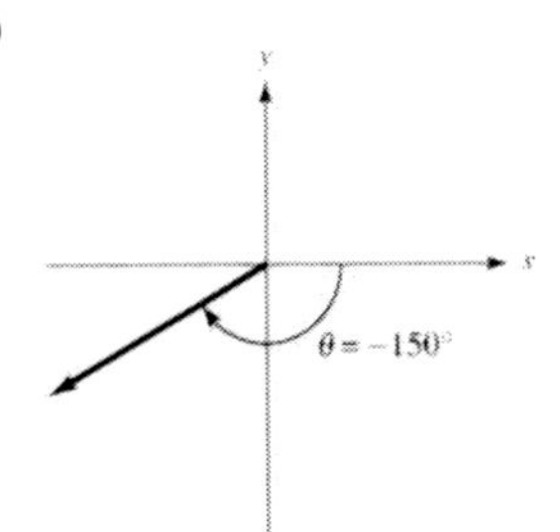

(b) $-150° + 360° = 210°$

(c) $(-150°)\dfrac{\pi}{180°} = -\dfrac{5}{6}\pi$ radians

(d) $\theta' = 30°$

(e) $\sin\theta = -\dfrac{1}{2}$ $\quad \csc\theta = -2$

$\cos\theta = -\dfrac{\sqrt{3}}{2}$ $\quad \sec\theta = -\dfrac{2}{\sqrt{3}} = -\dfrac{2\sqrt{3}}{3}$

$\tan\theta = \dfrac{\sqrt{3}}{3}$ $\quad \cot\theta = \dfrac{3}{\sqrt{3}} = \sqrt{3}$

2. $2.55 \text{ rad} = 2.55\left(\dfrac{180°}{\pi}\right) \approx 146.1°$

3. $\sec^2\theta = 1 + \tan^2\theta = 1 + \left(-\dfrac{12}{5}\right)^2 = \dfrac{169}{25}$

$\sec\theta = -\dfrac{13}{5}$ (since θ in Quandrant II)

$\cos\theta = -\dfrac{5}{13}$

4. $f(x) = 3 - 2\sin\pi x$

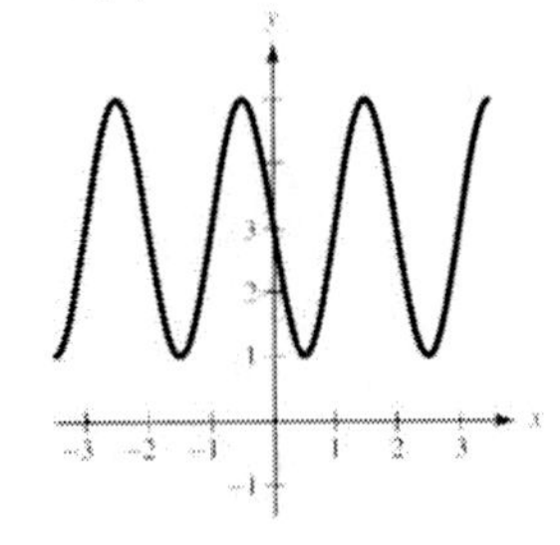

5. $f(x) = \tan(3x)$

Period: $\dfrac{\pi}{3}$

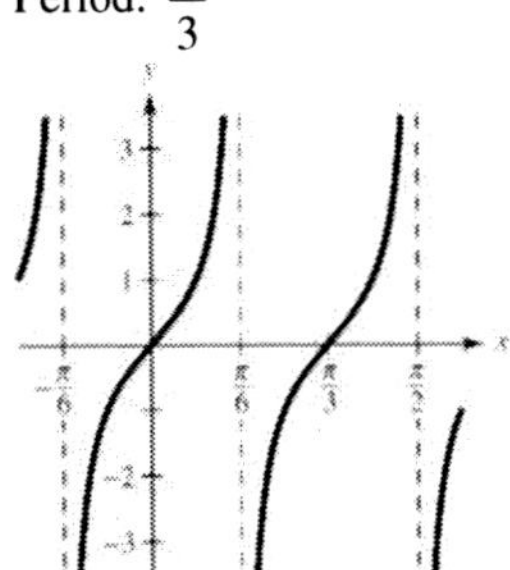

6. $f(x) = \dfrac{1}{2}\sec(x + \pi)$

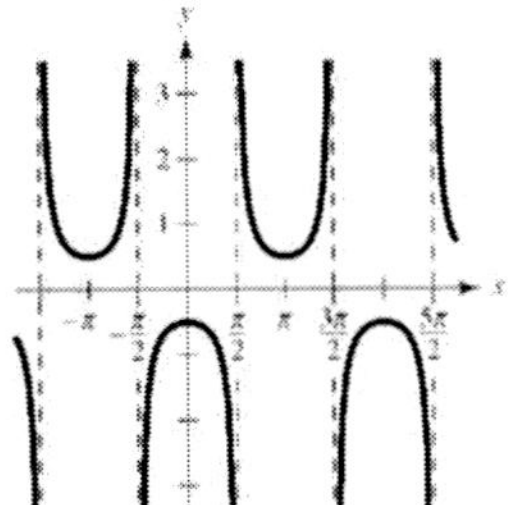

7. Amplitude: 3

Cosine curve reflected about the x-axis

Period: $2 \Rightarrow h(x) = -3\cos(\pi x)$

Answer: $a = -3,\ b = \pi,\ c = 0$

8. Let $\theta = \arctan\dfrac{3}{4}$.

$\tan\theta = \dfrac{3}{4}$

$\sin\left(\arctan\dfrac{3}{4}\right) = \dfrac{3}{5}$

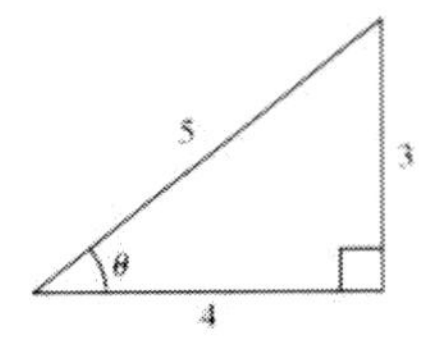

9. Let $\arcsin\left(-\dfrac{1}{2}\right) = \theta$.

$\sin\theta = -\dfrac{1}{2}$

$\theta = -\dfrac{\pi}{6}$

$\tan\left(-\dfrac{\pi}{6}\right) = -\dfrac{\sqrt{3}}{3}$

10. Let $\theta = \arctan 2x$.

$$\tan\theta = 2x$$

$$\sin(\arctan 2x) = \sin\theta = \frac{2x}{\sqrt{1+4x^2}}$$

11.
$$\frac{\sin\theta - 1}{\cos\theta} - \frac{\cos\theta}{\sin\theta - 1} = \frac{\sin^2\theta - 2\sin\theta + 1 - \cos^2\theta}{\cos\theta(\sin\theta - 1)}$$
$$= \frac{\sin^2\theta - 2\sin\theta + \sin^2\theta}{\cos\theta(\sin\theta - 1)}$$
$$= \frac{2\sin\theta(\sin\theta - 1)}{\cos\theta(\sin\theta - 1)} = 2\tan\theta$$

12. $\cot^2\alpha(\sec^2\alpha - 1) = \cot^2\alpha(\tan^2\alpha) = 1$

13.
$$\sin(x+y)\sin(x-y) = [\sin x\cos y + \cos x\sin y][\sin x\cos y - \sin y\cos x]$$
$$= \sin^2 x\cos^2 y - \sin^2 y\cos^2 x$$
$$= \sin^2 x(1 - \sin^2 y) - \sin^2 y(1 - \sin^2 x)$$
$$= \sin^2 x - \sin^2 x\sin^2 y - \sin^2 y + \sin^2 y\sin^2 x$$
$$= \sin^2 x - \sin^2 y$$

14.
$$\sin^2 x\cos^2 x = \frac{1}{4}(2\sin x\cos x)^2$$
$$= \frac{1}{4}(\sin 2x)^2$$
$$= \frac{1}{4}\cdot\frac{1-\cos 4x}{2}$$
$$= \frac{1}{8}(1 - \cos 4x)$$

15.
$$\sin^2 x + 2\sin x + 1 = 0$$
$$(\sin x + 1)^2 = 0$$
$$\sin x + 1 = 0$$
$$\sin x = -1$$
$$x = \frac{3\pi}{2} + 2n\pi$$

16.
$$3\tan\theta - \cot\theta = 0$$
$$3\tan\theta - \frac{1}{\tan\theta} = 0$$
$$3\tan^2\theta - 1 = 0$$
$$\tan\theta = \pm\frac{1}{\sqrt{3}} \Rightarrow \theta = \frac{\pi}{6} + n\pi, \frac{5\pi}{6} + n\pi$$

17. Graph $y = \cos^2 x - 5\cos x - 1$ on $[0, 2\pi)$.

Roots are $x \approx 1.7646, 4.5186$.

18.

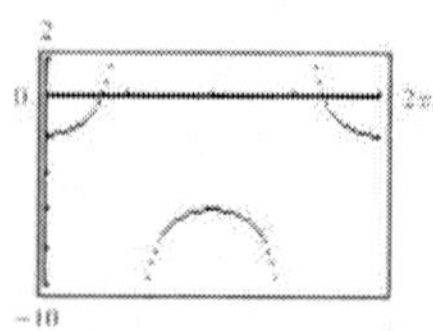

Zeros: $x \approx 1.047, 5.236$

Algebraically:
$$\frac{1+\sin x}{\cos x} + \frac{\cos x}{1+\sin x} = 4$$
$$\frac{1 + 2\sin x + \sin^2 x + \cos^2 x}{\cos x(1+\sin x)} = 4$$
$$\frac{2 + 2\sin x}{\cos x(1+\sin x)} = 4$$
$$\frac{2}{\cos x} = 4$$
$$\cos x = \frac{1}{2}$$
$$x = \frac{\pi}{3}, \frac{5\pi}{3}$$

19.

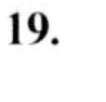

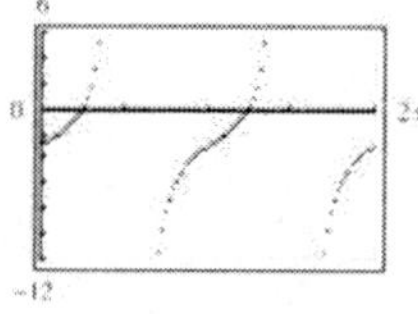

Zeros: $x \approx 0.785, 3.927$

Algebraically: $\tan^3 x - \tan^2 x + 3\tan x - 3 = 0$
$$\tan^2 x(\tan x - 1) + 3(\tan x - 1) = 0$$
$$(\tan^2 x + 3)(\tan x - 1) = 0$$
$$\tan x = 1 \Rightarrow x = \frac{\pi}{4}, \frac{5\pi}{4}$$

20. $\sin u = \frac{12}{13} \Rightarrow \cos u = \frac{5}{13} \Rightarrow \tan u = \frac{12}{5}$

$\cos v = \frac{3}{5} \Rightarrow \sin v = \frac{4}{5} \Rightarrow \tan v = \frac{4}{3}$

$$\tan(u - v) = \frac{\tan u - \tan v}{1 + \tan u\tan v} = \frac{(12/5) - (4/3)}{1 + (12/5)(4/3)} = \frac{16}{63}$$

21. $$\tan(2\theta) = \frac{2\tan\theta}{1 - \tan^2\theta} = \frac{2(1/2)}{1 - (1/4)} = \frac{4}{3}$$

22. $\sec^2\theta = \tan^2\theta + 1 = \frac{25}{9} \Rightarrow \sec\theta = -\frac{5}{3} \Rightarrow$

$\cos\theta = -\frac{3}{5}$ (Quadrant III)

$$\sin\frac{\theta}{2} = \pm\sqrt{\frac{1-\cos\theta}{2}} = \pm\frac{2}{\sqrt{5}} = \pm\frac{2\sqrt{5}}{5}$$

$\frac{\theta}{2}$ in Quadrant II: $\sin\frac{\theta}{2} = \frac{2\sqrt{5}}{5}$

23. $\cos 8x + \cos 4x = 2\cos\left(\frac{8x+4x}{2}\right)\cos\left(\frac{8x-4x}{2}\right)$

$= 2\cos 6x \cos 2x$

24. $\tan x\left(1-\sin^2 x\right) = \frac{\sin x}{\cos x}\cos^2 x = \sin x \cos x$

$= \frac{1}{2}\left(2\sin x\cos x\right) = \frac{1}{2}\sin 2x$

25. $\sin 3\theta \sin\theta = \frac{1}{2}\left[\cos(3\theta-\theta) - \cos(3\theta+\theta)\right]$

$= \frac{1}{2}\left(\cos 2\theta - \cos 4\theta\right)$

26. $\sin 3x \cos 2x = \frac{1}{2}\left(\sin(3x+2x) + \sin(3x-2x)\right)$

$= \frac{1}{2}\left(\sin 5x + \sin x\right)$

27. $\frac{2\cos 3x}{\sin 4x - \sin 2x} = \frac{2\cos 3x}{2\cos 3x \cdot \sin x} = \frac{1}{\sin x} = \csc x$

28. $\sin B = \frac{\sin A}{a} b = 0.2569 \Rightarrow B \approx 14.9°$

$C = 180° - 46° - 14.9° = 119.1°$

$c = \frac{a}{\sin A}\left(\sin C\right) \approx 17.0$

29. $A = 32°,\ b = 8,\ c = 10$

$a^2 = b^2 + c^2 - 2bc\cos A$

$= 64 + 100 - 2(8)(10)\cos 32°$

$= 28.3123 \Rightarrow a \approx 5.32$

$\sin B = \frac{b\sin A}{a} \approx 0.7967 \Rightarrow B \approx 52.82°$

$C = 180° - B - A = 95.18°$

30. $B = 180° - 24° - 101° = 55°$

$b = \frac{a}{\sin A}\sin B \approx 20.14$

$c = \frac{a}{\sin A}\sin C \approx 24.13$

31. $\cos A = \frac{b^2+c^2-a^2}{2bc} = 0.8982 \Rightarrow A \approx 26.1°$

$\cos B = \frac{a^2+c^2-b^2}{2bc} = 0.8355 \Rightarrow B \approx 33.3°$

$C = 180° - 26.1° - 33.3° = 120.6°$

32. $A = \frac{1}{2}bh = \frac{1}{2}19 \cdot 14\sin 82°$

≈ 131.7 sq. in.

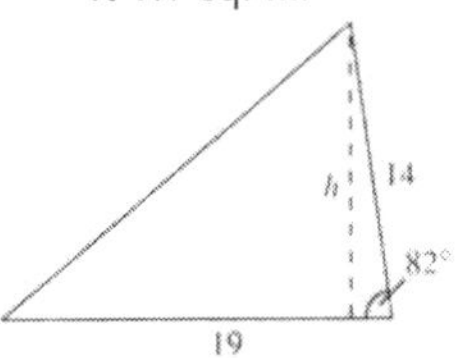

33. $s = \frac{12+16+18}{2} = 23$

Area $= \sqrt{s(s-a)(s-b(s-c))}$

$= \sqrt{23(11)(7)(5)} \approx 94.10$ sq. in.

34. $\mathbf{u} = \langle 3, 5\rangle = 3\mathbf{i} + 5\mathbf{j}$

35. $\mathbf{v} = \mathbf{i} - 2\mathbf{j}$

$\|\mathbf{v}\| = \sqrt{1+4} = \sqrt{5}$

Unit vector: $\frac{1}{\sqrt{5}}\mathbf{i} - \frac{2}{\sqrt{5}}\mathbf{j} = \left\langle \frac{\sqrt{5}}{5}, \frac{-2\sqrt{5}}{5}\right\rangle$

36. $\mathbf{u}\cdot\mathbf{v} = 3(1) + 4(-2) = -5$

37. $\mathbf{u}\cdot\mathbf{v} = 0$

$\langle 1, 2k\rangle . \langle 2, -1\rangle = 0$

$2 - 2k = 0$

$k = 1$

38. $\text{proj}_{\mathbf{v}}\mathbf{u} = \frac{8-10}{26}\langle 1, 5\rangle = \left\langle -\frac{1}{13}, -\frac{5}{13}\right\rangle = \mathbf{w}_1$

$\mathbf{w}_2 = \mathbf{u} - \mathbf{w}_1 = \langle 8, -2\rangle - \left\langle -\frac{1}{13}, -\frac{5}{13}\right\rangle$

$= \left\langle \frac{105}{13}, -\frac{21}{13}\right\rangle$

$\mathbf{u} = \mathbf{w}_1 + \mathbf{w}_2$

39. $|z| = 3\sqrt{2},\ \theta = \frac{3\pi}{4}:\ 3\sqrt{2}\left(\cos\frac{3\pi}{4} + i\sin\frac{3\pi}{4}\right)$

40. $6\sqrt{3}\left(-\frac{\sqrt{3}}{2} + \frac{1}{2}i\right) = -9 + 3\sqrt{3}i$

41. $\left[4\left(\cos 30° + i\sin 30°\right)\right]\left[6\left(\cos 120° + i\sin 120°\right)\right]$

$= 24\left(\cos(30° + 120°) + i\sin(30° + 120°)\right)$

$= 24\left(\cos 150° + i\sin 150°\right)$

$= 24\left(-\frac{\sqrt{3}}{2} + i\frac{1}{2}\right) = -12\sqrt{3} + 12i$

42. $2+i=\sqrt{5}(\cos\theta+i\sin\theta)$, where $\tan\theta=\dfrac{1}{2}$

$\Rightarrow \theta\approx 0.4636$

Square roots:

$$z_1=5^{1/4}\left(\cos\frac{\theta}{2}+i\sin\frac{\theta}{2}\right)$$

$$z_2=5^{1/4}\left(\cos\left(\frac{\theta+2\pi}{2}\right)+i\sin\left(\frac{\theta+2\pi}{2}\right)\right)$$

$z_1\approx 1.4533+0.3436i$

$z_2\approx -1.4553-0.3436i$

43. $1=1(\cos 0+i\sin 0)$

$$\cos\left(\frac{0+2\pi k}{3}\right)+i\sin\left(\frac{0+2\pi k}{3}\right),\ k=0,\ 1,\ 2$$

$k=0:\cos 0+i\sin 0=1$

$$k=1:\cos\frac{2\pi}{3}+i\sin\frac{2\pi}{3}=-\frac{1}{2}+\frac{\sqrt{3}}{2}i$$

$$k=2:\cos\frac{4\pi}{3}+i\sin\frac{4\pi}{3}=-\frac{1}{2}-\frac{\sqrt{3}}{2}i$$

44. $x^4=-625$

Four fourth roots of $-625=625(\cos\pi+i\sin\pi)$ are:

$$\sqrt[4]{625}\left(\cos\frac{\pi+2\pi k}{4}+i\sin\frac{\pi+2\pi k}{4}\right);k=0,\ 1,\ 2,\ 3$$

$$5\left(\cos\frac{\pi}{4}+i\sin\frac{\pi}{4}\right)$$

$$5\left(\cos\frac{3\pi}{4}+i\sin\frac{3\pi}{4}\right)$$

$$5\left(\cos\frac{5\pi}{4}+i\sin\frac{5\pi}{4}\right)$$

$$5\left(\cos\frac{7\pi}{4}+i\sin\frac{7\pi}{4}\right)$$

45. $\tan 18^\circ=\dfrac{h}{200}$

$\tan 16^\circ 45'=\dfrac{k}{200}$

Hence,

$f=h-k=200\tan 18^\circ-200\tan 16^\circ 45'$

$\approx 4.8\approx 5$ feet.

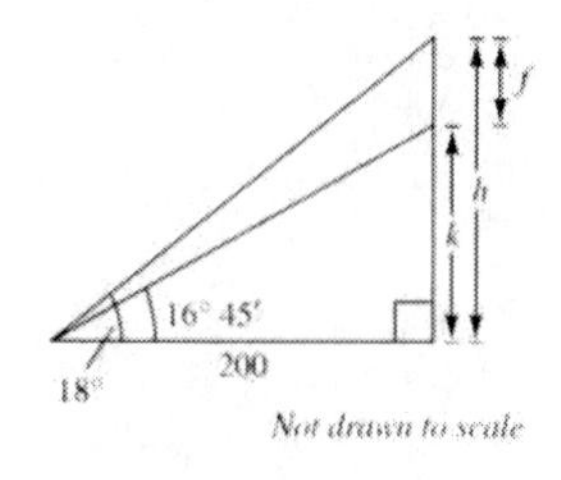

46. Since $d=a\sin\left(\dfrac{(2\pi)}{b}t\right)$ and given the maximum displacement is 7 inches, with a period of 8 seconds, $a=7$ and $\dfrac{2\pi}{b}=8\Rightarrow b=\dfrac{\pi}{4}$.

So, $d=7\sin\left(\dfrac{\pi}{4}t\right)$.

47.

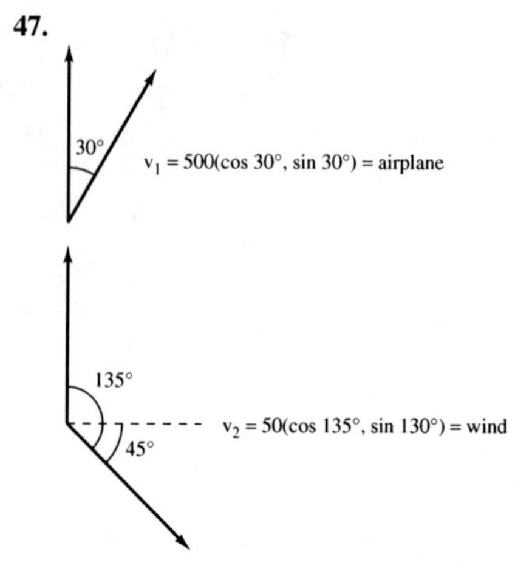

$\mathbf{v}=\mathbf{v}_1+\mathbf{v}_2=500\langle\cos 30^\circ,\ \sin 30^\circ\rangle+50\langle\cos 135^\circ,\ \sin 135^\circ\rangle$

$\mathbf{v}\approx\langle 397.7,\ 285.4\rangle$

$\|\mathbf{v}\|\approx\sqrt{(397.7)^2+(285.4)^2}\approx 489.45$ km/hr

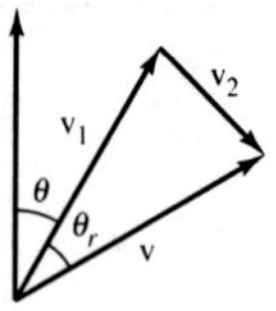

$$\theta_r=\tan^{-1}\left(\frac{285.4}{397.7}\right)\approx 35.66^\circ\Rightarrow\theta=90^\circ-\theta_r=54.34^\circ$$

The direction of the airplane is 54.34° at an airspeed relative to the ground of 489.45 km/hr.

48. $\cos A=\dfrac{60^2+125^2-100^2}{2(60)(125)}=0.615\Rightarrow A\approx 52.05^\circ$

$\cos B=\dfrac{100^2+125^2-60^2}{2(100)(125)}=0.881\Rightarrow B\approx 28.24$

Angle between vectors $=A+B\approx 80.3^\circ$

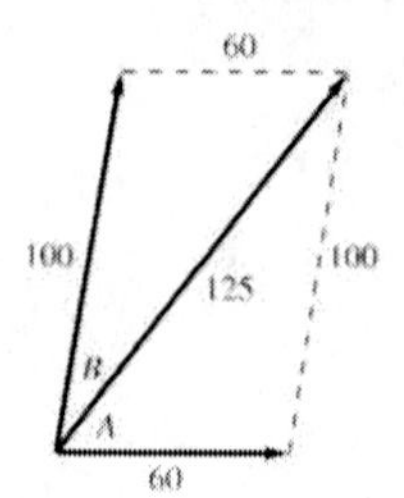

CHAPTER 7

Section 7.1

1. system, equations

3. substitution

5. break-even point

7. (a) $4(0)-(-3)\stackrel{?}{=}1$

$6(0)+(-3)\stackrel{?}{=}-6$

$3\neq 1$

$-3\neq -6$

No, $(0,\ -3)$ is not a solution.

(b) $4(-1)-(-5)\stackrel{?}{=}1$

$6(-1)+(-5)\stackrel{?}{=}6$

$1=1$

$-11\neq -6$

No, $(-1,\ -5)$ is not a solution.

(c) $4\left(-\frac{3}{2}\right)-(-3)\stackrel{?}{=}1$

$6\left(-\frac{3}{2}\right)+(3)\stackrel{?}{=}-6$

$-9\neq 1$

$-6=-6$

No, $\left(-\frac{3}{2},\ 3\right)$ is not a solution.

(d) $4\left(-\frac{1}{2}\right)-(-3)\stackrel{?}{=}1$

$6\left(-\frac{1}{2}\right)+(-3)\stackrel{?}{=}6$

$1=1$

$-6=-6$

Yes, $\left(-\frac{1}{2},\ -3\right)$ is a solution.

9. (a) $0\stackrel{?}{=}-2e^{-2}$

$3(-2)-0\stackrel{?}{=}2$

$0\neq -2e^{-2}$

$-6\neq 2$

No, $(-2,\ 0)$ is not a solution.

(b) $-2\stackrel{?}{=}-2e^{0}$

$3(0)-(-2)\stackrel{?}{=}2$

$-2=-2$

$2=2$

Yes, $(0,\ -2)$ is a solution.

(c) $-3\stackrel{?}{=}-2e^{0}$

$3(0)-(-3)\stackrel{?}{=}2$

$-3\neq -2$

$3\neq 2$

No, $(0,\ -3)$ is not a solution.

(d) $-5\stackrel{?}{=}-2e^{-1}$

$3(-1)-(-5)\stackrel{?}{=}2$

$-5\neq -2e^{-1}$

$2=2$

No, $(-1,\ -5)$ is not a solution.

11. $\begin{cases}2x+y=6 & \text{Equation 1}\\ -x+y=0 & \text{Equation 2}\end{cases}$

Solve for y in Equation 1: $y=6-2x$

Substitute for y in Equation 2: $-x+(6-2x)=0$

Solve for x: $-3x+6=0\Rightarrow x=2$

Back-substitute $x=2$: $y=6-2(2)=2$

Answer: $(2,\ 2)$

13. $\begin{cases}x-y=-4 & \text{Equation 1}\\ x^2-y=-2 & \text{Equation 2}\end{cases}$

Solve for y in Equation 1: $y=x+4$

Substitute for y in Equation 2: $x^2-(x+4)=-2$

Solve for x: $x^2-x-2=0$

$\Rightarrow(x+1)(x-2)=0$

$\Rightarrow x=-1,\ 2$

Back-substitute $x=-1$: $y=-1+4=3$

Back-substitute $x=2$: $y=2+4=6$

Answer: $(-1,\ 3),\ (2,\ 6)$

15. $\begin{cases} 3x + y = 2 & \text{Equation 1} \\ x^3 - 2 + y = 0 & \text{Equation 2} \end{cases}$

Solve for y in Equation 1: $y = 2 - 3x$

Substitute for y in Equation 2: $x^3 - 2 + (2 - 3x) = 0$

Solve for x: $x^3 - 3x = 0 \Rightarrow x(x^2 - 3) = 0 \Rightarrow x = 0, \pm\sqrt{3}$

Back-substitute: $x = 0: y = 2$

$x = \sqrt{3}: y = 2 - 3\sqrt{3}$

$x = -\sqrt{3}: y = 2 + 3\sqrt{3}$

Answer: $(0, 2), (\sqrt{3}, 2 - 3\sqrt{3}), (-\sqrt{3}, 2 + 3\sqrt{3})$

17. $\begin{cases} -\frac{7}{2}x - y = -18 & \text{Equation 1} \\ 8x^2 - 2y^3 = 0 & \text{Equation 2} \end{cases}$

Solve for x in Equation 1: $-\frac{7}{2}x = y - 18 \Rightarrow x = -\frac{2}{7}y + \frac{36}{7}$

Substitute for x in Equation 2: $8\left(-\frac{2}{7}y + \frac{36}{7}\right)^2 - 2y^3 = 0$

Solve for x: $-2y^3 + 8\left(\frac{4}{49}y^2 - \frac{144}{49}y + \frac{36^2}{49}\right) = 0$

$49y^3 - 16y^2 + 576y - 5184 = 0$

$(y - 4)(49y^2 + 180y + 1296) = 0$

Hence, $y = 4$ and $x = -\frac{2}{7}(4) + \frac{36}{7} = 4$.

Answer: $(4, 4)$

19. $\begin{cases} x - y = 0 & \text{Equation 1} \\ 5x - 3y = 10 & \text{Equation 2} \end{cases}$

Solve for y in Equation 1: $y = x$

Substitute for y in Equation 2: $5x - 3x = 10$

Solve for x: $2x = 10 \Rightarrow x = 5$

Back-substitute in Equation 1: $y = x = 5$

Answer: $(5, 5)$

21. $\begin{cases} 2x - y + 2 = 0 & \text{Equation 1} \\ 4x + y - 5 = 0 & \text{Equation 2} \end{cases}$

Solve for y in Equation 1: $y = 2x + 2$

Substitute for y in Equation 2:

$4x + (2x + 2) - 5 = 0$

Solve for x:

$4x + (2x + 2) - 5 = 0 \Rightarrow 6x - 3 = 0 \Rightarrow x = \frac{1}{2}$

Back-substitute $x = \frac{1}{2}$: $y = 2x + 2 = 2\left(\frac{1}{2}\right) + 2 = 3$

Answer: $\left(\frac{1}{2}, 3\right)$

23. $\begin{cases} 1.5x + 0.8y = 2.3 \Rightarrow 15x + 8y = 23 \\ 0.3x - 0.2y = 0.1 \Rightarrow 3x - 2y = 1 \end{cases}$

Solve for y in Equation 2: $-2y = 1 - 3x$

$y = \dfrac{3x - 1}{2}$

Substitute for y in Equation 1: $15x + 8\left(\dfrac{3x - 1}{2}\right) = 23$

$15x + 12x - 4 = 23$

$27x = 27$

$x = 1$

Then, $y = \dfrac{3x - 1}{2} = \dfrac{3(1) - 1}{2} = 1$. *Answer:* $(1, 1)$

25. $\begin{cases} \frac{1}{5}x + \frac{1}{2}y = 8 & \text{Equation 1} \\ x + y = 20 & \text{Equation 2} \end{cases}$

Solve for x in Equation 2: $x = 20 - y$

Substitute for x in Equation 1: $\frac{1}{5}(20 - y) + \frac{1}{2}y = 8$

Solve for y: $4 + \frac{3}{10}y = 8 \Rightarrow y = \frac{40}{3}$

Back-substitute $y = \frac{40}{3}$: $x = 20 - y$

$= 20 - \frac{40}{3} = \frac{20}{3}$

Answer: $\left(\frac{20}{3}, \frac{40}{3}\right)$

27. $\begin{cases} -\frac{5}{3}x + y = 5 & \text{Equation 1} \\ -5x + 3y = 6 & \text{Equation 2} \end{cases}$

Solve for y in Equation 1: $y = 5 + \frac{5}{3}x$

Substitute for y in Equation 2: $-5x + 3\left(5 + \frac{5}{3}x\right) = 6$

Solve for x: $-5x + 15 + 5x = 6$

$15 \neq 6$ Inconsistent

No solution

29. $\begin{cases} x + y = 18{,}000 & \text{Equation 1} \\ 0.04x + 0.06y = 1000 & \text{Equation 2} \end{cases}$

Solve for y in Equation 1: $y = 18{,}000 - x$

Substitute for y in Equation 2:

$0.04x + 0.06(18{,}000 - x) = 1000$

Solve for x: $0.04x + 1080 - 0.06x = 1000$

$-0.02x = -80$

$x = 4000$

Back-substitute: $y = 18{,}000 - 4000 = 14{,}000$

Answer: $(4000, 14{,}000)$

\$4000 at 4% and \$14,000 at 6%

31. $\begin{cases} x + y = 18{,}000 & \text{Equation 1} \\ 0.076x + 0.088y = 1542 & \text{Equation 2} \end{cases}$

Solve for y in Equation 1: $y = 18{,}000 - x$

Substitute for y in Equation 2:

$$0.076x + 0.088(18{,}000 - x) = 1542$$

Solve for x:

$$0.076x + 1584 - 0.088x = 1542$$
$$-0.012x = -42$$
$$x = 3500$$

Back-substitute: $y = 18{,}000 - 3500 = 14{,}500$

Answer: $(3500,\ 14{,}500)$

\$3500 at 7.6% and \$14,500 at 8.8%

33. $\begin{cases} x^2 - 2x + y = 8 & \text{Equation 1} \\ x - y = -2 & \text{Equation 2} \end{cases}$

Solve for y in Equation 2: $y = x + 2$

Substitute for y in Equation 1:

$x^2 - 2x + (x + 2) = 8$

$x^2 - x - 6 = 0$

$(x - 3)(x + 2) = 0$

$x = 3,\ -2$

$x = 3 \Rightarrow y = 5$

$x = -2 \Rightarrow y = 0$

Answer: $(3, 5), (-2, 0)$

35. $\begin{cases} 2x^2 - y = 1 & \text{Equation 1} \\ x - y = 2 & \text{Equation 2} \end{cases}$

Solve for y in Equation 2: $y = x - 2$

Substitute for y in Equation 1: $2x^2 - (x - 2) = 1$

$$2x^2 - x + 1 = 0$$

No real solution

37. $\begin{cases} x^3 - y = 0 & \text{Equation 1} \\ x - y = 0 & \text{Equation 2} \end{cases}$

Solve for y in Equation 2: $y = x$

Substitute for y in Equation 1: $x^3 - x = 0$

Solve for x:

$x(x - 1)(x + 1) = 0 \Rightarrow x = 0,\ 1,\ -1$

Back-substitute: $x = 0 \Rightarrow y = 0$

$x = 1 \Rightarrow y = 1$

$x = -1 \Rightarrow y = -1$

Answer: $(0, 0), (1, 1), (-1, -1)$

39. $\begin{cases} -x + 2y = 2 \\ 3x + y = 15 \end{cases}$

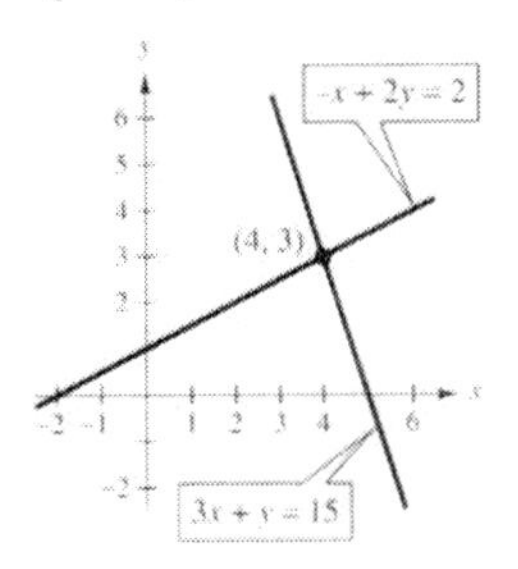

Point of intersection: $(4, 3)$

41. $\begin{cases} x - 3y = -2 \\ 5x + 3y = 17 \end{cases}$

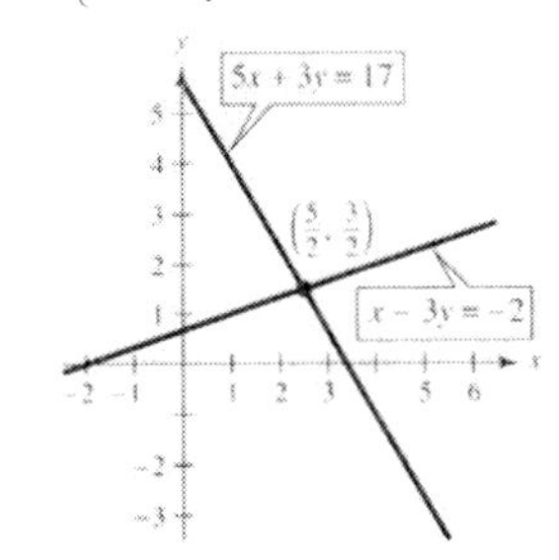

Point of intersection: $\left(\frac{5}{2}, \frac{3}{2}\right)$

43. $\begin{cases} x^2 + y = 1 \\ x + y = 2 \end{cases}$

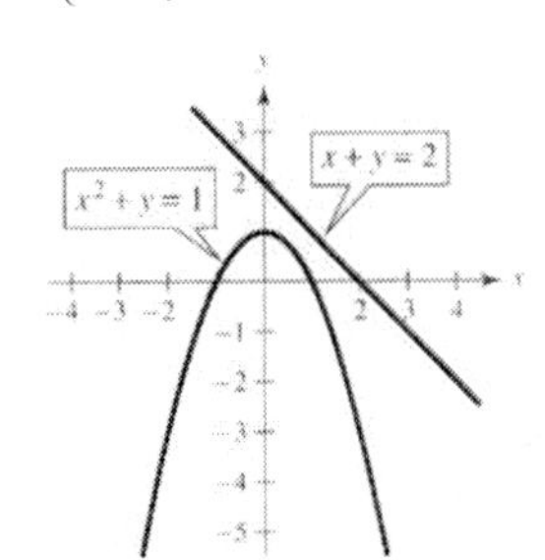

No solution

45. $\begin{cases} -x + y = 3 \Rightarrow y_1 = x + 3 \\ x^2 - 6x - 27 + y^2 = 0 \Rightarrow \quad y_2 = \sqrt{6x - x^2 + 27} \end{cases}$

$y_3 = -\sqrt{6x - x^2 + 27}$

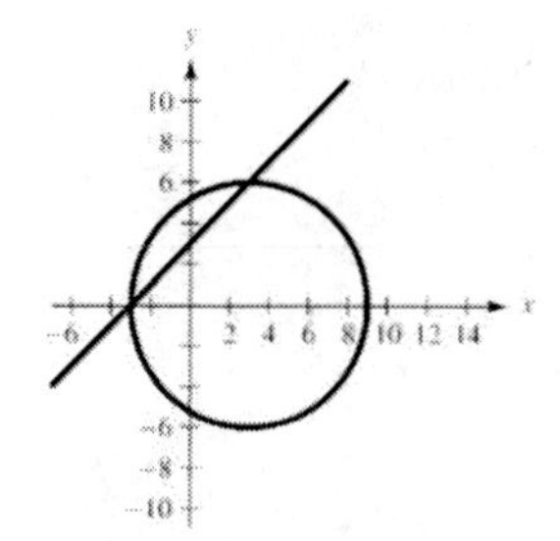

Points of intersection: $(-3, 0), (3, 6)$

47. $\begin{cases} 7x + 8y = 24 \Rightarrow y_1 = -\frac{7}{8}x + 3 \\ x - 8y = 8 \Rightarrow y_2 = \frac{1}{8}x - 1 \end{cases}$

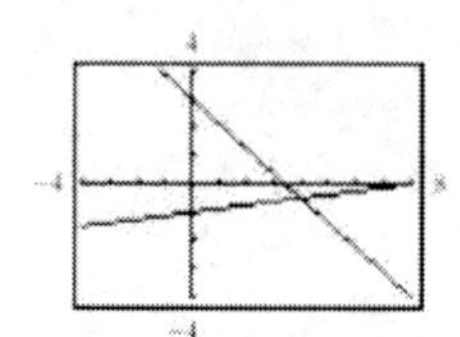

Points of intersection: $(4, -0.5)$

49. $x - y^2 = -1 \Rightarrow y^2 = x + 1 \Rightarrow y_1 = \sqrt{x + 1}$

$y_2 = -\sqrt{x + 1}$

$x - y = 5 \Rightarrow y_3 = x - 5$

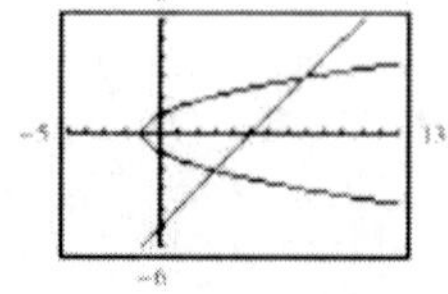

Points of intersection: $(8, 3), (3, -2)$

51. $x^2 + y^2 = 8 \Rightarrow y_1 = \sqrt{8 - x^2},\ y_2 = -\sqrt{8 - x^2}$

$y = x^2 \Rightarrow y_3 = x^2$

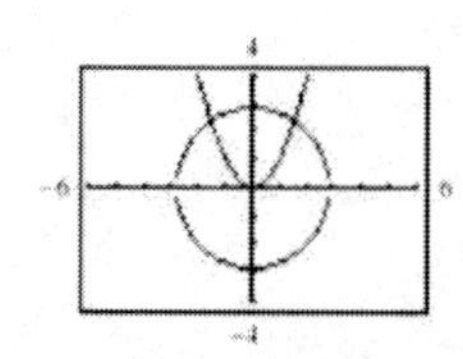

Points of intersection: $(1.540, 2.372), (-1.540, 2.372)$

53. $\begin{cases} y = e^x \\ x - y + 1 = 0 \Rightarrow y = x + 1 \end{cases}$

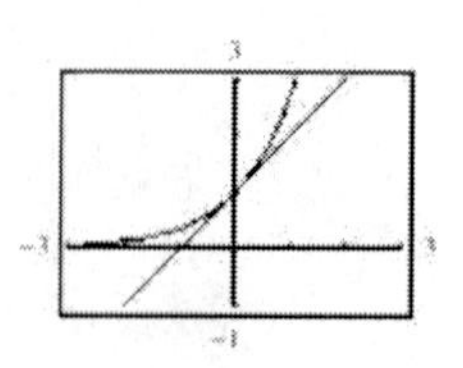

Point of intersection: $(0, 1)$

55. $\begin{cases} x + 2y = 8 \Rightarrow y_1 = 4 - x/2 \\ y = 2 + \ln x \Rightarrow y_2 = 2 + \ln x \end{cases}$

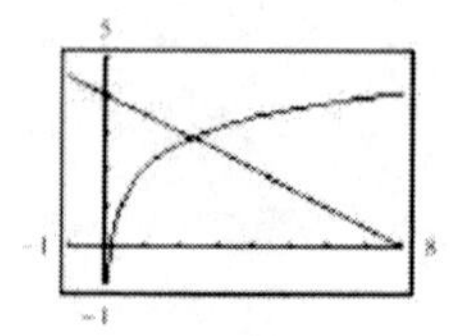

Point of intersection: $(2.318, 2.841)$

57. $\begin{cases} y = \sqrt{x} + 4 \\ y = 2x + 1 \end{cases}$

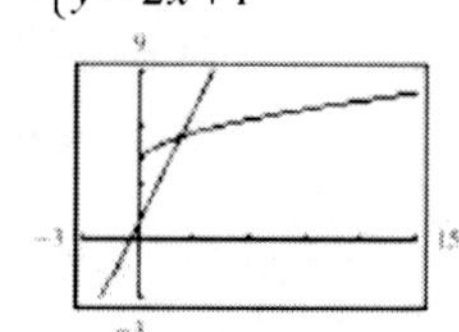

Point of intersection: $(2.25, 5.5)$

59. $\begin{cases} x^2 + y^2 = 169 \Rightarrow y_1 = \sqrt{169 - x^2} \text{ and} \\ \qquad y_2 = -\sqrt{169 - x^2} \\ x^2 - 8y = 104 \Rightarrow y_3 = \frac{1}{8}x^2 - 13 \end{cases}$

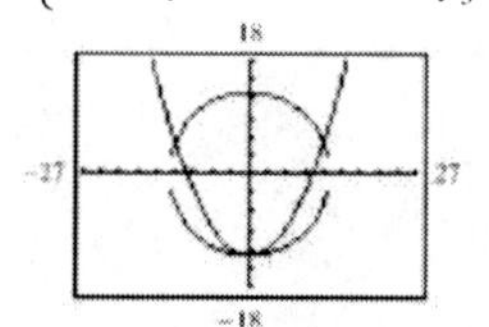

Points of intersection: $(0, -13), (\pm 12, 5)$

61. $\begin{cases} y = 2x & \text{Equation 1} \\ y = x^2 + 1 & \text{Equation 2} \end{cases}$

Substitute for y in Equation 2: $2x = x^2 + 1$

Solve for x:

$x^2 - 2x + 1 = (x-1)^2 = 0 \Rightarrow x = 1$

Back-substitute $x = 1$ in Equation 1: $y = 2x = 2$

Answer: $(1, 2)$

63. $\begin{cases} 3x - 7y + 6 = 0 & \text{Equation 1} \\ x^2 - y^2 = 4 & \text{Equation 2} \end{cases}$

Solve for y in Equation 1: $y = \dfrac{3x+6}{7}$

Substitute for y in Equation 2: $x^2 - \left(\dfrac{3x+6}{7}\right)^2 = 4$

Solve for x: $x^2 - \left(\dfrac{9x^2 + 36x + 36}{49}\right) = 4$

$$49x^2 - (9x^2 + 36x + 36) = 196$$

$$40x^2 - 36x - 232 = 0$$

$$10x^2 - 9x - 58 = 0 \Rightarrow x = \frac{9 \pm \sqrt{81 + 40(58)}}{20} \Rightarrow x = \frac{29}{10}, -2$$

Back-substitute $x = \dfrac{29}{10}$: $y = \dfrac{3x+6}{7} = \dfrac{3(29/10)+6}{7} = \dfrac{21}{10}$

Back-substitute $x = -2$: $y = \dfrac{3x+6}{7} = 0$

Answer: $\left(\frac{29}{10}, \frac{21}{10}\right), (-2, 0)$

65. $x^2 + y^2 = 1$

$x + y = 4$

Graphing $y_1 = \sqrt{1-x^2}$, $y_2 = -\sqrt{1-x^2}$ and $y_3 = 4 - x$, you see that there are no points of intersection.

No solution

67. $\begin{cases} y = 2x + 1 \\ y = \sqrt{x+2} \end{cases}$

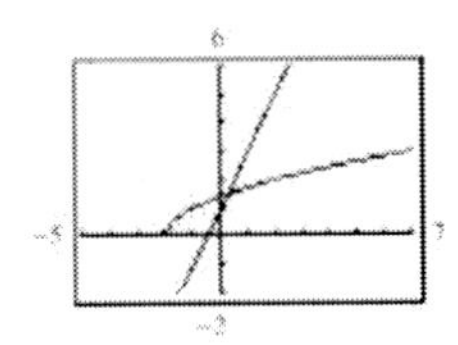

Point of intersection: $\left(\frac{1}{4}, \frac{3}{2}\right)$ or $(0.25, 1.5)$

69. $\begin{cases} y - e^{-x} = 1 \Rightarrow y = e^{-x} + 1 \\ y - \ln x = 3 \Rightarrow y = \ln x + 3 \end{cases}$

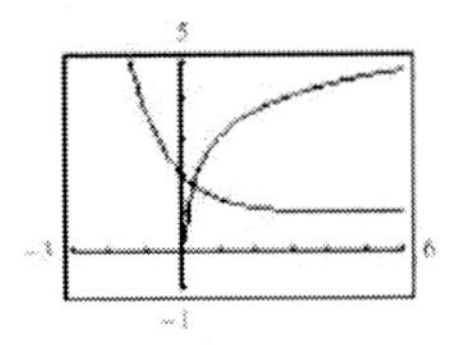

Point of intersection: $(0.287, 1.751)$

71. $\begin{cases} y = x^3 - 2x^2 + 1 & \text{Equation 1} \\ y = 1 - x^2 & \text{Equation 2} \end{cases}$

Substitute for y in Equation 2:

$x^3 - 2x^2 + 1 = 1 - x^2$

Solve for x: $x^3 - x^2 = 0$

$x^2(x-1) = 0 \Rightarrow x = 0, 1$

Back-substitute: $x = 0 \Rightarrow y = 1$

$x = 1 \Rightarrow y = 0$

Answer: $(0, 1), (1, 0)$

73. $\begin{cases} xy - 1 = 0 & \text{Equation 1} \\ 2x - 4y + 7 = 0 & \text{Equation 2} \end{cases}$

Solve for y in Equation 1: $y = \dfrac{1}{x}$

Substitute for y in Equation 2: $2x - 4\left(\dfrac{1}{x}\right) + 7 = 0$

Solve for x: $2x^2 - 4 + 7x = 0$

$\Rightarrow (2x-1)(x+4) = 0 \Rightarrow x = \dfrac{1}{2}, -4$

Back-substitute $x = \dfrac{1}{2}$: $y = \dfrac{1}{1/2} = 2$

Back-substitute $x = -4$: $y = \dfrac{1}{-4} = -\dfrac{1}{4}$

Answer: $\left(\dfrac{1}{2}, 2\right), \left(-4, -\dfrac{1}{4}\right)$

75. $C = 8650x + 250{,}000,\ R = 9950x$

$R = C$

$9950x = 8650x + 250{,}000$

$1300x = 250{,}000$

$x \approx 192$ units

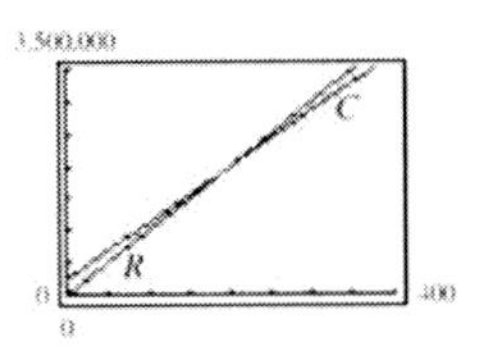

$R \approx \$1{,}910{,}400$

77. $C = 5.5\sqrt{x} + 10{,}000,\ R = 3.29x$

$$R = C$$
$$3.29x = 5.5\sqrt{x} + 10{,}000$$
$$3.29x - 10{,}000 = 5.5\sqrt{x}$$
$$10.8241x^2 - 65{,}800x + 100{,}000{,}000 = 30.25x$$
$$10.8241x^2 - 65{,}830.25x + 100{,}000{,}000 = 0$$
$$x \approx 3133 \text{ units}$$

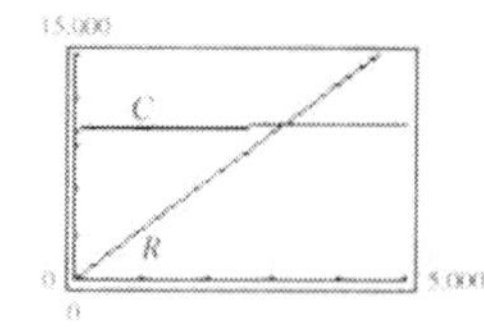

In order for the revenue to break even with the cost, 3133 units must be sold. $R = \$10{,}308$.

79. $2l + 2w = 30 \Rightarrow l + w = 15$

$$l = w + 3 \Rightarrow \quad (w + 3) + w = 15$$
$$2w = 12$$
$$w = 6$$

$l = w + 3 = 9$

Dimensions: 6 meters $\times$ 9 meters

81. $N = 360 - 24x$ Animated film

$N = 24 + 18x$ Horror film

(a)

Week x	1	2	3	4	5	6	7	8	9	10	11	12
Animated	336	312	288	264	240	216	192	168	144	120	96	72
Horror	42	60	78	96	114	132	150	168	186	204	222	240

(b) For $x = 8$, $N = 168$

(c) $360 - 24x = 24 + 18x$

$$336 = 42x$$
$$x = 8$$
$$N = 24 + 18(8) = 168$$

(d) The answers are the same.

(e) During week 8, the same numbe

rented.

83. (a) The total cost will be the sum of the variable cost plus the fixed cost (initial cost).

$C = 9.45x + 16{,}000$

The total revenue is the selling price times the number of units sold.

$R = 55.95x$

(b)

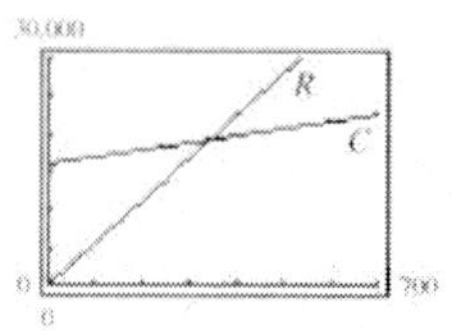

The break-even point is the point of intersection of the cost function and the revenue function.

$$55.95x = 9.45x + 16{,}000$$
$$46.5x = 16{,}000$$
$$x \approx 344 \text{ units}$$

85. $2l + 2w = 40 \Rightarrow l + w = 20 \Rightarrow w = 20 - l$

$lw = 96 \Rightarrow \quad l(20 - l) = 96$

$$20l - l^2 = 96$$
$$0 = l^2 - 20l + 96$$
$$0 = (l - 8)(l - 12)$$
$$l = 8 \text{ or } l = 12$$

$l = 12,\ w = 8$

If the length is supposed to be greater than the width, we have $l = 12$ miles and $w = 8$ miles.

87. (a) $\begin{cases} x + y = 20{,}000 & \text{Equation 1} \\ 0.055x + 0.075y = 1300 & \text{Equation 2} \end{cases}$

(b)

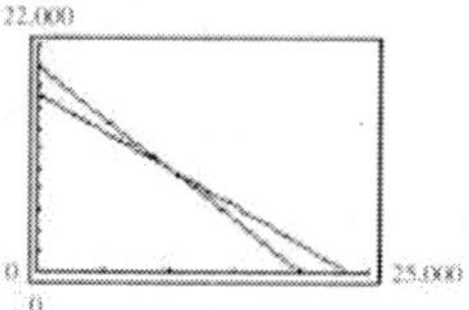

(c) Solve for y in Equation 1: $y = 20{,}000 - x$

$$0.055x + 0.075(20{,}000 - x) = 1300$$
$$0.055x + 1500 - 0.075x = 1300$$
$$-0.02x = -200$$
$$x = 10{,}000$$

Back-substitute: $y = 20{,}000 - 10{,}000 = 10{,}000$

To earn \$1300 in interest, \$10,000 should be invested in each fund, earning 5.5% and 7.5%.

89. (a)

t	Year	Arizona	Indiana
0	2000	5118	6080
1	2001	5289.9	6115.7
2	2002	5461.8	6151.4
3	2003	5633.7	6187.1
4	2004	5805.6	6222.8
5	2005	5977.5	6258.5
6	2006	6149.4	6294.2
7	2007	6321.3	6329.9
8	2008	6493.2	6365.6

(b) In 2008, the population of Arizona, 6,493,200, is greater than the population of Indiana, 6,365,600.

(c)

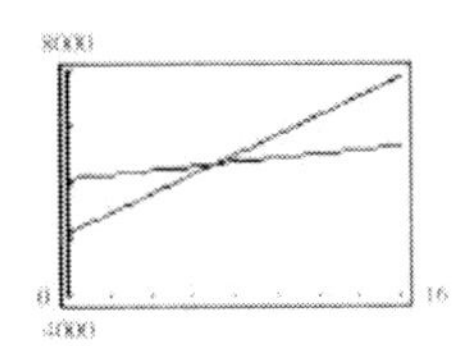

The point of intersection is approximately $(7.0, 6332.0)$.

(d) $\begin{cases} A = 171.9t + 5118 \\ I = 35.7t + 6080 \end{cases}$

Set $A = I \Rightarrow 171.9t + 5118 = 35.7t + 6080$

$$136.2t = 962$$

$$t = \frac{962}{136.2} \approx 7.06$$

$I = 35.7(7.06) + 6080 \approx 6332.04$

$(7.06, 6332.04)$

(e) At one point during 2007, the populations of Arizona and Indiana were equal.

91. False. You could solve for x first.

93. The system has no solution if you arrive at a false statement, such as $4 = 8$, or you have a quadratic equation with a negative discriminant, which would yield imaginary roots.

95. Answers will vary. For example,

(a) $\begin{cases} 3x + y = 3 \\ 3x + y = 5 \end{cases}$

(b) $\begin{cases} 3x + y = 4 \\ 2x + y = 2 \end{cases}$

(c) $\begin{cases} 6x + 3y = 9 \\ 2x + y = 3 \end{cases}$

97. (a)

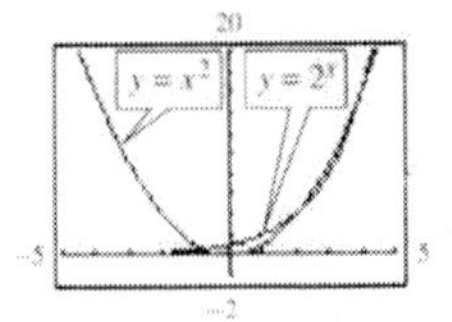

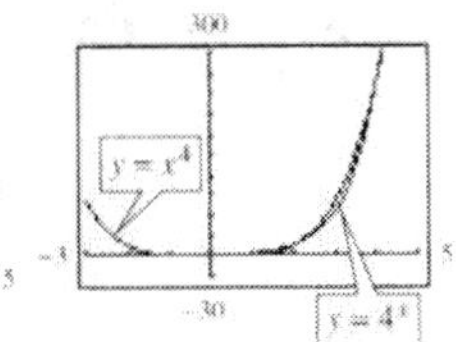

(b) Based on the graphs in part (a), it appears that for $b > 1$, there are three points of intersection for the graphs of $y = b^x$ and $y = x^b$ when b is an even number.

99. $(-2, 7), (5, 5)$

$$m = \frac{5 - 7}{5 - (-2)} = -\frac{2}{7}$$

$$y - 7 = -\frac{2}{7}(x - (-2))$$

$$y - 7 = -\frac{2}{7}x - \frac{4}{7}$$

$$y = -\frac{2}{7}x + \frac{45}{7}$$

101. $(6, 3), (10, 3)$

$$m = \frac{3 - 3}{10 - 6} = 0$$

The line is horizontal.

$y = 3$

103. $\left(\frac{3}{5}, 0\right), (4, 6)$

$$m = \frac{6 - 0}{4 - \frac{3}{5}} = \frac{6}{\frac{17}{5}} = \frac{30}{17}$$

$$y - 6 = \frac{30}{17}(x - 4)$$

$$y - 6 = \frac{30}{17}x - \frac{120}{17}$$

$$y = \frac{30}{17}x - \frac{18}{17}$$

105. $f(x) = \dfrac{5}{x - 6}$

Domain: all $x \neq 6$
Vertical asymptote: $x = 6$
Horizontal asymptote: $y = 0$

107. $f(x) = \dfrac{x^2 + 2}{x^2 - 16}$

Domain: all $x \neq \pm 4$
Vertical asymptotes: $x = \pm 4$
Horizontal asymptote: $y = 1$

109. $f(x) = \dfrac{x + 1}{x^2 + 1}$

Domain: all real numbers x
Horizontal asymptote: $y = 0$

Section 7.2

1. method, elimination

3. A system of linear equations with no solution is inconsistent.

5. A system of linear equations where the lines are coincident or identical is consistent.

7. $\begin{cases} 2x + y = 5 & \text{Equation 1} \\ x - y = 1 & \text{Equation 2} \end{cases}$

Add to eliminate y: $3x = 6 \Rightarrow x = 2$

Substitute $x = 2$ in Equation 2: $2 - y = 1 \Rightarrow y = 1$

Answer: $(2, 1)$

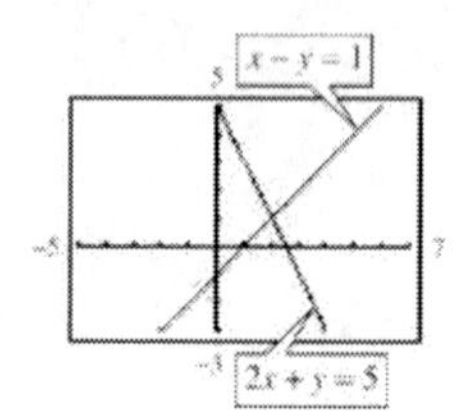

9. $\begin{cases} x + y = 0 & \text{Equation 1} \\ 3x + 2y = 1 & \text{Equation 2} \end{cases}$

Multiply Equation 1 by -2: $-2x - 2y = 0$

Add this to Equation 2 to eliminate y: $x = 1$

Substitute $x = 1$ in Equation 1: $1 + y = 0 \Rightarrow y = -1$

Answer: $(1, -1)$

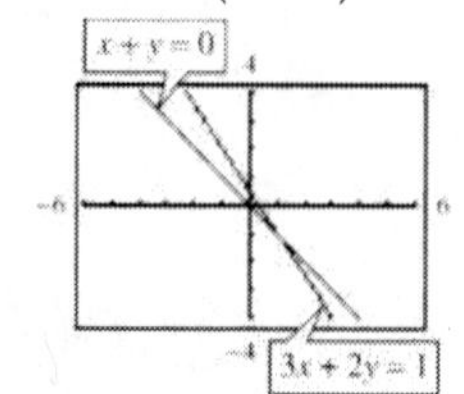

11. $\begin{cases} x - y = 2 & \text{Equation 1} \\ -2x + 2y = 5 & \text{Equation 2} \end{cases}$

Multiply Equation 1 by 2: $2x - 2y = 4$

Add this to Equation 2: $0 = 9$

There are no solutions.

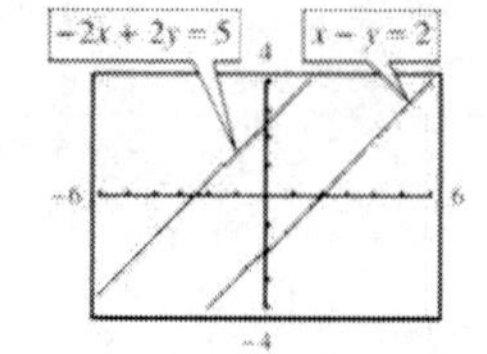

13. $\begin{cases} x + 2y = 3 & \text{Equation 1} \\ x - 2y = 1 & \text{Equation 2} \end{cases}$

Add to eliminate y:

$2x = 4$

$x = 2$

Substitute $x = 2$ into Equation 1:

$2 + 2y = 3$

$2y = 1$

$y = \frac{1}{2}$

Answer: $\left(2, \frac{1}{2}\right)$

15. $\begin{cases} 2x + 3y = 18 & \text{Equation 1} \\ 5x - y = 11 & \text{Equation 2} \end{cases}$

Multiply Equation 2 by 3: $15x - 3y = 33$

Add this to Equation 1 to eliminate y:

$17x = 51 \Rightarrow x = 3$

Substitute $x = 3$ in Equation 1:

$6 + 3y = 18 \Rightarrow y = 4$

Answer: $(3, 4)$

17. $\begin{cases} 3r + 2s = 10 & \text{Equation 1} \\ 2r + 5s = 3 & \text{Equation 2} \end{cases}$

Multiply Equation 1 by 2 and Equation 2 by -3:

$6r + 4s = 20$

$-6r - 15s = -9$

Add to eliminate r:

$-11s = 11 \Rightarrow s = -1$

Substitute $s = -1$ in Equation 1:

$3r + 2(-1) = 10 \Rightarrow r = 4$

Answer: $(4, -1)$

19. $\begin{cases} 5u + 6v = 24 & \text{Equation 1} \\ 3u + 5v = 18 & \text{Equation 2} \end{cases}$

Multiply Equation 1 by 3 and Equation 2 by (-5):

$15u + 18v = 72$

$-15u - 25v = -90$

Add to eliminate u: $-7v = -18 \Rightarrow v = \dfrac{18}{7}$

Substitute $v = \frac{18}{7}$ in Equation 2:

$3u + 5\left(\frac{18}{7}\right) = 18 \Rightarrow u = \frac{12}{7}$

Answer: $\left(\frac{12}{7}, \frac{18}{7}\right)$

21. $\begin{cases} 1.8x + 1.2y = 4 & \text{Equation 1} \\ 9x + 6y = 3 & \text{Equation 2} \end{cases}$

Multiply Equation 1 by (-5): $-9x - 6y = -20$

Add this to Equation 2: $0 = -17$

Inconsistent; no solution

23. $2x - 5y = 0$

$x - \ \ y = 3 \Rightarrow y = x - 3$

$2x - 5(x - 3) = 0$

$-3x = -15$

$x = 5, \ y = 2$

Matches (b).
One solution; consistent

25. $\left.\begin{array}{l} 2x - 5y = 0 \\ 2x - 3y = -4 \end{array}\right\} \begin{array}{l} -2y = 4 \\ y = -2, \ x = -5 \end{array}$

One solution; consistent
Matches (c).

27. $\begin{cases} 5x + 3y = 6 & \text{Equation 1} \\ 3x - \ \ y = 5 & \text{Equation 2} \end{cases}$

Multiply Equation 2 by 3 and add to eliminate y:

$5x + 3y = 6$

$9x - 3y = 15$

$14x = 21$

$x = \frac{3}{2}$

Substitute $x = \frac{3}{2}$ into Equation 1:

$5\left(\frac{3}{2}\right) + 3y = 6$

$\frac{15}{2} + 3y = 6$

$3y = -\frac{3}{2}$

$y = -\frac{1}{2}$

Answer: $\left(\frac{3}{2}, -\frac{1}{2}\right)$

29. $\begin{cases} \frac{2}{5}x - \frac{3}{2}y = \ \ 4 & \text{Equation 1} \\ \frac{1}{5}x - \frac{3}{4}y = -2 & \text{Equation 2} \end{cases}$

Multiply Equation 2 by -2 and add to Equation 1:

$0 = 8$

Inconsistent; no solution

31. $\begin{cases} \frac{3}{4}x + \ \ y = \frac{1}{8} & \text{Equation 1} \\ \frac{9}{4}x + 3y = \frac{3}{8} & \text{Equation 2} \end{cases}$

Multiply Equation 1 by -3: $-\frac{9}{4}x - 3y = -\frac{3}{8}$

Add this to Equation 2: $\qquad 0 = 0$

There are an infinite number of solutions.

The solutions consist of all $(x, \ y)$ satisfying

$\frac{3}{4}x + y = \frac{1}{8}$, or $6x + 8y = 1$.

33. $\begin{cases} \dfrac{x+3}{4} + \dfrac{y-1}{3} = 1 & \text{Equation 1} \\ 2x - y = 12 & \text{Equation 2} \end{cases}$

Multiply Equation 1 by 12 and Equation 2 by 4:

$\begin{cases} 3x + 4y = 7 \\ 8x - 4y = 48 \end{cases}$

Add to eliminate y: $11x = 55 \Rightarrow x = 5$

Substitute $x = 5$ into Equation 2:

$2(5) - y = 12 \Rightarrow y = -2$

Answer: $(5, \ -2)$

35. $\begin{cases} -5x + \ 6y = -3 & \text{Equation 1} \\ 20x - 24y = 12 & \text{Equation 2} \end{cases}$

Multiply Equation 2 by $\frac{1}{4}$ and add:

$-5x + 6y = -3$

$5x - 6y = 3$

$0 = 0$

There are an infinite number of solutions. All points on the line $-5x + 6y = -3$.

37. $\begin{cases} 2.5x - \ 3y = 1.5 & \text{Equation 1} \\ 10x - 12y = 6 & \text{Equation 2 multiplied by 5} \end{cases}$

Multiply Equation 1 by (-4):

$-10x + 12y = -6$

Add this to Equation 2 to eliminate x:

$0 = 0$

The solution set consist of all points lying on the line $10x - 12y = 6$.

All points on the line $5x - 6y = 3$.

39. $\begin{cases} 0.2x - 0.5y = -27.8 & \text{Equation 1} \\ 0.3x + 0.4y = \ \ 68.7 & \text{Equation 2} \end{cases}$

Multiply Equation 1 by 40 and Equation 2 by 50:

$\begin{cases} 8x - 20y = -1112 \\ 15x + 20y = \ \ 3435 \end{cases}$

Adding the equations eliminates y:

$23x = 2323 \Rightarrow x = 101$

Substitute $x = 101$ into Equation 1:

$8(101) - 20y = -1112 \Rightarrow y = 96$

Answer: $(101, \ 96)$

41. $\begin{cases} 0.05x - 0.03y = 0.21 & \text{Equation 1} \\ 0.07x + 0.02y = 0.16 & \text{Equation 2} \end{cases}$

Multiply Equation 1 by 200 and Equation 2 by 300:

$\begin{cases} 10x - 6y = 42 \\ 21x + 6y = 48 \end{cases}$

Add to eliminate y: $31x = 90$

$x = \frac{90}{31}$

Substitute $x = \frac{90}{31}$ in Equation 2:

$0.07\left(\frac{90}{31}\right) + 0.02y = 0.16$

$y = -\frac{67}{31}$

Answer: $\left(\frac{90}{31}, \ -\frac{67}{31}\right)$

43. Let $X = \frac{1}{x}$ and $Y = \frac{1}{y}$.

$$\begin{cases} X + 3Y = 2 & \text{Equation 1} \\ 4X - Y = -5 & \text{Equation 2} \end{cases}$$

Multiply Equation 1 by 4:

$$\begin{cases} 4X + 12Y = 8 \\ 4X - Y = -5 \end{cases}$$

Subtract to eliminate X:

$$13Y = 13 \Rightarrow Y = 1.$$

Hence,

$$X = 2 - 3Y = 2 - 3(1) = -1$$

$$x = \frac{1}{X} = -1,\ y = \frac{1}{Y} = 1$$

Answer: $(-1, 1)$

45. Let $X = \frac{1}{x}$ and $Y = \frac{1}{y}$.

$$\begin{cases} X + 2Y = 5 & \text{Equation 1} \\ 3X - 4Y = -5 & \text{Equation 2} \end{cases}$$

Multiply Equation 1 by 2:

$$\begin{cases} 2X + 4Y = 10 \\ 3X - 4Y = -5 \end{cases}$$

Adding the equations eliminates Y:

$$5X = 5 \Rightarrow X = 1.$$

Hence,

$$2Y = 5 - X = 4 \Rightarrow Y = 2.$$

$$x = \frac{1}{X} = 1,\ y = \frac{1}{Y} = \frac{1}{2}$$

Answer: $(1, \frac{1}{2})$

47. $$\begin{cases} 2x - 5y = 0 \Rightarrow y = \frac{2}{5}x \\ x - y = 3 \Rightarrow y = x - 3 \end{cases}$$

The system is consistent.

There is one solution, $(5, 2)$.

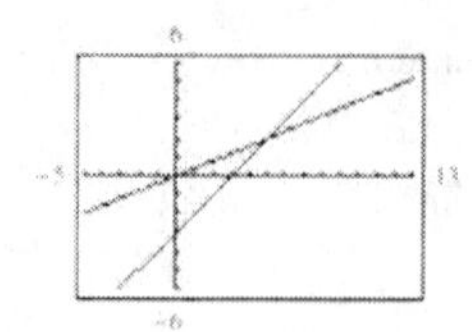

49. $$\begin{cases} \frac{3}{5}x - y = 3 \Rightarrow y = \frac{3}{5}x - 3 \\ -3x + 5y = 9 \Rightarrow y = \frac{1}{5}(3x + 9) = \frac{3}{5}x + \frac{9}{5} \end{cases}$$

The lines are parallel.
The system is inconsistent.

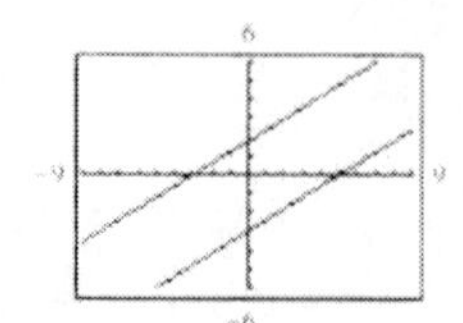

51. $$\begin{cases} 8x - 14y = 5 \Rightarrow y = (8x - 5)/14 = \frac{4}{7}x - \frac{5}{14} \\ 2x - 3.5y = 1.25 \Rightarrow y = (2x - 1.25)/3.5 = \frac{4}{7}x - \frac{5}{14} \end{cases}$$

The system is consistent. The solution set consists of all points on the line

$y = \frac{4}{7}x - \frac{5}{14}$, or $8x - 14y = 5$.

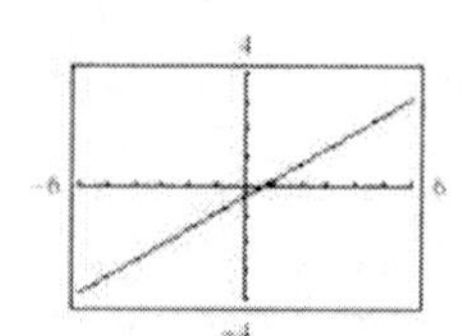

53. $$\begin{cases} 6y = 42 \Rightarrow y = 7 \\ 6x - y = 16 \Rightarrow y = 6x - 16 \end{cases}$$

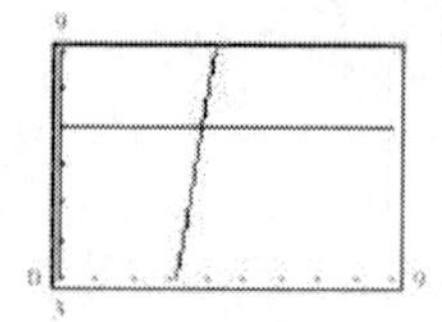

Answer: $(\frac{23}{6}, 7) \approx (3.833, 7)$

55. $$\begin{cases} \frac{3}{2}x - \frac{1}{5}y = 8 \Rightarrow y = 5(\frac{3}{2}x - 8) \\ -2x + 3y = 3 \Rightarrow y = \frac{1}{3}(3 + 2x) \end{cases}$$

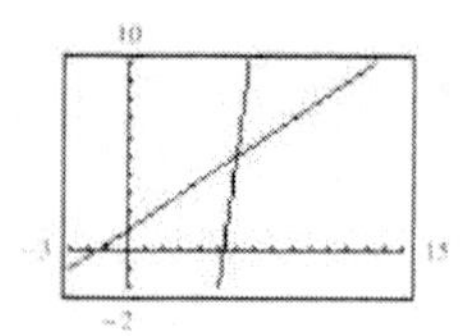

Answer: $(6, 5)$

57. $$\frac{1}{3}x + y = -\frac{1}{3} \Rightarrow y = -\frac{1}{3} - \frac{1}{3}x$$

$$5x - 3y = 7 \Rightarrow y = \frac{1}{3}(5x - 7)$$

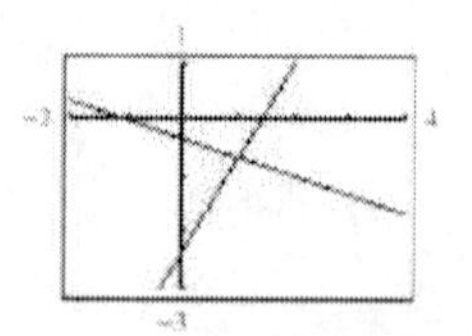

Answer: $(1, -0.667)$

59. $$\begin{cases} 0.5x + 2.2y = 9 \Rightarrow y = 1/2.2(9 - 0.5x) \\ 6x + 0.4y = -22 \Rightarrow y = 1/0.4(-22 - 6x) \end{cases}$$

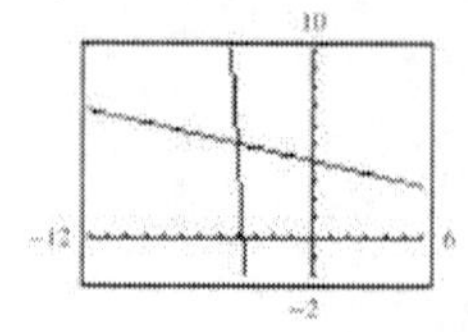

Answer: $(-4, 5)$

61. $\begin{cases} 3x - 5y = 7 & \text{Equation 1} \\ 2x + y = 9 & \text{Equation 2} \end{cases}$

Multiply Equation 2 by 5:

$10x + 5y = 45$

Add this to Equation 1:

$13x = 52 \Rightarrow x = 4$

Back-substitute $x = 4$ into Equation 2:

$2(4) + y = 9 \Rightarrow y = 1$

Answer: $(4, 1)$

63. $\begin{cases} y = 2x - 5 & \text{Equation 1} \\ y = 5x - 11 & \text{Equation 2} \end{cases}$

Set Equation 1 equal to Equation 2:

$2x - 5 = 5x - 11$

$-3x = -6$

$x = 2$

Substitute $x = 2$ into Equation 1:

$y = 2(2) - 5 = -1$

Answer: $(2, -1)$

65. $\begin{cases} x - 5y = 21 \\ 6x + 5y = 21 \end{cases}$

Adding the equations, $7x = 42 \Rightarrow x = 6$.

Back-substituting, $x - 5y = 6 - 5y = 21 \Rightarrow$

$-5y = 15 \Rightarrow y = -3$

Answer: $(6, -3)$

67. $\begin{cases} -5x + 9y = 13 & \text{Equation 1} \\ y = x - 4 & \text{Equation 2} \end{cases}$

Substitute Equation 2 into Equation 1:

$-5x + 9(x - 4) = 13$

$-5x + 9x - 36 = 13$

$4x = 49$

$x = \frac{49}{4}$

Substitute $x = \frac{49}{4}$ into Equation 2:

$y = \frac{49}{4} - 4 = \frac{33}{4}$

Answer: $\left(\frac{49}{4}, \frac{33}{4}\right)$

69. There are infinitely many systems that have the solution $(0, 8)$. One possible system is

$\begin{cases} x + y = 8 \\ -x + y = 8 \end{cases}$

71. There are infinitely many systems that have the solution $\left(3, \frac{5}{2}\right)$. One possible system:

$2(3) + 2\left(\frac{5}{2}\right) = 11 \Rightarrow 2x + 2y = 11$

$3 - 4\left(\frac{5}{2}\right) = -7 \Rightarrow x - 4y = -7$

73.

Demand $=$ Supply

$500 - 0.4x = 380 + 0.1x$

$-0.5x = -120$

$x = 240$ units

$p = 500 - 0.4(240) = \$404$

Answer: $(240, 404)$

75.

Demand $=$ Supply

$140 - 0.00002x = 80 + 0.00001x$

$60 = 0.00003x$

$x = 2{,}000{,}000$ units

$p = \$100.00$

Answer: $(2{,}000{,}000, 100)$

77. Let $x =$ the ground speed and $y =$ the wind speed.

$\begin{cases} 3.6(x - y) = 1800 & \text{Equation 1} & x - y = 500 \\ 3(x + y) = 1800 & \text{Equation 2} & x + y = 600 \end{cases}$

$2x = 1100$

$x = 550$

Substituting $x = 550$ in Equation 2:

$550 + y = 600$

$y = 50$

Answer: $x = 550$ mph, $y = 50$ mph

79. (a) $\begin{cases} A + C = 1175 & \text{Equation 1} \\ 5A + 3.5C = 5087.5 & \text{Equation 2} \end{cases}$

(b) Multiply Equation 1 by 5:

$5A + 5C = 5875$

$5A + 3.5C = 5087.5$

Subtracting eliminates A:

$1.5C = 787.5$

$C = 525$

Hence, $A = 1175 - 525 = 650$.

650 adult tickets and 525 child tickets

(c)

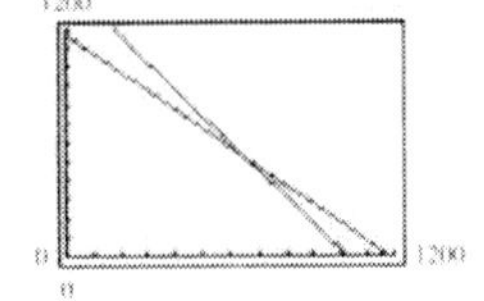

Let $C = y_1 = 1175 - x$

$C = y_2 = \frac{1}{3.5}(5087.5 - 5x)$

Point of intersection: $(A, C) = (650, 525)$

81. Let M = number of oranges and let R = number of grapefruits.

$$\begin{cases} M + R = 16 & \text{Equation 1} \\ 0.95M + 1.05R = 15.90 & \text{Equation 2} \end{cases}$$

Solving for R in Equation 1: $R = 16 - M$.
Substituting into Equation 2:

$$0.95M + 1.05(16 - M) = 15.9$$
$$0.95M + 16.8 - 1.05M = 15.9$$
$$0.9 = 0.1M$$
$$M = 9$$

Hence, $R = 16 - 9 = 7$.
9 oranges and 7 grapefruits

83. Let m = number of movies and let v = number of video games.

$$\begin{cases} m + v = 310 & \text{Equation 1} \\ 3m + 2.5v = 867.5 & \text{Equation 2} \end{cases}$$

Solve for m in Equation 1: $m = 310 - v$.
Substitute for m Equation 2:

$$3(310 - v) + 2.5v = 867.5$$
$$62.5 = 0.5v$$
$$v = 125$$
$$m = 310 - 125 = 185$$

185 movies and 125 video games

85. $\begin{cases} 5b + 10a = 20.2 \Rightarrow & -10b - 20a = -40.4 \\ 10b + 30a = 50.1 \Rightarrow & 10b + 30a = 50.1 \end{cases}$

$$10a = 9.7$$
$$a = 0.97$$
$$b = 2.1$$

Least squares regression line: $y = 0.97x + 2.1$

87. $\begin{cases} 5b + 10a = 2.7 & \text{Equation 1} \\ 10b + 30a = -19.6 & \text{Equation 2} \end{cases}$

Multiply Equation 1 by -2:

$$-10b - 20a = -5.4$$
$$10b + 30a = -19.6$$

Adding,

$$10a = -25.0$$
$$a = -2.5$$
$$b = (2.7 + 10(2.5))/5 = 5.54$$

$$y = -2.5x + 5.54$$

89. (a) $\begin{cases} 4b + 7a = 174 \Rightarrow & 28b + 49a = 1218 \\ 7b + 13.5a = 322 \Rightarrow & -28b - 54a = -1288 \end{cases}$

Adding, $-5a = -70 \Rightarrow a = 14$, $b = 19$.
Thus, $y = 14x + 19$

(b) Using a graphing utility, you obtain $y = 14x + 19$.

(c)

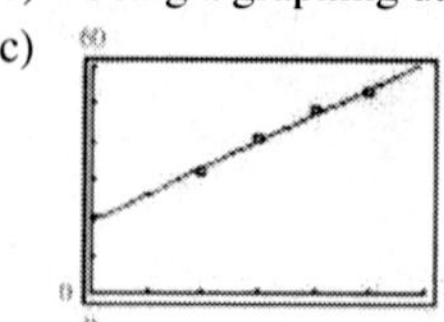

(d) If $x = 1.6$, $(160 \text{ pounds/acre})$,

$$y = 14(1.6) + 19 = 41.4 \text{ bushels per acre.}$$

91. True. A consistent linear system has either one solution or an infinite number of solutions.

93. False. At times, only a reasonable approximation is possible graphically.

95. (a) $\begin{cases} x + y = 10 & \text{Equation 1} \\ x + y = 20 & \text{Equation 2} \end{cases}$

Subtract Equation 2 from Equation 1: $0 = -10$
System is inconsistent $\Rightarrow$ no solution

(b) $\begin{cases} x + y = 3 & \text{Equation 1} \\ 2x + 2y = 6 & \text{Equation 2} \end{cases}$

Multiply Equation 1 by (-2): $-2x - 2y = -6$

Add this to Equation 2: $0 = 0$ (dependent)

The system has an infinite number of solutions.

97. $u \sin x + v \cos x = 0$
$u \cos x - v \sin x = \sec x$

Multiply the first equation by $\sin x$, the second by $\cos x$, and add the equations:

$$u \sin^2 x + u \cos^2 x = \sec x \cdot \cos x$$
$$u = 1$$

Hence, $v \cos x = -u \sin x = -\sin x$
$$v = -\tan x.$$

99. $-11 - 6x \geq 33$

$$-6x \geq 44$$
$$x \leq -\tfrac{44}{6} = -\tfrac{22}{3}$$

101. $|x - 8| < 10$

$$-10 < x - 8 < 10$$
$$-2 < x < 18$$

103. $2x^2 + 3x - 35 < 0$

$(2x-7)(x+5) < 0$

Critical numbers: $\frac{7}{2}, -5$.

Testing the three intervals,

$-5 < x < \frac{7}{2}$.

105. $\ln x + \ln 6 = \ln 6x$

107. $\log_9 12 - \log_9 x = \log_9 \frac{12}{x}$

109. $2\ln x - \ln(x+2) = \ln x^2 - \ln(x+2)$

$= \ln\left(\frac{x^2}{x+2}\right)$

111. Answers will vary. (Make a Decision)

Section 7.3

1. row-echelon

3. Gaussian

5. three-dimensional

7. A consistent system with exactly one solution is independent.

9. (a) $3(3)-(5)+(-3)\stackrel{?}{=}1$ Yes

$2(3) \quad -3(-3)\stackrel{?}{=}-14$ No

$5(5)+2(-3)\stackrel{?}{=}8$ No

No, $(3, 5, -3)$ is not a solution.

(b) $3(-1)-(0)+(4)\stackrel{?}{=}1$ Yes

$2(-1) \quad -3(4)\stackrel{?}{=}-14$ Yes

$5(0)+2(4)\stackrel{?}{=}8$ Yes

Yes, $(-1, 0, 4)$ is solution.

(c) $3(0)-(-1)+(3)\stackrel{?}{=}1$ No

$2(8) \quad -3(3)\stackrel{?}{=}-14$ No

$5(-1)+2(3)\stackrel{?}{=}8$ No

No, $(0, -1, 3)$ is not a solution.

(d) $(1)-(0)+(4)\stackrel{?}{=}1$ No

$2(1) \quad -3(4)\stackrel{?}{=}-14$ Yes

$5(0)+2(4)\stackrel{?}{=}8$ Yes

No, $(1, 0, 4)$ is not a solution.

11. (a) $4(0)+(1)-(1)\stackrel{?}{=}0$ Yes

$-8(0)-6(1)+(1)\stackrel{?}{=}-\frac{7}{4}$ No

$3(0)-(1) \stackrel{?}{=}-\frac{9}{4}$ No

No, $(0, 1, 1)$ is not a solution.

(b) $4\left(-\frac{3}{2}\right)+\left(\frac{5}{4}\right)-\left(-\frac{5}{4}\right)\stackrel{?}{=}0$ No

$-8\left(-\frac{3}{2}\right)-6\left(\frac{5}{4}\right)+\left(-\frac{5}{4}\right)\stackrel{?}{=}-\frac{7}{4}$ No

$3\left(-\frac{3}{2}\right)-\left(\frac{5}{4}\right) \stackrel{?}{=}-\frac{9}{4}$ No

No, $\left(-\frac{3}{2}, \frac{5}{4}, -\frac{5}{4}\right)$ is not a solution.

(c) $4\left(-\frac{1}{2}\right)+\left(\frac{3}{4}\right)-\left(-\frac{5}{4}\right)\stackrel{?}{=}0$ Yes

$-8\left(-\frac{1}{2}\right)-6\left(\frac{3}{4}\right)+\left(-\frac{5}{4}\right)\stackrel{?}{=}-\frac{7}{4}$ Yes

$3\left(-\frac{1}{2}\right)-\left(\frac{3}{4}\right) \stackrel{?}{=}-\frac{9}{4}$ Yes

Yes, $\left(-\frac{1}{2}, \frac{3}{4}, -\frac{5}{4}\right)$ is a solution.

(d) $4\left(-\frac{1}{2}\right)+\left(\frac{1}{6}\right)-\left(-\frac{3}{4}\right)\stackrel{?}{=}0$ No

$-8\left(-\frac{1}{2}\right)-6\left(\frac{1}{6}\right)+\left(-\frac{3}{4}\right)\stackrel{?}{=}-\frac{7}{4}$ No

$3\left(-\frac{1}{2}\right)-\left(\frac{1}{6}\right) \stackrel{?}{=}-\frac{9}{4}$ No

No, $\left(-\frac{1}{2}, \frac{1}{6}, -\frac{3}{4}\right)$ is not a solution.

13. $\begin{cases} 2x - y + 5z = 16 & \text{Equation 1} \\ y + 2z = 2 & \text{Equation 2} \\ z = 2 & \text{Equation 3} \end{cases}$

Back-substitute $z = 2$ into Equation 2:

$$y + 2(2) = 2$$
$$y = -2$$

Back-substitute $z = 2$ and $y = -2$ into Equation 1:

$$2x - (-2) + 5(2) = 16$$
$$2x = 4$$
$$x = 2$$

Answer: $(2, -2, 2)$

15. $\begin{cases} 2x - y - 3z = 10 & \text{Equation 1} \\ y + z = 12 & \text{Equation 2} \\ z = 2 & \text{Equation 3} \end{cases}$

Back-substitute $z = 2$ into Equation 2:

$$y + 2 = 12$$
$$y = 10$$

Back-substitute $y = 10$ and $z = 2$ into Equation 1:

$$2x + 10 - 3(2) = 10$$
$$2x = 6$$
$$x = 3$$

Answer: $(3, 10, 2)$

17. $\begin{cases} 4x - 2y + z = 8 & \text{Equation 1} \\ -y + z = 4 & \text{Equation 2} \\ z = 11 & \text{Equation 3} \end{cases}$

Back-substitute $z = 11$ into Equation 2:

$$-y + 11 = 4$$
$$-y = -7$$
$$y = 7$$

Back-substitute $y = 7$ and $z = 11$ into Equation 1:

$$4x - 2(7) + 11 = 8$$
$$4x - 14 + 11 = 8$$
$$4x = 11$$
$$x = \tfrac{11}{4}$$

Answer: $\left(\dfrac{11}{4}, 7, 11\right)$

19. $\begin{cases} x - 2y + 3z = 5 & \text{Equation 1} \\ -x + 3y - 5z = 4 & \text{Equation 2} \\ 2x \quad\quad - 3z = 0 & \text{Equation 3} \end{cases}$

Add Equation 1 to Equation 2.

$y - 2z = 9$ New Equation 2

This is the first step in putting the system in row-echelon from.

$\begin{cases} x - 2y + 3z = 5 \\ y - 2z = 9 \\ 2x \quad\quad - 3z = 0 \end{cases}$

21. $\begin{cases} x + y + z = 6 & \text{Equation 1} \\ 2x - y + z = 3 & \text{Equation 2} \\ 3x \quad\quad - z = 0 & \text{Equation 3} \end{cases}$

$\begin{cases} x + y + z = 6 & \\ -3y - z = -9 & (-2)\text{Eq. 1+Eq. 2} \\ -3y - 4z = -18 & (-3)\text{Eq. 1+ Eq. 3} \end{cases}$

$\begin{cases} x + y + z = 6 & \\ -3y - z = -9 & \\ -3z = -9 & (-1)\text{Eq. 2+ Eq. 3} \end{cases}$

$-3z = -9 \Rightarrow z = 3$

$-3y - 3 = -9 \Rightarrow y = 2$

$x + 2 + 3 = 6 \Rightarrow x = 1$

Answer: $(1, 2, 3)$

23. $\begin{cases} 2x \quad\quad + 2z = 2 & \text{Equation 1} \\ 5x + 3y \quad\quad = 4 & \text{Equation 2} \\ 3y - 4z = 4 & \text{Equation 3} \end{cases}$

$\begin{cases} x \quad\quad + z = 1 & \left(\frac{1}{2}\right)\text{Eq. 1} \\ 5x + 3y \quad\quad = 4 & \\ 3y - 4z = 4 & \end{cases}$

$\begin{cases} x \quad\quad + z = 1 & \\ 3y - 5z = -1 & (-5)\text{ Eq. 1 + Eq. 2} \\ 3y - 4z = 4 & \end{cases}$

$\begin{cases} x \quad\quad + z = 1 & \\ 3y - 5z = -1 & \\ z = 5 & (-1)\text{ Eq. 2 + Eq. 3} \end{cases}$

$3y - 5(5) = -1 \Rightarrow y = 8$

$x + 5 = 1 \Rightarrow x = -4$

Answer: $(-4, 8, 5)$

25. $\begin{cases} 4x + y - 3z = 11 & \text{Equation 1} \\ 2x - 3y + 2z = 9 & \text{Equation 2} \\ x + y + z = -3 & \text{Equation 3} \end{cases}$

$\begin{cases} x + y + z = -3 & \text{Interchange Equations} \\ 2x - 3y + 2z = 9 & \text{1 and 3.} \\ 4x + y - 3z = 11 & \end{cases}$

$\begin{cases} x + y + z = -3 & \\ -5y \quad\quad = 15 & (-2)\text{Eq. 1 + Eq. 2} \\ -3y - 7z = 23 & (-4)\text{Eq. 1 + Eq. 3} \end{cases}$

$y = -3 \Rightarrow -3(-3) - 7z = 23$

$\Rightarrow -7z = 14$

$\Rightarrow z = -2$

$x + (-3) + (-2) = -3 \Rightarrow x = 2$

Answer: $(2, -3, -2)$

27. $\begin{cases} 3x - 2y + 4z = 1 & \text{Equation 1} \\ x + y - 2z = 3 & \text{Equation 2} \\ 2x - 3y + 6z = 8 & \text{Equation 3} \end{cases}$

$\begin{cases} x + y - 2z = 3 & \text{Interchange} \\ 3x - 2y + 4z = 1 & \text{Equations 1 and 2.} \\ 2x - 3y + 6z = 8 \end{cases}$

$\begin{cases} x + y - 2z = 3 \\ -5y + 10z = -8 & -3\text{ Eq. }1 + \text{Eq. }2 \\ -5y + 10z = 2 & -2\text{ Eq. }1 + \text{Eq. }3 \end{cases}$

$\begin{cases} x + y - 2z = 3 \\ -5y + 10z = -8 \\ 0 = 10 & -\text{Eq. }2 + \text{Eq. }3 \end{cases}$

Inconsistent; no solution.

29. $\begin{cases} 3x + 3y + 5z = 1 & \text{Equation 1} \\ 3x + 5y + 9z = 0 & \text{Equation 2} \\ 5x + 9y + 17z = 0 & \text{Equation 3} \end{cases}$

$\begin{cases} 6x + 6y + 10z = 2 & 2\text{ Eq. }1 \\ 3x + 5y + 9z = 0 \\ 5x + 9y + 17z = 0 \end{cases}$

$\begin{cases} x - 3y - 7z = 2 & -\text{Eq. }3 + \text{Eq. }1 \\ 3x + 5y + 9z = 0 \\ 5x + 9y + 17z = 0 \end{cases}$

$\begin{cases} x - 3y - 7z = 2 \\ 14y + 30z = -6 & -3\text{ Eq. }1 + \text{Eq. }2 \\ 24y + 52z = -10 & -5\text{ Eq. }1 + \text{Eq. }3 \end{cases}$

$\begin{cases} x - 3y - 7z = 2 \\ 84y + 180z = -36 & 6\text{ Eq. }2 \\ 84y + 182z = -35 & 3.5\text{ Eq. }3 \end{cases}$

$\begin{cases} x - 3y - 7z = 2 \\ 84y + 180z = -36 \\ 2z = 1 & -\text{Eq. }2 + \text{Eq. }3 \end{cases}$

$2z = 1 \Rightarrow z = \frac{1}{2}$

$84y + 180\left(\frac{1}{2}\right) = -36 \Rightarrow y = -\frac{3}{2}$

$x - 3\left(-\frac{3}{2}\right) - 7\left(\frac{1}{2}\right) = 2 \Rightarrow x = 1$

Answer: $\left(1, -\frac{3}{2}, \frac{1}{2}\right)$

31. $\begin{cases} 3x - 3y + 6z = 6 & \text{Equation 1} \\ x + 2y - z = 5 & \text{Equation 2} \\ 5x - 8y + 13z = 7 & \text{Equation 3} \end{cases}$

$\begin{cases} x - y + 2z = 2 & \left(\frac{1}{3}\right)\text{Eq. }1 \\ x + 2y - z = 5 \\ 5x - 8y + 13z = 7 \end{cases}$

$\begin{cases} x - y + 2z = 2 \\ 3y - 3z = 3 & (-1)\text{Eq. }1 + \text{Eq. }2 \\ -3y + 3z = -3 & (-5)\text{Eq. }1 + \text{Eq. }3 \end{cases}$

$\begin{cases} x - y + 2z = 2 \\ y - z = 1 \\ 0 = 0 \end{cases}$

$\begin{cases} x + z = 3 \\ y - z = 1 \end{cases}$

Let $z = a$, then:

$y = a + 1$

$x = -a + 3$

Answer: $(-a + 3, a + 1, a)$

33. $\begin{cases} x - 2y + 3z = 4 & \text{Equation 1} \\ 3x - y + 2z = 0 & \text{Equation 2} \\ x + 3y - 4z = -2 & \text{Equation 3} \end{cases}$

$\begin{cases} x - 2y + 3z = 4 \\ 5y - 7z = -12 & -3\text{Eq. }1 + \text{Eq. }2 \\ 5y - 7z = -6 & -1\text{Eq. }1 + \text{Eq. }3 \end{cases}$

$\begin{cases} x - 2y + 3z = 4 \\ 5y - 7z = -12 \\ 0 = 6 & -\text{Eq. }2 + \text{Eq. }3 \end{cases}$

No solution; inconsistent.

35. $\begin{cases} x + 4z = 1 & \text{Equation 1} \\ x + y + 10z = 10 & \text{Equation 2} \\ 2x - y + 2z = -5 & \text{Equation 3} \end{cases}$

$\begin{cases} x + 4z = 1 \\ y + 6z = 9 & -\text{Eq. }1 + \text{Eq. }2 \\ -y - 6z = -7 & -2\text{Eq. }1 + \text{Eq. }3 \end{cases}$

$\begin{cases} x + 4z = 1 \\ y + 6z = 9 \\ 0 = 2 & \text{Eq. }2 + \text{Eq. }3 \end{cases}$

No solution; inconsistent.

37. $\begin{cases} x+2y+z=1 & \text{Equation 1} \\ x-2y+3z=-3 & \text{Equation 2} \\ 2x+y+z=-1 & \text{Equation 3} \end{cases}$

$\begin{cases} x+2y+z=1 \\ -4y+2z=-4 & (-1)\text{Eq. }1+\text{Eq. }2 \\ -3y-z=-3 & (-2)\text{Eq. }1+\text{Eq. }3 \end{cases}$

$\begin{cases} x+2y+z=1 \\ y-\frac{1}{2}z=1 & \left(-\frac{1}{4}\right)\text{Eq. }2 \\ 3y+z=3 & (-1)\text{Eq. }3 \end{cases}$

$\begin{cases} x+2y+z=1 \\ y-\frac{1}{2}z=1 \\ \frac{5}{2}z=0 & (-3)\text{Eq. }2+\text{Eq. }3 \end{cases}$

$z=0$

$y=1-0=1$

$x+2y+z=1 \Rightarrow x=1-2=-1$

Answer: $(-1, 1, 0)$

39. $\begin{cases} x-2y+5z=2 & \text{Equation 1} \\ 4x-z=0 & \text{Equation 2} \end{cases}$

$\begin{cases} x-2y+5z=2 \\ 8y-21z=-8 & -4\text{ Eq. }1+\text{Eq. }2 \end{cases}$

$\begin{cases} x-2y+5z=2 \\ y-\frac{21}{8}z=-1 & \frac{1}{8}\text{Eq. }2 \end{cases}$

$\begin{cases} x-\frac{1}{4}z=0 & 2\text{ Eq. }2+\text{Eq. }1 \\ y-\frac{21}{8}z=-1 \end{cases}$

Let $z=a$, then $y=\frac{21}{8}a-1$ and $x=\frac{1}{4}a$

Answer: $\left(\frac{1}{4}a, \frac{21}{8}a-1, a\right)$

41. $\begin{cases} 2x-3y+z=-2 & \text{Equation 1} \\ -4x+9y=7 & \text{Equation 2} \end{cases}$

$\begin{cases} 2x-3y+z=-2 \\ 3y+2z=3 & 2\text{ Eq. }1+\text{Eq. }2 \end{cases}$

$\begin{cases} 2x+3z=1 & \text{Eq.}2+\text{Eq.}1 \\ 3y+2z=3 \end{cases}$

Let $z=a$, then:

$y=-\dfrac{2}{3}a+1$

$x=-\dfrac{3}{2}a+\dfrac{1}{2}$

Answer: $\left(-\dfrac{3}{2}a+\dfrac{1}{2}, -\dfrac{2}{3}a+1, a\right)$

43. There are an infinite number of linear systems that have $(3, -4, 2)$ as their solution.

One possible system is:

$\begin{matrix} 3+(-4)+2=1 \\ 2(3)+(-4)+2=4 \\ 3+(-4)-3(2)=-7 \end{matrix} \Rightarrow \begin{cases} x+y+z=1 \\ 2x+y+z=4 \\ x+y-3z=-7 \end{cases}$

45. There are an infinite number of linear systems that have $\left(-6, -\frac{1}{2}, -\frac{7}{4}\right)$ as their solution.

One possible system is:

$\begin{matrix} -6+\left(-\frac{1}{2}\right)+2\left(-\frac{7}{4}\right)=-10 \\ -(-6)+12\left(-\frac{1}{2}\right)+8\left(-\frac{7}{4}\right)=-14 \\ -6+14\left(-\frac{1}{2}\right)-4\left(-\frac{7}{4}\right)=-6 \end{matrix} \Rightarrow \begin{cases} x+y+2z=-10 \\ -x+12y+8z=-14 \\ x+14y-4z=-6 \end{cases}$

47. $2x+3y+4z=12$

$(6, 0, 0), (0, 4, 0), (0, 0, 3), (4, 0, 1)$

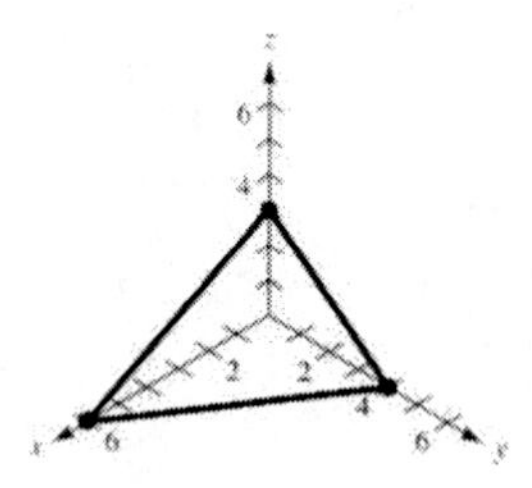

49. $2x+y+z=4$

$(2, 0, 0), (0, 4, 0), (0, 0, 4), (0, 2, 2)$

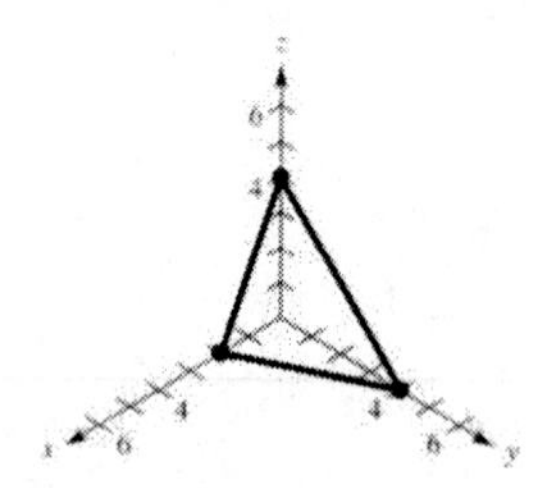

51. $\dfrac{7}{x^2-14x}=\dfrac{7}{x(x-14)}=\dfrac{A}{x}+\dfrac{B}{x-14}$

53. $\dfrac{12}{x^3-10x^2}=\dfrac{12}{x^2(x-10)}=\dfrac{A}{x}+\dfrac{B}{x^2}+\dfrac{C}{x-10}$

55. $\dfrac{4x^2+3}{(x-5)^3}=\dfrac{A}{(x-5)}+\dfrac{B}{(x-5)^2}+\dfrac{C}{(x-5)^3}$

57. $\dfrac{1}{x^2-1}=\dfrac{A}{x+1}+\dfrac{B}{x-1}$

$1=A(x-1)+B(x+1)=(A+B)x+(B-A)$

$\begin{cases} A+B=0 \\ -A+B=1 \end{cases}$

$2B=1 \Rightarrow B=\frac{1}{2} \Rightarrow A=-\frac{1}{2}$

$\dfrac{1}{x^2-1}=\dfrac{-1/2}{x+1}+\dfrac{1/2}{x-1}=\dfrac{1}{2}\left(\dfrac{1}{x-1}-\dfrac{1}{x+1}\right)$

59. $\dfrac{1}{x^2+x}=\dfrac{1}{x(x+1)}=\dfrac{A}{x}+\dfrac{B}{x+1}$

$1=A(x+1)+Bx=(A+B)x+A$

$\begin{cases} A+B=0 \\ \quad A=1 \Rightarrow B=-1 \end{cases}$

$\dfrac{1}{x^2-x}=\dfrac{1}{x}+\dfrac{-1}{x+1}=\dfrac{1}{x}-\dfrac{1}{x+1}$

61. $\dfrac{5-x}{2x^2+x-1}=\dfrac{5-x}{(2x-1)(x+1)}$

$=\dfrac{A}{2x-1}+\dfrac{B}{x+1}$

$5-x=A(x+1)+B(2x-1)$

$=(A+2B)x+(A-B)$

$\begin{cases} A+2B=-1 \quad \Rightarrow A=-1-2B \\ A-\ \ B=\ 5 \end{cases}$

$(-1-2B)-B=5 \Rightarrow B=-2$ and $A=3$

$\dfrac{5-x}{2x^2+x-1}=\dfrac{3}{2x-1}+\dfrac{-2}{x+1}$

63. $\dfrac{x^2+12x+12}{x^3-4x}=\dfrac{x^2+12x+12}{x(x-2)(x+2)}=\dfrac{A}{x}+\dfrac{B}{x+2}+\dfrac{C}{x-2}$

$x^2+12x+12=A(x+2)(x-2)+Bx(x-2)+Cx(x+2)$

$=(A+B+C)x^2+(-2B+2C)x+(-4A)$

$\begin{cases} A+B+C=1 \\ \quad -2B+2C=12 \\ -4A \qquad\quad =12 \Rightarrow A=-3 \end{cases}$

$\begin{cases} B+C=4 \\ -B+C=6 \end{cases}$

$2C=10 \Rightarrow C=5 \Rightarrow B=-1$

$\dfrac{x^2+12x+12}{x^3-4x}=\dfrac{-3}{x}+\dfrac{-1}{x+2}+\dfrac{5}{x-2}$

65. $\dfrac{4x^2+2x-1}{x^2(x+1)}=\dfrac{A}{x}+\dfrac{B}{x^2}+\dfrac{C}{x+1}$

$4x^2+2x-1=Ax(x+1)+B(x+1)+Cx^2$

$=(A+C)x^2+(A+B)x+B$

$\begin{cases} A \qquad +C=\ 4 \\ A+B \qquad =\ 2 \\ \quad B \qquad =-1 \end{cases}$

$B=-1 \Rightarrow A=3 \Rightarrow C=1$

$\dfrac{4x^2+2x-1}{x^2(x+1)}=\dfrac{3}{x}+\dfrac{-1}{x^2}+\dfrac{1}{x+1}$

67. $\dfrac{2x^2-x^2+x+5}{x^2+3x+2}=2x-7+\dfrac{18x+19}{(x+1)(x+2)}$

$\dfrac{18x+19}{(x+1)(x+2)}=\dfrac{A}{x+1}+\dfrac{B}{x+2}$

$18x+19=A(x+2)+B(x+1)$

$=(A+B)x+(2A+B)$

$\begin{cases} A+B=18 \\ 2A+B=19 \end{cases}$

$A=1 \Rightarrow B=17$

$\dfrac{2x^3-x^2+x+5}{x^2+3x+2}=2x-7+\dfrac{1}{x+1}+\dfrac{17}{x+2}$

69. $\dfrac{x^4}{(x-1)^3}=x+3+\dfrac{6x^2-8x+3}{(x-1)^3}$

$\dfrac{6x^2-8x+3}{(x-1)^3}=\dfrac{A}{x-1}+\dfrac{B}{(x-1)^2}+\dfrac{C}{(x-1)^3}$

$6x^2-8x+3=A(x-1)^2+B(x-1)+C$

$=Ax^2+(-2A+B)x+(A-B+C)$

$\begin{cases} A \qquad\quad =\ 6 \\ -2A+B \quad =-8 \\ \ A-B+C=\ 3 \end{cases}$

$A=6 \Rightarrow B=-8+2(6)=4 \Rightarrow C=3-6+4=1$

$\dfrac{x^4}{(x-1)^3}=\dfrac{6}{x-1}+\dfrac{4}{(x-1)^2}+\dfrac{1}{(x-1)^3}+x+3$

71. $\dfrac{x-12}{x(x-4)} = \dfrac{A}{x} + \dfrac{B}{x-4}$

$x - 12 = A(x-4) + Bx$

$\begin{cases} A + B = 1 \\ -4A = -12 \end{cases} \Rightarrow A = 3,\ B = -2$

$\dfrac{x-12}{x(x-4)} = \dfrac{3}{x} - \dfrac{2}{x-4}$

$y = \dfrac{x-12}{x(x-4)} \quad y = \dfrac{3}{x},\ y = -\dfrac{2}{x-4}$

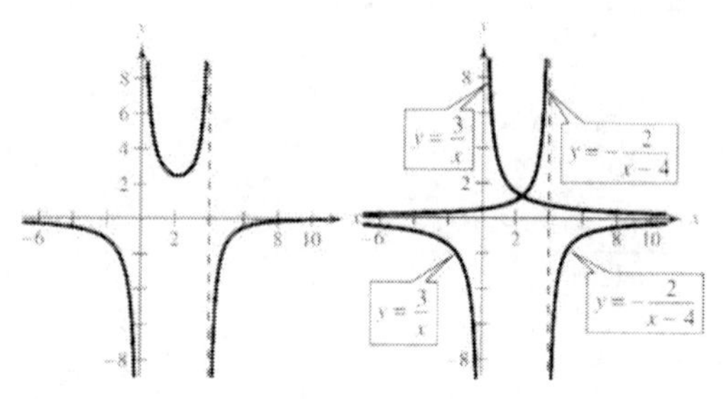

Vertical asymptotes: $x = 0$ and $x = 4$

Vertical asymptotes: $x = 0$ and $x = 4$

The combination of the vertical asymptotes of the terms of the decompositions are the same as the vertical asymptotes of the rational function.

73. $s = \frac{1}{2}at^2 + v_0t + s_0$

$(1, 128), (2, 80), (3, 0)$

$\begin{cases} 128 = \dfrac{1}{2}a + v_0 + s_0 \Rightarrow a + 2v_0 + 2s_0 = 256 \\ 80 = 2a + 2v_0 + s_0 \Rightarrow 2a + 2v_0 + s_0 = 80 \\ 0 = \dfrac{9}{2}a + 3v_0 + s_0 \Rightarrow 9a + 6v_0 + 2s_0 = 0 \end{cases}$

Solving the system, $a = -32, v_0 = 0, s_0 = 144$.

Thus, $s = \dfrac{1}{2}(-32)t^2 + (0)t + 144$

$= -16t^2 + 144.$

75. $s = \frac{1}{2}at^2 + v_0t + s_0$

$(1, 352), (2, 272), (3, 160)$

$\begin{cases} 352 = \frac{1}{2}a + v_0 + s_0 \Rightarrow a + 2v_0 + 2s_0 = 704 \\ 272 = 2a + 2v_0 + s_0 \Rightarrow 2a + 2v_0 + s_0 = 272 \\ 160 = \frac{9}{2}a + 3v_0 + s_0 \Rightarrow 9a + 6v_0 + 2s_0 = 320 \end{cases}$

Solving the system, $a = -32,\ v_0 = -32,\ s_0 = 400$.

Thus, $s = \frac{1}{2}(-32)t^2 - 32t + 400$

$= -16t^2 - 32t + 400.$

77. $y = ax^2 + bx + c$ passing through $(0, 0), (2, -2), (4, 0)$

$\begin{cases} (0,\ 0):\ 0 = 4a + 2b + c \Rightarrow c = -4a - 2b \\ (2,\ -2): -2 = 4a + 2b + c \Rightarrow -1 = 2a + b \\ (4,\ 0):\ 0 = 16a + 4b + c \Rightarrow 0 = 4a + b \end{cases}$

Answer: $a = \frac{1}{2},\ b = -2,\ c = 0$

The equation of the parabola is $y = \frac{1}{2}x^2 - 2x$.

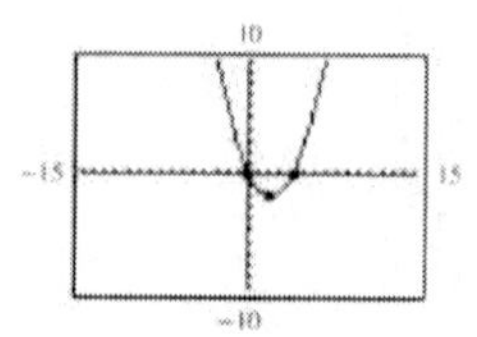

79. $y = ax^2 + bx + c$ passing through $(2, 0), (3, -1), (4, 0)$

$\begin{cases} (2,\ 0):\ 0 = 4a + 2b + c \Rightarrow c = -4a - 2b \\ (3,\ -1):\ -1 = 9a + 3b + c \Rightarrow -1 = 5a + b \\ (4,\ 0):\ 0 = 16a + 4b + c \Rightarrow 0 = 12a + 2b \end{cases}$

Answer: $a = 1, b = -6, c = 8$

The equation of the parabola is $y = x^2 - 6x + 8$.

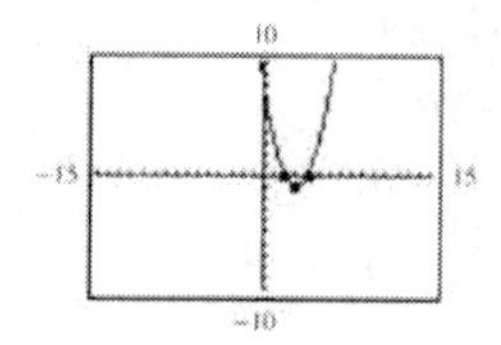

81. $x^2 + y^2 + Dx + Ey + F = 0$ passing through $(0, 0), (5, 5), (10, 0)$

$\begin{cases} (0, 0):\ F = 0 \Rightarrow F = 0 \\ (5, 5):\ 25 + 25 + 5D + 5E + F = 0 \Rightarrow 5D + 5E = -50 \\ (10, 0):\ 100 + 10D + F = 0 \Rightarrow 10D = -100 \end{cases}$

$10D = -100 \Rightarrow D = -10$

$5(-10) + 5E = -50 \Rightarrow E = 0$

The equation of the circle is $x^2 + y^2 - 10x = 0$.

To graph, solve for y.

$x^2 + y^2 - 10x = 0$

$y^2 = -x^2 + 10x$

$y = \pm\sqrt{-x^2 + 10x}$

Let $y_1 = \sqrt{-x^2 + 10x}$ and $y_2 = \sqrt{-x^2 + 10x}$.

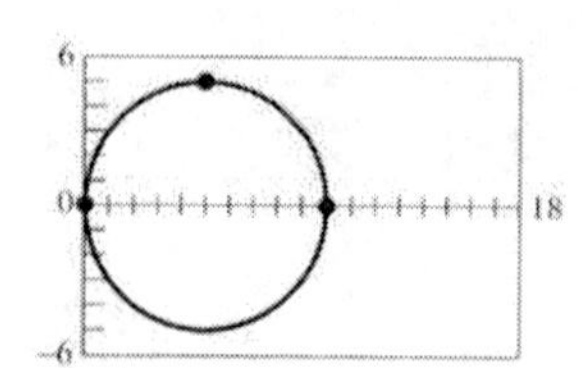

83. $x^2 + y^2 + Dx + Ey + F = 0$ passes through $(-3, -1)$, $(2, 4)$, $(-6, 8)$.

$(-3, -1)$: $10 - 3D - E + F = 0 \Rightarrow 10 = 3D + E - F$

$(2, 4)$: $20 + 2D + 4E + F = 0 \Rightarrow 20 = -2D - 4E - F$

$(-6, 8)$: $100 - 6D + 8E + F = 0 \Rightarrow 100 = 6D - 8E - F$

Answer: $D = 6$, $E = -8$, $F = 0$

The equation of the circle is $x^2 + y^2 + 6x - 8y = 0$. To graph, complete the squares first, then solve for y.

$$(x^2 + 6x + 9) + (y^2 - 8y + 16) = 0 + 9 + 16$$
$$(x+3)^2 + (y-4)^2 = 25$$
$$(y-4)^2 = 25 - (x+3)^2$$
$$y - 4 = \pm\sqrt{25 - (x+3)^2}$$
$$y = 4 \pm \sqrt{25 - (x+3)^2}$$

Let $y_1 = 4 + \sqrt{25 - (x+3)^2}$ and $y_2 = 4 - \sqrt{25 - (x+3)^2}$.

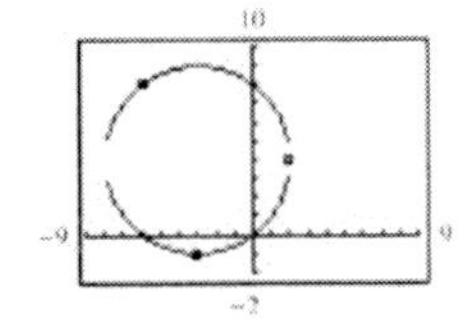

85. Let x = amount invested at 8%,
y = amount invested at 9%, and
z = amount invested at 10%.

$$\begin{cases} x + y + z = 775{,}000 \\ 0.08x + 0.09y + 0.1z = 67{,}500 \\ x - 4z = 0 \end{cases}$$

Solving the system, $x = \$300{,}000$, $y = \$400{,}000$, and $z = \$75{,}000$.

87. Let C = amount in certificates of deposit.
Let M = amount in muncipal bonds.
Let B = amount in blue-chip stocks.
Let G = amount in growth stocks.

$$\begin{cases} C + M + B + G = 500{,}000 \\ 0.03C + 0.05M + 0.08B + 0.1G = 0.05(500{,}000) \\ B + G = \frac{1}{4}(500{,}000) \end{cases}$$

The system has infinitely many solutions.
Let $G = s$, then $B = 125{,}000 - s$
$M = 187{,}500 - s$
$C = 187{,}500 + s$

Answer: $(187{,}500 + s,\ 187{,}500 - s,\ 125{,}000 - s,\ s)$

One possible solution: Let $s = \$100{,}000$.
Certification of deposit: \$287,500
Municipal bonds: \$87,500
Blue-chip stocks: \$25,000
Growth stocks: \$100,000

89. Let x = number of 1-point free throws.
Let y = number of 2-point baskets.
Let z = number of 3-point baskets.

$$\begin{cases} x + 2y + 3z = 53 \\ x - y = -4 \\ x - z = 3 \end{cases}$$

Solving the system, $x = 9$, $y = 13$, $z = 6$.
The University of Connecticut scored 9 free throws, 13 2-point baskets, and 6 3-point baskets to score a total of 53 points.

91. $\begin{cases} I_1 - I_2 + I_3 = 0 & \text{Equation 1} \\ 3I_1 + 2I_2 = 7 & \text{Equation 2} \\ 2I_2 + 4I_3 = 8 & \text{Equation 3} \end{cases}$

$$\begin{cases} I_1 - I_2 + I_3 = 0 \\ 5I_2 - 3I_3 = 7 & -3\text{ Eq. } 1 + \text{Eq. } 2 \\ 2I_2 + 4I_3 = 8 \end{cases}$$

$$\begin{cases} I_1 - I_2 + I_3 = 0 \\ 10I_2 - 6I_3 = 14 & 2\text{ Eq. } 2 \\ 10I_2 + 20I_3 = 40 & 5\text{ Eq. } 3 \end{cases}$$

$$\begin{cases} I_1 - I_2 + I_3 = 0 \\ 10I_2 - 6I_3 = 14 \\ 26I_3 = 26 & -\text{Eq. } 2 + \text{Eq. } 3 \end{cases}$$

$$26I_3 = 26 \Rightarrow I_3 = 1$$
$$10I_2 - 6(1) = 14 \Rightarrow I_2 = 2$$
$$I_1 - 2 + 1 = 0 \Rightarrow I_1 = 1$$

Answer: $I_1 = 1$ ampere, $I_2 = 2$ amperes, $I_3 = 1$ ampere

93. Least squares regression parabola through $(-4, 5)$, $(-2, 6)$, $(2, 6)$, $(4, 2)$

$$\begin{cases} 4c + 40a = 19 \\ 40b = -12 \\ 40c + 544a = 160 \end{cases}$$

Solving the system, $a = -\frac{5}{24}$, $b = -\frac{3}{10}$, and $c = \frac{41}{6}$.

Thus, $y = -\frac{5}{24}x^2 - \frac{3}{10}x + \frac{41}{6}$.

95. Least squares regression parabola through $(0, 0), (2, 2), (3, 6), (4, 12)$

$$\begin{cases} 4c + 9b + 29a = 20 \\ 9c + 29b + 99a = 70 \\ 29c + 99b + 353a = 254 \end{cases}$$

Solving the system, $a = 1$, $b = -1$, and $c = 0$.

Thus, $y = x^2 - x$.

97. (a) $\begin{cases} a(30)^2 + b(30) + c = 55 \\ a(40)^2 + b(40) + c = 105 \\ a(50)^2 + b(50) + c = 188 \end{cases}$

Solving the system, $a = 0.165$, $b = -6.55$, and $c = 103$.

$y = 0.165x^2 - 6.55x + 103$

(b)

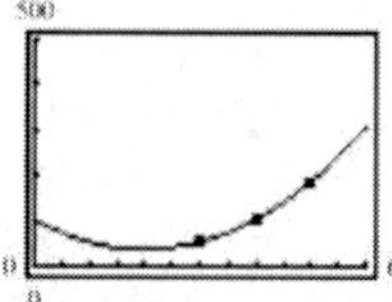

(c) For $x = 70$, $y = 453$ feet.

99. (a) $\dfrac{2000(4-3x)}{(11-7x)(7-4x)} = \dfrac{A}{(11-7x)} + \dfrac{B}{(7-4x)}, \; 0 \le x \le 1$

$2000(4-3x) = A(7-4x) + B(11-7x)$

$\begin{cases} -6000 = -4A - 7B \\ 8000 = 7A + 11B \end{cases} \Rightarrow \begin{matrix} A = -2000 \\ B = 2000 \end{matrix}$

$\dfrac{2000(4-3x)}{(11-7x)(7-4x)} = \dfrac{-2000}{11-7x} + \dfrac{2000}{7-4x}$

$= \dfrac{2000}{7-4x} - \dfrac{2000}{11-7x}$

(b) $y_1 = \dfrac{2000}{7-4x}$

$y_2 = \dfrac{2000}{11-7x}$

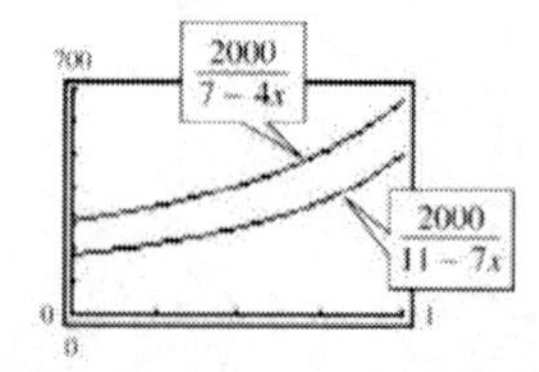

101. False, The coefficient of y in the second equation is not 1.

103. $A = -1 \Rightarrow B = 1$. No, the problem was not worked correctly. You must first divide the improper fraction.

105. No, they are not equivalent. In the second system, the constant in the second equation should be -11 and the coefficient of z in the third equation should be 2.

107. $\begin{cases} y + \lambda = 0 \\ x + \lambda = 0 \\ x + y - 10 = 0 \end{cases} \begin{matrix} \Rightarrow x = y = -\lambda \\ \\ \Rightarrow 2x - 10 = 0 \end{matrix}$

$x = 5$

$y = 5$

$\lambda = -5$

109. (a) $f(x) = x^3 + x^2 - 12x$

$= x(x^2 + x - 12) = x(x+4)(x-3)$

$\Rightarrow x = 0, \; -4, \; 3$

(b)

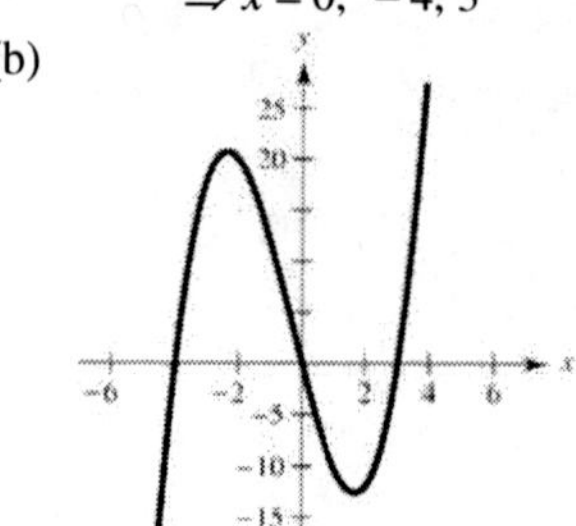

111. (a) 6

-3)

(b)

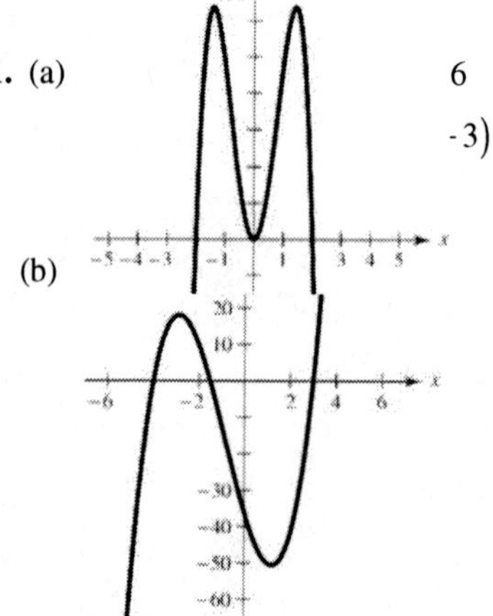

113. $4\sqrt{3}\tan\theta - 3 = 1$

$4\sqrt{3}\tan\theta = 4$

$\tan\theta = \dfrac{1}{\sqrt{3}} = \dfrac{\sqrt{3}}{3}$

$\theta = \dfrac{\pi}{6} + n\pi$, n is an integer

115. Answers will vary. (Make a Decision)

Section 7.4

1. matrix

3. Gauss-Jordan elimination

5. No, the augmented matrix for the system $\begin{cases} -2x+3y=5 \\ 6x+7y=4 \end{cases}$ is $\begin{bmatrix} -2 & 3 & \vdots & 5 \\ 6 & 7 & \vdots & 4 \end{bmatrix}$ is 2×3.

7. Since the matrix has one row and two columns, its dimension is 1×2.

9. Since the matrix has three rows and one column, its dimension is 3×1.

11. Since the matrix has two rows and two columns, its dimension is 2×2.

13. $\begin{cases} 4x-3y=-5 \\ -x+3y=12 \end{cases}$

$$\begin{bmatrix} 4 & -3 & \vdots & -5 \\ -1 & 3 & \vdots & 12 \end{bmatrix}$$

The dimension is 2×3.

15.

$$\begin{cases} x+10y-2z=2 \\ 5x-3y+4z=0 \\ 2x+y=6 \end{cases}$$

$$\begin{bmatrix} 1 & 10 & -2 & \vdots & 2 \\ 5 & -3 & 4 & \vdots & 0 \\ 2 & 1 & 0 & \vdots & 6 \end{bmatrix}$$

The dimension is 3×4.

17. $\begin{cases} 7x-5y+z=13 \\ 19x-8z=10 \end{cases}$

$$\begin{bmatrix} 7 & -5 & 1 & \vdots & 13 \\ 19 & 0 & -8 & \vdots & 10 \end{bmatrix}$$

The dimension is 2×4.

19. $\begin{bmatrix} 3 & 4 & \vdots & 9 \\ 1 & -1 & \vdots & -3 \end{bmatrix}$

$$\begin{cases} 3x+4y=9 \\ x-y=-3 \end{cases}$$

21. $\begin{bmatrix} 9 & 12 & 3 & \vdots & 0 \\ -2 & 18 & 5 & \vdots & 10 \\ 1 & 7 & -8 & \vdots & -4 \end{bmatrix}$

$$\begin{cases} 9x+12y+3z=0 \\ -2x+18y+5z=10 \\ x+7y-8z=-4 \end{cases}$$

23. Add -3 times Row 2 to Row 1.

25. Interchange Rows 1 and 2.

27. $\begin{bmatrix} 1 & 4 & 3 \\ 2 & 10 & 5 \end{bmatrix}$

$$-2R_1+R_2\rightarrow\begin{bmatrix} 1 & 4 & 3 \\ 0 & \boxed{2} & -1 \end{bmatrix}$$

29. $\begin{bmatrix} 1 & 1 & 4 & -1 \\ 3 & 8 & 10 & 3 \\ -2 & 1 & 12 & 6 \end{bmatrix}$

$$\begin{matrix} \\ -3R_1+R_2\rightarrow \\ 2R_1+R_3\rightarrow \end{matrix}\begin{bmatrix} 1 & 1 & 4 & -1 \\ 0 & 5 & \boxed{-2} & \boxed{6} \\ 0 & 3 & \boxed{20} & \boxed{4} \end{bmatrix}$$

$$\tfrac{1}{5}R_2\rightarrow\begin{bmatrix} 1 & 1 & 4 & -1 \\ 0 & 1 & -\frac{2}{5} & \frac{6}{5} \\ 0 & 3 & \boxed{20} & \boxed{4} \end{bmatrix}$$

31. (a) $\begin{bmatrix} -3 & 4 & \vdots & 22 \\ 6 & -4 & \vdots & -28 \end{bmatrix}$

(i) $R_2+R_1\rightarrow\begin{bmatrix} 3 & 0 & \vdots & -6 \\ 6 & -4 & \vdots & -28 \end{bmatrix}$

(ii) $-2R_1+R_2\rightarrow\begin{bmatrix} 3 & 0 & \vdots & -6 \\ 0 & -4 & \vdots & -16 \end{bmatrix}$

(iii) $-\frac{1}{4}R_2\rightarrow\begin{bmatrix} 3 & 0 & \vdots & -6 \\ 0 & 1 & \vdots & 4 \end{bmatrix}$

(iv) $\frac{1}{3}R_1\rightarrow\begin{bmatrix} 1 & 0 & \vdots & -2 \\ 0 & 1 & \vdots & 4 \end{bmatrix}$

Answer: $x=-2,\ y=4$

(b) $\begin{cases} -3x+4y=22 & \text{Equation 1} \\ 6x-4y=-28 & \text{Equation 2} \end{cases}$

Add Equation 1 and Equation 2 to eliminate y:

$$3x=-6$$
$$x=-2$$

Substitute $x=-2$ into Equation 1:

$$-3(-2)+4y=22$$
$$4y=16$$
$$y=4$$

Answer: $(-2, 4)$

(c) Answers will vary.

33. (i)

```
row+([A],2,1)→[B
]
       [[3 0  -6 ]
        [6 -4 -28]]
```

(ii)

```
*row+(-2,[B],1,2
)→[C]
       [[3 0  -6 ]
        [0 -4 -16]]
```

(iii)

```
*row(-1/4,[C],2)
→[D]
          [[3 0 -6]
           [0 1 4 ]]
```

(iv)

```
*row(1/3,[D],1)→
[E]
          [[1 0 -2]
           [0 1 4 ]]
```

35. $\begin{bmatrix} 1 & 0 & 0 & 0 \\ 0 & 1 & 1 & 5 \\ 0 & 0 & 0 & 0 \end{bmatrix}$

This matrix is in reduced row-echelon form.

37. $\begin{bmatrix} 3 & 0 & 3 & 7 \\ 0 & -2 & 0 & 4 \\ 0 & 0 & 1 & 5 \end{bmatrix}$

The first nonzero entries in rows one and two are not one. The matrix is not in row-echelon form.

39. $\begin{bmatrix} 1 & 0 & 0 & 1 \\ 0 & 1 & 0 & -1 \\ 0 & 0 & 0 & 2 \end{bmatrix}$

The first nonzero entry in row three is two, not one. The matrix is not in row-echelon form.

41. $\begin{bmatrix} 1 & 2 & 3 & 0 \\ -1 & 4 & 0 & -5 \\ 2 & 6 & 3 & 10 \end{bmatrix}$

$$\begin{matrix} \\ R_1 + R_2 \rightarrow \\ -2R_1 + R_3 \rightarrow \end{matrix} \begin{bmatrix} 1 & 2 & 3 & 0 \\ 0 & 6 & 3 & -5 \\ 0 & 2 & -3 & 10 \end{bmatrix}$$

$$\begin{matrix} \\ \frac{1}{6}R_2 \rightarrow \\ \\ \end{matrix} \begin{bmatrix} 1 & 2 & 3 & 0 \\ 0 & 1 & \frac{1}{2} & -\frac{5}{6} \\ 0 & 2 & -3 & 10 \end{bmatrix}$$

$$\begin{matrix} \\ \\ -2R_2 + R_3 \rightarrow \end{matrix} \begin{bmatrix} 1 & 2 & 3 & 0 \\ 0 & 1 & \frac{1}{2} & -\frac{5}{6} \\ 0 & 0 & -4 & \frac{35}{3} \end{bmatrix}$$

$$\begin{matrix} \\ \\ -\frac{1}{4}R_3 \rightarrow \end{matrix} \begin{bmatrix} 1 & 2 & 3 & 0 \\ 0 & 1 & \frac{1}{2} & -\frac{5}{6} \\ 0 & 0 & 1 & -\frac{35}{12} \end{bmatrix}$$

43. $\begin{bmatrix} 1 & -1 & -1 & 1 \\ 5 & -4 & 1 & 8 \\ -6 & 8 & 18 & 0 \end{bmatrix}$

$$\begin{matrix} \\ -5R_1 + R_2 \rightarrow \\ 6R_1 + R_3 \rightarrow \end{matrix} \begin{bmatrix} 1 & -1 & -1 & 1 \\ 0 & 1 & 6 & 3 \\ 0 & 2 & 12 & 6 \end{bmatrix}$$

$$\begin{matrix} \\ \\ -2R_2 + R_3 \rightarrow \end{matrix} \begin{bmatrix} 1 & -1 & -1 & 1 \\ 0 & 1 & 6 & 3 \\ 0 & 0 & 0 & 0 \end{bmatrix}$$

45. $\begin{bmatrix} 3 & 3 & 3 \\ -1 & 0 & -4 \\ 2 & 4 & -2 \end{bmatrix}$

$$\begin{matrix} \frac{1}{3}R_1 \rightarrow \\ \\ \\ \end{matrix} \begin{bmatrix} 1 & 1 & 1 \\ -1 & 0 & -4 \\ 2 & 4 & -2 \end{bmatrix}$$

$$\begin{matrix} \\ R_1 + R_2 \rightarrow \\ -2R_1 + R_3 \rightarrow \end{matrix} \begin{bmatrix} 1 & 1 & 1 \\ 0 & 1 & -3 \\ 0 & 2 & -4 \end{bmatrix}$$

$$\begin{matrix} -R_2 + R_1 \rightarrow \\ \\ -2R_2 + R_3 \rightarrow \end{matrix} \begin{bmatrix} 1 & 0 & 4 \\ 0 & 1 & -3 \\ 0 & 0 & 2 \end{bmatrix}$$

$$\begin{matrix} \\ \\ \frac{1}{2}R_3 \rightarrow \end{matrix} \begin{bmatrix} 1 & 0 & 4 \\ 0 & 1 & -3 \\ 0 & 0 & 1 \end{bmatrix}$$

$$\begin{matrix} -4R_3 + R_1 \rightarrow \\ 3R_3 + R_2 \rightarrow \\ \\ \end{matrix} \begin{bmatrix} 1 & 0 & 0 \\ 0 & 1 & 0 \\ 0 & 0 & 1 \end{bmatrix}$$

47. $\begin{bmatrix} -4 & 1 & 0 & 6 \\ 1 & -2 & 3 & -4 \end{bmatrix}$

$\begin{matrix} \to R_1 \to \\ \to R_2 \to \end{matrix} \begin{bmatrix} 1 & -2 & 3 & -4 \\ -4 & 1 & 0 & 6 \end{bmatrix}$

$4R_1 + R_2 \to \begin{bmatrix} 1 & -2 & 3 & -4 \\ 0 & -7 & 12 & -10 \end{bmatrix}$

$-\frac{1}{7}R_2 \to \begin{bmatrix} 1 & -2 & 3 & -4 \\ 0 & 1 & -\frac{12}{7} & \frac{10}{7} \end{bmatrix}$

$2R_2 + R_1 \to \begin{bmatrix} 1 & 0 & -\frac{3}{7} & -\frac{8}{7} \\ 0 & 1 & -\frac{12}{7} & \frac{10}{7} \end{bmatrix}$

49. $\begin{cases} x - 2y = 4 \\ \quad\; y = -3 \end{cases}$

$x = 2y + 4 = 2(-3) + 4 = -2$

Answer: $(x, y) = (-2, -3)$

51. $\begin{cases} x - y + 2z = 4 \\ \quad y - z = 2 \\ \qquad z = -2 \end{cases}$

$y - (-2) = 2$

$y = 0$

$x - 0 + 2(-2) = 4$

$x = 8$

Answer: $(8, 0, -2)$

53. $\begin{bmatrix} 1 & 0 & \vdots & 7 \\ 0 & 1 & \vdots & -5 \end{bmatrix}$

$x = 7$

$y = -5$

Answer: $(7, -5)$

55. $\begin{bmatrix} 1 & 0 & 0 & \vdots & -4 \\ 0 & 1 & 0 & \vdots & -8 \\ 0 & 0 & 1 & \vdots & 2 \end{bmatrix}$

$x = -4$

$y = -8$

$z = 2$

Answer: $(-4, -8, 2)$

57. $\begin{cases} x + 2y = 7 \\ 2x + y = 8 \end{cases}$

$\begin{bmatrix} 1 & 2 & \vdots & 7 \\ 2 & 1 & \vdots & 8 \end{bmatrix}$

$-2R_1 + R_2 \to \begin{bmatrix} 1 & 2 & \vdots & 7 \\ 0 & -3 & \vdots & -6 \end{bmatrix}$

$-\frac{1}{3}R_2 \to \begin{bmatrix} 1 & 2 & \vdots & 7 \\ 0 & 1 & \vdots & 2 \end{bmatrix}$

$y = 2$

$x + 2(2) = 7 \Rightarrow \quad x = 3$

Answer: $(-3, 2)$

59. $\begin{cases} -x + y = -22 \\ 3x + 4y = 4 \\ 4x - 8y = 32 \end{cases}$

$\begin{bmatrix} -1 & 1 & \vdots & -22 \\ 3 & 4 & \vdots & 4 \\ 4 & -8 & \vdots & 32 \end{bmatrix}$

$\begin{matrix} \\ 3R_1 + R_2 \to \\ 4R_1 + R_3 \to \end{matrix} \begin{bmatrix} -1 & 1 & \vdots & -22 \\ 0 & 7 & \vdots & -62 \\ 0 & -4 & \vdots & -56 \end{bmatrix}$

$\begin{matrix} \\ \to R_2 \to \\ \to R_3 \to \end{matrix} \begin{bmatrix} -1 & 1 & \vdots & -22 \\ 0 & -4 & \vdots & -56 \\ 0 & 7 & \vdots & -62 \end{bmatrix}$

$\begin{matrix} \\ -\frac{1}{4}R_2 \to \\ -7R_2 + R_3 \to \end{matrix} \begin{bmatrix} -1 & 1 & \vdots & -22 \\ 0 & 1 & \vdots & 14 \\ 0 & 0 & \vdots & -160 \end{bmatrix}$

No solution, inconsistent

61. $\begin{cases} 3x + 2y - z + w = 0 \\ x - y + 4z + 2w = 25 \\ -2x + y + 2z - w = 2 \\ x + y + z + w = 6 \end{cases}$

$\begin{bmatrix} 3 & 2 & -1 & 1 & \vdots & 0 \\ 1 & -1 & 4 & 2 & \vdots & 25 \\ -2 & 1 & 2 & -1 & \vdots & 2 \\ 1 & 1 & 1 & 1 & \vdots & 6 \end{bmatrix}$

$\begin{bmatrix} 1 & -1 & 4 & 2 & \vdots & 25 \\ 0 & 5 & -13 & -5 & \vdots & -75 \\ 0 & -1 & 10 & 3 & \vdots & 52 \\ 0 & 2 & -3 & -1 & \vdots & -19 \end{bmatrix}$

$\begin{bmatrix} 1 & -1 & 4 & 2 & \vdots & 25 \\ 0 & 1 & -10 & -3 & \vdots & -52 \\ 0 & 0 & 37 & 10 & \vdots & 185 \\ 0 & 0 & 17 & 5 & \vdots & 85 \end{bmatrix}$

$\begin{bmatrix} 1 & -1 & 4 & 2 & \vdots & 25 \\ 0 & 1 & -10 & -3 & \vdots & -52 \\ 0 & 0 & 37 & 10 & \vdots & 185 \\ 0 & 0 & 0 & \frac{15}{37} & \vdots & 0 \end{bmatrix}$

$w = 0,\ z = \frac{185}{37} = 5,\ y = -52 + 10(5) = -2$

$x = 25 + (-2) - 4(5) = 3$

Answer: $(3, -2, 5, 0)$

63. $\begin{cases} x \quad\quad - 3z = -2 \\ 3x + \ y - 2z = \ 5 \\ 2x + 2y + \ z = \ 4 \end{cases}$

$$\begin{bmatrix} 1 & 0 & -3 & \vdots & -2 \\ 3 & 1 & -2 & \vdots & 5 \\ 2 & 2 & 1 & \vdots & 4 \end{bmatrix}$$

$$\begin{matrix} \\ -3R_1 + R_2 \rightarrow \\ -2R_1 + R_3 \rightarrow \end{matrix} \begin{bmatrix} 1 & 0 & -3 & \vdots & -2 \\ 0 & 1 & 7 & \vdots & 11 \\ 0 & 2 & 7 & \vdots & 8 \end{bmatrix}$$

$$\begin{matrix} \\ \\ -2R_2 + R_3 \rightarrow \end{matrix} \begin{bmatrix} 1 & 0 & -3 & \vdots & -2 \\ 0 & 1 & 7 & \vdots & 11 \\ 0 & 0 & -7 & \vdots & -14 \end{bmatrix}$$

$$\begin{matrix} \\ \\ -\frac{1}{7}R_3 \rightarrow \end{matrix} \begin{bmatrix} 1 & 0 & -3 & \vdots & -2 \\ 0 & 1 & 7 & \vdots & 11 \\ 0 & 0 & 1 & \vdots & 2 \end{bmatrix}$$

$$\begin{matrix} 3R_3 + R_1 \rightarrow \\ -7R_3 + R_2 \rightarrow \\ \\ \end{matrix} \begin{bmatrix} 1 & 0 & 0 & \vdots & 4 \\ 0 & 1 & 0 & \vdots & -3 \\ 0 & 0 & 1 & \vdots & 2 \end{bmatrix}$$

Answer: $(4, -3, 2)$

65. $\begin{cases} x + y - 5z = 3 \\ x \quad\quad - 2z = 1 \\ 2x - y - z = 0 \end{cases}$

$$\begin{bmatrix} 1 & 1 & -5 & \vdots & 3 \\ 1 & 0 & -2 & \vdots & -1 \\ 2 & -1 & -1 & \vdots & 0 \end{bmatrix}$$

$$\begin{matrix} \\ -R_1 + R_2 \rightarrow \\ -2R_1 + R_3 \rightarrow \end{matrix} \begin{bmatrix} 1 & 1 & -5 & \vdots & 3 \\ 0 & -1 & 3 & \vdots & -2 \\ 0 & -3 & 9 & \vdots & -6 \end{bmatrix}$$

$$\begin{matrix} \\ \\ -3R_2 + R_3 \rightarrow \end{matrix} \begin{bmatrix} 1 & 1 & -5 & \vdots & 3 \\ 0 & -1 & 3 & \vdots & -2 \\ 0 & 0 & 0 & \vdots & 0 \end{bmatrix}$$

$$\begin{matrix} R_2 + R_1 \rightarrow \\ -R_2 \rightarrow \\ \\ \end{matrix} \begin{bmatrix} 1 & 0 & -2 & \vdots & 1 \\ 0 & 1 & -3 & \vdots & 2 \\ 0 & 0 & 0 & \vdots & 0 \end{bmatrix}$$

Let $z = a$, any real number

$y - 3a = 2 \Rightarrow y = 3a + 2$

$x - 2a = 1 \Rightarrow x = 2a + 1$

Answer: $(2a + 1, 3a + 2, a)$

67. $\begin{cases} -x + \ y - z = -14 \\ 2x - \ y + z = \ 21 \\ 3x + 2y + z = \ 19 \end{cases}$

$$\begin{bmatrix} -1 & 1 & -1 & \vdots & -14 \\ 2 & -1 & 1 & \vdots & 21 \\ 3 & 2 & 1 & \vdots & 19 \end{bmatrix}$$

$$\begin{matrix} \\ 2R_1 + R_2 \rightarrow \\ 3R_1 + R_3 \rightarrow \end{matrix} \begin{bmatrix} -1 & 1 & -1 & \vdots & -14 \\ 0 & 1 & -1 & \vdots & -7 \\ 0 & 5 & -2 & \vdots & -23 \end{bmatrix}$$

$$\begin{matrix} -R_1 \rightarrow \\ \\ -5R_2 + R_3 \rightarrow \end{matrix} \begin{bmatrix} 1 & -1 & 1 & \vdots & 14 \\ 0 & 1 & -1 & \vdots & -7 \\ 0 & 0 & 3 & \vdots & 12 \end{bmatrix}$$

$$\begin{matrix} \\ \\ \frac{1}{3}R_3 \rightarrow \end{matrix} \begin{bmatrix} 1 & -1 & 1 & \vdots & 14 \\ 0 & 1 & -1 & \vdots & -7 \\ 0 & 0 & 1 & \vdots & 4 \end{bmatrix}$$

$$\begin{matrix} -R_3 + R_1 \rightarrow \\ R_3 + R_2 \rightarrow \\ \\ \end{matrix} \begin{bmatrix} 1 & -1 & 0 & \vdots & 10 \\ 0 & 1 & 0 & \vdots & -3 \\ 0 & 0 & 1 & \vdots & 4 \end{bmatrix}$$

$$\begin{matrix} R_2 + R_1 \rightarrow \\ \\ \\ \end{matrix} \begin{bmatrix} 1 & 0 & 0 & \vdots & 7 \\ 0 & 1 & 0 & \vdots & -3 \\ 0 & 0 & 1 & \vdots & 4 \end{bmatrix}$$

Answer: $(7, -3, 4)$

69. $\begin{cases} 3x + 3y + 12z = \ 6 \\ x + \ y + \ 4z = \ 2 \\ 2x + 5y + 20z = 10 \\ -x + 2y + \ 8z = \ 4 \end{cases}$

$$\begin{bmatrix} 3 & 3 & 12 & \vdots & 6 \\ 1 & 1 & 4 & \vdots & 2 \\ 2 & 5 & 20 & \vdots & 10 \\ -1 & 2 & 8 & \vdots & 4 \end{bmatrix} \Rightarrow \begin{bmatrix} 1 & 0 & 0 & \vdots & 0 \\ 0 & 1 & 4 & \vdots & 2 \\ 0 & 0 & 0 & \vdots & 0 \\ 0 & 0 & 0 & \vdots & 0 \end{bmatrix}$$

Let $z = a$, any real number

$y = -4a + 2$

$x = 0$

Answer: $(0, -4a + 2, a)$

71. $\begin{cases} 2x + 10y + 2z = \ 6 \\ x + \ 5y + 2z = \ 6 \\ x + \ 5y + \ z = \ 3 \\ -3x - 15y - 3z = -9 \end{cases}$

$$\begin{bmatrix} 2 & 10 & 2 & \vdots & 6 \\ 1 & 5 & 2 & \vdots & 6 \\ 1 & 5 & 1 & \vdots & 3 \\ -3 & -15 & -3 & \vdots & -9 \end{bmatrix} \Rightarrow \begin{bmatrix} 1 & 5 & 0 & \vdots & 0 \\ 0 & 0 & 1 & \vdots & 3 \\ 0 & 0 & 0 & \vdots & 0 \\ 0 & 0 & 0 & \vdots & 0 \end{bmatrix}$$

$z = 3, y = a, x = -5a$

Answer: $(-5a, a, 3)$

73. (a) $z=-3;\ y=5(-3)+16=1; x=2(1)-(-3)-6=-1$

Answer: $(-1,\ 1,\ -3)$

(b) $z=-3,\ y=-3(-3)-8=1,\ x=-1+2(-3)+6=-1$

Answer: $(-1,\ 1,\ -3)$

Yes, the systems yield the same solutions.

75. (a) $z=8,\ y=7(8)-54=2, x=4(2)-5(8)+27=-5$

Answer: $(-5,\ 2, 8)$

(b) $z=8,\ y=-5(8)+42=2,\ x=6(2)-8+15=19$

Answer: $(19,\ 2, 8)$

No, solutions are different.

77. $f(x)=ax^2+bx+c$

$$\begin{cases} f(1)=a+b+c=8 \\ f(2)=4a+2b+c=13 \\ f(3)=9a+3b+c=20 \end{cases}$$

$$\begin{bmatrix} 1 & 1 & 1 & \vdots & 8 \\ 4 & 2 & 1 & \vdots & 13 \\ 9 & 3 & 1 & \vdots & 20 \end{bmatrix}$$

$$\begin{matrix} \\ -4R_1+R_2 \to \\ -9R_1+R_3 \to \end{matrix} \begin{bmatrix} 1 & 1 & 1 & \vdots & 8 \\ 0 & -2 & -3 & \vdots & -19 \\ 0 & -6 & -8 & \vdots & -52 \end{bmatrix}$$

$$\begin{matrix} \\ -\frac{1}{2}R_2 \to \\ -3R_2+R_3 \to \end{matrix} \begin{bmatrix} 1 & 1 & 1 & \vdots & 8 \\ 0 & 1 & \frac{3}{2} & \vdots & \frac{19}{2} \\ 0 & 0 & 1 & \vdots & 5 \end{bmatrix}$$

$$c=5$$
$$b+\tfrac{3}{2}(5)=\tfrac{19}{2} \Rightarrow \quad b=2$$
$$a+2+5=8 \Rightarrow \quad a=1$$

Answer: $y=x^2+2x+5$

79. $f(x)=ax^2+bx+c$

$$\begin{cases} f(1)=a+b+c=2 \\ f(-2)=4a-2b+c=11 \\ f(3)=9a+3b+c=16 \end{cases}$$

$$\begin{bmatrix} 1 & 1 & 1 & \vdots & 2 \\ 4 & -2 & 1 & \vdots & 11 \\ 9 & 3 & 1 & \vdots & 16 \end{bmatrix}$$

$$\begin{matrix} \\ -4R_1+R_2 \to \\ -9R_1+R_3 \to \end{matrix} \begin{bmatrix} 1 & 1 & 1 & \vdots & 2 \\ 0 & -6 & -3 & \vdots & 3 \\ 0 & -6 & -8 & \vdots & -2 \end{bmatrix}$$

$$\begin{matrix} \\ \\ (-1)R_2+R_3 \to \end{matrix} \begin{bmatrix} 1 & 1 & 1 & \vdots & 2 \\ 0 & -6 & -3 & \vdots & 3 \\ 0 & 0 & -5 & \vdots & -5 \end{bmatrix}$$

$$-5c=-5 \Rightarrow c=1$$
$$-6b-3c=3 \Rightarrow -6b=3+3=6 \Rightarrow b=-1$$
$$a+b+c=2 \Rightarrow a=2+1-1=2$$

Answer: $y=2x^2-x+1$

81. $f(x)=ax^2+bx+c$

$f(-2)=4a-2b+c=-15$

$f(-1)=\ a-\ b+c=\ 7$

$f(1)=\ a+\ b+c=-3$

Solving the system, $a=-9,\ b=-5,\ c=11$.

$f(x)=-9x^2-5x+11$

83. $f(x)=ax^3+bx^2+cx+d$

$f(-2)=-8a+4b-2c+d=\ -7$

$f(-1)=\ -a+\ b-\ c+d=\ 2$

$f(1)=\ a+\ b+\ c+d=\ -4$

$f(2)=\ 8a+4b+2c+d=\ -7$

Solving the system,
$a=1,\ b=-2,\ c=-4,\ d=1$.

$f(x)=x^3-2x^2-4x+1$

85. $$\begin{cases} I_1-\ I_2+\ I_3=0 \\ 2I_1+2I_2 \qquad =7 \\ \qquad 2I_2+4I_3=8 \end{cases}$$

$$\begin{bmatrix} 1 & -1 & 1 & \vdots & 0 \\ 2 & 2 & 0 & \vdots & 7 \\ 0 & 2 & 4 & \vdots & 8 \end{bmatrix}$$

$$\begin{matrix} \\ -2R_1+R_2 \to \\ \\ \end{matrix} \begin{bmatrix} 1 & -1 & 1 & \vdots & 0 \\ 0 & 4 & -2 & \vdots & 7 \\ 0 & 2 & 4 & \vdots & 8 \end{bmatrix}$$

$$\begin{matrix} \\ R_3 \to \\ R_2 \to \end{matrix} \begin{bmatrix} 1 & -1 & 1 & \vdots & 0 \\ 0 & 2 & 4 & \vdots & 8 \\ 0 & 4 & -2 & \vdots & 7 \end{bmatrix}$$

$$\begin{matrix} \\ \frac{1}{2}R_2 \to \\ \\ \end{matrix} \begin{bmatrix} 1 & -1 & 1 & \vdots & 0 \\ 0 & 1 & 2 & \vdots & 4 \\ 0 & 4 & -2 & \vdots & 7 \end{bmatrix}$$

$$\begin{matrix} \\ \\ -4R_2+R_3 \to \end{matrix} \begin{bmatrix} 1 & -1 & 1 & \vdots & 0 \\ 0 & 1 & 2 & \vdots & 4 \\ 0 & 0 & -10 & \vdots & -9 \end{bmatrix}$$

$$\begin{matrix} \\ \\ -\frac{1}{10}R_3 \to \end{matrix} \begin{bmatrix} 1 & -1 & 1 & \vdots & 0 \\ 0 & 1 & 2 & \vdots & 4 \\ 0 & 0 & 1 & \vdots & \frac{9}{10} \end{bmatrix}$$

$I_3=\frac{9}{10}$ amperes; $I_2+2\left(\frac{9}{10}\right)=4 \Rightarrow I_2=\frac{11}{5}$ amperes;
$I_1-\frac{11}{5}+\frac{9}{10}=0 \Rightarrow I_1=\frac{13}{10}$ amperes

87. Let x = number of \$1 bills.
Let y = number of \$5 bills.
Let z = number of \$10 bills.
Let w = number of \$20 bills.

$$\begin{cases} x + 5y + 10z + 20w = 95 \\ x + y + z + w = 26 \\ y - 4z = 0 \\ x - 2y = -1 \end{cases}$$

$$\left[\begin{array}{cccc|c} 1 & 5 & 10 & 20 & 95 \\ 1 & 1 & 1 & 1 & 26 \\ 0 & 1 & -4 & 0 & 0 \\ 1 & -2 & 0 & 0 & -1 \end{array}\right] \Rightarrow \left[\begin{array}{cccc|c} 1 & 0 & 0 & 0 & 15 \\ 0 & 1 & 0 & 0 & 8 \\ 0 & 0 & 1 & 0 & 2 \\ 0 & 0 & 0 & 1 & 1 \end{array}\right]$$

$x = 15$
$y = 8$
$z = 2$
$w = 1$

The server has 15 \$1 bills, 8 \$5 bills, 2 \$10 bills, and one \$20 bill.

89. $$\frac{8x^2}{(x-1)^2(x+1)} = \frac{A}{x+1} + \frac{B}{x-1} + \frac{C}{(x-1)^2}$$

$$8x^2 = A(x-1)^2 + B(x-1)(x+1) + C(x+1)$$
$$8x^2 = A(x^2 - 2x + 1) + B(x^2 - 1) + C(x+1)$$
$$8x^2 = (A+B)x^2 + (-2A + C)x + (A - B + C)$$

System of equations: $\begin{cases} A + B = 8 \\ -2A + C = 0 \\ A - B + C = 0 \end{cases}$

$$\left[\begin{array}{ccc|c} 1 & 1 & 0 & 8 \\ -2 & 0 & 1 & 0 \\ 1 & -1 & 1 & 0 \end{array}\right] \xrightarrow{\text{rref}} \left[\begin{array}{ccc|c} 1 & 0 & 0 & 2 \\ 0 & 1 & 0 & 6 \\ 0 & 0 & 1 & 4 \end{array}\right]$$

$A = 2,\ B = 6,\ C = 4$

$$\frac{8x^2}{(x-1)^2(x+1)} = \frac{2}{x+1} + \frac{6}{x-1} + \frac{4}{(x-1)^2}$$

91. (a) $f(t) = at^2 + bt + c$
$f(6) = 36a + 6b + c = 65.82$
$f(7) = 49a + 7b + c = 68.77$
$f(8) = 64a + 8b + c = 71.70$

$$\left[\begin{array}{ccc|c} 36 & 6 & 1 & 65.82 \\ 49 & 7 & 1 & 68.77 \\ 64 & 8 & 1 & 71.70 \end{array}\right] \Rightarrow \left[\begin{array}{ccc|c} 1 & 0 & 0 & -0.01 \\ 0 & 1 & 0 & 3.08 \\ 0 & 0 & 1 & 42.7 \end{array}\right]$$

$y = -0.01t^2 + 3.08t + 4.77$

(b)

(c) 2010:
$t = 10 \Rightarrow y = -0.01(10)^2 + 3.08(10) + 47.7 = \$77.50$
2015:
$t = 15 \Rightarrow y = -0.01(15)^2 + 3.08(15)^2 + 47.7 = \$91.65$

2020:
$t = 20 \Rightarrow y = -0.01(20)^2 + 3.08(20)^2 + 47.7 = \$105.50$

(d) Answers will vary.

93. (a) $$\begin{cases} x_1 + x_3 = 600 \\ x_1 - x_2 - x_4 = 0 \\ x_2 + x_5 = 500 \\ x_3 + x_6 = 600 \\ x_4 - x_6 + x_7 = 0 \\ x_5 + x_7 = 500 \end{cases}$$

$$\left[\begin{array}{ccccccc|c} 1 & 0 & 1 & 0 & 0 & 0 & 0 & 600 \\ 1 & -1 & 0 & -1 & 0 & 0 & 0 & 0 \\ 0 & 1 & 0 & 0 & 1 & 0 & 0 & 500 \\ 0 & 0 & 1 & 0 & 0 & 1 & 0 & 600 \\ 0 & 0 & 0 & 1 & 0 & -1 & 1 & 0 \\ 0 & 0 & 0 & 0 & 1 & 0 & 1 & 500 \\ 0 & 0 & 0 & 0 & 0 & 0 & 0 & 0 \end{array}\right]$$

$$\Rightarrow \left[\begin{array}{ccccccc|c} 1 & 0 & 0 & 0 & 0 & -1 & 0 & 0 \\ 0 & 1 & 0 & 0 & 0 & 0 & -1 & 0 \\ 0 & 0 & 1 & 0 & 0 & 1 & 0 & 600 \\ 0 & 0 & 0 & 1 & 0 & -1 & 1 & 0 \\ 0 & 0 & 0 & 0 & 1 & 0 & 1 & 500 \\ 0 & 0 & 0 & 0 & 0 & 0 & 0 & 0 \\ 0 & 0 & 0 & 0 & 0 & 0 & 0 & 0 \end{array}\right]$$

$x_1 - x_6 = 0$
$x_2 - x_7 = 0$
$x_3 + x_6 = 600$
$x_4 - x_6 + x_7 = 0$
$x_5 + x_7 = 500$
$x_6 = s$
$x_7 = t$

$x_1 = s,\ x_2 = t,\ x_3 = 600 - s,\ x_4 = s - t,\ x_5 = 500 - t,$
$x_6 = s,\ x_7 = t$

(b) When $x_6 = 0,\ x_7 = 0.$

$x_1 = 0,\ x_2 = 0,\ x_3 = 600,\ x_4 = 0,\ x_5 = 500,\ x_6 = 0,\ x_7 = 0$

(c) When $x_5 = 400,\ x_6 = 500.$
$x_1 = 500,\ x_2 = 100,\ x_3 = 100,\ x_4 = 400,\ x_5 = 400,$
$x_6 = 500,\ x_7 = 100$

95. True. If while in the process of using Gaussian elimination, a row is obtained with entries of zero except for the last row entry, then it can be concluded that the system is inconsistent.

97. $\begin{cases} x + 3z = -2 & \text{Equation 1} \\ y + 4z = 1 & \text{Equation 2} \end{cases}$

(Equation 1) + (Equation 2) $\rightarrow$ new Equation 1
(Equation 1) + 2(Equation 2) $\rightarrow$ new Equation 2
2(Equation 1) + (Equation 2) $\rightarrow$ new Equation 3

$$\begin{cases} x+\ y+\ 7z=-1 \\ x+2y+11z=\ 0 \\ 2x+y+10z=-3 \end{cases}$$

99. No. Answers will vary.

101. $f(x)=\dfrac{7}{-x-1}$

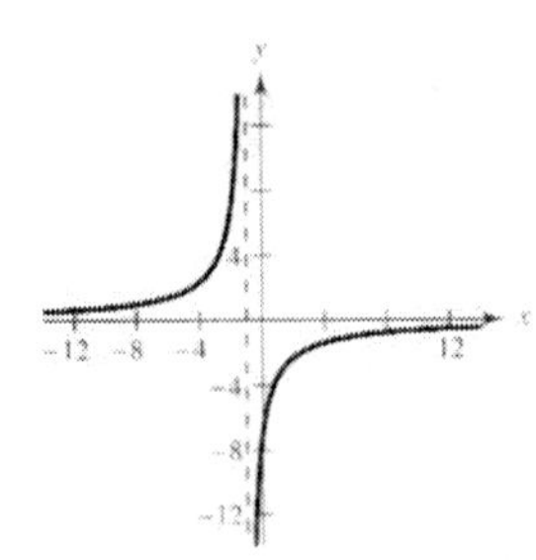

Vertical asymptote: $x=-1$
Horizontal asymptote: $y=0$

103. $f(x)=\dfrac{x^2-2x-3}{x-4}=x+2+\dfrac{5}{x-4}$

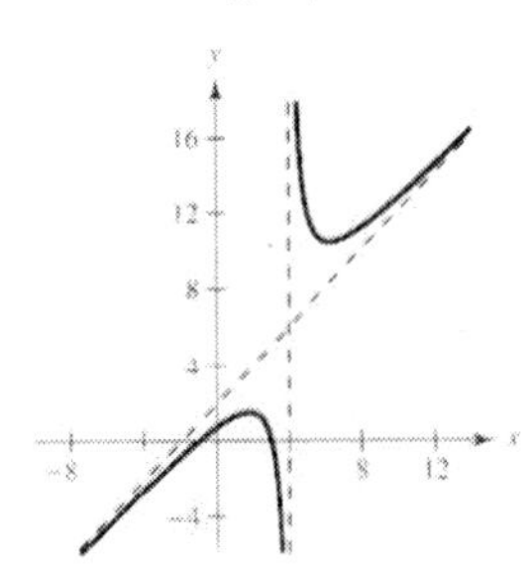

Vertical asymptote: $x=4$
Slant asymptote: $y=x+2$

Section 7.5

1. equal

3. zero, O

5. (a) iii (b) i (c) iv (d) v (e) ii

7. No. In general $AB\neq BA$; matrix multiplication is not commutative.

9. $x=5,\ y=-8$

11. $2x+7=5\Rightarrow x=-1$
$3y=12\Rightarrow y=4$
$3y-14=4\Rightarrow z=6$

13. (a) $A+B=\begin{bmatrix}5&-2\\3&1\end{bmatrix}+\begin{bmatrix}3&1\\-2&6\end{bmatrix}=\begin{bmatrix}8&-1\\1&7\end{bmatrix}$

(b) $A-B=\begin{bmatrix}5&-2\\3&1\end{bmatrix}-\begin{bmatrix}3&1\\-2&6\end{bmatrix}=\begin{bmatrix}2&-3\\5&-5\end{bmatrix}$

(c) $3A=3\begin{bmatrix}5&-2\\3&1\end{bmatrix}=\begin{bmatrix}3(5)&3(-2)\\3(3)&3(1)\end{bmatrix}=\begin{bmatrix}15&-6\\9&3\end{bmatrix}$

(d) $3A-2B=3\begin{bmatrix}5&-2\\3&1\end{bmatrix}-2\begin{bmatrix}3&1\\-2&6\end{bmatrix}$

$=\begin{bmatrix}15&-6\\9&3\end{bmatrix}+\begin{bmatrix}-6&-2\\4&-12\end{bmatrix}=\begin{bmatrix}9&-8\\13&-9\end{bmatrix}$

15. (a) $A+B=\begin{bmatrix}8&-1\\2&3\\-4&5\end{bmatrix}+\begin{bmatrix}1&6\\-1&-5\\1&10\end{bmatrix}=\begin{bmatrix}8+1&-1+6\\2-1&3-5\\-4+1&5+10\end{bmatrix}$

$=\begin{bmatrix}9&5\\1&-2\\-3&15\end{bmatrix}$

(b) $A-B=\begin{bmatrix}8&-1\\2&3\\-4&5\end{bmatrix}-\begin{bmatrix}1&6\\-1&-5\\1&10\end{bmatrix}=\begin{bmatrix}8-1&-1-6\\2+1&3+5\\-4-1&5-10\end{bmatrix}$

$=\begin{bmatrix}7&-7\\3&8\\-5&-5\end{bmatrix}$

(c) $3A=3\begin{bmatrix}8&-1\\2&3\\-4&5\end{bmatrix}=\begin{bmatrix}3(8)&3(-1)\\3(2)&3(3)\\3(-4)&3(5)\end{bmatrix}=\begin{bmatrix}24&-3\\6&9\\-12&15\end{bmatrix}$

(d) $3A-2B=\begin{bmatrix}24&-3\\6&9\\-12&15\end{bmatrix}-\begin{bmatrix}2&12\\-2&-10\\2&20\end{bmatrix}=\begin{bmatrix}22&-15\\8&19\\-14&-5\end{bmatrix}$

17. (a) $A+B=\begin{bmatrix}4 & 5 & -1 & 3 & 4\\ 1 & 2 & -2 & -1 & 0\end{bmatrix}+\begin{bmatrix}1 & 0 & -1 & 1 & 0\\ -6 & 8 & 2 & -3 & -7\end{bmatrix}$

$=\begin{bmatrix}4+1 & 5+0 & -1-1 & 3+1 & 4+0\\ 1-6 & 2+8 & -2+2 & -1-3 & 0-7\end{bmatrix}$

$=\begin{bmatrix}5 & 5 & -2 & 4 & 4\\ -5 & 10 & 0 & -4 & -7\end{bmatrix}$

(b) $A-B=\begin{bmatrix}4 & 5 & -1 & 3 & 4\\ 1 & 2 & -2 & -1 & 0\end{bmatrix}-\begin{bmatrix}1 & 0 & -1 & 1 & 0\\ -6 & 8 & 2 & -3 & -7\end{bmatrix}$

$=\begin{bmatrix}4-1 & 5-0 & -1+1 & 3-1 & 4-0\\ 1+6 & 2-8 & -2-2 & -1+3 & 0+7\end{bmatrix}$

$=\begin{bmatrix}3 & 5 & 0 & 2 & 4\\ 7 & -6 & -4 & 2 & 7\end{bmatrix}$

(c) $3A=3\begin{bmatrix}4 & 5 & -1 & 3 & 4\\ 1 & 2 & -2 & -1 & 0\end{bmatrix}$

$=\begin{bmatrix}3(4) & 3(5) & 3(-1) & 3(3) & 3(4)\\ 3(1) & 3(2) & 3(-2) & 3(-1) & 3(10)\end{bmatrix}$

$=\begin{bmatrix}12 & 15 & -3 & 9 & 12\\ 3 & 6 & -6 & -3 & 0\end{bmatrix}$

(d) $3A-2B=\begin{bmatrix}12 & 15 & -3 & 9 & 12\\ 3 & 6 & -6 & -3 & 0\end{bmatrix}-\begin{bmatrix}2 & 0 & -2 & 2 & 0\\ -12 & 16 & 4 & -6 & -14\end{bmatrix}$

$=\begin{bmatrix}10 & 15 & -1 & 7 & 12\\ 15 & -10 & -10 & 3 & 14\end{bmatrix}$

19. (a) $A+B$ is not possible.
(b) $A-B$ is not possible.

(c) $3A=3\begin{bmatrix}6 & 0 & 3\\ -1 & -4 & 0\end{bmatrix}=\begin{bmatrix}3(6) & 3(0) & 3(3)\\ 3(-1) & 3(-4) & 3(0)\end{bmatrix}$

$=\begin{bmatrix}18 & 0 & 9\\ -3 & -12 & 0\end{bmatrix}$

(d) $3A-2B$ is not possible.

21. $\begin{bmatrix}-5 & 0\\ 3 & -6\end{bmatrix}+\begin{bmatrix}7 & 1\\ -2 & -1\end{bmatrix}+\begin{bmatrix}-10 & -8\\ 14 & 6\end{bmatrix}=\begin{bmatrix}-5 & 0\\ 3 & -6\end{bmatrix}+\begin{bmatrix}-3 & -7\\ 12 & 5\end{bmatrix}$

$=\begin{bmatrix}-8 & -7\\ 15 & -1\end{bmatrix}$

23. $4\left(\begin{bmatrix}-4 & 0 & 1\\ 0 & 2 & 3\end{bmatrix}-\begin{bmatrix}2 & 1 & -2\\ 3 & -6 & 0\end{bmatrix}\right)=4\begin{bmatrix}-6 & -1 & 3\\ -3 & 8 & 3\end{bmatrix}$

$=\begin{bmatrix}-24 & -4 & 12\\ -12 & 32 & 12\end{bmatrix}$

25. $\frac{3}{7}\begin{bmatrix}2 & 5\\ -1 & -4\end{bmatrix}+6\begin{bmatrix}-3 & 0\\ 2 & 2\end{bmatrix}\approx\begin{bmatrix}-17.143 & 2.143\\ 11.571 & 10.286\end{bmatrix}$

27. $-1\begin{bmatrix}3.211 & 6.829\\ -1.004 & 4.914\\ 0.055 & -3.889\end{bmatrix}-1\begin{bmatrix}1.630 & -3.090\\ 5.256 & 8.335\\ -9.768 & 4.251\end{bmatrix}$

$=\begin{bmatrix}-4.841 & -3.739\\ -4.252 & -13.249\\ 9.713 & -0.362\end{bmatrix}$

29. $X=3A-2B=3\begin{bmatrix}-2 & -1\\ 1 & 0\\ 3 & -4\end{bmatrix}-2\begin{bmatrix}0 & 3\\ 2 & 0\\ -4 & -1\end{bmatrix}$

$=\begin{bmatrix}-6 & -3\\ 3 & 0\\ 9 & -12\end{bmatrix}-\begin{bmatrix}0 & 6\\ 4 & 0\\ -8 & -2\end{bmatrix}=\begin{bmatrix}-6 & -9\\ -1 & 0\\ 17 & -10\end{bmatrix}$

31. $X=-\frac{3}{2}A+\frac{1}{2}B=-\frac{3}{2}\begin{bmatrix}-2 & -1\\ 1 & 0\\ 3 & -4\end{bmatrix}+\frac{1}{2}\begin{bmatrix}0 & 3\\ 2 & 0\\ -4 & -1\end{bmatrix}=\begin{bmatrix}3 & 3\\ -\frac{1}{2} & 0\\ -\frac{13}{2} & \frac{11}{2}\end{bmatrix}$

33. A is 3×2 and B is $3\times 3\Rightarrow AB$ is not defined.

35. $AB=\begin{bmatrix}-1 & 6\\ -4 & 5\\ 0 & 3\end{bmatrix}\begin{bmatrix}2 & 3\\ 0 & 9\end{bmatrix}=\begin{bmatrix}-2 & 51\\ -8 & 33\\ 0 & 27\end{bmatrix}$

37. $AB=\begin{bmatrix}5 & 0 & 0\\ 0 & -8 & 0\\ 0 & 0 & 7\end{bmatrix}\begin{bmatrix}\frac{1}{5} & 0 & 0\\ 0 & -\frac{1}{8} & 0\\ 0 & 0 & \frac{1}{2}\end{bmatrix}=\begin{bmatrix}1 & 0 & 0\\ 0 & 1 & 0\\ 0 & 0 & \frac{7}{2}\end{bmatrix}$

39. $AB=\begin{bmatrix}5\\ 6\end{bmatrix}\begin{bmatrix}-3 & -1 & -5 & -9\end{bmatrix}=\begin{bmatrix}-15 & -5 & -25 & -45\\ -18 & -6 & -30 & -54\end{bmatrix}$

41. (a) $AB=\begin{bmatrix}1 & 2\\ 4 & 2\end{bmatrix}\begin{bmatrix}2 & -1\\ -1 & 8\end{bmatrix}=\begin{bmatrix}0 & 15\\ 6 & 12\end{bmatrix}$

(b) $BA=\begin{bmatrix}2 & -1\\ -1 & 8\end{bmatrix}\begin{bmatrix}1 & 2\\ 4 & 2\end{bmatrix}=\begin{bmatrix}-2 & 2\\ 31 & 14\end{bmatrix}$

(c) $A^2=AA=\begin{bmatrix}1 & 2\\ 4 & 2\end{bmatrix}\begin{bmatrix}1 & 2\\ 4 & 2\end{bmatrix}=\begin{bmatrix}9 & 6\\ 12 & 12\end{bmatrix}$

43. (a) $AB=\begin{bmatrix}3 & -1\\ 1 & 3\end{bmatrix}\begin{bmatrix}1 & -3\\ 3 & 1\end{bmatrix}=\begin{bmatrix}3-3 & -9-1\\ 1+9 & -3+3\end{bmatrix}$

$=\begin{bmatrix}0 & -10\\ 10 & 0\end{bmatrix}$

(b) $BA = \begin{bmatrix} 1 & -3 \\ 3 & 1 \end{bmatrix}\begin{bmatrix} 3 & -1 \\ 1 & 3 \end{bmatrix} = \begin{bmatrix} 3-3 & -1-9 \\ 9+1 & -3+3 \end{bmatrix}$

$= \begin{bmatrix} 0 & -10 \\ 10 & 0 \end{bmatrix}$

(c) $A^2 = AA = \begin{bmatrix} 3 & -1 \\ 1 & 3 \end{bmatrix}\begin{bmatrix} 3 & -1 \\ 1 & 3 \end{bmatrix} = \begin{bmatrix} 9-1 & -3-3 \\ 3+3 & -1+9 \end{bmatrix}$

$= \begin{bmatrix} 8 & -6 \\ 6 & 8 \end{bmatrix}$

45. (a) $AB = \begin{bmatrix} 7 \\ 8 \\ -1 \end{bmatrix}\begin{bmatrix} 1 & 1 & 2 \end{bmatrix} = \begin{bmatrix} 7 & 7 & 14 \\ 8 & 8 & 16 \\ -1 & -1 & -2 \end{bmatrix}$

(b) $BA = \begin{bmatrix} 1 & 1 & 2 \end{bmatrix}\begin{bmatrix} 7 \\ 8 \\ -1 \end{bmatrix} = \begin{bmatrix} 7+8-2 \end{bmatrix} = \begin{bmatrix} 13 \end{bmatrix}$

(c) A^2 is not defined.

47. $AB = \begin{bmatrix} 70 & -17 & 73 \\ 32 & 11 & 6 \\ 16 & -38 & 70 \end{bmatrix}$

49. $AB = \begin{bmatrix} 151 & 25 & 48 \\ 516 & 279 & 387 \\ 47 & -20 & 87 \end{bmatrix}$

51. $\left(\begin{bmatrix} 3 & 1 \\ 0 & -2 \end{bmatrix}\begin{bmatrix} 1 & 0 \\ -2 & 2 \end{bmatrix}\right)\begin{bmatrix} 1 & 0 \\ 2 & 4 \end{bmatrix} = \begin{bmatrix} 1 & 2 \\ 4 & -4 \end{bmatrix}\begin{bmatrix} 1 & 0 \\ 2 & 4 \end{bmatrix} = \begin{bmatrix} 5 & 8 \\ -4 & -16 \end{bmatrix}$

53. $\begin{bmatrix} 0 & 2 & -2 \\ 4 & 1 & 2 \end{bmatrix}\left(\begin{bmatrix} 4 & 0 \\ 0 & -1 \\ -1 & 2 \end{bmatrix} + \begin{bmatrix} -2 & 3 \\ -3 & 5 \\ 0 & -3 \end{bmatrix}\right)$

$= \begin{bmatrix} 0 & 2 & -2 \\ 4 & 1 & 2 \end{bmatrix}\begin{bmatrix} 2 & 3 \\ -3 & 4 \\ -1 & -1 \end{bmatrix} = \begin{bmatrix} -4 & 10 \\ 3 & 14 \end{bmatrix}$

55. $\begin{bmatrix} 1 & 2 & \vdots & 4 \\ 3 & 2 & \vdots & 0 \end{bmatrix}$

(a) $\begin{bmatrix} 1 & 2 \\ 3 & 2 \end{bmatrix}\begin{bmatrix} 2 \\ 1 \end{bmatrix} = \begin{bmatrix} 4 \\ 8 \end{bmatrix} \Rightarrow \begin{bmatrix} 2 \\ 1 \end{bmatrix}$ is not a solution.

(b) $\begin{bmatrix} 1 & 2 \\ 3 & 2 \end{bmatrix}\begin{bmatrix} -2 \\ 3 \end{bmatrix} = \begin{bmatrix} 4 \\ 0 \end{bmatrix} \Rightarrow \begin{bmatrix} -2 \\ 3 \end{bmatrix}$ is a solution.

(c) $\begin{bmatrix} 1 & 2 \\ 3 & 2 \end{bmatrix}\begin{bmatrix} -4 \\ 4 \end{bmatrix} = \begin{bmatrix} 4 \\ -4 \end{bmatrix} \Rightarrow \begin{bmatrix} -4 \\ 4 \end{bmatrix}$ is not a solution.

(d) $\begin{bmatrix} 1 & 2 \\ 3 & 2 \end{bmatrix}\begin{bmatrix} 2 \\ -3 \end{bmatrix} = \begin{bmatrix} -4 \\ 0 \end{bmatrix} \Rightarrow \begin{bmatrix} 2 \\ -3 \end{bmatrix}$ is not a solution.

57. $\begin{bmatrix} -2 & -3 & \vdots & -6 \\ 4 & 2 & \vdots & 20 \end{bmatrix}$

(a) $\begin{bmatrix} -2 & -3 \\ 4 & 2 \end{bmatrix}\begin{bmatrix} 3 \\ 0 \end{bmatrix} = \begin{bmatrix} -6 \\ 12 \end{bmatrix} \Rightarrow \begin{bmatrix} 3 \\ 0 \end{bmatrix}$ is not a solution.

(b) $\begin{bmatrix} -2 & -3 \\ 4 & 2 \end{bmatrix}\begin{bmatrix} 6 \\ -2 \end{bmatrix} = \begin{bmatrix} -6 \\ 20 \end{bmatrix} \Rightarrow \begin{bmatrix} 6 \\ -2 \end{bmatrix}$ is a solution.

(c) $\begin{bmatrix} -2 & -3 \\ 4 & 2 \end{bmatrix}\begin{bmatrix} -6 \\ 6 \end{bmatrix} = \begin{bmatrix} -6 \\ -12 \end{bmatrix} \Rightarrow \begin{bmatrix} -6 \\ 6 \end{bmatrix}$ is not a solution.

(d) $\begin{bmatrix} -2 & -3 \\ 4 & 2 \end{bmatrix}\begin{bmatrix} 4 \\ 2 \end{bmatrix} = \begin{bmatrix} -14 \\ 20 \end{bmatrix} \Rightarrow \begin{bmatrix} 4 \\ 2 \end{bmatrix}$ is not a solution.

59. (a) $A = \begin{bmatrix} -1 & 1 \\ -2 & 1 \end{bmatrix}, X = \begin{bmatrix} x_1 \\ x_2 \end{bmatrix}, B = \begin{bmatrix} 4 \\ 0 \end{bmatrix}$

(b) By Gauss-Jordan elimination on

$\begin{bmatrix} -1 & 1 & \vdots & 4 \\ -2 & 1 & \vdots & 0 \end{bmatrix}$

$\begin{array}{r} -R_1 \rightarrow \\ 2R_1 + R_2 \rightarrow \end{array} \begin{bmatrix} 1 & -1 & \vdots & -4 \\ 0 & -1 & \vdots & -8 \end{bmatrix}$

$\begin{array}{r} -R_2 + R_1 \rightarrow \\ -R_2 \rightarrow \end{array} \begin{bmatrix} 1 & 0 & \vdots & 4 \\ 0 & 1 & \vdots & 8 \end{bmatrix}$, we have

$x_1 = 4$ and $x_2 = 8$.

Answer: $X = \begin{bmatrix} 4 \\ 8 \end{bmatrix}$

61. (a) $A = \begin{bmatrix} -2 & -3 \\ 6 & 1 \end{bmatrix}, X = \begin{bmatrix} x_1 \\ x_2 \end{bmatrix}, B = \begin{bmatrix} -4 \\ -36 \end{bmatrix}$

(b) By Gauss-Jordan elimination on

$\begin{bmatrix} -2 & -3 & \vdots & -4 \\ 6 & 1 & \vdots & -36 \end{bmatrix}$

$3R_1 + R_2 \begin{bmatrix} -2 & -3 & \vdots & -4 \\ 0 & -8 & \vdots & -48 \end{bmatrix}$

$\left(-\frac{1}{8}\right)R_2 \begin{bmatrix} -2 & -3 & \vdots & -4 \\ 0 & 1 & \vdots & 6 \end{bmatrix}$

$3R_1 + R_2 \begin{bmatrix} -2 & 0 & \vdots & 14 \\ 0 & 1 & \vdots & 6 \end{bmatrix}$

$-\frac{1}{2}R_1 \begin{bmatrix} 1 & 0 & \vdots & -7 \\ 0 & 1 & \vdots & 6 \end{bmatrix}$, we have

$x_1 = -7$, $x_2 = 6$.

Answer: $X = \begin{bmatrix} -7 \\ 6 \end{bmatrix}$

63. (a) $A = \begin{bmatrix} 1 & -2 & 3 \\ -1 & 3 & -1 \\ 2 & -5 & 5 \end{bmatrix}, X = \begin{bmatrix} x_1 \\ x_2 \\ x_3 \end{bmatrix}, B = \begin{bmatrix} 9 \\ -6 \\ 17 \end{bmatrix}$

(b) By Gauss-Jordan elimination on

$$\begin{bmatrix} 1 & -2 & 3 & \vdots & 9 \\ -1 & 3 & -1 & \vdots & -6 \\ 2 & -5 & 5 & \vdots & 17 \end{bmatrix}$$

$$\begin{matrix} \\ R_1 + R_2 \to \\ -2R_1 + R_3 \to \end{matrix} \begin{bmatrix} 1 & -2 & 3 & \vdots & 9 \\ 0 & 1 & 2 & \vdots & 3 \\ 0 & -1 & -1 & \vdots & -1 \end{bmatrix}$$

$$\begin{matrix} 2R_2 + R_1 \to \\ \\ R_2 + R_3 \to \end{matrix} \begin{bmatrix} 1 & 0 & 7 & \vdots & 15 \\ 0 & 1 & 2 & \vdots & 3 \\ 0 & 0 & 1 & \vdots & 2 \end{bmatrix}$$

$$\begin{matrix} -7R_3 + R_1 \to \\ -2R_3 + R_2 \to \\ \\ \end{matrix} \begin{bmatrix} 1 & 0 & 0 & \vdots & 1 \\ 0 & 1 & 0 & \vdots & -1 \\ 0 & 0 & 1 & \vdots & 2 \end{bmatrix}, \text{ we have}$$

$x_1 = 1, x_2 = -1, x_3 = 2.$

$$Answer: X = \begin{bmatrix} 1 \\ -1 \\ 2 \end{bmatrix}$$

65. (a) $A = \begin{bmatrix} 1 & -5 & 2 \\ -3 & 1 & -1 \\ 0 & -2 & 5 \end{bmatrix}, X = \begin{bmatrix} x_1 \\ x_2 \\ x_3 \end{bmatrix}, B = \begin{bmatrix} -20 \\ 8 \\ -16 \end{bmatrix}$

(b) By Gauss-Jordan elimination on

$$\begin{bmatrix} 1 & -5 & 2 & \vdots & -20 \\ -3 & 1 & -1 & \vdots & 8 \\ 0 & -2 & 5 & \vdots & -16 \end{bmatrix}$$

$$3R_1 + R_2 \quad \begin{bmatrix} 1 & -5 & 2 & \vdots & -20 \\ 0 & -14 & 5 & \vdots & -52 \\ 0 & -2 & 5 & \vdots & -16 \end{bmatrix}$$

$$\begin{matrix} \\ R_3 \to \\ R_2 \to \end{matrix} \begin{bmatrix} 1 & -5 & 2 & \vdots & -20 \\ 0 & -2 & 5 & \vdots & -16 \\ 0 & -14 & 5 & \vdots & -52 \end{bmatrix}$$

$$-7R_2 + R_3 \quad \begin{bmatrix} 1 & -5 & 2 & \vdots & -20 \\ 0 & -2 & 5 & \vdots & -16 \\ 0 & 0 & -30 & \vdots & 60 \end{bmatrix}$$

$$-\tfrac{1}{30}R_3 \quad \begin{bmatrix} 1 & -5 & 2 & \vdots & -20 \\ 0 & -2 & 5 & \vdots & -16 \\ 0 & 0 & 1 & \vdots & -2 \end{bmatrix}$$

$$\begin{matrix} -2R_3 + R_1 \\ -5R_3 + R_2 \\ \\ \end{matrix} \begin{bmatrix} 1 & -5 & 0 & \vdots & -16 \\ 0 & -2 & 0 & \vdots & -6 \\ 0 & 0 & 1 & \vdots & -2 \end{bmatrix}$$

$$\left(-\tfrac{1}{2}\right)R_2 \quad \begin{bmatrix} 1 & -5 & 0 & \vdots & -16 \\ 0 & 1 & 0 & \vdots & 3 \\ 0 & 0 & 1 & \vdots & -2 \end{bmatrix}$$

$$5R_2 + R_1 \quad \begin{bmatrix} 1 & 0 & 0 & \vdots & -1 \\ 0 & 1 & 0 & \vdots & 3 \\ 0 & 0 & 1 & \vdots & -2 \end{bmatrix}, \text{ we have}$$

$x_1 = -1, x_2 = 3, x_3 = -2.$

$$Answer: X = \begin{bmatrix} -1 \\ 3 \\ -2 \end{bmatrix}$$

67. (a) $A(B+C) = \begin{bmatrix} 7 & -2 & 5 \\ -6 & 13 & -8 \\ 16 & 11 & -3 \end{bmatrix}$

(b) $AB + AC = \begin{bmatrix} 7 & -2 & 5 \\ -6 & 13 & -8 \\ 16 & 11 & -3 \end{bmatrix}$

The answers are the same.

69. (a) $(A+B)^2 = \begin{bmatrix} 26 & 11 & 0 \\ 11 & 20 & -3 \\ 11 & 14 & 0 \end{bmatrix}$

(b) $A^2 + AB + BA + B^2 = \begin{bmatrix} 26 & 11 & 0 \\ 11 & 20 & -3 \\ 11 & 14 & 0 \end{bmatrix}$

The answers are the same.

71. (a) $A(BC) = \begin{bmatrix} 25 & -34 & 28 \\ -53 & 34 & -7 \\ -76 & 30 & 21 \end{bmatrix}$

(b) $(AB)C = \begin{bmatrix} 25 & -34 & 28 \\ -53 & 34 & -7 \\ -76 & 30 & 21 \end{bmatrix}$

The answers are the same.

73. (a) and (b)

$$A + cB = \begin{bmatrix} 1 & 2 & -2 \\ -1 & 1 & 0 \end{bmatrix} + (-2)\begin{bmatrix} -1 & 4 & -1 \\ -2 & -1 & 0 \end{bmatrix}$$
$$= \begin{bmatrix} 1 & 2 & -2 \\ -1 & 1 & 0 \end{bmatrix} + \begin{bmatrix} 2 & -8 & 2 \\ 4 & 2 & 0 \end{bmatrix}$$
$$= \begin{bmatrix} 3 & -6 & 0 \\ 3 & 3 & 0 \end{bmatrix}$$

75. (a) and (b) $c(AB) \Rightarrow$ Not possible, A is a 2×3 matrix and B is a 2×3 matrix.

77. (a) and (b) $CA - BC \Rightarrow$ Not possible because CA is 3×3, and BC is 2×2.

79. (a) and (b)

$$cdA = (-2)(-3)\begin{bmatrix} 1 & 2 & -2 \\ -1 & 1 & 0 \end{bmatrix} = \begin{bmatrix} 6 & 12 & -12 \\ -6 & 6 & 0 \end{bmatrix}$$

81. $A = \begin{bmatrix} 2 & 0 \\ 4 & 5 \end{bmatrix}$

$$f(A) = A^2 - 5A + 2I = \begin{bmatrix} 2 & 0 \\ 4 & 5 \end{bmatrix}\begin{bmatrix} 2 & 0 \\ 4 & 5 \end{bmatrix} - 5\begin{bmatrix} 2 & 0 \\ 4 & 5 \end{bmatrix} + 2\begin{bmatrix} 1 & 0 \\ 0 & 1 \end{bmatrix}$$
$$= \begin{bmatrix} -4 & 0 \\ 8 & 2 \end{bmatrix}$$

83. $1.20\begin{bmatrix} 70 & 50 & 25 \\ 35 & 100 & 70 \end{bmatrix} = \begin{bmatrix} 84 & 60 & 30 \\ 42 & 120 & 84 \end{bmatrix}$

85. $0.90A = 0.90\begin{bmatrix} 100 & 120 & 60 & 40 \\ 140 & 160 & 200 & 80 \end{bmatrix}$
$$= \begin{bmatrix} 90 & 108 & 54 & 36 \\ 126 & 144 & 180 & 72 \end{bmatrix}$$

87. $BA = \begin{bmatrix} 3.50 & 6.00 \end{bmatrix}\begin{bmatrix} 125 & 100 & 75 \\ 100 & 175 & 125 \end{bmatrix}$
$$= \begin{bmatrix} \$1037.50 & \$1400 & \$1012.50 \end{bmatrix}$$

The entries in the last matrix BA represent the profit for both crops at each of the three outlets.

89. $ST = \begin{bmatrix} 1.0 & 0.5 & 0.2 \\ 1.6 & 1.0 & 0.2 \\ 2.5 & 2.0 & 0.4 \end{bmatrix}\begin{bmatrix} 12 & 10 \\ 9 & 8 \\ 6 & 5 \end{bmatrix} = \begin{bmatrix} \$17.70 & \$15.00 \\ \$29.40 & \$25.00 \\ \$50.40 & \$43.00 \end{bmatrix}$

The entries represent the labor cost for each boat size at each plant.

91. $P^2 = \begin{bmatrix} 0.6 & 0.1 & 0.1 \\ 0.2 & 0.7 & 0.1 \\ 0.2 & 0.2 & 0.8 \end{bmatrix}\begin{bmatrix} 0.6 & 0.1 & 0.1 \\ 0.2 & 0.7 & 0.1 \\ 0.2 & 0.2 & 0.8 \end{bmatrix}$
$$= \begin{bmatrix} 0.40 & 0.15 & 0.15 \\ 0.28 & 0.53 & 0.17 \\ 0.32 & 0.32 & 0.68 \end{bmatrix}$$

This product represents the proportion of changes in party affiliation after *two* elections.

93. True. To add two matrices, corresponding entries are added, therefore the matrices must have the same dimension.

95. $A + 2C$ is not possible. A and C are not of the same dimension.

97. AB is not possible. The number of columns of A does not equal the number of rows of B.

99. $BC - D$ is possible. The resulting dimension is 2×2.

101. $D(A - 3B)$ is possible. The resulting dimension is 2×3.

103. $(A + B)^2 = \begin{bmatrix} 1 & 0 \\ 2 & 1 \end{bmatrix}$

$$A^2 + 2AB + B^2 = \begin{bmatrix} 0 & 0 \\ 3 & 2 \end{bmatrix}$$

105. $(A + B)(A - B) = \begin{bmatrix} 3 & -2 \\ 4 & 3 \end{bmatrix}$

$$A^2 - B^2 = \begin{bmatrix} 2 & -2 \\ 5 & 4 \end{bmatrix}$$

107. $AC = \begin{bmatrix} 0 & 1 \\ 0 & 1 \end{bmatrix}\begin{bmatrix} 2 & 3 \\ 2 & 3 \end{bmatrix} = \begin{bmatrix} 2 & 3 \\ 2 & 3 \end{bmatrix}$

$$BC = \begin{bmatrix} 1 & 0 \\ 1 & 0 \end{bmatrix}\begin{bmatrix} 2 & 3 \\ 2 & 3 \end{bmatrix} = \begin{bmatrix} 2 & 3 \\ 2 & 3 \end{bmatrix}$$

$AC = BC$, but $A \neq B$.

109. (a) $A^2 = \begin{bmatrix} i & 0 \\ 0 & i \end{bmatrix}\begin{bmatrix} i & 0 \\ 0 & i \end{bmatrix} = \begin{bmatrix} -1 & 0 \\ 0 & -1 \end{bmatrix}$ and $i^2 = -1$

$$A^3 = A^2A = \begin{bmatrix} -1 & 0 \\ 0 & -1 \end{bmatrix}\begin{bmatrix} i & 0 \\ 0 & i \end{bmatrix} = \begin{bmatrix} -i & 0 \\ 0 & -i \end{bmatrix} \text{ and } i^3 = -i$$

$$A^4 = A^3A = \begin{bmatrix} -i & 0 \\ 0 & -i \end{bmatrix}\begin{bmatrix} i & 0 \\ 0 & i \end{bmatrix} = \begin{bmatrix} 1 & 0 \\ 0 & 1 \end{bmatrix} \text{ and } i^4 = 1$$

(b) $B^2 = \begin{bmatrix} 0 & -i \\ i & 0 \end{bmatrix}\begin{bmatrix} 0 & -i \\ i & 0 \end{bmatrix} = \begin{bmatrix} 1 & 0 \\ 0 & 1 \end{bmatrix}$.

The identity matrix

111. (a) $A = \begin{bmatrix} 0 & 2 \\ 0 & 0 \end{bmatrix}$, $B = \begin{bmatrix} 0 & 2 & 3 \\ 0 & 0 & 4 \\ 0 & 0 & 0 \end{bmatrix}$

(b) A^2 and B^3 are both are zero matrices.

(c) If A is 4×4, then A^4 will be the zero matrix.

(d) If A is $n \times n$, then A^n is the zero matrix.

113. $3 \ln 4 - \frac{1}{3}\ln(x^2 + 3) = \ln 4^3 - \ln(x^2 + 3)^{1/3}$

$$= \ln\left[\frac{64}{(x^2 + 3)^{1/3}}\right]$$

Section 7.6

1. inverse

3. No. A square matrix must be invertible or nonsingular to have an inverse.

5. $$AB = \begin{bmatrix} 2 & 1 \\ 5 & 3 \end{bmatrix}\begin{bmatrix} 3 & -1 \\ -5 & 2 \end{bmatrix} = \begin{bmatrix} 2(3)+1(-5) & 2(-1)+1(2) \\ 5(3)+3(-5) & 5(-1)+3(2) \end{bmatrix} = \begin{bmatrix} 1 & 0 \\ 0 & 1 \end{bmatrix}$$

$$BA = \begin{bmatrix} 3 & -1 \\ -5 & 2 \end{bmatrix}\begin{bmatrix} 2 & 1 \\ 5 & 3 \end{bmatrix} = \begin{bmatrix} 3(2)+(-1)(5) & 3(1)+(-1)(3) \\ -5(2)+2(5) & -5(1)+2(3) \end{bmatrix} = \begin{bmatrix} 1 & 0 \\ 0 & 1 \end{bmatrix}$$

7.

$$AB = \begin{bmatrix} 1 & 2 \\ 3 & 4 \end{bmatrix}\begin{bmatrix} -2 & 1 \\ \frac{3}{2} & -\frac{1}{2} \end{bmatrix} = \begin{bmatrix} -2+3 & 1-1 \\ -6+6 & 3-2 \end{bmatrix} = \begin{bmatrix} 1 & 0 \\ 0 & 1 \end{bmatrix}$$

$$BA = \begin{bmatrix} -2 & 1 \\ \frac{3}{2} & -\frac{1}{2} \end{bmatrix}\begin{bmatrix} 1 & 2 \\ 3 & 4 \end{bmatrix} = \begin{bmatrix} -2+3 & -4+4 \\ \frac{3}{2}-\frac{3}{2} & 3-2 \end{bmatrix} = \begin{bmatrix} 1 & 0 \\ 0 & 1 \end{bmatrix}$$

9. $$AB = \begin{bmatrix} 2 & -17 & 11 \\ -1 & 11 & -7 \\ 0 & 3 & -2 \end{bmatrix}\begin{bmatrix} 1 & 1 & 2 \\ 2 & 4 & -3 \\ 3 & 6 & -5 \end{bmatrix}$$

$$= \begin{bmatrix} 2-34+33 & 2-68+66 & 4+51-55 \\ -1+22-21 & -1+44-42 & -2-33+35 \\ 6-6 & 12-12 & -9+10 \end{bmatrix}$$

$$= \begin{bmatrix} 1 & 0 & 0 \\ 0 & 1 & 0 \\ 0 & 0 & 1 \end{bmatrix}$$

$$BA = \begin{bmatrix} 1 & 1 & 2 \\ 2 & 4 & -3 \\ 3 & 6 & -5 \end{bmatrix}\begin{bmatrix} 2 & -17 & 11 \\ -1 & 11 & -7 \\ 0 & 3 & -2 \end{bmatrix}$$

$$= \begin{bmatrix} 2-1 & -17+11+6 & 11-7-4 \\ 4-4 & -34+44-9 & 22-28+6 \\ 6-6 & -51+66-15 & 33-42+10 \end{bmatrix} = \begin{bmatrix} 1 & 0 & 0 \\ 0 & 1 & 0 \\ 0 & 0 & 1 \end{bmatrix}$$

11. $$[A \ \vdots \ I] = \left[\begin{array}{cc:cc} 2 & 0 & 1 & 0 \\ 0 & 3 & 0 & 1 \end{array}\right]$$

$$\begin{array}{l} \frac{1}{2}R_1 \rightarrow \\ \frac{1}{3}R_2 \rightarrow \end{array} \left[\begin{array}{cc:cc} 1 & 0 & \frac{1}{2} & 0 \\ 0 & 1 & 0 & \frac{1}{3} \end{array}\right] = [I \ \vdots \ A^{-1}]$$

$$A^{-1} = \begin{bmatrix} \frac{1}{2} & 0 \\ 0 & \frac{1}{3} \end{bmatrix} = \frac{1}{6}\begin{bmatrix} 3 & 0 \\ 0 & 2 \end{bmatrix}$$

13. $$[A \ \vdots \ I] = \left[\begin{array}{cc:cc} 1 & -2 & 1 & 0 \\ 2 & -3 & 0 & 1 \end{array}\right]$$

$$\begin{array}{l} \\ -2R_1 + R_2 \rightarrow \end{array} \left[\begin{array}{cc:cc} 1 & -2 & 1 & 0 \\ 0 & 1 & -2 & 1 \end{array}\right]$$

$$\begin{array}{l} 2R_2 + R_1 \rightarrow \\ \\ \end{array} \left[\begin{array}{cc:cc} 1 & 0 & -3 & 2 \\ 0 & 1 & -2 & 1 \end{array}\right]$$

$$A^{-1} = \begin{bmatrix} -3 & 2 \\ -2 & 1 \end{bmatrix}$$

15. $$A = \begin{bmatrix} 2 & 7 & 1 \\ -3 & -9 & 2 \end{bmatrix}$$

A has no inverse because it is not square.

17. $$[A \ \vdots \ I] = \left[\begin{array}{ccc:ccc} 1 & 1 & 1 & 1 & 0 & 0 \\ 3 & 5 & 4 & 0 & 1 & 0 \\ 3 & 6 & 5 & 0 & 0 & 1 \end{array}\right]$$

$$\begin{array}{l} \\ -3R_1 + R_2 \rightarrow \\ -3R_1 + R_3 \rightarrow \end{array} \left[\begin{array}{ccc:ccc} 1 & 1 & 1 & 1 & 0 & 0 \\ 0 & 2 & 1 & -3 & 1 & 0 \\ 0 & 3 & 2 & -3 & 0 & 1 \end{array}\right]$$

$$\begin{array}{l} -R_2 + R_1 \rightarrow \\ \frac{1}{2}R_2 \rightarrow \\ -3R_2 + R_3 \rightarrow \end{array} \left[\begin{array}{ccc:ccc} 1 & 0 & \frac{1}{2} & \frac{5}{2} & -\frac{1}{2} & 0 \\ 0 & 1 & \frac{1}{2} & -\frac{3}{2} & \frac{1}{2} & 0 \\ 0 & 0 & \frac{1}{2} & \frac{3}{2} & -\frac{3}{2} & 1 \end{array}\right]$$

$$\begin{array}{l} -R_3 + R_1 \rightarrow \\ -R_3 + R_2 \rightarrow \\ 2R_3 \rightarrow \end{array} \left[\begin{array}{ccc:ccc} 1 & 0 & 0 & 1 & 1 & -1 \\ 0 & 1 & 0 & -3 & 2 & -1 \\ 0 & 0 & 1 & 3 & -3 & 2 \end{array}\right]$$

$$= [I \ \vdots \ A^{-1}]$$

$$A^{-1} = \begin{bmatrix} 1 & 1 & -1 \\ -3 & 2 & -1 \\ 3 & -3 & 2 \end{bmatrix}$$

19. $$[A \ \vdots \ I] = \left[\begin{array}{ccc:ccc} 1 & 0 & 0 & 1 & 0 & 0 \\ 3 & 4 & 0 & 0 & 1 & 0 \\ 2 & 5 & 5 & 0 & 0 & 1 \end{array}\right]$$

$$\begin{array}{l} \\ -3R_1 + R_2 \rightarrow \\ -2R_1 + R_3 \rightarrow \end{array} \left[\begin{array}{ccc:ccc} 1 & 0 & 0 & 1 & 0 & 0 \\ 0 & 4 & 0 & -3 & 1 & 0 \\ 0 & 5 & 5 & -2 & 0 & 1 \end{array}\right]$$

$$\begin{array}{l} \\ \frac{1}{4}R_2 \rightarrow \\ \frac{1}{5}R_3 \rightarrow \end{array} \left[\begin{array}{ccc:ccc} 1 & 0 & 0 & 1 & 0 & 0 \\ 0 & 1 & 0 & -\frac{3}{4} & \frac{1}{4} & 0 \\ 0 & 1 & 1 & -\frac{2}{5} & 0 & \frac{1}{5} \end{array}\right]$$

$$\begin{array}{l} \\ \\ R_2 - R_3 \rightarrow \end{array} \left[\begin{array}{ccc:ccc} 1 & 0 & 0 & 1 & 0 & 0 \\ 0 & 1 & 0 & -\frac{3}{4} & \frac{1}{4} & 0 \\ 0 & 0 & -1 & -\frac{7}{20} & \frac{1}{4} & -\frac{1}{5} \end{array}\right]$$

$$-R_3 \rightarrow \left[\begin{array}{ccc:ccc} 1 & 0 & 0 & 1 & 0 & 0 \\ 0 & 1 & 0 & -\frac{3}{4} & \frac{1}{4} & 0 \\ 0 & 0 & 1 & \frac{7}{20} & -\frac{1}{4} & \frac{1}{5} \end{array}\right]$$

$$= \left[I \ \vdots \ A^{-1}\right]$$

$$A^{-1} = \begin{bmatrix} 1 & 0 & 0 \\ -\frac{3}{4} & \frac{1}{4} & 0 \\ \frac{7}{20} & -\frac{1}{4} & \frac{1}{5} \end{bmatrix}$$

21. $A^{-1} = \begin{bmatrix} -\frac{3}{2} & \frac{3}{2} & 1 \\ \frac{9}{2} & -\frac{7}{2} & -3 \\ -1 & 1 & 1 \end{bmatrix}$

23. $A^{-1} = \begin{bmatrix} -12 & -5 & -9 \\ -4 & -2 & -4 \\ -8 & -4 & -6 \end{bmatrix}$

25. $A^{-1} = \frac{5}{11}\begin{bmatrix} 0 & -4 & 2 \\ -22 & 11 & 11 \\ 22 & -6 & -8 \end{bmatrix}$

27. $A^{-1} = \begin{bmatrix} 1 & 0 & 1 & 0 \\ 0 & 1 & 0 & 1 \\ 2 & 0 & 1 & 0 \\ 0 & 1 & 0 & 2 \end{bmatrix}$

29. $\begin{bmatrix} 5 & 1 \\ -2 & -2 \end{bmatrix}^{-1} = \frac{1}{5(-2)-(-2)(1)}\begin{bmatrix} -2 & -1 \\ 2 & 5 \end{bmatrix}$

$$= \frac{1}{-8}\begin{bmatrix} -2 & -1 \\ 2 & 5 \end{bmatrix} = \begin{bmatrix} \frac{1}{4} & \frac{1}{8} \\ -\frac{1}{4} & -\frac{5}{8} \end{bmatrix}$$

31. $\begin{bmatrix} \frac{7}{2} & -\frac{3}{4} \\ \frac{1}{5} & \frac{4}{5} \end{bmatrix}^{-1} = \frac{1}{\left(\frac{7}{2}\right)\left(\frac{4}{5}\right)-\left(-\frac{3}{4}\right)\left(\frac{1}{5}\right)}\begin{bmatrix} \frac{4}{5} & \frac{3}{4} \\ -\frac{1}{5} & \frac{7}{2} \end{bmatrix}$

$$= \frac{20}{59}\begin{bmatrix} \frac{4}{5} & \frac{3}{4} \\ -\frac{1}{5} & \frac{7}{2} \end{bmatrix}$$

$$= \frac{1}{59}\begin{bmatrix} 16 & 15 \\ -4 & 70 \end{bmatrix}$$

33. $\begin{bmatrix} 2 & 3 \\ -1 & 5 \end{bmatrix}^{-1} = \frac{1}{2(5)-(3)(-1)}\begin{bmatrix} 5 & -3 \\ 1 & 2 \end{bmatrix} = \frac{1}{13}\begin{bmatrix} 5 & -3 \\ 1 & 2 \end{bmatrix} = \begin{bmatrix} \frac{5}{13} & -\frac{3}{13} \\ \frac{1}{13} & \frac{2}{13} \end{bmatrix}$

35. $\begin{bmatrix} 1 & 2 \\ -2 & 0 \end{bmatrix}^{-1} = \begin{bmatrix} 0 & -\frac{1}{2} \\ \frac{1}{2} & \frac{1}{4} \end{bmatrix} \Rightarrow k = 0$

37. $\begin{bmatrix} x \\ y \end{bmatrix} = \begin{bmatrix} -3 & 2 \\ -2 & 1 \end{bmatrix}\begin{bmatrix} 5 \\ 10 \end{bmatrix} = \begin{bmatrix} 5 \\ 0 \end{bmatrix}$

Answer: $(5, 0)$

39. $\begin{bmatrix} x \\ y \end{bmatrix} = \begin{bmatrix} -3 & 2 \\ -2 & 1 \end{bmatrix}\begin{bmatrix} 4 \\ 2 \end{bmatrix} = \begin{bmatrix} -8 \\ -6 \end{bmatrix}$

Answer: $(-8, -6)$

41. $\begin{bmatrix} x \\ y \\ z \end{bmatrix} = \begin{bmatrix} 1 & 1 & -1 \\ -3 & 2 & -1 \\ 3 & -3 & 2 \end{bmatrix}\begin{bmatrix} 0 \\ 5 \\ 2 \end{bmatrix} = \begin{bmatrix} 3 \\ 8 \\ -11 \end{bmatrix}$

Answer: $(3, 8, -11)$

43. $\begin{bmatrix} x_1 \\ x_2 \\ x_3 \\ x_4 \end{bmatrix} = \begin{bmatrix} -24 & 7 & 1 & -2 \\ -10 & 3 & 0 & -1 \\ -29 & 7 & 3 & -2 \\ 12 & -3 & -1 & 1 \end{bmatrix}\begin{bmatrix} 0 \\ 1 \\ -1 \\ 2 \end{bmatrix} = \begin{bmatrix} 2 \\ 1 \\ 0 \\ 0 \end{bmatrix}$

Answer: $(2, 1, 0, 0)$

45. $A = \begin{bmatrix} 3 & 4 \\ 5 & 3 \end{bmatrix}$

$$A^{-1} = \frac{1}{9-20}\begin{bmatrix} 3 & -4 \\ -5 & 3 \end{bmatrix}$$

$$\begin{bmatrix} x \\ y \end{bmatrix} = -\frac{1}{11}\begin{bmatrix} 3 & -4 \\ -5 & 3 \end{bmatrix}\begin{bmatrix} -2 \\ 4 \end{bmatrix} = -\frac{1}{11}\begin{bmatrix} -22 \\ 22 \end{bmatrix} = \begin{bmatrix} 2 \\ -2 \end{bmatrix}$$

Answer: $(2, -2)$

47. $A = \begin{bmatrix} -4 & 0.8 \\ 2 & -0.4 \end{bmatrix}$

$$A^{-1} = \frac{1}{1.6-1.6}\begin{bmatrix} -4 & -0.8 \\ -2 & -0.4 \end{bmatrix}$$

A^{-1} does not exist.
[The system actually has no solution.]

49. $A = \begin{bmatrix} -\frac{1}{4} & \frac{3}{8} \\ \frac{3}{2} & \frac{3}{4} \end{bmatrix}$

$$A^{-1} = \begin{bmatrix} -1 & \frac{1}{2} \\ 2 & \frac{1}{3} \end{bmatrix}$$

$$\begin{bmatrix} x \\ y \end{bmatrix} = A^{-1}b = \begin{bmatrix} -1 & \frac{1}{2} \\ 2 & \frac{1}{3} \end{bmatrix}\begin{bmatrix} -2 \\ -12 \end{bmatrix} = \begin{bmatrix} -4 \\ -8 \end{bmatrix}$$

Answer: $(-4, -8)$

51. $A = \begin{bmatrix} 4 & -1 & 1 \\ 2 & 2 & 3 \\ 5 & -2 & 6 \end{bmatrix}$

$A^{-1} = \frac{1}{55}\begin{bmatrix} 18 & 4 & -5 \\ 3 & 19 & -10 \\ -14 & 3 & 10 \end{bmatrix}$

$\begin{bmatrix} x \\ y \\ z \end{bmatrix} = \frac{1}{55}\begin{bmatrix} 18 & 4 & -5 \\ 3 & 19 & -10 \\ -14 & 3 & 10 \end{bmatrix}\begin{bmatrix} -5 \\ 10 \\ 1 \end{bmatrix}$

$= \frac{1}{55}\begin{bmatrix} -55 \\ 165 \\ 110 \end{bmatrix} = \begin{bmatrix} -1 \\ 3 \\ 2 \end{bmatrix}$

Answer: $(-1, 3, 2)$

53. $A = \begin{bmatrix} 5 & -3 & 2 \\ 2 & 2 & -3 \\ -1 & 7 & -8 \end{bmatrix}, B = \begin{bmatrix} 2 \\ 3 \\ 4 \end{bmatrix}$

A^{-1} does not exist.
The system actually has an infinite number of solutions of the form

$x = 0.3125t + 0.8125$

$y = 1.1875t + 0.6875$

$z = t$

where t is any real number.

55. $A = \begin{bmatrix} 7 & -3 & 0 & 2 \\ -2 & 1 & 0 & -1 \\ 4 & 0 & 1 & -2 \\ -1 & 1 & 0 & -1 \end{bmatrix}, B = \begin{bmatrix} 41 \\ -13 \\ 12 \\ -8 \end{bmatrix}$

$A^{-1} = \begin{bmatrix} 0 & -1 & 0 & 1 \\ -1 & -5 & 0 & 3 \\ -2 & -4 & 1 & -2 \\ -1 & -4 & 0 & 1 \end{bmatrix}$

$\begin{bmatrix} x \\ y \\ z \\ w \end{bmatrix} = \begin{bmatrix} 0 & -1 & 0 & 1 \\ -1 & -5 & 0 & 3 \\ -2 & -4 & 1 & -2 \\ -1 & -4 & 0 & 1 \end{bmatrix}\begin{bmatrix} 41 \\ -13 \\ 12 \\ -8 \end{bmatrix} = \begin{bmatrix} 5 \\ 0 \\ -2 \\ 3 \end{bmatrix}$

Answer: $(5, 0, -2, 3)$

57. $X = A^{-1}B = \frac{1}{11}\begin{bmatrix} 50 & -600 & -4 \\ -13 & 200 & 5 \\ -26 & 400 & -1 \end{bmatrix}\begin{bmatrix} 10,000 \\ 705 \\ 0 \end{bmatrix} = \begin{bmatrix} 7000 \\ 1000 \\ 2000 \end{bmatrix}$

Answer: \$7000 in AAA-rated bonds, \$1000 in A-rated bonds, and \$2000 in B-rated bonds

59. $X = A^{-1}B = \frac{1}{11}\begin{bmatrix} 50 & -600 & -4 \\ -13 & 200 & 5 \\ -26 & 400 & -1 \end{bmatrix}\begin{bmatrix} 12,000 \\ 835 \\ 0 \end{bmatrix} = \begin{bmatrix} 9000 \\ 1000 \\ 2000 \end{bmatrix}$

Answer: \$9000 in AAA-rated bonds, \$1000 in A-rated bonds, and \$2000 in B-rated bonds

61. $E_1 = 15, E_2 = 17$

$\begin{cases} 2I_1 + 4I_3 = 15 \\ I_2 + 4I_3 = 17 \\ I_1 + I_2 - I_3 = 0 \end{cases}$

$A = \begin{bmatrix} 2 & 0 & 4 \\ 0 & 1 & 4 \\ 1 & 1 & -1 \end{bmatrix}$

$A^{-1} = \frac{1}{14}\begin{bmatrix} 5 & -4 & 4 \\ -4 & 6 & 8 \\ 1 & 2 & -2 \end{bmatrix}$

$\begin{bmatrix} I_1 \\ I_2 \\ I_3 \end{bmatrix} = \frac{1}{14}\begin{bmatrix} 5 & -4 & 4 \\ -4 & 6 & 8 \\ 1 & 2 & -2 \end{bmatrix}\begin{bmatrix} 15 \\ 17 \\ 0 \end{bmatrix} = \begin{bmatrix} 0.5 \\ 3 \\ 3.5 \end{bmatrix}$

Answer: $I_1 = 0.5$ ampere, $I_2 = 3$ amperes, $I_3 = 3.5$ amperes

63. $A^{-1}\begin{bmatrix} 500 \\ 500 \\ 400 \end{bmatrix} = \begin{bmatrix} 100 \\ 100 \\ 100 \end{bmatrix}$

Answer: 100 bags for seedlings, 100 bags for general planting, and 100 bags for hardwood plants

65. (a) Let x = number of roses.
Let y = number of lilies.
Let z = number of irises.

$\begin{cases} x + y + z = 120 \\ 2.5x + 4y + 2z = 300 \\ x - 2y - 2z = 0 \end{cases}$

(b) $\underbrace{\begin{bmatrix} 1 & 1 & 1 \\ 2.5 & 4 & 2 \\ 1 & -2 & -2 \end{bmatrix}}_{A}\underbrace{\begin{bmatrix} x \\ y \\ z \end{bmatrix}}_{X} = \underbrace{\begin{bmatrix} 120 \\ 300 \\ 0 \end{bmatrix}}_{B}$

(c) $X = A^{-1}B = \begin{bmatrix} \frac{2}{3} & 0 & \frac{1}{3} \\ -\frac{7}{6} & \frac{1}{2} & -\frac{1}{12} \\ \frac{3}{2} & -\frac{1}{2} & -\frac{1}{4} \end{bmatrix}\begin{bmatrix} 120 \\ 300 \\ 0 \end{bmatrix} = \begin{bmatrix} 80 \\ 10 \\ 30 \end{bmatrix}$

80 roses, 10 lilies, 30 irises

67. True. $AA^{-1} = A^{-1}A = I$

69. Answers will vary. One possibility is if A is a 2×2 matrix given by $A = \begin{bmatrix} a & b \\ c & d \end{bmatrix}$, then A is invertible (A^{-1} exists) if and only if $ad - bc \neq 0$.

71. (a) Given

$$A = \begin{bmatrix} a_{11} & 0 \\ 0 & a_{12} \end{bmatrix}, A^{-1} = \begin{bmatrix} \frac{1}{a_{11}} & 0 \\ 0 & \frac{1}{a_{22}} \end{bmatrix}, a_{11} \neq 0, a_{22} \neq 0$$

Given

$$A = \begin{bmatrix} a_{11} & 0 & 0 \\ 0 & a_{22} & 0 \\ 0 & 0 & a_{33} \end{bmatrix}, A^{-1} = \begin{bmatrix} \frac{1}{a_{11}} & 0 & 0 \\ 0 & \frac{1}{a_{22}} & 0 \\ 0 & 0 & \frac{1}{a_{33}} \end{bmatrix}.$$

$a_{11}, a_{22}, a_{33} \neq 0$

(b) In general, the inverse of the diagonal matrix A is

$$\begin{bmatrix} \frac{1}{a_{11}} & 0 & 0 & \cdots & 0 \\ 0 & \frac{1}{a_{22}} & 0 & \cdots & 0 \\ 0 & 0 & \frac{1}{a_{33}} & \cdots & 0 \\ \vdots & \vdots & \vdots & \vdots & \vdots \\ 0 & 0 & 0 & \cdots & \frac{1}{a_{nn}} \end{bmatrix} \text{ (assuming } a_{ii} \neq 0\text{)}$$

73. $e^{2x} + 2e^x - 15 = (e^x + 5)(e^x - 3) = 0 \Rightarrow e^x = 3$

$\Rightarrow x = \ln 3 \approx 1.099$

75. $7 \ln 3x = 12$

$\ln 3x = \frac{12}{7}$

$3x = e^{12/7}$

$x = \frac{1}{3}e^{12/7} \approx 1.851$

77. Answers will vary. (Make a Decision)

Section 7.7

1. determinant

3. For a square matrix B, given the minor $M_{23} = 5$, the cofactor is $C_{23} = (-1)^{2+3} M_{23} = (-1)(5) = -5$.

5. $|4| = 4$

7. $\begin{vmatrix} 8 & 4 \\ 2 & 3 \end{vmatrix} = 8(3) - 4(2) = 24 - 8 = 16$

9. $\begin{vmatrix} 6 & 2 \\ -5 & 3 \end{vmatrix} = 6(3) - (2)(-5) = 18 + 10 = 28$

11. $\begin{vmatrix} -7 & 6 \\ \frac{1}{2} & 3 \end{vmatrix} = -7(3) - 6\left(\frac{1}{2}\right) = -21 - 3 = -24$

13. $\begin{vmatrix} 0.3 & 0.2 & 0.2 \\ 0.2 & 0.2 & 0.2 \\ -0.4 & 0.4 & 0.3 \end{vmatrix} = -0.002$

15. $\begin{bmatrix} 3 & 4 \\ 2 & -5 \end{bmatrix}$

(a) $M_{11} = -5$

$M_{12} = 2$

$M_{21} = 4$

$M_{22} = 3$

(b) $C_{11} = M_{11} = -5$

$C_{12} = -M_{12} = -2$

$C_{21} = -M_{21} = -4$

$C_{22} = M_{22} = 3$

17. $\begin{bmatrix} -4 & 6 & 3 \\ 7 & -2 & 8 \\ 1 & 0 & -5 \end{bmatrix}$

(a) $M_{11} = \begin{vmatrix} -2 & 8 \\ 0 & -5 \end{vmatrix} = 10$

$M_{12} = \begin{vmatrix} 7 & 8 \\ 1 & -5 \end{vmatrix} = -43$

$M_{13} = \begin{vmatrix} 7 & -2 \\ 1 & 0 \end{vmatrix} = 2$

$M_{21} = \begin{vmatrix} 6 & 3 \\ 0 & -5 \end{vmatrix} = -30$

$M_{22} = \begin{vmatrix} -4 & 3 \\ 1 & -5 \end{vmatrix} = 17$

$M_{23} = \begin{vmatrix} -4 & 6 \\ 1 & 0 \end{vmatrix} = -6$

$M_{31} = \begin{vmatrix} 6 & 3 \\ -2 & 8 \end{vmatrix} = 54$

$M_{32} = \begin{vmatrix} -4 & 3 \\ 7 & 8 \end{vmatrix} = -53$

$M_{33} = \begin{vmatrix} -4 & 6 \\ 7 & -2 \end{vmatrix} = -34$

(b) $C_{11} = (-1)^2 M_{11} = 10$

$C_{12} = (-1)^3 M_{12} = 43$

$C_{13} = (-1)^4 M_{13} = 2$

$C_{21} = (-1)^3 M_{21} = 30$

$C_{22} = (-1)^4 M_{22} = 17$

$C_{23} = (-1)^5 M_{23} = 6$

$C_{31} = (-1)^4 M_{31} = 54$

$C_{32} = (-1)^5 M_{32} = 53$

$C_{33} = (-1)^6 M_{33} = -34$

19. (a) $\begin{vmatrix} -3 & 2 & 1 \\ 4 & 5 & 6 \\ 2 & -3 & 1 \end{vmatrix} = -3\begin{vmatrix} 5 & 6 \\ -3 & 1 \end{vmatrix} - 2\begin{vmatrix} 4 & 6 \\ 2 & 1 \end{vmatrix} + \begin{vmatrix} 4 & 5 \\ 2 & -3 \end{vmatrix}$

$= -3(23) - 2(-8) - 22 = -75$

(b) $\begin{vmatrix} -3 & 2 & 1 \\ 4 & 5 & 6 \\ 2 & -3 & 1 \end{vmatrix} = -2\begin{vmatrix} 4 & 6 \\ 2 & 1 \end{vmatrix} + 5\begin{vmatrix} -3 & 1 \\ 2 & 1 \end{vmatrix} + 3\begin{vmatrix} -3 & 1 \\ 4 & 6 \end{vmatrix}$

$= -2(-8) + 5(-5) + 3(-22) = -75$

21. (a) $\begin{vmatrix} 6 & 0 & -3 & 5 \\ 4 & 13 & 6 & -8 \\ -1 & 0 & 7 & 4 \\ 8 & 6 & 0 & 2 \end{vmatrix} = -4\begin{vmatrix} 0 & -3 & 5 \\ 0 & 7 & 4 \\ 6 & 0 & 2 \end{vmatrix} + 13\begin{vmatrix} 6 & -3 & 5 \\ -1 & 7 & 4 \\ 8 & 0 & 2 \end{vmatrix}$

$-6\begin{vmatrix} 6 & 0 & -3 \\ -1 & 0 & 7 \\ 8 & 6 & 0 \end{vmatrix} - 8\begin{vmatrix} 6 & 0 & -3 \\ -1 & 0 & 7 \\ 8 & 6 & 0 \end{vmatrix}$

$= -4(-282) + 13(-298) - 6(-174)$

$-8(-234)$

$= 170$

(b) $\begin{vmatrix} 6 & 0 & -3 & 5 \\ 4 & 13 & 6 & -8 \\ -1 & 0 & 7 & 4 \\ 8 & 6 & 0 & 2 \end{vmatrix} = 0\begin{vmatrix} 4 & 6 & -8 \\ -1 & 7 & 4 \\ 8 & 0 & 2 \end{vmatrix} + 13\begin{vmatrix} 6 & -3 & 5 \\ -1 & 7 & 4 \\ 8 & 0 & 2 \end{vmatrix}$

$+0\begin{vmatrix} 6 & -3 & 5 \\ 4 & 6 & -8 \\ 8 & -0 & 2 \end{vmatrix} + 6\begin{vmatrix} 6 & -3 & 5 \\ 4 & 6 & -8 \\ -1 & 7 & 4 \end{vmatrix}$

$= 0 + 13(-298) + 0 + 6(674) = 170$

23. Expand along Column 3.

$\begin{vmatrix} 1 & 4 & -2 \\ 3 & 2 & 0 \\ -1 & 4 & 3 \end{vmatrix} = -2\begin{vmatrix} 3 & 2 \\ -1 & 4 \end{vmatrix} + 3\begin{vmatrix} 1 & 4 \\ 3 & 2 \end{vmatrix}$

$= -2(14) + 3(-10) = -58$

25. Expand along Row 2.

$\begin{vmatrix} 6 & 3 & -7 \\ 0 & 0 & 0 \\ 4 & -6 & 3 \end{vmatrix} = 0\begin{vmatrix} 3 & -7 \\ -6 & 3 \end{vmatrix} - 0\begin{vmatrix} 6 & -7 \\ 4 & 3 \end{vmatrix} + 0\begin{vmatrix} 6 & 3 \\ 4 & -6 \end{vmatrix} = 0$

27. Expand along Column 1.

$\begin{vmatrix} -1 & 2 & -5 \\ 0 & 3 & -4 \\ 0 & 0 & 3 \end{vmatrix} = -1\begin{vmatrix} 3 & -4 \\ 0 & 3 \end{vmatrix} - 0\begin{vmatrix} 2 & -5 \\ 0 & 3 \end{vmatrix} + 0\begin{vmatrix} 2 & -5 \\ 3 & -4 \end{vmatrix}$

$= -1(9) - 0(6) + 0(7) = -9$

29. Expand along Column 3.

$\begin{vmatrix} 2 & 6 & 6 & 2 \\ 2 & 7 & 3 & 6 \\ 1 & 5 & 0 & 1 \\ 3 & 7 & 0 & 7 \end{vmatrix} = 6\begin{vmatrix} 2 & 7 & 6 \\ 1 & 5 & 1 \\ 3 & 7 & 7 \end{vmatrix} - 3\begin{vmatrix} 2 & 6 & 2 \\ 1 & 5 & 1 \\ 3 & 7 & 7 \end{vmatrix}$

$= 6(-20) - 3(16) = -168$

31. Expand along Column 2.

$\begin{vmatrix} 3 & 2 & 4 & -1 & 5 \\ -2 & 0 & 1 & 3 & 2 \\ 1 & 0 & 0 & 4 & 0 \\ 6 & 0 & 2 & -1 & 0 \\ 3 & 0 & 5 & 1 & 0 \end{vmatrix} = -2\begin{vmatrix} -2 & 1 & 3 & 2 \\ 1 & 0 & 4 & 0 \\ 6 & 2 & -1 & 0 \\ 3 & 5 & 1 & 0 \end{vmatrix} = (-2)(-2)\begin{vmatrix} 1 & 0 & 4 \\ 6 & 2 & -1 \\ 3 & 5 & 1 \end{vmatrix}$

$= 4(103) = 412$

33. $\begin{vmatrix} 1 & -1 & 8 & 4 \\ 2 & 6 & 0 & -4 \\ 2 & 0 & 2 & 6 \\ 0 & 2 & 8 & 0 \end{vmatrix} = -336$

35. $\begin{vmatrix} 3 & -2 & 4 & 3 & 1 \\ -1 & 0 & 2 & 1 & 0 \\ 5 & -1 & 0 & 3 & 2 \\ 4 & 7 & -8 & 0 & 0 \\ 1 & 2 & 3 & 0 & 2 \end{vmatrix} = 410$

37. (a) $\begin{vmatrix} -1 & 0 \\ 0 & 3 \end{vmatrix} = -3$

(b) $\begin{vmatrix} 2 & 0 \\ 0 & -1 \end{vmatrix} = -2$

(c) $\begin{bmatrix} -1 & 0 \\ 0 & 3 \end{bmatrix}\begin{bmatrix} 2 & 0 \\ 0 & -1 \end{bmatrix} = \begin{bmatrix} -2 & 0 \\ 0 & -3 \end{bmatrix}$

(d) $\begin{vmatrix} -2 & 0 \\ 0 & -3 \end{vmatrix} = 6$; $|AB| = |A||B|$

39. (a) $\begin{vmatrix} -1 & 2 & 1 \\ 1 & 0 & 1 \\ 0 & 1 & 0 \end{vmatrix} = 2$

(b) $\begin{vmatrix} -1 & 0 & 0 \\ 0 & 2 & 0 \\ 0 & 0 & 3 \end{vmatrix} = -6$

(c) $\begin{bmatrix} -1 & 2 & 1 \\ 1 & 0 & 1 \\ 0 & 1 & 0 \end{bmatrix}\begin{bmatrix} -1 & 0 & 0 \\ 0 & 2 & 0 \\ 0 & 0 & 3 \end{bmatrix} = \begin{bmatrix} 1 & 4 & 3 \\ -1 & 0 & 3 \\ 0 & 2 & 0 \end{bmatrix}$

(d) $\begin{vmatrix} 1 & 4 & 3 \\ -1 & 0 & 3 \\ 0 & 2 & 0 \end{vmatrix} = -12;\ |AB| = |A||B|$

41. (a) $\begin{vmatrix} 6 & 4 & 0 & 1 \\ 2 & -3 & -2 & -4 \\ 0 & 1 & 5 & 0 \\ -1 & 0 & -1 & 1 \end{vmatrix} = -25$

(b) $\begin{vmatrix} 0 & -5 & 0 & -2 \\ -2 & 4 & -1 & -4 \\ 3 & 0 & 1 & 0 \\ 1 & -2 & 3 & 0 \end{vmatrix} = -220$

(c) $\begin{bmatrix} 6 & 4 & 0 & 1 \\ 2 & -3 & -2 & -4 \\ 0 & 1 & 5 & 0 \\ -1 & 0 & -1 & 1 \end{bmatrix}\begin{bmatrix} 0 & -5 & 0 & -2 \\ -2 & 4 & -1 & -4 \\ 3 & 0 & 1 & 0 \\ 1 & -2 & 3 & 0 \end{bmatrix} = \begin{bmatrix} -7 & -16 & -1 & -28 \\ -4 & -14 & -11 & 8 \\ 13 & 4 & 4 & -4 \\ -2 & 3 & 2 & 2 \end{bmatrix}$

(d) $\begin{vmatrix} -7 & -16 & -1 & -28 \\ -4 & -14 & -11 & 8 \\ 13 & 4 & 4 & -4 \\ -2 & 3 & 2 & 2 \end{vmatrix} = 5500;\ |AB| = |A||B|$

43. $\begin{vmatrix} w & x \\ y & z \end{vmatrix} = wz - xy$

$-\begin{vmatrix} y & z \\ w & x \end{vmatrix} = -(xy - wz) = wz - xy$

Thus, $\begin{vmatrix} w & x \\ y & z \end{vmatrix} = -\begin{vmatrix} y & z \\ w & x \end{vmatrix}$.

45. $\begin{vmatrix} w & x \\ y & z \end{vmatrix} = wz - xy$

$\begin{vmatrix} w & x + cw \\ y & z + cy \end{vmatrix} = w(z + cy) - y(x + cw) = wz - xy$

Thus, $\begin{vmatrix} w & x \\ y & z \end{vmatrix} = \begin{vmatrix} w & x + cw \\ y & z + cy \end{vmatrix}$.

47. $\begin{vmatrix} 1 & x & x^2 \\ 1 & y & y^2 \\ 1 & z & z^2 \end{vmatrix} = \begin{vmatrix} y & y^2 \\ z & z^2 \end{vmatrix} - \begin{vmatrix} x & x^2 \\ z & z^2 \end{vmatrix} + \begin{vmatrix} x & x^2 \\ y & y^2 \end{vmatrix}$

$= (yz^2 - y^2z) - (xz^2 - x^2z) + (xy^2 - x^2y)$

$= yz^2 - xz^2 - y^2z + x^2z + xy(y - x)$

$= z^2(y - x) - z(y^2 - x^2) + xy(y - x)$

$= z^2(y - x) - z(y - x)(y + x) + xy(y - x)$

$= (y - x)[z^2 - z(y + x) + xy]$

$= (y - x)[z^2 - zy - zx + xy]$

$= (y - x)[z^2 - zx - zy + xy]$

$= (y - x)[z(z - x) - y(z - x)]$

$= (y - x)(z - x)(z - y)$

49. $\begin{vmatrix} x & 2 \\ 1 & x \end{vmatrix} = 2$

$x^2 - 2 = 2$

$x^2 = 4$

$x = \pm 2$

51. $\begin{vmatrix} 2x & -3 \\ -2 & 2x \end{vmatrix} = 3$

$4x^2 - 6 = 3$

$4x^2 = 9$

$x^2 = \frac{9}{4}$

$x = \pm\frac{3}{2}$

53. $\begin{vmatrix} x & 1 \\ 2 & x - 2 \end{vmatrix} = -1$

$x^2 - 2x - 2 = -1$

$x^2 - 2x - 1 = 0$

$x = \dfrac{2 \pm \sqrt{4 + 4}}{2}$

$x = 1 \pm \sqrt{2}$

55. $\begin{vmatrix} x + 3 & 2 \\ 1 & x + 2 \end{vmatrix} = 0$

$(x + 3)(x + 2) - 2 = 0$

$x^2 + 5x + 4 = 0$

$(x + 4)(x + 1) = 0$

$x = -4,\ -1$

57. $\begin{vmatrix} 2x & 1 \\ -1 & x-1 \end{vmatrix} = x$

$$2x^2 - 2x + 1 = x$$
$$2x^2 - 3x + 1 = 0$$
$$(x-1)(2x-1) = 0$$
$$x = 1, \frac{1}{2}$$

59. $\begin{vmatrix} 1 & 2 & x \\ -1 & 3 & 2 \\ 3 & -2 & 1 \end{vmatrix} = 0$

$$1\begin{vmatrix} 3 & 2 \\ -2 & 1 \end{vmatrix} - 2\begin{vmatrix} -1 & 2 \\ 3 & 1 \end{vmatrix} + x\begin{vmatrix} -1 & 3 \\ 3 & -2 \end{vmatrix} = 0$$
$$7 - 2(-7) + x(-7) = 0$$
$$21 = 7x$$
$$x = 3$$

61. $\begin{vmatrix} 4u & -1 \\ -1 & 2v \end{vmatrix} = 8uv - 1$

63. $\begin{vmatrix} e^{2x} & e^{3x} \\ 2e^{2x} & 3e^{3x} \end{vmatrix} = 3e^{5x} - 2e^{5x} = e^{5x}$

65. $\begin{vmatrix} x & \ln x \\ 1 & \frac{1}{x} \end{vmatrix} = 1 - \ln x$

67. True. Expand along the row of zeros.

69. Answers will vary. Possible answer

$$A = \begin{bmatrix} 1 & 0 & -3 \\ 6 & -2 & 7 \\ 9 & 5 & -1 \end{bmatrix}, B = \begin{bmatrix} 3 & 1 & 5 \\ -8 & 1 & 0 \\ -7 & 6 & -2 \end{bmatrix}$$

$$A + B = \begin{bmatrix} 1+3 & 0+1 & -3+5 \\ 6-8 & -2+1 & 7+0 \\ 9+7 & 5+6 & -1-2 \end{bmatrix} = \begin{bmatrix} 4 & 1 & 2 \\ -2 & -1 & 7 \\ 2 & 11 & -3 \end{bmatrix}$$

Expand along Row 1 to find $|A + B|$.

$$|A + B| = 4\begin{vmatrix} -1 & 7 \\ 11 & -3 \end{vmatrix} - 1\begin{vmatrix} -2 & 7 \\ 2 & -3 \end{vmatrix} + 2\begin{vmatrix} -2 & -1 \\ 2 & 11 \end{vmatrix}$$
$$= 4(-74) - (-8) + 2(-20)$$
$$= -328$$

Expand along Row 1 to find $|A|$.

$$|A| = 1\begin{vmatrix} -2 & 7 \\ 5 & -1 \end{vmatrix} - 0\begin{vmatrix} 6 & 7 \\ 9 & -1 \end{vmatrix} + 3\begin{vmatrix} 6 & -2 \\ 9 & 5 \end{vmatrix}$$
$$= -33 - 0 - 3(48)$$
$$= -177$$

Expand along Column 3 to find $|B|$.

$$|B| = 5\begin{vmatrix} -8 & 1 \\ -7 & 6 \end{vmatrix} - 0\begin{vmatrix} 3 & 1 \\ -7 & 6 \end{vmatrix} - 2\begin{vmatrix} 3 & 1 \\ -8 & 1 \end{vmatrix}$$
$$= 5(-41) - 2(11)$$
$$= -227$$
$$|A| + |B| = -177 - 227 = -404$$

Therefore, $|A + B| \neq |A| + |B|$.

71. (a) $|A| = 6$

(b) $A^{-1} = \begin{bmatrix} \frac{1}{3} & -\frac{1}{3} \\ \frac{1}{3} & \frac{1}{6} \end{bmatrix}$

(c) $\det(A^{-1}) = \frac{1}{6}$

(d) In general, $\det(A^{-1}) = \frac{1}{\det A}$.

73. (a) $|A| = 2$

(b) $A^{-1} = \begin{bmatrix} -4 & -5 & 1.5 \\ -1 & -1 & 0.5 \\ -1 & -1 & 0 \end{bmatrix}$

(c) $\det(A^{-1}) = \frac{1}{2}$

(d) In general, $\det(A^{-1}) = \frac{1}{\det A}$.

75. (a) Columns 2 and 3 are interchanged.
(b) Rows 1 and 3 are interchanged.

77. (a) 3 is factored out of the first row of A.
(b) 2 is factored out of the first column of A and 4 is factored out of the second column of A.

79. (a) $\begin{bmatrix} 3 & -2 \\ 0 & 5 \end{bmatrix}; \begin{vmatrix} 3 & -2 \\ 0 & 5 \end{vmatrix} = 15$

(b) $\begin{bmatrix} 3 & -7 & 1 \\ 0 & -5 & -9 \\ 0 & 0 & 5 \end{bmatrix}; \begin{vmatrix} 3 & -7 & 1 \\ 0 & -5 & -9 \\ 0 & 0 & 5 \end{vmatrix} = -75$

(c) $\begin{bmatrix} 4 & 0 & 0 & 0 \\ 3 & -3 & 0 & 0 \\ 3 & 6 & 5 & 0 \\ 2 & -2 & 1 & 2 \end{bmatrix}; \begin{vmatrix} 4 & 0 & 0 & 0 \\ 3 & -3 & 0 & 0 \\ 3 & 6 & 5 & 0 \\ 2 & -2 & 1 & 2 \end{vmatrix} = -120$

The determinant of a triangular matrix is the product of the entries along the main diagonal.

81. $|AB| = |A||B|$

$$A = \begin{bmatrix} a_{11} & a_{12} \\ a_{21} & a_{22} \end{bmatrix},\ B = \begin{bmatrix} b_{11} & b_{12} \\ b_{21} & b_{22} \end{bmatrix}$$

$$|A| = a_{11}a_{22} - a_{21}a_{12},\ |B| = b_{11}b_{22} - b_{21}b_{12}$$

$$|A||B| = a_{11}a_{22}b_{11}b_{22} - a_{11}a_{22}b_{21}b_{12} - a_{21}a_{12}b_{11}b_{22} + a_{21}a_{12}b_{21}b_{12}$$

$$AB = \begin{bmatrix} a_{11} & a_{12} \\ a_{21} & a_{22} \end{bmatrix}\begin{bmatrix} b_{11} & b_{12} \\ b_{21} & b_{22} \end{bmatrix}$$

$$= \begin{bmatrix} a_{11}b_{11} + a_{12}b_{21} & a_{11}b_{12} + a_{12}b_{22} \\ a_{21}b_{11} + a_{22}b_{21} & a_{21}b_{12} + a_{22}b_{22} \end{bmatrix}$$

$$|AB| = (a_{11}b_{11} + a_{12}b_{21})(a_{21}b_{12} + a_{22}b_{22})$$

$$-(a_{21}b_{11} + a_{22}b_{21})(a_{11}b_{12} + a_{12}b_{22})$$

$$= a_{11}b_{11}a_{21}b_{12} + a_{11}b_{11}a_{22}b_{22} + a_{12}b_{21}a_{21}b_{12} + a_{12}b_{21}a_{22}b_{22}$$

$$- a_{21}b_{11}a_{11}b_{12} - a_{21}b_{11}a_{12}b_{22} - a_{22}b_{21}a_{11}b_{12} - a_{22}b_{21}a_{12}b_{22}$$

(*Note*: First and fifth terms and fourth and eighth terms when combined subtract out, leaving the second, third, sixth, and seventh terms.)

$$= a_{11}b_{11}a_{22}b_{22} + a_{12}b_{21}a_{21}b_{12} - a_{21}b_{11}a_{12}b_{22} - a_{22}b_{21}a_{11}b_{12}$$

Conclusion: $|AB| = |A||B|$

83. $x^2 - 3x + 2 = (x-2)(x-1)$

85. $4y^2 - 12y + 9 = (2y-3)^2$

87. $\begin{cases} 3x - 10y = 46 \\ x + \ \ y = -2 \end{cases}$

$y = -x - 2$

$3x - 10(-x-2) = 46$

$13x = 26$

$x = 2$

$y = -2 - 2 = -4$

Answer: $(2, -4)$

Section 7.8

1. Cramer's Rule

3. The area of the triangle with vertices $(x_1, y_1), (x_2, y_2), (x_3, y_3)$ can be found by:

$$Area = \pm\tfrac{1}{2}\begin{vmatrix} x_1 & y_1 & 1 \\ x_2 & y_2 & 1 \\ x_3 & y_3 & 1 \end{vmatrix},$$

so if the determinant is equal to -6, then multiply by $-\frac{1}{2}$ so yield an area of 3 square units.

5. Vertices: $(-2, 4), (2, 3), (-1, 5)$

$$\tfrac{1}{2}\begin{vmatrix} -2 & 4 & 1 \\ 2 & 3 & 1 \\ -1 & 5 & 1 \end{vmatrix} = \tfrac{1}{2}\left[-2\begin{vmatrix} 3 & 1 \\ 5 & 1 \end{vmatrix} - 4\begin{vmatrix} 2 & 1 \\ -1 & 1 \end{vmatrix} + \begin{vmatrix} 2 & 3 \\ -1 & 5 \end{vmatrix}\right]$$

$$= \tfrac{1}{2}\left[-2(-2) - 4(3) + 13\right] = \tfrac{1}{2}(5) = \tfrac{5}{2}$$

Area $= \frac{5}{2}$ square units

7. Vertices: $(0, \frac{1}{2}), (\frac{5}{2}, 0), (4, 3)$

$$\tfrac{1}{2}\begin{vmatrix} 0 & \frac{1}{2} & 1 \\ \frac{5}{2} & 0 & 1 \\ 4 & 3 & 1 \end{vmatrix} = \tfrac{1}{2}\left[-\tfrac{1}{2}\left(-\tfrac{3}{2}\right) + \tfrac{15}{2}\right] = \tfrac{1}{2}\left[\tfrac{33}{4}\right] = \tfrac{33}{8}$$

Area $= \frac{33}{8}$ square units

9. Vertices: $(-3, 2), (1, 2), (-1, -4), (-5, -4)$

$$\begin{vmatrix} -3 & 2 & 1 \\ 1 & 2 & 1 \\ -1 & -4 & 1 \end{vmatrix} = -3(6) - 2(2) + 1(-2) = -24$$

Area of rhombus $= |-24| = 24$ square units

11. $4 = \pm\frac{1}{2}\begin{vmatrix} -1 & 5 & 1 \\ -2 & 0 & 1 \\ x & 2 & 1 \end{vmatrix}$

$8 = \pm\left[(-1)(-2) - 5(-2-x) + 1(-4)\right]$

$8 = \pm\left[5x + 8\right]$

$5x + 8 = 8$ or $5x + 8 = -8$

$x = 0$ or $x = -\frac{16}{5}$

$x = 0, -\frac{16}{5}$

13. Points: $(3, -1), (0, -3), (12, 5)$

$$\begin{vmatrix} 3 & -1 & 1 \\ 0 & -3 & 1 \\ 12 & 5 & 1 \end{vmatrix} = 3(-8) + 12(2) = 0$$

The points are collinear.

15. Points: $\left(2, -\frac{1}{2}\right), (-4, 4), (6, -3)$

$$\begin{vmatrix} 2 & -\frac{1}{2} & 1 \\ -4 & 4 & 1 \\ 6 & -3 & 1 \end{vmatrix} = 2(7) + \tfrac{1}{2}(-10) + 1(-12)$$

$$= -3 \neq 0$$

The points are not collinear.

17. $\begin{vmatrix} 1 & -2 & 1 \\ x & 2 & 1 \\ 5 & 6 & 1 \end{vmatrix} = 0$

$$1(-4) + 2(x-5) + 1(6x-10) = 0$$
$$8x - 24 = 0$$
$$x = 3$$

19. $\begin{cases} -7x + 11y = -1 \\ 3x - 9y = 9 \end{cases}$

$$x = \frac{\begin{vmatrix} -1 & 11 \\ 9 & -9 \end{vmatrix}}{\begin{vmatrix} -7 & 11 \\ 3 & -9 \end{vmatrix}} = \frac{-90}{30} = -3$$

$$y = \frac{\begin{vmatrix} -7 & -1 \\ 3 & 9 \end{vmatrix}}{\begin{vmatrix} -7 & 11 \\ 3 & -9 \end{vmatrix}} = \frac{-60}{30} = -2$$

Answer: $(-3, -2)$

21. $\begin{cases} 3x + 2y = -2 \\ 6x + 4y = 4 \end{cases}$

$$\begin{vmatrix} 3 & 2 \\ 6 & 4 \end{vmatrix} = 12 - 12 = 0$$

Cramer's Rule cannot be used.
(In fact, the system is inconsistent.)

23. $\begin{cases} 4x - y + z = -5 \\ 2x + 2y + 3z = 10 \\ 5x - 2y + 6z = 1 \end{cases}$ $D = \begin{vmatrix} 4 & -1 & 1 \\ 2 & 2 & 3 \\ 5 & -2 & 6 \end{vmatrix} = 55$

$$x = \frac{\begin{vmatrix} -5 & -1 & 1 \\ 10 & 2 & 3 \\ 1 & -2 & 6 \end{vmatrix}}{55} = \frac{-55}{55} = -1,$$

$$y = \frac{\begin{vmatrix} 4 & -5 & 1 \\ 2 & 10 & 3 \\ 5 & 1 & 6 \end{vmatrix}}{55} = \frac{165}{55} = 3,$$

$$z = \frac{\begin{vmatrix} 4 & -1 & -5 \\ 2 & 2 & 10 \\ 5 & -2 & 1 \end{vmatrix}}{55} = \frac{110}{55} = 2$$

Answer: $(-1, 3, 2)$

25. (a) $\begin{cases} 3x + 3y + 5z = 1 \\ 3x + 5y + 9z = 2 \\ 5x + 9y + 17z = 4 \end{cases}$

$$\begin{cases} 3x + 3y + 5z = 1 \\ 2y + 4z = 1 \\ 4y + \frac{26}{3}z = \frac{7}{3} \end{cases}$$

$$\begin{cases} 3x + 3y + 5z = 1 \\ 2y + 4z = 1 \\ \frac{2}{3}z = \frac{1}{3} \end{cases}$$

$z = \frac{1}{2}$

$2y + 4\left(\frac{1}{2}\right) = 1 \Rightarrow y = -\frac{1}{2}$

$3x + 3\left(-\frac{1}{2}\right) + 5\left(\frac{1}{2}\right) = 1 \Rightarrow x = 0$

Answer: $\left(0, -\frac{1}{2}, \frac{1}{2}\right)$

(b) $D = \begin{vmatrix} 3 & 3 & 5 \\ 3 & 5 & 9 \\ 5 & 9 & 17 \end{vmatrix} = 4$

$$x = \frac{\begin{vmatrix} 1 & 3 & 5 \\ 2 & 5 & 9 \\ 4 & 9 & 17 \end{vmatrix}}{4} = 0$$

$$y = \frac{\begin{vmatrix} 3 & 1 & 5 \\ 3 & 2 & 9 \\ 5 & 4 & 17 \end{vmatrix}}{4} = -\frac{1}{2}$$

$$z = \frac{\begin{vmatrix} 3 & 3 & 1 \\ 3 & 5 & 2 \\ 5 & 9 & 4 \end{vmatrix}}{4} = \frac{1}{2}$$

Answer: $\left(0, -\frac{1}{2}, \frac{1}{2}\right)$

27. (a) $a = \frac{\begin{vmatrix} 15{,}489.6 & 1260 & 190 \\ 2422.0 & 190 & 30 \\ 398.8 & 30 & 5 \end{vmatrix}}{\begin{vmatrix} 8674 & 1260 & 190 \\ 1260 & 190 & 30 \\ 190 & 30 & 5 \end{vmatrix}} \approx -1.086$

$$b = \frac{\begin{vmatrix} 8674 & 15{,}489.6 & 190 \\ 1260 & 2422.0 & 30 \\ 190 & 398.8 & 5 \end{vmatrix}}{\begin{vmatrix} 8674 & 1260 & 190 \\ 1260 & 190 & 30 \\ 190 & 30 & 5 \end{vmatrix}} \approx 15.949$$

$$c = \frac{\begin{vmatrix} 8674 & 1260 & 15{,}489.6 \\ 1260 & 190 & 2422.0 \\ 190 & 30 & 398.8 \end{vmatrix}}{\begin{vmatrix} 8674 & 1260 & 190 \\ 1260 & 190 & 30 \\ 190 & 30 & 5 \end{vmatrix}} \approx 25.326$$

$y = -1.086t^2 + 15.949t + 25.326$

(b)

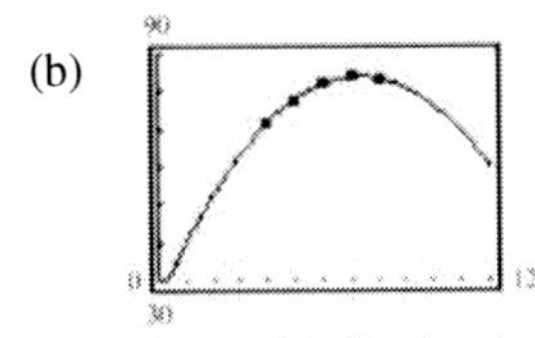

The model fits the data well.

29. (a) The uncoded row matrices are the rows of the 5×3 matrix:

$$\begin{matrix} \text{T} & \text{E} & \text{X} \\ \text{T} & & \text{M} \\ \text{E} & & \text{A} \\ \text{T} & & \text{W} \\ \text{O} & \text{R} & \text{K} \end{matrix} \begin{bmatrix} 20 & 5 & 24 \\ 20 & 0 & 13 \\ 5 & 0 & 1 \\ 20 & 0 & 23 \\ 15 & 18 & 11 \end{bmatrix}$$

(b) $$\begin{bmatrix} 20 & 5 & 24 \\ 20 & 0 & 13 \\ 5 & 0 & 1 \\ 20 & 0 & 23 \\ 15 & 18 & 11 \end{bmatrix} \begin{bmatrix} 1 & -1 & 0 \\ 1 & 0 & -1 \\ -6 & 2 & 3 \end{bmatrix} = \begin{bmatrix} -119 & 28 & 67 \\ -58 & 6 & 39 \\ -1 & -3 & 3 \\ -118 & 26 & 69 \\ -33 & 7 & 15 \end{bmatrix}$$

(c) *Answer:*

−119 28 67 −58 6 39 −1

−3 3 −118 26 69 −33 7 15

31. $\overset{\text{K}\ \ \text{E}\ \ \text{Y}}{[11\ \ 5\ \ 25]}\ \overset{\_\ \ \text{U}\ \ \text{N}}{[0\ \ 21\ \ 14]}\ \overset{\text{D}\ \ \text{E}\ \ \text{R}}{[4\ \ 5\ \ 18]}$

$\overset{\_\ \ \text{R}\ \ \text{U}}{[0\ \ 18\ \ 21]}\ \overset{\text{G}\ \ \_\ \ \_}{[7\ \ 0\ \ 0]}$

$$\begin{aligned} [11\ \ 5\ \ 25]A &= [1\ \ -43\ \ -108] \\ [0\ \ 21\ \ 14]A &= [49\ \ 91\ \ 91] \\ [4\ \ 5\ \ 18]A &= [1\ \ -29\ \ -73] \\ [0\ \ 18\ \ 21]A &= [33\ \ 42\ \ 15] \\ [7\ \ 0\ \ 0]A &= [7\ \ 14\ \ 14] \end{aligned}$$

Cryptogram:

1 −43 −108 49 91 91 1 −29 −73

33 42 15 7 14 14

33. $A = \begin{bmatrix} 1 & 2 \\ 3 & 5 \end{bmatrix}, A^{-1} = \begin{bmatrix} -5 & 2 \\ 3 & -1 \end{bmatrix}$

$$\begin{bmatrix} 11 & 21 \\ 64 & 112 \\ 25 & 50 \\ 29 & 53 \\ 23 & 46 \\ 40 & 75 \\ 55 & 92 \end{bmatrix} \begin{bmatrix} -5 & 2 \\ 3 & -1 \end{bmatrix} = \begin{bmatrix} 8 & 1 \\ 16 & 16 \\ 25 & 0 \\ 14 & 5 \\ 23 & 0 \\ 25 & 5 \\ 1 & 18 \end{bmatrix} \begin{matrix} \text{H} & \text{A} \\ \text{P} & \text{P} \\ \text{Y} & \\ \text{N} & \text{E} \\ \text{W} & \\ \text{Y} & \text{E} \\ \text{A} & \text{R} \end{matrix}$$

Message: HAPPY NEW YEAR

35. $A = \begin{bmatrix} 1 & 2 & 1 \\ -1 & 0 & 2 \\ 1 & -1 & -2 \end{bmatrix}, A^{-1} = \begin{bmatrix} \frac{2}{3} & 1 & \frac{4}{3} \\ 0 & -1 & -1 \\ \frac{1}{3} & 1 & \frac{2}{3} \end{bmatrix}$

$$\begin{bmatrix} 3 & 18 & 21 \\ 31 & 29 & 13 \\ -2 & -1 & 4 \\ -6 & 28 & 54 \\ -3 & 4 & 12 \\ 16 & 8 & 1 \\ 6 & 6 & 0 \\ 27 & -12 & -39 \\ 15 & -19 & -27 \\ 5 & 10 & 5 \end{bmatrix} \begin{bmatrix} \frac{2}{3} & 1 & \frac{4}{3} \\ 0 & -1 & -1 \\ \frac{1}{3} & 1 & \frac{2}{3} \end{bmatrix} = \begin{bmatrix} 9 & 6 & 0 \\ 25 & 15 & 21 \\ 0 & 3 & 1 \\ 14 & 20 & 0 \\ 2 & 5 & 0 \\ 11 & 9 & 14 \\ 4 & 0 & 2 \\ 5 & 0 & 22 \\ 1 & 7 & 21 \\ 5 & 0 & 0 \end{bmatrix}$$

Message: IF YOU CANT BE KIND BE VAGUE

37. True. Cramer's Rule requires that the determinant of the coefficient matrix be nonzero.

39. Answers will vary.

41.
$$\begin{aligned} y - 5 &= \frac{5-3}{-1-7}(x+1) = \frac{-1}{4}(x+1) \\ 4y - 20 &= -x - 1 \\ 4y + x &= 19 \\ x + 4y - 19 &= 0 \end{aligned}$$

43.
$$\begin{aligned} y + 3 &= \frac{-3+1}{3-10}(x-3) = \frac{2}{7}(x-3) \\ 7y + 21 &= 2x - 6 \\ 7y - 2x &= -27 \\ 2x - 7y - 27 &= 0 \end{aligned}$$

Chapter 7 Review Exercises

1. $\begin{cases} x + y = 2 \Rightarrow & y = 2 - x \\ x - y = 0 \Rightarrow & x - (2 - x) = 0 \end{cases}$

$$2x - 2 = 0$$
$$x = 1$$
$$y = 2 - 1 = 1$$

Answer: $(1, 1)$

3. $\begin{cases} 4x - y = 1 \\ 8x + y = 17 \Rightarrow y = 17 - 8x \end{cases}$

$$4x - (17 - 8x) = 1$$
$$4x - 17 + 8x = 1$$
$$12x = 18$$
$$x = \tfrac{3}{2}$$
$$y = 17 - 8\left(\tfrac{3}{2}\right) = 5$$

Answer: $\left(\tfrac{3}{2}, 5\right)$

5. $\begin{cases} 0.5x + y = 0.75 \Rightarrow y = -0.5x + 0.75 \\ 1.25x - 4.5y = -2.5 \end{cases}$

$$1.25x - 4.5(-0.5x + 0.75) = -2.5$$
$$1.25x + 2.25x - 3.375 = -2.5$$
$$3.5x = 0.875$$
$$x = 0.25$$
$$y = -0.5(0.25) + 0.75 = 0.625$$

Answer: $(0.25, 0.625)$

7. $\begin{cases} x^2 - y^2 = 9 \\ x - y = 1 \Rightarrow x = y + 1 \end{cases}$

$$(y+1)^2 - y^2 = 9$$
$$2y + 1 = 9$$
$$y = 4$$
$$x = 5$$

Answer: $(5, 4)$

9. $\begin{cases} y = 2x^2 \\ y = x^4 - 2x^2 \Rightarrow 2x^2 = x^4 - 2x^2 \end{cases}$

$$0 = x^4 - 4x^2$$
$$0 = x^2(x^2 - 4)$$
$$0 = x^2(x+2)(x-2)$$
$$x = 0, x = -2, x = 2$$
$$y = 0, y = 8, y = 8$$

Answer: $(0, 0), (-2, 8), (2, 8)$

11. $\begin{cases} 5x + 6y = 7 \Rightarrow y_1 = \frac{1}{6}(7 - 5x) \\ -x - 4y = 0 \Rightarrow y_2 = -\frac{x}{4} \end{cases}$

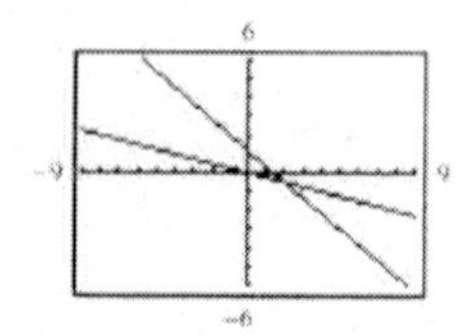

Point of intersection: $\left(2, -\tfrac{1}{2}\right)$

13. $\begin{cases} y^2 - 4x = 0 \Rightarrow & y^2 = 4x \Rightarrow y = \pm 2\sqrt{x} \\ x + y = 0 \Rightarrow & y = -x \end{cases}$

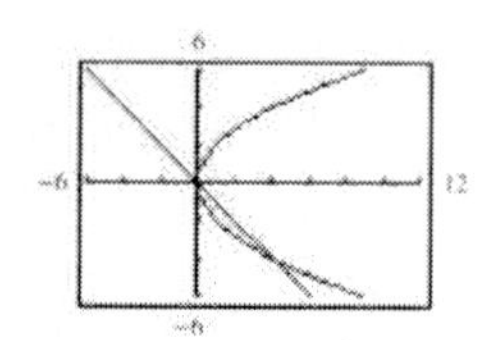

Points of intersection: $(0, 0), (4, -4)$

15. $\begin{cases} y = 3 - x^2 \\ y = 2x^2 + x + 1 \end{cases}$

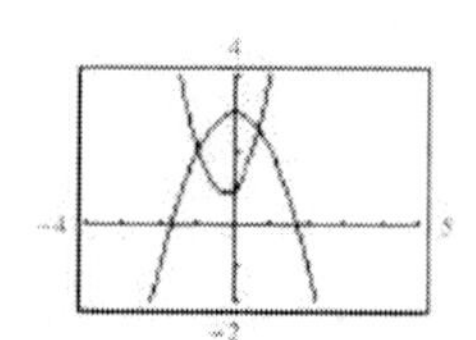

Points of intersection: $(0.67, 2.56), (-1, 2)$

17. $\begin{cases} y = 2(6 - x) \\ y = 2^{x-2} \end{cases}$

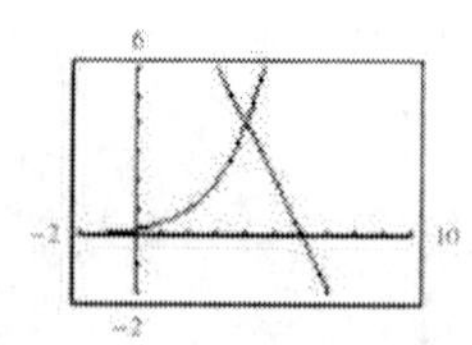

Point of intersection: $(4, 4)$

19. Let x = number of plants.

$\begin{cases} C = 6.43x + 5000 & \text{Cost equation} \\ R = 12.68x & \text{Revenue equation} \end{cases}$

$$R = C \Rightarrow 6.43x + 5000 = 12.68x$$
$$-6.25x = -5000$$
$$x = 800 \text{ plants}$$

21. $\begin{cases} 2l + 2w = 480 \\ l = 1.50w \end{cases}$

$$2(1.50w) + 2w = 480$$
$$5w = 480$$
$$w = 96$$
$$l = 144$$

The dimensions are 96×144 meters.

23. $\begin{cases} 2x - y = 2 \Rightarrow 16x - 8y = 16 \\ 6x + 8y = 39 \Rightarrow 6x + 8y = 39 \end{cases}$

$$22x = 55$$
$$x = \tfrac{55}{22} = \tfrac{5}{2}$$
$$y = 3$$

Answer: $\left(\frac{5}{2}, 3\right)$

25. $\begin{cases} 0.2x + 0.3y = 0.14 \\ 0.4x + 0.5y = 0.20 \end{cases}$

-2 times Equation 1 and add to Equation 2 to eliminate x:

$$-0.4x - 0.6y = -0.28$$
$$\underline{0.4x + 0.5y = 0.20}$$
$$-0.1y = -0.08$$
$$y = 0.8$$

Substitute $y = 0.8$ into Equation 1:

$$0.2x + 0.3(0.8) = 0.14$$
$$0.2x + 0.24 = 0.14$$
$$0.2x = -0.1$$
$$x = -0.5$$

Answer: $(-0.5, 0.8)$

27. $\begin{cases} 1/5x + 3/10y = 7/50 \Rightarrow 20x + 30y = 14 \Rightarrow 20x + 30y = 14 \\ 2/5x + 1/2y = 1/5 \Rightarrow 4x + 5y = 2 \Rightarrow -20x - 25y = -10 \end{cases}$

$$5y = 4$$
$$y = \tfrac{4}{5}$$
$$x = -\tfrac{1}{2}$$

Answer: $\left(-\frac{1}{2}, \frac{4}{5}\right)$ or $(-0.5, 0.8)$

29. $\begin{cases} 3x - 2y = 0 \Rightarrow 3x - 2y = 0 \\ 3x + 2(y + 5) = 10 \Rightarrow 3x + 2y = 0 \end{cases}$

$$6x = 0$$
$$x = 0$$
$$y = 0$$

Answer: $(0, 0)$

31. $\begin{cases} 1.25x - 2y = 3.5 \Rightarrow 5x - 8y = 14 \\ 5x - 8y = 14 \Rightarrow -5x + 8y = -14 \end{cases}$

$$0 = 0$$

Infinite number of solutions

Let $y = a$, then $5x - 8a = 14 \Rightarrow x = \frac{14}{5} + \frac{8}{5}a$.

Answer: $\left(\frac{14}{5} + \frac{8}{5}a, a\right)$

33. $\begin{cases} 3x + 2y = 0 \Rightarrow y = -\frac{3}{2}x \\ x - y = 4 \Rightarrow y = x - 4 \end{cases}$

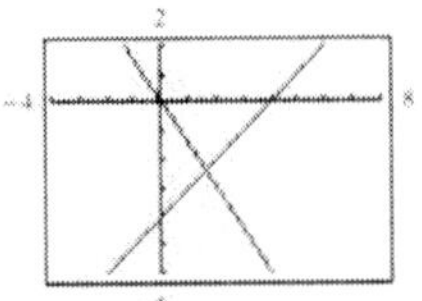

Consistent

Answer: $(1.6, -2.4)$

35. $\begin{cases} \frac{1}{4}x - \frac{1}{5}y = 2 \Rightarrow y = \frac{5}{4}x - 10 \\ -5x + 4y = 8 \Rightarrow y = \frac{1}{4}(8 + 5x) = \frac{5}{4}x + 2 \end{cases}$

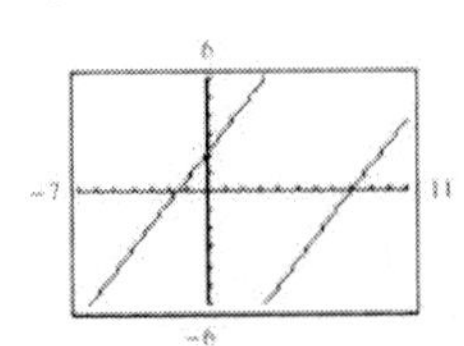

Inconsistent; lines are parallel.

37. $\begin{cases} 2x - 2y = 8 \Rightarrow y = x - 4 \\ 4x + 1.5y = -5.5 \Rightarrow y = \frac{8}{3}x + \frac{11}{3} \end{cases}$

Consistent

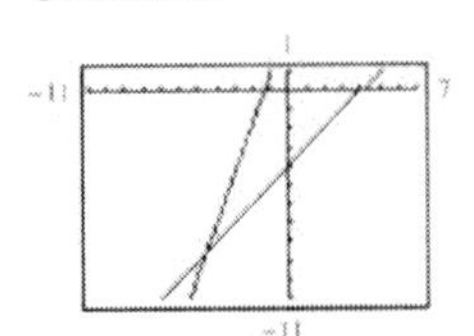

Answer: $(-4.6, -8.6)$

39. Demand = Supply

$$37 - 0.0002x = 22 + 0.00001x$$
$$15 = 0.00021x$$
$$x = \frac{500{,}000}{7}, \; p = \frac{159}{7}$$

Point of equilibrium: $\left(\dfrac{500{,}000}{7}, \dfrac{159}{7}\right)$

41. Let $x =$ speed of the slower plane.

Let $y =$ speed of the faster plane.

Then, distance of first plane + distance of second plane = 275 miles.

(rate of first plane)(time) + (rate of second plane)(time) = 275 miles.

$$\begin{cases} x\left(\frac{40}{60}\right) + y\left(\frac{40}{60}\right) = 275 \\ y = x + 25 \end{cases}$$

$$\frac{2}{3}x + \frac{2}{3}(x+25) = 275$$
$$4x + 50 = 825$$
$$4x = 775$$
$$x = 193.75 \text{ mph}$$
$$y = x + 25 = 218.75 \text{ mph}$$

43.
$$\begin{cases} x - 4y + 3z = 3 \\ -y + z = -1 \\ z = -5 \end{cases}$$

Substitute $z = -5$ into Equation 2:

$$-y + (-5) = -1$$
$$-y = 4$$
$$y = -4$$

Substitute $y = -4$ and $z = -5$ into Equation 1:

$$x - 4(-4) + 3(-5) = 3$$
$$x + 16 - 15 = 3$$
$$x = 2$$

Answer: $(2, -4, -5)$

45.
$$\begin{cases} x + 3y - z = 13 \\ 2x - 5z = 23 \\ 4x - y - 2z = 14 \end{cases}$$

$$\begin{cases} x + 3y - z = 13 \\ -6y - 3z = -3 & -2\text{ Eq.1} + \text{Eq.2} \\ -13y + 2z = -38 & -4\text{ Eq.1} + \text{Eq.3} \end{cases}$$

$$\begin{cases} x + 3y - z = 13 \\ -6y - 3z = -3 \\ \frac{17}{2}z = -\frac{63}{2} & -\frac{13}{6}\text{ Eq.2} + \text{Eq.3} \end{cases}$$

$$\frac{17}{2}z = -\frac{63}{2} \Rightarrow z = -\frac{63}{17}$$
$$-6y - 3\left(-\frac{63}{17}\right) = -3 \Rightarrow y = \frac{40}{17}$$
$$x + 3\left(\frac{40}{17}\right) - \left(-\frac{63}{17}\right) = 13 \Rightarrow x = \frac{38}{17}$$

Answer: $\left(\frac{38}{17}, \frac{40}{17}, -\frac{63}{17}\right)$

47.
$$\begin{cases} x - 2y + z = -6 \\ 2x - 3y = -7 \\ -x + 3y - 3z = 11 \end{cases}$$

$$\begin{cases} x - 2y + z = -6 \\ y - 2z = 5 & -2\text{ Eq.1} + \text{Eq. 2} \\ y - 2z = 5 & \text{Eq.1} + \text{Eq. 3} \end{cases}$$

$$\begin{cases} x - 2y + z = -6 \\ y - 2z = 5 \\ 0 = 0 & -\text{Eq. 2} + \text{Eq. 3} \end{cases}$$

Let $z = a$, then $y = 2a + 5$.

$$x - 2(2a+5) + a = -6$$
$$x - 3a - 10 = -6$$
$$x = 3a + 4$$

Answer: $(3a+4, 2a+5, a)$, where a is any real number

49.
$$\begin{cases} x - 2y + 3z = -5 \\ 2x + 4y + 5z = 1 \\ x + 2y + z = 0 \end{cases}$$

$$\begin{cases} x - 2y + 3z = -5 \\ 8y - z = 11 & -2\text{ Eq.1} + \text{Eq. 2} \\ 4y - 2z = 5 & -\text{Eq.1} + \text{Eq. 3} \end{cases}$$

$$\begin{cases} x - 2y + 3z = -5 \\ 4y - 2z = 5 & \text{Interchange Eq.2 and Eq.3} \\ 3z = 1 & -2\text{Eq. 2} + \text{Eq. 3} \end{cases}$$

$$z = \frac{1}{3}$$
$$4y = 5 + 2\left(\frac{1}{3}\right) = \frac{17}{3} \Rightarrow y = \frac{17}{12}$$
$$x = 2\left(\frac{17}{12}\right) - 3\left(\frac{1}{3}\right) - 5 = -\frac{19}{6}$$

Answer: $\left(-\frac{19}{6}, \frac{17}{12}, \frac{1}{3}\right)$

51.
$$\begin{cases} 5x - 12y + 7z = 16 \\ 3x - 7y + 4z = 9 \end{cases}$$

3 times Eq.1 and (-5) times Eq. 2:

$$\begin{cases} 15x - 36y + 21z = 48 \\ -15x + 35y - 20z = -45 \end{cases}$$

Adding, $-y + z = 3 \Rightarrow y = z - 3$.

$$5x - 12(z-3) + 7z = 16$$
$$5x - 5z + 36 = 16$$
$$5x = 5z - 20$$
$$x = z - 4$$

Let $z = a$, then $x = a - 4$ and $y = a - 3$.

Answer: $(a-4, a-3, a)$, where a is any real number

53.

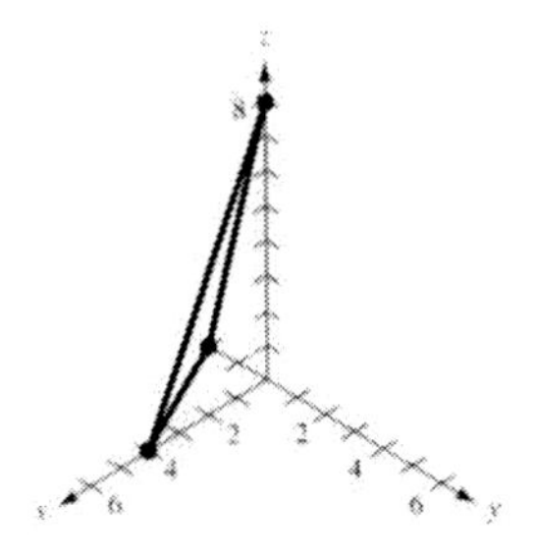

$(4, 0, 0), (0, -2, 0), (0, 0, 8), (1, 0, 6)$

55. $\frac{4-x}{x^2+6x+8} = \frac{A}{x+2} + \frac{B}{x+4}$

$4-x = A(x+4) + B(x+2)$

$= (A+B)x + (4A+2B)$

$\begin{cases} A + B = -1 \\ 4A + 2B = 4 \end{cases} \Rightarrow A = 3, B = -4$

$\frac{4-x}{x^2+6x+8} = \frac{3}{x+2} - \frac{4}{x+4}$

57. $\frac{x^2+2x}{x^3-x^2+x-1} = \frac{A}{x-1} + \frac{Bx+C}{x^2+1}$

$x^2+2x = A(x^2+1) + (Bx+C)(x-1)$

$= (A+B)x^2 + (C-B)x + (A-C)$

$\begin{cases} A+B=1 \\ -B+C=2 \Rightarrow A = \frac{3}{2}, B = -\frac{1}{2}, C = \frac{3}{2} \\ A-C=0 \end{cases}$

$\frac{x^2+2x}{x^3-x^2+x-1} = \frac{3/2}{x-1} + \frac{-(1/2)x+3/2}{x^2+1} = \frac{1}{2}\left(\frac{3}{x-1} - \frac{x-3}{x^2+1}\right)$

59. $\frac{x^2+3x-3}{x^3+2x^2+x+2} = \frac{x^2+3x-3}{(x+2)(x^2+1)}$

$= \frac{A}{x+2} + \frac{Bx+C}{x^2+1}$

$x^2+3x-3 = A(x^2+1) + (Bx+C)(x+2)$

$\begin{cases} A + B = 1 \\ 2B + C = 3 \\ A + 2C = -3 \end{cases}$

$\begin{cases} A + B = 1 \\ 2B + C = 3 \\ -B + 2C = -4 \end{cases}$

$\begin{cases} A + B = 1 \\ B - 2C = 4 \\ 5C = -5 \end{cases}$

$C = -1, B = 2(-1) + 4 = 2, A = 1 - 2 = -1$

$\frac{x^2+3x-3}{x^3+2x^2+x+2} = \frac{-1}{x+2} + \frac{2x-1}{x^2+1}$

61. $y = ax^2 + bx + c$

$(-1, -4) \Rightarrow a - b + c = -4$

$(1, -2) \Rightarrow a + b + c = -2$

$(2, 5) \Rightarrow 4a + 2b + c = 5$

Solving the system,
$a = 2, b = 1, c = -5.$

$y = 2x^2 + x - 5$

63. Let x = number of par-3 holes.
Let y = number of par-4 holes.
Let z = number of par-5 holes.

$\begin{cases} x + y + z = 18 & \text{Equation 1} \\ y - 2z = 2 & \text{Equation 2} \\ x - z = 0 & \text{Equation 3} \end{cases}$

$\begin{cases} x + y + z = 18 \\ y - 2z = 2 \\ -y - 2z = -18 & -\text{Eq. 1} + \text{Eq. 3} \end{cases}$

$\begin{cases} x + y + z = 18 \\ y - 2z = 2 \\ -4z = -16 & \text{Eq. 2} + \text{Eq. 3} \end{cases}$

$-4z = -16 \Rightarrow z = 4$

$y - 2(4) = 2 \Rightarrow y = 10$

$x + 10 + 4 = 18 \Rightarrow x = 4$

Answer: 4 par–3 holes, 10 par–4 holes, 4 par–5 holes

65. Dimension 3×1

67. Dimension 1×1

69. $\begin{cases} 6x - 7y = 11 \\ -2x + 5y = -1 \end{cases}$

$\begin{bmatrix} 6 & -7 & \vdots & 11 \\ -2 & 5 & \vdots & -1 \end{bmatrix}$

71. $\begin{cases} 8x - 7y + 4z = 12 \\ 3x - 5y + 2z = 20 \\ 5x + 3y - 3z = 26 \end{cases}$

$\begin{bmatrix} 8 & -7 & 4 & \vdots & 12 \\ 3 & -5 & 2 & \vdots & 20 \\ 5 & 3 & -3 & \vdots & 26 \end{bmatrix}$

73. $\begin{bmatrix} 5 & 1 & 7 & \vdots & -9 \\ 4 & 2 & 0 & \vdots & 10 \\ 9 & 4 & 2 & \vdots & 3 \end{bmatrix}$

$\begin{cases} 5x + y + 7z = -9 \\ 4x + 2y = 10 \\ 9x + 4y + 2z = 3 \end{cases}$

75.
$$\begin{bmatrix} 0 & 1 & 1 \\ 1 & 2 & 3 \\ 2 & 2 & 2 \end{bmatrix}$$

$$\begin{matrix} R_1 + R_2 \to \\ -R_1 + R_2 \to \\ -2R_1 + R_3 \to \end{matrix} \begin{bmatrix} 1 & 3 & 4 \\ 0 & -1 & -1 \\ 0 & -4 & -6 \end{bmatrix}$$

$$\begin{matrix} 3R_2 + R_1 \to \\ -R_2 \to \\ -4R_2 + R_3 \to \end{matrix} \begin{bmatrix} 1 & 0 & 1 \\ 0 & 1 & 1 \\ 0 & 0 & -2 \end{bmatrix}$$

77. $\begin{bmatrix} 3 & -2 & 1 & 0 \\ 4 & -3 & 0 & 1 \end{bmatrix} \Rightarrow \begin{bmatrix} 1 & 0 & 3 & -2 \\ 0 & 1 & 4 & -3 \end{bmatrix}$

79. $\begin{bmatrix} 1.5 & 3.6 & 4.2 \\ 0.2 & 1.4 & 1.8 \\ 2.0 & 4.4 & 6.4 \end{bmatrix} \Rightarrow \begin{bmatrix} 1 & 0 & 0 \\ 0 & 1 & 0 \\ 0 & 0 & 1 \end{bmatrix}$

81. $\begin{cases} 5x + 4y = 2 \\ -x + y = -22 \end{cases}$

$$\left[\begin{array}{cc:c} 5 & 4 & 2 \\ -1 & 1 & -22 \end{array}\right]$$

$$\begin{matrix} 4R_2 + R_1 \to \\ R_1 + R_2 \to \end{matrix} \left[\begin{array}{cc:c} 1 & 8 & -86 \\ 0 & 9 & -108 \end{array}\right]$$

$9y = -108$

$y = -12$

$x = -8(-12) - 86 = 10$

Answer: $(10, -12)$

83. $\begin{cases} 0.3x - 0.1y = -0.13 \\ 0.2x - 0.3y = -0.25 \end{cases}$

$$\left[\begin{array}{cc:c} 0.3 & -0.1 & -0.13 \\ 0.2 & -0.3 & -0.25 \end{array}\right]$$

$$\begin{matrix} 10R_1 \to \\ 10R_2 \to \end{matrix} \left[\begin{array}{cc:c} 3 & -1 & -1.3 \\ 2 & -3 & -2.5 \end{array}\right]$$

$$\begin{matrix} \frac{1}{3}R_1 \to \\ 2R_1 + (-3)R_2 \to \end{matrix} \left[\begin{array}{cc:c} 1 & -\frac{1}{3} & -\frac{13}{30} \\ 0 & 7 & \frac{49}{10} \end{array}\right]$$

$$\begin{matrix} \\ \frac{1}{7}R_2 \to \end{matrix} \left[\begin{array}{cc:c} 1 & -\frac{1}{3} & -\frac{13}{30} \\ 0 & 1 & \frac{7}{10} \end{array}\right]$$

$$\frac{1}{3}R_2 + R_1 \to \left[\begin{array}{cc:c} 1 & 0 & -\frac{1}{5} \\ 0 & 1 & \frac{7}{10} \end{array}\right]$$

$x = -\frac{1}{5},\ y = \frac{7}{10}$

Answer: $\left(-\frac{1}{5}, \frac{7}{10}\right)$

85. $\begin{cases} 2x + 3y + 3z = 3 \\ 6x + 6y + 12z = 13 \\ 12x + 9y - z = 2 \end{cases}$

$$\left[\begin{array}{ccc:c} 2 & 3 & 3 & 3 \\ 6 & 6 & 12 & 13 \\ 12 & 9 & -1 & 2 \end{array}\right]$$

$$\begin{matrix} \\ -3R_1 + R_2 \to \\ -6R_1 + R_3 \to \end{matrix} \left[\begin{array}{ccc:c} 2 & 3 & 3 & 3 \\ 0 & -3 & 3 & 4 \\ 0 & -9 & -19 & -16 \end{array}\right]$$

$$\begin{matrix} R_2 + R_1 \to \\ \\ -3R_2 + R_3 \to \end{matrix} \left[\begin{array}{ccc:c} 2 & 0 & 6 & 7 \\ 0 & -3 & 3 & 4 \\ 0 & 0 & -28 & -28 \end{array}\right]$$

$$\begin{matrix} \frac{1}{2}R_1 \to \\ -\frac{1}{3}R_2 \to \\ -\frac{1}{28}R_3 \to \end{matrix} \left[\begin{array}{ccc:c} 1 & 0 & 3 & \frac{7}{2} \\ 0 & 1 & -1 & -\frac{4}{3} \\ 0 & 0 & 1 & 1 \end{array}\right]$$

$z = 1$

$y - 1 = -\frac{4}{3} \Rightarrow y = -\frac{1}{3}$

$x + 3(1) = \frac{7}{2} \Rightarrow x = \frac{1}{2}$

Answer: $\left(\frac{1}{2}, -\frac{1}{3}, 1\right)$

87. $\begin{cases} x + 2y - z = 1 \\ \quad y + z = 0 \end{cases}$

$$\left[\begin{array}{ccc:c} 1 & 2 & -1 & 1 \\ 0 & 1 & 1 & 0 \end{array}\right] \Rightarrow \left[\begin{array}{ccc:c} 1 & 0 & -3 & 1 \\ 0 & 1 & 1 & 0 \end{array}\right]$$

$y = -z$

$x = 1 + 3z$

Answer: $(1 + 3a, -a, a)$, a is a real number

89. $\begin{cases} -x + y + 2z = 1 \\ 2x + 3y + z = -2 \\ 5x + 4y + 2z = 4 \end{cases}$

$$\begin{bmatrix} -1 & 1 & 2 & \vdots & 1 \\ 2 & 3 & 1 & \vdots & -2 \\ 5 & 4 & 2 & \vdots & 4 \end{bmatrix}$$

$$\begin{matrix} -R_1 \rightarrow \\ 2R_1 + R_2 \rightarrow \\ 5R_1 + R_3 \rightarrow \end{matrix} \begin{bmatrix} 1 & -1 & -2 & \vdots & -1 \\ 0 & 5 & 5 & \vdots & 0 \\ 0 & 9 & 12 & \vdots & 9 \end{bmatrix}$$

$$\begin{matrix} \\ \frac{1}{5}R_2 \rightarrow \\ \\ \end{matrix} \begin{bmatrix} 1 & -1 & -2 & \vdots & -1 \\ 0 & 1 & 1 & \vdots & 0 \\ 0 & 9 & 12 & \vdots & 9 \end{bmatrix}$$

$$\begin{matrix} R_2 + R_1 \rightarrow \\ \\ -9R_2 + R_3 \rightarrow \end{matrix} \begin{bmatrix} 1 & 0 & -1 & \vdots & -1 \\ 0 & 1 & 1 & \vdots & 0 \\ 0 & 0 & 3 & \vdots & 9 \end{bmatrix}$$

$$\begin{matrix} \\ \\ \frac{1}{3}R_3 \rightarrow \end{matrix} \begin{bmatrix} 1 & 0 & -1 & \vdots & -1 \\ 0 & 1 & 1 & \vdots & 0 \\ 0 & 0 & 1 & \vdots & 3 \end{bmatrix}$$

$$\begin{matrix} R_3 + R_1 \rightarrow \\ -R_3 + R_2 \rightarrow \\ \\ \end{matrix} \begin{bmatrix} 1 & 0 & 0 & \vdots & 2 \\ 0 & 1 & 0 & \vdots & -3 \\ 0 & 0 & 1 & \vdots & 3 \end{bmatrix}$$

$x = 2,\ y = -3,\ z = 3$

Answer: $(2, -3, 3)$

91. $\begin{cases} x + y + 2z = 4 \\ x - y + 4z = 1 \\ 2x - y + 2z = 1 \end{cases} \begin{bmatrix} 1 & 1 & 2 & \vdots & 4 \\ 1 & -1 & 4 & \vdots & 1 \\ 2 & -1 & 2 & \vdots & 1 \end{bmatrix} \Rightarrow \begin{bmatrix} 1 & 0 & 0 & \vdots & 1 \\ 0 & 1 & 0 & \vdots & 2 \\ 0 & 0 & 1 & \vdots & \frac{1}{2} \end{bmatrix}$

Answer: $(1, 2, \frac{1}{2})$

93. $\begin{cases} x + 2y - z = 3 \\ x - y - z = -3 \\ 2x + y + 3z = 10 \end{cases}$

$$\begin{bmatrix} 1 & 2 & -1 & \vdots & 3 \\ 1 & -1 & -1 & \vdots & -3 \\ 2 & 1 & 3 & \vdots & 10 \end{bmatrix}$$

$$\begin{matrix} \\ R_1 - R_2 \rightarrow \\ 2R_1 - R_3 \rightarrow \end{matrix} \begin{bmatrix} 1 & 2 & -1 & \vdots & 3 \\ 0 & 3 & 0 & \vdots & 6 \\ 0 & 3 & -5 & \vdots & -4 \end{bmatrix}$$

$$\begin{matrix} \\ \frac{1}{3}R_2 \rightarrow \\ R_2 - R_3 \rightarrow \end{matrix} \begin{bmatrix} 1 & 2 & -1 & \vdots & 3 \\ 0 & 1 & 0 & \vdots & 2 \\ 0 & 0 & 5 & \vdots & 10 \end{bmatrix}$$

$$\begin{matrix} R_1 - 2R_2 \rightarrow \\ \\ \frac{1}{5}R_3 \rightarrow \end{matrix} \begin{bmatrix} 1 & 0 & -1 & \vdots & -1 \\ 0 & 1 & 0 & \vdots & 2 \\ 0 & 0 & 1 & \vdots & 2 \end{bmatrix}$$

$$\begin{matrix} R_1 + R_3 \rightarrow \\ \\ \\ \end{matrix} \begin{bmatrix} 1 & 0 & 0 & \vdots & 1 \\ 0 & 1 & 0 & \vdots & 2 \\ 0 & 0 & 1 & \vdots & 2 \end{bmatrix}$$

$x = 1,\ y = 2,\ z = 2$

Answer: $(1, 2, 2)$

95. $\begin{bmatrix} 1 & 2 & -1 & \vdots & 7 \\ 0 & -1 & -1 & \vdots & 4 \\ 4 & 0 & -1 & \vdots & 16 \end{bmatrix} \Rightarrow \begin{bmatrix} 1 & 0 & 0 & \vdots & 3 \\ 0 & 1 & 0 & \vdots & 0 \\ 0 & 0 & 1 & \vdots & -4 \end{bmatrix}$

Answer: $(3, 0, -4)$

97. $\begin{bmatrix} 3 & -1 & 5 & -2 & \vdots & -44 \\ 1 & 6 & 4 & -1 & \vdots & 1 \\ 5 & -1 & 1 & 3 & \vdots & -15 \\ 0 & 4 & -1 & -8 & \vdots & 58 \end{bmatrix} \Rightarrow \begin{bmatrix} 1 & 0 & 0 & 0 & \vdots & 2 \\ 0 & 1 & 0 & 0 & \vdots & 6 \\ 0 & 0 & 1 & 0 & \vdots & -10 \\ 0 & 0 & 0 & 1 & \vdots & -3 \end{bmatrix}$

Answer: $(2, 6, -10, -3)$

99. $x = 12$
$y = -7$

101. $x + 3 = 5x - 1 \Rightarrow x = 1$
$-4y = -44 \Rightarrow y = 11$
$y + 5 = 16 \Rightarrow y = 11$
$6x = 6 \Rightarrow x = 1$
Answer: $x = 1,\ y = 11$

103. $A = \begin{bmatrix} 7 & 3 \\ -1 & 5 \end{bmatrix},\ B = \begin{bmatrix} 10 & -20 \\ 14 & -3 \end{bmatrix}$

(a) $A + B = \begin{bmatrix} 7 & 3 \\ -1 & 5 \end{bmatrix} + \begin{bmatrix} 10 & -20 \\ 14 & -3 \end{bmatrix} = \begin{bmatrix} 17 & -17 \\ 13 & 2 \end{bmatrix}$

(b) $A - B = \begin{bmatrix} 7 & 3 \\ -1 & 5 \end{bmatrix} - \begin{bmatrix} 10 & -20 \\ 14 & -3 \end{bmatrix} = \begin{bmatrix} -3 & 23 \\ -15 & 8 \end{bmatrix}$

(c) $2A = 2\begin{bmatrix} 7 & 3 \\ -1 & 5 \end{bmatrix} = \begin{bmatrix} 14 & 6 \\ -2 & 10 \end{bmatrix}$

(d) $A + 3B = \begin{bmatrix} 7 & 3 \\ -1 & 5 \end{bmatrix} + 3\begin{bmatrix} 10 & -20 \\ 14 & -3 \end{bmatrix} = \begin{bmatrix} 37 & -57 \\ 41 & -4 \end{bmatrix}$

105. $A = \begin{bmatrix} 6 & 0 & 7 \\ 5 & -1 & 2 \\ 3 & 2 & 3 \end{bmatrix},\ B = \begin{bmatrix} 0 & 5 & 1 \\ -4 & 8 & 6 \\ 2 & -1 & 1 \end{bmatrix}$

(a) $A + B = \begin{bmatrix} 6 & 0 & 7 \\ 5 & -1 & 2 \\ 3 & 2 & 3 \end{bmatrix} + \begin{bmatrix} 0 & 5 & 1 \\ -4 & 8 & 6 \\ 2 & -1 & 1 \end{bmatrix} = \begin{bmatrix} 6 & 5 & 8 \\ 1 & 7 & 8 \\ 5 & 1 & 4 \end{bmatrix}$

(b) $A - B = \begin{bmatrix} 6 & 0 & 7 \\ 5 & -1 & 2 \\ 3 & 2 & 3 \end{bmatrix} - \begin{bmatrix} 0 & 5 & 1 \\ -4 & 8 & 6 \\ 2 & -1 & 1 \end{bmatrix} = \begin{bmatrix} 6 & -5 & 6 \\ 9 & -9 & -4 \\ 1 & 3 & 2 \end{bmatrix}$

(c) $2A = 2\begin{bmatrix} 6 & 0 & 7 \\ 5 & -1 & 2 \\ 3 & 2 & 3 \end{bmatrix} = \begin{bmatrix} 12 & 0 & 14 \\ 10 & -2 & 4 \\ 6 & 4 & 6 \end{bmatrix}$

(d) $A + 3B = \begin{bmatrix} 6 & 0 & 7 \\ 5 & -1 & 2 \\ 3 & 2 & 3 \end{bmatrix} + 3\begin{bmatrix} 0 & 5 & 1 \\ -4 & 8 & 6 \\ 2 & -1 & 1 \end{bmatrix} = \begin{bmatrix} 6 & 15 & 10 \\ -7 & 23 & 20 \\ 9 & -1 & 6 \end{bmatrix}$

107. $\begin{bmatrix} 2 & 1 & 0 \\ 0 & 5 & -4 \end{bmatrix} - 3\begin{bmatrix} 5 & 3 & -6 \\ 0 & -2 & 5 \end{bmatrix} = \begin{bmatrix} 2 & 1 & 0 \\ 0 & 5 & -4 \end{bmatrix}$

$-\begin{bmatrix} 15 & 9 & -18 \\ 0 & -6 & 15 \end{bmatrix}$

$= \begin{bmatrix} -13 & -8 & 18 \\ 0 & 11 & -19 \end{bmatrix}$

109. $-\begin{bmatrix} 8 & -1 \\ -2 & 4 \end{bmatrix} - 5\begin{bmatrix} -2 & 0 \\ 3 & -1 \end{bmatrix} + \begin{bmatrix} 7 & -8 \\ 4 & 3 \end{bmatrix}$

$= \begin{bmatrix} -8 & 1 \\ 2 & -4 \end{bmatrix} - \begin{bmatrix} -10 & 0 \\ 15 & -5 \end{bmatrix} + \begin{bmatrix} 7 & -8 \\ 4 & 3 \end{bmatrix}$

$= \begin{bmatrix} 9 & -7 \\ -9 & 4 \end{bmatrix}$

111. $3\begin{bmatrix} 8 & -2 & 5 \\ 1 & 3 & -1 \end{bmatrix} + 6\begin{bmatrix} 4 & -2 & -3 \\ 2 & 7 & 6 \end{bmatrix} = \begin{bmatrix} 48 & -18 & -3 \\ 15 & 51 & 33 \end{bmatrix}$

113. $X = 3A - 2B$

$$X = 3\begin{bmatrix} -4 & 0 \\ 1 & -5 \\ -3 & 2 \end{bmatrix} - 2\begin{bmatrix} -2 & 2 \\ -2 & 1 \\ 4 & 4 \end{bmatrix}$$

$$X = \begin{bmatrix} -12 & 0 \\ 3 & -15 \\ -9 & 6 \end{bmatrix} + \begin{bmatrix} 4 & -4 \\ 4 & -2 \\ -8 & -8 \end{bmatrix} = \begin{bmatrix} -8 & -4 \\ 7 & -17 \\ -17 & -2 \end{bmatrix}$$

115. $3X + 2A = B$

$3X = B - 2A$

$$3X = \begin{bmatrix} -2 & 2 \\ -2 & 1 \\ 4 & 4 \end{bmatrix} - 2\begin{bmatrix} -4 & 0 \\ 1 & -5 \\ -3 & 2 \end{bmatrix} = \begin{bmatrix} 6 & 2 \\ -4 & 11 \\ 10 & 0 \end{bmatrix}$$

$$X = \frac{1}{3}\begin{bmatrix} 6 & 2 \\ -4 & 11 \\ 10 & 0 \end{bmatrix} = \begin{bmatrix} 2 & \frac{2}{3} \\ -\frac{4}{3} & \frac{11}{3} \\ \frac{10}{3} & 0 \end{bmatrix}$$

117. $\begin{bmatrix} 1 & 2 \\ 5 & -4 \\ 6 & 0 \end{bmatrix}\begin{bmatrix} 6 & -2 & 8 \\ 4 & 0 & 0 \end{bmatrix}$

$$= \begin{bmatrix} 1(6)+2(4) & 1(-2)+2(0) & 1(8)+2(0) \\ 5(6)+(-4)(4) & 5(-2)+(-4)(0) & 5(8)+(-4)(0) \\ 6(6)+(0)(4) & 6(-2)+(0)(0) & 6(8)+(0)(0) \end{bmatrix}$$

$$= \begin{bmatrix} 14 & -2 & 8 \\ 14 & -10 & 40 \\ 36 & -12 & 48 \end{bmatrix}$$

119. $\begin{bmatrix} 6 & -5 & 7 \end{bmatrix}\begin{bmatrix} -1 \\ 4 \\ 8 \end{bmatrix} = \begin{bmatrix} 30 \end{bmatrix}$

121. $\begin{bmatrix} 4 & 1 \\ 11 & -7 \\ 12 & 3 \end{bmatrix}\begin{bmatrix} 3 & -5 & 6 \\ 2 & -2 & -2 \end{bmatrix} = \begin{bmatrix} 14 & -22 & 22 \\ 19 & -41 & 80 \\ 42 & -66 & 66 \end{bmatrix}$

123. $\begin{bmatrix} 2 & 1 \\ 6 & 0 \end{bmatrix}\left(\begin{bmatrix} 4 & 2 \\ -3 & 1 \end{bmatrix} + \begin{bmatrix} -2 & 4 \\ 0 & 4 \end{bmatrix}\right) = \begin{bmatrix} 2 & 1 \\ 6 & 0 \end{bmatrix}\begin{bmatrix} 2 & 6 \\ -3 & 5 \end{bmatrix}$

$$= \begin{bmatrix} 2(2)+1(-3) & 2(6)+1(5) \\ 6(2)+0 & 6(6)+0 \end{bmatrix}$$

$$= \begin{bmatrix} 1 & 17 \\ 12 & 36 \end{bmatrix}$$

125. (a) $AB = \begin{bmatrix} 40 & 64 & 52 \\ 60 & 82 & 76 \\ 76 & 96 & 84 \end{bmatrix}\begin{bmatrix} 3.25 & 0.25 \\ 3.35 & 0.30 \\ 3.49 & 0.35 \end{bmatrix}$

$$= \begin{bmatrix} 525.88 & 47.40 \\ 734.94 & 66.20 \\ 861.76 & 77.20 \end{bmatrix}$$

The entries are the dairy mart's sales and profits on milk for Friday, Saturday, and Sunday.

(b) The dairy mart's profit for Friday through Sunday is the sum of the entries in the second column of AB or \$190.80.

127. $AB = \begin{bmatrix} -4 & -1 \\ 7 & 2 \end{bmatrix}\begin{bmatrix} -2 & -1 \\ 7 & 4 \end{bmatrix} = \begin{bmatrix} 1 & 0 \\ 0 & 1 \end{bmatrix}$; $BA = I_2$

129. $AI_2 = \left[\begin{array}{cc:cc} -6 & 5 & 1 & 0 \\ -5 & 4 & 0 & 1 \end{array}\right]$

$$\begin{array}{l} -\frac{1}{6}R_1 \to \\ -\frac{1}{5}R_2 \to \end{array}\left[\begin{array}{cc:cc} 1 & -\frac{5}{6} & -\frac{1}{6} & 0 \\ 1 & -\frac{4}{5} & 0 & -\frac{1}{5} \end{array}\right]$$

$$-R_1 + R_2 \to \left[\begin{array}{cc:cc} 1 & -\frac{5}{6} & -\frac{1}{6} & 0 \\ 0 & \frac{1}{30} & \frac{1}{6} & -\frac{1}{5} \end{array}\right]$$

$$25R_2 + R_1 \to \left[\begin{array}{cc:cc} 1 & 0 & 4 & 0 \\ 0 & \frac{1}{30} & \frac{1}{6} & -\frac{1}{5} \end{array}\right]$$

$$30R_2 \to \left[\begin{array}{cc:cc} 1 & 0 & 4 & 0 \\ 0 & 1 & 5 & -6 \end{array}\right] = I_2A^{-1}$$

$$A^{-1} = \begin{bmatrix} 4 & 0 \\ 5 & -6 \end{bmatrix}$$

131. $AI_3 = \left[\begin{array}{ccc|ccc} 2 & 0 & 3 & 1 & 0 & 0 \\ -1 & 1 & 1 & 0 & 1 & 0 \\ 2 & -2 & 1 & 0 & 0 & 1 \end{array}\right]$

$$\begin{array}{r} \frac{1}{2}R_1 \rightarrow \\ R_1 + 2R_2 \rightarrow \\ R_1 - R_3 \rightarrow \end{array} \left[\begin{array}{ccc|ccc} 1 & 0 & \frac{3}{2} & \frac{1}{2} & 0 & 0 \\ 0 & 2 & 5 & 1 & 2 & 0 \\ 0 & 2 & 2 & 1 & 0 & -1 \end{array}\right]$$

$$\begin{array}{r} \\ \frac{1}{2}R_2 \rightarrow \\ R_2 - R_3 \rightarrow \end{array} \left[\begin{array}{ccc|ccc} 1 & 0 & \frac{3}{2} & \frac{1}{2} & 0 & 0 \\ 0 & 1 & \frac{5}{2} & \frac{1}{2} & 1 & 0 \\ 0 & 0 & 3 & 0 & 2 & 1 \end{array}\right]$$

$$\begin{array}{r} \\ \\ \frac{1}{3}R_3 \rightarrow \end{array} \left[\begin{array}{ccc|ccc} 1 & 0 & \frac{3}{2} & \frac{1}{2} & 0 & 0 \\ 0 & 1 & \frac{5}{2} & \frac{1}{2} & 1 & 0 \\ 0 & 0 & 1 & 0 & \frac{2}{3} & \frac{1}{3} \end{array}\right]$$

$$\begin{array}{r} R_1 - \frac{3}{2}R_3 \rightarrow \\ R_2 - \frac{5}{2}R_3 \\ \\ \end{array} \left[\begin{array}{ccc|ccc} 1 & 0 & 0 & \frac{1}{2} & -1 & -\frac{1}{2} \\ 0 & 1 & 0 & \frac{1}{2} & -\frac{2}{3} & \frac{-5}{6} \\ 0 & 0 & 1 & 0 & \frac{2}{3} & \frac{1}{3} \end{array}\right] = I_3A^{-1}$$

$$A^{-1} = \begin{bmatrix} \frac{1}{2} & -1 & -\frac{1}{2} \\ \frac{1}{2} & -\frac{2}{3} & -\frac{5}{6} \\ 0 & \frac{2}{3} & \frac{1}{3} \end{bmatrix}$$

133. $\begin{bmatrix} 2 & 6 \\ 3 & -6 \end{bmatrix}^{-1} = \begin{bmatrix} \frac{1}{5} & \frac{1}{5} \\ \frac{1}{10} & -\frac{1}{15} \end{bmatrix}$

135. $\begin{bmatrix} 1 & 2 & 0 \\ -1 & 1 & 1 \\ 0 & -1 & 0 \end{bmatrix}^{-1} = \begin{bmatrix} 1 & 0 & 2 \\ 0 & 0 & -1 \\ 1 & 1 & 3 \end{bmatrix}$

137. $\begin{bmatrix} -7 & 2 \\ -8 & 2 \end{bmatrix}^{-1} = \frac{1}{(-7)(2)-(2)(-8)}\begin{bmatrix} 2 & -2 \\ 8 & -7 \end{bmatrix} = \begin{bmatrix} 1 & -1 \\ 4 & -\frac{7}{2} \end{bmatrix}$

139. $\begin{bmatrix} -1 & 10 \\ 2 & 20 \end{bmatrix}^{-1} = \frac{1}{(-1)(20)-(10)(2)}\begin{bmatrix} 20 & -10 \\ -2 & -1 \end{bmatrix}$

$$= \frac{1}{-40}\begin{bmatrix} 20 & -10 \\ -2 & -1 \end{bmatrix}$$

$$= \begin{bmatrix} -\frac{1}{2} & \frac{3}{4} \\ \frac{1}{20} & \frac{1}{40} \end{bmatrix}$$

141. $\begin{bmatrix} -1 & 4 \\ 2 & -7 \end{bmatrix}^{-1} = \begin{bmatrix} 7 & 4 \\ 2 & 1 \end{bmatrix}$

$$\begin{bmatrix} x \\ y \end{bmatrix} = \begin{bmatrix} 7 & 4 \\ 2 & 1 \end{bmatrix}\begin{bmatrix} 8 \\ -5 \end{bmatrix} = \begin{bmatrix} 36 \\ 11 \end{bmatrix}$$

Answer: $(36, 11)$

143. $\begin{bmatrix} 3 & 2 & -1 \\ 1 & -1 & 2 \\ 5 & 1 & 1 \end{bmatrix}^{-1} = \begin{bmatrix} -1 & -1 & 1 \\ 3 & \frac{8}{3} & -\frac{7}{3} \\ 2 & \frac{7}{3} & -\frac{5}{3} \end{bmatrix}$

$$\begin{bmatrix} x \\ y \\ z \end{bmatrix} = \begin{bmatrix} -1 & -1 & 1 \\ 3 & \frac{8}{3} & -\frac{7}{3} \\ 2 & \frac{7}{3} & -\frac{5}{3} \end{bmatrix}\begin{bmatrix} 6 \\ -1 \\ 7 \end{bmatrix} = \begin{bmatrix} 2 \\ -1 \\ -2 \end{bmatrix}$$

Answer: $(2, -1, -2)$

145. $\begin{bmatrix} 1 & 2 & 1 & -1 \\ 2 & 1 & 1 & 1 \\ 1 & -1 & -3 & 0 \\ 0 & 0 & 1 & 1 \end{bmatrix}^{-1} = \begin{bmatrix} -1 & \frac{4}{3} & -\frac{2}{3} & -\frac{7}{3} \\ 2 & -\frac{5}{3} & \frac{4}{3} & \frac{11}{3} \\ -1 & 1 & -1 & -2 \\ 1 & -1 & 1 & 3 \end{bmatrix}$

$$\begin{bmatrix} x \\ y \\ z \\ w \end{bmatrix} = \begin{bmatrix} -1 & \frac{4}{3} & -\frac{2}{3} & -\frac{7}{3} \\ 2 & -\frac{5}{3} & \frac{4}{3} & \frac{11}{3} \\ -1 & 1 & -1 & -2 \\ 1 & -1 & 1 & 3 \end{bmatrix}\begin{bmatrix} -2 \\ 1 \\ 0 \\ 1 \end{bmatrix} = \begin{bmatrix} 1 \\ -2 \\ 1 \\ 0 \end{bmatrix}$$

Answer: $(1, -2, 1, 0)$

147. $\begin{cases} x + 2y = -1 \\ 3x + 4y = -5 \end{cases}$

$$\begin{bmatrix} 1 & 2 \\ 3 & 4 \end{bmatrix}^{-1} = \begin{bmatrix} -2 & 1 \\ \frac{3}{2} & -\frac{1}{2} \end{bmatrix} \Rightarrow \begin{bmatrix} x \\ y \end{bmatrix} = \begin{bmatrix} -2 & 1 \\ \frac{3}{2} & -\frac{1}{2} \end{bmatrix}\begin{bmatrix} -1 \\ -5 \end{bmatrix} = \begin{bmatrix} -3 \\ 1 \end{bmatrix}$$

$x = -3,\ y = 1$

Answer: $(-3, 1)$

149. $\begin{cases} -3x - 3y - 4z = 2 \\ \qquad\quad y + z = -1 \\ 4x + 3y + 4z = -1 \end{cases}$

$$\begin{bmatrix} -3 & -3 & -4 \\ 0 & 1 & 1 \\ 4 & 3 & 4 \end{bmatrix}^{-1} = \begin{bmatrix} 1 & 0 & 1 \\ 4 & 4 & 3 \\ 4 & -3 & -3 \end{bmatrix} \Rightarrow \begin{bmatrix} x \\ y \\ z \end{bmatrix} = \begin{bmatrix} 1 & 0 & 1 \\ 4 & 4 & 3 \\ -4 & -3 & -3 \end{bmatrix}\begin{bmatrix} 2 \\ -1 \\ -1 \end{bmatrix} = \begin{bmatrix} 1 \\ 1 \\ -2 \end{bmatrix}$$

$x = 1,\ y = 1,\ z = -2$

Answer: $(1, 1, -2)$

151. $\begin{vmatrix} 8 & 5 \\ 2 & -4 \end{vmatrix} = 8(-4) - 2(5) = -42$

153. $\begin{vmatrix} 50 & -30 \\ 10 & 5 \end{vmatrix} = 50(5) - (-30)(10) = 550$

155. $A = \begin{bmatrix} 2 & -1 \\ 7 & 4 \end{bmatrix}$

(a) $M_{11} = 4 \quad M_{21} = -1$

$M_{12} = 7 \quad M_{22} = 2$

(b) $C_{11} = 4 \quad C_{21} = 1$

$C_{12} = -7 \quad C_{22} = 2$

157. $A = \begin{bmatrix} 3 & 2 & -1 \\ -2 & 5 & 0 \\ 1 & 8 & 6 \end{bmatrix}$

(a) $M_{11} = \begin{vmatrix} 5 & 0 \\ 8 & 6 \end{vmatrix} = 30,\ M_{12} = \begin{vmatrix} -2 & 0 \\ 1 & 6 \end{vmatrix} = -12,$

$M_{13} = \begin{vmatrix} -2 & 5 \\ 1 & 8 \end{vmatrix} = -21$

$M_{21} = \begin{vmatrix} 2 & -1 \\ 8 & 6 \end{vmatrix} = 20,\ M_{22} = \begin{vmatrix} 3 & -1 \\ 1 & 6 \end{vmatrix} = 19,$

$M_{23} = \begin{vmatrix} 3 & 2 \\ 1 & 8 \end{vmatrix} = 22$

$M_{31} = \begin{vmatrix} 2 & -1 \\ 5 & 0 \end{vmatrix} = 5,\ M_{32} = \begin{vmatrix} 3 & -1 \\ -2 & 0 \end{vmatrix} = -2,$

$M_{33} = \begin{vmatrix} 3 & 2 \\ -2 & 5 \end{vmatrix} = 19$

(b) $C_{11} = 30,\ C_{12} = 12,\ C_{13} = -21$

$C_{21} = -20,\ C_{22} = 19,\ C_{23} = -22$

$C_{31} = 5,\ C_{32} = 2,\ C_{33} = 19$

159. $\begin{vmatrix} -2 & 4 & 1 \\ -6 & 0 & 2 \\ 5 & 3 & 4 \end{vmatrix} = 6\begin{vmatrix} 4 & 1 \\ 3 & 4 \end{vmatrix} - 2\begin{vmatrix} -2 & 4 \\ 5 & 3 \end{vmatrix}$

$= 6(13) - 2(-26) = 130$

161. $\begin{vmatrix} -2 & 0 & 0 \\ 2 & -1 & 0 \\ -1 & 1 & 3 \end{vmatrix} = (-2)\begin{vmatrix} -1 & 0 \\ 1 & 3 \end{vmatrix} = (-2)\left[(-1)(3) - (1)(0)\right] = 6$

(Expansion along Row 1)

163. $\begin{vmatrix} 1 & 0 & -2 \\ 0 & 1 & 0 \\ -2 & 0 & 1 \end{vmatrix} = 1\begin{vmatrix} 1 & -2 \\ -2 & 1 \end{vmatrix} = 1(1) - (-2)(-2) = 1 - 4 = -3$

(Expansion along Row 2)

165. $\begin{vmatrix} 3 & 0 & -4 & 0 \\ 0 & 8 & 1 & 2 \\ 6 & 1 & 8 & 2 \\ 0 & 3 & -4 & 1 \end{vmatrix} = 3\begin{vmatrix} 8 & 1 & 2 \\ 1 & 8 & 2 \\ 3 & -4 & 1 \end{vmatrix} + (-4)\begin{vmatrix} 0 & 8 & 2 \\ 6 & 1 & 2 \\ 0 & 3 & 1 \end{vmatrix}$

(Expansion along Row 1)

$= 3\left[8(8 - (-8)) - 1(1 - 6) + 2(-4 - 24)\right] - 4\left[0 - 6(8 - 6) + 0\right]$

$= 3\left[128 + 5 - 56\right] - 4\left[-12\right]$

$= 279$

167. $(1, 0), (5, 0), (5, 8)$

$\frac{1}{2}\begin{vmatrix} 1 & 0 & 1 \\ 5 & 0 & 1 \\ 5 & 8 & 1 \end{vmatrix} = \frac{1}{2}(32) = 16$

Area = 16 square units

169. $\left(\frac{1}{2}, 1\right), \left(2, -\frac{5}{2}\right), \left(\frac{3}{2}, 1\right)$

$\frac{1}{2}\begin{vmatrix} \frac{1}{2} & 1 & 1 \\ 2 & -\frac{5}{2} & 1 \\ \frac{3}{2} & 1 & 1 \end{vmatrix} = \frac{1}{2}\left(\frac{7}{2}\right) = \frac{7}{4}$

Area = $\frac{7}{4}$ square units

171. $(2, 4), (5, 6), (4, 1)$

$\frac{1}{2}\begin{vmatrix} 2 & 4 & 1 \\ 5 & 6 & 1 \\ 4 & 1 & 1 \end{vmatrix} = \frac{1}{2}(-13)$

Area = $\frac{13}{2}$ square units

173. $(-2, -1), (4, 9), (-2, -9), (4, 1)$

The figure is rhombus.

$\begin{vmatrix} -2 & -1 & 1 \\ 4 & 9 & 1 \\ -2 & -9 & 1 \end{vmatrix} = -48$

Area = 48 square units

175. $\begin{vmatrix} -1 & 7 & 1 \\ 3 & -9 & 1 \\ -3 & 15 & 1 \end{vmatrix} = (-1)\begin{vmatrix} -9 & 1 \\ 15 & 1 \end{vmatrix} - (7)\begin{vmatrix} 3 & 1 \\ -3 & 1 \end{vmatrix} + (1)\begin{vmatrix} 3 & -9 \\ -3 & 15 \end{vmatrix}$

$= 0$

Collinear

177. $x = \dfrac{\begin{vmatrix} 5 & 2 \\ 1 & 1 \end{vmatrix}}{\begin{vmatrix} 1 & 2 \\ -1 & 1 \end{vmatrix}} = \dfrac{3}{3} = 1$

$y = \dfrac{\begin{vmatrix} 1 & 5 \\ -1 & 1 \end{vmatrix}}{\begin{vmatrix} 1 & 2 \\ -1 & 1 \end{vmatrix}} = \dfrac{6}{3} = 2$

Answer: $(1, 2)$

179. $x = \dfrac{\begin{vmatrix} 6 & -2 \\ -23 & 3 \end{vmatrix}}{\begin{vmatrix} 5 & -2 \\ -11 & 3 \end{vmatrix}} = \dfrac{-28}{-7} = 4$

$y = \dfrac{\begin{vmatrix} 5 & 6 \\ -11 & -23 \end{vmatrix}}{\begin{vmatrix} 5 & -2 \\ -11 & 3 \end{vmatrix}} = \dfrac{-49}{-7} = 7$

Answer: $(4, 7)$

181. $x = \dfrac{\begin{vmatrix} -11 & 3 & -5 \\ -3 & -1 & 1 \\ 15 & -4 & 6 \end{vmatrix}}{\begin{vmatrix} -2 & 3 & -5 \\ 4 & -1 & 1 \\ -1 & -4 & 6 \end{vmatrix}} = \dfrac{-14}{14} = -1$

$y = \dfrac{\begin{vmatrix} -2 & -11 & -5 \\ 4 & -3 & 1 \\ -1 & 15 & 6 \end{vmatrix}}{14} = \dfrac{56}{14} = 4$

$z = \dfrac{\begin{vmatrix} -2 & 3 & -11 \\ 4 & -1 & -3 \\ -1 & -4 & 15 \end{vmatrix}}{14} = \dfrac{70}{14} = 5$

Answer: $(-1, 4, 5)$

183. $\begin{vmatrix} 1 & -3 & 2 \\ 2 & 2 & -3 \\ 1 & -7 & 8 \end{vmatrix} = 20$

$|A_1| = 0$

$|A_2| = -48$

$|A_3| = -52$

$x = 0,\ y = -\frac{48}{20} = -2.4,\ z = -\frac{52}{20} = -2.6$

Answer: $(0, -2.4, -2.6)$

185. (a) $\begin{cases} x - 3y + 2z = 5 \\ 2x + y - 4z = -1 \\ 2x + 4y + 2z = 3 \end{cases}$

$\begin{cases} x - 3y + 2z = 5 \\ 7y - 8z = -11 \\ 10y - 2z = -7 \end{cases}$

$\begin{cases} x - 3y + 2z = 5 \\ 7y - 8z = -11 \\ \frac{66}{7}z = \frac{61}{7} \end{cases}$

$z = \dfrac{61}{66}$

$7y - 8\left(\dfrac{61}{66}\right) = -11 \Rightarrow y = -\dfrac{17}{33}$

$x - 3\left(-\dfrac{17}{33}\right) + 2\left(\dfrac{61}{66}\right) = 5 \Rightarrow x = \dfrac{53}{33}$

Answer: $\left(\dfrac{53}{33}, -\dfrac{17}{33}, \dfrac{61}{66}\right)$

(b) $D = \begin{vmatrix} 1 & -3 & 2 \\ 2 & 1 & -4 \\ 2 & 4 & 2 \end{vmatrix} = 66$

$x = \dfrac{\begin{vmatrix} 5 & -3 & 2 \\ -1 & 1 & -4 \\ 3 & 4 & 2 \end{vmatrix}}{66} = \dfrac{106}{66} = \dfrac{53}{33}$

$y = \dfrac{\begin{vmatrix} 1 & 5 & 2 \\ 2 & -1 & -4 \\ 2 & 3 & 2 \end{vmatrix}}{66} = \dfrac{-34}{66} = \dfrac{-17}{33}$

$z = \dfrac{\begin{vmatrix} 1 & -3 & 5 \\ 2 & 1 & -1 \\ 2 & 4 & 3 \end{vmatrix}}{66} = \dfrac{61}{66}$

Answer: $\left(\dfrac{53}{33}, -\dfrac{17}{33}, \dfrac{61}{66}\right)$

187. L O O K _ O U T _ B E L O W

(a) $\begin{bmatrix} 12 & 15 & 15 \end{bmatrix} \begin{bmatrix} 11 & 0 & 15 \end{bmatrix} \begin{bmatrix} 21 & 20 & 0 \end{bmatrix} \begin{bmatrix} 2 & 5 & 12 \end{bmatrix}$

$\begin{bmatrix} 15 & 23 & 0 \end{bmatrix}$

(b) $\begin{bmatrix} 12 & 15 & 15 \end{bmatrix} A = \begin{bmatrix} -21 & 6 & 0 \end{bmatrix}$

$\begin{bmatrix} 11 & 0 & 15 \end{bmatrix} A = \begin{bmatrix} -68 & 8 & 45 \end{bmatrix}$

$\begin{bmatrix} 21 & 20 & 0 \end{bmatrix} A = \begin{bmatrix} 102 & -42 & -60 \end{bmatrix}$

$\begin{bmatrix} 2 & 5 & 12 \end{bmatrix} A = \begin{bmatrix} -53 & 20 & 21 \end{bmatrix}$

$\begin{bmatrix} 15 & 23 & 0 \end{bmatrix} A = \begin{bmatrix} 99 & -30 & -69 \end{bmatrix}$

Cryptogram:

−21 6 0 −68 8 45 102 −42 −60 −53 20 21 99 −30 −60

189. $A^{-1} = \begin{bmatrix} \frac{1}{2} & -\frac{1}{4} & \frac{1}{4} \\ -\frac{1}{4} & -\frac{3}{8} & -\frac{1}{8} \\ -\frac{1}{2} & -\frac{1}{4} & \frac{1}{4} \end{bmatrix}$

$\begin{bmatrix} 32 & -46 & 37 \\ 9 & -48 & 15 \\ 3 & -14 & 10 \\ -1 & -6 & 2 \\ -8 & -22 & -3 \end{bmatrix} \begin{bmatrix} \frac{1}{2} & -\frac{1}{4} & \frac{1}{4} \\ -\frac{1}{4} & -\frac{3}{8} & -\frac{1}{8} \\ -\frac{1}{2} & -\frac{1}{4} & \frac{1}{4} \end{bmatrix} = \begin{bmatrix} 9 & 0 & 23 \\ 9 & 12 & 12 \\ 0 & 2 & 5 \\ 0 & 2 & 1 \\ 3 & 11 & 0 \end{bmatrix} \begin{matrix} \text{I} & \text{–} & \text{W} \\ \text{I} & \text{L} & \text{L} \\ \text{–} & \text{B} & \text{E} \\ \text{–} & \text{B} & \text{A} \\ \text{C} & \text{K} & \text{–} \end{matrix}$

Message: I WILL BE BACK

191. $A^{-1} = \begin{bmatrix} \frac{2}{3} & \frac{1}{3} & \frac{1}{3} \\ -\frac{1}{3} & \frac{1}{3} & \frac{1}{3} \\ \frac{1}{6} & \frac{1}{3} & -\frac{1}{6} \end{bmatrix}$

$\begin{bmatrix} 21 & -11 & 14 \\ 29 & -11 & -18 \\ 32 & -6 & -26 \\ 31 & -19 & -12 \\ 10 & 6 & 26 \\ 13 & -11 & -2 \\ 37 & 28 & -8 \\ 5 & 13 & 36 \end{bmatrix} \begin{bmatrix} \frac{2}{3} & \frac{1}{3} & \frac{1}{3} \\ -\frac{1}{3} & \frac{1}{3} & \frac{1}{3} \\ \frac{1}{6} & \frac{1}{3} & -\frac{1}{6} \end{bmatrix} = \begin{bmatrix} 20 & 8 & 1 \\ 20 & 0 & 9 \\ 19 & 0 & 13 \\ 25 & 0 & 6 \\ 9 & 14 & 1 \\ 12 & 0 & 1 \\ 14 & 19 & 23 \\ 5 & 18 & 0 \end{bmatrix} \begin{matrix} \text{T} & \text{H} & \text{A} \\ \text{T} & \text{–} & \text{I} \\ \text{S} & \text{–} & \text{M} \\ \text{Y} & \text{–} & \text{F} \\ \text{I} & \text{N} & \text{A} \\ \text{L} & \text{–} & \text{A} \\ \text{N} & \text{S} & \text{W} \\ \text{E} & \text{R} & \text{–} \end{matrix}$

Message: THAT IS MY FINAL ANSWER

193. (a) $\begin{cases} 4b + 20a = 70.3 \\ 20b + 120a = 357 \end{cases}$

$$b = \frac{\begin{vmatrix} 70.3 & 20 \\ 357 & 120 \end{vmatrix}}{\begin{vmatrix} 4 & 20 \\ 20 & 120 \end{vmatrix}} = 16.2$$

$$a = \frac{\begin{vmatrix} 4 & 70.3 \\ 20 & 357 \end{vmatrix}}{\begin{vmatrix} 4 & 20 \\ 20 & 120 \end{vmatrix}} = 0.275$$

$y = 0.275t + 16.2$

(b)

(c) and (d) Let $y = 20$ and solve for t.

$$20 = 0.275t + 16.2$$
$$3.8 = 0.275t$$
$$t \approx 13.8 \Rightarrow 2013$$

195. True. Expansion by Row 3 gives

$$\begin{vmatrix} a_{11} & a_{12} & a_{13} \\ a_{21} & a_{22} & a_{23} \\ a_{31}+c_1 & a_{32}+c_2 & a_{33}+c_3 \end{vmatrix}$$

$$= (a_{31}+c_1)\begin{vmatrix} a_{12} & a_{13} \\ a_{22} & a_{23} \end{vmatrix} - (a_{32}+c_2)\begin{vmatrix} a_{11} & a_{13} \\ a_{21} & a_{23} \end{vmatrix} + (a_{33}+c_3)\begin{vmatrix} a_{11} & a_{12} \\ a_{21} & a_{22} \end{vmatrix}$$

$$= a_{31}\begin{vmatrix} a_{12} & a_{13} \\ a_{22} & a_{23} \end{vmatrix} - a_{32}\begin{vmatrix} a_{11} & a_{13} \\ a_{21} & a_{23} \end{vmatrix} + a_{33}\begin{vmatrix} a_{11} & a_{12} \\ a_{21} & a_{22} \end{vmatrix} + c_1\begin{vmatrix} a_{12} & a_{13} \\ a_{22} & a_{23} \end{vmatrix}$$

$$- c_2\begin{vmatrix} a_{11} & a_{13} \\ a_{21} & a_{22} \end{vmatrix} + c_3\begin{vmatrix} a_{11} & a_{12} \\ a_{21} & a_{22} \end{vmatrix}$$

$$= \begin{vmatrix} a_{11} & a_{12} & a_{13} \\ a_{21} & a_{22} & a_{23} \\ a_{31} & a_{32} & a_{33} \end{vmatrix} + \begin{vmatrix} a_{11} & a_{12} & a_{13} \\ a_{21} & a_{22} & a_{23} \\ c_1 & c_2 & c_3 \end{vmatrix}$$

Note: Expand each of these matrices by Row 3 to see the previous step.

197. A square $n \times n$ matrix A has an inverse A^{-1} if $\det(A) \neq 0$.

Chapter 7 Test

1. $x - y = 6 \Rightarrow y = x - 6.$ Then

$3x + 5(x-6) = 2 \Rightarrow$

$$8x = 32 \Rightarrow x = 4,\ y = 4 - 6 = -2.$$

Answer: $(4,\ -2)$

2. $y = x - 1 = (x-1)^3 \Rightarrow x = 1$ or

$1 = (x-1)^2 = x^2 - 2x + 1 \Rightarrow x^2 - 2x = 0.$

Thus, $x = 1$ or $x(x-2) = 0 \Rightarrow x = 0,\ 1,\ 2.$

Answer: $(0,\ -1),\ (1,\ 0),\ (2,\ 1)$

3. $x - y = 3 \Rightarrow y = x - 3 \Rightarrow 4x - (x-3)^2 = 7$

$$4x - (x^2 - 6x + 9) = 7$$
$$x^2 - 10x + 16 = 0$$
$$(x-2)(x-8) = 0$$

$x = 2,\ 8$

Answer: $(2,\ -1)(8,\ 5)$

4. $\begin{cases} 2x + 5y = -11 & \text{Equation 1} \\ 5x - y = 19 & \text{Equation 2} \end{cases}$

$-\frac{5}{2}$ times Eq.1 added to Eq. produces

$-\frac{27}{2}y = \frac{93}{2} \Rightarrow y = -\frac{31}{9}.$

Then $2x + 5\left(-\frac{31}{9}\right) = -11 \Rightarrow x = \frac{28}{9}.$

Answer: $\left(\frac{28}{9},\ -\frac{31}{9}\right)$

5. $\begin{cases} 3x - 2y + z = 0 \\ 6x + 2y + 3z = -2 \\ 3x - 4y + 5z = 5 \end{cases}$

$\begin{cases} 3x - 2y + z = 0 \\ 6y + z = -2 \\ -2y + 4z = 5 \end{cases}$

$\begin{cases} 3x - 2y + z = 0 \\ y - 2z = -\frac{5}{2} \\ 13z = 13 \end{cases}$

$z = 1$

$y = 2(1) - \frac{5}{2} = -\frac{1}{2}$

$x = \frac{1}{3}\left(-1 + 2\left(-\frac{1}{2}\right)\right) = \frac{1}{3}(-2) - \frac{2}{3}$

Answer: $\left(-\frac{2}{3},\ -\frac{1}{2},\ 1\right)$

6. $\begin{cases} x-4y-z=3 \\ 2x-5y+z=0 \\ 3x-3y+2z=-1 \end{cases}$

$\begin{cases} x-4y-z=3 \\ 3y+3z=-6 \\ 9y+5z=-10 \end{cases}$

$\begin{cases} x-4y-z=3 \\ y+z=-2 \\ -4z=8 \end{cases}$

$z=-2$

$y=-2-(-2)=0$

$x=3+4(0)+(-2)=1$

Answer: $(1, 0, -2)$

7. $6=a(0)^2+b(0)+c \Rightarrow c=6$

$2=a(-2)^2+b(-2)+c$

$\frac{9}{2}=a(3)^2+b(3)+c$

Hence, $\begin{cases} 4a-2b+6=2 \text{ or } 2a-b=-2 \\ 9a+3b+6=\frac{9}{2} \text{ or } 9a+3b=-\frac{3}{2}. \end{cases}$

Solving this system for a and b, you obtain $a=-\frac{1}{2}$, $b=1$. Thus, $y=-\frac{1}{2}x^2+x+6$.

8. $\frac{5x-2}{(x-1)^2}=\frac{A}{x-1}+\frac{B}{(x-1)^2}$

$(5x-2)=A(x-1)+B=Ax+(-A+B)$

$\begin{cases} A=5 \\ -A+B=-2 \Rightarrow B=3 \end{cases}$

$\frac{5x-2}{(x-1)^2}=\frac{5}{x-1}+\frac{3}{(x-1)^2}$

9. $\frac{x^3+x^2+x+2}{x^4+x^2}=\frac{A}{x}+\frac{B}{x^2}+\frac{Cx+D}{x^2+1}$

$x^3+x^2+x+2=Ax(x^2+1)+B(x^2+1)+(Cx+D)x^2$
$=(A+C)x^3+(B+D)x^2+Ax+B$

$\begin{cases} A+C=1 \\ B+D=1 \\ A=1 \\ B=2 \end{cases}$

$A=1,\ C=0,\ B=2,\ D=-1$

$\frac{x^3+x^2+x+2}{x^4+x^2}=\frac{1}{x}+\frac{2}{x^2}-\frac{1}{x^2+1}$

10. $\begin{bmatrix} 2 & 1 & 2 & \vdots & 4 \\ 2 & 2 & 0 & \vdots & 5 \\ 2 & -1 & 6 & \vdots & 2 \end{bmatrix}$ row reduces to

$\begin{bmatrix} 1 & 0 & 2 & \vdots & 1.5 \\ 0 & 1 & -2 & \vdots & 1 \\ 0 & 0 & 0 & \vdots & 0 \end{bmatrix}.$

Infinite number of solutions. Let $z=a,\ y=2a+1,\ x=1.5-2a.$

Answer: $(1.5-2a,\ 1+2a,\ a)$, where a is any real number

11. $\begin{bmatrix} 2 & 3 & 1 & \vdots & 10 \\ 2 & -3 & -3 & \vdots & 22 \\ 4 & -2 & 3 & \vdots & -2 \end{bmatrix}$ row reduces to

$\begin{bmatrix} 1 & 0 & 0 & \vdots & 5 \\ 0 & 1 & 0 & \vdots & 2 \\ 0 & 0 & 1 & \vdots & -6 \end{bmatrix}.$

Answer: $(5, 2, -6)$

12. (a) $A-B=\begin{bmatrix} 1 & 0 & 4 \\ -7 & -6 & -1 \\ 0 & 4 & 0 \end{bmatrix}$

(b) $3A=\begin{bmatrix} 15 & 12 & 12 \\ -12 & -12 & 0 \\ 3 & 6 & 0 \end{bmatrix}$

(c) $3A-2B=\begin{bmatrix} 7 & 4 & 12 \\ -18 & -16 & -2 \\ 1 & 10 & 0 \end{bmatrix}$

(d) $AB=\begin{bmatrix} 36 & 20 & 4 \\ -28 & -24 & -4 \\ 10 & 8 & 2 \end{bmatrix}$

13. $\begin{bmatrix} -2 & 2 & 3 & \vdots & 1 & 0 & 0 \\ 1 & -1 & 0 & \vdots & 0 & 1 & 0 \\ 0 & 1 & 4 & \vdots & 0 & 0 & 1 \end{bmatrix}$ reduces to

$\begin{bmatrix} 1 & 0 & 0 & \vdots & -\frac{4}{3} & -\frac{5}{3} & 1 \\ 0 & 1 & 0 & \vdots & -\frac{4}{3} & -\frac{8}{3} & 1 \\ 0 & 0 & 1 & \vdots & \frac{1}{3} & \frac{2}{3} & 0 \end{bmatrix}.$

$A^{-1}=\begin{bmatrix} -\frac{4}{3} & -\frac{5}{3} & 1 \\ -\frac{4}{3} & -\frac{8}{3} & 1 \\ \frac{1}{3} & \frac{2}{3} & 0 \end{bmatrix}$

$A^{-1}\begin{bmatrix} 7 \\ -5 \\ -1 \end{bmatrix}=\begin{bmatrix} -2 \\ 3 \\ -1 \end{bmatrix}$

Answer: $(-2, 3, -1)$

14. $\begin{vmatrix} -25 & 18 \\ 6 & -7 \end{vmatrix} = (-25)(-7) - 6(18) = 67$

15. $\det(A) = \begin{vmatrix} 4 & 0 & 3 \\ 1 & -8 & 2 \\ 3 & 2 & 2 \end{vmatrix}$

$= 4(-16-4) - 0 + 3(2+24)$

$= -80 + 78 = -2$

16. $\begin{vmatrix} x_1 & y_1 & 1 \\ x_2 & y_2 & 1 \\ x_3 & y_3 & 1 \end{vmatrix} = \begin{vmatrix} -1 & 1 & 1 \\ 4 & 11 & 1 \\ -1 & -5 & 1 \end{vmatrix}$

$= -1(16) - 1(4+1) + 1(-20+11)$

$= -16 - 5 - 9 = -30$

Area $= |-30| = 30$ square units

17. $x = \dfrac{\begin{vmatrix} 3 & -2 \\ -1 & 4 \end{vmatrix}}{\begin{vmatrix} 2 & -2 \\ 1 & 4 \end{vmatrix}} = \dfrac{10}{10} = 1$

$y = \dfrac{\begin{vmatrix} 2 & 3 \\ 1 & -1 \end{vmatrix}}{\begin{vmatrix} 2 & -2 \\ 1 & 4 \end{vmatrix}} = \dfrac{-5}{10} = -\dfrac{1}{2}$

Answer: $\left(1, -\dfrac{1}{2}\right)$

18. Upper left: $400 + x_2 = x_1$

Upper right: $x_1 + x_3 = x_4 + 600$

Lower left: $300 = x_2 + x_3 + x_5$

Lower right: $x_5 + x_4 = 100$

$$\begin{cases} x_1 - x_2 = 400 \\ x_1 + x_3 - x_4 = 600 \\ x_2 + x_3 + x_5 = 300 \\ x_4 + x_5 = 100 \end{cases}$$

Solving the system:

$$\begin{bmatrix} 1 & -1 & 0 & 0 & 0 & \vdots & 400 \\ 1 & 0 & 1 & -1 & 0 & \vdots & 600 \\ 0 & 1 & 1 & 0 & 1 & \vdots & 300 \\ 0 & 0 & 0 & 1 & 1 & \vdots & 100 \end{bmatrix} \rightarrow \begin{bmatrix} 1 & 0 & 1 & 0 & 1 & \vdots & 700 \\ 0 & 1 & 1 & 0 & 1 & \vdots & 300 \\ 0 & 0 & 0 & 1 & 1 & \vdots & 100 \\ 0 & 0 & 0 & 0 & 0 & \vdots & 0 \end{bmatrix}$$

Letting $x_3 = a$ and $x_5 = b$ be real numbers, we have:

$x_5 = b$

$x_4 = 100 - b$

$x_3 = a$

$x_2 = 300 - a - b$

$x_1 = 700 - b - a$

CHAPTER 8

Section 8.1

1. terms

3. index, upper limit, lower limit

5. (a) A finite sequence is a function whose domain is the set of the first n positive integers.

(b) An infinite sequence is a function whose domain is the set of positive integers.

7. $a_n = 2n + 5$

$a_1 = 2(1) + 5 = 7$

$a_2 = 2(2) + 5 = 9$

$a_3 = 2(3) + 5 = 11$

$a_4 = 2(4) + 5 = 13$

$a_5 = 2(5) + 5 = 15$

9. $a_n = 3^n$

$a_1 = 3^1 = 3$

$a_2 = 3^2 = 9$

$a_3 = 3^3 = 27$

$a_4 = 3^4 = 81$

$a_5 = 3^5 = 243$

11. $a_n = \left(-\frac{1}{2}\right)^n$

$a_1 = \left(-\frac{1}{2}\right)^1 = -\frac{1}{2}$

$a_2 = \left(-\frac{1}{2}\right)^2 = \frac{1}{4}$

$a_3 = \left(-\frac{1}{2}\right)^3 = -\frac{1}{8}$

$a_4 = \left(-\frac{1}{2}\right)^4 = \frac{1}{16}$

$a_5 = \left(-\frac{1}{2}\right)^5 = -\frac{1}{32}$

13. $a_n = \frac{n+1}{n}$

$a_1 = \frac{1+1}{1} = 2$

$a_2 = \frac{2+1}{2} = \frac{3}{2}$

$a_3 = \frac{3+1}{3} = \frac{4}{3}$

$a_4 = \frac{4+1}{4} = \frac{5}{4}$

$a_5 = \frac{5+1}{5} = \frac{6}{5}$

15. $a_n = \frac{n}{n^2+1}$

$a_1 = \frac{1}{1^2+1} = \frac{1}{2}$

$a_2 = \frac{2}{2^2+1} = \frac{2}{5}$

$a_3 = \frac{3}{3^2+1} = \frac{3}{10}$

$a_4 = \frac{4}{4^2+1} = \frac{4}{17}$

$a_5 = \frac{5}{5^2+1} = \frac{5}{26}$

17. $a_n = \frac{1+(-1)^n}{n}$

(a) $a_1 = 0$

$a_2 = 1$

$a_3 = 0$

$a_4 = 0.5$

$a_5 = 0$

(b) $a_1 = \frac{1+(-1)}{1} = 0$

$a_2 = \frac{1+1}{2} = \frac{2}{2} = 1$

$a_3 = \frac{1+(-1)}{3} = 0$

$a_4 = \frac{1+1}{4} = \frac{2}{4} = \frac{1}{2}$

$a_5 = \frac{1+(-1)}{5} = 0$

19. $a_n = 1 - \frac{1}{2^n}$

(a) $a_1 = 0.5$

$a_2 = 0.75$

$a_3 = 0.875$

$a_4 = 0.9375$

$a_5 \approx 0.9688$

(b) $a_1 = 1 - \frac{1}{2^1} = \frac{1}{2}$

$a_2 = 1 - \frac{1}{2^2} = 1 - \frac{1}{4} = \frac{3}{4}$

$a_3 = 1 - \frac{1}{2^3} = \frac{7}{8}$

$a_4 = 1 - \frac{1}{2^4} = \frac{15}{16}$

$a_5 = 1 - \frac{1}{2^5} = \frac{31}{32}$

21. $a_n = \frac{1}{n^{3/2}}$

(a) $a_1 = 1$

$a_2 \approx 0.3536$

$a_3 \approx 0.1925$

$a_4 = 0.125$

$a_5 \approx 0.0894$

(b) $a_1 = \frac{1}{1} = 1$

$a_2 = \frac{1}{2^{3/2}}$

$a_3 = \frac{1}{3^{3/2}}$

$a_4 = \frac{1}{4^{3/2}} = \frac{1}{8}$

$a_5 = \frac{1}{5^{3/2}}$

23. $a_n = \frac{(-1)^n}{n^2}$

(a) $a_1 = -1$

$a_2 = 0.25$

$a_3 \approx -0.1111$

$a_4 = 0.0625$

$a_5 = -0.04$

(b) $a_1 = \frac{-1}{1} = -1$

$a_2 = \frac{1}{4}$

$a_3 = -\frac{1}{9}$

$a_4 = \frac{1}{16}$

$a_5 = -\frac{1}{25}$

25. $a_n = (2n-1)(2n+1)$

(a) $a_1 = 3$

$a_2 = 15$

$a_3 = 35$

$a_4 = 63$

$a_5 = 99$

(b) $a_1 = (1)(3) = 3$

$a_2 = (3)(5) = 15$

$a_3 = (5)(7) = 35$

$a_4 = (7)(9) = 63$

$a_5 = (9)(11) = 99$

27. $a_n = 2(3n-1) + 5$

n	1	2	3	4	5	6	7	8	9	10
a_n	9	15	21	27	33	39	45	51	57	63

29. $a_n = 1 + \frac{n+1}{n}$

n	1	2	3	4	5	6	7	8	9	10
a_n	3	2.5	2.33	2.25	2.2	2.17	2.14	2.13	2.11	2.1

31. $a_n = (-1)^n + 1$

n	1	2	3	4	5	6	7	8	9	10
a_n	0	2	0	2	0	2	0	2	0	2

33. $a_{10} = \frac{10^2}{10^2 + 1} = \frac{100}{101}$

35. $a_{25} = (-1)^{25}[3(25) - 2] = -73$

37. $a_6 = \frac{2^6}{2^6 + 1} = \frac{64}{65}$

39. 1, 4, 7, 10, 13, ...

$a_n = 1 + (n-1)3 = 3n - 2$

$a_n = 4n - 1$

41. 0, 3, 8, 15, 24, ⋯

$a_n = n^2 - 1$

43. $\frac{2}{3}, \frac{3}{4}, \frac{4}{5}, \frac{5}{6}, \frac{6}{7}, \ldots$

$a_n = \frac{n+1}{n+2}$

45. $\frac{1}{2}, \frac{-1}{4}, \frac{1}{8}, \frac{-1}{16}, \ldots$

$a_n = \frac{(-1)^{n+1}}{2^n}$

47. $1+\frac{1}{1}, 1+\frac{1}{2}, 1+\frac{1}{3}, 1+\frac{1}{4}, 1+\frac{1}{5}, \dots$

$a_n = 1 + \frac{1}{n}$

49. $1, \frac{1}{2}, \frac{1}{6}, \frac{1}{24}, \frac{1}{120}, \dots$

$a_n = \frac{1}{n!}$

51. $1, 3, 1, 3, 1, 3, \dots$

$a_n = 2 + (-1)^n$

53. $a_1 = 28,\ a_{k+1} = a_k - 4$

$a_1 = 28$
$a_2 = a_1 - 4 = 28 - 4 = 24$
$a_3 = a_2 - 4 = 24 - 4 = 20$
$a_4 = a_3 - 4 = 20 - 4 = 16$
$a_5 = a_4 - 4 = 16 - 4 = 12$

55. $a_1 = 3,\ a_{k+1} = 2(a_k - 1)$

$a_1 = 3$
$a_2 = 2(a_1 - 1) = 2(3 - 1) = 4$
$a_3 = 2(a_2 - 1) = 2(4 - 1) = 6$
$a_4 = 2(a_3 - 1) = 2(6 - 1) = 10$
$a_5 = 2(a_4 - 1) = 2(10 - 1) = 18$

57. $a_0 = 1, a_1 = 3,\ a_k = a_{k-2} + a_{k-1}$

$a_0 = 1$
$a_1 = 3$
$a_2 = a_0 + a_1 = 1 + 3 = 4$
$a_3 = a_1 + a_2 = 3 + 4 = 7$
$a_4 = a_2 + a_3 = 4 + 7 = 11$

59. $a_1 = 6,\ a_{k+1} = a_k + 2$

$a_1 = 6$
$a_2 = a_1 + 2 = 6 + 2 = 8$
$a_3 = a_2 + 2 = 8 + 2 = 10$
$a_4 = a_3 + 2 = 10 + 2 = 12$
$a_5 = a_4 + 2 = 12 + 2 = 14$

In general, $a_n = 2n + 4$.

61. $a_1 = 81,\ a_{k+1} = \frac{1}{3}a_k$

$a_1 = 81$
$a_2 = \frac{1}{3}a_1 = \frac{1}{3}(81) = 27$
$a_3 = \frac{1}{3}a_2 = \frac{1}{3}(27) = 9$
$a_4 = \frac{1}{3}a_3 = \frac{1}{3}(9) = 3$
$a_5 = \frac{1}{3}a_4 = \frac{1}{3}(3) = 1$

In general, $a_n = 81\left(\frac{1}{3}\right)^{n-1} = 81(3)\left(\frac{1}{3}\right)^n = \frac{243}{3^n}$.

63. $a_n = \frac{1}{n!}$

(a) $a_0 = 1$
$a_1 = 1$
$a_2 = 0.5$
$a_3 \approx 0.1667$
$a_4 \approx 0.0417$

(b) $a_0 = \frac{1}{0!} = 1$
$a_1 = \frac{1}{1!} = 1$
$a_2 = \frac{1}{2!} = \frac{1}{2}$
$a_3 = \frac{1}{3!} = \frac{1}{6}$
$a_4 = \frac{1}{4!} = \frac{1}{24}$

65. $a_n = \frac{n!}{2n+1}$

(a)
$a_0 = 1$
$a_1 \approx 0.3333$
$a_2 \approx 0.4$
$a_3 \approx 0.8571$
$a_4 \approx 2.6667$

(b)
$a_0 = \frac{0!}{1} = 1$
$a_1 = \frac{1!}{2+1} = \frac{1}{3}$
$a_2 = \frac{2!}{4+1} = \frac{2}{5}$
$a_3 = \frac{3!}{6+1} = \frac{6}{7}$
$a_4 = \frac{4!}{8+1} = \frac{24}{9} = \frac{8}{3}$

67. $a_n = \frac{(-1)^{2n}}{(2n)!}$

(a) $a_0 = 1$

$a_1 = 0.5$

$a_2 \approx 0.0417$

$a_3 \approx 0.0014$

$a_4 \approx 0.0000248$

(b) $a_0 = \frac{(-1)^0}{0!} = 1$

$a_1 = \frac{(-1)^2}{2!} = \frac{1}{2}$

$a_2 = \frac{(-1)^4}{4!} = \frac{1}{24}$

$a_3 = \frac{(-1)^6}{6!} = \frac{1}{720}$

$a_4 = \frac{(-1)^8}{8!} = \frac{1}{40,320}$

69. $\frac{2!}{4!} = \frac{2!}{4 \cdot 3 \cdot 2!} = \frac{1}{12}$

71. $\frac{12!}{4!8!} = \frac{12 \cdot 11 \cdot 10 \cdot 9 \cdot 8!}{4!8!}$

$= \frac{12 \cdot 11 \cdot 10 \cdot 9}{4 \cdot 3 \cdot 2} = 495$

73. $\frac{(n+1)!}{n!} = \frac{(n+1)n!}{n!} = n+1$

75. $\frac{(2n-1)!}{(2n+1)!} = \frac{(2n-1)!}{(2n+1)(2n)(2n-1)!}$

$= \frac{1}{2n(2n+1)}$

77. $a_n = \frac{8}{n+1}$

$a_n \to 0$ as $n \to \infty$

$a_1 = 4,\ a_{10} = \frac{8}{11}$

Matches graph (c).

79. $a_n = 4(0.5)^{n-1}$

$a_n \to 0$ as $n \to \infty$

$a_1 = 4,\ a_{10} \approx 0.008$

Matches graph (d).

81. $a_n = \frac{2}{3}n$

83. $a_n = 16(-0.5)^{n-1}$

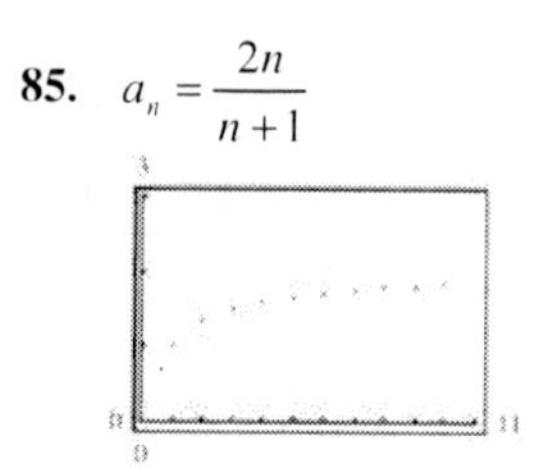

85. $a_n = \frac{2n}{n+1}$

87. $\sum_{i=1}^{5}(2i+1) = (2+1)+(4+1)+(6+1)+(8+1)+(10+1) = 35$

89. $\sum_{k=1}^{4} 10 = 10+10+10+10 = 40$

91. $\sum_{i=0}^{4} i^2 = 0^2 + 1^2 + 2^2 + 3^2 + 4^2 = 30$

93. $\sum_{k=0}^{3} \frac{1}{k^2+1} = \frac{1}{1} + \frac{1}{1+1} + \frac{1}{4+1} + \frac{1}{9+1} = \frac{9}{5}$

95. $\sum_{i=1}^{4}\left[(i-1)^2 + (i+1)^3\right] = \left[(0)^2 + (2)^3\right] + \left[(1)^2 + (3)^3\right]$

$+ \left[(2)^2 + (4)^3\right] + \left[(3)^2 + (5)^3\right] = 238$

97. $\sum_{i=1}^{4} 2^i = 2^1 + 2^2 + 2^3 + 2^4 = 30$

99. $\sum_{j=1}^{6}(24 - 3j) = 81$

101. $\sum_{k=0}^{4} \frac{(-1)^k}{k+1} = \frac{47}{60}$

103. $\frac{1}{3(1)} + \frac{1}{3(2)} + \frac{1}{3(3)} + \cdots + \frac{1}{3(9)} = \sum_{i=1}^{9} \frac{1}{3i} \approx 0.94299$

105. $\left[2\left(\frac{1}{8}\right)+3\right] + \left[2\left(\frac{2}{8}\right)+3\right] + \left[2\left(\frac{3}{8}\right)+3\right] + \cdots + \left[2\left(\frac{8}{8}\right)+3\right]$

$= \sum_{i=1}^{8}\left[2\left(\frac{i}{8}\right)+3\right] = 33$

107. $3 - 9 + 27 - 81 + 243 - 729 = \sum_{i=1}^{6} (-1)^{i+1} 3^i = -546$

109. $\frac{1}{1^2} - \frac{1}{2^2} + \frac{1}{3^2} - \frac{1}{4^2} + \cdots - \frac{1}{20^2} = \sum_{i=1}^{20} \frac{(-1)^{i+1}}{i^2} \approx 0.82128$

111. $\frac{1}{4} + \frac{3}{8} + \frac{7}{16} + \frac{15}{32} + \frac{31}{64} = \sum_{j=1}^{5} \frac{2^j - 1}{2^{j+1}} = \frac{129}{64} = 2.015625$

113. $\sum_{i=1}^{4} 5\left(\frac{1}{2}\right)^i = 4.6875 = \frac{75}{16}$

115. $\sum_{n=1}^{3} 4\left(-\frac{1}{2}\right)^n = -1.5 = -\frac{3}{2}$

117. (a) $\sum_{i=1}^{4} 6\left(\frac{1}{10}\right)^i = 6\left(\frac{1}{10}\right) + 6\left(\frac{1}{10}\right)^2 + 6\left(\frac{1}{10}\right)^3 + 6\left(\frac{1}{10}\right)^4$

$= 0.6666$

$= \frac{3333}{5000}$

(b) $\sum_{i=1}^{\infty} 6\left(\frac{1}{10}\right)^i = 6[0.1 + 0.01 + 0.001 + \cdots]$

$= 6[0.111\ldots]$

$= 0.666\ldots$

$= \frac{2}{3}$

119. (a) $\sum_{k=1}^{4} \left(\frac{1}{10}\right)^k = \frac{1}{10} + \frac{1}{100} + \frac{1}{1000} + \frac{1}{10{,}000}$

$= 0.1111$

$= \frac{1111}{10{,}000}$

(b) $\sum_{k=1}^{\infty} \left(\frac{1}{10}\right)^k = 0.1 + 0.01 + 0.001 + \cdots$

$= 0.111\ldots$

$= \frac{1}{9}$

121. $A_n = 5000\left(1 + \frac{0.03}{4}\right)^n, \; n = 1, 2, 3, \ldots$

(a) $A_1 = 5000\left(1 + \frac{0.03}{4}\right)^1 = \$5037.50$

$A_2 \approx \$5075.28 \quad A_3 \approx \$5113.35$

$A_4 \approx \$5151.70 \quad A_5 \approx \$5190.33$

$A_6 \approx \$5229.26 \quad A_7 \approx \$5268.48$

$A_8 \approx \$5307.99$

(b) $A_{40} \approx \$6741.74$

123. $a_n = 296.477n^2 - 469.11n + 3606.2$

$a_1 \approx \$3433.6$

$a_2 \approx \$3853.9$

$a_3 \approx \$4867.2$

$a_4 \approx \$6473.4$

$a_5 \approx \$8672.6$

$a_6 \approx \$11{,}464.7$

$a_7 \approx \$14{,}489.8$

$a_8 \approx \$18{,}827.8$

Approximate revenue $\approx \$72.443$ million

Actual revenue $\approx \$72.443$ million

125. True by the Properties of Sums.

127. $a_0 = 1, \; a_1 = 1, \; a_{k+2} = a_{k+1} + a_k; \; b_n = \frac{a_{n+1}}{a_n}, \; n > 0$

$a_0 = 1$

$a_1 = 1$

$a_2 = 1 + 1 = 2$

$a_3 = 2 + 1 = 3$

$a_4 = 3 + 2 = 5$

$a_5 = 5 + 3 = 8$

$a_6 = 8 + 5 = 13$

$a_7 = 13 + 8 = 21$

$a_8 = 21 + 13 = 34$

$a_9 = 34 + 21 = 55$

$a_{10} = 55 + 34 = 89$

$a_{11} = 89 + 55 = 144$

$b_0 = \frac{1}{1} = 1$

$b_1 = \frac{2}{1} = 2$

$b_2 = \frac{3}{2}$

$b_3 = \frac{5}{3}$

$b_4 = \frac{8}{5}$

$b_5 = \frac{13}{8}$

$b_6 = \frac{21}{13}$

$b_7 = \frac{34}{21}$

$b_8 = \frac{55}{34}$

$b_9 = \frac{89}{55}$

129. $a_n = \frac{(1 + \sqrt{5})^n - (1 - \sqrt{5})^n}{2^n \sqrt{5}}$

$a_1 = \frac{(1 + \sqrt{5})^1 - (1 - \sqrt{5})^1}{2^1 \sqrt{5}} = 1$

$a_2 = 1, \; a_3 = 2$

$a_4 = 3, \; a_5 = 5$

131. $a_{n+1} = \dfrac{\left(1+\sqrt{5}\right)^{n+1} - \left(1-\sqrt{5}\right)^{n+1}}{2^{n+1}\sqrt{5}}$

$a_{n+2} = \dfrac{\left(1+\sqrt{5}\right)^{n+2} - \left(1-\sqrt{5}\right)^{n+2}}{2^{n+2}\sqrt{5}}$

133. $a_n = \dfrac{x^n}{n!}$

$a_1 = \dfrac{x}{1} = x$

$a_2 = \dfrac{x^2}{2!} = \dfrac{x^2}{2}$

$a_3 = \dfrac{x^3}{3!} = \dfrac{x^3}{6}$

$a_4 = \dfrac{x^4}{4!} = \dfrac{x^4}{24}$

$a_5 = \dfrac{x^5}{5!} = \dfrac{x^5}{120}$

135. $a_n = \dfrac{(-1)^n x^{2n+1}}{2n+1}$

$a_1 = \dfrac{(-1)x^{2+1}}{2+1} = \dfrac{-x^3}{3}$

$a_2 = \dfrac{(-1)^2 x^{2(2)+1}}{2(2)+1} = \dfrac{x^5}{5}$

$a_3 = \dfrac{(-1)^3 x^{2(3)+1}}{2(3)+1} = -\dfrac{x^7}{7}$

$a_4 = \dfrac{(-1)^4 x^{2(4)+1}}{2(4)+1} = \dfrac{x^9}{9}$

$a_5 = \dfrac{(-1)^5 x^{2(5)+1}}{2(5)+1} = \dfrac{-x^{11}}{11}$

137. $a_n = \dfrac{(-1)^n x^{2n}}{(2n)!}$

$a_1 = -\dfrac{x^2}{2}$

$a_2 = \dfrac{x^4}{4!} = \dfrac{x^4}{24}$

$a_3 = \dfrac{-x^6}{6!} = -\dfrac{x^6}{720}$

$a_4 = \dfrac{x^8}{8!} = \dfrac{x^8}{40,320}$

$a_5 = \dfrac{-x^{10}}{10!} = -\dfrac{x^{10}}{3,628,800}$

139. $a_n = \dfrac{(-1)^n x^n}{n!}$

$a_1 = \dfrac{(-1)x}{1!} = -x$

$a_2 = \dfrac{(-1)^2 x^2}{2!} = \dfrac{x^2}{2}$

$a_3 = \dfrac{-x^3}{3!} = -\dfrac{x^3}{6}$

$a_4 = \dfrac{x^4}{4!} = \dfrac{x^4}{24}$

$a_5 = \dfrac{-x^5}{5!} = -\dfrac{x^5}{120}$

141. $a_n = \dfrac{(-1)^{n+1}(x+1)^n}{n!}$

$a_1 = x+1$

$a_2 = -\dfrac{(x+1)^2}{2}$

$a_3 = \dfrac{(x+1)^3}{6}$

$a_4 = -\dfrac{(x+1)^4}{24}$

$a_5 = \dfrac{(x+1)^5}{120}$

143. $a_n = \dfrac{1}{2n} - \dfrac{1}{2n+2}$

$a_1 = \dfrac{1}{2} - \dfrac{1}{4} = \dfrac{1}{4}$

$a_2 = \dfrac{1}{4} - \dfrac{1}{6} = \dfrac{1}{12}$

$a_3 = \dfrac{1}{6} - \dfrac{1}{8} = \dfrac{1}{24}$

$a_4 = \dfrac{1}{8} - \dfrac{1}{10} = \dfrac{1}{40}$

$a_5 = \dfrac{1}{10} - \dfrac{1}{12} = \dfrac{1}{60}$

nth partial sum $= \left(\dfrac{1}{2} - \dfrac{1}{4}\right) + \left(\dfrac{1}{4} - \dfrac{1}{6}\right) + \cdots + \left(\dfrac{1}{2n} - \dfrac{1}{2n+2}\right)$

$= \dfrac{1}{2} - \dfrac{1}{2n+2}$

145. $a_n = \frac{1}{n+1} - \frac{1}{n+2}$

$a_1 = \frac{1}{2} - \frac{1}{3} = \frac{1}{6}$

$a_2 = \frac{1}{3} - \frac{1}{4} = \frac{1}{12}$

$a_3 = \frac{1}{4} - \frac{1}{5} = \frac{1}{20}$

$a_4 = \frac{1}{5} - \frac{1}{6} = \frac{1}{30}$

$a_5 = \frac{1}{6} - \frac{1}{7} = \frac{1}{42}$

$n\text{th partial sum} = \left(\frac{1}{2} - \frac{1}{3}\right) + \left(\frac{1}{3} - \frac{1}{4}\right) + \cdots + \left(\frac{1}{n+1} - \frac{1}{n+2}\right)$

$= \frac{1}{2} - \frac{1}{n+2}$

147. Yes. If the sequence is finite and the terms are integer terms, then the sum can be found.

149. (a) $A - B = \begin{bmatrix} 8 & 1 \\ -3 & 7 \end{bmatrix}$

(b) $2B - 3A = \begin{bmatrix} -22 & -7 \\ 3 & -18 \end{bmatrix}$

(c) $AB = \begin{bmatrix} 18 & 9 \\ 18 & 0 \end{bmatrix}$

(d) $BA = \begin{bmatrix} 0 & 6 \\ 27 & 18 \end{bmatrix}$

151. (a) $A - B = \begin{bmatrix} -3 & -7 & 4 \\ 4 & 4 & 1 \\ 1 & 4 & 3 \end{bmatrix}$

(b) $2B - 3A = \begin{bmatrix} 8 & 17 & -14 \\ -12 & -13 & -9 \\ -3 & -15 & -10 \end{bmatrix}$

(c) $AB = \begin{bmatrix} -2 & 7 & -16 \\ 4 & 42 & 45 \\ 1 & 23 & 48 \end{bmatrix}$

(d) $BA = \begin{bmatrix} 16 & 31 & 42 \\ 10 & 47 & 31 \\ 13 & 22 & 25 \end{bmatrix}$

Section 8.2

1. $a_n = a_1 + (n-1)d$

3. A sequence is arithmetic when the differences between consecutive terms are the same or constant, which is known as the common difference.

5. $10, 8, 6, 4, 2, \ldots$

Arithmetic sequence, $d = -2$

7. $3, \frac{5}{2}, 2, \frac{3}{2}, 1, \ldots$

Arithmetic sequence, $d = -\frac{1}{2}$

9. $3.7, 4.3, 4.9, 5.5, 6.1, \ldots$

Arithmetic sequence, $d = 0.6$

11. $a_n = 8 + 13n$

$a_1 = 8 + 13(1) = 21$

$a_2 = 8 + 13(2) = 34$

$a_3 = 8 + 13(3) = 47$

$a_4 = 8 + 13(4) = 60$

$a_5 = 8 + 13(5) = 73$

Arithmetic sequence, $d = 13$

13. $a_n = \frac{1}{n+1}$

$a_1 = \frac{1}{1+1} = \frac{1}{2}$

$a_2 = \frac{1}{2+1} = \frac{1}{3}$

$a_3 = \frac{1}{3+1} = \frac{1}{4}$

$a_4 = \frac{1}{4+1} = \frac{1}{5}$

$a_5 = \frac{1}{5+1} = \frac{1}{6}$

Not an arithmetic sequence

15. $a_n = 150 - 7n$

$a_1 = 150 - 7(1) = 143$

$a_2 = 150 - 7(2) = 136$

$a_3 = 150 - 7(3) = 129$

$a_4 = 150 - 7(4) = 122$

$a_5 = 150 - 7(5) = 115$

Arithmetic sequence, $d = -7$

17. $a_n = 3 + 2(-1)^n$

$a_1 = 3 + 2(-1)^1 = 1$

$a_2 = 3 + 2(-1)^2 = 5$

$a_3 = 3 + 2(-1)^3 = 1$

$a_4 = 3 + 2(-1)^4 = 5$

$a_5 = 3 + 2(-1)^5 = 1$

Not an arithmetic sequence

19. $a_n = (-1)^n$

$a_1 = (-1)^1 = -1$

$a_2 = (-1)^2 = 1$

$a_3 = (-1)^3 = -1$

$a_4 = (-1)^4 = 1$

$a_5 = (-1)^5 = -1$

Not an arithmetic sequence

21. $a_1 = 1,\ d = 3$

$a_n = a_1 + (n-1)d = 1 + (n-1)(3) = 3n - 2$

23. $a_1 = 100,\ d = -8$

$a_n = a_1 + (n-1)d$

$= 100 + (n-1)(-8) = 108 - 8n$

25. $4,\ \frac{3}{2},\ -1,\ -\frac{7}{2},\ldots,d = -\frac{5}{2}$

$a_n = a_1 + (n-1)d = 4 + (n-1)\left(-\frac{5}{2}\right) = \frac{13}{2} - \frac{5}{2}n$

27. $a_1 = 5, a_4 = 15$

$a_4 = a_1 + 3d \Rightarrow 15 = 5 + 3d \Rightarrow d = \frac{10}{3}$

$a_n = a_1 + (n-1)d = 5 + (n-1)\left(\frac{10}{3}\right) = \frac{10}{3}n + \frac{5}{3}$

29. $a_3 = 94, a_6 = 85$

$a_6 = a_3 + 3d \Rightarrow 85 = 94 + 3d \Rightarrow d = -3$

$a_1 = a_3 - 2d \Rightarrow a_1 = 94 - 2(-3) = 100$

$a_n = a_1 + (n-1)d$

$= 100 + (n-1)(-3) = 103 - 3n$

31. $a_1 = 5, d = 6$

$a_1 = 5$

$a_2 = 5 + 6 = 11$

$a_3 = 11 + 6 = 17$

$a_4 = 17 + 6 = 23$

$a_5 = 23 + 6 = 29$

33. $a_1 = 10, d = -12$

$a_1 = -10$

$a_2 = -10 - 12 = -22$

$a_3 = -22 - 12 = -34$

$a_4 = -34 - 12 = -46$

$a_5 = -46 - 12 = -58$

35. $a_8 = 26, a_{12} = 42$

$26 = a_8 = a_1 + (n-1)d = a_1 + 7d$

$42 = a_{12} = a_1 + (n-1)d = a_1 + 11d$

Answer: $d = 4,\ a_1 = -2$

$a_1 = -2$

$a_2 = -2 + 4 = 2$

$a_3 = 2 + 4 = 6$

$a_4 = 6 + 4 = 10$

$a_5 = 10 + 4 = 14$

37. $a_3 = 19, a_{15} = -1.7$

$a_{15} = a_3 + 12d$

$-1.7 = 19 + 12d \Rightarrow d = -1.725$

$a_3 = a_1 + 2d \quad \Rightarrow 19 = a_1 + 2(-1.725)$

$\Rightarrow a_1 = 22.45$

$a_2 = a_1 - 1.725 = 20.725$

$a_3 = 19$

$a_4 = 19 - 1.725 = 17.275$

$a_5 = 17.275 - 1.725 = 15.55$

39. $a_1 = 15,\ a_{k+1} = a_k + 4$

$a_2 = a_1 + 4 = 15 + 4 = 19$

$a_3 = 19 + 4 = 23$

$a_4 = 23 + 4 = 27$

$a_5 = 27 + 4 = 31$

$d = 4,\ a_n = 11 + 4n$

41. $a_1 = \frac{3}{4},\ a_{k+1} = -\frac{1}{10} + a_k$

$a_2 = -\frac{1}{10} + \frac{3}{5} = \frac{5}{10} = \frac{1}{2}$

$a_3 = -\frac{1}{10} + \frac{1}{2} = \frac{4}{10} = \frac{2}{5}$

$a_4 = -\frac{1}{10} + \frac{2}{5} = \frac{3}{10}$

$a_5 = -\frac{1}{10} + \frac{3}{10} = \frac{1}{5}$

$d = -\frac{1}{10}$

$a_n = \frac{7}{10} - \frac{1}{10}n$

43. $a_1 = 5,\ a_2 = 11 \Rightarrow d = 6$

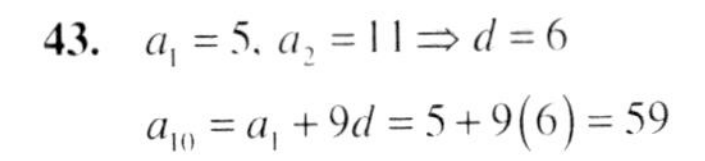

$a_{10} = a_1 + 9d = 5 + 9(6) = 59$

45. $a_1 = 4.2,\ a_2 = 6.6 \Rightarrow d = 2.4$

$a_7 = a_1 + 6d = 4.2 + 6(2.4) = 18.6$

47. $a_n = 15 - \frac{3}{2}n$

49. $a_n = 0.4n - 2$

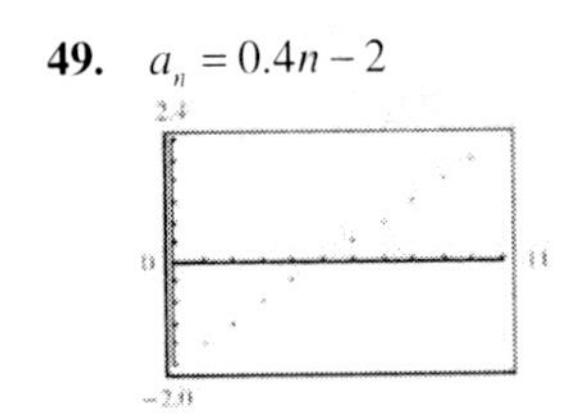

51. $a_n = 4n - 5$

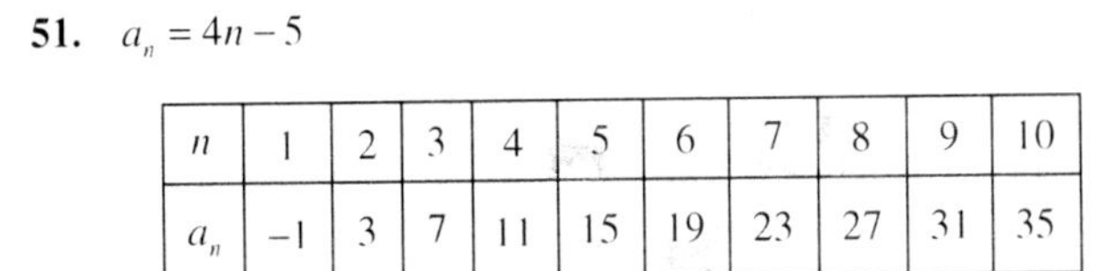

n	1	2	3	4	5	6	7	8	9	10
a_n	−1	3	7	11	15	19	23	27	31	35

53. $a_n = 20 - \frac{3}{4}n$

n	1	2	3	4	5	6	7	8	9	10
a_n	19.25	18.5	17.75	17	16.25	15.5	14.75	14	13.25	12.5

55. $a_n = 1.5 + 0.05n$

n	1	2	3	4	5	6	7	8	9	10
a_n	1.55	1.6	1.65	1.7	1.75	1.8	1.85	1.9	1.95	2.0

57. $S_{10} = \frac{10}{2}(2 + 20) = 110$

59. $S_5 = \frac{5}{2}(-1 + (-9)) = -25$

61. $S_{100} = \sum_{n=1}^{100} n = 1 + 2 + \cdots + 99 + 100 = \frac{100}{2}(1 + 100) = 5050$

63. $S_{131} = \frac{131}{2}(-100 + 30) = -4585$

65. $8, 20, 32, 44, \ldots,\ n = 10$

$a_1 = 8,\ a_2 = 20 \Rightarrow d = 12$

$a_{10} = a_1 + 9d = 8 + 9(12) = 116$

$S_{10} = \frac{10}{2}(8 + 116) = 620$

67. $0.5, 1.3, 2.1, 2.9, \ldots,\ n = 10$

$a_1 = 0.5,\ a_2 = 1.3 \Rightarrow d = 0.8$

$a_{10} = a_1 + 9d = 0.5 + 9(0.8) = 7.7$

$S_{10} = \frac{10}{2}(a_1 + a_{10}) = 5(0.5 + 7.7) = 41$

69. $a_1 = 100,\ a_{25} = 220,\ n = 25$

$S_{25} = \frac{25}{2}(a_1 + a_{25}) = 12.5(100 + 220) = 4000$

71. $a_1 = 1,\ a_{50} = 50,\ n = 50$

$\sum_{n=1}^{50} n = \frac{50}{2}(1 + 50) = 1275$

73. $\sum_{n=11}^{30} n - \sum_{n=1}^{10} n = \frac{20}{2}(11 + 30) - \frac{10}{2}(1 + 10)$

$= 410 - 55 = 355$

75. $\sum_{n=1}^{500}(n + 8) = \frac{500}{2}(9 + 508) = 129{,}250$

77. $\sum_{n=1}^{20}(2n + 1) = 440$

79. $\sum_{n=1}^{100}\frac{n + 1}{2} = 2575$

81. $\sum_{i=1}^{60}\left(250 - \frac{2}{5}i\right) = 14{,}268$

83. $a_1 = 14,\ a_{18} = 31$

$S_{18} = \frac{18}{2}(14 + 31) = 405$ bricks

85. $a_1 = 30{,}000$

$a_2 = 30{,}000 + 5000 = 35{,}000$

$d = 5000$

$a_5 = 30{,}000 + 4(5000) = 50{,}000$

$S_5 = \frac{5}{2}(30{,}000 + 50{,}000) = \$200{,}000$

87. (a) $a_n = 0.84n + 14.9$

(b)

Year	2001	2002	2003	2004
Sales (in billions of dollars)	15.7	16.6	17.4	18.3

Year	2005	2006	2007	2008
Sales (in billions of dollars)	19.1	19.9	20.8	21.6

The model fits the data will.

(c) $a_n = 0.84n + 14.9$

$a_1 = 15.74,\ a_8 = 21.62$

$S_8 = \frac{8}{2}(15.74 + 21.62) = \$149.44$ billion

(d) $a_9 = 22.46,\ a_{16} = 28.34$

$$S_8 = \frac{8}{2}(22.46 + 28.34) = \$203.2 \text{ billion}$$

Answers will vary.

89. True. If the nth term of an arithmetic sequence is a_n and the common difference is d, the $(n+1)$th term can be found using the recursion formula, $a_{n+1} = a_n + d$.

91. $a_1 = x$

$a_2 = x + 2x = 3x$

$a_3 = 3x + 2x = 5x$

$a_4 = 7x$

$a_5 = 9x$

$a_6 = 11x$

$a_7 = 13x$

$a_8 = 15x$

$a_9 = 17x$

$a_{10} = 19x$

93. $a_{20} = a_1 + 19(3) = a_1 + 57$

$$S = \frac{n}{2}(a_1 + a_{20})$$

$$= \frac{20}{2}(a_1 + (a_1 + 57)) = 650$$

$$10(2a_1 + 57) = 650$$

$$20a_1 = 80$$

$$a_1 = 4$$

95. $S = \frac{n}{2}((a_1 + 5) + (a_n + 5))$

$$= \frac{n}{2}(a_1 + a_2 + 10)$$

$$= \frac{n}{2}(a_1 + a_2) + 5n$$

$$= S_n + 5_n$$

97. Gauss might have done the following:

$1 + 2 + 3 + \cdots + 99 + 100 = x$

$100 + 99 + \cdots + 2 + 1 = x$

Adding: $101 + 101 + \cdots + 101 + 101 = 2x$

$$100(101) = 2x \Rightarrow x = \frac{100(101)}{2} = 5050$$

In general, $1 + 2 + \cdots + n = \frac{n(n+1)}{2}$.

99. $S = \frac{n(n+1)}{2} = \frac{200(201)}{2} = 20{,}100$

101. $S = 1 + 3 + 5 + \cdots + 101$

$$= (1 + 2 + 3 + \cdots + 101) - (2 + 4 + \cdots + 100)$$

$$= \frac{101(102)}{2} - 2\left(\frac{50(51)}{2}\right)$$

$$= 5151 - 2550 = 2601$$

103. $\begin{bmatrix} 2 & -1 & 7 & \vdots & -10 \\ 3 & 2 & -4 & \vdots & 17 \\ 6 & -5 & 1 & \vdots & -20 \end{bmatrix}$ row reduces to

$$\begin{bmatrix} 1 & 0 & 0 & \vdots & 1 \\ 0 & 1 & 0 & \vdots & 5 \\ 0 & 0 & 1 & \vdots & -1 \end{bmatrix}.$$

Answer: $(1, 5, -1)$

105. Answers will vary. (Make a Decision)

Section 8.3

1. geometric, common

3. geometric series

5. In order for an infinite geometric series to have a sum, $|r| < 1$ and $S = \sum_{i=0}^{\infty} a_1 r^i = \frac{a_1}{1-r}$.

7. 5, 15, 45, 135,...

Geometric sequence

$r = 3$

9. 6, 18, 30, 42,...

Not a geometric sequence

(**Note:** It is an arithmetic sequence with $d = 12$.)

11. $1, -\frac{1}{2}, \frac{1}{4}, -\frac{1}{8}, \ldots$

Geometric sequence

$r = -\frac{1}{2}$

13. $\frac{1}{8}, \frac{1}{4}, \frac{1}{2}, 1, \ldots$

Geometric sequence

$r = 2$

15. $1, \frac{1}{2}, \frac{1}{3}, \frac{1}{4}, \ldots$

Not a geometric sequence

17. $a_1 = 6,\ r = 3$

$a_2 = 6(3) = 18$

$a_3 = 18(3) = 54$

$a_4 = 54(3) = 162$

$a_5 = 162(3) = 486$

19. $a_1 = 1,\ r = \frac{1}{2}$

$a_2 = 1\left(\frac{1}{2}\right) = \frac{1}{2}$

$a_3 = \frac{1}{2}\left(\frac{1}{2}\right) = \frac{1}{4}$

$a_4 = \frac{1}{4}\left(\frac{1}{2}\right) = \frac{1}{8}$

$a_5 = \frac{1}{8}\left(\frac{1}{2}\right) = \frac{1}{16}$

21. $a_1 = 5,\ r = -\frac{1}{10}$

$a_2 = 5\left(-\frac{1}{10}\right) = -\frac{1}{2}$

$a_3 = \left(-\frac{1}{2}\right)\left(-\frac{1}{10}\right) = \frac{1}{20}$

$a_4 = \frac{1}{20}\left(-\frac{1}{10}\right) = -\frac{1}{200}$

$a_5 = \left(-\frac{1}{200}\right)\left(-\frac{1}{10}\right) = \frac{1}{2000}$

23. $a_1 = 1,\ r = e$

$a_2 = 1(e) = e$

$a_3 = (e)(e) = e^2$

$a_4 = (e^2)(e) = e^3$

$a_5 = (e^3)(e) = e^4$

25. $a_1 = 64,\ a_{k+1} = \frac{1}{2}a_k$

$a_2 = \frac{1}{2}(64) = 32$

$a_3 = \frac{1}{2}(32) = 16$

$a_4 = \frac{1}{2}(16) = 8$

$a_5 = \frac{1}{2}(8) = 4$

$r = \frac{1}{2},\ a_n = 64\left(\frac{1}{2}\right)^{n-1} = 128\left(\frac{1}{2}\right)^n$

27. $a_1 = 9,\ a_{k+1} = 2a_k$

$a_2 = 2(9) = 18$

$a_3 = 2(18) = 36$

$a_4 = 2(36) = 72$

$a_5 = 2(72) = 144$

$r = 2,\ a_n = \left(\frac{9}{2}\right)2^n = 9\left(2^{n-1}\right)$

29. $a_1 = 6,\ a_{k+1} = -\frac{3}{2}a_k$

$a_2 = -\frac{3}{2}(6) = -9$

$a_3 = -\frac{3}{2}(-9) = \frac{27}{2}$

$a_4 = -\frac{3}{2}\left(\frac{27}{2}\right) = -\frac{81}{4}$

$a_5 = -\frac{3}{2}\left(-\frac{81}{4}\right) = \frac{243}{8}$

$r = -\frac{3}{2},\ a_n = 6\left(-\frac{3}{2}\right)^{n-1} = -4\left(-\frac{3}{2}\right)^n$

31. $a_1 = 6,\ r = \frac{1}{3},\ n = 12$

$a_n = a_1 r^{n-1}$

(a) $a_{12} = 6\left(\frac{1}{3}\right)^{12-1} = 6\left(\frac{1}{3}\right)^{11} \approx 0.000034$

(b) $a_{12} = 6\left(\frac{1}{3}\right)^{11} = 6\left(\frac{1}{177{,}147}\right) = \frac{2}{59{,}049}$

33. $a_1 = 8,\ r = -\frac{4}{3},\ n = 7$

$a_n = a_1 r^{n-1}$

(a) $a_7 = 8\left(-\frac{4}{3}\right)^{7-1} = 8\left(-\frac{4}{3}\right)^6 \approx 44.949$

(b) $a_7 = 8\left(-\frac{4}{3}\right)^6 = \frac{32{,}768}{729}$

35. $a_1 = -1,\ r = 3,\ n = 6$

$a_n = a_1 r^{n-1}$

(a) $a_6 = (-1)(3)^{6-1} = (-1)(3)^5 = -243$

(b) $a_6 = (-1)(3)^5 = -243$

37. $a_1 = 500,\ r = -1.02,\ n = 14$

$a_n = a_1 r^{n-1}$

(a) $a_{14} = 500(-1.02)^{14-1} = 500(-1.02)^{13} \approx -646.803$

(b) $a_{14} = 500(-1.02)^{13} \approx -646.803$

39. $7, 21, 63, \ldots$

$r = 3$

$a_n = 7(3)^{n-1}$

$a_9 = 7(3)^{9-1} = 45{,}927$

41. $5, 30, 180, \ldots$

$r = \frac{30}{5} = 6$

$a_n = 5(6)^{n-1}$

$a_{10} = 5(6)^{10-1} = 50{,}388{,}480$

43. $a_1 = 4,\ a_4 = \frac{1}{2},\ n = 10$

$a_4 = a_1 r^3$

$\frac{1}{2} = 4r^3$

$\frac{1}{8} = r^3$

$\frac{1}{2} = r$

$a_n = a_1 r^{n-1}$

$a_{10} = 4\left(\frac{1}{2}\right)^9 = \frac{1}{128}$

45. $a_2 = -18,\ a_5 = \frac{2}{3},\ n = 6$

$a_5 = a_2 r^3$

$\frac{2}{3} = -18r^3$

$-\frac{1}{27} = r^3$

$-\frac{1}{3} = r$

$a_6 = a_5 r$

$a_6 = \left(\frac{2}{3}\right)\left(-\frac{1}{3}\right) = -\frac{2}{9}$

47. $a_n = 12(-0.75)^{n-1}$

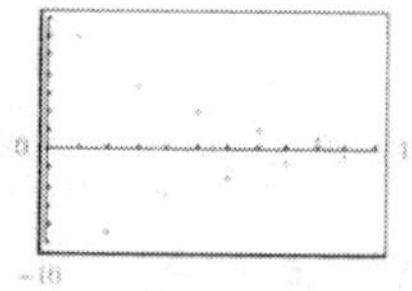

49. $a_n = 2(1.3)^{n-1}$

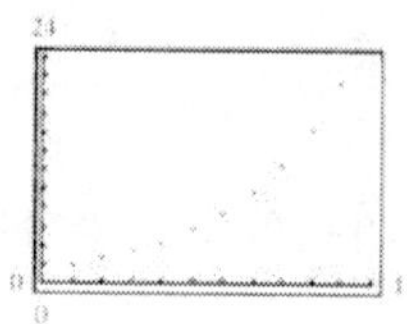

51. $8,\ -4,\ 2,\ -1,\ \frac{1}{2}, \ldots$

$S_1 = 8$

$S_2 = 8 + (-4) = 4$

$S_3 = 8 + (-4) + 2 = 6$

$S_4 = 8 + (-4) + 2 + (-1) = 5$

$S_5 = 8 + (-4) + 2 + (-1) + \frac{1}{2} = \frac{11}{2}$

53. $\sum_{n=1}^{\infty} 16\left(-\frac{1}{2}\right)^{n-1}$

n	1	2	3	4	5	6	7	8	9	10
S_n	16	24	28	30	31	31.5	31.75	31.875	31.9375	31.96875

55. $\sum_{n=1}^{9} 2^{n-1} \Rightarrow a_1 = 1,\ r = 2$

$S_9 = \frac{1(1-2^9)}{1-2} = 511$

57. $\sum_{i=1}^{7} 64\left(-\frac{1}{2}\right)^{i-1} \Rightarrow a_1 = 64,\ r = -\frac{1}{2}$

$S_7 = 64\left[\frac{1-(-1/2)^7}{1-(-1/2)}\right] = \frac{128}{3}\left[1-\left(-\frac{1}{2}\right)^7\right] = 43$

59. $\sum_{n=0}^{20} 3\left(\frac{3}{2}\right)^n = \sum_{n=1}^{21} 3\left(\frac{3}{2}\right)^{n-1} \Rightarrow a_1 = 3,\ r = \frac{3}{2}$

$S_{21} = 13\left[\frac{1-(3/2)^{21}}{1-(3/2)}\right]$

$= -6\left[1-\left(\frac{3}{2}\right)^{21}\right] \approx 29{,}921.31$

61. $\sum_{i=1}^{10} 8\left(-\frac{1}{4}\right)^{i-1} \Rightarrow a_1 = 8,\ r = -\frac{1}{4}$

$S_{10} = 8\left[\frac{1-(-1/4)^{10}}{1-(-1/4)}\right] = \frac{32}{5}\left[1-\left(-\frac{1}{4}\right)^{10}\right] \approx 6.4$

63. $\sum_{n=0}^{5} 300(1.06)^n = \sum_{n=1}^{6} 300(1.06)^{n-1} \Rightarrow a_1 = 300,\ r = 1.06$

$S_6 = 300\left(\frac{1-(1.06)^6}{1-1.06}\right) \approx 2092.60$

65. $5 + 15 + 45 + \cdots + 3645$

$r = 3$ and $3645 = 5(3)^{n-1} \Rightarrow n = 7$

$\sum_{n=1}^{7} 5(3)^{n-1}$

67. $2-\frac{1}{2}+\frac{1}{8}-\cdots+\frac{1}{2048}$

$r=-\frac{1}{4}$ and $\frac{1}{2048}=2\left(-\frac{1}{4}\right)^{n-1} \Rightarrow n=7$

$$\sum_{n=1}^{7} 2\left(-\frac{1}{4}\right)^{n-1}$$

69. $a_1=10,\ r=\frac{4}{5}$

$$\sum_{n=0}^{\infty} 10\left(\frac{4}{5}\right)^n=\frac{a_1}{1-r}=\frac{10}{1-\frac{4}{5}}=50$$

71. $a_1=5,\ r=-\frac{1}{2}$

$$\sum_{n=0}^{\infty} 5\left(-\frac{1}{2}\right)^n=\frac{a_1}{1-r}=\frac{5}{1-\left(-\frac{1}{2}\right)}=\frac{5}{\left(\frac{3}{2}\right)}=\frac{10}{3}$$

73. $\sum_{n=1}^{\infty} 2\left(\frac{7}{3}\right)^{n-1}$ does not have a finite sum $\left(\frac{7}{3}>1\right)$.

75. $a_1=10,\ r=0.11$

$$\sum_{n=0}^{\infty} 10(0.11)^n=\frac{a_1}{1-r}=\frac{10}{1-0.11}=\frac{10}{0.89}$$
$$=\frac{1000}{89}\approx 11.236$$

77. $a_1=-3,\ r=-0.9$

$$\sum_{n=0}^{\infty} -3(-0.9)^n=\frac{a_1}{1-r}=\frac{-3}{1-(-0.9)}$$
$$=\frac{-3}{1.9}=-\frac{30}{19}\approx -1.579$$

79. $9+6+4+\frac{8}{3}+\cdots=\sum_{n=0}^{\infty} 9\left(\frac{2}{3}\right)^n$

$$=\frac{9}{1-\frac{2}{3}}=\frac{9}{\frac{1}{3}}=27$$

81. $3-1+\frac{1}{3}-\frac{1}{9}+\cdots=\sum_{n=0}^{\infty} 3\left(-\frac{1}{3}\right)^n=\frac{a_1}{1-r}=\frac{3}{1-(-1/3)}$

$$=3\left(\frac{3}{4}\right)=\frac{9}{4}$$

83. $0.\overline{36}=\sum_{n=0}^{\infty} 0.36(0.01)^n$

$$=\frac{0.36}{1-0.01}=\frac{0.36}{0.99}=\frac{36}{99}=\frac{4}{11}$$

85. $1.2\overline{5}=1.2+\sum_{n=0}^{\infty} 0.05(0.1)^n$

$$=\frac{6}{5}+\frac{0.05}{1-0.1}$$
$$=\frac{6}{5}+\frac{0.05}{0.9}$$
$$=\frac{6}{5}+\frac{5}{90}=\frac{113}{90}$$

87. $8+16+32+64+\cdots$

The series is geometric, $r=\frac{a_2}{a_1}=\frac{16}{8}=2$.

Because the first term is $a_1=8$, the nth term is $a_n=a_1r^{n-1}=8(2)^{n-1}$.

The sum of the first 15 terms is

$$S_{15}=\sum_{n=1}^{15} 8(2)^{n-1}=8\left(\frac{1-2^{15}}{1-2}\right)=262{,}136.$$

89. $90+30+10+\frac{10}{3}+\cdots$

The series is geometric, $r=\frac{a_2}{a_1}=\frac{30}{90}=\frac{1}{3}$. Because the first term is $a_1=90$, the nth term is

$$a_n=a_1r^{n-1}=90\left(\frac{1}{3}\right)^{n-1}.$$

The sum of the first 15 terms is

$$S_{15}=\sum_{n=1}^{15} 90\left(\tfrac{1}{3}\right)^{n-1}=90\left[\frac{1-\left(\frac{1}{3}\right)^{15}}{1-\left(\frac{1}{3}\right)}\right]=135.$$

91. $\sum_{n=1}^{\infty} 6n$

The series is arithmetic with common difference $d=6$. Because the first term is $a_1=6$ and the 15th term is $a_{15}=90$, the sum of the first 15 terms is

$$S_{15}=\sum_{n=1}^{15} 6n=\frac{15}{2}(6+90)=720.$$

93. $\sum_{n=0}^{\infty} 6(0.8)^n$

The series is geometric with common ratio $r=0.8$.

$$\sum_{n=0}^{\infty} 6(0.8)^n=\sum_{n=1}^{\infty} 6(0.8)^{n-1}$$

The sum of the first 15 terms is $S_{15}=\sum_{n=1}^{15} 6(0.8)^{n-1}=6\left(\frac{1-0.8^{15}}{1-0.8}\right)\approx 28.944.$

95. $A = P\left(1+\frac{r}{n}\right)^{nt} = 1000\left(1+\frac{0.03}{n}\right)^{n(10)}$

(a) $n = 1$: $A = 1000(1+0.03)^{10} \approx \$1343.92$

(b) $n = 2$: $A = 1000\left(1+\frac{0.03}{2}\right)^{2(10)} \approx \$1346.86$

(c) $n = 4$: $A = 1000\left(1+\frac{0.03}{4}\right)^{4(10)} \approx \$1348.35$

(d) $n = 12$: $A = 1000\left(1+\frac{0.03}{12}\right)^{12(10)} \approx \$1349.35$

(e) $n = 365$: $A = 1000\left(1+\frac{0.03}{365}\right)^{365(10)} \approx \$1349.84$

97. Let $N = 12t$ be the total number of deposits.

$$A = P\left(1+\frac{r}{12}\right) + P\left(1+\frac{r}{12}\right)^2 + \cdots + P\left(1+\frac{r}{12}\right)^N$$

$$= \left(1+\frac{r}{12}\right)\left[P + P\left(1+\frac{r}{12}\right) + \cdots + P\left(1+\frac{r}{12}\right)^{N-1}\right]$$

$$= P\left(1+\frac{r}{12}\right)\sum_{n=1}^{N}\left(1+\frac{r}{12}\right)^{n-1}$$

$$= P\left(1+\frac{r}{12}\right)\frac{1-\left(1+\frac{r}{12}\right)^N}{1-\left(1+\frac{r}{12}\right)}$$

$$= P\left(1+\frac{r}{12}\right)\left(-\frac{12}{r}\right)\left[1-\left(1+\frac{r}{12}\right)^N\right]$$

$$= P\left(\frac{12}{r}+1\right)\left[-1+\left(1+\frac{r}{12}\right)^N\right]$$

$$= P\left[\left(1+\frac{r}{12}\right)^N - 1\right]\left(1+\frac{12}{r}\right)$$

$$= P\left[\left(1+\frac{r}{12}\right)^{12t} - 1\right]\left(1+\frac{12}{r}\right)$$

99. $P = \$50$, $r = 7\%$, $t = 20$ years

(a) Compounded monthly:

$$A = 50\left[\left(1+\frac{0.07}{12}\right)^{12(20)} - 1\right]\left(1+\frac{12}{0.07}\right) \approx \$26,198.27$$

(b) Compounded continuously:

$$A = \frac{50e^{0.07/12}\left(e^{0.07(20)}-1\right)}{e^{0.07/12}-1} \approx \$26,263.88$$

101. $P = 100$, $r = 5\% = 0.05$, $t = 40$

(a) Compounded monthly:

$$A = 100\left[\left(1+\frac{0.05}{12}\right)^{12(40)} - 1\right]\left(1+\frac{12}{0.05}\right) \approx \$153,237.86$$

(b) Compounded continuously:

$$A = \frac{100e^{0.05/12}\left(e^{0.05(40)}-1\right)}{e^{0.05/12}-1} \approx \$153,657.02$$

103. First shaded area: $\frac{16^2}{4}$

Second shaded area: $\frac{16^2}{4} + \frac{1}{2}\cdot\frac{16^2}{4}$

Third shaded area: $\frac{16^2}{4} + \frac{1}{2}\frac{16^2}{4} + \frac{1}{4}\frac{16^2}{4}$, etc.

Total area of shaded region:

$$\frac{16^2}{4}\sum_{n=0}^{5}\left(\frac{1}{2}\right)^n = 64\left[\frac{1-(1/2)^6}{1-1/2}\right] = 128\left(1-\left(\frac{1}{2}\right)^6\right) = 126 \text{ square inches}$$

105. (a) $a_0 = 70$ degrees

$a_1 = 0.8(70) = 56$ degrees

$\vdots$

$a_n = (0.8)^n(70)$

(b) $a_6 = (0.8)^6(70) \approx 18.35°\text{F}$

$a_{12} = (0.8)^{12}(70) \approx 4.81°\text{F}$

(c)

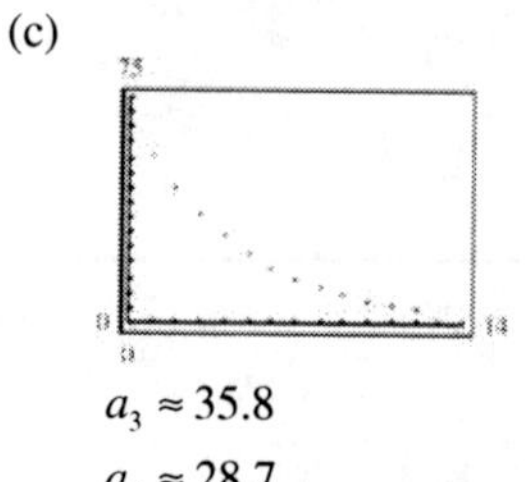

$a_3 \approx 35.8$

$a_4 \approx 28.7$

Thus, the water freezes between 3 and 4 hours, about 3.5 hours.

107. (a) $a_n \approx 1269.10(1.006)^n$

(b) The population of China is growing at a rate of about 0.6% per year.

(c) $2015 \Rightarrow n = 15$

$a_{15} = 1269.10(1.006)^{15} \approx 1388.2$ million.

This value is close to the value predicted by the U.S. Census Bureau.

(d) $a_n = 1.35$ billion $\Rightarrow a_n = 1350$ million

$$1350 = 1269.10(1.006)^n$$

$$\frac{1350}{1269.10} = (1.006)^n$$

$$\ln\left(\frac{1350}{1269.10}\right) = n\ln(1.006)$$

$$\frac{\ln\left(\frac{1350}{1269.10}\right)}{\ln 1.006} = n$$

$$n \approx 10.33$$

So, sometime during the year 2010 China's population will reach 1.35 billion people.

109. The ball falls 6 feet, then rebounds to a height of $\frac{3}{4}(6) = 4.5$ feet, then falls 4.5 feet, then rebounds $\frac{3}{4}(4.5) = 3.375$ feet, and so on. The terms of this geometric series can be written as

$6 + (4.5)(2) + (3.375)(2) + \cdots$ or

$6 + 9 + \frac{3}{4}(9) + \left(\frac{3}{4}\right)^2(9) + \cdots$ or $6 + \sum_{n=1}^{\infty} 9\left(\frac{3}{4}\right)^{n-1}$

So, the sum is $6 + \frac{9}{1 - \frac{3}{4}} = 42$ feet.

(**Note:** Other possible solutions close yield 42 feet as an answer.)

111. True. If the sequence is geometric and the common ratio is $r = 1$, then all terms are equal to a_1, $a_n = a_1r^{n-1} \Rightarrow a_n = a_1$. If the sequence is arithmetic with common difference $d = 0$, then all terms are equal to a_1, $a_n = a_1 + (n-1)d \Rightarrow a_n = a_1$. Therefore both sequences have the same terms, $a_n = a_1$.

113. $a_1 = 3$, $r = \frac{x}{2}$

$$a_2 = 3\left(\frac{x}{2}\right) = \frac{3x}{2}$$

$$a_3 = \frac{3x}{2}\left(\frac{x}{2}\right) = \frac{3x^2}{4}$$

$$a_4 = \frac{3x^2}{4}\left(\frac{x}{2}\right) = \frac{3x^3}{8}$$

$$a_5 = \frac{3x^3}{8}\left(\frac{x}{2}\right) = \frac{3x^4}{16}$$

115. $a_1 = 100$, $r = e^x$, $n = 9$

$$a_n = a_1r^{n-1}$$

$$a_9 = 100\left(e^x\right)^8 = 100e^{8x}$$

117. (a) $f(x) = 6\left[\frac{1 - 0.5^x}{1 - 0.5}\right]$

$$\sum_{n=0}^{\infty} 6\left(\frac{1}{2}\right)^n = \frac{6}{1 - 1/2} = 12$$

The horizontal asymptote of $f(x)$ is $y = 12$. This corresponds to the sum of the series.

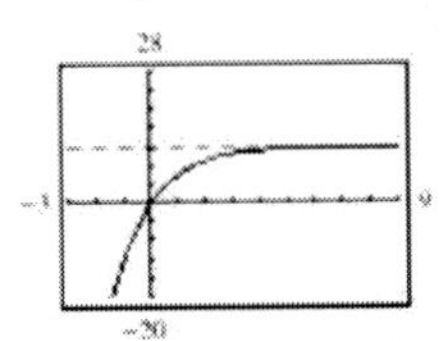

(b) $f(x) = 2\left[\frac{1 - 0.8^x}{1 - 0.8}\right]$

$$\sum_{n=0}^{\infty} 2\left(\frac{4}{5}\right)^n = \frac{2}{1 - 4/5} = 10$$

The horizontal asymptote of $f(x)$ is $y = 10$. This corresponds to the sum of the series.

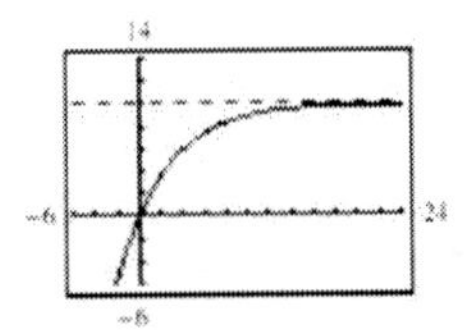

119. To use the first two terms of a geometric series to find the nth term, first divide the second term by the first term, to obtain the common ratio. The nth term is the first term multiplied by the common ratio raised to the $(n-1)th$ power.

$$r = \frac{a_2}{a_1},\ a_n = a_1r^{n-1}$$

121. $\text{Time} = \frac{\text{Distance}}{\text{Speed}} = \frac{200}{50} + \frac{200}{42} = 200\left(\frac{92}{2100}\right)$ hours

$$\text{Speed} = \frac{\text{Distance}}{\text{Time}} = \frac{400}{200(92/2100)} = \frac{2(2100)}{92} \approx 45.65 \text{ mph}$$

123. $\det\begin{bmatrix} -1 & 3 & 4 \\ -2 & 8 & 0 \\ 2 & 5 & -1 \end{bmatrix} = 4(-10 - 16) - 1(-8 + 6)$

$$= -104 + 2 = -102$$

125. Answers will vary. (Make a Decision)

Section 8.4

1. ${}_nC_r$ or $\binom{n}{r}$

3. Both the Binomial Theorem and Pascal's Triangle can be used to find binomial coefficients.

5. ${}_7C_5 = \dfrac{7!}{2!5!} = \dfrac{7\cdot6\cdot5!}{2\cdot5!} = \dfrac{42}{2} = 21$

7. ${}_{20}C_{15} = \dfrac{20!}{15!5!} = \dfrac{20\cdot19\cdot18\cdot17\cdot16}{5\cdot4\cdot3\cdot2\cdot1} = 15{,}504$

9. ${}_{14}C_1 = \dfrac{14!}{13!1!} = \dfrac{14\cdot13!}{13!} = 14$

11. $\binom{12}{0} = {}_{12}C_0 = \dfrac{12!}{0!12!} = 1$

13. $\binom{10}{4} = {}_{10}C_4 = \dfrac{10!}{6!4!} = \dfrac{10\cdot9\cdot8\cdot7}{4\cdot3\cdot2\cdot1} = 210$

15. $\binom{100}{98} = {}_{100}C_{98} = \dfrac{100!}{98!2!} = \dfrac{100\cdot99}{2\cdot1} = 4950$

17. ${}_{41}C_{36} = 749{,}398$

19. ${}_{50}C_{48} = 1225$

21. ${}_{250}C_2 = 31{,}125$

23.
$$\begin{aligned}(x+2)^4 &= {}_4C_0x^4 + {}_4C_1x^3(2) + {}_4C_2x^2(2)^2 \\ &\quad + {}_4C_3x(2)^3 + {}_4C_4(2)^4 \\ &= x^4 + 8x^3 + 24x^2 + 32x + 16\end{aligned}$$

25.
$$\begin{aligned}(a+3)^3 &= {}_3C_0a^3 + {}_3C_1a^2(3) + {}_3C_2a(3)^2 + {}_3C_3(3)^3 \\ &= a^3 + 3a^2(3) + 3a(3)^2 + (3)^3 \\ &= a^3 + 9a^2 + 27a + 27\end{aligned}$$

27.
$$\begin{aligned}(y-4)^3 &= {}_3C_0y^3 - {}_3C_1y^2(4)^2 + {}_3C_2y(4)^2 - {}_3C_3(4)^3. \\ &= y^3 - 12y^2 + 48y - 64\end{aligned}$$

29.
$$\begin{aligned}(x+y)^5 &= {}_5C_0x^5 + {}_5C_1x^4y + {}_5C_2x^3y^2 + {}_5C_3x^2y^3 \\ &\quad + {}_5C_4xy^4 + {}_5C_5y^5 \\ &= x^5 + 5x^4y + 10x^3y^2 + 10x^2y^3 + 5xy^4 + y^5\end{aligned}$$

31.
$$\begin{aligned}(r+3s)^6 &= {}_6C_0(r)^6 + {}_6C_1(r)^5(3s)^1 + {}_6C_2(r)^4(3s)^2 + {}_6C_3(r)^3(3s)^3 + {}_6C_2(r)^2(3s)^4 + {}_6C_1(r)^1(3s)^5 + {}_6C_6(3s)^6 \\ &= r^6 + 18r^5s + 135r^4s^2 + 540r^3s^3 + 1215r^2s^4 + 1458rs^5 + 729s^6\end{aligned}$$

33.
$$\begin{aligned}(x-y)^5 &= {}_5C_0x^5 - {}_5C_1x^4y + {}_5C_2x^3y^2 - {}_5C_3x^2y^3 + {}_5C_4xy^4 - {}_5C_5y^5 \\ &= x^5 - 5x^4y + 10x^3y^2 - 10x^2y^3 + 5xy^4 - y^5\end{aligned}$$

35. $\left(1-4x\right)^3 = {}_3C_0 1^3 - {}_3C_1 1^2\left(4x\right) + {}_3C_2 1\left(4x\right)^2 - {}_3C_3\left(4x\right)^3$

$= 1 - 3\left(4x\right) + 3\left(4x\right)^2 - \left(4x\right)^3$

$= 1 - 12x + 48x^2 - 64x^3$

37. $\left(x^2+2\right)^4 = {}_4C_0\left(x^2\right)^4 + {}_4C_1\left(x^2\right)^3\left(2\right) + {}_4C_2\left(x^2\right)^2 2^2 + {}_4C_3\left(x^2\right)2^3 + {}_4C_4\left(2\right)^4$

$= x^8 + 8x^6 + 24x^4 + 32x^2 + 16$

39. $\left(x^2-5\right)^5 = {}_5C_0\left(x^2\right)^5 - {}_5C_1\left(x^2\right)^4\left(5\right) + {}_5C_2\left(x^2\right)^3\left(5^2\right) - {}_5C_3\left(x^2\right)^2\left(5^3\right) + {}_5C_4\left(x^2\right)\left(5\right)^4 - {}_5C_5\left(5^5\right)$

$= x^{10} - 25x^8 + 250x^6 - 1250x^4 + 3125x^2 - 3125$

41. $\left(x^2+y^2\right)^4 = {}_4C_0\left(x^2\right)^4 + {}_4C_1\left(x^2\right)^3\left(y^2\right) + {}_4C_2\left(x^2\right)^2\left(y^2\right)^2 + {}_4C_3\left(x^2\right)\left(y^2\right)^3 + {}_4C_4\left(y^2\right)^4$

$= x^8 + 4x^6y^2 + 6x^4y^4 + 4x^2y^6 + y^8$

$= x^{12} + 6x^{10}y^2 + 15x^8y^4 + 20x^6y^6 + 15x^4y^8 + 6x^2y^{10} + y^{12}$

43. $\left(x^3-y\right)^6 = {}_6C_0\left(x^3\right)^6 - {}_6C_1\left(x^3\right)^5 y + {}_6C_2\left(x^3\right)^4 y^2 - {}_6C_3\left(x^3\right)^3 y^3 + {}_6C_4\left(x^3\right)^2 y^4 - {}_6C_5\left(x^3\right)y^5 + {}_6C_6 y^6$

$= x^{18} - 6x^{15}y + 15x^{12}y^2 - 20x^9y^3 + 15x^6y^4 - 6x^3y^5 + y^6$

45. $\left(\frac{1}{x}+y\right)^5 = {}_5C_0\left(\frac{1}{x}\right)^5 + {}_5C_1\left(\frac{1}{x}\right)^4 y + {}_5C_2\left(\frac{1}{x}\right)^3 y^2 + {}_5C_3\left(\frac{1}{x}\right)^2 y^3 + {}_5C_4\left(\frac{1}{x}\right)y^4 + {}_5C_5 y^5$

$= \frac{1}{x^5} + \frac{5y}{x^4} + \frac{10y^2}{x^3} + \frac{10y^3}{x^2} + \frac{5y^4}{x} + y^5$

47. $\left(\frac{2}{x}-y\right)^4 = {}_4C_0\left(\frac{2}{x}\right)^4 - {}_4C_1\left(\frac{2}{x}\right)^3 y + {}_4C_2\left(\frac{2}{x}\right)^2 y^2 + {}_4C_3\left(\frac{2}{x}\right)y^3 + {}_4C_4 y^4$

$= \frac{16}{x^4} - \frac{32}{x^3}y + \frac{24}{x^2}y^2 - \frac{8}{x}y^3 + y^4$

49. $\left(4x-1\right)^3 - 2\left(4x-1\right)^4 = \left(64x^3 - 48x^2 + 12x - 1\right) - 2\left(256x^4 - 256x^3 + 96x^2 - 16x + 1\right)$

$= -512x^4 + 576x^3 - 240x^2 + 44x - 3$

51. $2\left(x-3\right)^4 + 5\left(x-3\right)^2 = 2\left[x^4 - 4\left(x^3\right)\left(3\right) + 6\left(x^2\right)\left(3^2\right) - 4\left(x\right) + \left(3^3\right) + 3^4\right] + 5\left[x^2 - 2\left(x\right)\left(3\right) + 3^2\right]$

$= 2\left(x^4 - 12x^3 + 54x^2 - 108x + 81\right) + 5\left(x^2 - 6x + 9\right)$

$= 2x^4 - 24x^3 + 113x^2 - 246x + 207$

53. $-3\left(x-2\right)^3 - 4\left(x+1\right)^6 = \left(-3x^3 + 18x^2 - 36x + 24\right) - \left(4x^6 + 24x^5 + 60x^4 + 80x^3 + 60x^2 + 24x + 4\right)$

$= -4x^6 - 24x^5 - 60x^4 - 83x^3 - 42x^2 - 60x + 20$

55. $\left(x+8\right)^{10}$, $n = 4$

${}_{10}C_3 x^{10-3}\left(8\right)^3 = 120x^7\left(512\right) = 61{,}440x^7$

57. $\left(x-6y\right)^5$, $n = 3$

${}_5C_2 x^{5-2}\left(-6y\right)^2 = 10x^3\left(36\right)y^2 = 360x^3y^2$

59. $\left(4x+3y\right)^9$, $n = 8$

${}_9C_7\left(4x\right)^{9-7}\left(3y\right)^7 = 36\left(16\right)x^2\left(3^7\right)y^7$

$= 1{,}259{,}712x^2y^7$

61. $\left(10x-3y\right)^{12}$, $n = 10$

${}_{12}C_9\left(10x\right)^{12-9}\left(-3y\right)^9 = 220\left(10\right)^3\left(-3\right)^9 x^3y^9$

$= -4{,}330{,}260{,}000x^3y^9$

63. $\left(x+3\right)^{12}$, ax^5

${}_{12}C_7 x^5\left(3\right)^7 = 1{,}732{,}104x^5$

$a = 1{,}732{,}104$

65. The term involving x^8y^2 in the expansion of $\left(x-2y\right)^{10}$ is ${}_{10}C_2 x^8\left(-2y\right)^2 = \frac{10!}{2!8!}\cdot 4x^8y^2 = 180x^8y^2$.

The coefficient is 180.

67. The term involving x^6y^3 in the expansion of $(3x-2y)^9$ is

$${}_9C_3(3x)^6(-2y)^3 = 84(3)^6(-2)^3x^6y^3$$
$$= -489{,}888x^6y^3.$$

The coefficient is –489,888.

69. The term involving x^8y^6 in the expansion of $(x^2+y)^{10}$ is ${}_{10}C_6(x^2)^4y^6 = 210x^8y^6$. The coefficient is 210.

71. 4th entry of 7th row: ${}_7C_4 = 35$

73. 5th entry of 6th row: ${}_6C_5 = 6$

75. 4th row of Pascal's Triangle: 1 4 6 4 1

$$(3t-2v)^4 = 1(3t)^4 - 4(3t)^3(2v) + 6(3t)^2(2v)^2 - 4(3t)(2v)^3 + 1(2v)^4$$
$$= 81t^4 - 216t^3v + 216t^2v^2 - 96tv^3 + 16v^4$$

77. 5th row of Pascal's Triangle: 1 5 10 10 5 1

$$(2x-3y)^5 = 1(2x)^5 - 5(2x)^4(3y) + 10(2x)^3(3y)^2 - 10(2x)^2(3y)^3 + 5(2x)(3y)^4 - (3y)^5$$
$$= 32x^5 - 240x^4y + 720x^3y^2 - 1080x^2y^3 + 810xy^4 - 243y^5$$

79. 5th row of Pascal's Triangle: 1 5 10 10 5 1

$$(x+2y)^5 = 1(x)^5 + 5(x)^4(2y) + 10(x)^3(2y)^2 + 10(x)^2(2y)^3 + 5(x)(2y)^4 + 1(2y)^5$$
$$= x^5 + 10x^4y + 40x^3y^2 + 80x^2y^3 + 80xy^4 + 32y^5$$

81. $$(\sqrt{x}+5)^3 = (\sqrt{x})^3 + 3(\sqrt{x})^2(5) + 3(\sqrt{x})(5)^2 + (5)^3$$
$$= x\sqrt{x} + 15x + 75\sqrt{x} + 125$$

83. $$(x^{2/3}-y^{1/3})^3 = (x^{2/3})^3 - 3(x^{2/3})^2(y^{1/3}) + 3(x^{2/3})(y^{1/3})^2 - (y^{1/3})^3$$
$$= x^2 - 3x^{4/3}y^{1/3} + 3x^{2/3}y^{2/3} - y$$

85. $$\frac{f(x+h)-f(x)}{h} = \frac{(x+h)^3 - x^3}{h}$$
$$= \frac{x^3 + 3x^2h + 3xh^2 + h^3 - x^3}{h}$$
$$= \frac{h(3x^2 + 3xh + h^2)}{h}$$
$$= 3x^2 + 3xh + h^2,\ h \neq 0$$

87. $$\frac{f(x+h)-f(x)}{h} = \frac{(x+h)^6 - x^6}{h}$$
$$= \frac{(x^6 + 6x^5h + 15x^4h^2 + 20x^3h^3 + 15x^2h^4 + 6xh^5 + h^6) - x^6}{h}$$
$$= \frac{h(6x^5 + 15x^4h + 20x^3h^2 + 15x^2h^3 + 6xh^4 + h^5)}{h}$$
$$= 6x^5 + 15x^4h + 20x^3h^2 + 15x^2h^3 + 6xh^4 + h^5,\ h \neq 0$$

89. $$\frac{f(x+h)-f(x)}{h} = \frac{\sqrt{x+h}-\sqrt{x}}{h}$$
$$= \frac{\sqrt{x+h}-\sqrt{x}}{h}\cdot\frac{\sqrt{x+h}+\sqrt{x}}{\sqrt{x+h}+\sqrt{x}}$$
$$= \frac{(x+h)-x}{h\left[\sqrt{x+h}+\sqrt{x}\right]}$$
$$= \frac{1}{\sqrt{x+h}+\sqrt{x}},\ h \neq 0$$

91. $$(1+i)^4 = {}_4C_0 1^4 + {}_4C_1(1)^3 i + {}_4C_2(1)^2 i^2 + {}_4C_3 1\cdot i^3 + {}_4C_4 i^4$$
$$= 1 + 4i - 6 - 4i + 1$$
$$= -4$$

93. $$(4+i)^4 = {}_4C_0(4)^4 + {}_4C_1(4^3)i + {}_4C_2(4^2)(i^2) + {}_4C_3(4)(i^3) + {}_4C_4 i^4$$
$$= 256 + 256i - 96 - 16i + 1$$
$$= 161 + 240i$$

95. $$(2-3i)^6 = {}_6C_0 2^6 - {}_6C_1 2^5(3i) + {}_6C_2 2^4(3i)^2 - {}_6C_3 2^3(3i)^3 + {}_6C_4 2^2(3i)^4 - {}_6C_5 2(3i)^5 + {}_6C_6(3i)^6$$
$$= 64 - 576i - 2160 + 4320i + 4860 - 2916i - 729$$
$$= 2035 + 828i$$

97. $$(5+\sqrt{-16})^3 = (5+4i)^3$$
$$= {}_3C_0 5^3 + {}_3C_1 5^2(4i) + {}_3C_2 5(4i)^2 + {}_3C_3(4i)^3$$
$$= 125 + 300i - 240 - 64i$$
$$= -115 + 236i$$

99. $$(4+\sqrt{3}i)^4 = {}_4C_0 4^4 + {}_4C_1 4^3\sqrt{3}i + {}_4C_2 4^2(\sqrt{3}i)^2 + {}_4C_3 4(\sqrt{3}i)^3 + {}_4C_4(\sqrt{3}i)^4$$
$$= 256 + 256\sqrt{3}i - 288 - 48\sqrt{3}i + 9$$
$$= -23 + 208\sqrt{3}i$$

101. $$\left(-\frac{1}{2}+\frac{\sqrt{3}}{2}i\right)^3 = \frac{1}{8}(-1+\sqrt{3}i)^3$$
$$= \frac{1}{8}\left[{}_3C_0(-1)^3 + {}_3C_1(-1)^2(\sqrt{3}i) + {}_3C_2(-1)(\sqrt{3}i)^2 + {}_3C_3(\sqrt{3}i)^3\right]$$
$$= \frac{1}{8}\left[-1 + 3\sqrt{3}i + 9 - 3\sqrt{3}i\right]$$
$$= 1$$

103. $\left(\frac{1}{4}-\frac{\sqrt{3}}{4}i\right)^3 = {}_3C_0\left(\frac{1}{3}\right)^3 - {}_3C_1\left(\frac{1}{3}\right)^2\left(\frac{\sqrt{3}}{3}i\right) + {}_3C_2\left(\frac{1}{3}\right)\left(\frac{\sqrt{3}}{3}i\right)^2 - {}_3C_3\left(\frac{\sqrt{3}}{3}i\right)^3 = \frac{1}{64} - \frac{3\sqrt{3}}{64}i - \frac{9}{64} + \frac{3\sqrt{3}}{64} = -\frac{1}{8}$

105. $(1.02)^8 = (1+0.02)^8$

$$\begin{aligned}
&= 1 + 8(0.02) + 28(0.02)^2 + 56(0.02)^3 + 70(0.02)^4 + 56(0.02)^5 + 28(0.02)^6 + 8(0.02)^7 + (0.02)^8 \\
&= 1 + 0.16 + 0.0112 + 0.000448 + \cdots \\
&\approx 1.172
\end{aligned}$$

107. $(2.99)^{12}$

$$\begin{aligned}
&= (3-0.01)^{12} \\
&= 3^{12} - 12(3)^{11}(0.01) + 66(3)^{10}(0.01)^2 - 220(3)^9(0.01)^3 + 495(3)^8(0.01)^4 - 792(3)^7(0.01)^5 + 924(3)^6(0.01)^6 \\
&\quad -792(3)^5(0.01)^7 + 495(3)^4(0.01)^8 - 220(3)^3(0.01)^9 + 66(3)^2(0.01)^{10} - 12(3)(0.01)^{11} + (0.01)^{12} \\
&\approx 510{,}568.785
\end{aligned}$$

109. g is shifted four units to the left of f.

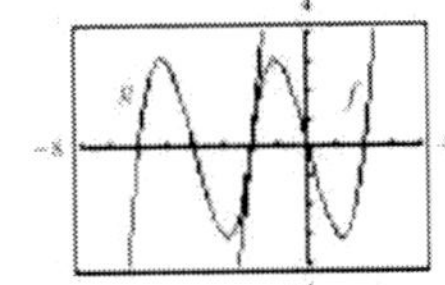

$$\begin{aligned}
f(x) &= x^3 - 4x \\
g(x) &= f(x+4) \\
&= (x+4)^3 - 4(x+4) \\
&= x^3 + 12x^2 + 48x + 64 - 4x - 16 \\
&= x^3 + 12x^2 + 44x + 48
\end{aligned}$$

111. $f(x) = (1-x)^3$

$g(x) = 1 - 3x$

$h(x) = 1 - 3x + 3x^2$

$p(x) = 1 - 3x + 3x^2 - x^3 = f(x)$

Since $p(x)$ is the expansion of $f(x)$, they have the same graph.

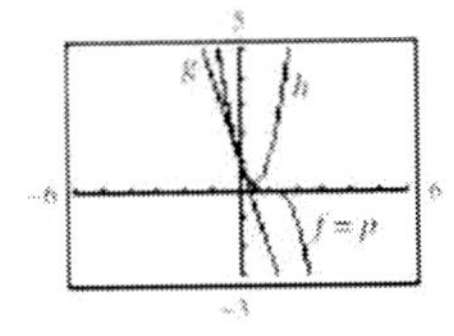

113. ${}_7C_4\left(\frac{1}{2}\right)^4\left(\frac{1}{2}\right)^3 = 35\left(\frac{1}{16}\right)\left(\frac{1}{8}\right) \approx 0.273$

115. ${}_8C_4\left(\frac{1}{3}\right)^4\left(\frac{2}{3}\right)^4 = 70\left(\frac{1}{81}\right)\left(\frac{16}{81}\right) \approx 0.171$

117. (a) $f(t) = 0.044t^2 + 0.44t + 8.3$

$$\begin{aligned}
g(t) &= f(t+10) \\
&= 0.44(t+10)^2 + 0.44(t+10) + 8.3 \\
&= 0.44t^2 + 1.32t + 17.1
\end{aligned}$$

(b)

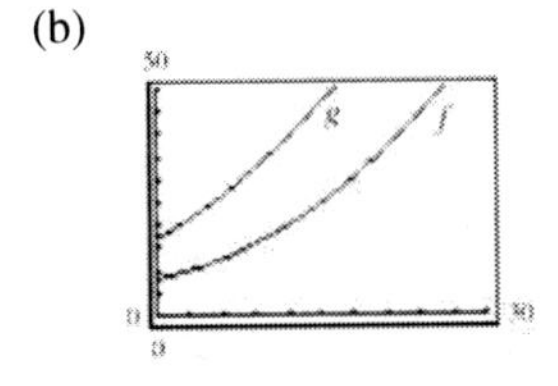

119. True. The coefficients from the Binomial Theorem can be used to find the numbers in Pascal's Triangle.

121. False. The x^4y^8 term is

${}_{12}C_4x^4(-2y)^8 = 495x^4(-2)^8y^8 = 126{,}720x^4y^8$.

[**Note:** 7920 is the coefficient of x^8y^4.]

123. $(n+1)$ terms

125. (a) Second term of $(2x-3y)^5$ is

$5(2x)^4(-3y)^1 = -240x^4y$.

(b) Fourth term of $\left(\frac{1}{2}x + 7y\right)^6$ is

${}_6C_3\left(\frac{1}{2}x\right)^3(7y)^3 = 857.5x^3y^3$.

127.
$$\begin{aligned}
{}_nC_{n-r} &= \frac{n!}{[n-(n-r)]!(n-r)!} \\
&= \frac{n!}{r!(n-r)!} = \frac{n!}{(n-r)!r!} = {}_nC_r
\end{aligned}$$

129. $_nC_r + {_nC_{r-1}} = \dfrac{n!}{(n-r)!r!} + \dfrac{n!}{(n-r+1)!(r-1)!}$

$= \dfrac{n!(n-r+1)}{(n-r)!r!(n-r+1)} + \dfrac{n!r}{(n-r+1)!(r-1)!r}$

$= \dfrac{n!(n-r+1)}{(n-r+1)!r!} + \dfrac{n!r}{(n-r+1)!r!}$

$= \dfrac{n!(n-r+1+r)}{(n-r+1)!r!}$

$= \dfrac{n!(n+1)}{(n-r+1)!r!}$

$= \dfrac{(n+1)!}{(n+1-r)!r!} = {_{n+1}C_r}$

131. $A = \begin{bmatrix} -1 & -4 \\ 1 & 2 \end{bmatrix}$

$A^{-1} = \dfrac{1}{(-1)(2)-(1)(-4)} \begin{bmatrix} 2 & 4 \\ -1 & -1 \end{bmatrix}$

$= \dfrac{1}{2}\begin{bmatrix} 2 & 4 \\ -1 & -1 \end{bmatrix} = \begin{bmatrix} 1 & 2 \\ -\frac{1}{2} & -\frac{1}{2} \end{bmatrix}$

Section 8.5

1. Fundamental Counting Principle

3. A permutation is an ordering of n elements.

5. Odd integers: 1, 3, 5, 7, 9, 11, 13, 15
8 ways

7. Prime integers: 1, 3, 5, 7, 11, 13
6 ways

9. Sum is 20:
$5+15, 6+14, 7+13, 8+12, 9+11, 10+10, 11+9,$
$12+8, 13+7, 14+6, 15+5$
11 ways

11. Distinct integers whose sum is 20:
$5+15, 6+14, 7+13, 8+12, 9+11, 11+9, 12+8, 13+7, 14+6,$
$15+5$
10 ways

13. Amplifiers: 4 choices
Compact disc players: 6 choices
Speakers: 5 choices
Total: $4 \cdot 6 \cdot 5 = 120$ ways

15. $2^{10} = 1024$ ways

17. (a) $9 \cdot 10 \cdot 10 = 900$ numbers
(b) $9 \cdot 9 \cdot 8 = 648$ numbers
(c) $9 \cdot 10 \cdot 2 = 180$ numbers

19. $2(8 \cdot 10 \cdot 10)(10 \cdot 10 \cdot 10 \cdot 10) = 16{,}000{,}000$ numbers

21. (a) $26^3 + 26^3 = 35{,}152$
(b) There are $2 \cdot 25^3$ possibilities that don't have Q.
Hence, $2 \cdot 26^3 - 2 \cdot 25^3 = 3902$ have at least one Q.

23. (a) $10^5 = 100{,}000$ zip codes
(b) There are $2 \cdot 10^4 = 20{,}000$ zip codes beginning with a one or a two.

25. (a) $6 \cdot 5 \cdot 4 \cdot 3 \cdot 2 \cdot 1 = 720$ ways
(b) $6 \cdot 1 \cdot 4 \cdot 1 \cdot 2 \cdot 1 = 48$ ways

27. $_4P_4 = \dfrac{4!}{(4-4)!} = \dfrac{4!}{0!} = 24$

29. $_8P_3 = \dfrac{8!}{(8-3)!} = \dfrac{8!}{5!} = 8 \cdot 7 \cdot 6 = 336$

31. $_5P_4 = \dfrac{5!}{(5-4)!} = \dfrac{5!}{1!} = 120$

33. $_{20}P_6 = 27{,}907{,}200$

35. $_{120}P_4 = 197{,}149{,}680$

37. $5! = 120$ ways

39. $9! = 362{,}880$ ways

41. $_{12}P_4 = \dfrac{12!}{8!}$
$= 12 \cdot 11 \cdot 10 \cdot 9$
$= 11{,}880$ ways

43. $37 \cdot 37 \cdot 37 = 50{,}653$ lock combinations

45.

ABCD	BACD	CABD	DABC
ABDC	BADC	CADB	DACB
ACBD	BCAD	CBAD	DBAC
ACDB	BCDA	CBDA	DBCA
ADBC	BDAC	CDAB	DCAB
ADCB	BDCA	CDBA	DCBA

47. $\dfrac{7!}{2!1!3!1!} = \dfrac{7!}{2!3!} = \dfrac{7 \cdot 6 \cdot 5 \cdot 4}{2 \cdot 1} = 420$

49. $\dfrac{7!}{2!1!1!1!1!1!} = \dfrac{7!}{2!}$
$= 7 \cdot 6 \cdot 5 \cdot 4 \cdot 3$
$= 2520$

51. $_5C_2 = \frac{5!}{2!3!} = \frac{5 \cdot 4}{2} = 10$

53. $_4C_1 = \frac{4!}{1!3!} = 4$

55. $_{25}C_0 = \frac{25!}{0!25!} = 1$

57. $_{20}C_5 = 15{,}504$

59. $_{42}C_5 = 850{,}668$

61. AB, AC, AD, AE, AF, BC, BD, BE, BF, CD, CE, CF, DE, DF, EF

$_6C_2 = 15$ ways

63. $_{100}C_{15} = \frac{100!}{85!15!} = 2.53 \times 10^{17}$ committees

65. $_{59}C_5 \cdot {_{39}C_1} = \frac{59!}{54!5!} \cdot \frac{39!}{38!1!}$

$= 5{,}006{,}386 \cdot 39$

$= 195{,}249{,}054$ ways

67. There are 27 good sets and 3 defective sets.

(a) $_{27}C_4 = 17{,}550$ ways

(b) $_{27}C_2 \cdot {_3C_2} = 351 \cdot 3 = 1053$ ways

(c)

$_{27}C_4 + {_{27}C_3} \cdot {_3C_1} + {_{27}C_2} \cdot {_3C_2} = 17{,}550 + 8775 + 1053$

$= 23{,}378$ ways

69. Select 2 jacks: $_4C_2 = 6$

Select 3 aces: $_4C_3 = 4$

Total: $6 \cdot 4 = 24$ ways

71. There are 7 administrators, 12 faculty, and 25 students.

$_7C_1 \cdot {_{12}C_3} \cdot {_{25}C_2} = 7 \cdot 220 \cdot 300 = 462{,}000$ committees

73. $_5C_2 - 5 = 10 - 5 = 5$ diagonals

75. $_8C_2 - 8 = 28 - 8 = 20$ diagonals

77. $14 \cdot {_nP_3} = {_{n+2}P_4}$

Note: $n \geq 3$ for this to be defined.

$$14\left[\frac{n!}{(n-3)!}\right] = \frac{(n+2)!}{(n-2)!}$$

$14n(n-1)(n-2) = (n+2)(n+1)n(n-1)$ (We can divide here by $n(n-1)$ since $n \neq 0$, $n \neq 1$.)

$$14n - 28 = n^2 + 3n + 2$$

$$0 = n^2 - 11n + 30$$

$$0 = (n-5)(n-6)$$

$$n = 5 \text{ or } n = 6$$

79. $_nP_4 = 10 \cdot {_{n-1}P_3}$

$$\frac{n!}{(n-4)!} = 10\frac{(n-1)!}{(n-4)!}$$

$$n! = 10(n-1)!$$

$$n = 10$$

81. $_{n+1}P_3 = 4 \cdot {_nP_2}$

$$\frac{(n+1)!}{(n-2)!} = 4\frac{n!}{(n-2)!}$$

$$(n+1)! = 4n!$$

$$n = 3$$

83. $4 \cdot {_{n+1}P_2} = {_{n+2}P_3}$

$$4\frac{(n+1)!}{(n-1)!} = \frac{(n+2)!}{(n-1)!}$$

$$4(n+1)! = (n+2)!$$

$$n = 2$$

85. False. Because order does not matter, it is an example of a combination.

87. $_{100}P_{80} \approx 3.836 \times 10^{139}$

This number is too large for some calculators to evaluate.

89. The symbol $_nP_r$ means the number of ways to choose and order r elements out of a set of n elements.

91. $_nP_{n-1} = \frac{n!}{(n-(n-1))!} = \frac{n!}{1!} = \frac{n!}{0!} = {_nP_n}$

93. $_nC_{n-1} = \frac{n!}{[n-(n-1)]!(n-1)!}$

$$= \frac{n!}{(1)!(n-1)!}$$

$$= \frac{n!}{(n-1)!1!} = {_nC_1}$$

95. $\log_2(x-3) = 5$

$$2^5 = x - 3$$

$$2^5 + 3 = x$$

$$x = 35$$

97. $x = \frac{\begin{vmatrix} -14 & 3 \\ 2 & -2 \end{vmatrix}}{\begin{vmatrix} -5 & 3 \\ 7 & -2 \end{vmatrix}} = \frac{22}{-11} = -2$

$y = \frac{\begin{vmatrix} -5 & -14 \\ 7 & 2 \end{vmatrix}}{\begin{vmatrix} -5 & 3 \\ 7 & -2 \end{vmatrix}} = \frac{88}{-11} = -8$

Answer: $(-2, -8)$

Section 8.6

1. sample space

3. mutually exclusive

5. The probability of an event E is given by $P(E) = \frac{n(E)}{n(S)}$, where $0 \le P(E) \le 1$.

7. The probability of a certain event is $P(E) = 1$.

9. $\{(H, 1), (H, 2), (H, 3) (H, 4), (H, 5), (H, 6), (T, 1), (T, 2) (T, 3), (T, 4), (T, 5), (T, 6)\}$

11. $\{ABC, ACB, BAC, BCA, CAB, CBA\}$

13. $E = \{HTT, THT, TTH\}$

$$P(E) = \frac{n(E)}{n(S)} = \frac{3}{8}$$

15. $E = \{HHH, HHT, HTH, HTT, THH, THT, TTH\}$

$$P(E) = \frac{n(E)}{n(S)} = \frac{7}{8}$$

17. $E = \{K, K, K, K, Q, Q, Q, Q, J, J, J, J\}$

$$P(E) = \frac{n(E)}{n(S)} = \frac{12}{52} = \frac{3}{13}$$

19. $E = 3$ spade face cards

$$P(E) = \frac{n(E)}{n(S)} = \frac{3}{52}$$

21. $E = \{(1, 5), (2, 4), (3, 3), (4, 2), (5, 1)\}$

$$P(E) = \frac{n(E)}{n(S)} = \frac{5}{36}$$

23. not $E = \{(5, 6), (6, 5), (6, 6)\}$

$$n(E) = n(S) - n(\text{not } E) = 36 - 3 = 33$$

$$P(E) = \frac{n(E)}{n(S)} = \frac{33}{36} = \frac{11}{12}$$

25. $E = 66$ ticket holders less than 19 years old

$$P(E) = \frac{n(E)}{n(S)} = \frac{66}{2200} = \frac{3}{100}$$

27. $E = 792$ ticket holders 19 to 39 years old

$$P(E) = \frac{n(E)}{n(S)} = \frac{792}{2200} = \frac{9}{25}$$

29. $P(E) = \frac{{}_3C_2}{{}_6C_2} = \frac{3}{15} = \frac{1}{5}$

31. $P(E) = \frac{{}_4C_2}{{}_6C_2} = \frac{6}{15} = \frac{2}{5}$

33. $P(E') = 1 - P(E) = 1 - 0.75 = 0.25$

35. $P(E') = 1 - P(E) = 1 - \frac{2}{3} = \frac{1}{3}$

37. $P(E) = 1 - P(E') = 1 - p = 1 - 0.12 = 0.88$

39. $P(E) = 1 - P(E') = 1 - \frac{13}{20} = \frac{7}{20}$

41. $E =$ The card is a club or a king.
Event $A =$ club

Event $B =$ king

$$P(E) = P(A) + P(B) - P(A \cap B)$$
$$= \frac{13}{52} + \frac{4}{52} - \frac{1}{52} = \frac{16}{52} = \frac{4}{13}$$

43. $E =$ the card is a face card or 2

Event $A =$ face card

Event $B = 2$

$$P(E) = P(A) + P(B) - P(A \cap B)$$
$$= \frac{12}{52} + \frac{4}{52} - \frac{0}{52} = \frac{16}{52} = \frac{4}{13}$$

45. $E = 20$ to 21 years old

$$P(E) = \frac{n(E)}{n(S)} = \frac{18}{32} = \frac{9}{16}$$

47. $E =$ older than 21 years old

$$P(E) = \frac{n(E)}{n(S)} = \frac{3}{32}$$

49. $E =$ all three numbers are even

$$P(E) = \frac{n(E)}{n(S)} = \frac{5 \cdot 5 \cdot 5}{10 \cdot 10 \cdot 10} = \frac{125}{1000} = \frac{1}{8}$$

51. $E =$ two numbers are less than 5 and the other number is 10

$$P(E) = \frac{n(E)}{n(S)} = \frac{4 \cdot 4 \cdot 1}{10 \cdot 10 \cdot 10} = \frac{16}{1000} = \frac{2}{125}$$

53. $p + p + 2p = 1$

$p = 0.25$

Taylor: $0.50 = \frac{1}{2}$, Moore: $0.25 = \frac{1}{4}$, Perez: $0.25 = \frac{1}{4}$

55. (a) 20.22 million people with advanced degree

(b) E = Bachelor's degree or higher

$P(E) = 0.191 + 0.103 = 0.294$

(c) E = high school degree or some post secondary education

$$P(E) = 0.312 + 0.172 + 0.088 + 0.191 + 0.103 = 0.8651$$

57. (a) Number of ways to arrange the digits $= 5! = 120$

Probability of guessing each digit $= \frac{1}{120}$

(b) Number of ways to arrange the digits if first digit is known $= 4! = 24$

Probability of guessing other 4 digits $= \frac{1}{24}$

59. (a) $\frac{{}_9C_4}{{}_{12}C_4} = \frac{126}{495} = \frac{14}{55}$ (4 good units)

(b) $\frac{({}_9C_2)({}_3C_2)}{{}_{12}C_4} = \frac{108}{495} = \frac{12}{55}$ (2 good units)

(c) $\frac{({}_9C_3)({}_3C_1)}{{}_{12}C_4} = \frac{252}{495} = \frac{28}{55}$ (3 good units)

At least 2 good units: $\frac{12}{55} + \frac{28}{55} + \frac{14}{55} = \frac{54}{55}$

61. $(0.32)^2 = 0.1024$

63. (a) $\left(\frac{1}{5}\right)^6 = \frac{1}{15{,}625}$

(b) $\left(\frac{4}{5}\right)^6 = \frac{4096}{15{,}625} = 0.262144$

(c) $1 - 0.262144 = 0.737856 = \frac{11{,}529}{15{,}625}$

65. (a) If the *center* of the coin falls within the circle of radius $d/2$ around a vertex, the coin will cover the vertex.

$$P(\text{coin covers a vertex}) = \frac{\text{Area in which coin may fall so that it covers a vertex}}{\text{Total area}} = \frac{n\left[\pi\left(\frac{d}{2}\right)^2\right]}{nd^2} = \frac{1}{4}\pi$$

(b) Experimental results will vary.

67. True. $P(E) + P(E') = 1$

69. (a) As you consider successive people with distinct birthdays, the probabilities must decrease to take into account the birth dates already used. Since the birth dates of people are independent events, multiply the respective probabilities of distinct birthdays.

(b) $\frac{365}{365} \cdot \frac{364}{365} \cdot \frac{363}{365} \cdot \frac{362}{365}$

(c) $P_1 = \frac{365}{365} = 1$

$$P_2 = \frac{365}{365} \cdot \frac{364}{365} = \frac{364}{365} P_1 = \frac{365 - (2-1)}{365} P_1$$

$$P_3 = \frac{365}{365} \cdot \frac{364}{365} \cdot \frac{363}{365} = \frac{363}{365} P_2 = \frac{365 - (3-1)}{365} P_2$$

$$P_n = \frac{365}{365} \cdot \frac{364}{365} \cdot \frac{363}{365} \cdots \frac{365-(n-1)}{365} = \frac{365-(n-1)}{365} P_{n-1}$$

(d) Q_n is the probability that the birthdays are *not* distinct, which is equivalent to at least 2 people having the same birthday.

(e)

n	10	15	20	23	30	40	50
P_n	0.88	0.75	0.59	0.49	0.29	0.11	0.03
Q_n	0.12	0.25	0.41	0.51	0.71	0.89	0.97

(f) 23; see the table in part (e).

71. $P(A) = 0.76$ and $P(B) = 0.58$

(a) A and B cannot be mutually exclusive, because $P(A) + P(B) = 0.76 + 0.58 = 1.34 > 1$.

(b) A' and B' can be mutually exclusive, because $A' = 1 - 0.76 = 0.24$ and $B' = 1 - 0.58 = 0.42 \Rightarrow A' + B' = 0.24 + 0.42 = 0.66 < 1$.

(c) $0.76 \le P(A \cup B) \le 1$, where $P(A \cup B) = P(A) + P(B) - P(A \cap B)$.

73. ${}_6C_2 = \frac{6!}{4!2!} = \frac{6 \cdot 5 \cdot 4!}{4!2} = 15$

75. ${}_{11}C_8 = \frac{11!}{8!3!} = \frac{11 \cdot 10 \cdot 9 \cdot 8!}{8!6} = 165$

Chapter 8 Review Exercises

1. $a_n = \dfrac{2^n}{2^n+1}$

$a_1 = \dfrac{2^1}{2^1+1} = \dfrac{2}{3}$

$a_2 = \dfrac{2^2}{2^2+1} = \dfrac{4}{5}$

$a_3 = \dfrac{2^3}{2^3+1} = \dfrac{8}{9}$

$a_4 = \dfrac{2^4}{2^4+1} = \dfrac{16}{17}$

$a_5 = \dfrac{2^5}{2^5+1} = \dfrac{32}{33}$

3. Common difference is 5.

$a_n = 5n,\ n = 1, 2, \ldots$

5. Denominators are successive odd numbers.

$a_n = \dfrac{2}{2n-1},\ n = 1, 2, 3, \ldots$

7. $a_1 = 9,\ a_{k+1} = a_k - 4$

$a_2 = a_1 - 4 = 9 - 4 = 5$

$a_3 = 5 - 4 = 1$

$a_4 = 1 - 4 = -3$

$a_5 = -3 - 4 = -7$

9. $\dfrac{18!}{20!} = \dfrac{18!}{20 \cdot 19 \cdot 18!}$

$= \dfrac{1}{20 \cdot 19} = \dfrac{1}{380}$

11. $\dfrac{(n+1)!}{(n-1)!} = \dfrac{(n+1)n(n-1)!}{(n-1)!} = n(n+1)$

13. $\displaystyle\sum_{i=1}^{6} 5 = 6(5) = 30$

15. $\displaystyle\sum_{j=1}^{4} \frac{6}{j^2} = \frac{6}{1^2} + \frac{6}{2^2} + \frac{6}{3^2} + \frac{6}{4^2}$

$= 6 + \dfrac{3}{2} + \dfrac{2}{3} + \dfrac{3}{8} = \dfrac{205}{24}$

17. $\displaystyle \frac{1}{2(1)} + \frac{1}{2(2)} + \frac{1}{2(3)} + \cdots + \frac{1}{2(20)} = \sum_{k=1}^{20} \frac{1}{2k}$

≈ 1.799

19. $\displaystyle \frac{1}{2} + \frac{2}{3} + \frac{3}{4} + \cdots + \frac{9}{10} = \sum_{k=1}^{9} \frac{k}{k+1} \approx 7.071$

21. (a) $\displaystyle\sum_{k=1}^{4} \frac{5}{10^k} = \frac{5}{10} + \frac{5}{100} + \frac{5}{1000} + \frac{5}{10{,}000}$

$= 0.5 + 0.05 + 0.005 + 0.0005$

$= 0.5555 = \dfrac{1111}{2000}$

(b) $\displaystyle\sum_{k=1}^{\infty} \frac{5}{10^k} = \frac{5}{10} \sum_{k=0}^{\infty} \frac{1}{10^k} = \frac{5}{10} \cdot \frac{1}{1 - 1/10} = \frac{5}{10} \cdot \frac{10}{9} = \frac{5}{9}$

23. (a) $\displaystyle\sum_{k=1}^{4} 2(0.5)^k = 2(0.5) + 2(0.5)^2 + 2(0.5)^3 + 2(0.5)^4$

$= 1.875 = \dfrac{15}{8}$

(b) $\displaystyle\sum_{k=1}^{\infty} 2(0.5)^k = 2(0.5)\frac{1}{1-0.5} = 2$

25. $a_n = 3000\left(1 + \dfrac{0.02}{4}\right)^n$

(a) $a_1 = 3000\left(1 + \dfrac{0.02}{4}\right)^1 = \$3015$

$a_2 = 3000\left(1 + \dfrac{0.02}{4}\right)^2 = \$3030.08$

$a_3 = 3000\left(1 + \dfrac{0.02}{4}\right)^3 = \$3045.23$

$a_4 = 3000\left(1 + \dfrac{0.02}{4}\right)^4 = \$3060.45$

$a_5 = 3000\left(1 + \dfrac{0.02}{4}\right)^5 = \$3075.75$

$a_6 = 3000\left(1 + \dfrac{0.02}{4}\right)^6 = \$3091.13$

$a_7 = 3000\left(1 + \dfrac{0.02}{4}\right)^7 = \$3106.59$

$a_8 = 3000\left(1 + \dfrac{0.02}{4}\right)^8 = \$3122.12$

(b) $a_{40} = 3000\left(1 + \dfrac{0.02}{4}\right)^{40} = \$3662.38$

27. Arithmetic

$d = 3 - 5 = -2$

29. Arithmetic

$d = 1 - \dfrac{1}{2} = \dfrac{1}{2}$

31. $a_1 = 3,\ d = 4$

$a_1 = 3$
$a_2 = 3 + 4 = 7$
$a_3 = 7 + 4 = 11$
$a_4 = 11 + 4 = 15$
$a_5 = 15 + 4 = 19$

33. $a_4 = 10,\ a_{10} = 28$

$a_{10} = a_4 + 6d$
$28 = 10 + 6d$
$18 = 6d$
$3 = d$
$a_1 = a_4 - 3d$
$a_1 = 10 - 3(3)$
$a_1 = 1$
$a_2 = 1 + 3 = 4$
$a_3 = 4 + 3 = 7$
$a_4 = 7 + 3 = 10$
$a_5 = 10 + 3 = 13$

35. $a_1 = 35,\ a_{k+1} = a_k - 3$

$a_1 = 35$
$a_2 = a_1 - 3 = 35 - 3 = 32$
$a_3 = a_2 - 3 = 32 - 3 = 29$
$a_4 = a_3 - 3 = 29 - 3 = 26$
$a_5 = a_4 - 3 = 26 - 3 = 23$
$a_n = 35 + (n-1)(-3) = 38 - 3n,\ d = -3$

37. $a_n = 100 + (n-1)(-3) = 103 - 3n$

$$\sum_{n=1}^{25}(103 - 3n) = \sum_{n=1}^{25}103 - 3\sum_{n=1}^{25}n = 25(103) - 3\left[\frac{(25)(26)}{2}\right] = 1600$$

39. $$\sum_{j=1}^{10}(2j - 3) = 2\sum_{j=1}^{10}j - \sum_{j=1}^{10}3$$
$$= 2\left[\frac{10(11)}{2}\right] - 10(3) = 80$$

41. The sum of the first 50 positive multiples of 5

$$S_{50} = \sum_{n=1}^{50}5n = 5 + \cdots + 250 = \frac{50}{2}(5 + 250)$$
$$= 6375$$

43. Starting salary: \$36,000 with a salary increase of \$2250 per year

$a_n = 36{,}000 + (n-1)(2250)$

(a) $a_5 = 36{,}000 + (5-1)(2250) = \$45{,}000$

(b) $$S_5 = \sum_{n=1}^{5}(36{,}000 + (n-1)(2250))$$
$$= 36{,}000 + \cdots + 45{,}000$$
$$= \frac{5}{2}(36{,}000 + 45{,}000) = \$202{,}500$$

45. Geometric

$r = 2$

47. Geometric

$r = -\frac{1}{3}$

49. $a_1 = 4,\ r = -\frac{1}{4}$

$a_1 = 4$
$a_2 = 4\left(-\frac{1}{4}\right) = -1$
$a_3 = -1\left(-\frac{1}{4}\right) = \frac{1}{4}$
$a_4 = \frac{1}{4}\left(-\frac{1}{4}\right) = -\frac{1}{16}$
$a_5 = -\frac{1}{16}\left(-\frac{1}{4}\right) = \frac{1}{64}$

51. $a_1 = 9,\ a_3 = 4$

$a_3 = a_1r^2$
$4 = 9r^2$
$\frac{4}{9} = r^2 \Rightarrow r = \pm\frac{2}{3}$

$a_1 = 9$
$a_2 = 9\left(\frac{2}{3}\right) = 6$
$a_3 = 6\left(\frac{2}{3}\right) = 4$
$a_4 = 4\left(\frac{2}{3}\right) = \frac{8}{3}$
$a_5 = \frac{8}{3}\left(\frac{2}{3}\right) = \frac{16}{9}$

or

$a_1 = 9$
$a_2 = 9\left(-\frac{2}{3}\right) = -6$
$a_3 = -6\left(-\frac{2}{3}\right) = 4$
$a_4 = 4\left(-\frac{2}{3}\right) = -\frac{8}{3}$
$a_5 = -\frac{8}{3}\left(-\frac{2}{3}\right) = \frac{16}{9}$

53. $a_1 = 120,\ a_{k+1} = \frac{1}{3}a_k$

$a_1 = 120$
$a_2 = \frac{1}{3}(120) = 40$
$a_3 = \frac{1}{3}(40) = \frac{40}{3}$
$a_4 = \frac{1}{3}\left(\frac{40}{3}\right) = \frac{40}{9}$
$a_5 = \frac{1}{3}\left(\frac{40}{9}\right) = \frac{40}{27}$
$a_n = 120\left(\frac{1}{3}\right)^{n-1},\ r = \frac{1}{3}$

55. (a) and (b)

$$a_1 = 16,\ a_2 = -8,\ n = 6$$
$$r = \frac{a_2}{a_1} = \frac{-8}{16} = -\frac{1}{2}$$
$$a_n = a_1 r^{n-1}$$
$$a_6 = 16\left(-\frac{1}{2}\right)^{6-1} = 16\left(-\frac{1}{2}\right)^5 = -\frac{16}{32} = -\frac{1}{2}$$

57. $\sum_{i=1}^{7} 2^{i-1} = \frac{1-2^7}{1-2} = 127$

59. $\sum_{n=1}^{7} (-4)^{n-1} = \frac{1-(-4)^7}{1-(-4)}$
$= 3277$

61. $\sum_{i=1}^{\infty} 4\left(\frac{7}{8}\right)^{i-1} = \sum_{i=0}^{\infty} 4\left(\frac{7}{8}\right)^{i} = \frac{4}{1-7/8} = 32$

63. $\sum_{k=1}^{\infty} 4\left(\frac{2}{3}\right)^{k-1} = \sum_{k=0}^{\infty} 4\left(\frac{2}{3}\right)^{k} = \frac{4}{1-2/3} = 12$

65. Initial value: \$130,000
Each year, value depreciates 0.30 of previous year's value

(a) $a_t = 0.70\ a_{t-1}$ or $a_0 = 130{,}000$

$$a_1 = 130{,}000(0.70)$$
$$a_2 = 130{,}000(0.70)^2$$
$$\vdots$$
$$a_t = 130{,}000(0.7)^t$$

(b) $a_5 = 130{,}000(0.7)^5 = \$21{,}849.10$

67. ${}_{10}C_8 = \frac{10!}{8!2!} = \frac{10\cdot 9\cdot 8!}{8!2} = 45$

69. $\binom{9}{4} = {}_9C_4 = \frac{9!}{4!5!} = \frac{9\cdot 8\cdot 7\cdot 6\cdot 5!}{4\cdot 3\cdot 2\cdot 5!} = 126$

71. $(x+5)^4 = {}_4C_0x^4 + {}_4C_1x^3(5) + {}_4C_2x^2(5)^2 + {}_4C_3x(5)^3 + {}_4C_4(5)^4$
$= x^4 + 20x^3 + 150x^2 + 500x + 625$

73. $(a-4b)^5 = {}_5C_0a^5 - {}_5C_1a^4(4b) + {}_5C_2a^3(4b)^2 - {}_5C_3a^2(4b)^3$
$+ {}_5C_4a(4b)^4 - {}_5C_5(4b)^5$
$= a^5 - 20a^4b + 160a^3b^2 - 640a^2b^3$
$+ 1280ab^4 - 1024b^5$

75. The 4th number in the 6th row is ${}_6C_3 = 20$.

77. The 5th number in the 8th row is ${}_8C_4 = 70$.

79. Two numbers from 1 to 15 whose sum is 12 (without replacement)

10 ways: $1+11$
$2+10$
$3+9$
$4+8$
$5+7$
$7+5$
$8+4$
$9+3$
$10+2$
$11+1$

81. (a) $(4)(3)(6)(3) = 216$ schedules
(b) $(2)(3)(6)(3) = 108$ schedules
(c) $(2)(3)(2)(3) = 36$ schedules

83. ${}_{12}P_{10} = \frac{12!}{2!} = 239{,}500{,}800$

85. $\frac{8!}{2!2!2!1!1!} = \frac{8!}{8} = 7! = 5040$ permutations

87. ${}_{n+1}P_2 = 4\cdot {}_nP_1$

$$\frac{(n+1)!}{(n-1)!} = 4\cdot\frac{n!}{(n-1)!}$$
$$(n+1)! = 4\cdot n!$$
$$n = 3$$

89. ${}_8C_6 = \frac{8!}{2!6!} = \frac{8\cdot 7}{2} = 28$

91. $12! = 479{,}001{,}600$ ways

93. $\frac{10}{10}\cdot\frac{1}{9} = \frac{1}{9}$

95. (a) $\frac{208}{500} = 0.416$
(b) $\frac{400}{500} = 0.8$
(c) $\frac{37}{500} = 0.074$

97. True.
$\frac{(n+2)!}{n!} = \frac{(n+2)(n+1)n!}{n!} = (n+2)(n+1)$

99. (a) Each term is obtained by adding the same constant (common difference) to the preceding term.
(b) Each term is obtained by multiplying the same constant (common ratio) by the preceding term.

CHAPTER 9

Section 9.1

1. conic section

3. circle, center

5. The standard equation of a circle is given by $(x-h)^2+(y-k)^2=r^2$, where the point (h, k) is the center of the circle and the radius is r.

7. $$(x-0)^2+(y-0)^2=(4)^2$$
$$x^2+y^2=16$$

9. $$\text{Radius } =\sqrt{(3-1)^2+(7-0)^2}$$
$$=\sqrt{4+49}=\sqrt{53}$$
$$(x-h)^2+(y-k)^2=r^2$$
$$(x-3)^2+(y-7)^2=53$$

11. $\text{Diameter}=2\sqrt{7} \Rightarrow \text{radius}=\sqrt{7}$
$$(x-h)^2+(y-k)^2=r^2$$
$$(x+3)^2+(y+1)^2=7$$

13. $x^2+y^2=49$

Center: $(0, 0)$

Radius: 7

15. $(x+2)^2+(y-7)^2=16$

Center: $(-2, 7)$

Radius: 4

17. $(x-1)^2+y^2=15$

Center: $(1, 0)$

Radius: $\sqrt{15}$

19. $$\frac{1}{4}x^2+\frac{1}{4}y^2=1$$
$$x^2+y^2=4$$

Center: $(0, 0)$

Radius: 2

21. $$\frac{4}{3}x^2+\frac{4}{3}y^2=1$$
$$x^2+y^2=\frac{3}{4}$$

Center: $(0, 0)$

Radius: $\dfrac{\sqrt{3}}{2}$

23. $$(x^2-2x+1)+(y^2+6y+9)=-9+1+9$$
$$(x-1)^2+(y+3)^2=1$$

Center: $(1, -3)$

Radius: 1

25. $$4(x^2+3x+\tfrac{9}{4})+4(y^2-6y+9)=-41+9+36$$
$$4(x+\tfrac{3}{2})^2+4(y-3)^2=4$$
$$(x+\tfrac{3}{2})^2+(y-3)^2=1$$

Center: $(-\tfrac{3}{2}, 3)$

Radius: 1

27. $$x^2=16-y^2$$
$$x^2+y^2=16$$

Center: $(0, 0)$

Radius: 4

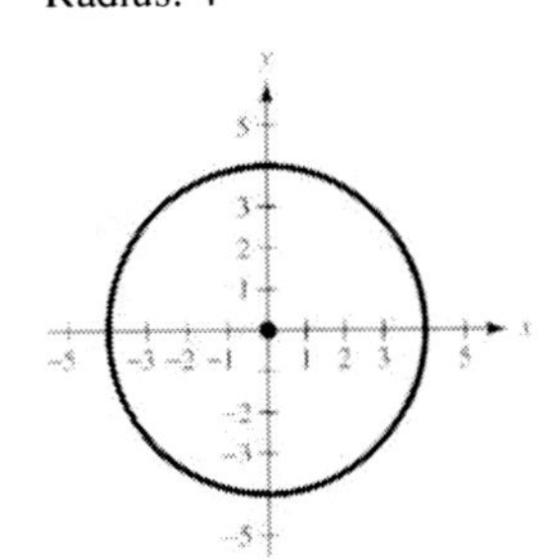

29. $$x^2+4x+y^2+4y-1=0$$
$$(x^2+4x+4)+(y^2+4y+4)=1+4+4$$
$$(x+2)^2+(y+2)^2=9$$

Center: $(-2, -2)$

Radius: 3

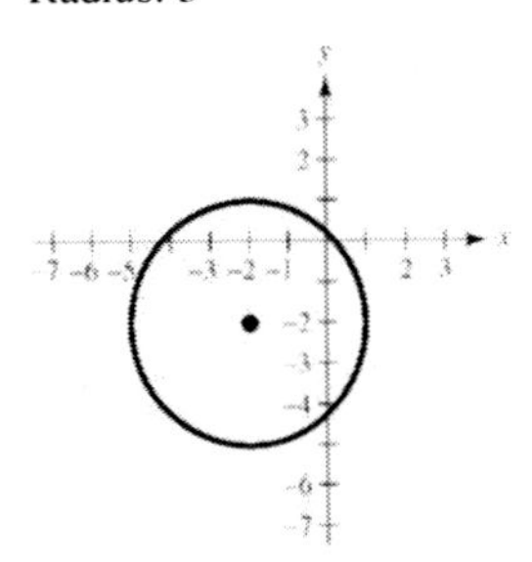

31. $$x^2 - 14x + y^2 + 8y + 40 = 0$$
$$(x^2 - 14x + 4) + (y^2 + 8y + 16) = -40 + 49 + 16$$
$$(x-7)^2 + (y+4)^2 = 25$$

Center: $(7, -4)$

Radius: 5

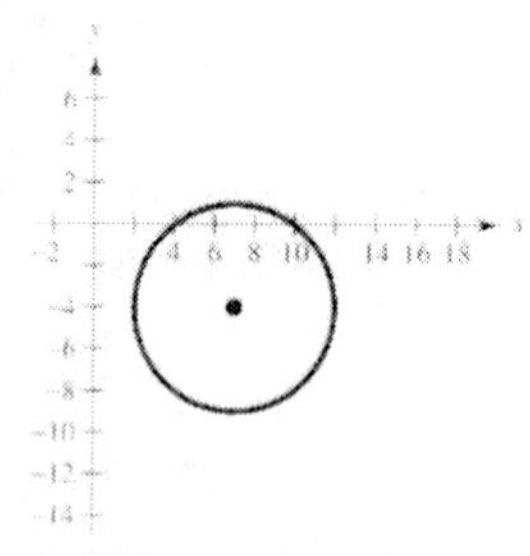

33. $$x^2 + 2x + y^2 - 35 = 0$$
$$(x^2 + 2x + 1) + y^2 = 35 + 1$$
$$(x+1)^2 + y^2 = 36$$

Center: $(-1, 0)$

Radius: 6

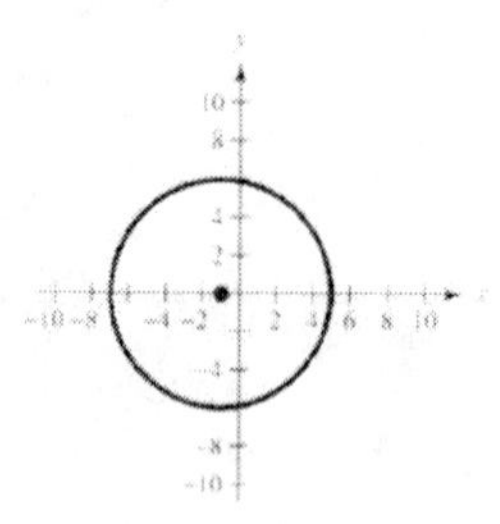

35. y-intercepts: $(0-2)^2 + (y+3)^2 = 9$
$$4 + (y+3)^2 = 9$$
$$(y+3)^2 = 5$$
$$y = -3 \pm \sqrt{5}$$
$$\left(0, -3 \pm \sqrt{5}\right)$$

x-intercepts: $(x-2)^2 + (0+3)^2 = 9$
$$(x-2)^2 = 0$$
$$x = 2$$
$$(2, 0)$$

37. y-intercepts: Let $x = 0$.
$$y^2 - 6y - 27 = 0$$
$$y^2 - 6y + 9 = 27 + 9$$
$$(y-3)^2 = 36$$
$$y - 3 = \pm 6$$
$$y = 9, -3$$
$$(0, 9), (0, -3)$$

x-intercepts: Let $y = 0$.
$$x^2 - 2x - 27 = 0$$
$$x^2 - 2x + 1 = 27 + 1$$
$$(x-1)^2 = 28$$
$$x - 1 = \pm\sqrt{28}$$
$$x = 1 \pm 2\sqrt{7}$$
$$\left(1 \pm 2\sqrt{7}, 0\right)$$

39. y-intercepts: $(0-6)^2 + (y+3)^2 = 16$
$$(y+3)^2 = 16 - 36$$
$$= -20$$

No solution

No y-intercepts

x-intercepts: $(x-6)^2 + (0+3)^2 = 16$
$$(x-6)^2 = 7$$
$$x - 6 = \pm\sqrt{7}$$
$$x = 6 \pm \sqrt{7}$$
$$\left(6 \pm \sqrt{7}, 0\right)$$

41. (a) Radius: 52; center: $(0, 0)$

$x^2 + y^2 = 52^2 \Rightarrow x^2 + y^2 = 2704$

(b) The distance from $(-40, -30)$ or 40 miles west and 30 miles south is
$$d = \sqrt{(0+40)^2 + (0+30)^2}$$
$$= \sqrt{40^2 + 30^2}$$
$$= \sqrt{2500} = 50 \text{ miles.}$$

Therefore, you would have felt the earthquake.

(c) $x^2 + y^2 = 52^2$

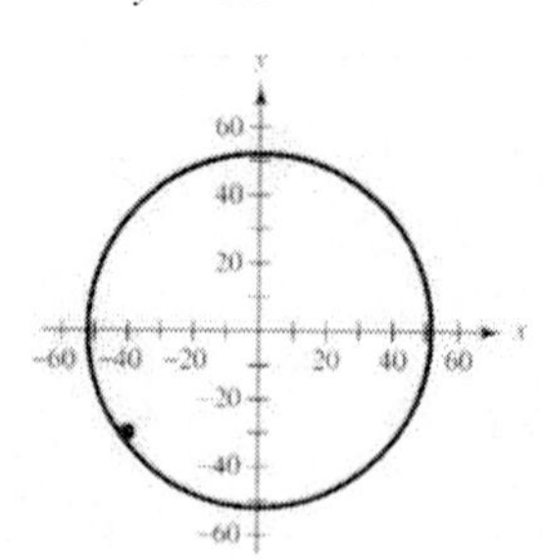

You were $52 - 50 = 2$ miles from the outer boundary.

43. $y^2 = -4x$

Vertex: $(0, 0)$

Opens to the left since p is negative.
Matches graph (e).

45. $x^2 = -8y$

Vertex: $(0, 0)$

Opens downward since p is negative.
Matches graph (d).

47. $(y-1)^2 = 4(x-3)$

Vertex: $(3, 1)$

Opens to the right since p is positive.
Matches graph (a).

49. Vertex: $(0, 0) \Rightarrow h = 0, k = 0$

Graph opens upward.

$x^2 = 4py$

Point on graph: $(3, 6)$

$3^2 = 4p(6)$

$9 = 24p$

$\frac{3}{8} = p$

Thus, $x^2 = 4\left(\frac{3}{8}\right)y \Rightarrow y = \frac{2}{3}x^2$
$\Rightarrow x^2 = \frac{3}{2}y.$

51. Vertex: $(0, 0) \Rightarrow h = 0, k = 0$

Focus: $\left(0, -\frac{3}{2}\right) \Rightarrow p = -\frac{3}{2}$

$(x-h)^2 = 4p(y-k)$

$x^2 = 4\left(-\frac{3}{2}\right)y$

$x^2 = -6y$

53. Vertex: $(0, 0) \Rightarrow h = 0, k = 0$

Focus: $(-2, 0) \Rightarrow p = -2$

$(y-k)^2 = 4p(x-h)$

$y^2 = 4(-2)x$

$y^2 = -8x$

55. Vertex: $(0, 0) \Rightarrow h = 0, k = 0$

Directrix: $y = 1 \Rightarrow p = -1$

$(x-h)^2 = 4p(y-k)$

$x^2 = 4(-1)y$

$x^2 = -4y$

57. Vertex: $(0, 0) \Rightarrow h = 0, k = 0$

Directrix: $x = 2 \Rightarrow p = -2$

$y^2 = 4px$

$y^2 = -8x$

59. Vertex: $(0, 0) \Rightarrow h = 0, k = 0$

Horizontal axis and passes through the point $(4, 6)$

$(y-k)^2 = 4p(x-h)$

$(y-0)^2 = 4p(x-0)$

$y^2 = 4px$

$6^2 = 4p(4)$

$36 = 16p \Rightarrow p = \frac{9}{4}$

$y^2 = 4\left(\frac{9}{4}\right)x$

$y^2 = 9x$

61. $y = \frac{1}{2}x^2$

$x^2 = 2y = 4\left(\frac{1}{2}\right)y;\ p = \frac{1}{2}$

Vertex: $(0, 0)$

Focus: $\left(0, \frac{1}{2}\right)$

Directrix: $y = -\frac{1}{2}$

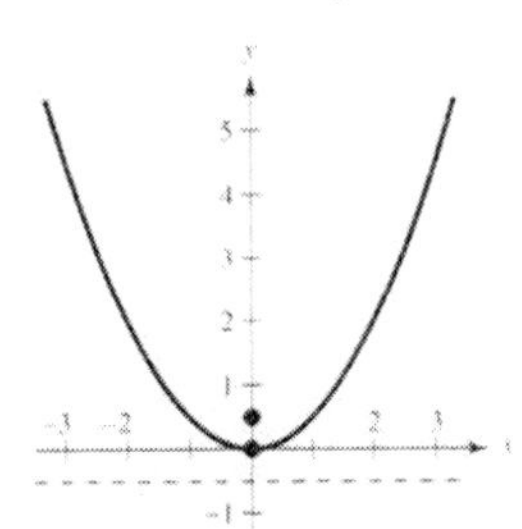

63. $y^2 = -6x$

$y^2 = 4\left(-\frac{3}{2}\right)x;\ p = -\frac{3}{2}$

Vertex: $(0, 0)$

Focus: $\left(\frac{3}{-2}, 0\right)$

Directrix: $x = \frac{3}{2}$

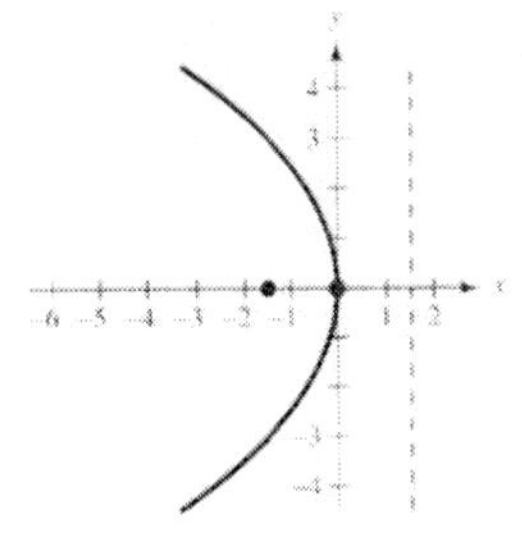

65. $x^2 + 6y = 0$

$$x^2 = -6y$$

$$x^2 = 4\left(-\tfrac{3}{2}\right)y$$

Vertex: $(0, 0)$

Focus: $\left(0, -\tfrac{3}{2}\right)$

Directrix: $y = \tfrac{3}{2}$

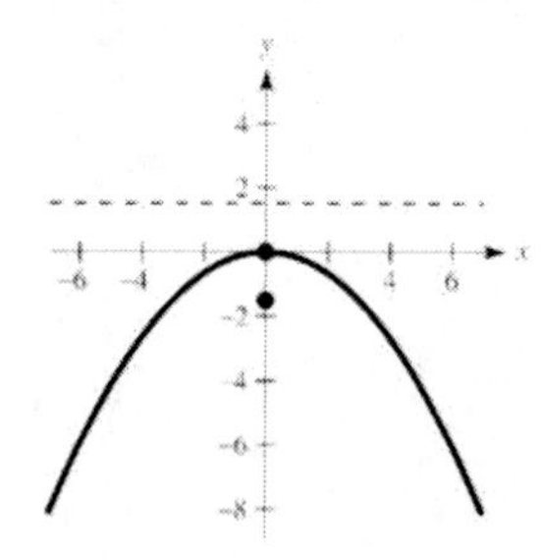

67. $(x+1)^2 + 8(y+2) = 0$

$$(x+1)^2 = -8(y+2)$$

$$(x+1)^2 = 4(-2)(y+2)$$

Vertex: $(-1, -2)$

Focus: $(-1, -4)$

Directrix: $y = 0$

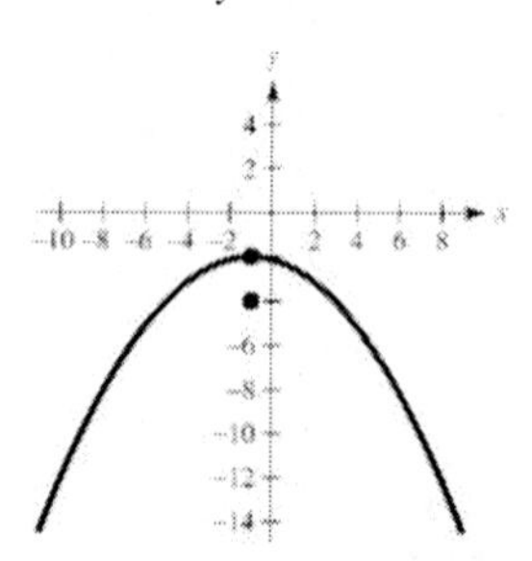

69. $y^2 + 6y + 8x + 25 = 0$

$$(y+3)^2 = 4(-2)(x+2);\ p = -2$$

Vertex: $(-2, -3)$

Focus: $(-4, -3)$

Directrix: $x = 0$

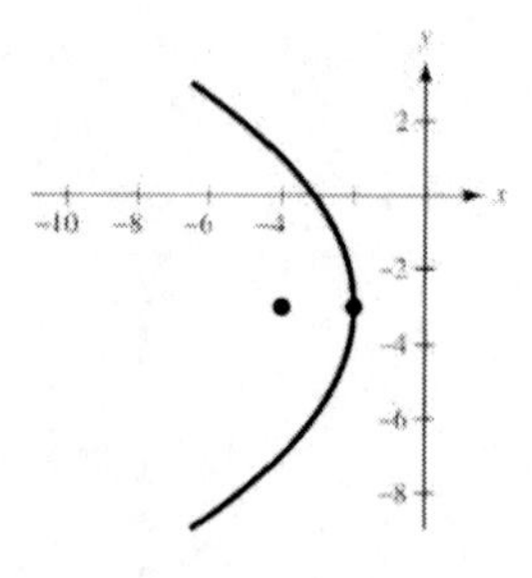

71. $\left(x+\tfrac{3}{2}\right)^2 = 4(y-2) \Rightarrow h = -\tfrac{3}{2},\ k = 2,\ p = 1$

Vertex: $\left(-\tfrac{3}{2}, 2\right)$

Focus: $\left(-\tfrac{3}{2}, 2+1\right) = \left(-\tfrac{3}{2}, 3\right)$

Directrix: $y = 1$

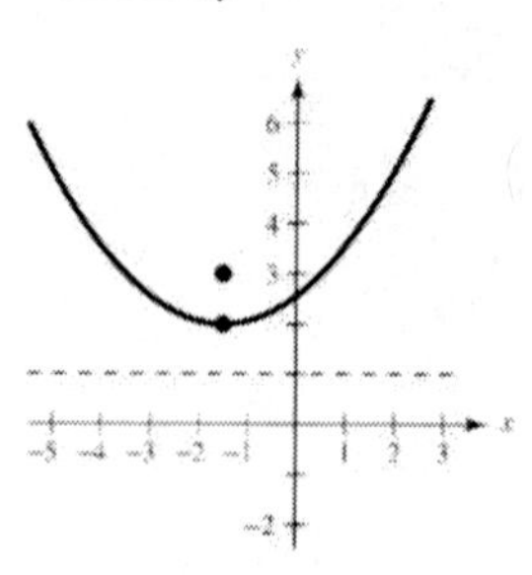

73. $y = \tfrac{1}{4}\left(x^2 - 2x + 5\right)$

$$4y - 4 = (x-1)^2$$

$$(x-1)^2 = 4(1)(y-1)$$

$h = 1,\ k = 1,\ p = 1$

Vertex: $(1, 1)$

Focus: $(1, 2)$

Directrix: $y = 0$

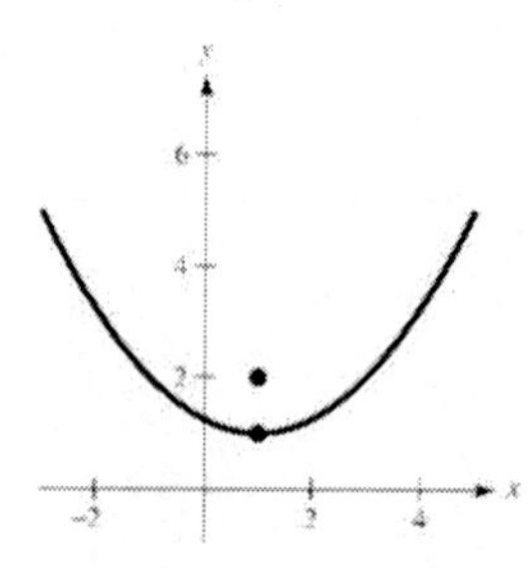

75. $x^2 + 4x + 6y - 2 = 0$

$$x^2 + 4x + 4 = -6y + 2 + 4 = -6y + 6$$

$$(x+2)^2 = -6(y-1)$$

$$(x+2)^2 = 4\left(-\tfrac{3}{2}\right)(y-1)$$

Vertex: $(-2, 1)$

Focus: $\left(-2, 1-\tfrac{3}{2}\right) = \left(-2, -\tfrac{1}{2}\right)$

Directrix: $y = \tfrac{5}{2}$

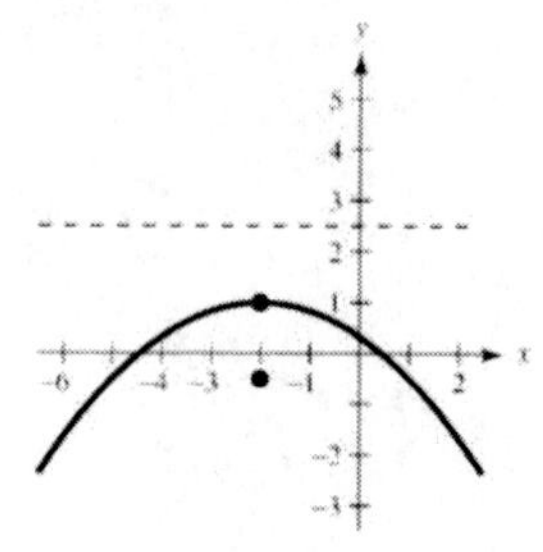

77. $y^2 + x + y = 0$

$y^2 + y + \frac{1}{4} = -x + \frac{1}{4}$

$\left(y + \frac{1}{2}\right)^2 = 4\left(-\frac{1}{4}\right)\left(x - \frac{1}{4}\right)$

$h = \frac{1}{4},\ k = -\frac{1}{2}, p = -\frac{1}{4}$

Vertex: $\left(\frac{1}{4},\ -\frac{1}{2}\right)$

Focus: $\left(0,\ -\frac{1}{2}\right)$

Directrix: $x = \frac{1}{2}$

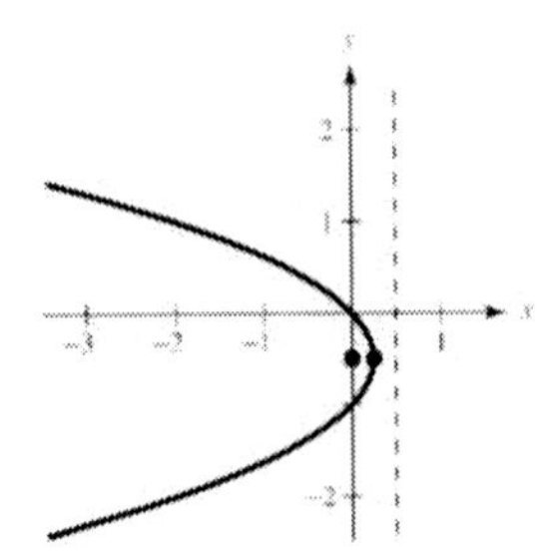

79. Vertex: $(3,\ 1) \Rightarrow h = 3,\ k = 1$

Opens downward

Passes through: $(2,\ 0),\ (4, 0)$

$$y = -(x-2)(x-4)$$
$$= -x^2 + 6x - 8$$
$$= -(x-3)^2 + 1$$
$$(x-3)^2 = -(y-1)$$

81. Vertex: $(-4,\ 0) \Rightarrow h = -4,\ k = 0$

Opens to the right

Passes through: $(0,\ 4)$

$$(y-0)^2 = 4p(x+4)$$
$$y^2 = 4p(x+4)$$
$$(4)^2 = 4p(0+4)$$
$$16 = 16p$$
$$1 = p$$
$$y^2 = 4(x+4)$$

83. Vertex: $(-2,\ 0) \Rightarrow h = -2,\ k = 0$

Opens to the right

Focus: $\left(-\frac{3}{2},\ 0\right)$

$$\frac{1}{2} = p$$
$$y^2 = 4\left(\frac{1}{2}\right)(x+2)$$
$$y^2 = 2(x+2)$$

85. Vertex: $(5,\ 2)$

Focus: $(3,\ 2)$

Horizontal axis: $p = 3 - 5 = -2$

$$(y-2)^2 = 4(-2)(x-5)$$
$$(y-2)^2 = -8(x-5)$$

87. Vertex: $(0,\ 4)$

Directrix: $y = 2$

Vertical axis

$p = 4 - 2 = 2$

$$(x-0)^2 = 4(2)(y-4)$$
$$x^2 = 8(y-4)$$

89. Focus: $(2,\ 2)$

Directrix: $x = -2$

Horizontal axis

Vertex: $(0,\ 2)$

$p = 2 - 0 = 2$

$$(y-2)^2 = 4(2)(x-0)$$
$$(y-2)^2 = 8x$$

91. $y^2 - 8x = 0$ and $x - y + 2 = 0$

$y^2 = 8x$ $\qquad$ $y_3 = x + 2$

$y_1 = \sqrt{8x}$

$y_2 = -\sqrt{8x}$

The point of tangency is $(2,\ 4)$.

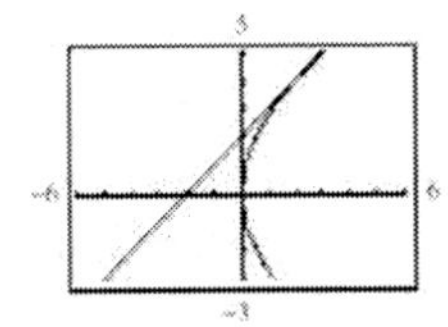

93. $x^2 = 2y,\ (4,\ 8),\ p = \dfrac{1}{2},$ focus: $\left(0,\ \dfrac{1}{2}\right)$

$$d_1 = \frac{1}{2} - b$$
$$d_2 = \sqrt{(4-0)^2 + \left(8 - \frac{1}{2}\right)^2} = \frac{17}{2}$$
$$d_1 = d_2 \Rightarrow \frac{1}{2} - b = \frac{17}{2} \Rightarrow b = -8$$
$$m = \frac{8-(-8)}{4-0} = 4$$

Tangent line: $y = 4x - 8 \Rightarrow 4x - y - 8 = 0$

Let $y = 0 \Rightarrow x = 2 \Rightarrow x$-intercept $(2,\ 0)$.

95. $y = -2x^2 \Rightarrow x^2 = -\frac{1}{2}y = 4\left(-\frac{1}{8}\right)y$

$\Rightarrow p = -\frac{1}{8}$

Focus: $\left(0, -\frac{1}{8}\right)$

$d_1 = \frac{1}{8} + b$

$d_2 = \sqrt{(-1-0)^2 + \left(-2 + \frac{1}{8}\right)^2} = \frac{17}{8}$

$d_1 = d_2 \Rightarrow \frac{1}{8} + b = \frac{17}{8} \Rightarrow b = 2$

$m = \frac{-2-2}{-1-0} = 4$

Tangent line: $y = 4x + 2 \Rightarrow 4x - y + 2 = 0$

Let $y = 0 \Rightarrow x = -\frac{1}{2} \Rightarrow x$-intercept $\left(-\frac{1}{2}, 0\right)$.

97. (a) $x^2 = 4py$

$32^2 = 4p\left(\frac{1}{12}\right)$

$1024 = \frac{1}{3}p$

$3072 = p$

$x^2 = 4(3072)y$

$y = \frac{x^2}{12,288}$

(b) $\frac{1}{24} = \frac{x^2}{12,288}$

$\frac{12,288}{24} = x^2$

$512 = x^2$

$x \approx 22.6$ feet

99. (a) $x^2 = 4py,\ p = \frac{3}{2}$

$x^2 = 4\left(\frac{3}{2}\right)y = 6y$

$\left(\text{or } y^2 = 6x\right)$

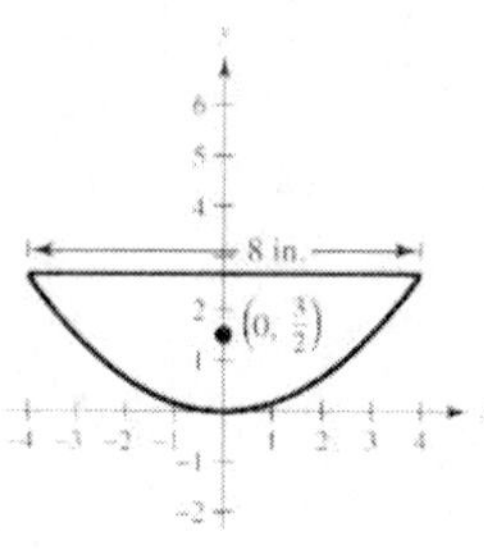

(b) When $x = 4$,

$6y = 16$

$y = \frac{16}{6} = \frac{8}{3}$.

Depth: $\frac{8}{3} \approx 2.67$ inches

101. (a)

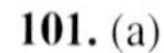

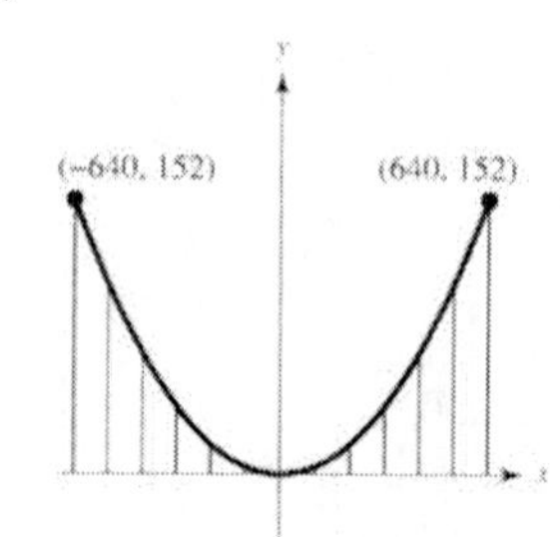

(b) $x^2 = 4py$

$640^2 = 4p(152)$

$p = \frac{12,800}{19}$

$y = \frac{19}{51,200}x^2$

(c)

x	0	200	400	500	500
y	0	14.84	59.38	92.77	133.59

103. Vertex: $(0, 0)$

$y^2 = 4px$

Point: $(1000, 800)$

$800^2 = 4p(1000) \Rightarrow p = 160$

$y^2 = 4(160)x$

$y^2 = 640x$

105. (a) $x^2 = -\frac{v^2}{16}(y - s)$

$x^2 = -\frac{(28)^2}{16}(y - 100)$

$x^2 = -49(y - 100)$

(b) The ball bits the ground when $y = 0$.

$x^2 = -49(0 - 100)$

$x^2 = 4900$

$x = 70$

The ball travels 70 feet.

107. The slope of the line joining $(3, -4)$ and the center is $-\frac{4}{3}$. The slope of the tangent line at $(3, -4)$ is $\frac{3}{4}$. Thus, the tangent line is:

$y + 4 = \frac{3}{4}(x - 3)$

$4y + 16 = 3x - 9$

$3x - 4y = 25$.

109. The slope of the line joining $\left(2, -2\sqrt{2}\right)$ and the center is $\left(-2\sqrt{2}\right)/2 = -\sqrt{2}$. The slope of the tangent line is $1/\sqrt{2} = \sqrt{2}/2$. Thus, the tangent line is:

$$y + 2\sqrt{2} = \frac{\sqrt{2}}{2}(x - 2)$$
$$2y + 4\sqrt{2} = \sqrt{2}x - 2\sqrt{2}$$
$$\sqrt{2}x - 2y = 6\sqrt{2}.$$

111. False. The center is $(0, -5)$.

113. False. A circle is a conic section.

115. True. The vertix is the closest point to the directrix or focus.

117. True. If the vertex and focus of a parabola are on a horizontal line, then the directrix of the parabola is a vertical line.

119. The graph of $x^2 + y^2 = 0$ is a single point, $(0, 0)$.

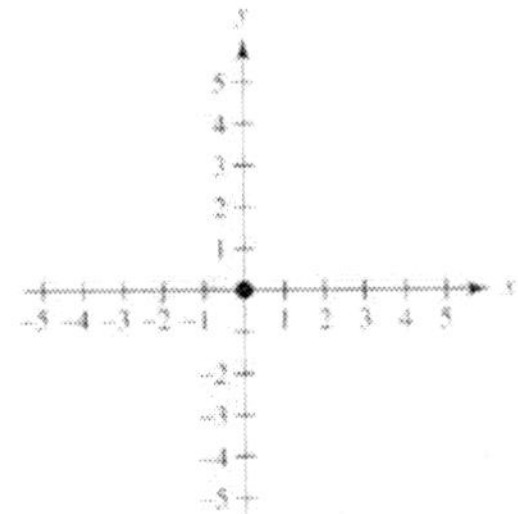

The plane intersects the double-napped cone at the vertices of the cones.

121. $(y+1)^2 = 2(x-2)$

For the lower half of the parabola,

$$y + 1 = -\sqrt{2(x-2)}$$
$$y = -1 - \sqrt{2(x-2)}.$$

123. $f(x) = 2x^2 + 3x$

Relative minimum: $(-0.75, -1.13)$

125. $f(x) = x^5 - 3x - 1$

Relative minimum: $(0.88, -3.11)$

Relative maximum: $(-0.88, 1.11)$

Section 9.2

1. ellipse, foci

3. minor axis

In Exercises 5–8, use the equation of the ellipse $\dfrac{x^2}{2^2} + \dfrac{y^2}{8^2} = 1.$

5. The major axis is vertical since $8 > 2$.

7. The length of the minor axis is $2b$ or $2(2) = 4$ units.

9. $\dfrac{x^2}{4} + \dfrac{y^2}{9} = 1$

Center: $(0, 0)$

$a = 3, b = 2$

Vertical major axis
Matches graph (b).

11. $\dfrac{(x-2)^2}{16} + (y+1)^2 = 1$

Center: $(2, -1)$

$a = 4, b = 1$

Horizontal major axis
Matches graph (a).

13. $\dfrac{x^2}{64} + \dfrac{y^2}{9} = 1$

Center: $(0, 0)$

$a = 8, b = 3,$

$c = \sqrt{64 - 9} = \sqrt{55}$

Vertices: $(\pm 8, 0)$

Foci: $\left(\pm\sqrt{55}, 0\right)$

$e = \dfrac{c}{a} = \dfrac{\sqrt{55}}{8}$

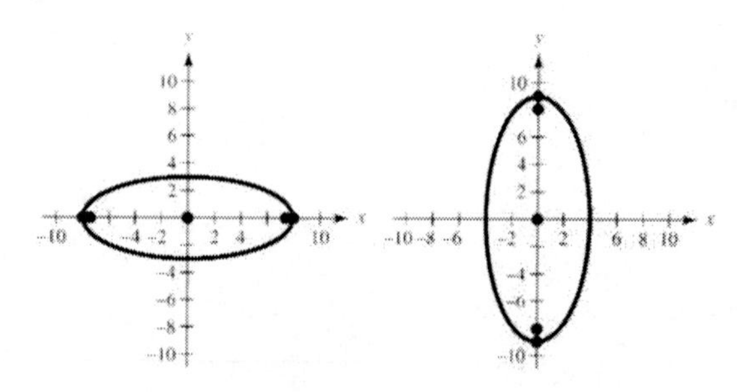

15. $\frac{(x-4)^2}{16}+\frac{(y+1)^2}{25}=1$

Center: $(4,\ -1)$

$a=5,\ b=4,\ c=3$

Vertices: $(4,\ -1\pm 5)\Rightarrow(4,\ -6),\ (4,4)$

Foci: $(4,\ -1\pm 3)\Rightarrow(4,\ -4),\ (4,2)$

$e=\frac{c}{a}=\frac{3}{5}$

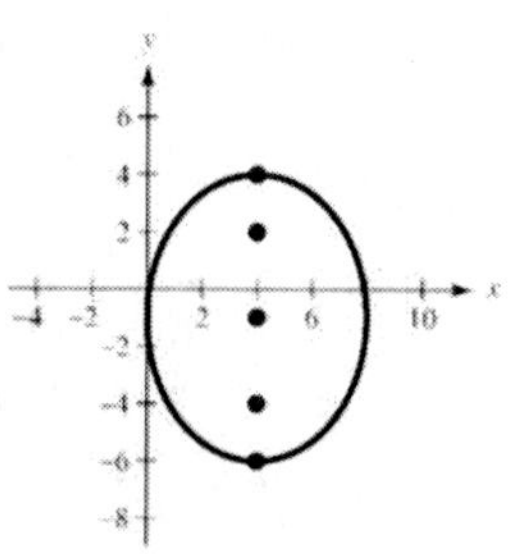

17. $\frac{(x+5)^2}{\frac{9}{4}}+(y-1)^2=1$

Center: $(-5,\ 1)$

$a=\frac{3}{2},\ b=1,\ c=\sqrt{\frac{9}{4}-1}=\frac{\sqrt{5}}{2}$

Foci: $\left(-5+\frac{\sqrt{5}}{2},\ 1\right),\ \left(-5-\frac{\sqrt{5}}{2},\ 1\right)$

Vertices:

$\left(-5+\frac{3}{2},\ 1\right)=\left(-\frac{7}{2},\ 1\right),\ \left(-5-\frac{3}{2},\ 1\right)=\left(-\frac{13}{2},\ 1\right)$

$e=\frac{\sqrt{5}/2}{3/2}=\frac{\sqrt{5}}{3}$

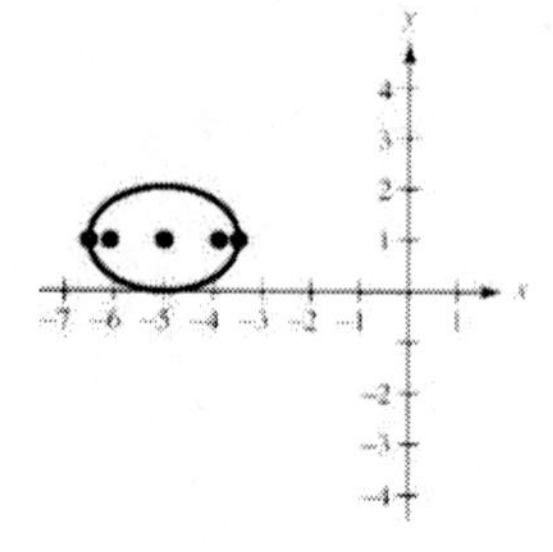

19. Center: $(0,\ 0)\Rightarrow h=0,\ k=0$

$a=4,\ b=2$

Vertical major axis

$\frac{x^2}{4}+\frac{y^2}{16}=1$

21. Center: $(0,\ 0)\Rightarrow h=0,\ k=0$

$a=3,\ c=2\Rightarrow b=\sqrt{9-4}=\sqrt{5}$

Horizontal major axis

$\frac{x^2}{9}+\frac{y^2}{5}=1$

23. Center: $(0,\ 0)\Rightarrow h=0,\ k=0$

Foci: $(\pm 5,\ 0)\Rightarrow c=5$

Horizontal major axis: 14 units $\Rightarrow a=7$

$$a^2=b^2+c^2$$
$$7^2=b^2+5^2$$
$$24=b^2$$
$$2\sqrt{6}=b$$
$$\frac{(x-h)^2}{a^2}+\frac{(y-k)^2}{b^2}=1$$
$$\frac{x^2}{49}+\frac{y^2}{24}=1$$

25. Vertices: $(0,\ \pm 5)\Rightarrow a=5$

Center: $(0,\ 0)\Rightarrow h=0,\ k=0$

Vertical major axis

$$\frac{(x-h)^2}{b^2}+\frac{(y-k)^2}{a^2}=1$$
$$\frac{x^2}{b^2}+\frac{y^2}{25}=1$$

Point: $(4,\ 2)$

$$\frac{4^2}{b^2}+\frac{2^2}{25}=1$$
$$\frac{16}{b^2}=1-\frac{4}{25}=\frac{21}{25}$$
$$400=21b^2$$
$$\frac{400}{21}=b^2$$
$$\frac{x^2}{\frac{400}{21}}+\frac{y^2}{25}=1$$
$$\frac{21x^2}{400}+\frac{y^2}{25}=1$$

27. Center: $(2, 3) \Rightarrow h = 2, k = 3$

$a = 3, b = 1$

Vertical major axis

$$\frac{(x-h)^2}{b^2} + \frac{(y-k)^2}{a^2} = 1$$

$$\frac{(x-2)^2}{1} + \frac{(y-3)^2}{9} = 1$$

29. Center: $(4, 2) \Rightarrow h = 4, k = 2$

$a = 4, b = 1 \Rightarrow c = \sqrt{16-1} = \sqrt{15}$

Horizontal major axis

$$\frac{(x-4)^2}{16} + \frac{(y-2)^2}{1} = 1$$

31. Center: $(0, 4) \Rightarrow h = 0, k = 4$

$c = 4, a = 18 \Rightarrow b^2 = a^2 - c^2 = 324 - 16 = 308$

Vertical major axis

$$\frac{x^2}{308} + \frac{(y-4)^2}{324} = 1$$

33. Vertices: $(3, 1), (3, 9) \Rightarrow a = 4$

Center: $(3, 5) \Rightarrow h = 3, k = 5$

Minor axis of length $6 \Rightarrow b = 3$

Vertical major axis

$$\frac{(x-h)^2}{b^2} + \frac{(y-k)^2}{a^2} = 1$$

$$\frac{(x-3)^2}{9} + \frac{(y-5)^2}{16} = 1$$

35. Center: $(0, 4) \Rightarrow h = 0, k = 4$

Vertices: $(-4, 4), (4, 4) \Rightarrow a = 4$

$a = 2c \Rightarrow 4 = 2c \Rightarrow c = 2$

$2^2 = 4^2 - b^2 \Rightarrow b^2 = 12$

Horizontal major axis

$$\frac{(x-h)^2}{a^2} + \frac{(y-k)^2}{b^2} = 1$$

$$\frac{x^2}{16} + \frac{(y-4)^2}{12} = 1$$

37. (a) $x^2 + 9y^2 = 36$

$$\frac{x^2}{36} + \frac{y^2}{4} = 1$$

(b) $a = 6, b = 2, c = \sqrt{36-4} = \sqrt{32} = 4\sqrt{2}$

Center: $(0, 0)$

Vertices: $(\pm 6, 0)$

Foci: $(\pm 4\sqrt{2}, 0)$

$$e = \frac{c}{a} = \frac{4\sqrt{2}}{6} = \frac{2\sqrt{2}}{3}$$

(c)

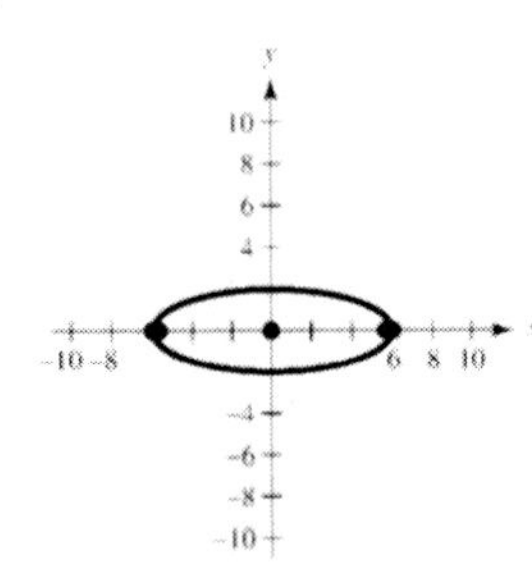

39. (a) $49x^2 + 4y^2 - 196 = 0$

$$49x^2 + 4y^2 = 196$$

$$\frac{x^2}{4} + \frac{y^2}{49} = 1$$

(b) $a = 7, b = 2, c = \sqrt{49-4} = \sqrt{45} = 3\sqrt{5}$

Center: $(0, 0)$

Vertices: $(0, \pm 7)$

Foci: $(0, \pm 3\sqrt{5})$

$$e = \frac{c}{a} = \frac{3\sqrt{5}}{7}$$

(c)

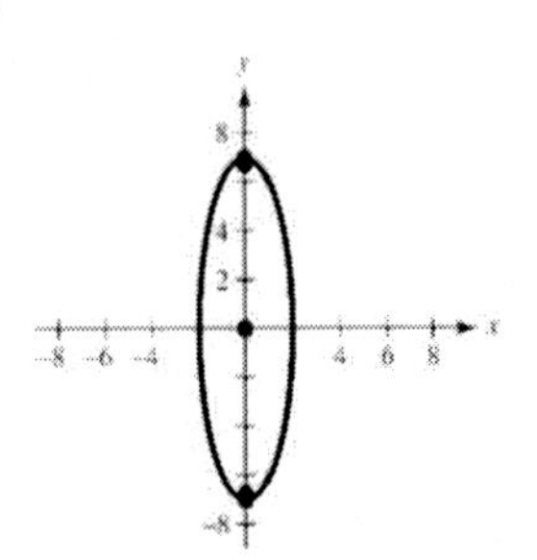

41. (a) $9x^2 + 4y^2 + 36x - 24y + 36 = 0$

$$9(x^2 + 4x + 4) + 4(y^2 - 6y + 9) = -36 + 36 + 36$$

$$\frac{(x+2)^2}{4} + \frac{(y-3)^2}{9} = 1$$

(b) $a = 3, b = 2, c = \sqrt{5}$

Center: $(-2, 3)$

Foci: $(-2, 3 \pm \sqrt{5})$

Vertices: $(-2, 6), (-2, 0)$

$$e = \frac{c}{a} = \frac{\sqrt{5}}{3}$$

(c)

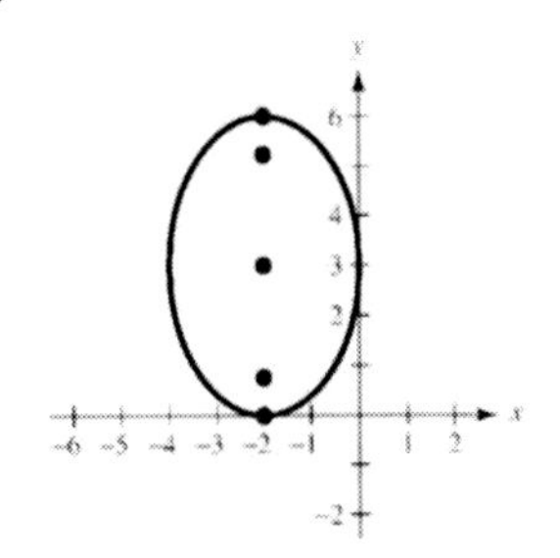

43. (a)

$$6x^2 + 2y^2 + 18x - 10y + 2 = 0$$

$$6\left(x^2 + 3x + \frac{9}{4}\right) + 2\left(y^2 - 5y + \frac{25}{4}\right) = -2 + \frac{27}{2} + \frac{25}{2}$$

$$6\left(x + \frac{3}{2}\right)^2 + 2\left(y - \frac{5}{2}\right)^2 = 24$$

$$\frac{\left(x + \frac{3}{2}\right)^2}{4} + \frac{\left(y - \frac{5}{2}\right)^2}{12} = 1$$

(b) $a = 2\sqrt{3},\ b = 2,\ c = 2\sqrt{2}$

Center: $\left(-\frac{3}{2}, \frac{5}{2}\right)$

Foci: $\left(-\frac{3}{2}, \frac{5}{2} \pm 2\sqrt{2}\right)$

Vertices: $\left(-\frac{3}{2}, \frac{5}{2} \pm 2\sqrt{3}\right)$

$e = \frac{c}{a} = \frac{\sqrt{2}}{\sqrt{3}} = \frac{\sqrt{6}}{3}$

(c)

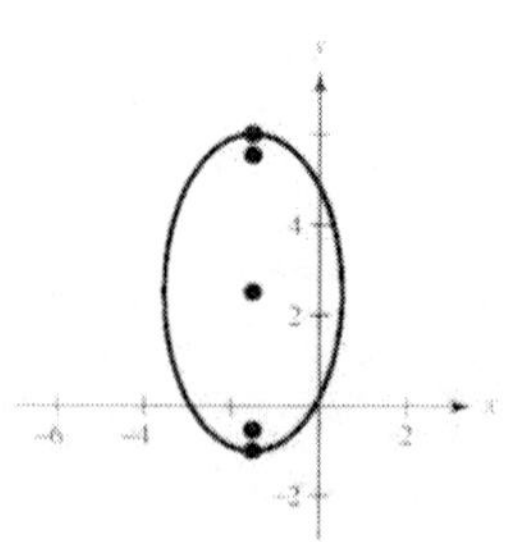

45. (a)

$$16x^2 + 25y^2 - 32x + 50y + 16 = 0$$

$$16\left(x^2 - 2x + 1\right) + 25\left(y^2 + 2y + 1\right) = -16 + 16 + 25$$

$$\frac{(x-1)^2}{\frac{25}{16}} + (y+1)^2 = 1$$

(b) $a = \frac{5}{4},\ b = 1,\ c = \frac{3}{4}$

Center: $(1, -1)$

Foci: $\left(\frac{7}{4}, -1\right), \left(\frac{1}{4}, -1\right)$

Vertices: $\left(\frac{9}{4}, -1\right), \left(-\frac{1}{4}, -1\right)$

$e = \frac{c}{a} = \frac{3}{5}$

(c)

47. (a)

$$12x^2 + 20y^2 - 12x + 40y - 37 = 0$$

$$12\left(x^2 - 1 + \frac{1}{4}\right) + 20\left(y^2 + 2y + 1\right) = 37 + 3 + 20$$

$$12\left(x - \frac{1}{2}\right)^2 + 20(y+1)^2 = 60$$

(b) $a = \sqrt{5},\ b = \sqrt{3},\ c = \sqrt{5-3} = \sqrt{2}$

Center: $\left(\frac{1}{2}, -1\right)$

Vertices: $\left(\frac{1}{2} \pm \sqrt{5}, -1\right)$

Foci: $\left(\frac{1}{2} \pm \sqrt{2}, -1\right)$

$e = \frac{c}{a} = \frac{\sqrt{2}}{\sqrt{5}} = \frac{\sqrt{10}}{5}$

(c)

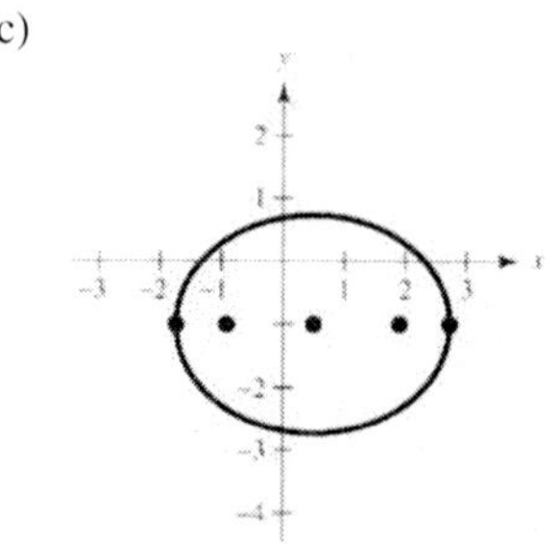

49. $\frac{x^2}{4} + \frac{y^2}{9} = 1$

$a = 3,\ b = 2,$

$c = \sqrt{9-4} = \sqrt{5}$

$e = \frac{c}{a} = \frac{\sqrt{5}}{3}$

51.

$$x^2 + 9y^2 - 10x + 36y + 52 = 0$$

$$\left(x^2 - 10x + 25\right) + 9\left(y^2 + 4y + 4\right) = -52 + 25 + 36$$

$$(x-5)^2 + 9(y+2)^2 = 9$$

$$\frac{(x-5)^2}{9} + \frac{(y+2)^2}{1} = 1$$

$a = 3,\ b = 1,\ c = \sqrt{9-1} = 2\sqrt{2}$

$e = \frac{c}{a} = \frac{2\sqrt{2}}{3}$

53. Vertices: $(\pm 5, 0)$, $a = 5$, $h = 0$, $k = 0$

Eccentricity: $e = \dfrac{3}{5} = \dfrac{c}{a}$

$$\frac{3}{5} = \frac{c}{5}$$

$$3 = c$$

$$b^2 - a^2 - c^2 = 25 - 9 = 16$$

$$\frac{x^2}{a^2} + \frac{y^2}{b^2} = 1$$

$$\frac{x^2}{25} + \frac{y^2}{16} = 1$$

55. Foci: $(\pm 3, 0) \Rightarrow c = 3$

Eccentricity: $e = \dfrac{4}{5}$

$$e = \frac{c}{a} \Rightarrow \frac{4}{5} = \frac{3}{a}$$

$$4a = 15$$

$$a = \tfrac{15}{4}$$

$$a^2 = b^2 + c^2$$

$$\frac{225}{16} = b^2 + 9$$

$$\frac{81}{16} = b^2$$

The center is the midpoint between the foci or $(0, 0)$ and the major axis is horizontal.

$$\frac{x^2}{a^2} + \frac{y^2}{b^2} = 1$$

$$\frac{x^2}{\frac{225}{16}} + \frac{y^2}{\frac{81}{16}} = 1$$

57. Let $\dfrac{x^2}{a^2} + \dfrac{y^2}{b^2} = 1$ be the equation of the ellipse. Then $b = 2$ and $a = 3 \Rightarrow c^2 = a^2 - b^2 = 9 - 4 = 5$. Thus, the tacks are placed at $(\pm\sqrt{5}, 0)$. The string has a length of $2a = 6$ feet.

59. Center: $(0, 0)$, $e = 0.97$

$$2a = 35.88 \Rightarrow a = 17.94 \Rightarrow a^2 \approx 321.84$$

$$e = \frac{c}{a} \Rightarrow 0.97 = \frac{c}{17.94} \Rightarrow c = 17.4018$$

$$c^2 = a^2 - b^2 \Rightarrow b^2 = a^2 - c^2 \approx 19.02$$

Ellipse: $\dfrac{x^2}{321.84} + \dfrac{y^2}{19.02} = 1$

61. $a + c = 947 + 6378 = 7325$

$a - c = 228 + 6378 = 6606$

$2a = 13,931$

$a = 6965.5$

$c = 7325 - 6965.5$

$= 359.5$

$e = \dfrac{c}{a} \approx 0.0516$

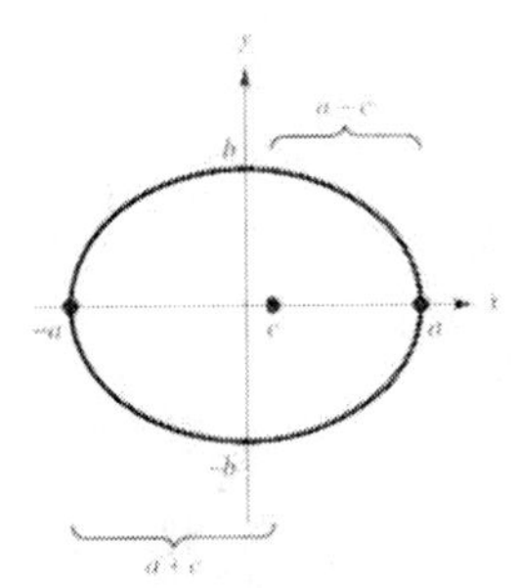

63. $\dfrac{x^2}{4} + \dfrac{y^2}{1} = 1$

$a = 2$, $b = 1$, $c = \sqrt{3}$

Points on the ellipse: $(\pm 2, 0)$, $(0, \pm 1)$

Length of latus recta: $\dfrac{2b^2}{a} = 1$

Additional points: $\left(\sqrt{3}, \pm\dfrac{1}{2}\right)$, $\left(-\sqrt{3}, \pm\dfrac{1}{2}\right)$

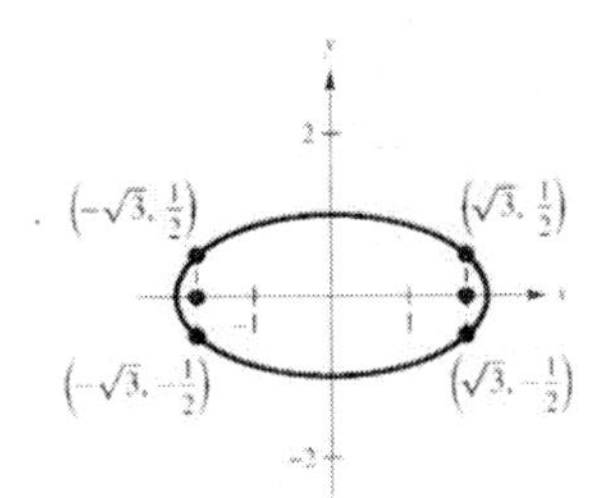

65. $9x^2 + 4y^2 = 36$

$$\frac{x^2}{4} + \frac{y^2}{9} = 1$$

Points on the ellipse: $(\pm 2, 0)$, $(0, \pm 3)$

Length of latus recta: $\dfrac{2b^2}{a} = \dfrac{2 \cdot 2^2}{3} = \dfrac{8}{3}$

Additional points: $\left(\pm\dfrac{4}{3}, -\sqrt{5}\right)$, $\left(\pm\dfrac{4}{3}, \sqrt{5}\right)$

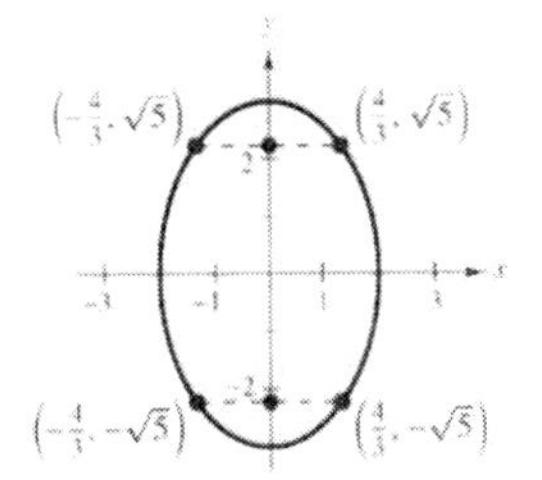

67. True. If $e \approx 1$, then the ellipse is elongated, not circular.

69. $\dfrac{x^2}{328} + \dfrac{y^2}{327} = 1,\ a = 328,\ b = 327$

$$a^2 = b^2 + c^2$$
$$328 = 327 + c^2$$
$$1 = c^2$$
$$\sqrt{1} = c$$
$$1 = c$$

The eccentricity is $e = \dfrac{c}{a} = \dfrac{1}{\sqrt{328}} \approx 0.055.$

Because the eccentricity is close to 0, the ellipse is nearly circular.

71. (a) The length of the string is $2a$.

(b) The path is an ellipse because the sum of the distances from the two thumbtacks is always the length of the string, that is, it is constant.

73. Center: $(6,\ 2)$

Foci: $(2,\ 2),\ (10,\ 2) \Rightarrow c = 4$

$(a+c)+(a-c) = 2a = 36 \Rightarrow a = 18$

$b^2 = a^2 - b^2 \Rightarrow b = \sqrt{18^2 - 16} = \sqrt{308}$

Horizontal major axis

$$\frac{(x-6)^2}{324} + \frac{(y-2)^2}{308} = 1$$

75. Arithmetic: $d = -11$

77. Geometric: $r = 2$

79. $\displaystyle\sum_{n=0}^{6} 3^n = 1093$

Section 9.3

1. hyperbola

3. $Ax^2 + Bxy + Cy^2 + Dx + Ey + F = 0$

5. The equations $\dfrac{(x-h)^2}{a^2} - \dfrac{(y-k)^2}{b^2} = 1$ and $\dfrac{x^2}{a^2} - \dfrac{y^2}{b^2} = 1$ have a horizontal transverse axis because the x-term is the positive fraction in standard form. Similarly, the equations $\dfrac{(y-k)^2}{a^2} - \dfrac{(x-h)^2}{b^2} = 1$ and $\dfrac{y^2}{a^2} - \dfrac{x^2}{b^2} = 1$ have a vertical transverse axis because the y-term is the positive fraction in standard form.

7. Center: $(0,\ 0)$

$a = 3,\ b = 5,\ c = \sqrt{34}$

Vertical transverse axis

Matches graph (b).

9. Center: $(1,\ 0)$

$a = 4,\ b = 2$

Horizontal transverse axis

Matches graph (a).

11. $x^2 - y^2 = 1$

$a = 1,\ b = 1,\ c = \sqrt{2}$

Center: $(0,\ 0)$; Vertices: $(\pm 1,\ 0)$

Foci: $(\pm\sqrt{2},\ 0)$

Asymptotes: $y = \pm\dfrac{b}{a}x = \pm x$

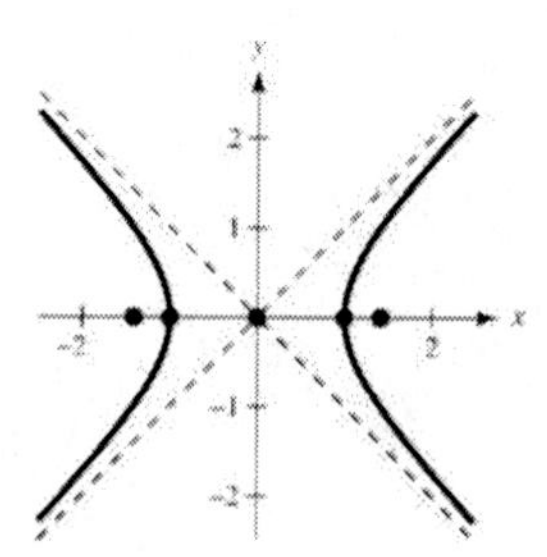

13. $\dfrac{y^2}{1} - \dfrac{x^2}{4} = 1$

$a = 1,\ b = 2,\ c = \sqrt{5}$

Center: $(0,\ 0)$

Vertices: $(0,\ \pm 1)$; Foci: $(0,\ \pm\sqrt{5})$

Asymptotes: $y = \pm\dfrac{a}{b}x = \pm\dfrac{1}{2}x$

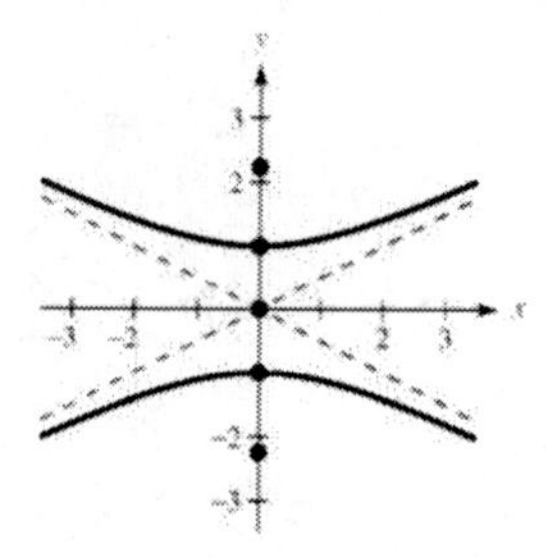

15. $\dfrac{y^2}{25} - \dfrac{x^2}{81} = 1$

$a = 5,\ b = 9,\ c = \sqrt{a^2 + b^2} = \sqrt{106}$

Center: $(0,\ 0)$

Vertices: $(0,\ \pm 5)$

Foci: $\left(0,\ \pm\sqrt{106}\right)$

Asymptotes: $y = \pm\dfrac{a}{b}x = \pm\dfrac{5}{9}x$

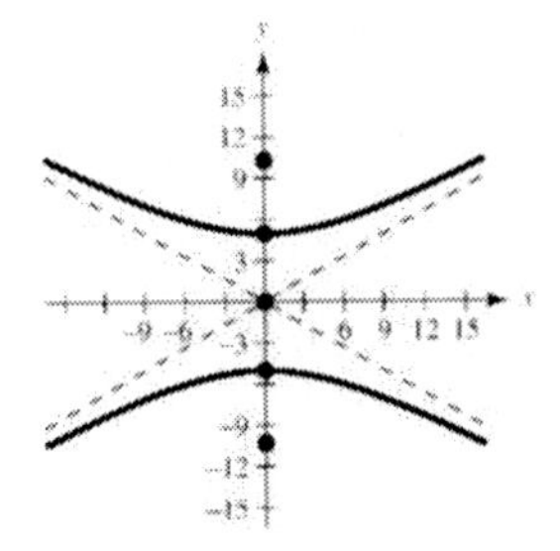

17. $\dfrac{(x-1)^2}{4} - \dfrac{(y+2)^2}{1} = 1$

$a = 2,\ b = 1,\ c = \sqrt{5}$

Center: $(1,\ -2)$

Vertices: $(-1,\ -2),\ (3,\ -2)$

Foci: $\left(1 \pm \sqrt{5},\ -2\right)$

Asymptotes: $y = k \pm \dfrac{b}{a}(x-h) = -2 \pm \dfrac{1}{2}(x-1)$

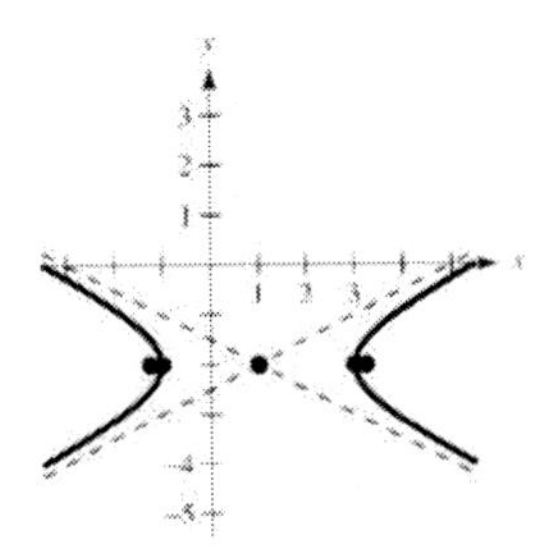

19. $\dfrac{(y+5)^2}{\frac{1}{9}} - \dfrac{(x-1)^2}{\frac{1}{4}} = 1$

$a = \dfrac{1}{3},\ b = \dfrac{1}{2},\ c = \sqrt{\dfrac{1}{9} + \dfrac{1}{4}} = \dfrac{\sqrt{13}}{6}$

Center: $(1,\ -5)$

Vertices: $\left(1,\ -5 \pm \dfrac{1}{3}\right) \Rightarrow \left(1,\ -\dfrac{16}{3}\right),\ \left(1,\ -\dfrac{14}{3}\right)$

Foci: $\left(1,\ -5 \pm \dfrac{\sqrt{13}}{6}\right)$

Asymptotes: $y = k \pm \dfrac{a}{b}(x-h)$

$y = -5 \pm \dfrac{2}{3}(x-1)$

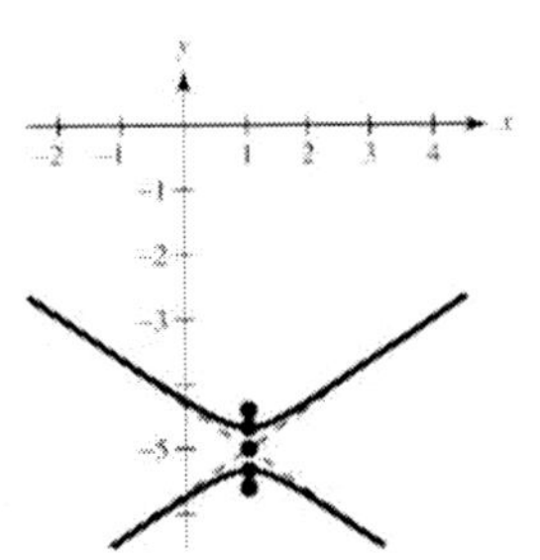

21. (a) $4x^2 - 9y^2 = 36$

$\dfrac{x^2}{9} - \dfrac{y^2}{4} = 1$

(b) Center: $(0,\ 0)$

$a = 3,\ b = 2,\ c = \sqrt{9+4} = \sqrt{13}$

Vertices: $(\pm 3,\ 0)$

Foci: $\left(\pm\sqrt{13},\ 0\right)$

Asymptotes: $y = \pm\dfrac{b}{a}x = \pm\dfrac{2}{3}x$

(c)

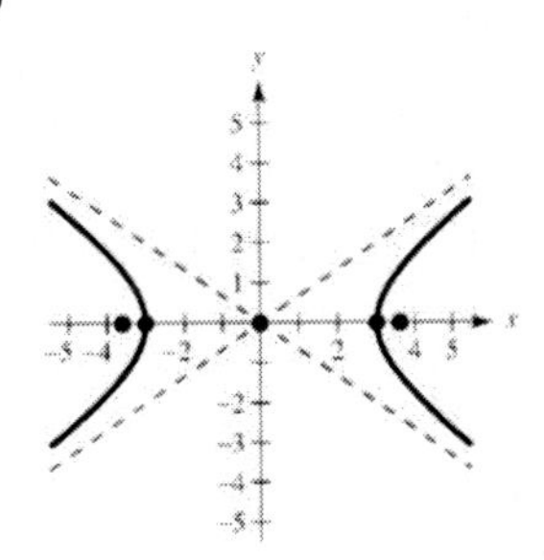

23. (a) $2x^2 - 3y^2 = 6$

$\dfrac{x^2}{3} - \dfrac{y^2}{2} = 1$

(b) $a = \sqrt{3},\ b = \sqrt{2},\ c = \sqrt{5}$

Center: $(0,\ 0)$

Vertices: $\left(\pm\sqrt{3},\ 0\right)$

Foci: $\left(\pm\sqrt{5},\ 0\right)$

Asymptotes: $y = \pm\sqrt{\dfrac{2}{3}}x$

$= \pm\dfrac{\sqrt{6}}{3}x$

(c)

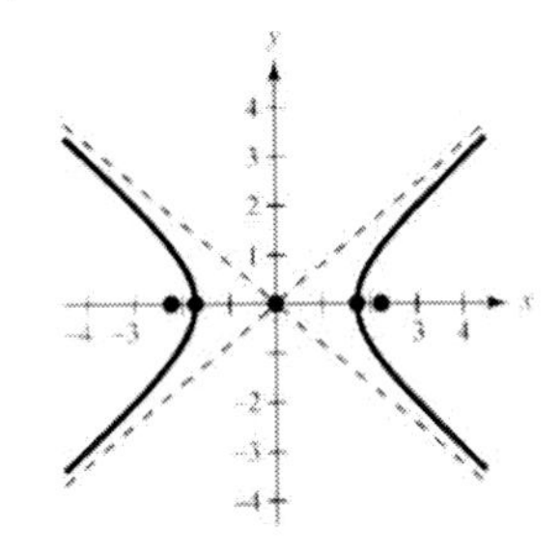

25. (a)
$$9x^2 - y^2 - 36x - 6y + 18 = 0$$
$$9(x^2 - 4x + 4) - (y^2 + 6y + 9) = -18 + 36 - 9$$
$$\frac{(x-2)^2}{1} - \frac{(y+3)^2}{9} = 1$$

(b) $a = 1,\ b = 3,\ c = \sqrt{10}$

Center: $(2,\ -3)$

Vertices: $(1,\ -3),\ (3,\ -3)$

Foci: $(2 \pm \sqrt{10},\ -3)$

Asymptotes: $y = k \pm \frac{b}{a}(x - h) = -3 \pm 3(x - 2)$

(c)

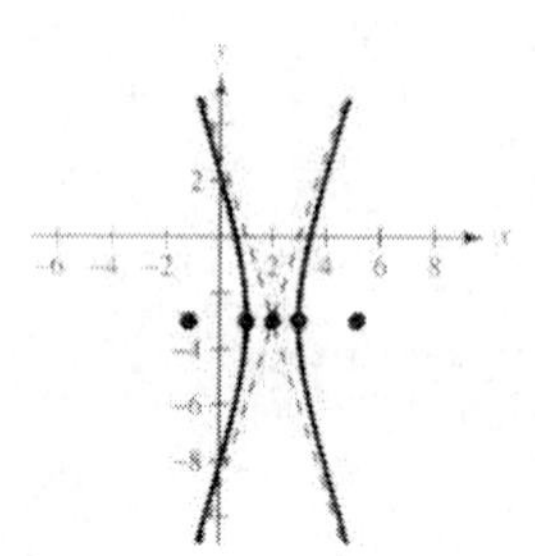

27. (a)
$$x^2 - 9y^2 + 2x - 54y - 80 = 0$$
$$(x^2 + 2x + 1) - 9(y^2 + 6y + 9) = 80 + 1 - 81$$
$$(x+1)^2 - 9(y+3)^2 = 0$$
$$y + 3 = \pm\tfrac{1}{3}(x + 1)$$

(b) Degenerate conic is two lines intersecting at $(-1,\ -3)$.

(c)

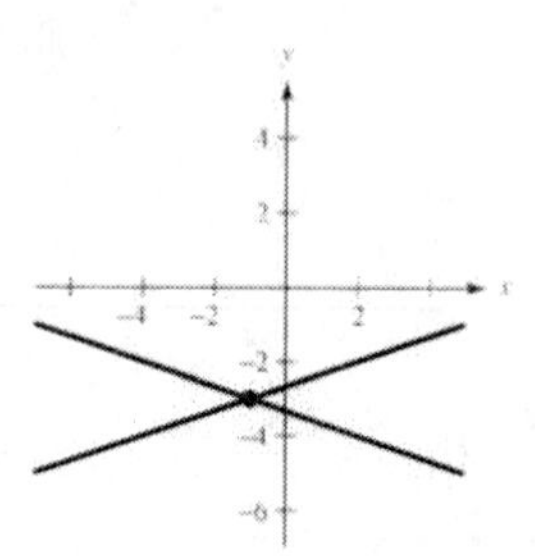

29. (a)
$$9y^2 - x^2 + 2x + 54y + 62 = 0$$
$$9(y^2 + 6y + 9) - (x^2 - 2x + 1) = -62 - 1 + 81$$
$$\frac{(y+3)^2}{2} - \frac{(x-1)^2}{18} = 1$$

(b) $a = \sqrt{2},\ b = 3\sqrt{2},\ c = 2\sqrt{5}$

Center: $(1,\ -3)$

Vertices: $(1,\ -3 \pm \sqrt{2})$

Foci: $(1,\ -3 \pm 2\sqrt{5})$

Asymptotes: $y = k \pm \frac{a}{b}(x - h) = -3 \pm \frac{1}{3}(x - 1)$

(c)

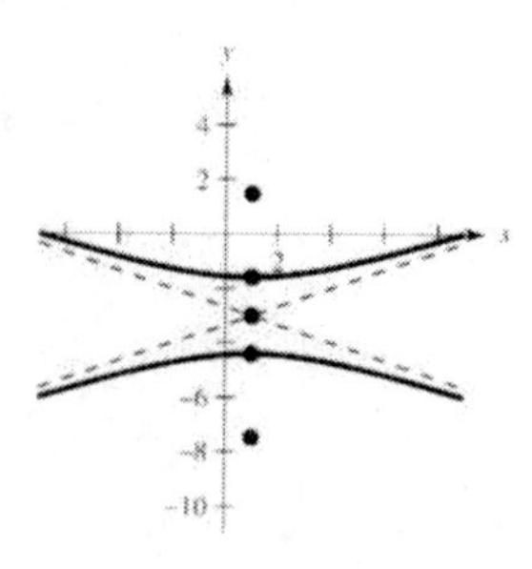

31. Vertices: $(0,\ \pm 2) \Rightarrow a = 2$

Foci: $(0,\ \pm 4) \Rightarrow c = 4$

$b^2 = c^2 - a^2 = 16 - 4 = 12$

Center: $(0,\ 0)$

$$\frac{y^2}{a^2} - \frac{x^2}{b^2} = 1$$
$$\frac{y^2}{4} - \frac{x^2}{12} = 1$$

33. Vertices: $(\pm 1,\ 0) \Rightarrow a = 1$

Asymptotes:

$y = \pm 5x \quad \Rightarrow \frac{b}{a} = 5$

$\Rightarrow b = 5$

Center: $(0,\ 0)$

$$\frac{x^2}{a^2} - \frac{y^2}{b^2} = 1$$
$$\frac{x^2}{1} - \frac{y^2}{25} = 1$$

35. Foci: $(0,\ \pm 8) \Rightarrow c = 8$

Asymptotes: $y = \pm 4x \Rightarrow \frac{a}{b} = 4 \Rightarrow a = 4b$

Center: $(0,\ 0) = (h,\ k)$

$c^2 = a^2 + b^2 \Rightarrow 64 = 16b^2 + b^2$

$\frac{64}{17} = b^2 \Rightarrow a^2 = \frac{1024}{17}$

$$\frac{y^2}{a^2} - \frac{x^2}{b^2} = 1$$
$$\frac{y^2}{1024/17} - \frac{x^2}{64/17} = 1$$
$$\frac{17y^2}{1024} - \frac{17x^2}{64} = 1$$

37. Vertices: $(2, 0), (6, 0) \Rightarrow a = 2$

Foci: $(0, 0), (8, 0) \Rightarrow c = 4$

$b^2 = c^2 - a^2 \Rightarrow 16 - 4 = 12$

Center: $(4, 0) = (h, k)$

$$\frac{(x-h)^2}{a^2} - \frac{(y-k)^2}{b^2} = 1$$

$$\frac{(x-4)^2}{4} - \frac{y^2}{12} = 1$$

39. Vertices: $(4, 1), (4, 9) \Rightarrow a = 4$

Foci: $(4, 0), (4, 10) \Rightarrow c = 5$

$b^2 = c^2 - a^2 \Rightarrow 25 - 16 = 9$

Center: $(4, 5) = (h, k)$

$$\frac{(y-k)^2}{a^2} - \frac{(x-h)^2}{b^2} = 1$$

$$\frac{(y-5)^2}{16} - \frac{(x-4)^2}{9} = 1$$

41. Vertices: $(2, 3), (2, -3) \Rightarrow a = 3$

Solution point: $(0, 5)$

Center: $(2, 0) = (h, k)$

$$\frac{(y-k)^2}{a^2} - \frac{(x-h)^2}{b^2} = 1$$

$$\frac{y^2}{9} - \frac{(x-2)^2}{b^2} = 1$$

$$b^2 = \frac{9(x-2)^2}{y^2 - 9}$$

$$= \frac{9(-2)^2}{25 - 9} = \frac{36}{16} = \frac{9}{4}$$

$$\frac{y^2}{9} - \frac{(x-2)^2}{\frac{9}{4}} = 1$$

43. Vertices: $(0, 4), (0, 0)$

Center: $(0, 2), a = 2$

$$\frac{(y-k)^2}{a^2} - \frac{(x-h)^2}{b^2} = 1$$

$$\frac{(y-2)^2}{4} - \frac{x^2}{b^2} = 1$$

Passes through $(\sqrt{5}, -1)$

$$\frac{(-1-2)^2}{4} - \frac{5}{b^2} = 1$$

$$\frac{9}{4} - 1 = \frac{5}{b^2}$$

$$b^2 = 4 \Rightarrow b = 2$$

$$\frac{(y-2)^2}{4} - \frac{x^2}{4} = 1$$

45. Vertices: $(1, 2), (3, 2) \Rightarrow a = 1$

Center: $(2, 2) = (h, k)$

Asymptotes: $y = x, y = 4 - x$

$\frac{b}{a} = 1 \Rightarrow b = 1$

$$\frac{(x-h)^2}{a^2} - \frac{(y-k)^2}{b^2} = 1$$

$$\frac{(x-2)^2}{1} - \frac{(y-2)^2}{1} = 1$$

47. Vertices: $(0, 2), (6, 2) \Rightarrow a = 3$

Asymptotes: $y = \frac{2}{3}x, y = 4 - \frac{2}{3}x$

$\frac{b}{a} = \frac{2}{3} \Rightarrow b = 2$

Center: $(3, 2) = (h, k)$

$$\frac{(x-h)^2}{a^2} - \frac{(y-k)^2}{b^2} = 1$$

$$\frac{(x-3)^2}{9} - \frac{(y-2)^2}{4} = 1$$

49. F_1: Friend's location $(-10{,}560, 0)$

F_2: Your location $(10{,}560, 0)$

$P(x, y)$: Location of lightning strike

$(1100)(18) = 19{,}800$

$$\frac{x^2}{a^2} - \frac{y^2}{b^2} = 1$$

$c = 10{,}560,\ a = \frac{19{,}800}{2} = 9900 \Rightarrow a^2 = 98{,}010{,}000$

$b^2 = c^2 - a^2 = 13{,}503{,}600$

$$\frac{x^2}{98{,}010{,}000} - \frac{y^2}{13{,}503{,}600} = 1$$

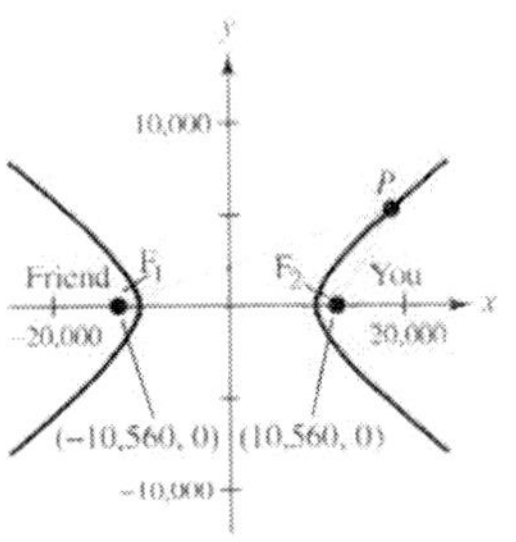

51. (a) $\dfrac{x^2}{a^2} - \dfrac{y^2}{b^2} = 1$

$a = 1$; $(2, 9)$ is on the curve, so

$$\frac{4}{1} - \frac{81}{b^2} = 1 \Rightarrow \frac{81}{b^2} = 3$$

$$\Rightarrow b^2 = \frac{81}{3} \Rightarrow b = 3\sqrt{3}.$$

$$\frac{x^2}{1} - \frac{y^2}{27} = 1, \ -9 \le y \le 9$$

(b) Because each unit is $\frac{1}{2}$ foot, 4 inches is $\frac{2}{3}$ of a unit. The base is 9 units from the origin, so

$$y = 9 - \frac{2}{3} = 8\frac{1}{3}.$$

When $y = \dfrac{25}{3}$,

$$x^2 = 1 + \frac{(25/3)^2}{27} \Rightarrow x \approx 1.88998.$$

So the width is $2x \approx 3.779956$ units, or 22.68 inches, or 1.88998 feet.

53. Center: $(0, 0)$

Focus: $(24, 0)$

$$b^2 = c^2 - a^2 = 24^2 - a^2 = 576 - a^2$$

$$\frac{x^2}{a^2} - \frac{y^2}{576 - a^2} = 1$$

$$\frac{24^2}{a^2} - \frac{24^2}{576 - a^2} = 1$$

$$\frac{576}{a^2} - \frac{576}{576 - a^2} = 1$$

$$576(576 - a^2) - 576a^2 = a^2(576 - a^2)$$

$$a^4 - 1728a^2 + 331{,}776 = 0$$

$$a \approx \pm 38.83 \text{ or } a \approx \pm 14.83$$

Since $a < c$ and $c = 24$, we choose $a = 14.83$. The vertex is approximately at $(14.83, 0)$.

[**Note:** By the Quadratic Formula, the exact value of a is $a = 12(\sqrt{5} - 1)$.]

55. $9x^2 + 4y^2 - 18x + 16y - 119 = 0$

$A = 9, C = 4$

$AC = 36 > 0$, Ellipse

57. $16x^2 - 9y^2 + 32x + 54y - 209 = 0$

$A = 16, C = -9$

$AC = 16(-9) < 0$, Hyperbola

59. $y^2 + 12x + 4y + 28 = 0$

$C = 1, A = 0$

$AC = 0$, Parabola

61. $x^2 + y^2 + 2x - 6y = 0$

$A = C = 1$, Circle

63. $x^2 - 6x - 2y + 7 = 0$

$A = 1, C = 0, D = -6, E = -2, F = 7$

$AC = 0 \Rightarrow$ Parabola

65. $xy + 4 = 0$

$B^2 - 4AC = 1 \Rightarrow$ The graph is a hyperbola.

$$\cot 2\theta = \frac{A - C}{B} = 0 \Rightarrow \theta \Rightarrow 45^\circ$$

Matches graph (e).

67. $-2x^2 + 3xy + 2y^2 + 3 = 0$

$$B^2 - 4AC = (3)^2 - 4(-2)(2)$$

$= 25 \Rightarrow$ The graph is a hyperbola.

$$\cot 2\theta = \frac{A - C}{B} = -\frac{4}{3} \Rightarrow \theta \approx -18.43^\circ$$

Matches graph (f).

69. $3x^2 + 2xy + y^2 - 10 = 0$

$$B^2 - 4AC = (2)^2 - 4(3)(1) = -8$$

The graph is an ellipse or circle.

$$\cot 2\theta = \frac{A - C}{B} = 1 \Rightarrow \theta = 22.5^\circ$$

Matches graph (d).

71. $\theta = 90^\circ$; Point: $(0, 3)$

$x = x'\cos\theta - y'\sin\theta$	$y = x'\sin\theta + y'\cos\theta$
$0 = x'\cos 90^\circ - y'\sin 90^\circ$	$3 = x'\sin 90^\circ + y'\cos 90^\circ$
$0 = y'$	$3 = x'$

Thus, $(x', y') = (3, 0)$.

73. $xy + 1 = 0, A = 0, B = 1, C = 0$

$$\cot 2\theta = \frac{A - C}{B} = 0 \Rightarrow 2\theta = \frac{\pi}{2} \Rightarrow \theta = \frac{\pi}{4}$$

$$x = x'\cos\frac{\pi}{4} - y'\sin\frac{\pi}{4} = x'\left(\frac{\sqrt{2}}{2}\right) - y'\left(\frac{\sqrt{2}}{2}\right) = \frac{x' - y'}{\sqrt{2}}$$

$$y = x'\sin\frac{\pi}{4} + y'\cos\frac{\pi}{4} = x'\left(\frac{\sqrt{2}}{2}\right) + y'\left(\frac{\sqrt{2}}{2}\right) = \frac{x' + y'}{\sqrt{2}}$$

$$xy + 1 = 0$$

$$\left(\frac{x' - y'}{\sqrt{2}}\right)\left(\frac{x' + y'}{\sqrt{2}}\right) + 1 = 0$$

$$\frac{(y')^2}{2} - \frac{(x')^2}{2} = 1, \text{ Hyperbola}$$

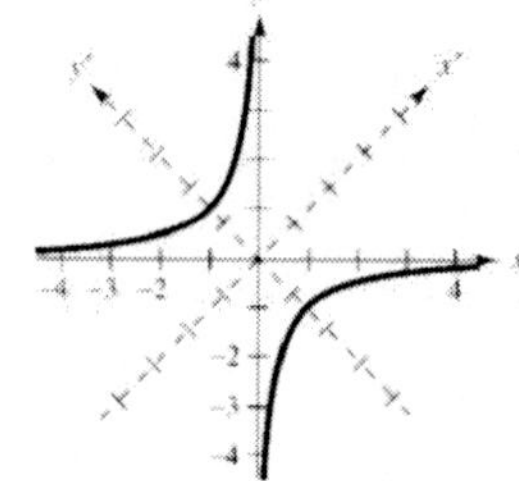

$$x = x'\cos\frac{\pi}{4} - y'\sin\frac{\pi}{4} = x'\left(\frac{\sqrt{2}}{2}\right) - y'\left(\frac{\sqrt{2}}{2}\right) = \frac{x' - y'}{\sqrt{2}}$$

$$y = x'\sin\frac{\pi}{4} + y'\cos\frac{\pi}{4} = x'\left(\frac{\sqrt{2}}{2}\right) + y'\left(\frac{\sqrt{2}}{2}\right) = \frac{x' + y'}{\sqrt{2}}$$

$$xy - 2 = 0$$

$$\left(\frac{x' - y'}{\sqrt{2}}\right)\left(\frac{x' + y'}{\sqrt{2}}\right) - 2 = 0$$

$$\frac{(x')^2 - (y')^2}{2} = 2$$

$$\frac{(x')^2}{4} - \frac{(y')^2}{4} = 1, \text{ Hyperbola}$$

75. $x^2 - 4xy + y^2 + 1 = 0$

$A = 1,\ B = -4,\ C = 1$

$$\cot 2\theta = \frac{A - C}{B} = 0 \Rightarrow 2\theta = \frac{\pi}{2} \Rightarrow \theta = \frac{\pi}{4}$$

$$x = x'\cos\frac{\pi}{4} - y'\sin\frac{\pi}{4} = x'\left(\frac{\sqrt{2}}{2}\right) - y'\left(\frac{\sqrt{2}}{2}\right) = \frac{\sqrt{2}}{2}(x' - y')$$

$$y = x'\sin\frac{\pi}{4} + y'\cos\frac{\pi}{4} = x'\left(\frac{\sqrt{2}}{2}\right) + y'\left(\frac{\sqrt{2}}{2}\right) = \frac{\sqrt{2}}{2}(x' + y')$$

$$x^2 - 4xy + y^2 + 1 = 0$$

$$\left[\frac{\sqrt{2}}{2}(x' - y')\right]^2 - 4\left[\frac{\sqrt{2}}{2}(x' - y')\frac{\sqrt{2}}{2}(x' + y')\right] + \left[\frac{\sqrt{2}}{2}(x' + y')\right]^2 + 1 = 0$$

$$\frac{1}{2}(x')^2 - x'y' + \frac{1}{2}(y')^2 - 2\left[(x')^2 - (y')^2\right] + \frac{1}{2}(x')^2 + x'y' + \frac{1}{2}(y')^2 + 1 = 0$$

$$-(x')^2 + 3(y')^2 = -1$$

$$(x')^2 - \frac{(y')^2}{1/3} = 1; \text{ Hyperbola}$$

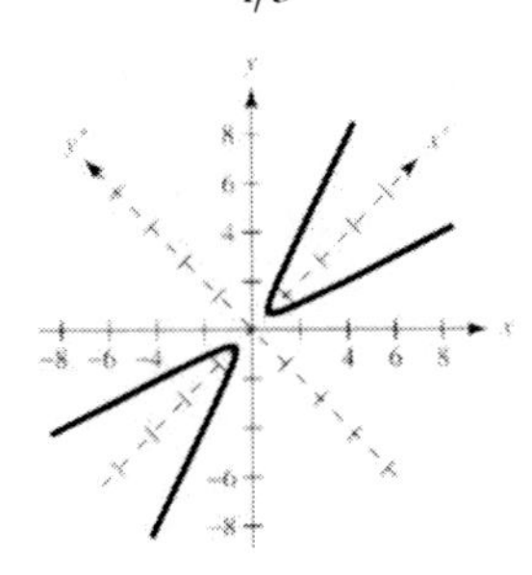

77. $5x^2 - 6xy + 5y^2 - 12 = 0$

$A = 5,\ B = -6,\ C = 5$

$$\cot 2\theta = \frac{A - C}{B} = 0 \Rightarrow 2\theta = \frac{\pi}{2} \Rightarrow \theta = \frac{\pi}{4}$$

$$x = x'\cos\frac{\pi}{4} - y'\sin\frac{\pi}{4} = \frac{\sqrt{2}}{2}(x' - y')$$

$$y = x'\sin\frac{\pi}{4} + y'\cos\frac{\pi}{4} = \frac{\sqrt{2}}{2}(x' + y')$$

$$5x^2 - 6xy + 5y^2 - 12 = 0$$

$$5\left[\frac{\sqrt{2}}{2}(x'-y')\right]^2-6\left[\frac{\sqrt{2}}{2}(x'-y')\frac{\sqrt{2}}{2}(x'+y')\right]+5\left[\frac{\sqrt{2}}{2}(x'+y')\right]^2=12$$

$$\frac{5}{2}(x')^2-5x'y'+\frac{5}{2}(y')^2-3(x')^2+3(y')^2+5x'y'+\frac{5}{2}(y')^2=12$$

$$2(x')^2+8(y')^2=12$$

$$\frac{(x')^2}{6}+\frac{(y')^2}{3/2}=1, \text{ Ellipse}$$

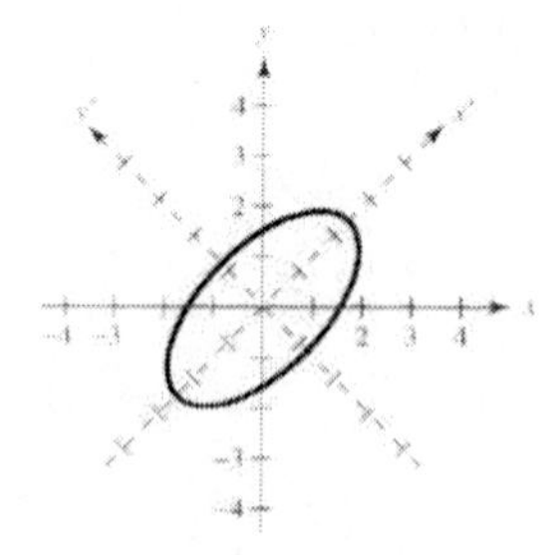

79. $3x^2-2\sqrt{3}xy+y^2+2x+2\sqrt{3}y=0$

$A=3,\ B=-2\sqrt{3},\ C=1$

$\cot 2\theta=\frac{A-C}{B}=-\frac{1}{\sqrt{3}}\Rightarrow\theta=60°$

$$x=x'\cos 60°-y'\sin 60°=x'\left(\frac{1}{2}\right)-y'\left(\frac{\sqrt{3}}{2}\right)=\frac{x'-\sqrt{3}y'}{2}$$

$$y=x'\sin\theta+y'\cos\theta=\frac{\sqrt{3}x'+y'}{2}\quad 3x^2-2\sqrt{3}xy+y^2+2x+2\sqrt{3}y=0$$

$$3\left(\frac{x'-\sqrt{3}y'}{2}\right)^2-2\sqrt{3}\left(\frac{x'-\sqrt{3}y'}{2}\right)\left(\frac{\sqrt{3}x'+y'}{2}\right)+\left(\frac{\sqrt{3}x'+y'}{2}\right)^2+2\left(\frac{x'-\sqrt{3}y}{2}\right)+2\sqrt{3}\left(\frac{\sqrt{3}x'+y'}{2}\right)=0$$

$$\frac{3(x')^2}{4}-\frac{6\sqrt{3}x'y'}{4}+\frac{9(y')^2}{4}-\frac{6(x')^2}{4}+\frac{4\sqrt{3}x'y'}{4}+\frac{6(y')^2}{4}+\frac{3(x')^2}{4}+\frac{2\sqrt{3}x'y'}{4}+\frac{(y')^2}{4}+x'-\sqrt{3}y'+3x'+\sqrt{3y'}=0$$

$$4(y')^2+4x'=0$$

$$x'=-(y')^2, \text{ Parabola}$$

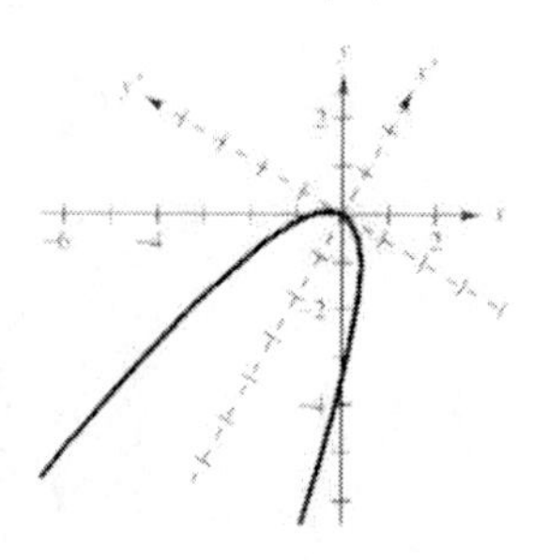

81. $x^2 + 3xy + y^2 = 20$

$\cot 2\theta = \frac{A-C}{B} = \frac{1-1}{3} = 0 \Rightarrow \frac{\pi}{4} = 45°$

Solve for y in terms of x:

$$y^2 + 3xy = 20 - x^2$$

$$y^2 + 3xy + \frac{9x^2}{4} = 20 - x^2 + \frac{9x^2}{4}$$

$$\left(y + \frac{3}{2}x\right)^2 = 20 + \frac{5x^2}{4} = \frac{80 + 5x^2}{4}$$

$$y = -\frac{3}{2}x \pm \frac{\sqrt{80 + 5x^2}}{2}$$

Graph $y_1 = -\frac{3x}{2} + \frac{\sqrt{80 + 5x^2}}{2}$ and

$y_2 = -\frac{3x}{2} - \frac{\sqrt{80 + 5x^2}}{2}$.

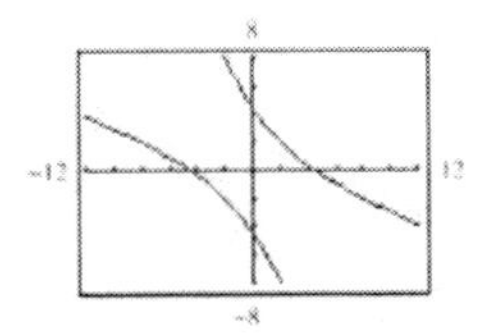

83. $17x^2 + 32xy - 7y^2 = 75$

$\cot 2\theta = \frac{A-C}{B} = \frac{17+7}{32} = \frac{24}{32} = \frac{3}{4} \Rightarrow \theta \approx 26.57°$

Solve for y in terms of x by completing the square.

$$-7y^2 + 32xy = -17x^2 + 75$$

$$y^2 - \frac{32}{7}xy = \frac{17}{7}x^2 - \frac{75}{7}$$

$$y^2 - \frac{32}{7}xy + \frac{256}{49}x^2 = \frac{119}{49}x^2 - \frac{525}{49} + \frac{256}{49}x^2$$

$$\left(y - \frac{16}{7}x\right)^2 = \frac{375x^2 - 525}{49}$$

$$y = \frac{16}{7}x \pm \sqrt{\frac{375x^2 - 525}{49}}$$

$$y = \frac{16x \pm 5\sqrt{15x^2 - 21}}{7}$$

Graph $y_1 = \frac{16x + 5\sqrt{15x^2 - 21}}{7}$ and

$y_2 = \frac{16x - 5\sqrt{15x^2 - 21}}{7}$.

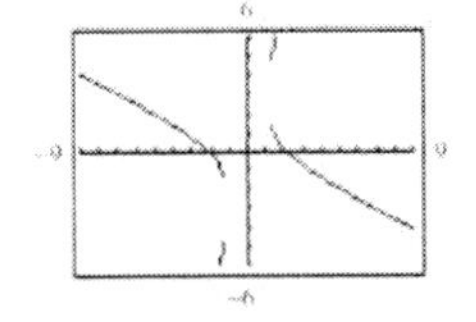

85. $y^2 - 16x^2 = 0$

$$y^2 = 16x^2$$

$$y = \pm 4x$$

Two interesting lines

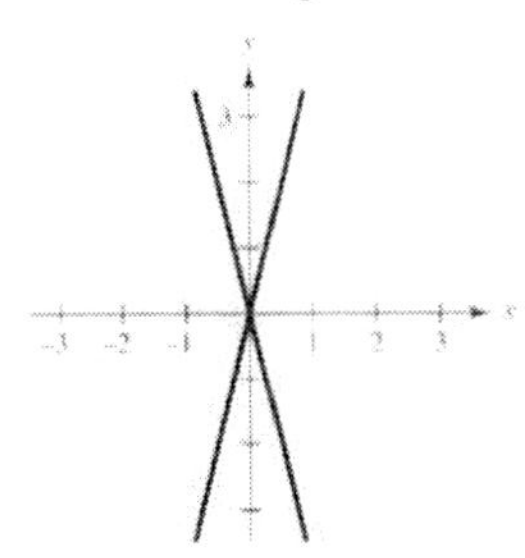

87. $x^2 + 2xy + y^2 - 1 = 0$

$$(x + y)^2 - 1 = 0$$

$$(x + y)^2 = 1$$

$$x + y = \pm 1$$

$$y = -x \pm 1$$

Two parallel lines

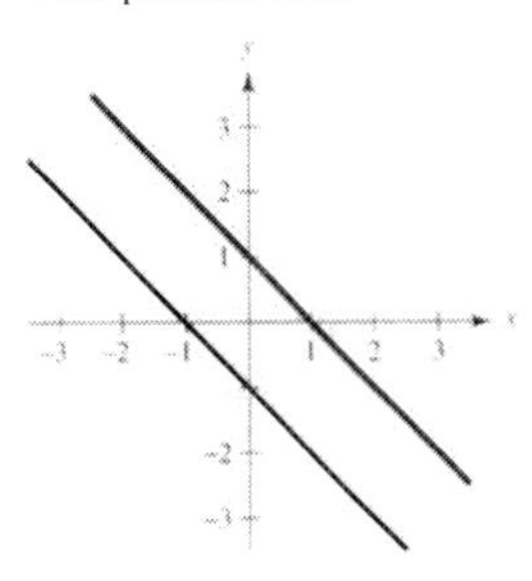

89. True. $e = \frac{c}{a} = \frac{\sqrt{a^2 + b^2}}{a}$

91. False. For example,

$x^2 - y^2 - 2x + 2y = 0$

$(x-1)^2 - (y-1)^2 = 0$

is the graph of two intersecting lines.

93. False. The coefficients of the x^2- and y^2-terms become A' and C', respectively.

95. The asymptotes pass through the corners of the rectangle.

97. Center: $(6, 2)$

Horizontal transverse axis

Foci at $(2, 2)$ and $(10, 2) \Rightarrow c = 4$

$(c + a) - (c - a) = 6 \Rightarrow a = 3$

$b^2 = c^2 - a^2 = 16 - 9 = 7$

$$\frac{(x-6)^2}{9} - \frac{(y-2)^2}{7} = 1$$

99. If $A = C \neq 0$, then by completing the square you obtain a circle.
If $A = 0$ and $C \neq 0$, then $Cy^2 + Dx + Ey + F = 0$ is a parabola (complete the square). Same for $A \neq 0$ and $C = 0$.
If $AC > 0$, then both A and C are positive (or both negative). By completing the square you obtain an ellipse.
If $AC < 0$, then A and C have opposite signs. You obtain a hyperbola.

101. $(x^3 - 3x^2) - (6 - 2x - 4x^2) = x^3 + x^2 + 2x - 6$

103.
$$\begin{array}{r|rrrr} -2 & 1 & 0 & -3 & 4 \\ & & -2 & 4 & -2 \\ \hline & 1 & -2 & 1 & 2 \end{array}$$
$$\frac{x^2 - 3x + 4}{x + 2} = x^2 - 2x + 1 + \frac{2}{x + 2}$$

105. $x^3 - 16x = x(x^2 - 16) = x(x - 4)(x + 4)$

107. $2x^3 - 24x^2 + 72x = 2x(x^2 - 12x + 36)$
$= 2x(x - 6)^2$

109. $16x^3 + 54 = 2(8x^3 + 27)$
$= 2(2x + 3)(4x^2 - 6x + 9)$

111. $f(x) = |x + 3|$

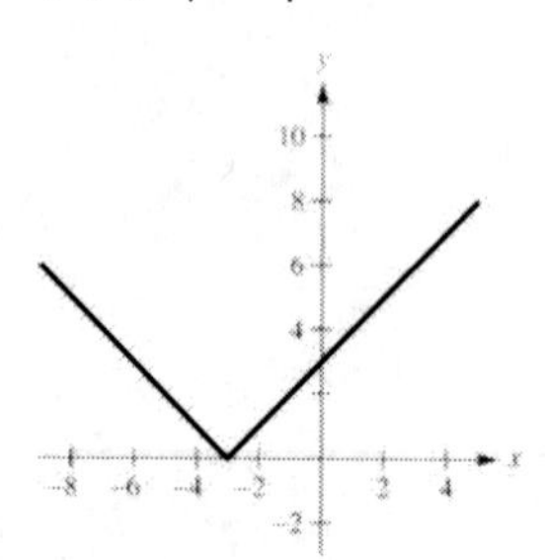

113. $g(x) = \sqrt{4 - x^2}$

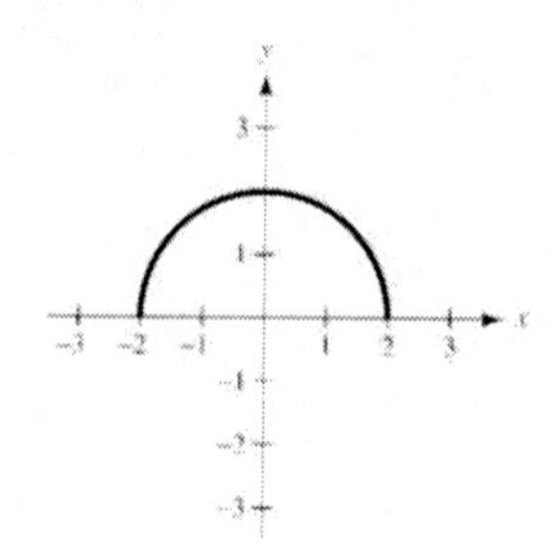

115. $h(t) = -(t - 2)^3 + 3$

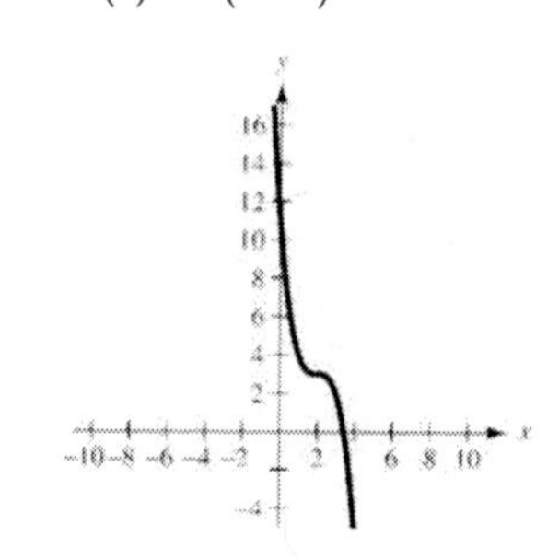

117. $f(t) = [\![t - 5]\!] + 1$

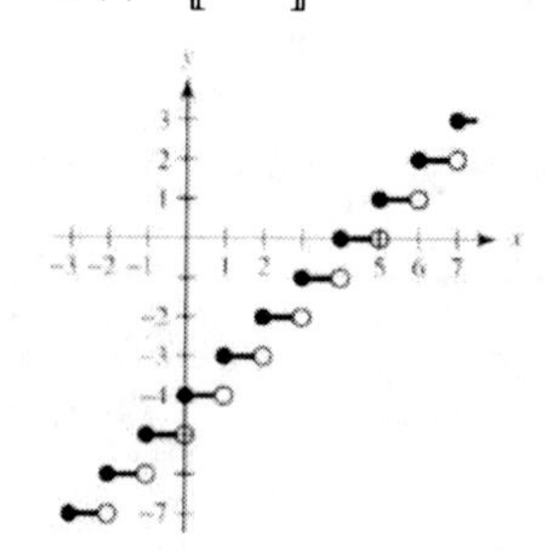

Section 9.4

1. plane curve, parametric equations, parameter

3. Given a set of parametric equations, the process of finding a corresponding rectangular equation is called eliminating the parameter.

5. $x = t$
$y = t + 2$
$y = x + 2$, line
Matches (c).

7. $x = \sqrt{t}$
$y = t$
$y = x^2$, parabola, $x \geq 0$
Matches (b).

9. $x = \sqrt{t}$, $y = 2 - t$

(a)

t	0	1	2	3	4
x	0	1	$\sqrt{2}$	$\sqrt{3}$	2
y	2	1	0	−1	−2

(b)

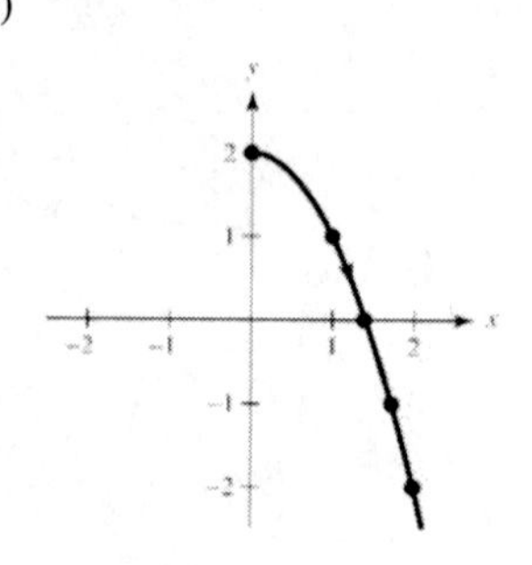

The curve starts at $(0, 2)$ and moves along the right half of the parabola.

(c)

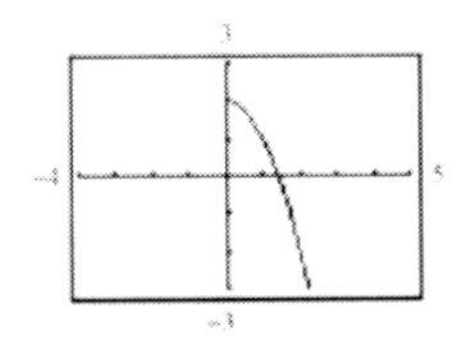

(d) $y = 2 - t = 2 - x^2$, parabola
The graph is an entire parabola rather than just the right half.

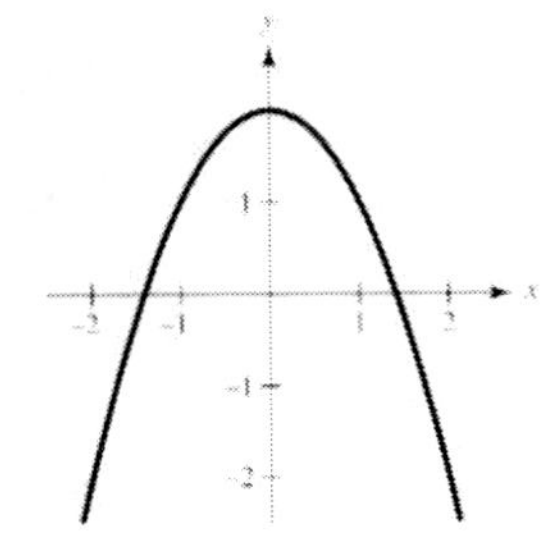

11. The graph opens upward, contains $(1, 0)$, and is oriented left to right. Matches (b).

13. $x = t, y = -4t$

$y = -4x$

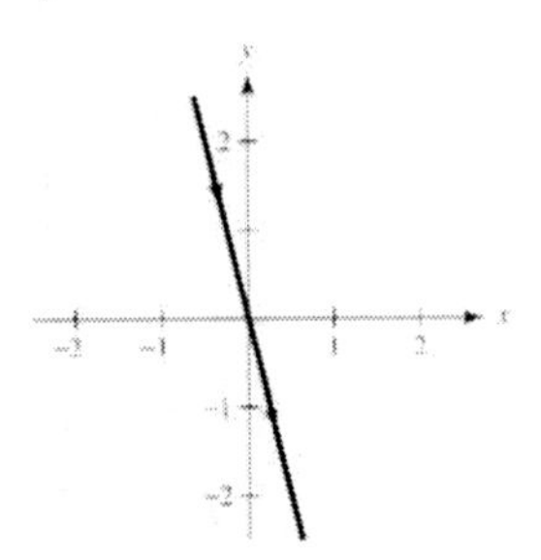

15. $x = 3t - 3,\ y = 2t + 1$

$t = \dfrac{x+3}{3}$

$y = 2\left(\dfrac{x+3}{3}\right) + 1$

$y = \dfrac{2}{3}x + 3$

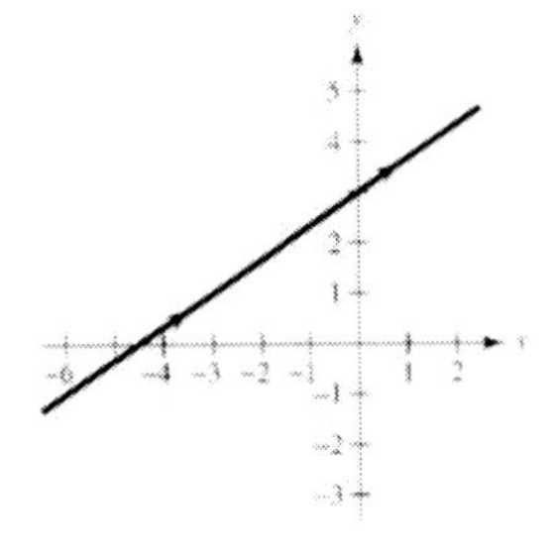

17. $x = \dfrac{1}{4}t,\ y = t^2$

$y = (4x)^2$

$y = 16x^2$

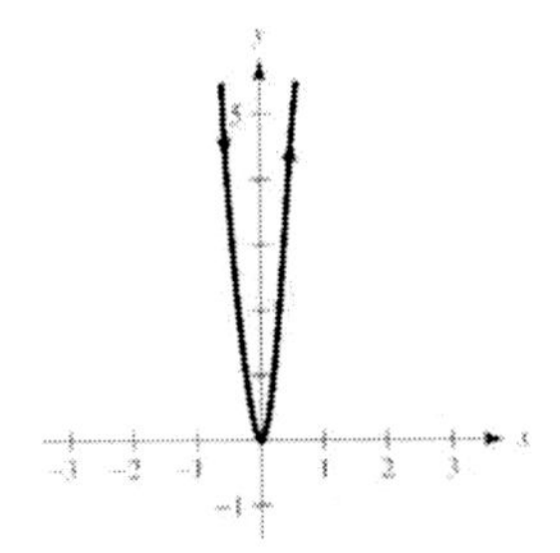

19. $x = t + 2,\ y = t^2$

$t = x - 2$

$y = (x - 2)^2$

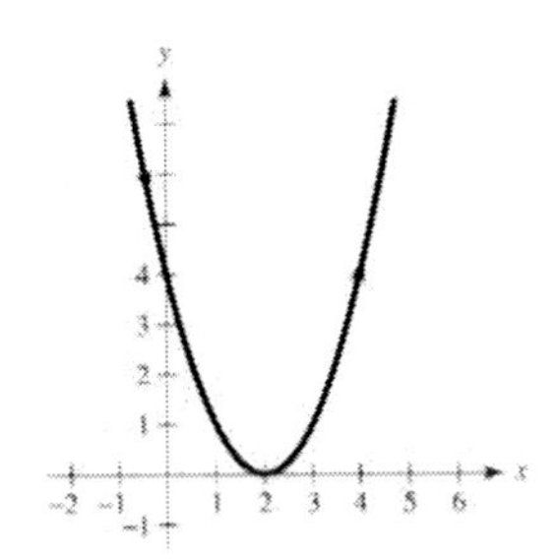

21. $x = 2t,\ y = |t - 2|$

$t = \dfrac{x}{2} \Rightarrow y = |t - 2|$

$= \left|\dfrac{x}{2} - 2\right|$

$= \dfrac{1}{2}|x - 4|$

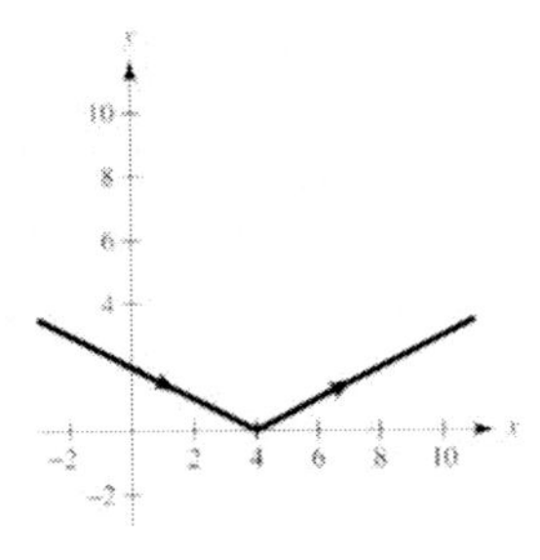

23. $x = 2\cos\theta,\ y = 3\sin\theta$

$$\left(\frac{x}{2}\right)^2 = \cos^2\theta,\ \left(\frac{y}{3}\right)^2 = \sin^2\theta$$

$$\frac{x^2}{4} + \frac{y^2}{9} = \cos^2\theta + \sin^2\theta = 1$$

$$\frac{x^2}{4} + \frac{y^2}{9} = 1, \text{ ellipse}$$

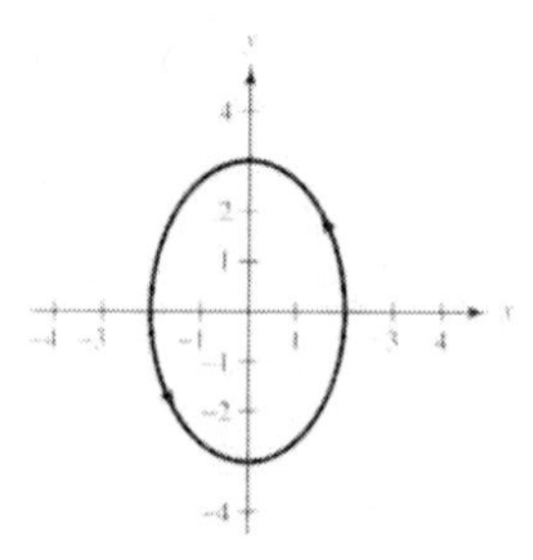

25. $x = e^{-t} \Rightarrow \dfrac{1}{x} = e^t$

$$y = e^{3t} \Rightarrow y = \left(e^t\right)^3$$

$$y = \left(\frac{1}{x}\right)^3$$

$$y = \frac{1}{x^3},\ x > 0,\ y > 0$$

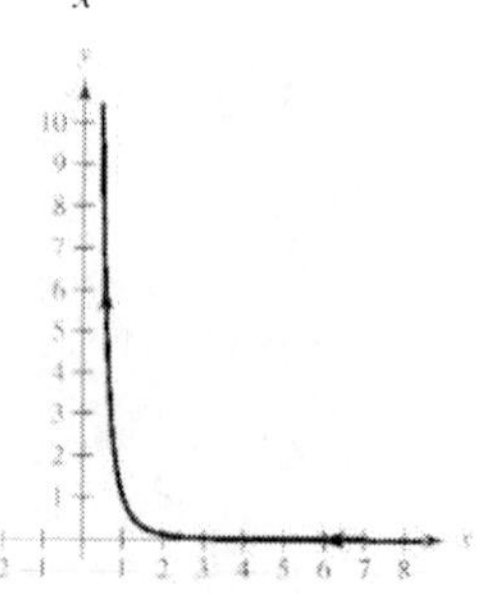

27. $x = t^3 \Rightarrow x^{1/3} = t$

$$y = 3\ln t \Rightarrow y = \ln t^3$$

$$y = \ln\left(x^{1/3}\right)^3$$

$$y = \ln x$$

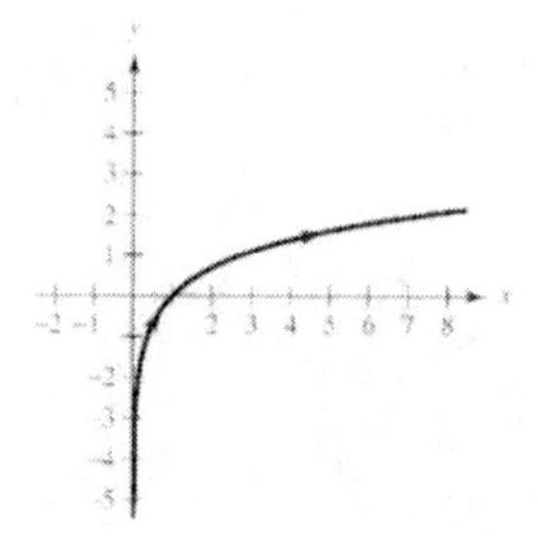

29. $x = 4 + 3\cos\theta,\ y = -2 + \sin\theta$

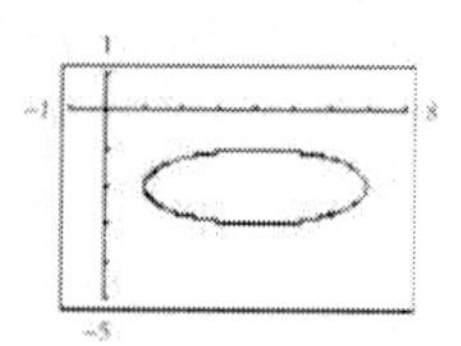

31. $x = 4\sec\theta,\ y = 2\tan\theta$

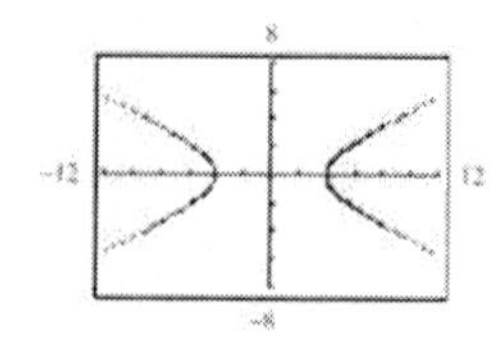

33. $x = \dfrac{t}{2}$

$$y = \ln\left(t^2 + 1\right)$$

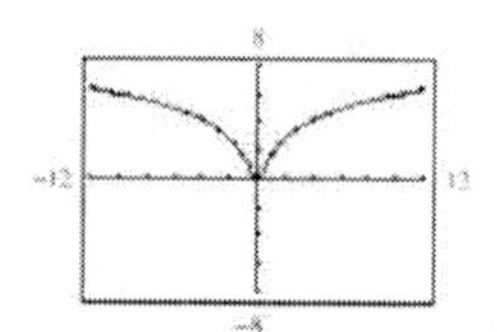

35. Each curve represents a portion of the line $y = 2x + 1$.

(a) $x = t$

$y = 2t + 1$

Domain: $-\infty < x < \infty$

Orientation: Left to right

(b) $x = \cos\theta$

$y = 2\cos\theta + 1$

Domain: $-1 \le x \le 1$

Orientation: Depends on θ

(c) $x = e^{-t}$

$y = 2e^{-t} + 1$

Domain: $0 < x < \infty$

Orientation: Right to left

(d) $x = e^t$

$y = 2e^t + 1$

Domain: $0 < x < \infty$

Orientation: Left to right

37. $t = \dfrac{(x - x_1)}{(x_2 - x_1)}$

$$y = y_1 + \left(\frac{x - x_1}{x_2 - x_1}\right)(y_2 - y_1)$$

$$y - y_1 = \left(\frac{y_2 - y_1}{x_2 - x_1}\right)(x - x_1)$$

39. $x = h + a\cos\theta$

$y = k + b\sin\theta$

$\dfrac{x-h}{a} = \cos\theta,\ \dfrac{y-k}{b} = \sin\theta$

$\dfrac{(x-h)^2}{a^2} + \dfrac{(y-k)^2}{b^2} = 1$

41. Line through (x_1, y_1) and (x_2, y_2):

$x = x_1 + t(x_2 - x_1)$

$y = y_1 + t(y_2 - y_1)$

Line though $(3, 1)$ and $(-2, 6)$:

$x = 3 + t(-2-3) \Rightarrow x = 3 - 5t$

$y = 1 + t(6-1) \Rightarrow y = 1 + 5t$

43. Ellipse: $x = h + a\cos\theta$

$y = k + b\sin\theta$

Vertices: $(\pm 5, 0)$; Foci: $(\pm 3, 0)$

The center is the midpoint between the vertices or $(0, 0)$. In addition, the distance from the center to a vertex is $a = 5$. Also, the distance from the center to a focus is $c = 3$.

$a^2 = b^2 + c^2$

$25 = b^2 + 9$

$16 = b^2$

$4 = b$

$x = 0 + 5\cos\theta \Rightarrow x = 5\cos\theta$

$y = 0 + 4\sin\theta \Rightarrow y = 4\sin\theta$

45. $y = 5x - 3$

(a) Let $t = x$: $x = t$

$y = 5t - 3$

(b) Let $t = 2 - x$: $x = 2 - t$

$y = 5(2-t) - 3 \Rightarrow y = -5t + 7$

47. $y = \dfrac{1}{x}$

(a) Let $t = x$: $x = t$

$y = \dfrac{1}{t}$

(b) Let $t = 2 - x$: $x = 2 - t$

$y = \dfrac{1}{2-t}$

49. $y = 6x^2 - 5$

(a) Let $t = x$: $x = t$

$y = 6t^2 - 5$

(b) Let $t = 2 - x$: $x = 2 - t$

$y = 6(2-t)^2 - 5$

$= 6t^2 - 24t + 19$

51. $y = e^x$

(a) Let $t = x$: $x = t$

$y = e^t$

(b) Let $t = 2 - x$: $x = 2 - t$

$y = e^{2-t}$

53. $x = 2\cot\theta,\ y = 2\sin^2\theta$

55. $x = \theta + \sin\theta,\ y = 1 - \cos\theta$

57. Matches (b).

59. Matches (d).

61. $x = (v_0\cos\theta)t$ and $y = h + (v_0\sin\theta)t - 16t^2$

(a) $v_0 = 100$ mph ≈ 146.67 ft/sec, $h = 3$

$x = (146.67\cos\theta)t$

$y = 3 + (146.67\sin\theta)t - 16t^2$

(b) $x = (146.67\cos 15^\circ)t$

$y = 3 + (146.67\sin 15^\circ)t - 16t^2$

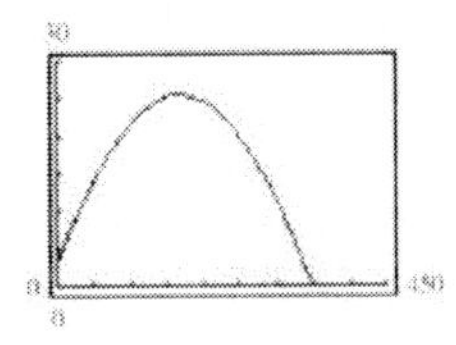

The ball travels about 347 feet. So, it is not a home run.

(c) $x = (146.67\cos 23^\circ)t$

$y = 3 + (146.67\sin 15^\circ)t - 16t^2$

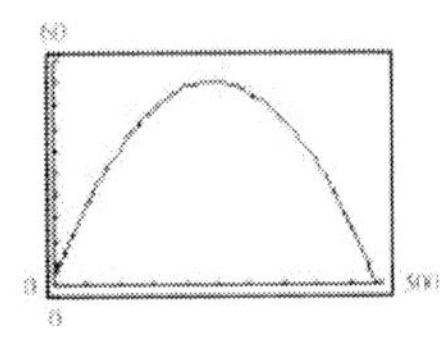

The ball travels about 490 feet. So, it is a home run.

(d) Trying several values for θ, you'll find that a minimum angle of about 19° is required for the hit to be a home run.

63. True

$x = t$ first set

$y = t^2 + 1 = x^2 + 1$

$x = 3t$ second set

$y = 9t^2 + 1 = (3t)^2 + 1 = x^2 + 1$

65. False. For example, $x = t^2$ and $y = t$ does not represent y as a function of x.

67. The graph is the same, but the orientation is reversed.

69. $f(-x) = \dfrac{4(-x)^2}{(-x)^2 + 1} = \dfrac{4x^2}{x^2 + 1} = f(x)$

Symmetric about the y-axis
Even function

71. $y = e^x \neq e^{-x}$; $e^{-x} \neq -e^x$

No symmetry
Neither even nor odd

Section 9.5

1. pole

3. The rectangular coordinates (x, y) and the polar coordinates (r, θ) are related by the following equations:

$x = r\cos\theta$ and $\tan\theta = \dfrac{y}{x}$

$y = r\sin\theta$ $\quad r^2 = x^2 + y^2$

5. Polar coordinates: $\left(4, \dfrac{\pi}{2}\right)$

$x = 4\cos\left(\dfrac{\pi}{2}\right) = 0$

$y = 4\sin\left(\dfrac{\pi}{2}\right) = 4$

Rectangular coordinates: $(0, 4)$

7. Polar coordinates: $\left(-1, \dfrac{5\pi}{4}\right)$

$x = -1\cos\left(\dfrac{5\pi}{4}\right) = \dfrac{\sqrt{2}}{2}$

$y = -1\sin\left(\dfrac{5\pi}{4}\right) = \dfrac{\sqrt{2}}{2}$

Rectangular coordinates: $\left(\dfrac{\sqrt{2}}{2}, \dfrac{\sqrt{2}}{2}\right)$

9.

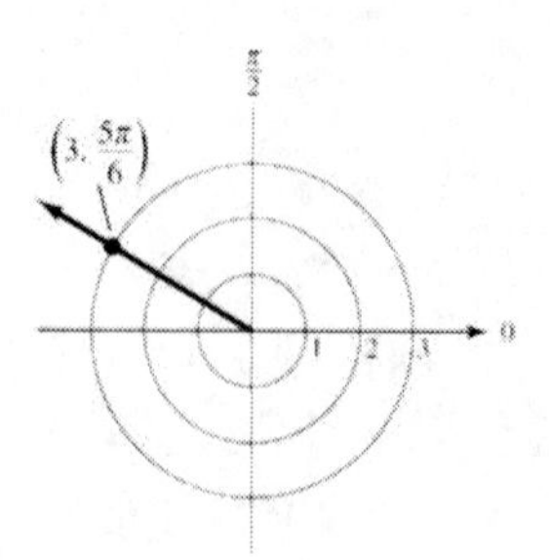

Three additional representations:

$\left(3, \dfrac{5\pi}{6} - 2\pi\right) = \left(3, -\dfrac{7\pi}{6}\right)$

$\left(-3, \dfrac{5\pi}{6} + \pi\right) = \left(-3, \dfrac{11\pi}{6}\right)$

$\left(-3, \dfrac{5\pi}{6} - \pi\right) = \left(-3, -\dfrac{\pi}{6}\right)$

11.

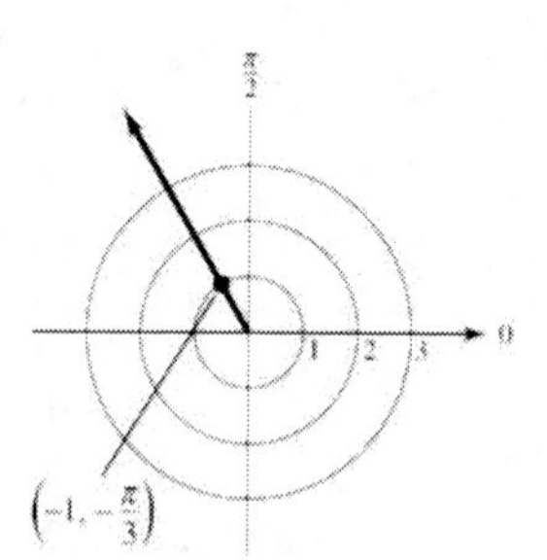

Three additional representations:

$\left(-1, -\dfrac{\pi}{3} + 2\pi\right) = \left(-1, \dfrac{5\pi}{3}\right)$

$\left(1, -\dfrac{\pi}{3} + \pi\right) = \left(1, \dfrac{2\pi}{3}\right)$

$\left(1, -\dfrac{\pi}{3} - \pi\right) = \left(1, -\dfrac{4\pi}{3}\right)$

13.

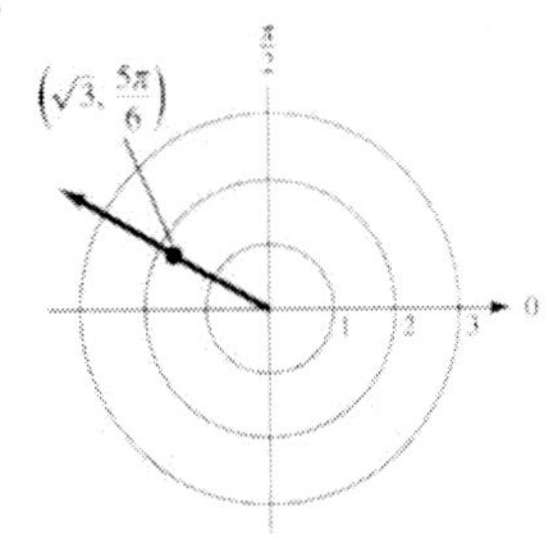

Three additional representations:

$$\left(\sqrt{3}, \frac{5\pi}{6} - 2\pi\right) = \left(\sqrt{3}, -\frac{7\pi}{6}\right)$$

$$\left(-\sqrt{3}, \frac{5\pi}{6} + \pi\right) = \left(-\sqrt{3}, \frac{11\pi}{6}\right)$$

$$\left(-\sqrt{3}, \frac{5\pi}{6} - \pi\right) = \left(-\sqrt{3}, -\frac{\pi}{6}\right)$$

15.

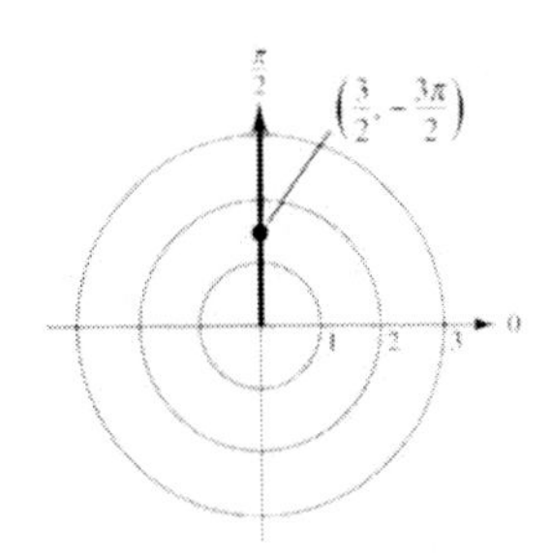

Three additional representations:

$$\left(\frac{3}{2}, -\frac{3\pi}{2} + 2\pi\right) = \left(\frac{3}{2}, \frac{\pi}{2}\right)$$

$$\left(-\frac{3}{2}, \frac{-3\pi}{2} + 3\pi\right) = \left(-\frac{3}{2}, \frac{3\pi}{2}\right)$$

$$\left(-\frac{3}{2}, \frac{-3\pi}{2} + \pi\right) = \left(-\frac{3}{2}, -\frac{\pi}{2}\right)$$

17. Polar coordinates: $\left(4, -\frac{\pi}{3}\right)$

$$x = 4\cos\left(-\frac{\pi}{3}\right) = 2$$

$$y = 4\sin\left(-\frac{\pi}{3}\right) = -2\sqrt{3}$$

Rectangular coordinates: $\left(2, -2\sqrt{3}\right)$

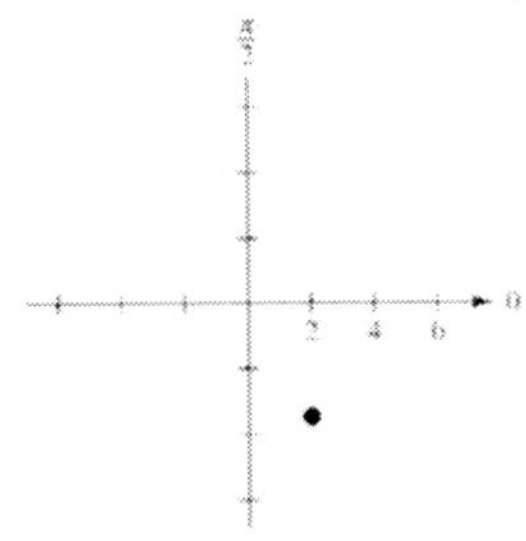

19. Polar coordinates: $\left(-1, \frac{-3\pi}{4}\right)$

$$x = -1\cos\left(\frac{-3\pi}{4}\right) = \frac{\sqrt{2}}{2}$$

$$y = -1\sin\left(\frac{-3\pi}{4}\right) = \frac{\sqrt{2}}{2}$$

Rectangular coordinates: $\left(\frac{\sqrt{2}}{2}, \frac{\sqrt{2}}{2}\right)$

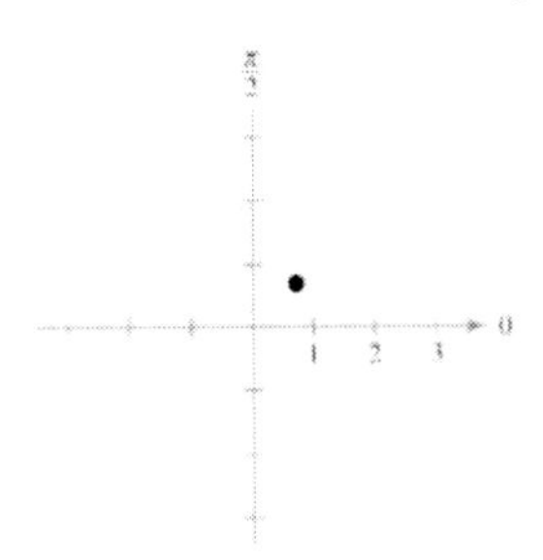

21. Polar coordinates: $\left(0, -\frac{7\pi}{6}\right)$ (origin!)

$$x = 0\cos\left(-\frac{7\pi}{6}\right) = 0$$

$$y = 0\sin\left(-\frac{7\pi}{6}\right) = 0$$

Rectangular coordinates: $(0, 0)$

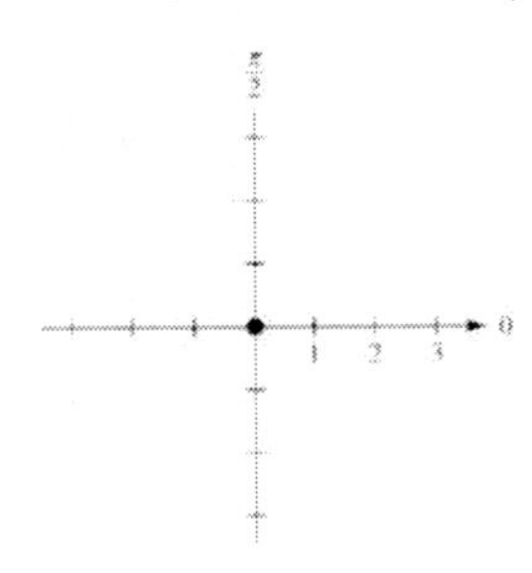

23. Polar coordinates: $\left(\sqrt{2}, 2.36\right)$

$$x = \sqrt{2}\cos(2.36) \approx -1.004$$

$$y = \sqrt{2}\sin(2.36) \approx 0.996$$

Rectangular coordinates: $(-1.004, 0.996)$

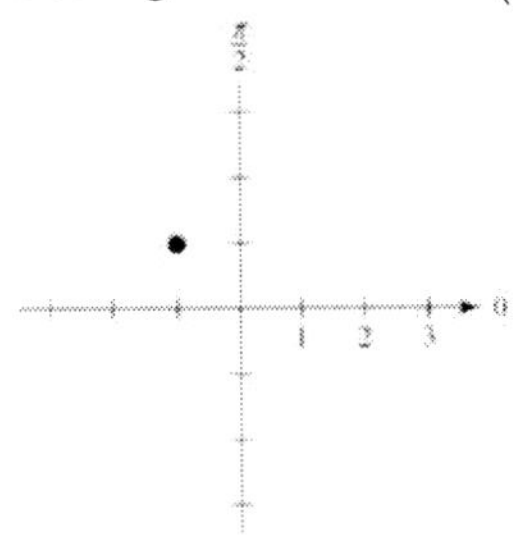

25. Polar coordinates: $(-5, -2.36)$

$x = -5\cos(-2.36) \approx 3.549$

$y = -5\sin(-2.36) \approx 3.522$

Rectangular coordinates: $(3.549, 3.522)$

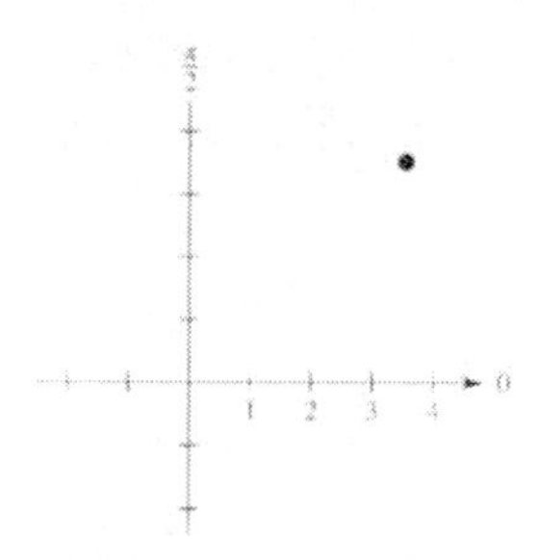

27. $(r, \theta) = \left(2, \dfrac{2\pi}{9}\right) \Rightarrow (x, y) = (1.53, 1.29)$

29. $(r, \theta) = (-4.5, 1.3) \Rightarrow (x, y) = (-1.20, -4.34)$

31. $(r, \theta) = (2.5, 1.58) \Rightarrow (x, y) = (-0.02, 2.50)$

33. $(r, \theta) = (-4.1, -0.5) \Rightarrow (x, y) = (-3.60, 1.97)$

35. Rectangular coordinates: $(-7, 0)$

$r = 7,\ \tan\theta = 0,\ \theta = 0$

Polar coordinates: $(7, \pi), (-7, 0)$

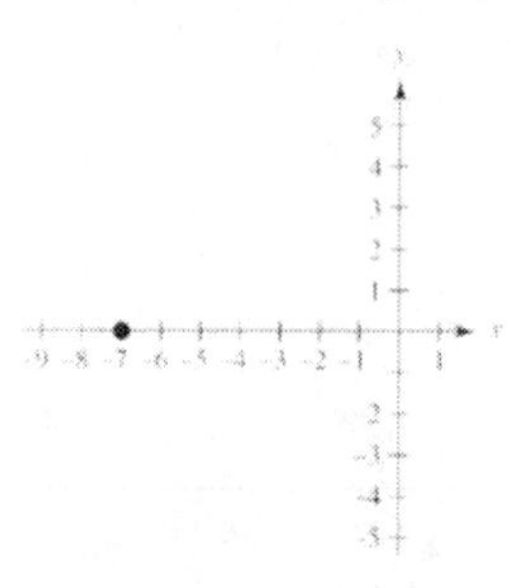

37. Rectangular coordinates: $(1, 1)$

$r = \sqrt{2},\ \tan\theta = 1,\ \theta = \pi/4$

Polar coordinates: $\left(\sqrt{2}, \dfrac{\pi}{4}\right), \left(-\sqrt{2}, \dfrac{5\pi}{4}\right)$

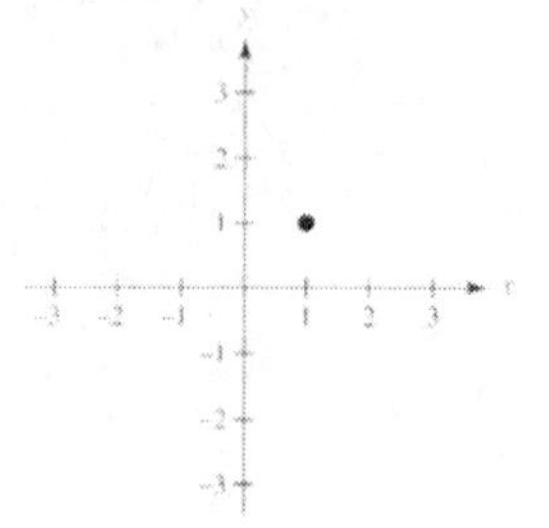

39. Rectangular coordinates: $(-3, 4)$

$r = \sqrt{(-3)^2 + (4)^2} = \sqrt{9+16} = 5$

$\tan\theta = \dfrac{4}{-3},\ \theta \approx 2.214$

Polar coordinates: $(5, 2.214), (-5, 5.356)$

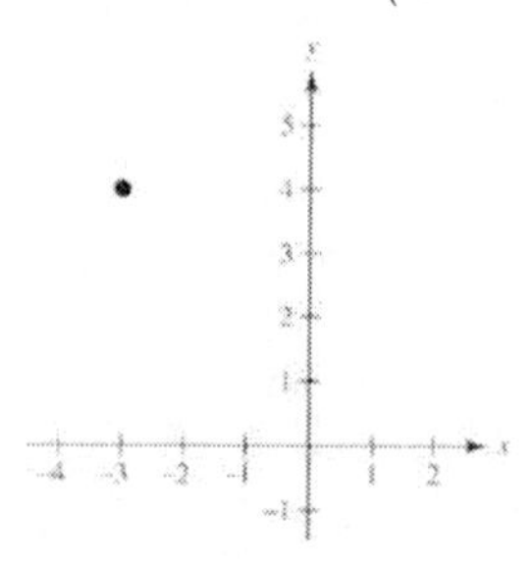

41. Rectangular coordinates: $(-\sqrt{3}, -\sqrt{3})$

$r = \sqrt{3+3} = \sqrt{6},\ \tan\theta = 1,\ \theta = \dfrac{\pi}{4}$

Polar coordinates: $\left(\sqrt{6}, \dfrac{5\pi}{4}\right), \left(-\sqrt{6}, \dfrac{\pi}{4}\right)$

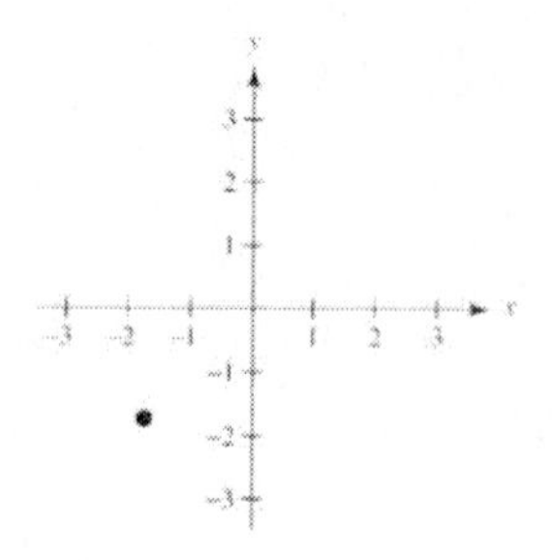

43. $(x, y) = (6, 9)$

$r = \sqrt{6^2 + 9^2} = \sqrt{117} \approx 10.8$

$\tan\theta = \dfrac{9}{6} = \dfrac{3}{2} \Rightarrow \theta \approx 0.983$

Polar coordinates: $(10.8, 0.983), (-10.8, 4.124)$

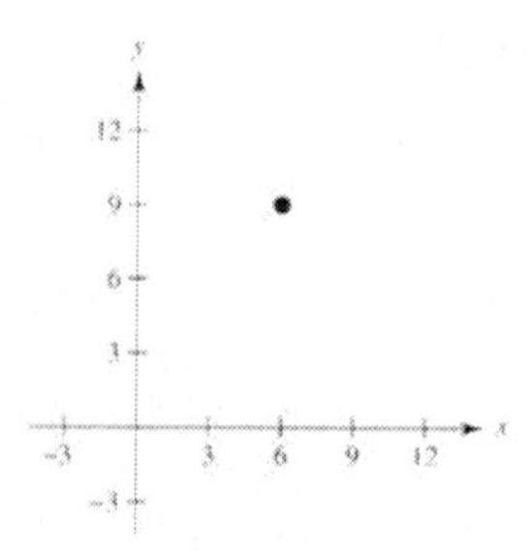

45. $(x, y) = (3, -2) \Rightarrow r = \sqrt{3^2 + (-2)^2} = \sqrt{3} \approx 3.61$

$\theta = \arctan\left(-\frac{2}{3}\right) \approx -0.588$

$(r, \theta) \approx (3.61, -0.588)$

47. $(x, y) = (\sqrt{3}, 2) \Rightarrow \quad r = \sqrt{3 + 2^2} = \sqrt{7} \approx 2.65$

$$\theta = \arctan\left(\frac{2}{\sqrt{3}}\right) \approx 0.857$$

$(r, \theta) \approx (2.65, 0.857)$

49. $(x, y) = \left(\frac{5}{2}, \frac{4}{3}\right) \Rightarrow r = \sqrt{\left(\frac{5}{2}\right)^2 + \left(\frac{4}{3}\right)^2}$

$$= \frac{17}{6} \approx 2.83$$

$$\theta = \arctan\left(\frac{\frac{4}{3}}{\frac{5}{2}}\right)$$

$$\approx 0.490$$

$(r, \theta) \approx (2.83, 0.490)$

51. $x^2 + y^2 = 9$

$$r^2 = 9$$
$$r = 3$$

53. $y = 4$

$$r\sin\theta = 4$$
$$r = 4\csc\theta$$

55. $x = 8$

$$r\cos\theta = 8$$
$$r = 8\sec\theta$$

57. $3x - y + 2 = 0$

$$3(r\cos\theta) - (r\sin\theta) + 2 = 0$$
$$r(3\cos\theta - \sin\theta) = -2$$
$$r = -\frac{2}{3\cos\theta - \sin\theta}$$
$$= \frac{2}{\sin\theta - 3\cos\theta}$$

59. $xy = 4$

$$(r\cos\theta)(r\sin\theta) = 4$$
$$r^2\cos\theta\sin\theta = 4$$
$$r^2(2\cos\theta\sin\theta) = 8$$
$$r^2\sin 2\theta = 8$$
$$r^2 = 8\csc 2\theta$$

61. $(x^2 + y^2)^2 = 9(x^2 - y^2)$

$$(r^2)^2 = 9(r^2\cos^2\theta - r^2\sin^2\theta)$$
$$r^2 = 9(\cos^2\theta - \sin^2\theta)$$
$$r^2 = 9\cos(2\theta)$$

63. $x^2 + y^2 - 6x = 0$

$$r^2 - 6r\cos\theta = 0$$
$$r^2 = 6r\cos\theta$$
$$r = 6\cos\theta$$

65. $x^2 + y^2 - 2ax = 0$

$$r^2 - 2ar\cos\theta = 0$$
$$r(r - 2a\cos\theta) = 0$$
$$r = 2a\cos\theta$$

67. $y^2 = x^3$

$$(r\sin\theta)^2 = (r\cos\theta)^3$$
$$\sin^2\theta = r\cos^3\theta$$
$$r = \frac{\sin^2\theta}{\cos^3\theta}$$
$$= \tan^2\theta\sec\theta$$

69. $r = 4\sin\theta$

$$r^2 = 4r\sin\theta$$
$$r^2 - 4r\sin\theta = 0$$
$$x^2 + y^2 - 4y = 0$$

71. $\theta = \frac{2\pi}{3}$

$$\tan\theta = \tan\frac{2\pi}{3}$$
$$\frac{y}{x} = -\sqrt{3}$$
$$y = -\sqrt{3}x$$

73. $\theta = \frac{5\pi}{6}$

$$\tan\theta = \tan\frac{5\pi}{6} = \frac{y}{x}$$
$$\frac{-\sqrt{3}}{3} = \frac{y}{x}$$
$$y = -\frac{\sqrt{3}}{3}x$$

75. $\theta = \frac{\pi}{2}$, vertical line

$$x = 0$$

77. $r = 4$

$$r^2 = 16$$
$$x^2 + y^2 = 16$$

79. $r = -3\csc\theta$

$$r\sin\theta = -3$$
$$y = -3$$

81.
$$r^2 = \cos\theta$$
$$r^3 = r\cos\theta$$
$$\left(x^2+y^2\right)^{3/2} = x$$
$$x^2+y^2 = x^{2/3}$$
$$\left(x^2+y^2\right)^3 = x^2$$

83.
$$r = 2\sin 3\theta$$
$$r = 2\left(3\sin\theta - 4\sin^3\theta\right)$$
$$r^4 = 6r^3\sin\theta - 8r^3\sin^3\theta$$
$$\left(x^2+y^2\right)^2 = 6\left(x^2+y^2\right)y - 8y^3$$
$$\left(x^2+y^2\right)^2 = 6x^2y - 2y^3$$

85.
$$r = \frac{1}{1-\cos\theta}$$
$$r - r\cos\theta = 1$$
$$\sqrt{x^2+y^2} - x = 1$$
$$x^2+y^2 = 1+2x+x^2$$
$$y^2 = 2x+1$$

87.
$$r = \frac{6}{2-3\sin\theta}$$
$$r\left(2-3\sin\theta\right) = 6$$
$$2r = 6+3r\sin\theta$$
$$2\left(\pm\sqrt{x^2+y^2}\right) = 6+3y$$
$$4\left(x^2+y^2\right) = \left(6+3y\right)^2$$
$$4x^2+4y^2 = 36+36y+9y^2$$
$$4x^2-5y^2-36y-36 = 0$$

89. The graph of $r = 6$ is a circle centered at the origin with a radius of 6 units.
$$r = 6$$
$$r^2 = 36$$
$$x^2+y^2 = 36$$

91.
$$\theta = \frac{\pi}{4}$$
$$\tan\theta = \tan\frac{\pi}{4} = 1 = \frac{y}{x}$$
$$y = x$$

The graph is the line $y = x$, which makes an angle of $\theta = \frac{\pi}{4}$ with the positive x-axis.

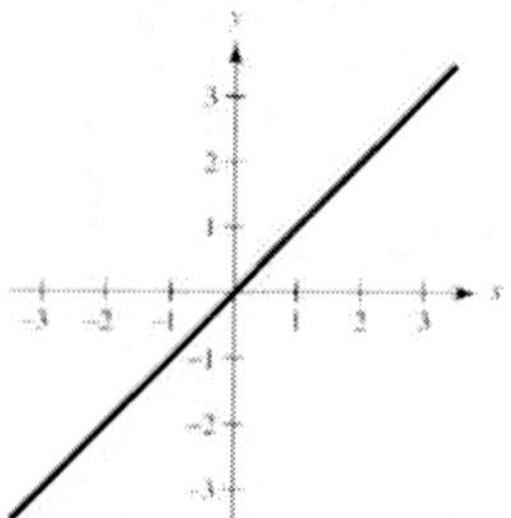

93.
$$r = 3\sec\theta$$
$$r\cos\theta = 3$$
$$x = 3$$
$$x-3 = 0$$

Vertical line through $(3, 0)$

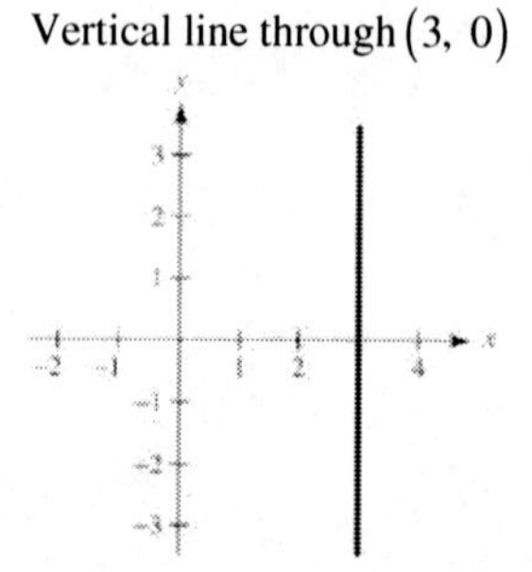

95. True. The distances from the origin are the same.

97. (a) $(r_1, \theta_1) = (x_1, y_1)$, where $x_1 = r_1\cos\theta_1$ and $y_1 = r_1\sin\theta_1$.

$(r_2, \theta_2) = (x_2, y_2)$, where $x_2 = r_2\cos\theta_2$ and $y_2 = r_2\sin\theta_2$.

Then $x_1^2 + y_1^2 = r_1^2\cos^2\theta_1 + r_1^2\sin^2\theta_1 = r_1^2$ and $x_2^2+y_2^2 = r_2^2$. Thus,

$$d = \sqrt{(x_1-x_2)^2+(y_1-y_2)^2}$$
$$= \sqrt{x_1^2 - 2x_1x_2 + x_2^2 + y_1^2 - 2y_1y_2 + y_2^2}$$
$$= \sqrt{\left(x_1^2+y_1^2\right)+\left(x_2^2+y_2^2\right)-2\left(x_1x_2+y_1y_2\right)}$$
$$= \sqrt{r_1^2+r_2^2-2\left(r_1r_2\cos\theta_1\cos\theta_2+r_1r_2\sin\theta_1\sin\theta_2\right)}$$
$$= \sqrt{r_1^2+r_2^2-2r_1r_2\cos\left(\theta_1-\theta_2\right)}.$$

(b) If $\theta_1 = \theta_2$, the points are on the same line through the origin. In this case,

$$d = \sqrt{r_1^2+r_2^2-2r_1r_2\cos(0)} = \sqrt{\left(r_1-r_2\right)^2} = \left|r_1-r_2\right|.$$

(c) If $\theta_1 - \theta_2 = 90°$, $d = \sqrt{r_1^2+r_2^2}$, the Pythagorean Theorem.

(d) For instance, $\left(3, \frac{\pi}{6}\right), \left(4, \frac{\pi}{3}\right)$ gives $d \approx 2.053$ and $\left(-3, \frac{7\pi}{6}\right), \left(-4, \frac{4\pi}{3}\right)$ give $d \approx 2.053$.

99.
$$r = \cos\theta + 3\sin\theta$$
$$r = \frac{x}{r} + \frac{3y}{r}$$
$$r^2 = x + 3y$$
$$x^2 + y^2 = x + 3y$$
$$x^2 - x + y^2 - 3y = 0$$
$$\left(x - \frac{1}{2}\right)^2 + \left(y - \frac{3}{2}\right)^2 = \frac{5}{2}$$

The graph is a circle, with center $\left(\frac{1}{2}, \frac{3}{2}\right)$ and radius $\frac{\sqrt{10}}{2}$.

101. $\cos A = \frac{b^2 + c^2 - a^2}{2bc} = \frac{19^2 + 25^2 - 13^2}{2(19)(25)} = 0.86$

$A \approx 30.7°$

$\cos B = \frac{a^2 + c^2 - b^2}{2ac} = \frac{13^2 + 25^2 - 19^2}{2(13)(25)} = 0.66615$

$B \approx 48.2°$

$C \approx 180° - 30.7° - 48.2° \approx 101.1°$

103. $B = 180° - 56° - 38° = 86°$

$$\frac{a}{\sin A} = \frac{c}{\sin C} \Rightarrow a = \frac{c\sin A}{\sin C} = \frac{12\sin(56°)}{\sin(38°)} \approx 16.16$$

$$\frac{b}{\sin B} = \frac{c}{\sin C} \Rightarrow b = \frac{c\sin B}{\sin C} = \frac{12\sin(86°)}{\sin(38°)} \approx 19.44$$

Section 9.6

1. convex limaçon

3. lemniscate

5. The graph of a polar equation is symmetric with respect to the line $\theta = \frac{\pi}{2}$ its because replacing (r, θ) with $(r, \pi - \theta)$ or $(-r, -\theta)$ yields an equivalent equation.

7. $r = 3\cos 2\theta$ is a rose curve.

9. $r^2 = 9\cos 2\theta$ is a lemniscate.

11. $r = 6\sin 2\theta$ is a rose curve.

13. The graph is symmetric about the line $\theta = \pi/2$, and passes through $(r, \theta) = (3, 3\pi/2)$. Matches (a).

15. The graph has four leaves. Matches (c).

17. $r = 5 + 4\cos\theta$

$\theta = \frac{\pi}{2}$: $-r = 5 + 4\cos(-\theta)$

$-r = 5 + 4\cos\theta$

Not an equivalent equation

$r = 5 + 4\cos(\pi - \theta)$

$r = 5 + 4(\cos\pi\cos\theta + \sin\pi\sin\theta)$

$r = 5 - 4\cos\theta$

Not an equivalent equation

Polar axis: $r = 5 + 4\cos(-\theta)$

$r = 5 + 4\cos\theta$

Equivalent equation

Pole: $-r = 5 + 4\cos\theta$

Not an equivalent equation

$r = 5 + 4\cos(\pi + \theta)$

$r = 5 + 4\cos(\cos\pi\cos\theta - \sin\pi\sin\theta)$

$r = 5 - 4\cos\theta$

Not an equivalent equation

Answer: Symmetric with respect to the polar axis

19. $r = \frac{2}{1 + \sin\theta}$

$\theta = \frac{\pi}{2}$: $r = \frac{2}{1 + \sin(\pi - \theta)}$

$r = \frac{2}{1 + \sin\pi\cos\theta - \cos\pi\sin\theta}$

$r = \frac{2}{1 + \sin\theta}$

Equivalent equation

Polar axis: $r = \frac{2}{1 + \sin(-\theta)}$

$r = \frac{2}{1 - \sin\theta}$

Not an equivalent equation

$-r = \frac{2}{1 + \sin(\pi - \theta)}$

$-r = \frac{2}{1 + \sin\pi\cos\theta - \cos\pi\sin\theta}$

$-r = \frac{2}{1 + \sin\theta}$

Not an equivalent equation

Pole: $-r = \frac{2}{1 + \sin\theta}$

Not an equivalent equation

$r = \frac{2}{1 + \sin(\pi + \theta)}$

$r = \frac{2}{1 + \sin\pi\cos\theta + \cos\pi\sin\theta}$

$r = \frac{2}{1 - \sin\theta}$

Not an equivalent equation

Answer: Symmetric with respect to $\theta = \frac{\pi}{2}$

21. $r = 6\sin\theta$

$\theta = \frac{\pi}{2}$: $-r = 6\sin(-\theta)$

$r = 6\sin\theta$

Equivalent equation

Polar axis: $r = 6\sin(-\theta)$

$r = -6\sin(\theta)$

Not an equivalent equation

$-r = 6\sin(\pi - \theta)$

$-r = 6(\sin\pi\cos\theta - \cos\pi\sin\theta)$

$-r = 6\sin\theta$

Not an equivalent equation

Pole: $-r = 6\sin\theta$

Not an equivalent equation

$r = 6\sin(\pi + \theta)$

$r = -6\sin\theta$

Not an equivalent equation

Answer: Symmetric with respect to $\theta = \frac{\pi}{2}$

23. $r^2 = 16\sin 2\theta$

$\theta = \frac{\pi}{2}$: $(-r)^2 = 16\sin(2(-\theta))$

$r^2 = -16\sin 2\theta$

Not an equivalent equation

$r^2 = 16\sin(2(\pi - \theta))$

$r^2 = 16\sin(2\pi - 2\theta)$

$r^2 = -16\sin 2\theta$

Not an equivalent equation

Polar axis: $r^2 = 16\sin(2(-\theta))$

$r^2 = -16\sin 2\theta$

Not an equivalent equation

$(-r)^2 = 16\sin(2(\pi - \theta))$

$r^2 = -16\sin 2\theta$

Not an equivalent equation

Pole: $(-r)^2 = 16\sin(2\theta)$

$r^2 = 16\sin 2\theta$

Equivalent equation

Answer: Symmetric with respect to pole

25. $r = 5$

Circle

27. $r = 3\sin\theta$

Symmetric with respect to $\theta = \frac{\pi}{2}$

Circle with radius of $\frac{3}{2}$

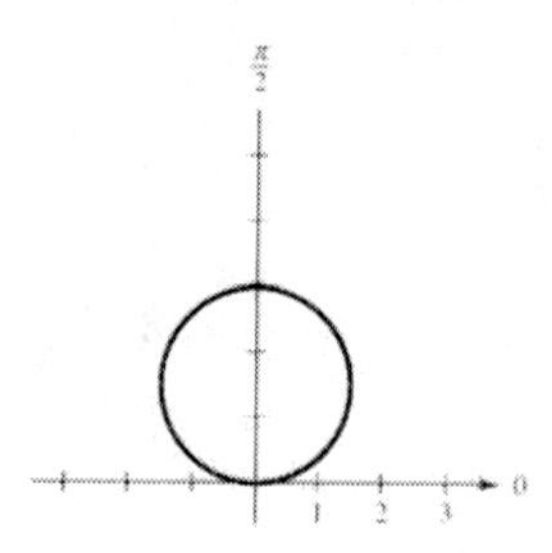

29. $r = 3(1 - \cos\theta)$

Cardioid

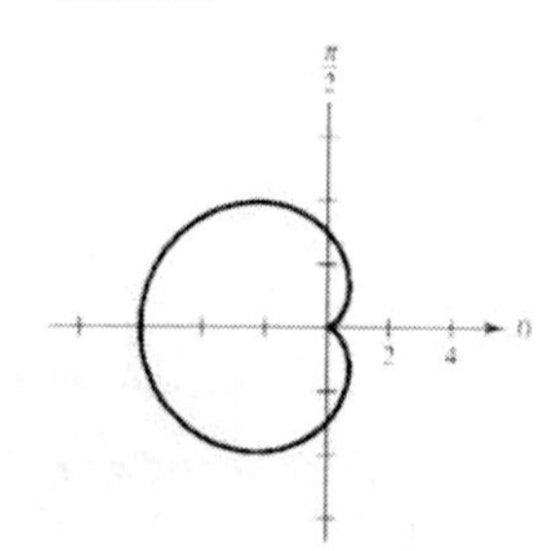

31. $r = 3 - 4\cos\theta$

Limaçon with inner loop

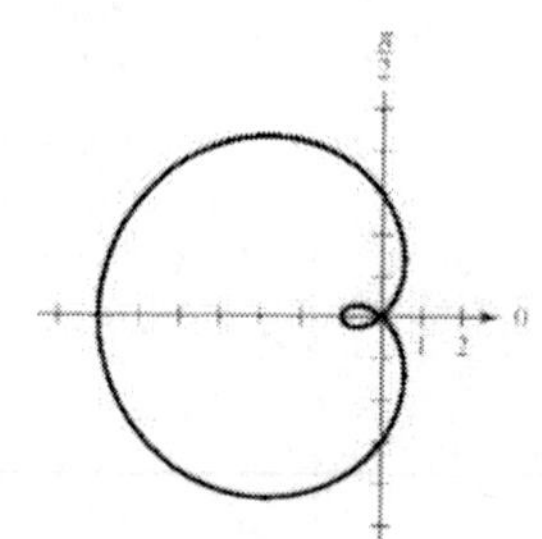

33. $r = 4 + 5\sin\theta$

Limaçon with inner loop

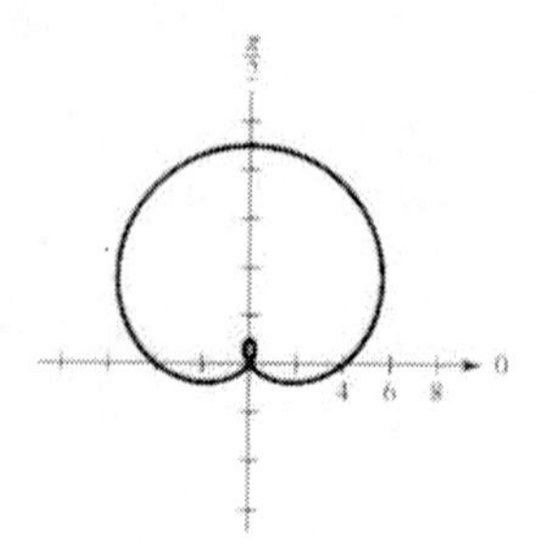

35. $r = 5\cos 3\theta$

3-petal rose curve

Symmetry: Polar axis

$$r = 5\cos 3\theta$$
$$r = 5\cos(3(-\theta))$$
$$r = 5\cos(-3\theta)$$
$$r = 5\cos 3\theta$$

Equivalent equation

Zeros: $r = 0$

$$5\cos 3\theta = 0$$
$$\cos 3\theta = 0$$
$$3\theta = \frac{\pi}{2}, \frac{3\pi}{2}, \frac{5\pi}{2}$$
$$\theta = \frac{\pi}{6}, \frac{\pi}{2}, \frac{5\pi}{6}$$

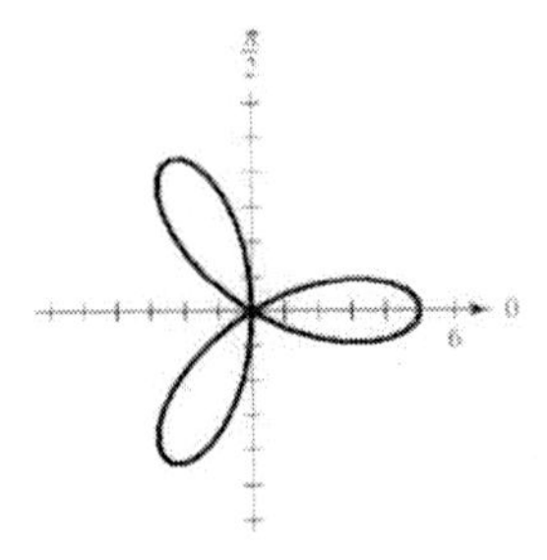

37. $r = 7\sin 2\theta$

4-petal rose curve

Symmetry: $\theta = \frac{\pi}{2}$, polar axis, and the pole

$$r = 7\sin 2\theta$$
$$\theta = \frac{\pi}{2}: \; r = 7\sin(2(\pi - \theta))$$
$$r = 7\sin(2\pi - 2\theta)$$
$$r = 7(\sin 2\pi \cos 2\theta - \cos 2\pi \sin 2\theta)$$
$$r = 7\sin 2\theta$$

Equivalent equation

Polar axis: $-r = 7\sin(2(\pi - \theta))$

$$-r = 7\sin(2\pi - 2\theta)$$
$$-r = 7(\sin 2\pi \cos 2\theta - \cos 2\pi \sin 2\theta)$$
$$-r = -7\sin 2\theta$$
$$r = 7\sin 2\theta$$

Equivalent equation

Pole: $r = 7\sin(2(\pi + \theta))$

$$r = 7\sin(2\pi + 2\theta)$$
$$r = 7(\sin 2\pi \cos 2\theta + \cos 2\pi \sin 2\theta)$$
$$r = 7\sin 2\theta$$

Equivalent equation

Zeros: $r = 0$

$$7\sin 2\theta = 0$$
$$\sin 2\theta = 0$$
$$2\theta = 0, \pi, 2\pi, 3\pi, 4\pi$$
$$\theta = 0, \frac{\pi}{2}, \pi, \frac{3\pi}{2}, 2\pi$$

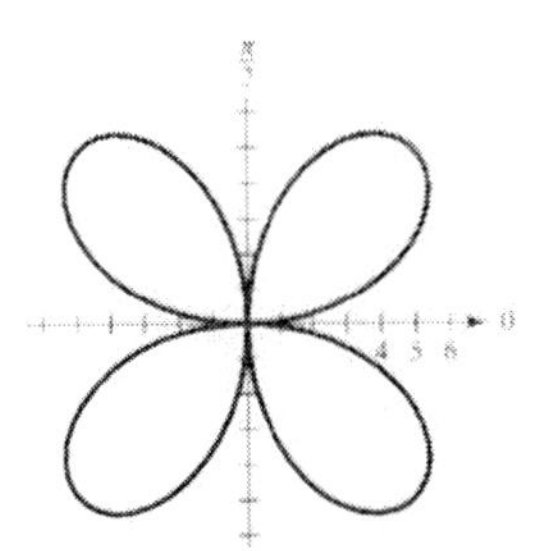

41. $r = 2(5 - \sin\theta)$

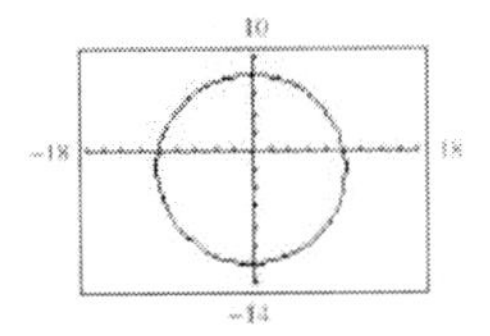

$0 \le \theta < 2\pi$

43. $r = \dfrac{3}{\sin\theta - 2\cos\theta}, \; 0 \le \theta \le \dfrac{\pi}{2}$

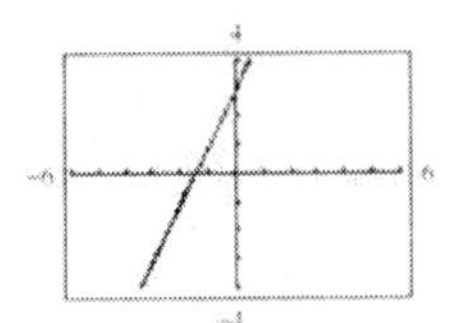

45. $r^2 = 4\cos 2\theta$

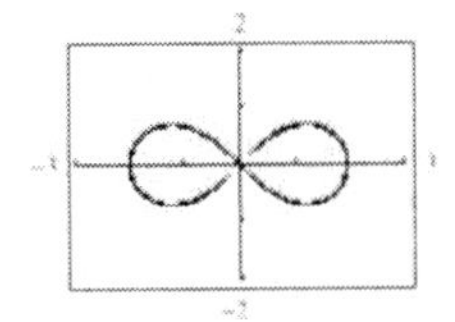

$-2\pi \le \theta \le 2\pi$

47. $r = 8\sin\theta\cos^2\theta$

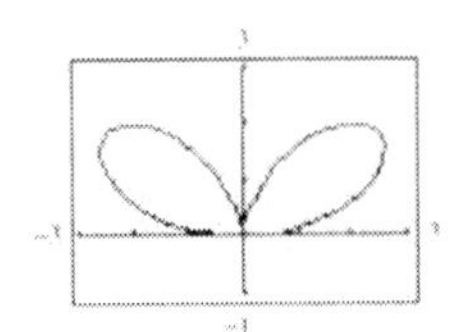

$0 \le \theta \le \pi$

49. $r = 2\csc\theta + 6$

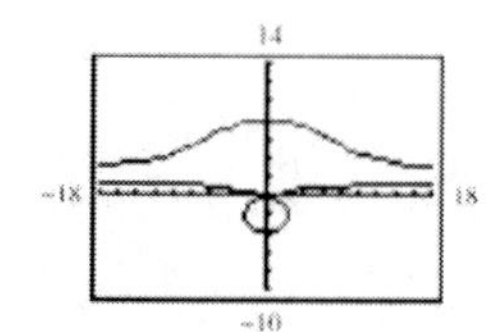

$0 \le \theta < 2\pi$

51. $r = e^{2\theta}$

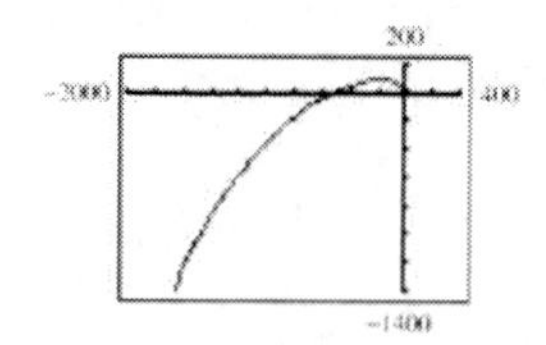

$0 \le \theta \le 2\pi$

53. $r = 3 - 4\cos\theta$

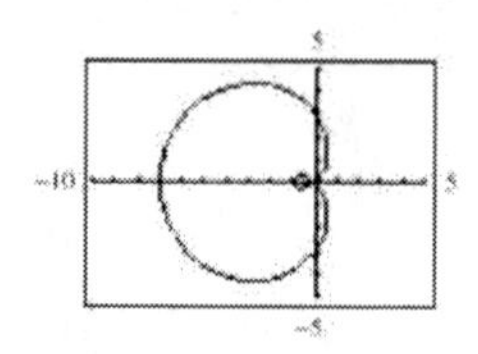

$0 \le \theta < 2\pi$

55. $r = 2\cos\left(\dfrac{3\theta}{2}\right)$, $0 \le \theta < 4\pi$

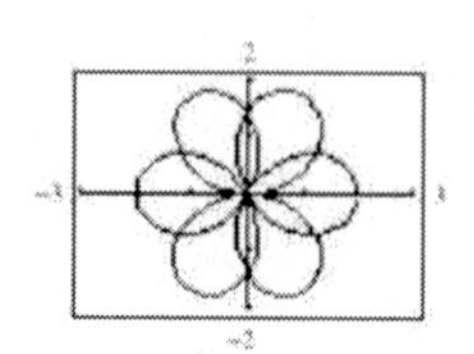

57. $r^2 = 16\sin 2\theta$

$r = \pm\sqrt{16\sin 2\theta} = \pm 4\sqrt{\sin 2\theta}$

Graph both equations using $0 \le \theta < \dfrac{\pi}{2}$.

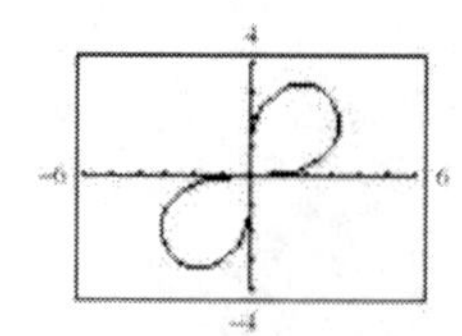

59. $r = 2 - \sec\theta$

From the graph, as $y \to \pm\infty$, $x \to -1$. So, $x = -1$ is an asymptote.

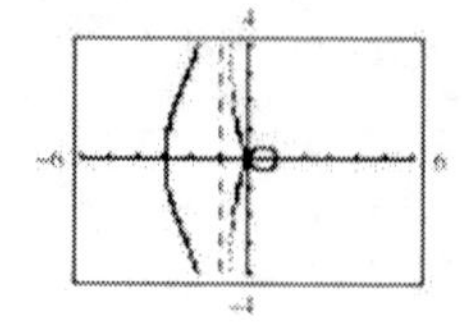

61. $r = \dfrac{3}{\theta}$

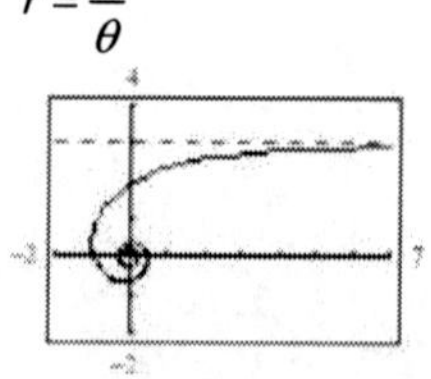

From the graph, as $x \to \infty$, $y \to 3$. So, $y = 3$ is a horizontal asymptote.

63. True. It has five petals.

65. The graph of $r = f(\theta)$ is rotated about the pole through an angle ϕ. Let (r, θ) be any point on the graph of $r = f(\theta)$. Then $(r, \theta + \phi)$ is rotated through the angle ϕ, and since $r = f((\theta + \phi)\phi) = f(\theta)$, it follows that $(r, \theta + \phi)$ is on the graph of $r = f(\theta - \phi)$.

67. (a) $r = 2 - \sin\left(\theta - \dfrac{\pi}{4}\right)$

$= 2 - \dfrac{\sqrt{2}}{2}(\sin\theta - \cos\theta)$

(b) $r = 2 - \sin\left(\theta - \dfrac{\pi}{2}\right)$

$= 2 + \cos\theta$

(c) $r = 2 - \sin(\theta - \pi)$

$= 2 + \sin\theta$

(d) $r = 2 - \sin\left(\theta - \dfrac{3\pi}{2}\right)$

$= 2 - \cos\theta$

69. $r = 2 + k\sin\theta$

$k = 0$: $r = 2$, circle

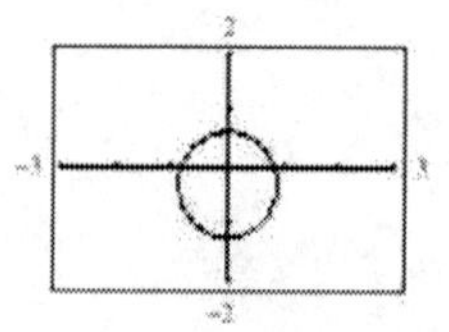

$k = 1$: $r = 2 + \sin\theta$, convex limaçon

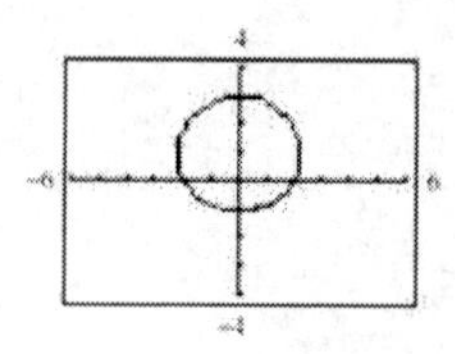

$k = 2$: $r = 2 + 2\sin\theta$, cardioid

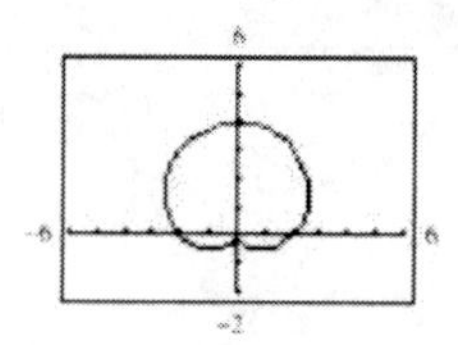

$k = 3$: $r = 2 + 3\sin\theta$, limaçon with inner loop

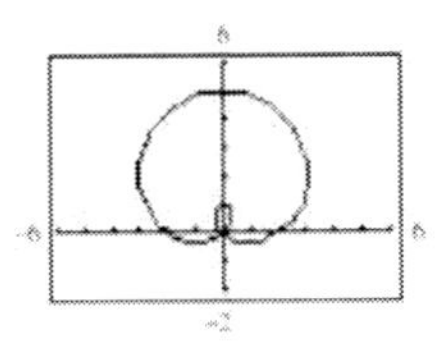

71. $y = \dfrac{x^2 - 9}{x + 1}$

$\dfrac{x^2 - 9}{x + 1} = 0$

$x^2 - 9 = 0$

$x^2 = 9$

$x = \pm 3$

73. $y = 5 - \dfrac{3}{x - 2}$

$5 - \dfrac{3}{x - 2} = 0$

$5 = \dfrac{3}{x - 2}$

$5(x - 2) = 3$

$5x - 10 = 3$

$5x = 13$

$x = \dfrac{13}{5}$

Section 9.7

1. conic

3. An equation of the form $r = \dfrac{ep}{1 + e\cos\theta}$ corresponds to a conic with a vertical directrix.

5. $r = \dfrac{2e}{1 + e\cos\theta}$

(a) Parabola
(b) Ellipse
(c) Hyperbola

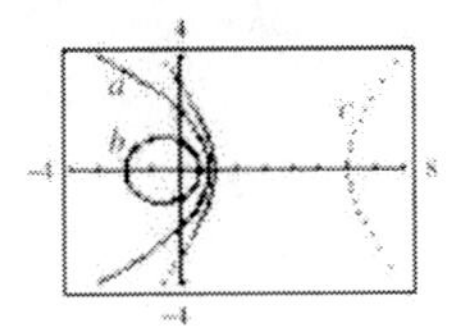

7. $r = \dfrac{2e}{1 - e\sin\theta}$

(a) Parabola
(b) Ellipse
(c) Hyperbola

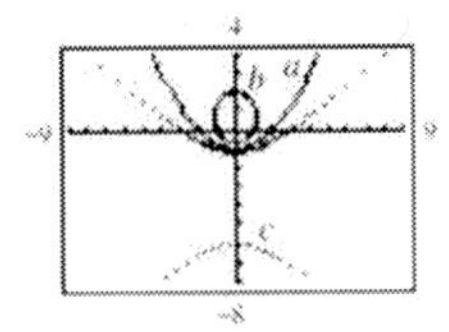

9. $r = \dfrac{4}{1 - \cos\theta}$

$e = 1 \Rightarrow$ parabola

Vertical directrix to left of pole

Matches (b).

11. $r = \dfrac{3}{2 + \cos\theta} = \dfrac{3/2}{1 + (1/2)\cos\theta}$

$e = \dfrac{1}{2} \Rightarrow$ ellipse

Vertical directrix to right of pole

Matches (f).

13. $r = \dfrac{3}{1 + 2\sin\theta}$

$e = 2 \Rightarrow$ hyperbola

Horizontal directrix above the pole.

Matches (d).

15. $r = \dfrac{3}{1 - \cos\theta}$

$e = 1 \Rightarrow$ parabola

17. $r = \dfrac{4}{4 - \cos\theta} = \dfrac{1}{1 - (1/4)\cos\theta}$

$e = \dfrac{1}{4} \Rightarrow$ ellipse

19. $r = \dfrac{8}{4 + 3\sin\theta} = \dfrac{2}{1 + (3/4)\sin\theta}$

$e = \dfrac{3}{4} \Rightarrow$ ellipse

21. $r = \dfrac{6}{2 + \sin\theta} = \dfrac{(1/2)(6)}{1 + (1/2)\sin\theta}$

$e = \dfrac{1}{2} \Rightarrow$ ellipse

23. $r = \dfrac{3}{4 - 8\cos\theta} = \dfrac{3/4}{1 - 2\cos\theta}$

$e = 2 \Rightarrow$ hyperbola

25. $r = \dfrac{-5}{1-\sin\theta}$

$e = 1 \Rightarrow$ parabola

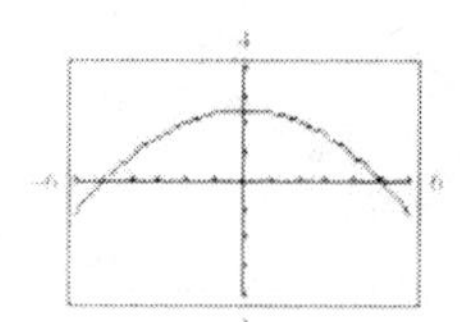

27. $r = \dfrac{14}{14+17\sin\theta} = \dfrac{1}{1+(17/14)\sin\theta}$

$e = \dfrac{17}{14} \Rightarrow$ hyperbola

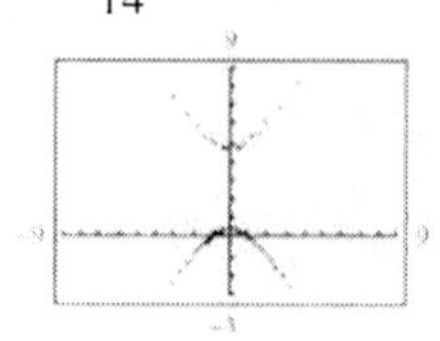

29. $r = \dfrac{3}{-4+2\cos\theta}$

$= \dfrac{-3/4}{1-(1/2)\cos\theta}$

$e = \dfrac{1}{2} \Rightarrow$ ellipse

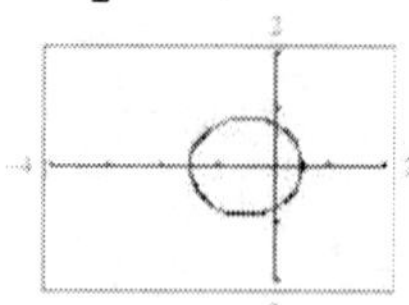

31. $r = \dfrac{3}{1-\cos(\theta - \pi/4)}$

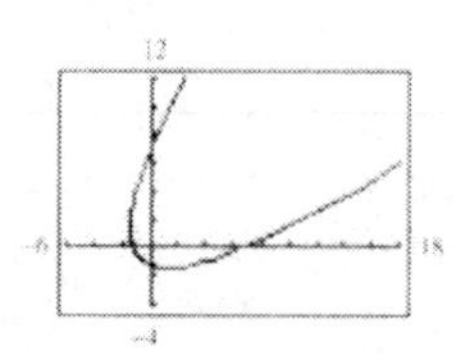

33. $r = \dfrac{4}{4-\cos(\theta + 3\pi/4)}$

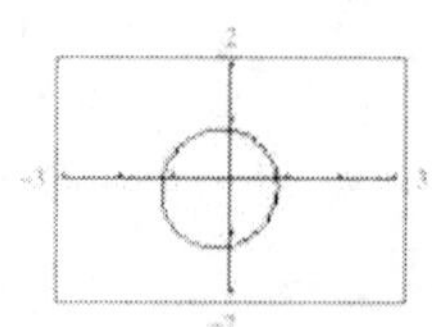

35. $r = \dfrac{8}{4+3\sin(\theta + \pi/6)}$

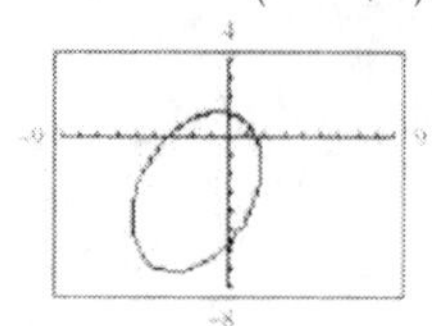

37. $e = 1,\ x = -1,\ p = 1$

Vertical directrix to the left of the pole

$r = \dfrac{1(1)}{1-1\cos\theta} = \dfrac{1}{1-\cos\theta}$

39. $e = \dfrac{1}{2},\ y = 1,\ p = 1$

Horizontal directrix above the pole

$r = \dfrac{(1/2)(1)}{1+(1/2)\sin\theta} = \dfrac{1}{2+\sin\theta}$

41. $e = 2,\ x = 1,\ p = 1$

Vertical directrix to the right of the pole

$r = \dfrac{2(1)}{1+2\cos\theta} = \dfrac{2}{1+2\cos\theta}$

43. Vertex: $\left(1,\ -\dfrac{\pi}{2}\right) \Rightarrow e = 1,\ p = 2$

Horizontal directrix below the pole

$r = \dfrac{1(2)}{1-1\sin\theta} = \dfrac{2}{1-\sin\theta}$

45. Vertex: $(5,\ \pi) \Rightarrow e = 1,\ p = 10$

Vertical directrix to left of pole

$r = \dfrac{1(10)}{1-1\cos\theta} = \dfrac{10}{1-\cos\theta}$

47. Center: $(4,\ \pi),\ c = 4,\ a = 6,\ e = \dfrac{2}{3}$

Vertical directrix to the right of the pole

$r = \dfrac{(2/3)p}{1+(2/3)\cos\theta} = \dfrac{2p}{3+2\cos\theta}$

$2 = \dfrac{2p}{3+2\cos 0} = \dfrac{2p}{5} \Rightarrow p = 5$

$r = \dfrac{10}{3+2\cos\theta}$

49. Center: $(8,\ 0),\ c = 8,\ a = 12,\ e = \dfrac{c}{a} = \dfrac{2}{3}$

Vertical directrix to left of pole

$r = \dfrac{(2/3)p}{1-(2/3)\cos\theta} = \dfrac{2p}{3-2\cos\theta}$

$20 = \dfrac{2p}{3-2} = 2p \Rightarrow p = 10$

$r = \dfrac{20}{3-2\cos\theta}$

51. Center: $(5, 0)$, $c = 5$, $a = 3$, $e = \frac{5}{3}$

Vertical directrix to the right of the pole

$$r = \frac{(5/3)p}{1+(5/3)\cos\theta}$$

$$2 = \frac{(5/3)p}{1+(5/3)\cos\theta} \Rightarrow p = \frac{16}{5}$$

$$r = \frac{\frac{5}{3}\left(\frac{16}{5}\right)}{1+\frac{5}{3}\cos\theta} = \frac{16/3}{1+5/3\cos\theta}$$

$$r = \frac{16}{3+5\cos\theta}$$

53. When $\theta = 0$, $r = c + a = ea + a = a(1+e)$.

Therefore,

$$a(1+e) = \frac{ep}{1-e\cos 0}$$

$$a(1+e)(1-e) = ep$$

$$a(1-e^2) = ep.$$

Thus, $r = \dfrac{ep}{1-e\cos\theta} = \dfrac{(1-e^2)a}{1-e\cos\theta}$.

In Exercises 55 – 58, use the following result from Exercises 53 and 54:

$$r = \frac{(1-e^2)a}{1-e\cos\theta}$$

Perihelion: $r = a(1-e)$

Aphelion: $r = a(1+e)$

55. Earth: $a = 9.295\times10^7$, $e = 0.0167$

$$r = \frac{\left(1-(0.0167)^2\right)9.295\times10^7}{1-0.0167\cos\theta} = \frac{9.2930\times10^7}{1-0.0167\cos\theta}$$

Perihelion: $r = 9.295\times10^7(1-0.0167)$

$= 9.1404\times10^7$ miles

Aphelion: $r = 9.295\times10^7(1+0.0167)$

$= 9.4508\times10^7$ miles

57. Venus: $a = 6.7283\times10^7$; $e = 0.0068$

$$r = \frac{\left(1-(0.0068)^2\right)6.7283\times10^7}{1-0.0068\cos\theta} = \frac{6.7280\times10^7}{1-0.0068\cos\theta}$$

Perihelion: $r = 6.7283\times10^7(1-0.0068)$

$= 6.6781\times10^7$ miles

Aphelion: $r = 6.7283\times10^7(1+0.0068)$

$= 6.7695\times10^7$ miles

59. $a = 4.498\times10^9$, $e = 0.0086$, Neptune

$a = 5.906\times10^9$, $e = 0.2488$, Pluto

(a) Neptune: $r = \dfrac{(1-0.0086^2)4.498\times10^9}{1-0.0086\cos\theta} = \dfrac{4.4977\times10^9}{1-0.0086\cos\theta}$

Pluto: $r = \dfrac{(1-0.2488^2)5.906\times10^9}{1-0.2488\cos\theta} = \dfrac{5.5404\times10^9}{1-0.2488\cos\theta}$

(b) Neptune:

Perihelion:

$4.498\times10^9\times(1-0.0086) \approx 4.4593\times10^9$ km

Aphelion:

$4.498\times10^9(1+0.0086) \approx 4.5367\times10^9$ km

Pluto:

Perihelion:

$5.906\times10^9\times(1-0.2488) \approx 4.4366\times10^9$ km

Aphelion:

$5.906\times10^9(1+0.2488) \approx 7.3754\times10^9$ km

(c)

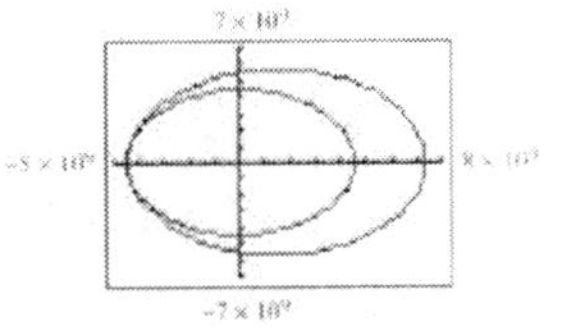

(d) Yes. Pluto is closer to the sun for just a very short time. Pluto was considered the ninth planet because its mean distance from the sun is larger than that of Neptune.

(e) Although the graphs intersect, the orbits do not. So, the planets won't collide.

61. $r = \dfrac{4}{-3-3\sin\theta} = \dfrac{-4/3}{1+\sin\theta}$

False. The directrix is below the pole.

63. True. Both $r = \dfrac{ex}{1-e\cos\theta}$ and $r = \dfrac{e(-x)}{1+e\cos\theta}$ for $e > 1$, represent the same hyperbola.

65.

$$\frac{x^2}{a^2}+\frac{y^2}{b^2} = 1$$

$$\frac{r^2\cos^2\theta}{a^2}+\frac{r^2\sin^2\theta}{b^2} = 1$$

$$\frac{r^2\cos^2\theta}{a^2}+\frac{r^2(1-\cos^2\theta)}{b^2} = 1$$

$$r^2b^2\cos^2\theta + r^2a^2 - r^2a^2\cos^2\theta = a^2b^2$$

$$r^2(b^2-a^2)\cos^2\theta + r^2a^2 = a^2b^2$$

For an ellipse, $b^2 - a^2 = -c^2$. Hence,

$$-r^2c^2\cos^2\theta + r^2a^2 = a^2b^2$$

$$-r^2\left(\frac{c}{a}\right)^2\cos^2\theta + r^2 = b^2,\ e = \frac{c}{a}$$

$$-r^2e^2\cos^2\theta + r^2 = b^2$$

$$r^2\left(1 - e^2\cos^2\theta\right) = b^2$$

$$r^2 = \frac{b^2}{1 - e^2\cos^2\theta}.$$

67. $\frac{x^2}{169} + \frac{y^2}{144} = 1$

$a = 13,\ b = 12,\ c = 5,\ e = \frac{5}{13}$

$$r^2 = \frac{144}{1 - (25/169)\cos^2\theta} = \frac{24,336}{169 - 25\cos^2\theta}$$

69. $\frac{x^2}{25} + \frac{y^2}{16} = 1$

$a = 5,\ b = 4,\ c = 3,\ e = \frac{3}{5}$

$$r^2 = \frac{b^2}{1 - e^2\cos^2\theta} = \frac{16}{1 - (9/25)\cos^2\theta} = \frac{400}{25 - 9\cos^2\theta}$$

71. Hyperbola $\Rightarrow$ One focus: $(5,\ 0) \Rightarrow c = 5$

Vertices: $(4,\ 0),\ (4,\ \pi) \Rightarrow a = 4$

$c^2 = a^2 + b^2 \Rightarrow 25 = 16 + b^2 \Rightarrow b^2 = 9$

$$\frac{x^2}{a^2} - \frac{y^2}{b^2} = 1 \Rightarrow r^2 = \frac{-b^2}{1 - e^2\cos^2\theta}$$

$$r^2 = \frac{-(3)^2}{1 - \left(\frac{5}{4}\right)^2\cos^2\theta}$$

$$r^2 = \frac{-9}{1 - \frac{25}{16}\cos^2\theta}$$

$$r^2 = \frac{-144}{16 - 25\cos^2\theta}$$

$$r^2 = \frac{144}{25\cos^2\theta - 16}$$

73. $r = \frac{4}{1 - 0.4\cos\theta}$

Vertical directrix to left of pole

(a) $e = 0.4 \Rightarrow$ ellipse

(b) $r = \frac{4}{1 + 0.4\cos\theta}$

Vertical directrix to right of pole

Graph is reflected in line $\theta = \pi/2$.

$r = \frac{4}{1 - 0.4\sin\theta}$

Horizontal directrix below pole

$90°$ rotation counterclockwise

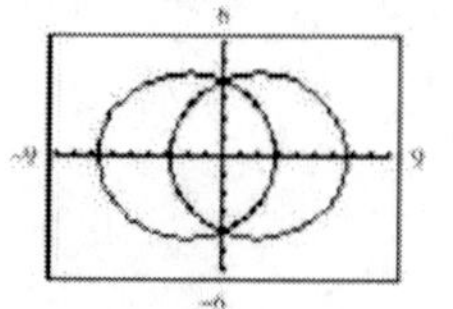

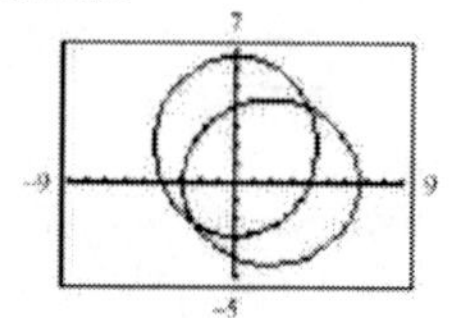

75. $r = a\sin\theta + b\cos\theta$

$r^2 = ar\sin\theta + br\cos\theta$

$x^2 + y^2 = ay + bx$

Circle

77. $\cos(u + v) = \cos u\cos v - \sin u\sin v$

$$= \frac{4}{5}\left(\frac{1}{\sqrt{2}}\right) - \left(-\frac{3}{5}\right)\left(-\frac{1}{\sqrt{2}}\right)$$

$$= \frac{1}{5\sqrt{2}} = \frac{\sqrt{2}}{10}$$

79. $\sin(u - v) = \sin u\cos v - \sin v\cos u$

$$= \left(-\frac{3}{5}\right)\left(\frac{1}{\sqrt{2}}\right) - \left(-\frac{1}{\sqrt{2}}\right)\left(\frac{4}{5}\right)$$

$$= \frac{1}{5\sqrt{2}} = \frac{\sqrt{2}}{10}$$

Chapter 9 Review Exercises

1. Hyperbola

3. Radius $= \sqrt{(-3 - 0)^2 + (-4 - 0)^2}$

$= \sqrt{9 + 16} = \sqrt{25} = 5$

$x^2 + y^2 = 25$

5. Radius $= \frac{1}{2}\sqrt{(5 - (-1))^2 + (6 - 2)^2}$

$= \frac{1}{2}\sqrt{36 + 16} = \frac{1}{2}\sqrt{52} = \sqrt{13}$

Center $= \left(\frac{5 + (-1)}{2},\ \frac{6 + 2}{2}\right) = (2,\ 4)$

$(x - 2)^2 + (y - 4)^2 = 13$

7. $\frac{1}{2}x^2 + \frac{1}{2}y^2 = 18$

$x^2 + y^2 = 36$

Center: $(0,\ 0)$

Radius: 6

9. $$16x^2 + 16y^2 - 16x + 24y - 3 = 0$$
$$16\left(x^2 - x + \tfrac{1}{4}\right) + 16\left(y^2 + \tfrac{3}{2}y + \tfrac{9}{16}\right) = 3 + 4 + 9$$
$$16\left(x - \tfrac{1}{2}\right)^2 + 16\left(y + \tfrac{3}{4}\right)^2 = 16$$
$$\left(x - \tfrac{1}{2}\right)^2 + \left(y + \tfrac{3}{4}\right)^2 = 1$$

Center: $\left(\frac{1}{2}, -\frac{3}{4}\right)$

Radius: 1

11. $$\left(x^2 + 4x + 4\right) + \left(y^2 + 6y + 9\right) = 3 + 4 + 9$$
$$(x + 2)^2 + (y + 3)^2 = 16$$

Center: $(-2, -3)$

Radius: 4

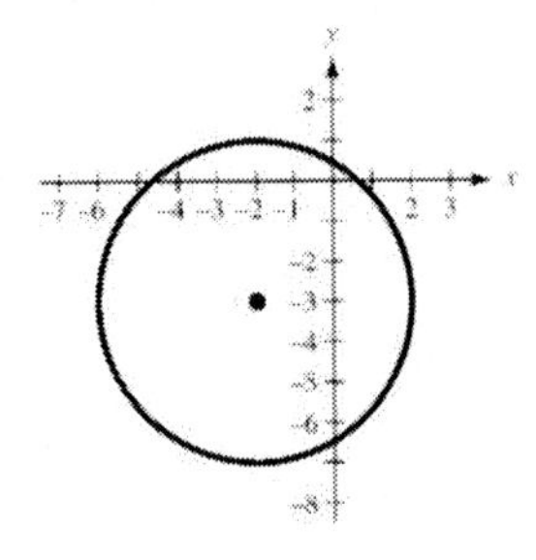

13. x-intercepts: $(x - 3)^2 + (0 + 1)^2 = 7$
$$(x - 3)^2 = 6$$
$$x - 3 = \pm\sqrt{6}$$
$$x = 3 \pm \sqrt{6}$$
$$\left(3 \pm \sqrt{6}, 0\right)$$

y-intercepts: $(0 - 3)^2 + (y + 1)^2 = 7$

$(y + 1)^2 = -2$, impossible

No y-intercepts

15. $4x - y^2 = 0$

$y^2 = 4(1)x,\ p = 1$

Vertex: $(0, 0)$

Focus: $(1, 0)$

Directrix: $x = -1$

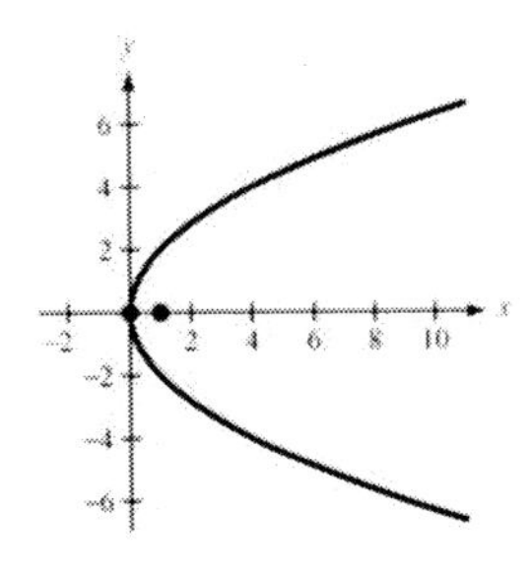

17. $\frac{1}{2}y^2 + 18x = 0$
$$\tfrac{1}{2}y^2 = -18x$$
$$y^2 = -36x = 4(-9)x,\ p = -9$$

Vertex: $(0, 0)$

Focus: $(-9, 0)$

Directrix: $x = 9$

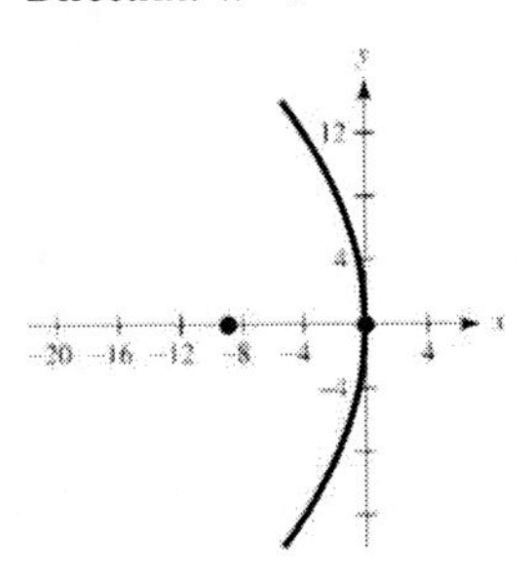

19. Vertex: $(0, 0)$, focus $(4, 0)$

$h = 0,\ k = 0,\ p = 4$

Horizontal axis
$$y^2 = 4px$$
$$= 4(4)(x)$$
$$= 16x$$

21. Vertex: $(-6, 4)$

Passes through $(0, 0)$

Vertical axis
$$(x + 6)^2 = 4p(y - 4)$$
$$(0 + 6)^2 = 4p(0 - 4)$$
$$36 = -16p$$
$$-\tfrac{9}{4} = p$$
$$(x + 6)^2 = 4\left(-\tfrac{9}{4}\right)(y - 4)$$
$$(x + 6)^2 = -9(y - 4)$$

23. $x^2 = -2y = 4\left(-\dfrac{1}{2}\right)y,\ p = -\dfrac{1}{2}$

Focus: $\left(0, -\dfrac{1}{2}\right)$
$$d_1 = \frac{1}{2} + b$$
$$d_2 = \sqrt{(2 - 0)^2 + \left(-2 + \frac{1}{2}\right)^2} = \frac{5}{2}$$
$$d_1 = d_2 \Rightarrow \frac{1}{2} + b = \frac{5}{2} \Rightarrow b = 2$$

Slope of tangent line: $\dfrac{b + 2}{0 - 2} = \dfrac{4}{-2} = -2$

Equation: $y + 2 = -2(x - 2)$
$$y = -2x + 2$$

x-intercept: $(1, 0)$

25. $x^2 = 4p(y-12)$

$(4, 10)$ on curve:

$16 = 4p(10-12) = -8p \Rightarrow p = -2$

$x^2 = 4(-2)(y-12) = -8y + 96$

$y = \dfrac{-x^2 + 96}{8}$

$y = 0$ if $x^2 = 96 \Rightarrow x = 4\sqrt{6} \Rightarrow$ width is $8\sqrt{6}$ meters.

27. $\dfrac{x^2}{4} + \dfrac{y^2}{16} = 1$

$a = 4,\ b = 2,\ c = \sqrt{16-4} = \sqrt{12} = 2\sqrt{3}$

Center: $(0, 0)$

Vertices: $(0, \pm 4)$

Foci: $\left(0, \pm 2\sqrt{3}\right)$

$$\text{Eccentricity} = \frac{c}{a} = \frac{2\sqrt{3}}{4} = \frac{\sqrt{3}}{2}$$

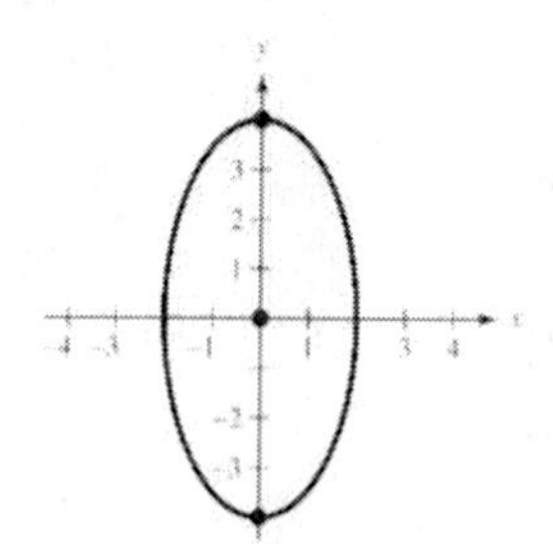

29. $\dfrac{(x+1)^2}{25} + \dfrac{(y-2)^2}{49} = 1$

$a = 7,\ b = 5,\ c = \sqrt{49-25} = 2\sqrt{6}$

Center: $(-1, 2)$

Vertices: $(-1, 9)$, $(-1, -5)$

Foci: $\left(-1, 2 \pm 2\sqrt{6}\right)$

Eccentricity $e = \dfrac{c}{a}$

$e = \dfrac{2\sqrt{6}}{7}$

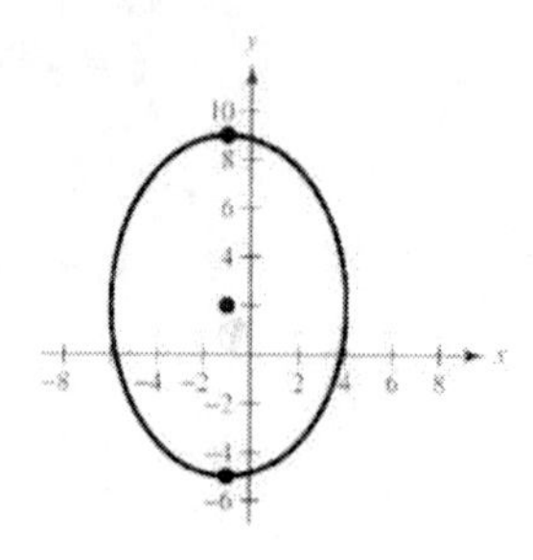

31. (a) $16\left(x^2 - 2x + 1\right) + 9\left(y^2 + 8y + 16\right) = -16 + 16 + 144$

$16(x-1)^2 + 9(y+4)^2 = 144$

$\dfrac{(x-1)^2}{9} + \dfrac{(y+4)^2}{16} = 1$

(b) Center: $(1, -4)$

$a = 4,\ b = 3,\ c = \sqrt{16-9} = \sqrt{7}$

Vertices: $(1, 0)$, $(1, -8)$

Foci: $\left(1, -4 \pm \sqrt{7}\right)$

$e = \dfrac{c}{a} = \dfrac{\sqrt{7}}{4}$

(c)

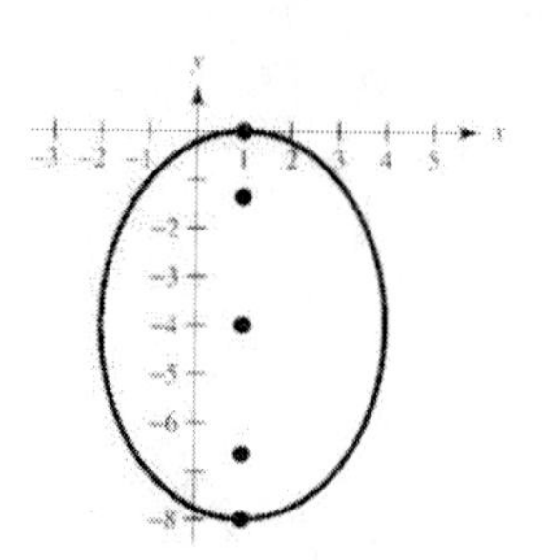

33. (a) $3\left(x^2 + 4x + 4\right) + 8\left(y^2 - 14y + 49\right) = -403 + 12 + 392$

$3(x+2)^2 + 8(y-7)^2 = 1$

$\dfrac{(x+2)^2}{1/3} + \dfrac{(y-7)^2}{1/8} = 1$

(b) Center: $(-2, 7)$

$a = \dfrac{\sqrt{3}}{3},\ b = \dfrac{\sqrt{2}}{4}$

$c^2 = a^2 - b^2 = \dfrac{1}{3} - \dfrac{1}{8} = \dfrac{5}{24} \Rightarrow c = \dfrac{\sqrt{30}}{12}$

Vertices: $\left(-2 \pm \dfrac{\sqrt{3}}{3}, 7\right)$

Foci: $\left(-2 \pm \dfrac{\sqrt{30}}{12}, 7\right)$

$e = \dfrac{c}{a} = \dfrac{\sqrt{30}/12}{\sqrt{3}/3} = \dfrac{\sqrt{10}}{4}$

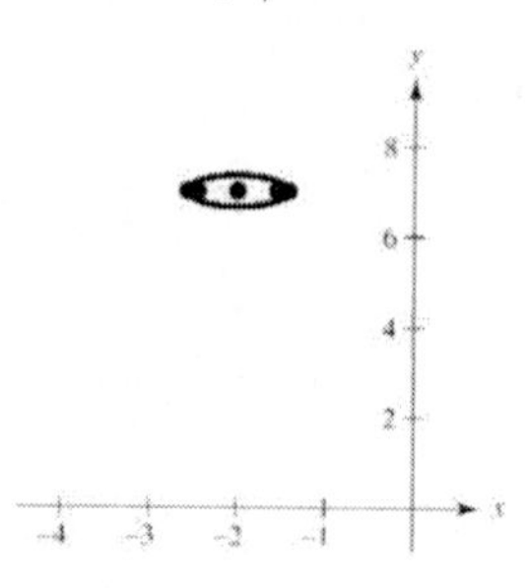

35. Vertices: $(\pm 5, 0)$

Foci: $(\pm 4, 0)$

$a = 5, c = 4 \Rightarrow b = 3,$

$$\frac{x^2}{25} + \frac{y^2}{9} = 1$$

37. Vertices: $(-3, 0), (7, 0)$

Foci: $(0, 0), (4, 0)$

Horizontal major axis

Center: $(2, 0)$

$a = 5, c = 2,$

$b = \sqrt{25-4} = \sqrt{21}$

$$\frac{(x-h)^2}{a^2} + \frac{(y-k)^2}{b^2} = 1$$

$$\frac{(x-2)^2}{25} + \frac{y^2}{21} = 1$$

39. $a = 5, b = 4, c = \sqrt{a^2 - b^2} = \sqrt{25-16} = 3$

The foci should be placed 3 feet on either side of the center and have the same height as the pillars.

41. $a - c = 1.3495 \times 10^9$

$a + c = 1.5045 \times 10^9$

Adding, $2a = 2.854 \times 10^9 \Rightarrow a = 1.427 \times 10^9$. Then

$c = 1.5045 \times 10^9 - 1.427 \times 10^9 = 0.0775 \times 10^9$

$e = \frac{c}{a} \approx 0.0543.$

43. Vertices: $(\pm 4, 0)$; foci: $(\pm 6, 0)$

$$\frac{x^2}{a^2} - \frac{y^2}{b^2} = 1$$

$a = 4$

$c^2 = a^2 + b^2 \Rightarrow 36 = 16 + b^2$

$\Rightarrow b = \sqrt{20} = 2\sqrt{5}$

$$\frac{x^2}{16} - \frac{y^2}{20} = 1$$

45. Foci: $(0, 0), (8, 0) \Rightarrow c = 4$

Center: $(4, 0)$

Asymptotes:

$y = \pm 2(x-4) \Rightarrow \frac{b}{a} = 2 \Rightarrow b = 2a$

$c^2 = a^2 + b^2$

$16 = a^2 + (2a)^2 = 5a^2 \Rightarrow a = \frac{4}{\sqrt{5}}, b = \frac{8}{\sqrt{5}}$

$$\frac{(x-4)^2}{16/5} - \frac{y^2}{64/5} = 1$$

47. (a) $5y^2 - 4x^2 = 20$

$$\frac{y^2}{4} - \frac{x^2}{5} = 1$$

(b) $a = 2, b = \sqrt{5},$

$c = \sqrt{4+5} = 3$

Center: $(0, 0)$

Vertices: $(0, \pm 2)$

Foci: $(0, \pm 3)$

$e = \frac{c}{a} = \frac{3}{2}$

(c)

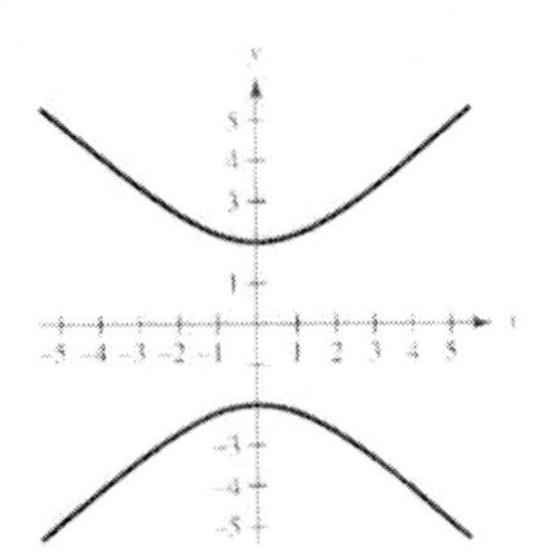

49. (a) $9(x^2 - 2x + 1) - 16(y^2 + 2y + 1) = 151 + 9 - 16$

$9(x-1)^2 - 16(y+1)^2 = 144$

$$\frac{(x-1)^2}{16} - \frac{(y+1)^2}{9} = 1$$

(b) Center: $(1, -1)$, $a = 4, b = 3, c = 5$

Vertices: $(5, -1), (-3, -1)$

Foci: $(6, -1), (-4, -1)$

$e = \frac{c}{a} = \frac{5}{4}$

(c)

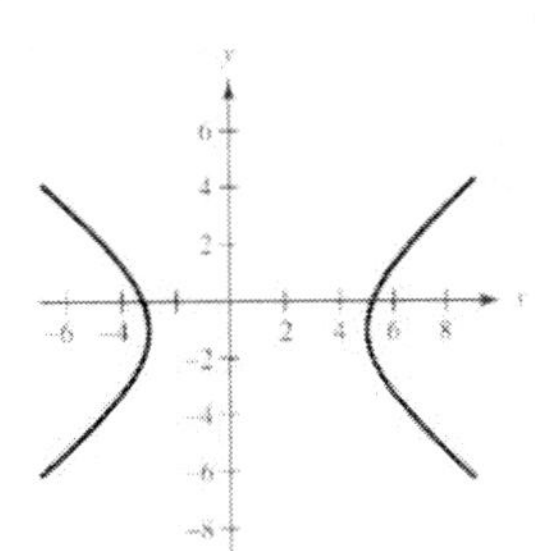

51. (a) $(y^2 - 2y + 1) - 4(x^2 + 12x + 36) = -59 + 1 - 144$

$(y-1)^2 - 4(x+6)^2 = -202$

$$\frac{(x+6)^2}{(101/2)} - \frac{(y-1)^2}{202} = 1$$

(b) Center: $(-6, 1)$

$a^2 = \frac{101}{2}, b^2 = 202, c^2 = \frac{101}{2} + 202 = \frac{505}{2}$

Vertices: $\left(-6 \pm \sqrt{\frac{101}{2}},\ 1\right)$

Foci: $\left(-6 \pm \sqrt{\frac{505}{2}},\ 1\right)$

$e = \frac{c}{a} = \frac{\sqrt{505}}{\sqrt{101}} = \sqrt{5}$

(c)

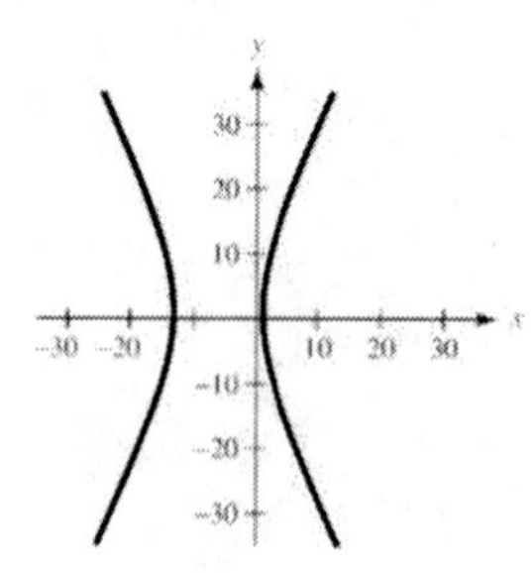

53. $d_2 - d_1 = 186{,}000(0.0005)$

$2a = 93$

$a = 46.5$

$c = 100$

$b = \sqrt{c^2 - a^2}$

$\frac{x^2}{a^2} - \frac{y^2}{b^2} = 1$

$x = 60 \Rightarrow y^2 = b^2\left(\frac{x^2}{a^2} - 1\right) = \left(100^2 - 46.5^2\right)\left(\frac{60^2}{46.5^2} - 1\right)$

$\approx 5211.57 \Rightarrow y \approx 72.2$

72.2 miles north

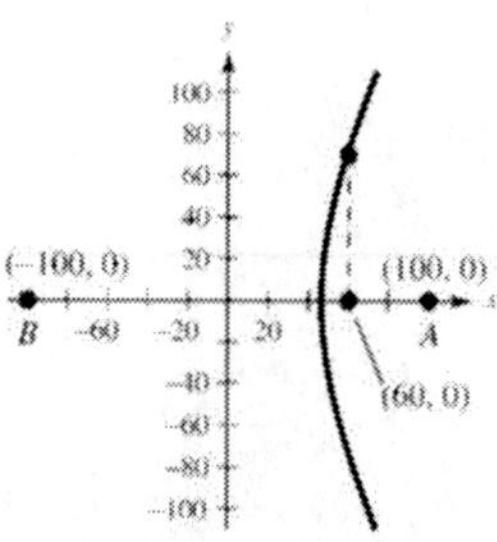

55. $3x^2 + 2y^2 - 12x + 12y + 29 = 0$

$AC = 3(2) = 6 > 0$

Ellipse

57. $5x^2 - 2y^2 + 10x - 4y + 17 = 0$

$AC = 5(-2) = -10 < 0$

Hyperbola

59. $xy - 3 = 0$

$A = 0,\ B = 1,\ C = 0$

$\cot 2\theta = \frac{A - C}{B} = 0 \Rightarrow \theta = \frac{\pi}{4}$

$x = \frac{\sqrt{2}}{2}(x' - y'),\ y = \frac{\sqrt{2}}{2}(x' + y')$

$xy = 3$

$\frac{\sqrt{2}}{2}(x' - y')\frac{\sqrt{2}}{2}(x' + y') = 3$

$\frac{1}{2}(x')^2 - \frac{1}{2}(y')^2 = 3$

$\frac{(x')^2}{6} - \frac{(y')^2}{6} = 1$

Hyberbola

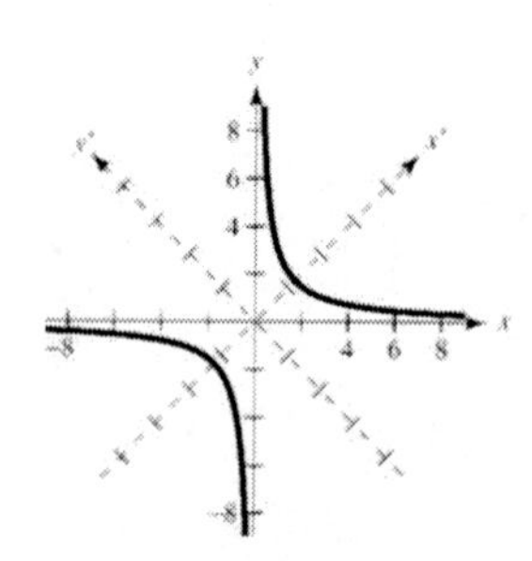

61. $5x^2 - 2xy + 5y^2 + 12 = 0$

$A = 5,\ B = -2,\ C = 5$

$\cot 2\theta = 0 \Rightarrow \theta = \frac{\pi}{4}$

$x = \frac{\sqrt{2}}{2}(x' - y'), y = \frac{\sqrt{2}}{2}(x' + y')$

$5x^2 - 2xy + 5y^2 = 12$

$$5\left[\frac{\sqrt{2}}{2}(x' - y')\right]^2 - 2\left[\frac{\sqrt{2}}{2}(x' - y')\right]\left[\frac{\sqrt{2}}{2}(x' + y')\right] + 5\left[\frac{\sqrt{2}}{2}(x' + y')\right]^2 = 12$$

$$5\left[\frac{1}{2}(x')^2 - x'y' + \frac{1}{2}(y')^2\right] - (x')^2 + (y')^2 + 5\left[\frac{1}{2}(x')^2 + x'y' + \frac{1}{2}(y')^2\right] = 12$$

$$4(x')^2 + 6(y')^2 = 12$$

$$\frac{(x')^2}{3} + \frac{(y')^2}{2} = 1$$

Ellipse

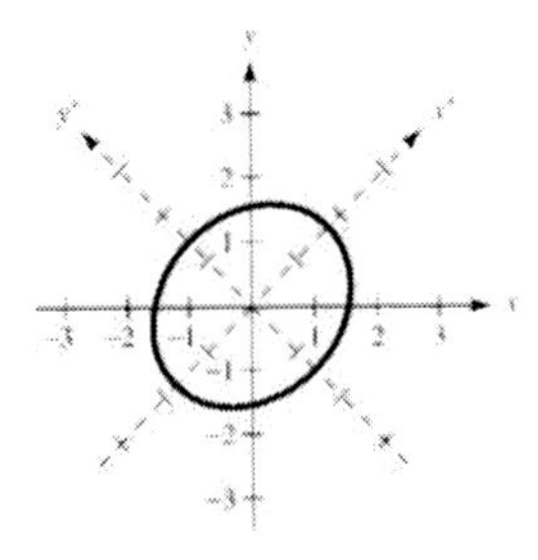

63. $x = 3t - 2,\ y = 7 - 4t$

t	−2	−1	0	1	2	3
x	−8	−5	−2	1	4	7
y	15	11	7	3	−1	−5

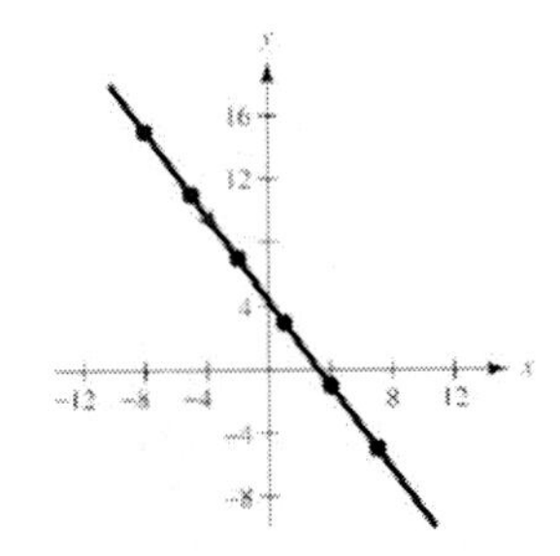

65. $x = 2t$

$y = 4t$

Let $t = \dfrac{x}{2}$.

$$y = 4\left(\frac{x}{2}\right)$$

$$y = 2x$$

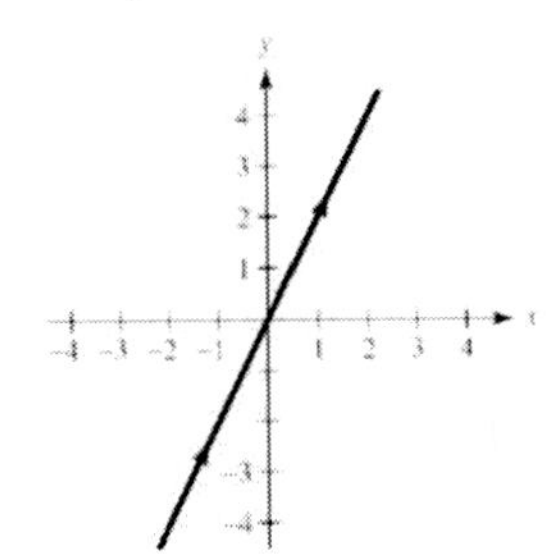

67. $x = t^2 + 2,\ y = 4t^2 - 3$

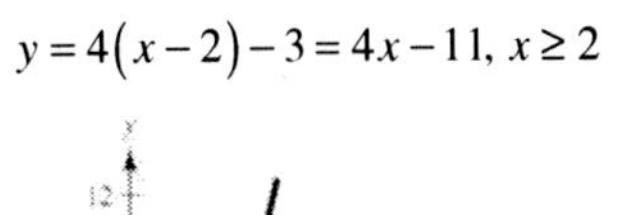

$t^2 = x - 2 \Rightarrow$

$y = 4(x - 2) - 3 = 4x - 11,\ x \ge 2$

69. $x = t^3,\ y = \dfrac{1}{2}t^2$

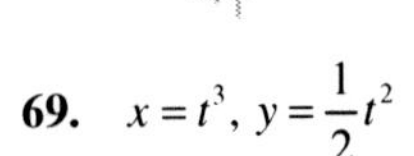

$t = x^{1/3} \Rightarrow y = \dfrac{1}{2}x^{2/3}$

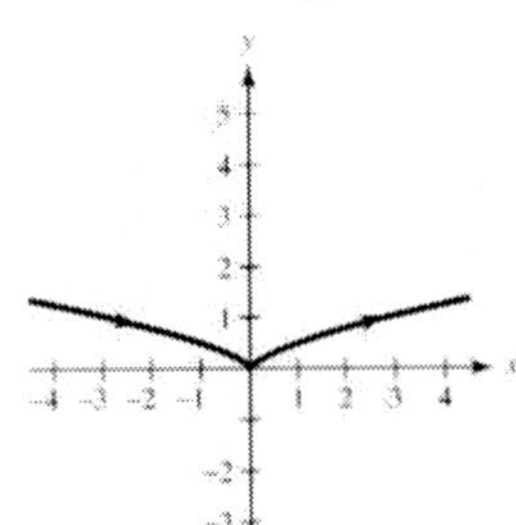

71. $x = \sqrt[3]{t}$

$y = t$

$t = x^3 \Rightarrow y = t = x^3$

$y = x^3$

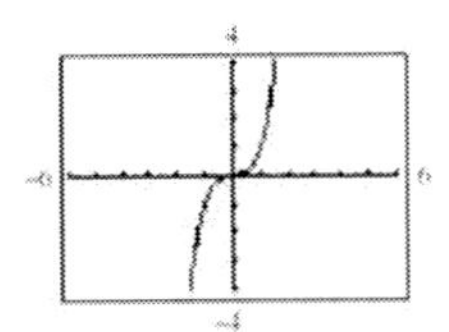

73. $x = \frac{1}{t}$

$y = t$

$y = t = \frac{1}{x}$

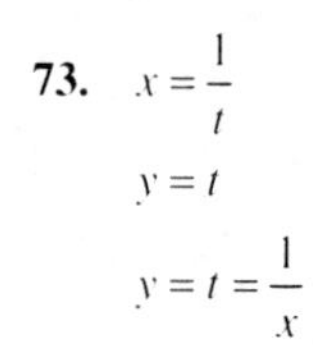

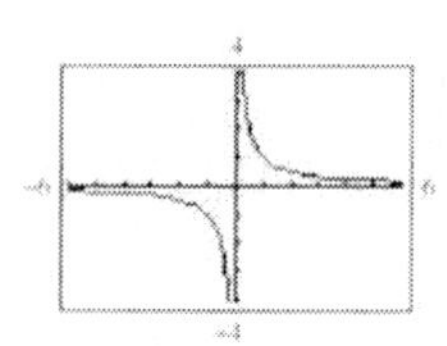

75. $x = 2t$

$y = 4t$

$y = 2(2t) = 2x$

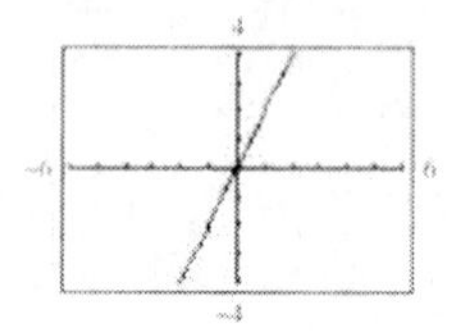

77. $x = 1 + 4t$

$y = 2 - 3t$

$t = \frac{x-1}{4} \Rightarrow y = 2 - 3\left(\frac{x-1}{4}\right) = 2 - \frac{3}{4}x + \frac{3}{4}$

$y = \frac{11}{4} - \frac{3}{4}x$

$3x + 4y - 11 = 0$

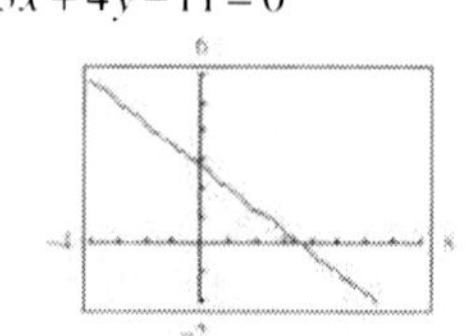

79. $x = 3$

$y = t$

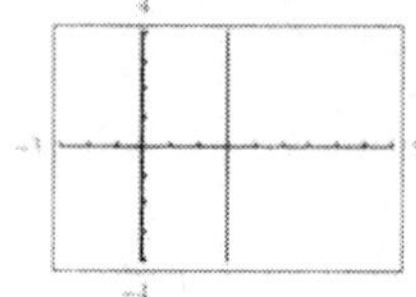

81. $x = 6\cos\theta,\ y = 6\sin\theta$

$\cos\theta = \frac{x}{6},\ \sin\theta = \frac{y}{6}$

$\frac{x^2}{36} + \frac{y^2}{36} = 1$

$x^2 + y^2 = 36$

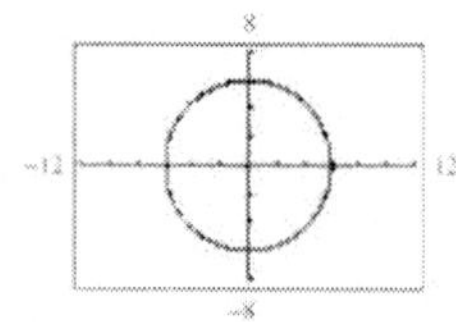

83. $y = 6x + 2$

(a) Let $t = x$.

$x = t$

$y = 6t + 2$

(b) Let $t = 1 - x$.

$x = 1 - t$

$y = 6(1-t) + 2 \Rightarrow y = -6t + 8$

85. $y = x^2 + 2$

(a) Let $t = x$.

$x = t$

$y = t^2 + 2$

(b) Let $t = 1 - x$.

$x = 1 - t$

$y = (1-t)^2 + 2 \Rightarrow y = t^2 - 2t + 3$

87. $x = x_1 + t(x_2 - x_1) = 3 + t(8 - 3) = 5t + 3$

$y = y_1 + t(y_2 - y_1) = 5 + t(0 - 0) = 5$

or $x = t,\ y = 5$

89. $x = x_1 + t(x_2 - x_1)$

$= -1 + t[10 - (-1)] = 11t - 1$

$y = y_1 + t(y_2 - y_1) = 6 + t(0 - 6) = -6t + 6$

91. $(90, 4)$ is on the curve:

$90 = 0.82v_0 t \Rightarrow v_0 = \frac{90}{0.82}t$

$4 = 7 + 0.57\left[\frac{90}{0.82t}\right]t - 16t^2 \Rightarrow$

$16t^2 = 3 + \frac{0.57(90)}{0.82} \Rightarrow t \approx 2.024$

Hence, $v_0 \approx \frac{90}{0.82(2.024)} \approx 54.23$ ft/sec.

93. From Exercise 92:

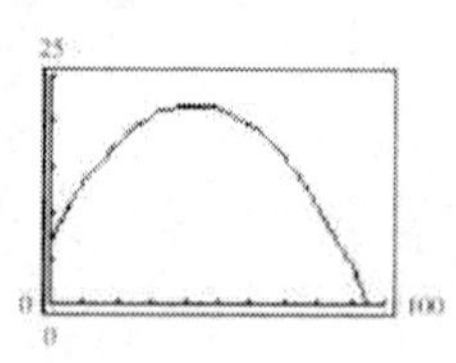

The maximum height is approximately 21.9 feet for $t \approx 0.97$.

95.

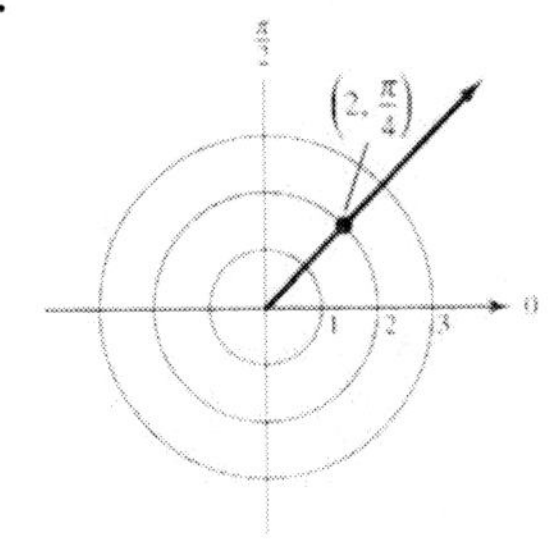

Three additional representations:

$$\left(2, \frac{\pi}{4} - 2\pi\right) = \left(2, -\frac{7\pi}{4}\right)$$

$$\left(-2, \frac{\pi}{4} + \pi\right) = \left(-2, \frac{5\pi}{4}\right)$$

$$\left(-2, \frac{\pi}{4} - \pi\right) = \left(-2, -\frac{3\pi}{4}\right)$$

97.

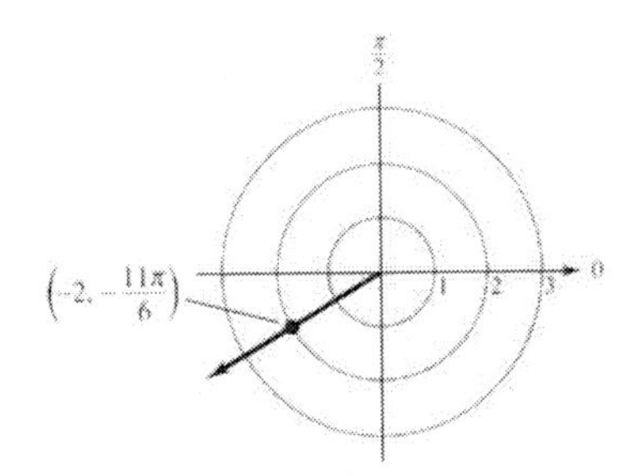

Three additional representations:

$$\left(-2, -\frac{11\pi}{6} + 2\pi\right) = \left(-2, \frac{\pi}{6}\right)$$

$$\left(2, -\frac{11\pi}{6} + 3\pi\right) = \left(2, \frac{7\pi}{6}\right)$$

$$\left(2, -\frac{11\pi}{6} + \pi\right) = \left(2, -\frac{5\pi}{6}\right)$$

99.

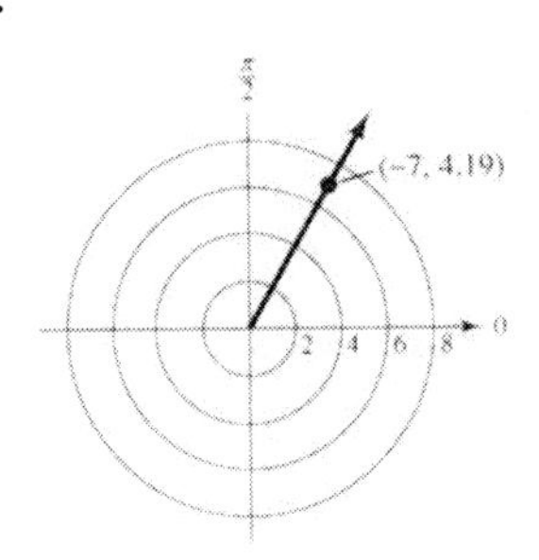

Three additional representations:

$(-7, 4.19 - 2\pi) \approx (-7, -2.09)$

$(7, 4.19 - 3\pi) \approx (7, -5.23)$

$(7, 4.19 - \pi) \approx (7, 1.05)$

101. $(r, \theta) = \left(5, -\frac{7\pi}{6}\right)$

$$x = r\cos\theta = 5\left(-\frac{\sqrt{3}}{2}\right) = -\frac{5\sqrt{3}}{2}$$

$$y = r\sin\theta = 5\left(\frac{1}{2}\right) = \frac{5}{2}$$

$$(x, y) = \left(-\frac{5\sqrt{3}}{2}, \frac{5}{2}\right)$$

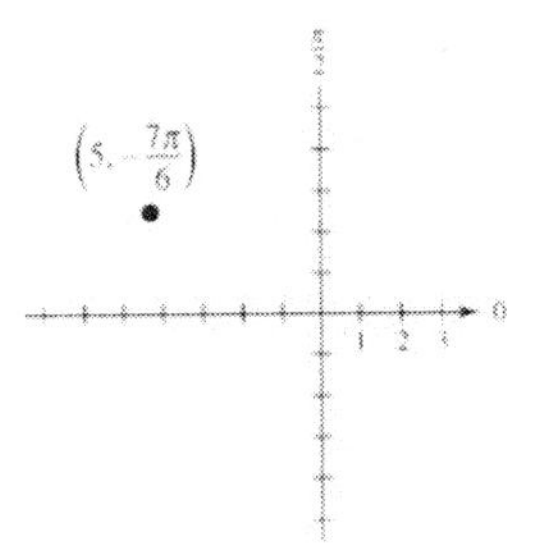

103. $\left(2, -\frac{5\pi}{3}\right)$

$$x = r\cos\theta = 2\left(\frac{1}{2}\right) = 1$$

$$y = r\sin\theta = 2\left(\frac{\sqrt{3}}{2}\right) = \sqrt{3}$$

$$(x, y) = \left(1, \sqrt{3}\right)$$

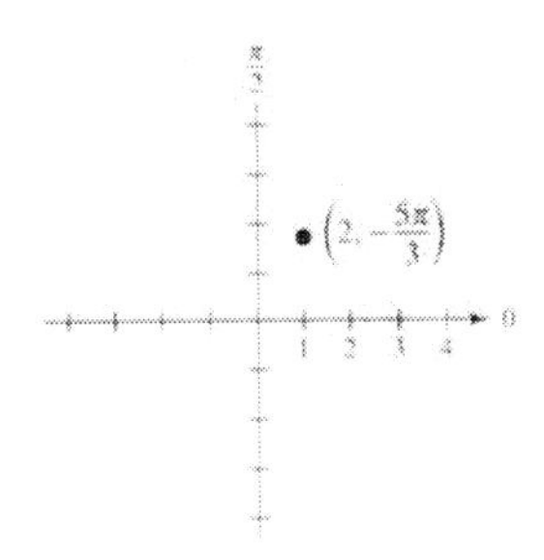

105. $(r, \theta) = \left(3, \frac{3\pi}{4}\right)$

$$(x, y) = \left(3\cos\frac{3\pi}{4}, 3\sin\frac{3\pi}{4}\right) = \left(\frac{-3\sqrt{2}}{2}, \frac{3\sqrt{2}}{2}\right)$$

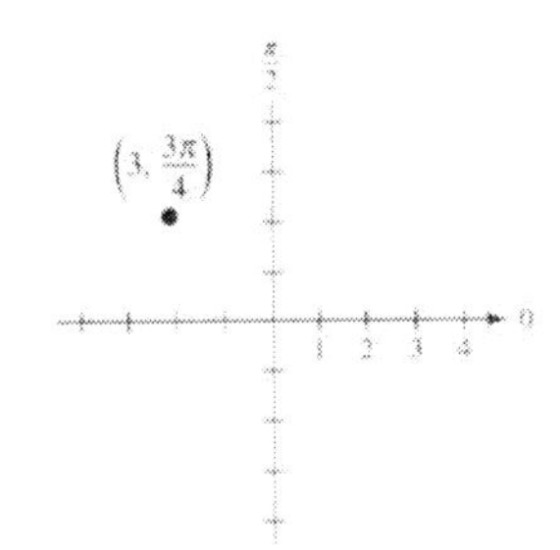

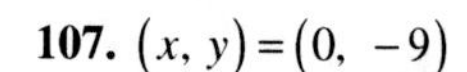

107. $(x, y) = (0, -9)$

$r = 9$, $\tan\theta$ undefined, $\theta = \dfrac{\pi}{2}$

$(r, \theta) = \left(9, \dfrac{3\pi}{2}\right), \left(-9, \dfrac{\pi}{2}\right)$

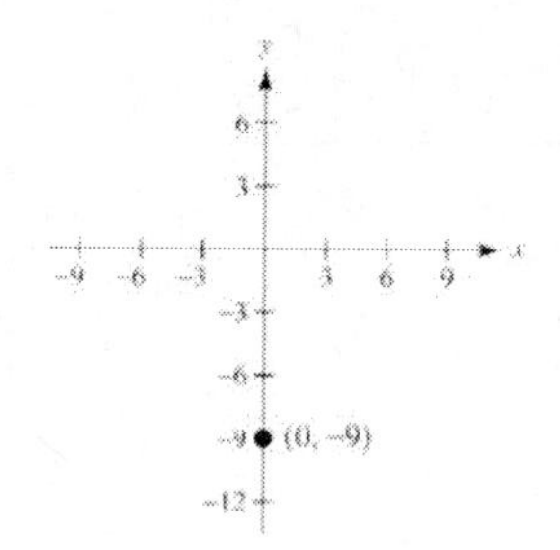

109. $(x, y) = (5, -5)$

$r = 5\sqrt{2}$, $\tan\theta = -1 \Rightarrow \theta = \dfrac{3\pi}{4}$

$(r, \theta) = \left(5\sqrt{2}, \dfrac{7\pi}{4}\right), \left(-5\sqrt{2}, \dfrac{3\pi}{4}\right)$

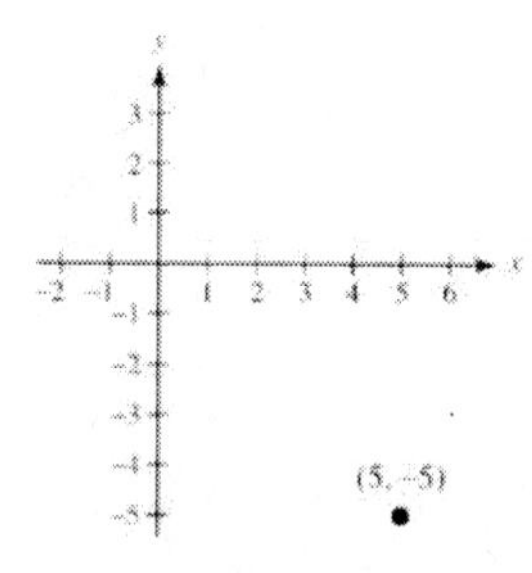

111. $x^2 + y^2 = 81$

$r^2 = 81$

$r = 9$

113. $x^2 + y^2 - 4x = 0$

$r^2 - 4r\cos\theta = 0$

$r = 4\cos\theta$

115. $xy = 5$

$(r\cos\theta)(r\sin\theta) = 5$

$r^2 = 5\csc\theta\sec\theta$

117. $4x^2 + y = 1$

$4(r\cos\theta)^2 + (r\sin\theta)^2 = 1$

$4r^2\cos^2\theta + r^2(1 - \cos^2\theta) = 1$

$r^2\left[3\cos^2\theta + 1\right] = 1$

$r^2 = \dfrac{1}{3\cos^2\theta + 1}$

119. $r = 5$

$x^2 + y^2 = 5^2 = 25$

121. $r = 3\cos\theta$

$r^2 = 3r\cos\theta$

$x^2 + y^2 = 3x$

123. $r^2 = \cos 2\theta$

$r^2 = 1 - 2\sin^2\theta$

$r^4 = r^2 - 2r^2\sin^2\theta$

$\left(x^2 + y^2\right)^2 = x^2 + y^2 - 2y^2$

$\left(x^2 + y^2\right)^2 - x^2 + y^2 = 0$

125. $\theta = \dfrac{5\pi}{6}$

$\tan\theta = \dfrac{y}{x} = -\dfrac{1}{\sqrt{3}}$

$y = -\dfrac{\sqrt{3}}{3}x$, line

127. $r = 5$, circle

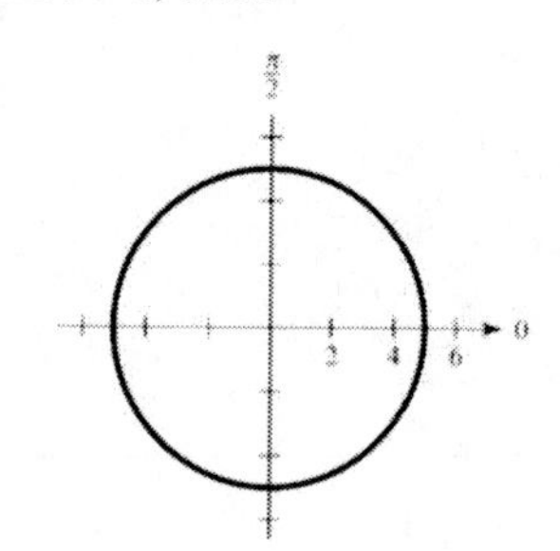

129. $\theta = \dfrac{\pi}{2}$, y-axis

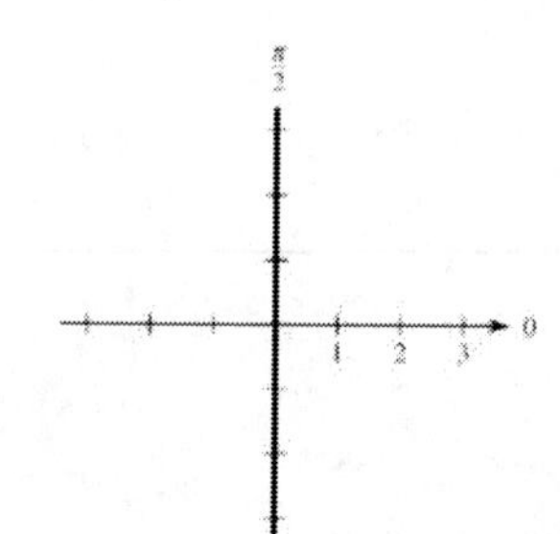

131. $r = 5\cos\theta$, circle

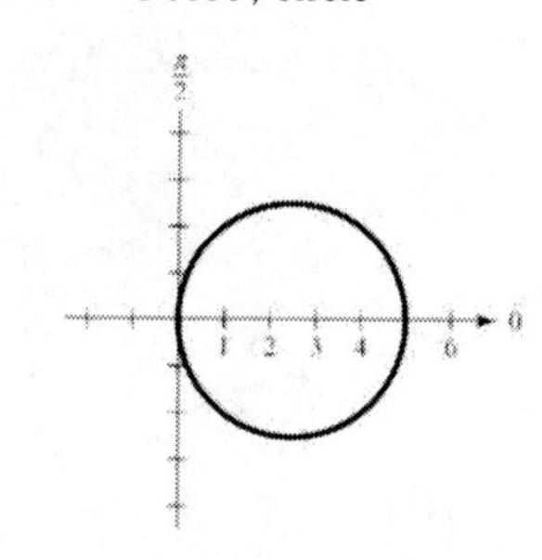

133. $r = 5 + 4\cos\theta$

Dimpled limaçon

Symmetric with respect to polar axis

r is maximum at $\theta = 0$: $(r, \theta) = (9, 0)$

$r \neq 0$ (No zeros)

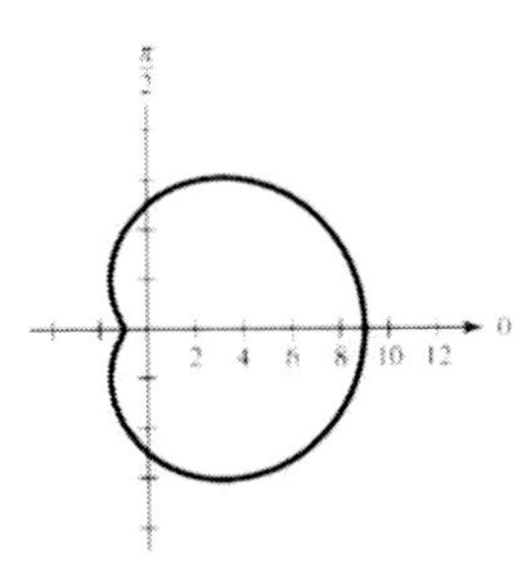

135. $r = 3 - 5\sin\theta$

Limaçon with loop

Symmetric: line $\theta = \dfrac{\pi}{2}$

Maximum $|r|$-value: $|r| = 8$ when $\theta = \dfrac{3\pi}{2}$

Zeros: $r = 0$ when $\theta \approx 0.6435, 2.4981 \left(\sin\theta = \dfrac{3}{5}\right)$

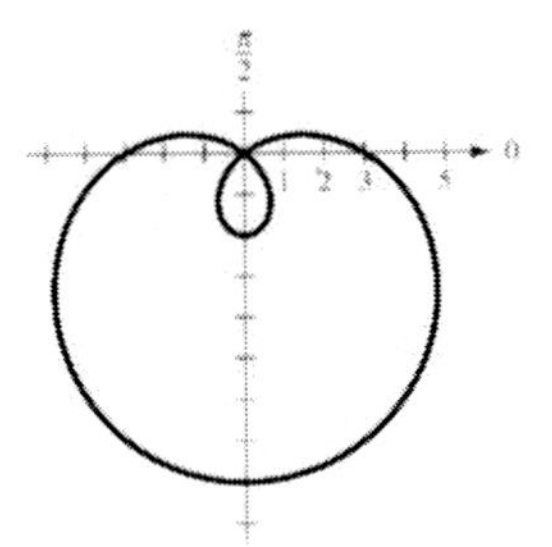

137. $r = -3\cos 2\theta,\ 0 \le \theta \le 2\pi$

Four-leaved rose curve

Symmetric with respect to $\theta = \dfrac{\pi}{2}$, polar axis, and pole

The value of $|r|$ is a maximum (3) at

$\theta = 0, \dfrac{\pi}{2}, \pi, \dfrac{3\pi}{2}$.

$r = 0$ for $\theta = \dfrac{\pi}{4}, \dfrac{3\pi}{4}, \dfrac{5\pi}{4}, \dfrac{7\pi}{4}$

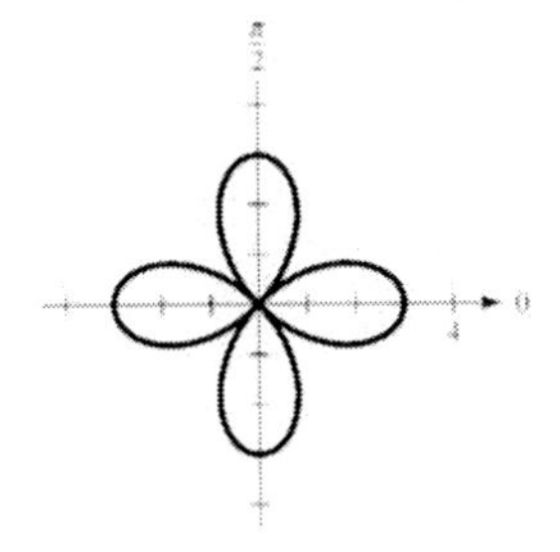

139. $r^2 = 5\sin 2\theta$

Lemniscate

Symmetry with respect to pole

Maximum $|r|$-value: $\sqrt{5}$ when $\theta = \dfrac{\pi}{4}, \dfrac{5\pi}{4}$

Zeros: $r = 0$ when $\theta = 0, \dfrac{\pi}{2}, \pi, \dfrac{3\pi}{2}$

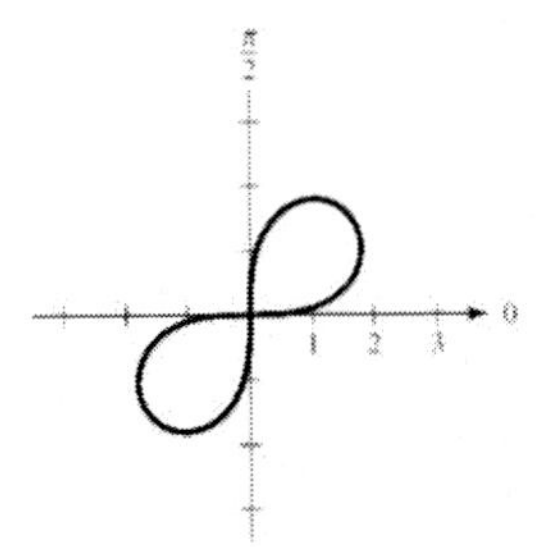

141. $r = \dfrac{1}{1 + 2\sin\theta}$

$e = 2 > 1 \Rightarrow$ hyperbola

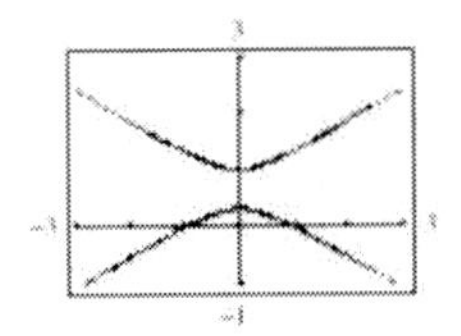

143. $r = \dfrac{4}{5 - 3\cos\theta}$

$= \dfrac{4/5}{1 - (3/5)\cos\theta}$

$e = \dfrac{3}{5} < 1 \Rightarrow$ ellipse

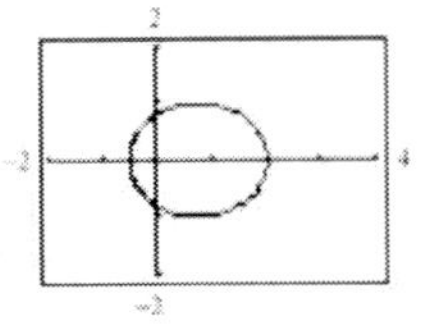

145. $r = \dfrac{5}{6 + 2\sin\theta}$

$= \dfrac{5/6}{1 + (1/3)\sin\theta}$

$e = \dfrac{1}{3} < 1 \Rightarrow$ ellipse

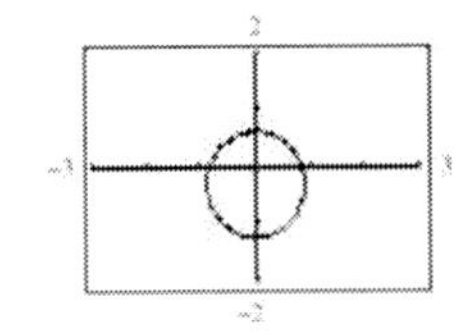

147. Parabola, vertex: $(2, \pi)$

$$e = 1$$

$$r = \frac{4}{1 - \cos\theta}$$

149. Ellipse: $r = \dfrac{ep}{1 - e\cos\theta}$

Vertices: $(5, 0), (1, \pi) \Rightarrow a = 3$

One focus: $(0, 0) \Rightarrow c = 2$

$$e = \frac{c}{a} = \frac{2}{3}$$

$$5 = \frac{2/3\,p}{1 - (2/3)\cos\theta} \Rightarrow p = \frac{5}{2}$$

$$r = \frac{(2/3)(5/2)}{1 - (2/3)\cos\theta} = \frac{5/3}{1 - (2/3)\cos\theta} = \frac{5}{3 - 2\cos\theta}$$

151. $e = 0.093$

Use $r = \dfrac{ep}{1 - e\cos\theta}$.

$$2a = \frac{0.093p}{1 - 0.093\cos\theta} + \frac{0.093p}{1 - 0.093\cos\pi}$$

$$= 0.1876p$$

$$= 3.05 \Rightarrow p \approx 16.258,\ ep \approx 1.512$$

$$r = \frac{1.512}{1 - 0.093\cos\theta}$$

Perihelion: $\dfrac{1.512}{1 + 0.093} \approx 1.383$ astronomical units

Aphelion: $\dfrac{1.512}{1 - 0.093} \approx 1.667$ astronomical units

153. False. The y^4-term is not second degree.

155. (a) Vertical translation
(b) Horizontal translation
(c) Reflection in the y-axis
(d) Vertical shrink

157. The orientation of the graph would be reversed.

Chapter 9 Test

1. $y^2 = 8x = 2(4)x$

Parabola

Vertex: $(0, 0)$

Focus: $(2, 0)$

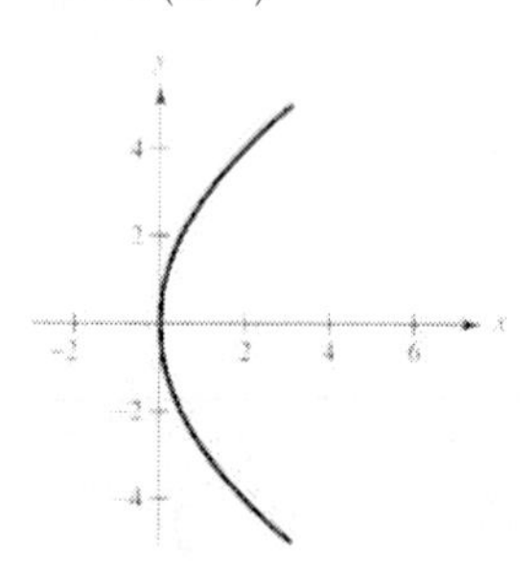

2. $y^2 - 4x + 4 = 0$

$$y^2 = 4x - 4$$

$$y^2 = 4(x - 1)$$

Parabola

Vertex: $(1, 0)$

Focus: $(2, 0)$

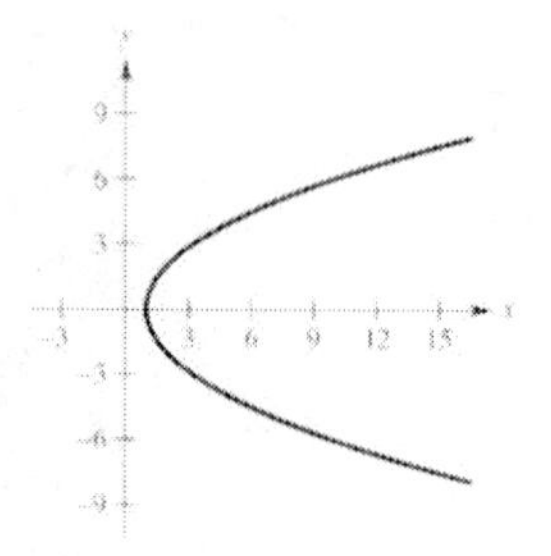

3.

$$x^2 - 4y^2 - 4x = 0$$

$$x^2 - 4x + 4 - 4y^2 = 4$$

$$(x-2)^2 - 4y^2 = 4$$

$$\frac{(x-2)^2}{4} - y^2 = 1$$

Hyperbola

Center: $(2, 0)$

$a = 2,\ b = 1,\ c = \sqrt{5}$

Vertices: $(0, 0), (4, 0)$

Foci: $(2 \pm \sqrt{5}, 0)$

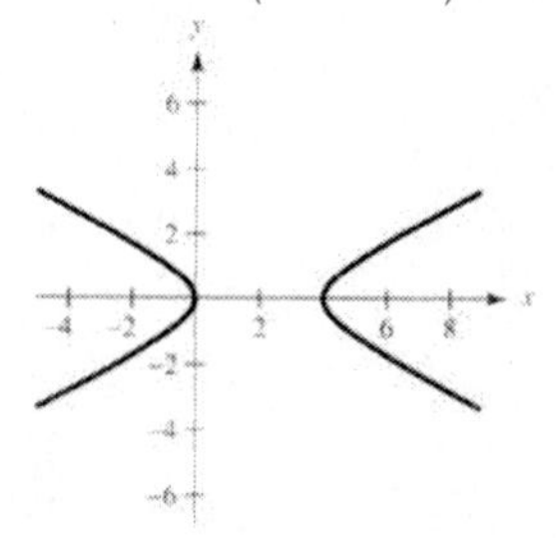

4. Vertex: $(6, -2)$, $p = 2$

$(y+2)^2 = 4(2)(x-6)$

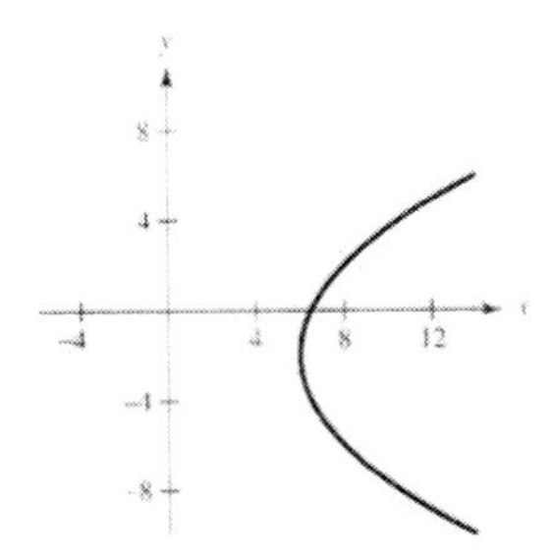

5. Center: $(-6, 3)$

$a = 7, b = 4$

$$\frac{(x+6)^2}{16} + \frac{(y-3)^2}{49} = 1$$

6. $a = 3, \frac{3}{2} = \frac{a}{b} \Rightarrow b = 2$

$$\frac{y^2}{9} - \frac{x^2}{4} = 1$$

7. $x^2 - \frac{y^2}{4} = 1$

$\frac{y^2}{4} = x^2 - 1$

$y = \pm 2\sqrt{x^2 - 1}$

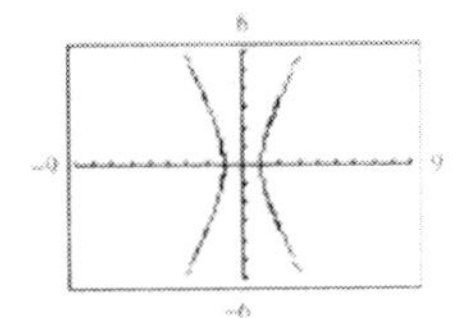

8. (a) $\cot 2\theta = \frac{A-C}{B} = \frac{1-1}{6} = 0 \Rightarrow \theta = \frac{\pi}{4}$ or $45°$

(b) $B^2 - 4AC = 36 - 4 = 32 > 0 \Rightarrow$ Hyperbola

$y^2 + 6xy + (x^2 - 6) = 0$

$$y = \frac{-6x \pm \sqrt{36x^2 - 4(x^2 - 6)}}{2}$$

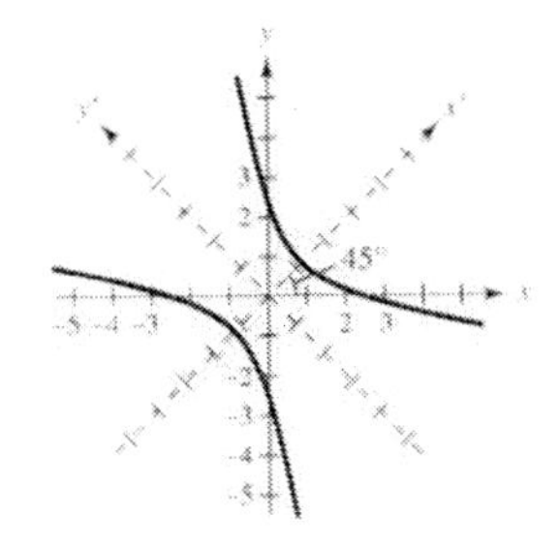

9. $x = t^2 - 6$

$y = \frac{1}{2}t - 1 \Rightarrow t = 2(y+1)$

$x = \left[2(y+1)\right]^2 - 6 = 4y^2 + 8y - 2$

$x = 4y^2 + 8y - 2$ or $(y+1)^2 = \frac{1}{4}(x+6)$

Parabola

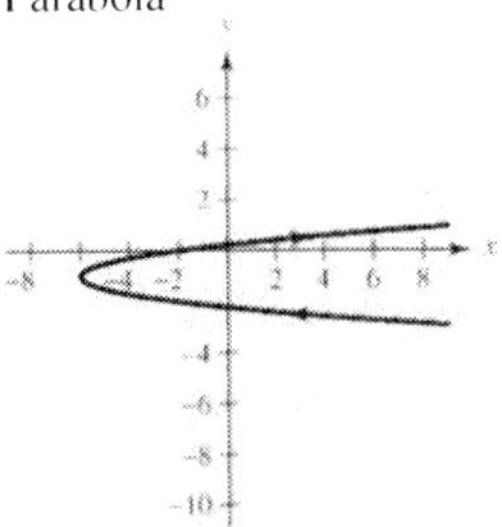

10. $x = \sqrt{t^2 + 2}$

$y = \frac{t}{4} \Rightarrow t = 4y$

$x = \sqrt{16y^2 + 2}$

$x^2 = 16y^2 + 2$

Right-hand portion of hyperbola

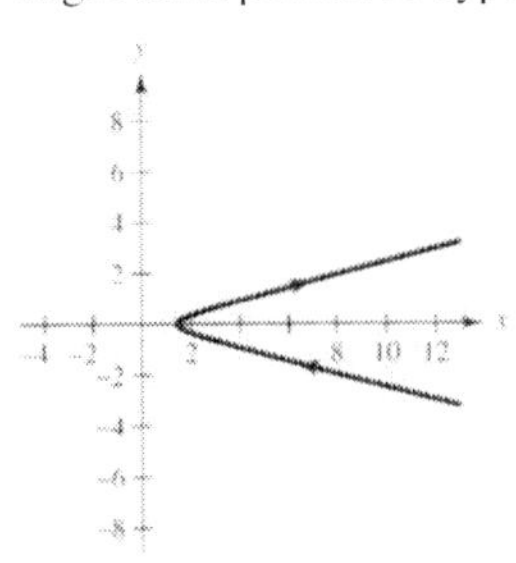

11. $x = 2 + 3\cos\theta$

$y = 2\sin\theta$

$$\left(\frac{x-2}{3}\right)^2 + \left(\frac{y}{2}\right)^2 = 1$$

$$\frac{(x-2)^2}{9} + \frac{y^2}{4} = 1$$

Ellipse

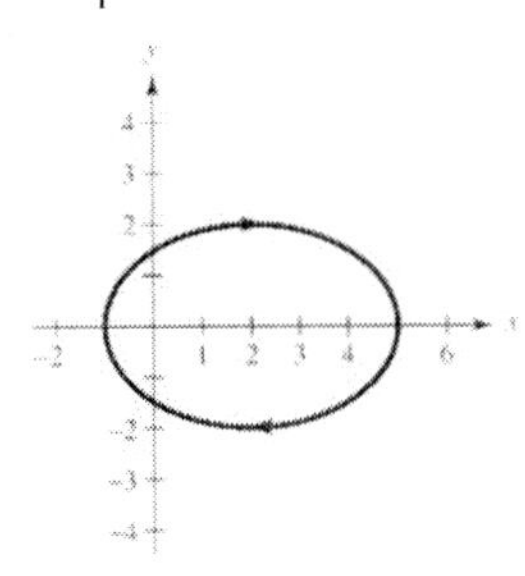

12. $4x + y = 7$

(a) Let $t = x$.

$x = t$

$y = -4t + 7$

(b) Let $t = 2 - x$.

$x = 2 - t$

$y = -4(2 - t) + 7 \Rightarrow y = 4t - 1$

13. $y = \dfrac{3}{x}$

(a) Let $t = x$.

$x = t$

$y = \dfrac{3}{t}$

(b) Let $t = 2 - x$.

$x = 2 - t$

$y = \dfrac{3}{2 - t}$

14. $y = x^2 + 10$

(a) Let $t = x$.

$x = t$

$y = t^2 + 10$

(b) Let $t = 2 - x$.

$x = 2 - t$

$y = (2 - t)^2 + 10 \Rightarrow y = t^2 - 4t + 14$

15. $\left(-2, \dfrac{5\pi}{6}\right)$

$x = r\cos\theta \Rightarrow x = -2\cos\dfrac{5\pi}{6} = \sqrt{3}$

$y = r\sin\theta \Rightarrow y = -2\sin\dfrac{5\pi}{6} = -1$

Answer: $\left(\sqrt{3}, -1\right)$

16. $(x, y) = (2, -2)$, $r = \sqrt{8} = 2\sqrt{2}$, $\theta = \dfrac{7\pi}{4}$

$(r, \theta) = \left(2\sqrt{2}, \dfrac{7\pi}{4}\right) = \left(2\sqrt{2}, -\dfrac{\pi}{4}\right) = \left(-2\sqrt{2}, \dfrac{3\pi}{4}\right)$

17. $x^2 + y^2 - 3x = 0$

$r^2 - 3r\cos\theta = 0$

$r^2 = 3r\cos\theta$

$r = 3\cos\theta$

18. $r = 2\sin\theta$

$r^2 = 2r\sin\theta$

$x^2 + y^2 = 2y$

$x^2 + y^2 - 2y + 1 = 1$

$x^2 + (y - 1)^2 = 1$

19. $r = 2 + 3\sin\theta$

Limaçon with inner loop

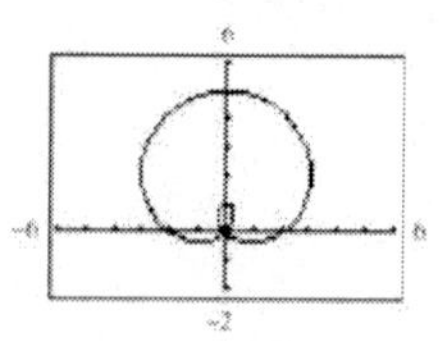

20. $r = \dfrac{1}{1 - \cos\theta}$

$e = 1 \Rightarrow$ Parabola

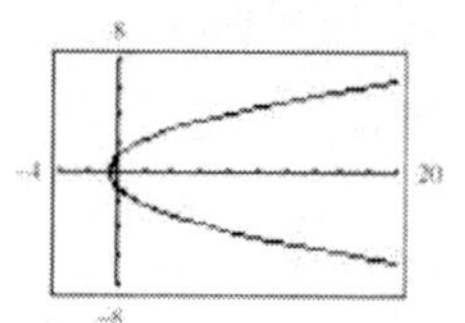

21. $r = \dfrac{4}{2 + 3\sin\theta} = \dfrac{2}{1 + \frac{3}{2}\sin\theta}$

$e = \dfrac{3}{2} \Rightarrow$ Hyperbola

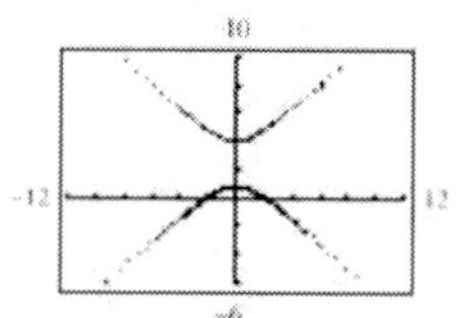

22. $r = \dfrac{ep}{1 + e\sin\theta} = \dfrac{\left(\frac{1}{4}\right)(4)}{1 + \left(\frac{1}{4}\right)\sin\theta}$

$r = \dfrac{4}{4 + \sin\theta} = \dfrac{1}{1 + \left(\frac{1}{4}\right)\sin\theta}$

23. $r = \dfrac{ep}{1 + e\sin\theta} = \dfrac{\left(\frac{5}{4}\right)(2)}{1 + \left(\frac{5}{4}\right)\sin\theta} = \dfrac{10}{4 + 5\sin\theta}$

24. $r = 8\cos\theta$

The maximum value of $|r|$ occurs when $|\cos 3\theta| = 1$.

Hence, the maximum is $8 = |r|$.

$r = 0 \Rightarrow \cos 3\theta = 0$

$\Rightarrow 3\theta = \dfrac{\pi}{2} + n\pi$

$\Rightarrow \theta = \dfrac{\pi}{6} + \dfrac{n\pi}{3}$

On the interval $0 < \theta \le \pi$, $\theta = \dfrac{\pi}{6}, \dfrac{\pi}{2}, \dfrac{5\pi}{6}$.

Cumulative Test for Chapters 7–9

1. $\begin{bmatrix} -1 & -3 & \vdots & 5 \\ 4 & 2 & \vdots & 10 \end{bmatrix}$ row reduces to $\begin{bmatrix} 1 & 0 & \vdots & 4 \\ 0 & 1 & \vdots & -3 \end{bmatrix}$.

Answer: $(4, -3)$

2. $2x - y^2 = 0$

$x - y = 4 \Rightarrow x = y + 4$

$2(y+4) - y^2 = 0$

$y^2 - 2y - 8 = 0$

$(y-4)(y+2) = 0$

$y = 4,\ x = 8$

$y = -2,\ x = 2$

Answer: $(2, -2), (8, 4)$

3. $\begin{bmatrix} 2 & -3 & 1 & \vdots & 13 \\ -4 & 1 & -2 & \vdots & -6 \\ 1 & -3 & 3 & \vdots & 12 \end{bmatrix}$ row reduces to

$\begin{bmatrix} 1 & 0 & 0 & \vdots & \frac{3}{5} \\ 0 & 1 & 0 & \vdots & -4 \\ 0 & 0 & 1 & \vdots & -\frac{1}{5} \end{bmatrix}$.

Answer: $\left(\frac{3}{5}, -4, -\frac{1}{5}\right)$

4. $\begin{bmatrix} 1 & -4 & 3 & \vdots & 5 \\ 5 & 2 & -1 & \vdots & 1 \\ -2 & -8 & 0 & \vdots & 30 \end{bmatrix}$ row reduces to

$\begin{bmatrix} 1 & 0 & 0 & \vdots & 1 \\ 0 & 1 & 0 & \vdots & -4 \\ 0 & 0 & 1 & \vdots & -4 \end{bmatrix}$.

Answer: $(1, -4, -4)$

5. $3A - B = \begin{bmatrix} -7 & -10 & -16 \\ -6 & 18 & 9 \\ -12 & 16 & 7 \end{bmatrix}$

6. $5A + 3B = \begin{bmatrix} -18 & 15 & -14 \\ 28 & 11 & 34 \\ -20 & 52 & -1 \end{bmatrix}$

7. $AB = \begin{bmatrix} 3 & -31 & 2 \\ 22 & 18 & 6 \\ 52 & -40 & 14 \end{bmatrix}$

8. $BA = \begin{bmatrix} 5 & 36 & 31 \\ -36 & 12 & -36 \\ 16 & 0 & 18 \end{bmatrix}$

9. (a) $\begin{bmatrix} 1 & 2 & -1 \\ 3 & 7 & -10 \\ -5 & -7 & -15 \end{bmatrix}^{-1} = \begin{bmatrix} -175 & 37 & -13 \\ 95 & -20 & 7 \\ 14 & -3 & 1 \end{bmatrix}$

(b) $\det(A) = \begin{vmatrix} 1 & 2 & -1 \\ 3 & 7 & -10 \\ -5 & -7 & -15 \end{vmatrix}$

$= 1(-105 - 70) - 2(-45 - 50) - 1(-21 + 35)$

$= -175 + 190 - 14 = 1$

10. $\begin{vmatrix} 0 & 0 & 1 \\ 6 & 2 & 1 \\ 8 & 10 & 1 \end{vmatrix} = 44 \Rightarrow$ Area $= \frac{1}{2}(44) = 22$ sq. units

11. (a) $a_1 = \dfrac{(-1)^{1+1}}{2(1)+3} = \dfrac{1}{5}$

$a_2 = -\dfrac{1}{7}$

$a_3 = \dfrac{1}{9}$

$a_4 = -\dfrac{1}{11}$

$a_5 = \dfrac{1}{13}$

(b) $a_1 = 3(2)^{1-1} = 3$

$a_2 = 6$

$a_3 = 12$

$a_4 = 24$

$a_5 = 48$

12. $\displaystyle\sum_{k=1}^{6} (7k - 2) = \frac{7(6)(7)}{2} - 2(6) = 135$

13. $\displaystyle\sum_{k=1}^{4} \frac{2}{k^2 + 4} = \frac{2}{1+4} + \frac{2}{4+4} + \frac{2}{9+4} + \frac{2}{16+4}$

$\approx \dfrac{47}{52}$

14. $\displaystyle\sum_{n=0}^{10} 9\left(\frac{3}{4}\right)^n = 9\left(\frac{1 - \left(\frac{3}{4}\right)^{11}}{1 - \left(\frac{3}{4}\right)}\right) \approx 34.4795$

15. $\displaystyle\sum_{n=0}^{50} 100\left(-\frac{1}{2}\right)^n = \sum_{n=1}^{51} 100\left(-\frac{1}{2}\right)^{n-1}$

$= 100\dfrac{1 - \left(-\frac{1}{2}\right)^{51}}{1 - \left(-\frac{1}{2}\right)}$

$\approx \dfrac{2}{3}(100) \approx 66.67$

16. $\sum_{n=0}^{\infty} 3\left(-\frac{3}{5}\right)^n = \frac{3}{1-\left(-\frac{3}{5}\right)} = \frac{3}{\frac{8}{5}} = \frac{15}{8}$

17. $\sum_{n=1}^{\infty} 5(-0.02)^n = 5(-0.02)\frac{1}{1-(-0.02)}$

$= \frac{-0.1}{1.02} = -\frac{5}{51}$

18. $4 - 2 + 1 - \frac{1}{2} + \frac{1}{4} - \cdots = \sum_{n=0}^{\infty} 4\left(-\frac{1}{2}\right)^n$

$= \frac{4}{1-\left(-\frac{1}{2}\right)} = \frac{8}{3}$

19. (a) ${}_{20}C_{18} = \frac{20!}{(20-18)!18!} = \frac{20!}{2!18!} = 190$

(b) $\binom{20}{2} = \frac{20!}{(20-2)!2!} = \frac{20}{18!2!} = 190$

20. $(x+3)^4 = x^4 + 12x^3 + 54x^2 + 108x + 81$

21. $(2x+y^2)^5 = 32x^5 + 80x^4y^2 + 80x^3y^4 + 40x^2y^6 + 10xy^8 + y^{10}$

22. $(x-2y)^6 = x^6 - 12x^5y + 60x^4y^2 - 160x^3y^3 + 240x^2y^4 - 192xy^5 + 64y^6$

23. $(3a-4b)^8 = 6561a^8 - 69{,}698a^7b + 326{,}592a^6b^2 - 870{,}912a^5b^3$

$+ 1{,}451{,}520a^4b^4 - 1{,}548{,}288a^3b^5 + 1{,}032{,}192a^2b^6$

$- 393{,}216ab^7 + 65{,}536b^8$

24. Permutations of L, I, O, N, S

${}_5P_5 = \frac{5!}{(5-5)!} = \frac{5!}{0!} = 120$

25. Permutations of S, E, A, B, E, E, S

${}_7P_7 = \frac{7!}{(7-7)!} = \frac{7!}{0!} = 5040$

However, because there are 3! permutations of three letter Es, and 2! permutations of two letter Ss, the number of distinguishable permutations is:

$\frac{7!}{3!2!} \Rightarrow \frac{5040}{6 \cdot 2} = 420$ permutations.

26. Permutations of B, O, B, B, L, E, H, E, A, D

${}_{10}P_{10} = \frac{10!}{0!} = 3{,}628{,}800$

However, because there are 3! permutations of three letter Bs, and 2! permutations of two letter Es, the number of distinguishable permutations is:

$\frac{10!}{3!2!} = \frac{3{,}628{,}800}{6 \cdot 2} = 302{,}400$ permutations.

27. Permutations of I, N, T, U, I, T, I, O, N

${}_9P_9 = \frac{9!}{0!} = 9! = 362{,}880$

However, because there are 3! permutations of three letter Is, 2! permutations of two letter Ts, and 2! permutations of two letter Ns, the number of distinguishable permutations is:

$\frac{9!}{3!2!2!} = 15{,}120$ permutations.

28. Hyperbola with center $(5, -3)$

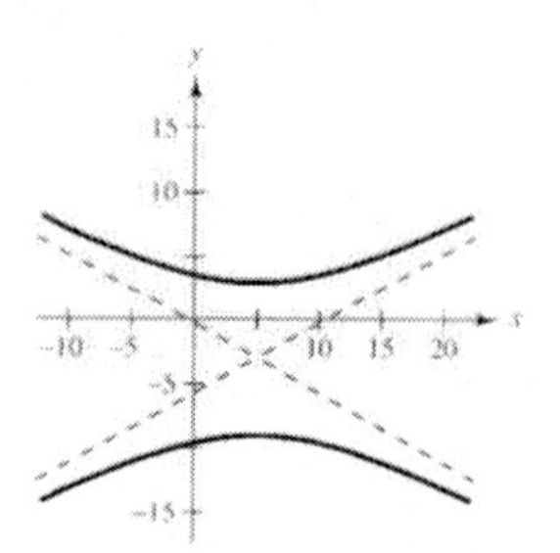

29. Ellipse with center $(2, -1)$

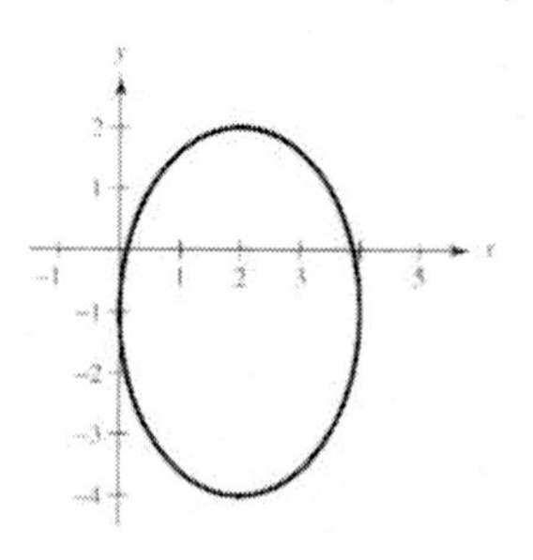

30. Hyperbola with center $(0, 0)$

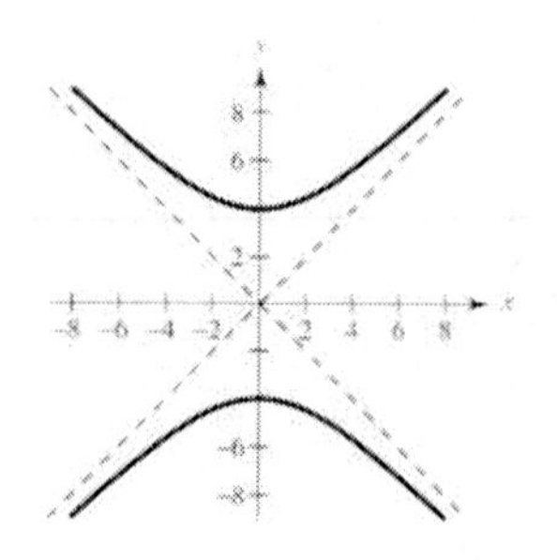

31. $(x^2 - 2x + 1) + (y^2 - 4y + 4) = -1 + 1 + 4$

$(x-1)^2 + (y-2)^2 = 4$

Circle

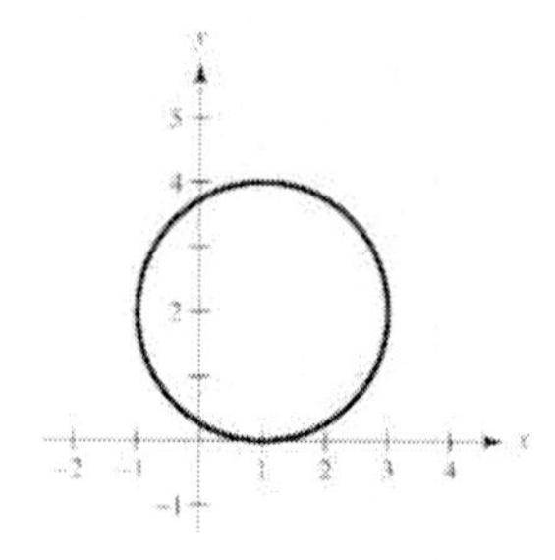

32. $(x-2)^2 = 4p(y-3)$

$(0, 0): 4 = 4p(-3) \Rightarrow p = -\frac{1}{3}$

$(x-2)^2 = 4\left(-\frac{1}{3}\right)(y-3)$

$(x-2)^2 = -\frac{4}{3}(y-3)$

33. Center: $(1, 4)$

$a = 5, b = 2$

$$\frac{(x-1)^2}{25} + \frac{(y-4)^2}{4} = 1$$

34. Center: $(0, -4)$; $a = 2$

$$\frac{(y+4)^2}{4} - \frac{x^2}{b^2} = 1$$

$(4, 0): 4 - \frac{16}{b^2} = 1 \Rightarrow \frac{16}{b^2} = 3 \Rightarrow b^2 = \frac{16}{3}$

$$\frac{(y+4)^2}{4} - \frac{x^2}{\frac{16}{3}} = 1$$

35. $B^2 - 4AC = 16 - 8 = 8 \Rightarrow$ Hyperbola

$\cot 2\theta = \frac{1-2}{-4} = \frac{1}{4} \Rightarrow \theta \approx 38°$

Graph as:

$2y^2 - 4xy + (x^2 - 6) = 0$

$$y = \frac{4x \pm \sqrt{16x^2 - 8(x^2 - 6)}}{4}$$

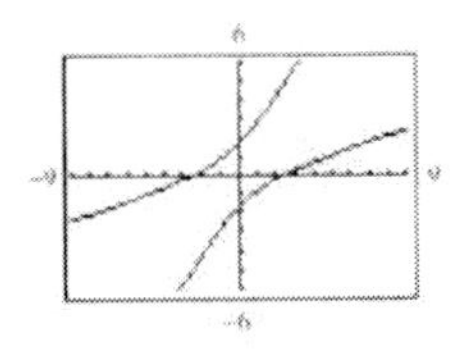

36. $x = 2t + 1, y = t^2$

(a), (b)

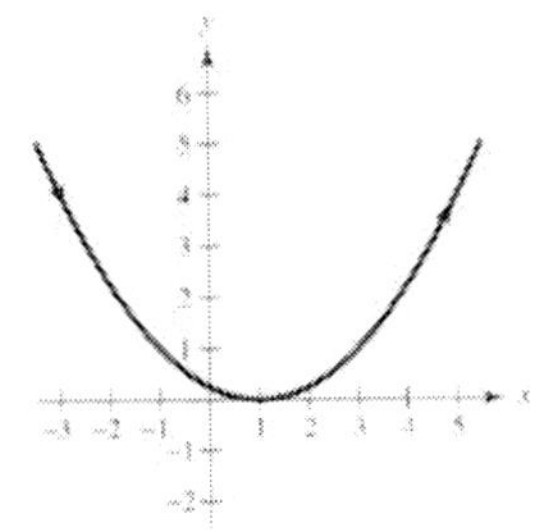

(c) $t = \frac{x-1}{2} \Rightarrow y = \left(\frac{x-1}{2}\right)^2 = \frac{1}{4}(x-1)^2$

37. $x = \cos\theta, y = 2\sin^2\theta$

(a), (b)

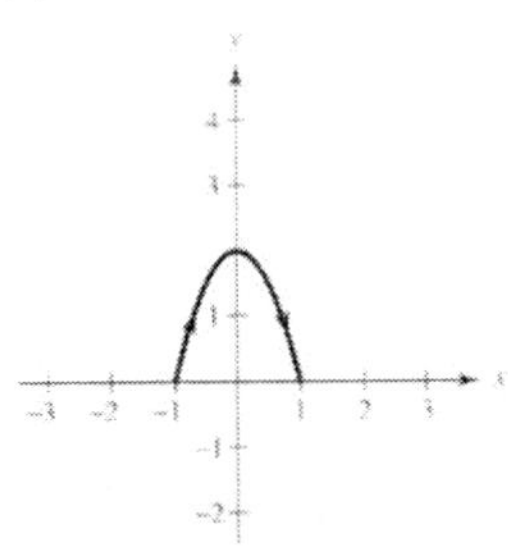

(c) $y = 2\sin^2\theta$

$= 2(1 - \cos^2\theta)$

$= 2(1 - x^2), \ -1 \le x \le 1$

38. $x = 4 \ln t, y = \frac{1}{2}t^2$

(a), (b)

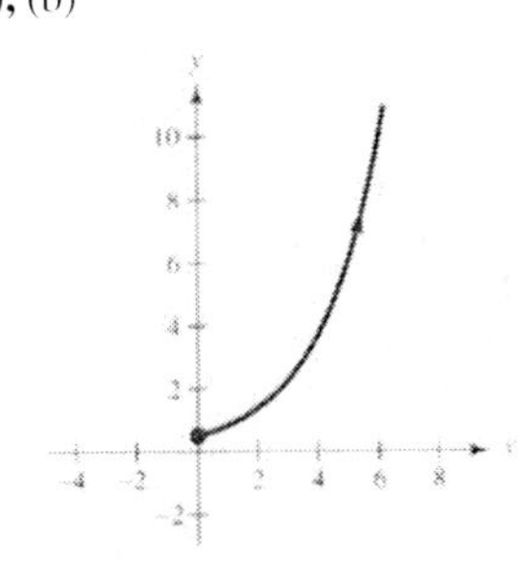

(c) $t = e^{x/4} \Rightarrow y = \frac{1}{2}e^{x/2}, x \ge 0$

39. $y = 3x - 2$

(a) Let $t = x$.

$x = t$

$y = 3t - 2$

(b) Let $t = \frac{x}{2}$.

$x = 2t$

$y = 3(2t) - 2 \Rightarrow y = 6t - 2$

40. $x^2 - y = 16$

(a) Let $t = x$.

$x = t$

$y = t^2 - 16$

(b) Let $t = \frac{x}{2}$.

$x = 2t$

$y = (2t)^2 - 16 \Rightarrow y = 4t^2 - 16$

41. $y = \frac{2}{x}$

(a) Let $t = x$.

$x = t$

$y = \frac{2}{t}$

(b) Let $t = \frac{x}{2}$.

$x = 2t$

$y = \frac{2}{2t} \Rightarrow y = \frac{1}{t}$

42. $y = \frac{e^{2x}}{e^{2x}+1}$

(a) Let $t = x$.

$x = t$

$y = \frac{e^{2t}}{e^{2t}+1}$

(b) Let $t = \frac{x}{2}$.

$x = 2t$

$y = \frac{e^{2(2t)}}{e^{2(2t)}+1} \Rightarrow y = \frac{e^{4t}}{e^{4t}+1}$

43. $(r, \theta) = \left(8, \frac{5\pi}{6}\right)$

$\left(8, -\frac{7\pi}{6}\right), \left(-8, -\frac{\pi}{6}\right), \left(-8, \frac{11\pi}{6}\right)$

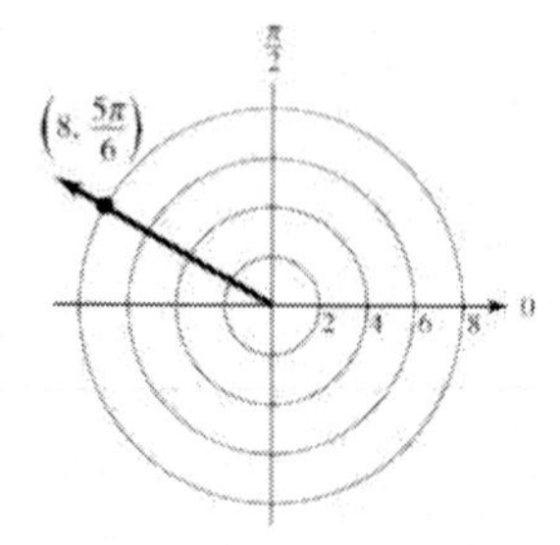

44. $(r, \theta) = \left(5, -\frac{3\pi}{4}\right)$

$\left(5, \frac{5\pi}{4}\right), \left(-5, \frac{\pi}{4}\right), \left(-5, -\frac{7\pi}{4}\right)$

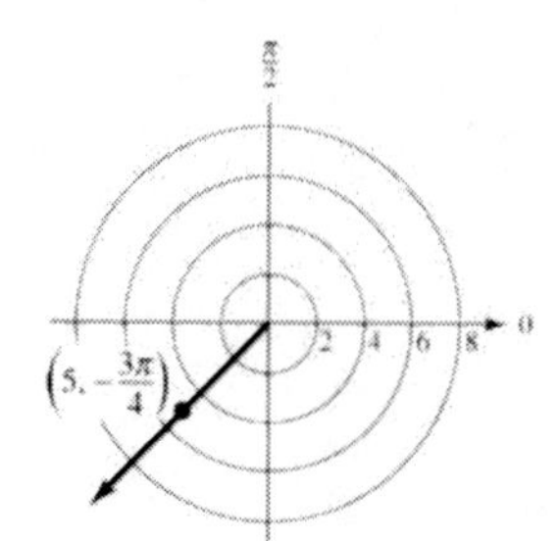

45. $(r, \theta) = \left(-2, \frac{5\pi}{4}\right)$

$\left(-2, -\frac{3\pi}{4}\right), \left(2, \frac{\pi}{4}\right), \left(2, -\frac{7\pi}{4}\right)$

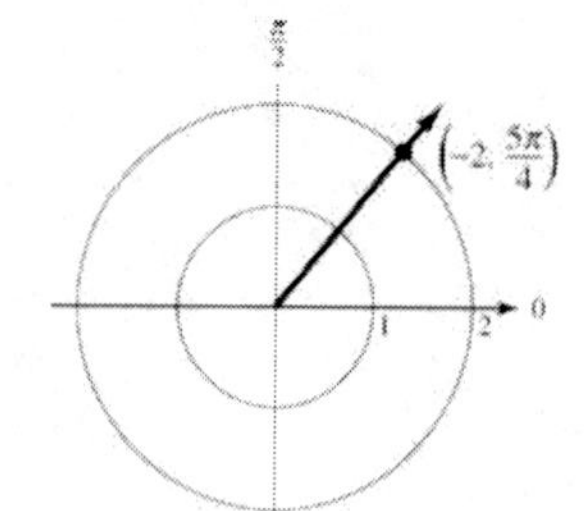

46. $(r, \theta) = \left(-3, -\frac{11\pi}{6}\right)$

$\left(-3, \frac{\pi}{6}\right), \left(3, \frac{7\pi}{6}\right), \left(3, -\frac{5\pi}{6}\right)$

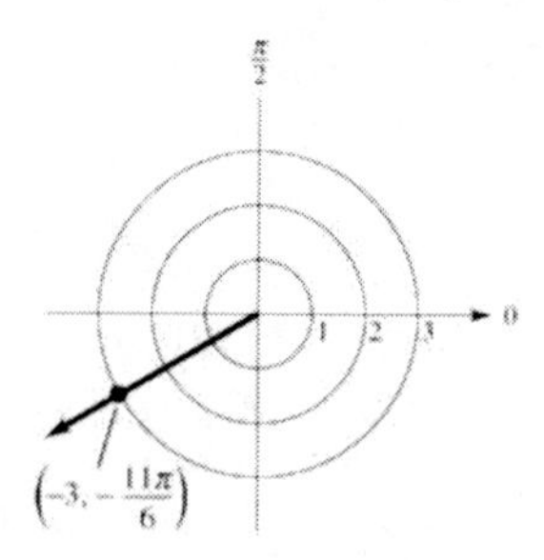

47.

$$4x + 4y + 1 = 0$$
$$4r\cos\theta + 4r\sin\theta + 1 = 0$$
$$r\left[4\cos\theta + 4\sin\theta\right] = -1$$
$$r = \frac{-1}{4\cos\theta + 4\sin\theta}$$

48.

$$r = 4\cos\theta$$
$$r^2 = 4r\cos\theta$$
$$x^2 + y^2 = 4x$$
$$x^2 + y^2 - 4x = 0$$
$$x^2 - 4x + 4 + y^2 = 4$$
$$(x-2)^2 + y^2 = 4$$

49.

$$r = \frac{2}{4 - 5\cos\theta}$$
$$4r - 5r\cos\theta = 2$$
$$4\left(x^2 + y^2\right)^{1/2} - 5x = 2$$
$$16\left(x^2 + y^2\right) = (5x + 2)^2 = 25x^2 + 20x + 4$$
$$16y^2 + 9x^2 - 20x = 4$$

50. $r = -\dfrac{\pi}{6}$, Circle

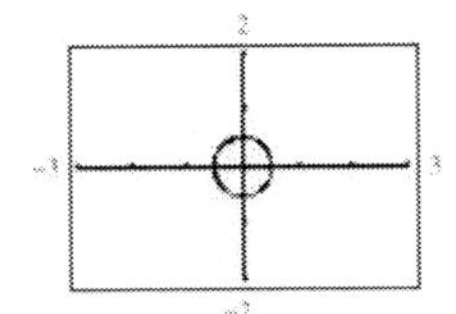

51. $r = 3 - 2\sin\theta$
Limaçon

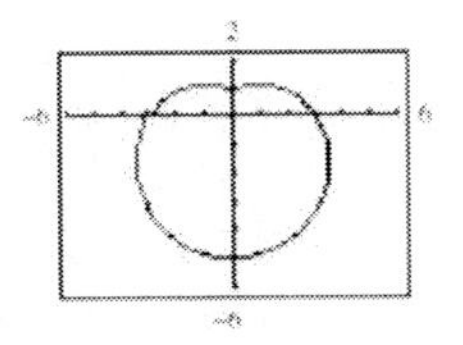

52. $r = 2 + 5\cos\theta$
Limaçon with inner loop

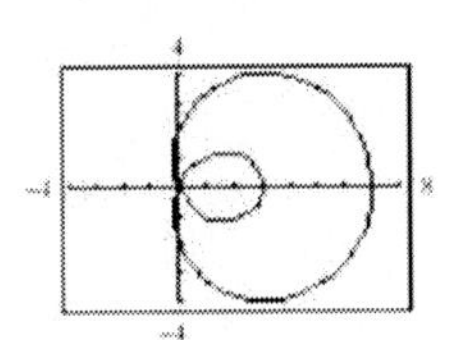

53. $32{,}500 + 32{,}500(1.05) + \cdots + 32{,}500(1.05)^{14}$

$$= \sum_{n=1}^{15} 32{,}500(1.05)^{n-1}$$

$$= 32{,}500\left(\frac{1-1.05^{15}}{1-1.05}\right)$$

$$\approx \$701{,}303.32$$

54. There are two ways to select the first digit (4 or 5), and two ways for the second digit. Hence, $p = \dfrac{1}{4}$.

55. Let $y = -ax^2 + 16$.

$$(6,\ 14):\ 14 = -a(6)^2 + 16$$

$$36a = 2$$

$$a = \tfrac{1}{18}$$

$$y = -\tfrac{1}{18}x^2 + 16$$

$$y = 0:\ 16 = \tfrac{1}{18}x^2$$

$$x^2 = 288$$

$$x = 12\sqrt{2}$$

Width: $24\sqrt{2}$ meters

CHAPTER 10

Section 10.1

1. three-dimensional

3. Distance Formula

5. surface, space

7. The standard equation of a sphere with radius r and center (h, k, j) is given by
$(x-h)^2+(y-k)^2+(z-j)^2=r^2$.

9. $A(-1, 4, 3)$, $B(1, 3, -2)$, $C(-3, 0, -2)$

11. $A(-2, -1, 4)$, $B(3, -2, 0)$, $C(-2, 2, -3)$

13.

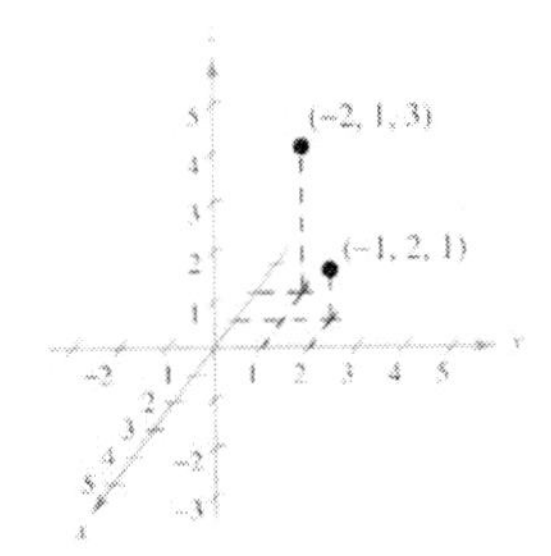

15.

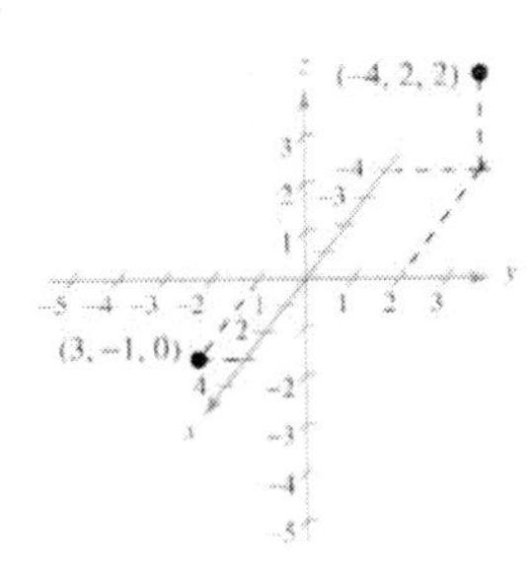

17.

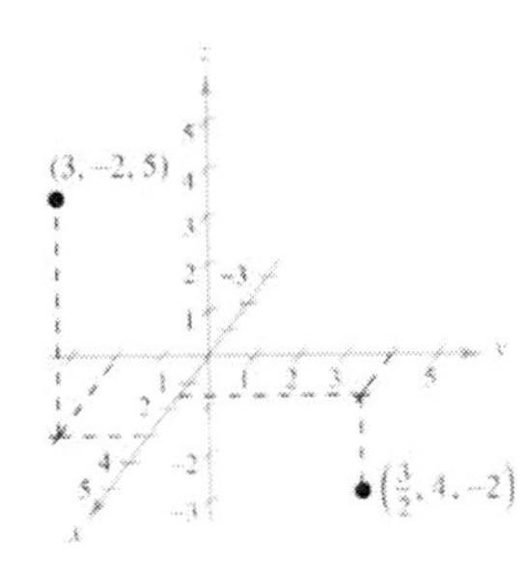

19. $x=-3,\ y=3,\ z=5 \Rightarrow (-3, 3, 5)$

21. $x=11,\ y=0,\ z=0 \Rightarrow (11, 0, 0)$

23. Octant V

25. Octants I, II, III, IV
(above the xy-plane)

27. Octants II, IV, VI, VIII

29. $d=\sqrt{(7-3)^2+(4-2)^2+(8-(-5))^2}$
$=\sqrt{4^2+2^2+13^2}$
$=\sqrt{16+4+169}$
$=\sqrt{189}$
$=3\sqrt{21}\approx 13.748$ units

31. $d=\sqrt{(5-0)^2+(2-0)^2+(6-0)^2}$
$=\sqrt{25+4+36}=\sqrt{65}\approx 8.062$ units

33. $d=\sqrt{[6-(-1)]^2+[0-4]^2+[-9-(-2)]^2}$
$=\sqrt{7^2+4^2+7^2}$
$=\sqrt{49+16+49}$
$=\sqrt{114}$
≈ 10.677 units

35. $d=\sqrt{(1-0)^2+[0-(-2)]^2+(-10-0)^2}$
$=\sqrt{1+4+100}$
$=\sqrt{105}\approx 10.247$ units

37. $d_1=\sqrt{(0-0)^2+(0-4)^2+(2-0)^2}=\sqrt{20}=2\sqrt{5}$
$d_2=\sqrt{(0-(-2))^2+(0-5)^2+(2-2)^2}=\sqrt{29}$
$d_3=\sqrt{(-2-0)^2+(5-4)^2+(2-0)^2}=3$
$d_1^2+d_3^2=20+9=29=d_2^2$

39. $d_1=\sqrt{(2-0)^2+(2-0)^2+(1-0)^2}=\sqrt{9}=3$
$d_2=\sqrt{(2-2)^2+(-4-2)^2+(4-1)^2}=\sqrt{45}=3\sqrt{5}$
$d_3=\sqrt{(2-0)^2+(-4-0)^2+(4-0)^2}=\sqrt{36}=6$
$d_1^2+d_3^2=9+36=45=d_2^2$

41. $d_1=\sqrt{(5-1)^2+(-1+3)^2+(2+2)^2}=\sqrt{16+4+16}=\sqrt{36}=6$
$d_2=\sqrt{(5+1)^2+(-1-1)^2+(2-2)^2}=\sqrt{36+4}=\sqrt{40}=2\sqrt{10}$
$d_3=\sqrt{(-1-1)^2+(1+3)^2+(2+2)^2}=\sqrt{4+16+16}=\sqrt{36}=6$
$d_1=d_3$, Isosceles triangle

43. $d_1=\sqrt{(8-4)^2+(1+1)^2+(2+2)^2}=\sqrt{36}=6$
$d_2=\sqrt{(8-2)^2+(1-3)^2+(2-2)^2}=\sqrt{40}=2\sqrt{10}$
$d_3=\sqrt{(4-2)^2+(-1-3)^2+(-2-2)^2}=\sqrt{36}=6$
$d_1=d_3$, Isosceles triangle

45. Midpoint: $\left(\frac{3-3}{2}, \frac{-6+4}{2}, \frac{10+4}{2}\right)=(0, -1, 7)$

47. Midpoint: $\left(\frac{6+(-4)}{2}, \frac{-2+2}{2}, \frac{5+7}{2}\right) = (1, 0, 6)$

49. Midpoint: $\left(\frac{-2+7}{2}, \frac{8-4}{2}, \frac{10+2}{2}\right) = \left(\frac{5}{2}, 2, 6\right)$

51. Center: $(3, 2, 4) \Rightarrow h = 3, k = 2, j = 4$

Radius: $r = 4$

$$(x-h)^2 + (y-k)^2 + (z-j)^2 = r^2$$
$$(x-3)^2 + (y-2)^2 + (z-4)^2 = 4^2$$
$$(x-3)^2 + (y-2)^2 + (z-4)^2 = 16$$

53. Center: $(-1, 2, 0) \Rightarrow h = -1, k = 2, j = 0$

Radius: $r = \sqrt{3}$

$$(x-h)^2 + (y-k)^2 + (z-j)^2 = r^2$$
$$(x-(-1))^2 + (y-2)^2 + (z-0)^2 = \left(\sqrt{3}\right)^2$$
$$(x+1)^2 + (y-2)^2 + z^2 = 3$$

55. Center: $(0, 4, 3) \Rightarrow h = 0, k = 4, j = 3$

Radius: $r = 4$

$$(x-h)^2 + (y-k)^2 + (z-j)^2 = r^2$$
$$(x-0)^2 + (y-4)^2 + (z-3)^2 = 4^2$$
$$x^2 + (y-4)^2 + (z-3)^2 = 16$$

57. Center: $(-3, 7, 5) \Rightarrow h = -3, k = 7, j = 5$

$$\text{Radius} = \frac{\text{Diameter}}{2} = 5$$
$$(x-h)^2 + (y-k)^2 + (z-j)^2 = r^2$$
$$(x+3)^2 + (y-7)^2 + (z-5)^2 = 5^2 = 25$$

59. Center: $\left(\frac{3+0}{2}, \frac{0+0}{2}, \frac{0+6}{2}\right) = \left(\frac{3}{2}, 0, 3\right) \Rightarrow$

$h = \frac{3}{2}, k = 0, j = 3$

$$\text{Radius: } \sqrt{\left(3-\frac{3}{2}\right)^2 + (0-0)^2 + (0-3)^2}$$
$$= \sqrt{\frac{9}{4} + 9}$$
$$= \sqrt{\frac{45}{4}}$$
$$(x-h)^2 + (y-k)^2 + (z-j)^2 = r^2$$
$$\left(x-\frac{3}{2}\right)^2 + (y-0)^2 + (z-3)^2 = \frac{45}{4}$$

61.
$$x^2 + y^2 + z^2 - 6x = 0$$
$$(x^2 - 6x + 9) + y^2 + z^2 = 9$$
$$(x-3)^2 + y^2 + z^2 = 9$$

Center: $(3, 0, 0)$

Radius: 3

$$x^2 + \left(y^2 - 9y + \frac{81}{4}\right) + z^2 = \frac{81}{4}$$
$$x^2 + \left(y - \frac{9}{2}\right)^2 + z^2 = \frac{81}{4}$$

Center: $\left(0, \frac{9}{2}, 0\right)$

Radius: $\frac{9}{2}$

63.
$$x^2 + y^2 + z^2 - 4x + 2y = 0$$
$$(x^2 - 4x + 4) + (y^2 + 2y + 1) + z^2 = 4 + 1$$
$$(x-2)^2 + (y+1)^2 + z^2 = 5$$

Center: $(2, -1, 0)$

Radius: $\sqrt{5}$

65.
$$x^2 + y^2 + z^2 - 4x + 2y - 6z + 10 = 0$$
$$(x^2 - 4x + 4) + (y^2 + 2y + 1) + (z^2 - 6z + 9) = -10 + 4 + 1 + 9$$
$$(x-2)^2 + (y+1)^2 + (z-3)^2 = 4$$

Center: $(2, -1, 3)$

Radius: 2

67.
$$x^2 + y^2 + z^2 + 4x - 8z + 19 = 0$$
$$(x^2 + 4x + 4) + y^2 + (z^2 - 8z + 16) = -19 + 4 + 16$$
$$(x+2)^2 + y^2 + (z-4)^2 = 1$$

Center: $(-2, 0, 4)$

Radius: 1

69.
$$9x^2 + 9y^2 + 9z^2 - 18x - 6y - 72z + 73 = 0$$
$$x^2 + y^2 + z^2 - 2x - \frac{2}{3}y - 8z = -\frac{73}{9}$$
$$(x^2 - 2x + 1) + \left(y^2 - \frac{2}{3}y + \frac{1}{9}\right) + (z^2 - 8z + 16) = -\frac{73}{9} + 1 + \frac{1}{9} + 16$$
$$(x-1)^2 + \left(y - \frac{1}{3}\right)^2 + (z-4)^2 = 9$$

Center: $\left(1, \frac{1}{3}, 4\right)$

Radius: 3

71. $$4x^2 + 4y^2 + 4z^2 - 8x + 16y - 1 = 0$$

$$4(x^2 - 2x + 1) + 4(y^2 + 4y + 4) + 4z^2 = 4 + 16 + 1$$

$$(x-1)^2 + (y+2)^2 + z^2 = \frac{21}{4}$$

Center: $(1, -2, 0)$

Radius: $\dfrac{\sqrt{21}}{2}$

73. $$9x^2 - 6x + 9y^2 + 18y + 9z^2 = -1$$

$$x^2 - \frac{2}{3}x + \frac{1}{9} + y^2 + 2y + 1 + z^2 = -\frac{1}{9} + \frac{1}{9} + 1$$

$$\left(x - \frac{1}{3}\right)^2 + (y+1)^2 + z^2 = 1$$

Center: $\left(\dfrac{1}{3}, -1, 0\right)$

Radius: 1

75. xz-trace $(y = 0): (x-1)^2 + z^2 = 36$, Circle

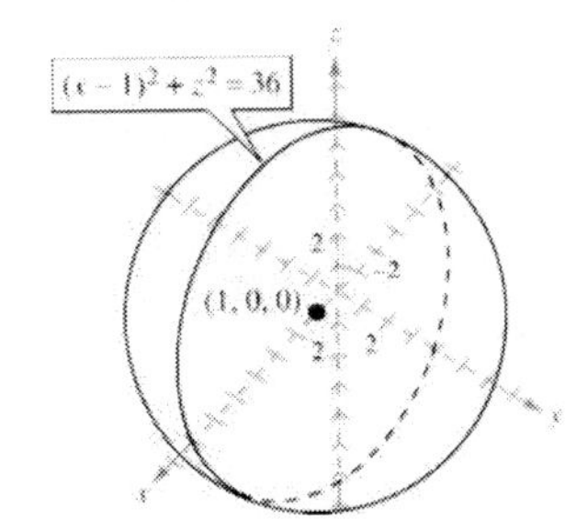

77. yz-trace$(x = 0): (y-3)^2 + z^2 = 9 - 4 = 5$, Circle

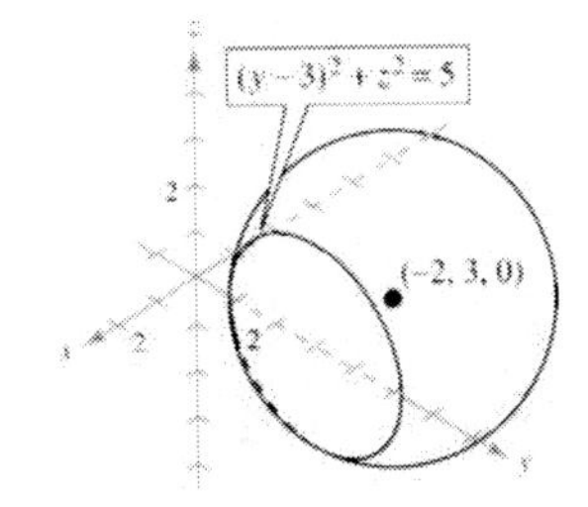

79. $$(x^2 - 2x + 1) + y^2 + (z^2 - 4z + 4) = -1 + 1 + 4$$

$$(x-1)^2 + y^2 + (z-2)^2 = 4$$

yz-trace $(x = 0): y^2 + (z-2)^2 = 3$, Circle

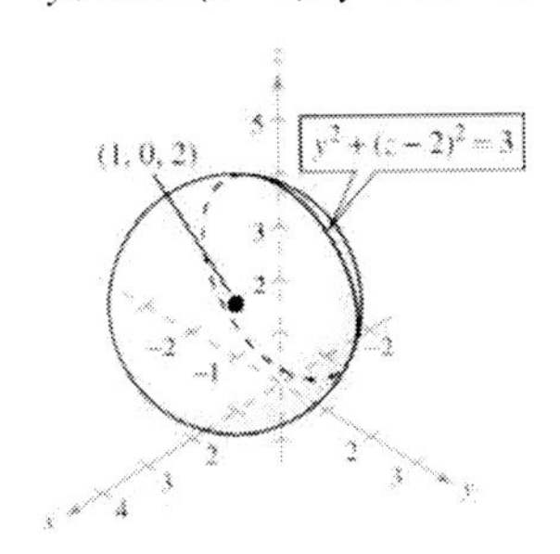

81. $$x^2 + y^2 + z^2 - 6x - 8y - 10z + 46 = 0$$

$$x^2 - 6x + y^2 - 8y + (z^2 - 10z + 25) = -46 + 25$$

$$x^2 - 6x + y^2 - 8y + (z-5)^2 = -21$$

$$z_1 = 5 + \sqrt{-21 - x^2 + 6x - y^2 + 8y}$$

$$z_2 = 5 - \sqrt{-21 - x^2 + 6x - y^2 + 8y}$$

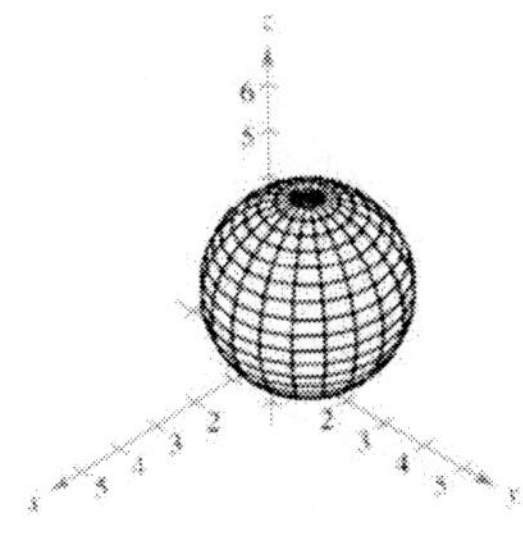

83. $$4x^2 + 4y^2 + 4z^2 - 8x - 16y + 8x - 25 = 0$$

$$x^2 + y^2 + z^2 - 2x - 4y + 2z = \frac{25}{4}$$

$$x^2 - 2x + y^2 - 4y + (z^2 + 2z + 1) = \frac{25}{4} + 1$$

$$x^2 - 2x + y^2 - 4y + (z+1)^2 = \frac{29}{4}$$

$$z_1 = -1 + \sqrt{\frac{29}{4} - x^2 + 2x - y^2 + 4y}$$

$$z_2 = -1 - \sqrt{\frac{29}{4} - x^2 + 2x - y^2 + 4y}$$

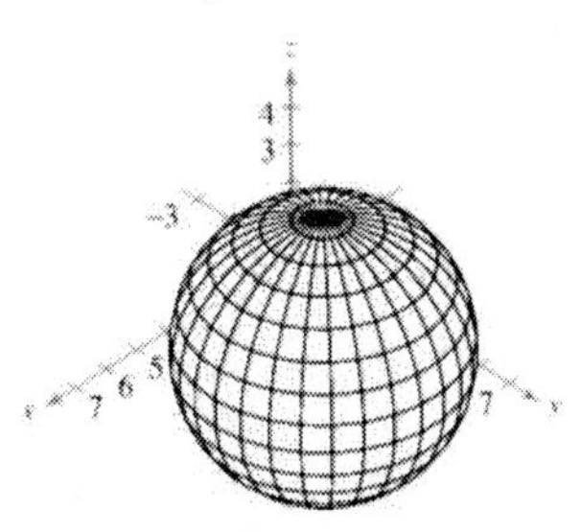

85. The length of each side is 3.
Thus, $(x, y, z) = (3, 3, 3)$.

87. $d = 205 \Rightarrow r = \dfrac{205}{2} = 102.5$

$$x^2 + y^2 + z^2 = \left(\frac{205}{2}\right)^2$$

89. False. x is the directed distance from the yz-plane to P.

91. In the xy-plane, the z-coordinate is 0.
In the xz-plane, the y-coordinate is 0.
In the yz-plane, the x-coordinate is 0.

93. $x_m = \dfrac{x_1 + x_2}{2} \Rightarrow x_2 = 2x_m - x_1$

Similarly for y_2 and z_2,

$(x_2, y_2, z_2) = (2x_m - x_1, 2y_m - y_1, 2z_m - z_1)$.

95. $v^2 + 3v + \frac{9}{4} = 2 + \frac{9}{4}$

$$\left(v + \frac{3}{2}\right)^2 = \frac{17}{4}$$

$$v + \frac{3}{2} = \pm\frac{\sqrt{17}}{2}$$

$$v = -\frac{3}{2} \pm \frac{\sqrt{17}}{2}$$

97. $x^2 - 5x + \frac{25}{4} = -5 + \frac{25}{4}$

$$\left(x - \frac{5}{2}\right)^2 = \frac{5}{4}$$

$$x - \frac{5}{2} = \pm\frac{\sqrt{5}}{2}$$

$$x = \frac{5}{2} \pm \frac{\sqrt{5}}{2}$$

99. $4y^2 + 4y = 9$

$$y^2 + y + \frac{1}{4} = \frac{9}{4} + \frac{1}{4}$$

$$\left(y + \frac{1}{2}\right)^2 = \frac{10}{4}$$

$$y + \frac{1}{2} = \pm\frac{\sqrt{10}}{2}$$

$$y = -\frac{1}{2} \pm \frac{\sqrt{10}}{2}$$

Section 10.2

1. zero

3. parallel

5. The magnitude (or length) of $\mathbf{v} = \langle v_1, v_2, v_3 \rangle$ is $\|\mathbf{v}\| = \sqrt{v_1^2 + v_2^2 + v_3^2}$.

7. (a) $\mathbf{v} = \langle 4-4, 2-2, 4-0 \rangle = \langle 0, 0, 4 \rangle$

(b)

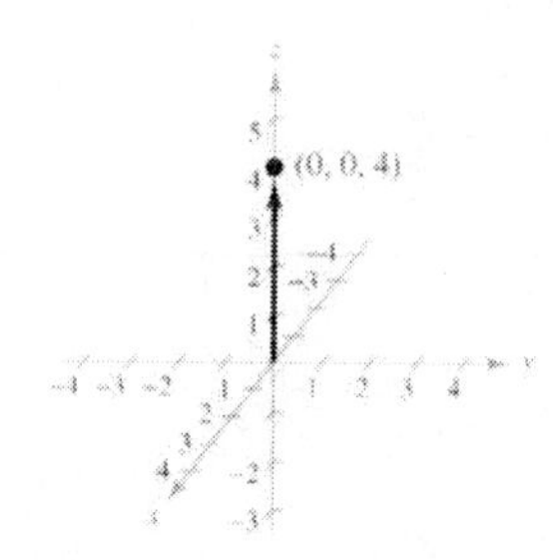

9. (a) $\mathbf{v} = \langle 0-2, 3-0, 2-1 \rangle = \langle -2, 3, 1 \rangle$

(b)

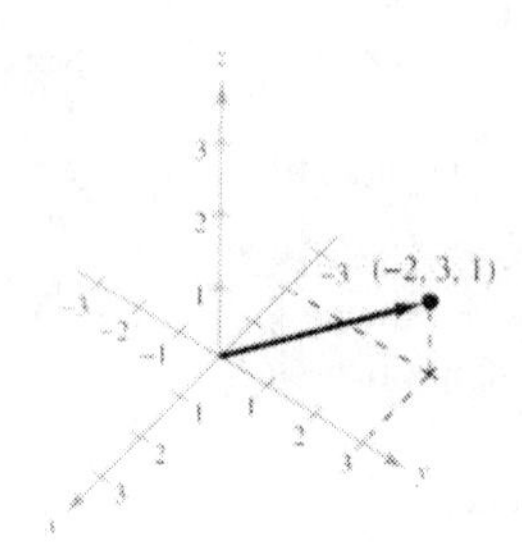

11. (a) $\mathbf{v} = \langle 1-(-6), -1-4, 3-(-2) \rangle$
$= \langle 7, -5, 5 \rangle$

(b) $\|\mathbf{v}\| = \sqrt{7^2 + (-5)^2 + 5^2}$
$= \sqrt{49 + 25 + 25}$
$= \sqrt{99}$
$= 3\sqrt{11}$

(c) $\frac{\mathbf{v}}{\|\mathbf{v}\|} = \frac{1}{3\sqrt{11}}\langle 7, -5, 5 \rangle = \frac{\sqrt{11}}{33}\langle 7, -5, 5 \rangle$

13. (a) $\mathbf{v} = \langle 1-(-1), 4-2, -4-(-4) \rangle = \langle 2, 2, 0 \rangle$

(b) $\|\mathbf{v}\| = \sqrt{2^2 + 2^2 + 0^2} = \sqrt{8} = 2\sqrt{2}$

(c) Unit vector: $\frac{1}{2\sqrt{2}}\langle 2, 2, 0 \rangle = \left\langle \frac{\sqrt{2}}{2}, \frac{\sqrt{2}}{2}, 0 \right\rangle$

15. $\mathbf{v} = \langle 1, 1, 3 \rangle$

(a) $2\mathbf{v} = \langle 2, 2, 6 \rangle$

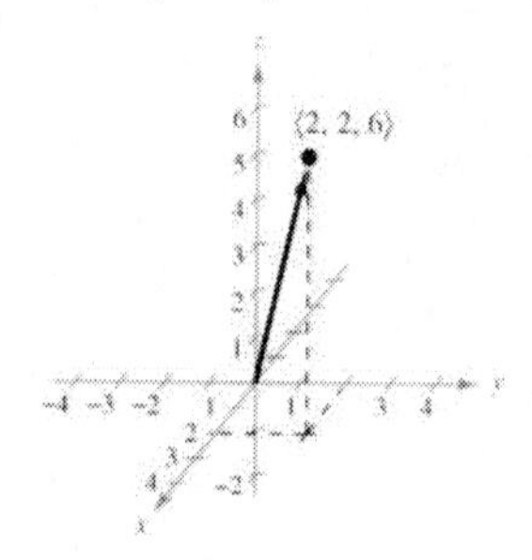

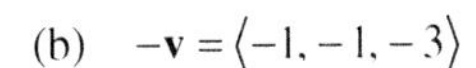

(b) $-\mathbf{v} = \langle -1, -1, -3 \rangle$

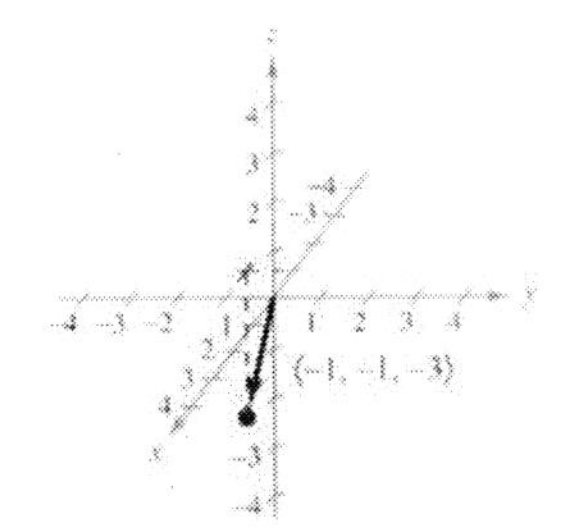

(c) $\frac{3}{2}\mathbf{v} = \left\langle \frac{3}{2}, \frac{3}{2}, \frac{9}{2} \right\rangle$

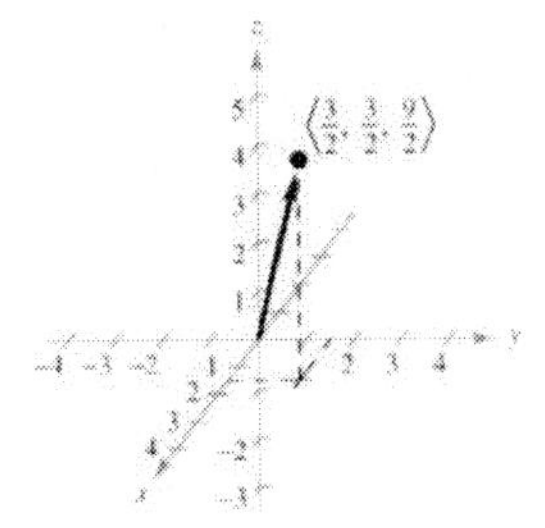

(d) $0\mathbf{v} = \langle 0, 0, 0 \rangle$

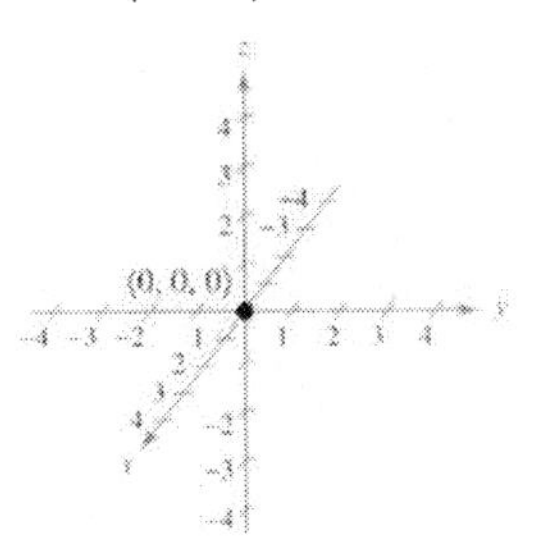

17. $\mathbf{v} = 2\mathbf{i} + 2\mathbf{j} - \mathbf{k}$

(a) $2\mathbf{v} = 4\mathbf{i} + 4\mathbf{j} - 2\mathbf{k}$

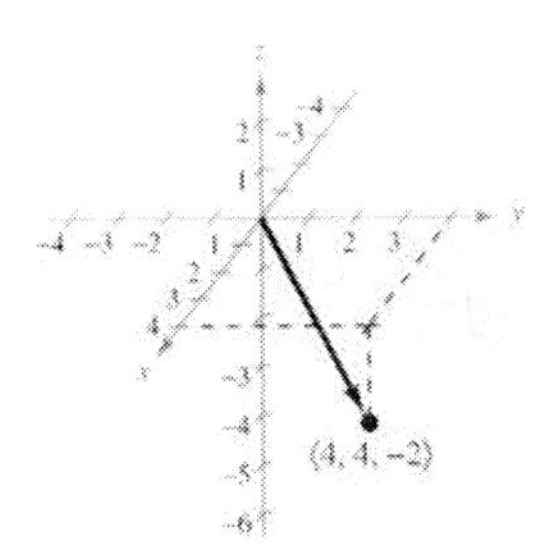

(b) $-\mathbf{v} = -2\mathbf{i} - 2\mathbf{j} + \mathbf{k}$

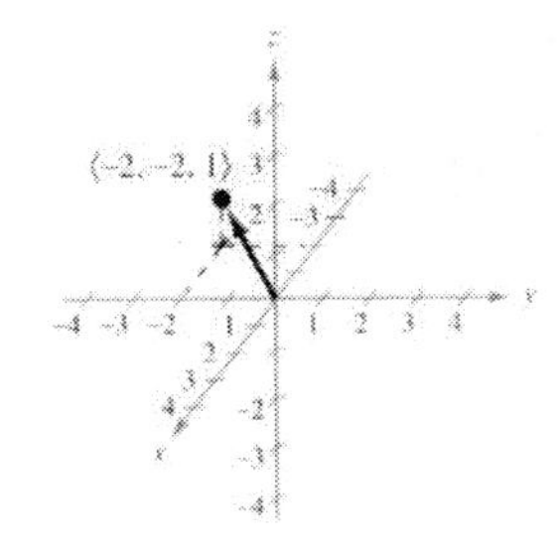

(c) $\frac{5}{2}\mathbf{v} = 5\mathbf{i} + 5\mathbf{j} - \frac{5}{2}\mathbf{k}$

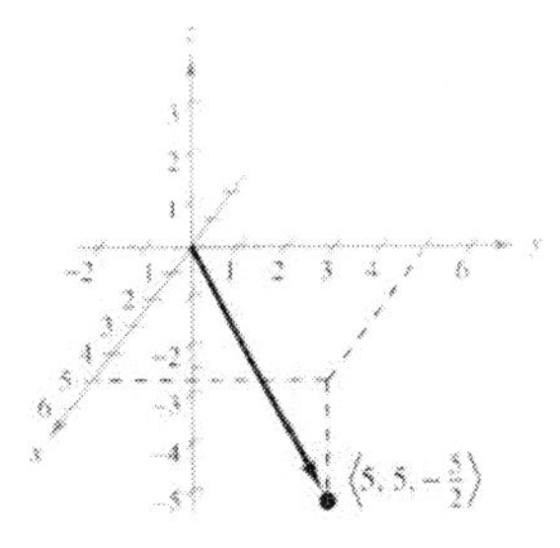

(d) $0\mathbf{v} = \mathbf{0}$

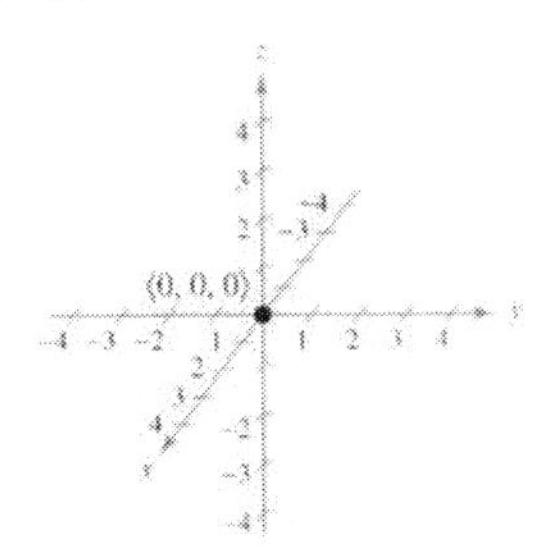

19. $\mathbf{z} = \mathbf{u} - 2\mathbf{v} = \langle -1, 3, 2 \rangle - 2\langle 1, -2, -2 \rangle = \langle -3, 7, 6 \rangle$

21. $2\mathbf{z} - 4\mathbf{u} = \mathbf{w} \Rightarrow \mathbf{z} = \frac{1}{2}(4\mathbf{u} + \mathbf{w})$

$$= \frac{1}{2}\left(4\langle -1, 3, 2 \rangle + \langle 5, 0, -5 \rangle\right)$$

$$= \left\langle \frac{1}{2}, 6, \frac{3}{2} \right\rangle$$

23. $\mathbf{z} = 2\langle -1, 3, 2 \rangle - 3\langle 1, -2, -2 \rangle + \frac{1}{2}\langle 5, 0, -5 \rangle$

$$= \left\langle -\frac{5}{2}, 12, \frac{15}{2} \right\rangle$$

25. $\|\mathbf{v}\| = \|\langle 7, 8, 9 \rangle\| = \sqrt{49 + 64 + 49} = \sqrt{162} = 9\sqrt{2}$

27. $\|\mathbf{v}\| = \sqrt{1^2 + (-2)^2 + 4^2} = \sqrt{21}$

29. $\|\mathbf{v}\| = \sqrt{2^2 + (-4)^2 + 1^2} = \sqrt{21}$

31. $\mathbf{v} = \langle 1 - 1, 0 - (-3), -1 - 4 \rangle = \langle 0, 3, -5 \rangle$

$\|\mathbf{v}\| = \sqrt{0 + 3^2 + (-5)^2} = \sqrt{34}$

33. $\|\mathbf{u}\| = \sqrt{5^2 + (-12)^2} = \sqrt{169} = 13$

(a) $\frac{1}{13}(5\mathbf{i} - 12\mathbf{k})$

(b) $-\frac{1}{13}(5\mathbf{i} - 12\mathbf{k})$

35. (a) $\dfrac{\mathbf{u}}{\|\mathbf{u}\|} = \dfrac{\langle 8, 3, -1\rangle}{\sqrt{74}}$

$= \dfrac{1}{\sqrt{74}}(8\mathbf{i} + 3\mathbf{j} - \mathbf{k}) = \dfrac{\sqrt{74}}{74}\langle 8, 3, -1\rangle$

(b) $-\dfrac{1}{\sqrt{74}}(8\mathbf{i} + 3\mathbf{j} - \mathbf{k}) = -\dfrac{\sqrt{74}}{74}\langle 8, 3, -1\rangle$

37. $6\mathbf{u} - 4\mathbf{v} = 6\langle -1, 3, 4\rangle - 4\langle 5, 4.5, -6\rangle$

$= \langle -6, 18, 24\rangle + \langle -20, -18, 24\rangle$

$= \langle -26, 0, 48\rangle$

39. $\mathbf{u} + \mathbf{v} = \langle -1, 3, 4\rangle + \langle 5, 4.5, -6\rangle = \langle 4, 7.5, -2\rangle$

$\|\mathbf{u} + \mathbf{v}\| = \sqrt{4^2 + 7.5^2 + (-2)^2} = \dfrac{1}{2}\sqrt{305} \approx 8.73$

41. $\mathbf{u} \cdot \mathbf{v} = \langle 4, 4, -1\rangle . \langle 2, -5, -8\rangle$

$= 8 - 20 + 8 = -4$

43. $\mathbf{u} \cdot \mathbf{v} = \langle 2, -5, 3\rangle \cdot \langle 9, 3, -1\rangle$

$= 18 - 15 - 3 = 0$

45. $\cos\theta = \dfrac{\mathbf{u} \cdot \mathbf{v}}{\|\mathbf{u}\|\|\mathbf{v}\|} = \dfrac{-8}{\sqrt{8}\sqrt{25}} \Rightarrow \theta \approx 124.45°$

47. $\cos\theta = \dfrac{\mathbf{u} \cdot \mathbf{v}}{\|\mathbf{u}\|\|\mathbf{v}\|} = \dfrac{-120}{\sqrt{1700}\sqrt{73}} \Rightarrow \theta \approx 109.92°$

49. $-\dfrac{3}{2}\mathbf{v} = -\dfrac{3}{2}\langle 8, -4, -10\rangle = \langle -12, 6, 15\rangle = \mathbf{u} \Rightarrow$ parallel

51. $\mathbf{u} \cdot \mathbf{v} = (0)(8) + (1)(-4) + (6)(-10) = -64 \Rightarrow$ neither

53. $\mathbf{u} \cdot \mathbf{v} = -4 + 3 + 1 = 0$

Orthogonal

55. $\mathbf{v} = \langle 7 - 5, 3 - 4, -1 - 1\rangle = \langle 2, -1, -2\rangle$

$\mathbf{u} = \langle 4 - 7, 5 - 3, 3 - (-1)\rangle = \langle -3, 2, 4\rangle$

Since $\mathbf{u}$ and $\mathbf{v}$ are not parallel, the points are not collinear.

57. $\mathbf{v} = \langle -1 - 1, 2 - 3, 5 - 2\rangle = \langle -2, -1, 3\rangle$

$\mathbf{u} = \langle 3 - (-1), 4 - 2, -1 - 5\rangle = \langle 4, 2, -6\rangle$

Since $\mathbf{u} = -2\mathbf{v}$, the points are collinear.

59. The vector $\langle 1, 2, 0\rangle$ joining $(1, 2, 0)$ and $(0, 0, 0)$ is perpendicular to the vector $\langle -2, 1, 0\rangle$ joining $(-2, 1, 0)$ and $(0, 0, 0)$:

$\langle 1, 2, 0\rangle \cdot \langle -2, 1, 0\rangle = -2 + 2 = 0$

The triangle is a right triangle.

61. The three sides of the triangle are given by the vectors:

$\mathbf{u} = \langle -2, 4, -2\rangle$

$\mathbf{v} = \langle -3, 5, -4\rangle$

$\mathbf{w} = \langle -1, 1, -2\rangle$

$\mathbf{u} \cdot \mathbf{v} = 34 > 0$

$\mathbf{u} \cdot \mathbf{w} = 10 > 0$

$\mathbf{v} \cdot \mathbf{w} = 16 > 0$

The triangle has three acute angles, so the triangle is an acute triangle.

63. $\mathbf{v} = \langle 2, -4, 7\rangle = \langle q_1 - 1, q_2 - 5, q_3 - 0\rangle \Rightarrow$

$$\left.\begin{array}{l} 2 = q_1 - 1 \\ -4 = q_2 - 5 \\ 7 = q_3 \end{array}\right\} \Rightarrow \left.\begin{array}{l} q_1 = 3 \\ q_2 = 1 \\ q_3 = 7 \end{array}\right\} \Rightarrow \text{Terminal point is } (3, 1, 7).$$

65. $\mathbf{v} = \left\langle 4, \dfrac{3}{2}, -\dfrac{1}{4}\right\rangle = \left\langle q_1 - 2, q_2 - 1, q_3 + \dfrac{3}{2}\right\rangle$

$4 = q_1 - 2 \Rightarrow q_1 = 6$

$\dfrac{3}{2} = q_2 - 1 \Rightarrow q_2 = \dfrac{5}{2}$

$-\dfrac{1}{4} = q_3 + \dfrac{3}{2} \Rightarrow q_3 = -\dfrac{7}{4}$

Terminal point: $\left(6, \dfrac{5}{2}, -\dfrac{7}{4}\right)$

67.

$\|c\mathbf{u}\| = 3$

$\|c\langle 1, 2, 3\rangle\| = 3$

$\sqrt{c^2\left(1^2 + 2^2 + 3^2\right)} = 3$

$\sqrt{14c^2} = 3$

$14c^2 = 9$

$c^2 = \dfrac{9}{14}$

$c = \pm\sqrt{\dfrac{9}{14}} = \dfrac{\pm 3\sqrt{14}}{14}$

69. $\mathbf{v} = \langle q_1, q_2, q_3\rangle$

Since $\mathbf{v}$ lies in the yz-plane, $q_1 = 0$. Since $\mathbf{v}$ makes an angle of 45°, $|q_2| = |q_3|$. Finally, $\|\mathbf{v}\| = 4$ implies that $q_2{}^2 + q_3{}^2 = 16$. Thus,

$q_2 = q_3 = 2\sqrt{2}$ and $\mathbf{v} = \langle 0, 2\sqrt{2}, 2\sqrt{2}\rangle$, or

$q_2 = 2\sqrt{2}$ and $q_3 = -2\sqrt{2}$ and $\mathbf{v} = \langle 0, 2\sqrt{2}, -2\sqrt{2}\rangle$.

71.

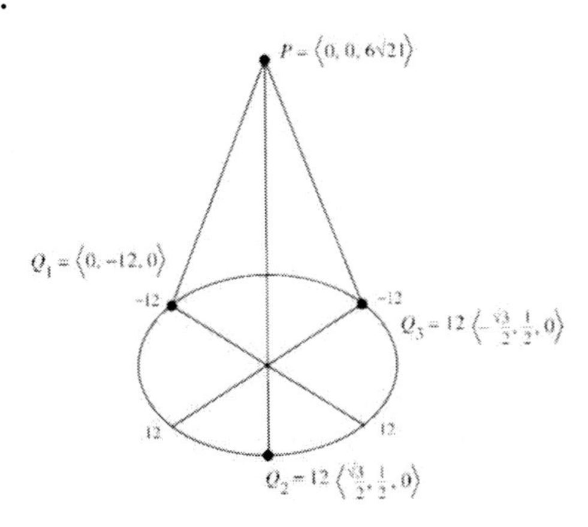

$Q_1 = (0, -12, 0)$

$Q_2 = (6\sqrt{3}, 6, 0)$

$Q_3 = (-6\sqrt{3}, 6, 0)$

$P = (0, 0, 6\sqrt{21})$

$\overline{PQ_1} = \langle 0, -12, -6\sqrt{21} \rangle$

$\overline{PQ_2} = \langle 6\sqrt{3}, 6, -6\sqrt{21} \rangle$

$\overline{PQ_3} = \langle -6\sqrt{3}, 6, -6\sqrt{21} \rangle$

Let $\mathbf{F}_1, \mathbf{F}_2, \mathbf{F}_3$ be tension on each chain.

Since $\|\mathbf{F}_1\| = \|\mathbf{F}_2\| = \|\mathbf{F}_3\|$, there exists a constant c such that

$\mathbf{F}_1 = c\langle 0, -12, -6\sqrt{21} \rangle$

$\mathbf{F}_2 = c\langle 6\sqrt{3}, 6, -6\sqrt{21} \rangle$

$\mathbf{F}_3 = c\langle -6\sqrt{3}, 6, -6\sqrt{21} \rangle.$

The total force is $-10k = \mathbf{F}_1 + \mathbf{F}_2 + \mathbf{F}_3$.
The vertical component k satisfies

$-10 = -18\sqrt{21}c \Rightarrow c = \dfrac{5}{9\sqrt{21}}.$

So,

$\mathbf{F}_1 = \dfrac{5}{9\sqrt{21}} = \langle 0, -12, -6\sqrt{21} \rangle = \left\langle 0, \dfrac{-20}{3\sqrt{21}}, \dfrac{-10}{3} \right\rangle$

$\mathbf{F}_2 = \dfrac{5}{9\sqrt{21}}\langle 6\sqrt{3}, 6, -6\sqrt{21} \rangle = \left\langle \dfrac{10}{3\sqrt{7}}, \dfrac{10}{3\sqrt{21}}, \dfrac{-10}{3} \right\rangle$

$\mathbf{F}_3 = \dfrac{5}{9\sqrt{21}}\langle -6\sqrt{3}, 6, -6\sqrt{21} \rangle = \left\langle \dfrac{-10}{3\sqrt{7}}, \dfrac{10}{3\sqrt{21}}, \dfrac{-10}{3} \right\rangle$

$\|\mathbf{F}_1\| = \|\mathbf{F}_2\| = \|\mathbf{F}_3\| \approx 3.64$ pounds

73. True. $\cos\theta = 0 \Rightarrow \theta = 90°$

75. This set is a sphere.
$(x - x_1)^2 + (y - y_1)^2 + (z - z_2)^2 = 16$

Section 10.3

1. cross product

3. $\|\mathbf{u}\|\|\mathbf{v}\|\sin\theta$

5. Given $\mathbf{u} \times \mathbf{v} = \langle 1, 0, -2 \rangle$, $\mathbf{v} \times \mathbf{u} = -(\mathbf{u} \times \mathbf{v}) = \langle -1, 0, 2 \rangle$.

7. $\mathbf{u} \times \mathbf{v} = \begin{vmatrix} \mathbf{i} & \mathbf{j} & \mathbf{k} \\ 1 & -1 & 0 \\ 0 & 1 & -1 \end{vmatrix} = \mathbf{i} + \mathbf{j} + \mathbf{k} = \langle 1, 1, 1 \rangle$

$(\mathbf{u} \times \mathbf{v}) \cdot \mathbf{u} = \langle 1, 1, 1 \rangle \cdot \langle 1, -1, 0 \rangle = 0$

$(\mathbf{u} \times \mathbf{v}) \cdot \mathbf{v} = \langle 1, 1, 1 \rangle \cdot \langle 0, 1, -1 \rangle = 0$

9. $\mathbf{u} \times \mathbf{v} = \begin{vmatrix} \mathbf{i} & \mathbf{j} & \mathbf{k} \\ 3 & -2 & 5 \\ 0 & -1 & 1 \end{vmatrix} = 3\mathbf{i} - 3\mathbf{j} - 3\mathbf{k} = \langle 3, -3, -3 \rangle$

$(\mathbf{u} \times \mathbf{v}) \cdot \mathbf{u} = \langle 3, -3, -3 \rangle \cdot \langle 3, -2, 5 \rangle = 0$

$(\mathbf{u} \times \mathbf{v}) \cdot \mathbf{v} = \langle 3, -3, -3 \rangle \cdot \langle 0, -1, 1 \rangle = 0$

11. $\mathbf{u} \times \mathbf{v} = \begin{vmatrix} \mathbf{i} & \mathbf{j} & \mathbf{k} \\ -10 & 0 & 6 \\ 7 & 0 & 0 \end{vmatrix} = 42\mathbf{j} = \langle 0, 42, 0 \rangle$

$(\mathbf{u} \times \mathbf{v}) \cdot \mathbf{u} = \langle 0, 42, 0 \rangle \cdot \langle -10, 0, 6 \rangle = 0$

$(\mathbf{u} \times \mathbf{v}) \cdot \mathbf{v} = \langle 0, 42, 0 \rangle \cdot \langle 7, 0, 0 \rangle = 0$

13. $\mathbf{u} \times \mathbf{v} = \begin{vmatrix} \mathbf{i} & \mathbf{j} & \mathbf{k} \\ 6 & 2 & 1 \\ 1 & 3 & -2 \end{vmatrix} = -7\mathbf{i} + 13\mathbf{j} + 16\mathbf{k} = \langle -7, 13, 16 \rangle$

$(\mathbf{u} \times \mathbf{v}) \cdot \mathbf{u} = \langle -7, 13, 16 \rangle \cdot \langle 6, 2, 1 \rangle = 0$

$(\mathbf{u} \times \mathbf{v}) \cdot \mathbf{v} = \langle -7, 13, 16 \rangle \cdot \langle 1, 3, -2 \rangle = 0$

15. $\mathbf{u} \times \mathbf{v} = \begin{vmatrix} \mathbf{i} & \mathbf{j} & \mathbf{k} \\ 2 & 4 & 3 \\ -1 & 3 & -2 \end{vmatrix} = -17\mathbf{i} + \mathbf{j} + 10\mathbf{k} = \langle -17, 1, 10 \rangle$

$(\mathbf{u} \times \mathbf{v}) \cdot \mathbf{u} = \langle -17, 1, 10 \rangle \cdot \langle 2, 4, 3 \rangle = 0$

$(\mathbf{u} \times \mathbf{v}) \cdot \mathbf{v} = \langle -17, 1, 10 \rangle \cdot \langle -1, 3, -2 \rangle = 0$

17. $\mathbf{u}\times\mathbf{v}=\begin{vmatrix}\mathbf{i} & \mathbf{j} & \mathbf{k}\\ \frac{1}{2} & -\frac{2}{3} & 1\\ -\frac{3}{4} & 1 & \frac{1}{4}\end{vmatrix}=-\frac{7}{6}\mathbf{i}-\frac{7}{8}\mathbf{j}=\left\langle -\frac{7}{6},-\frac{7}{8},0\right\rangle$

$(\mathbf{u}\times\mathbf{v})\cdot\mathbf{u}=\left\langle -\frac{7}{6},-\frac{7}{8},0\right\rangle\cdot\left\langle \frac{1}{2},-\frac{2}{3},1\right\rangle=0$

$(\mathbf{u}\times\mathbf{v})\cdot\mathbf{v}=\left\langle -\frac{7}{6},-\frac{7}{8},0\right\rangle\cdot\left\langle -\frac{3}{4},1,\frac{1}{4}\right\rangle=0$

19. $\mathbf{u}\times\mathbf{v}=\begin{vmatrix}\mathbf{i} & \mathbf{j} & \mathbf{k}\\ 0 & 0 & 6\\ -1 & 3 & 1\end{vmatrix}=18\mathbf{i}-6\mathbf{j}=\langle -18,-6,0\rangle$

$(\mathbf{u}\times\mathbf{v})\cdot\mathbf{v}=\langle -18,-6,0\rangle\cdot\langle 0,0,6\rangle=0$

$(\mathbf{u}\times\mathbf{v})\cdot\mathbf{v}=\langle -18,-6,0\rangle\cdot\langle -1,3,1\rangle=0$

21. $\mathbf{u}\times\mathbf{v}=\begin{vmatrix}\mathbf{i} & \mathbf{j} & \mathbf{k}\\ -1 & 0 & 1\\ 0 & 1 & -2\end{vmatrix}=-\mathbf{i}-2\mathbf{j}-\mathbf{k}=\langle -1,-2,-1\rangle$

$(\mathbf{u}\times\mathbf{v})\cdot\mathbf{u}=\langle -1,-2,-1\rangle\cdot\langle -1,0,1\rangle=0$

$(\mathbf{u}\times\mathbf{v})\cdot\mathbf{v}=\langle -1,2,-1\rangle\cdot\langle 0,1,-2\rangle=0$

23. $\mathbf{u}\times\mathbf{v}=\begin{vmatrix}\mathbf{i} & \mathbf{j} & \mathbf{k}\\ 2 & 4 & 3\\ 0 & -2 & 1\end{vmatrix}=\langle 10,-2,-4\rangle$

25. $\mathbf{u}\times\mathbf{v}=\begin{vmatrix}\mathbf{i} & \mathbf{j} & \mathbf{k}\\ 1 & -2 & 4\\ -4 & 2 & -1\end{vmatrix}=-6\mathbf{i}-15\mathbf{j}-6\mathbf{k}$

27. $\mathbf{u}\times\mathbf{v}=\begin{vmatrix}\mathbf{i} & \mathbf{j} & \mathbf{k}\\ 6 & -5 & 1\\ \frac{1}{2} & -\frac{3}{4} & \frac{2}{10}\end{vmatrix}=-\frac{1}{4}\mathbf{i}-\frac{7}{10}\mathbf{j}-2\mathbf{k}$

29. $\mathbf{u}\times\mathbf{v}=\begin{vmatrix}\mathbf{i} & \mathbf{j} & \mathbf{k}\\ 2 & -3 & 4\\ 0 & -1 & 1\end{vmatrix}=\mathbf{i}-2\mathbf{j}-2\mathbf{k}$

$\|\mathbf{u}\times\mathbf{v}\|=\sqrt{(1)^2+(-2)^2+(-2)^2}=\sqrt{9}=3$

$\text{Unit vector}=\frac{\mathbf{u}\times\mathbf{v}}{\|\mathbf{u}\times\mathbf{v}\|}=\frac{1}{3}(\mathbf{i}-2\mathbf{j}-2\mathbf{k})=\frac{1}{3}\mathbf{i}-\frac{2}{3}\mathbf{j}-\frac{2}{3}\mathbf{k}$

31. $\mathbf{u}\times\mathbf{v}=\begin{vmatrix}\mathbf{i} & \mathbf{j} & \mathbf{k}\\ 3 & 1 & 0\\ 0 & 1 & 1\end{vmatrix}=\mathbf{i}-3\mathbf{j}+3\mathbf{k}$

$\|\mathbf{u}\times\mathbf{v}\|=\sqrt{19}$

$\text{Unit vector}=\frac{\mathbf{u}\times\mathbf{v}}{\|\mathbf{u}\times\mathbf{v}\|}=\frac{1}{\sqrt{19}}(\mathbf{i}-3\mathbf{j}+3\mathbf{k})$

$=\frac{\sqrt{19}}{19}(1,-3,3)$

33. $\mathbf{u}\times\mathbf{v}=\begin{vmatrix}\mathbf{i} & \mathbf{j} & \mathbf{k}\\ -3 & 2 & -5\\ \frac{1}{2} & -\frac{3}{4} & \frac{1}{10}\end{vmatrix}=\left\langle -\frac{71}{20},-\frac{11}{5},\frac{5}{4}\right\rangle$

Consider the parallel vector $\langle -71,-44,25\rangle=\mathbf{w}$.

$\|\mathbf{w}\|=\sqrt{71^2+44^2+25^2}=\sqrt{7602}$

$\text{Unit vector}=\frac{\mathbf{u}\times\mathbf{v}}{\|\mathbf{w}\|}=\frac{1}{\sqrt{7602}}\langle -71,-44,25\rangle$

$=\frac{\sqrt{7602}}{7602}\langle -71,-44,25\rangle$

35. $\mathbf{u}\times\mathbf{v}=\begin{vmatrix}\mathbf{i} & \mathbf{j} & \mathbf{k}\\ 1 & 1 & -1\\ 1 & 1 & 1\end{vmatrix}=2\mathbf{i}-2\mathbf{j}$

$\|\mathbf{u}\times\mathbf{v}\|=2\sqrt{2}$

$\text{Unit vector}=\frac{\mathbf{u}\times\mathbf{v}}{\|\mathbf{u}\times\mathbf{v}\|}=\frac{1}{2\sqrt{2}}(2\mathbf{i}-2\mathbf{j})$

$=\frac{1}{\sqrt{2}}\mathbf{i}-\frac{1}{\sqrt{2}}\mathbf{j}=\frac{\sqrt{2}}{2}\mathbf{i}-\frac{\sqrt{2}}{2}\mathbf{j}$

37. $\mathbf{u}\times\mathbf{v}=\begin{vmatrix}\mathbf{i} & \mathbf{j} & \mathbf{k}\\ 0 & 0 & 1\\ 1 & 0 & 1\end{vmatrix}=\mathbf{j}$

$\text{Area}=\|\mathbf{u}\times\mathbf{v}\|=\|\mathbf{j}\|=1$ square unit

39. $\mathbf{u}\times\mathbf{v}=\begin{vmatrix}\mathbf{i} & \mathbf{j} & \mathbf{k}\\ 3 & 4 & 6\\ 2 & -1 & 5\end{vmatrix}=26\mathbf{i}-3\mathbf{j}-11\mathbf{k}$

$\text{Area}=\|\mathbf{u}\times\mathbf{v}\|=\sqrt{26^2+(-3)^2+(-11)^2}$

$=\sqrt{806}$ square units

41. $\mathbf{u}\times\mathbf{v}=\begin{vmatrix}\mathbf{i} & \mathbf{j} & \mathbf{k}\\ 4 & 4 & -6\\ 0 & 4 & 6\end{vmatrix}=48\mathbf{i}-24\mathbf{j}+16\mathbf{k}$

$\text{Area}=\|\mathbf{u}\times\mathbf{v}\|=\sqrt{(48)^2+(-24)^2+(16)^2}$

$=\sqrt{3136}=56$ square units

43. (a) $\overrightarrow{AB} = \langle 3-2, 1-(-1), 2-4\rangle = \langle 1, 2, -2\rangle$ is parallel to

$\overrightarrow{DC} = \langle 0-(-1), 5-3, 6-8\rangle = \langle 1, 2, -2\rangle$.

$\overrightarrow{AD} = \langle -3, 4, 4\rangle$ is parallel to $\overrightarrow{BC} = \langle -3, 4, 4\rangle$.

(b) $$\overrightarrow{AB}\times\overrightarrow{AD} = \begin{vmatrix} \mathbf{i} & \mathbf{j} & \mathbf{k} \\ 1 & 2 & -2 \\ -3 & 4 & 4 \end{vmatrix} = \langle 16, 2, 10\rangle$$

$$\text{Area} = \left\|\overrightarrow{AB}\times\overrightarrow{AD}\right\| = \sqrt{16^2+2^2+10^2} = \sqrt{360} = 6\sqrt{10} \text{ square units}$$

(c) $\overrightarrow{AB}\cdot\overrightarrow{AD} = \langle 1, 2, -2\rangle\cdot\langle -3, 4, 4\rangle$

$\neq 0 \Rightarrow$ not a rectangle

45. $\mathbf{u} = \langle 1-0, 2-0, 3-0, \rangle = \langle 1, 2, 3\rangle$

$\mathbf{v} = \langle -3-0, 0-0, 0-0\rangle = \langle -3, 0, 0\rangle$

$$\mathbf{u}\times\mathbf{v} = \begin{vmatrix} \mathbf{i} & \mathbf{j} & \mathbf{k} \\ 1 & 2 & 3 \\ -3 & 0 & 0 \end{vmatrix} = \langle 0, -9, 6\rangle$$

$$\text{Area} = \frac{1}{2}\|\mathbf{u}\times\mathbf{v}\| = \frac{1}{2}\sqrt{81+36} = \frac{3}{2}\sqrt{13}$$

47. $\mathbf{u} = \langle -2-2, \ -2-3, \ 0-(-5)\rangle = \langle -4, \ -5, \ 5\rangle$

$\mathbf{v} = \langle 3-2, \ 0-3, \ 6-(-5)\rangle = \langle 1, \ -3\ , \ 11\rangle$

$$\mathbf{u}\times\mathbf{v} = \begin{vmatrix} \mathbf{i} & \mathbf{j} & \mathbf{k} \\ -4 & -5 & 5 \\ 1 & -3 & 11 \end{vmatrix} = \langle -40, \ 49, \ 17\rangle$$

$$\text{Area} = \frac{1}{2}\|\mathbf{u}\times\mathbf{v}\| = \frac{1}{2}\sqrt{(-40)^2+49^2+17^2} = \frac{1}{2}\sqrt{4290} \text{ square units}$$

49. $$\mathbf{u}\cdot(\mathbf{v}\times\mathbf{w}) = \begin{vmatrix} 2 & 3 & 3 \\ 4 & 4 & 0 \\ 0 & 0 & 4 \end{vmatrix} = 2(16)-3(16)+3(0) = -16$$

51. $$\mathbf{u}\cdot(\mathbf{v}\times\mathbf{w}) = \begin{vmatrix} 2 & 3 & 1 \\ 1 & -1 & 0 \\ 4 & 3 & 1 \end{vmatrix} = 2(-1)-3(1)+1(7) = 2$$

53. $$\mathbf{u}\cdot(\mathbf{v}\times\mathbf{w}) = \begin{vmatrix} 1 & 1 & 0 \\ 0 & 1 & 1 \\ 1 & 0 & 1 \end{vmatrix} = 1+1 = 2$$

Volume $= |\mathbf{u}\cdot(\mathbf{v}\times\mathbf{w})| = 2$ cubic units

55. $$\mathbf{u}\cdot(\mathbf{v}\times\mathbf{w}) = \begin{vmatrix} 0 & 2 & 2 \\ 0 & 0 & -2 \\ 3 & 0 & 2 \end{vmatrix} = 0-2(6)+2(0) = -12$$

Volume $= |\mathbf{u}\cdot(\mathbf{v}\times\mathbf{w})| = 12$ cubic units

57. $\mathbf{u} = \langle 4, 0, 0\rangle, \ \mathbf{v} = \langle 0, \ -2, \ 3\rangle, \ \mathbf{w} = \langle 0, \ 5, \ 3\rangle$

$$\mathbf{u}\cdot(\mathbf{v}\times\mathbf{w}) = \begin{vmatrix} 4 & 0 & 0 \\ 0 & -2 & 3 \\ 0 & 5 & 3 \end{vmatrix} = 4(-21) = -84$$

Volume $= |-84| = 84$ cubic units

59. $$\mathbf{V}\times\mathbf{F} = \begin{vmatrix} \mathbf{i} & \mathbf{j} & \mathbf{k} \\ 0 & -\frac{1}{2}\cos 40^\circ & -\frac{1}{2}\sin 40^\circ \\ 0 & 0 & -p \end{vmatrix} = \left(\frac{p}{2}\cos 40^\circ\right)\mathbf{i}$$

(a) $T(p) = \|\mathbf{V}\times\mathbf{F}\| = \frac{p}{2}\cos 40^\circ$

(b)

p	15	20	25	30	35	40	45
T	5.75	7.66	9.58	11.49	13.41	15.32	17.24

61. True. The cross product is defined for vectors in three-dimensional space.

63. If $\mathbf{u}$ and $\mathbf{v}$ are orthogonal, then $\sin\theta = 1$ and hence, $\|\mathbf{u}\times\mathbf{v}\| = \|\mathbf{u}\|\|\mathbf{v}\|\sin\theta = \|\mathbf{u}\|\|\mathbf{v}\|$.

65. $$\mathbf{u}\times\mathbf{v} = \begin{vmatrix} \mathbf{i} & \mathbf{j} & \mathbf{k} \\ \cos\alpha & \sin\alpha & 0 \\ \cos\beta & \sin\beta & 0 \end{vmatrix} = (\cos\alpha\sin\beta - \sin\alpha\cos\beta)\mathbf{k}$$

Area of triangle formed by the unit vectors $\mathbf{u}$ and $\mathbf{v}$ is

$\frac{1}{2}(\text{base})(\text{height}) = \frac{1}{2}(1)\sin(\alpha-\beta)$.

The area is also given by

$\frac{1}{2}\|\mathbf{u}\times\mathbf{v}\| = \frac{1}{2}|\cos\alpha\sin\beta - \sin\alpha\cos\beta|$.

Notice that $\cos\alpha\sin\beta - \sin\alpha\cos\beta$ is negative.

Thus, $\sin(\alpha-\beta) = \sin\alpha\cos\beta - \cos\alpha\sin\beta$.

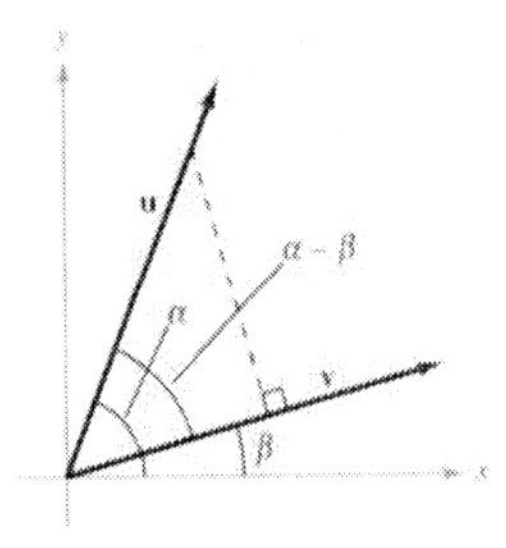

67. $\cos 480^\circ = \cos 120^\circ = -\frac{1}{2}$

Section 10.4

1. direction

3. For two distinct planes is three-space with normal vectors $\mathbf{n}_1$ and $\mathbf{n}_2$, $\mathbf{n}_1 \cdot \mathbf{n}_2 = 0$. The planes are perpendicular.

5. The coordinate plane parallel to the plane $x = 2$ is the yz-plane.

7. $x = x_1 + at = 0 + t$
   $y = y_1 + bt = 0 + 2t$
   $z = z_1 + ct = 0 + 3t$
   (a) Parametric equations: $x = t, y = 2t, z = 3t$
   (b) Symmetric equations: $\frac{x}{1} = \frac{y}{2} = \frac{z}{3}$

9.

$$x = x_1 + at = -4 + \frac{1}{2}t, y = y_1 + bt = 1 + \frac{4}{3}t, z = z_1 + ct = 0 - t$$

   (a) Parametric equations:
   $x = -4 + \frac{1}{2}t, y = 1 + \frac{4}{3}t, z = -t$
   Equivalently: $x = -4 + 3t$, $y = 1 + 8t$, $z = -6t$
   (b) Symmetric equations: $\frac{x+4}{3} = \frac{y-1}{8} = \frac{z}{-6}$

11. $x = x_1 + at = 2 + 2t$
   $y = y_1 + bt = -3 - 3t$
   $z = z_1 + ct = 5 + t$
   (a) Parametric equation:
   $x = 2 + 2t, y = -3 - 3t, z = 5 + t$
   (b) Symmetric equations: $\frac{x-2}{2} = \frac{y+3}{-3} = z - 5$

13. $\mathbf{v} = \langle 1-2, 4-0, -3-2 \rangle = \langle -1, 4, -5 \rangle$
   Point: (2, 0, 2)
   (a) $x = 2 - t, y = 4t, z = 2 - 5t$
   (b) $\frac{x-2}{-1} = \frac{y}{4} = \frac{z-2}{-5}$

15. $\mathbf{v} = \langle 1-(-3), -2-8, 16-15 \rangle = \langle 4, -10, 1 \rangle$
   Point: (−3, 8, 15)
   (a) $x = -3 + 4t, y = 8 - 10t, z = 15 + t$
   (b) $\frac{x+3}{4} = \frac{y-8}{-10} = \frac{z-15}{1}$

17. $\mathbf{v} = \langle -1-3, 1-1, 5-2 \rangle = \langle -4, 0, 3 \rangle$
   Point: (3, 1, 2)
   (a) $x = 3 - 4t, y = 1, z = 2 + 3t$
   (b) $\frac{x-3}{-4} = \frac{z-2}{3}, y = 1$
   Not possible

19. $\mathbf{v} = \left\langle 1 + \frac{1}{2}, -\frac{1}{2} - 2, 0 - \frac{1}{2} \right\rangle = \left\langle \frac{3}{2}, -\frac{5}{2}, -\frac{1}{2} \right\rangle$
   or $\langle 3, -5, -1 \rangle$
   Point: $\left(-\frac{1}{2}, 2, \frac{1}{2} \right)$
   (a) $x = -\frac{1}{2} + 3t, y = 2 - 5t, z = \frac{1}{2} - t$
   (b) $\frac{x + \frac{1}{2}}{3} = \frac{y-2}{-5} = \frac{z - \frac{1}{2}}{-1}$

21. $x = 2t, y = 2 + t, z = 1 + \frac{1}{2}t$

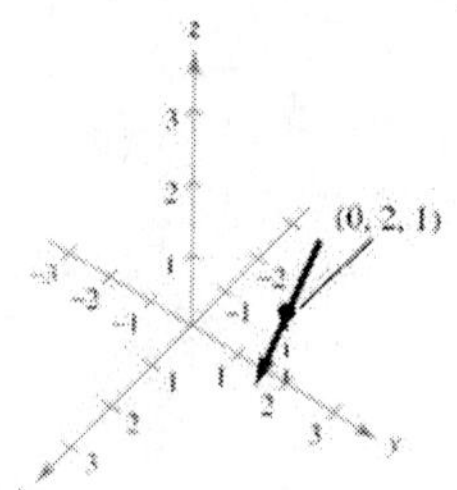

23. $a(x - x_1) + b(y - y_1) + c(z - z_1) = 0$
   $1(x-2) + 0(y-1) + 0(z-2) = 0$
   $x - 2 = 0$

25. $-2(x-5) + 1(y-6) - 2(z-3) = 0$
   $-2x + y - 2z + 10 = 0$

27. $\mathbf{n} = \langle -1, -2, 1 \rangle$
   $-1(x-2) - 2(y-0) + 1(z-0) = 0$
   $-x - 2y + z + 2 = 0$

29. $\mathbf{u} = \langle 1-0, 2-0, 3-0\rangle = \langle 1, 2, 3\rangle$

$\mathbf{v} = \langle -2-0, 3-0, 3-0\rangle = \langle -2, 3, 3\rangle$

$$\mathbf{n} = \mathbf{u}\times\mathbf{v} = \begin{vmatrix} \mathbf{i} & \mathbf{j} & \mathbf{k} \\ 1 & 2 & 3 \\ -2 & 3 & 3 \end{vmatrix} = \langle -3, -9, 7\rangle$$

$$\begin{aligned} -3(x-0) - 9(y-0) + (z-0) &= 0 \\ -3x - 9y + 7z &= 0 \\ 3x + 9y - 7z &= 0 \end{aligned}$$

31. $\mathbf{u} = \langle 3-2, 4-3, 2+2\rangle = \langle 1, 1, 4\rangle$

$\mathbf{v} = \langle 1-2, -1-3, 0+2\rangle = \langle -1, -4, 2\rangle$

$$\mathbf{n} = \mathbf{u}\times\mathbf{v} = \begin{vmatrix} \mathbf{i} & \mathbf{j} & \mathbf{k} \\ 1 & 1 & 4 \\ -1 & -4 & 2 \end{vmatrix} = \langle 18, -6, -3\rangle$$

$$\begin{aligned} 18(x-2) - 6(y-3) - 3(z+2) &= 0 \\ 18x - 6y - 3z - 24 &= 0 \\ 6x - 2y - z - 8 &= 0 \end{aligned}$$

33. $\mathbf{n} = \mathbf{j}$: $0(x-2) + 1(y-5) + 0(z-3) = 0$

$$y - 5 = 0$$

35. $\langle 0-(-1), 2-(-2), 4-0\rangle = \langle 1, 4, 4\rangle$ and $\langle 1, 0, 0\rangle$ are parallel to the plane.

$$\mathbf{n} = \begin{vmatrix} \mathbf{i} & \mathbf{j} & \mathbf{k} \\ 1 & 4 & 4 \\ 1 & 0 & 0 \end{vmatrix} = \langle 0, 4, -4\rangle$$

$$\begin{aligned} 0(x-0) + 4(y-2) - 4(z-4) &= 0 \\ 4y - 4z + 8 &= 0 \\ y - z + 2 &= 0 \end{aligned}$$

37. $\langle -1-2, 1-2, -1-1\rangle = \langle -3, -1, -2\rangle$ and $\langle 2, -3, 1\rangle$ are parallel to plane.

$$\mathbf{n} = \begin{vmatrix} \mathbf{i} & \mathbf{j} & \mathbf{k} \\ -3 & -1 & -2 \\ 2 & -3 & 1 \end{vmatrix} = \langle -7, -1, 11\rangle$$

$$\begin{aligned} -7(x-2) - 1(y-2) + 11(z-1) &= 0 \\ -7x - y + 11z + 5 &= 0 \end{aligned}$$

39. $\mathbf{v} = \langle 0, 0, 1\rangle$ and $P = (2, 3, 4)$

$x = 2$

$y = 3$

$z = 4 + t$

41. $\mathbf{v} = \langle 3, 2, -1\rangle$ and $P = (2, 3, 4)$

$x = 2 + 3t$

$y = 3 + 2t$

$z = 4 - t$

43. $\mathbf{v} = \langle 2, -1, 3\rangle$ and $P = (5, -3, -4)$

$x = 5 + 2t$

$y = -3 - t$

$z = -4 + 3t$

45. $\mathbf{v} = \langle -1, 1, 1\rangle$ and $P = (2, 1, 2)$

$x = 2 - t$

$y = 1 + t$

$z = 2 + t$

47. $\mathbf{n}_1 = \langle 5, -3, 1\rangle$, $\mathbf{n}_2 = \langle 1, 4, 7\rangle$

$\mathbf{n}_1 \cdot \mathbf{n}_2 = 5 - 12 + 7 = 0$; orthogonal

49. $\mathbf{n}_1 = \langle 2, 0, -1\rangle$, $\mathbf{n}_2 = \langle 4, 1, 8\rangle$

$\mathbf{n}_1 \cdot \mathbf{n}_2 = 8 - 8 = 0$; orthogonal

51. (a) $\mathbf{n}_1 = \langle 3, -4, 5\rangle$, $\mathbf{n}_2 = \langle 1, 1, -1\rangle$; normal vectors to planes

$$\cos\theta = \frac{|\mathbf{n}_1 \cdot \mathbf{n}_2|}{\|\mathbf{n}_1\|\|\mathbf{n}_2\|} = \frac{|-6|}{\sqrt{50}\sqrt{3}} = \frac{6}{\sqrt{150}} \Rightarrow \theta \approx 60.67^\circ$$

(b) $3x - 4y + 5z = 6$ Equation 1

$x + y - z = 2$ Equation 2

(-3) times Equation 2 added to Equation 1 gives

$$-7y + 8z = 0$$

$$y = \frac{8}{7}z.$$

Substituting back into Equation 2,

$$x = 2 - y + z = 2 - \frac{8}{7}z + z = 2 - \frac{1}{7}z.$$

Letting $t = z/7$, you obtain $x = 2 - t$, $y = 8t$, $z = 7t$.

53. (a) $\mathbf{n}_1 = \langle 1, 1, -1\rangle$, $\mathbf{n}_2 = \langle 2, -5, -1\rangle$; normal vectors to planes

$$\cos\theta = \frac{|\mathbf{n}_1 \cdot \mathbf{n}_2|}{\|\mathbf{n}_1\|\|\mathbf{n}_2\|} = \frac{|-2|}{\sqrt{3}\sqrt{30}} = \frac{2}{\sqrt{90}} \Rightarrow \theta \approx 77.83^\circ$$

(b) $x + y - z = 0$ Equation 1

$2x - 5y - z = 1$ Equation 2

(-2) times Equation 1 added to Equation 2 gives

$$-7y = z = 1$$

$$y = \frac{z-1}{7}.$$

Substituting back into Equation 1,

$$x = z - y = z - \frac{z-1}{7} = \frac{6z}{7} + \frac{1}{7} = \frac{1}{7}(6z+1).$$

Letting $z = t$, $x = \frac{6t+1}{7}$, $y = \frac{t-1}{7}$. Equivalently, let $y = t$, $z = 7t+1$ and $x = 6t+1$.

55. $x + 2y + 3z = 6$

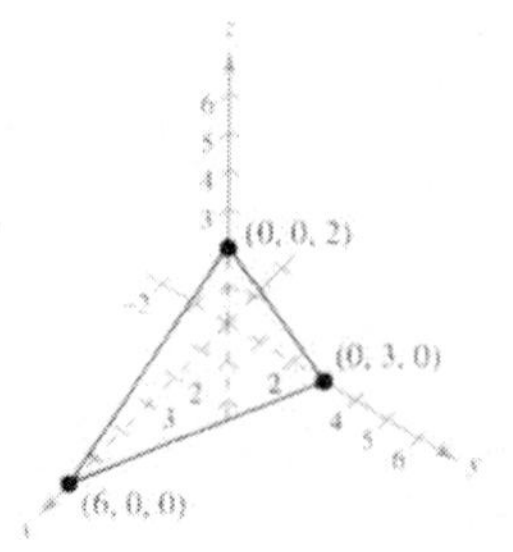

57. $x + 2y = 4$

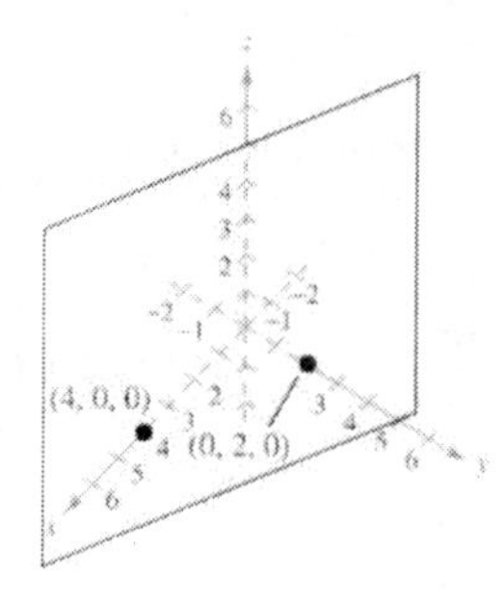

59. $3x + 2y - z = 6$

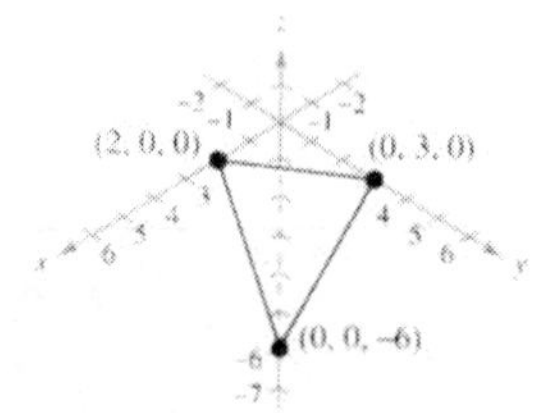

61. $D = \frac{|\overrightarrow{PQ} \cdot \mathbf{n}|}{\|\mathbf{n}\|}$

$P = (1, 0, 0)$ on plane, $Q = (0, 0, 0)$,

$\mathbf{n} = \langle 8, -4, 1\rangle$, $\overrightarrow{PQ} = \langle -1, 0, 0\rangle$

$$D = \frac{|\langle -1, 0, 0\rangle \cdot \langle 8, -4, 1\rangle|}{\sqrt{64+16+1}} = \frac{|-8|}{\sqrt{81}} = \frac{8}{9}$$

63. $D = \frac{|\overrightarrow{PQ} \cdot \mathbf{n}|}{\|\mathbf{n}\|}$

$P = (2, 0, 0)$ on plane, $Q = (4, -2, -2)$,

$\mathbf{n} = \langle 2, -1, 1\rangle$, $\overrightarrow{PQ} = \langle 2, -2, -2\rangle$

$$D = \frac{|\langle 2, -2, -2\rangle \cdot \langle 2, -1, 1\rangle|}{\sqrt{6}} = \frac{4}{\sqrt{6}} = \frac{2\sqrt{6}}{3}$$

65. The normal vector to plane STP: $(0, 0, 0)$, $(2, 2, 12)$, and $(10, 0, 0)$ is given by $\overline{ST} = \mathbf{v}_1 = \langle 2, 2, 12\rangle$ and $\overline{SP} = \mathbf{v}_2 = \langle 10, 0, 0\rangle$.

$$\mathbf{v}_1 \times \mathbf{v}_2 = \begin{vmatrix} \mathbf{i} & \mathbf{j} & \mathbf{k} \\ 2 & 2 & 12 \\ 10 & 0 & 0 \end{vmatrix} = \langle 0, 120, -20\rangle$$

$$\mathbf{n}_1 = \langle 0, 6, -1\rangle$$

The normal vector to plane STQ: $(0, 0, 0)$, $(2, 2, 12)$, and $(10, 10, 0)$ is given by

$\overline{ST} = \mathbf{u}_1 = \langle 2, 2, 12\rangle$ and $\overline{SQ} = \mathbf{u}_2 = \langle 10, 10, 0\rangle$.

$$\mathbf{u}_1 \times \mathbf{u}_2 = \begin{vmatrix} \mathbf{i} & \mathbf{j} & \mathbf{k} \\ 2 & 2 & 12 \\ 10 & 10 & 0 \end{vmatrix} = \langle -120, 0, 20\rangle$$

$$\mathbf{n}_2 = \langle -6, 0, 1\rangle$$

The angle θ between two adjacent sides is given by

$$\cos\theta = \frac{|\mathbf{n}_1 \cdot \mathbf{n}_2|}{\|\mathbf{n}_1\|\,\|\mathbf{n}_2\|} = \frac{|-1|}{\sqrt{37}\sqrt{37}} = \frac{1}{37}$$

$$\theta \approx 88.45^\circ$$

67. False. Lines that do not intersect and are not in the same plane may not be parallel.

69. The error in writing symmetric equations of the line that passes through $(1, 2, 6)$ and is parallel to $\mathbf{v} = \langle 3, 5, 4\rangle$ is that the point and the vector were reversed and should be $\frac{x-1}{3} = \frac{y-2}{5} = \frac{z-6}{4}$.

71. The lines are parallel:

$$-\frac{3}{2}(10, 18, 20) = \langle -15, 27, -30\rangle.$$

Chapter 10 Review Exercises

1. (a) and (b)

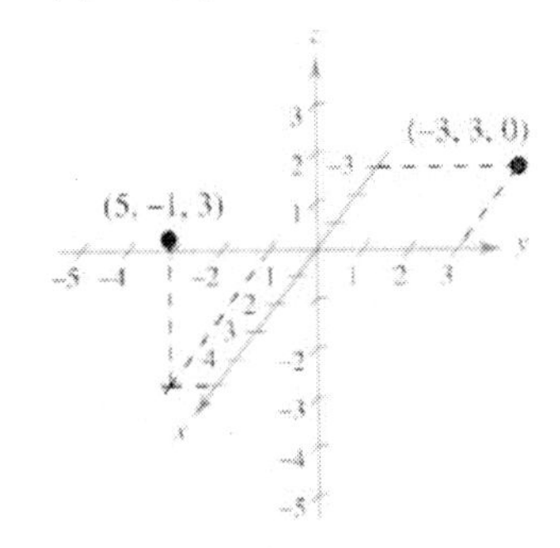

3. $x = -5,\ y = 3,\ z = 0 \Rightarrow (-5, 3, 0)$

5. $d = \sqrt{(x_2 - x_1)^2 + (y_2 - y_1)^2 + (z_2 - z_1)^2}$

$= \sqrt{(5-4)^2 + (2-0)^2 + (1-6)^2} = \sqrt{1+4+25} = \sqrt{30}$

7. $d_1 = \sqrt{(3-0)^2 + (-2-3)^2 + (0-2)^2} = \sqrt{9+25+4} = \sqrt{38}$

$d_2 = \sqrt{(0-0)^2 + (5-3)^2 + (-3-2)^2} = \sqrt{4+25} = \sqrt{29}$

$d_3 = \sqrt{(0-3)^2 + (5-(-2))^2 + (-3-0)^2} = \sqrt{9+49+9} = \sqrt{67}$

$d_1^2 + d_2^2 = 38 + 29 = 67 = d_3^2$

9. Midpoint: $\left(\dfrac{x_1+x_2}{2}, \dfrac{y_1+y_2}{2}, \dfrac{z_1+z_2}{2}\right)$

$= \left(\dfrac{-2+2}{2}, \dfrac{3+(-5)}{2}, \dfrac{2+(-4)}{2}\right)$

$= (0, -1, -1)$

11. Midpoint: $\left(\dfrac{x_1+x_2}{2}, \dfrac{y_1+y_2}{2}, \dfrac{z_1+z_2}{2}\right)$

$= \left(\dfrac{10+(-8)}{2}, \dfrac{6+(-4)}{2}, \dfrac{-12+(-6)}{2}\right)$

$= (1, 1, -9)$

13. $(x-2)^2 + (y-3)^2 + (z-5)^2 = 1$

15. Radius: 6

$(x-1)^2 + (y-5)^2 + (z-2)^2 = 36$

17. $(x^2 - 4x + 4) + (y^2 - 6y + 9) + z^2 = -4 + 4 + 9$

$(x-2)^2 + (y-3)^2 + z^2 = 9$

Center: (2, 3, 0)

Radius: 3

19. (a) xz-trace $(y = 0)$: $x^2 + z^2 = 7$, circle

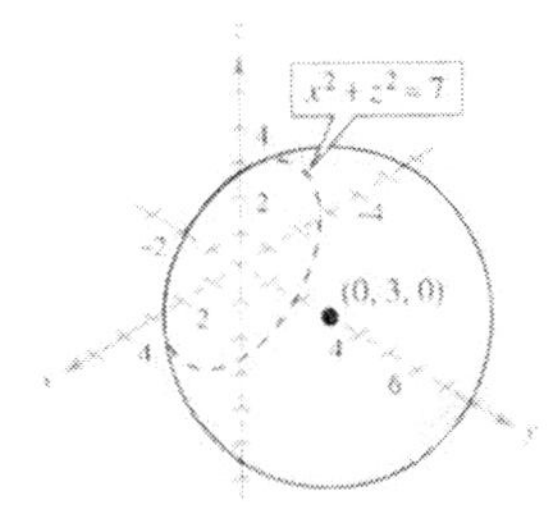

(b) yz-trace $(x = 0)$: $(y-3)^2 + z^2 = 16$, circle

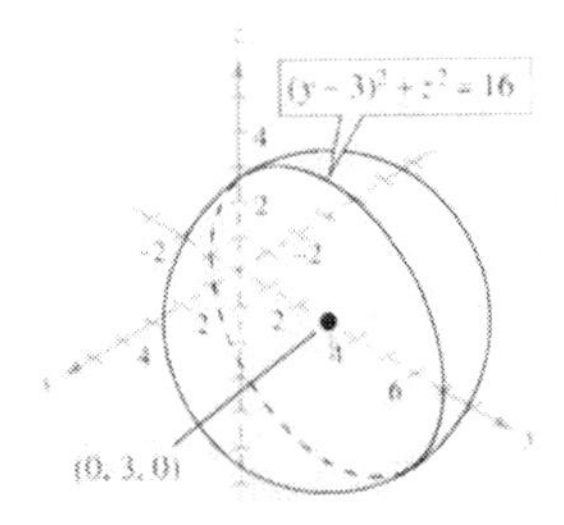

21. (a) $\mathbf{v} = \langle 3-2, 3-(-1), 0-3\rangle = \langle 1, 4, -3\rangle$

(b) $\|\mathbf{v}\| = \sqrt{(1)^2 + (4)^2 + (-3)^2} = \sqrt{1+16+9} = \sqrt{26}$

(c) $\mathbf{u} = \dfrac{\mathbf{v}}{\|\mathbf{v}\|} = \dfrac{1}{\sqrt{26}}\langle 1, 4, -3\rangle = \dfrac{\sqrt{26}}{26}\langle 1, 4, -3\rangle$

23. (a) $\mathbf{v} = \langle -3-7, 2-(-4), 10-3\rangle = \langle -10, 6, 7\rangle$

(b) $\|\mathbf{v}\| = \sqrt{(-10)^2 + 6^2 + 7^2} = \sqrt{185}$

(c) $\mathbf{u} = \dfrac{\mathbf{v}}{\|\mathbf{v}\|} = \dfrac{1}{\sqrt{185}}\langle -10, 6, 7\rangle = \dfrac{\sqrt{185}}{185}\langle -10, 6, 7\rangle$

25. $\mathbf{u}\cdot\mathbf{v} = -1(0) + 4(-6) + 3(5) = -9$

27. $\mathbf{u}\cdot\mathbf{v} = 2(1) - 1(0) + 1(-1) = 1$

29. $\cos\theta = \dfrac{\mathbf{u}\cdot\mathbf{v}}{\|\mathbf{u}\|\ \|\mathbf{v}\|} = \dfrac{2-2+0}{\|\sqrt{5}\|\ \|\sqrt{6}\|} = 0$

$\theta = 90^\circ$, **u** and **v** are orthogonal.

31. $\cos\theta = \dfrac{\mathbf{u}\cdot\mathbf{v}}{\|\mathbf{u}\|\ \|\mathbf{v}\|} = \dfrac{-4-4+8}{\|\sqrt{40}\|\ \|\sqrt{7}\|} = 0$

$\theta = 90^\circ$, **u** and **v** are orthogonal.

33. $\mathbf{u}\cdot\mathbf{v} = 7(-1) + (-2)(4) + 3(5) = 0$

Orthogonal

35. $\mathbf{u}\cdot\mathbf{v} = 30 + 15 - 45 = 0$

Orthogonal

37. First two points: $\mathbf{u} = \langle -3, 4, 1 \rangle$

Last two points: $\mathbf{v} = \langle 0, -2, 6 \rangle$

Since $\mathbf{u} \neq c\mathbf{v}$, the points are not collinear.

39. First two points: $\langle 4, -2, -10 \rangle$

First and third points: $\langle 2, -1, -5 \rangle$

Since $\langle 4, -2, -10 \rangle = 2\langle 2, -1, -5 \rangle$, the three points are collinear.

41. Let **a**, **b**, and **c** be the three force vectors determined by $A(0, 10, 10)$, $B(-4, -6, 10)$, and $C(4, -6, 10)$.

$$\mathbf{a} = \|\mathbf{a}\| \frac{\langle 0, 10, 10 \rangle}{10\sqrt{2}} = \|\mathbf{a}\| \left\langle 0, \frac{1}{\sqrt{2}}, \frac{1}{\sqrt{2}} \right\rangle$$

$$\mathbf{b} = \|\mathbf{b}\| \frac{\langle -4, -6, 10 \rangle}{\sqrt{152}} = \|\mathbf{b}\| \left\langle \frac{-2}{\sqrt{38}}, \frac{-3}{\sqrt{38}}, \frac{5}{\sqrt{38}} \right\rangle$$

$$\mathbf{c} = \|\mathbf{c}\| \frac{\langle 4, -6, 10 \rangle}{\sqrt{152}} = \|\mathbf{c}\| \left\langle \frac{2}{\sqrt{38}}, \frac{-3}{\sqrt{38}}, \frac{5}{\sqrt{38}} \right\rangle$$

Must have $\mathbf{a} + \mathbf{b} + \mathbf{c} = 300\mathbf{k}$. Thus,

$$\frac{-2}{\sqrt{38}}\|\mathbf{b}\| + \frac{2}{\sqrt{38}}\|\mathbf{c}\| = 0$$

$$\frac{1}{\sqrt{2}}\|\mathbf{a}\| - \frac{3}{\sqrt{38}}\|\mathbf{b}\| - \frac{3}{\sqrt{38}}\|\mathbf{c}\| = 0$$

$$\frac{1}{\sqrt{2}}\|\mathbf{a}\| + \frac{5}{\sqrt{38}}\|\mathbf{b}\| + \frac{5}{\sqrt{38}}\|\mathbf{c}\| = 300.$$

From the first equation, $\|\mathbf{b}\| = \|\mathbf{c}\|$. From the second equation, $\frac{1}{\sqrt{2}}\|\mathbf{a}\| = \frac{6}{\sqrt{38}}\|\mathbf{b}\|$. From the third equation, $\frac{1}{\sqrt{2}}\|\mathbf{a}\| = 300 - \frac{10}{\sqrt{38}}\|\mathbf{b}\|$. Thus,

$$\frac{6}{\sqrt{38}}\|\mathbf{b}\| = 300 - \frac{10}{\sqrt{38}}\|\mathbf{b}\| \Rightarrow \frac{16}{\sqrt{38}}\|\mathbf{b}\| = 300 \text{ and}$$

$$\|\mathbf{b}\| = \|\mathbf{c}\| = \frac{75\sqrt{38}}{4} \approx 115.58 \text{ pounds.}$$

Finally, $\|\mathbf{a}\| = \sqrt{2}\left(\frac{6}{\sqrt{38}}\right)\left(\frac{75\sqrt{38}}{4}\right)$

$$= \frac{225\sqrt{2}}{2} \approx 159.10 \text{ pounds.}$$

43. $\mathbf{u} \times \mathbf{v} = \begin{vmatrix} \mathbf{i} & \mathbf{j} & \mathbf{k} \\ -2 & 8 & 2 \\ 1 & 1 & -1 \end{vmatrix} = -10\mathbf{i} - 10\mathbf{k} = \langle -10, 0, -10 \rangle$

45. $\mathbf{u} \times \mathbf{v} = \begin{vmatrix} \mathbf{i} & \mathbf{j} & \mathbf{k} \\ -3 & 2 & -5 \\ 10 & -15 & 2 \end{vmatrix} = \langle -71, -44, 25 \rangle$

$\|\mathbf{u} \times \mathbf{v}\| = \sqrt{7602}$

Unit vector: $\frac{1}{\sqrt{7602}}\langle -71, -44, 25 \rangle$

47. $A(2, -1, 1)$, $B(5, 1, 4)$, $C(0, 1, 1)$, $D(3, 3, 4)$

(a) $\overrightarrow{AB} = \langle 3, 2, 3 \rangle$

$\overrightarrow{CD} = \langle 3, 2, 3 \rangle$

(b) $\overrightarrow{AC} = \langle -2, 2, 0 \rangle$

$$\overrightarrow{AB} \times \overrightarrow{AC} = \begin{vmatrix} \mathbf{i} & \mathbf{j} & \mathbf{k} \\ 3 & 2 & 3 \\ -2 & 2 & 0 \end{vmatrix} = \langle -6, -6, 10 \rangle$$

$$\text{Area} = \|\overrightarrow{AB} \times \overrightarrow{AC}\| = \sqrt{(-6)^2 + (-6)^2 + (10)^2}$$
$$= \sqrt{36 + 36 + 100} = \sqrt{172} = 2\sqrt{43} \text{ square units}$$

(c) $\overrightarrow{AB} \cdot \overrightarrow{AC} = -6 + 4 + 0 = -2 \neq 0 \Rightarrow$ not a rectangle

49. The parallelogram is determined by the three vectors with initial point $(0, 0, 0)$.

$\mathbf{u} = \langle 3, 0, 0 \rangle$, $\mathbf{v} = \langle 2, 0, 5 \rangle$, $\mathbf{w} = \langle 0, 5, 1 \rangle$

$$\mathbf{u} \cdot (\mathbf{v} \times \mathbf{w}) = \begin{vmatrix} 3 & 0 & 0 \\ 2 & 0 & 5 \\ 0 & 5 & 1 \end{vmatrix} = -75$$

Volume $= |-75| = 75$ cubic units

51. $\mathbf{v} = \langle 9-3, 11-0, 6-2 \rangle = \langle 6, 11, 4 \rangle$

Point: $(3, 0, 2)$

(a) $x = 3 + 6t$, $y = 11t$, $z = 2 + 4t$

(b) $\frac{x-3}{6} = \frac{y}{11} = \frac{z-2}{4}$

53. $\mathbf{v} = \langle 3+1, 6-3, -1-5 \rangle = \langle 4, 3, -6 \rangle$, point: $(-1, 3, 5)$

(a) $x = -1 + 4t$, $y = 3 + 3t$, $z = 5 - 6t$

(b) $\frac{x+1}{4} = \frac{y-3}{3} = \frac{z-5}{-6}$

55. Use $2\mathbf{v} = \langle -4, 5, 2 \rangle$, point: $(0, 0, 0)$.

(a) $x = -4t$, $y = 5t$, $z = 2t$

(b) $\frac{x}{-4} = \frac{y}{5} = \frac{z}{2}$

57. $\mathbf{u} = \langle 5, 0, 2 \rangle$, $\mathbf{v} = \langle 2, 3, 8 \rangle$

$$\mathbf{u} \times \mathbf{v} = \begin{vmatrix} \mathbf{i} & \mathbf{j} & \mathbf{k} \\ 5 & 0 & 2 \\ 2 & 3 & 8 \end{vmatrix} = \langle -6, -36, 15 \rangle$$

$\mathbf{n} = \langle 2, 12, -5 \rangle$

$$a(x - x_0) + b(y - y_0) + c(z - z_0) = 0$$
$$2(x - 0) + 12(y - 0) - 5(z - 0) = 0$$
$$2x + 12y - 5z = 0$$

59. $\mathbf{n} = \mathbf{k}$, normal vector

$$0(x-5)+0(y-3)+1(z-2)=0$$
$$z-2=0$$

61. $3x-2y+3z=6$

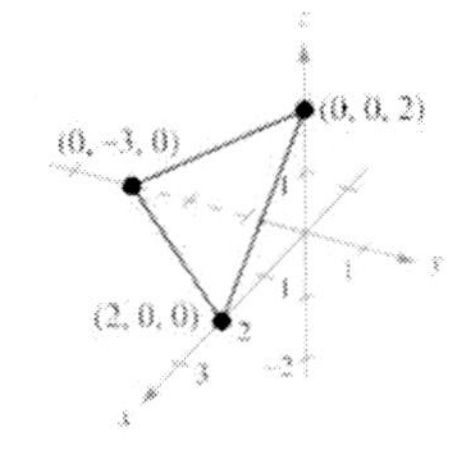

63. $2x-3z=6$

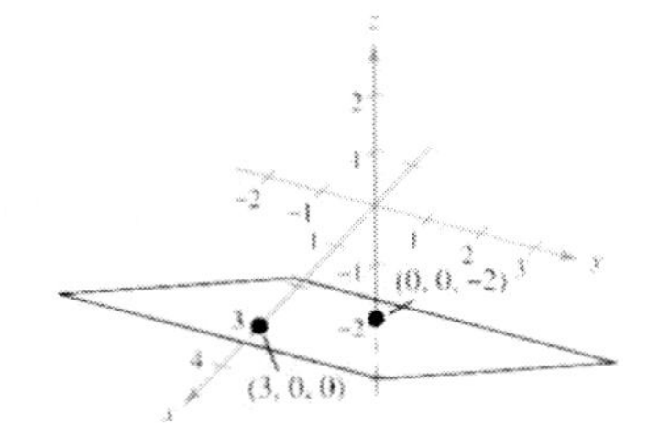

65. $\mathbf{n} = \langle 2, -20, 6\rangle$, $P = (0, 0, 1)$ in plane,

$Q = (2, 3, 10)$, $\overrightarrow{PQ} = \langle 2, 3, 9\rangle$

$$D = \frac{|\overrightarrow{PQ}\cdot\mathbf{n}|}{\|\mathbf{n}\|} = \frac{|-2|}{\sqrt{440}} = \frac{1}{\sqrt{110}} = \frac{\sqrt{110}}{110} \approx 0.0953$$

67. $\mathbf{n} = \langle 1, -10, 3\rangle$, $P = (2, 0, 0)$ in plane,

$Q = (0, 0, 0)$, $\overrightarrow{PQ} = \langle -2, 0, 0\rangle$

$$D = \frac{|\overrightarrow{PQ}\cdot\mathbf{n}|}{\|\mathbf{n}\|} = \frac{|-2|}{\sqrt{1+100+9}} = \frac{2}{\sqrt{110}}$$
$$= \frac{2\sqrt{110}}{110} = \frac{\sqrt{110}}{55} \approx 0.191$$

69. False. $\mathbf{u}\times\mathbf{v} = -(\mathbf{v}\times\mathbf{u})$

71.
$$\mathbf{u}\cdot\mathbf{u} = \langle 3, -2, 1\rangle\cdot\langle 3, -2, 1\rangle$$
$$= 9+4+1$$
$$= 14$$
$$= \|\mathbf{u}\|^2$$

73. $\mathbf{u}\cdot(\mathbf{v}+\mathbf{w}) = \langle 3, -2, 1\rangle\cdot\langle 1, -2, -1\rangle = 6$

$\mathbf{u}\cdot\mathbf{v} + \mathbf{u}\cdot\mathbf{w} = 11 + (-5) = 6$

75.
$$\mathbf{u}\times\mathbf{v} = \begin{vmatrix} \mathbf{i} & \mathbf{j} & \mathbf{k} \\ u_1 & u_2 & u_3 \\ v_1 & v_2 & v_3 \end{vmatrix}$$
$$= (u_2v_3 - u_3v_2)\mathbf{i} - (u_1v_3 - u_3v_1)\mathbf{j} + (u_1v_2 - u_2v_1)\mathbf{k}$$

Chapter 10 Test

1. (a)–(c)

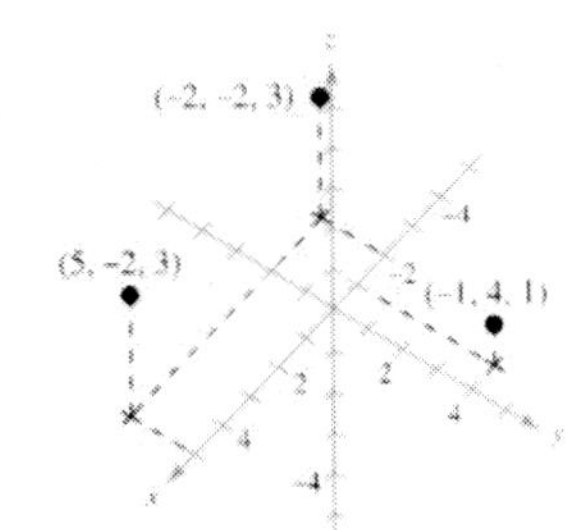

2.
$$AB = \sqrt{(8-6)^2+(-2-4)^2+(5+1)^2} = \sqrt{76}$$
$$AC = \sqrt{(8+4)^2+(-2-3)^2+(5-0)^2} = \sqrt{144+25+25} = \sqrt{194}$$
$$BC = \sqrt{(6+4)^2+(4-3)^2+(-1-0)^2} = \sqrt{100+1+1} = \sqrt{102}$$

No. $\left(\sqrt{76}\right)^2 + \left(\sqrt{102}\right)^2 \neq \left(\sqrt{194}\right)^2$

3. $\text{Midpoint} = \left(\dfrac{8+6}{2}, \dfrac{-2+4}{2}, \dfrac{5-1}{2}\right) = (7, 1, 2)$

4.
$$\text{Diameter} = \sqrt{(8-6)^2+(-2-4)^2+(5+1)^2}$$
$$= \sqrt{4+36+36} = \sqrt{76}$$

$\text{Radius} = \sqrt{19}$

$(x-7)^2+(y-1)^2+(z-2)^2 = 19$

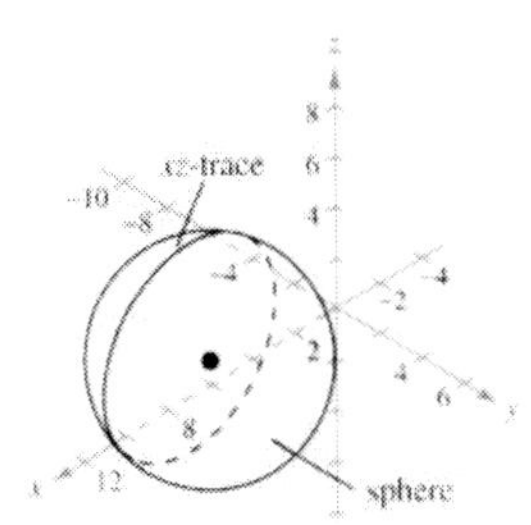

5. $\mathbf{v} = \langle 4-2, 4-(-1), -7-3\rangle = \langle 2, 5, -10\rangle$

$\|\mathbf{v}\| = \sqrt{2^2+5^2+(-10)^2} = \sqrt{129}$

6. $\mathbf{v} = \langle 3-6, -3-2, 8-0\rangle = \langle -3, -5, 8\rangle$

$\|\mathbf{v}\| = \sqrt{(-3)^2+(-5)^2+8^2} = \sqrt{98} = 7\sqrt{2}$

7. $\mathbf{u} = \langle 6-8,\ 4-(-2),\ -1-5\rangle = \langle -2,\ 6,\ -6\rangle$

$\mathbf{v} = \langle -4-8,\ 3-(-2),\ 0-5\rangle = \langle -12,\ 5,\ -5\rangle$

8. (a) $\|\mathbf{v}\| = \sqrt{(-12)^2 + 5^2 + (-5)^2} = \sqrt{194}$

(b) $\mathbf{u}\cdot\mathbf{v} = (-2)(-12) + 6(5) + (-6)(-5) = 84$

(c) $\mathbf{u}\times\mathbf{v} = \begin{vmatrix} \mathbf{i} & \mathbf{j} & \mathbf{k} \\ -2 & 6 & -6 \\ -12 & 5 & -5 \end{vmatrix} = \langle 0,\ 62,\ 62\rangle$

9. $\cos\theta = \dfrac{\mathbf{u}\cdot\mathbf{v}}{\|\mathbf{u}\|\ \|\mathbf{v}\|} = \dfrac{84}{\sqrt{76}\sqrt{194}} \approx 0.6918$

$\theta \approx 46.23°$ or 0.8068 radian

10. $\mathbf{v} = \langle 6-8,\ 4+2,\ -1-5\rangle = \langle -2,\ 6,\ -6\rangle$

(a) $x = 8-2t,\ y = -2+6t,\ z = 5-6t$

(b) $\dfrac{x-8}{-2} = \dfrac{y+2}{6} = \dfrac{z-5}{-6}$

11. $\mathbf{u}\cdot\mathbf{v} = 0-2-6 \neq 0$ and $\mathbf{u} \neq c\mathbf{v} \Rightarrow$ neither

12. $\mathbf{u} = -3\mathbf{i} + 2\mathbf{j} - \mathbf{k}$

$\mathbf{v} = \mathbf{i} + \mathbf{j} - \mathbf{k}$

$\mathbf{u}\cdot\mathbf{v} = (-3)(1) + (2)(1) + (-1)(-1) = 0$

u and **v** are orthogonal.

13. First two points: $\mathbf{v} = \langle 4,\ 8,\ -2\rangle$

Last two points: $\mathbf{w} = \langle 4,\ 8,\ -2\rangle$

Opposite sides are parallel and equal length.

Adjacent sides: **v** and $\mathbf{u} = \langle 1,\ -3,\ 3\rangle$

Area $= \|\mathbf{u}\times\mathbf{v}\|$

$\mathbf{u}\times\mathbf{v} = \begin{vmatrix} \mathbf{i} & \mathbf{j} & \mathbf{k} \\ 1 & -3 & 3 \\ 4 & 8 & -2 \end{vmatrix} = \langle -18,\ 14,\ 20\rangle$

$\|\mathbf{u}\times\mathbf{v}\| = \sqrt{18^2 + 14^2 + 20^2} = 2\sqrt{230} \approx 30.33$ square units

14. $\mathbf{u} = \langle 0,\ 8,\ -1\rangle,\ \mathbf{v} = \langle 4,\ 5,\ -4\rangle$

$\mathbf{n} = \mathbf{u}\times\mathbf{v} = \begin{vmatrix} \mathbf{i} & \mathbf{j} & \mathbf{k} \\ 0 & 8 & -1 \\ 4 & 5 & -4 \end{vmatrix} = \langle -27,\ -4,\ -32\rangle$

$$\begin{aligned} -27(x+3) - 4(y+4) - 32(z-2) &= 0 \\ -27x - 4y - 32z - 33 &= 0 \\ 27x + 4y + 32z + 33 &= 0 \end{aligned}$$

15. Let $A(0,\ 0,\ 5)$ be the vertex.

$\mathbf{u} = \overline{AD} = \langle 4,\ 0,\ 0\rangle,\ \mathbf{v} = \overline{AB} = \langle 0,\ 10,\ 0\rangle,$

$\mathbf{w} = \overline{AE} = \langle 0,\ 1,\ -5\rangle$

$\mathbf{u}\cdot(\mathbf{v}\times\mathbf{w}) = \begin{vmatrix} 4 & 0 & 0 \\ 0 & 10 & 0 \\ 0 & 1 & -5 \end{vmatrix} = 4(-50) = -200$

Volume $= |-200| = 200$ cubic units

16. $3x + 6y + 2z = 18$

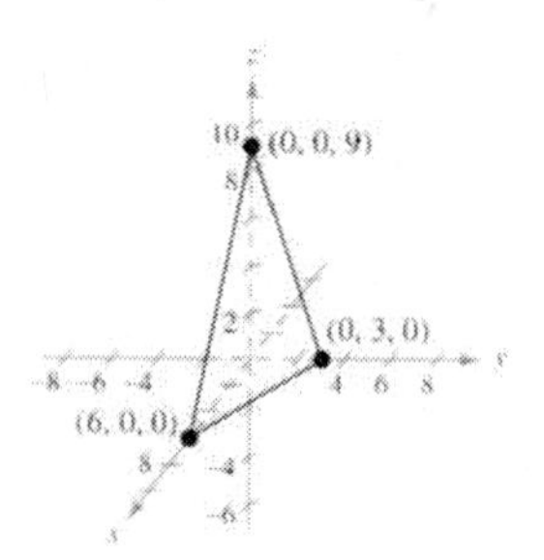

17. $5x - y - 2z = 10$

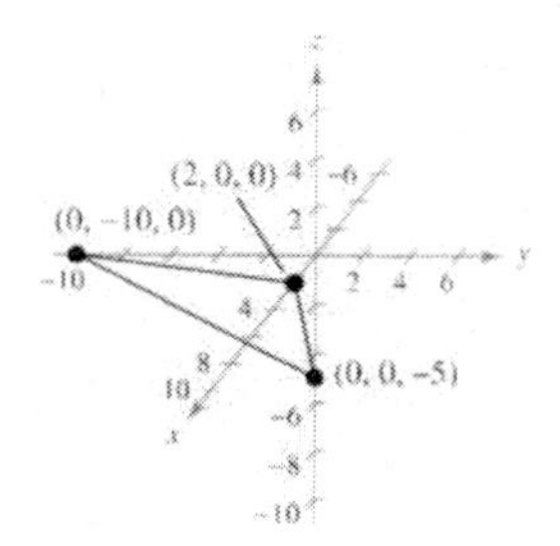

18. $\mathbf{n} = \langle 3,\ -2,\ 1\rangle,\ Q = (2,\ -1,\ 6),\ P = (0,\ 0,\ 6)$ in plane,

$\overline{PQ} = \langle 2,\ -1,\ 0\rangle$

$D = \dfrac{|\overline{PQ}\cdot\mathbf{n}|}{\|\mathbf{n}\|} = \dfrac{|8|}{\sqrt{14}} = \dfrac{4\sqrt{14}}{7}$

CHAPTER 11

Section 11.1

1. limit

3. $\lim_{x\to 0} 3 = 3$ since $\lim_{x\to c} b = b$.

5. (a)

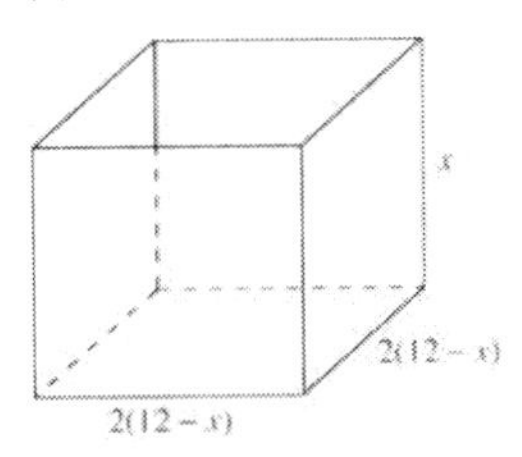

(b) $V = l \cdot w \cdot h$

$= (24 - 2x)(24 - 2x)x$

$= [2(12 - x)][2(12 - x)]x$

$= 4(12 - x)^2 x$

$= 4x(12 - x)^2$

(c)

x	3	3.5	3.9	4
V	972	1011.5	1023.5	1024

x	4.1	4.5	5
V	1023.5	1012.5	980

$\lim_{x\to 4} V = 1024$

(d)

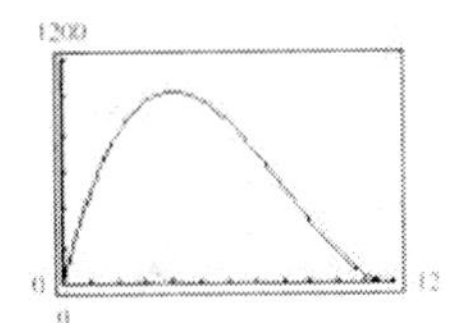

7. $\lim_{x\to 2}(5x + 4) = 14$

x	1.9	1.99	1.999	2	2.001	2.01	2.1
$f(x)$	13.5	13.95	3.995	14	14.005	14.05	14.5

The limit is reached.

9. $\lim_{x\to -1} \dfrac{x+1}{x^2 - x - 2} = -\dfrac{1}{3}$

x	−1.1	−1.01	−1.001	−1.0
$f(x)$	−0.3226	−0.3322	−0.3332	Error

x	−0.999	−0.99	−0.9
$f(x)$	−0.3334	−0.3344	−0.3348

The limit is not reached.

11. $\lim_{x\to 0} \dfrac{\tan x}{2x} = \dfrac{1}{2}$

x	−0.1	−0.01	−0.001	0	0.001	0.01	0.1
$f(x)$	0.5017	0.50002	0.5000002	Error	0.5000002	0.50002	0.5017

The limit is not reached.

13. $\lim_{x\to -1} \dfrac{x^2 - 1}{x + 1} = -2$

x	−1.1	−1.01	−1.001	−1
$f(x)$	−2.1	−2.01	−2.001	Error

x	−0.999	−0.99	−0.9
$f(x)$	−1.999	−1.99	−1.9

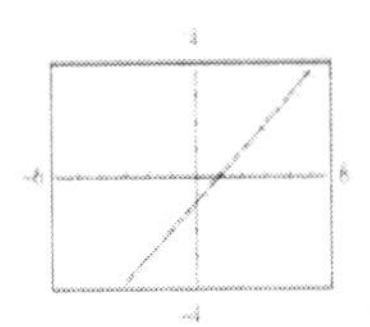

15. $\lim_{x\to 1} \dfrac{x - 1}{x^2 + 2x - 3} = \dfrac{1}{4}$

x	0.9	0.99	0.999	1.0
$f(x)$	0.2564	0.2506	0.2501	Error

x	1.001	1.01	1.1
$f(x)$	0.2499	0.2494	0.2439

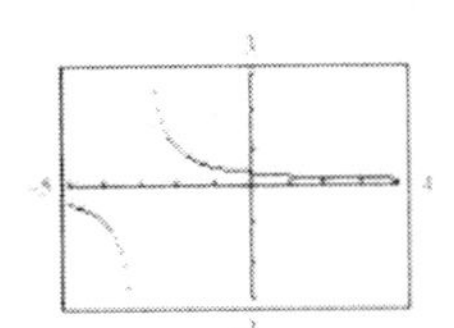

17. $\lim_{x\to 0} \dfrac{\sqrt{x+5} - \sqrt{5}}{x} \approx 0.2236 \left(\text{Actual limit is } \dfrac{1}{2\sqrt{5}}.\right)$

x	−0.1	−0.01	−0.001	0	0.001	0.01	0.1
$f(x)$	0.2247	0.2237	0.2236	Error	0.2236	0.2235	0.2225

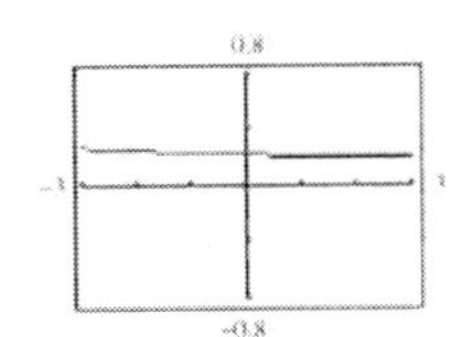

19. $\lim_{x \to -4} \frac{[x/(x+2)] - 2}{x+4} = \frac{1}{2}$

x	-4.1	-4.01	-4.001	-4.0	-3.999	-3.99	-3.9
$f(x)$	0.4762	0.4975	0.4998	Error	0.5003	0.5025	0.5263

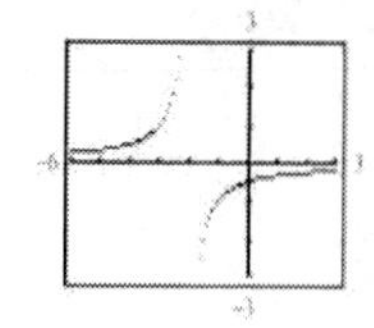

21. $\lim_{x \to 0} \frac{\sin x}{x} = 1$

x	-0.1	-0.01	-0.001	0
$f(x)$	0.9983	0.99998	0.9999998	Error

x	0.001	0.01	0.1
$f(x)$	0.9999998	0.99998	0.9983

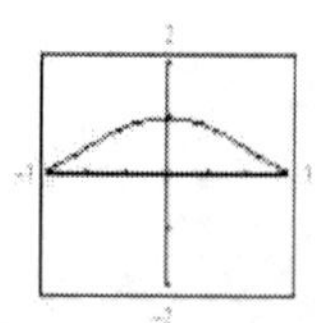

23. $\lim_{x \to 0} \frac{\sin^2 x}{x} = 0$

x	-0.1	-0.01	-0.001	0
$f(x)$	-0.0997	-0.0100	-0.0010	Error

x	0.001	0.01	0.1
$f(x)$	0.0010	0.0100	0.0997

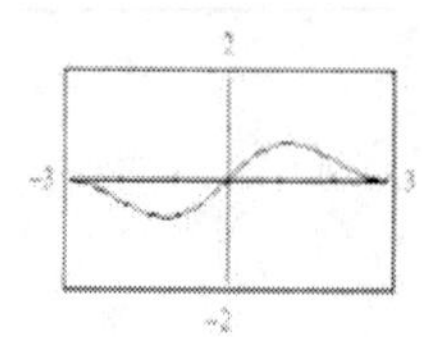

25. $\lim_{x \to 0} \frac{e^{2x} - 1}{2x} = 1.0$

x	-0.1	-0.01	-0.001	0	0.001	0.01	0.1
$f(x)$	0.9063	0.9901	0.9990	Error	1.0010	1.0101	1.1070

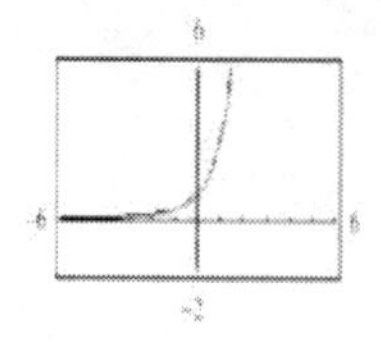

27. $\lim_{x \to 2} \frac{\ln(2x-3)}{x-2} = 2$

x	1.9	1.99	1.999	2
$f(x)$	2.2314	2.0203	2.002	Error

x	2.001	2.01	2.1
$f(x)$	1.998	1.9803	1.8232

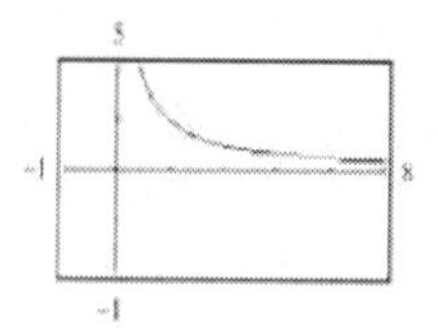

29. $f(x) = \begin{cases} 3, & x \neq 2 \\ 1, & x = 2 \end{cases}$

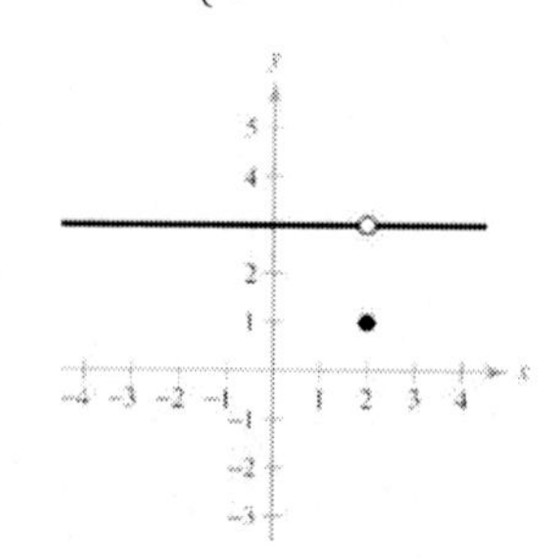

$\lim_{x \to 2} f(x) = 3$

31. $f(x) = \begin{cases} 2x+1, & x < 2 \\ x+3 & x \geq 2 \end{cases}$

The limit exists as x approaches 2:

$\lim_{x \to 2} f(x) = 5$

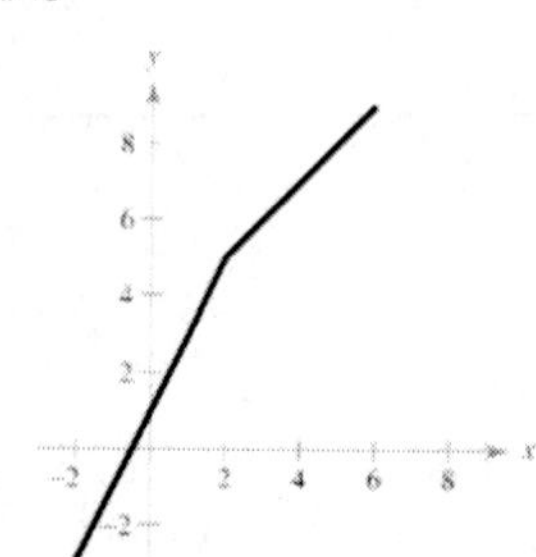

33. $\lim_{x \to 0}(2 - x^2) = 2$

35. $\lim_{x \to -2} \frac{|x+2|}{x+2}$ does not exist. $f(x) = \frac{|x+2|}{x+2}$ equals -1 to the left of -2, and equals 1 to the right of -2.

37. The limit does not exist because $f(x)$ does not approach a real number as x approaches 1.

39. The limit does not exist because $f(x)$ oscillates between 2 and -2.

41. $\lim_{x\to 0} \frac{5}{2+e^{\frac{1}{x}}}$ does not exist.

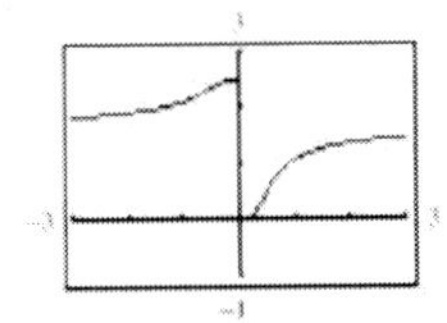

43. $\lim_{x\to 0} \cos\frac{1}{x}$ does not exist.

The graph oscillates between -1 and 1.

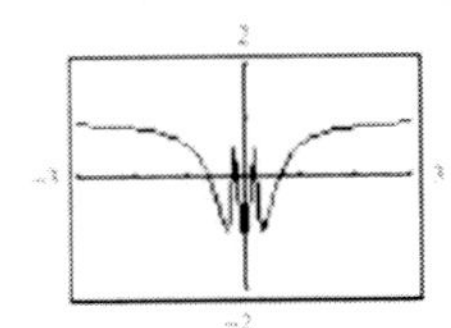

45. $\lim_{x\to 4} \frac{\sqrt{x+3}-1}{x-4}$ does not exist.

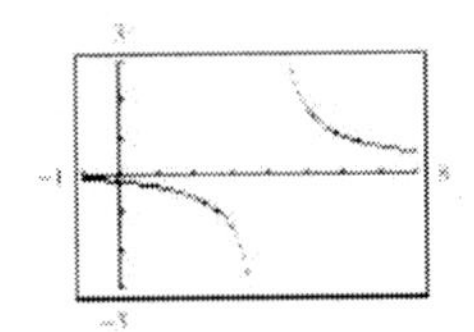

47. $\lim_{x\to 4} \ln(x+3) \approx 1.946$ (Exact limit is ln 7.)

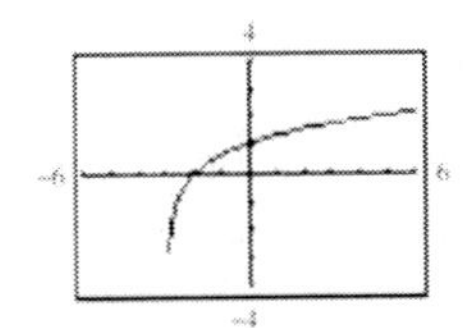

49. $\lim_{x\to c} f(x) = 4, \ \lim_{x\to c} g(x) = 8$

(a) $\lim_{x\to c}\left[-2g(x)\right] = -2(8) = -16$

(b) $\lim_{x\to c}\left[f(x)+g(x)\right] = 4+8 = 12$

(c) $\lim_{x\to c}\frac{f(x)}{g(x)} = \frac{4}{8} = \frac{1}{2}$

(d) $\lim_{x\to c}\sqrt{f(x)} = \sqrt{4} = 2$

51. (a) $\lim_{x\to 2} f(x) = 2^3 = 8$

(b) $\lim_{x\to 2} g(x) = \frac{\sqrt{2^2+5}}{2\left(2^2\right)} = \frac{3}{8}$

(c) $\lim_{x\to 2}\left[f(x)g(x)\right] = 8\left(\frac{3}{8}\right) = 3$

(d) $\lim_{x\to 2}\left[g(x)-f(x)\right] = \frac{3}{8} - 8 = -\frac{61}{8}$

53. $\lim_{x\to 5}\left(10-x^2\right) = 10-5^2 = -15$

55. $\lim_{x\to -3}(2x^2+4x+1) = 2(-3)^2+4(-3)+1 = 7$

57. $\lim_{x\to -3}\frac{3x}{x^2+1} = \frac{3(-3)}{(-3)^2+1} = -\frac{9}{10}$

59. $\lim_{x\to -2}\frac{5x+3}{2x-9} = \frac{5(-2)+3}{2(-2)-9} = \frac{-7}{-13} = \frac{7}{13}$

61. $\lim_{x\to -1}\sqrt{x+2} = \sqrt{-1+2} = 1$

63. $\lim_{x\to 3} e^x = e^3 \approx 20.0855$

65. $\lim_{x\to \pi}\sin 2x = \sin 2\pi = 0$

67. $\lim_{x\to \frac{1}{2}}\arcsin x = \arcsin\frac{1}{2} = \frac{\pi}{6} \approx 0.5236$

69. True. If $f(x)$ approaches a different number from the right side of c than if approaches from left side of c, the limit does not exist.

71. (a) and (b) Answers will vary.

73. (a) No. The limit may or may not exist, and if it does exist, it may not equal 4.

(b) No. $f(2)$ may or may not exist, and if $f(2)$ exists, it may not equal 4.

75. $\frac{5-x}{3x-15} = \frac{5-x}{-3(5-x)} = -\frac{1}{3}, \ x \neq 5$

77. $\frac{15x^2+7x-4}{15x^2+x-2} = \frac{(3x-1)(5x+4)}{(3x-1)(5x+2)}$

$= \frac{5x+4}{5x+2}, \ x \neq \frac{1}{3}$

79. $\frac{x^2+27}{x^2+x-6} = \frac{(x+3)\left(x^2-3x+9\right)}{(x+3)(x-2)}$

$= \frac{x^2-3x+9}{x-2}, \ x \neq -3$

Section 11.2

1. dividing out technique

3. The rationalizing technique can be used to find a limit for an expressing involving radicals.

5. $g(x)=\dfrac{-2x^2+x}{x},\ g_2(x)=-2x+1$

(a) $\lim\limits_{x\to 0} g(x)=1$

(b) $\lim\limits_{x\to -1} g(x)=3$

(c) $\lim\limits_{x\to -2} g(x)=5$

7. $g(x)=\dfrac{x^3-x}{x-1},\ g_2(x)=x^2+x=x(x+1)$

(a) $\lim\limits_{x\to 1} g(x)=2$

(b) $\lim\limits_{x\to -1} g(x)=0$

(c) $\lim\limits_{x\to 0} g(x)=0$

9. $$\lim_{x\to 6}\frac{x-6}{x^2-36}=\lim_{x\to 6}\frac{x-6}{(x-6)(x+6)}=\lim_{x\to 6}\frac{1}{x+6}=\frac{1}{12}$$

11. $$\lim_{x\to 2}\frac{x^2-x-2}{x-2}=\lim_{x\to 2}\frac{(x-2)(x+1)}{x-2}=\lim_{x\to 2}(x+1)=2+1=3$$

13. $$\lim_{x\to -1}\frac{1-2x-3x^2}{x+1}=\lim_{x\to -1}\frac{(x+1)(1-3x)}{1+x}=\lim_{x\to -1}(1-3x)=4$$

15. $$\lim_{t\to 2}\frac{t^3-8}{t-2}=\lim_{t\to 2}\frac{(t-2)(t^2+2t+4)}{t-2}=\lim_{t\to 2}(t^2+2t+4)=4+4+4=12$$

17. $$\lim_{x\to 2}\frac{x^5-32}{x-2}=\lim_{x\to 2}\frac{(x-2)(x^4+2x^3+4x^2+8x+16)}{x-2}=\lim_{x\to 2}(x^4+2x^3+4x^2+8x+16)=80$$

(Note: To factor x^5-32, divide x^5-32 by $x-2$.)

19. $$\lim_{x\to -4}\frac{x^2+x-12}{x^2+6x+8}=\lim_{x\to -4}\frac{(x+4)(x-3)}{(x+4)(x+2)}=\lim_{x\to -4}\frac{x-3}{x+2}=\frac{-7}{-2}=\frac{7}{2}$$

21. $$\lim_{x\to -1}\frac{x^3+2x^2-x-2}{x^3+4x^2-x-4}=\lim_{x\to -1}\frac{(x-1)(x+1)(x+2)}{(x-1)(x+1)(x+4)}=\lim_{x\to -1}\frac{(x-1)(x+2)}{(x-1)(x+4)}=\frac{(-2)(1)}{(-2)(3)}=\frac{1}{3}$$

23. $$\lim_{y\to 0}\frac{\sqrt{5+y}-\sqrt{5}}{y}=\lim_{y\to 0}\frac{\sqrt{5+y}-\sqrt{5}}{y}\cdot\frac{\sqrt{5+y}+\sqrt{5}}{\sqrt{5+y}+\sqrt{5}}=\lim_{y\to 0}\frac{(5+y)-5}{y\left(\sqrt{5+y}+\sqrt{5}\right)}=\lim_{y\to 0}\frac{1}{\sqrt{5+y}+\sqrt{5}}=\frac{1}{2\sqrt{5}}=\frac{\sqrt{5}}{10}$$

25. $$\lim_{x\to -3}\frac{\sqrt{x+7}-2}{x+3}=\lim_{x\to -3}\frac{\sqrt{x+7}-2}{x+3}\cdot\frac{\sqrt{x+7}+2}{\sqrt{x+7}+2}=\lim_{x\to -3}\frac{(x+7)-4}{(x+3)\left(\sqrt{x+7}+2\right)}=\lim_{x\to -3}\frac{1}{\sqrt{x+7}+2}=\frac{1}{4}$$

27. $$\lim_{x\to 0}\frac{1/(x+1)-1}{x}=\lim_{x\to 0}\frac{1-(1+x)}{(x+1)x}=\lim_{x\to 0}\frac{-1}{1+x}=-1$$

29. $$\lim_{x\to 0}\frac{\frac{1}{x+4}-\frac{1}{4}}{x}=\lim_{x\to 0}\frac{\frac{4-(x+4)}{4(x+4)}}{x}=\lim_{x\to 6}\frac{-x}{4(x+4)}\cdot\frac{1}{x}=\lim_{x\to 0}\frac{-1}{4(x+4)}=-\frac{1}{16}$$

31. $$\lim_{x\to\frac{\pi}{2}}\frac{1-\sin x}{\cos x}=\lim_{x\to\frac{\pi}{2}}\frac{1-\sin x}{\cos x}\cdot\frac{1+\sin x}{1+\sin x}=\lim_{x\to\frac{\pi}{2}}\frac{1-\sin^2 x}{\cos x(1+\sin x)}=\lim_{x\to\frac{\pi}{2}}\frac{\cos^2 x}{\cos x(1+\sin x)}=\lim_{x\to\frac{\pi}{2}}\frac{\cos x}{1+\sin x}=0$$

33. $$\lim_{x\to 0}\frac{\cos 2x}{\cot 2x}=\lim_{x\to 0}\frac{\cos 2x}{(\cos 2x)/\sin(2x)}=\lim_{x\to 0}\sin 2x=0$$

35. $\lim_{x\to\frac{\pi}{2}} \frac{\sin x - 1}{x} = \frac{1-1}{\frac{\pi}{2}} = 0$

37. $\lim_{x\to 0} \frac{e^{2x-1}}{x} \approx 2.000$

$\left(\text{Note: } \lim_{x\to 0} \frac{e^{2x-1}}{x} = 2\right)$

x	−0.1	−0.01	−0.001
$f(x)$	1.813	1.980	1.998

x	0	0.001	0.01	0.1
$f(x)$	Error	2.002	2.020	2.214

39. $\lim_{x\to 0} \frac{\sqrt{2x+1}-1}{x} \approx 1.000$ $\left(\text{Note: } \lim_{x\to 0} \frac{\sqrt{2x+1}-1}{x} = 1\right)$

x	−0.1	−0.01	−0.001
$f(x)$	1.056	1.005	1.001

x	0	0.001	0.01	0.1
$f(x)$	Error	0.9995	0.995	0.954

41. $\lim_{x\to 0}(1-x)^{2/x} \approx 0.135$ $\left(\text{Note: } \lim_{x\to 0}(1-x)^{2/x} = e^{1/2}\right)$

x	−0.1	−0.01	−0.001
$f(x)$	0.149	0.137	0.135

x	0	0.001	0.01	0.1
$f(x)$	Error	0.135	0.134	0.122

43. $\lim_{x\to 0} \frac{\sin 2x}{x} = 2$

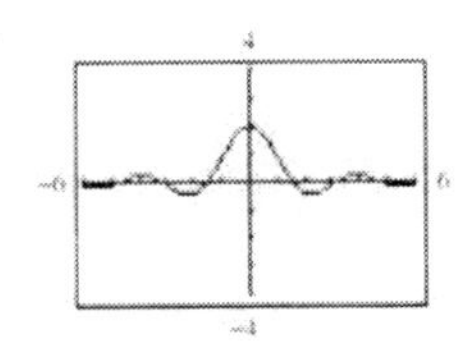

45. $\lim_{x\to 0} \frac{\tan x}{x} = 1$

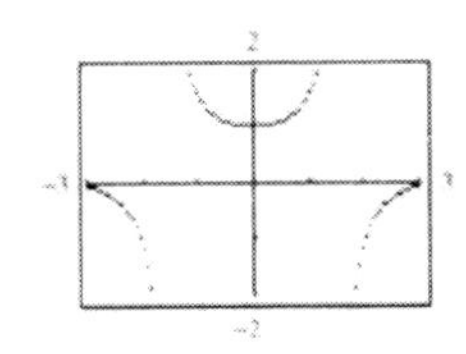

47. $\lim_{x\to 1} \frac{1-\sqrt[3]{x}}{1-x} = \frac{1}{3} \approx 0.333$

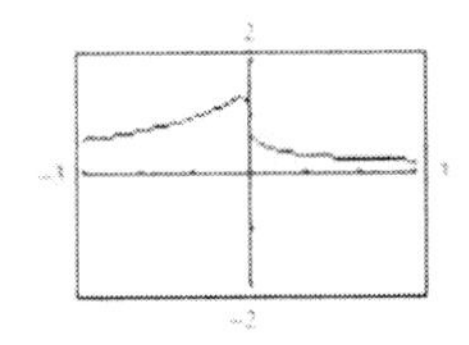

49. $f(x) = \frac{|x-6|}{x-6}$

$\lim_{x\to 6^+} f(x) = 1$

$\lim_{x\to 6^-} f(x) = -1$

Limit does not exist.

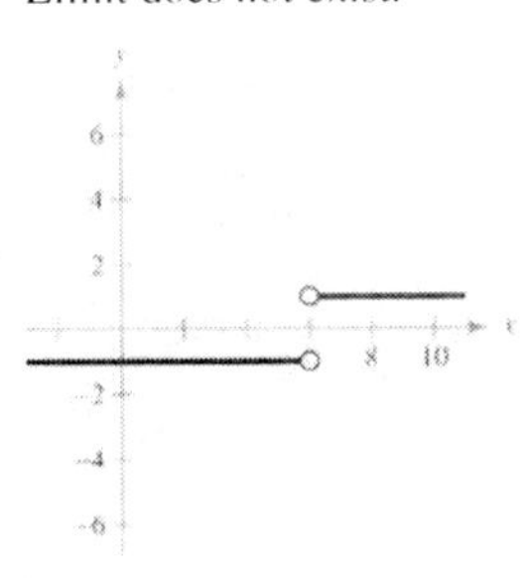

51. $f(x) = \frac{1}{x^2+1}$

$$\lim_{x\to 1^-} \frac{1}{x^2+1} = \lim_{x\to 1^+} \frac{1}{x^2+1} = \lim_{x\to 1} \frac{1}{x^2+1} = \frac{1}{2}$$

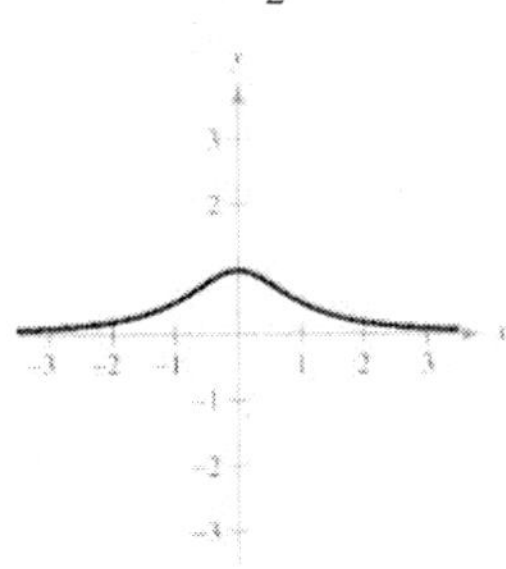

53. $\lim_{x\to 2^-} f(x) = 2 - 1 = 1$

$\lim_{x\to 2^+} f(x) = 2(2) - 3 = 1$

$\lim_{x\to 2} f(x) = 1$

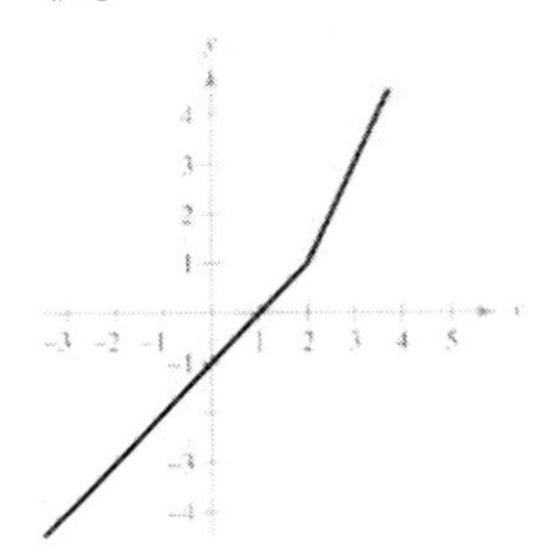

55. $f(x) = \begin{cases} 4 - x^2, x \le 1 \\ 3 - x, x > 1 \end{cases}$

$\lim_{x \to 1^-} f(x) = 4 - 1 = 3$

$\lim_{x \to 1^+} f(x) = 3 - 1 = 2$

$\lim_{x \to 1} f(x)$ dose not exist.

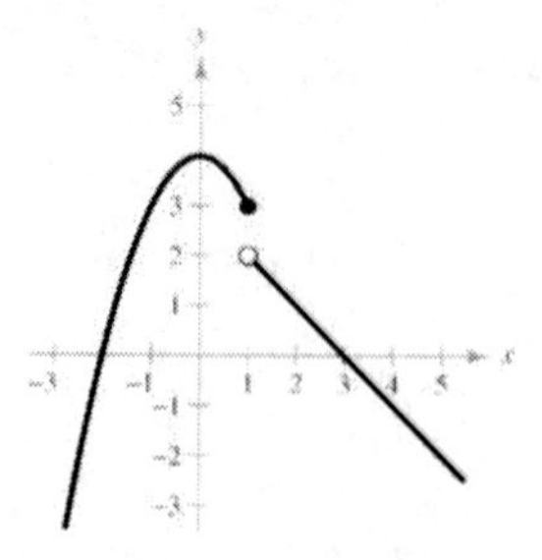

57. $f(x) = \dfrac{x - 1}{x^2 - 1}$

(a)

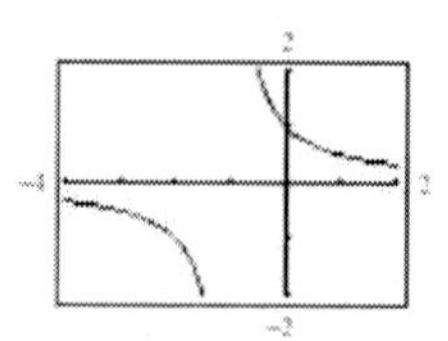

Graphically, $\lim_{x \to 1^-} \dfrac{x - 1}{x^2 - 1} = \dfrac{1}{2}$.

(b)

x	0.5	0.9	0.99	0.999	1
$f(x)$	0.6667	0.5263	0.5025	0.5003	Error

Numerically, $\lim_{x \to 1^-} \dfrac{x - 1}{x^2 - 1} = \dfrac{1}{2}$.

(c) Algebraically, $\lim_{x \to 1^-} \dfrac{x - 1}{x^2 - 1} = \lim_{x \to 1^-} \dfrac{x - 1}{(x - 1)(x + 1)}$

$= \lim_{x \to 1^-} \dfrac{1}{x + 1} = \dfrac{1}{2}$.

59. $f(x) = \dfrac{4 - \sqrt{x}}{x - 16}$

(a)

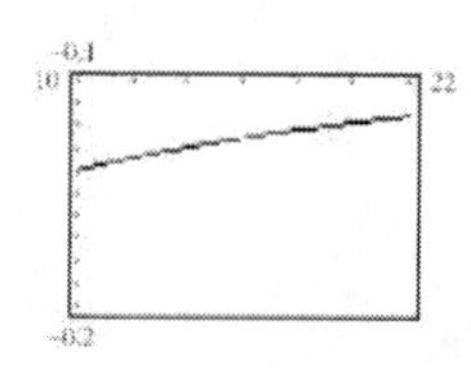

Graphically, $\lim_{x \to 16^+} \dfrac{4 - \sqrt{x}}{x - 16} = -\dfrac{1}{8}$.

(b)

x	16	16.001	16.01	16.1	16.5
$f(x)$	Error	−0.1250	−0.1250	−0.1248	−0.1240

Numerically, $\lim_{x \to 16^+} \dfrac{4 - \sqrt{x}}{x - 16} = -0.125$.

(c) Algebraically, $\lim_{x \to 16^+} \dfrac{4 - \sqrt{x}}{x - 16} = \lim_{x \to 16^+} \dfrac{4 - \sqrt{x}}{\left(\sqrt{x} - 4\right)\left(\sqrt{x} + 4\right)}$

$= \lim_{x \to 16^+} \dfrac{-1}{\sqrt{x + 4}} = \dfrac{-1}{4 + 4} = -\dfrac{1}{8}$.

61.

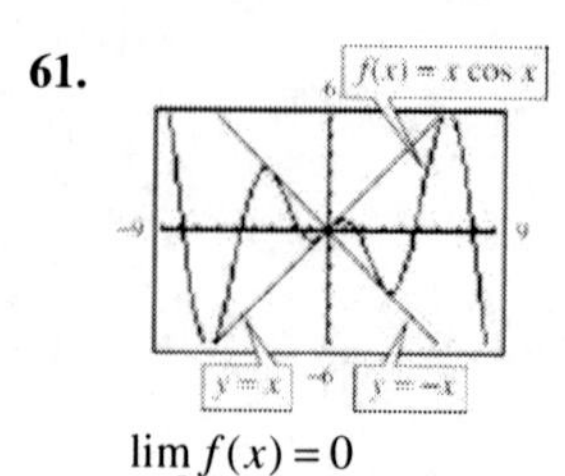

$\lim_{x \to 0} f(x) = 0$

63.

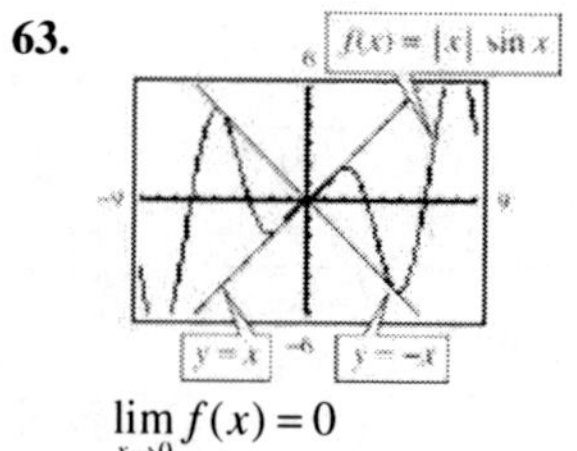

$\lim_{x \to 0} f(x) = 0$

65.

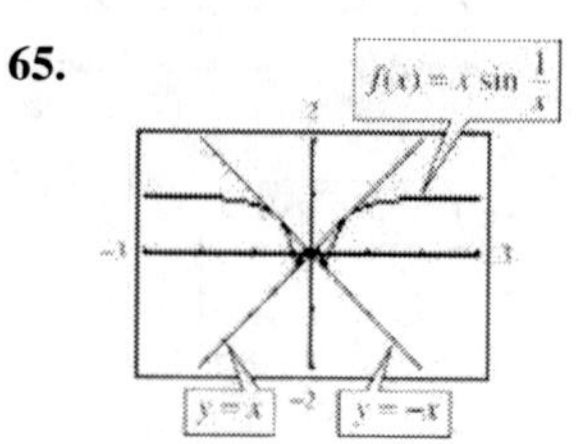

$\lim_{x \to 0} f(x) = 0$

67. (a) Can be evaluated by direct substitution:
$\lim_{x \to 0} x^2 \sin x^2 = 0^2 \sin 0^2 = 0$

(b) Cannot be evaluated by direct substitution:
$\lim_{x \to 0} \dfrac{\sin x^2}{x^2} = 1$

69. $\lim_{t \to 1} \dfrac{(-16(1) + 128) - (-16t^2 + 128)}{1 - t} = \lim_{t \to 1} \dfrac{16t^2 - 16}{1 - t}$

$= \lim_{t \to 1} \dfrac{16(t - 1)(t + 1)}{1 - t}$

$= \lim_{t \to 1} -16(t + 1)$

$= -32 \dfrac{\text{ft}}{\text{sec}}$

71. Answers will vary. As $t \to 2$ from the left, $f(t) \to 39.00$. As $t \to 2$ from the right, $f(t) \to 46.80$.

73. (a)

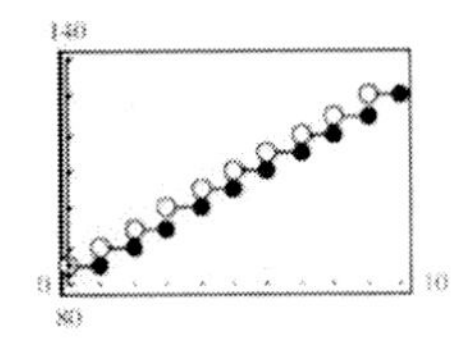

(b)

x	5	5.3	5.4	5.5	5.6	5.7	6
$C(x)$	105	110	110	110	110	110	110

$\lim_{x \to 5.5} C(x) = 110$

(c)

x	4	4.5	4.9	5	5.1	5.5	6
$C(x)$	100	105	105	105	110	110	110

$\lim_{x \to 5^-} C(x) = 105$

$\lim_{x \to 5^+} C(x) = 110$

So, $\lim_{x \to 5} C(x)$ does not exist.

75.
$$\lim_{h \to 0} \frac{f(x+h) - f(x)}{h} = \lim_{h \to 0} \frac{3(x+h) - 1 - (3x-1)}{h} = \lim_{h \to 0} \frac{3x + 3h - 1 - 3x + 1}{h} = \lim_{h \to 0} \frac{3h}{h} = 3$$

77.
$$\lim_{h \to 0} \frac{f(x+h) - f(x)}{h} = \lim_{h \to 0} \frac{\sqrt{x+h} - \sqrt{x}}{h} \cdot \left(\frac{\sqrt{x+h} + \sqrt{x}}{\sqrt{x+h} + \sqrt{x}}\right) = \lim_{h \to 0} \frac{(x+h) - x}{h\left(\sqrt{x+h} + \sqrt{x}\right)} = \lim_{h \to 0} \frac{1}{\sqrt{x+h} + \sqrt{x}} = \frac{1}{2\sqrt{x}}$$

79.
$$\lim_{h \to 0} \frac{f(x+h) - f(x)}{h} = \lim_{h \to 0} \frac{((x+h)^2 - 3(x+h)) - (x^2 - 3x)}{h} = \lim_{h \to 0} \frac{x^2 + 2xh + h^2 - 3x - 3h - x^2 + 3x}{h} = \lim_{h \to 0} \frac{2xh + h^2 - 3h}{h} = \lim_{h \to 0} (2x + h - 3) = 2x - 3$$

81.
$$\lim_{h \to 0} \frac{f(x+h) - f(x)}{h} = \lim_{h \to 0} \frac{1/(x+h+2) - 1/(x+2)}{h} = \lim_{h \to 0} \frac{(x+2) - (x+h+2)}{h(x+h+2)(x+2)} = \lim_{h \to 0} \frac{-h}{h(x+h+2)(x+2)} = \lim_{h \to 0} \frac{-1}{(x+h+2)(x+2)} = -\frac{1}{(x+2)^2}$$

83. True. See text page 761.

85. Many answers possible. Sample answers given.

(a)

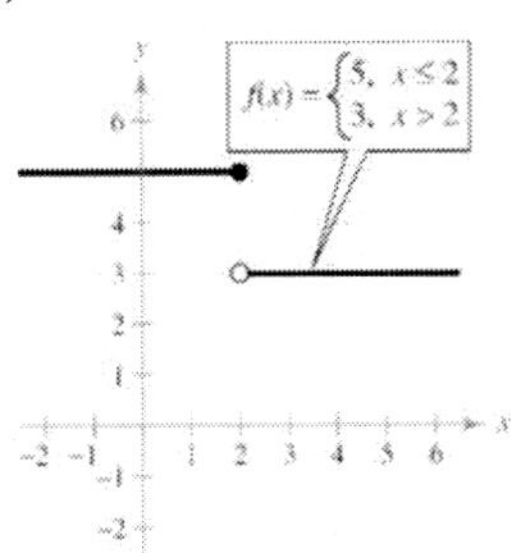

(b)

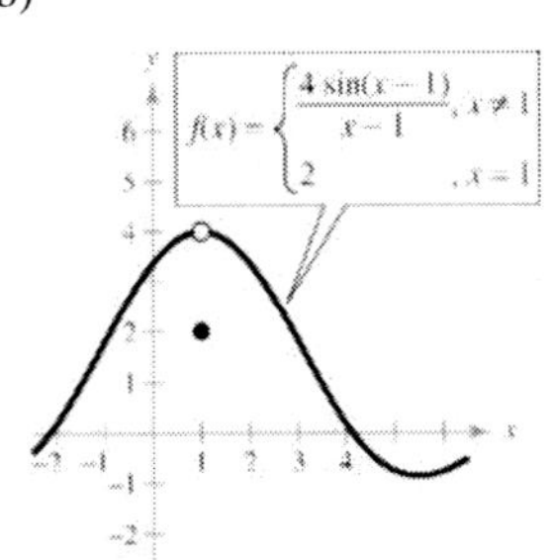

87. Answers will vary.

89. $r = \frac{3}{1 + \cos\theta}$, $e = 1$

Parabola

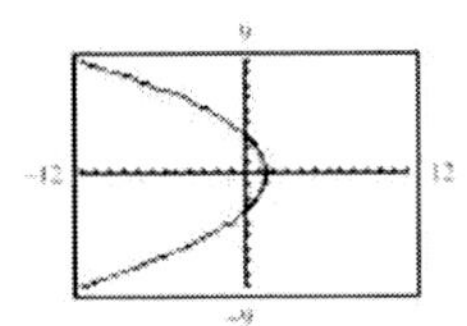

91. $r = \frac{9}{2 + 3\cos\theta} = \frac{\frac{9}{2}}{1 + \left(\frac{3}{2}\right)\cos\theta}$, $e = \frac{3}{2}$

Hyperbola

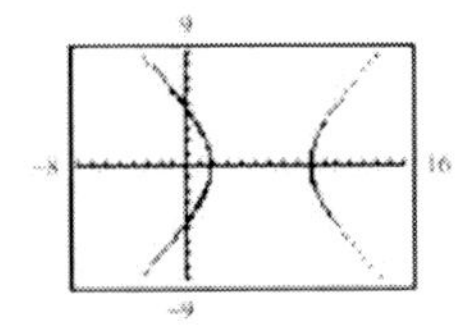

93. $\langle 7, -2, 3\rangle \cdot \langle -1, 4, 5\rangle = -7 - 8 + 15$

$= 0 \Rightarrow$ orthogonal

95. $\langle 2, -3, 1\rangle \cdot \langle -2, 2, 2\rangle = -8 \neq 0$

Not multiples of each other; neither parallel nor orthogonal

Section 11.3

1. Calculus

3. secant line

5. The slope of a graph at a point is equal to the slope of the tangent line at that point. So, the slope of the graph of f at the point $(1, 5)$ is 2.

7. Slope is 0 at (x, y).

9. Slope is $\frac{1}{2}$ at (x, y).

11.
$$m_{sec} = \frac{g(3+h) - g(3)}{h} = \frac{(3+h)^2 - 4(3+h) - (-3)}{h} = \frac{h^2 + 2h}{h}$$
$$m = \lim_{h\to 0} \frac{h^2 + 2h}{h} = \lim_{h\to 0} \frac{h(h+2)}{h} = \lim_{h\to 0}(h+2) = 2$$

13.
$$m_{sec} = \frac{g(1+h) - g(1)}{h} = \frac{5 - 2(1+h) - 3}{h} = \frac{-2h}{h}$$
$$m = \lim_{h\to 0} \frac{-2h}{h} = -2$$

15.
$$m_{sec} = \frac{g(2+h) - g(2)}{h} = \frac{[4/(2+h)] - 2}{h} = \frac{4 - 2(2+h)}{(2+h)h} = \frac{-2}{2+h},\ h \neq 0$$
$$m = \lim_{h\to 0}\left(\frac{-2}{2+h}\right) = -1$$

17.
$$m_{sec} = \frac{h(9+k) - h(9)}{k} = \frac{\sqrt{9+k} - 3}{k} \cdot \frac{\sqrt{9+k} + 3}{\sqrt{9+k} + 3} = \frac{(9+k) - 9}{k\left[\sqrt{9+k} + 3\right]} = \frac{1}{\sqrt{9+k} + 3},\ k \neq 0$$
$$m = \lim_{k\to 0} \frac{1}{\sqrt{9+k} + 3} = \frac{1}{6}$$

19.
$$m_{sec} = \frac{f(x+h) - f(x)}{h} = \frac{\left[4 - (x+h)^2\right] - \left(4 - x^2\right)}{h} = \frac{4 - x^2 - 2xh - h^2 - 4 + x^2}{h} = \frac{-2xh - h^2}{h} = -2x - h,\ h \neq 0$$
$$m = \lim_{h\to 0}(-2x - h) = -2x$$

(a) At $(0, 4)$, $m = -2(0) = 0$.

(b) At $(-2, 0)$, $m = -2(-2) = 4$.

21.
$$m_{sec} = \frac{f(x+h) - f(x)}{h} = \frac{\frac{1}{x+h+4} - \frac{1}{x+4}}{h} = \frac{(x+4) - (x+4+h)}{(x+h+4)(x+4)(h)} = \frac{-h}{(x+h+4)(x+4)h} = \frac{-1}{(x+h+4)(x+4)},\ h \neq 0$$
$$m = \lim_{h\to 0} \frac{-1}{(x+h+4)(x+4)} = -\frac{1}{(x+4)^2}$$

(a) At $\left(0, \frac{1}{4}\right)$, $m = \frac{-1}{(0+4)^2} = -\frac{1}{16}$.

(b) At $\left(-2, \frac{1}{2}\right)$, $m = \frac{-1}{(-2+4)^2} = -\frac{1}{4}$.

23.
$$m_{sec} = \frac{f(x+h) - f(x)}{h} = \frac{\sqrt{x+h-1} - \sqrt{x-1}}{h} = \frac{\sqrt{x+h-1} - \sqrt{x-1}}{h} \cdot \frac{\sqrt{x+h-1} + \sqrt{x-1}}{\sqrt{x+h-1} + \sqrt{x-1}}$$
$$= \frac{(x+h-1) - (x-1)}{h\left(\sqrt{x+h-1} + \sqrt{x-1}\right)} = \frac{h}{h\left(\sqrt{x+h-1} + \sqrt{x-1}\right)} = \frac{1}{\sqrt{x+h-1} + \sqrt{x-1}}$$

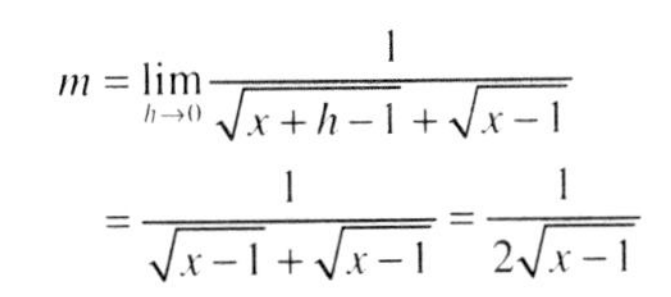

$$m = \lim_{h \to 0} \frac{1}{\sqrt{x+h-1} + \sqrt{x-1}}$$

$$= \frac{1}{\sqrt{x-1} + \sqrt{x-1}} = \frac{1}{2\sqrt{x-1}}$$

(a) At $(2, 1)$, $m = \dfrac{1}{2\sqrt{2-1}} = \dfrac{1}{2}$.

(b) At $(10, 3)$, $m = \dfrac{1}{2\sqrt{10-1}} = \dfrac{1}{2\sqrt{9}} = \dfrac{1}{6}$.

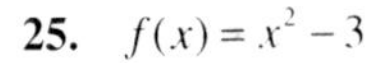

25. $f(x) = x^2 - 3$

Tangent line at $(1, -2)$: $y = 2x - 4$

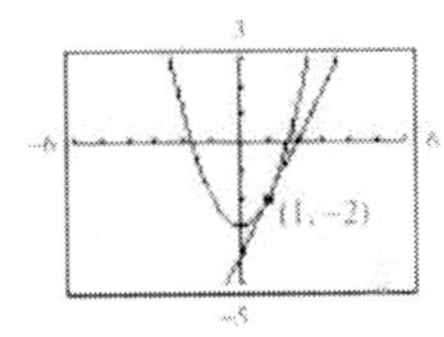

Slope at $(1, -2) = 2$

27. $f(x) = \sqrt{2-x}$

Tangent line at $(1, 1)$: $y = -\dfrac{1}{2}x + \dfrac{3}{2}$

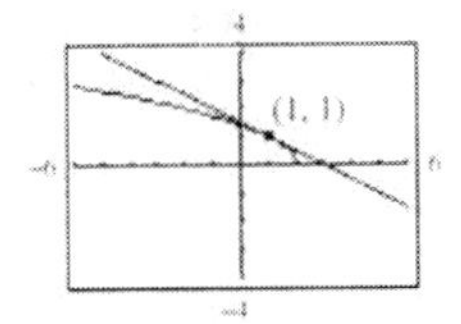

Slope at $(1, 1) = -\dfrac{1}{2}$

29. $f(x) = \dfrac{4}{x+1}$

Tangent line at $(1, 2)$: $y = -x + 3$

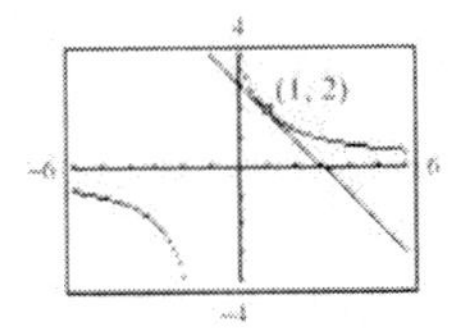

Slope at $(1, 2) = -1$.

31. $$f'(x) = \lim_{h \to 0} \frac{f(x+h) - f(x)}{h}$$

$$= \lim_{h \to 0} \frac{\left[4 - 3(x+h)^2\right] - (4 - 3x^2)}{h}$$

$$= \lim_{h \to 0} \frac{-3(x^2 + 2xh + h^2) + 3x^2}{h}$$

$$= \lim_{h \to 0} \frac{-6xh - 3h^2}{h} = \lim_{h \to 0}(-6x - 3h) = -6x$$

33. $$f'(x) = \lim_{h \to 0} \frac{f(x+h) - f(x)}{h} = \lim_{h \to 0} \frac{5-5}{h} = 0$$

35. $$g'(x) = \lim_{h \to 0} \frac{g(x+h) - g(x)}{h}$$

$$= \lim_{h \to 0} \frac{\left[9 - \frac{1}{3}(x+h)\right] - \left[9 - \frac{1}{3}x\right]}{h}$$

$$= \lim_{h \to 0} \frac{-\frac{1}{3}h}{h} = -\frac{1}{3}$$

37. $$f'(x) = \lim_{h \to 0} \frac{f(x+h) - f(x)}{h}$$

$$= \lim_{h \to 0} \frac{\dfrac{1}{(x+h)^2} - \dfrac{1}{x^2}}{h}$$

$$= \lim_{h \to 0} \frac{x^2 - (x^2 + 2xh + h^2)}{(x+h)^2 x^2 h}$$

$$= \lim_{h \to 0} \frac{-2x - h}{(x+h)^2 x^2} = -\frac{2x}{x^4} = -\frac{2}{x^3}$$

39. $$f'(x) = \lim_{h \to 0} \frac{f(x+h) - f(x)}{h}$$

$$= \lim_{h \to 0} \frac{\sqrt{x+h-4} - \sqrt{x-4}}{h} \cdot \frac{\sqrt{x+h-4} + \sqrt{x-4}}{\sqrt{x+h-4} + \sqrt{x-4}}$$

$$= \lim_{h \to 0} \frac{(x+h-4) - (x-4)}{h\left[\sqrt{x+h-4} + \sqrt{x-4}\right]}$$

$$= \lim_{h \to 0} \frac{1}{\sqrt{x+h-4} + \sqrt{x-4}}$$

$$= \frac{1}{2\sqrt{x-4}}$$

41. $$f'(x) = \lim_{h \to 0} \frac{f(x+h) - f(x)}{h}$$

$$= \lim_{h \to 0} \frac{\dfrac{1}{\sqrt{x+h-9}} - \dfrac{1}{\sqrt{x-9}}}{h} \cdot \frac{\dfrac{1}{\sqrt{x+h-9}} + \dfrac{1}{\sqrt{x-9}}}{\dfrac{1}{\sqrt{x+h-9}} + \dfrac{1}{\sqrt{x-9}}}$$

$$= \lim_{h \to 0} \frac{\dfrac{1}{(x+h-9)} - \dfrac{1}{(x-9)}}{h\left[\dfrac{1}{\sqrt{x+h-9}} + \dfrac{1}{\sqrt{x-9}}\right]}$$

$$= \lim_{h \to 0} \frac{(x-9) - (x+h-9)}{h(x+h-9)(x-9)\left[\dfrac{1}{\sqrt{x+h-9}} + \dfrac{1}{\sqrt{x-9}}\right]}$$

$$= \lim_{h \to 0} \frac{-1}{(x+h-9)(x-9)\left[\dfrac{1}{\sqrt{x+h-9}} + \dfrac{1}{\sqrt{x-9}}\right]}$$

$$= \frac{-1}{(x-9)^2\left[\dfrac{2}{\sqrt{x-9}}\right]} = -\frac{1}{2(x-9)^{\frac{3}{2}}}$$

43. (a) $m_{\text{sec}} = \dfrac{f(2+h)-f(2)}{h}$

$= \dfrac{(2+h)^2 - 1 - 3}{h}$

$= \dfrac{4 + 4h + h^2 - 4}{h}$

$= 4 + h,\ h \neq 0$

$m = \lim_{h \to 0}(4+h) = 4$

(b) $y - 3 = 4(x - 2)$

$y = 4x - 5$

(c)

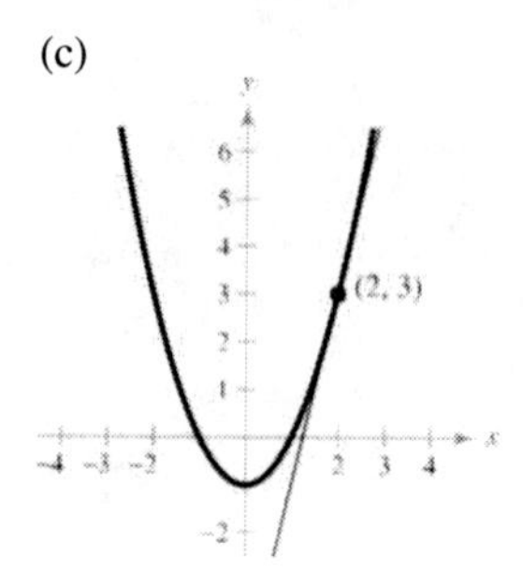

45. (a) $m_{\text{sec}} = \dfrac{f(1+h)-f(1)}{h}$

$= \dfrac{(1+h)^3 - 2(1+h) + 1}{h}$

$= \dfrac{1 + 3h + 3h^2 + h^3 - 2 - 2h + 1}{h}$

$= \dfrac{h^3 + 3h^2 + h}{h} = h^2 + 3h + 1,\ h \neq 0$

$m = \lim_{h \to 0}(h^2 + 3h + 1) = 1$

(b) $y + 1 = 1(x - 1)$

$y = x - 2$

(c)

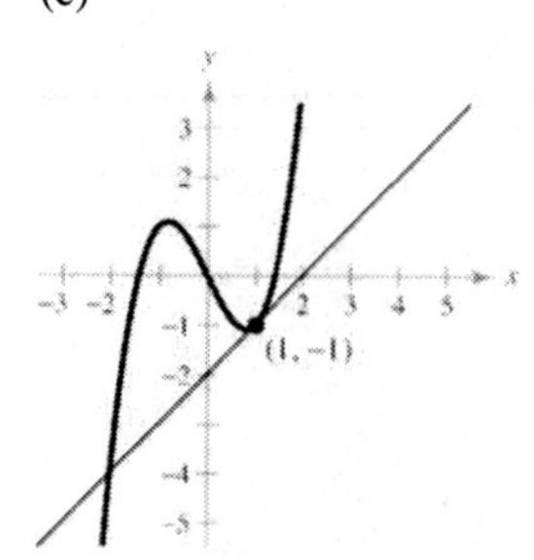

47. (a) $m_{\text{sec}} = \dfrac{f(3+h)-f(3)}{h} = \dfrac{\sqrt{3+h+1} - 2}{h}$

$= \dfrac{\sqrt{4+h} - 2}{h} \cdot \dfrac{\sqrt{4+h} + 2}{\sqrt{4+h} + 2}$

$= \dfrac{4 + h - 4}{h\left[\sqrt{4+h} + 2\right]}$

$= \dfrac{1}{\sqrt{4+h} + 2}$

$m = \lim_{h \to 0} \dfrac{1}{\sqrt{4+h} + 2} = \dfrac{1}{4}$

(b) $y - 2 = \dfrac{1}{4}(x - 3)$

$y = \dfrac{1}{4}x + \dfrac{5}{4}$

(c)

49. (a) $m_{\text{sec}} = \dfrac{f(-4+h) - f(-4)}{h}$

$= \dfrac{\dfrac{1}{-4+h+5} - 1}{h}$

$= \dfrac{1 + 4 - h - 5}{h(-4+h+5)} = \dfrac{-1}{h+1},\ h \neq 0$

$m = \lim_{h \to 0} \dfrac{-1}{h+1} = -1$

(b) $y - 1 = -1(x + 4)$

$y = -x - 3$

(c)

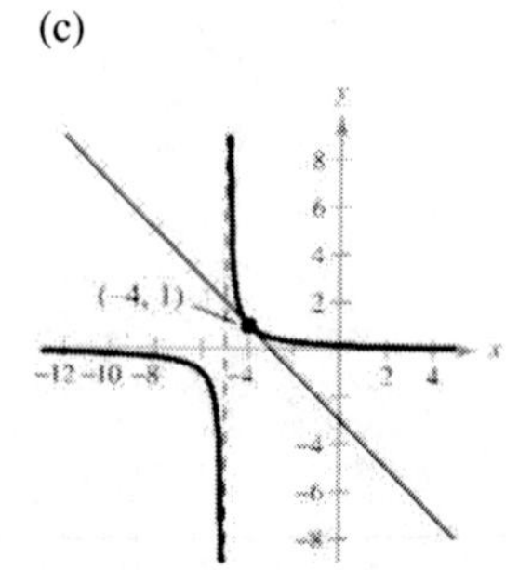

51.

x	−2	−1.5	−1	−0.5
$f(x)$	2	1.125	0.5	0.125
$f'(x)$	−2	−1.5	−1	−0.5

x	0	0.5	1	1.5	2
$f(x)$	0	0.125	0.5	1.125	2
$f'(x)$	0	0.5	1	1.5	2

$f(x) = \dfrac{1}{2}x^2$

$f'(x) = x$

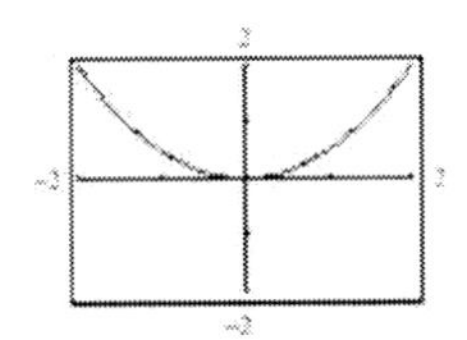

They appear to be the same.

53.

x	-2	-1.5	-1	-0.5	0
$f(x)$	1	1.225	1.414	1.581	1.732
$f'(x)$	0.5	0.408	0.354	0.316	0.289

x	0.5	1	1.5	2
$f(x)$	1.871	2	2.121	2.236
$f'(x)$	0.267	0.25	0.236	0.224

$f(x) = \sqrt{x+3}$

$f'(x) = \dfrac{1}{2\sqrt{x+3}}$

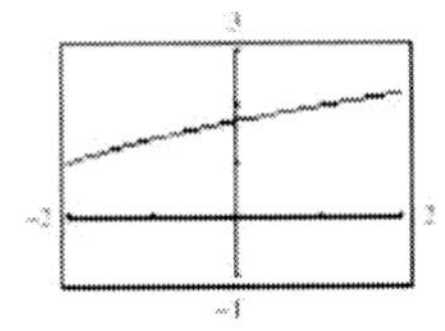

They appear to be the same.

55. $f'(x) = \lim_{h\to 0}\dfrac{f(x+h)-f(x)}{h}$

$= \lim_{h\to 0}\dfrac{[(x+h)^2 - 4(x+h)+3]-[x^2-4x+3]}{h}$

$= \lim_{h\to 0}\dfrac{(x^2+2xh+h^2-4x-4h+3)-(x^2-4x+3)}{h}$

$= \lim_{h\to 0}\dfrac{2xh+h^2-4h}{h} = \lim_{h\to 0} 2x+h-4 = 2x-4$

$f'(x) = 0 = 2x-4 \Rightarrow x = 2$

f has a horizontal tangent at $(2, -1)$.

57. $f'(x) = \lim_{h\to 0}\dfrac{f(x+h)-f(x)}{h}$

$= \lim_{h\to 0}\dfrac{[3(x+h)^3 - 9(x+h)]-[3x^3-9x]}{h}$

$= \lim_{h\to 0}\dfrac{3(x^3+3x^2h+3xh^2+h^3)-9x-9h-3x^3+9x}{h}$

$= \lim_{h\to 0}\dfrac{9x^2h+9xh^2+3h^3-9h}{h}$

$= \lim_{h\to 0}(9x^2+9xh+3h-9) = 9x^2-9$

$f'(x) = 0 = 9x^2-9 = 9(x+1)(x-1) \Rightarrow x = \pm 1$

f has horizontal tangents at $(1, -6)$ and $(-1, 6)$.

59. $f'(x) = 4x^3 - 4x = 0$

$4x(x-1)(x+1) = 0$

$x = 0, 1, -1$

f has horizontal tangents at $(0, 0)$, $(1, -1)$, and $(-1, -1)$.

61. $f'(x) = -2\sin x + 1 = 0$

$\sin x = \dfrac{1}{2}$

$x = \dfrac{\pi}{6}, \dfrac{5\pi}{6}$

f has horizontal tangents at $\left(\dfrac{\pi}{6}, \sqrt{3}+\dfrac{\pi}{6}\right)$ and $\left(\dfrac{5\pi}{6}, \dfrac{5\pi}{6}-\sqrt{3}\right)$.

63. $f'(x) = x^2e^x + 2xe^x = 0$

$xe^x(x+2) = 0$

$x = 0, -2$

f has horizontal tangents at $(0, 0)$ and $(-2, 4e^{-2})$.

65. $f'(x) = \ln x + 1 = 0$

$\ln x = -1$

$x = e^{-1}$

f has a horizontal tangent at $\left(e^{-1}, -e^{-1}\right)$.

67. (a) $y = -0.41t^2 + 54.7t + 8529$

(b)

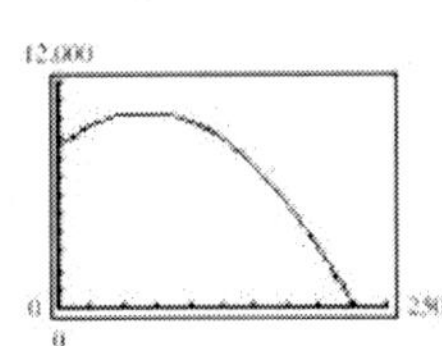

The slope at $t = 20$ is approximately 38. The population is increasing by approximately 38,000 people per year in 2000.

(c) $y' = \lim_{h\to 0}\dfrac{f(t+h)-f(t)}{h}$

$y' = \lim_{h\to 0}\dfrac{\left(-0.41(t+h)^2 + 54.7(t+h) + 8529 - \left(-0.41t^2+54.7t+8529\right)\right)}{h}$

$y' = \lim_{h\to 0}\dfrac{\left(-0.41t^2 - 0.82th - 0.41h^2 + 54.7t + 54.7h + 8529 + 0.41t^2 - 54.7t - 8529\right)}{h}$

$y' = \lim_{h\to 0}\dfrac{-0.082th - 0.41h^2 + 54.7h}{h}$

$y' = \lim_{h\to 0}\dfrac{h\left(-0.82t - 0.41h + 54.7\right)}{h}$

$y' = \lim_{h\to 0}\left(-0.82t - 0.41h + 54.7\right)$

$y' = -0.82t + 54.7$

y' at $t = 20$: $m = 38.3$

(d) Answers will vary.

69. (a) $V = \frac{4}{3}\pi r^3$

$$V'(r) = \lim_{h\to 0}\frac{V(r+h)-V(r)}{h}$$
$$= \lim_{h\to 0}\frac{\left(\frac{4}{3}\right)\pi(r+h)^3 - \left(\frac{4}{3}\right)\pi r^3}{h}$$
$$= \lim_{h\to 0}\left(\frac{4}{3}\pi\right)\frac{r^3+3r^2h+3rh^2+h^3-r^3}{h}$$
$$= \lim_{h\to 0}\frac{4}{3}\pi(3r^2+3rh+h^2) = 4\pi r^2$$

(b) $V'(4) = 4\pi(4)^2 \approx 201.06$

(c) Cubic inches per inch; The derivative is a formula for rate of change.

71. $s(t) = -16t^2 + 64t + 80$

(a) Using the limit definition, $s'(t) = -32t + 64$.

(b) $s(0) = 80$, $s(3) = 128$

$$\text{Average rate of change} = \frac{128-80}{3} = \frac{48}{3} = 16\tfrac{\text{ft}}{\text{sec}}$$

(c) $s'(t) = -32t + 64 = 0 \Rightarrow t = 2$ seconds

Answers will vary.

(d) $s(t) = -16t^2 + 64t + 80$

$= 0 \Rightarrow t = 5$ seconds

$s'(5) = -32(5) + 64 = -9$ ft/sec

(e)

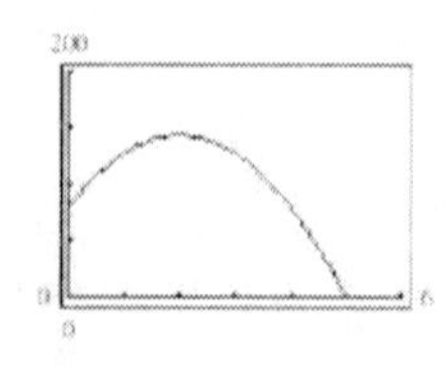

73. True. The slope is $2x$, which is different for all x.

75. Matches (b).

(Derivative is always positive, but decreasing.)

77. Matches (d).
(Derivative is -1 for $x < 0$, 1 for $x > 0$.)

79. Answers will vary.
Sample answer: The graph shown has a positive slope.

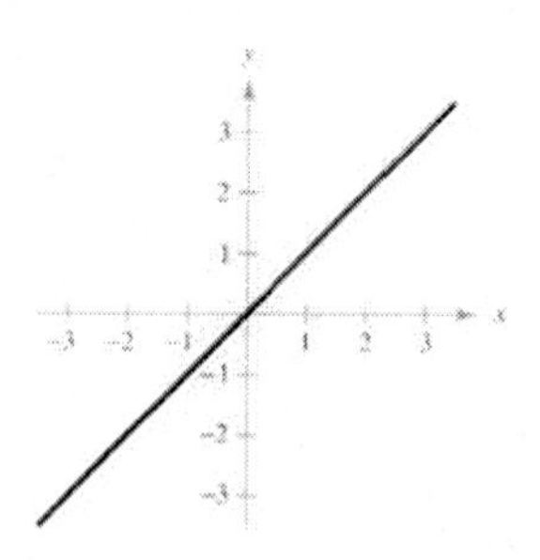

81. Given the following: $f'(x) < 0$ for $x < 1$

$f'(x) > 0$ for $x > 1$

$f'(1) = 0$

Answers will vary: Sample answer:

$f'(x) < 0$ for $x < 1 \rightarrow f$ is decreasing on $(-\infty, 1)$

$f'(x) > 0$ for $x > 1 \rightarrow f$ is increasing on $(1, \infty)$

$f'(1) = 0 \rightarrow x = 1$ is a critical number of f and

$(1, f(1))$ is a relative minimum of f.

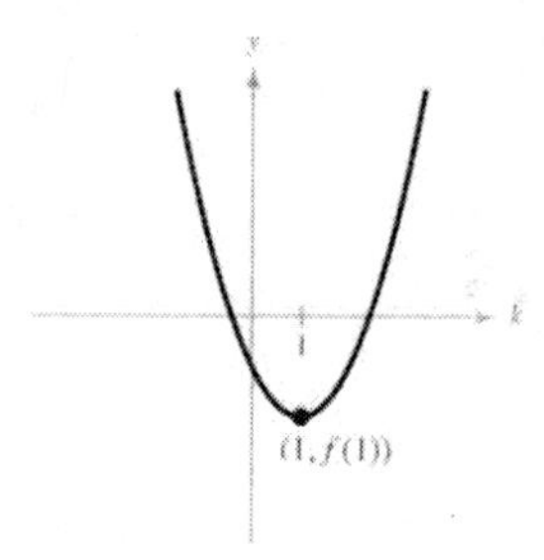

83. $f(x) = \dfrac{x-2}{x^2-4x+3} = \dfrac{x-2}{(x-3)(x-1)}$

Intercepts: $(2, 0)$, $\left(0, -\frac{2}{3}\right)$

Vertical asymptotes: $x = 1$, $x = 3$

Horizontal asymptote: $y = 0$

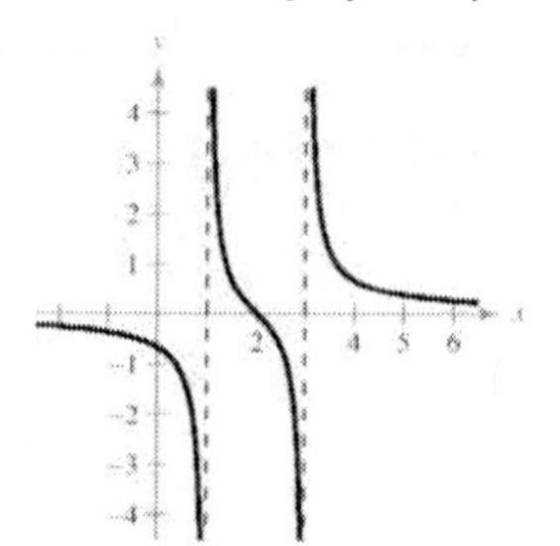

Section 11.4

1. If the line $y = 5$ is a horizontal asymptote of the graph of a function f, then $\lim_{x\to\infty} f(x) = 5$.

3. converge

5. Intercept: $(0, 0)$

Horizontal asymptote: $y = 4$

Matches (c).

7. Horizontal asymptote: $y = 4$

Vertical asymptote: $x = 0$

Matches (d).

9. Vertical asymptotes: $x = \pm 1$
Horizontal asymptote: $y = 1$
Matches (f).

11. Vertical asymptote: $x = 2$
Horizontal asymptote: $y = -2$
Matches (h).

13. $\lim\limits_{x\to\infty} \dfrac{3}{x^2} = 0$

15. $\lim\limits_{x\to\infty} \dfrac{3+x}{3-x} = -1$

17. $\lim\limits_{x\to-\infty} \dfrac{5x-2}{6x+1} = \dfrac{5}{6}$

19. $\lim\limits_{x\to-\infty} \dfrac{4x^2-3}{2-x^2} = \dfrac{4}{-1} = -4$

21. $\lim\limits_{t\to\infty} \dfrac{t^2}{t+3}$ does not exist.

23. $\lim\limits_{t\to\infty} \dfrac{4t^2+3t-1}{3t^2+2t-5} = \dfrac{4}{3}$

25. $\lim\limits_{y\to-\infty} \dfrac{3+8y-4y^2}{3-y-2y^2} = \dfrac{-4}{-2} = 2$

27. $\lim\limits_{x\to-\infty} \dfrac{-(x^2+3)}{(2-x)^2} = \lim\limits_{x\to-\infty} \dfrac{-x^2-3}{x^2-4x+4} = -1$

29. $\lim\limits_{x\to\infty} \left[\dfrac{x}{(x+1)^2} - 4\right] = 0 - 4 = -4$

31. $\lim\limits_{t\to\infty} \left(\dfrac{1}{3t^2} - \dfrac{5t}{t+2}\right) = 0 - 5 = -5$

33. $f(x) = \dfrac{3x}{x-1}$

(a)

x	10^0	10^1	10^2	10^3
$f(x)$	Error	-3.33	-3.03	-3.003

x	10^4	10^5	10^6
$f(x)$	-3.0003	-3.00003	-3.000003

$\lim\limits_{x\to\infty} \dfrac{3x}{x-1} = -3$

(b)

$\lim\limits_{x\to\infty} \dfrac{3x}{x-1} = -3$

(c) $\lim\limits_{x\to\infty} \dfrac{3x}{x-1} = -3$

35. $f(x) = \dfrac{2x}{1-x^2}$

(a)

x	10^0	10^1	10^2	10^3
$f(x)$	Error	-0.202	-0.0200	-0.002

x	10^4	10^5	10^6
$f(x)$	-0.0002	-0.00002	-0.000002

$\lim\limits_{x\to\infty} \dfrac{2x}{1-x^2} = 0$

(b)

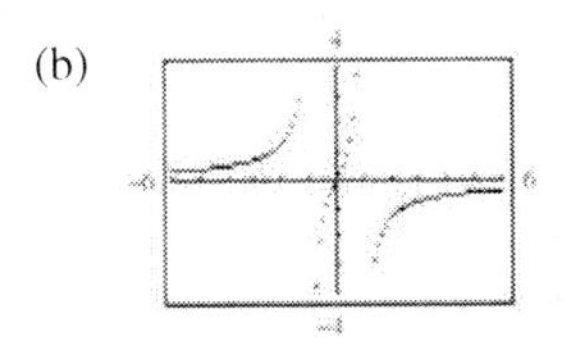

$\lim\limits_{x\to\infty} \dfrac{2x}{1-x^2} = 0$

(c) $\lim\limits_{x\to\infty} \dfrac{2x}{1-x^2} = 0$

37. $f(x) = 1 - \dfrac{3}{x^2}$ $\left(\text{or } f(x) = \dfrac{x^2-3}{x^2}\right)$

(a)

x	10^0	10^1	10^2	10^3
$f(x)$	-2	0.97	0.9997	0.999997

x	10^4	10^5	10^6
$f(x)$	0.99999997	0.9999999997	1

$\lim\limits_{x\to\infty} \left(1 - \dfrac{3}{x^2}\right) = 1$

(b)

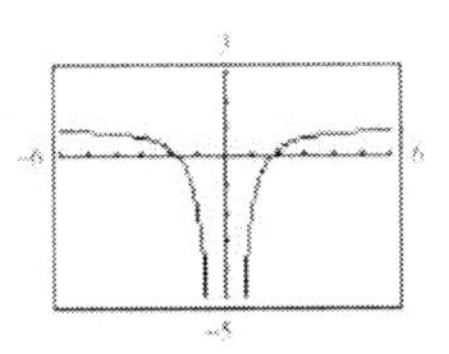

$\lim\limits_{x\to\infty} \left(1 - \dfrac{3}{x^2}\right) = 1$

(c) $\lim\limits_{x\to\infty} \left(1 - \dfrac{3}{x^2}\right) = 1 \Rightarrow \lim\limits_{x\to\infty} 1 - \lim\limits_{x\to\infty} \dfrac{3}{x^2} = 1 - 0 = 1$

39. $f(x) = x - \sqrt{x^2 + 2}$

(a)

x	10^0	10^1	10^2	10^3
$f(x)$	–0.7321	–0.0995	–0.0100	–0.0010

x	10^4	10^5	10^6
$f(x)$	-1.0×10^{-5}	-1.0×10^{-5}	-1.0×10^{-6}

$\lim_{x\to\infty} f(x) = 0$

(b)

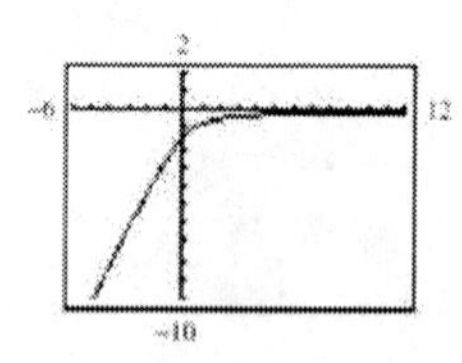

$\lim_{x\to\infty} f(x) = 0$

41. $f(x) = 3\left(2x - \sqrt{4x^2 + x}\right)$

(a)

x	10^0	10^1	10^2	10^3
$f(x)$	–0.7082	–0.7454	–0.7495	–0.74995

x	10^4	10^5	10^6
$f(x)$	–0.749995	–0.7499995	–0.75

$\lim_{x\to\infty} f(x) = -0.75$

(b)

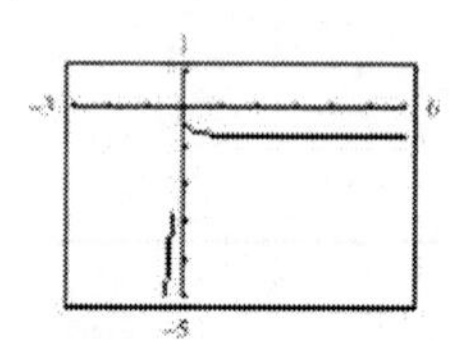

$\lim_{x\to\infty} f(x) = -0.75$

43. $a_n = \dfrac{n+1}{n^2+1}$

$1, \frac{3}{5}, \frac{2}{5}, \frac{5}{17}, \frac{3}{13}$

$\lim_{n\to\infty} \dfrac{n+1}{n^2+1} = 0$

45. $a_n = \dfrac{n}{2n+1}$

$\frac{1}{3}, \frac{2}{5}, \frac{3}{7}, \frac{4}{9}, \frac{5}{11}$

$\lim_{n\to\infty} \dfrac{n}{2n+1} = \dfrac{1}{2}$

47. $a_n = \dfrac{n^2}{3n+2}$

$\frac{1}{5}, \frac{1}{2}, \frac{9}{11}, \frac{8}{7}, \frac{25}{17}$

$\lim_{n\to\infty} \dfrac{n^2}{3n+2}$ does not exist.

49. $a_n = \dfrac{(n+1)!}{n!}$

2, 3, 4, 5, 6

$\lim_{n\to\infty} \dfrac{(n+1)!}{n!} = \lim_{x\to\infty}(n+1)$ does not exist.

51. $a_n = \dfrac{(-1)^n}{n}$

$-1, \frac{1}{2}, -\frac{1}{3}, \frac{1}{4}, -\frac{1}{5}$

$\lim_{n\to\infty} \dfrac{(-1)^n}{n} = 0$

53.

n	10^0	10^1	10^2	10^3
a_n	2	1.55	1.505	1.5005

$\lim_{n\to\infty} a_n = 1.5$

$$a_n = \frac{1}{n}\left(n + \frac{1}{n}\left[\frac{n(n+1)}{2}\right]\right) = 1 + \frac{1}{n^2}\left[\frac{n^2+n}{2}\right] = 1 + \frac{n^2+n}{2n^2}$$

$\lim_{n\to\infty} a_n = 1 + \dfrac{1}{2} = \dfrac{3}{2}$

55. $a_n = \dfrac{10}{n^3}\left[\dfrac{n(n+1)(3n+1)}{6}\right]$

n	10^0	10^1	10^2	10^3
a_n	13.33	5.683	5.06683	5.0066683

$\lim_{n\to\infty} a_n = 5$

$$a_n = \frac{10}{n^3}\left[\frac{3n^3+4n^2+n}{6}\right]$$

$$= \left(5 + \frac{20}{3n} + \frac{5}{3n^2}\right)$$

$\lim_{n\to\infty} a_n = 5 + 0 = 5$

57. (a) Average cost $= \overline{C} = \dfrac{C}{x} = 13.50 + \dfrac{45{,}750}{x}$

(b) $\overline{C}(100) = \$471$

$\overline{C}(1000) = \$59.25$

(c) $\lim_{x\to\infty} C(x) = 13.50$

As more units are produced, the fixed costs (45,750) become less dominant.

59. $R(t)=\dfrac{61.018t^2+1260.64}{0.0578t^2+1}$

(a)

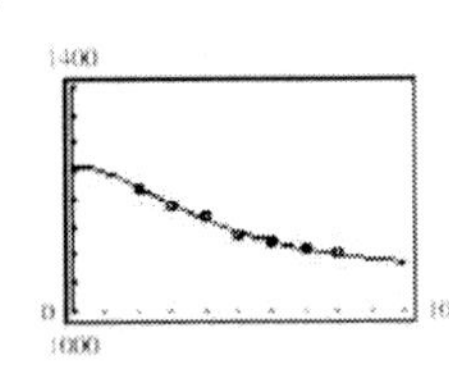

The model fits the data well

(b) $2009: R(9)=1092 \Rightarrow 1{,}092{,}000$ reserves

$2010: R(10)=1086 \Rightarrow 1{,}086{,}000$ reserves

(c) $\lim_{t\to\infty}\dfrac{61.018t^2+1260.64}{0.0578t^2+1}=\dfrac{61.018}{0.0578}=1056$

$\Rightarrow 1{,}056{,}000$ reserves

As the time passes, the number of United States military reserve personnel approaches 1,056,000.

(d) Answers will vary.

61. False. $f(x)=\dfrac{x^2+1}{1}$ does not have a horizontal asymptote.

63. True. See page 784 in the text.

65. Answers will vary. Sample answer:

Let $f(x)=x^2$ and $g(x)=x^2$. Then

$\lim_{x\to\infty}x^2=\infty$ and $\lim_{x\to\infty}\left[f(x)-g(x)\right]=0.$

67. Converges to 0

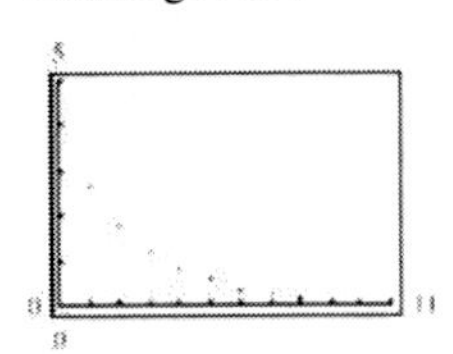

69. Diverges

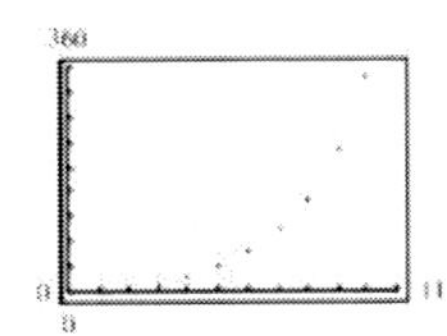

71. The correct answer is $\lim_{x\to\infty}\dfrac{1-2x-x^2}{4x^2+1}=\dfrac{-1}{4}.$

73. $y=x^4$

(a) $f(x)=(x+3)^4$

(b) $f(x)=x^4-1$

(c) $f(x)=-2+x^4$

(d) $f(x)=\dfrac{1}{2}(x-4)^4$

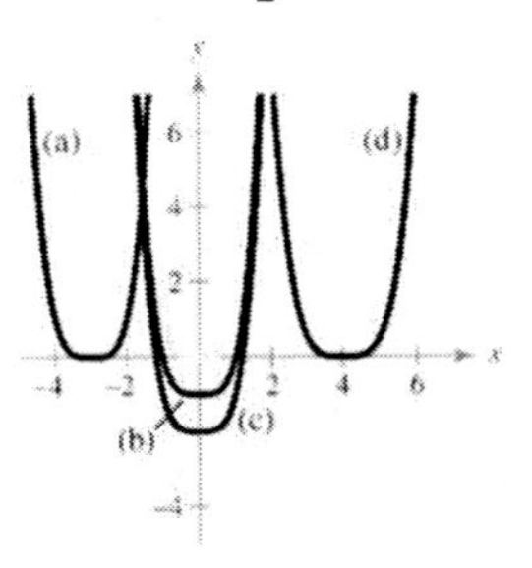

75. $\sum_{i=1}^{6}(2i+3)=5+7+9+11+13+15=60$

77. $\sum_{k=1}^{10}15=10(15)=150$

Section 11.5

1. $\dfrac{n(n+1)}{2}$

3. Better approximations of the area of the shaded region are found by using more rectangles, therefore 100 rectangles of equal width would obtain a better approximation than that of 10 rectangles of equal width.

5. $\sum_{i=1}^{60}7=7(60)=420$

7. $\sum_{i=1}^{20}i^3=\dfrac{20^2(21)^2}{4}=44{,}100$

9. $\sum_{k=1}^{20}(k^3+2)=\dfrac{20^2(21)^2}{4}+2(20)$

$=44{,}100+40$

$=44{,}140$

11. $\sum_{j=1}^{25}(j^2+j)=\dfrac{25(26)(51)}{6}+\dfrac{25(26)}{2}=5850$

13. (a) $S(n)=\sum_{i=1}^{n}\dfrac{i^3}{n^4}=\dfrac{1}{n^4}\left[\dfrac{n^2(n+1)^2}{4}\right]$

$=\dfrac{n^2+2n+1}{4n^2}$

(b)

n	10^0	10^1	10^2	10^3	10^4
$S(n)$	1	0.3025	0.255025	0.25050025	0.25005

(c) $\lim_{n\to\infty}S(n)=\dfrac{1}{4}$

15. (a) $S(n) = \sum_{i=1}^{n} \frac{3}{n^3}(1+i^2)$

$= \frac{3}{n^3}\left[n + \frac{n(n+1)(2n+1)}{6}\right]$

$= \frac{3}{n^2} + \frac{6n^2+9n+3}{6n^2} = \frac{2n^2+3n+7}{2n^2}$

(b)

n	10^0	10^1	10^2	10^3	10^4
$S(n)$	6	1.185	1.0154	1.0015	1.00015

(c) $\lim_{n\to\infty} S(n) = 1$

17. (a) $S(n) = \sum_{i=1}^{n}\left(\frac{i^2}{n^3} + \frac{2}{n}\right)\left(\frac{1}{n}\right)$

$= \frac{1}{n}\left[\frac{n(n+1)(2n+1)}{6n^3} + \frac{2n}{n}\right]$

$= \frac{1}{6n^3}(2n^2+3n+1) + \frac{2}{n}$

$= \frac{14n^2+3n+1}{6n^3}$

(b)

n	10^0	10^1	10^2	10^3	10^4
$S(n)$	3	0.2385	0.02338	0.00233	0.0002333

(c) $\lim_{n\to\infty} S(n) = 0$

19. $f(x) = x+4$, $[-1, 2]$, $n = 6$, width $= \frac{1}{2}$

Intervals: $\left[-1, -\frac{1}{2}\right], \left[-\frac{1}{2}, 0\right], \left[0, \frac{1}{2}\right],$

$\left[\frac{1}{2}, 1\right], \left[1, \frac{3}{2}\right], \left[\frac{3}{2}, 2\right]$

The height is obtained by evaluating f at the right-hand endpoint of each interval.

Area $\approx \frac{1}{2}[3.5 + 4.45 + 5 + 5.5 + 6]$

$= 14.25$ square units

21. $f(x) = \frac{1}{4}x^3$, $[0, 2]$, $n = 8$, width $= \frac{1}{4}$

Intervals: $\left[0, \frac{1}{4}\right], \left[\frac{1}{4}, \frac{1}{2}\right], \left[\frac{1}{2}, \frac{3}{4}\right], \left[\frac{3}{4}, 1\right],$

$\left[1, \frac{5}{4}\right], \left[\frac{5}{4}, \frac{3}{2}\right], \left[\frac{3}{2}, \frac{7}{4}\right], \left[\frac{7}{4}, 2\right]$

The height is obtained by evaluating f at the right-hand endpoint of each interval.

Area $\approx \sum_{i=1}^{8} f\left(\frac{i}{4}\right)\left(\frac{1}{4}\right) = \sum_{i=1}^{8} \frac{1}{4}\left(\frac{i}{4}\right)^3\left(\frac{1}{4}\right)$

$= 1.265625$ square units

23. Width of each rectangle is $\frac{12}{n}$. The height is

$f\left(\frac{12}{n}i\right) = -\frac{1}{3}\left(\frac{12}{n}i\right) + 4.$

$A \approx \sum_{i=1}^{n}\left[-\frac{1}{3}\left(\frac{12i}{n}\right) + 4\right]\left(\frac{12}{n}\right)$

Note: Exact area is 24.

n	4	8	20	50
Approximate area	18	21	22.8	23.52

25. The width of each rectangle is $\frac{3}{n}$. The height is

$f\left(\frac{3i}{n}\right) = \frac{1}{9}\left(\frac{3i}{n}\right)^3.$

$A \approx \sum_{i=1}^{n} \frac{1}{9}\left(\frac{3i}{n}\right)^3\left(\frac{3}{n}\right)$

n	4	8	20	50
Approximate area	3.52	2.85	2.48	2.34

27. $f(x) = 2x + 5$, $[0, 4]$

The width of each rectangle is $\frac{4}{n}$. The height is

$f\left(\frac{4i}{n}\right) = 2\left(\frac{4i}{n}\right) + 5 = \frac{8i}{n} + 5.$

$A \approx \sum_{i=1}^{n}\left(\frac{8i}{n} + 5\right)\left(\frac{4}{n}\right)$

n	4	8	20	50	100	∞
Area	40	38	36.8	36.32	36.16	36

$A \approx \sum_{i=1}^{n}\left(\frac{8i}{n} + 5\right)\left(\frac{4}{n}\right) = \sum_{i=1}^{n}\left(\frac{20}{n} + \frac{32}{n^2}i\right)$

$= \frac{20}{n}(n) + \frac{32}{n^2}\left(\frac{n(n+1)}{2}\right) = 20 + 16\left(\frac{n^2+n}{n^2}\right)$

$A = \lim_{n\to\infty}\left[20 + 16\left(\frac{n^2+n}{n^2}\right)\right] = 20 + 16 = 36$

$A \approx \sum_{i=1}^{n}\left(\frac{48i}{n^2} + \frac{4}{n}\right) = \frac{48}{n^2}\left(\frac{n(n+1)}{2}\right) + \frac{4}{n}(n)$

$= 24\left(\frac{n^2+n}{n^2}\right) + 4$

$A = \lim_{n\to\infty}\left[24\left(\frac{n^2+n}{n^2}\right) + 4\right] = 28$

29. $f(x) = 9 - x^2,\ [0, 2]$

The width of each rectangle is $\frac{2}{n}$. The height is

$$f\left(\frac{2i}{n}\right) = 9 - \left(\frac{2i}{n}\right)^2 = 9 - \frac{4i^2}{n^2}.$$

$$A \approx \sum_{i=1}^{n}\left(9 - \frac{4i^2}{n^2}\right)\left(\frac{2}{n}\right)$$

n	4	8	20	50	100	∞
Area	14.25	14.8125	15.13	15.2528	15.2932	$\frac{46}{3}$

$$A \approx \sum_{i=1}^{n}\left(\frac{18}{n} - \frac{8i^2}{n^3}\right)$$

$$= \left(\frac{18}{n}\right)n - \frac{8}{n^3}\left(\frac{n(n+1)(2n+1)}{6}\right)$$

$$= 18 - \frac{4}{3}\left(\frac{n(n+1)(2n+1)}{n^3}\right)$$

$$A = \lim_{n\to\infty}\left[18 - \frac{4}{3}\left(\frac{n(n+1)(2n+1)}{n^3}\right)\right] = 18 - \frac{8}{3} = \frac{46}{3}$$

31. $f(x) = \frac{1}{2}x + 4,\ [-1, 3]$

The width of each rectangle is $\frac{4}{n}$. The height is

$$f\left(-1 + \frac{4i}{n}\right) = \frac{1}{2}\left(-1 + \frac{4i}{n}\right) + 4 = \frac{7}{2} + \frac{2i}{n}.$$

$$A \approx \sum_{i=1}^{n}\left(\frac{7}{2} + \frac{2i}{n}\right)\left(\frac{4}{n}\right)$$

n	4	8	20	50	100	∞
Area	19	18.5	18.2	18.08	18.04	18

$$A \approx \sum_{i=1}^{n}\left(\frac{14}{n} + \frac{8i}{n^2}\right) = \left(\frac{14}{n}\right)n + \frac{8}{n^2}\left(\frac{n(n+1)}{2}\right)$$

$$A = \lim_{n\to\infty}\left[14 + \frac{4}{n^2}\left(\frac{n(n+1)}{2}\right)\right] = 14 + 4 = 18$$

33. $A \approx \sum_{i=1}^{n} f\left(\frac{i}{n}\right)\left(\frac{1}{n}\right)$

$$= \sum_{i=1}^{n}\left[4\left(\frac{i}{n}\right) + 1\right]\left(\frac{1}{n}\right)$$

$$= \frac{1}{n}\sum_{i=1}^{n}\left[\frac{4}{n}i + 1\right]$$

$$= \frac{1}{n}\left[\frac{4}{n}\frac{n(n+1)}{2} + n\right]$$

$$= \frac{1}{n}\left[2(n+1) + n\right]$$

$$= \frac{3n+2}{n}$$

$$A = \lim_{n\to\infty}\frac{3n+2}{n} = 3 \text{ square units}$$

35. $A \approx \sum_{i=1}^{n} f\left(\frac{3}{n}i\right)\left(\frac{3}{n}\right)$

$$= \sum_{i=1}^{n}\left[-\left(\frac{3}{n}i\right) + 4\right]\left(\frac{3}{n}\right)$$

$$= \frac{3}{n}\sum_{i=1}^{n}\left[-\frac{3}{n}i + 4\right]$$

$$= \frac{3}{n}\left[-\frac{3}{n}\cdot\frac{n(n+1)}{2} + 4n\right]$$

$$= \frac{3}{n}\left[-\frac{3}{2}n - \frac{3}{2} + 4n\right]$$

$$= \frac{3}{n}\left(\frac{5}{2}n - \frac{3}{2}\right)$$

$$= \frac{15}{2} - \frac{9}{2n}$$

$$A = \lim_{n\to\infty}\left[\frac{15}{2} - \frac{9}{2n}\right] = \frac{15}{2} \text{ square units}$$

37. $A \approx \sum_{i=1}^{n} f\left(\frac{4}{n}i\right)\frac{4}{n}$

$$= \sum_{i=1}^{n}\left[16 - \left(\frac{4}{n}i\right)^2\right]\frac{4}{n}$$

$$= \frac{4}{n}\sum_{i=1}^{n}\left[16 - \frac{16}{n^2}i^2\right]$$

$$= \frac{4}{n}\left[16n - \frac{16}{n^2}\cdot\frac{n(n+1)(2n+1)}{6}\right]$$

$$= \frac{4}{n}\left[16n - \frac{16}{n^2}\left(\frac{2n^3 + 3n^2 + n}{6}\right)\right]$$

$$= \frac{4}{n}\left[16n - \frac{16}{3}n - 8 - \frac{8}{3n}\right]$$

$$= \frac{4}{n}\left[\frac{32}{3}n - 8 - \frac{8}{3n}\right]$$

$$= \frac{128}{3} - \frac{32}{n} - \frac{32}{3n^2}$$

$$A = \lim_{n\to\infty}\left(\frac{128}{3} - \frac{32}{n} - \frac{32}{3n^2}\right) = \frac{128}{3} \text{ square units}$$

39. $A \approx \sum_{i=1}^{n} f\left(\frac{i}{n}\right)\frac{1}{n}$

$$= \sum_{i=1}^{n}\left[1-\left(\frac{i}{n}\right)^3\right]\frac{1}{n}$$

$$= \sum_{i=1}^{n}\left[\left(1-\frac{i^3}{n^3}\right)\frac{1}{n}\right]$$

$$= \frac{1}{n}\left[n - \frac{1}{n^3}\cdot\frac{n^2(n+1)^2}{4}\right]$$

$$= \frac{1}{n}\left[n - \frac{1}{n^3}\cdot\frac{n^2(n^2+2n+1)}{4}\right]$$

$$= \frac{1}{n}\left[n - \frac{n}{4} - \frac{1}{2} - \frac{1}{4n}\right] = \frac{1}{n}\left(\frac{3}{4}n - \frac{1}{2} - \frac{1}{4n}\right)$$

$$= \frac{3}{4} - \frac{1}{2n} - \frac{1}{4n^2}$$

$$A = \lim_{n\to\infty}\left(\frac{3}{4} - \frac{1}{2n} - \frac{1}{4n^2}\right) = \frac{3}{4} \text{ square unit}$$

41. $A \approx \sum_{i=1}^{n} g\left(\frac{i}{n}\right)\left(\frac{1}{n}\right)$

$$= \sum_{i=1}^{n}\left[2\left(\frac{i}{n}\right) - \left(\frac{i}{n}\right)^3\right]\left(\frac{1}{n}\right)$$

$$= \frac{1}{n}\sum_{i=1}^{n}\left[\frac{2}{n}i - \frac{1}{n^3}i^3\right]$$

$$= \frac{1}{n}\left[\frac{2}{n}\cdot\frac{n(n+1)}{2} - \frac{1}{n^3}\cdot\frac{n^2(n+1)^2}{4}\right]$$

$$= \frac{n+1}{n} - \frac{(n+1)^2}{4n^2}$$

$$A = \lim_{n\to\infty}\left[\frac{n+1}{n} - \frac{(n+1)^2}{4n^2}\right]$$

$$= 1 - \frac{1}{4} = \frac{3}{4} \text{ square unit}$$

43. $A \approx \sum_{i=1}^{n} f\left(\frac{6}{n}i\right)\frac{6}{n}$

$$= \sum_{i=1}^{n}\left[\left(\frac{6}{n}i\right)^2 + 4\left(\frac{6}{n}i\right)\right]\frac{6}{n}$$

$$= \sum_{i=1}^{n}\left[\frac{36}{n^2}i^2 + \frac{24}{n}i\right]\frac{6}{n}$$

$$= \frac{6}{n}\left[\frac{36}{n^2}\cdot\frac{n(n+1)(2n+1)}{6} + \frac{24}{n}\cdot\frac{n(n+1)}{2}\right]$$

$$= \frac{6}{n}\left[\frac{36}{n^2}\cdot\frac{2n^3+3n^2+n}{6} + \frac{24}{n}\cdot\frac{n^2+n}{2}\right]$$

$$= \frac{6}{n}\left[12n + 18 + \frac{6}{n} + 12n + 12\right]$$

$$= \frac{6}{n}\left(24n + 30 + \frac{6}{n}\right)$$

$$= 144 + \frac{180}{n} + \frac{36}{n^2}$$

$$A = \lim_{n\to\infty}\left[144 + \frac{180}{n} + \frac{36}{n^2}\right] = 144 \text{ square units}$$

45. $y = (-3.0\cdot 10^{-6})x^3 + 0.002x^2 - 1.05x + 400$

Note that $y = 0$ when $x = 500$.

Area $\approx$ 105,208.33 square feet $\approx$ 2.4153 acres

47. True. See Formula 2, page 789 of the text.

49. Area is approximately a triangle of base 2 and height 3.

Area ≈ 4; (c)

Chapter 11 Review Exercises

1. $f(x) = 6x - 1$

$\lim_{x\to 3}(6x-1) = 17$

The limit can be reached.

x	2.9	2.99	2.999	3	3.001	3.01	3.1
$f(x)$	16.4	16.94	16.994	17	17.006	17.06	17.6

3. $f(x) = \frac{1-e^{-x}}{x}$

$\lim_{x\to 0}\frac{1-e^{-x}}{x} = 1$

The limit cannot be reached.

x	−0.1	−0.01	−0.001	0	0.001
$f(x)$	1.0517	1.0050	1.0005	Error	0.9995

x	0.01	0.1
$f(x)$	0.9950	0.9516

5. $\lim_{x\to 1}(3-x)=2$

7. $\lim_{x\to -3}\frac{|x+3|}{x+3}$ does not exist, the limits from the left and from the right do not agree. The one-sided limits are not equal.

9. $\lim_{x\to c} f(x)=2,\ \lim_{x\to c} g(x)=5$

(a) $\lim_{x\to c}\left[f(x)\right]^3=(2)^3=8$

(b) $\lim_{x\to c}\left[3f(x)-g(x)\right]=3(2)-5=1$

(c) $\lim_{x\to c}\left[f(x)g(x)\right]=(2)(5)=10$

(d) $\lim_{x\to c}\frac{f(x)}{g(x)}=\frac{2}{5}$

11. $\lim_{x\to 4}\left(\frac{1}{2}x+3\right)=\frac{1}{2}(4)+3=5$

13. $\lim_{t\to 3}\frac{t^2+1}{t}=\frac{9+1}{3}=\frac{10}{3}$

15. $\lim_{x\to -2}\sqrt[3]{4x}=(-8)^{\frac{1}{3}}=-2$

17. $\lim_{x\to \pi}\sin 3x=\sin 3\pi=0$

19. $\lim_{x\to -1}e^{-x}=e^{-(-1)}=e$

21. $\lim_{x\to -1/2}\arcsin x=\arcsin\left(-\frac{1}{2}\right)=-\frac{\pi}{6}$

23. $\lim_{t\to -2}\frac{t+2}{t^2-4}=\lim_{t\to -2}\frac{t+2}{(t+2)(t-2)}$

$=\lim_{t\to -2}\frac{1}{t-2}=-\frac{1}{4}$

25. $\lim_{x\to 5}\frac{x-5}{x^2+5x-50}=\lim_{x\to 5}\frac{x-5}{(x-5)(x+10)}$

$=\lim_{x\to 5}\frac{1}{x+10}=\frac{1}{15}$

27. $\lim_{x\to 1}\frac{x^2+7x-8}{x^2-3x+2}=\lim_{x\to 1}\frac{(x+8)(x-1)}{(x-2)(x-1)}$

$=\lim_{x\to 1}\left(\frac{x+8}{x-2}\right)=\frac{9}{-1}=-9$

29. $\lim_{x\to -1}\frac{1/(x+2)-1}{x+1}=\lim_{x\to -1}\frac{1-(x+2)}{(x+2)(x+1)}$

$=\lim_{x\to -1}\frac{-(x+1)}{(x+2)(x+1)}$

$=\lim_{x\to -1}\frac{-1}{(x+2)}=-1$

31. $\lim_{u\to 0}\frac{\sqrt{4+u}-2}{u}=\lim_{u\to 0}\frac{\sqrt{4+u}-2}{u}\cdot\frac{\sqrt{4+u}+2}{\sqrt{4+u}+2}$

$=\lim_{u\to 0}\frac{(4+u)-4}{u\left(\sqrt{4+u}+2\right)}$

$=\lim_{u\to 0}\frac{1}{\sqrt{4+u}+2}=\frac{1}{4}$

33. (a)

$\lim_{x\to 3}\frac{x-3}{x^2-9}=\frac{1}{6}$

(b)

x	2.9	2.99	3	3.01	3.1
y_1	0.1695	0.1669	Error	0.1664	0.1639

$\lim_{x\to 3}\frac{x-3}{x^2-9}\approx 0.17$

35. (a)

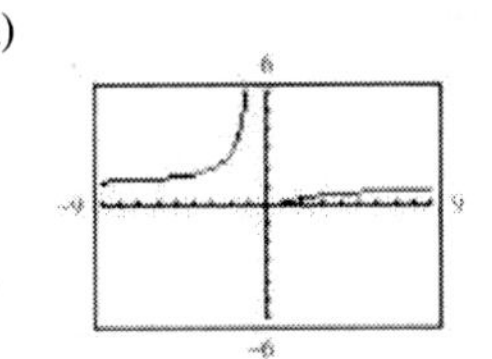

$\lim_{x\to 0}e^{-2/x}$ does not exist.

(b) Answers will vary.

x	–0.1	–0.01	–0.001	0
y_1	4.85 E 8	7.2 E 86	Error	Error

x	0.001	0.01	0.1
y_1	0	1 E –87	2.1 E –9

$\lim_{x\to 0}e^{-2/x}$ does not exist.

37. (a)

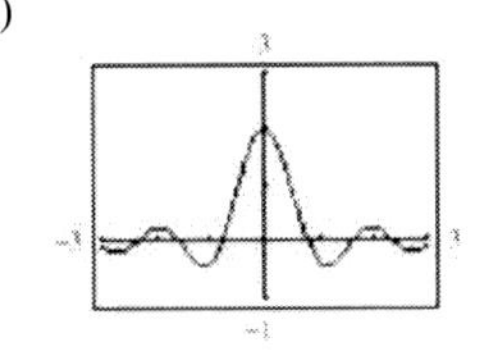

$\lim_{x\to 0}\frac{\sin 4x}{2x}=2$

(b)

x	-0.1	-0.01	-0.001	0	0.001
y_1	1.9471	1.9995	1.999995	error	1.999995

x	0.01	0.1
y_1	1.995	1.9471

$$\lim_{x \to 0} \frac{\sin 4x}{2x} \approx 2$$

39. (a)

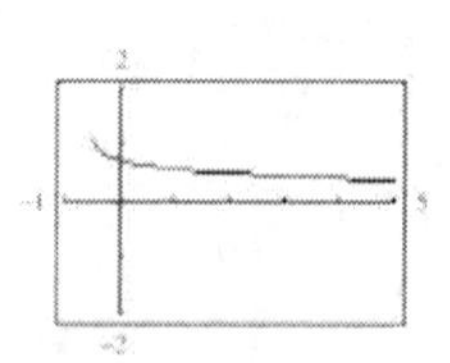

$$\lim_{x \to 1^+} \frac{\sqrt{2x+1} - \sqrt{3}}{x-1} \approx 0.577$$

$$\left(\text{Exact value: } \frac{\sqrt{3}}{3}\right)$$

(b)

x	1.1	1.01	1.001	1.0001
y_1	0.5680	0.5764	0.5773	0.5773

$$\lim_{x \to 1^+} \frac{\sqrt{2x+1} - \sqrt{3}}{x-1} \approx 0.577$$

41. $\lim_{x \to 3^+} f(x) = 1$

$\lim_{x \to 3^-} f(x) = -1$

So, $\lim_{x \to 3} \frac{|x-3|}{x-3}$ does not exist.

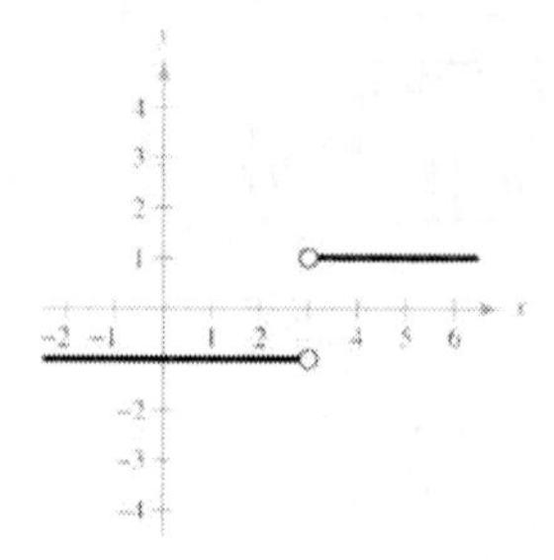

43. $\lim_{x \to 2} \frac{2}{x^2 - 4}$ does not exist because $f(x)$ decreases without bound as $x \to 2$ from the left and $f(x)$ increases without bound as $x \to 2$ from the right.

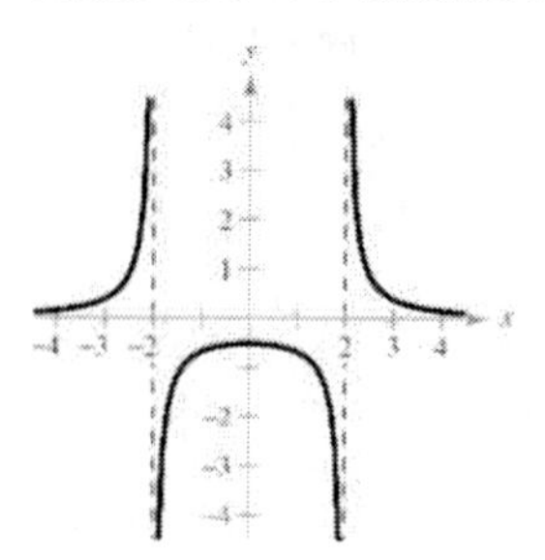

45. $\lim_{x \to 5^-} \frac{|x-5|}{x-5} = -1$

$\lim_{x \to 5^+} \frac{|x-5|}{x-5} = 1$

So, $\lim_{x \to 5} \frac{|x-5|}{x-5}$ does not exist.

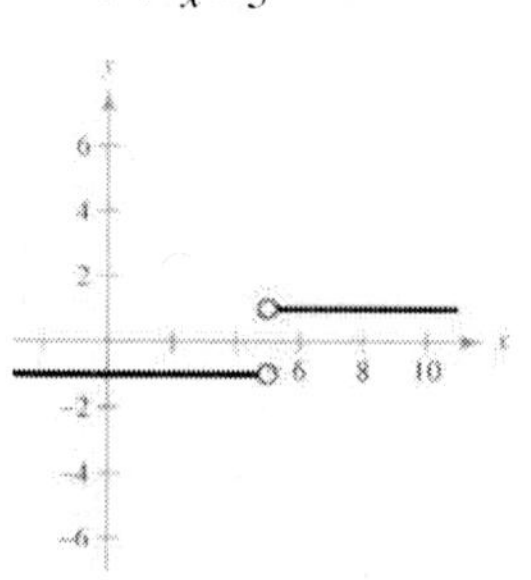

47. $\lim_{x \to 2^-} f(x) = 3$

$\lim_{x \to 2^+} f(x) = 1$

So, $\lim_{x \to 2} f(x)$ does not exist.

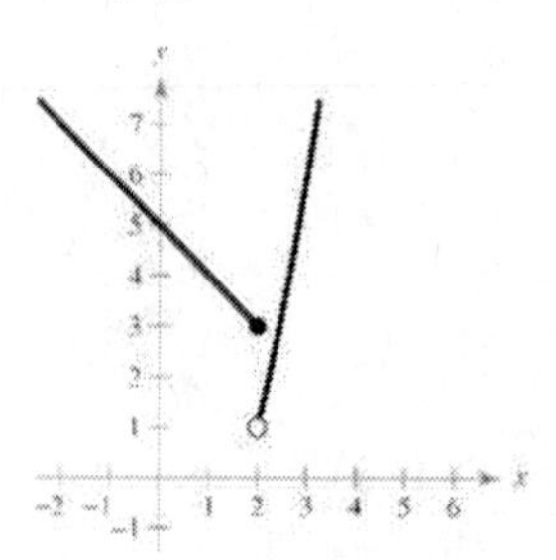

49.
$$\lim_{h \to 0} \frac{f(x+h) - f(x)}{h}$$
$$= \lim_{h \to 0} \frac{3(x+h) - (x+h)^2 - (3x - x^2)}{h}$$
$$= \lim_{h \to 0} \frac{3x + 3h - x^2 - 2xh - h^2 - 3x + x^2}{h}$$
$$= \lim_{h \to 0} \frac{3h - 2xh - h^2}{h}$$
$$= \lim_{h \to 0} (3 - 2x - h) = 3 - 2x$$

51. Slope ≈ 2

Answers will vary.

53. $f(x) = x^2 - 2x$

Tangent line at $(2, 0)$: $y = 2x - 4$

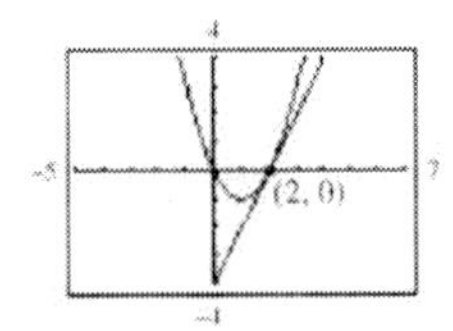

Slope at $(2, f(2)) = 2$

55. 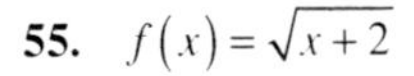 $f(x) = \sqrt{x+2}$

Tangent line at $(2, 2)$: $y = \frac{1}{4}x + \frac{3}{2}$

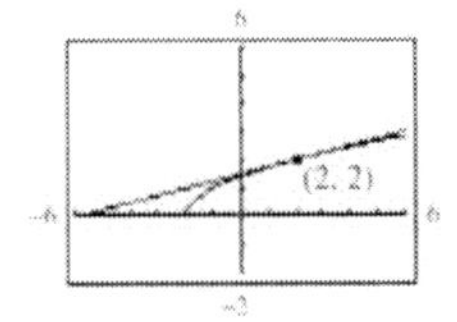

Slope at $(2, f(2)) = \frac{1}{4}$

57. $f(x) = \dfrac{6}{x-4}$

Tangent line at $(2, -3)$: $y = -\frac{3}{2}x$

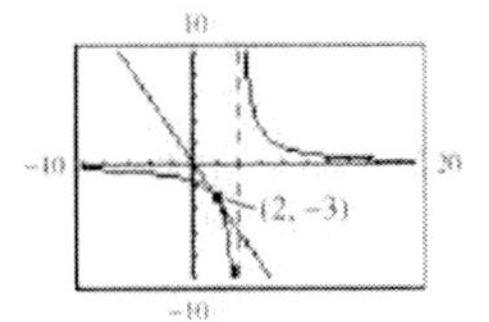

Slope at $(2, f(2)) = -\frac{3}{2}$

59.
$$\begin{aligned}
m &= \lim_{h\to 0}\frac{f(x+h)-f(x)}{h}\\
&= \lim_{h\to 0}\frac{(x+h)^2 - 4(x+h) - (x^2 - 4x)}{h}\\
&= \lim_{h\to 0}\frac{x^2 + 2xh + h^2 - 4x - 4h - x^2 - 4x}{h}\\
&= \lim_{h\to 0}\frac{2xh + h^2 - 4h}{h}\\
&= \lim_{h\to 0}(2x + h - 4) = 2x - 4
\end{aligned}$$

(a) At $(0, 0)$, $m = 2(0) - 4 = -4$.

(b) At $(5, 5)$, $m = 2(5) - 4 = 6$.

61.
$$\begin{aligned}
m &= \lim_{h\to 0}\frac{f(x+h)-f(x)}{h} = \lim_{h\to 0}\frac{\frac{4}{x+h-6} - \frac{4}{x-6}}{h}\\
&= \lim_{h\to 0}\frac{4(x-6) - 4(x+h-6)}{(x+h-6)(x-6)h}\\
&= \lim_{h\to 0}\frac{-4h}{(x+h-6)(x-6)h}\\
&= \lim_{h\to 0}\frac{-4}{(x+h-6)(x-6)} = -\frac{4}{(x-6)^2}
\end{aligned}$$

(a) At $(7, 4)$, $m = \dfrac{-4}{(7-6)^2} = -4$.

(b) At $(8, 2)$, $m = \dfrac{-4}{(8-6)^2} = -1$.

63. $f'(x) = \lim_{h\to 0}\dfrac{f(x+h)-f(x)}{h} = \lim_{h\to 0}\dfrac{5-5}{h} = 0$

65.
$$\begin{aligned}
h'(x) &= \lim_{k\to 0}\frac{h(x+k)-h(x)}{k}\\
&= \lim_{k\to 0}\frac{\left[5 - \frac{1}{2}(x+k)\right] - \left[5 - \frac{1}{2}x\right]}{k}\\
&= \lim_{k\to 0}\frac{-\frac{1}{2}k}{k} = -\frac{1}{2}
\end{aligned}$$

67.
$$\begin{aligned}
g'(x) &= \lim_{h\to 0}\frac{g(x+h)-g(x)}{h}\\
&= \lim_{h\to 0}\frac{2(x+h)^2 - 1 - (2x^2 - 1)}{h}\\
&= \lim_{h\to 0}\frac{2x^2 + 4xh + 2h^2 - 2x^2}{h}\\
&= \lim_{h\to 0}(4x + 2h)\\
&= 4x
\end{aligned}$$

69.
$$\begin{aligned}
f'(t) &= \lim_{h\to 0}\frac{f(t+h)-f(t)}{h}\\
&= \lim_{h\to 0}\frac{\sqrt{t+h+5} - \sqrt{t+5}}{h}\cdot\frac{\sqrt{t+h+5} + \sqrt{t+5}}{\sqrt{t+h+5} + \sqrt{t+5}}\\
&= \lim_{h\to 0}\frac{(t+h+5) - (t+5)}{h\left(\sqrt{t+h+5} + \sqrt{t+5}\right)}\\
&= \lim_{h\to 0}\frac{1}{\sqrt{t+h+5} + \sqrt{t+5}}\\
&= \frac{1}{2\sqrt{t+5}}
\end{aligned}$$

71. $g'(s) = \dfrac{g(s+h) - g(s)}{h}$

$= \lim_{h\to 0} \dfrac{\dfrac{4}{s+h+5} - \dfrac{4}{s+5}}{h}$

$= \lim_{h\to 0} \dfrac{4s + 20 - 4s - 4h - 20}{(s+h+5)(s+5)h}$

$= \lim_{h\to 0} \dfrac{-4h}{(s+h+5)(s+5)h}$

$= \lim_{h\to 0} \dfrac{-4}{(s+h+5)(s+5)}$

$= -\dfrac{4}{(s+5)^2}$

73. $g'(x) = \lim_{h\to 0} \dfrac{g(x+h) - g(x)}{h}$

$= \lim_{h\to 0} \dfrac{\dfrac{1}{\sqrt{x+h+4}} - \dfrac{1}{\sqrt{x+4}}}{h}$

$= \lim_{h\to 0} \dfrac{\sqrt{x+4} - \sqrt{x+h+4}}{h\sqrt{x+h+4}\sqrt{x+4}} \cdot \dfrac{\sqrt{x+4} + \sqrt{x+h+4}}{\sqrt{x+4} + \sqrt{x+h+4}}$

$= \lim_{h\to 0} \dfrac{(x+4) - (x+h+4)}{h\sqrt{x+h+4}\sqrt{x+4}\left[\sqrt{x+4} + \sqrt{x+h+4}\right]}$

$= \lim_{h\to 0} \dfrac{-1}{\sqrt{x+h+4}\sqrt{x+4}\left[\sqrt{x+4} + \sqrt{x+h+4}\right]}$

$= \dfrac{-1}{(x+4)2\sqrt{x+4}}$

$= -\dfrac{1}{2(x+4)^{\frac{3}{2}}}$

75. (a) $m_{sec} = \dfrac{f(0+h) - f(0)}{h}$

$= \dfrac{2h^2 - 1 - (-1)}{h}$

$= \dfrac{2h^2}{h}$

$= 2h,\ h \neq 0$

$m = \lim_{h\to 0} 2h = 0$

(b) $y + 1 = 0(x - 0)$

$y = -1$

(c)

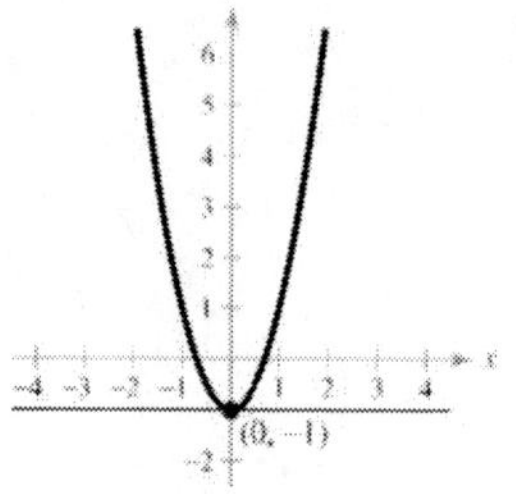

77. (a) $m_{sec} = \dfrac{f(-1+h) - f(-1)}{h}$

$= \dfrac{(-1+h)^3 + 1 - (-1+1)}{h}$

$= \dfrac{h^3 - 3h^2 + 3h - 1 + 1}{h}$

$= \dfrac{h^3 - 3h^2 + 3h}{h}$

$= h^2 - 3h + 3,\ h \neq 0$

$m = \lim_{h\to 0}(h^2 - 3h + 3) = 3$

(b) $y - 0 = 3(x + 1)$

$y = 3x + 3$

(c)

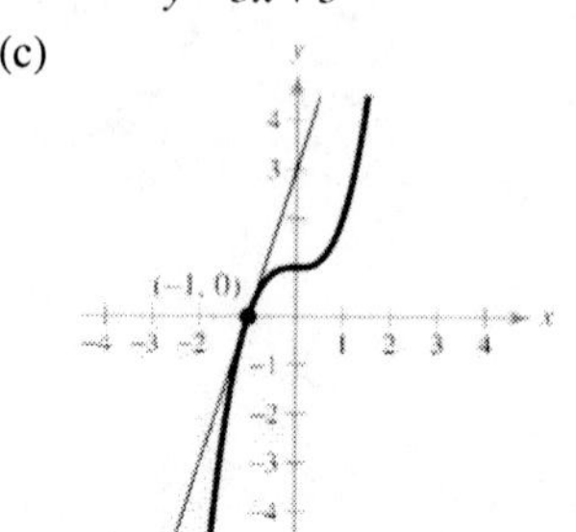

79. $\lim_{x\to\infty} \dfrac{4x}{2x - 3} = \dfrac{4}{2} = 2$

81. $\lim_{x\to-\infty} \dfrac{2x}{x^2 - 25} = 0$

83. $\lim_{x\to\infty} \dfrac{x^2}{2x+3}$ does not exist.

85. $\lim_{x\to\infty}\left[\dfrac{x}{(x-2)^2} + 3\right] = 0 + 3 = 3$

87. $a_n = \dfrac{2n - 3}{5n + 4}$

$a_1 = -\dfrac{1}{9}$

$a_2 = \dfrac{1}{14}$

$a_3 = \dfrac{3}{19}$

$a_4 = \dfrac{5}{24}$

$a_5 = \dfrac{7}{29}$

$\lim_{n\to\infty} a_n = \dfrac{2}{5}$

89. $a_n = \frac{(-1)^n}{n^3}$

$a_1 = -1$

$a_2 = \frac{1}{8}$

$a_3 = -\frac{1}{27}$

$a_4 = \frac{1}{64}$

$a_5 = -\frac{1}{125}$

$\lim_{n\to\infty} a_n = 0$

91. $a_n = \frac{1}{2n^2}\left[3 - 2n(n+1)\right] = \frac{3}{2n^2} - \frac{n+1}{n}$

$a_1 = -\frac{1}{2}$

$a_2 = -\frac{9}{8}$

$a_3 = -\frac{7}{6}$

$a_4 = -\frac{37}{32}$

$a_5 = -\frac{57}{50}$

$\lim_{n\to\infty} a_n = 0 - 1 = -1$

93. (a) $\sum_{i=1}^{n}\left(\frac{4i^2}{n^2} - \frac{i}{n}\right)\frac{1}{n} = \frac{4}{n^3}\sum_{i=1}^{n} i^2 - \frac{1}{n^2}\sum_{i=1}^{n} i$

$= \frac{4}{n^3}\frac{n(n+1)(2n+1)}{6} - \frac{1}{n^2}\frac{n(n+1)}{2}$

$= \frac{4n(n+1)(2n+1) - 3n^2(n+1)}{6n^3}$

$= \frac{n(n+1)(8n+4-3n)}{6n^3}$

$= \frac{(n+1)(5n+4)}{6n^2}$

(b)

n	10^0	10^1	10^2	10^3	10^4
$s(n)$	3	0.99	0.8484	0.8348	0.8335

(c) $\lim_{n\to\infty} S(n) = \frac{5}{6}$

95. $f(x) = 4 - x$, $[0, 3]$, $n = 6$, width $= \frac{1}{2}$

Intervals:

$\left[0, \frac{1}{2}\right], \left[\frac{1}{2}, 1\right], \left[1, \frac{3}{2}\right], \left[\frac{3}{2}, 2\right], \left[2, \frac{5}{2}\right], \left[\frac{5}{2}, 3\right]$

The height is obtained by evaluating f at the right-hand endpoint of each interval.

Area $\approx \sum_{i=1}^{6} f\left(\frac{1}{2}i\right)\left(\frac{1}{2}\right) = \sum_{i=1}^{6}\left(4 - \frac{1}{2}i\right)\left(\frac{1}{2}\right)$

$= \frac{1}{2}\left(\sum_{i=1}^{6} 4 - \frac{1}{2}\sum_{i=1}^{6} i\right)$

$= \frac{1}{2}\left(24 - \frac{1}{2}(21)\right)$

$= 6.75$ square units

97. $f(x) = \frac{1}{4}x^2$, $b - a = 4 - 0 = 4$

$A \approx \sum_{i=1}^{n} f\left(\frac{4i}{n}\right)\left(\frac{4}{n}\right)$

$= \sum_{i=1}^{n} \frac{1}{4}\left(\frac{4i}{n}\right)^2\left(\frac{4}{n}\right)$

$= \frac{1}{n}\sum_{i=1}^{n} \frac{16}{n^2} i^2$

$= \frac{16}{n^3}\frac{n(n+1)(2n+1)}{6}$

$= \frac{8(n+1)(2n+1)}{3n^2}$

n	4	8	20	50
Approximate area	7.5	6.375	5.74	5.4944

$\left(\text{Exact area is } \frac{16}{3} \approx 5.33.\right)$

99. $A = \lim_{n\to\infty}\sum_{i=1}^{n}\left(10 - \frac{10i}{n}\right)\left(\frac{10}{n}\right)$

$= \lim_{n\to\infty}\left[\frac{100}{n}\sum_{i=1}^{n} 1 - \frac{100}{n^2}\sum_{i=1}^{n} i\right]$

$= \lim_{n\to\infty}\left[\frac{100}{n}(n) - \frac{100}{n^2}\left(\frac{n(n+1)}{2}\right)\right]$

$= \lim_{n\to\infty}\left[100 - 50\frac{n(n+1)}{n^2}\right]$

$= 100 - 50 = 50$ square units, exact area

101. $A \approx \sum_{i=1}^{n} f\left(\frac{3}{n}i\right)\frac{3}{n}$

$$= \sum_{i=1}^{n}\left[\left(\frac{3}{n}i\right)^2 + 4\right]\frac{3}{n}$$

$$= \frac{3}{n}\sum_{i=1}^{n}\left(\frac{9}{n^2}i^2 + 4\right)$$

$$= \frac{3}{n}\left[\frac{9}{n^2}\frac{n(n+1)(2n+1)}{6} + 4n\right]$$

$$= \frac{3}{n}\left[\frac{9}{n^2}\cdot\frac{2n^3+3n^2+n}{6} + 4n\right]$$

$$= \frac{3}{n}\left[3n + \frac{9}{2} + \frac{3}{2n} + 4n\right]$$

$$= 21 + \frac{27}{2n} + \frac{9}{2n^2}$$

$$A = \lim_{n\to\infty}\left(21 + \frac{27}{2n} + \frac{9}{2n^2}\right) = 21 \text{ square units}$$

103. $A \approx \sum_{i=1}^{n} f\left(\frac{4}{n}i\right)\frac{4}{n}$

$$= \sum_{i=1}^{n}\left[\left(\frac{4}{n}i\right)^3 + 1\right]\frac{4}{n}$$

$$= \frac{4}{n}\sum_{i=1}^{n}\left(\frac{64}{n^3}i^3 + 1\right)$$

$$= \frac{4}{n}\left[\frac{64}{n^3}\cdot\frac{n^2(n+1)^2}{4} + n\right]$$

$$= \frac{4}{n}\left[\frac{64}{n^3}\frac{n^2(n^2+2n+1)}{4} + n\right]$$

$$= \frac{4}{n}\left[16n + 32 + \frac{16}{n} + n\right]$$

$$= \frac{4}{n}\left(17n + 32 + \frac{16}{n}\right)$$

$$= 68 + \frac{128}{n} + \frac{64}{n^2}$$

$$A = \lim_{n\to\infty}\left(68 + \frac{128}{n} + \frac{64}{n^2}\right) = 68 \text{ square units}$$

105. (a) $y = (-3.376\times10^{-7})x^3 + (3.753\times10^{-4})x^2 - 0.168x + 132$

(b)

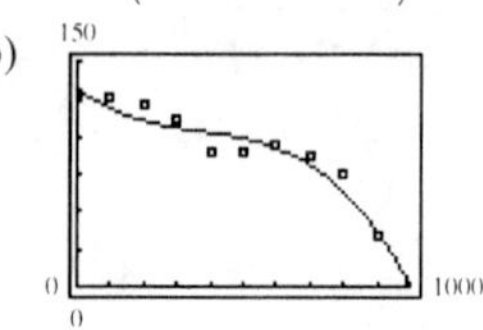

(c) Area $\approx$ 88,868 square feet

Answers will vary.

107. True. The derivative of f at a point is the slope of the tangent line to the graph at that point, therefore at $(z, f(z))$ the derivative is $f'(z)$.

Chapter 11 Test

1. $\lim_{x\to-2}\frac{x^2-1}{2x} = \frac{(-2)^2-1}{2(-2)} = -\frac{3}{4}$

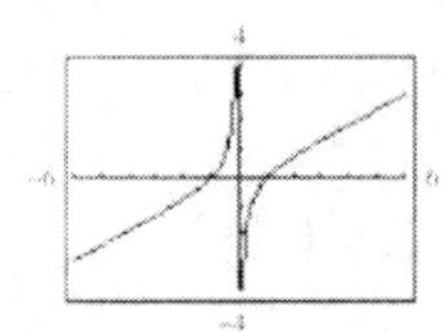

Limit is -0.75.

2.

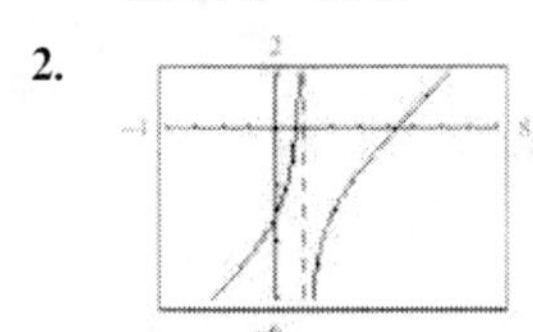

The limit does not exist.

$\lim_{x\to1}\frac{-x^2+5x-3}{1-x}$ does not exist.

$x = 1$ is a vertical asymptote.

3.

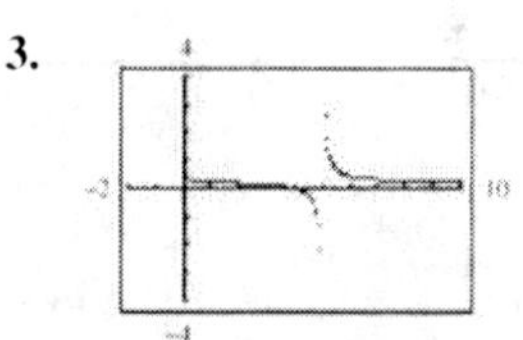

$\lim_{x\to5}\frac{\sqrt{x}-2}{x-5}$ does not exist.

4. $f(x) = \frac{\sin 3x}{x}$

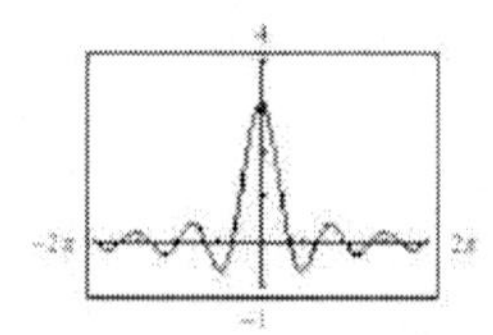

$\lim_{x\to0}\frac{\sin 3x}{x} = 3$

x	−0.1	−0.01	−0.001
$f(x)$	2.9552	2.9996	2.999996

x	0	0.001	0.01	0.1
$f(x)$	Error	2.999996	2.9996	2.9552

$$\lim_{x\to 0}\frac{\sin 3x}{x} = 3$$

5. $f(x) = \dfrac{e^{2x}-1}{x}$

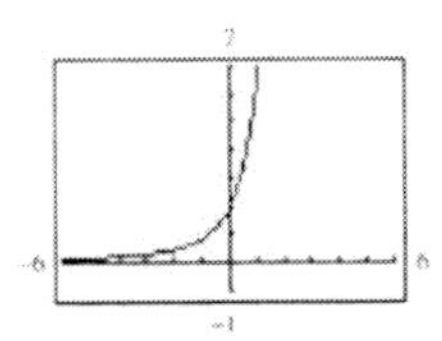

$$\lim_{x\to 0}\frac{e^{2x}-1}{x} = 2$$

x	−0.1	−0.01	−0.001	0	0.001	0.01	0.1
$f(x)$	1.8127	1.9801	1.9980	Error	2.0020	2.0201	2.2140

$$\lim_{x\to 0}\frac{e^{2x}-1}{x} = 2$$

6. (a)
$$\frac{f(x+h)-f(x)}{h} = \frac{3(x+h)^2-5(x+h)-2-(3x^2-5x-2)}{h}$$
$$= \frac{3x^2+6xh+3h^2-5h-3x^2}{h}$$
$$= 6x+3h-5$$
$$f'(x) = \lim_{h\to 0}[6x+3h-5] = 6x-5$$
$$f'(2) = 6(2)-5 = 7$$

(b)
$$\frac{f(x+h)-f(x)}{h} = \frac{\left[2(x+h)^3+6(x+h)\right]-\left[2x^3-6x\right]}{h}$$
$$= \frac{2x^3+6x^2h+6xh^2+2h^3+6x+6h-2x^3-6x}{h}$$
$$= \frac{6x^2h+6xh^2+2h^3+6h}{h}$$
$$= 6x^2+6xh+2h^2+6,\ h\neq 0$$
$$f'(x) = \lim_{h\to 0}\left[6x^2+6xh+2h^2+6\right] = 6x^2+6$$
$$f'(-1) = 6(-1)^2+6 = 12$$

7. $f(x) = 3-\dfrac{2}{5}x$
$$f'(x) = \lim_{h\to 0}\frac{f(x+h)-f(x)}{h}$$
$$= \lim_{h\to 0}\frac{\left[3-\frac{2}{5}(x+h)\right]-\left(3-\frac{2}{5}x\right)}{h}$$
$$= \lim_{h\to 0}\frac{3-\frac{2}{5}x-\frac{2}{5}h-3+\frac{2}{5}x}{h}$$
$$= \lim_{h\to 0}\frac{-\frac{2}{5}h}{h}$$
$$= \lim_{h\to 0}\left(-\frac{2}{5}\right) = -\frac{2}{5}$$

8. $f(x) = 2x^2+4x-1$
$$f'(x) = \lim_{h\to 0}\frac{f(x+h)-f(x)}{h}$$
$$= \lim_{h\to 0}\frac{2(x+h)^2+4(x+h)-1-\left[2x^2+4x-1\right]}{h}$$
$$= \lim_{h\to 0}\frac{2x^2+4xh+2h^2+4h-2x^2}{h}$$
$$= \lim_{h\to 0}(4x+2h+4) = 4x+4$$

9. $f(x) = \dfrac{1}{x+1}$
$$f'(x) = \lim_{h\to 0}\frac{f(x+h)-f(x)}{h}$$
$$= \lim_{h\to 0}\frac{\frac{1}{x+h+1}-\frac{1}{x+1}}{h}$$
$$= \lim_{h\to 0}\frac{\frac{(x+1)-(x+h+1)}{(x+h+1)(x+1)}}{h}$$
$$= \lim_{h\to 0}\frac{-h}{(x+h+1)(x+1)}\cdot\frac{1}{h}$$
$$= \lim_{h\to 0}-\frac{1}{(x+h+1)(x+1)} = -\frac{1}{(x+1)^2}$$

10. $\displaystyle\lim_{x\to\infty}\frac{6}{5x-1} = 0$

11. $\displaystyle\lim_{x\to\infty}\frac{1-3x^2}{x^2-5} = -3$

12. $\displaystyle\lim_{x\to-\infty}\frac{x^2}{3x+2}$ does not exist.

$f(x) = \dfrac{x^2}{3x+2}$ decreases without bound as $x\to-\infty$.

13. $0, \frac{3}{4}, \frac{14}{19}, \frac{12}{17}, \frac{36}{53}$

$\displaystyle\lim_{n\to\infty} a_n = \frac{1}{2}$

14. $0, 1, 0, \frac{1}{2}, 0$

$\lim_{n\to\infty} a_n = 0$

15. Width of each rectangle: $\frac{1}{2}$

Heights: $8, \frac{15}{2}, 6, \frac{7}{2}$

Area $\approx \frac{1}{2}\left[8 + \frac{15}{2} + 6 + \frac{7}{2}\right] = \frac{25}{2}$ square units

16. Width: $\dfrac{4}{n}$, Height: $f\left(-2 + \dfrac{4i}{n}\right) = \left(-2 + \dfrac{4i}{n}\right) + 2$

$$= \frac{4i}{n}$$

$$A \approx \sum_{i=1}^{n}\left(\frac{4i}{n}\right)\left(\frac{4}{n}\right) = \frac{16}{n^2}\sum_{i=1}^{n} i = \frac{16}{n^2}\frac{n(n+1)}{2}$$

$$A = \lim_{n\to\infty}\frac{16}{n^2}\cdot\frac{n(n+1)}{2} = 8 \text{ square units}$$

17.
$$A \approx \sum_{i=1}^{n} f\left(\frac{2}{n}i\right)\frac{2}{n}$$

$$= \sum_{i=1}^{n}\left[7 - \left(\frac{2}{n}i\right)^2\right]\frac{2}{n}$$

$$= \frac{2}{n}\sum_{i=1}^{n}\left[7 - \frac{4}{n^2}i^2\right]$$

$$= \frac{2}{n}\left[7n - \frac{4}{n^2}\frac{n(n+1)(2n+1)}{6}\right]$$

$$= \frac{2}{n}\left[7n - \frac{4}{n^2}\cdot\frac{2n^3 + 3n^2 + n}{6}\right]$$

$$= \frac{2}{n}\left[7n - \frac{4}{3}n - 2 - \frac{2}{3n}\right]$$

$$= \frac{2}{n}\left(\frac{17}{3}n - 2 - \frac{2}{3n}\right)$$

$$= \frac{34}{3} - \frac{4}{n} - \frac{4}{3n^2}$$

$$A = \lim_{n\to\infty}\left(\frac{34}{3} - \frac{4}{n} - \frac{4}{3n^2}\right) = \frac{34}{3} \text{ square units}$$

18. (a) $y = 8.79x^2 - 6.2x - 0.4$

(b) Velocity = Derivative = $17.58x - 6.2$

At $x = 5$, velocity ≈ 81.7 ft/sec.

Cumulative Test for Chapters 10 and 11

1. $(-6, 1, 2)$

2. $(0, -5, 0)$

3.
$$d = \sqrt{(4-(-2))^2 + (-5-3)^2 + (1-(-6))^2}$$
$$= \sqrt{36 + 64 + 49}$$
$$= \sqrt{149}$$

4. $d_1 = 3, d_2 = 4, d_3 = \sqrt{4^2 + 3^2} = 5$

$d_1^{\,2} + d_2^{\,2} = d_3^{\,2}$

5. Midpoint: $\left(\dfrac{3-5}{2}, \dfrac{4+0}{2}, \dfrac{-1+2}{2}\right) = \left(-1, 2, \dfrac{1}{2}\right)$

6. Center $= \left(\dfrac{0+4}{2}, \dfrac{0+4}{2}, \dfrac{0+8}{2}\right) = (2, 2, 4)$

Radius $= \sqrt{2^2 + 2^2 + 4^2} = \sqrt{24}$

$(x-2)^2 + (y-2)^2 + (z-4)^2 = 24$

7. xy-trace: $(z = 0)$

$(x-2)^2 + (y+1)^2 = 4$, Circle

yz-trace: $(x = 0)$

$4 + (y+1)^2 + z^2 = 4$ or $(y+1)^2 + z^2 = 0$, Point

$(0, -1, 0)$, Point

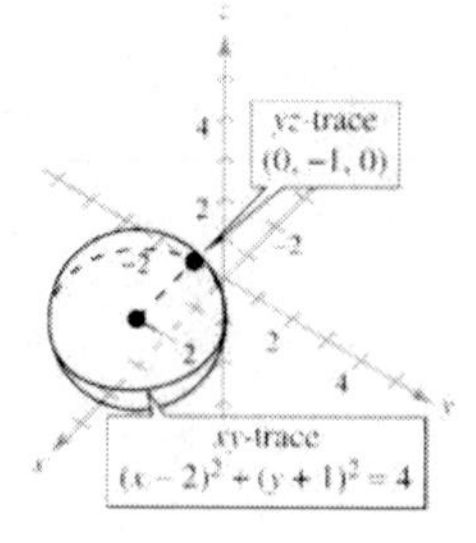

8. $\mathbf{u}\cdot\mathbf{v} = \langle 2, -6, 0\rangle\cdot\langle -4, 5, 3\rangle$

$= -8 - 30 = -38$

$$\mathbf{u}\times\mathbf{v} = \begin{vmatrix} \mathbf{i} & \mathbf{j} & \mathbf{k} \\ 2 & -6 & 0 \\ -4 & 5 & 3 \end{vmatrix} = \langle -18, -6, -14\rangle$$

9. $\mathbf{u}\cdot\mathbf{v} \neq 0$, $\mathbf{u} \neq c\mathbf{v} \Rightarrow$ neither

10. $\mathbf{u}\cdot\mathbf{v} = -8 - 12 + 20 = 0 \Rightarrow$ orthogonal

11. $3\mathbf{u} = \langle -3, 18, -9\rangle = -\mathbf{v} \Rightarrow$ parallel

12. $\overline{DA} = \langle 0, 2, 0\rangle$, $\overline{DC} = \langle 2, 1, 0\rangle$, $\overline{DH} = \langle 0, 0, 3\rangle$

$$\begin{vmatrix} 0 & 2 & 0 \\ 2 & 1 & 0 \\ 0 & 0 & 3 \end{vmatrix} = 12 \text{ cubic units}$$

13. (a) Vector is $\langle 5+2, 8-3, 25-0\rangle = \langle 7, 5, 25\rangle$.

$x = -2 + 7t,\ y = 3 + 5t,\ z = 25t$

(b) $\dfrac{x+2}{7} = \dfrac{y-3}{5} = \dfrac{z}{25}$

14. $\mathbf{v} = \langle 2, -4, 1\rangle$ and $P = (-1, 2, 0)$

$x = -1 + 2t$

$y = 2 - 4t$

$z = t$

15. $\mathbf{u} = \langle -2, 3, 0\rangle$, $\mathbf{v} = \langle 5, 8, 25\rangle$

$$\mathbf{u}\times\mathbf{v} = \begin{vmatrix} \mathbf{i} & \mathbf{j} & \mathbf{k} \\ -2 & 3 & 0 \\ 5 & 8 & 25 \end{vmatrix} = \langle 75, 50, -31\rangle$$

Normal to plane
Plane: $75x + 50y - 31z = 0$

16.

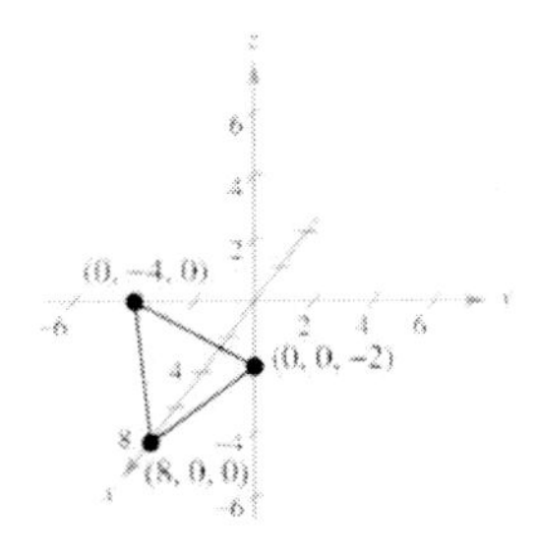

17. $\mathbf{n} = \langle 2, -5, 1\rangle$, $Q = (0, 0, 25)$, $P = (0, 0, 10)$ in plane, $\overline{PQ} = \langle 0, 0, 15\rangle$

$$D = \frac{|\overline{PQ}\cdot\mathbf{n}|}{\|\mathbf{n}\|} = \frac{15}{\sqrt{30}} = \frac{\sqrt{30}}{2} \approx 2.74$$

18. Normal to plane containing:
$(-1, -1, 3)$, $(0, 0, 0)$ and $(2, 0, 0)$ is

$$\langle -1, -1, 3\rangle \times \langle 2, 0, 0\rangle = \begin{vmatrix} \mathbf{i} & \mathbf{j} & \mathbf{k} \\ -1 & -1 & 3 \\ 2 & 0 & 0 \end{vmatrix} = \langle 0, 6, 2\rangle \text{ or } \mathbf{n}_1 = \langle 0, 3, 1\rangle$$

Normal to front face is:

$$\langle 1, -1, 3\rangle \times \langle 0, 2, 0\rangle = \begin{vmatrix} \mathbf{i} & \mathbf{j} & \mathbf{k} \\ 1 & -1 & 3 \\ 0 & 2 & 0 \end{vmatrix} = \langle -6, 0, 2\rangle \text{ or } \mathbf{n}_2 = \langle -3, 0, 1\rangle$$

Angle between sides:

$$\cos\theta = \frac{|\mathbf{n}_1\cdot\mathbf{n}_2|}{\|\mathbf{n}_1\|\|\mathbf{n}_2\|} = \frac{1}{\sqrt{10}\sqrt{10}} = \frac{1}{10} \Rightarrow \theta \approx 84.26^\circ$$

19. $\lim\limits_{x\to 4}(5x - x^2) = 5(4) - 4^2 = 4$

20. $\lim\limits_{x\to -2^+}\dfrac{x+2}{(x+2)(x-1)} = \lim\limits_{x\to -2^+}\dfrac{1}{x-1} = -\dfrac{1}{3}$

21. $\lim\limits_{x\to 7}\dfrac{x-7}{(x-7)(x+7)} = \lim\limits_{x\to 7}\dfrac{1}{x+7} = \dfrac{1}{14}$

22. $$\lim_{x\to 0}\frac{\sqrt{x+4}-2}{x}\cdot\frac{\sqrt{x+4}+2}{\sqrt{x+4}+2} = \lim_{x\to 0}\frac{(x+4)-4}{x\left(\sqrt{x+4}+2\right)}$$

$$= \lim_{x\to 0}\frac{1}{\sqrt{x+4}+2} = \frac{1}{2+2} = \frac{1}{4}$$

23. $\lim\limits_{x\to 4^-}\dfrac{|x-4|}{x-4} = -1$

24. $\lim\limits_{x\to 0}\sin\left(\dfrac{\pi}{x}\right)$ does not exist.

25. $$\lim_{x\to 0}\frac{\frac{1}{x+3}-\frac{1}{3}}{x} = \lim_{x\to 0}\frac{\frac{3-(x+3)}{3(x+3)}}{x}$$

$$= \lim_{x\to 0}\frac{-x}{3(x+3)}\cdot\frac{1}{x}$$

$$= \lim_{x\to 0} -\frac{1}{3(x+3)} = -\frac{1}{9}$$

26. $$\lim_{x\to 0}\frac{\sqrt{x+16}-4}{x}\cdot\frac{\sqrt{x+16}+4}{\sqrt{x+16}+4} = \lim_{x\to 0}\frac{x+16-16}{x\left(\sqrt{x+16}+4\right)}$$

$$= \lim_{x\to 0}\frac{x}{x\left(\sqrt{x+16}+4\right)}$$

$$= \lim_{x\to 0}\frac{1}{\sqrt{x+16}+4} = \frac{1}{8}$$

27. $$\lim_{x\to 2^-}\frac{x-2}{x^2-4} = \lim_{x\to 2^-}\frac{x-2}{(x-2)(x+2)}$$

$$= \lim_{x\to 2^-}\frac{1}{x+2} = \frac{1}{4}$$

28. $$f'(x) = \lim_{h\to 0}\frac{f(x+h)-f(x)}{h}$$

$$= \lim_{h\to 0}\frac{4-(x+h)^2-(4-x^2)}{h}$$

$$= \lim_{h\to 0}\frac{4-x^2-2xh-h^2-4+x^2}{h}$$

$$= \lim_{h\to 0}\frac{-2xh-h^2}{h}$$

$$= \lim_{h\to 0}(-2x-h) = -2x, \text{ Slope}$$

At $(0, 4)$, $m = 0$.

29. $f(x)=\sqrt{x+3}$

$$m=\lim_{h\to 0}\frac{f(x+h)-f(x)}{h}$$
$$=\lim_{h\to 0}\frac{\sqrt{x+h+3}-\sqrt{x+3}}{h}\cdot\frac{\sqrt{x+h+3}+\sqrt{x+3}}{\sqrt{x+h+3}+\sqrt{x+3}}$$
$$=\lim_{h\to 0}\frac{(x+h+3)-(x+3)}{h\left[\sqrt{x+h+3}+\sqrt{x+3}\right]}$$
$$=\lim_{h\to 0}\frac{1}{\sqrt{x+h+3}+\sqrt{x+3}}=\frac{1}{2\sqrt{x+3}}$$

At $(-2, 1)$, $m=\frac{1}{2}$.

30. $f(x)=\frac{1}{x+3}$

$$m=\lim_{h\to 0}\frac{f(x+h)-f(x)}{h}$$
$$=\lim_{h\to 0}\frac{\frac{1}{x+h+3}-\frac{1}{x+3}}{h}$$
$$=\lim_{h\to 0}\frac{(x+3)-(x+h+3)}{h(x+h+3)(x+3)}$$
$$=\lim_{h\to 0}\frac{-1}{(x+h+3)(x+3)}$$
$$=-\frac{1}{(x+3)^2}$$

At $\left(1, \frac{1}{4}\right)$, $m=-\frac{1}{16}$.

31. $f'(x)=\lim_{h\to 0}\frac{f(x+h)-f(x)}{h}$

$$=\lim_{h\to 0}\frac{(x+h)^2-(x+h)-(x^2-x)}{h}$$
$$=\lim_{h\to 0}\frac{x^2+2xh+h^2-x-h-x^2+x}{h}$$
$$=\lim_{h\to 0}\frac{2xh+h^2-h}{h}$$
$$=\lim_{h\to 0}(2x+h-1)=2x-1, \text{ Slope}$$

At $(1, 0)$, $m=1$.

32. $\lim_{x\to\infty}\frac{x^3}{x^2-9}$ does not exist.

Function increases without bound.

33. $\lim_{x\to\infty}\frac{3-7x}{x+4}=-7$

34. $\lim_{x\to\infty}\frac{3x^2+1}{x^2+4}=3$

35. $\lim_{x\to\infty}\frac{2x}{x^2+3x-2}=0$

36. $\lim_{x\to\infty}\frac{3-x}{x^2+1}=0$

37. $\lim_{x\to\infty}\frac{3+4x-x^3}{2x^2+3}$

does not exist. Function decrease without bound.

38. $\sum_{i=1}^{50}(1-i^2)=50-\frac{50(51)(101)}{6}=-42,875$

39. $\sum_{k=1}^{20}(3k^2-2k)=3\frac{20(21)(41)}{6}-2\frac{20(21)}{2}$

$=8610-420=8190$

40. $\sum_{i=1}^{40}(12+i^3)=12(40)+\frac{40^2(41)^2}{4}$

$=480+672,400=672,880$

41. $f(x)=2x$, $[0, 3]$, $n=6$, width $=\frac{1}{2}$

Intervals: $\left[0, \frac{1}{2}\right], \left[\frac{1}{2}, 1\right], \left[1, \frac{3}{2}\right], \left[\frac{3}{2}, 2\right], \left[2, \frac{5}{2}\right], \left[\frac{5}{2}, 3\right]$

The height is obtained by evaluating f at the right-hand endpoint of each interval.

$$\text{Area}\approx\sum_{i=1}^{6}f\left(\frac{1}{2}i\right)\left(\frac{1}{2}\right)=\sum_{i=1}^{6}\left(2\cdot\frac{1}{2}i\right)\left(\frac{1}{2}\right)$$
$$=\frac{1}{2}\sum_{i=1}^{6}i$$
$$=\frac{1}{2}(21)$$
$$=\frac{21}{2}\text{ square units}$$

42. $f(x)=5-\frac{1}{2}x^2$, $[0, 2]$, $n=4$, width $=\frac{1}{2}$

Intervals: $\left[0, \frac{1}{2}\right], \left[\frac{1}{2}, 1\right], \left[1, \frac{3}{2}\right], \left[\frac{3}{2}, 2\right]$

The height is obtained by evaluating f at the right-hand endpoint of each interval.

$$\text{Area}\approx\sum_{i=1}^{4}f\left(\frac{1}{2}i\right)\left(\frac{1}{2}\right)=\sum_{i=1}^{4}\left[5-\frac{1}{2}\left(\frac{1}{2}i\right)^2\right]\left(\frac{1}{2}\right)$$
$$=\frac{1}{2}\sum_{i=1}^{4}\left(5-\frac{1}{8}i^2\right)$$
$$=\frac{1}{2}\left(\sum_{i=1}^{4}5-\frac{1}{8}\sum_{i=1}^{4}i^2\right)$$
$$=\frac{1}{2}\left[20-\frac{1}{8}(30)\right]$$
$$=8.125\text{ square units}$$

43. $f(x) = \frac{1}{4}(x+1)^2$, $[0, 2]$, $n = 4$, width $= \frac{1}{2}$

Intervals: $\left[0, \frac{1}{2}\right], \left[\frac{1}{2}, 1\right], \left[1, \frac{3}{2}\right], \left[\frac{3}{2}, 2\right]$

The height is obtained by evaluating f at the right-hand endpoint of each interval.

$$\text{Area} \approx \sum_{i=1}^{4} f\left(\frac{1}{2}i\right)\left(\frac{1}{2}\right) = \sum_{i=1}^{4}\left[\frac{1}{4}\left(\frac{1}{2}i+1\right)^2\right]\left(\frac{1}{2}\right)$$

$$= \frac{1}{8}\sum_{i=1}^{4}\left(\frac{1}{2}i+1\right)^2$$

$$= \frac{1}{8}\left(\frac{43}{2}\right)$$

$$= 2.6875 \text{ square units}$$

44. $f(x) = \frac{1}{x^2+1}$, $[-1, 1]$, $n = 8$, width $= \frac{1}{2}$

$$\text{Area} \approx \frac{1}{4}\left[\frac{1}{1+\left(-\frac{3}{4}\right)^2} + \frac{1}{1+\left(-\frac{1}{2}\right)^2} + \frac{1}{1+\left(-\frac{1}{4}\right)^2} + \frac{1}{1+0} + \frac{1}{1+\left(\frac{1}{4}\right)^2} + \frac{1}{1+\left(\frac{1}{2}\right)^2} + \frac{1}{1+\left(\frac{3}{4}\right)^2} + \frac{1}{1+1^2}\right]$$

$$\approx \frac{1}{4}\left[2(0.64) + 2(0.8) + 2(0.941176) + 1 + \frac{1}{2}\right]$$

$$\approx 1.566 \text{ square units}$$

45. $f(x) = x + 2$, $[0, 1]$

The width of each rectangle is $1/n$. The height is

$$f\left(\frac{i}{n}\right) = \frac{i}{n} + 2.$$

$$A \approx \sum_{i=1}^{n}\left[\frac{i}{n}+2\right]\frac{1}{n} = \sum_{i=1}^{n}\left(\frac{1}{n^2}i + \frac{2}{n}\right)$$

$$A = \lim_{n\to\infty}\left[\frac{1}{n^2}\frac{n(n+1)}{2} + \frac{2}{n}(n)\right]$$

$$= \frac{1}{2} + 2 = \frac{5}{2} \text{ square units}$$

46. $f(x) = x^2 + 1$, $[0, 4]$

The width of each rectangle is $4/n$. The height is

$$f\left(\frac{4i}{n}\right) = \left(\frac{4i}{n}\right)^2 + 1.$$

$$A \approx \sum_{i=1}^{n}\left(\frac{16i^2}{n^2}+1\right)\left(\frac{4}{n}\right) = \sum_{i=1}^{n}\left(\frac{64}{n^3}i^2 + \frac{4}{n}\right)$$

$$A = \lim_{n\to\infty}\left[\frac{64}{n^3}\frac{n(n+1)(2n+1)}{6} + \frac{4}{n}(n)\right]$$

$$= \frac{64}{3} + 4 = \frac{76}{3} \text{ square units}$$

47. $f(x) = 4 - x^2$, $[0, 2]$

The width of each rectangle is $2/n$. The height is

$$f\left(\frac{2i}{n}\right) = 4 - \left(\frac{2i}{n}\right)^2.$$

$$A \approx \sum_{i=1}^{n}\left[4 - \frac{4i^2}{n^2}\right]\left(\frac{2}{n}\right) = \sum_{i=1}^{n}\left(\frac{8}{n} - \frac{8i^2}{n^3}\right)$$

$$A = \lim_{n\to\infty}\left[\frac{8}{n}(n) - \frac{8}{n^3}\frac{n(n+1)(2n+1)}{6}\right]$$

$$= 8 - \frac{8}{3} = \frac{16}{3} \text{ square units}$$

48. Width: $\frac{1}{n}$, Height: $f\left(\frac{i}{n}\right) = 1 - \left(\frac{i}{n}\right)^3$

$$A = \sum_{i=1}^{n}\left(1 - \left(\frac{i}{n}\right)^3\right)\left(\frac{1}{n}\right) = \frac{1}{n}\sum_{i=1}^{n} 1 - \frac{1}{n^4}\sum_{i=1}^{n} i^3$$

$$= \frac{1}{n}(n) - \frac{1}{n^4}\left[\frac{n^2(n+1)^2}{4}\right]$$

$$A = \lim_{n\to\infty}\left[1 - \frac{1}{n^4}\left(\frac{n^2(n+1)^2}{4}\right)\right] = 1 - \frac{1}{4} = \frac{3}{4} \text{ square units}$$

APPENDIX C
Review of Graphs, Equations, and Inequalities

APPENDIX C
Review of Graphs, Equations, and Inequalities

Appendix C.1 The Cartesian Plane

1. Cartesian

3. Midpoint Formula

5. iii

6. vi

7. i

8. iv

9. v

10. ii

11. A: $(2, 6)$, B: $(-6, -2)$, C: $(4, -4)$, D: $(-3, 2)$

13.

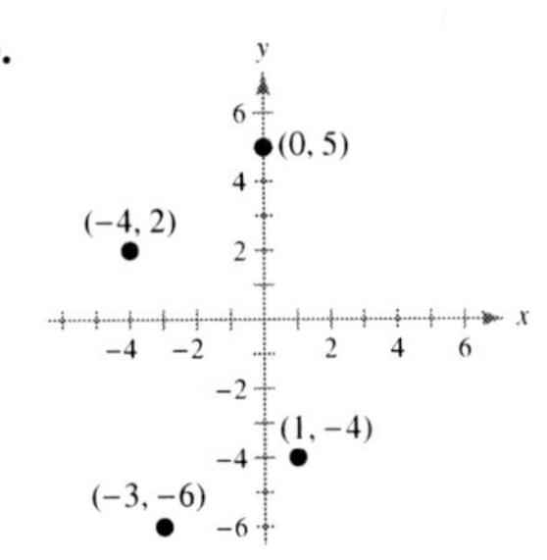

15.

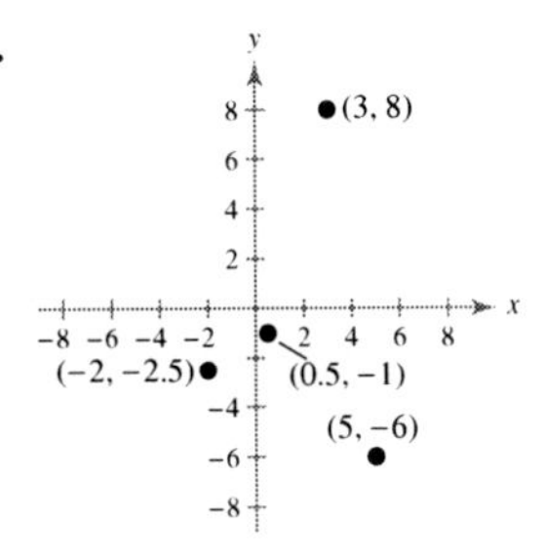

17. $(-5, 4)$

19. $(0, -6)$

21. $x > 0 \Rightarrow$ The point lies in Quadrant I or in Quadrant IV.

$y < 0 \Rightarrow$ The point lies in Quadrant III or in Quadrant IV.

$x > 0$ and $y < 0 \Rightarrow (x, y)$ lies in Quadrant IV.

23. $x = -4 \Rightarrow x$ is negative $\Rightarrow$ The point lies in Quadrant II or in Quadrant III.

$y > 0 \Rightarrow$ The point lies in Quadrant I or in Quadrant II.

$x = -4$ and $y > 0 \Rightarrow (x, y)$ lies in Quadrant II.

25. $y < -5 \Rightarrow y$ is negative $\Rightarrow$ The point lies in Quadrant III or in Quadrant IV.

27. If $-y > 0$ then $y < 0$.

$x < 0 \Rightarrow$ The point lies in Quadrant II or in Quadrant III.

$y < 0 \Rightarrow$ The point lies in Quadrant III or in Quadrant IV.

$x < 0$ and $y < 0 \Rightarrow (x, y)$ lies in Quadrant III.

29. If $xy > 0$, then either x and y are both positive, or both negative. Hence, (x, y) lies in either Quadrant I or Quadrant III.

31. $(6, -3)$, $(6, 5)$

$$d = \sqrt{(6-6)^2 + (5-(-3))^2} = \sqrt{64} = 8$$

33. $(-3, -1)$, $(2, -1)$

$$d = \sqrt{\left(2-(-3)^2\right) + \left(-1-(-1)^2\right)} = \sqrt{25} = 5$$

35. $$d = \sqrt{\left(3-(-2)^2\right) + (-6-6)^2} = \sqrt{5^2 + (-12)^2}$$
$$= \sqrt{25 + 144} = \sqrt{169} = 13$$

37. $\left(\frac{1}{2}, \frac{4}{3}\right)$, $(2, -1)$

$$d = \sqrt{\left(\frac{1}{2} - 2\right)^2 + \left(\frac{4}{3} + 1\right)^2}$$
$$= \sqrt{\frac{9}{4} + \frac{49}{9}}$$
$$= \sqrt{\frac{277}{36}} = \frac{\sqrt{277}}{6} \approx 2.77$$

39. $(-4.2, 3.1), (-12.5, 4.8)$

$$\begin{aligned} d &= \sqrt{(-4.2 + 12.5)^2 + (3.1 - 4.8)^2} \\ &= \sqrt{68.89 + 2.89} \\ &= \sqrt{71.78} \approx 8.47 \end{aligned}$$

41. (a) The distance between $(1, 1)$ and $(4, 1)$ is 3.

The distance between $(4, 1)$ and $(4, 5)$ is 4.

The distance between $(1, 1)$ and $(4, 5)$ is

$$\begin{aligned} \sqrt{(4 - 1)^2 + (5 - 1)^2} &= \sqrt{9 + 16} \\ &= \sqrt{25} = 5. \end{aligned}$$

(b) $3^2 + 4^2 = 9 + 16 = 25 = 5^2$

43. (a) The distance between $(-1, 1)$ and $(9, 1)$ is 10.

The distance between $(9, 1)$ and $(9, 4)$ is 3.

The distance between $(-1, 1)$ and $(9, 4)$ is

$$\begin{aligned} \sqrt{(9 - (-1))^2 + (4 - 1)^2} &= \sqrt{100 + 9} \\ &= \sqrt{109}. \end{aligned}$$

(b) $10^2 + 3^2 = 109 = \left(\sqrt{109}\right)^2$

45. Find the distances between pairs of points.

$$\begin{aligned} d_1 &= \sqrt{(4 - 2)^2 + (0 - 1)^2} = \sqrt{5} \\ d_2 &= \sqrt{(4 + 1)^2 + (0 + 5)^2} = \sqrt{50} \\ d_3 &= \sqrt{(2 + 1)^2 + (1 + 5)^2} = \sqrt{45} \end{aligned}$$

$$\left(\sqrt{5}\right)^2 + \left(\sqrt{45}\right)^2 = \left(\sqrt{50}\right)^2$$

Because $d_1^2 + d_3^2 = d_2^2$, the triangle is a right triangle.

47. $d_1 = \sqrt{(1 - 3)^2 + (-3 - 2)^2} = \sqrt{4 + 25} = \sqrt{29}$

$d_2 = \sqrt{(3 + 2)^2 + (2 - 4)^2} = \sqrt{25 + 4} = \sqrt{29}$

$d_3 = \sqrt{(1 + 2)^2 + (-3 - 4)^2} = \sqrt{9 + 49} = \sqrt{58}$

$d_1 = d_2$. Triangle is isosceles.

49. Find the distances between pairs of points.

$$\begin{aligned} d_1 &= \sqrt{(0 - 2)^2 + (9 - 5)^2} = \sqrt{4 + 16} = \sqrt{20} = 2\sqrt{5} \\ d_2 &= \sqrt{(-2 - 0)^2 + (0 - 9)^2} = \sqrt{4 + 81} = \sqrt{85} \\ d_3 &= \sqrt{(0 - (-2))^2 + (-4 - 0)^2} = \sqrt{4 + 16} = \sqrt{20} = 2\sqrt{5} \\ d_4 &= \sqrt{(0 - 2)^2 + (-4 - 5)^2} = \sqrt{4 + 81} = \sqrt{85} \end{aligned}$$

Opposite sides have equal lengths of $2\sqrt{5}$ and $\sqrt{85}$, so the figure is a parallelogram.

51. First show that the diagonals are equal in length.

$$\begin{aligned} d_1 &= \sqrt{(0 - (-3))^2 + (8 - 1)^2} = \sqrt{9 + 49} = \sqrt{58} \\ d_2 &= \sqrt{(2 - (-5))^2 + (3 - 6)^2} = \sqrt{49 + 9} = \sqrt{58} \end{aligned}$$

Now use the Pythagorean Theorem to verify that at least one angle is 90° (and, hence, they are all right angles).

$$\begin{aligned} d_3 &= \sqrt{(0 - (-5))^2 + (8 - 6)^2} = \sqrt{25 + 4} = \sqrt{29} \\ d_4 &= \sqrt{(-3 - (-5))^2 + (1 - 6)^2} = \sqrt{4 + 25} = \sqrt{29} \end{aligned}$$

Thus, $d_3^2 + d_4^2 = d_1^2$.

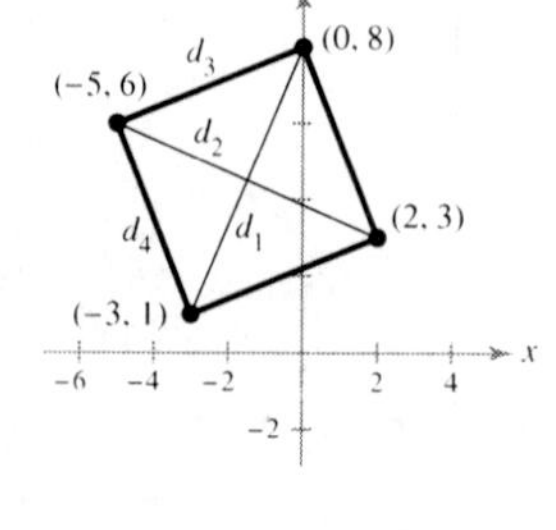

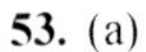

53. (a)

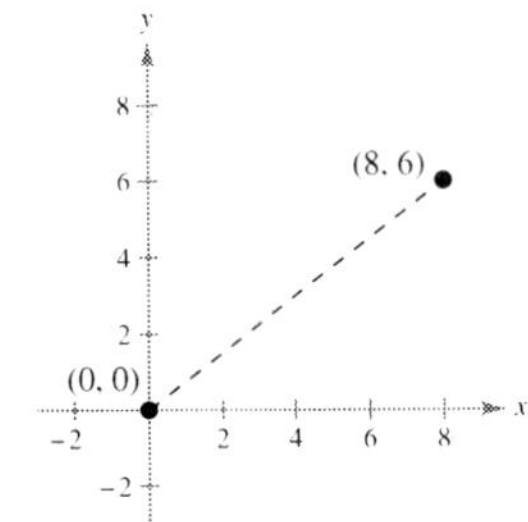

(b) $d = \sqrt{(8 - 0)^2 + (6 - 0)^2}$

$= \sqrt{64 + 36} = 10$

(c) $\left(\dfrac{8 + 0}{2}, \dfrac{6 + 0}{2}\right) = (4, 3)$

55. (a)

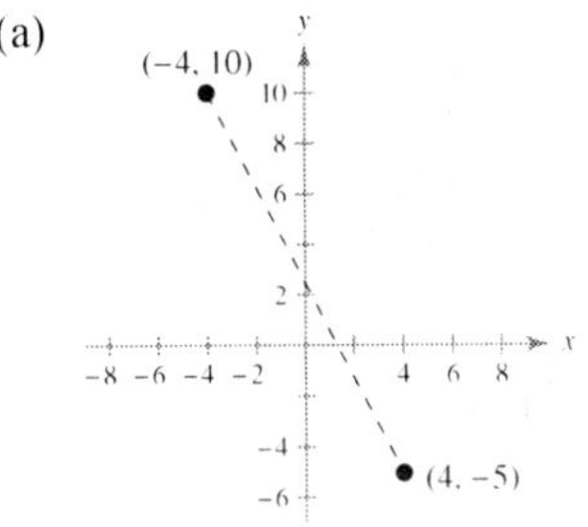

(b) $d = \sqrt{(4 + 4)^2 + (-5 - 10)^2}$

$= \sqrt{64 + 225} = 17$

(c) $\left(\dfrac{4 - 4}{2}, \dfrac{-5 + 10}{2}\right) = \left(0, \dfrac{5}{2}\right)$

57. (a)

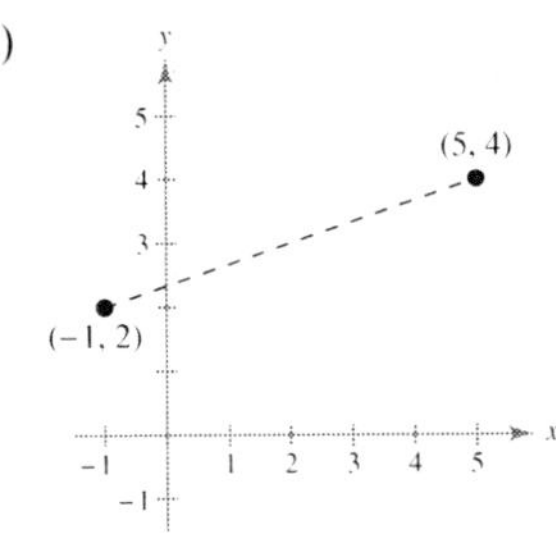

(b) $d = \sqrt{(5 + 1)^2 + (4 - 2)^2}$

$= \sqrt{36 + 4} = \sqrt{40} = 2\sqrt{10}$

(c) $\left(\dfrac{-1 + 5}{2}, \dfrac{2 + 4}{2}\right) = (2, 3)$

59. (a)

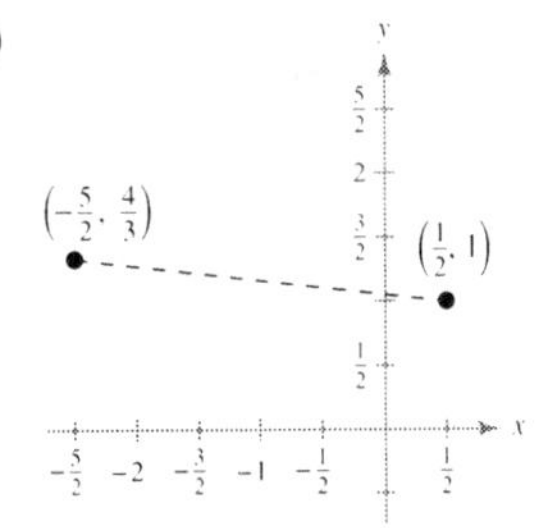

(b) $d = \sqrt{\left(\dfrac{1}{2} + \dfrac{5}{2}\right)^2 + \left(1 - \dfrac{4}{3}\right)^2}$

$d = \sqrt{9 + \dfrac{1}{9}} = \dfrac{\sqrt{82}}{3}$

(c) $\left(\dfrac{-\frac{5}{2} + \frac{1}{2}}{2}, \dfrac{\frac{4}{3} + 1}{2}\right) = \left(-1, \dfrac{7}{6}\right)$

61. (a)

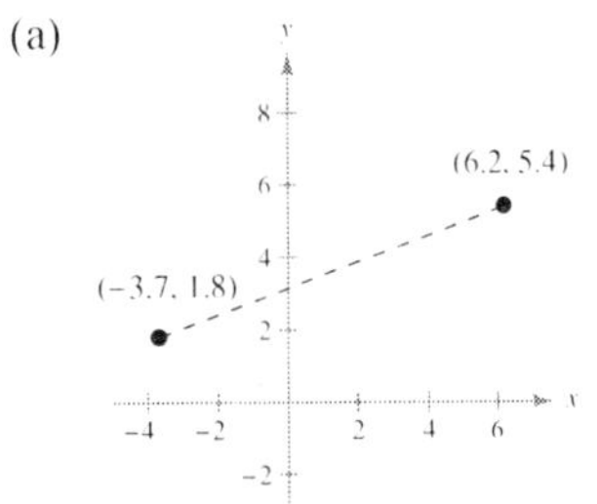

(b) $d = \sqrt{(6.2 + 3.7)^2 + (5.4 - 1.8)^2}$

$= \sqrt{98.01 + 12.96} = \sqrt{110.97}$

(c) $\left(\dfrac{6.2 - 3.7}{2}, \dfrac{5.4 + 1.8}{2}\right) = (1.25, 3.6)$

63. Calculate the midpoint:

$\left(\dfrac{2003 + 2009}{2}, \dfrac{287 + 942}{2}\right) = (2006, 614.5)$

The sales in 2006 are \$614.5 million.

65. $(x - 0)^2 + (y - 0)^2 = 5^2$

$x^2 + y^2 = 25$

67. $(x - 2)^2 + (y + 1)^2 = 4^2$

$(x - 2)^2 + (y + 1)^2 = 16$

69. $(x + 1)^2 + (y - 2)^2 = r^2$

$(0 + 1)^2 + (0 - 2)^2 = r^2 \Rightarrow r^2 = 5$

$(x + 1)^2 + (y - 2)^2 = 5$

71. $r = \frac{1}{2}\sqrt{(6-0)^2 + (8-0)^2} = \frac{1}{2}\sqrt{100} = 5$

Center: $\left(\frac{0+6}{2}, \frac{0+8}{2}\right) = (3, 4)$

$(x-3)^2 + (y-4)^2 = 25$

73. Because the circle is tangent to the x-axis, the radius is 1.

$(x+2)^2 + (y-1)^2 = 1$

75. The center is the midpoint of one of the diagonals of the square.

Center: $\left(\frac{7+(-1)}{2}, \frac{-2+10}{2}\right) = (3, -6)$

The radius is one half the length of a side of the square.

Radius: $\frac{1}{2}(7-(-1)) = 4$

Circle: $(x-3)^2 + (y+6)^2 = 16$

77. $(x-2)^2 + (y+1)^2 = 16$

79. 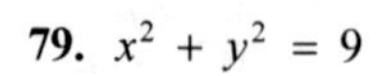

$x^2 + y^2 = 9$

Center: $(0, 0)$

Radius: 3

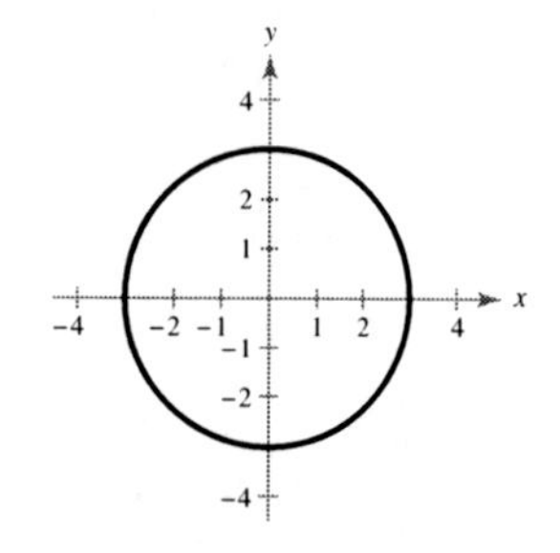

81. Center: $(1, -3)$

Radius: 2

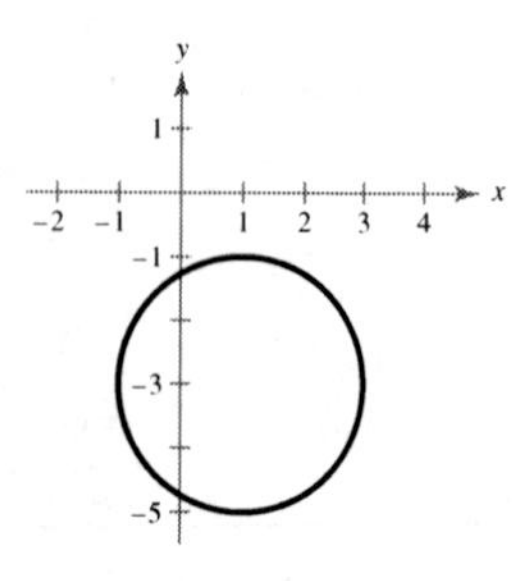

83. Center: $\left(\frac{1}{2}, \frac{1}{2}\right)$

Radius: $\frac{3}{2}$

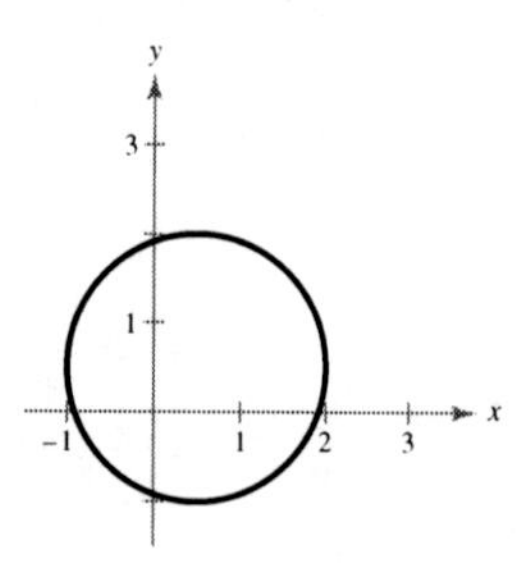

85. The x-coordinates are increased by 2, and the y-coordinates are increased by 5.

Old vertex	*Shifted vertex*
$(-1, -1)$	$(1, 4)$
$(-2, -4)$	$(0, 1)$
$(2, -3)$	$(4, 2)$

87.

Old vertex	*Shifted vertex*
$(0, 2)$	$(-1, 5)$
$(-3, 5)$	$(-4, 8)$
$(-5, 2)$	$(-6, 5)$
$(-2, -1)$	$(-3, 2)$

89. $d = \sqrt{(45-10)^2 + (40-15)^2}$

$= \sqrt{35^2 + 25^2}$

$= \sqrt{1850}$

$= 5\sqrt{74}$

≈ 43 yards

91. Let $(0, 0)$ represent the point of departure, Naples, and $(120, 150)$ represent the destination, Rome.

$d = \sqrt{(120-0)^2 + (150-0)^2}$

$= \sqrt{36{,}900} \approx 192.1$ km

93. False. The polygon could be a rhombus. For example, consider the points $(4, 0)$, $(0, 6)$, $(-4, 0)$ and $(0, -6)$.

95. The y-coordinate of a point on the x-axis is 0.
The x-coordinate of a point on the y-axis is 0.

97. Since $x_m = \dfrac{x_1 + x_2}{2}$ and $y_m = \dfrac{y_1 + y_2}{2}$ we have:

$$2x_m = x_1 + x_2 \qquad 2y_m = y_1 + y_2$$
$$2x_m - x_1 = x_2 \qquad 2y_m - y_1 = y_2$$

So, $(x_2, y_2) = (2x_m - x_1, 2y_m - y_1)$.

(a) $(x_2, y_2) = (2x_m - x_1, 2y_m - y_1) = (2(4) - 1, 2(-1) - (-2)) = (7, 0)$

(b) $(x_2, y_2) = (2x_m - x_1, 2y_m - y_1) = (2(2) - (-5), 2(4) - 11) = (9, -3)$

99. Use the Midpoint Formula to prove the diagonals of the parallelogram bisect each other.

$$\left(\frac{b + a}{2}, \frac{c + 0}{2}\right) = \left(\frac{a + b}{2}, \frac{c}{2}\right)$$
$$\left(\frac{a + b + 0}{2}, \frac{c + 0}{2}\right) = \left(\frac{a + b}{2}, \frac{c}{2}\right)$$

Appendix C.2 Graphs of Equations

1. solution point

3. intercepts

5. $y = \sqrt{x + 4}$

(a) $(0, 2)$: $2 \stackrel{?}{=} \sqrt{0 + 4}$
$2 = 2$ ✓

Yes, the point *is* on the graph.

(b) $(12, 4)$: $4 \stackrel{?}{=} \sqrt{12 + 4}$
$4 = \sqrt{16}$ ✓

Yes, the point *is* on the graph.

7. $y = 4 - |x - 2|$

(a) $(1, 5)$: $5 \stackrel{?}{=} 4 - |1 - 2|$
$5 \neq 4 - 1$

No, the point *is not* on the graph.

(b) $(1.2, 3.2)$: $3.2 \stackrel{?}{=} 4 - |1.2 - 2|$
$3.2 \stackrel{?}{=} 4 - |-0.8|$
$3.2 \stackrel{?}{=} 4 - 0.8$
$3.2 \stackrel{?}{=} 3.2$ ✓

Yes, the point *is* on the graph.

9. $x^2 + y^2 = 20$

(a) $(3, -2)$: $3^2 + (-2)^2 \stackrel{?}{=} 20$
$9 + 4 \stackrel{?}{=} 20$
$13 \neq 20$

No, the point *is not* on the graph.

(b) $(-4, 2)$: $(-4)^2 + 2^2 \stackrel{?}{=} 20$
$16 + 4 \stackrel{?}{=} 20$
$20 = 20$

Yes, the point *is* on the graph.

11. $y = \frac{3}{2}x - 1$

x	-2	0	$\frac{2}{3}$	1	2
y	-4	-1	0	$\frac{1}{2}$	2
Solution point	$(-2, -4)$	$(0, -1)$	$\left(\frac{3}{2}, 0\right)$	$\left(1, \frac{1}{2}\right)$	$(2, 2)$

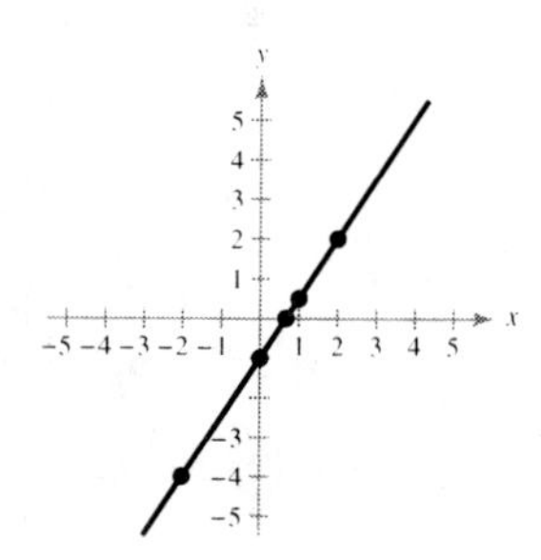

13. $y = x^2 - 2x$

x	-1	0	1	2	3
y	3	0	-1	0	3
Solution point	$(-1, 3)$	$(0, 0)$	$(1, -1)$	$(2, 0)$	$(3, 3)$

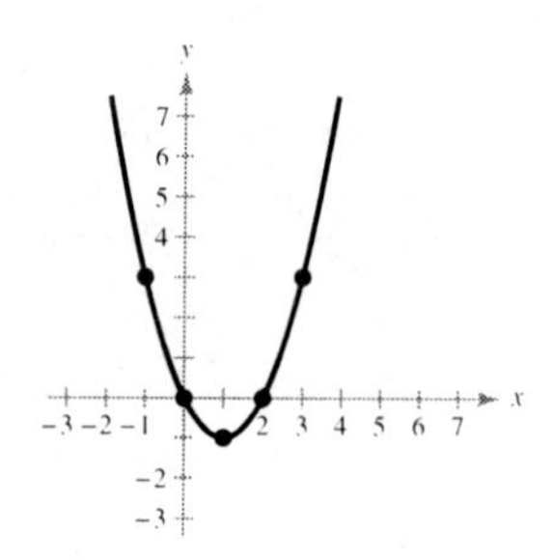

15. $y = 2\sqrt{x}$ has one intercept $(0, 0)$.

Matches graph (b).

16. $y = 4 - x^2$ has intercepts $(0, 4)$, $(2, 0)$, and $(-2, 0)$.

Matches graph (d).

17. $y = \sqrt{9 - x^2}$ has intercepts $(0, 3)$, $(-3, 0)$ and $(3, 0)$.

Matches graph (c).

19. $y = -4x + 1$

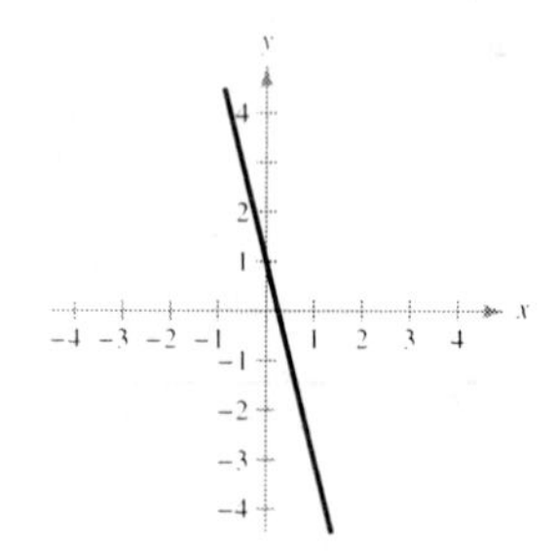

21. $y = 2 - x^2$

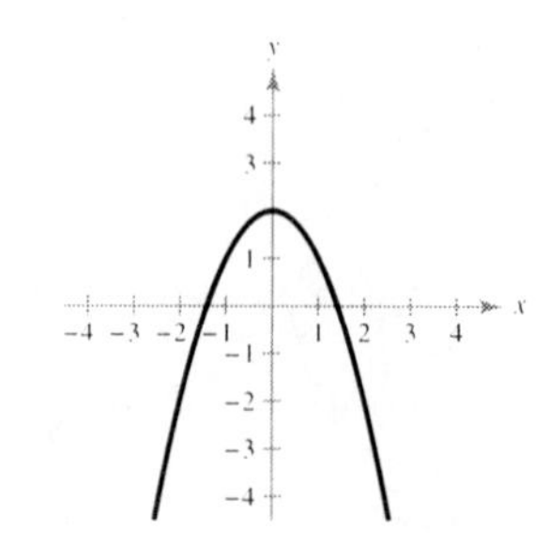

23. $y = x^2 - 3x$

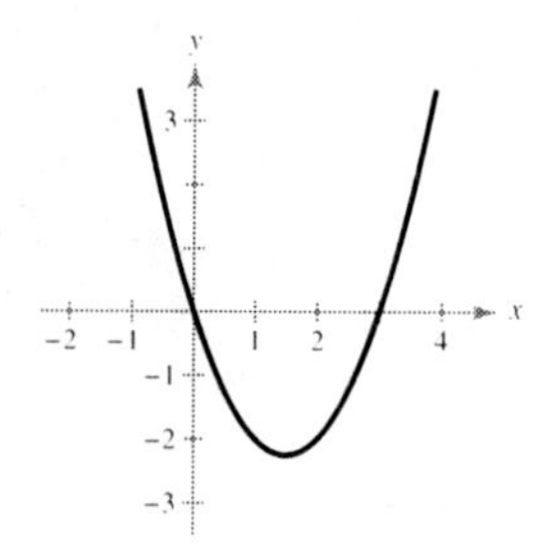

25. $y = x^3 + 2$

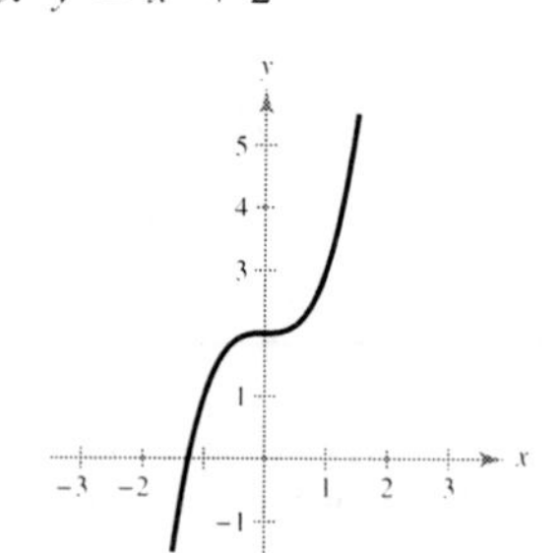

27. $y = \sqrt{x - 3}$

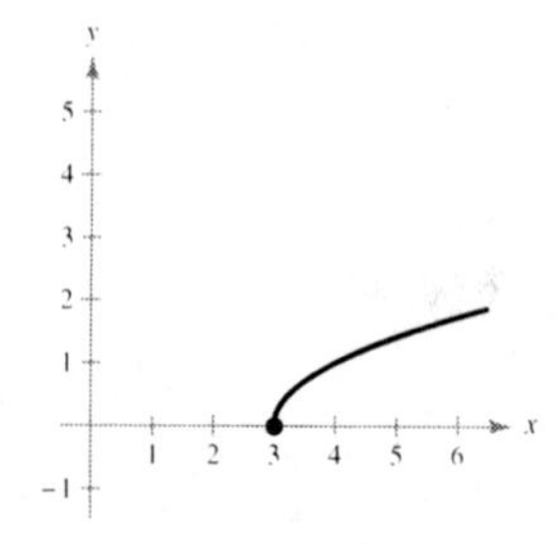

29. $y = |x - 2|$

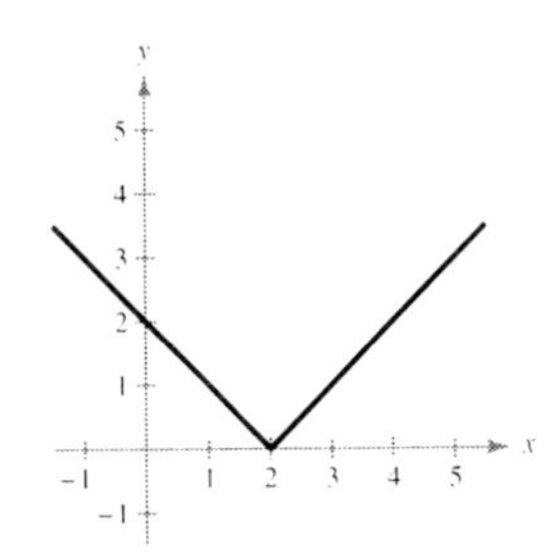

31. $x = y^2 - 1$

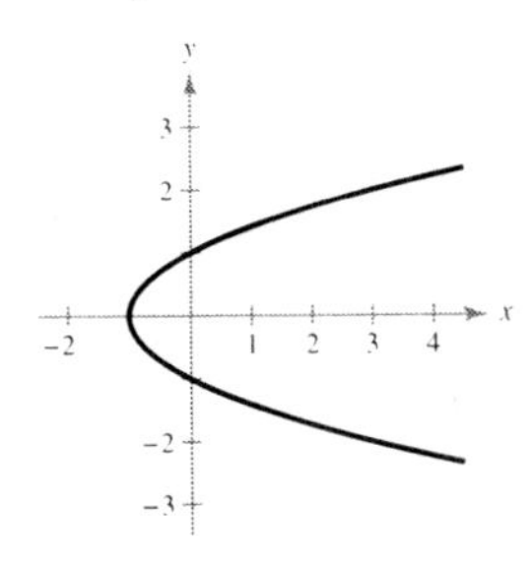

33. $y = x - 7$

Intercepts: $(0, -7), (7, 0)$

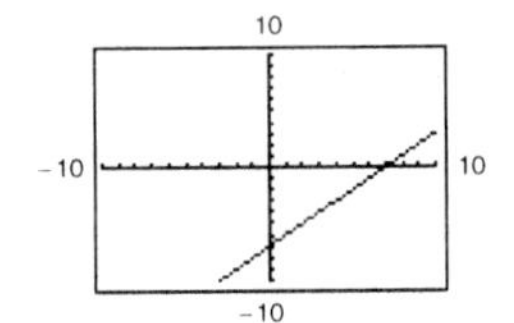

35. $y = 3 - \frac{1}{2}x$

Intercepts: $(6, 0), (0, 3)$

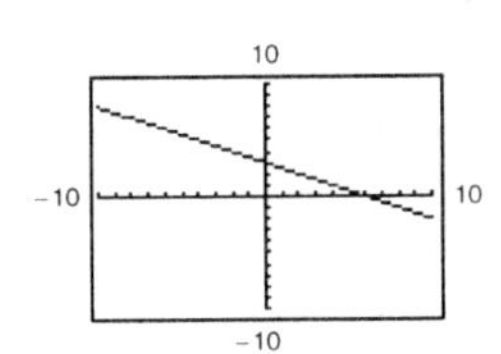

37. $y = \dfrac{2x}{x - 1}$

Intercepts: $(0, 0)$

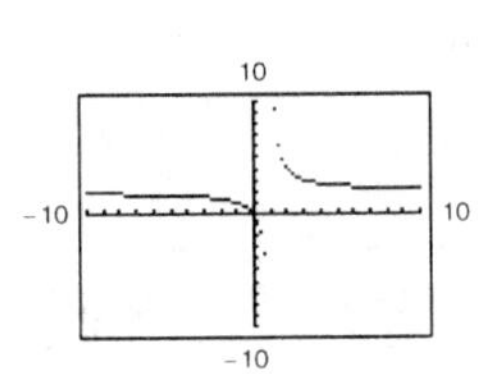

39. $y = x\sqrt{x + 3}$

Intercepts: $(0, 0), (-3, 0)$

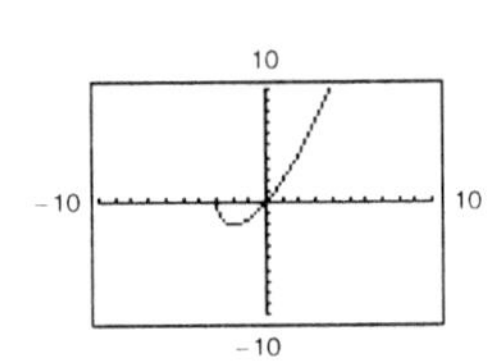

41. $y = \sqrt[3]{x - 8}$

Intercepts: $(8, 0), (0, -2)$

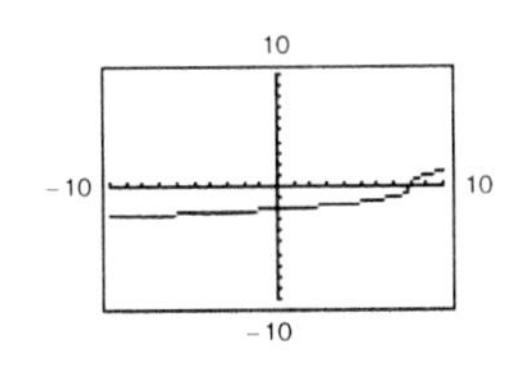

43. $y = x^2 - 4x + 3$

Intercepts: $(3, 0), (1, 0), (0, 3)$

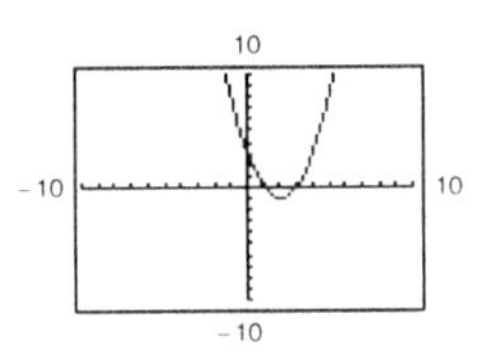

45. $y = x^2(x - 4) + 4x$

$= x^3 - 4x^2 + 4x$

Intercepts: $(0, 0), (2, 0)$

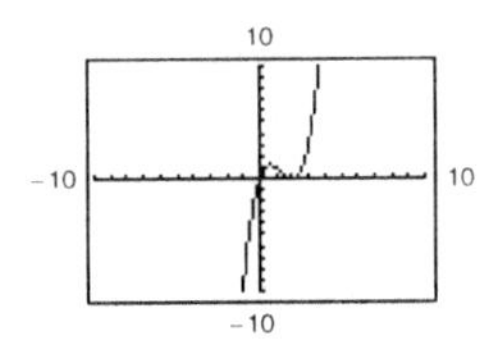

47. $y = -10x + 50$

Range/Window

Xmin = −10
Xmax = 10
Xscl = 2
Ymin = −50
Ymax = 100
Yscl = 25

49. $y_1 = \frac{1}{4}(x^2 - 8)$

$y_2 = \frac{1}{4}x^2 - 2$

The graphs are identical.
The Distributive Property is illustrated.

51. $y_1 = \frac{1}{5}\left[10(x^2 - 1)\right]$

$y_2 = 2(x^2 - 1)$

The graphs are identical.
The Associative Property of Multiplication is illustrated.

53. $y = \sqrt{5 - x}$

(a) $(3, y) \approx (3, 1.41)$

(b) $(x, 3) \approx (-4, 3)$

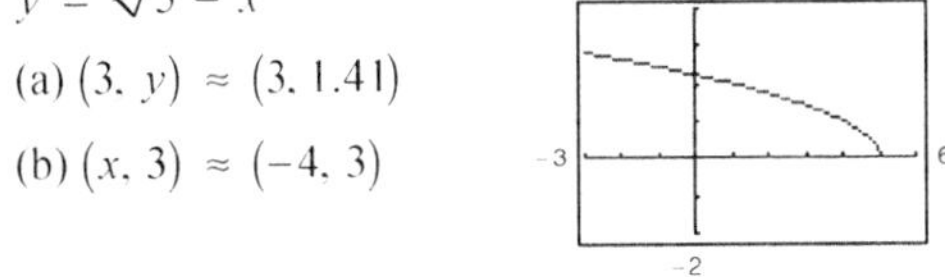

55. $y = x^5 - 5x$

(a) $(-0.5, y) \approx (-0.5, 2.47)$

(b) $(x, -4) = (1, -4)$ or $(x, -4) \approx (-1.65, -4)$

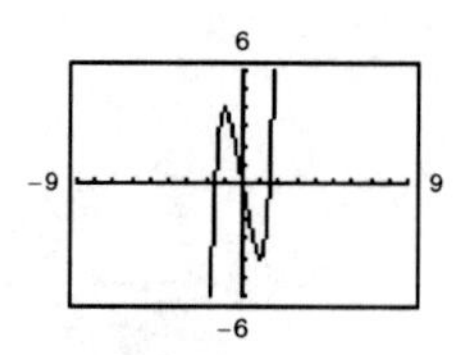

57. $x^2 + y^2 = 16$

$$y^2 = 16 - x^2$$

$$y = \pm\sqrt{16 - x^2}$$

Use $y_1 = \sqrt{16 - x^2}$

$y_2 = -\sqrt{16 - x^2}$

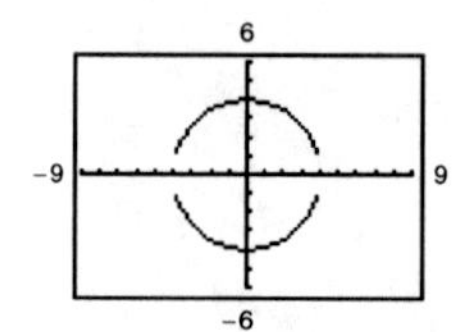

59. $(x - 1)^2 + (y - 2)^2 = 9$

$$(y - 2)^2 = 9 - (x - 1)^2$$

$$y - 2 = \pm\sqrt{9 - (x - 1)^2}$$

$$y = 2 \pm \sqrt{9 - (x - 1)^2}$$

Use $y_1 = 2 + \sqrt{9 - (x - 1)^2}$

$y_2 = 2 - \sqrt{9 - (x - 1)^2}$

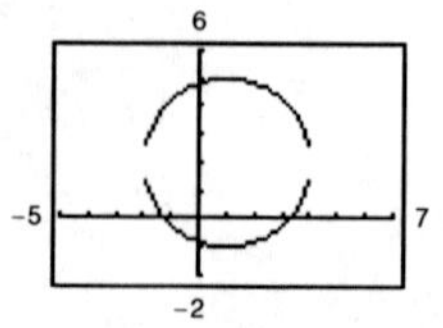

61. $(x - 1)^2 + (y - 2)^2 = 25$

(a) $(1 - 1)^2 + (2 - 2)^2 = 0 \neq 25$ No

(b) $(-2 - 1)^2 + (6 - 2)^2 = 9 + 16 = 25$ Yes

(c) $(5 - 1)^2 + (-1 - 2)^2 = 16 + 9 = 25$ Yes

(d) $(0 - 1)^2 + \left(2 + 2\sqrt{6} - 2\right)^2 = 1 + 24 = 25$ Yes

63. (a) $y = 225{,}000 - 20{,}000t,\ 0 \le t \le 8$

Window

$X_{min} = 0$

$X_{max} = 8$

$X_{scl} = 1$

$Y_{min} = 60{,}000$

$Y_{max} = 230{,}000$

$Y_{scl} = 10{,}000$

230,000

0 8

60,000

(b) When $t = 5.8$, $y = 109{,}000$. Algebraically, $225{,}000 - 20{,}000(5.8) = \$109{,}000$.

(c) When $t = 2.35$, $y = 178{,}000$. Algebraically, $225{,}000 - 20{,}000(2.35) = \$178{,}000$.

65. (a)

Year	2000	2001	2002	2003	2004	2005	2006	2007	2008
Median sales price (in thousands of dollars)	150.9	151.7	159.3	171.3	185.0	198.0	207.7	211.6	207.2

The model fits the data well.

(b)

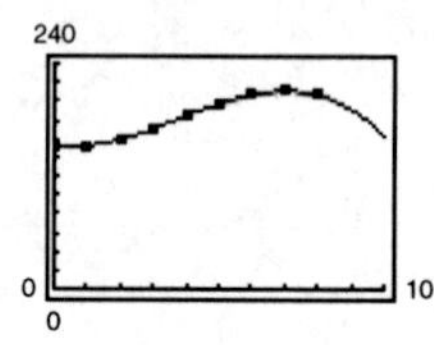

The model fits the data well.

(c) 2012: \$55,231; 2014: −\$136,682; No. There is a large decrease in the price over a short period of time, leading to negative values.

(d) 2000, 2010

67. False, $y = 1 - x^2$ has two x-intercepts, $(1, 0)$ and $(-1, 0)$. Also, $y = x^2 = 1$ has no x intercepts.

69. Option 1: $w_1 = 3000 + 0.07x$

Option 2: $w_2 = 3400 + 0.05x$

(x is the amount of sales)

$$\begin{aligned} w_1 &= w_2 \\ 3000 + 0.07x &= 3400 + 0.05x \\ 0.02x &= 400 \\ x &= 20{,}000 \end{aligned}$$

If sales equal 20,000, the options are equivalent. For sales less than 20,000, choose option 2. For sales greater than 20,000, choose option 1.

Appendix C.3 Solving Equations Algebraically and Graphically

1. equation

3. extraneous

5. $\dfrac{5}{2x} - \dfrac{4}{x} = 3$

(a) $\dfrac{5}{2(-1/2)} - \dfrac{5}{4(-1/2)} \stackrel{?}{=} 3$

$3 = 3$

$x = -\dfrac{1}{2}$ *is* a solution.

(b) $\dfrac{5}{2(4)} - \dfrac{4}{4} \stackrel{?}{=} 3$

$-\dfrac{3}{8} \neq 3$

(c) $\dfrac{5}{2(0)} - \dfrac{4}{0}$ is undefined

$x = 0$ *is not* solution.

(d) $\dfrac{5}{2(1/4)} - \dfrac{4}{1/4} \stackrel{?}{=} 3$

$-6 \neq 3$

7. $3 + \dfrac{1}{x+2} = 4$

(a) $3 + \dfrac{1}{(-1)+2} \stackrel{?}{=} 4$

$4 = 4$

$x = -1$ *is* a solution.

(b) $3 + \dfrac{1}{(-2)+2} = 3 + \dfrac{1}{0}$ is undefined.

$x = -2$ *is not* a solution.

(c) $3 + \dfrac{1}{0+2} \stackrel{?}{=} 4$

$\dfrac{7}{2} \neq 4$

$x = 0$ *is not* a solution.

(d) $3 + \dfrac{1}{5+2} \stackrel{?}{=} 4$

$\dfrac{22}{7} = 4$

$x = 5$ *is not* a solution.

9. $\dfrac{\sqrt{x+4}}{6} + 3 = 4$

(a) $\dfrac{\sqrt{-3+4}}{6} + 3 \stackrel{?}{=} 4$

$\dfrac{19}{6} \neq 4$

$x = -3$ *is not* a solution.

(b) $\dfrac{\sqrt{0+4}}{6} + 3 \stackrel{?}{=} 4$

$\dfrac{10}{3} \neq 4$

$x = 0$ *is not* a solution.

(c) $\dfrac{\sqrt{21+4}}{6} + 3 \stackrel{?}{=} 4$

$\dfrac{23}{6} \neq 4$

$x = 21$ *is not* a solution.

(d) $\dfrac{\sqrt{32+4}}{6} + 3 \stackrel{?}{=} 4$

$4 = 4$

$x = 32$ *is* a solution.

11. $2(x-1) = 2x - 2$ is an *identity* by the Distributive Property. It is true for all real values of x.

13. $x^2 - 8x + 5 = (x-4)^2 - 11$ is an *identity* since

$$\begin{aligned}(x-4)^2 - 11 &= x^2 - 8x + 16 - 11\\ &= x^2 - 8x + 5.\end{aligned}$$

15. $3 + \dfrac{1}{x+1} = \dfrac{4x}{x+1}$ is *conditional*. There are real values of x which the equation is not true.

17.
$$\begin{aligned}3x - 5 &= 2x + 7\\ 3x - 2x &= 7 + 5\\ x &= 12\end{aligned}$$

19.
$$\begin{aligned}4y + 2 - 5y &= 7 - 6y\\ -y + 2 &= 7 - 6y\\ 6y - y &= 7 - 2\\ 5y &= 5\\ y &= 1\end{aligned}$$

21.
$$\begin{aligned}3(y-5) &= 3 + 5y\\ 3y - 15 &= 3 + 5y\\ -18 &= 2y\\ y &= -9\end{aligned}$$

23.
$$\begin{aligned}\frac{x}{5} - \frac{x}{2} &= 3\\ \frac{2x - 5x}{10} &= 3\\ -3x &= 30\\ x &= -10\end{aligned}$$

25.
$$\begin{aligned}\tfrac{3}{2}(z+5) - \tfrac{1}{4}(z+24) &= 0\\ 4\left(\tfrac{3}{2}\right)(z+5) - 4\left(\tfrac{1}{4}\right)(z+24) &= 4(0)\\ 6(z+5) - (z+24) &= 0\\ 6z + 30 - z - 24 &= 0\\ 5z &= -6\\ z &= -\frac{6}{5}\end{aligned}$$

27.
$$\begin{aligned}\frac{2(z-4)}{5} + 5 &= 10z\\ \frac{(2z-8)+25}{5} &= 10z\\ 2z + 17 &= 50z\\ 17 &= 48z\\ z &= \frac{17}{48}\end{aligned}$$

29.
$$\begin{aligned}\frac{100-4u}{3} &= \frac{5u+6}{4} + 6\\ 12\left(\frac{100-4u}{3}\right) &= 12\left(\frac{5u+6}{4}\right) + 12(6)\\ 4(100-4u) &= 3(5u+6) + 72\\ 400 - 16u &= 15u + 18 + 72\\ -31u &= -310\\ u &= 10\end{aligned}$$

31.
$$\begin{aligned}\frac{5x-4}{5x+4} &= \frac{2}{3}\\ 3(5x-4) &= 2(5x+4)\\ 15x - 12 &= 10x + 8\\ 5x &= 20\\ x &= 4\end{aligned}$$

33.
$$\begin{aligned}\frac{1}{x-3} + \frac{1}{x+3} &= \frac{10}{x^2-9}\\ \frac{(x+3)+(x-3)}{x^2-9} &= \frac{10}{x^2-9}\\ 2x &= 10\\ x &= 5\end{aligned}$$

35.
$$\begin{aligned}\frac{7}{2x+1} - \frac{8x}{2x-1} &= -4\\ 7(2x+1) - 8x(2x+1) &= -4(2x+1)(2x-1)\\ 14x + 7 - 16x^2 - 8x &= -16x^2 + 4\\ 6x &= 11\\ x &= \frac{11}{6}\end{aligned}$$

37.
$$\begin{aligned}\frac{1}{x} + \frac{2}{x-5} &= 0\\ 1(x-5) + 2x &= 0\\ 3x - 5 &= 0\\ 3x &= 5\\ x &= \frac{5}{3}\end{aligned}$$

39.
$$\begin{aligned}\frac{3}{x(x-3)} + \frac{4}{x} &= \frac{1}{x-3}\\ 3 + 4(x-3) &= x\\ 3 + 4x - 12 &= x\\ 3x &= 9\\ x &= 3\end{aligned}$$

A check reveals that $x = 3$ is an extraneous solution, so there is no solution.

41. $y = x - 5$

Let $y = 0$: $0 = x - 5 \Rightarrow x = 5 \Rightarrow (5, 0)$ x-intercept

Let $x = 0$: $y = 0 - 5 \Rightarrow y = -5 \Rightarrow (0, -5)$ y-intercept

43. $y = x^2 + x - 2$

Let $y = 0$: $(x^2 + x - 2) = (x + 2)(x - 1) = 0 \Rightarrow x = -2, 1 \Rightarrow (-2, 0), (1, 0)$ x-intercepts

Let $x = 0$: $y = 0^2 + 0 - 2 = -2 \Rightarrow (0, -2)$ y-intercept

45. $y = x\sqrt{x + 2}$

Let $y = 0$: $0 = x\sqrt{x + 2} \Rightarrow x = 0, -2 \Rightarrow (0, 0), (-2, 0)$ x-intercepts

Let $x = 0$: $y = 0\sqrt{0 + 2} = 0 \Rightarrow (0, 0)$ y-intercept

47. $xy = 4$

If $x = 0$, then $0y = 0 = 4$, which is impossible. Similarly, $y = 0$ is impossible.

Hence there are no intercepts.

49. $y = |x - 2| - 4$

Let $y = 0$: $|x - 2| - 4 = 0 \Rightarrow |x - 2| = 4 \Rightarrow x = -2, 6 \Rightarrow (-2, 0), (6, 0)$ x-intercepts

Let $x = 0$: $|0 - 2| - 4 = |-2| - 4 = 2 - 4 = -2 = y \Rightarrow (0, -2)$ y-intercept

51. $xy - 2y - x + 1 = 0$

Let $y = 0$:

$-x + 1 = 0 \Rightarrow x = 1 \Rightarrow (1, 0)$ x-intercept

Let $x = 0$:

$-2y + 1 = 0 \Rightarrow y = \frac{1}{2} \Rightarrow \left(0, \frac{1}{2}\right)$ y-intercept

53.

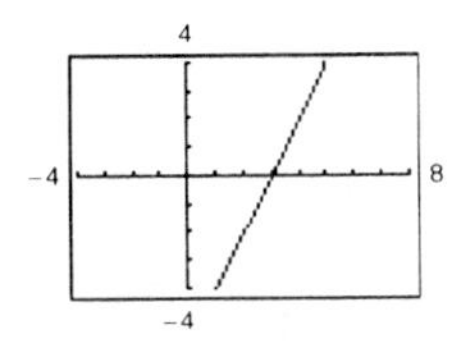

$$y = 0 = 2(x - 1) - 4$$
$$= 2x - 2 - 4$$
$$= 2x - 6 \Rightarrow 2x = 6 \Rightarrow x = 3$$

$(3, 0)$

55.

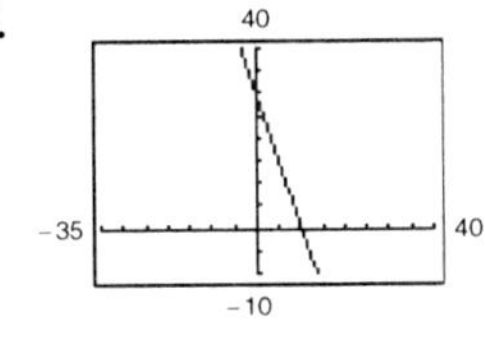

$$y = 0 = 20 - (3x - 10)$$
$$= 20 - 3x + 10$$
$$= 30 - 3x \Rightarrow 3x = 30 \Rightarrow x = 10$$

$(10, 0)$

57. $f(x) = 5(4 - x)$

$$5(4 - x) = 0$$
$$4 - x = 0$$
$$x = 4$$

59. $f(x) = x^3 - 6x^2 + 5x$

$$x^3 - 6x^2 + 5x = 0$$
$$x(x^2 - 6x + 5) = 0$$
$$x(x - 5)(x - 1) = 0$$
$$x = 0, 5, 1$$

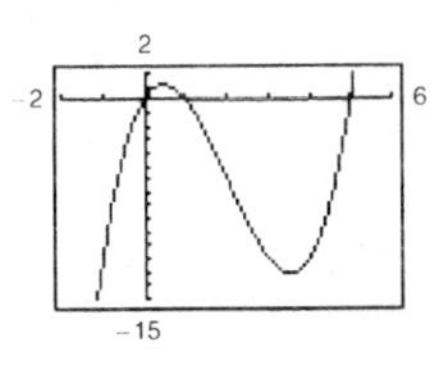

61. $f(x) = \dfrac{x + 2}{3} - \dfrac{x - 1}{5} - 1$

$$\frac{x + 2}{3} - \frac{x - 1}{5} - 1 = 0$$
$$5(x + 2) - 3(x - 1) - 15 = 0$$
$$2x = 2$$
$$x = 1$$

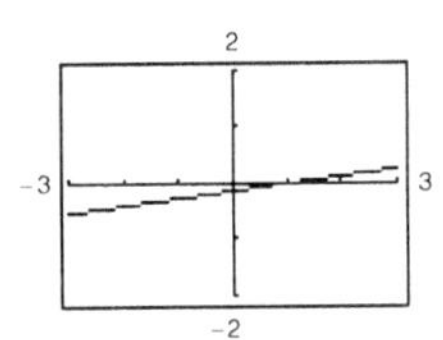

63. $2.7x - 0.4x = 1.2$

$2.3x = 1.2$

$x = \frac{1.2}{2.3} \approx 0.522$

$f(x) = 2.7x - 0.4x - 1.2 = 0$

$x \approx 0.522$

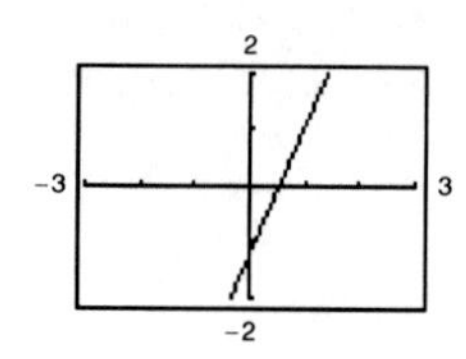

65. $25(x - 3) = 12(x + 2) - 10$

$25x - 75 = 12x + 24 - 10$

$13x - 89 = 0$

$x = \frac{89}{13}$

$f(x) = 25(x - 3) - 12(x + 2) + 10 = 0$

$x = 6.846$

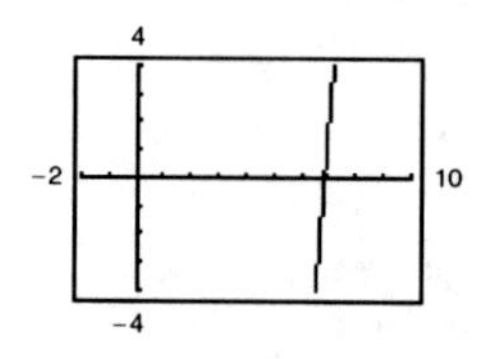

67. $\frac{3x}{2} + \frac{1}{4}(x - 2) = 10$

$\frac{6x}{4} + \frac{x}{4} = 10 + \frac{1}{2}$

$\frac{7x}{4} = \frac{21}{2}$

$x = 6$

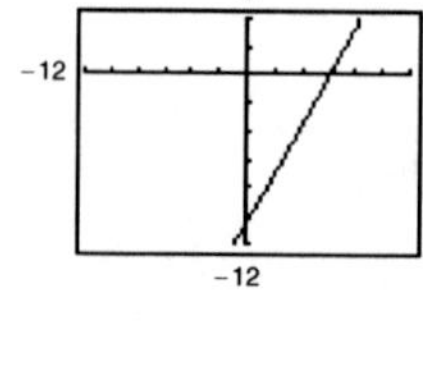

$f(x) = \frac{3x}{2} + \frac{1}{4}(x - 2) - 10 = 0$

$x = 6.0$

69. $x^3 - x + 4 = 0$

$x \approx -1.796$

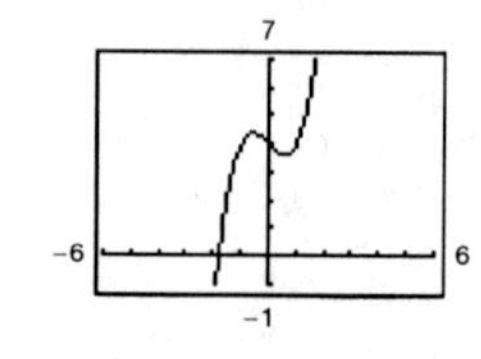

71. $\frac{2x}{3} = 10 - \frac{24}{x}$

$\frac{2x}{3}(3x) = 10(3x) - \frac{24}{x}(3x)$

$2x^2 = 30x - 72$

$2x^2 - 30x + 72 = 0$

$x^2 - 15x + 36 = 0$

$(x - 3)(x - 12) = 0$

$x = 3, 12$

$f(x) = \frac{2x}{3} - 10 + \frac{24}{x}$

$x = 3, 12$

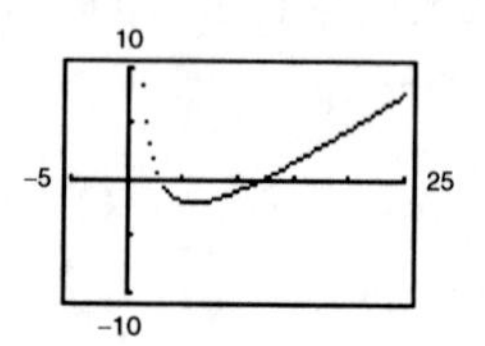

73. $\frac{2}{x + 2} - \frac{4}{x - 2} = 5$

$3(x - 2) - 4(x + 2) = 5(x + 2)(x - 2)$

$3x - 6 - 4x - 8 = 5(x^2 - 4)$

$0 = 5x^2 + x - 6$

$0 = (x - 1)(5x + 6)$

$x = 1, -\frac{6}{5}$

$f(x) = \frac{3}{x + 2} - \frac{4}{x - 2} - 5$

$= 0$

$x = 1.0, -1.2$

75. $(x + 2)^2 = x^2 - 6x + 1$

$x^2 + 4x + 4 = x^2 - 6x + 1$

$10x = -3$

$x = -\frac{3}{10}$

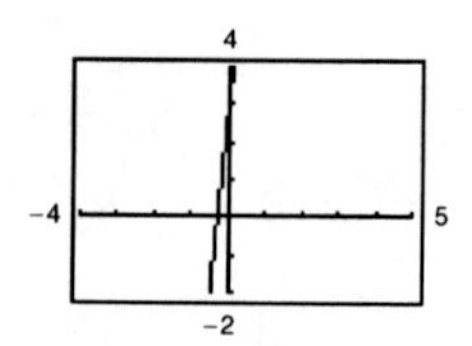

$f(x) = (x + 2)^2 - x^2 + 6x - 1$

$x = -\frac{3}{10}$

77. $2x^3 - x^2 - 18x + 9 = 0$

$x = -3.0, 0.5, 3.0$

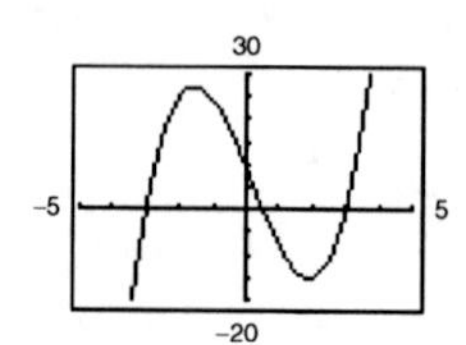

79. $x^4 = 2x^3 + 1$

$x^4 - 2x^3 - 1 = 0$

$x \approx -0.717, 2.107$

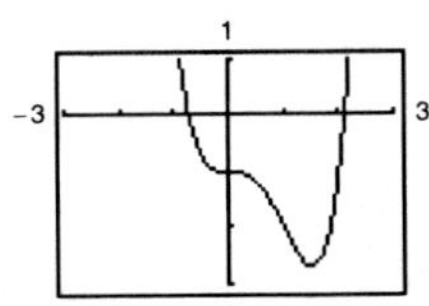

81. $\dfrac{2}{x+2} = 3$

$\dfrac{2}{x+2} - 3 = 0$

$x = -\dfrac{4}{3}$

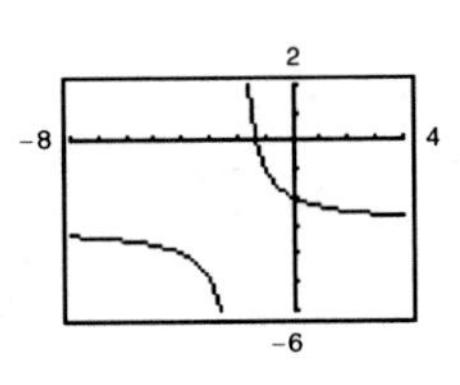

83. $|x - 3| = 4$

$|x - 3| - 4 = 0$

$x = -1, 7$

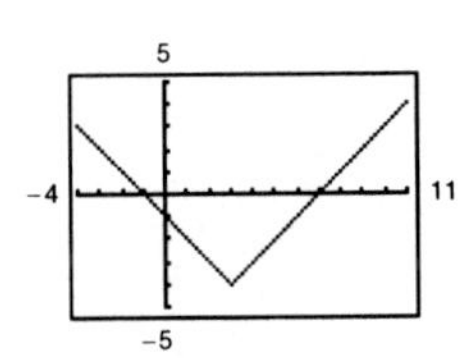

85. $\sqrt{x-2} = 3$

$\sqrt{x-2} - 3 = 0$

$x = 11$

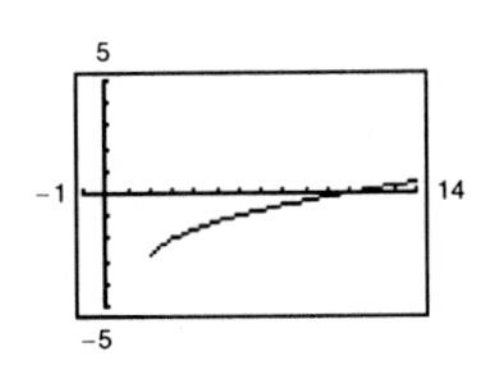

87. $y = 2 - x$

$y = 2x - 1$

$2 - x = 2x - 1$

$3 = 3x$

$x = 1, y = 2 - 1 = 1$

$(x, y) = (1, 1)$

89. $x - y = -4 \Rightarrow y = x + 4$

$x^2 - y = -2 \Rightarrow y = x^2 + 2$

$x^2 + 2 = x + 4$

$x^2 - x - 2 = 0$

$(x - 2)(x + 1) = 0$

$x = 2, y = 6$

$x = -1, y = 3$

$(2, 6), (-1, 3)$

91. $y = x^2 - x + 1$

$y = x^2 + 2x + 4$

$x^2 - x + 1 = x^2 + 2x + 4$

$-3 = 3x$

$x = -1$

$y = (-1)^2 - (-1) + 1 = 3$

$(x, y) = (-1, 3)$

93. $y = 9 - 2x$

$y = x - 3$

$(4, 1)$

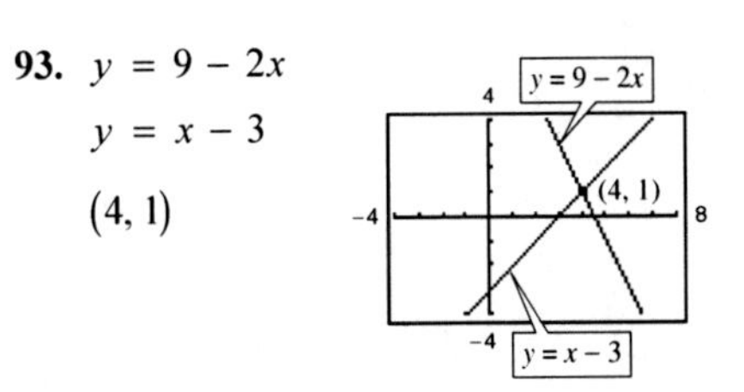

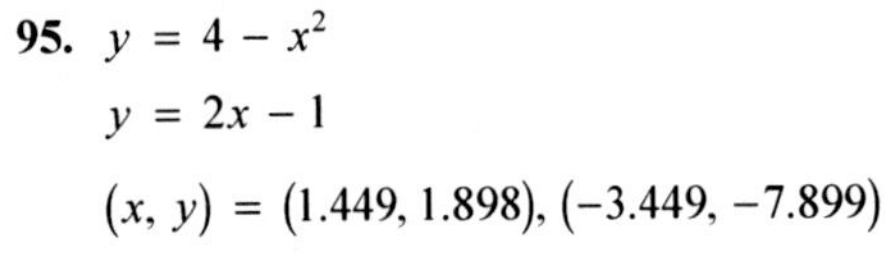

95. $y = 4 - x^2$

$y = 2x - 1$

$(x, y) = (1.449, 1.898), (-3.449, -7.899)$

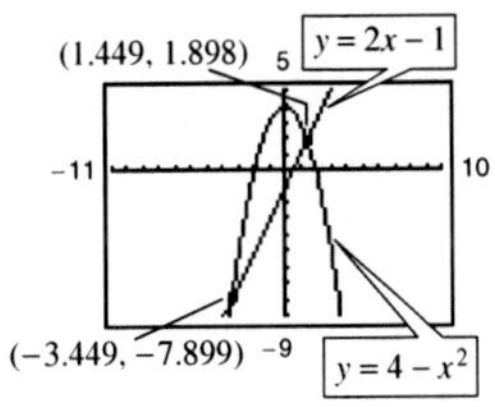

97. $y = 2x^2$

$y = x^4 - 2x^2$

$(x, y) = (0, 0), (2, 8), (-2, 8)$

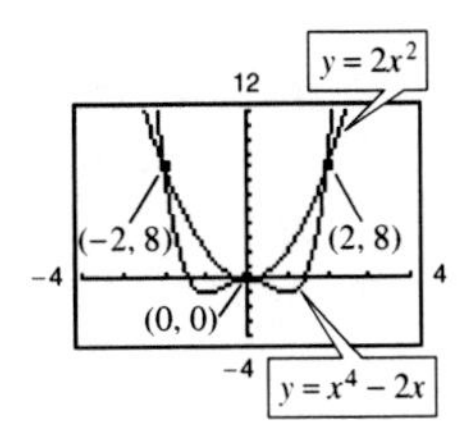

99. $6x^2 + 3x = 0$

$3x(2x + 1) = 0$

$3x = 0$ or $2x + 1 = 0$

$x = 0$ or $x = -\frac{1}{2}$

101. $x^2 - 2x - 8 = 0$

$(x - 4)(x + 2) = 0$

$x - 4 = 0$ or $x + 2 = 0$

$x = 4$ or $x = -2$

103. $3 + 5x - 2x^2 = 0$

$(3 - x)(1 + 2x) = 0$

$3 - x = 0$ or $1 + 2x = 0$

$x = 3$ or $x = -\frac{1}{2}$

105. $x^2 + 4x = 12$

$x^2 + 4x - 12 = 0$

$(x + 6)(x - 2) = 0$

$x + 6 = 0$ or $x - 2 = 0$

$x = -6$ or $x = 2$

107. $(x + a)^2 - b^2 = 0$

$[(x + a) + b][(x + a) - b] = 0$

$x + a + b = 0 \Rightarrow x = -a - b$

$x + a - b = 0 \Rightarrow x = -a + b$

109. $x^2 = 49$

$x = \pm\sqrt{49} = \pm 7$

111. $(x - 12)^2 = 16$

$x - 12 = \pm\sqrt{16} = \pm 4$

$x = 12 \pm 4$

$x = 16, 8$

113. $(3x - 1)^2 + 6 = 0$

$(3x - 1)^2 = -6$

$3x - 1 = \pm\sqrt{-6} = \pm\sqrt{6}i$

$x = \frac{1}{3} \pm \frac{\sqrt{6}}{3}i \approx 0.33 \pm 0.82i$

115. $(2x - 1)^2 = 12$

$2x - 1 = \pm\sqrt{12} = \pm\sqrt{3}$

$2x = 1 \pm 2\sqrt{3}$

$x = \frac{1}{2} \pm \sqrt{3}$

$x \approx 2.23, -1.23$

117. $(x - 7)^2 = (x + 3)^2$

$x - 7 = \pm(x + 3)$

$x - 7 = x + 3$, impossible

$x - 7 = -(x + 3) \Rightarrow 2x = 4$

$\Rightarrow x = 2$

119. $x^2 + 4x = 32$

$x^2 + 4x + 4 = 32 + 4$

$(x + 2)^2 = 36$

$x + 2 = \pm 6$

$x = -2 \pm 6$

$x = -8, 4$

121. $x^2 + 6x + 2 = 0$

$x^2 + 6x = -2$

$x^2 + 6x + 3^2 = -2 + 3^2$

$(x + 3)^2 = 7$

$x + 3 = \pm\sqrt{7}$

$x = -3 \pm \sqrt{7}$

123. $9x^2 - 18x + 3 = 0$

$x^2 - 2x + \frac{1}{3} = 0$

$x^2 - 2x = -\frac{1}{3}$

$x^2 - 2x + 1^2 = -\frac{1}{3} + 1^2$

$(x - 1)^2 = \frac{2}{3}$

$x - 1 = \pm\sqrt{\frac{2}{3}}$

$x = 1 \pm \sqrt{\frac{2}{3}}$

$x = 1 \pm \frac{\sqrt{6}}{3}$

125. $-6 + 2x - x^2 = 0$

$x^2 - 2x + 1 = -6 + 1$

$(x - 1)^2 = -5$

$x - 1 = \pm\sqrt{-5}$

$= \pm\sqrt{5}i$

$x = 1 \pm \sqrt{5}i$

127. $2x^2 + 5x - 8 = 0$

$$x^2 + \frac{5}{2}x - 4 = 0$$

$$x^2 + \frac{5}{2}x + \frac{25}{16} = 4 + \frac{25}{16}$$

$$\left(x + \frac{5}{4}\right)^2 = \frac{89}{16}$$

$$x + \frac{5}{4} = \pm\frac{\sqrt{89}}{4}$$

$$x = \frac{-5}{4} \pm \frac{\sqrt{89}}{4}$$

129. $-x^2 + 2x + 2 = 0$

$$x = \frac{-b \pm \sqrt{b^2 - 4ac}}{2a}$$

$$= \frac{-2 \pm \sqrt{2^2 - 4(-1)(2)}}{2(-1)}$$

$$= \frac{-2 \pm 2\sqrt{3}}{-2} = 1 \pm \sqrt{3}$$

131. $x^2 + 8x - 4 = 0$

$$x = \frac{-b \pm \sqrt{b^2 - 4ac}}{2a}$$

$$= \frac{-8 \pm \sqrt{8^2 - 4(1)(-4)}}{2(1)}$$

$$= \frac{-8 \pm 4\sqrt{5}}{2} = -4 \pm 2\sqrt{5}$$

133. $x^2 + 3x + 8 = 0$

$$x = \frac{-3 \pm \sqrt{9 - 4(8)}}{2}$$

$$= \frac{-3 \pm \sqrt{-23}}{2}$$

$$= -\frac{3}{2} \pm \frac{\sqrt{23}i}{2}$$

135. $28x - 49x^2 = 4$

$$49x^2 + 28x - 4 = 0$$

$$x = \frac{-b \pm \sqrt{b^2 - 4ac}}{2a}$$

$$= \frac{-28 \pm \sqrt{28^2 - 4(-49)(-4)}}{2(-49)}$$

$$= \frac{-28 \pm 0}{-98} = \frac{2}{7}$$

137. $4x^2 + 16x + 17 = 0$

$$x = \frac{-b \pm \sqrt{b^2 - 4ac}}{2a}$$

$$= \frac{-16 \pm \sqrt{16^2 - 4(-49)(-4)}}{2(4)}$$

$$= \frac{-16 \pm \sqrt{-16}}{8}$$

$$= \frac{-16 \pm 4i}{8}$$

$$= -2 \pm \frac{1}{2}i$$

139. $x^2 - 2x - 1 = 0$

$$x^2 - 2x = 1$$

$$x^2 - 2x + 1^2 = 1 + 1^2$$

$$x - 1 = \pm\sqrt{2}$$

$$x = 1 \pm \sqrt{2}$$

141. $(x + 3)^2 = 81$

$$x + 3 = \pm 9$$

$$x + 3 = 9 \text{ or } x + 3 = -9$$

$$x = 6 \text{ or } \quad x = -12$$

143. $x^2 - 14x + 49 = 0$

$$(x - 7)^2 = 0$$

$$x = 7$$

145. $x^2 - x - \frac{11}{4} = 0$

$$x^2 - x + \frac{1}{4} = \frac{11}{4} + \frac{1}{4}$$

$$\left(x - \frac{1}{2}\right)^2 = 3$$

$$x - \frac{1}{2} = \pm\sqrt{3}$$

$$x = \frac{1}{2} \pm \sqrt{3}$$

$$x = \frac{1}{2} \pm \sqrt{3}$$

147. $(x + 1)^2 = x^2$

$$(x + 1)^2 - x^2 = 0$$

$$(x + 1 - x)(x + 1 + x) = 0$$

$$2x + 1 = 0$$

$$2x = -1$$

$$x = -\frac{1}{2}$$

149.
$$4x^4 - 16x^2 = 0$$
$$4x^2(x^2 - 4) = 0$$
$$4x^2(x - 2)(x + 2) = 0$$
$$x = 0, \pm 2$$

151.
$$5x^3 + 30x^2 + 45 = 0$$
$$5(x^2 + 6x + 9) = 0$$
$$5x(x + 3)^2 = 0$$
$$5x = 0 \Rightarrow x = 0$$
$$x + 3 = 0 \Rightarrow x = -3$$

153.
$$4x^4 - 18x^2 = 0$$
$$2x^2(2x^2 - 9) = 0$$
$$2x^2 = 0 \Rightarrow x = 0$$
$$2x^2 - 9 = 0 \Rightarrow x \pm\frac{3\sqrt{2}}{2}$$

155.
$$x^3 - 3x^2 - x + 3 = 0$$
$$x^2(x - 3) - (x - 3) = 0$$
$$(x - 2)(x^2 - 1) = 0$$
$$(x - 3)(x + 1)(x - 1) = 0$$
$$x - 3 = 0 \Rightarrow x = 3$$
$$x + 1 = 0 \Rightarrow x = -1$$
$$x - 1 = 0 \Rightarrow x = 1$$

157.
$$x^4 - 4x^2 + 3 = 0$$
$$(x^2 - 3)(x^2 - 1) = 0$$
$$(x + \sqrt{3})(x - \sqrt{3})(x + 1)(x - 1) = 0$$
$$x + \sqrt{3} = 0 \Rightarrow x = -\sqrt{3}$$
$$x - \sqrt{3} = 0 \Rightarrow x = \sqrt{3}$$
$$x + 1 = 0 \Rightarrow x = -1$$
$$x - 1 = 0 \Rightarrow x = 1$$

159.
$$4x^4 - 65x^2 + 16 = 0$$
$$(4x^2 - 1)(x^2 - 16) = 0$$
$$(2x + 1)(2x - 1)(x + 4)(x - 4) = 0$$
$$2x + 1 = 0 \Rightarrow x = -\tfrac{1}{2}$$
$$2x - 1 = 0 \Rightarrow x = \tfrac{1}{2}$$
$$x + 4 = 0 \Rightarrow x = -4$$
$$x - 4 = 0 \Rightarrow x = 4$$

161.
$$\frac{1}{t^2} + \frac{8}{t} + 15 = 0$$
$$1 + 8t + 15t^2 = 0$$
$$(1 + 3t)(1 + 5t) = 0$$
$$1 + 3t = 0 \Rightarrow t = -\frac{1}{3}$$
$$1 + 5t = 0 \Rightarrow t = -\frac{1}{5}$$

163.
$$6\left(\frac{s}{s + 1}\right)^2 + 5\left(\frac{s}{s + 1}\right) - 6 = 0$$

Let $u = \dfrac{s}{s + 1}$.
$$6u^2 + 5u - 6 = 0$$
$$(3u - 2)(2u + 3) = 0$$
$$3u - 2 = 0 \Rightarrow u = \frac{2}{3}$$
$$2u + 3 = 0 \Rightarrow u = -\frac{3}{2}$$
$$\frac{s}{s + 1} = \frac{2}{3} \Rightarrow s = 2$$
$$\frac{s}{s + 1} = -\frac{3}{2} \Rightarrow s = -\frac{3}{5}$$

165.
$$2x + 9\sqrt{x} - 5 = 0$$
$$(2\sqrt{x} - 1)(\sqrt{x} + 5) = 0$$
$$\sqrt{x} = \tfrac{1}{2} \Rightarrow x = \tfrac{1}{4}$$
$(\sqrt{x} = -5$ is not possible.)

Note: You can see graphically that there is only one solution.

167.
$$\sqrt{x - 10} - 4 = 0$$
$$\sqrt{x - 10} = 4$$
$$x - 10 = 6$$
$$x = 26$$

169.
$$\sqrt{x + 1} - 3x = 1$$
$$\sqrt{x + 1} = 3x + 1$$
$$x + 1 = 9x^2 + 6x + 1$$
$$0 = 9x^2 + 5x$$
$$0 = x(9x + 5)$$
$$x = 0$$
$9x + 5 = 0 \Rightarrow x = -\tfrac{5}{9}$, extraneous

171. $\sqrt[3]{2x+1} + 8 = 0$

$$\sqrt[3]{2x+1} = -8$$
$$2x + 1 = -512$$
$$2x = -513$$
$$x = -\tfrac{513}{2} = -256.5$$

173. $\sqrt{x} - \sqrt{x-5} = 1$

$$\sqrt{x} = 1 + \sqrt{x-5}$$
$$\left(\sqrt{x}\right)^2 = \left(1 + \sqrt{x-5}\right)^2$$
$$x = 1 + 2\sqrt{x-5} + x - 5$$
$$4 = 2\sqrt{x-5}$$
$$2 = \sqrt{x-5}$$
$$4 = x - 5$$
$$9 = x$$

175. $(x-5)^{2/3} = 16$

$$x - 5 = \pm 16^{3/2}$$
$$x - 5 = \pm 64$$
$$x = 69, -59$$

177. $3x(x-1)^{1/2} + 2(x-1)^{3/2} = 0$

$$(x-1)^{1/2}\left[3x + 2(x-1)\right] = 0$$
$$(x-1)^{1/2}(5x-2) = 0$$
$$(x-1)^{1/2} = 0 \Rightarrow x - 1 = 0 \Rightarrow x = 1$$

$5x - 2 = 0 \Rightarrow x = \frac{2}{5}$ which is extraneous.

179.

$$\frac{1}{x} - \frac{1}{x+1} = 3$$
$$x(x+1)\frac{1}{x} - x(x+1)\frac{1}{x+1} = x(x+1)(3)$$
$$x + 1 - x = 3x(x+1)$$
$$1 = 3x^2 + 3x$$
$$0 = 3x^2 + 3x - 1;\ a = 3,\ b = 3,\ c = -1$$
$$x = \frac{-3 \pm \sqrt{(3)^2 - 4(3)(-1)}}{2(3)} = \frac{-3 \pm \sqrt{21}}{6}$$

181.

$$x = \frac{3}{x} + \frac{1}{2}$$
$$(2x)(x) = (2x)\left(\frac{3}{x}\right) + (2x)\left(\frac{1}{2}\right)$$
$$2x^2 = 6 + x$$
$$2x^2 - x - 6 = 0$$
$$(2x+3)(x-2) = 0$$
$$2x + 3 = 0 \Rightarrow x = -\frac{3}{2}$$
$$x - 2 = 0 \Rightarrow x = 2$$

183. $|2x - 1| = 5$

$$2x - 1 = 5 \Rightarrow x = 3$$
$$-(2x-1) = 5 \Rightarrow x = -2$$

185. $|x| = x^2 + x - 3$

$$x = x^2 + x - 3$$
$$x^2 - 3 = 0$$
$$x = \pm\sqrt{3}$$

OR

$$-x = x^2 + x - 3$$
$$x^2 + 2x - 3 = 0$$
$$(x-1)(x+3) = 0$$
$$x - 1 = 0 \Rightarrow x = 1$$
$$x + 3 = 0 \Rightarrow x = -3$$

Only $x = \sqrt{3}$, and $x = -3$ are solutions to the original equation. $x = -\sqrt{3}$ and $x = 1$ are extraneous. Note that the graph of $y = x^2 + x - 3 - |x|$ has two x-intercepts.

187. $y = x^3 - 2x^2 - 3x$

(a)

(b) x-intercepts: $(-1, 0), (0, 0), (3, 0)$

(c) $0 = x^3 - 2x^2 - 3x$

$$0 = x(x+1)(x-3)$$
$$x = 0$$
$$x + 1 = 0 \Rightarrow x = -1$$
$$x - 3 = 0 \Rightarrow x = 3$$

(d) The x-intercepts are the same as the solutions.

189. $y = \sqrt{11x - 30} - x$

(a)

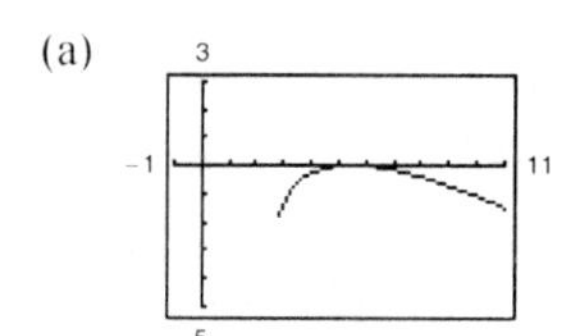

(b) x-intercepts: $(5, 0), (6, 0)$

(c)
$$0 = \sqrt{11x - 30} - x$$
$$x = \sqrt{11x - 30}$$
$$x^2 = 11x - 30$$
$$x^2 - 11x + 30 = 0$$
$$(x - 5)(x - 6) = 0$$
$$x - 5 = 0 \Rightarrow x = 5$$
$$x - 6 = 0 \Rightarrow x = 6$$

(d) The x-intercepts and the solutions are the same.

191. $y = \dfrac{1}{x} - \dfrac{4}{x - 1} - 1$

(a)

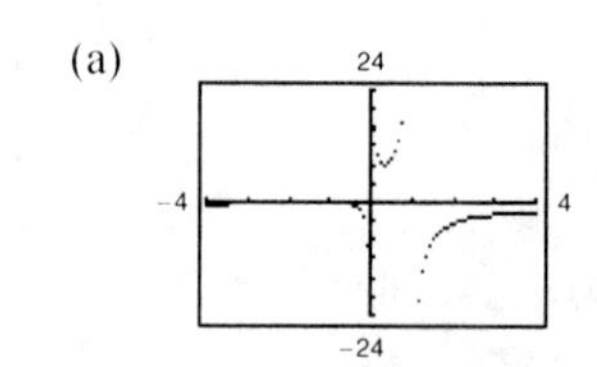

(b) x-intercepts: $(-1, 0)$

(c)
$$0 = \frac{1}{x} - \frac{4}{x - 1} - 1$$
$$0 = (x - 1) - 4x - x(x - 1)$$
$$0 = x - 1 - 4x - x^2 + x$$
$$0 = -x^2 - 2x - 1$$
$$0 = x^2 + 2x + 1$$
$$x + 1 = 0 \Rightarrow x = -1$$

(d) The x-intercepts and the solutions are the same.

193. $y = |x + 1| - 2$

(a)

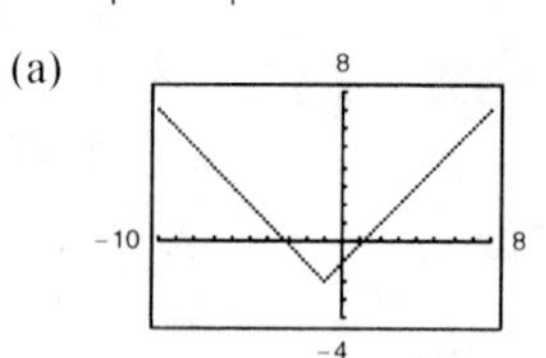

(b) x-intercepts: $(1, 0), (-3, 0)$

(c)
$$0 = |x + 1| - 2$$
$$2 = |x + 1|$$
$$x + 1 = 2 \text{ or } -(x + 1) = 2$$
$$x = 1 \text{ or } \quad -x - 1 = 2$$
$$-x = 3$$
$$x = -3$$

(d) The x-intercepts and the solutions are the same.

195. (a)

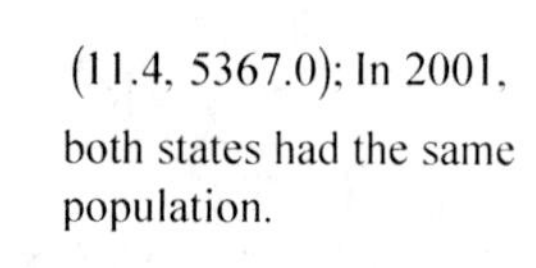

(11.4, 5367.0); In 2001, both states had the same population.

(b)
$$51.1t + 4785 = 162.0t + 3522$$
$$1263 = 110.9t$$
$$11.4 \approx t$$
$$M = A \approx 5367.0$$

The point (11.4, 5367.0) indicates the year, 2001, in which the two populations were the same, about 5367.0 thousand.

(c) The slope indicates the change in population per year. Arizona's population is growing faster.

(d) Maryland: 6,011,400; Arizona: 7,410,000

Answers will vary.

197. $C = 0.45x^2 - 1.65x + 50.75, 10 \le x \le 25$

(a)

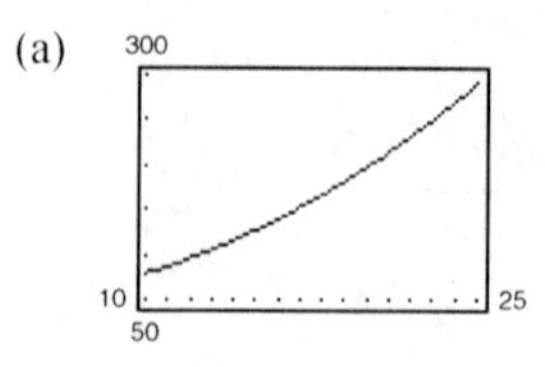

(b) If $C = 150$, then $x = 16.797$ degrees.

(c) If temperature is increased 10° to 20°, then C increases from 79.25 to 197.75, a factor of 2.5.

199. False. Two linear equations could have an infinite number of points of intersection. For example, $x + y = 1$ and $2x + 2y = 2$.

201. $2x - 5c = 10 + 3c - 3x,\ x = 3$

$2(3) - 5c = 10 + 3c - 3(3)$

$6 - 5c = 1 + 3c$

$5 = 8c$

$c = \frac{5}{8}$

203. (a) $ax^2 + bx = 0$

$x(ax + b) = 0$

$x = 0$

$x = -b/a$

(b) $ax^2 - ax = 0$

$ax(x - 1) = 0$

$x = 0$

$x = 1$

Appendix C.4 Solving Inequalities Algebraically and Graphically

1. double

3. $x \le -a,\ x \ge a$

5. no

7. $x < 3$

Matches (f).

8. $x \ge 5$

Matches (a).

9. $-3 < x \le 4$

Matches (d).

10. $0 \le x \le \frac{9}{2}$

Matches (b).

11. $-1 \le x \le \frac{5}{2}$

Matches (e).

12. $-1 < x < \frac{5}{2}$

Matches (c)

13. (a) $x = 3$

$5(3) - 12 \overset{?}{>} 0$

$3 > 0$

Yes, $x = 3$ is a solution.

(b) $x = -3$

$5(-3) - 12 \overset{?}{>} 0$

$-27 \not> 0$

No, $x = -3$ is not a solution.

(c) $x = \frac{5}{2}$

$5\left(\frac{5}{2}\right) - 12 \overset{?}{>} 0$

$\frac{1}{2} > 0$

Yes, $x = \frac{5}{2}$ is a solution.

(d) $x = \frac{3}{2}$

$5\left(\frac{3}{2}\right) - 12 \overset{?}{>} 0$

$-\frac{9}{2} \not> 0$

No, $x = \frac{3}{2}$ is not a solution.

15. $-1 < \dfrac{3 - x}{2} \le 1$

(a) $x = -1$

$-1 \overset{?}{<} \dfrac{3 - (-1)}{2} \overset{?}{\le} 1$

$-1 < \dfrac{4}{2} \not\le 1$

No, $x = 0$ is not a solution.

(b) $x = \sqrt{5}$

$-1 \overset{?}{<} \dfrac{3 - \sqrt{5}}{2} \overset{?}{\le} 1$

$-1 < 0.382 \le 1$

Yes, $x = \sqrt{5}$ is a solution.

(c) $x = 1$

$-1 \overset{?}{<} \dfrac{3 - 1}{2} \overset{?}{\le} 1$

$-1 < 1 \le 1$

Yes, $x = 1$ is a solution.

(d) $x = 5$

$-1 \overset{?}{<} \dfrac{3 - 5}{2} \le 1$

$-1 \not< -1 \le 1$

No, $x = 5$ is not a solution.

17.
$$-10x < 40$$
$$-\tfrac{1}{10}(-10x) > -\tfrac{1}{10}(40)$$
$$x > -4$$

19. $4x + 7 < 3 + 2x$
$$2x < -4$$
$$x < -2$$

21. $4(x + 1) < 2x + 3$
$$4x + 4 < 2x + 3$$
$$2x < -1$$
$$x < -\tfrac{1}{2}$$

23. $\tfrac{3}{4}x - 6 \le x - 7$
$$1 \le \tfrac{1}{4}x$$
$$4 \le x$$
$$x \ge 4$$

25.
$$1 < 2x + 3 < 9$$
$$-2 < 2x < 6$$
$$-1 < x < 3$$

27.
$$-8 \le -3(x - 2) < 13$$
$$-8 \le 1 - 3x + 6 < 13$$
$$-8 \le -3x + 7 < 13$$
$$-15 \le -3x < 6$$
$$5 \ge x > -2 \Rightarrow -2 < x \le 5$$

29. $-4 < \dfrac{2x - 3}{3} < 4$
$$-12 < 2x - 3 < 12$$
$$-9 < 2x < 15$$
$$-\tfrac{9}{2} < x < \tfrac{15}{2}$$

31. $5 - 2x \ge 1$
$$-2 \ge -4$$
$$x \le 2$$

33. $3(x + 1) < x + 7$
$$3x + 3 < x + 7$$
$$2x < 4$$
$$x < 2$$

35.

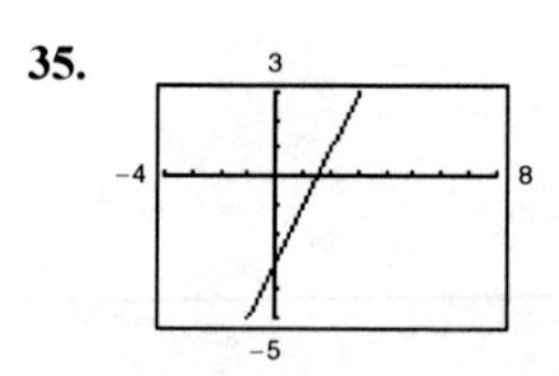

Using the graph, (a) $y \ge 1$ for $x \ge 2$ and (b) $y \le 0$ for $x \le \tfrac{3}{2}$.

Algebraically:

(a)
$$y \ge 1$$
$$2x - 3 \ge 1$$
$$2x \ge 4$$
$$x \ge 2$$

(b)
$$y \le 0$$
$$2x - 3 \le 0$$
$$2x \le 3$$
$$x \le \tfrac{3}{2}$$

37.

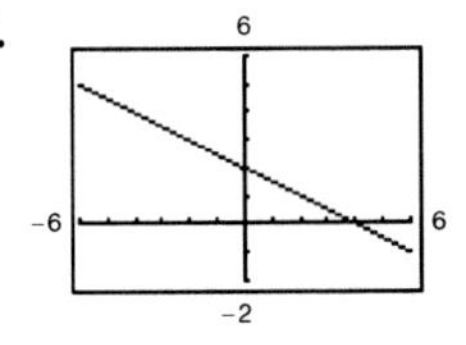

Using the graph, (a) $0 \le y \le 3$ for $-2 \le x \le 4$, and (b) $y \ge 0$ for $x \le 4$.

Algebraically:

(a) $$0 \le y \le 3$$
$$-1 \le -\tfrac{1}{2}x + 2 \le 3$$
$$-2 \le -\tfrac{1}{2}x \le 1$$
$$4 \ge x \ge -2$$

(b) $$y \ge 0$$
$$-\tfrac{1}{2}x + 2 \ge 0$$
$$2 \ge \tfrac{1}{2}x$$
$$4 \ge x$$

39. $|5x| > 10$

$$5x < 10 \text{ or } 5x > 10$$
$$x < -2 \text{ or } x > 2$$

41. $|x - 7| < 6$

$$-6 < x - 7 < 6$$
$$1 < x < 13$$

43. $|x + 14| + 3 > 17$

$$|x + 14| > 14$$
$$x + 14 < -14 \text{ or } x + 14 > 14$$
$$x < -28 \text{ or } x > 0$$

45. $10|1 - 2x| < 5$

$$|1 - 2x| < \tfrac{1}{2}$$
$$-\tfrac{1}{2} < 1 - 2x < \tfrac{1}{2}$$
$$-\tfrac{3}{2} < -2x < -\tfrac{1}{2}$$
$$\tfrac{3}{4} > x > \tfrac{1}{4}$$
$$\tfrac{1}{4} < x < \tfrac{3}{4}$$

47. $y = |x - 3|$

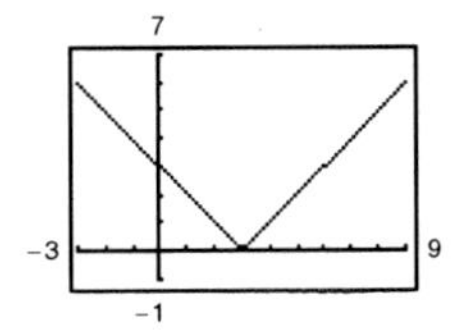

Graphically, (a) $y \le 2$ for $1 \le x \le 5$ and (b) $y \ge 4$ for $x \le -1$ or $x \ge 7$.

Algebraically:

(a) $$y \le 2$$
$$|x - 3| \le 2$$
$$-2 \le x - 3 \le 2$$
$$1 \le x \le 5$$

(b) $$y \ge 4$$
$$|x - 3| \ge 4$$
$$x - 3 \le -4 \text{ or } x - 3 \ge 4$$
$$x \le -1 \text{ or } x \ge 7$$

49. The midpoint of the interval $[-3, 3]$ is 0. The interval represents all the real numbers x no more than three units from 0.

$$|x - 0| \le 3$$
$$|x| \le 3$$

51. The midpoint of the interval $[-3, 3]$ is 0. The two intervals represent all numbers x more than three units from 0.

$$|x - 0| > 3$$
$$|x| > 3$$

53. All real numbers within 10 units of 7

$$|x - 7| \le 10$$

55. All real numbers at least five units from 3

$$|x - 3| \ge 5$$

57. $$x^2 - 4x - 5 > 0$$
$$(x - 5)(x + 1) > 0$$

Critical numbers: $-1, 5$

Testing the intervals $(-\infty, -1)$, $(-1, 5)$ and $(5, \infty)$, we have $x^2 - 4x - 5 > 0$ on $(-\infty, -1)$ and $(5, \infty)$.

Similarly, $x^2 - 4x - 5 < 0$ on $(-1, 5)$.

59. $2x^2 - 4x - 3 = 0$

$$x = \frac{4 \pm \sqrt{16 + 24}}{4} = 1 \pm \frac{\sqrt{10}}{2}$$

Entirely negative: $\left(1 - \frac{\sqrt{10}}{2}, 1 + \frac{\sqrt{10}}{2}\right) \approx (-0.581, 2.581)$

Entirely positive: $\left(-\infty, 1 - \frac{\sqrt{10}}{2}\right) \cup \left(1 + \frac{\sqrt{10}}{2}, \infty\right)$

61. $x^2 - 4x + 5 > 0$ for all x. There are no critical numbers because $x^2 - 4x + 5 \neq 0$. The only test interval is $(-\infty, \infty)$.

63.
$$(x + 2)^2 < 25$$
$$x^2 + 4x + 4 < 25$$
$$x^2 + 4x - 21 < 0$$
$$(x + 7)(x - 3) < 0$$

Critical numbers: $x = -7, x = 3$

Test intervals: $(-\infty, -7), (-7, 3), (3, \infty)$

Test: Is $(x + 7)(x - 3) < 0$?

Solution set: $(-7, 3)$

65.
$$x^2 + 4x + 4 \geq 9$$
$$x^2 + 4x - 5 \geq 0$$
$$(x + 5)(x - 1) \geq 0$$

Critical numbers: $x = -5, x = 1$

Test intervals: $(-\infty, -5), (-5, 1), (1, \infty)$

Test: Is $(x + 5)(x - 1) \geq 0$?

Solution set: $(-\infty, -5] \cup [1, \infty)$

67.
$$x^3 - 4x \geq 0$$
$$x(x + 2)(x - 2) \geq 0$$

Critical numbers: $x = 0, x = \pm 2$

Test intervals: $(-\infty, -2), (-2, 0), (0, 2), (2, \infty)$

Test: Is $x(x + 2)(x - 2) \geq 0$?

Solution set: $[-2, 0] \cup [2, \infty)$

69.
$$2x^3 + 5x^2 - 6x - 9 > 0$$
$$(x + 1)(x + 3)(2x - 3) > 0$$

Critical numbers: $-3, -1, \frac{3}{2}$

Testing the four intervals, we see that $2x^3 + 5x^2 - 6x - 9 > 0$ on $(-3, -1)$ and $\left(\frac{3}{2}, \infty\right)$.

71.
$$x^3 - 3x^2 - x + 3 > 0$$
$$(x + 1)(x - 1)(x - 3) > 0$$

Critical numbers: $-1, 1, 3$

Testing the four intervals, we see that $x^3 - 3x^2 - x + 3 > 0$ on $(-1, 1)$ and $(3, \infty)$.

73. $3x^2 - 11x + 16 \leq 0$

Since $b^2 - 4ac = -71 < 0$, there are no real solutions to $3x^2 - 11x + 16 = 0$. In fact, $3x^2 - 11x + 16 > 0$ for all x.

75. $x^2 + 3x + 8 > 0$

There are no critical numbers because $x^2 + 3x + 8 \neq 0$. The solution is all real numbers x.

77. (a) $f(x) = g(x)$ when $x = 1$.

(b) $f(x) \geq g(x)$ when $x \geq 1$.

(c) $f(x) > g(x)$ when $x > 1$.

79. $y = -x^2 + 2x + 3$

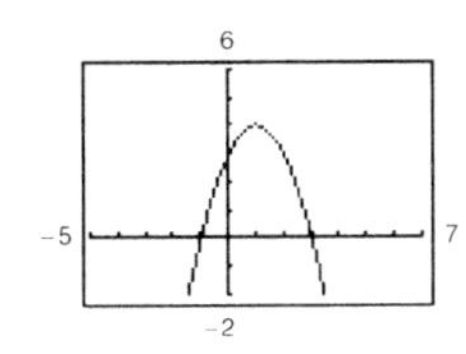

(a) $y \le 0$ when $x \le -1$ or $x \ge 3$.

(b) $y \ge 3$ when $0 \le x \le 2$ or $x \ge 3$.

Algebraically,

$$-x^2 + 2x + 3 \le 0$$
$$x^2 - 2x - 3 \ge 0$$
$$(x - 3)(x + 1) \ge 0$$

Critical numbers: $x = -1, x = 3$

Testing the intervals $(-\infty, -1)$, $(-, 3)$, and $(3, \infty)$, you obtain $x \le -1$ or $x \ge 3$.

$$-x^2 + 2x + 3 \ge 3$$
$$-x^2 + 2x \ge 0$$
$$x^2 - 2x \le 0$$
$$x(x - 2) \le 0$$

Critical numbers: $x = 0, x = 2$

Testing the intervals $(-\infty, 0)$, $(0, 2)$, and $(2, \infty)$, you obtain $0 \le x \le 2$.

81. $\dfrac{1}{x} - x > 0$

$$\frac{1 - x^2}{x} > 0$$

Critical numbers: $x = 0, x = \pm 1$

Test intervals: $(-\infty, -1)$, $(-1, 0)$, $(0, 1)$, $(1, \infty)$

Test: Is $\dfrac{1 - x^2}{x} > 0$?

Solution set: $(-\infty, -1) \cup (0, 1)$

83. $\dfrac{x + 6}{x + 1} - 2 < 0$

$$\frac{x + 6 - 2(x + 1)}{x + 1} < 0$$
$$\frac{4 - x}{x + 1} < 0$$

Critical numbers: $x = -1, x = 4$

Test intervals: $(-\infty, -1)$, $(-1, 4)$, $(4, \infty)$

Test: is $\dfrac{4 - x}{x + 1} < 0$?

Solution set: $(-\infty, -1) \cup (4, \infty)$

85. $y = \dfrac{3x}{x - 2}$

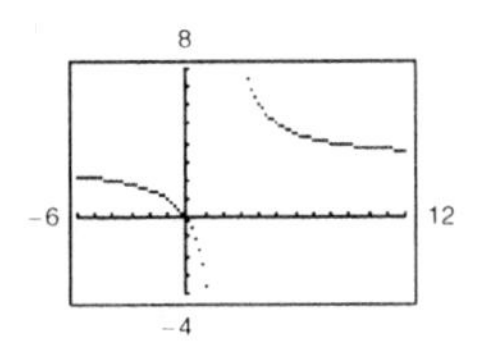

(a) $y \le 0$ when $y \le x < 2$.

(b) $y \ge 6$ when $2 < x \le 4$.

87. $\sqrt{x - 5}$

Need: $x - 5 \ge 0$

$x \ge 5$

Domain: $[5, \infty)$

89. $\sqrt[3]{6 - x}$

Domain: all real x

91. $\sqrt{x^2 - 4}$

Need: $x^2 - 4 \ge 0$

$(x + 2)(x - 2) \ge 0$

$x \le -2$ or $x \ge 2$

Domain: $(-\infty, -2] \cup [2, \infty)$

93. (a) $P(t) = 2000$

This occurs at the point of intersection, $t \approx 4$, or 2004.

(b) Less than two million: $P(t) < 2000$

This occurs for $t < 4$, or before 2004.

Greater than two million: $P(t) > 2000$

This occurs for $t > 4$, or after 2004.

95. (a) $s = -16t^2 + v_0t + s_0$

$s = -16t^2 + 160t$

$s = 16t(10 - t)$

$s = 0$ when $t = 10$ seconds.

(b) $s = -16t^2 + 160t > 384$

$16t^2 - 160t + 384 < 0$

$16(t - 6)(t - 4) < 0$

$s > 384$ when $4 < t < 6$.

97. (a)

(b) $20 < D < 25$ for $10.1 < t < 14.3$, or between 2000 and 2004

(c) $20 < D < 25$

$20 < 0.0510t^2 - 0.045t + 15.25 < 25$

To solve these inequalities, find the critical numbers.

$-0.0510t^2 + 0.045t + 4.75 = 0$

$$t = \frac{-0.045 \pm \sqrt{(0.045)^2 - 4(-0.0510)(4.75)}}{2(-0.0510)}$$

$$= \frac{-0.045 \pm \sqrt{0.971025}}{-0.102}$$

Because $0 < t < 18$, select the negative sign, $t \approx 10.1$. Hence, $20 < D$ for $10.1 < t$. Similarly, $D < 25$ for $t < 14.2749$.

99. $91.88t + 331.7 \geq 900$

$91.88t \geq 568.3$

$t \geq 6.19$

In the year 2006, there were at least 900 Bed Bath & Beyond Stores.

101. $91.88t + 331.7 = 30.22t + 404.0$

$61.66t = 72.3$

$t \approx 1.17$

In 2001, there were the same number of Bed Bath & Beyond stores as Williams-Sonoma stores.

103. When $t = 2$, $v \approx 333$ vibrations per second.

105. When $200 \leq v \leq 400$, $1.2 < t < 2.4$.

107. False. If $-10 \leq x \leq 8$, then $10 \geq -x$ and $-x \geq -8$.

109. The polynomial. $f(x) = (x - a)(x - b)$ is zero at $x = a$ and $x = b$.

111. (iv) $a < b$

(ii) $2a < 2b$

(iii) $2a < a + b < 2b$

(i) $a < \frac{a + b}{2} < b$

CPSIA information can be obtained
at www.ICGtesting.com
Printed in the USA
FFOW01n0018240915
17154FF

9 781285 867779